R. F. Albrecht, C. R. Reeves,
and N. C. Steele (eds.)

Artificial Neural Nets and Genetic Algorithms

Proceedings of the International Conference
in Innsbruck, Austria, 1993

Springer-Verlag Wien New York

Dr. Rudolf F. Albrecht
Institut für Informatik
Universität Innsbruck, Innsbruck, Austria

Dr. Colin R. Reeves
Division of Statistics and Operational Research

Dr. Nigel C. Steele
Division of Mathematics
School of Mathematical and Information Sciences
Coventry University, Coventry, U.K.

Printed on acid-free paper

With 503 Figures

ISBN-13:978-3-211-82459-7 e-ISBN-13:978-3-7091-7533-0
DOI: 10.1007/978-3-7091-7533-0

Preface

This Conference has as its themes two areas of research which have their origins in mathematical models constructed in order to gain understanding of important natural processes. By focussing on the process *models* rather than the processes themselves, significant new computational techniques have evolved which have found application in a large number of diverse fields. Some of this diversity is reflected in the topics which are the subjects of contributions to the conference programme.

The high profile enjoyed, or perhaps suffered by the field of neural network research has been a mixed blessing, leading to both under- and over-statement of their capabilities and potential applications. It is believed in many quarters that Minsky's conviction on the limited learning capabilities of perceptron structures, and the subsequent publicity received seriously diminished the scope of research into Artificial Intelligence. Equally, the widespread current awareness of the existence of artificial neural network based techniques may lead to their use in unnecessary circumstances, when the power is wasted and a simpler approach would be both adequate and more informative.

Amongst the papers accepted for presentation, there are contributions reporting theoretical developments in the design of neural networks, and in the management of their learning. There are also a number of contributions reflecting applications to speech recognition tasks, control of industrial processes as well as to credit scoring, and so on. In each case, participants will be eager to learn of the expected advantages of the approach over more conventional techniques.

In contrast, the development of genetic algorithms seems to have been more measured, and although it is interesting to note that a `hype warning' was recently posted on the electronically-distributed GA List Digest, these developments have so far avoided the glare of destructive publicity. It is fair to say that in both fields the over-blown claims for their capabilities have in the main not come from active researchers, but from those with a product to sell. As the papers in these proceedings demonstrate, the concern in the research community is to come to a better understanding of the procedures, and of where they need to be improved.

For example, we have several methodological papers which consider how GAs can be improved using an experimental approach, as well as by hybridizing with other useful techniques such as tabu search. The closely related area of classifier systems also receives a significant amount of coverage as we try to find better ways for their implementation. Further, while there are many papers which explore ways in which GAs can be applied to real problems, nearly all involve some understanding of the context in order to apply the GA paradigm more successfully. That this can indeed be done is evidenced by the range of applications covered in these proceedings. The important area of parallel implementations has not been neglected either, and of course, given the twin-theme nature of this conference, it is not surprising that there are several descriptions of ways in which GAs have been used for configuring and training neural networks.

It is our hope that the many stimulating papers collected in these proceedings will be of interest to the now extensive research community in both neural networks and genetic algorithms, and further advance our goals of better understanding and more successfully applying these emerging technologies.

February 8, 1993

Rudolf F. Albrecht
Innsbruck

Nigel C. Steele
Coventry

Colin R. Reeves
Coventry

Contents

GeneticAlgoritms

Artificial Neural Networks & Genetic Algorithms

ANNGA 93

International Conference on Artificial Neural Networks & Genetic Algorithms
Innsbruck/Austria, April 14-16, 1993

Programme Committee

R Albrecht, Institute of Informatics, University of Innsbruck, A-6020 Innsbruck, Austria.

K Warwick, Department of Cybernetics, University of Reading, Whiteknights, PO Box 22, Reading, Berkshire, RG8 2AY, UK.

H Saxén , Heat Engineering Laboratory, Department of Chemical Engineering, Abo Akademi, Bishopsgatan 8, SF-20500 Abo, Finland.

C Reeves, Division of Statistics and OR, Coventry University, Priory Street, Coventry CV1 5FB, UK.

E Hines, Department of Engineering Science, University of Warwick, Coventry CV4 7AL,UK.

C Peterson, Department of Theoretical Physics, University of Lund, Sölvegatan 14A, S-22362 Lund, Sweden.

G Smith, School of Information Systems, University of East Anglia, Norwich NR4 7TJ, UK.

M Dorigo, Milano/Berkeley, International Computer Science Institute, 1947 Center Street,Suite 600, Berkeley, CA 94704 - 1105, USA.

A Starita, Dipartimento di Informatica, Universita di Pisa, Corso Italia 40, I-56125 Pisa, Italy.

M Christodoulou, Director Laboratory of Automation, Technical University of Crete, Department of Electronic and Computer Engineering, GR-73100 Chania, Crete, Greece.

S Beaty, Department of Mechanical Engineering, Colorado State University, Fort Collins, CO80523, USA.

D Fielder, Vice President for Research and Graduate Programs, Office of the President, Georgia Institute of Technology, Atlanta, GA30332, USA.

A Sperduti, Dipartimento di Informatica, Universita di Pisa, Corso Italia 40, I-56100 Pisa, Italy.

D Würz, IPS, CLU B3, ETH-Center, CH-8092 Zürich, Switzerland.

J Bernasconi, Asea Brown Boveri Research Center, CH-5405 Baden-Dättwil, Switzerland.

V Bhavsar, Faculty of Computer Science, University of New Brunswick, PO Box 4400, Fredericton, New Brunswick E3B 5A3, Canada.

Organising Committee

Prof R Albrecht, Innsbruck, Chairman.

Prof N C Steele, Coventry.

Dr H Druckmüller, Innsbruck.

A Zuderell, Innsbruck.

Workshop Summary

Recognising that the Conference would attract participants from two distinct areas, initial arrangements were made to hold a pre-conference workshop to serve as a general introduction to both of the Conference themes.

The scope of the workshop was extended to cover additional topics so that participants already working in the fields could also benefit by joining later sessions.

Overall, the workshop was designed to cover two major areas namely pattern recognition and associated problems, and optimization.

In the first of these areas, the development of neural computing has made a significant contribution in the past decade, and two sessions are devoted to the exposition of some of the basic methodology, in both software and hardware applications.

Techniques for optimization have also made considerable recent progress, and feed-back neural networks and genetic algorithms have been at the leading edge of some of these developments. These are covered in two further sessions.

A more detailed description of the plan of the sessions is given below.

1 An Introduction to Neural Networks

Nigel Steele
Coventry University

This session, as the opening session of the workshop, is aimed at those with limited knowledge of the field of Artificial Neural Networks and is not intended for experienced practitioners.

An introduction to the field is given, covering the basic concepts of network structure and network learning. Attention is focussed mainly on the multi-layer perceptron network, with learning by error back-propagation.

Ideas on improving learning performance are discussed, and some 'hands-on' experience incoporated.

Some alternative network structures are introduced, and the session ends with a discussion of the capabilities of trained networks for pattern recognition and classification tasks.

2 Hardware Implementation of Neural Nets

Kevin Warwick
University of Reading

In this session some of the principles and practice of implementing ANNs in hardware are presented. Following an introduction to some hardware analogues of ANNs, the basis and structure of some digital and analogue artificial networks are described.

Issues involved in their implementation are discussed, along with examples and case studies drawn from a wide range of applications including image processing, speech recognition and robotics.

Finally, some ways in which hardware ANNs can be expected to develop are outlined.

3 Optimization with NNs

Carsten Petersen
Lund University

An introduction to the use of feed-back artificial neural networks (ANNs) for obtaining good approximate solutions to combinatorial optimization problems is given, assuming no previous knowledge in the field. This approach maps the problems onto energy functions in terms of neurons encoding different solutions. The energies are then minimized by iteratively solving so-called mean field equations. In this settling process the system encounters phase transitions. A systematic prescription is given for estimating the phase transition temperatures in advance, which facilitates an automatized choice of optimal parameters. This MFT procedure avoids to a large extent unwanted local minima. It differs from existing methods since the system 'feels' its way towards good solutions rather than fully or partly exploring different possible solutions. This methodology is illustrated for both binary (Ising) and K-valued (Potts) neurons using graph partition and traveling salesman problems (TSP) as examples. Also, a realistic high school scheduling problem is dealt with in some detail using Potts encoding. These problems are all characterized by equality constraints. The formalism can also be modified to deal with inequality constraints. This is illustrated with the knapsack problem.

Closely related to the Potts neural approach is the deformable templates method, where template coordinates replace binary decision neuronic elements as independent variables. In problems of geometrical nature like TSP and track finding this method is more economical with its fewer degrees of freedom.

4 Optimization with GAs

Colin Reeves
Coventry University

Genetic algorithms (GAs) have aroused intense interest in the past few years because of their flexibility and versatility in solving problems which traditional methods of optimization find difficult. Reported applications cover the fields of Operational Research, Engineering, Biology, Medicine, Control Theory and Robotics - to mention only a few areas.

In this session, the basic concepts of Genetic algorithms are introduced, using a worked example for emphasis. Some of the many ways in which GAs have been modified and extended are then reviewed. GAs need no simplifying assumptions of linearity, continuity etc., and thus can solve highly complex real-world problems, some of which are discussed, including problems of optimization, sequencing, design and control.

Research in GAs is a burgeoning field, but despite their very real successes, they also have some disadvantages. We look at some of these, and suggest ways in which they might be overcome. Chief among these may well be the development of hybrid procedures incorporating such techniques as Tabu Search; a brief introduction to the latter topic is also included.

The class of refractory neural nets

A. CLEMENTI, M. DI IANNI
Dipartimento di Scienze dell'Informazione,
Università di Roma "La Sapienza",
via Salaria 113, 00198 Roma, Italy
P.MENTRASTI
Dipartimento di Matematica,
Università di Roma "La Sapienza",
P.le Aldo Moro 2, 00185 Roma, Italy

Abstract. We introduce the absolute refractory behaviour into the formal neuron model. While a probabilistic approach to such a refractory model has yet been attempted, in this paper, a deterministic analysis is realized. A first result consists in showing a not expensive algorithm to transform each refractory net into an equivalent not refractory one. Such a result is then exploited to obtain an upper bound to the computational complexity of two classical problems: the reachability and stabilization problems.They find their principal motivations in control and learning theories whenever the necessity to a priori determine the lenght of both transients and limit cycles arises. Finally, we prove that, when the connection matrices of nets are symmetric, the complementary problem of stabilization is NP-complete and reachability is P-complete.

1 Introduction

The relevance of neural nets consists in their parallel evolution modality, in the structural and functional homogeneity of their computational units and in the fact that they represent one of the most meaningful formalization of some of the main characteristics of biological neurons: their on/off behaviour and the dependence of their activity on the "sum" of the actions generated by connected neurons. What the studied neuron models do not consider from a computational point of view is an important characteristic of physiologic neurons, that is their impossibility to receive input solicitations after they have entered the state on (absolute refractory state).

A probabilistic approach to such a refractory model has yet been attempted [1, 2]. In this paper, a deterministic analysis of some classical computational problems from the point of view of complexity theory is realized. Problems under examination deal with the capability of the net to reach an equilibrium point starting from any initial configuration (stabilization problem) or to reach a given configuration Y starting from X (reachability problem). The problems just defined find their principal motivations in control and learning theories whenever the necessity to a priori determine the lengths of both transients and limit cycles arises.

A first result of this paper consists in showing that each refractory net can be transformed by a not expensive algorithm into an equivalent not refractory one. This result is then exploited to establish an upper bound to the complexity of the problems just stated. Furtherly, we show that, when symmmetrical connections between neurons exist, the complementary problem of stabilization is NP-complete, that is every problem in NP (i.e. solvable in polynomial time by a nondeterministic Turing machine) is polynomially reducible to it; in other words, unless P=NP, it does not exist a polynomial-time deterministic algorithm to solve it. Finally we prove that the reachability problem is P-complete, that is it is in P (i.e. it is solvable in polynomial time by a deterministic Turing machine) but a polylogarithmic-time parallel algorithm using a polynomial number of processors, which solves it, is unlikely to exist.

The organization of the paper is straightforward: in §2 the necessary formal definitions of net models and problems are given; in §3 the reduction from a refractory net to a not refractory one is shown and in §3.1, 3.2 the complexity of the above problems is exactly determined.

2 Net models and problems

Argument of this paper is formal neuron which is a computation unit defined by abstracting some peculiar characteristics from physiologic neurons, that is their on/off behaviour and the dependence of their activity from the "sum" of the actions generated by the connected neurons. Among the proposed models, we focus our attention on two of them both coming from the classical model by Mc Culloch & Pitt ([3]).

A threshold binary neuron (in short, *tb-neuron*) is a 4-tuple $\langle In, P, O, th\rangle$ where:

- $In : \mathbf{N} \rightarrow \{0,1\}^n$ represents its n input lines which at each instant may be on (i.e. 1) or off (i.e. 0); the jth component of In will be denoted as In_j;
- $P = \langle p_1, \ldots, p_n\rangle$, $p_i \in \mathbf{Z}$ is the weight associated with the i-th input line;
- $O : \mathbf{N} \rightarrow \{0,1\}$ is the state-output function: $O(t)$ represents both the state and the output of the neuron at time t;
- $th \in \mathbf{Z}$ is the threshold of the neuron.

As usual, each *tb*-neuron operates in discrete time steps. Its activation modality is governed by the following relation:

$$O(t+1) = HS(\sum_{j=1}^{n} p_j \cdot In_j(t) - th). \quad (1)$$

where HS is the Heavyside function. that is, $HS(x) = 1$ if $x \geq 0$, 0 otherwise.

The neuron model just defined does not consider an important characteristic of physiologic neurons: their impossibility to receive input solicitations for a fixed amount of time after they have entered the state on. Formally, this fact can be modelled by defining the *refractory tb-neuron* (in short *rtb-neuron*), a *tb*-neuron having one absolute refractory step with activation modality governed by the following relation (see [4, 5]):

$$O(t+1) = HS(\sum_{j=1}^{n} p_j In_j(t) - th)(1 - O(t)). \quad (2)$$

We are now going to define a neural net as a set of neurons which are usually of the same type (*tb* or *rtb*). Given a set $I = \{1, \ldots, n\}$ of neurons, a neural net N is a triple $\langle I, W, Th\rangle$ where $W \in \mathbf{Z}^{n \times n}$ is a matrix such that W_{ij} is the weight of the connection from the output line of neuron j to the input line of neuron i and Th is the threshold array. The array $X(t) = \langle x_1(t), \ldots, x_n(t)\rangle$ where $x_i(t)$ is the state of neuron i at instant $t \in \mathbf{N}$ is called *configuration* of N at time t.

Let N be a net; an iterative dynamical system [6] $U = \langle X, S\rangle$ can be associated with it, where:

- $X = \{0,1\}^n$ is the space of states of the system;
- $S : \{0,1\}^n \rightarrow \{0,1\}^n$ is the state transition function determined by the equation (with parallel evolution mode):
  $x_i(t+1) = HS(\sum_{j=1}^{n} W_{ij} x_j(t) - Th_i)$ when *tb*-neurons are considered,
  $x_i(t+1) = HS(\sum_{j=1}^{n} W_{ij} x_j(t) - Th_i)[1 - x_i(t)]$ when *rtb*-neurons are considered.

The *trajectory* of U with initial configuration X_0 is defined as $Tr_X(t) = S^t(X)$, the composition of S t times. Since $\{0,1\}^n$ is finite, every trajectory is *definitively periodic*, that is for each initial configuration X_0 there exist $t', c \in \mathbf{N}$ such that $Tr_X(t) = Tr_X(t+c)$ for each $t \geq t'$. We call *limit cycle* of a trajectory the sequence of states which repeats itself and *transient* the sequence of different states before the cycle. The *order* of the limit cycle is the number of different states appearing in it. If the limit cycle has order one, it is an *equilibrium point* (a fixed point of S).

Two iterative dynamical systems $\langle X, S\rangle$, $\langle X', S'\rangle$ are said *isomorphic* if there exists a bijection $\phi : X \rightarrow X'$ such that $\phi(S(x)) = S'(\phi(x))$ for each $x \in X$.

The previous definitions can be easily extended to neural nets; hence in the following we shall usually speak about limit cycles and equilibrium points of nets instead of dynamical systems. In particular, two nets having isomorphic dynamical systems will be said *equivalent.* It is not hard to show that, if two nets are equivalent, the lengths of their transients and the orders of their limit cycles are the same.

Two different problems, related to trajectories, arise naturally which have the following definitions:

- the LIMIT CYCLE (in short LC) problem consists in deciding whether, given a neural net, an initial configuration exists which ends in a limit cycle of order at least 2;
- the REACHABILITY (in short CREP) problem consists in deciding whether, given a neural net and two configurations X and Y, the net starting from X reaches Y.

Together with the LC problem it is often considered its complementary, denoted as STABILIZATION.

Both of the two problems previously defined have been proved to be PSPACE-complete [7] in the general case. However, when only *tb*-neurons are considered and the matrix W is symmetric, the order of the limit cycles is not greater than two [8], while the length l of the transient of each trajectory is bounded by [9, 10]:

$$l \leq \frac{1}{2}\left(\sum_{i,k=1}^{n} |W(i,k)| + \sum_{i=1}^{n} |T(i)|\right). \qquad (3)$$

For the sake of brevity the class of nets with symmetric matrix W will be denoted as *Sym*. In [7] the last stated results have been exploited to show the NP-completeness of LC in *Sym* and the P-completeness of CREP in *Sym* when the weights and thresholds are restricted to values bounded by some polynomial function in n.

3 Refractory nets

In this section the problems previously defined are studied when the nets are made up of *rtb*-neurons only. Such kinds of nets are called *refractory nets.* Let us denote as *RSym* the class of refractory nets $N = \langle W, Th\rangle$ in which W is symmetric and as RN_k the class of nets in *RSym* such that $max_{1\leq i,j\leq n}(|W_{i,j}|) \leq kn^k$.

Observe that, since an *rtb*-neuron cannot remain in state 1, every refractory net eventually admits only the state in which all neurons are set to 0 (*quiescent state*) as equilibrium point. However an interesting relation exists between the refractory and the not refractory models which is stated by the following theorem:

Theorem 3.1 *Let $N = \langle W, Th\rangle$ be a net in RSym. Then it is always possible to define an equivalent net $N' = \langle W', Th'\rangle \in Sym$.*

Proof. The net $N' = \langle W', Th'\rangle$ is defined as follows

$$W'_{i,j} = \begin{cases} W_{i,j} & \text{if } i \neq j \\ -(1 + \sum_{k=1}^{n} |W_{i,k}|) & \text{otherwise} \end{cases}$$

$$Th'_i = Th_i \quad i = 1, \ldots, n$$

It is not hard to prove that N' is in *Sym* and is equivalent to N [11]. □

By the definition of equivalence, it is obvious that the converse of the previous theorem does not come true.

The above theorem allows to extend to *RSym* the properties proved for *Sym*. In particular, nets in *RSym* admit only cycles of order not greater than two and the length of each transient is bounded by the relation (3). In particular, this implies that LC is in NP when restricted to *Rsym*, while CREP is in P for instances restricted to the class RN_k. In short such problems restricted to the defined classes, will be denoted respectively as RSYM-LC and K-CREP. Notice that STABILIZATION reduces in this case to ask whether every trajectory ends in the quiescent state.

3.1 The RSYM-LC *problem*

In this section the NP-completeness of the RSYM-LC problem will be shown. The proof is based on a reduction from a slightly different version

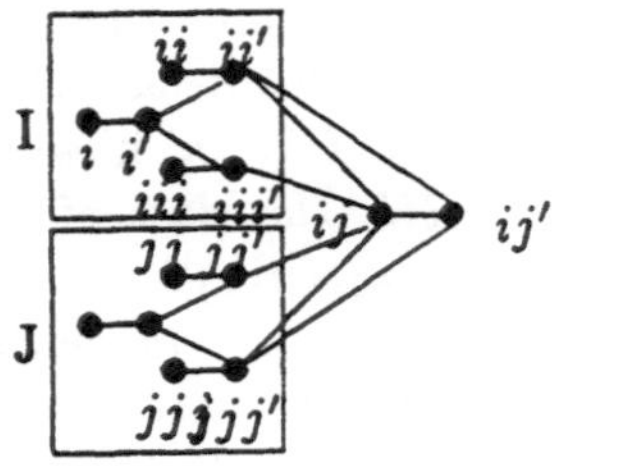

Figure 1: The subnet representing nodes i, j and the edge (i, j)

of the Travelling Salesman problem (in short TSP), a well known NP-complete problem [12]. In the original version it consists in deciding, given a complete undirected graph $G = (V, E)$, a function $d : E \rightarrow \mathbf{N}$ (the *distance* function) and a constant k, if G contains a tour of size at most k passing through each node exactly once. The version we consider, referred to as MAX-TSP, differs from the previous one in its requiring a tour of size *at least* k. It is not hard to show a polynomial reduction from TSP to MAX-TSP: in fact, if $\langle G, d, k\rangle$ is an instance of TSP, the corresponding instance of MAX-TSP is $\langle G, d', k'\rangle$ with $d'((i,j)) = max_{(h,l)\in E}(d((h,l))) - d((i,j)) + 1$ and $k' = n(d_M + 1) - k$, where $d_M = max_{(i,j)\in E}(d(i,j))$. This proves the NP-completeness of MAX-TSP.

We are now able to prove the following result:

Theorem 3.2 *The problem* RSYM-LC *is NP-complete.*

Proof. The inclusion of RSYM-LC in NP follows from theorem 3.1.

With every instance $\langle G, d, k\rangle$ of MAX-TSP, we associate a net $RF = \langle W, T\rangle \in RSym$ which admits at least one limit cycle with order greater than 1 if and only if G contains a tour of size at least k. If G contains n nodes, then RF is built up of $6n + n(n-1) + 4$ neurons and is shown in figures 1 and 2. Let $w \geq 1$ be a constant, for each node $u_i \in V(G)$ there are three pairs of neurons in RF denoted as

- i and i' which have thresholds, respectively, $w/16$ and $w/4$;

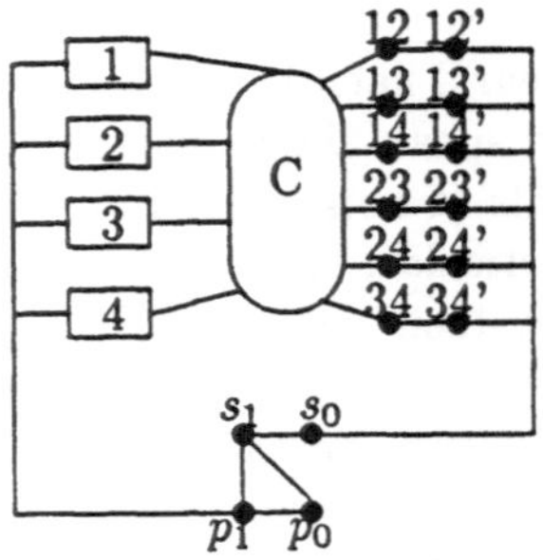

Figure 2: A schematic view of the net corresponding to 4-nodes graph. Rectangles stand for nodes as described in Figure 1

- ii and ii', which have thresholds , respectively, $w/2$ and $2w$;
- iii and iii' which have thresholds, respectively, $w/2 - w/8$ and $3w$.

Further, for each edge $(u_i, u_j) \in E$, the net contains the pair of neurons ij and ij' having thresholds $4w + 1$ and $d(i,j) + (4w + 1)$. Finally, there are two pairs of neurons named p_0, p_1 and s_0, s_1; in turn, their thresholds are $nw/32$, $w/64$, $\frac{n(n-1)}{2}d_M + k + 1$ and $\frac{n(n-1)}{2}d_M + w/128 + 1$.

The weights are listed in tables 1 and 2.

First of all, we observe that an equivalent definition of MAX-TSP is the following: deciding if there exists a subset $E' \subseteq E$ of size n such that for each $i = 1, \ldots, n$ exactly two edges incident on i are included in E' and $\sum_{e\in E'} d(e) \geq k$.

If $\langle G, d, k\rangle$ admits answer yes the net RF admits a limit cycle with order two. In fact, consider the following initial configuration ($t = 0$): the state of the neuron ij is 1 if and only if (u_i, u_j) belongs to E', the state of all neurons i and ii is 1, the state of the neurons p_1 and s_0 is 1. The state of all the other neurons is 0. It is not hard to see that at each even instant this configuration repeats itself.

On the contrary, suppose that $\langle G, d, k\rangle$ admits answer no. Two cases are possible: any subset $E' \subseteq E$ either is not a tour or $\sum_{e\in E'} d(e) < k$. Let C_0 be an initial configuration: if $x_{ij}(t_0) = 1$ and $x_{ij'}(t_0) = 1$ for some $i \neq j$, at time $t_1 = t_0+1$ their states become 0 and so remain in every successive step. Thus, after step t_1 the state of the neurons ij and ij' ($i \neq j$) corresponds to a subset E' of E: if $x_{ij}(t) = 1$ or $x_{ij'}(t) = 1$, $t \geq 1$, the edge (i,j) is in E', otherwise it is in $E - E'$.

1. if E' is not a tour two cases may occur:
   I. there exists a node u_1 incident to at most one edge in E'. In this case $x_{11'}(t) = 0$ for $t \geq 2$. This implies: $x_{11}(t) = 0$ for $t \geq 3$, $x_{1'}(t) = 0$ for $t \geq 4$ and $x_1(t) = 0$ for $t \geq 5$. So at some step $t' \leq 6$ the state of p_0 becomes 0 and so remains in all successive instants. It is not hard to see that two further steps are sufficient to set to 0 the states of s_0 and s_1. By the definition of the weights, s_0 is now able to set to 0 the state of all neurons;

   II. there exists a node incident to more than two edges in E'. With a procedure similar to the previous one it is possible to show that, in a constant number of steps, the net reaches the quiescent equilibrium point;
2. if E' is a tour but its size is less than k then, after at most two steps, s_0 reaches the state 0 and so remains in all successive instants. As before, by the definition of weights and thresholds, s_0 sets to 0 the state of all neurons.

The assert is thus completely proved. □

3.2 The CREP *problem*

As stated at the end of section 3 the CREP problem admits a polynomial algorithm when restricted to the classes RN_k for each $k \in N$. Let us now prove the P-completeness of the problem (for $k \geq 2$) with respect to the LOGSPACE reducibility. As a consequence, such a problem is candidate to be an "innerently sequential problem", that is it does not admit any efficient parallel algorithm unless P=NC, where NC⊆P is the class of efficiently parallelizable (i.e. solvable in polylogarithmic time with a polynomial number of processors on a PRAM) problems. In the next theorem a reduction from a well known P-complete problem, i.e. MONOTONE CIRCUIT VALUE ([13]), in short MCV, is shown. The MCV problem is defined as follows: given a monotone boolean circuit C realizing a boolean function $f(x_1, \ldots, x_m)$ and an assignment V to the $x_1, \ldots, x_m$, is $f(V) = 1$?[1]

Theorem 3.3 *The problem problem* CREP *in* RN_2 *is P-complete.*

Proof. Without loss of generality, only layered synchronous monotone circuits having gates realizing the boolean functions $\wedge, \vee, \neg$ are considered. Our goal is to transform each instance $\langle C, V\rangle$ of MCV into an instance $\langle RF, X, Y\rangle$ of CREP so that the instance of the former problem admits answer yes if and only if the corresponding instance of the latter admits answer yes.

Let us assume that the gates of C are partitioned into the layers $C_1, \ldots, C_h$ where C_1 consists of the input gates $x_1, \ldots, x_m$ and C_h consists of the output gate y. Since $\wedge, \vee$ and $\neg$ are linear separable boolean functions, C can be interpreted as a layered neural network $RF' = \langle W', t'\rangle$ which computes, in $k-1$ steps, the value $f(V)$ for each assignment V. In general, W' is not symmetric. To obtain the symmetry we consider $W" = W' + W'^T$ where W'^T is the transpose of W'. Since the circuit is unidirectional, in the new network $RF" = \langle W", t'\rangle$ the neuron y computes at step $h-1$ the same value computed by the corresponding neuron in RF'.

We want to associate to each instance $\langle C, V\rangle$ of MCV an instance $\langle RF, X, Y\rangle$ of CREP, so that $f(V) = 1$ if and only if Y is reachable from X in exactly $h-1$ steps. However, if Y is not reached within this time, Y is not to be reached in any future instant. To make this possible, let us introduce a new network L which goal consists in counting exactly k steps and successively in

[1] A monotone boolean circuit is a directed acyclic graph, the nodes of which are logic gates, among which we distinguish n input gates and one output gate.

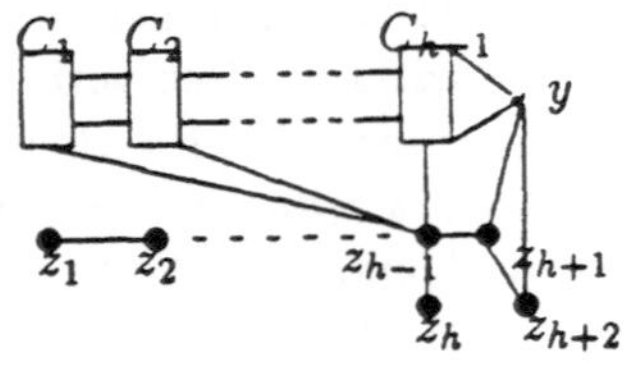

Figure 3: The net corresponding to a monotone circuit

setting off the network RF". L is made up of $h+2$ neurons, denoted as $z_1, \ldots z_{h+2}$, connected as shown in table 3, where $1 \le i \le h-2$. The thresholds are defined as follows: $t(z_i) = n^2 + h - i$, $1 \le i \le h-2$, $t(z_i) = 1$, $h-1 \le i \le h+2$.

RF is the connection between RF" and L; the weights of connections between z_{h-1} and every neuron of RF" (except y) and between y and both z_{h+1} and z_{h+2} are set to $-n$ (see Figure 3).

When starting from the initial configuration $X = \langle V, 0, \ldots, 0, 1, 0, \ldots, 0\rangle$, where 1 is the state referred to the neuron z_1, after exactly $h-1$ steps RF reaches the configuration $Y' = \langle 0, \ldots, 0, f(V), 0, \ldots, 0, 1, 1, 0\rangle$, where the state $f(V)$ of to the neuron y and the pair of states 1 is referred to the pair of neurons z_{k+1} and z_{k+2}. Independently from the value $f(V)$, RF in the step k ends in the limit cycle in which the two pairs of neurons z_{k-1}, z_k and z_{k+1}, z_{k+2} alternate their states and all the other neurons are in the state 0. So if Y is choosed $\langle 0, \ldots, 0, 1, 0, \ldots, 0, 1, 1, 0\rangle$ it is not hard to verify that $f(V) = 1$ if and only if RF, starting from X, reaches Y. □

4 Conclusions

We have shown that the behaviour of every refractory net can be "simulated" by a not refractory one. Thus, a qualitative analysis of refractory nets can be performed by exploiting results concerning not refractory ones (as, for instance, the order of limit cycles and the lenght of transients) and we can think the class of refractory nets as a restricted version of the not refractory ones. Is this an effective restriction? In other words, we ask whether the new proposed model has the same computational capacity as the old one or not. The complexity results obtained in this paper give a partial answer to the previous question, in the sense that refractory nets maintain the capability to model "hard" decisional problems. On the other hand, till now it is not clear how to define a new suitable equivalence relation between neural nets so to make possible the inverse simulation.

However we want to observe that, to make the behaviour of formal neurons closer to the behaviour of biological ones, it is meaningful to introduce a more natural definition of refractority. In fact, after an instant of absolute refractority, a real neuron remains in a state of *relative* refractority, in the sense that the threshold of the neuron comes back to its normal value gradually. A simulation of such a model has been attempted by using a cellular-automata oriented machine, the CAM-6 (see [14]), which has led to conclude that relative refractory nets show dynamical behaviour basically more complex than the model studied in this paper. It is still an open problem to determine the exact computational complexity of LC and CREP problems in this case.

Other possible directions for future research are represented by the study of net models containing different formal neuron models.

References

[1] W.A. Little, "The existence of persistent states in the brain", Mathematical Biosciences, 19, 101, 1974.

[2] A. Giordano, "Activity in neural nets with locality constraints", Proc. of third italian conference in theoretical computer science, World scientific, Mantova (Italy), November 1989

[3] W.S. Mc Culloch and W. Pitts, *A logical calculus of the ideas immanent in nervous activity*, Bull. Math. Biophys., 5, 115,1943.

[4] E.R. Caianiello, "Outline of a theory of thought-process and thinking machines", J. of Theoretical Biology, 1, 204, 1961.

[5] E.R. Caianiello, A. de Luca and L.M. Ricciardi, "Reverberation and control of neural networks", Kybernetik, 4, 10, 1967.

[6] E. Bienestock, F. Fogelman Soulie and G. Weiesbuch, (EDs), *Disordered systems and biological organization*, Springer-Verlag 1985.

[7] P. Campadelli, "Reti di neuroni: un approccio computazionale allo studio del sistema nervoso", PhD thesis in Computer Science, Universita' degli Studi di Milano e Torino, 1988.

[8] F. Fogelman Soulie, "Lyapunov functions and their use in automata networks", in: E. Bienestock, F. Fogelman Soulie and G. Weiesbuch, (EDs), *Disordered systems and biological organization*, Springer-Verlag 1985.

[9] F. Fogelman Soulie, E. Goles and G. Weiesbuch, "Transient length in sequential iteration of threshold functions", Disc. Appl. Math., 6, 95, 1983.

[10] F. Fogelman Soulie, E. Goles and D. Pellegrin, "Decreasing energy functions as a tool for studying threshold networks", Disc. Appl. Math., 12, 261, 1985.

[11] A. Clementi and P. Mentrasti, "Cellular automata and neural networks: links and computational problems", Proc. of third Workshop on parallel architectures and neural nets, Vietri (Italy), 1990.

[12] M.R. Garey and D.S. Johnson, *Computers and intractability: a guide to the theory of NP-completeness*, Freeman 1979.

[13] L.M. Goldschager, "The monotone and planar circuit value problems are LOGSPACE-complete for P", SIGACT News, 9, 2, 25-29, 1977.

[14] T. Toffoli and N. Margolus, *Cellular automata machines*, MIT Press, Cambridge, 1987.

	i	i'	ii	ii'	iii	iii'	ij	ij'
i	0	$w/16$	0	0	0	0	0	0
i'	$w/16$	0	$w/4$	$w/4$	$-w/8$	$-w/8$	0	0
ii	0	$w/4$	0	$w/2$	0	0	0	0
ii'	0	$w/4$	$w/2$	0	0	0	w	w
iii	0	$-w/8$	0	0	0	$w/2$	0	0
iii'	0	$-w/8$	0	0	$w/2$	0	w	0
ij	0	0	0	w	0	w	0	$4w+1$
ij'	0	0	0	w	0	0	$4w+1$	0
s_0	0	0	0	0	0	0	0	$(4w+2)d(i,j)$
s_1	0	0	0	0	0	0	0	0
p_0	$w/32$	$w/32$	0	0	0	0	0	0
p_1	0	0	0	0	0	0	0	0

Table 1

	s_0	s_1	p_0	p_1
i	0	0	$hw/32$	0
i'	0	0	$hw/32$	0
ii	0	0	0	0
ii'	0	0	0	0
iii	0	0	0	0
iii'	0	0	0	0
ij	0	0	0	0
ij'	$(4w+2)d(i,j)$	0	0	0
s_0	0	$(4w+2)d_M+1$	0	0
s_1	$(4w+2)d_M+1$	0	$w/132$	$w/132$
p_0	0	$w/132$	0	$w/64$
p_1	0	$w/132$	$w/64$	0

Table 2

	z_{i-1}	z_i	...	z_{h-1}	z_h	z_{h+1}	z_{h+2}
z_{i-1}	0	n^2+h-i	...	0	0	0	0
z_i	n^2+h-i	0	...	0	0	0	0
z_{h-1}	0	0	...	0	n^2+1	$n+1$	0
z_h	0	0	...	n^2+1	0	0	0
z_{h+1}	0	0	...	$n+1$	0	0	$n+1$
z_{h+2}	0	0	...	0	0	$n+1$	0

Table 3

The Boltzmann ECE Neural Network: A Learning Machine for Estimating Unknown Probability Distributions

ELIAS B. KOSMATOPOULOS AND MANOLIS A. CHRISTODOULOU*

Technical University of Crete, GREECE

Abstract—This paper treats the following problem: consider an ergodic signal source $\mathcal{S}$. Suppose that each time the source transmitts a multidimensional signal x according to an unknown ergodic probability distribution with density $p(x)$. Then the problem is to estimate the unknown density $p(x)$. The problem is solved via a Reccurent High-Order Neural Network (RHONN) and is based on the Energy Coordinates Equivalence (ECE) principle proposed by the authors. In the proposed method the signals are considered to be the states of a stochastic gradient dynamical system (Langevin s.d.e.) after it converges (in a stochastic manner). Then the (unknown) system is identified using ECE neural networks. After the learning procedure converges, the energy function of the ECE neural network is the estimate of the unknown probability distribution.

Keywords—Recurrent high order neural networks, Boltzmann machines, Energy coordinates equivalence, Langevin stochastic dynamical systems, Unknown probability distribution estimation, Stability, Stochastic system identification.

1 Introduction

The problem of estimating unknown probability distributions is very interesting and important in many areas of science and engineering (physics, chemistry, telecommunications, industrial engineering, information theory, pattern recognition, etc). Moreover learning and intelligent machines must be able to estimate the probability distributions of external processes in order to be able to recognize and predict them. This is why the problem of estimating unknown probabilities is of great interest in the neural networks science.

The simplest neural network for estimating unknown probability distributions is the so-called self-organizing map or competitive learning neural network (Kohonen, 1988, 1990; Grossberg, 1969, 1982, 1987). In these cases, whenever a signal x is input to the neural network, the neurons compete and the neuron (or group of neurons in the multidimensional case) closest to the incoming signal is the winner one. Then the state of the winner neuron (and perhaps of some of its neighborhoods) is adjusted using a simple linear correction learning law:

$$w_i(t+1) = w_i(t) + e(t)[w_i(t) - x_i(t)]$$

where w_i denotes the i-th neuron state and $e(\cdot)$ an exponential decaying function of time. It has been proved (Kohonen, 1988; Clark & Ravishankar, 1990; Kosko, 1991b; Kong & Kosko, 1990) that competitive neural networks partition the state-space into classes D_i; the state of the i-th neuron (group of neurons) converge in the probabilistic centroid of the class D_i. In this case, if the number of classes is sufficiently large we can estimate the unknown probability distribution using the simple rule (Kosko, 1991b; Kong & Kosko, 1990):

$$p(V) = \frac{n_V}{N}$$

where $p(V)$ is the probability the signals x to belong in the volume $V \subset \Re^n$, n_V is the number of synaptic vectors w_i in V and N is the total number of synaptic vectors.

Although such a technique is quite efficient and very easy to be implemented in the case where N is

*The authors are with Dept. of Electronic & Computer Engineering, Technical University of Crete,73100 Chania, Crete, GREECE.

small, it can not be utilized in cases where we desire an explicit mathematical formula for the unknown probability ditribution. Also, the number of synaptic weights cannot in practice be as large as desired due to hardware or software limitations, and hence the accuracy of such methods in estimating unknown probability distributions is not always the desired.

A second technique for estimating unknown probability distributions is the so-called Boltzmann machine (Ackley *et al.*, 1985; Amari *et al.*, 1992). In this case a recurrent neural network is used whose neurons take either the value 1 (excited) or the value 0 (quiescent) depending on a probability distribution which is a function of the network energy. The learning in Boltzmann machines is performed in two phases and in general it requires many training examples in order to give reliable results (in fact the number of examples must be infinite). On the other hand the Boltzmann machine can realize (estimate) only a small amount of probability distributions (Amari *et al.*, 1992). This happens because to the network energy function is quadratic while there are high-order terms needed in order to be able to realize any probability distribution.

The above problem can be overcome if we add higher-order connections in the Boltzmann neural networks (Amari, 1991). In this case the neural network can realize any probability distribution. However efficient learning laws are not known for training high-order Boltzmann machines like that proposed in Amari (1991).

In this paper we propose a completely different technique for estimating unknown probability distributions. In fact this technique is the same with that of Energy Coordinates Equivalence (ECE) principle (Kosmatopoulos *et al.*, 1992c; Kosmatopoulos & Christodoulou, 1992b, 1992c, 1992e, 1992f). The basic principle used is this: if we consider a stochastic dynamical system of the Langevin form

$$dx = -\nabla V(x)dt + d\xi \qquad (1.1)$$

where $V(x)$ is a positive definite scalar "energy" function, $\nabla V(x)$ is its gradient with respect to x, and ξ is a Wienner process (i.e. $d\xi/dt$ is a zero-mean Gaussian white noise process), then the probability distribution of x converges to an ergodic probability distribution with density $p(x) = a \cdot exp(V(x)/b)$ where a and b are constants. Conversely, for any probability distribution with density $p(x)$ there is a dynamical system of the form (1.1) such that the probability distribution of the states x converge to the distribution with density $p(x)$. These two facts lead us to reformulate the problem of estimating unknown probability distributions into the problem of approximating the system (1.1). What we need is an appropriate dynamical system able to approximate (1.1). Such a system is the Reccurrent High-Order Neural Network (RHONN) since it is a gradient system and it is able to approximate any dynamical system (Kosmatopoulos & Christodoulou, 1992a, 1992c, Kosmatopoulos *et al.*, 1992a, 1992b, 1992c). Moreover RHONNs are structurally stable systems (Kosko, 1990, 1991a; Kosmatopoulos & Christodoulou, 1992d) which simply means that they remain stable even if there is some disturbance in their dynamics.

An appropriate RHONN model that can be used for estimation of unknown distribution probabilities is the *Energy Coordinates Equivalent (ECE)* RHONN; ECE neural networks approximate the energy of the unknown processes rather than their dynamics. Knowledge of the energy of a process means that not only the dynamics of the process are known, but also its properties (Kosmatopoulos & Christodoulou, 1992c, 1992e, 1992f; Kosmatopoulos *et al.*, 1992c). This very powerful property of ECE networks has found great applicability in learning of the input/output behaviour and the qualitative properties of dynamical systems (Kosmatopoulos & Christodoulou, 1992c; Kosmatopoulos *et al.*, 1992c), in control of unknown dynamical systems (Kosmatopoulos & Christodoulou, 1992f) in storage, filtering and recognition of spatiotemporal patterns which are affected by multiplicative noise (Kosmatopoulos & Christodoulou, 1992b, 1992e) and in learning vector quantization (Kosmatopoulos & Christodoulou, 1992b). The ECE neural networks are also shown to fit well in estimation of unknown probability distributions since estimation of unknown probability distribution is equivalent to the approximation of the energy of a stochastic dynamical system.

2 Probability Distributions and Additive Stochastic Dynamical Systems

In this section we present some preliminary results that will be utilized in the next sections. Consider a probability distribution with density $p(x)$ where $x \in \mathcal{X} \subset \Re^n$ is an n-dimensional vector. Hereafter this paper we assume that $\mathcal{X}$ is a compact subset of $\Re^n$. The following result is nessecary for the rest of the paper.

Theorem 2.1 *Consider a probability distribution with density $p(x)$. Then there exist M, m_k, J_k, $d_j(k)$ such that*

$$\log(p(x)) = -\sum_{k=1}^{M} m_k \prod_{j\in J_k} s_j^{d_j(k)} + \psi \tag{2.1}$$

where

- $\{J_1, J_2, \ldots, J_M\}$ *is a collection of M not-ordered subsets of* $\{1, 2, \ldots, n\}$,
- m_k *are real parameters*
- $d_j(k)$ *are integers,* $d_j(k) \geq 0$
- ψ *a "normalizing" constant depending on the particular selection of J_k, m_k and $d_i(k)$*

and the vector $[s_1, s_2, \ldots, s_n]^T$ is defined as

$$s_i = S(x_i), \qquad i = 1, \ldots, n \tag{2.2}$$

where $S(\cdot)$ is an invertible, smooth, and monotone increasing function. □

For simplicity, in this paper we use a simple linear function in place of $S(\cdot)$. More presicely we assume that

$$S(x_i) = x_i \tag{2.3}$$

Of course the results of this paper are also valid, if instead of (2.3), we use another invertible, smooth, and monote increasing function, e.g. a sigmoidal.

The next Theorem states that a stochastic gradient dynamical system, can realize any probability distribution. In other words the next Theorem states that for any ergodic probability distribution $p(x)$ there exists a stochastic dynamical system whose states are governed by a probability distribution that converges to the desired one $p(x)$.

Theorem 2.2 *Consider the Langevin stochastic differential equation*

$$dx = -\nabla V(x)dt + d\xi \tag{2.4}$$

where $V : \mathcal{X} \rightarrow \Re^+$ and ξ is a scalar Wienner process with $E\{\dot{\xi}\} = 0$, $E\{\dot{\xi}(t)\dot{\xi}(\tau)\} = \sqrt{2}\delta(t-\tau)$. Then

(i)

$$\lim_{t\rightarrow\infty} p(x;t) = ae^{-V(x)} \tag{2.5}$$

where a is a positive "normalizing" constant.

(ii) *For any probability distribution with density $p(x)$ there exists a dynamical system of the form (2.4) such that*

$$\lim_{t\rightarrow\infty} p(x;t) = p(x) = ae^{-V(x)} \tag{2.6}$$

where

$$V(x) = \log(a) + \sum_{k=1}^{M} m_k \prod_{j\in J_k} s_j^{d_j(k)} \tag{2.7}$$

□

In the sequel we make (for simplicity) the following assumption.

Assumption 2.3 *In the sequel of this paper we assume that $a = 1$ and hence $\log(a) = 0$.* □

All the results can be easily extended in the case where $a \neq 1$. However the assumption that $a = 1$ makes the analysis less complicated. In fact the constant a is given by

$$a = \frac{1}{\int_{\mathcal{X}} \exp\{-V(x)\}d\mathcal{X}}$$

We close this section by analyzing the stability properties of a class of stochastic dynamical systems. This class is defined in the next Definition.

Definition 2.4 *A stochastic dynamical system is said to be an* additive stochastic dynamical system *iff it is of the following form*

$$dx = f(x)dt + gd\xi \tag{2.8}$$

where $f(\cdot)$ is a smooth random vector field and $g \in \Re$. In other words, a stochastic dynamical system is said to be additive if its diffusion coefficient g is indepented of the state vector x. □

Additive stochastic dynamical systems possess nice properties that the general Ito ones of the form $dx = f(x)dt + g(x)d\xi$ do not possess. In the sequel we follow the stability analysis of additive stochastic dynamical systems as proposed by Kosko (1990, 1991a) in the context of neural networks. The following Lemma is needed in the next sections

Lemma 2.5 *Consider the additive stochastic dynamical system*

$$dx = f(x)dt + gd\xi \Leftrightarrow \dot{x} = f(x) + g\dot{\xi}$$

If there exists a scalar positive definite smooth function $y = V(x)$, $V : \mathcal{X} \rightarrow \Re_+$ such that

$$\dot{E}\{y\} = E\{\nabla V^T(x)f(x)\} \leq 0 \tag{2.9}$$

then

$$\lim_{t\rightarrow\infty} E\{x\} = x^*$$

where x^* is a constant point in $\mathcal{X}$ (equilibrium) and moreover x^* is such that $V(x^*) = 0$ and $f(x^*) = 0$. Moreover

$$\lim_{t\to\infty} E\{\dot{x}\} = 0$$

Here $E\{\cdot\}$ denotes expectation. □

3 The Boltzmann ECE Neural Network

The neural network that is used for the estimation of the unknown probability distribution is of the form (Dempo *et al.*, 1991; Kosko, 1990; Kosmatopoulos & Christodoulou, 1992a, 1992b, 1992c, 1992d, 1992e, 1992f; Kosmatopoulos *et al.*, 1992a, 1992b, 1992c):

$$\dot{z}_i = -\sum_{k=1}^{L} w_{ik} \frac{d_i(k)}{y_i} \prod_{j\in I_k} y_j^{d_j(k)} + \zeta_i \qquad (3.1)$$

where

- $\{I_1, I_2, \ldots, I_L\}$ is a collection of L not-ordered subsets of $\{1, 2, \ldots, n\}$,
- w_{ik} are the synaptic weights of the neural network
- $d_j(k)$ are integers, $d_j(k) \geq 0$,
- z_i is the state of the i-th neuron, $i = 1, 2, \ldots, n$,
- $\dot{z}_i = dz_i/dt$

and $[y_1, y_2, \ldots, y_n]^T$ is the vector consisting of all inputs to each neuron, i.e.

$$y_i = S(zi), \qquad i = 1, \ldots, n \qquad (3.2)$$

where $S(\cdot)$ has been defined in the previous section. The parameter ζ_i denotes the i-th *fast synaptic weight* of the neural network, and its meaning will be made clear in the sequel.

Note that (Cohen & Grossberg 1983; Dempo *at al.*, 1991; Kosmatopoulos & Christodoulou, 1992a, 1992b, 1992c, 1992c, 1992d, 1992e, 1992f; Kosmatopoulos *et al.*, 1992a, 1992b, 1992c) in the case where the synaptic weights of the neural network are *symmetric*, that is

$$w_{ik} = w_{jk} = w_k, \quad \forall i, j \qquad (3.3)$$

the neural network (3.1) can be written similarly to the (2.4) form. More presicely if we set

$$L(z) = \sum_{k=1}^{L} w_{ik} \prod_{j\in I_k} y_j^{d_j(k)} \qquad (3.4)$$

then the neural network (3.1) can be written as

$$\dot{z} = \nabla L(z) + \zeta \Leftrightarrow dz = \nabla L(z)dt + d\eta$$

where obviously $\dot{\eta} = \zeta$. Of course here the parameters ζ_i do not denote necessarily Gaussian processes.

The following Theorem is a simple corollary of Theorem 2.1.

Theorem 3.1 *Consider the neural network (3.1). Then for any probability distribution with density $p(x)$ there exists a neural network of the form (3.1) such that*

$$\log(p(z)) = \log(a) - L(z) \qquad (3.5)$$

where $L(\cdot)$ is given by (3.4). □

The main problem, therefore, is that, given the observations x which are governed by the unknown probability distribution $p(x)$, how one can find the appropriate synaptic weights w_{ik}, the collections I_k and the powers $d_i(k)$ such that (3.5) holds. If this is achieved then obviously the problem of estimation of unknown probability distributions has been solved. In this paper, we will concentrate our attention in the optimal selection of the synaptic weights w_{ik}. The selection of the optimal collections I_k and the powers $d_i(k)$ will not be treated in this paper since, in general, the more collections I_k we have the better approximation we do. Hence we assume that the number of collections I_k is sufficiently large in order to make the neural network able to approximate the unknown probability distributions sufficiently close.

Remark 3.2 We mention here that although theoretically a problem that is solvable with a network of a given number of high-order connections, it can be also solved by a larger network which imbeds the smaller one, in practice things are different. This is true because bigger number of high-order connections means larger searching space, more complexity and danger in overfitting the data[1]. Although the problem of deciding the optimal number of high-order connections is an open problem for further research, we a usually overcome it by the trial-and-error method. ◇

Although there exists a variety of ECE learning laws that can be implemented in the above neural

[1]The latest case is known in the classification theory as the "tuning to the noise" problem and it occurs when the number of free parameters is large relative to the training data.

network we will concentrate our attention to the following

$$\dot{w}_{ik} = -\gamma \sum_{i=1}^{n} \left\{ \sum_{k=1}^{L} \frac{d_i(k)}{y_i} \prod_{j\in I_k} y_j^{d_j(k)} [x_i - z_i] \right\}$$
$$\dot{\zeta}_i = -\gamma(x_i - z_i) \tag{3.6}$$

where γ is a positive design scalar.

Remark 3.3 Note that the learning laws (3.6) do not destroy the symmetry property (3.3). This is very important since in the case where the symmetry property (3.3) does not hold the neural network (3.1) is not a gradient system anymore and hence Theorem 3.1 does not hold. ◇

4 Analysis of the Boltzmann ECE Neural Networks: An ECE Theoretical Point of view

In this section we analyze the capabilities of neural network (3.1) whose synaptic weights are adjusted according to (3.6). Before doing so, we need some basic elements of the ECE theory. Of course it is not possible to describe all the principles of the ECE theory in this section. However we can describe only the concepts that will be needed in this paper. An interesting reader is referred to Kosmatopoulos & Christodoulou, (1992b, 1992c, 1992e, 1992f; Kosmatopoulos *et al.*, 1992c) for more details.

The fundamental pronciple of ECE theory is this: consider that there is an unknown system that we wish to learn its input/output behaviour and qualitative properties. Then if we construct appropriate energy coordinates of this unknown system, we can use a RHONN which will approximate - via learning - the energy coordinates of the unknown system by the RHONN's energy coordinates. In the case where the RHONN is described by the simplified version (3.1) its i-th energy coordinate is defined as follows

$$^{i}L(z) = -\sum_{k=1}^{L} w_{ik} \frac{d_i(k)}{y_i} \prod_{j\in I_k} y_j^{d_j(k)} + \zeta_i \tag{4.1}$$

The next step is to construct appropriate learning laws in order to (i) approximate the unknown system behaviour as close as possible and (ii) force the synaptic weights to converge to the optimal values. In Kosmatopoulos & Christodoulou, (1992c), Kosmatopoulos *et al.* (1992c) we provide with learning laws satisfy the aforementioned conditions (i) and (ii). We note that the learning laws (3.6) are simple cases of the above ones.

After this brief and quite simplified description of the ECE principle we are ready to proceed to the main contributions of this paper. As we have already seen the unknown probability distribution can be realized by the dynamical system (2.4) where $V(x)$ is given by (2.7). Consider now the following Lemma.

Lemma 4.1 *Consider the neural network (3.1) and a dynamical system of the form (2.4) where $V(\cdot)$ is given by (2.7). Consider that Assumption 2.3 holds. Assume that the regressor terms $\frac{d_i(k)}{y_i}\prod_{j\in I_k} y_j^{d_j(k)}$ are sufficiently rich in the sense that for any positive δ, there exist positive reals α_{ik} and β_{ik}, such that*

$$0 \le \alpha_{ik} \le \int_t^{t+\delta} \left[\frac{d_i(k)}{y_i} \prod_{j\in I_k} y_j^{d_j(k)} \right]^2 dt \le \beta_{ik}$$

Then

$$\lim_{t\to\infty} E\left\{ [\dot{x}_i - \dot{z}_i]^2 \right\} = 0 \tag{4.2}$$

and

$$\lim_{t\to\infty} E\left\{ [w_{ik} - m_k]^2 \right\} = 0 \tag{4.3}$$

where $E\{\cdot\}$ denotes the expectation. □

Let us now return to the case where the neural network (3.1) whose weights are adjusted according to (3.6) is used for the estimation of an unknown probability $p(x)$. The following Lemma is introduced.

Lemma 4.2 *Consider a signal source that transmitts randomly signals x according to an ergodic but unknown probability with density $p(x)$. Suppose that Assumption 2.3 holds. Then*

$$\lim_{t\to\infty} E\left\{ [w_{ik} - w_{ik}^*]^2 \right\} = 0 \tag{4.4}$$

i.e. the synaptic weights w_{ik} converge (in a stochastic manner) to some constant values. Moreover,

$$\lim_{t\to\infty} E\{\dot{z}_i\} = 0 \tag{4.5}$$

□

The next Theorem demonstrates the capability of the neural network (3.1) whose weights are adjusted according to (3.6) to estimate any unknown ergodic probability distribution with density $p(x)$. We remind that from Theorem 2.1 we have that $p(x)$ can be written as

$$\log(p(x)) = -\sum_{k=1}^{M} m_k \prod_{j\in J_k} s_j^{d_j(k)}$$

Theorem 4.3 *Consider a signal source that transmitts randomly signals x according to an ergodic but unknown probability distribution with density $p(x)$. Suppose that Assumption 2.3 holds. Then*

$$\lim_{t\to\infty} E\left\{\left[p(x) - a \cdot e^{-L(x)}\right]^2\right\} = 0 \qquad (4.6)$$

Moreover

$$\lim_{t\to\infty} E\left\{[w_{ik} - m_k]^2\right\} = 0 \qquad (4.7)$$

□

5 Conclusions

This paper shows that there exists a one-to-one correspondence between ergodic probability distributions on the one hand, and stochastic dynamical systems that are additive and gradient, on the other. The keyword for such a correspondence is the *energy coordinates*. In other words knowledge of the energy coordinates of an additive gradient stochastic system leads to the knowledge of the underlying probability distribution, and vice versa. Hence, the basic concept of *Energy Coordinates Equivalence* proposed by the authors previously seems to be valid not only in the case where one wishes to design a learning machine that is able to learn the input/output behaviour and the qualitative properties of unknown dynamical systems as originally proposed in Kosmatopoulos & Christodoulou (1992c) and Kosmatopoulos *et al.* (1992c), but also in the case where we wish to design unknown probability estimators or spatiotemporal pattern recognizers (Kosmatopoulos & Christodoulou, 1992a, 1992e). Moreover *the same learning laws that have been proposed in Kosmatopoulos & Christodoulou (1992c) and Kosmatopoulos et al. (1992c) for learning the behaviour and qualitative properties of dynamical systems, are used here for unknown probability estimation.*

The importance of the above facts is multi-fold:

1. Gradient RHONN neural networks are able to perform different tasks. Pattern recognition as well as dynamical system approximation and control, spatiotemporal pattern recognition, unknown probability estimation and centroid (learning vector quantization) estimation problems can be solved using gradient RHONNs.
2. The concept of utilization of the neural networks *energy* seems to be a common denominator in most of neural machines. For instance associative memory neural networks use this concept in order to store the desired patterns as the - possible local - minima of the energy of the neural network; similar conditions hold in simulated annealing neural networks as well as statistical associative memories (see e.g. Kamp and Hasler, 1990) and competitive and self-organizing networks. In ECE neural networks this concept is generalized: while in associative memory neural networks the aim is to approximate the minima of the energy of the unknown process with those of the neural network, in the ECE case, the neural network energy approximates the energy of the unknown process[3]. Note that in this sense Boltzmann machines resemble ECE ones; ECE networks as well as Boltzmann machines approximate the energy function and not only the equilibria of the unknown process. However Boltzmann machines can be applied only in cases of estimating unknown probability distributions, while the ECE ones are more general-purpose.
3. At last we mention that the present work seems to be a bridge between dynamical systems theory, information and signal processing theory, and neurophysiology. RHONNs of the form (3.1) or of the more general form given by Dempo *et al.* (1991), Kosmatopoulos and Christodoulou (1992a-1992f) and Kosmatopoulos *et al.*, (1992b, 1992c) which are generalized cases of the Cohen-Grossberg networks (Cohen & Grossberg, 1983; Kosko 1990, 1991a), seem to descibe quite accuratelly the electrochemical, molecular and other processes taking place in the mammalian nervous system. On the other hand, these neural networks, although they are dynamical systems, are able to realize any probability distribution, as shown in this paper, and to represent and compute regular languages as shown by Miller *et al.* (1991).

References

[1] Ackley, D. H., Hinton, G. E., & Sejnowski, T. J. (1985). A learning algorithm for Boltzmann machines. *Cognitive Science*, **9**, 147-169.

[2] Amari, S. (1991). Dualistic geometry of the manifold of higher-order neurons. *Neural Networks*, **4**, 443-451.

[3]Here the term "process" means either pattern source, probability distribution or dynamical system.

[3] Amari, S., Kurata, K., & Nagaoka, H. (1992). Information geometry of Boltzmann machines. *IEEE Transactions on Neural Networks*, **3**, 260-2271.

[4] Clark, D. M., & Ravishankar K., (1990). A convergence theorem for Grossberg learning. *Neural Networks*, **3**, 87-92.

[5] Cohen, M. A., & Grossberg, S. (1983) Absolute stability of global pattern formation and parallel memory storage by competitive neural networks. *IEEE Transactions on Systems, Man, and Cybernetics*, **13**, 815-826.

[6] Dempo, A., Farotimi, O., & Kailath, T., (1991). High-order absolutely stable neural networks. *IEEE Transactions on Circuits and Systems*, **38**, 57-65.

[7] Grossberg, S. (1969). On learning and energy-entropy dependence in recurrent and nonrecurrent signed networks. *J. Statist. Phys.*, **1**, 319-350.

[8] Grossberg, S., (1982). *Studies of Mind and Brain*. Boston: Reidel.

[9] Grossberg, S. (1987). Competitive learning: From interactive activation to adaptive resonance. *Cognitive Science*, **11**, 121-134.

[10] Kamp, Y. & Hasler, M. (1990). *Recursive Neural Networks for Associative Memory*, J. Wiley & Sons.

[11] Kohonen, T. (1988) *Self-Organization and Associative Memory*. 2nd. ed., New York: Springer-Verlag, 1988.

[12] Kohonen, T. (1990). The self-organizing map. *Proceedings of the IEEE*, **78**, 1464-1480.

[13] Kong, S.-G., & Kosko, B. (1990). Differential competitive learning for centroid estimation and phoneme recognition. *IEEE Transactions on Neural Networks*, **2**, 118-124.

[14] Kosko, B. (1990). Unsupervised learning in noise. *IEEE Transactions on Neural Networks*, **1**, 44-57.

[15] Kosko, B. (1991a). Structural stability of unsupervised learning in feedback neural networks. *IEEE Transactions on Automatic Control*, **36**, 758-792.

[16] Kosko, B. (1991b). Stochastic Competitive Learning. *IEEE Transactions on Neural Networks*, **2**, 522-529.

[17] Kosmatopoulos, E. B., & Christodoulou, M. A. (1992a). Dynamical distributed Neural Networks for Nonlinear System Identification. *Neural Network World*, **3-4**, 241-267.

[18] Kosmatopoulos, E. B., & Christodoulou, M. A. (1992b). ECE neural networks for learning, recognizing, and filtering of spatiotemporal patterns. *IFAC Workshop on Mutual Impact of Computing Power and Control Theory*, Prague, Czechoslovakia.

[19] Kosmatopoulos, E. B., & Christodoulou, M. A. (1992c). "Energy coordinates equivalent neural networks for nonlinear system identification," *Symposium on Implicit and Nonlinear Systems*, Texas, USA.

[20] Kosmatopoulos, E. B., & Christodoulou, M. A. (1992d). Stability analysis of recurrent high order neural networks. Submitted to *IEEE Transactions on Circuits and Systems.*

[21] Kosmatopoulos, E. B., & Christodoulou, M. A. (1992e). Stability, Filtering, and Learning Properties of Competitive ECE Neural Networks. Submitted to *IEEE Transactions on Systems, Man, and Cybernetics.*

[22] Kosmatopoulos, E. B., & Christodoulou, M. A. (1992f). Control of Unknown Nonlinear Systems via ECE Neural Networks. Submitted to *IEEE Transactions on Automatic Control.*

[23] Kosmatopoulos, E. B., Ioannou, P. A., & Christodoulou, M. A. (1992a). Identification of nonlinear systems using new dynamic neural network structures. Submitted to *IEEE Conference on Decision and Control 1992.*

[24] Kosmatopoulos, E. B., Polycarpou, M. M., Christodoulou, M. A., & Ioannou, P. A. (1992b) High-order neural network structures for identification of dynamical systems. Submitted to *IEEE Transactions on Neural Networks.*

[25] Kosmatopoulos, E. B., Polycarpou, M. M., Christodoulou, M. A., & Ioannou, P. A. (1992c). Learning via energy coordinates equivalence. *In preparation.*

[26] Miller, M.I., Roysam, B., Smith, K.R., & O'Sullivan, J.A. (1991). Representing and computing regular languages on massivelly parallel networks. *IEEE Transactions on Neural Networks*, **2**, 56-72.

THE FUNCTIONAL INTRICACY OF NEURAL NETWORKS A MATHEMATICAL STUDY

Starkermann Rudolf, Dr. sc. techn., DSc., PhD
Grabemattweg 14
CH-4553 Niederrohrdorf, Switzerland

ABSTRACT

The motive of the essay is to provide awareness of the **functional** complexity of our life-structure and to emphasize the incomprehensibility of this polymorphic mightiness. The consideration is based on the assumption that neurons with axons, dendrites, and synapses form closed functional loops and that axons spread via multiple dendrits and synapses to other neurons, forming in this way highly entangled, constantly operating networks. Two different **generalized** patterns of neural structures are investigated. The generalization is needed for the purpose of mathematical formulization. In one form of interaction each axon spreads to all other neurons; in the second form there are two bilateral paths of interaction between individual neuron-loops. In both cases a mathematical formula allows to illustrate the doubtless limit of perception of functional comportment of networks, although the investigation hereinconsiders only the architectural structure. With an illustrated examples it is referred to the similarity of the structure of the brain of mammals and the structure of the information networks of technical multi-controlled installations. Thus, it might seem that the basic structure of handling information is universal in nature.

INTRODUCTION

Panta rhei (Heraklit von Ephesus): Everything is in constant motion. - Life is automatic, it is self-sustained, self-controlling, and in a continuous flow. If this were not the case, life could not exist and proceed. Self-control is achieved by feedback linkages. It is the human's endeavour to know how matter functions. But unless the behaviour of a piece of matter can be described mathematically or measured physically with sufficient accuracy, the matter's behaviour is not **really known**. In most cases only a narrative prospect of what it does is expressed.

What is meant by **really known**? To **fire a rocket** is a social, or political, or military term. To **calculate the trajectory of a rocket, including the initial conditions of its path and calculating the time and the location when it hits the ground** is knowing what the rocket does in every instant after it was fired. - To look after the **heating of a room** is a social or ethical term. To **calculate the temperature-time function** is a mathematical-physical term. It means knowing the pattern of the temperature at every instant whilst heating or cooling of the room is performed. Such knowing reveals itself by solving the set of interconnected differential equations which describe the functional behaviour of all individual parts of a composite, including all initial conditions and all signals which enter a system and disturb its behaviour. Then and only then is the system's comportment **really** known.

With the assumption in mind that **knowing** is the time-functional behaviour of a system in the broadest and most accurate sense, the essay wants to emphasize that **the functioning of the brain of a living being cannot be understood by a brain of a living being**. Acknowledging that the **single** feedback loop is already an extremely complex mathematical and psycho-social concept defying thorough understanding and interpretation, it is demonstrated how tremendously the functional complexity grows as the number of interacting loops and the number of pathways in the interaction network increase. The **number of loops** which can be traced within a system shall be called its **complexity.** Thus, the complexity of a single loop is **one**.

STRUCTURE I

It was found that there is an analogy between the

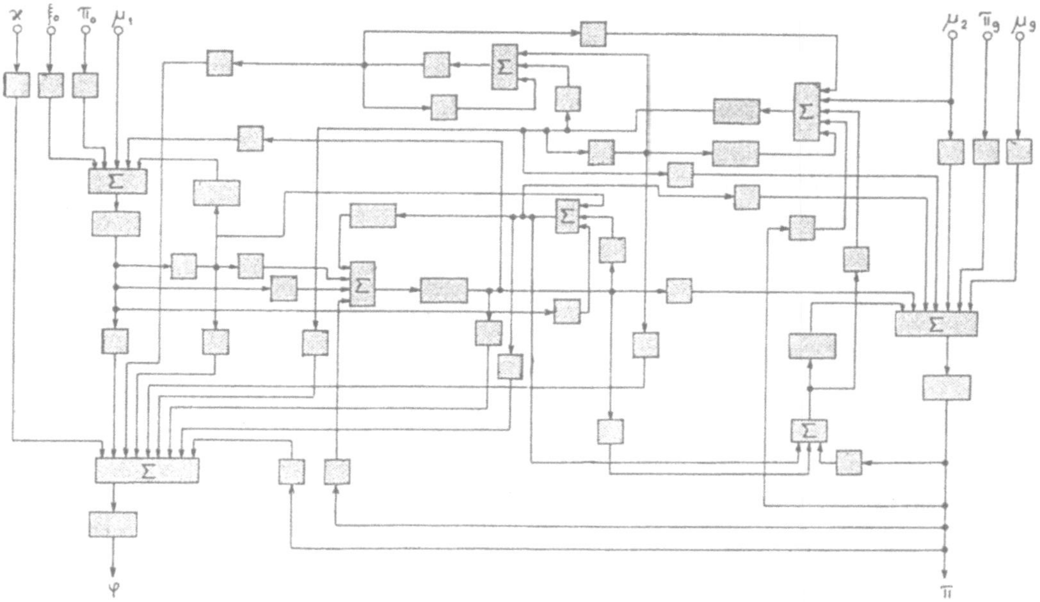

Fig. 1: Structure of an industrial turbine plant.

μ_1, μ_2 = Process input variables, the output variables of the system of controllers

φ, π, = Process output variables, feedback signals to the contollers

κ, β_0, π_0, π_g, μ_g = Disturbance variables, they penetrate the system at random

structure of technical automatic control systems and the architectural structure of the brain of mammals. **Fig. 1** shows the structure of a turbine of which two variables are to be controlled automatically, the frequency φ and the extraction pressure π (The differential equations to structure the system were linearized). The control system is not shown in Fig. 1, only the system to be controlled, the turbine, is displayed.

It can be seen that information signals are firstly transformed (shaded blocks) and then collected in a summing point Σ. After their summation the one outlet signal becomes transformed again and then splits into different channels of the same signal. These equal signals become transformed anew and then travel to different other summing points. The repetitive element of the structure **Fig. 1** is shown with **Fig. 2**.

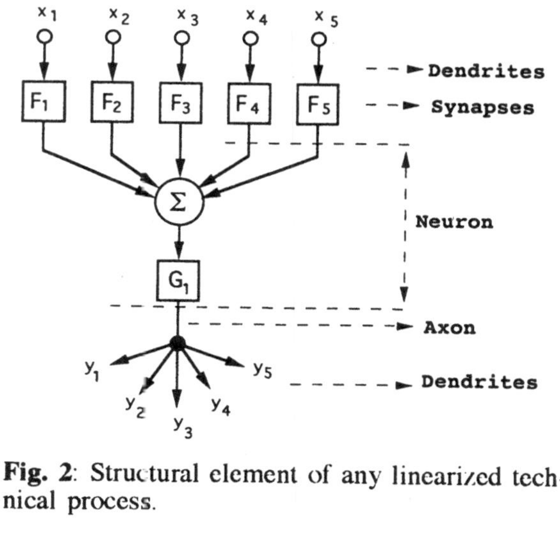

Fig. 2: Structural element of any linearized technical process.

The analog structure of the brain is given with **Fig. 3**, the neuron N with its dendrites D, the synapses S, and the axon A as the output. The average number of inputs D into a neuron is said to be 1000, but there shall be neurons with up to 100 000 inputs. **Fig. 4** shows part of a brain of four neurons. One loop is heavily traced, the loop

N_1-N_2. Such a loop has the character of a **feed-back device**.

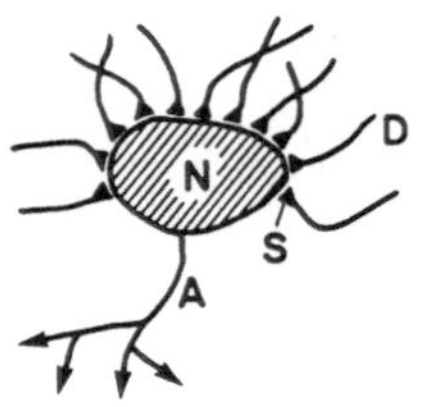

Fig. 3: The structural element of the mammal's brain.

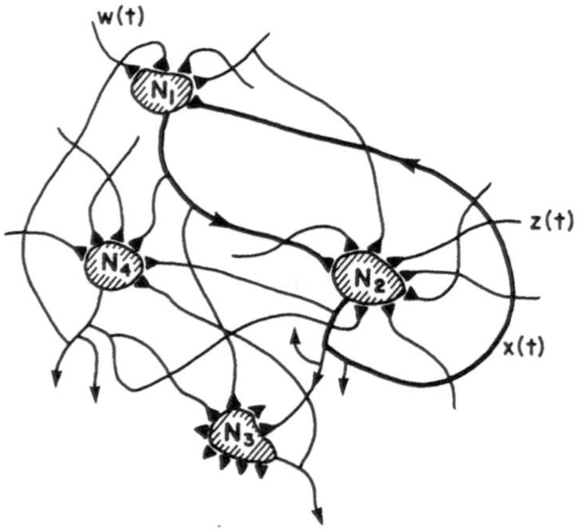

Fig. 4: The net–work of 4 elements of **Fig. 3**.

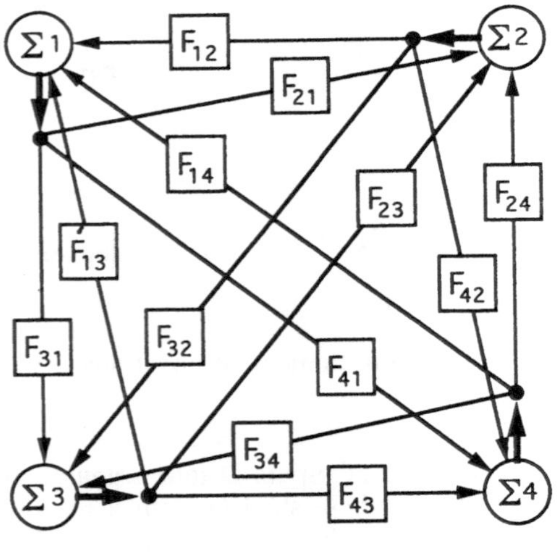

Fig. 5: The schematic brain of four neurons.

Now, in counting the possible loops to find the magnitude of complexity, a brain of four neurons and one of six neurons are investigated.

In **Fig. 5** the summing points Σ1, Σ2, Σ3 and Σ4 signify the neurons of a four neuron brain. The structure contains the following 20 loops:

1. $\Sigma 1-F_{21}-\Sigma 2-F_{12}-\Sigma 1$
2. $\Sigma 1-F_{21}-\Sigma 2-F_{32}-\Sigma 3-F_{13}-\Sigma 1$
3. $\Sigma 1-F_{21}-\Sigma 2-F_{32}-\Sigma 3-F_{43}-\Sigma 4-F_{14}-\Sigma 1$
4. $\Sigma 1-F_{21}-\Sigma 2-F_{42}-\Sigma 4-F_{14}-\Sigma 1$
5. $\Sigma 1-F_{21}-\Sigma 2-F_{42}-\Sigma 4-F_{34}-\Sigma 3-F_{13}-\Sigma 1$
6. $\Sigma 1-F_{31}-\Sigma 3-F_{13}-\Sigma 1$
7. $\Sigma 1-F_{31}-\Sigma 3-F_{23}-\Sigma 2-F_{12}-\Sigma 1$
8. $\Sigma 1-F_{31}-\Sigma 3-F_{23}-\Sigma 2-F_{42}-\Sigma 4-F_{14}-\Sigma 1$
9. $\Sigma 1-F_{31}-\Sigma 3-F_{43}-\Sigma 4-F_{14}-\Sigma 1$
10. $\Sigma 1-F_{31}-\Sigma 3-F_{43}-\Sigma 4-F_{24}-\Sigma 2-F_{12}-\Sigma 1$
11. $\Sigma 1-F_{41}-\Sigma 4-F_{14}-\Sigma 1$
12. $\Sigma 1-F_{41}-\Sigma 4-F_{24}-\Sigma 2-F_{12}-\Sigma 1$
13. $\Sigma 1-F_{41}-\Sigma 4-F_{24}-\Sigma 2-F_{32}-\Sigma 3-F_{13}-\Sigma 1$
14. $\Sigma 1-F_{41}-\Sigma 4-F_{34}-\Sigma 3-F_{13}-\Sigma 1$
15. $\Sigma 1-F_{41}-\Sigma 4-F_{34}-\Sigma 3-F_{23}-\Sigma 2-F_{12}-\Sigma 1$
16. $\Sigma 2-F_{32}-\Sigma 3-F_{23}-\Sigma 2$
17. $\Sigma 2-F_{32}-\Sigma 3-F_{43}-\Sigma 4-F_{24}-\Sigma 2$
18. $\Sigma 2-F_{42}-\Sigma 4-F_{24}-\Sigma 2$
19. $\Sigma 2-F_{42}-\Sigma 4-F_{34}-\Sigma 3-F_{23}-\Sigma 2$
20. $\Sigma 3-F_{43}-\Sigma 4-F_{34}-\Sigma 3$

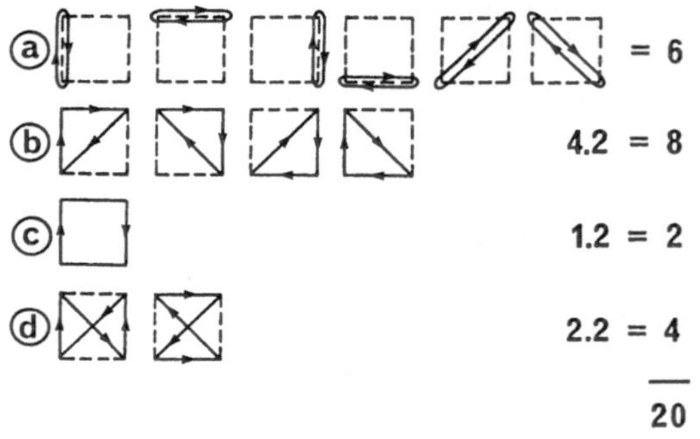

Fig. 6: The possible twenty loops of **Fig. 5**.

They are illustrated in **Fig. 6**. Thus, the complexity of this four neuron brain is **20**.

It has to be noticed that in an arrangement of 3 or more neurons there are two possible loops which can be traced, one going around in the clockwise, the other in the counterclockwise direction. They both have to be counted because they are built with different transfer functions (different dendrites).

Fig. 7 lists all the possible 409 loops which can be traced in a six-neuron "brain". The potential of complexit is already on a level of incomprehensibility. The complexity is **409**.

The formula to calculate the number A_n of all possible loops in a brain of n neurons in generalized interconnectedness – what is indeed hypothetical and in reality a chaos – is:

$$A_n = \sum_{i=1}^{n-1} \frac{n!\,(n-i)!}{(i-1)!\,(n+1-i)!},\ (n \geq 2) =$$

$$\sum_{i=1}^{n-1} \binom{n}{i-1}(n-1)!\ ,\ (n \geq 2) \qquad (1)$$

The derivation of the formula is in [1].

Table I is the list of all loops A_n which can be traced in a system of generally interconnected neurons n from 2 to 18.

The study of the **functional** behaviour of one loop (n = 2) in the technical area requires the handling of complex numbers, of system dynamics, of physics, and of linear algebra.

A short calculation results in the following astonishment: Taking a brain of 30 neurons which are generaly interconnected, i.e., each output axon spreads via dendrites to all other 29 neurons, then there can be traced so many loops that if per loop a square millimeter for a carpet were used, the carpet would cover the surface of the earth 50 000 000 000 times. If one layer had a heigh of 0.1 mm, the 5×10^{10} layers would form a stratum of 5000 km thickness around the globe. And an ant has already 60 000 neurons in its tiny brain, whereas the human being possesses about 1 000 000 000 000 000 such cells!

Table I: Number of possible loops, A_n, in a generalized interconnectedness of n neurons from 2 to 18.

n	A_n
2	1
3	5
4	20
5	84
6	409
7	2 365
8	16 064
9	125 664
10	1 112 073
11	10 976 173
12	119 481 284
13	1 421 542 628
14	18 384 340 113
15	255 323 504 917
16	3 809 950 976 992
17	60 683 990 530 208
18	1 027 542 662 034 897

STRUCTURE II

There is no doubt that the **generalized** structure as shown above is not real, that it means too much of a chaos, because each brain is very specifically structured.

Let's assume now that of loops of two neurons each loop has two mutual paths of communication with every other loop, as **Fig. 8** illustrates:

Each unit in **Fig. 8** has its own self control in the form of a feedback signal. The interactions S_{12} and S_{21} form a forward action, whereas the two paths, V_{12} and V_{21}, perform a mutual control of each other unit's doing, a mutual observation. V_{12} and V_{21} form a crossed feedback constellation.

Fig. 9 illustrates the net of four such units, simplified as a line diagram. The entanglement becomes obvious.

The formula to calculate all possible loop-tracks B_n of such structural agglomerations is:

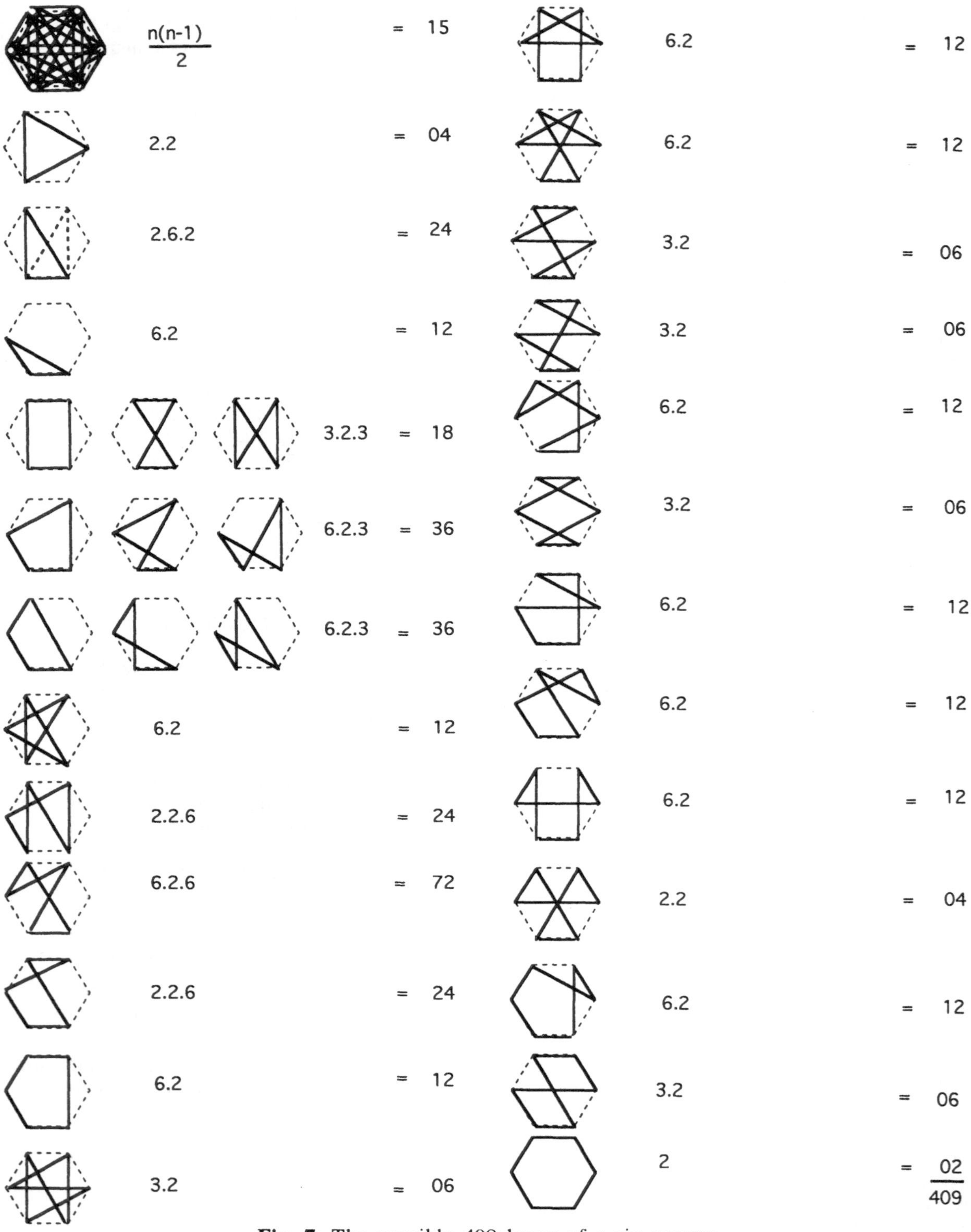

Fig. 7: The possible 409 loops of a six–neuron "brain".

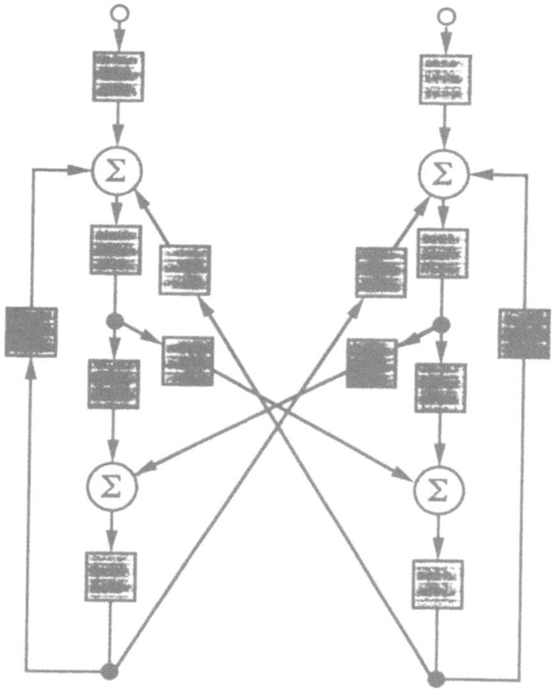

Fig. 8: Structure of two neuron clusters, called **units**, with two pairs of mutual interaction.

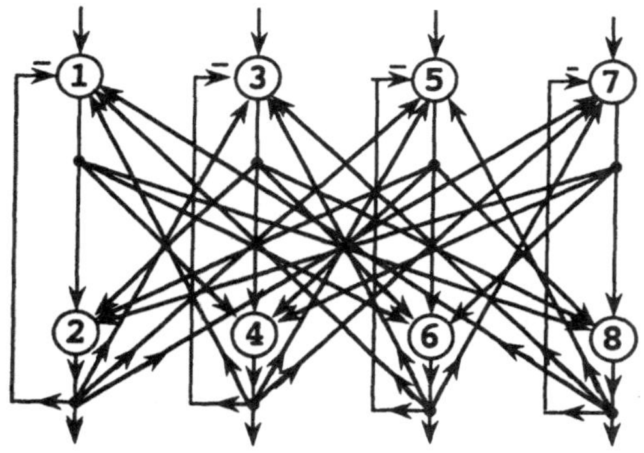

Fig. 9: The generalized structure of four units which have mutual interaction and mutual observation.

$$B_n = \sum_{i=1}^{n} \left\{ \frac{\left[\frac{n!}{(n-i)!} \right]^2}{i} \right\} = (n!)^2 \sum_{i=1}^{n} \frac{1}{[(n-i)!]^2 i} = \sum_{i=1}^{n} \frac{\left[\binom{n}{i} i! \right]^2}{i} \qquad (2)$$

Table II gives the calculation of B_n up to 15 units.

Table II: Number of possible loops, B_n, of the Structure II.

n	L(CD)
1	1
2	6
3	39
4	424
5	7 905
6	227 766
7	9 324 511
8	512 970 144
9	36 452 217 969
10	3 247 721 402 870
11	354 391 641 042 791
12	46 474 986 465 907 176
13	7 210 847 466 760 853 809
14	1 306 387 103 146 257 800 774
15	273 269 900 360 634 449 732 895

The formula (2) gives numbers which are about the square of the numbers of formula (1).

CONCLUSION

Once the knowledge of the **functioning** of neurons is available, a further step in the investigation of the behaviour of neural networks will be possible. But it has to be mentioned that the understanding of one single loop in its functional comportment demands substantial physical and mathematical cognizance. Two neurons can form a feedback loop. But perhaps there are already feedback loops within the neuron. Perhaps one single neuron is already a world in itself, and the brain is like a galaxy of worlds. Therefore, efforts to become able to understand our own identity are highly in vain. Even whilst knowledge increases, little by little, the social complexity, which sends its signals in every brain and changes the content of neurons, grows in an explosive way. Ironically already established knowledge is continuously lost through death of individuals and must be reestablished be the young. To finish: When it comes to investigate our own brain, the statement may be justified: **An identity cannot perceive itself**. My estimation is that the brain which can perceive the behaviour of another, smaller brain has to be at least on a complexity level which is 10^9 higher than the brain to be investigated.

Reference

[1] Starkermann R., "*The Structural Complexity of Continuously Functioning Multi-Goal Systems*", Proceedings of the IVth International Conference on Mathematical Modelling, Zurich, Switzerland, Aug. 15–17, **1983**, pp. 97–102. Pergamon Press, ISBM 0–08–030156–8.

Evolving Neural Feedforward Networks

Heinrich Braun

Institut für Logik, Komplexität
und Deduktionssysteme
Universität Karlsruhe
braun@ira.uka.de

Joachim Weisbrod

Institut für Programmstrukturen
und Datenorganisation
Universität Karlsruhe
weisbrod@ira.uka.de

Abstract

For many practical problem domains the use of neural networks has led to very satisfactory results. Nevertheless the choice of an appropriate, problem specific network architecture still remains a very poorly understood task. Given an actual problem, one can choose a few different architectures, train the chosen architectures a few times and finally select the architecture with the best behaviour. But, of course, there may exist totally different and much more suited topologies. In this paper we present a genetic algorithm driven network generator that evolves neural feedforward network architectures for specific problems. Our system ENZO[1] optimizes both the network topology and the connection weights at the same time, thereby saving an order of magnitude in necessary learning time. Together with our new concept to solve the crucial neural network problem of permuted internal representations this approach provides an efficient and successfull crossover operator. This makes ENZO very appropriate to manage the large networks needed in application oriented domains. In experiments with three different applications our system generated very successful networks. The generated topologies possess distinct improvements referring to network size, learning time, and generalization ability.

1 Introduction

The imitation of biological mechanisms works well in the case of both neural networks and genetic algorithms. Therefore the idea to follow the biological paradigm and to optimize neural networks (NN) using genetic algorithms (GA) is very obvious. Actually, the optimization of network topologies according to some explicit performance criteria seems to be a task that is predestinated to the use of evolutionary methods.

In our work we show that it is possible to generate appropriate problem specific feedforward network architectures by simultaneously optimizing both network topology and connection weights using genetic algorithms. Our network generator ENZO represents such a system, that has been successfully employed to accomplish the needs of different tasks. Using ENZO we tested our algorithm with three different applications: (1) a pattern recognition problem, (2) the emulation of the kinematics of a backing up truck derived from [1], and (3) the endgame of the two-player game *Nine Men's Morris* [2], employing a (23-12-10-10)–topology, a (3-20-3)–topology and a (120-60-20-2-1)–topology, respectively. So at least the last one of them is much more complex than problems considered in publications on comparable topics ([3], [4], [5], [6], [7], [8], [9], [10], [11], [12], [13], [14], [15]).

Arguing with Miller et.al. [9] the search space of possible network topologies is infinitely large, not differentiable, complex, noisy, deceptive and multimodal. These attributes make random, enumerative, gradient descent or heuristic search methods unpracticable. Genetic algorithms, however, represent a search method that is able to manage the demands of the examined search space [16]. Accordingly, there has been a lot of work concerning the use of genetic methods in order to evolve problem specific network architectures.

On the other hand there have been attempts to achieve robuster learning techniques by using genetic methods ([12], [13]), the most promising of them be-

[1] Evolutiver Netzwerk–Optimierer

```
while evolution do
(1)  select one rsp. two parents               selection
(2)  generate one offspring                    mutation rsp. crossover
(3)  evaluate offspring                        learning
                                               & evaluation
(4)  insert offspring in population            survival of the fittest
     and delete the last population element
```

Figure 1: Genetic algorithm skeleton

ing combinations of standard backpropagation (BP) and GA ([3], [14]). But from our point of view there is no reason to strictly separate topology and weight optimization. Our network generator ENZO successfully hybridizes both optimization processes, additionally establishing powerful mechanisms to both improve the optimization process and to save learning time.

Genetic topology optimization methods can be divided into two classes relating to their phenotype–genotype mapping: there are *strong* and *weak* representations. In strong representation schemes each gene of the genotype's genstring is interpreted as an individual connection between two units of the represented network. So the length of the genstring is equivalent to the number of potential connections allowed by the represented architecture.

In weak representation schemes the genes correspond to more abstract network properties. Examples for such weak encodings can be found in [6], [7], or [15]. We agree with Miller et.al. [9], that weak schemes may be useful for 'capturing the architectural regularities of large networks rather efficiently'. But their application also requires a much more detailed knowledge about both genetic and neural mechanisms. For this reason in our work we preferred a strong encoding.

Interesting explorations using strong representation schemes are described in [8], [9], or [14], for instance. But in all of these papers the resulting algorithms are only evaluated with very 'small' applications like the XOR-problem or the 2-bit-adder and it seems no trivial conclusion to generalize the results of these experiments to real-world applications.

2 Our approach

2.1 Basic algorithm

Our main design decisions were influenced by the desire to create a network generator, that was both easy to control and able to handle 'large', real-world applications. For this reason we selected a strong representation scheme, meaning that every gene of the genotype relates to exactly one connection of the represented network. Therefore, the set of possible connections is fixed and the genetic algorithm searches for an optimal topology using a subset of these connections. However, a gene of our genstring has to encode more than just the two states ⟨*existing & learnable*⟩ and ⟨*not existing*⟩. Especially for applications with complex networks like the *Nine Men's Morris* example mentioned above it is often necessary or at least profitable to fix some network properties a priori. Examples for such properties are fixed weights (⟨*existing & not learnable*⟩) or connections, that are linked together in order to have the same weights (⟨ *existing & linked to* ...⟩). Our system was to be able to consider a priori arrangements like that.

ENZO uses the same scheme as already proposed by Braun ([17], [18]) for finding the optimal solution of large travelling salesman problems (see figure 1).

Our algorithm briefly works as follows (see figure 2): Given the population size S, the initial connection density P_I, and the specification of the 'maximal' topology, ENZO generates a start population pop of S different networks, each of them using about $100 \cdot P_I\%$ of the total number of connections allowed by the specified network architecture: For every initial network, each of its potential connections is established with the given probability P_I. These nets are trained, evaluated and sorted according to their determined fitness values. These fitness values can incorporate any design criteria judged important for the given network application. Arbitrary linear combinations of interesting optimization criteria (e.g. successful training, generalization ability, or network size) may be applied.

Now ENZO starts to create offsprings using crossover and/or mutation. With a polynomial bias α

```
PROC ENZO-algorithm;
pop:=Generate_Start_Population(S, P_I, maximal topology);
repeat
    net1:=Selection(pop,α);
    offspring:=copy(net1);
    if (crossover requested) then
        net2:=Selection(pop,α);
        offspring:=Crossover(net1,net2,P_C,F_BP);
    fi
    if (mutation requested) then
        Mutation(offspring,P_M,F_BP);
    fi
    Training(offspring);
    Evaluation(offspring);
    Insertion(offspring,pop);
until (best element satisfies);
```

Figure 2: ENZO — basic algorithm

preferring population individuals with a higher ranking ENZO selects one or two parent networks. Considering the selected network(s) our network generator creates one offspring.

To recombine two parent networks net_1 and net_2 our crossover operator checks for each potential connection, whether this connection is used by the two parents. If the connection exists twice, the emerging offspring receives this connection, too. If the connection is only used by one parent net, the offspring gets a chance P_C to obtain this connection. Mutation takes place by changing the state of each potential connection with a given probability P_M.

At last the new network is evaluated and inserted into the population according to its fitness value, whereby removing the population element with lowest fitness. This means, ENZO does not consider generations, i.e. individuals with high fitness values may live very long.

The explanations above describe only the fundamental skeleton underlying our algorithm. In the following section we will take a closer look at some of the decisive components embedded into this skeleton.

2.2 Crucial mechanisms

The crucial obstacle to efficient evolution driven network generation are *permuted internal representations*. That is, in order to solve the given task two successfully trained networks may extract the same features and use the same internal representations, but distribute these internal representations in a totally different way among their hidden neurons — the contributions of the hidden neurons to the overall solution may be internally permuted. Therefore this problem is also referred to as the phenomenon of different *structural mappings* coding the same *functional mapping*.

This may cause significant problems to the crossover operator, because two parents successfully solving the given task with almost identical functional mappings may use totally different structural mappings. Applying the crossover operator to such parents will create an offspring with partly doubled and partly missing internal representations. That means, this offspring is unlikely to achieve good results.

We encounter the danger of generating poor offsprings when recombining such parents by introducing *connection specific distance coefficients*. In our network specification we assign an additional attribute L_{ij} to each possible connection C_{ij} between unit j and unit i representing the connection's length. These coefficients may be induced by an imaginary layout of the network on the 2–dimensional plane, for instance.

Considering these connection lengths L_{ij} our system successfully tries to prevent permuted internal representations by preferring for each functional mapping the structural mapping with the shortest amount of connection lengths. Every decision concerning the insertion or deletion of the actual connection C_{ij} is influenced by its length attribute L_{ij}, i.e. every possible connection obeys its own special probabilities $P_{M_{ij}} = f(\mathrm{P_M}, L_{ij})$ and $P_{C_{ij}} = f(P_C, L_{ij})$.

Making long connections less probable than short ones is both biologically plausible and sensible under hardware aspects.

The second important mechanism introduced is what we call *reduced weight transmission*. This mechanism justifies our former statement of ENZO representing a hybrid system. ENZO combines genetic topology optimization with genetic learning.

Especially in application oriented domains with large networks computation time becomes the main problem. In our algorithm offsprings not only receive topological properties from their parents, but also knowledge. This is achieved by the following mechanism: If an offspring's connection C_{ij} is established, its weight W_{ij} receives a fraction $F_{BP} \cdot W'_{ij}$ of the relating parent weight W'_{ij} instead of a random value[2]. Therefore a new offspring's learning process does not start somewhere in weight space, but most likely in the neighbourhood of the expected optimum. This approach does not only save learning time, but also decreases the danger of getting stuck in a poor local optimum.

3 Experimental results

Analyzing experiments with three different applications we were able to justify our algorithm including its particular and crucial ideas. The complete results can be found in [19].

(1) Digit recognition

First we examined a digit recognition problem described in [20].

In order to classify seven distorted sets of the trained digits a (23-12-10-10)–topology was used. While in [20] no perfect network could be found[3], ENZO evolved lots of them. As optimization criterion we simply used the number of false classifications with the seven test sets, i.e. 70 distorted digits. To compare ENZO's performance with standard backpropagation we determined the quotient of necessary learning epochs divided by the number of obtained perfect nets. For standard BP we found (examining 6,000 networks):

$$\frac{6,000\ [nets] \cdot 220\ [\frac{epochs}{net}]}{21\ [perfect\ nets]} \approx 63,000\ [\frac{epochs}{perfect\ net}]$$

[2] F_{BP} is another user specified parameter in [0, 1].

[3] *Perfect* means no false classification with the given test sets.

To get comparable data using our network generator we started 6 runs, each of them having a population size of 25 networks and computing 1,000 offsprings:

$$\frac{(6 \cdot 25\ [nets] \cdot 200\ [\frac{epochs}{initial\ net}])}{141\ [perfect\ nets]} \qquad \text{(initialization)}$$

$$+ \frac{(6 \cdot 1,000\ [nets] \cdot 110\ [\frac{epochs}{offspring}])}{141\ [perfect\ nets]} \qquad \text{(evolution)}$$

$$\approx \quad 5,000\ [\frac{epochs}{perfect\ net}]$$

Notice that the 141 perfect nets are derived from a total amount of 150 evolved networks, namely 6 runs with a population size of 25 each.

The formulas above also show the success of our concept of *weight transmission* from ancestors to offsprings by evidently reducing the necessary training epochs (from around 200 for initial population elements down to 110 for offsprings).

The importance of our *distance coefficients* when using crossover could be significantly supported by experiments neglecting this distance criterion. Table 1 shows relating results[4]: compared to our normal crossover we obtained only 22 perfect nets instead of 119, reaching a mean population fitness of 1.86 instead of 0.25. The last column of table 1 indicates that without the connection specific L_{ij}'s more then half of the created offsprings could not even be successfully trained, i.e. they did not reach the given training criterion within the allowed 1,000 training epochs.

Table 2 contains additional results related to our concept of *weight transmission*[5] Different values for the weight reduction factor F_{BP} show, that ENZO can transmit both too much and too few knowledge from parents to offsprings. We found, that the optimum value for the parameter F_{BP} depends very much on the complexity of the given application.

Without *weight transmission*, i.e. $F_{BP} = 0$, ENZO evolved no perfect network at all.

(2) Kinematics of the truck backer–upper

Our second application was derived from [1]. A (3-20-3)–topology was used to emulate the kinematics

[4] In each case 6 runs creating 300 offsprings with a population size $S = 25$.

[5] Again, in each case 6 runs creating 300 offsprings with a population size $S = 25$.

	mean fitness:	perfect nets:	training epochs:	successfully trained nets:
with L_{ij}	0.25	110	~ 71	100%
without L_{ij}	1.86	22	~ 600	45%

Table 1: Results with and without *connection specific distance coefficients* L_{ij} (crossover and mutation).

	mean fitness:	perfect nets:	training epochs:	inserted offsprings:
$F_{BP} = 0.5$	1.91	1	~ 70	27%
$F_{BP} = 0.7$	0.75	68	~ 110	31%
$F_{BP} = 0.9$	1.24	24	~ 85	25%

Table 2: Results for different weight transmission factors F_{BP} (just mutation without crossover).

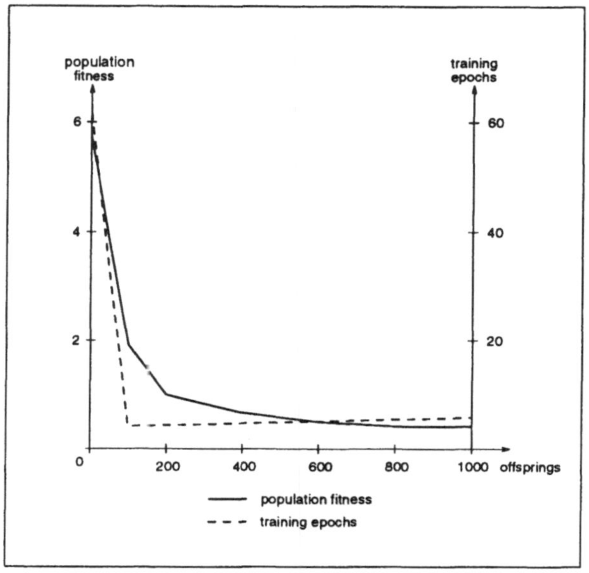

Figure 3: ENZO and 'Sokrates' (crossover and mutation).

of a backing up truck within a delimited range of situations. Despite of being very easy to learn (about 20 training epochs) this application could confirm our conclusions drawn from the digit recognition task.

Most of the experiments examined with this application were run to find good choices for the newly established parameters. Moreover, our results sustained the often published opinion that in the given domain mutation alone is a powerful mechanism to achieve satisfying performance (cf. [12], [14], or [15]).

(3) Nine Men's Morris

Finally we tested our system with a very hard problem requiring a very large network architecture. The task was to learn a scoring function for the endgame of the two-player game *Nine Men's Morris*. The so called 'Sokrates' net introduced in [21] has gained a high level performance, defeating most human opponents. In order to train the net the employed (60-30-10-1)-topology is doubled, leading to a (120-60-20-2-1)-topology. At this point ENZO's capability to consider a priori specifications determined by the user is required, because during training the two subnets have to remain identical, i.e. corresponding weights have to be linked together in order to be identically adjusted. With a training set consisting of about 90 pairs of endgame positions the network is to create a scoring function, that can generalize as good as possible onto the about 60,000 essentially different endgame configurations.

We started ENZO with the following parameter values: population size $S = 30$, initial connection density $P_I = 0.7$, crossover probability $P_C = 0.6$, mutation propability $P_M = 0.01$, relearning factor $F_{BP} = 0.7$. We let ENZO generate $1,000$ offsprings requiring about one week of computation time on a SUN 4-SLC workstation. Figure 3 shows results of one run using our crossover and mutation operators. It shows that the number of training epochs is reduced from 50 at the start to almost constantly 5 epochs in the evolution phase, as a consequence of *weight transmission*.

In order to compare differently trained 'Sokrates' nets in [2] the authors introduce the *average improvement* as a measure for the network's playing performance. Besides being used as fitness criterion this measure also allows a comparison between our best evolved net and the handcrafted original 'Sokrates' net. Table 3 shows an increase of the *average improvement* from 0.826 to 1.218, i.e. an performance enhancement of 47%. Additionally, our net used only 43% of its potential connections. At this point we have to emphasize, that the data used for evaluating the fitness of a given network architecture within ENZO's evolution phase and the data used for evaluating the evolved network's performance (as shown in table 3)

	Scoring improvement per move (1,000 moves)							Average	Moves to	
	2 =opt.	0	−2	−4	−6	−8	≤ −10	improvement	remis	loose
original 'Sokrates'	81.7%	5.4%	3.0%	2.4%	1.3%	1.6%	3.6%	0.826	0.5%	0.3%
ENZO	86.1%	4.1%	2.6%	2.1%	1.3%	1.2%	2.2%	1.218	0.2%	0.4%

Table 3: Generalization ability of the original 'Sokrates' net and our best evolved net

	Scoring improvement per move (1,000 moves)							Average	Moves to	
	2 =opt.	0	−2	−4	−6	−8	≤ −10	improvement	remis	loose
1. training	82.4%	5.0%	3.1%	1.6%	2.8%	1.6%	3.2%	0.939	0.1%	0.4%
2. training	83.2%	5.6%	2.9%	1.8%	2.2%	1.4%	2.6%	1.068	0.1%	0.4%
3. training	85.5%	2.6%	2.4%	2.2%	2.0%	1.8%	3.4%	0.971	0.1%	0.4%
4. training	83.5%	4.2%	2.5%	1.5%	2.2%	1.2%	4.6%	0.778	0.0%	0.3%
5. training	82.6%	5.8%	2.6%	1.6%	1.9%	1.0%	4.5%	0.826	0.1%	0.3%
6. training	84.9%	4.6%	2.2%	1.3%	1.7%	1.5%	3.4%	1.022	0.1%	0.3%
7. training	84.0%	5.0%	2.7%	1.1%	1.8%	1.5%	3.7%	0.922	0.1%	0.2%
8. training	81.7%	6.5%	2.4%	1.4%	2.0%	1.8%	3.9%	0.835	0.1%	0.2%
average values	*83.5%*	*4.9%*	*2.6%*	*1.5%*	*2.0%*	*1.5%*	*3.7%*	*0.920*	*0.1%*	*0.3%*

Table 4: Generalization ability of our best re–trained topology

were totally disjoint, of course.

In order to test how well suited the evolved topologies were, we reset the weights of our best performing network and re–trained the net eight times. Table 4 shows the performance properties of the eight re–trained networks. In average we get an increase of the *average improvement* from 0.826 to 0.920, i.e. an enhancement of performance of 11%. That means, using just the evolved topology (employing only 43% of the connections) distinctly improves the average performance. But on the other hand our mechanism of *weight transmission* turns out to be crucial, since we are far away from getting the performance of the best evolved net: comparing the best of the eight nets with the handcrafted original 'Sokrates' net we find an increase from 0.826 to 1.068, i.e./ an enhancement of performance of 29% instead of 47% for the best evolved net (see table 3).

This last experiment strongly confirms our expectation, that our network generator is able to evolve problem specific network topologies with evidently improved performance properties.

4 Conclusion and future directions

The essential virtues of our genetic algorithm consist (1) in the new combination of the parental properties when merging the parents' genes (crossover with *connection specific distance coefficients*) and (2) in speeding up the learning process by inheriting knowledge from the parents (*weight transmission*).

By solving the problem of permuted internal representations we can propose a successfull crossover operator and by hybridizing genetic topology optimization and genetic learning we can introduce a very efficient network generator, ENZO.

For assessing application oriented problems fast learning is crucial, because this is the most time consuming part of the genetic algorithm. In ENZO learning is speeded up by more than an order of magnitude using the *reduced weight transmission* heuristic, e.g. in application (3) shrinking the learning time on average to just 9% compared to learning from the scratch (random starting weights). Moreover, learning from the scratch means gradient descent to an 'average' local minimum whereas our weight transmission mechanism strongly biases the descent to a good local minimum. In application (1), for instance, the probability for generating a perfectly trained net is increased by factor 6 during evolution (compared to learning form the scratch). Combining both effects the average time for generating a perfect net is decreased by factor 20 in application (1) just due to the reduced weight transmission heuristic.

By examining experiments with our applications we were able to justify both our basic design decisions and some particular crucial details. With applications (1) and (2) being mainly employed to find good choices for our newly established parameters, application (3) represented a hard touchstone for the overall performance of our algorithm. On one hand there was a really 'large' network to be managed and on the other hand the network consisted of two equal sub–networks leading to specific a priori restrictions on the network specification. Our system evolved networks impressively surpassing our best handcrafted networks by 47% in performance while using only 43% connections. These results were achieved by an evolution process that required about a week of computation time (SUN 4-SLC workstation).

As already mentioned above, ENZO generates a completely specified neural network. Obviously, it is an interesting question, whether an appropriate problem specific topology is evolved. Using just the topology of the best network and training from scratch we could validate that the generated topologies possess three main advantages:

- smaller network size (e.g. application (3): generated network's size is about 43% compared to the handcrafted network)
- shorter training time (e.g. application (3): on average only 35% of the training epochs needed for the handcrafted network)
- higher generalization capability (e.g. application (3): performance is increased by 29%)

But, of course, networks with the evolved topology but re–trained weights do not reach the performance of the evolved networks due to the synergetic effect of evolving both topology and connection weights.

From our point of view with applications becoming more complex our proposed crossover operator will surpass evidently the pure use of mutation. In addition the possibility to consider a priori specifications determined by the network designer surely is essential for such real world applications.

References

[1] D. Nguyen and B. Widrow, *The truck backer-upper*, Proc. Internat. Network Conf. (Kluwer Academic Publishers, Dordrecht, 1990)

[2] H. Braun, J. Feulner, V. Ullrich, *Learning strategies for solving the problem of planning using backpropagation*, in: Proc. Fourth Intern. Conf. Neural Networks (Nimes, 1991)

[3] R.K. Belew, J. McInerney, N.N. Schraudolph, *Evolving networks: using the genetic algorithm with connectionist learning*, CSE Technical Report #CS90-174 (University of California, San Diego)

[4] D.J. Chalmers, *The evolution of learning: an experiment in genetic connectionism*, in: Proc. of the 1990 Connectionist Models Summer School (Morgan-Kaufmann, San Mateo, CA, 1990)

[5] D.B. Fogel, L.J. Fogel, V.W. Porto, *Evolving neural networks*, in: Biological Cybernetics 63 (Springer, Berlin, 1990)

[6] P.J.B. Hancock, L.S. Smith, *GANNET: Genetic design of a neural net for face recognition*, in: Parallel Problem Solving from Nature (Springer, Berlin, 1990)

[7] S. Harp, T. Samad, A. Guha, *Towards the genetic synthesis of neural networks*, in: Proc. Third Internat. Conf. Genetic Algorithms (Morgan Kaufmann, San Mateo, CA, 1989)

[8] K.U. Hoeffgen, H.P. Siemon, A. Ultsch, *Genetic improvements of feedforward nets for approximating functions*, in: Parallel Problem Solving from Nature (Springer, Berlin, 1990)

[9] G. Miller, P. Todd, S. Hedge, *Designing neural networks using genetic algorithms*, in: Proc. Third Internat. Conf. Genetic Algorithms (Morgan Kaufmann, San Mateo, CA, 1989)

[10] H. Muehlenbein, *Limitations of multi-layer perceptron networks – steps towards genetic neural networks*, in: Parallel Computing 14 (1990)

[11] W. Schiffmann, M. Joost, R. Werner, *Performance evaluation of evolutionary created neural network topologies*, in: Parallel Problem Solving from Nature (Springer, Berlin, 1990)

[12] M. Scholz, *A learning strategy for neural networks based on a modified evolutionary strategy* , in: Parallel Problem Solving from Nature (Springer, Berlin, 1990)

[13] D. Whitley, T. Hanson, *Optimizing neural networks using faster, more accurate genetic search*, in: Proc. Third Internat. Conf. Genetic Algorithms (Morgan Kaufmann, San Mateo, CA, 1989)

[14] D. Whitley, T. Starkweather, C. Bogart, *Genetic algorithms and neural networks: optimizing connections and connectivity*, in: Parallel Computing 14 (1990)

[15] D. Whitley, S. Dominic, R. Das, *Genetic reinforcement learning with multilayer neural networks*, in: Proc. Fourth Internat. Conf. Genetic Algorithms (Morgan Kaufmann, San Mateo, CA, 1991)

[16] J.H. Holland, *Adaption in natural and artificial systems*, (Ann Arbor: University of Michigan Press, 1975)

[17] H. Braun, *Massiv parallele Algorithmen fuer kombinatorische Optimierungsprobleme und ihre Implementierung auf einem Parallelrechner*, Dissertation TH Karlsruhe, Fakultaet fuer Informatik, 1990

[18] H. Braun, *On solving traveling salesman problems by genetic algor ithms*, in: Parallel Problem Solving from Nature, LNCS 496 (Berlin, 1991)

[19] J. Weisbrod, *Einsatz Genetischer Algorithmen zur Optimierung der Topologie mehrschichtiger Feedforward-Netzwerke*, Diplomarbeit TH Karlsruhe, Fakultät für Informatik (1992)

[20] J. Weisbrod, *Untersuchung der Einsatzmöglichkeiten Neuronaler Netze zur visuellen Mustererkennung gestanzter Ziffern*, Studienarbeit TH Karlsruhe, Fakultät für Elektrotechnik (1989)

[21] V. Ullrich, *Erlernen von Spielstrategien für Mühle durch Neuronale Netze*, Diplomarbeit TH Karlsruhe, Fakultät für Informatik (1991)

An Example of Neural Code: Neural Trees Implemented by LRAAMs

Alessandro SPERDUTI
(sperduti@icsi.berkeley.edu)
ICSI, 1947 Center St., Suite 600,
Berkeley, CA 94704, USA

Antonina STARITA
(starita@di.unipi.it)
Dipartimento di Informatica, Corso Italia 40,
56125 Pisa, Italy

Abstract

In this paper we discuss a general method which allows to implement a Neural Tree in a high-order recurrent network, the Executor, using an extension of the RAAM model, the Labeling RAAM model (LRAAM). Neural Trees and LRAAMs are briefly reviewed and the Executor defined. A Neural Tree is encoded by a LRAAM, and the decoding part of the LRAAM used to control the dynamics of the Executor. The main aspect of this kind of technique is that the weights of the LRAAM can be considered as a *neural code* implementing the Neural Tree on the Executor. An example of the method is presented for the 8 bits parity problem.

1 Introduction

Recently, several papers have addressed the question of how complex data structures, such as lists, stacks, trees and so on, could be implemented by a neural network. One proposal was the Recurrent Auto-Associative Memory (RAAM) of Pollack [11]. It allows to represent variable-sized recursive data structures, such as trees and lists, in a fixed-width patterns. The model uses backpropagation and an encoder network to develop compressed representations of the components of the structures to be represented. The same process is then repeated, in a recursive fashion, over the compressed representations of the components, until a single distributed pattern of activation is obtained for each structure. An extention of the RAAM, the Labeling RAAM (LRAAM), has been proposed in [19]. A LRAAM allows to store a label for each component of the structure to be represented, so to generate encoded representations of labeled graphs.

In this paper we show how LRAAMs and multiplicative connections [4] can be used to develop a general method allowing to implement a Neural Tree (NT) in a high-order recurrent network. This seems particularly important because Neural Trees have been recently proposed as a fast learning method in classification tasks. A NT is composed by a set of perceptrons (or neural networks) whose output is used to route the input pattern to the perceptron (or network) which is able to classify it, according to the tree structure generated by learning. Usually, the tree structure is stored and managed using classical symbolic data structures and programming. Therefore, NTs are decision trees [1] where the splitting of the data for each node, i.e., the classification of a pattern according to some feature, is performed by a perceptron [18] or a more complex neural network [14].

In Section 2 we review the definition of NT and show how to encode a NT in a LRAAM, i.e., to build up an Executor Network. The structure of the Executor Network is also discussed. An example of neural code for the 8 bits parity problem is given in Section 3.

2 Neural Trees Implemented by LRAAMs

2.1 Labeling RAAM

Labeling RAAM (LRAAM) differs from the RAAM because allows to encode labeled structures. The general structure of the encoder network for a LRAAM is shown in Figure 1.

The idea is to obtain a compressed representation (hidden layer activation) of a node of a labeled graph by allocating a part of the input (output) of an Encoder network to represent the label and the rest to represent one or more pointers. The compressed representation is then used as pointer to the node. In order to allows the use of such compressed representations, the part of the input (output) layer which represents a pointer must be of the same dimension

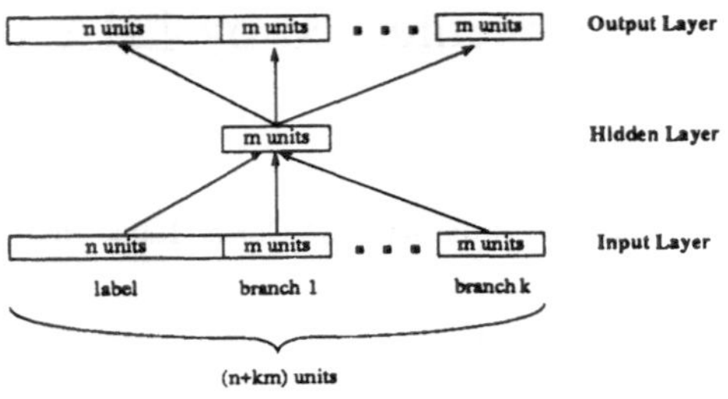

Figure 1: The encoder network for a general LRAAM.

of the hidden layer. A part of the label is used to indicate if the pointers are void or not.

Labeled graphs can be easily encoded using a LRAAM. It suffice to represent each node of the graph as a record with one field for the label and one field for each pointer to a connected node. The pointers need to be only logical pointers, since their actual values will be the patterns of hidden activation of the encoder network. At the beginning of learning their values are set at random. A graph will be represented by a list of such records, and such list will be the training set for the LRAAM.

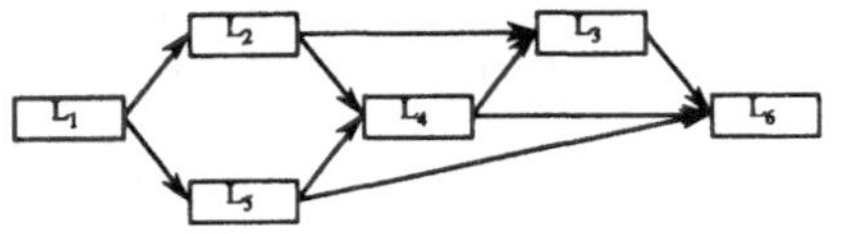

Figure 2: Example of graph.

For example, the graph shown in Figure 2 would be encoded as follows:

input		hidden		output
$(L_1\ P_{n2}\ P_{n5})$	$\rightarrow$	$P_{n1}(t)$	$\rightarrow$	$(L'_1(t)\ P'_{n2}(t)\ P'_{n5}(t))$
$(L_2\ P_{n3}\ P_{n4})$	$\rightarrow$	$P_{n2}(t)$	$\rightarrow$	$(L'_2(t)\ P'_{n3}(t)\ P'_{n4}(t))$
$(L_3\ P_{n6}\ nil)$	$\rightarrow$	$P_{n3}(t)$	$\rightarrow$	$(L'_3(t)\ P'_{n6}(t)\ nil'(t))$
$(L_4\ P_{n6}\ P_{n3})$	$\rightarrow$	$P_{n4}(t)$	$\rightarrow$	$(L'_4(t)\ P'_{n6}(t)\ P'_{n3}(t))$
$(L_5\ P_{n4}\ P_{n6})$	$\rightarrow$	$P_{n5}(t)$	$\rightarrow$	$(L'_5(t)\ P'_{n4}(t)\ P'_{n6}(t))$
$(L_6\ nil\ nil)$	$\rightarrow$	$P_{n6}(t)$	$\rightarrow$	$(L'_6(t)\ nil'(t)\ nil'(t))$

where L_i and P_{ni} are respectively the label and the pointer to the i-th node and t represents the time, or epoch, of training. If the backpropagation algorithm converges to perfect learning, i.e., the total error goes to zero, it can be stated that:

$$
\begin{array}{lll}
L_1 = L'_1 & L_2 = L'_2 & L_3 = L'_3 \\
L_4 = L'_4 & L_5 = L'_5 & L_6 = L'_6 \\
P_{n2} = P'_{n2} & P_{n3} = P'_{n3} & P_{n4} = P'_{n4} \\
P_{n5} = P'_{n5} & P_{n6} = P'_{n6} &
\end{array}
$$

Once the training is complete, the patterns of activations representing pointers can be used to retrieve information. Thus, for example, if the activity of the hidden units of the encoder is clamped to P_{n4}, the output of the encoder is (L_4, P_{n6}, P_{n3}), allowing further retrieval of information by decoding P_{n6} or P_{n3}, and so on.

2.2 Neural Trees

Neural Trees have been recently proposed as a fast learning method in classification tasks. They are decision trees ([1, 17, 3]) where the splitting of the data for each node, i.e., the classification of a pattern according to some features, is performed by a perceptron ([18]) or a more complex neural network ([14, 15, 16]). After learning, each node at every level of the tree corresponds to an exclusive subset of the training data and the leaf nodes of the tree completely partition the training set. In the operative mode, the internal nodes route the input pattern to the appropriate leaf node which represents the class of it. An example of how a binary neural tree splits the training set is shown in Figure 3. On the left side of the picture, it is shown the tree structure obtained by the learning procedure. The training set is shown on the top of the root and the succesive splitting of it at each tree node is shown. The 2D patterns belong to two different classes (grey or white) and their distribution on the plane is shown graphically. On the right side of the picture, the cuts generated on the patterns space by the four perceptrons used in the tree's nodes are reported.

A sketch of the learning and classification algorithms for a binary tree is reported. Either supervised ([18, 14, 15, 16, 9]) and unsupervised ([10, 6]) splitting of the data have been proposed. In the following we will see an example of supervised method, where the splitting is performed by a perceptron. More complex splitting rules can be obtained by using a multi-layer network instead of a simple perceptron. An algorithm for trees with valence higher than two can be found in ([14]).

One advantage of the neural tree approach is that the tree structure is constructed dynamically during learning and not linked to a static structure like a standard feed-forward network. Moreover, it allows incremental learning, since subtrees can be added as

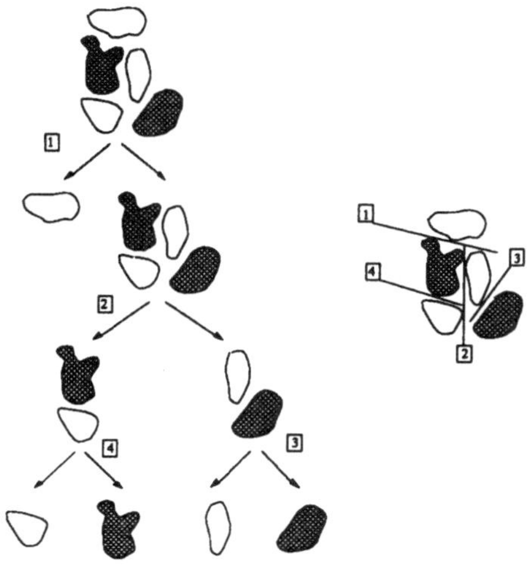

Figure 3: An example of neural tree.

well as deleted to recognize new classes of patterns or to improve the system performance.

Usually, the tree structure is stored and managed using classical symbolic data structures and programming. The joint use of multiplicative connections, *FRDs* and LRAAMs allows to implement a NT in a full neural architecture, giving rise to an example of *neural program*. The procedure to obtain this neural architecture is composed of two main steps:

1. Encode the NT in a LRAAM, where the labels are the weights of the discriminators associated to the tree nodes;
2. Load the weights of the LRAAM in the *Executor Network*.

In the following sections we discuss how to encode a NT in a LRAAM and we define the *Executor Network*.

2.3 Encoding a NT in a LRAAM

A NT can be encoded in a LRAAM in a very natural fashion. In fact, a NT is defined by the tree structure, the set of discriminators associated to each internal node of the tree and the set of labels associated to the leaves of the tree. A neural discriminator, once the type of activation functions used by it is fixed, can be defined by its set of weights. Thus, a NT can be represented by a labeled tree, where the label of each internal node is any predefined sequence of the appropriate set of weights and the label of a leaf is the label of one of the classes discriminated by the NT. This is quite the kind of structure that a LRAAM is able to encode.

The Learning Algorithm for Binary NT

Let $\{(\vec{I}_k, L_k)\}$ be the training set, where $\vec{I}_k$ is a real valued feature vector of dimension n and L_k the label of the associated class, $k = 1, \cdots, P$. Name $T(t)$ the current training set for node (t, D_t, r, l), where t is the ID of the node, D_t is the name of the discriminator associated to the node if the node is nonterminal or it is the label of a class if the node is a leaf, r the ID of the right child and l the ID of the left child.
The learning algorithm for each node t can be resumed in the following steps:

1. Set $T(root) = \{(\vec{I}_k, L_k)\}$;
2. If $T(t) = \emptyset$ then stop;
3. Train the discriminator D_t to split $T(t)$ in two sets: $T_0(t)$ and $T_1(t)$;
4. (a) If $T_0(t)$ contains only patterns of class C_j, $j \in [1, \cdots, P]$, then add the node (t_l, C_j, NIL, NIL) with training set $T(t_l) = \emptyset$, otherwise add the node $(t_l, D_{t_l}, t_{l_r}, t_{l_l})$ with training set $T(t_l) = T_0(t)$;
   (b) If $T_1(t)$ contains only patterns of class C_j, $j \in [1, \cdots, P]$, then add the node (t_r, C_j, NIL, NIL) with training set $T(t_r) = \emptyset$, otherwise add the node $(t_r, D_{t_r}, t_{r_r}, t_{r_l})$ with training set $T(t_r) = T_1(t)$;
5. repeat the same algorithm for each child node from step 2.

Some difficulties can arise from the fact that the weights of a discriminator are not binary (or bipolar) or outside the range $[-1, 1]$. This normalization problem can be solved by using linear units at the output layer of the LRAAM.

It is interesting to note that this kind of representation may optimize the number of parameters of the NT. Infact, if the dimension of the input patterns is far less than the number of nodes of the tree, linear redundancy in the weight space is present. This redundancy is automatically reduced by the LRAAM, since each pointer is a compressed version of the weights of a discriminator.

An alternative representation of the NT, which stresses this last aspect and solves the normaliza-

The Classification Algorithm for Binary NT

The class of a pattern I_k can be established by the following algorithm:

1. Set $t = root$;
2. If the node t is a leaf, classify I_k by the label $L_k = D_t$ and stop, otherwise:
   (a) If $D_t(I_k) = 0$ then set $t = t_l$;
   (b) If $D_t(I_k) = 1$ then set $t = t_r$;
3. go to step 2.

---

tion problem as well, can be obtained by compressing apart the weights of the discriminators and then using them as labels. In this way, however, it is not easy to control the precision of the encoding, since small errors in the representation of the labels in the LRAAM can lead to more serious errors in the results obtained by decoding them. In some cases, where the precision of the set of weights in not very important, this procedure can result to be more advantageous than the direct representation because of the reduced dimension of the labels.

2.4 The Executor Network

Once the NT has been codified in a LRAAM, the problem is how to implement the search on the tree, given a pattern in input. The idea is to integrate the decoding part of a general LRAAM network and a discriminator in the same network. In order to do that, multiplicative connections are used.

Multiplicative connections (or sigma-pi connections) were first proposed by Feldman and Ballard [4] and occasionally used by various authors [13, 7, 21, 12, 8, 5, 20]. Durbin and Rumelhart ([2]) have recently used units with multiplicative connections in an extension of the back-propagation algorithm.

Unlike the classic thresholded linear unit, the output of a sigma-pi unit (i.e., a unit using sigma-pi connections) is computed as a sum of contributions from a set of independent multiplicative clusters of input weights:

$$y = f(\sum_j w_j x_j), \tag{1}$$

where

$$x_j = \prod_h^k v_h I_h \tag{2}$$

is the product of k inputs I_h with their weights v_h within cluster j, and w_j is the weight on cluster j as whole.

Actually, we use a limited version of the multiplicative connections which allows the activation of one unit to control the strength of interconnection between two other units:

$$o_i(t) = f(\sum_j z_{ij}(t) I_j(t)), \tag{3}$$

$$z_{ij}(t) = w_{ij} c_{ij}(t), \tag{4}$$

where w_{ij} is a constant and $c_{ij}(t)$ is a dynamical control signal which gates the connection ij at time t.

The idea is to define a network, the Executor, which uses the outputs of the LRAAM as control signals in order to load the 'right' set of weights on a discriminator.

An example of Executor for a binary NT, with discriminators implemented by perceptrons, is shown in Figure 4.

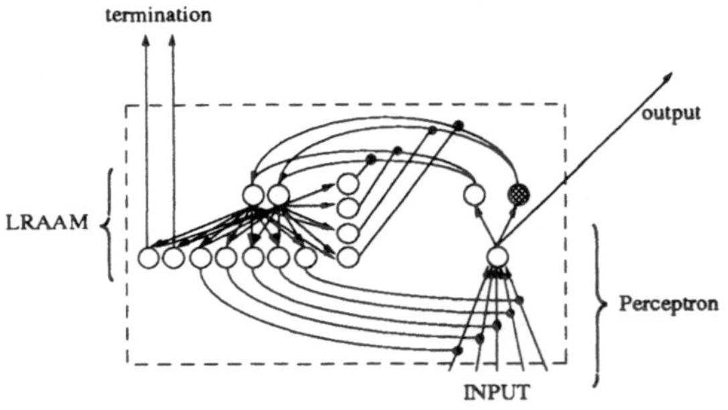

Figure 4: Example of Executor for a binary neural tree and perceptron discriminators.

In this case, the Executor contains a perceptron with the same topology of the discriminators used in the NT, but with multiplicative input connections. The weights of the connections are loaded by the output of a decoding network whose weights are defined by the decoding part of the LRAAM representing the NT. The output of the perceptron is transformed in a localized binary code, i.e., 1 is transformed in $(1,0)$ and -1 in $(0,1)$, by a couple of neurons with appropriate biases and activation functions. The output of these nodes is then feeded back to the input units of the decoder through multiplicative connections which are set by the output of the pointer's fields of the decoder. The input units of the decoder are linear units.

At the beginning of the classification task the pointer to the root of the NT, as codified by the LRAAM, is feeded into the input units of the decoder. After decoding of the pointer, the connections of the perceptron are loaded with the weights of the root discriminator and the connections outgoing from the localized representation of the perceptron output are setted with the pointers to the left and right children of the root. The pointer to the left or right child is then propagated to the input of the decoder, according to the output of the discriminator. The process is then automatically repeated until a leaf is reached. The termination condition is given by the bits of the label which codify the null pointer conditions.

Note that, if the number of classes is two and the same output convention for the discriminators is used, then the labels of the leaves can be removed from the tree since the output of the perceptron can be used instead.

2.5 LRAAM as Neural Program

The use of the decoder part of the LRAAM in the Executor gives rise to the concept of *neural program*. In fact, the weights of the LRAAM can be considered as a distributed neural code implementing a given NT. If the connections of the decoder in the Executor are multiplicative as well, then different NT can be *loaded and executed* on the same network once their encoding have been produced by a LRAAM. This kind of neural code has several interesting aspects. Firstly, it is entirely generated by learning, since both the parameters of NTs and LRAAMs are discovered by learning. Secondly, NTs of different size and depth can be encoded in LRAAMs of equal size. Lastly, the code is an analog code, with a continuos metric.

Actually, a NT is implemented not only by the parameters of the corresponding LRAAM, but also by the pointer to the root of the NT. Thus, the implementation of a NT via a LRAAM gives a well definite semantics in terms of a nonlinear system (LRAAM) with initial condition given by the pointer (generated by the LRAAM) to the root of the NT.

3 An Example of Neural Code

In this section, we will give an example of *neural code*. The problem at hand is the 8 bits parity problem, and it was solved by the neural tree shown in Figure 5.

The neural tree was synthetized using the Binary NT algorithm with perceptrons as discriminators and supervised learning. Each discriminator tried to classify correctly the highest number of patterns, so to reduce the total number of nodes in the tree. The technique employed to obtain this kind of discriminators was a mix between a variant of the perceptron algorithm (*Pocket*) proposed by Sirat and Nadal([18]) and the standard delta rule with L_1 norm suggested by Sankar and Mammone ([14]). Since we found the first technique not efficient in the first stages of the training and the second one in the final stages, we used the delta rule with L_1 norm for the first 100 epochs of the training and then switched to the perceptron procedure. The perceptron procedure was runned for at most 200 epochs and the obtained perceptron used as splitting rule. The consistency of the activation function between the two phases of the training was preserved by using a symmetrical sigmoid for the delta rule, so that the hard limiter used by the perceptron procedure could be obtained through an infinite steepness. The total number of epochs employed to built the tree was just below 2800, since only the training of the nodes with less than two leaf children reached the maximum number of epochs allowed for each node. The final neural tree was not robust with respect to perturbations in the parameters. In fact, due to the density of the patterns involved in the 8 bits parity problem, the hyperplanes defined by the weights of the perceptrons and used by the tree nodes as splitting rules, was very sensitive to noise. Small perturbations in the weights of a perceptron may change the disposition of the hyperplane in the input space and consequently modify the splitting rule.

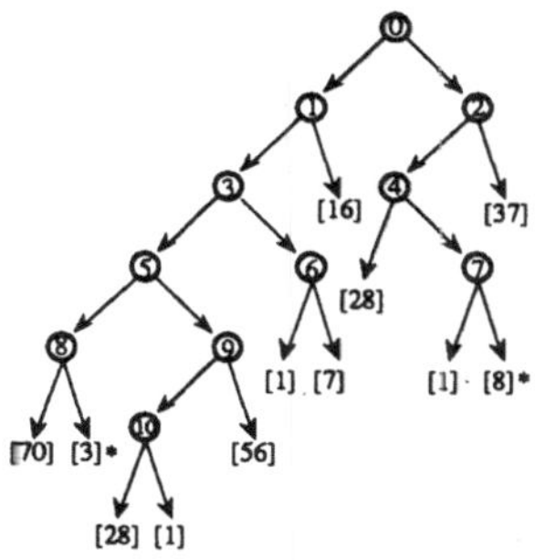

Figure 5: A neural tree solving the 8 bits parity problem. The patterns responding with 1 at each node are routed to the right child of the node. The number of patterns routed to each leaf is reported.

In spite of this sensitivity to perturbation of the weights, the neural tree can be encoded by a relatively small LRAAM with good results. The *neural code* for the tree was obtained by a six hidden nodes LRAAM with linear output nodes. The linear output nodes were used to obtain a more accurate approximation for the weights of the perceptrons which were used as labels. In this case, the normalization problem was not present, since the weights were normalized by the training procedure for the perceptron. The tollerance ν was setted to 0.05 and the training stopped after 13016 epochs. The representations obtained for the neural tree nodes are shown in Figure 6.

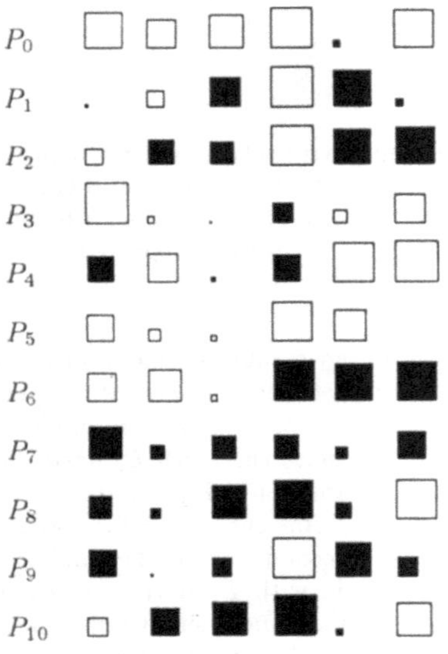

Figure 6: Representations obtained for the neural tree nodes.

The decoded labels presented an error higher than 0.05. A slight reduction of this error was obtained by using the decoded neural tree weights as training set for a one layer network with linear units and target vectors equal to the desired neural tree weights. The network was trained for 3000 epochs and then absorbed in the LRAAM, since the output units of the LRAAM were also linear units. The modulus of the error for each weight of the neural tree for the original and modified LRAAMs are shown in Figure 7.

The weights of the decoder part of the modified LRAAM was then used as *neural code* in a suitable sized Executor Network. The weights (without biases) of the decoder are shown in Figure 8.

The neural tree obtained with this procedure was not perfect, since two patterns were classified erroneously in the leaves marked by an asterisk (one pattern for each leaf) in Figure 5. Moreover, a pattern on the root was routed on the right branch of the tree instead of the left, but subsequently classified in the right manner. The errors are due to the approximations introduced in the weights by the LRAAM and a

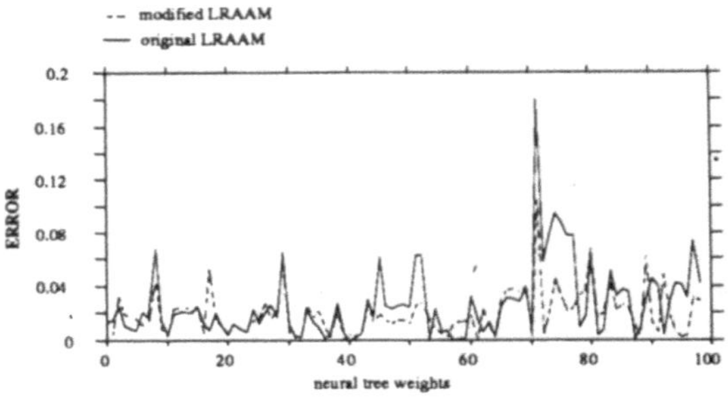

Figure 7: The modulus of the error for the decoded neural tree weights obtained by the original and modified LRAAM.

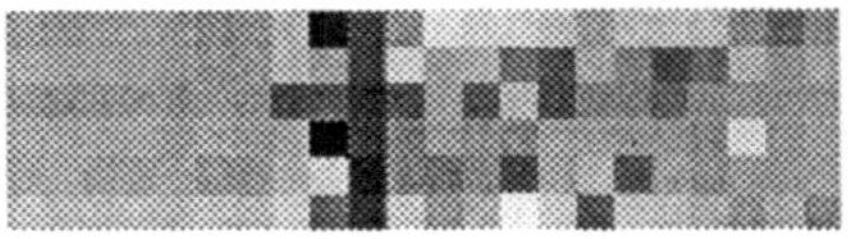

Figure 8: The *neural code* implementing the neural tree.

bigger LRAAM will be able to recover from them. It is however interesting to note that with 161 parameters (the weights of the decoder of the LRAAM) it is possible to have a full connectionist implementation of both the splitting rules and the routing process (classification) of the neural tree with a 0.78% error rate. Note that, the standard implementation of the neural tree must use 99 parameters only to implement the splitting rules.

4 Conclusion

In this paper we have discussed a general method to implement any Neural Tree in a high-order recurrent network (Executor) exploiting a constrained version of the π-connections. The idea is to use a LRAAM to encode the Neural Tree and then to load the set of weights defining the LRAAM in the Executor in order to simulate the classification process of the Neural Tree.

The advantage of this approach is to bring back into the neural network framework an hybrid technique, so that no external support to the network is needed. Moreover, due to the encoding characteristics of the LRAAMs, the number of parameters defining the Neural Tree is reduced.

However, the more interesting result of this approach is the synthesis of the *neural code* implementing the Neural Tree, which is the first step towards a treatment of neural networks closer to the computer science point of view. This change in the prospective can be better appreciated if it is considered the possibility to have a meta-network which changes dynamically the neural code executed by the Executor, according to the actual computational needs and context.

The final aim is the extension of this method to a very structured framework, where learning and programming could be melted in a single neural entity in a very efficient manner.

References

[1] L. Breiman, J. Friedman, R. Olshen, and C. Stone. *Classification and Regression Trees.* Wadsworth International Group, 1984.

[2] R. Durbin and D. E. Rumelhart. Product units: a computationally powerful and biologically plausible extension to backpropagation networks. *Neural Computation*, 1:133, 1989.

[3] L. Atlas et al. A performance comparison of trained multilayer perceptrons and trained classification trees. *Proceedings of the IEEE*, 78:1614–1619, 1992.

[4] J. A. Feldman and D. H. Ballard. Connectionist models and their propertirs. *Cognitive Science*, 6:205–254, 1982.

[5] C. L. Giles and T. Maxwell. Learning, invariance, and generalization in high-order neural networks. *Applied Optics*, 26:4972–4978, 1987.

[6] T. Li, L. Fang, and A. Jennings. Structurally adaptive self-organizing neural trees. In *International Joint Conference on Neural Networks*, pages 329–334, 1992.

[7] J. L. McClelland. Putting knowledge in its place. *Cognitive Science*, 9:113–146, 1985.

[8] B. W. Mel. *Connectionist Robot Motion Planning.* Perspectives in Artificial Intelligence. Academic Press, 1990.

[9] M. P. Perrone. A soft-competitive splitting rule for adaptive tree- structured neural networks. In *International Joint Conference on Neural Networks*, pages 689–693, 1992.

[10] M. P. Perrone and N. Intrator. Unsupervised splitting rules for neural tree classifiers. In *International Joint Conference on Neural Networks*, pages 820–825, 1992.

[11] J. B. Pollack. Recursive distributed representations. *Artificial Intelligence*, 46(1-2):77–106, 1990.

[12] J. B. Pollack. The induction of dynamical recognizers. *Machine Learning*, 7:227–252, 1991.

[13] D. E. Rumelhart and J. L. McClelland. *Parallel Distributed Processing: Explorations in the Microstructure of Cognition.* MIT Press, 1986.

[14] A. Sankar and R. Mammone. *Neural Tree Networks*, pages 281–302. Neural Networks: Theory and Applications. Academic Press, 1991.

[15] A. Sankar and R. Mammone. Optimal pruning of neural tree networks for improved generalization. In *International Joint Conference on Neural Networks*, pages 219–224, 1991.

[16] A. Sankar and R. Mammone. Speaker independent vowel recognition using neural tree networks. In *International Joint Conference on Neural Networks*, pages 809–814, 1991.

[17] I. K. Sethi. Entropy nets: From decision trees to neural networks. *Proceeding of the IEEE*, 78:1605–1613, 1990.

[18] J. A. Sirat and J-P. Nadal. Neural trees: a new tool for classification. *Network*, 1:423–438, 1990.

[19] A. Sperduti. *Optimization and Functional Reduced Descriptors in Neural Networks.* PhD thesis, Department of Computer Science, University of Pisa, 1993.

[20] S. S. Venkatesh and P. Baldi. Programmed interactions in higher-order neural networks: Maximal capacity. *Journal of Complexity*, 7:316–337, 1991.

[21] D. J. Volper and S. E. Hampson. Quadratic function nodes: use, structure, and training. *Neural Networks*, 3:93–108, 1990.

Kolmogorov's Theorem: From Algebraic Equations and Nomography to Neural Networks.

Alexander Kovačec *
Departamento de Matemática
Universidade de Coimbra
3000 Coimbra, Portugal
e-mail: contact coauthor

Bernardete Ribeiro
Laboratório de Informática e Sistemas, UC
Qta. Boavista, LT 1-1
3000 Coimbra, Portugal
e-mail: bribeiro@uc.pt.

Abstract

We trace the developments around Hilbert's thirteenth problem back to questions concerning algebraic equations.

Its solution, namely Kolmogorov's superposition theorem of 1956, is stated in an elaborate form and its relation with neural nets is explained. A detailed proof allows to initiate discussions concerning implementability.

We address individuals interested to form an opinion about the hotly debated applicability of the superposition theorem but also the philosophically inclined readers that want to learn the background of a mathematical problem with an eventful history, and who, by studying its proof will get a sense of the difference between construction and existence in mathematics.

0. Introduction

In 1900, at an international congress in Paris, David Hilbert, then already recognized as a very influential mathematician, see Reid [1] for a vivid portrait, posed the following as the thirteenth of a list of twenty three problems, devised to challenge mathematicians in the century to come [1].

> To prove that the equation of 7-th degree
>
> $$x^7 + ax^3 + bx^2 + cx + 1 = 0 \qquad (1)$$
>
> cannot be solved by means of superposition of continuous functions of only two arguments

In Section 1 we indicate the considerations that led to this conjecture, Section 2 states Kolmogorov's refutation in an elaborate form and discusses its potential relevance for neural networks, Section 3 gives a detailed proof of the theorem and Section 4 presents afterthoughts on the proof and its variants. In particular we initiate a discussion on implementability. Unfortunately neither of the two controversial papers by Girosi and Poggio [2] and by Kurkova [3] of whose existence we recently learned was constructively available for us within time but it is certain that in the context of our work they should be consulted.

Formulations and proofs of the theorem as we know it nowadays are the work of many authors, in particular Lorentz, Sprecher, Fridman, Arnold, and Kahane, while to Vituškin we owe important work on on the limits of extensibility of the theorem. The discovery of the relevance of the superposition theorem to neural networks is due to Hecht-Nielsen.

1. Tschirnhausen's transformations and Nomography

An extrordinary influential problem in mathematics was for centuries that of finding formulae for the solution of equations of the form

$$a_0x^n + a_1x^{n-1} + \ldots a_{n-1}x + a_n = 0. \qquad (2)$$

*Supported by JNICT Project PBIC/C/CEN 1129
[1] For brevity we slightly edited unimportant details

The history of this problem is well documented in pertinent books and among the developments we describe only those important for us.

Bring (1786) and Jerrard (1832) (see [4, Vol 1, p.274]) elaborated on a device first conceived by Tschirnhausen (1683). They showed that by simple algebraic transformations equations of the form (2) can always be brought into equivalent forms in which the coefficient of the highest power (x^n), as well as the constant is 1, and in which the $(n-1)$st, $(n-2)$nd and $(n-3)$rd powers do not figure at all. Thus e.g. as a first step of such a simplification substitute in (2) x by $y - \frac{a_1}{a_0 n}$, expand to write the resulting expression as a polynomial in y, and divide by a_0. The new polynomial has leading coefficient 1, and the coefficient of the $(n-1)$st power vanishes. The roots of the original polynomial are recovered from those of the new polynomial in an obvious way: by adding $\frac{a_1}{a_0 n}$. The Bring - Jerrard -work implies for the equation of the seventh degree exactly the equation (1). The equation of seventh degree is the first one in which there figure in spite of simplification at least three coefficients. It was also known since Niels H. Abel's fundamental insights of 1826, that (1), as a general equation of degree≥ 5 will not be solvable by means of finite expressions involving only the elementary arithmetic operations and root signs. This led mathematicians to reinforce the search for other means of solving equations of the higher degree; notably Hermite and Klein made useful contributions.

Hilbert's particular conjecture seems to have been stimulated also by two other facts to which he refers in his lecture. Firstly from that he had convinced himself that there are analytic functions of three variables not representable by the superposition of *analytic* functions of only two variables, and, secondly by the 1899 appearance of Maurice d'Ocagnes voluminous *Traité de Nomographie*, a work in which the contemporary knowledge on solution of equations by graphical means is presented;

Let us consider two simple examples of nomography (see [5]): First consider the equation $uvw = q$. In the chart of Figure 1a, q is determined for the values $u = 3, v = 1, w = 1.5$ by means of following the dotted lines, by first calculating (uv) and then taking that value and multiplying it with w; a process that is represented as exactly an instance of the superposition(=composition) of continuous functions of two variables: $uvw = \text{mult}(\text{mult}(u, v), w)$, where $\text{mult}(.,.)$ denotes the binary multiplication function. As an other example, Figure 1b shows an alignment chart for evaluation of the function $(x, y) \mapsto r = \sqrt{x^2 + y^2}$.

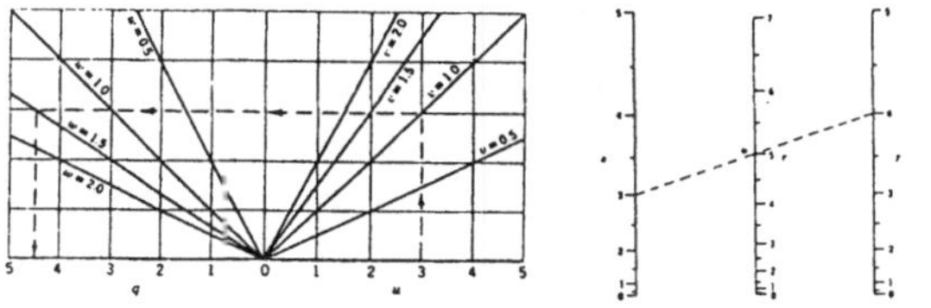

Fig 1. Finding (a) uvw and (b) $r = \sqrt{x^2 + y^2}$.

These particularly simple examples (using no curved lines) must suffice here to see with Hilbert that "it is evident how a great class of [continuous] functions of three and more variables can be represented by the principles of nomography, namely, all those functions which can be generated by taking first a [continuous] function of two variables, then by substituting each of the variables by [continuous] functions of two variables etc. ", However, in view of his experience with analytic functions he conjectured quite naturally that the solutions to the equation of the seventh degree would not belong to the class of functions so representable.

2. Refutation of Hilbert's Conjecture: Kolmogorov's Theorem and its Impact for Neural Nets.

Occupied with the topic for several years, in 1956, A. N. Kolmogorov, famous also for work in Fourier series, probability theory, classical mechanics, logics, and turbulence theory among others, refuted Hilbert's concrete conjecture by proving a very general superpositon theorem. He allegedly said, [6, p.40], that this was the technically most difficult work he ever had done. A refined form of this theorem reads as follows:

Theorem. *Let $\lambda_1, \ldots, \lambda_n$ be any rationally independent numbers. Then there exist $2n+1$ continuous increasing functions $\phi_1, \ldots, \phi_{2n+1}$: $[0,1] \mapsto I\!R$ such that for each continuous function $f : [0,1]^n \to I\!R$ there exists a continuous function $g_f : I\!R \to I\!R$ for which*

$$f(x_1, \ldots, x_n) = \sum_{k=1}^{2n+1} g_f(\lambda_1 \phi_k(x_1) + \ldots \lambda_n \phi_k(x_n)).$$

Since the proof for general n introduces no interesting new ideas, we discuss the theorem for $n = 2$ in a precise and rigorous form but as concrete as we are able to. For this reason we choose $1, \sqrt{2}$ as rationally independent constants. We shall prove:

There exist five continuous realvalued increasing functions $\phi_1, \ldots, \phi_5 : [0,1] \to [0,1]$ such that for every continuous $f : [0,1]^2 \to I\!R$ there exists a continuous real valued function $g = g_f$ defined on $[0, 1+\sqrt{2}]$ such that

$$f(x,y) = \sum_{k=1}^{5} g_f(\phi_k(x) + \sqrt{2}\phi_k(y)). \qquad (3)$$

The fact that apparently neither the functions ϕ_j, nor the function g_f whose existence is stated above can be given in a nice explicit way, let alone by formulas, not surprisingly turns out as the major obstacle for real world applications. Still, Hecht Nielsen's discovery relating the theorem with neural nets is important as it shows that our quest for approximation of functions by networks is sound.

In fact, assume devices represented by Figure 2 which translate the inputs t or $t_1, \ldots, t_n$, respectively, into outputs $h(t)$ and $\sum_{i=1}^n t_i$,

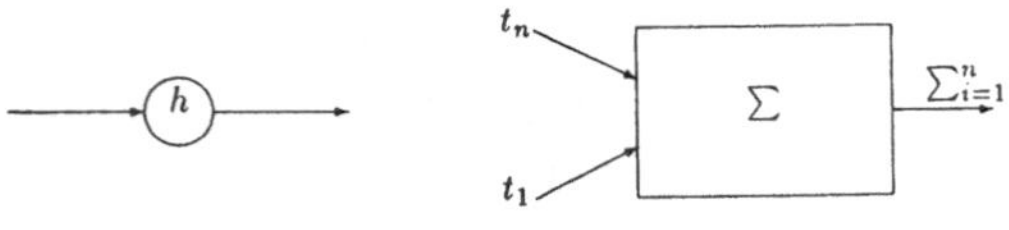

Fig 2. Representations for $h(t)$ and $\sum_{i=1}^n t_i$.

Then we see readily that the net of Figure 3 represents the function $f(x,y)$ as indicated.

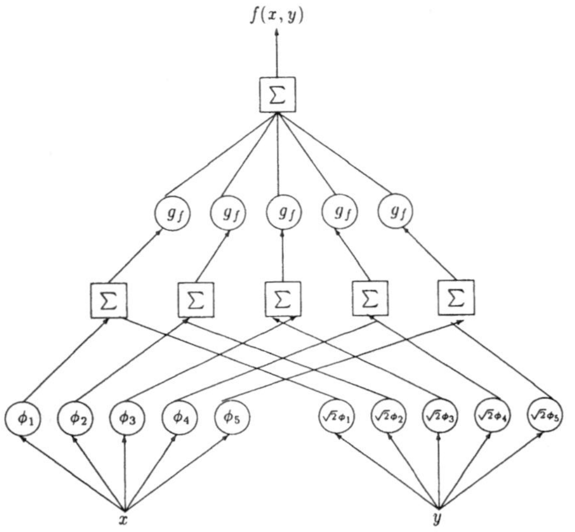

Fig 3. Topology of a Kolmogorov net for $f(x,y)$.

Note that the functions ϕ_i are independent of f. This implies for neural computing that a two variable function function $f(x,y)$ can be learned by learning the one variable function g_f.

3. A proof of Kolmogorov's theorem.

In this section we present a proof for $n = 2$ variables of the superposition theorem in detail and with an algorithmic flavour. The leading ideas will be apparent for every one acquainted with a moderate knowledge of real analysis. It is only towards the end of the proof that mathematical rigour will require some more subtle additional facts. These can be learned from Boas' elementary and delightful booklet [7].

We shall denote by $C(X)$ the set of all real valued continuous functions defined on a compact subset X of Euclidean spaces $\mathbb{R}$ or $\mathbb{R}^2$. For an $f \in C(X)$ we define as usual the (uniform) norm of f by $\|f\| = \max_{x \in X} |f(x)|$.

Assume then the following statement to hold true:

Lemma 1. *There exist five continuous functions*

$\Phi_1, \ldots, \Phi_5 : [0,1]^2 \to [0, 1+\sqrt{2}]$ *such that for every continuous function* $f : [0,1]^2 \to \mathbb{R}$ *there exists a continuous function* $\tilde{g}_f : [0, 1+\sqrt{2}] \to \mathbb{R}$ *of norm* $\|\tilde{g}_f\| \leq \frac{1}{7}$ *such that:*

$$\|f - \sum_{k=1}^{5} \|f\| \, \tilde{g}_f(\Phi_k(\cdot,\cdot))\| \leq \frac{6}{7}\|f\|. \qquad (4)$$

In the following algorithm KOLMO we assume ready accessibility to the functions $\Phi_1, \ldots, \Phi_5$ like in Lemma 1. The algorithm accepts as input a continuous $f : [0,1]^2 \to \mathbb{R}$; its relevant output is a one variable function g.

KOLMO(f)
  $g = 0$;
  for $j = 0$ **until** ∞
    begin
      $f := f - \|f\| \sum_{k=1}^{5} \tilde{g}_f(\Phi_k(\cdot,\cdot))$;
      $g := g + \|f\|\tilde{g}_f$;
    end

Disregard for a moment that KOLMO has an infinite **for**-loop. We claim that g is continuous and satisfies

$$f(x,y) = \sum_{k=1}^{5} g(\Phi_k(x,y)) \qquad (5).$$

To see this note that KOLMO produces in its fourth line inductively a sequence $\{f_j\}_{j=0}^{\infty}$ of functions which by Lemma 1 are continuous and satisfy $\|f_j\| \leq \frac{6}{7}\|f_{j-1}\| \leq (\frac{6}{7})^j\|f\|$ $(j \geq 1)$. By means of sequence $\{f_j\}_{j=0}^{\infty}$, we can write the returned g as

$$g = g_f = \sum_{j=0}^{\infty} \|f_j\|\tilde{g}_{f_j},$$

a function which by $\|\tilde{g}_{f_j}\| \leq \frac{1}{7}$ and elementary theorems of real analysis ([7 ,p.102]) exists and is continuous. Finally we have

$$\begin{aligned}\sum_{k=1}^{5} g(\Phi_k(\cdot,\cdot)) &= \sum_{j=0}^{\infty} \|f_j\|\tilde{g}_{f_j}(\Phi_k(\cdot,\cdot)) \\ &= \sum_{j=0}^{\infty}(f_j - f_{j+1}) = f,\end{aligned}$$

that is, we have presented f as a superposition of a function g_f, five function $\Phi_1, \ldots, \Phi_5$ being independent of f, and a sum which is trivially superposition of four additions. Replacing the ∞-symbol in the algorithm by, say 100 executions of the **for**-loop, we would exit with a function g such that $\|\sum g(\Phi_k(\cdot,\cdot)) - f\| < 10^{-10}\|f\|$. This should suffice for most practical purposes as a good substitute for the exact g_f.

We continue the proof in a top down manner. After a preparatory combinatorial consideration in numbers (1) and (2) below, (3) will show how to construct, assuming suitable Φ_i, a function $\tilde{g}_f$ so as to satisfy equation (4) of Lemma 1. To prove the existence of functions $\Phi_1, \ldots, \Phi_5$ as claimed in Lemma 1, and to show that they can be written as linear combinations of one-variable continuous and increasing functions $\phi_j : [0,1] \to [0,1]$, namely in the form $\Phi_j(x,y) = \phi_j(x) + \sqrt{2}\phi_j(y)$, so as to satisfy equation (3), will be the aim of numbers (4),(5),(6). It is in order to remark that apparently all proofs of the superposition theorem have as ingredients variations, to be discussed later, of the ideas containded in (1)-(4). These variations seem, given the current state of art concerning its implementability, insignificant.

1. Let N be a natural number, but assume, say $N \geq 10$, and define for $i = 1,2,3,4,5$ the sets $R_i := [0,1] \setminus \bigcup_{0 \leq s \leq N \; s \equiv i-1(5)} [\frac{s}{N}, \frac{s+1}{N}]$. For example, R_2 is obtained from the interval $[0,1]$ by removing $[\frac{1}{N}, \frac{2}{N}], [\frac{5}{N}, \frac{6}{N}], [\frac{10}{N}, \frac{11}{N}], \ldots$. Thus, each R_i is a union of about $N/5$ disjoint subintervals of $[0,1]$, the *i-intervals*, see Figure 4a for the case $N = 20$. The cartesian product $R_i \times R_i$ consists of about $(N/5)^2$ disjoint *i-rectangles*, see Figure 4b. We enumerate them somehow by $R_{i1}, R_{i2}, \ldots, R_{im}$, where $m \approx (N/5)^2$. (All rectangles not intersecting the boundary of $[0,1]^2$ are

in fact squares.)

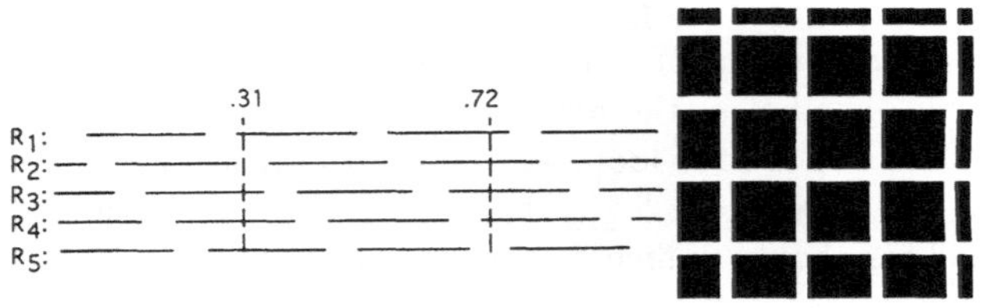

Fig 4. with sligthly different scales
(a): Subsets $R_1, \ldots, R_5$ of $[0,1]$, (b): $R_2 \times R_2$.

2. Take an arbitrary point $(x,y) \in [0,1]^2$. x lies in at least 4 of the sets R_i, y as well. The intersection of two 4–subsets of $\{1,2,3,4,5\}$ contains at least 3 elements. It follows that every $(x,y) \in [0,1]^2$ lies in *at least three* of the cartesian products $R_1 \times R_1, \ldots, R_5 \times R_5$. (If $N = 20$, then the point $(.31, .72)$, e.g. lies in the intersection of a 1-rectangle, a 3-rectangle, and a 4-rectangle.)

The considerations in (1),(2) are valid for any N. We shall define N in dependence of the normed function $\dot{f} := f/\|f\|$ (for which $\|\dot{f}\| = 1$). The rectangles and i–intervals we are speaking henceforth are those pertaining to that N.

3. The g to be constructed in this number is a $\tilde{g}_f$ as it figures in Lemma 1. Choose N so large that for all $(x,y) \in [0,1]^2$ the following holds: If $\dot{f}(x,y) > \frac{1}{7}$ then $\dot{f}|R > 0$ for each rectangle R to which (x,y) belongs. If $\dot{f}(x,y) < -\frac{1}{7}$ then $\dot{f}|R < 0$ for each rectangle to which (x,y) belongs. The coice of such N is possible by uniform continuity of f. In (4) we will show that the Φ_i can be chosen such that each Φ_i is constant on each of the i–rectangles and such that $(i,r) \neq (j,s)$ implies $\Phi_i(R_{ir}) \neq \Phi_j(R_{js})$. Assuming this for the moment we define a continuous function $g \in C([0, 1+\sqrt{2}])$ of norm $\|g\| \leq \frac{1}{7}$ such that for all i, r in question,

$$g(\Phi_i(R_{ir})) = \begin{cases} 1/7 & \text{iff } \dot{f}|R_{ir} > 0 \\ -1/7 & \text{iff } \dot{f}|R_{ir} < 0 \end{cases} .$$

(In general the function g will be, though continuous, quite "wild" even if f is "nice". In the executing KOLMO one has to associate a $\tilde{g}_{f_j}$ to functions which are partly smaller, partly larger than 0. For such f_j one would obtain a $\tilde{g}_{f_j}$ probably wilder than something like in Figure 5.

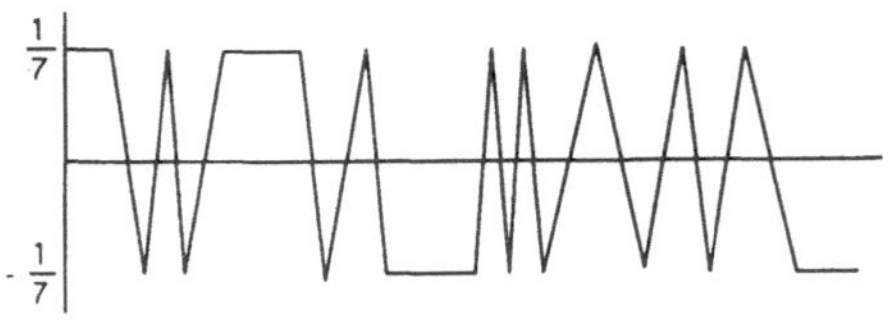

Fig 5. A continuous function of norm $\leq \frac{1}{7}$.)

Let

$$E(x,y) = \dot{f}(x,y) - \sum_{i=1}^{5} g(\Phi_i(x,y))$$

and keep in mind that we have $-1 \leq \dot{f}(x,y) \leq 1$ and $-\frac{1}{7} \leq g(\Phi_i(x,y)) \leq \frac{1}{7}$ $(i = 1, \ldots, 5)$ for any $(x,y) \in [0,1]^2$. We claim that $|E(x,y)| \leq \frac{6}{7}$:
Case 1. $\dot{f}(x,y) > \frac{1}{7}$. Since (x,y) belongs to at least three rectangles on each of which $\dot{f}$ (i.e. f) is positive, there are at least three indices i for which $g(\Phi_i(x,y)) = \frac{1}{7}$ and thus it follows that

$$-\frac{4}{7} = \frac{1}{7} - \frac{3}{7} - \frac{2}{7} \leq E(x,y) \leq 1 - \frac{3}{7} + \frac{2}{7} = \frac{6}{7}.$$

Case 2. $\dot{f}(x,y) < -\frac{1}{7}$. Then (x,y) belongs to at least three rectangles on each of which f is negative. Similarly to Case 1 it follows that

$$\frac{-6}{7} = -1 + \frac{3}{7} - \frac{2}{7} \leq E(x,y) \leq \frac{-1}{7} + \frac{3}{7} + \frac{2}{7} = \frac{4}{7}.$$

Case 3. $|\dot{f}(x,y)| \leq \frac{1}{7}$. Then we can still make the estimate

$$\frac{-6}{7} = -\frac{1}{7} - \frac{5}{7} \leq E(x,y) \leq \frac{1}{7} + \frac{5}{7} = \frac{6}{7}.$$

Our claim follows directly from these estimates.

4. Denote by $\mathcal{I}$ the class of all continuous, increasing functions $\phi : [0,1] \to [0,1]$. With the end of constructing the functions $\Phi_j(x,y)$ in the desired form, assume $\phi_1, \ldots, \phi_5$ to be five functions in $\mathcal{I}$ with two additional properties: Namely (i) that ϕ_1 is constant and equal to a certain rational number along each of the 1–intervals, ϕ_2 is constant and equal to a certain rational number on each of the 2–intervals

et cetera; Figure 6 gives an idea of how such a function looks like in case $N = 20$.

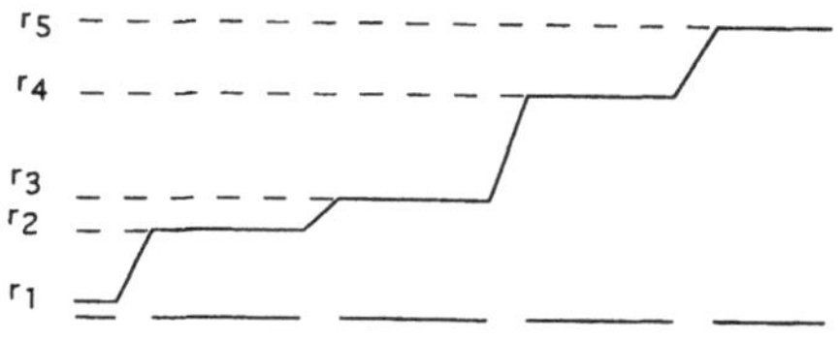

Fig 6. An example for a function ϕ_2 on R_2.

By the infinitude of the rationals we can assume (ii) for the ϕ_i also that the chosen sets of rationals are mutually disjoint; more precisely that for $i \neq j$, we have $\phi_i(R_i) \cap \phi_j(R_j) = \emptyset$.

Having thus fixed the quintuple $(\phi_1, \ldots, \phi_5) \in \mathcal{I}^5$, we define Φ_i as $\Phi_i(x, y) := \phi_i(x) + \sqrt{2}\phi_i(y)$, and claim that these are as required in (3).

In fact, since the ϕ_i are constant on each of the i–intervals, the constantness of $\Phi_i(x, y)$ on each of the i–rectangles is immediate. Furthermore, choose (x, y) in an i–rectangle and (x', y') in a j–rectangle. Then necessarily $\Phi_i(x, y) \neq \Phi_j(x', y')$, since otherwise the fraction $(\phi_i(x) - \phi_j(x'))/(\phi_i(y) - \phi_j(y'))$ of two rationals would be equal to the irrational number $\sqrt{2}$. This proves our claim.

Substituting the functions $\Phi_i(x, y)$ into the estimates of (3) we can say:

For all continuous functions $\dot{f} : [0, 1]^2 \to \mathbb{R}$ *of norm* $\|\dot{f}\| = 1$, *we can* constructively *find a quintuple* $(\phi_1, \ldots, \phi_5) \in \mathcal{I}^5$ *and a function* $g \in C[0, 1]$ *of norm* $\|g\| \leq \frac{1}{7}$ *such that for all* $(x, y) \in [0, 1]^2$, *we have*

$$|\dot{f}(x, y) - \sum_{k=1}^{5} g(\phi_k(x) + \sqrt{2}\phi_k(y))| \leq \frac{6}{7}.$$

This statement has (after multiplication with $\|f\|$) almost the form as required in Lemma 2.1, with the essential exception that the quintuple $(\phi_1, \ldots, \phi_5)$ depends on f. We show now that there exists a quintuple which is suitable for *any* f. It is at this point where more subtle mathematics creeps in. To simplify notation let $E(f, g, \phi_1, \ldots, \phi_5)(x, y) = f(x, y) - \sum_{k=1}^{5} \|f\| g(\phi_k(x) + \sqrt{2}\phi_k(y))$.

5. We consider then given f, an entirety U_f which, by $\frac{6}{7} < \frac{7}{8}$, will contain in particular the quintuple of the italicized statement above. Namely put

U_f := Set of *all* quintuples $(\phi_1, \ldots \phi_5) \in \mathcal{I}^5$ with the property that there exists a $g \in C[0, 1 + \sqrt{2}]$ of norm $\|g\| \leq \frac{1}{7}$, such that $\|E(\dot{f}, g, \phi_1, \ldots, \phi_5)\| < \frac{7}{8}$.

It is clear from the continuity of f, g, that, if $(\phi_1, \ldots, \phi_5) \in U_f$ and if $(\phi'_1, \ldots, \phi'_5) \in \mathcal{I}^5$ is "near" to $(\phi_1, \ldots, \phi_5)$, then $(\phi'_1, \ldots, \phi'_5)$ will also belong to U_f. In mathematical terms, we would say that U_f is *open*. Next fix an arbitrary quintuple $(\phi'_1, \ldots, \phi'_5)$ in $\mathcal{I}^5$. Then, since arbitrarily near to any real there lies a rational, by the construction procedure of (3) and (4), in which there is an enourmous liberty in the choice of N (all large enough N will do) and the rational image sets $\phi_i(R_i)$, we can arrange a quintuple $(\phi_1, \ldots, \phi_5)$ of functions in U_f such that ϕ_i is near to ϕ'_i for $i = 1, \ldots, 5$. In mathematical terms this says that U_f is *dense*.

(Separate carefully the two $N's$ in question. Once the existence of the universal quintuple is proven below, one needs for a step in the algorithm to find a $\tilde{g}_f$ only the N of (3). The N we were speaking of in the last paragraph was used to give an idea of how density of U_f can be established. Its simply saying that any increasing function can be approximated by a quasi step function as in Figure 6 if sufficiently many steps are used.)

6. To the ideas of (5) one can attach precise mathematical meaning. One endows $\mathcal{I}^5$ with a suitable metric, say $d((\phi_1, \ldots, \phi_5), (\phi'_1, \ldots, \phi'_5)) = (\sum_{i=1}^{5} \|\phi_i - \phi'_i\|^2)^{\frac{1}{2}}$ and finds that " U_f is open and dense in the metric space $(\mathcal{I}^5, d)$". Furthermore, if we take a Cauchy-sequence of quintuples in $(\mathcal{I}^5, d)$ then the limit quintuple in $(C[0, 1])^5$ exists and obviously each component of this quintuple is an increasing function. This shows: $\mathcal{I}^5$ is a complete

metric space.

Let us cite a famous theorem (see [7, p. 62]) in the following form.

Baire-Category theorem. *A complete metric space is a Baire-space. In particular any* countable *intersection of open dense sets is dense, thus nonempty.*

We apply this theorem as follows. Choose by the Weierstrass approximation theorem a countable family $\{h_j\}_{j=0}^{\infty}$ of continous functions, dense on the unit sphere of $(C[0,1], \|\cdot\|)$. Baire's theorem gives us that there exists a $(\phi_1, \ldots, \phi_5)$ in the intersection of the U_{h_j}. Thus for each j there exists $g \in C([0, 1+\sqrt{2}])$ of norm $\|g\| \leq \frac{1}{7}$ such that $\|h_j(x,y) - \sum_{k=1}^{5} g(\phi_k(x) + \sqrt{2}\phi_k(y))\| < \frac{7}{8}$.

Finally let $f \in C([0,1]^2)$ be arbitrary. Then, $f = \|f\|\dot{f}$ and by the triangle inequality, for any $\epsilon > 0$ there exists a j such that for suitable g, $|E(f, g, \phi_1, \ldots, \phi_5)(x,y)| \leq |f(x,y) - \|f\|h_j(x,y)| + |E(\|f\|h_j, g, \phi_1, \ldots, \phi_5)(x,y)| \leq \|f\|(\frac{7}{8} + \epsilon)$.

This holding for any $\epsilon > 0$, we find in conclusion the desired proof of Lemma 1 with functions $\Phi_k(x,y) = \phi_k(x) + \sqrt{2}\phi_k(y)$ and the proof of Kolmogorov's theorem is complete. ∎

4. Discussion of Proof and Relevance of Kolmogorov's Theorem.

All proofs employ ideas similar to the rectangle- idea mentioned in numbers (1),(2) and aim to construct functions Φ_i that satisfy the non-equalities ($\neq$) mentioned in (3). Likewise, the proofs available to us make use of estimates for first approximations like given in Lemma 1. There are small variations on how to construct the function g in (2) or to choose the N in (1) as a function of f. Some authors require that f oscillates within a rectangle not more than a certain small bound (instead of remaining merely sign constant as we did).

The use of the Baire category theorem whose proof necessitates a form of the axiom of choice (Beppo Levi 1902 and E. Zermelo 1904) has its origin in Kahane's proof [8]. In the second part of the nineteenth century such an argument would have been bitterly disputed because of its non-constructivity. Superficial reflections raise, because of the density of $\bigcap_{j=0}^{\infty} U_{h_j}$ in $\mathcal{I}^5$, the hope that for almost every quintuple in $\mathcal{I}^5$ the associated quintuple of functions $\Phi_k(x,y) = \phi_k(x) + \sqrt{2}\phi_k(y)$ should do quite well when used in the algorithm outlined at the beginning of Section 2. Can these claims be substantiated? - see the problem below.

Lorentz makes constructive efforts to find the functions ϕ_i. He gradually approximates the exact functions by defining them on more and more points. In his notation, letting $\alpha_{qi}^k = i \cdot 10^{-k+1} - (2q-1) \cdot 10^{-k}$ $\beta_{qi}^k = i \cdot 10^{-k+1} - (2q-9) \cdot 10^{-k}$, $(k = 1, 2, \ldots)$ and extending this definition for $k = 0$ by $\alpha_{qi}^k = 0$ $\beta_{qi}^k = 1$ The functions ϕ_q are defined inductively. Given a q, in the k−th step the values of ϕ_q are given on the points $\alpha_{qi}^k, \beta_{qi}^k$, as $i, j = 0, 1, \ldots, 10^{k-1}$, this however in dependence of values of ϕ_q on all points $\alpha_{qi}^l, \beta_{qi}^l$, existing for all $l < k$ in a subtle way. The process is never completed. It is easy to see that all points $\alpha_{qi}^k, \beta_{qi}^k$ are different for different index triples and given q as i, k vary over admitted indices, the set of all these points is a dense subset A_q of $[0,1]$. The functions ϕ_q are the unique extensions to $[0,1]$ of their semi-constructive definitions on A_q. An illustration of other aspects of his proof by figures is in [9] and can serve well as additional illustrations to our paper.

It is adequate to posit a word of warning: work of Vituškin (see reference [12] of reference Lorentz (1976) of [9]) implies that there exists even a polynomial (!) $f(x,y)$ not representable in form of superpositions in question using *differentiable* ϕ_k throughout. Vituškin has also negative results concerning stability questions. Roughly speaking, if f_1, f_2 are near to each other, then g_{f_1}, g_{f_2} need not be near to each other.

For those that are in spite of the difficulties interested in implementing the superposition theorem in this or that form we pose the following question, assuming a machine that could work with real valued functions defined on the whole continuum $[0,1]$, which of course is not the case.

Having Kahane's density result in mind, take any quintuple $(\phi_1, \ldots, \phi_5) \in \mathcal{I}^5$ "at random" [2] and some function $f_0 = f \in C[0,1]$.

Problem. *Will the machine be able to construct many of the functions f_j in the algorithm of Section 2?*

5. Conclusion

In this article we presented the historic background of the superposition theorem. This history is interesting since it illustrates once more, with a significant though less known example, that questions in pure mathematics can have unforeseen relevance in applications.

We presented a version of the proof of the superposition theorem detailed and readable enough to be understandable in its principal ideas by non-mathematicians. In particular we made an effort concerning algorithmic aspects.

In this connection a question is posed whose pursuit seems natural for gaining more insights concerning implementability.

Remark. The following bibliography is kept at a minimum and comprises with preference papers as yet more rarely cited in connection with the superposition theorem. They can serve as pointers however to a broad spectrum of what has relevance for the subject.

References

[1] Reid C, *'Hilbert'*, Springer, Berlin 1970.

[2] Girosi F and Poggio J, *'Representation properties of networks: Kolmogorov's theorem is irrelevant'*, Neural Computation 1, 456-469, 1989.

[3] Kurkova V, *'Kolmogorov's Theorem is relevant'*, Neural Computation 3, 617-622, 1991.

[4] Burnside W S and Panton A W, *'The Theory of Equations'*, Reprint of the seventh edition of 1912, Dover 1960.

[5] Levens A S, *'Nomography'*, John Wiley, New York, 1959.

[6] *'Andreij Nikolajevic Kolmogorov'*, Obituary. Bull. London. Math. Soc. 22, 1, 37-68, 1990.

[7] Boas R P, *'A Primer of Real Functions'*, third edition, Carus Mathematical Monograph 13, The Mathematical Association of America, 1981.

[8] Kahane J P, *'Sur le Théorème de Superposition de Kolmogorov'*, J. Approximation Theory, 13, 229-234, 1975.

[9] Cotter N E and Guillerm T J, *'The CMAC and a Theorem of Kolmogorov'*, Neural Networks, 5, 2, 221-228, 1992.

[2] it is not clear what that means

Output Zeroing within a Hopfield Network

David William Pearson
Laboratoire d'Electronique d'Automatique et d'Informatique
Centre des Systèmes de Production
Ecole des Mines d'Alès
6, Avenue de Clavières
30319 Alès, France

Abstract

In this paper we investigate a specific problem from the control theory literature, that of zeroing a system output, from the point of view of a neural network. For this we consider functions of the neural network states as defining a system output. In particular we are concerned with a continuous network of Hopfield type which could, in theory, be manufactured with available electrical components. Our aim is to impose a specific dynamics on a network by calculating the synaptic weights directly, without requiring training. Hence when a network is initialised in certain states we can be confident that the functions defining the output will remain sufficiently close to zero. We use (nonlinear) geometrical methods in our analysis and reliable numerical methods for our computations.

Introduction

Neural networks are sometimes treated as "black boxes" in that a network is trained for a particular application, hence the problems encountered tend to be practical and application dependent. We are looking from another point of view. In other words we ask which structures of synaptic weights will give rise to particular trajectories in a network, regardless of the application.

In this preliminary phase of our investigations we are only concerned with continuous networks of Hopfield type. We remark that our aim is to impose a certain dynamics on a network without resort to training, in this respect our work is similar to [1]. However our methods are perhaps more geometric in flavour and we address a specific problem to be found in the literature on (nonlinear) control theory [2]. In fact the mixture of problems and geometrical methods found in [2] plus the more theoretical aspects, to be found in [3] and [4], are some of the main driving forces behind our investigations.

At the same time as carrying out a theoretical analysis of this problem we believe it is important to take into consideration the numerical computation aspects. In particular one of the primary aims of this research is to lead to a network realised in hardware and so we need to consider robustness problems caused by hardware tolerances etc. Hence we always use reliable numerical methods such as those found in [5] for all our computations.

1. Computing Element Equations

We choose a standard normalised computing element as described in [6], where the output of each computing element is a function $x^i : \mathbb{R} \to \mathbb{R}$ determined by the differential equation

$$\dot{x}^i + x^i = \sum_{j=1}^{n} u^i_j \theta(x^j) + c^i \qquad (1.1)$$

where $\dot{x}^i := \frac{dx^i}{dt}$, $U := [u^i_j]$ is the matrix of synaptic weights, c^i is a constant and there are n computing elements. The function θ for this paper will be

$$\theta(x) = \frac{1}{1+e^{-x}} - \frac{1}{2} \quad \in C^{\omega} \qquad (1.2)$$

we remove the usual parameters from the function θ [6] in order to simplify notation. The state of the network at any time t is then provided by the state vector $\mathbf{x}^*(t) = [x^1(t), \dots, x^n(t)]$.

Considering all the equations of type (1.1) we see that the state of the network at any time is governed by the flow of the analytic vector field

$$\mathbf{f} = \sum_{i=1}^{n} \sum_{j=1}^{n} (-x^i + u^i_j \theta(x^j) + c^i) \partial_{x^i} \qquad (1.3)$$

where $\partial_{x^i} := \frac{\partial}{\partial x^i}$ is a standard shorthand notation ([3],[4]). Hence if the network is initialised at time t=0 with an initial state vector $\mathbf{x}_0$ then the state of the network at time t will be given by the flow of (1.3) passing through the point $\mathbf{x}_0$ ([2],[3],[4])

$$\Phi_t(\mathbf{x}_0) := \exp(t\mathbf{f})\mathbf{x}_0 \quad \mathbf{x}_0 = \mathbf{x}(0) \qquad (1.4)$$

2. Output Zeroing and Symmetry Groups

Consider once more the vector field (1.3), we think of this as determining the state of a dynamical system with each u^i_j multiplying a "feedback" term in the form $\theta(x^j)\partial_{x^i}$. Now suppose that we are given a set of functions $h_i : \mathbb{R}^n \rightarrow \mathbb{R}$, $h_i \in \mathbf{C}^\infty$ (or $\mathbf{C}^\omega$), i=1, ... ,m with m<n, which we regard as describing the "output" of the system

$$\mathbf{y}(t) := \mathbf{h}(\mathbf{x})(t) = \begin{bmatrix} h_1(\mathbf{x}(t)) \\ \cdots \\ h_m(\mathbf{x}(t)) \end{bmatrix} \qquad (2.1)$$

The object is to compute a set of synaptic weights to ensure that $\mathbf{y}(t)=\mathbf{0}$ for all t in a sufficiently small neighbourhood of the origin, for a given set of functions $\mathbf{h}$.

2.1 Definition

For $\mathbf{h}(\mathbf{x})=\mathbf{0}$, a symmetry group of the system is a local group of transformations, **G**, with the property that **G** transforms solutions of the system to other solutions. Thus if $\mathbf{S}:=\{\mathbf{x}\subset\mathbb{R}^n : \mathbf{h}(\mathbf{x})=\mathbf{0}\}$ then if whenever $\mathbf{x}\in\mathbf{S}$ and $\gamma\in\mathbf{G}$ is such that $\gamma\mathbf{x}$ is defined then $\gamma\mathbf{x}\in\mathbf{S}$ and **S** is said to be **G**-invariant [4]. In our case of course **G** will be the 1-parameter group Φ_t (1.4) with infinitesimal generator **f** (1.3).

2.2 Definition

The system of equations $\mathbf{h}(\mathbf{x})=\mathbf{0}$ is said to be of maximal rank if the Jacobian matrix $\frac{\partial h_i}{\partial x^j}$ is of full rank at every solution **x** of the system.

2.3 Theorem

If the system $\mathbf{h}(\mathbf{x})=\mathbf{0}$ has maximal rank then Φ_t is a symmetry group of the system if and only if

$$\mathbf{f}(h_i)(\mathbf{x}) = 0 \ , \ i=1, \dots, m \qquad (2.2)$$

the proof of this theorem can be found in [4].

3. Calculation of Synaptic Weights

In the previous section we have introduced the infinitesimal criterion (2.2). In this section we will investigate how to calculate the synaptic weights in order to achieve output zeroing for two particular classes of output functions. We begin with a very simple but important class.

3.1 Proposition

If the output functions are linear and satisfy the maximal rank condition, then the synaptic weights can always be calculated in order to make Φ_t a symmetry group of **S**, where **S** is defined in definition 2.1.

Proof.

Let

$$h_k(\mathbf{x}) = \sum_{i=1}^{n} \alpha^k_i x^i - \beta^k \qquad (3.1)$$

then using (1.3) we see that condition (2.2) requires

$$\sum_{i=1}^{n}\sum_{j=1}^{n} \alpha^k_i(-x^i+u^i_j\theta(x^j)+c^i) = 0 \ , \ k=1, \dots ,m \qquad (3.2)$$

Now if the c^i are calculated such that

$$\sum_{j=1}^{n} \alpha^k_i c^i = \beta^k \ , \ k=1, \dots ,m \qquad (3.3)$$

then making use of (3.1) we see that condition (3.2) is equivalent to

$$-h_k(\mathbf{x}) + \sum_{i=1}^{n}\sum_{j=1}^{n} \alpha^k_i u^i_j\theta(x^j) = 0 \ , \ k=1, \dots ,m$$

but $h_k(\mathbf{x})=0$ for $\mathbf{x}\in\mathbf{S}$ and so we simply need the synaptic weights to satisfy the equations

$$\sum_{i=1}^{n}\sum_{j=1}^{n} \alpha^k_i u^i_j\theta(x^j) = 0 \ , \ k=1, \dots ,m \qquad (3.4)$$

Now it is easy to see that conditions (3.3) and (3.4) can be satisfied by solving the matrix equations

$$A\mathbf{c} = \begin{bmatrix} \alpha^1_1 \dots \alpha^1_n \\ \dots\dots\dots \\ \alpha^m_1 \dots \alpha^m_n \end{bmatrix} \begin{bmatrix} c^1 \\ \dots \\ c^n \end{bmatrix} = \mathbf{b} = \begin{bmatrix} \beta^1 \\ \dots \\ \beta^m \end{bmatrix} \qquad (3.5)$$

and

$$AU = A\begin{bmatrix} u_1^1 \dots u_n^1 \\ \dots\dots\dots \\ u_1^n \dots u_n^n \end{bmatrix} = 0 \tag{3.6}$$

The m×n matrix A has rank=m where m<n because **h** has maximal rank hence equation (3.5) has a solution and in equation (3.6) the columns of U can always be chosen from ker(A) (the kernel of A) which has dimension n-m, hence the proposition is proved. We note that this solution allows a certain amount of flexibility in the choice of the matrix U and we go on to indicate one way in which this flexibility may be exploited in section 4.

The second class of output functions that we consider is $\mathbf{C}^\infty$, this is not quite as "neat and tidy" as the linear case however a general approach to the problem can still be found. The first thing to be done is to invoke Frobenius's theorem ([2],[4]) in order to change the coordinate system to one that is more convenient. Let

$$z^i = \psi^i(\mathbf{x}) = h_i(\mathbf{x}) \ , \ i=1,\dots,m \tag{3.7}$$

and

$$z^i = \psi^i(\mathbf{x}) \ , \ i=m+1,\dots,n \tag{3.8}$$

Now if $\mathbf{x}\in\mathbf{S}$ then $z^i=0$ in (3.7) and Frobenius's theorem tells us that we can always choose the extra coordinates (3.8) so that the Jacobian matrix of the transformation $\Psi:=[\psi^i]$ has rank n within a neighbourhood of a point of interest and such that $z^i=0$ in (3.8) as well at this point of interest. In our case "point of interest" will mean where $\mathbf{x}\in\mathbf{S}$ and $\mathbf{x}$ is an equilibrium point where the vector field (1.3) vanishes.

Under the change of coordinates (3.7) and (3.8) the representation of the vector field becomes

$$\mathbf{f} = \sum_{i=1}^{n}\sum_{j=1}^{n} \left(f^j(\Psi^{-1}(\mathbf{z}))\frac{\partial\psi^i}{\partial x^j}(\Psi^{-1}(\mathbf{z}))\right)\partial_{z^i} \tag{3.9}$$

where

$$f^i(\mathbf{x}) := -x^i + c^i + \sum_{k=1}^{n} u_k^i\theta(x^k)$$

and combining (2.1) and (3.7) we have

$$y^i = h_i(\mathbf{x}) = z^i \ , \ i=1,\dots,m \tag{3.10}$$

Thus in the new coordinates defined by the transformation Ψ the infinitesimal condition of theorem (2.3) given by (2.2) becomes simply

$$\mathbf{f}(h_k)(\Psi^{-1}(\mathbf{z})) = \sum_{i=1}^{n} f^i(\Psi^{-1}(\mathbf{z}))\frac{\partial\psi^k}{\partial x^i}(\Psi^{-1}(\mathbf{z})) \tag{3.11}$$

We select a point $\mathbf{x}_e\in\mathbf{S}$ and impose the condition $\mathbf{f}_{|\mathbf{x}_e}=\mathbf{0}$, we will discuss this later, and choose the transformation Ψ in (3.7) and (3.8) such that $\mathbf{x}_e=\Psi^{-1}(\mathbf{0})$. Now let $\mathbf{Z}$ be an open neighbourhood of the origin in $\mathbb{R}^n$, then equation (3.11) can be approximated for $\mathbf{z}\subset\mathbf{Z}$ by ignoring the second order terms and rearranged into the form

$$\mathbf{f}(h_k)(\Psi^{-1}(\mathbf{z})) = \sum_{i=0}^{n} \alpha_i^k(\mathbf{p})z^i \tag{3.12}$$

where $z^0:=1$ and $\mathbf{p}\subset\mathbb{R}^{n^2}\times\mathbb{R}^n$ is a point in the parameter space of synaptic weights u_j^i and constant terms c^i. Thus we see that if $\mathbf{f}(h_k)(\Psi^{-1}(\mathbf{z}))=0$ is to hold for $\mathbf{z}\subset\mathbf{Z}$ then we require

$$\alpha_i^k(\mathbf{p}) = 0 \ , \ i=0,\dots,n \ \ k=1,\dots,m \tag{3.13}$$

which provides us with m(n+1) equations in n(n+1) unknowns. As mentioned earlier, just after equation (3.11), we also have to consider how to impose $\mathbf{x}_e$ as an equilibrium point. In fact this is simple because from (1.3) $\mathbf{f}_{|\mathbf{x}_e}=\mathbf{0}$ implies

$$-x_e^i + c^i + \sum_{k=1}^{n} u_k^i\theta(x_e^k) = 0 \ , \ i=1,\dots,n \tag{3.14}$$

which provides us with a further n equations to add to (3.13). The general situation will be that of an underdetermined system of equations formed by (3.13) and (3.14), hence defining a specific subspace of the parameter space where solutions may be found. This is of importance in the next section.

4. Equlibrium Points and Stability

Let us assume that the conditions of proposition (3.1) are satisfied, then an orthonormal basis for ker(A) can be calculated by QR or SVD factorisation [5]. Thus if $Q=[Q_0,Q_1]$ is an n×n orthogonal matrix with the n-m columns of Q_1 spanning ker(A) then (3.6) can be satisfied by letting

$$U = Q_1P \tag{4.1}$$

for some (n-m)×n matrix P. In this section we investigate how to calculate a matrix P as in (4.1) in order to fix an equilibrium point of **f** and to ensure that the Jacobian matrix of **f** evaluated at the equilibrium point is stable, implying that all its eigenvalues lie in the left half complex plane.

Let the chosen equilibrium point be $\mathbf{x_e}$, then

$$\mathbf{df}(\mathbf{x_e}) = -I + U\mathbf{d}\theta(\mathbf{x_e}) \tag{4.2}$$

where **df** denotes the Jacobian matrix and

$$\mathbf{d}\theta(\mathbf{x_e}) = \text{diag}\,[\theta'(x_e^1)\,,\,\dots\,,\,\theta'(x_e^n)]$$

with $\theta'(x) := \dfrac{d\theta(x)}{dx}$.

From (4.1) we can prove that :-

i. we require $\mathbf{x_e} - \mathbf{c} \subset \text{im}(Q_1)$ (im(.) denotes image)
ii. m of the eigenvalues of $\mathbf{df}(\mathbf{x_e})$ are fixed with a value of -1.

For the proof of (i) let $\mathbf{f} = -\mathbf{x}+\mathbf{c}+U\theta(\mathbf{x})$ denote the vector field (1.3) and note that if $\mathbf{x_e}$ is an equilibrium point of **f** then $\mathbf{x_e} - \mathbf{c} = U\theta(\mathbf{x_e})$ and from (4.1)

$$Q_1P\theta(\mathbf{x_e}) = \mathbf{x_e} - \mathbf{c} \tag{4.3}$$

a solution exists for equation (4.3) iff $\mathbf{x_e} - \mathbf{c} \subset \text{im}(Q_1)$.

To prove (ii) let $G:=\mathbf{d}\theta(\mathbf{x_e})$ (note that G is diagonal), then taking the transpose of (4.2) and using (4.1) we see that

$$\begin{bmatrix} Q_0^* \\ Q_1^* \end{bmatrix}[-I+GP^*Q_1^*][Q_0,Q_1]=\begin{bmatrix} -I_m & \dots \\ 0 & -I_{n-m}+Q_1^*GP^* \end{bmatrix} \tag{4.4}$$

The matrix on the right hand side of (4.4) is block upper triangular with I_m denoting the m×m identity matrix, hence m of the eigenvalues of $\mathbf{d}\theta(\mathbf{x_e})$ will be fixed at -1 whatever the choice of P in (4.1).

In order to ensure the stability of $\mathbf{d}\theta(\mathbf{x_e})$ we simply have to calculate a matrix K such that all the elements of the spectrum of -I+K (denoted $\lambda(-I+K)$) lie strictly in the left half complex plane and then calculate a P such that

$$Q_1^*GP^* = K$$

which will always be possible because Q_1^* has rank n-m by construction, hence we can calculate

$$P^* = G^{-1}Q_1K \tag{4.5}$$

Now consider the calculation of P to fix an equilibrium point and ensure the stability at the point. It is obviously not necessary to have a diagonal matrix K in (4.5), however if K is chosen to be diagonal then we can extract some easily computed conditions to satisfy both the equilibrium point and the stability. The first condition is clearly

$$K = \text{diag}\,[\mu_i]\ ,\ \mu_i \in \mathbb{R}\ ,\ -\infty < \mu_i < 1 \tag{4.6}$$

to ensure the stability. Secondly assume that $\mathbf{x_e} - \mathbf{c}$ satisfies condition (i) above, then using (4.3) and substituting (4.5) for P we have

$$Q_1KQ_1^*G^{-1}\theta(\mathbf{x_e}) = \mathbf{x_e} - \mathbf{c}$$

because $\mathbf{x_e} - \mathbf{c} \subset \text{im}(Q_1)$ and Q is orthogonal this equation reduces to

$$KQ_1^*G^{-1}\theta(\mathbf{x_e}) = Q_1^*(\mathbf{x_e} - \mathbf{c}) \tag{4.7}$$

Now let

$$\mu_i = \frac{(Q_1^*(\mathbf{x_e} - \mathbf{c}))^i}{(Q_1^*G^{-1}\theta(\mathbf{x_e}))^i} \tag{4.8}$$

where $(.)^i$ denotes the ith component, then clearly equation (4.7) is satisfied and hence (4.3), also if the μ_i satisfy (4.6) then the stability is guaranteed in a neighbourhood of the equilibrium point.

In the case of C^∞ nonlinear output functions as usual we have to work a little harder in order to guarantee the stability in a neighbourhood of the equilibrium point, however we can discuss some general methods. These are based on Gerschgorin's theorems [5] and amount to the fact that if the elements of the matrix U in (4.2) satisfy the constraints

$$\sum_{j=1}^{n} |u_j^i \theta'(x_e^j)| < 1 \quad , \; i=1,\dots,n \tag{4.9}$$

then all the eigenvalues of $\mathbf{df}(\mathbf{x}_e)$ will lie in the left half complex plane.

It is obvious from (4.9) that we must use the underdetermination associated with equations (3.13) and (3.14) in order to calculate a U with a "suitably small" norm. From (3.13) and (3.14) we will have to solve the equation

$$A\mathbf{p} = \mathbf{b} \tag{4.10}$$

where A is a $(n(m+1)+m)\times(n(n+1))$ matrix with unknown rank. Assuming that a solution exists to equation (4.10) we would like to find one with minimal norm, hence tending to make the elements of U small because they form the first n^2 components of $\mathbf{p}$. Under these conditions a minimum norm solution is supplied with the aid of the SVD (Singular Value Decomposition) [5], where we calculate

$$A = V\Sigma W^* \tag{4.11}$$

for orthogonal matrices $V=[V_0,V_1]$, $W=[W_0,W_1]$ of compatible dimensions and $\Sigma=[\Sigma_0,0]$ with $\Sigma_0=\mathrm{diag}[\sigma_i]$ where the σ_i are the singular values arranged in nonincreasing order, $i=1,\dots,r$ and V_0,W_0 of column dimension r where r is the rank of A. A minimum norm solution is then given by

$$\mathbf{p} = W_0\Sigma_0^{-1}V_0^*\mathbf{b} \tag{4.12}$$

Clearly calculating $\mathbf{p}$ from (4.12) will not guarantee that the conditions (4.9) will be satisfied. If the resulting $\mathbf{df}(\mathbf{x}_e)$ is unstable after the calculation of $\mathbf{p}$ then one could consider adding constraints of the form

$$u_j^i = \tau_j^i \quad , \; 1 \le i,j \le n \tag{4.13}$$

for some small $\tau_j^i \in \mathbb{R}$. Obviously adding constraints of the form (4.13) could result in the overall system of equations (3.13), (3.14) and (4.13) becoming overdetermined. This being the case one can still use the SVD and (4.12) to determine a least squares solution, however the accuracy of the resulting equilibrium point and the closeness to zero of the output functions may be affected. Hence a comprimise between accuracy and stability should be sought.

5. Examples

In this final section we present some examples in order to clarify the above theory and introduce some of the associated numerical problems.

5.1 Example 1

For a network of dimension 4 we will use the following 3 linear output functions

$$h_1(\mathbf{x}) = x^1+2x^2+3x^3+4x^4$$

$$h_2(\mathbf{x}) = x^1-x^3$$

$$h_3(\mathbf{x}) = x^2+x^3$$

it is easily verified that this system has rank 3 and that

$$\mathbf{S} = \mathrm{span}\left\{\begin{bmatrix}-2\\2\\-2\\1\end{bmatrix}\right\}$$

Now select an equilibrium point $\mathbf{x}_e^*=[-0.5,0.5,-0,5,0.25]$, from equation (3.5) the vector $\mathbf{b}$ is zero and so $\mathbf{c}$ must be an element of ker(A) hence condition (i) just after equation (4.2) is trivially satisfied.

We used the QR decomposition [5] to calculate the orthogonal matrix $[Q_0,Q_1]$ associated with (4.1), in this particular case Q_1 has column dimension 1 and we let

$\mathbf{c}=-0.5Q_1=[-0.27735,0.27735,-0.27735,0.13867]^*$

this resulted in a value $\mu_1=0.42828$ when calculated using (4.8) and so stability is assured around $\mathbf{x}_e$. Finally combining (4.1) and (4.5) the matrix U was calculated as

$$\begin{bmatrix} 0.56075 & -0.56075 & 0.56075 & 0.26769 \\ -0.56075 & 0.56075 & -0.56075 & 0.26769 \\ 0.56075 & -0.56075 & 0.56075 & -0.26769 \\ -0.28037 & 0.28037 & -0.28037 & 0.13385 \end{bmatrix}$$

and it can be verified that

$\lambda(\mathbf{df}(\mathbf{x}_e)) = \{-1,-1,-1,-0.57172\}$ as expected.

5.2 Example 2

In this example we examine a 2-dimensional network with output function

$$h(\mathbf{x}) = x^1x^2 - \alpha$$

and for the new coordinates (3.7), (3.8) we choose

$$z^1 = x^1x^2 - \alpha$$

$$z^2 = x^2 - \beta$$

and the equilibrium point will be $\mathbf{x}_e^* = [\frac{\alpha}{\beta}, \beta]$. Now

$$\mathbf{x} = \Psi^{-1}(\mathbf{z}) = \begin{bmatrix} \frac{z^1+\alpha}{z^2+\beta} \\ z^2+\beta \end{bmatrix}$$

and

$$d\Psi|_{\mathbf{x}=\Psi^{-1}(\mathbf{z})} = \begin{bmatrix} z^2+\beta & \frac{z^1+\alpha}{z^2+\beta} \\ 0 & 1 \end{bmatrix}$$

which is easily seen to have rank 2 for $\mathbf{z}$ in a sufficiently small neighbourhood of the origin. From (3.11) we calculate

$$\mathbf{f}(h)(\Psi^{-1}(\mathbf{z})) = -2(z^1+\alpha)+c^1(z^2+\beta)+c^2\left(\frac{z^1+\alpha}{z^2+\beta}\right)$$
$$+(z^2+\beta)\left(u_1^1\theta\left(\frac{z^1+\alpha}{z^2+\beta}\right)+u_2^1\theta(z^2+\beta)\right)$$
$$+\left(\frac{z^1+\alpha}{z^2+\beta}\right)\left(u_1^2\theta\left(\frac{z^1+\alpha}{z^2+\beta}\right)+u_2^2\theta(z^2+\beta)\right)$$

The function θ in (1.2) can be expanded about the origin, after taking this approximation and ignoring terms of second order and above we are left with

$$\mathbf{f}(h)(\Psi^{-1}(\mathbf{z})) = -2(z^1+\alpha)+c^1(z^2+\beta)+c^2\left(\frac{\alpha}{\beta}+\frac{z^1}{\beta}-\frac{\alpha z^2}{\beta}\right)$$
$$+\frac{u_1^1}{4}\left(\frac{\alpha z^2}{\beta}+z^1+\alpha-\alpha z^2\right)+\frac{u_2^1}{4}(2\beta z^2+\beta^2)$$
$$+\frac{u_1^2}{4}\left(\frac{\alpha^2}{\beta^2}+2\frac{\alpha z^1}{\beta^2}-2\frac{\alpha^2 z^2}{\beta^2}\right)+\frac{u_2^2}{4}\left(\frac{\alpha z^2}{\beta}+z^1+\alpha-\alpha z^2\right)$$

which is seen to be in the form (3.12). Thus conditions (3.13) reduce to

$$-2\alpha + c^1\beta + \frac{c^2\alpha}{\beta} + \frac{u_1^1\alpha}{4} + \frac{u_2^1\beta^2}{4} + \frac{u_1^2\alpha^2}{4\beta^2} + \frac{u_2^2\alpha}{4} = 0$$

$$-2 + \frac{c^2}{\beta} + \frac{u_1^1}{4} + \frac{u_1^2\alpha}{2\beta^2} + \frac{u_2^2}{4} = 0$$

$$c^1 - \frac{c^2\alpha}{\beta} + \frac{u_1^1(\alpha-\alpha\beta)}{4\beta} + \frac{u_2^1\beta}{2} - \frac{u_1^2\alpha^2}{2\beta^2} + \frac{u_2^2(\alpha-\alpha\beta)}{4\beta} = 0$$

and these equations are added to

$$\frac{\alpha}{\beta} = c^1 + u_1^1\theta\left(\frac{\alpha}{\beta}\right) + u_2^1\theta(\beta)$$

$$\beta = c^2 + u_1^2\theta\left(\frac{\alpha}{\beta}\right) + u_2^2\theta(\beta)$$

hence producing in total 5 equations in the 6 unknowns $\{u_1^1, u_2^1, u_1^2, u_2^2, c^1, c^2\}$. Choosing $\alpha=0.2$ and $\beta=0.4$ the SVD (4.11) was calculated and produced the following set of singular values

$\sigma_1 = 2.87891$
$\sigma_2 = 1.49201$
$\sigma_3 = 0.15870$
$\sigma_4 = 0.11439$
$\sigma_5 = 0.00001$

which indicates clearly that the "numerical rank" of the coefficient matrix is 4. Thus setting the rank to 4 in (4.12) we calculate

$$\mathbf{c} = \begin{bmatrix} 0.35041 \\ 0.20966 \end{bmatrix}$$

and

$$U = \begin{bmatrix} 1.53728 & -0.40175 \\ 1.94560 & -0.49788 \end{bmatrix}$$

which results in an actual equilibrium point

$\mathbf{x}_e^* = [0.49870\ ,\ 0.39838]$ and

$\lambda(\mathbf{df}(\mathbf{x}_e)) = \{-0.99614\ ,\ -0.76221\}$.

Hence the resulting network is stable and has a point of equilibrium sufficiently close to the one required.

Some results from this example are shown in figures 1 and 2. In figure 1 the solid line represents the 1-surface defined by the output function h with α=0.2, whilst the dotted line is an actual system trajectory. In figure 2 we see the value of the output as a function of time.

Conclusions

In this paper we have demonstrated how a neural network can be used to solve a specific problem encountered in the control theory literature. Various conditions have been found which, if satisfied, guarantee a solution and we have illustrated how to calculate a matrix of synaptic weights in order to achieve the required behaviour. An important point of our methodology is that we compute the synaptic weights directly without requiring any learning phases.

In our future research we will further investigate the conditions to be placed on the output functions, in particular of course we would like to find necessary conditions for a solution. More research needs to be carried out on the numerical computation side and at the same time we will be addressing more problems from the control theory literature.

References

[1] Cohen,M.A. '*The Construction of Arbitrary Stable Dynamics in Nonlinear Neural Networks*' Neural Networks, 5, 1, pp(83-103), 1992.

[2] Isidori,A. '*Nonlinear Control Systems*' 2nd edition, Springer-Verlag, 1989.

[3] Gallot,S., Hulin,D. and Lafontaine,J. '*Riemannian Geometry*' Springer-Verlag, 1987.

[4] Olver,P.J. '*Applications of Lie Groups to Differential Equations*' Springer-Verlag, 1986.

[5] Golub,G.H. and Van Loan,C.F. '*Matrix Computations*' North Oxford Academic, 1983.

[6] Hopfield,J.J. '*Neurons with graded response have collective computational properties like those of two-state neurons*' Proc. Natl. Acad. Sci. USA, Biophysics, 81, pp(3088-3092), 1984.

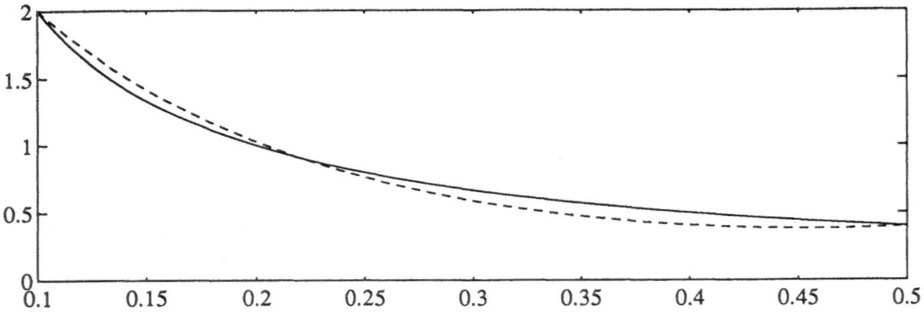

Figure 1

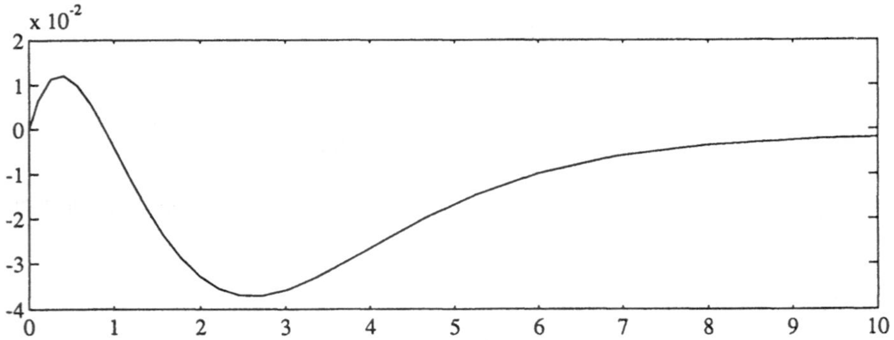

Figure 2

Evolving Recurrent Neural Networks

Kristian Lindgren[1], Anders Nilsson[1], Mats G. Nordahl[2], Ingrid Råde[1]

[1] Institute of Physical Resource Theory
Chalmers University of Technology
S-412 96 Göteborg
Sweden

[2] Santa Fe Institute
1660 Old Pecos Trail, Suite A
Santa Fe, New Mexico 87501
USA

Abstract

An evolutionary algorithm which allows entities to increase and decrease in complexity during the evolutionary process is applied to recurrent neural networks. Recognition of various regular languages provides a suitable set of test problems.

1 Introduction

A variety of applications of evolutionary methods to neural networks can be found in the literature (for a review see, e.g., [1]). Genetic algorithms have been used on their own as learning algorithms (e.g., [2, 3, 4]); hybrid approaches have for example used genetic algorithms to evolve a suitable network structure [5] and to evolve initial states for back-propagation [6]. The combination of a global evolutionary search with a local optimization algorithm that utilizes gradient information may prove to be a useful technique.

Almost all these applications have been to feed-forward networks. Recurrent networks form another area where evolutionary methods could be useful, in particular since learning algorithms similar to back-propagation [7, 8, 9] are less useful for fully recurrent nets due to excessive computational requirements. Evolutionary approaches to recurrent networks have not been extensively explored (see, e.g., [10, 11, 12] for a few exceptions). In [13] we introduced an algorithm where networks were allowed to grow during the evolutionary process, thereby allowing successive refinements of solutions to problems. This algorithm was in particular applied to the problem of recognizing strings from a regular language. In this contribution this algorithm is further explored; in particular we investigate its generalization ability.

An important feature of our approach is thus the use of representations of variable size, where entities that initially are very simple may become more complex during the course of evolution. In most genetic algorithm applications, representations in terms of fixed length binary strings have been used. Some exceptions can be found, e.g., the genetic programming paradigm of Koza [14], where the evolving entities are LISP programs (representable as trees), and crossover may interchange subtrees of varying size. Another example is given by the "messy genetic algorithms" introduced by Goldberg [15], where however an underlying fixed size representation is used. In the framework of more conventional learning algorithms several schemes where networks may grow and shrink have been introduced, in particular for the feed-forward case (e.g., [16, 17]), but also for recurrent networks, though with rather severe structure restrictions [18]. The importance of starting out with a small search space that can expand gradually has in particular been emphasized by Elman [19] in the context of neural network models of language learning.

Our approach was inspired by certain models of coevolution (e.g., [20]), where the interactions between species were modeled using game theory, and the genomes (corresponding to strategies in the game) were allowed to grow through neutral mutations analogous to gene duplication. In the simulations of coevolving strategies for the noisy iterated Prisoner's dilemma in [20], an initial state of simple strategies evolved towards greater complexity, and revealed the existence of pure strategies depending on the actions of both players during the two previous rounds which are evolutionarily stable, noise resistant, and achieve a nearly perfect score in the game.

Possible applications include time series prediction, and more generally problems involving sequences where correlations over longer distances have to be taken into account. We are presently exploring applications to protein secondary structure prediction. In this paper, however, only regular language inference will be considered (a regular language is a set of strings that a computational device with a finite number of internal states can recognize; we refer the reader to [21] for an introduction to automata theory). This provides us with a test domain where the difficulty of the problems to be solved can be varied in a systematic way (see [13] for further discussion).

2 The algorithm

In the simulations discussed below, the neurons s_i take values in the discrete set $\Sigma = \{0, 1\}$. This is useful for the problem of learning a regular language, since a network of continuous valued neurons need not correspond to a finite state device. Examples of embeddings of Turing machines in networks of a finite number of real valued units can be found, e.g., in [22] (see also [23], where computation universal piece-wise linear maps of the unit square were constructed; these could be viewed as networks of linear and threshold units). An approximate finite state description could in this case be extracted by partitioning the phase space, but the resulting finite automaton might depend sensitively on the partition, and no finite limit need exist as the partition is refined. In some cases, such as that of learning natural languages, it might be a distinct advantage to have a search space which includes both regular and more complex languages [24]; when we *a priori* restrict ourselves to regular languages this is not likely to be the case. The present algorithm could however be applied in the continuous case with minor modifications.

The connections w_{ij} can be either real valued or restricted to a few values (+1, 0, and −1); both cases will be investigated below. To each node i in the network is associated a threshold t_i, and the dynamics is given by simultaneously updating all units according to

$$\begin{aligned} s_i(t+1) &= 1 \quad \text{if} \sum_j w_{ij} s_j(t) > t_i \\ &= 0 \quad \text{otherwise.} \end{aligned} \qquad (1)$$

Input strings are fed sequentially to the network via a designated input node s_1. When the end of an input string is reached, 0's are fed to the input node for t_{accept} time steps. If the output node s_N is 1 at that point the string is accepted, otherwise it is rejected. The parameter t_{accept} is chosen independently for each network as the optimum in an interval $[0, .., t_{max}]$.

The adaptive moves allowed in the evolutionary process are of three kinds: connection mutations, threshold mutations, and node mutations, which change the size of the network by adding or deleting nodes (the present algorithm differs from that of [13] through the inclusion of evolving thresholds). In the discrete case, connection mutations change a non-zero weight to 0 with probability p_1 and flip its sign otherwise ($p_1 = 0.8$ in the simulations), while a zero connection is changed to +1 or −1 with equal probability. With real valued weights, a non-zero connection is removed with probability p_2 ($p_2 = 0.5$ was used), and otherwise replaced by a value (independent of the old one) drawn from a Gaussian distribution with zero average and standard deviation σ ($\sigma = 1.0$ below). A non-existent connection is added by randomly generating its value using the same distribution.

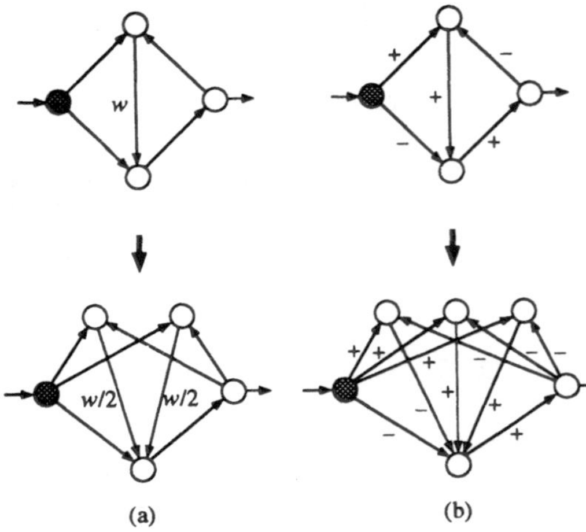

Figure 1: The action of a neutral mutation for (a) real-valued connections, (b) connections taking values ±1.

In the discrete case, thresholds will be restricted to the values +1, 0, or −1. The mutations on thresholds then operate in exactly the same way as the mutations on connections. In the continuous case, threshold mutations always introduce a new randomly chosen threshold.

Nodes are added to the network through neutral mutations that split one node into several without affecting the function of the network. In the case of weights with values in $\{+1, 0, -1\}$, two new nodes with identical incoming and mutually cancelling outgoing connections are added to the network, see Figure 1a. In the continuous case, several different types of neutral mutations are possible. In addition to mutations similar to Figure 1a, one could also consider mutations where the result is that the outputs from two nodes (instead of three) add up to the output from the node under consideration. In the simulations, we have used the procedure of Figure 1b, where inputs are replicated and outputs split equally. A promising but significantly more complex approach would be to instead let the mutation act on the inputs of a node, e.g., to add nodes to split the support of a certain node into two parts, allowing the specificity of individual neurons to increase. Node mutations may also remove nodes from the network; the fraction of size-decreasing node mutations was set to 0.75 in the simulations.

No crossover operator was used. We are not aware of any operator that would give a reasonable level of correlation between the fitness of parents and offspring [25] in the recurrent network case, where the repeated iteration of the network may amplify small differences, resulting in a more rugged fitness landscape. One approach with some promise could be to encourage modularity as proposed by Angeline and Pollack [26]. However, allowing the genome to grow during the course of evolution appears to serve the purpose of allowing solutions to gradually become more refined. This process will be investigated in more detail elsewhere [27].

The number of mutations in each update of an individual was chosen according to a discretized Gaussian distribution with average n_{mut} and standard deviation σ_{mut} (the values (1.0, 1.0) were used in the discrete case, and (2.0, 2.0) in the continuous case). The ratio between weight, threshold, and node mutations was 1:50:50. A population size of 20 was used. The reproduction step was based on ranking; the probability of removing a certain individual from the population was taken to be inversely proportional to the rank. The fitness of an individual was defined as the difference of the score on the training set and a small penalty term proportional to the size of the network.

3 Learning regular languages

We have applied evolving recurrent networks to regular language inference. A regular language is a set of finite strings that can be recognized by a computational device with a finite number of internal states, i.e., a finite automaton. A very simple example is given by the language $L = \{0, 1, 00, 01, 10, 000, 001, 010, 100, 101, 0000, 0001, \ldots\}$ of all finite strings without any occurrences of the substring 11 (the automaton accepting this language appears in Figure 2c). Regular language inference stands for the problem of guessing the language given a finite sample, a sample which could consist either of examples and counterexamples (positive and negative data), or only of strings in the language (positive data only). Here we consider only the simpler case of positive and negative data.

Regular language inference is an example of a problem where it is essential to allow the evolving entities to change in size. For good generalization one would like to find a minimal description consistent with the training set (for noisy data one would rather consider a trade-off between accuracy and description size based, e.g., on the minimum description length principle [28]). Finding the minimal finite automaton consistent with a set of positive and negative examples is known to be NP-complete [29]. Some heuristic algorithms for regular language inference can be found in the computer science literature, e.g., [30, 31], but it could still be worthwhile to explore the power of evolutionary methods in this context. An evolutionary approach, which involves a search of the space of hypotheses, naturally emphasizes generalization ability rather than the speed of the algorithm, and could be a useful complement to these heuristic algorithms.

Several attempts have also been made to use learning algorithms for recurrent neural networks for regular language inference, see, e.g., [32, 33, 24, 34, 35]). In most of these cases the size and structure of the network must be determined in advance. This restriction is avoided in our approach, where the network structure and the connections evolve simultaneously.

An interesting aspect of this problem is that one in fact attempts to generalize from a finite sample to an infinite set (with a finite description, the finite automaton). This makes the problem qualitatively different from that of supervised learning for feed-forward networks. Finding an appropriate measure of generalization in this situation is not entirely straight-forward; we have measured generalization by counting the fraction of correctly classified strings up to a certain length ($l = 14$ below). A less arbitrary measure would be the topological entropy ($\lim_{n\to\infty} \log N(n)/n$, where $N(n)$ is the number of strings of length n in the set) for the infinite set of incorrectly classified strings.

4 Results

Let us first consider the results for the relatively simple languages whose accepting automata are shown in Figure 2. Table 1 summarizes the simulation results for these five languages. In all cases averages over 100 runs (with different training data in each run, half of them positive and half negative) are shown. Training data were presented in two stages, where in the second stage a group of somewhat longer strings were added to the training set. The table shows the median of the total time required to train the network together with the average fraction of correctly clas-

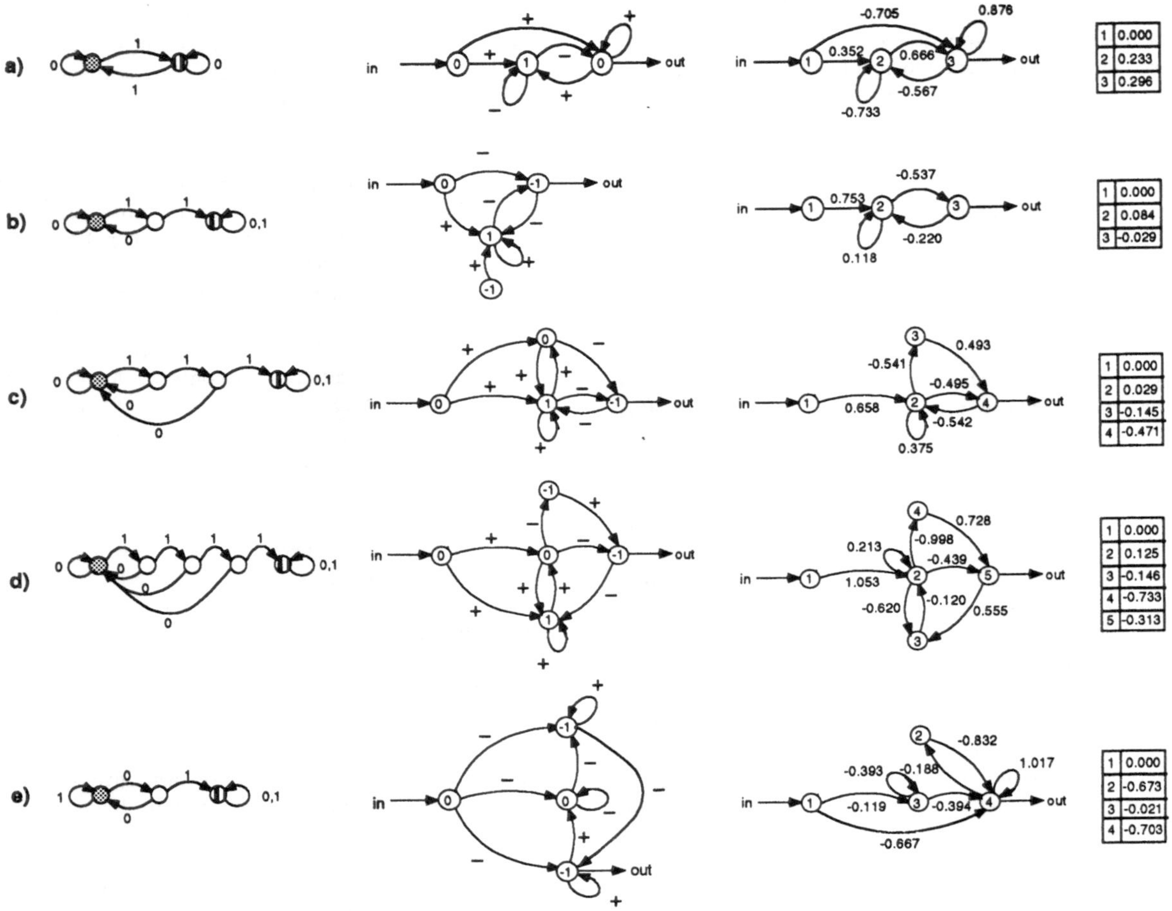

Figure 2: The first column shows the finite automata for the test languages: (a) parity, (b) $10^{2n-1}1$ forbidden, (c) 11 forbidden, (d) 111 forbidden, and (e) 1111 forbidden (start nodes are grey; striped nodes are forbidden final nodes). The second column gives the smallest networks discovered in the discrete case; the third column the smallest networks found with real weights (the values of the thresholds are shown to the right of the network).

sified strings of length ≤ 14, first for networks with weights in $\{+1, 0, -1\}$, then for networks with real valued weights.

The first of these languages (Figure 2a) consists of strings of even parity (strings containing an even number of 1's). Not surprisingly, parity turns out to be a significantly simpler problem in a sequential representation than for a feed-forward network. Of the problems studied, parity showed the shortest learning time in the discrete case with a median of 5682 individual updates. In the rest of our examples, we only consider automata where all nodes are allowed final nodes. The positive examples are then generated by a stochastic process obtained by chosing equal transition probabilities at nodes with two options for the next symbol. We have first considered a sequence of three finite complement languages (languages defined by a finite list of excluded blocks) where the blocks 11, 111, or 1111 are forbidden. The corresponding automata are shown in Figure 2b – 2d. While the time required for training increases significantly for the larger automata, reasonable generalization behavior is observed in all cases.

In the final example, Figure 2e, odd runs of 0's, i.e., finite blocks of the form $10^{2n-1}1$, are forbidden. This is the simplest conceivable example of a regular language which is not finite complement. For comparison, the parameters of the training set were chosen identically to Figure 2b. These automata might appear rather similar. However, sequences from the language corresponding to Figure 2e contain correlations over significantly longer distances than those

Problem	Fig.	training string length average	deviation	training set size	number of updates	general-ization	updates (real)	gen. (real)
Parity	2a	4	2	20	5682	0.995	60490	0.965
		6	2	20				
No 11	2b	3	2	15	18628	0.986	8669	0.991
		5	2	15				
No 111	2c	4	2	20	47281	0.971	187541	0.991
		6	2	20				
No 1111	2d	5	2	20	303869	0.960	272359	0.908
		7	2	20				
No $10^{2n+1}1$	2e	3	2	15	70267	0.965	166078	0.943
		5	2	15				

Table 1: For the languages whose accepting automata are shown in Fig. 2, the table shows the average and standard deviation of the training set string length for the first and second stages, the size of the training set, and the median number of individual updates needed and generalization fraction for networks with discrete and real weights, respectively.

Automaton	Fig.	training set parameters $(< l >, \sigma)_1$	size_1	$(< l >, \sigma)_2$	size_2	updates	general-ization
20,21		(3,2)	15	(5,2)	15	57280	0.997
220,131	3a	(3,2)	15	(4,3)	15	131230	0.809
030, 312	3b	(4,2)	15	(5,3)	15	86136	0.966
311,210	3c	(3,2)	20	(4,3)	20	91036	0.962
033,221	3d	(4,2)	15	(5,3)	15	372958	0.883
130,231	3e	(4,2)	15	(5,3)	15	626776	0.895
032,213		(3,2)	15	(5,2)	15	≈ 550000	0.868
130,221		(3,2)	15	(5,2)	15	76414	0.816
0141,4123	3f	(4,2)	20	(5,3)	20	63058	0.924
3044,2412	3g	(4,2)	20	(5,3)	20	759614	0.932
0133,4142	3h	(4,2)	20	(5,3)	20	≈ 650000	0.895
2430,1013	3i	(5,2)	20	(6,3)	20	≈ 600000	0.868
1420,3431		(4,2)	20	(6,2)	20	416098	0.819
4311,2042		(4,2)	20	(6,2)	20	69086	0.975
2311,0042		(4,2)	20	(6,2)	20	369150	0.964
02251,45203	3j	(6,2)	25	(7,3)	25	≥ 600000	0.879
45230,04011	3k	(6,2)	25	(7,3)	25	≥ 600000	0.956
04533,22144	3l	(6,2)	25	(7,3)	25	≥ 600000	0.816
23250,41001		(5,2)	25	(7,3)	25	≥ 600000	0.926
00454,23012		(5,2)	25	(7,3)	25	≥ 600000	0.960

Table 2: Results for a group of 20 randomly generated automata. The table shows the automaton (the notation is described in the text), the parameters of the training set: average $< l >$ and deviation σ of string length, and number of strings in stage 1 and 2, the median number of updates needed, and the average degree of generalization (fraction of strings of $l \leq 14$ classified correctly). Only discrete weights were considered.

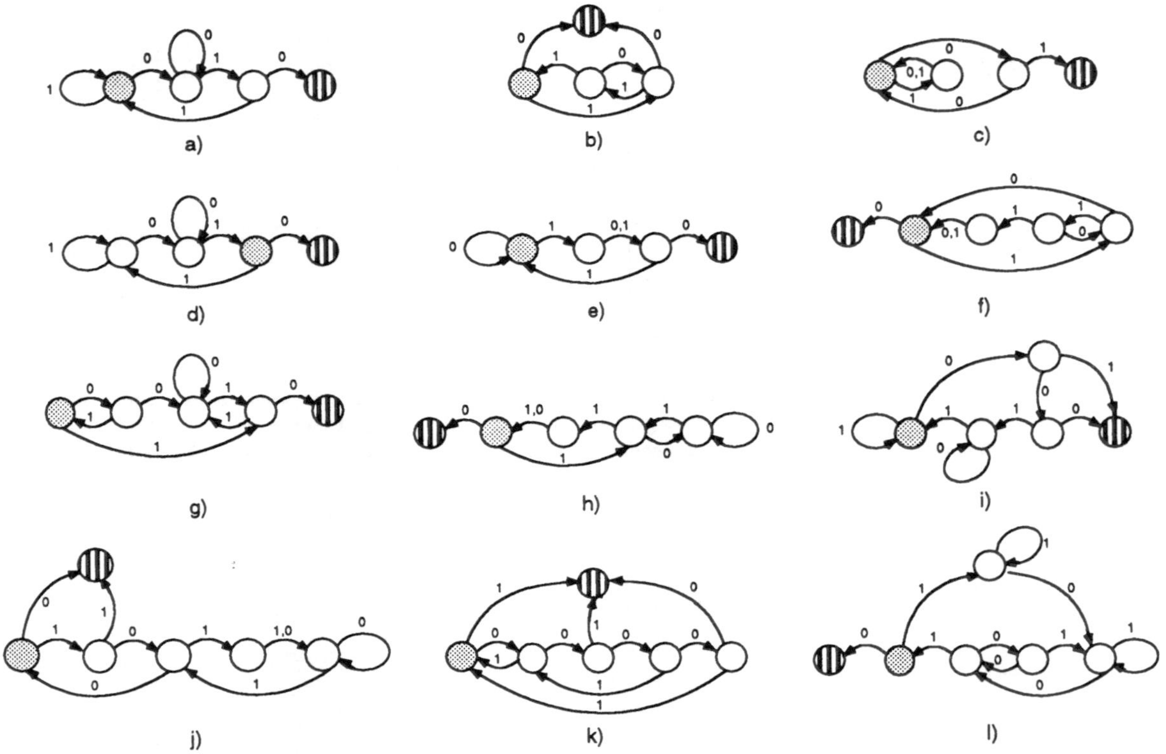

Figure 3: A selection of the randomly generated finite automata listed in Table 2.

of Figure 2b [13]. This fact is reflected in somewhat worse learning behavior for the automaton of Figure 2e when equal length training data are used. Examples of minimal network solutions found in each of these cases are also shown in Figure 2, both for discrete and real valued weights (a 3 node solution does exist for the case where 11 is excluded, but was not discovered in these runs).

Table 2 contains the results for a larger set of randomly generated finite automata with up to 5 states. Some of these are shown in Figure 3. In the first column of Table 2, automata are specified by giving two strings of integers $a(0)$, $a(1)$, where the integer $a(0)_i$ at position i gives the state to which there is a transition labeled by 0 from state i, unless $a(0)_i = 0$, in which case no such transition exists. States are denoted $1 \ldots N$; $a(1)$ represents transitions labeled by 1 in the same way. As an example, the automaton of Figure 2c would in this way be denoted $a(0)$, $a(1) =$ 111,230. Only the case of discrete weights is included in Table 2. At least 50 runs were performed for each automaton.

It is interesting to try to develop a better understanding of the generalization behavior, and how that depends on properties of the problem and the training set. We have measured how the generalization ability scales with the size of the training set for some of the automata in Table 2 and 3. The size of the training set is estimated as the product of the number of strings and the average length; as before generalization is measured in terms of the fraction of correctly classified strings of length $l \leq 14$.

Figures 4a and 4b show the results for two different automata with 3 states. The automaton of Figure 4a shows quite nice generalization behavior, in the sense that relatively little data gives a fairly high degree of generalization. However, this is somewhat misleading, since the automaton of Figure 4a has a rather low topological entropy (0.406), i.e., only a very small fraction of all strings are allowed. This means that even a network which only captures this fact, and nothing else of the structure of the automaton, can get a high generalization score. The automaton in Figure 4b has a larger fraction of allowed strings, but the remaining bias towards forbidden strings could still explain why the generalization ability initially appears to decrease with training set size. Beyond the the

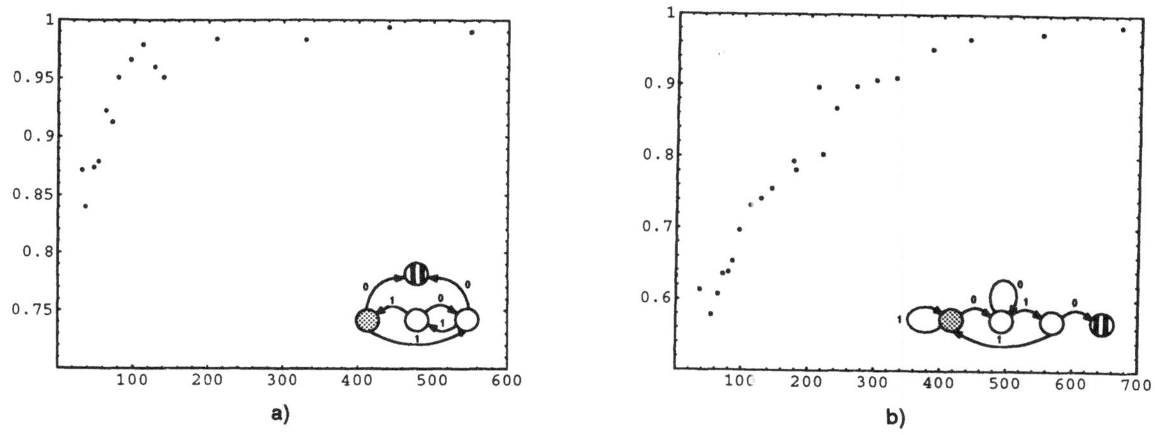

Figure 4: Generalization as a function of the training set size for two 3-state finite automata.

regime of very small training sets, the generalization error appears to decrease exponentially for both automata (for discussions for the feed-forward net case of exponentially decreasing generalization error versus worst-case bounds based on the Vapnik-Chervonenkis dimension see, e.g., [36, 37]).

5 Conclusions

We have trained recurrent neural networks to recognize various regular languages using an evolutionary algorithm where networks are allowed to grow and shrink during the evolutionary process. The generalization behavior in the case of generalization to an infinite set, e.g., a regular language, provides several interesting problems for further theoretical and experimental study.

Acknowledgements

We are grateful to the Santa Fe Institute, the Erna and Victor Hasselblad Foundation, and the Swedish Natural Science Research Council for support.

References

[1] Schaffer, J. D., Whitley, D., and Eshelman, J. L., *'Combinations of genetic algorithms and neural networks: a survey of the state of the art'*, in COGANN-92, International Workshop on Combinations of Genetic Algorithms and Neural Networks, pp. 1–37, IEEE Computer Society Press, Los Alamitos CA, 1992.

[2] Montana D. J. and Davis, L., *'Training feedforward neural networks using genetic algorithms'*, in Proceedings of the 11th Joint International Conference on Artificial Intelligence, pp. 762–767, 1989.

[3] Whitley, D. and Hanson, T., *'Optimizing neural networks using faster, more accurate genetic search'*, in Proceedings of the Third International Conference on Genetic Algorithms, pp. 391–396, Morgan Kauffman, San Mateo, CA, 1989.

[4] Whitley, D., Dominic, S. and Das, R., *'Genetic reinforcement learning with multilayer neural networks'*, in Proceedings of the Fourth International Conference on Genetic Algorithms, pp. 562–569, Morgan Kauffman, San Mateo CA, 1991.

[5] Miller, G. F., Todd, P. and Hedge, S. U., *'Designing neural networks using genetic algorithms'*, in Proceedings of the Third International Conference on Genetic Algorithms, pp. 379–384, Morgan Kauffman, San Mateo, CA, 1989.

[6] Belew, R. K., McInerney, J. and Schraudolph, N. N., *'Evolving networks: Using the genetic algorithm with connectionist learning'*, in Artificial Life II, pp. 511–547, Addison-Wesley, Redwood City CA, 1991.

[7] Elman, J. L., *'Finding structure in time'*, Cognitive Science, **14**, 179, 1990.

[8] Williams, R. J. and Zipser, D., *'A learning algorithm for continually running fully recurrent neural networks'*, Neural Computation, **1**, 270, 1989.

[9] Schmidhuber, J., *'A fixed size storage $O(n^3)$ time complexity learning algorithm for fully recurrent continually running networks'*, Neural Computation, **4**, 243, 1992.

[10] Bergmann, A., *'Self-organization by simulated evolution'*, in 1989 Lectures in Complex Systems,

pp. 455–463, (Addison-Wesley, Redwood City CA, 1990).

[11] de Garis, H., *'Steerable GenNETS: the genetic programming of steerable behavior in GenNETS'*, in Towards a practice of autonomous systems, Proceedings of the First European Conference on Artificial Life, pp. 272–281, MIT Press, Cambridge, MA, 1992.

[12] Torreele, J., *'Temporal processing with recurrent networks: An evolutionary approach'*, in Proceedings of the Fourth International Conference on Genetic Algorithms, pp. 555–561, Morgan Kauffman, San Mateo, CA, 1991.

[13] Lindgren, K., Nilsson, A., Nordahl. M. G., and Råde, I., *'Regular language inference using evolving neural networks'*, in COGANN-92, International Workshop on Combinations of Genetic Algorithms and Neural Networks, pp. 75–86, IEEE Computer Society Press, Los Alamitos CA, 1992.

[14] Koza, J. R., *'Genetic programming: On the programming of computers by means of natural selection'*, MIT Press, 1992.

[15] Goldberg, D. E., Deb, K. and Korb, B., *'Messy genetic algorithms: Motivation, analysis, and first results'*, Complex Systems, **3**, 493, 1988.

[16] Fahlman, S. E. and Lebiere, C., *'The cascade-correlation learning architecture'*, in Advances in Neural Information Processing Systems, Vol. 2, pp. 524–532, Morgan Kauffman, San Mateo CA, 1990.

[17] Frean, M., *'The upstart algorithm: A method for constructing and training feedforward neural networks'*, Neural Computation, **2**, 198, 1990.

[18] Fahlman, S. E., *'The recurrent cascade correlation architecture'*, technical report CMU-CS-91-100, School of Computer Science, Carnegie Mellon University, 1991.

[19] Elman, J. L., *'Incremental learning, or the importance of starting small'*, CRL Technical Report 9101, University of California, San Diego, 1991.

[20] Lindgren, K., *'Evolutionary phenomena in simple dynamics'*, in Artificial Life II, pp. 295–312, Addison-Wesley, Redwood City CA, 1991.

[21] Hopcroft, J. E. and Ullman, J. D., *'Introduction to Automata Theory, Languages, and Computation'*, Addison-Wesley, Reading MA, 1979.

[22] Siegelmann, H. T. and Sontag, E. D., *'On the computational power of neural nets'*, technical report SYCON-91-11, Rutgers Center for Systems and Control, 1991.

[23] Moore, C., *'Unpredictability and undecidability in dynamical systems'*, Physical Review Letters', **64**, 2354, 1990.

[24] Pollack, J. B., *'The induction of dynamical recognizers'*, Machine Learning, **7**, 227, 1991.

[25] Manderick, B., de Weger, M., and Spiessens, P., *The genetic algorithm and the structure of the fitness landscape*, in Proceedings of the Fourth International Conference on Genetic Algorithms, pp. 143–150, Morgan Kauffman, San Mateo CA, 1991.

[26] Angeline, P. J. and Pollack, J. B., *'Coevolving high-level representations'*, preprint.

[27] Lindgren, K., Nilsson, A., Nordahl. M. G., and Råde, I., in preparation.

[28] Rissanen, J., *'Stochastic Complexity in Statistical Inquiry'*, World Scientific, Singapore, 1989.

[29] Gold, E. M., *'Complexity of automaton identification from given data'*, Information and Control, **37**, 302, 1978.

[30] Biermann, A. W. and Feldman, J. A., *'On the synthesis of finite-state machines from samples of their behavior'*, IEEE Trans. Comput., **C-21**, 592, 1972.

[31] Miclet, L., *Grammatical inference*, in Syntactic and Structural Pattern Recognition Theory and Applications, World Scientific, Singapore, 1990.

[32] Cleeremans, A., Servan-Schreiber, D. and McClelland, J. L., *'Finite state automata and simple recurrent networks'*, Neural Computation, **1**, 372, 1989.

[33] Smith, A. W. and Zipser, D., *'Learning sequential structure with the real-time recurrent learning algorithm'*, International Journal of Neural Systems, **1**, 125, 1989.

[34] Giles, C. L., Miller, C. B., Chen, D., Chen, H. H., Sun, G. Z. and Lee, Y. C., *'Learning and extracting finite state automata with second-order recurrent neural networks'*, Neural Computation, **4**, 393, 1992.

[35] Watrous, R. C. and Kuhn, G. M., *'Induction of finite-state languages using second order recurrent networks'*, Neural Computation, **4**, 406, 1992.

[36] Schwarz, D. B., Samalam, V. K., Solla, S. A., and Denker, J. S., *'Exhaustive learning'*, Neural Computation, **2**, 374, 1990.

[37] Cohn, D. and Tesauro, G., *'How tight are the Vapnik-Chervonenkis bounds?'*, Neural Computation, **4**, 249, 1992.

Speeding Up Back Propagation by Partial Evaluation

Henrik Friborg Jacobsen[1], Carsten Krogh Gomard[2], Peter Sestoft[3]

Abstract

We automatically specialize a general Back Propagation learning algorithm to a particular network topology, obtaining a specialized learning algorithm which is faster than the general one.

The automatic specialization is done by a partial evaluator for a subset of the imperative programming language C.

1 Introduction

The training of a neural network may require many CPU-hours, or even CPU-days. In this paper we consider the training algorithm "Back Propagation" and find that the program transformation technique known as *partial evaluation* yields speedups of between 24 per cent and 51 per cent. We have tested the techniques on small nets (the XOR-problem) as well as large nets with many learning patterns (recognition of handwritten digits).

The key to the speedup is that the general Back Propagation algorithm performs much unnecessary repeated computation. Each time a pattern is presented and the weights are adjusted accordingly throughout the net, much testing on the *topology* of the net is done — even though it is known not to have changed since the last pattern processed.

Partial evaluation is a fully automatic program transformation technique, which given the general Back Propagation algorithm and a concrete net topology, automatically constructs a *dedicated* or *specialized* learning algorithm that works *only* for that topology. The advantage is that the specialized learning algorithm is more efficient than the general one.

[1]DIKU, Department of Computer Science, University of Copenhagen, Universitetsparken 1, DK-2100 Copenhagen Ø, Denmark.

[2]Computer Resources International, Bregnerødvej 144, DK-3460 Birkerød, Denmark. E-mail: cgo@csd.cri.dk. Most of this work was carried out while at DIKU.

[3]Department of Computer Science, Technical University of Denmark, Building 344, DK-2800 Lyngby, Denmark. E-mail: sestoft@id.dth.dk.

Outline

In Section 2.1 we describe partial evaluation. In Section 3 we give the structure of a general Back Propagation algorithm and point out what improvement (specialization) can be expected from partial evaluation. The effects of partial evaluation will be illustrated by considering the transformation of a small fragment of the algorithm. Finally, in Section 4 we comment on the sizes and runtimes of the involved programs.

2 Specialized programs

2.1 Partial evaluation

Assume we are given a program p that takes two pieces of input data d_1 and d_2. Denote the answer that is produced when p is run on d_1 and d_2 by ans. A short notation for this is:

$$p(d_1, d_2) = ans$$

A *residual program* for p with respect to d_1 is a program p_{d_1} such that the following holds for all d_2:

$$p_{d_1}(d_2) = ans$$

That is, the program p_{d_1} takes only one input (namely d_2) but produces the same result as would the original program p when applied to the two inputs simultaneously. In other words, p_{d_1} is a version of p, *specialized* to the value d_1 of its first argument.

For a simple example consider the following C-program that computes m+n:

```
int add(int m, int n)
{
   int sum;
   sum = n;
L: if (m)
      { sum = sum + 1;
        m = m - 1;
        goto L;
      }
   return sum;
}
```

Now suppose that we were given a concrete value for m, say 2. A residual program for add with respect to m = 2 would be a specialized program add_2 such as:

```
int add2(int n)
{
   int sum;
   sum = n;
   sum = sum + 1;
   sum = sum + 1;
   return sum;
}
```

This residual program is produced by simply executing all the statements depending only on `m` (such as the assignment `m = m-1;`) and by leaving all other statements unaltered to appear in the residual program (such as the assignment `sum = sum+1;`). Note that the assignment "`sum = sum+1;`" appears twice in the residual program, since with $m = 2$, two iterations of the loop will be necessary no matter the value of `n`.

A program that does partial evaluation (that is, one that produces residual programs) is called a *partial evaluator*. Calling the partial evaluator *pe*, the correctness criterion for a partial evaluator is that for any given program p and any data d_1:

$$pe(p, d_1) = p_{d_1}$$

such that for all d_2

$$p_{d_1}(d_2) = p(d_1, d_2)$$

2.2 Two steps can be faster than one

If d_1 is available at an earlier time than d_2, it may be reasonable to partially evaluate p with respect to d_1 to obtain p_{d_1} while "waiting" for d_2. But if d_1 and d_2 are supplied simultaneously, could there be any reason to use the more complicated two-step procedure (first generate p_{d_1}, then run it on d_2) instead of executing p on (d_1, d_2) in one step?

The answer is *yes*, and the reason is efficiency! An analogy: it is often faster to compile and run a program (a two-step procedure) than to execute the program using an interpreter (a one-step procedure).[4]

Loosely speaking, two-step evaluation may be faster if the second argument changes more frequently than the first argument within the program. Consider the case of the general Back Propagation algorithm, which has four input parameters:

1. A description of the network topology
2. Various parameters (learning rate, momentum, *etc.*)
3. The current learning pattern
4. The current values for weights and biases

In the algorithm, neither the topology of the net (item 1) nor the other parameters describing the net (item 2) change *at all*, whereas the patterns, weights and biases (items 3 and 4) change rapidly during execution of the algorithm. By partially evaluating the general Back Propagation algorithm with respect to the topology and parameters (items 1 and 2), all computations depending on these two inputs will be done once and for all instead of being repeated for each new pattern and each new set of weights. Section 3 will show the result of partially evaluating a fragment of the algorithm.

[4]This analogy is actually very close, as partial evaluators are often used to compile and even to generate stand-alone compilers [5].

2.3 How is partial evaluation done?

In this extended summary we shall not get into *how* partial evaluation is actually done. We emphasize that the techniques employed are fully automatic and that all experiments reported here are run on the computer.

Early partial evaluators operated on LISP-style functional programs (see for instance [3, 8]), or on simple imperative languages, intended mostly for theoretical use [4, 5]. More recently, considerable prog ss has been made in partial evaluation of practically useful subsets of imperative languages.

For example, partial evaluators constructed by Andersen [1, 2] and by Jacobsen [6] handle subsets of the programming language C which include arrays, pointers, and recursive procedures.

The book by Jones, Gomard, and Sestoft [7] also contains a chapter by Andersen on partial evaluation of C.

3 Application to back propagation

We have used a quite standard Back Propagation algorithm which we will not reproduce in its full length in this summary. Below we show the effects of partial evaluation on a fragment of the algorithm. The library function `logistic` computes

$$\texttt{logistic}(x) = \frac{1}{1 + e^{-x}}$$

```
for (X = no_of_inputs; X < no_of_units; X++ )
    { /* Compute activity for
         hidden and output units */
      if ( has_bias[ X ] )
        netinput = bias[ X ];
      else
        netinput = 0.0;

      for (X2 = 0; X2 < X; X2++)
        if ( has_weight[ X2 ][ X ] )
          netinput = netinput
                     + (activity[X2]
                        * weight[X2][X]);

      activity[ X ] = logistic( netinput );
      ...
    }
```

What is interesting about this piece of code is that much of the computation depends only on the topology of the net and not on the results of the training done so far. For a concrete example let us assume that we are given the following topology parameters:

```
no_of_inputs        = 2
no_of_units         = 5
has_bias[X]         = 1, for X ∈ {2,3,4},
has_bias[X]         = 0, otherwise
has_weight[X,X2]    = 1,
  for (X,X2) ∈ {(0,2),(0,3),(1,2),(1,3),(2,4),(3,4)}
has_weight[X,X2]    = 0, otherwise
```

Given the general algorithm fragment and the above topology data, our partial evaluator produced the following residual program which can be viewed as a specialized algorithm:

```
netinput = bias_2;
netinput = netinput + activity_0 + weight_0_2;
netinput = netinput + activity_1 + weight_1_2;
akt_2 = logistic( netinput );
netinput = bias_3;
netinput = netinput + activity_0 + weight_0_3;
netinput = netinput + activity_1 + weight_1_3;
akt_3 = logistic( netinput );
netinput = bias_4;
netinput = netinput + activity_2 + weight_2_4;
netinput = netinput + activity_3 + weight_3_4;
akt_4 = logistic( netinput );
```

Note the following

1. All testing on the topology of the net has disappeared.
2. Accordingly, the arrays holding topology information have disappeared.
3. The arrays holding weights and activities have been replaced by a set of individual variables.
4. The loops have been unrolled.

Obviously, this specialized algorithm will process the learning patterns faster than the original algorithm. In the next section we report statistics from our experiments.

In terms of Section 2.1, the general algorithm fragment corresponds to the program p, the topology data corresponds to the known input parameter d_1, our partial evaluator is pe, and the specialized algorithm shown above is p_{d_1}.

4 Speedup measurements

We specialized the general algorithm to the topology of a net with 5 units and 6 weights used to solve the XOR-problem. There were 4 learning patterns, iterated 3000 times. The results are in the table below:

Runtime, original algorithm	5.46 *sec.*
Runtime, specialized algorithm	2.66 *sec.*
Speedup	51 %

In a considerably larger experiment, we used a net with 382 units and 7240 weights to recognize hand written digits. There were 570 different learning patterns, iterated 100 times. Results:

Runtime, original algorithm	14181 *sec.*
Runtime, specialized algorithm	10774 *sec.*
Speedup	24 %

The speedup is not as large as for the XOR-problem, but still, a runtime consumption of three hours is preferable to a runtime consumption of four hours.

References

[1] L.O. Andersen. C program specialization. Technical Report 92/14, DIKU, University of Copenhagen, Denmark, May 1992.

[2] L.O. Andersen. Partial evaluation of C and automatic compiler generation (extended abstract). In U. Kastens and P. Pfahler, editors, *Compiler Construction, Paderborn, Germany, October 1992. (Lecture Notes in Computer Science, vol. 641)*, pages 251–257. Springer-Verlag, 1992.

[3] L. Beckman et al. A partial evaluator, and its use as a programming tool. *Artificial Intelligence*, 7(4):319–357, 1976.

[4] M.A. Bulyonkov. Polyvariant mixed computation for analyzer programs. *Acta Informatica*, 21:473–484, 1984.

[5] C.K. Gomard and N.D. Jones. Compiler generation by partial evaluation: a case study. *Structured Programming*, 12:123–144, 1991.

[6] H.F. Jacobsen. Speeding up the back-propagation algorithm by partial evaluation. DIKU Student Project 90-10-13, 32 pages. DIKU, University of Copenhagen. (In Danish), October 1990.

[7] N.D. Jones, C.K. Gomard, and P. Sestoft. *Partial Evaluation and Automatic Program Generation.* Prentice-Hall, 1993. To appear.

[8] N.D. Jones, P. Sestoft, and H. Søndergaard. Mix: A self-applicable partial evaluator for experiments in compiler generation. *Lisp and Symbolic Computation*, 2(1):9–50, 1989.

A New Min-Max Optimisation Approach for Fast Learning Convergence of Feed-Forward Neural Networks

A. Chella, A. Gentile, F. Sorbello, A. Tarantino

DIE - Department of Electrical Engineering, University of Palermo
Viale delle Scienze, 90128 Palermo Italy
Tel.: +39-91-6566111, Fax.: +39-91-488452, E-Mail: <sorbello@vlsipa.cres.it>

Abstract.- *One of the most critical aspect for a wide use of neural networks to real world problems is related to the learning process which is known to be computational expensive and time consuming.*

In this paper we propose a new approach to the problem of the learning process based on optimisation point of view. The developed algorithm is a minimax method based on a combination of the quasi-Newton and the Steepest descent methods: it was previously successfully applied in other areas and shows a faster convergence rate when compared with the classical learning rules.

The optimum point is reached by minimising the maximum of the error functions of the network without requiring any tuning of internal parameters.

Moreover, the proposed algorithm allows to obtain useful information about the size of the initial values of the weights by simple observations on the Gramian matrix associated to the network.

The algorithm has been tested on several wide-spread benchmarks. The proposed algorithm shows superior properties than the backpropagation either in terms of convergence rate and in terms of easiness of use; its performances are highly competitive when compared with the other learning methods available in literature.

Significant simulation results are also reported in the paper.

1. Neural network learning

A neural network can be described by a suitable network function

$$N(\mathbf{w},\mathbf{r}) = [O_1(\mathbf{w},\mathbf{r}), O_2(\mathbf{w},\mathbf{r}), \cdots, O_n(\mathbf{w},\mathbf{r})]^T \quad (1)$$

where $N(\mathbf{w},\mathbf{r})$ is an n-dimensional vector valued functions defined for any weight vector **w** and any input vector **r** and $O_i(\mathbf{w},\mathbf{r})$ i=1..n is the i-th output unit of the network.

The learning process of a neural network consists in finding a network function $N(\mathbf{w},\mathbf{r})$ which approximates a desired target function $T(\mathbf{r})$, defined over the same input space as the network function.

In order to estimate $T(\mathbf{r})$ on its domain T, the learning process requires $N(\mathbf{w},\mathbf{r})=T(\mathbf{r})$ on a subset of T, the so called training-set **R**.

The usual way to perform this task consist in
a) defining an error function associated to the network;
b) developing a technique to minimise this error function.

Under this point of view, the learning problem is a classic problem in optimisation.

A very large number of classical optimisation techniques have been tested on the learning problem. Most of them are iterative in nature and a general optimisation technique can be described as follows [1]:

let w_k be the weight vector at the k-th iteration, $\mathbf{d}_k$ the search direction and a_k the step along the same direction.

```
for (k=0; evaluate(wk) != CONVERGED;++k) {
dk=determine_search_direction ();
ak=determine_step ();
wk+1=wk+ak*dk
}
```

Optimisation algorithms vary in the way they determine a and **d**.

2. Application of the min-max algorithm to the learning process

A supervised learning rule can be revisited as an optimisation problem consisting in finding a set of weights which minimise the error function:

$$E(\mathbf{w}) = \sum_j \left(T_j - O_j(\mathbf{w})\right)^2 \quad (2)$$

This work has been supported by MURST (Ministero della Università e della Ricerca Tecnologica) and CNR (Consiglio Nazionale delle Ricerche) where **w** is a k-dimensional vector of all the weights associated to the network, T_j is the teaching input and $O_j(w)$ is the j-th output of the network.

Let's consider a feed forward neural network with n outputs. The error function E_p, associated to each input sample, can be computed as:

$$E_p = \sum_{j=1}^{n}\left(O_{pj} - t_{pj}\right)^2 \qquad p = 1,q \qquad (3)$$

where O_{pj} is the j-th component of the output vector and q is the number of training inputs.

The general statement of the minimax problem could be formulate as follows:

$$\min_{w}\{E\} = \min_{w}\left\{\max_{p}\{E_p\}\right\} \qquad (4)$$

where the network global error E is defined as the maximum error E_p with respect to all the input samples. More precisely, the maximum error value should be zero, but, actually the process will stop when the error is lower than a fixed quantity ε, depending on the particular problem at hand.

It is worth noting that, at the end of the training process, all the components E_p of the error are lower than ε (the network behaves in the same way for every input patterns) while this does not always happen by choosing, for instance, $E = \left(\sum_p E_p{}^2\right)^{\frac{1}{2}}$.

A full description of the general theory, on which the proposed algorithm is based, can be found in [2]. Some basic information are reported in order to allow the reader to understand the algorithm.

The function E is supposed to fulfil mild regularity conditions; namely, it is supposed to be continuous in R^k together with its first derivatives, although many examples are showed that these properties are not always indispensable.

An optimal point w° is supposed to exist in R^k. Bandler [3] has shown that the following conditions hold at the optimum point :

$$\sum_{p=1}^{r} u_p \nabla E_p^o = 0$$

$$u_p \geq 0 \ , \ p = 1,r \ ; \ r \leq q \qquad (5)$$

where the $E_p{}^o$'s represent the so called "equal maxima functions", the functions of **w** giving rise to r equal maxima with respect to the teaching input . If the functions $E_p{}^o$'s are convex in R^k then the above conditions are also sufficient.

A strategy to find the minimax optimum consists in find the equal maxima functions $E_p{}^o$ at the current point **w** and selecting one of those directions of R^k suitable for lowering simultaneously all the functions E_p.

In this section, $\nabla E_p{}^o$ is supposed to be non-zero irrespective of p, therefore, $n \geq 2$.

Using a vector notation, $\mathbf{C} \in \mathbf{R}^k$ is one of the above mentioned descent direction and the following inequalities must be satisfied:

$$\nabla E_p^o \cdot \mathbf{C} < 0 \qquad p = 1,r \qquad (6)$$

Let's define the sensitivity matrix **S** associated to the network as the matrix

$$\mathbf{S} = \begin{bmatrix} \nabla E_1 \\ \nabla E_2 \\ \vdots \\ \nabla E_r \end{bmatrix} \qquad (7)$$

whose rows are the components of the gradients of the equal maxima functions.

At each point $w \neq w^o$, a direction **C** satisfying the rel. (6) can be derived by solving the following equation set

$$\mathbf{S} \cdot \mathbf{C} = \mathbf{D} \qquad (8)$$

where **D** is an appropriate column vector whose components are all negative. In particular, $D_i = -f_i$, i=1,r.

It is possible to show [2] that at minimax point the non-homogeneous set (8) is inconsistent and vice-versa.

It is worth stressing that rel. (6) offer a way to determine some descent directions, others than those one(s) deriving from the set (8). These directions are easily detected by scanning the elements of two matrices derived from the sensitivity matrix S:

a) the Gramian matrix G

$$\mathbf{G} = \begin{bmatrix} \nabla E_1 \cdot \nabla E_1 & \nabla E_1 \cdot \nabla E_2 & \cdots & \nabla E_1 \cdot \nabla E_r \\ \cdots & \cdots & \cdots & \cdots \\ \nabla E_r \cdot \nabla E_1 & \nabla E_r \cdot \nabla E_2 & \cdots & \nabla E_r \cdot \nabla E_r \end{bmatrix} = \mathbf{S} \cdot \mathbf{S}^T \qquad (9)$$

b) the normalised Gramian matrix H which is the matrix of the normalised gradients

$$\frac{\nabla E_p}{|\nabla E_p|} \qquad p = 1,r \qquad (10).$$

The properties of these matrices are fully detailed in [2].

The algorithm is described below:

1. select a stop criterion $\varepsilon > 0$, an equal maxima measure δ and a maximum number of epochs (iterations) I;
2. at the current point **w** in the weights space, detect the set of the equal maxima functions $E_p = E_p(\mathbf{w}), \quad p = 1,r$, each differing from the maximum error value by less than δ and order them in descendent order;
3. IF r=1 THEN the direction opposite to the one of the gradient of the maximum-valued function is calculated and selected as descent direction $\mathbf{C_S}$; go to step 6.;
4. IF $r \geq 2$ THEN compute the components of the matrices **H** and **G**;

4.1 IF any of the terms of matrix H is -1 THEN EXIT (no descent direction exists);

4.2 IF r=2 THEN select the direction (if any) of the sum of the projections of the gradients on that plane of R^k where they are nearly aligned -descent direction $\mathbf{C_c}$; go to step 5.;

4.2.1 ELSE scan both matrix **H** and **G**;

4.2.1.1 IF a gradient has a positive inner product with the others, THEN select the corresponding direction as descent direction $\mathbf{C_A}$ and go to step 5.;

4.2.1.2 ELSE

4.2.1.2.1 IF the sum of all the gradients (either normalised or not) has positive inner product with all the gradients, THEN select the sum direction as the descent direction $\mathbf{C_B}$ and go to step 5;

4.2.1.2.2 ELSE select as descent direction $\mathbf{C_G}$ the Gauss direction

$$C_G = \frac{S^T \cdot G^{-1} \cdot D}{\left|S^T \cdot G^{-1} \cdot D\right|} \tag{11}$$

and go to step 6;

5. Select the Gauss direction as descent one as in step 4.2.1.2.2;

5.1 Between the found directions ($\mathbf{C_A}$ or $\mathbf{C_B}$ or $\mathbf{C_C}$) and $\mathbf{C_G}$, select the final descent direction $\mathbf{C_S}$ as that one giving rise to:

$$\max_z \left\{ \min_p \left\{ \nabla E_p^0 \cdot C_z \right\} \right\} \tag{12}$$

where z varies with the directions and p=1,r;

6. Compute a new point in the weight space w' performing an unidimensional optimisation based on the Golden Section.
7. IF any of the stop criterion (listed below) is satisfied THEN STOP;

7.1. ELSE go to step 2.

Stop conditions:

- sensitivity matrix **S** presents a null row;
- the difference between the minimum **w** point and w' is less than the prefixed quantity ε;
- the maximum number of epochs I is reached.

The strength of the algorithm mainly resides in lowering the maximum error by changing only an appropriate subset of weights. This subset is detected by scanning the Gramian matrix associated to the equal maxima error functions. Consequently, working on a reduced set of error functions (r<q) makes the method less expensive in terms of both storage and computations.

3. Experimental results

As known, the orography of the error surface presents a strong dependence from the specific problem at hand [4, 5] : for this reason, in this early experimental phase, the algorithm has been tested on a set of benchmarks showing a wide enough variety of shapes.

Moreover, the problem of scaling has been taken into account, in order to evaluate the performance of the method on even larger networks.

For all the benchmarks implemented, an effective strategy has got ready to choose the range of equal maxima δ, based on the ratio between, the greatest error function and the others.

Problem		Range	K	Trials	ε
Encoder	3-2-3	10	17	25	10^{-4}
	4-2-4	10	22	25	10^{-4}
	8-3-8	5	59	25	10^{-1}
	10-5-10	1	115	25	10^{-1}
Xor	2-2-1	10	9	25	10^{-4}
Parity	3-3-1	10	16	25	10^{-4}
	4-4-1	5	25	25	10^{-2}
	5-5-1	1	36	25	10^{-2}
Multiplexer	3-3-1	10	16	25	10^{-4}
Character Recognition	64-7-10	0.1	535	25	10^{-1}

Table 1

This strategy allows the user to avoid the annoying tuning of a variety of internal parameters, presents in most of the algorithms available.

As known, the choice of the range of initial network weights is an annoyng problem. In all the trials made, we successed in finding an optimal range for initial weights. This information is taken from simple observations on Gramian (or on Sensitivity matrix) which presents one or more rows of zeros when the range is too large. It is then possible to find the optimum by decreasing the range.

Table 1 shows the number of trials made for each test; K is the dimension of the weights space and ε is the minimum error value accepted. For all the networks implemented a sigmoidal function has been used as a transfer function of the units. The initial values of the weights are chosen randomly in [-Range,Range].

The results obtained for the tests employed, in terms of minimum, average and maximum number of epochs needed to convergence, are reported in Table 2.

Problem		Epochs		
		Max	Average	Min
Encoder	3-2-3	31	22	10
	4-2-4	48	35	14
	8-3-8	64	42	30
	10-5-10	26	22	19
Xor	2-2-1	13	8	4
Parity	3-3-1	20	14	8
	4-4-1	22	15	10
	5-5-1	25	17	11
Multiplexer	3-3-1	15	9	6
Character Recognition	64-7-10	49	27	20

Table 2

These first comparisons are quite encouraging.

Fahlman [5] obtained for the encoder 10-5-10, with a sigmoid as activation function, the same average number of epochs but the maximum value was 72 against 26 obtained with the proposed algorithm.

The performance of the proposed algorithm in the encoder problem, are also better than those reported by Kramer-Vincentelli in [1] with the Polak-Ribiere method, for which they have obtained the best results. This superior convergence speedup is even more evident with reference to the XOR task: the results reported in table 1 are clearly better than those obtained by Jacobs using his Delta-Bar-Delta learning rule [7].

We attempted also some experiments on character recognition problem, limiting this first approach to 8x8 patterns.

The chosen architecture is a multilayer perceptron 64-7-10 with sigmoidal units.

It is worth noting that the results are obtained by training the network with a very small set, in which only one representative of each digit is fed to the network. After training, the network has been tested by feeding it with noisy patterns. Each of these patterns is obtained by corrupting each digit with a certain percentage of pixels randomly generated.

Figure 1 shows the minimum, maximum and average errors obtained by averaging the results over 25 trials and using 10% noise level.

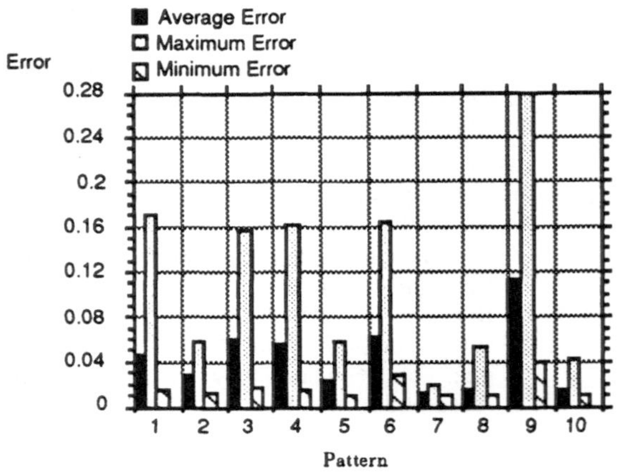

Figure 1

Figure 2 shows the trend of minimum, maximum and average errors versus the noise level. These values are obtained on the whole training set by averaging over 25 trials.

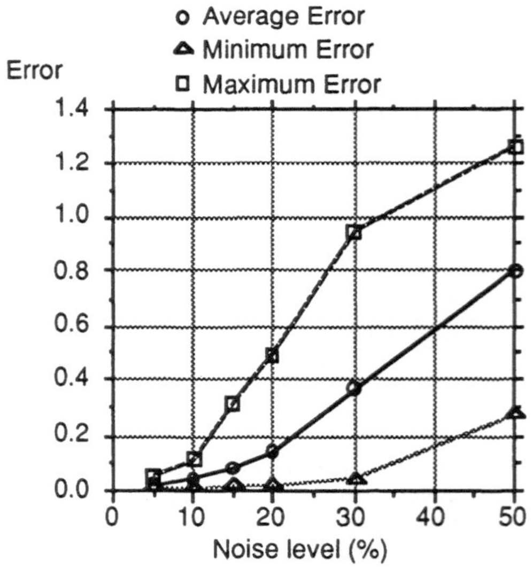

Figure 2

It is worth noting that an high recognition degree has been obtained with a noise level less than 20%.

Obviously, the performances of the network are reduced by increasing the noise level.

Conclusions

These first experimental results provide support for the good quality of the minimax approach to the learning process, when compared with the other optimisation methods developed in these years.

The proposed algorithm also presents easiness of use, due to the automatic tuning of internal parameters, and it is well-suited to parallel implementation.

The experimental phase showed the capability of the algorithm to provide useful information about the initial range of weights, giving a track to define a strategy fit to solve the annoying problem of the choice of the starting point of the network.

The first approach to character recognition showed also the capability of the algorithm to deal larger problems.

At present we are estimating the performances of the algorithm in character recognition using 16x16 bit-map from real world zip-codes.

References

[1] Kramer A H and A. Sangiovanni-Vincentelli *'Efficient Parallel Learning Algorithm for Neural Networks '*, in D. S. Touretzky (ed.), Advances in Neural Information Processing Systems, Morgan Kaufmann, 1989.

[2] Di Maio B and Sorbello F *'A Simple Algorithm for Min-Max Network Optimization '*, Alta Frequenza, 5, vol. LVII, 1988.

[3] Bandler J W *'Conditions for Minimax Optimum '*, IEEE Trans. on CT, July 1971

[4] Chella A and Sorbello F *'Fast Convergence of Neural Networks by Application of a New Min-Max Algorithm '*,in I. Alexander, J. Taylor (eds.), Artificial Neural Networks, 2, North-Holland Pubblisher, 1992

[5] Widrow B M Lehar A *'30 Year of Adaptive Neural Networks: Perceprton, Madaline, and Backpropagation '*, Proceedings of the IEEE, 78,9, 1990.

[6] Fahlman S E *'Faster Learning Variations on Backpropagation: an Empirical Study '*, in D.S. Touretzky G Hinton and Sejnowski (eds.), Proc. 1988 Connectionist Models Summer School, Morgan Kaufmann, 1988.

[7] Jacobs R A *'Increased Rates of Convergence Through Learning Rate Adaptation '*, Neural Networks, I, 1988.

Evolution of neural net architectures by a hierarchical grammar-based genetic system

Christian Jacob and Jan Rehder
Lehrstuhl für Programmiersprachen
Universität Erlangen-Nürnberg
Martensstraße 3, W-8520 Erlangen, Germany
Email: *jacob@informatik.uni-erlangen.de*

Abstract

We present a hierarchically structured system for the evolution of connectionist systems. Our approach is exemplified by evolution paradigms for neural network topologies and weights. Our descriptions of a network's connectivity are based on context-free grammars which are used to characterize signal flow from input to output neurons. Evolution of a simple control task gives a first impression about the capabilities of this approach.

1 Introduction

The design process for problem dependent neural network models usually consists of the following four stages: selection of a problem domain, selection or development of a suitable network architecture, choosing a learning algorithm for adjusting network specific parameters due to the problem domain, and testing of the network performance according to objective performance measures.

Although this sounds like a recipe for straight forward neural network development, the problem of designing application specific neural networks remains difficult due to the fact that the diverse phases of network design are interdependent and a great deal of experience is needed to choose suitable parameter values (Which architecture? Which connectivity? Which learning rule? ...).

Let us have a brief look at the problem of finding a suitable network architecture for a specific problem domain. This task often depends on the researcher's skill and experience to choose the proper architectural constraints for the net (More than one layer? Feed-forward connections only? Recurrent network?). In many cases, however, suitable net topologies only evolve from the process of supervised or unsupervised learning. With feed-forward networks trained by the error-backpropagation method the number of hidden layers and units may vary according to some performance criterions; with self-organizing networks not only the number of units needed is unknown, but also their connectivity structure. As Miller, Todd and Hegde [1] express it, "the network design stage remains something of a black art". The same is true for the selection of proper neuron activation functionality as well as learning rules for adjusting network weights and connectivity.

However, the "black art of network design" is not as "black" as it seems. Often the network designer has a rough idea about which neural network models could be tried to solve, say, classification, motion control or feature mapping tasks. In most cases there are some "rules of thumb" for special parameter values like the number of hidden units, the learning rate, the neuron activation function etc. So why not use this experience to enhance evolutional development of problem specific neural network models?

Nature solves part of the design problems of natural nervous systems through evolution mechanisms. To put it very simply, the network development process is naturally done in two stages: First, a coarse connectivity structure within specialized networks is evolved; the information about this structuring process is contained in genetic strings and has evolved through genetic mechanisms. Second, fine tuning of the network is done through neurological mechanisms controlled by environmental input signals, which cause the network to "learn" and adjust its performance to specialized tasks. Natural genetic coding, however, is much more complex and hierarchically organized than currently used bit-string codings applied in most of the genetic algorithm systems.

With these ideas in mind we want to develop an evolutional system which supports the different neural network design phases as described above. With "evolutional system" we mean a hybrid system capable of using genetic or evolutional as well as neural mechanisms for adaptation and learning.

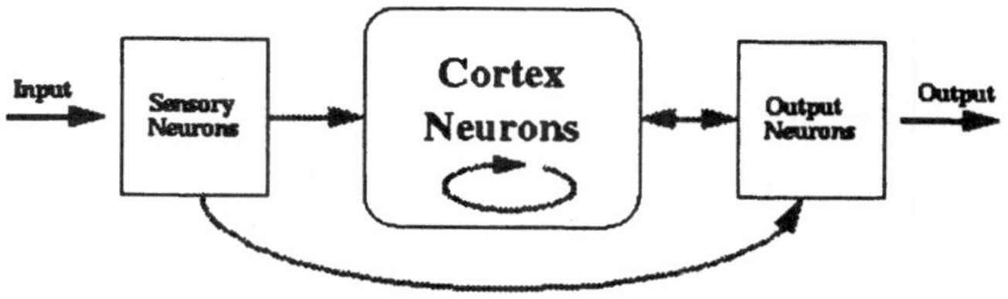

Figure 1: Neural net model

First of all we are proposing a coarse structure for an evolutional system for problemspecific neural network design. Secondly, we give a short description of our grammar-based approach for evolving neural net connectivity. The outlined system is currently being implemented and will serve as a basis for a hybrid system to get a better understanding when to switch from genetic search and adaptation methods to neurally inspired parameter adaptation and learning as well as to examine the use of higher-order codings and evolutional operators for a broad range of problem domains.

2 Related work

Several articles have been published concerning neural network design with the help of genetic algorithms. Most of the articles focus on genetic optimization techniques of neural net connectivity for the specialized class of feed-forward networks trained by error backpropagation algorithms [2], [3], [5]. In [5] and [2] two neural network development systems are described in detail, where genetic operators and search are used to evolve architectures for feed forward networks which are then trained through (modified versions of) backpropagation algorithms and evaluated by testing the resulting networks on predefined test data. In [3] neural learning is replaced by a genetic algorithm, leaving the fine-tuning or optimization of weight values to the genetic operators. Koza and Rice [6] evolve net connectivity and weights simultaneously by using a LISP S-expression coding. Bornholdt and Graudenz [7] let their GA-system operate on linked list data structures coding the net structure as well as the weights.

3 Hierarchical neural network design: A brief overview

We use an approach similar to [7] and [5] to evolve neural net connectivity, single neuron functionality and connection values. The network model used (see figure 1) consists of a predefined set of input and output neurons and a set of cortex neurons to be evolved. The input neurons are connected in feedforward direction only, i.e. inputs can be passed to either hidden neurons or output neurons directly. The cortex neurons are connected either to hidden or output neurons.

The functionality of the hidden neurons is depicted in figure 2: A summation function collects the neuron's incoming signals; this function might be a weighted summation of the inputs o_i, a Sigma-Pi-function or another appropriate input processing function. The summation value is processed by an activation function (linear, sigmoid, radial basis etc.) resulting in an internal activity of the neuron, which is taken by the output function (identity, linear threshold etc.) to compute an externally visible output value o_p that can be passed to other neurons.

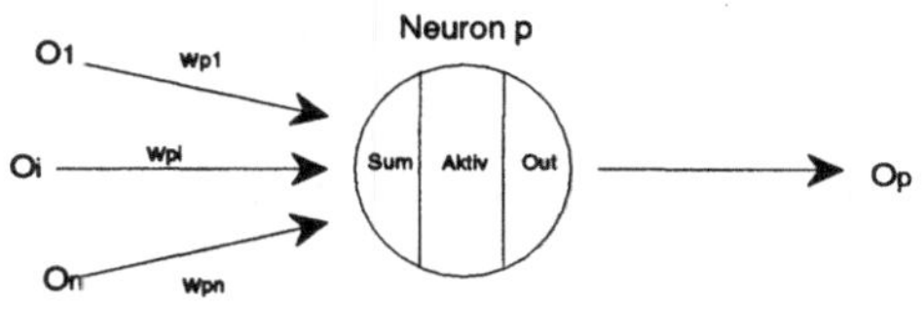

Figure 2: Model of a cortex neuron

The design process for a neural network then has to evolve a connectivity structure for the input, cortex and output neurons, a set of functional parameters defining the functionality of the hidden neurons, and a set of weight values for all connections. For each of these three, partly interdependent phases we

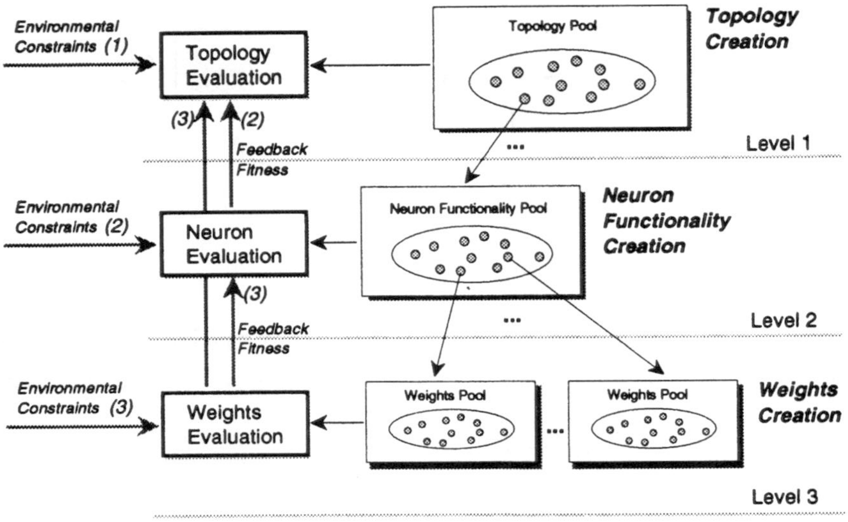

Figure 3: Design Hierarchy

use different string populations and codings. So the string length is usually rather short, and the population sizes can be kept small without giving up too much of a population's diversity. Thus, search spaces for the genetic algorithms are reduced. The results of the structured evolution and evalutation processes can be interpreted more easily by a human supervisor.

The evolution process for a neural network could be described as in figure 3. At each level there is a creation module which evolves a pool of competing chromosomes (= strings of coded parameters) and an evaluation module testing chromosome fitness via problemspecific constraints. For each level of pools there are two fitness values to be taken into account for each chromosome: environmental constraints define a "coarse" fitness value, whereas a kind of fitness fine tuning can be achieved by using "feedback" fitness values from levels below. At each level specialized parameter representations ("chromosomes") are used to generate new chromosome populations from previous ones. Strings and respective feedback fitness values are the only interface between the different evolution levels. Figure 4 outlines the coarse structure of an evolution module serving as a building block for the hierarchical evolution system.

A brief example might explain these ideas. Suppose we want to evolve feedforward networks for pattern classification. So the constraints for net topology (level 1) will accept only chromosomes describing feedforward connectivity, and will restrict the number of neurons, the maximal path length from input to output neurons or the connection density. Neuron functionality (level 2) could be constrained to activation functions that settle to zero for large absolute values. As a last constraint the weights (level 3) might be restricted to real values from the interval between zero and one.

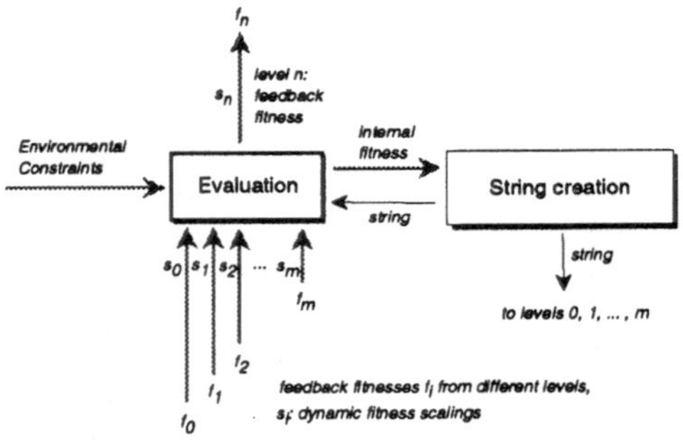

Figure 4: Structure of a string evolution module

With this structured network design it is easy to alter the sequence of modules and the number of modules used to evolve and optimize problemspecific net architectures. Furthermore, genetic modules may be replaced by, e.g., neurally inspired algorithms: Topology evolution might be done by a self-organizing process controlled by a set of input signals to the network. Neuron functionality can be tuned by a learning rule adapting, say, the threshold value for the

Topology	::=	*Path PathList**
PathList	::=	";" *Path*
Path	::=	*InputNeuron NeuronList OutputNeuron*
NeuronList	::=	(*CortexNeuron* | *OutputNeuron*)*
InputNeuron	::=	i, where $i \in$ *Sensory*
OutputNeuron	::=	o, where $o \in$ *Output*
CortexNeuron	::=	k, where $k \in$ *Cortex*

Figure 5: Production rules P of grammar G for net topology description

internal activation function. Finally, weight changes may be achieved with error-correcting rules (Hebb, generalized delta etc).

4 Net topology evolution: A grammar-based approach

Concerning the problem domain of net topology evolution we had especially two ideas in mind:

First of all, we want a parameter coding that is easily interpretable by human experts without the need to explain complicated decoding algorithms. This means that the structures we want to evolve should be rather close to (easily understandable) formal descriptions of the problem domain. [1]

Secondly, we wanted our system to behave much like a human expert trying to systematically improve network performance, although the evolution system should be far more efficient and persistent. With these operators the evolution processes should be much more intelligible, especially for people not interested in a deep understanding of evolution system tuning. This is an important prerequisite for an evolution system which shall be used not only for toy problems, but for "real world applications". [2]

In the following sections we will exemplify the basics of our approach to net topology evolution; this is only one level in our evolution system but the same ideas are applicable to the other levels as well. We will describe the coding and the operators used in evolving net topologies for the neural net model depicted in figure 1.

[1] Impressive results have been obtained by Koza [6] with the genetic programming paradigm based on the evolution of LISP-S-expressions. A more general grammar-based approach and some more arguments for the use of higher-order, problem-dependent codings have been presented by Antonisse [8].

[2] It seems to be a general problem of genetic or evolutionary systems that control parameter tuning remains a very difficult problem. This is especially true for the domain of neural network optimization. But this only means that the "black art problem" only has been shifted from one domain to another.

4.1 Grammar-based genetic parameter coding

In the current version of our network design system we use a contextfree grammar to describe and evolve strings that represent net connectivity structures. [3]

Our (contextfree) grammar $G := (N, T, S, P)$ is charcterized by

- non-terminals $N = \{$ *Topology*, *Path*, *PathList*, *NeuronList*, *InputNeuron*, *CortexNeuron*, *OutputNeuron* $\}$,
- terminals $T = \{\text{";"}\} \cup Sensory \cup Cortex \cup Output$,
- a startsymbol $S = Topology$ and
- production rules P as specified in figure 5. [4]

We refer to n input neurons, m output neurons and a previously undefined number of cortex neurons by the following sets of symbols:

$$\begin{aligned} Sensory &:= \{i_1, i_2, ..., i_n\} \\ Output &:= \{o_1, o_2, ..., o_m\} \\ Cortex &:= \{1, 2, 3, ...\} \end{aligned}$$

Note that the numbers n and m of sensory and output neurons, respectively, are predefined due to the problem domain, whereas the number of cortex neurons has to be evolved.

[3] A first advantage of the grammar approach is that the rules of the grammar can be used to generate strings which then automatically belong to the language $\mathcal{L}(G)$ of the grammar. Within the current prototype implementation strings are not modified by operators referring directly to the given grammar G; this will be done in future versions of the evolution system. Closure with respect to $\mathcal{L}(G)$ is only assured by appropriate definitions of the string operators.

[4] The non-terminals on the left side of the "::="-sign can be replaced by the strings on the right side. The string creation process begins with the startsymbol. A string t is defined to be in the language $\mathcal{L}(G)$ of G if t can be created from startsymbol S by a finite number of applications of the production rules P, and if t only consists of terminal symbols from T.

Each string generated by grammar G produces a list of paths from input to output neurons. As we will see even loops or recurrent connections can be modelled by this grammar. The following example string is produced by the grammar:

$$w = i_1 1223 o_1 \; ; \; i_1 3 o_1 1 o_2 \; ; \; i_2 21 o_2$$

The input neuron set is $Sensory := \{i_1, i_2\}$, the cortex neuron set is $Cortex := \{1,2,3\}$, and the set of output neurons is $Output := \{o_1, o_2\}$. Three paths describe the connection structure from input neurons over cortex neurons to output neurons (see figure 6):

$$\begin{aligned} p_1 &= i_1 \rightarrow 1 \rightarrow 2 \rightarrow 2 \rightarrow 3 \rightarrow o_1 \\ p_2 &= i_1 \rightarrow 3 \rightarrow o_1 \rightarrow 1 \rightarrow o_2 \\ p_3 &= i_2 \rightarrow 2 \rightarrow 1 \rightarrow o_2 \end{aligned}$$

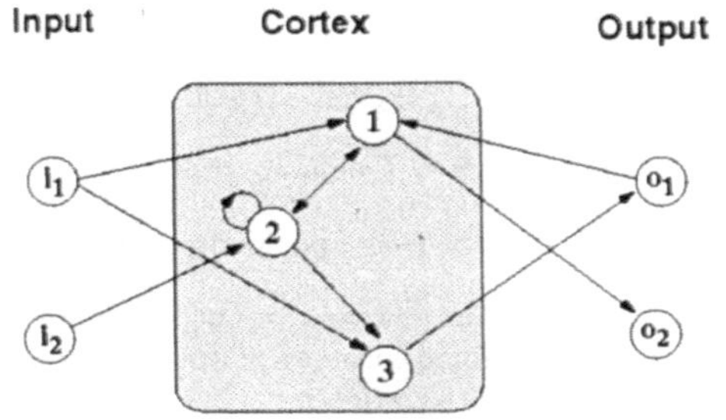

Figure 6: Example network produced by grammar G

As each path starts with an input and ends with an output neuron it is guaranteed that input signals eventually reach some of the output neurons and that there are no useless neurons which do not lie on any path from input to output neurons. Thus, all cortex neurons take part in the calculation of the output signals. If cortex neuron 3 had no output connection then it would be useless as it would only receive signals but never pass its own signals to other neurons; this effect is automatically prevented by the grammar.

The path descriptions can be decoded easily into a network (like in figure 6) by mapping the paths into an adjacency matrix; with this matrix at hand duplicate parts of the path description are eliminated (as e.g. the edge from neuron 1 to neuron o_2 in path p_2 and p_3). As we will see in the following sections, redundancy of the connectivity coding is essential for the evolution process.

4.2 Mutation on grammar-based chromosomes

Mutation on the grammar-based chromosomes should assure that we do not leave the language $\mathcal{L}(G)$ of the grammar G, i.e. we want closed operators on the strings.

Let $w = p_1; p_2; ...; p_n$ be a path concatenation, and let w' be the path resulting from applying the following mutation operators.

1. Create a new path p and insert p into w at position k:

$$w' = p_1; p_2; ...; p_{k-1}; p; p_k; ...; p_n$$

2. Remove a path p_k from w:

$$w' = p_1; p_2; ...; p_{k-1}; p_{k+1}; ...; p_n$$

3. Select a path $p_k = i_k c_{k1} c_{k2} ... c_{km} o_k$ from w and insert a new neuron $c \in Cortex \cup Output$ anywhere between i_k and o_k. Neuron c can be one of the neurons still available or can be an additional cortex neuron:

$$p'_k = i_k c_{k1} ... c_{kl-1} c c_{kl} ... c_{km} o_k$$

$$w' = p_1; p_2; ...; p_{k-1}; p'_k; p_{k+1}; ...; p_n$$

4. Select a path $p_k = i_k c_{k1} ... c_{kl-1} c_{kl} c_{kl+1} ... c_{km} o_k$ from w and remove a neuron c_{kl}:

$$p'_k = i_k c_{k1} ... c_{kl-1} c_{kl+1} ... c_{km} o_k$$

$$w' = p_1; p_2; ...; p_{k-1}; p'_k; p_{k+1}; ...; p_n$$

It is easy to see that we do not leave the grammar language $\mathcal{L}(G)$ when we apply these operators. In our implementation we use these mutation operators in the following way:

- Operators (1) and (2) are responsible for global search in the path description domain. That is why they are used with rather low probability rates. [5]
- An "edge-add" operator selects a path from the chromosome, duplicates the path, introduces and deletes neurons through operators (3) and (4) on the duplication string, and then inserts the mutated path into the chromosome. This introduces duplicate connectivity descriptions but extends the topological structure very smoothly because

[5] The effects of operators (1) and (2) are comparable to the mutation operators used for genetic algorithms on bit-strings.

it integrates existing single neurons into existing paths; due to redundant coding this does not disturb the overall connection structures very much but may just introduce a signal path necessary to solve an input-output mapping. Degree of redundancy is restricted by limiting the path lengths as well as the number of paths per string.

- To remove single edges we use operator (4).

4.3 Crossover on grammar-based chromosomes

The crossover operator picks out two chromosomes $chrom_1$ and $chrom_2$ and selects two crossover points k_1, k_2 and l_1, l_2 within each chromosome, respectively. The crossover points must lie between the paths, i.e. at the locations of the path separators ";". Then the path lists of $chrom_1$ and $chrom_2$ between the crossover points – the chromosomes are treated as ring structures – are exchanged resulting in new chromosomes $chrom_1'$ and $chrom_2'$: [6]

$$\begin{aligned} chrom_1 &= p_1; \dots; p_{k_1-1}; \underline{p_{k_1}; \dots; p_{k_2}}; p_{k_2+1} \dots; p_n \\ chrom_2 &= q_1; \dots; q_{l_1-1}; \underline{q_{l_1}; \dots; q_{l_2}}; q_{l_2+1} \dots; q_m \end{aligned}$$

$$\begin{aligned} chrom_1' &= p_1; \dots; p_{k_1-1}; \underline{q_{l_1}; \dots; q_{l_2}}; p_{k_2+1} \dots; p_n \\ chrom_2' &= q_1; \dots; q_{l_1-1}; \underline{p_{k_1}; \dots; p_{k_2}}; q_{l_2+1} \dots; q_m \end{aligned}$$

5 Some implementation details

5.1 Simulation system

Currently we have implemented levels 1 and 3 of the design hierarchy (see fig. 3). [7] The topology and weights creation modules can be distributed over a network of workstations communicating via socket interfaces. The topology module creates path descriptions as referred to in the last section and sends these strings in the compressed form of connectivity matrices to an array of subprocesses. These weights creation processes work independently from each other; they evaluate each topology string they have been sent by generating a pool of weights settings for the received topology. Each weights setting together with the topology then describes a fixed network structure which now can be evaluated for a predefined test environment (e.g. a pattern classification or parameter control task). For a fixed number of generations a genetic algorithm [8] for weights evolution then tries to find an optimal weights setting within the given environment. Finally, a weights string evolves which lets the network solve its task in an optimal way, and a fitness value serving as a performance measure for this network structure is returned to the topology module. These fitness values then control the evolution process at level 1.

5.2 First simulation results

For our first test experiment we defined a control task similar to the well-known pole-balancing task. A small ball thrown on a seesaw at random position and with random initial speed has to be balanced to the seesaw's centre. The control task is said to be successfully solved whenever the ball comes to rest near the seesaw hinge (see fig. 7).

The seesaw is controlled by networks as depicted in fig. 8 consisting of three input neurons and a single output neuron. The suitable number of cortex neurons has to be evolved. The input neurons take the current seesaw angle, the velocity of the ball and the position of the ball, accordingly. The output neuron controls the seesaw's delta angle.

5.2.1 A simple test experiment

Evaluation of a population of networks – all with the same topology structure – is performed as follows:

1. *Evaluate each weight setting:*
   Select initial position and speed of the ball.
   Let the network control the ball for a fixed number of cycles.
   To measure the network's performance distinguish three cases:
   (a) *The ball is tossed from the seesaw.*
   (b) *The ball is still on the seesaw but does not come to rest.*
   (c) *The network succeeds to bring the ball to rest; the ball's distance to the seesaw's centre is taken into account.*
2. *Perform selection and other GA-operators for a predefined number of generations on the weights setting population*
3. *Re-evaluate the best evolved weights settings:*
   In order to calculate a normalized fitness value, select initial test positions and speeds of the ball for a

[6] Without loss of generality: $k_1 \le k_2$, $l_1 \le l_2$ and $d_1 := (k_2 - k_1) < (l_2 - l_1) =: d_2$

[7] The functional properties of the processing elements (neurons) remain fixed during the evolution processes.

[8] We use binary as well as floating point representations for the weight values.

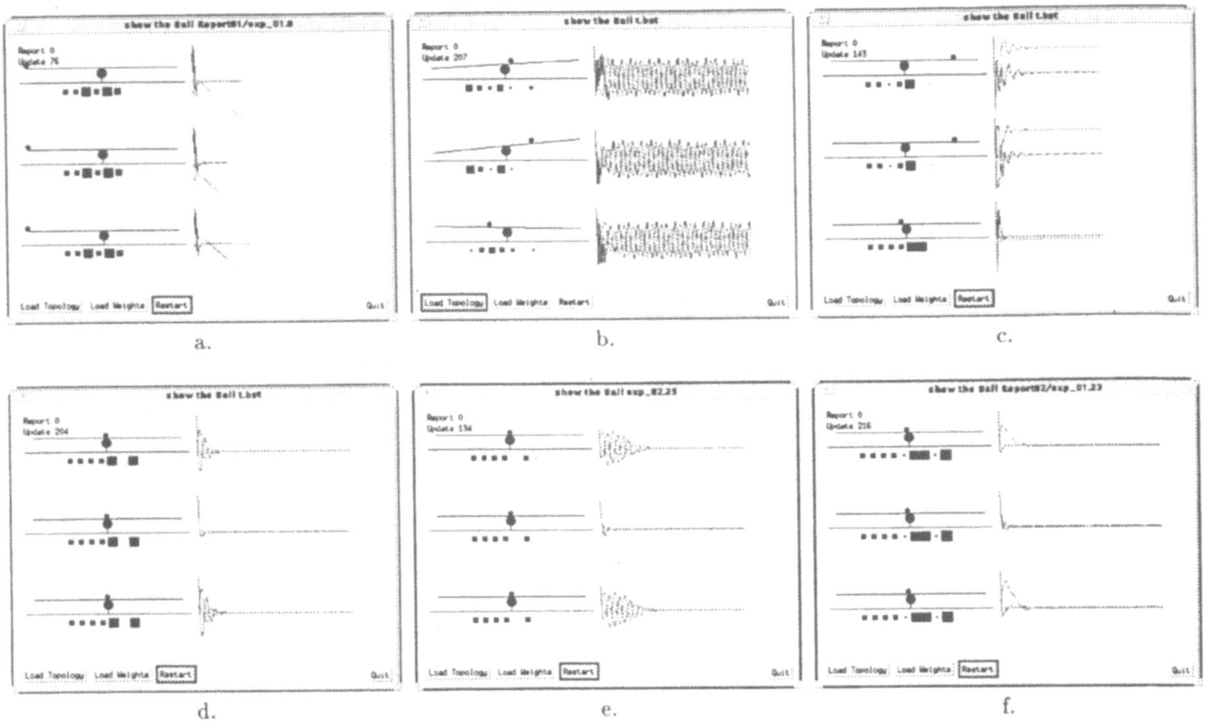

Figure 7: Performance of evolved control networks

fixed number of times. Let the network control the ball for a fixed number of test cycles and measure its performance.

Return the mean of these fitnesses as the final fitness value for the champion weight setting and the current topology. [9]

5.2.2 Evolved networks

Figure 7 gives a brief overview of control networks evolved. Each picture shows a seesaw (with its control net depicted below) for three different initial settings for the position and speed of the ball. The graphs on the right side plot the ball's position and speed and the seesaw's angle for every update cycle. Interestingly enough, the evolution process has developed very different strategies to control the ball: The worst strategy is not to react at all; finally the ball falls off the seesaw (a). Other networks try to keep the ball on the seesaw by periodically balancing (b); mostly the ball never comes to rest. Very smooth reaction of the seesaw sometimes succeeds in stopping the ball from rolling, however, the ball rests far from the seesaw's centre (c). The best strategies force the ball to the centre by periodically balancing (d,e) – some networks need a long time for that – or by immediately slowing down the ball and carefully pushing it to the centre (f).

[9]This ensures that the returned fitness values for the different topologies are comparable, if for the calculation of the final fitness value the same test intervals and cycles are used.

6 Conclusion and further research

Up to now we only have some first and very simple simulation results which are promising, however. The hierachical approach has to be tested for more complex problems. Furthermore, it has to be investigated in how much the network design hierarchy helps in evolving proper net architectures according to a number of constraints and performance measures. Last but not least, the system is still missing a "grammar frontend" which generically produces string representations and genetic operators from a given grammar specification.

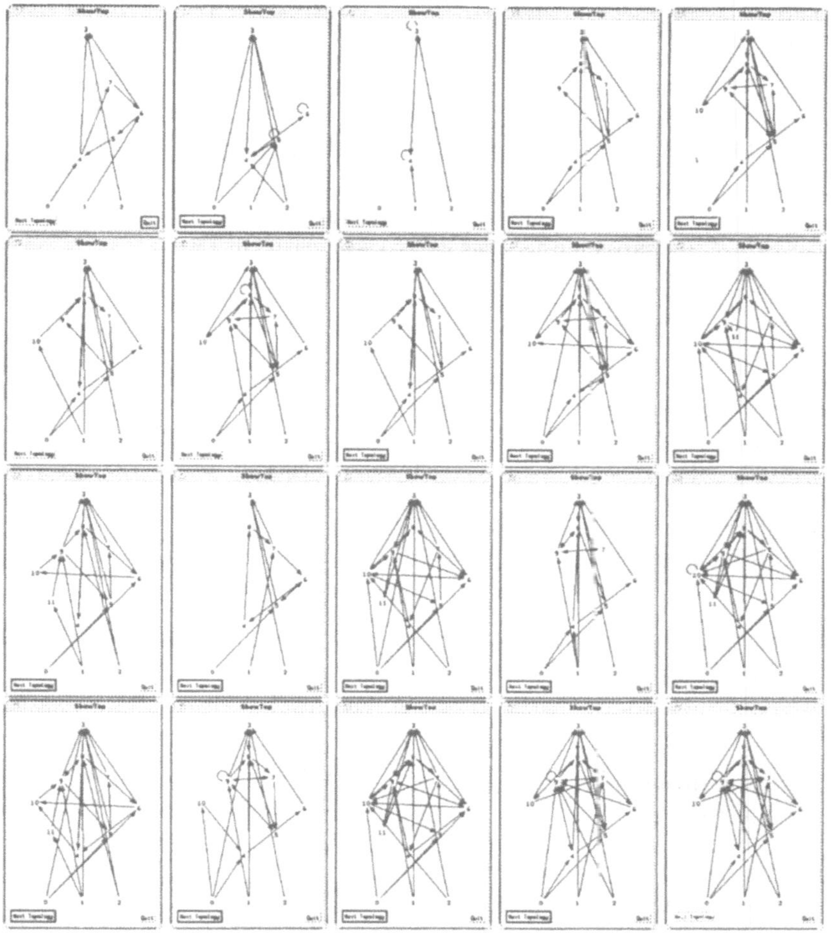

Figure 8: A highly fit generation of evolved topologies for the "seesaw" control task

References

[1] Miller, Hedges and Todd, *Designing Neural Networks using Genetic Algorithms*, Proceedings of the International Conference on Genetic Algorithms, 1989.

[2] Schiffmann W, Joost M and Werner R, *Performance Evaluation of Evolutionarily Created Neural Network Topologies*, in: Schwefel H P and M"anner R (eds), Parallel Problem Solving from Nature, Berlin, 1991.

[3] Whitley D, Starkweather T and Bogart C, *Genetic algorithms and neural networks: optimizing connections and connectivity*, Parallel Computing, vol 14, 1990.

[4] Hughes M, *Genetic Algorithm Workbench Documentation*, Cambridge, 1989.

[5] Harp S A, Samad T, *Genetic Synthesis of Neural Network Architecture*, in: Davis L (ed), Handbook of Genetic Algorithms, New York, 1991.

[6] Koza J R, Rice J P, *Genetic generation of both weights and architecture for a neural network*, in: Proceedings of the International Joint Conference on Neural Networks, 1991.

[7] Bornholdt S, Graudenz D, *General Asymmetric Neural Networks and Structure Desgin by Genetic Algorithms*, Neural Networks, vol. 5, 1992.

[8] Antonisse H J, *A Grammer-Based Genetic Algorithm*, in: Rawlins G (ed), Foundations of Genetic Algorithms, San Mateo, 1991.

Interactive Classification through Neural Networks

Mohamed Daoudi, Denis Hamad and Jack-Gérard Postaire

Centre d'automatique de Lille
Université des Sciences et Technologies de Lille
59655 Villeneuve d'Ascq Cedex, France.

Abstract- In this paper, we present a simple and fast way to provide the operator a plane representation of multidimensional data through neural networks for interactive classification. The superiority of humans over automatic clustering procedures comes from their ability in recognising cluster structures in a two dimensional space, even in the presence of outliers between the clusters, of bridging clusters and of all kinds of irrelevant details in the data points distribution. When giving the operator the interactive means which will help him to isolate clusters of two dimensional points, this visualisation becomes base of a clustering procedure where the operator doesn't loose his grip on the data he is analysing.

Index terms- On-line interactive classification, dimensionality reduction, neural network, back-propagation, multivariate data analysis, cluster analysis.

I. THE ROLE OF THE HUMAN OPERATOR IN PATTERN CLASSIFICATION.

The aim of pattern classification is to discover, in a population of multidimensional data, the presence of clusters which regroup similar data. In general, the data are characterised by a set of attributes which constitute the multidimensional observations. For convenience, the data are usually represented as points in a multidimensional space and the distance between those points is a classical measure of similarity between the data they represent.

When no a priori information about the data is available, i.e. when working in an unsupervised context, the analyst usually calls for clustering algorithms to separate the set of multidimensional observations into groups or clusters which share some property of similarity. Some clustering schemes are based on the optimisation of a criterion which takes into account the distances between and within the clusters [1]. Others are based on the analysis of the underlying probability density function of the available observations in order to extract its modes [2]. In most cases, when using such algorithms, the operator's part is reduced to adjusting the parameters which condition the performance of the procedures [3][4].

An approach where the analyst gets much more involved consists in representing the multidimensional data in a space of lower dimension than that of the original data space. Numerous methods are available to map multidimensional data as points in a plane while preserving similarity relationships between data elements. The most common among them is certainly the principal components analysis technique [5].

If it can be assumed that clusters on the plane are images of multidimensional clusters in the original data space, the interest of such a plane representation is to involve the outstanding visual perception capacities of human beings in the clustering process. Hence, provided the procedure used to map the N-dimensional observations X_q, q=1, 2, ..., Q, onto a two dimensional space do not change too much their relative positions, a human observer can usefully analyse graphic displays without conscious use of any analytical model of clusters, or any mathematical decision rule. It is hoped that any information lost in the dimensionality reduction process is fully compensated by the benefits associated with the integration of the human operator for organising the data [6][7].

The superiority of humans over automatic clustering procedures comes from their ability in recognising cluster structures in a two dimensional space, even in the presence of outliers between the clusters, of bridging clusters and of all kinds of irrelevant details in the data points distribution.

The aim of this paper is to present a simple and fast way to provide the operator a plane representation of multidimensional data through neural networks for interactive classification. After a quick presentation of the architecture of neural networks and training algorithms in section 2, it's shown how this technique of parallel processing of information provides a two dimensional representation of data in section 3.

When giving the operator the interactive means which will help him to isolate clusters of two dimensional points, this visualisation becomes base of a clustering procedure where the operator doesn't loose his grip on the data he is analysing (Section 4). The results we have obtained with artificially generated data show the quality discrimination between classes within same multidimensional samples which have been submitted to analysis (Section 5).

II. ARCHITECTURE OF THE MULTILAYER NEURAL NETWORK.

The network architecture considered here is of the type described in Rumelhart, Hinton, and Williams [8], [9], [10], [11]. Multilayer networks are made of an input layer which receives available information, one layer of output units, and one or several hidden layers.

In classical structures, the only connections allowed in this type of network go from layer to layer and from the input to the output, loops and connections within layer are not allowed.

Each unit output is camputed as follows. First a weighted sum of the activation levels O_h, $h = 1,2,...$ H of the H units connected to unit number i is computed as :

$$S_i = \sum_{h=1}^{H} w_{h,i} \cdot o_h$$

where $w_{h,i}$ is the weight of the connection from unit h to unit i. The output of unit number i is finally a function of $f(S_i)$, where f is generally a sigmoid function defined by :

$$f(S_i) = \frac{1}{1+e^{-S_i}}$$

In order to take advantage of such networks to represent multidimensional data on a plane, it is necessary to specify more clearly their architecture.

The input layer receives the information carried by the very observations to be classified. Let $X = [X_1, X_2, \cdots, X_q, \cdots, X_N]^T$ be the Q available N dimensional observations where $X_q = [x_{1,q}, \cdots, x_{n,q}, \cdots, x_{N,q}]^T$

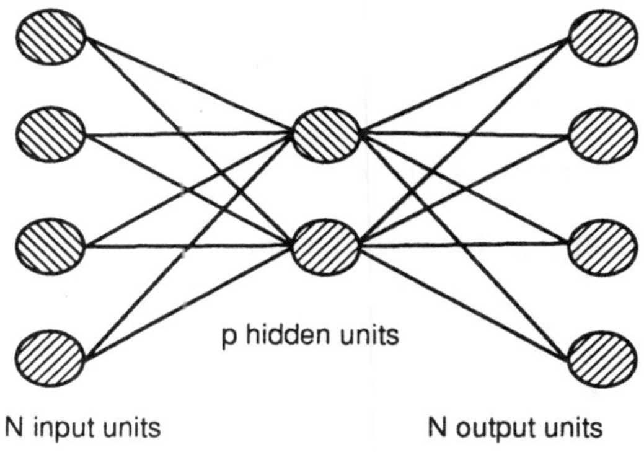

Figure 1. The network.

The input layer of the network is made of N units I_n, n=1, ..., N, such that unit I_n is solicited by attribute $x_{n,q}$, of observations

X_q when this one is presented to the network. Because of homogeneity constraints which will appear in the training phase, the output layer is also composed of N units, denoted O_n, n=1, 2, ..., N. Finally, a hidden layer made of only two units, brings all the information necessary for a plane representation of the data (Figure 1).

The training phase during which the network learns the structure of the data is of upmost importance. As any prototype with known classification is available, the network is trained in an unsupervised mode, using a back propagation technique in an auto associative mode. The Q available observations are successively presented to the network input. The weights of the connections are iteratively modified in order to obtain, at the output of the network, responses as close as possible to its inputs, for all the available observations, in order to minimise E.

To be more specific, the problem consists in minimising the mean square error E such as :

$$E = \sum_{q=1}^{Q} \left\| Y_q - X_q \right\|^2 \text{ where } Y_q = \left[y_{1,q}, \cdots, y_{n,q}, \cdots y_{N,q} \right]^T$$

is the output vector of the network.

During the training phase, the connection weight adjustment is performed by means of the generalised delta rule [10].

III. The mapping process.

Cottrell et al. have used linear units in conjunction with the technique of auto-association to perform image compression [12]. In the auto-association mode, which is called also auto-encoding or identity mapping [10], the network is modified iteratively until its output become close to its inputs. If the network uses hidden units, it will provide an efficient ways of compressing the information contained in the input patterns. An analysis of linear auto-association has been provided by Bourlard and Kamp [13] based on singular value decomposition of matrices (SVD). Another analysis of the behaviour of E has been published by Baldi and Hornik [14], [15]. The main result is that E has a unique local and global minimum corresponding to an orthogonal projection onto the subspace spanned by the first principal eigenvectors of a covariance matrix associated to the training patterns.

However, for most solutions found by running a gradient descent algorithm on the function E, the sub-space found by gradient descent algorithm is different from the sub-space found by the Principal Component Analysis. This is due to the fact that the backward errors used in the learning process [10] are distributed over all the units of the hidden layer.

The back-propagation approach is simple. It can be applied to non-linear networks and to a variety of problems without having any detailed a priori knowledge about their structure or mathematical properties of the optimal solution. We have used neural network with three layers : an input layer composed of N inputs units, and output layer composed of N outputs units, and a hidden layer composed of two units.

In the training phase, all the Q available observations are sequentially presented at the input of the network. Its weight connections are iteratively modified until the difference in mean square sense between the input and the output becomes smaller than a given threshold.

After this learning phase, the Q observations are presented one by one to the network. The two outputs of the hidden layer are considered as the two coordinates of the image in plane of the multidimensional input data. The data can be easily displayed so that this procedure can be used for interactive classification.

IV. Interactive classification.

Although the idea of using the outputs of the hidden layer is not new, it provides a useful interface for computer interactive classification of multidimensional data.

When called by the user, pop-up menus for data manipulation are displayed into utility boxes. The work space which is dominant in term of screen area, is a two dimensional space in which the user does the work of clustering the data. It is a large window used to display the image data points as dark spots on a bright background. All the available data points are mapped onto this two dimensional window.

All the items of the main menu are somewhat associated to relatively standard commands in computer interactive system, since the software has been developed under the X-Window environment. However, the "Classification Menu", which takes full advantage of the proposed two-dimensional mapping technique, deserves a particular attention.

A classical solution for identifying two dimensional clusters is to enclose them by boundaries, which can be drawn on the screen by means of the mouse. In order to release the operator from this tedious task, we propose another clustering strategy where the analyst selects the centres of the clusters by means of the mouse. Once the centres are clicked on the display, the well known ISODATA procedure is used to cluster the data [16].

Thanks to a zooming procedure, each cluster can appear full screen. A more thorough observation of each cluster displayed individually on the whole screen can reveal subclusters which could not be as obvious when the whole set is displayed. The user, by means of a coarse to fine strategy, can discover substructures which are not discernible in the display of the whole data set.

Once classification is achieved, each cluster is displayed in a different colour in order to allow the analyst to evaluate visually the quality of the results.

The statistical characteristics of each cluster as well as the assignation of the observations to each of them are available on request.

V. EXPERIMENTAL RESULTS AND DISCUSSION.

To provide some insight into the behaviour of the interactive system, and to present the main facilities of it, some results are reported, using an artificially generated data set.

The data are computer generated 4-dimensional gaussian random vectors. They consist of 1000 observations drawn from each of the two distributions with the statistical parameters given in table 1.

The results of classification obtained are compared with the Bayes theoretical minimum error rate achievable when the true statistics of the data are known. They demonstrate the ability of the procedure to cluster data of high dimensionality, even in a non trivial situation with overlapping clusters (Figure 2).

These results show that even someone who does not have any software system background can use the interactive system to classify multidimensional observations in an unsupervised context (Figure 3).

	Mean vector	Covariance matrix	Number of data points
Cluster 1	2 2 2 2	1 0 0 0 0 1 0 0 0 0 1 0 0 0 0 1	500
Cluster 2	0 0 0 0	1 0 0 0 0 1 0 0 0 0 1 0 0 0 0 1	500

Theoretic Bayes error rate : 2.5 %

Table 1. Cluster statistics.

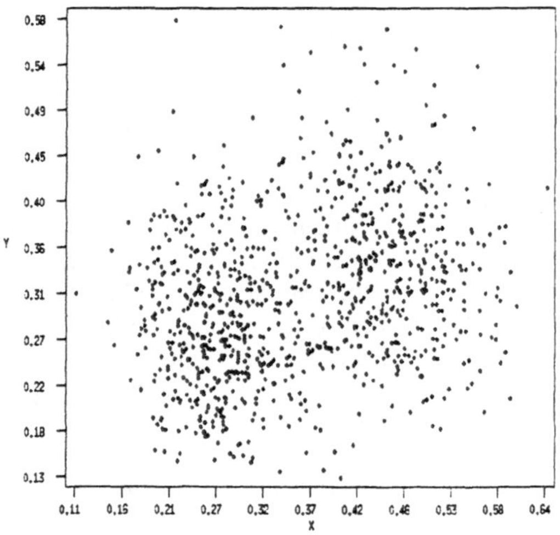

Figure 2. Outputs of hidden layer units

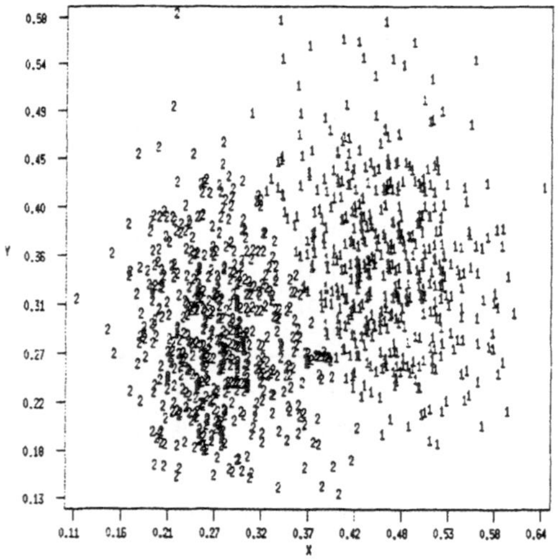

Figure 3. Clustering by ISODATA algorithm.

	Mean vector	Covariance matrix			
Cluster 1	2.02	0.96	0.04	0.08	0.02
	2.00	0.04	1.14	-0.01	-0.01
	1.98	0.08	-0.01	0.92	0.05
	1.91	0.02	-0.11	0.05	1.01
Cluster 2	0.04	1.24	0.21	0.14	0.14
	0.14	0.04	1.10	0.13	0.20
	0.01	0.14	0.13	1.11	0.22
	0.13	0.14	0.20	0.22	1.20

Table 2. Estimation of cluster statistics

	Cluster 1	Cluster 2
Cluster 1	461	13
Cluster 2	39	487

Actual error rate : 5.2%

Table 3. Confusion matrix

REFERENCES.

[1] Fukunaga, K., Koontz, W.L.G (1970) A criterion and an algorithm for grouping data. IEEE Trans. Comput., Vol. 21, pp. 171-178.

[2] Postaire, J.-G., Touzani, A. (1989) Mode boundary detection by relaxation for cluster analysis. Pattern recognition, Vol. 22, pp. 477-490.

[3] Devijver, P.A., Kittler, J. (1982) Pattern recognition : A statistical approach. Prentice Hall, Englewood Cliffs, N.J.

[4] Duda, R.O., Hart , P.E. (1973) Pattern classification and scene analysis. Wiley, New York.

[5] Chien, Y., (1978) Interactive pattern recognition. New York : Marcel Dekker.

[6] Sammon, J.W. (1970) Interactive pattern analysis and classification, IEEE Trans. Comput. Vol. C-19, pp 594-616.

[7] Fukunaga, K., Mantock, J. M. (1982) A non parametric two-dimensional display for classification. IEEE Trans. Pattern Anal. Machine Intell., Vol. PAMI-4, pp. 427-436.

[8] Le Cun, Y. (1986) Learning process in asymmetric threshold network. Disorder systems and biological Organisation, E. Bienenstock, F. Fogelman Soulie, and G. Weibush (Eds), Berlin : Springer.

[9] Lippman, R.P. (1987) An introduction to computing with neural nets. IEEE ASSP Magazine, pp 4-22

[10] Rumelhart, D.E., Hinton, G.E. and Williams, R.J. (1986) Learning internal representation by error propagation Parallel Distributed Processing : Explorations in the micro structures of cognition, Vol. 1, pp. 318-362, MIT Press, Cambridge, Mass.

[11] Wasserman, P.D. (1989) Neural computing Theory and Practise. VNR, New York.

[12] Cottrel, G.W., Munro, P.W. and Zipser, D. (1988) Image compression by back propagation : a demonstration of extensional programming. In : Advances in cognitive Science, Vol. 2, Sharkey, N. E. ed. Norwood, NJ Abbex.

[13] Boulard, H. Kamp, Y.(1988) Auto-association by the multilayer perceptrons and singular value decomposition. Biological cybernetics, Vol. 59, pp 291-294.

[14] Baldi, P., Hornik, K. (1989) Neural networks and principal component analysis : learning from examples without local minima. Neural Networks, Vol. 2, N° 1, pp 53-58.

[15] Baldi, P., Hornik, K. (1991) Back-propagation and unsupervised Learning in Linear Networks. Technical report, Jet Propulsion Laboratory and Division of Biology, California Institute of Technology.

[16] Ball, G. H., Hall, J. D. (1967) A clustering technique for summarising data. Behaviour Sci, Vol. 12, pp 153-155.

Visualization of Neural Network Operation for Improving the Performance Optimization Process

Adrian G. Williamson
& Richard D. Thomas

N219 Division of Computing
Coventry University
Priory Street
Coventry CV1 5FB
UK
email: csx122@uk.ac.cov.cch

Abstract

This paper considers the implementation of a neural network development environment, and describes a prototype that has been produced. The prototype has been targeted at pattern classifier applications, where human support can be beneficial. The initial aim of the system is to provide support for generating a neural network solution to a particular pattern classification problem. This takes the form of supporting the supervised training of a neural network. The support is designed to speed up the training process, by providing the operator with rapid indications of network performance, for the current configuration. The development environment currently supports the Neocognitron neural network paradigm, which is a versatile pattern classifier. The operation of the layers of this network can be presented visually to the operator, as an aid to setting up. A major consideration is the use of Graphical User Interfaces, and the use of X Windows and Motif for the development environment. In conclusion, the results of the work are considered, and future developments outlined.

1.0 INTRODUCTION

Recent work identifies the need to address more practical issues in the use of neural network pattern classifiers[1]. In particular, the optimization of training methods for pattern classifiers is of key importance. Optimization in the field of pattern recognition generally means improved recognition rates. It is envisaged that the provision of an interactive neural network development environment will considerably enhance the effectiveness of such a network application, by introducing new schemes for optimization.

It is suggested that the neocognitron proposed by Fukushima[2][3][4][5], is particularly suitable for an interactive development environment. This form of multi-layer network is based on the visual cortex, and as such has strong links with visualization. In order to classify a pattern, the network reconstructs a target pattern from an appropriate set of sub-patterns. The design of this network allows meaningful visual information to be extracted from each of the network layers.

The training process for the Neocognitron involves the definition of the target set, the refinement of the constituent sub-patterns and the setting of the remaining network parameters which control the recognition process. The objective of the training process is to produce a network which recognizes the target patterns, despite shifts in position, distortion or unwanted interference. The continual refinement of the network settings alters the extent to which these imperfect source patterns can be recognized. The aim of the interactive environment is to provide a rapid development path to a robust pattern classifier.

A prototype has been built which implements a graphical environment for training a supervised Neocognitron. Methods for the management and revision of all network parameters are provided, with an emphasis on Graphical User Interface (GUI) solutions.

2.0 THE NEOCOGNITRON

The Neocognitron is a proven network paradigm, which has only modest hardware requirements. The Neocognitron may be implemented as either a supervised or unsupervised network. It is particularly suited to visualization, because of the high visual content of its output. Unlike many techniques, the supervisor can use all the layer outputs to control the training process and derive an optimum network solution.

The main network structure consists of two major types of neuron or cell. Simple cells, (S cells), act as feature extractors. Complex cells, (C cells), tolerate positional errors and image rotation. The network is assembled from alternate layers of S and C cells. The linking of the layers uses the concept of connection regions, which respond more strongly at the centre of the region, than at the periphery. This simulates the selectivity of animal vision. The final layer of C cells is the output classification. The network forms a hierarchy of feature extraction, from simple line-segments to complete objects at the output stage. Distortion is tolerated at each C layer, which improves the overall robustness of the recognition process.

The layers are linked together by various types of synapse. These are excitatory and inhibitory connections, which can have either fixed or variable weights. If the weights are variable, then they start at zero, and are increased during training. In addition to the S and C cells there are inhibitory cells, (V cells). There is a V cell associated with each S cell. They have the same receptive field as the S cell and their inhibitory output is connected to input of their sister S cell. This inhibitory input adds selectivity to the S cell.

Training exists in both supervised and unsupervised forms. The self-organizing unsupervised network has a more complex structure, and training times are unpredictable. For these reasons supervised training was used in the current implementation, despite the volume of training patterns required. Earlier work[2] rejected this as biologically implausible, however optimization is more straightforward using the supervised method. Although it is easier to implement, the operator needs a greater understanding of the system.

3.0 IMPLEMENTATION

3.1 Configuration

It is apparent that the absolute dimensions and constant settings for the network are unpredictable. The alteration of dimensions and constants within the network influences many aspects of its operation. For example, creating too large a network may make real-time operation impossible, whilst modifying the constants can alter behaviour of the network significantly. A facility for altering these settings, without causing a lengthy re-compilation, needs to be provided.

The solution used in the prototype was a simple constraints definition language. A complete definition of the network settings is entered into a text file. Each relevant aspect, such as the total number of layers, is preceded by a descriptive textual flag. Checking syntax is simple, because the definition must be entered in a specific order. For example, the number of layers in the network always appears at the beginning of a file, and is always followed by a full description the first layer, then a full description of the second, and so on. Further syntax checking is possible, because each flag has a number and format type for the arguments which follow. A file not complying with these simple rules will generate a syntax error.

Semantic checking is more complex. Checking that there are as many layers as expected, is straightforward. The major difficulty arises in deciding if two consecutive layers 'fit' together. There are dimension constraints associated with the connection of two consecutive layers. There are also constraints associated with the definition of connection regions. There is no easy way of specifying how the connection regions attach to their respective layers. The interim solution is to provide semantic rules which apply to some, but not all variations of the network structure. For a better solution, a comprehensive set of rules would have to be considered.

A complete configuration file is parsed and checked. The details are extracted and used to initialize the network. However, deciding exactly what is contained within this configuration file is a process of continual refinement. Currently a standard configuration contains dimensions, constants and connection region definitions, as previously inferred. In addition to this, references to training patterns for each individual S layer exist. The file also contains combinations of S planes as inputs to one particular C plane, to minimize network size, but this is usually Neocognitron specific.

This solution is adequate for the problem of defining and altering network definitions. However, improvements could be made to the way items are entered. Rather than simple text entry, which gives no help or structure to the definition, an intelligent form might be provided. The form would help with full semantic checking, and allow the pre-emption of certain settings.

A prototype intelligent form was constructed using a simple 'wire-frame' GUI, in order to remove some of the problems connected with simple text entry.

3.2 Supervised training

Initially, the aim of training the network is to produce a network which recognizes all the target patterns. Then the operation of the network is refined, to allow the recognition of noisy and distorted patterns. The degree of noise and distortion tolerated may vary with the application. The Neocognitron example used, uses a supervised training strategy, which means that there will be a large number of training patterns required.

The supervised training module uses the information in the configuration file to train the network. The network begins as a network of nodes, with rules governing how these nodes may be connected together. The result of training, is a network of inter-connected nodes. The inter-connections have varying strengths. The lesser the connection strength, the greater the resistance offered to signals wishing to pass through it. Saving these connections weights is equivalent to saving the network state. When a network is fully trained the file containing these weights, together with the original configuration file, can be combined to create a trained network. This trained network can then be further tested, and either applied or refined.

The current method of training is via a simple batch job. The operator enters the command to train the network. They must then wait until the training process is complete, before any analysis of results can take place. Training is a time consuming process and one which may need to be repeated frequently to produce a robust network. Continuous diagnostic readouts would be a considerable drain on network processing resources, and are of questionable benefit. It is possible that the extra processing involved would slow operation down considerably, depending on the granularity of the information being analyzed. The current prototype concentrates on the more important issue of a valid training process. An extension of the current approach may well include displaying diagnostic information.

As an alternative to a more informative training process, the operator using a multitasking system is able to continue with other tasks, whilst the network is training. However, any further changes made to network settings during this time will have no effect on the training process currently in operation, which could prove misleading. This issue needs to be tackled by the final development system.

3.3 The Interactive Control Panel

The Graphical User Interface (GUI) is used to evaluate the effectiveness of the network in recognizing a range of test patterns. Test patterns are designed to probe the extremities of distortion and noise toleration. Displaying the reactions to test patterns, aids in network refinement. From conclusions drawn by Fukushima[3][5], C cells are chosen for display, providing a more condensed representation of the reaction information. An experienced operator can quickly interpret these outputs and decide what steps to take to improve performance.

The arrangement of the control panel is shown in Figure 1. The icon in the upper left hand corner is used for selecting test pattern files. The icon to the right is used to leave the application. The centre display area contains the outputs from the various C cells which comprise the network. Attention normally centres around those cells with the strongest response. Scale bars on the top of each layer control the lower bound of cell response intensity. The Neocognitron's ability to tolerate distortion and missing portions of a pattern, mean that it may respond weakly to patterns which seem to be inappropriate. Raising the display threshold allows the operator to keep the amount of visible information to a sensible minimum.

During refinement it is helpful to know what patterns have been used to train a layer. The correlation between C layer outputs and training patterns is not always direct, so a method for identifying the relationship between the two is also suggested. Activating the mouse button inside one of the layer containers, displays a window containing the training patterns for that layer. Clicking the mouse on one of these training patterns, explains its relationship to the associated C layer.

Upon initiation, the GUI constructs the aspects of the display which remain constant despite changes in the network configuration. Once these static aspects are complete, then the configuration file is examined, and the configuration dependent GUI settings are added.

Additional interface panels give assistance for setting up the configuration file, as shown in Figure 2. These are at an early stage of development, but attempt to give the operator rapid and informed control over the fundamental network set up. They have been used successfully during system development, and now await refinement in line with the appropriate conclusions drawn from system use.

At the outset, there was concern that the combined demand of updating the network and the display, would cause a significant time lapse between loading a test pattern and seeing the results on the screen. This proved to be unfounded and the two activities were performed acceptably in unison, although speed of operation may still be increased. Tasks can be divided neatly between two machines on a local area network (LAN). The display management takes place on the operator's console, whilst neural network calculations are performed elsewhere. This is a feature offered by the X Windows package, used to construct the GUI.

Currently the control panel does not support network training. This is a significant feature, which is under development. The initial problem is how to utilize the time required for training. It may be possible to minimize the extent of re-training required, if the alterations made to settings can be closely monitored and interpreted. Uniting all the operations connected with network construction, under one control centre, may provide such a mechanism. Such features will need to be supported by the central organization of data.

3.4 Application

Anyone wishing to use the network has first to construct an initial framework of settings. Then they must refine the controlling factors until the required performance has been achieved. The performance of the network can be measured by the number of test patterns which can be successfully recognized. The aim is to provide a network which recognizes all the 'test patterns' provided. The maximum possible size of the network depends on the combination of memory and processor power available.

The starting framework involves several choices by the operator. A set of target patterns for the network is the first logical step. From this an initial set of sub-patterns, used to train the intermediate network layers, must be created. The network dimensions can then be decided upon as well as settings for the constants involved in traversal calculations. There are guide-lines for creating these settings[2][5], however, arriving at the final values will involve many iterations.

Once the network has been successfully trained to recognize the target patterns, then a set of test patterns must be designed to test the robustness of the result. These patterns will determine the absolute tolerance of the network, and help to distinguish between a distorted shape and a blob.

The prototype was trained to recognize a set of simple geometric shapes. Optimized settings for characters and digits had already been suggested[3][5]. Although there were only five shapes involved, a sizable number of training patterns were required to recognize heavily distorted and incomplete images. The number of training patterns does not increase exponentially as the number of target patterns increase. On the contrary, if there is a large contingent of shared shapes, as in character recognition, then the increase in training patterns to cope with extra shapes, is kept to a minimum.

The controlling factor in determining the overall size of a network, is the size of the input layer. Although the number of target patterns does have a significant effect, increasing the size of the input layer by just two units can degrade system performance to unacceptable levels. This is caused by the non linear relationship between increases in input layer size, and the overall number of nodes subsequently required within the system[4]. The hardware in use was a network of monochrome DEC 5000 MIPS machines, all with 12 Mb of memory. A network with an input layer of 19x19 units was easily managed, whilst increasing this to 21x21 units, became almost intolerably slow. This prohibited the use of realistically sized images as test patterns. For example, an A4 image, scanned at 300 dpi, is approximately 3200x2490 units in size. For practical use, within current hardware restrictions, the method would probably be used in conjunction with a preprocessor.

The network produced has a surprising tolerance for misshapen images. However, recognition has not yet reached the performance sought. Much of the research time to date has been spent in trying to explore avenues to facilitate improvement, rather than exploiting any one completely. As such, several possibilities for further improvement can be suggested, on the basis of the test case.

4.0 CONCLUSIONS

The following section uses findings from the implementation of the prototype, to formulate several suggested improvements. Using an identified problem, or several related problems as a basis, methods of overcoming these perceived inadequacies have been proposed. In this way, the prototype cycle has helped to identify, and possibly solve, a number of problems.

Probably the most important conclusion from this work, is need for the full integration of information storage. This is important in this system. The main concern is not to reduce storage of repeated information, although this is worthwhile, but to ensure that any changes to settings are passed throughout the system. Changing some settings has a wide-ranging and almost invisible effect. This needs to be controlled, by the intelligent management of data in a central data store.

The entry of patterns and their corresponding descriptions, is an involved and time-consuming task. The prototype demonstrated the need for increased control and organization during this process. It is a target for the future, to integrate a pattern management tool into the planned data repository, so it too can be simplified and wasted effort removed.

X Windows is based on the client-server model. On networked systems, this is generally utilized to distribute the work load between several machines. Although this technique was applied in the running of the software, it was not fully investigated. Further work will reveal greater opportunities for the use of parallelism in the operation of the development environment.

Several observations have indicated the need for a complex set of rules to combat the problems identified. One such case is the verification of network syntax needed to ensure data is entered correctly. Another is a minimized re-training path for the network. These factors would tend to suggest the need for a Knowledge Base System to control some of the more complex transactions taking place. This will play an important underpinning role in the overall plan for the training environment.

The reason to opt for a GUI environment was mainly to produce visualization of results. It was hoped this would lead to better and quicker understanding, thus simplifying and accelerating the achievement of optimum settings for a network.

It is the goal of many GUIs to provide simplified methods for data entry. That is, they should be easier to use than the non-graphical counter-part. Without this, a GUI would simply be a prettier and less portable version of the data entry system which it replaces. When defining the neural network constraints it was found that the textual definition, whilst adequate, could benefit from improvement. The simple test form provided, indicated that ease of operation was increased. The database forms style, which this method provided, also had the added benefit of reducing the chance of error, and removing the need to enter data fields which could be calculated. It is suggested that this can be expanded further, by providing a set of bespoke data entry mechanisms for defining relationships and values. The need for an interactive graphical method for entering the nodes and links between successive layers in a network, is a prime example.

One of the most difficult problems, is how to best use the time required to re-train the network. The whole process of performance optimization relies on the speedy appraisal of new settings. Alterations to the network settings will not be affected, until the network is re-trained. It is certain that training needs to be executed as quickly as possible. This may take the form of a minimized re-train. It may be possible to produce an economic re-training path, based on the current state of the network and the set of proposed alterations. Thus, decreasing effort wasted by re-training parts of the network where it is unnecessary. This technique would require a complex set of rules to extract the fastest possible re-training route. This solution may be supported by a Knowledge Base System.

The current system relies on one type of neural network for its findings. It is believed that the particular paradigm chosen is, however, very informative. Not only in the specific factors concerned with improving its own performance, but also factors important in other pattern recognition systems, or network environments.

The prototype will now be developed into a wider support environment, using the conclusions drawn from this preliminary study.

REFERENCES

[1] R.P. Lippmann, *"Pattern Classification Using Neural Networks"*, IEEE Communications Magazine, 47-64, Nov 1989.

[2] K. Fukushima, S. Miyake, *"Neocognitron: a new algorithm for pattern recognition tolerant of deformations and shifts in position"*, Pattern Recognition, Vol 15, 455-469, 1982.

[3] K. Fukushima, S. Miyake, T. Ito, *"Neocognitron: a neural network model for a mechanism of visual pattern recognition"*, IEEE Transactions on Systems, Man and Cybernetics, SMC-13:826-834, 1983.

[4] K. Fukushima, *"A Neural Network for Visual Pattern Recognition"*, Proceedings of the IEEE International Conference on Neural Networks, 65-75, 1988.

[5] K. Fukushima, N. Wake, *"Handwritten Alphanumeric Character Recognition by the Neocognitron"*, IEEE Transactions on Neural Networks, Vol 2, No 3, 355-365, May 1991.

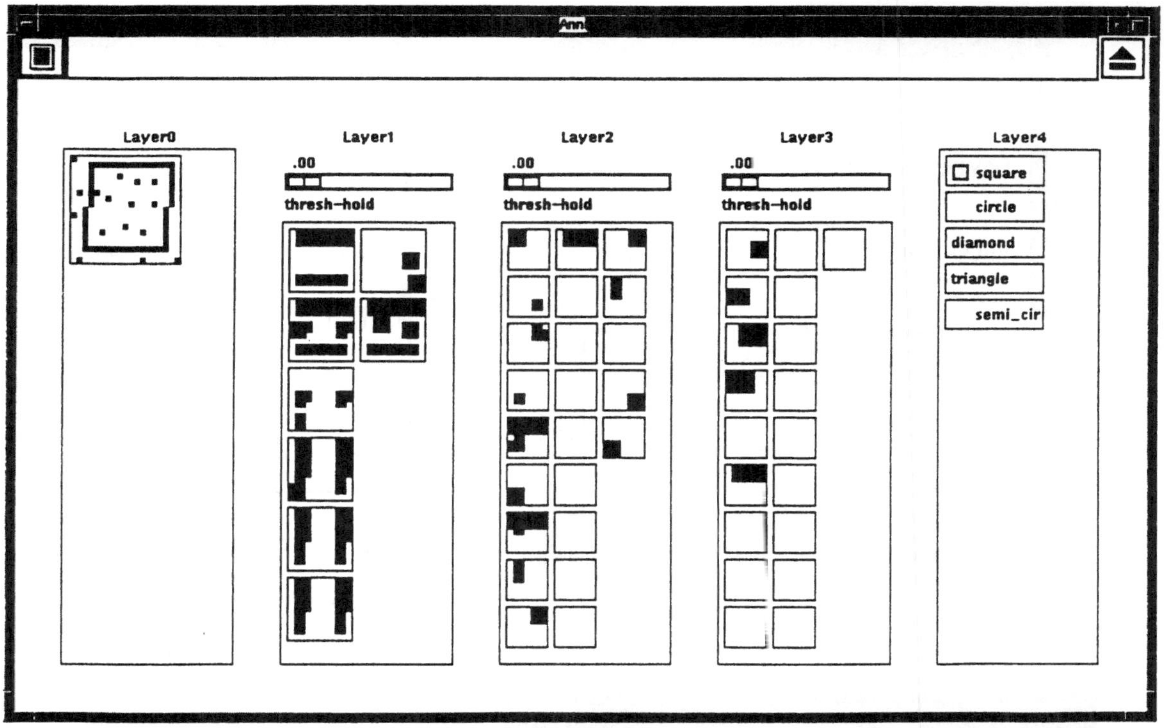

Figure 1: The Neocognitron Control Panel

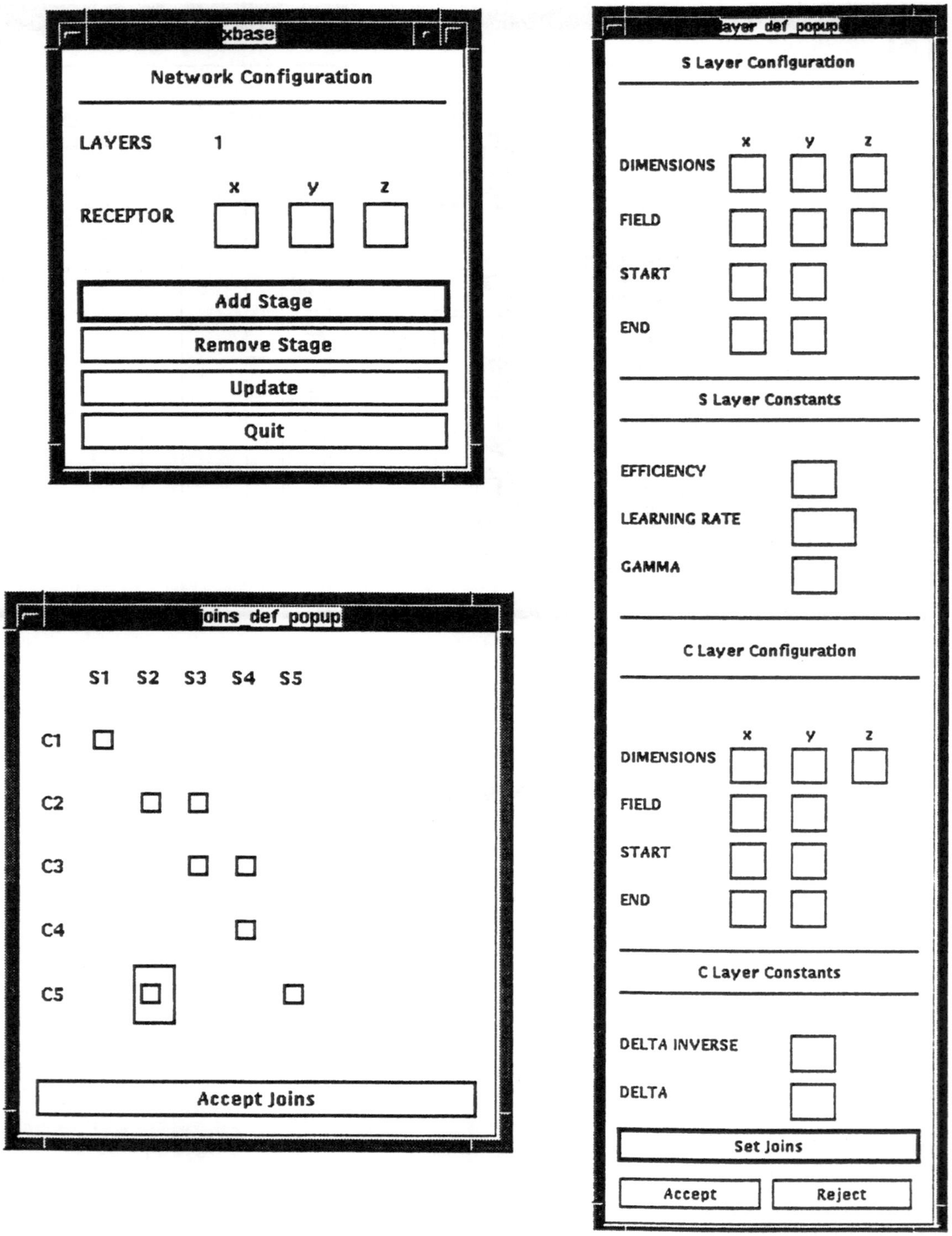

Figure 2: The Configuration Dialog

Identification of Nonlinear Systems Using Dynamical Neural Networks: A Singular Perturbation Approach

George A. Rovithakis and Manolis A. Christodoulou

Dept. of Electronic & Computer Engineering
Technical University of Crete
73100 Chania, Crete, GREECE

Abstract

In this paper, we employ singular perturbation analysis to examine the stability and robustness properties of a dynamical neural network identifier. Various cases that lead to modeling errors are taken into consideration. Input-output stability theory is used to assure convergence and internal stability of the identifier. Not all the plant states are assumed to be available for measurment.

1 Introduction

Adaptive control of dynamical systems has been an active area of research since the 1960's. At the beginning, research was focused on the problem of controlling and identifying linear time invariant plants. But it was not until the last decade, that the problem was solved for both continuous and discrete LTI systems [1]-[3].

Recent advances in nonlinear control theory and in particular feedback linearization techniques [4, 5], have inspired the development of adaptive control schemes for nonlinear plants [6]-[9]. A common assumption made in the above works, is that either all or part of the system dynamics are known. Although sometimes it is quite realistic, it constraints considerably the applications field. Furthermore, the most general problem of controlling a totally unknown plant, cannot even be discussed under these control schemes and assumptions.

An obvious solution to overcome the problem, is to introduce identification techniques in the control algorithm. The problem of identification consists of choosing an appropriate identification model and adjust its parameters such that the response of the model to an input signal approximates the response of the plant under the same input. It has been clear that in control systems theory, a mathematical description of a plant is often a prerequisite to analysis and controller design.

Furthermore, it is well known that global stability properties of model reference adaptive systems [10], are guaranteed under the "matching assumption", that the model order is not lower than that of the unknown plant. This restrictive assumption is likely to be violated in applications. Hence, it is important to determine the stability and robustness properties of adaptive schemes with respect to modeling errors.

Ioannou and Kokotovic [11], assumed a separation of time scales between the modeled and unmodeled phenomena and examined the performance of various types of identifiers and adaptive observers, when the order of the model is equal to the slow part of the plant.

Taylor *et al.* [6] and Kanelakopoulos *et al.* [7], examined the stability and robustness properties of nonlinear systems with parametric and dynamic uncertainties. In their work, the true plant is allowed to be of higher order due to unmodeled dynamics. Only the states appearing in the reduced order model were assumed to be available for measurment. Hence, their work appeared as an extension of the robustness analysis of Ioannou and Kokotovic [12], to nonlinear adaptive control.

Recently, neural networks appeared as powerful tools for learning static and dynamic highly nonlinear systems. Due to their massive parallelism, very fast adaptability and inherent approximation capabilities, neural networks have concentrated a lot of research, especially in the area of identification and control. In the literature one can find interesting works dealing with the subject [13],[14], the feasibility of which has been demonstrated through simulations.

Narendra *et al.* [15], originally formulated the problem. They proposed dynamic backpropagation schemes, which are static backpropagation neural networks, connected either in series or in parallel with linear dynamical systems.

However, in all works mentioned above it was assumed that all states of the unknown plant were available for measurment and furthermore, they lack theoretical verification of the results that simulations provide.

In this paper, we employ singular perturbation analysis, to examine the stability and robustness properties of a dynamical neural network identifier. We are concerned with all cases that may lead to modeling errors, (ie- parametric, dynamic, parametric and dynamic) and prove stability and convergence. To design the neural network identifier, we first consider only parametric uncertainties and build a learning law such that the identifier is input-output stable. At the same time the theory of input-output stability also guarantees internal stability as well. The next step is to introduce dynamic uncertainties and examine how stability is affected by them. In the algorithm, not all plant states are assumed to be available for measurment.

2 Problem Formulation and the Dynamic Neural Network Identifier

In this section we consider the problem of identifying a continuous time nonlinear dynamical system of the form

$$\dot{x} = f(x) + g(x)u \tag{2.1}$$

where $x \in \mathcal{M}$, a smooth manifold, the inputs $u \in \mathcal{U} \subset \Re^n$, where $\mathcal{U}$ is the class of admissible inputs, f is a smooth vector called the drift term, g is a matrix with columns g_i , $i = 1, 2, \ldots, n$ $g = [g_1\ g_2\ \ldots\ g_n]$, where g_i are smooth vectorfields and $x(0) = x_0$ is the initial condition in $\mathcal{M}$. In examining this problem , we will impose the following assumptions on the system to be approximated.

(A1) Given a class $\mathcal{U}$ of admissible inputs, then for any $u \in \mathcal{U}$ and any finite initial condition, the state trajectories are uniformly bounded for any finite $T > 0$. Hence, $| x(T) | < \infty$.

□

(A2) f, g are continuous with respect to their arguments and satisfy a local Lipschitz condition so that the solution $x(t)$ to the differential equation (2.1) is unique for any finite initial condition and $u \in \mathcal{U}$.

□

The above assumptions are required to guarantee the existence and uniqueness of solution of (2.1), for any finite initial condition and $u \in \mathcal{U}$. In order to identify the nonlinear dynamical system (2.1), we employ dynamical neural networks. Dynamic neural networks are recurrent, fully interconnected nets, containing dynamical elements in their neurons. Therefore, they are described by the following set of differential equations

$$\dot{\hat{x}} = A\hat{x} + BWS(x) + BW_{n+1}S'(x)u \tag{2.2}$$

where $\hat{x} \in \mathcal{M}$, the inputs $u \in \mathcal{U} \subset \Re^n$, W is a $n \times n$ matrix of synaptic weights, A, B are $n \times n$ diagonal matrices with elements the scalars a_i, b_i for all $i = 1, 2, 3, \ldots, n$ and W_{n+1} is a $n \times n$ diagonal matrix of synaptic weights of the form $W_{n+1} = diag[w_{1\ n+1} w_{2\ n+1} \ldots w_{n\ n+1}]$. Finally $S(x)$ is a vector-valued analytic function and $S'(x)$ is a $n \times n$ diagonal matrix, with elements $s(x_i)$, smooth, (at least twice differentiable), monotone increasing functions which are usually represented by sigmoids of the form

$$s(x_i) = \frac{k}{1 + e^{-lx_i}}$$

for all $i = 1, 2, 3, \ldots, n$ where k, l are parameters representing the bound, (k) and slope, (l), of sigmoid's curvature.

3 Parametric Uncertainty

Let us first assume that an exact model of the plant is available, (i.e we have no modeling error). The purpose of this section is to find a learning law that guarantees stability of the neural network, plus convergence of its output and weights to a desired value. Since we have only parametric uncertainties, we can assume that there exists weight values $W^\star$, $W^\star_{n+1}$ such that the system (2.1) is competely described by a neural network of the form

$$\dot{x} = Ax + BW^\star S(x) + BW^\star_{n+1} S'(x)u \tag{3.1}$$

where all matrices are as defined earlier.
Define the error between the identifier states and the real system states as

$$e = \hat{x} - x$$

Then from (2.2) and (3.1) we can obtain the error equation

$$\dot{e} = Ae + B\tilde{W}S(x) + B\tilde{W}_{n+1}S'(x)u \quad (3.2)$$

where

$$\tilde{W} = W - W^{\star}$$

$$\tilde{W}_{n+1} = W_{n+1} - W^{\star}_{n+1}$$

The Lyapunov synthesis method is used to derive stable adaptive laws. Therefore consider the Lyapunov function candidate

$$\begin{aligned} \mathcal{V}(e, \tilde{W}, \tilde{W}_{n+1}) &= \frac{1}{2}e^T Pe + \frac{1}{2}tr\{\tilde{W}^T\tilde{W}\} \\ &+ \frac{1}{2}tr\{\tilde{W}^T_{n+1}\tilde{W}_{n+1}\} \end{aligned}$$

where $P > 0$ is chosen to satisfy the Lyapunov equation

$$PA + A^T P = -I$$

Observe that since A is a diagonal matrix, P can be chosen to be a diagonal matrix too; simplifying, in this way, the calculations.
Differentiating $\mathcal{V}$ along the solution of (3.2) we obtain

$$\begin{aligned} \dot{\mathcal{V}} &= \frac{1}{2}(\dot{e}^T Pe + e^T P\dot{e}) + tr\{\dot{\tilde{W}}^T \tilde{W}\} \\ &+ tr\{\dot{\tilde{W}}^T_{n+1}\tilde{W}_{n+1}\} \end{aligned}$$

or

$$\begin{aligned} \dot{\mathcal{V}} &= \frac{1}{2}(-e^T e + S^T(x)\tilde{W}^T BPe) \\ &+ \frac{1}{2}(u^T S'(x)\tilde{W}_{n+1}BPe) \\ &+ \frac{1}{2}[(S^T(x)\tilde{W}^T BPe)^T] \\ &+ \frac{1}{2}[(u^T S'(x)\tilde{W}_{n+1}BPe)^T)] \\ &+ tr\{\dot{\tilde{W}}^T \tilde{W}\} + tr\{\dot{\tilde{W}}^T_{n+1}\tilde{W}_{n+1}\} \end{aligned}$$

Now
since $S^T(x)\tilde{W}^T BPe$, $u^T S'(x)\tilde{W}_{n+1}BPe$ are scalars

$$S^T(x)\tilde{W}^T BPe = (S^T(x)\tilde{W}^T BPe)^T$$

$$u^T S'(x)\tilde{W}_{n+1}BPe = (u^T S'(x)\tilde{W}_{n+1}BPe)^T$$

Therefore, $\dot{\mathcal{V}}$ becomes

$$\begin{aligned} \dot{\mathcal{V}} &= -\frac{1}{2}e^T e + S^T(x)\tilde{W}^T BPe \\ &+ u^T S'(x)\tilde{W}_{n+1}BPe + tr\{\dot{\tilde{W}}^T \tilde{W}\} \\ &+ tr\{\dot{\tilde{W}}^T_{n+1}\tilde{W}_{n+1}\} \end{aligned} \quad (3.3)$$

Hence, if we choose

$$tr\{\dot{\tilde{W}}^T \tilde{W}\} = -S^T(x)\tilde{W}^T BPe \quad (3.4)$$

$$tr\{\dot{\tilde{W}}^T_{n+1}\tilde{W}_{n+1}\} = -u^T S'(x)\tilde{W}_{n+1}BPe \quad (3.5)$$

then (2.6) becomes

$$\dot{\mathcal{V}} = -\frac{1}{2}e^T e \quad (3.6)$$

or

$$\dot{\mathcal{V}} = -\frac{1}{2}\|e\|^2 \leq 0 \quad (3.7)$$

From equations (3.4) and (3.5) we obtain learning laws in an element form as

$$\begin{aligned} \dot{w}_{ij} &= -b_i p_i s(x_j)e_i \\ \dot{w}_{in+1} &= -b_i s(x_i)p_i u_i e_i \end{aligned}$$

for all $i, j = 1, 2, 3, \ldots, n$.
Now we can prove the following theorem.

Theorem 3.1 *Consider the identification scheme (3.2). The learning law*

$$\begin{aligned} \dot{w}_{ij} &= -b_i p_i s(x_j)e_i \\ \dot{w}_{in+1} &= -b_i s(x_i)p_i u_i e_i \end{aligned}$$

for all $i, j = 1, 2, 3, \ldots, n$ guarantees the following properties

- $e, \hat{x}, W, W_{n+1} \in L_\infty, \quad e \in L_2$
- $\lim_{t\to\infty} e(t) = 0, \quad \lim_{t\to\infty} \dot{\tilde{W}}(t) = 0,$
- $\lim_{t\to\infty} \dot{\tilde{W}}_{n+1}(t) = 0$

□

Proof: We have shown that using the learning law

$$\begin{aligned} \dot{w}_{ij} &= -b_i p_i s(x_j)e_i \\ \dot{w}_{in+1} &= -b_i s(x_i)p_i u_i e_i \end{aligned}$$

for all $i, j = 1, 2, 3, \ldots, n$ $\dot{\mathcal{V}}$ becomes

$$\dot{\mathcal{V}} = -\frac{1}{2}\|e\|^2 \leq 0$$

Hence, $\mathcal{V} \in L_\infty$, which implies $e, \tilde{W}, \tilde{W}_{n+1} \in L_\infty$. Furthermore, $\hat{x} = e + x$ is also bounded. Since $\mathcal{V}$ is a non-increasing function of time and bounded from below, the $\lim_{t\to\infty} \mathcal{V} = \mathcal{V}_\infty$ exists. Therefore by integrating $\dot{\mathcal{V}}$ from 0 to ∞ we have

$$\int_0^\infty \|e\|^2 dt = 2[\mathcal{V}(0) - \mathcal{V}_\infty] < \infty$$

which implies that $e \in L_2$. By definition the sigmoid functions $s(x_i)$, $i = 1, 2, \ldots n$ are bounded for all x and by assumption all inputs to the neural network are also bounded. Hence from (3.2) we have that $\dot{e} \in L_\infty$. Since $e \in L_2 \bigcap L_\infty$ and $\dot{e} \in L_\infty$, using Barbalat's Lemma [17], we conclude that $\lim_{t\to\infty} e(t) = 0$. Now using the boundedness of u, $S(x)$, $S'(x)$ and the convergence of $e(t)$ to zero, we have that $\dot{W}$, $\dot{W}_{n+1}$ also converges to zero.

□

Remark 3.2 Under the assumptions of Theorem 2.1, we cannot conclude anything about the convergence of the weights to their optimal values. In order to guarantee convergence, $S(x)$, $S'(x)$, u need to satisfy a persistancy of excitation condition. A signal $z(t) \in \Re^n$ is persistently exciting in $\Re^n$ if there exist positive constants β_0, β_1, T such that

$$\beta_0 I \leq \int_t^{t+T} z(\tau) z^T(\tau) d\tau \leq \beta_1 I \quad \forall t \geq 0$$

However, such a condition cannot be verified a priori since $S(x)$ and $S'(x)$ are nonlinear functions of the state x.

□

4 Unmodeled Dynamics Present

In the previous section we assumed that there exist weight values $W^\star$, $W^\star_{n+1}$ such that a nonlinear dynamical system can be completely described by a neural network of the form

$$\dot{x} = Ax + BW^\star S(x) + BW^\star_{n+1} S'(x) u$$

where all matrices are as defined previously. It is well known however, that the model is of lower order than the plant, due to the unmodeled dynamics present in the plant. In the following, we extend our theory within the framework of singular perturbations, to include the case where dynamic uncertainties are present. For more details concerning singular perturbation theory, the interested reader is refered to [16]. Now we can assume that the unknown plant can be completely described by

$$\begin{aligned} \dot{x} &= Ax + BW^\star S(x) + BW^\star_{n+1} S'(x) u \\ &\quad + F(x, W, W_{n+1}) A_0^{-1} B_0 W_0 u + F(x, W, W_{n+1}) z \\ \mu \dot{z} &= A_0 z + B_0 W_0 u \quad , \quad z \in \Re^r \end{aligned} \tag{4.1}$$

where z is the state of the unmodeled dynamics and $\mu > 0$ a small singular perturbation scalar. If we define the error between the identifier states and the real system states as

$$e = \hat{x} - x$$

then from (2.2) and (4.1) we obtain the error equation

$$\begin{aligned} \dot{e} &= Ae + B\tilde{W} S(x) + B\tilde{W}_{n+1} S'(x) u \\ &\quad - F(x, W, W_{n+1}) A_0^{-1} B_0 W_0 u - F(x, W, W_{n+1}) z \\ \mu \dot{z} &= A_0 z + B_0 W_0 u \quad , \quad z \in \Re^r \end{aligned} \tag{4.2}$$

where
$F(x, W, W_{n+1})$, $B_0 W_0 u$, $B\tilde{W} S(x)$, $B\tilde{W}_{n+1} S'(x) u$, are bounded and differentiable with respect to their arguments for every $\tilde{w} \in B_{\tilde{w}}$ a ball in $\Re^{n \times n}$, $\tilde{w}_{n+1} \in B_{\tilde{w}_{n+1}}$ a ball in $\Re^n$ and all $x \in B_x$ a ball in $\Re^n$. We further assume that the unmodeled dynamics are asymptotically stable for all $x \in B_x$. In other words we assume that there exist a constant $\nu > 0$ such that

$$Re\ \lambda\{A_0\} \leq -\nu < 0$$

Note that $\dot{z}$ is large since μ is small and hence the unmodeled dynamics are fast. For a singular perturbation from $\mu > 0$ to $\mu = 0$ we obtain

$$z = A_0^{-1} B_0 W_0 U$$

Since the unmodeled dynamics are asymptotically stable the existence of A_0^{-1} is assured. As it is well known from singular perturbation theory, we express the state z as

$$z = h(x,\eta) + \eta \qquad (4.3)$$

where $h(x,\eta)$ is defined as the quasi-steady-state of z and η as its fast transient. In our case

$$h(x,\eta) = A_0^{-1} B_0 W_0 u$$

Substituting (4.3) into (4.2) we obtain the singularly perturbed model as

$$\begin{aligned} \dot{e} &= Ae + B\tilde{W}S(x) + B\tilde{W}_{n+1}S'(x)u \\ &\quad -F(x,W,W_{n+1})\eta \\ \mu\dot{\eta} &= A_0\eta - \mu\dot{h}(e,\tilde{W},\tilde{W}_{n+1},\eta,u) \qquad (4.4) \end{aligned}$$

where we define

$$\begin{aligned} \dot{h}(e,\tilde{W},\tilde{W}_{n+1},\eta,u) &= \frac{\partial h}{\partial e}\dot{e} + \frac{\partial h}{\partial \tilde{W}}\dot{\tilde{W}} \\ &\quad + \frac{\partial h}{\partial \tilde{W}_{n+1}}\dot{\tilde{W}}_{n+1} \\ &\quad + \frac{\partial h}{\partial u}\dot{u} \end{aligned}$$

Notice however, that in control case, u is a function of $e, \tilde{W}, \tilde{W}_{n+1}$ therefore making $\dot{h}(e,\tilde{W},\tilde{W}_{n+1},\eta,u)$ to be equal to

$$\begin{aligned} \dot{h}(e,\tilde{W},\tilde{W}_{n+1},\eta,u) &= \frac{\partial h}{\partial e}\dot{e} + \frac{\partial h}{\partial \tilde{W}}\dot{\tilde{W}} \\ &\quad + \frac{\partial h}{\partial \tilde{W}_{n+1}}\dot{\tilde{W}}_{n+1} \end{aligned}$$

Remark 4.1
$F(x,W,W_{n+1})A_0^{-1}B_0W_0u$, $F(x,W,W_{n+1})z$ in (4.1) can be viewed as correction terms in the input vectorfields and in the drift term of

$$\dot{x} = Ax + BW^{\star}S(x) + BW_{n+1}^{\star}S'(x)u$$

in the sense that the unknown system can now be described by a neural network plus the correction terms.

□

5 Robustness of the Neural Network Identifier

In this subsection we proceed to the stability analysis of the neural network identification scheme, when the unmodeled dynamics considered in the previous subsection are present. Before proceeding any further we need to prove the following Lemma.

Lemma 5.1 *It is true that $\dot{h}$ is bounded by*

$$\|\dot{h}(e,\tilde{W},\tilde{W}_{n+1},\eta,u)\| \le \rho_1\|e\| + \rho_2\|\eta\|$$

provided that the following inequalities hold

$$\begin{aligned} \|h_w\dot{\tilde{W}}\| &\le k_0\|e\| \\ \|h_{w_{n+1}}\dot{\tilde{W}}_{n+1}\| &\le k_1\|e\| \\ \|h_eB\tilde{W}_{n+1}S'(x)u\| &\le k_2\|e\| \\ \|h_eB\tilde{W}S(x)\| &\le k_3\|e\| \\ \|h_eF(x,\tilde{W},\tilde{W}_{n+1})\| &\le \rho_2 \\ \|h_eAe\| &\le k_4\|e\| \\ \|h_u\dot{u}\| &\le k_5\|e\| \end{aligned}$$

and

$$\rho_1 = k_0 + k_1 + k_2 + k_3 + k_4 + k_5$$

□

Proof: Differentiating $h(e,\tilde{W},\tilde{W}_{n+1},\eta,u)$ we obtain

$$\begin{aligned} \dot{h}(e,\tilde{W},\tilde{W}_{n+1},\eta,u) &= h_e\dot{e} + h_{\tilde{W}}\dot{\tilde{W}} + h_{\tilde{W}_{n+1}}\dot{\tilde{W}}_{n+1} \\ &\quad + h_u\dot{u} \end{aligned}$$

or

$$\begin{aligned} \dot{h} &= h_e(Ae + B\tilde{W}S(x) + B\tilde{W}_{n+1}S'(x)u) \\ &\quad - h_e(F(x,W,W_{n+1})\eta) + h_{\tilde{W}}\dot{\tilde{W}} \\ &\quad + h_{\tilde{W}_{n+1}}\dot{\tilde{W}}_{n+1} + h_u\dot{u} \end{aligned}$$

Therefore

$$\begin{aligned} \|\dot{h}\| &\le \|h_eAe\| + \|h_eB\tilde{W}S(x)\| \\ &\quad + \|h_eB\tilde{W}_{n+1}S'(x)u\| \\ &\quad + \|h_eF(x,W,W_{n+1})\eta\| + \|h_{\tilde{W}}\dot{\tilde{W}}\| \\ &\quad + \|h_{\tilde{W}_{n+1}}\dot{\tilde{W}}_{n+1}\| + \|h_u\dot{u}\| \\ &\le k_4\|e\| + k_3\|e\| + k_2\|e\| \\ &\quad + \|h_eF(x,W,W_{n+1})\|\|\eta\| \\ &\quad + k_0\|e\| + k_1\|e\| + k_5\|e\| \\ &\le k_4\|e\| + k_3\|e\| + k_2\|e\| + \rho_2\|\eta\| \\ &\quad + k_0\|e\| + k_1\|e\| + k_5\|e\| \end{aligned}$$

Hence,

$$\dot{h}(e,\tilde{W},\tilde{W}_{n+1},\eta,u) \le \rho_1\|e\| + \rho_2\|\eta\|$$

which concludes the proof.

□

We are now able to prove the following Theorem

Theorem 5.2 *The equilibrium of the singularly perturbed model is asymptotically stable for all*

$$\mu \in (0, \frac{1}{c_1c_2 + 2c_3})$$

and an estimate of its region of attraction is

$$S = \{e,\tilde{W},\tilde{W}_{n+1},\eta : \mathcal{V}(e,\tilde{W},\tilde{W}_{n+1},\eta) \le c\}$$

where c is the largest constant such that the set $\{e,\tilde{W},\tilde{W}_{n+1} : \mathcal{V}(e,\tilde{W},\tilde{W}_{n+1},0) \le c\}$ *is contained to* $B_e \times B_w \times B_{w_{n+1}}$. *Furthermore, the following properties are guaranteed*

- $e, \hat{X}, \eta, W, W_{n+1} \in L_\infty, \quad e, \eta \in L_2$
- $\lim_{t\to\infty} e(t) = 0, \ \lim_{t\to\infty} \eta(t) = 0$
- $\lim_{t\to\infty} \dot{\tilde{W}}(t) = 0, \quad \lim_{t\to\infty} \dot{\tilde{W}}_{n+1}(t) = 0$

□

Proof: Lets take the Lyapunov function candidate

$$\begin{aligned}\mathcal{V}(e,\tilde{W},\tilde{W}_{n+1},\eta) &= \frac{1}{2}c_1e^TPe + \frac{1}{2}c_2\eta^TP_0\eta \\ &+\frac{1}{2}c_1tr\{\tilde{W}^T\tilde{W}\} \\ &+\frac{1}{2}c_1tr\{\tilde{W}_{n+1}^T\tilde{W}_{n+1}\}\end{aligned}$$

where $P, P_0 > 0$ are chosen to satisfy the Lyapunov equation

$$PA + A^TP = -I$$

$$P_0A_0 + A_0^TP_0 = -I$$

Observe that $\mathcal{V}$ is a weighted sum of a slow and a fast part.Taking the time derivative of $\mathcal{V}$ and using the learning law

$$\begin{aligned}\dot{w}_{ij} &= -b_ip_is(x_j)e_i \\ \dot{w}_{in+1} &= -b_is(x_i)p_iu_ie_i\end{aligned}$$

for all $i = 1,2,\ldots,n$
we obtain, as in a previous subsection that

$$\begin{aligned}\dot{\mathcal{V}} &= -\frac{c_1}{2}\|e\|^2 - \frac{c_2}{2\mu}\|\eta\|^2 \\ &-c_1e^TPF(x,W,W_{n+1})\eta \\ &-c_2\eta^TP_0\dot{h}(e,\tilde{W},\tilde{W}_{n+1},\eta,u) \\ &\le -\frac{c_1}{2}\|e\|^2 - \frac{c_2}{2\mu}\|\eta\|^2 \\ &+\|c_1e^TPF(x,W,W_{n+1})\eta\| \\ &+\|c_2\eta^TP_0\dot{h}(e,\tilde{W},\tilde{W}_{n+1},\eta,u)\|\end{aligned}$$

and by employing Lemma (5.1) we obtain

$$\begin{aligned}\dot{\mathcal{V}} &\le -\frac{c_1}{2}\|e\|^2 - \frac{c_2}{2\mu}\|\eta\|^2 \\ &+c_1\|e^TPF(x,W,W_{n+1})\|\|\eta\| \\ &+c_2\|\eta P_0\|(\rho_1\|e\| + \rho_2\|\eta\|)\end{aligned}$$

which finally takes the form

$$\dot{\mathcal{V}} \le -\frac{c_1}{2}\|e\|^2 - c_2(\frac{1}{2\mu} - c_3)\|\eta\|^2 + c_1c_2\|e\|\|\eta\| \tag{5.1}$$

provided that the following inequalities hold

$$\begin{aligned}\|PF(x,W,W_{n+1})\| &\le c_2 \\ \|P_0\|\rho_1 &\le c_1 \\ \|P_0\|\rho_2 &\le c_3\end{aligned}$$

Therefore

$$\dot{\mathcal{V}} \le -\begin{bmatrix}\|e\| & \|\eta\|\end{bmatrix}\begin{bmatrix}\frac{c_1}{2} & -\frac{c_1c_2}{2} \\ -\frac{c_1c_2}{2} & c_2(\frac{1}{2\mu} - c_3)\end{bmatrix} \times \begin{bmatrix}\|e\| \\ \|\eta\|\end{bmatrix} \tag{5.2}$$

The 2×2 matrix in (5.3) is positive definite, when

$$\mu < \frac{1}{c_1c_2 + 2c_3}$$

Then $\dot{\mathcal{V}}$ is negative semidefinite. Since $\dot{\mathcal{V}} \le 0$ we conclude that $\mathcal{V} \in L_\infty$, which implies $e,\eta,\tilde{W},\tilde{W}_{n+1} \in L_\infty$.Furthermore, $\hat{x} = e + x$, $W = \tilde{W} + W^*$, $W_{n+1} = \tilde{W}_{n+1} + W_{n+1}^*$ are also bounded. Since $\mathcal{V}$ is a non-increasing function of time and bounded from below, the $\lim_{t\to\infty} \mathcal{V} = \mathcal{V}_\infty$ exists. Therefore by integrating $\dot{\mathcal{V}}$ from 0 to ∞ we have

$$\frac{c_1}{2}\int_0^\infty \|e\|^2dt + c_2(\frac{1}{2\mu} - c_3)\int_0^\infty \|\eta\|^2dt$$

$$-c_1c_2\int_0^\infty \|e\|\|\eta\|dt \le [\mathcal{V}(0) - \mathcal{V}_\infty] < \infty$$

which implies that $e, \eta \in L_2$.Furthermore

$$\begin{aligned}\dot{e} &= Ae + B\tilde{W}S(x) + B\tilde{W}_{n+1}S'(x)u \\ &\quad -F(x, W, W_{n+1})\eta \\ \mu\dot{\eta} &= A_0\eta - \mu\dot{h}(e, \tilde{W}, \tilde{W}_{n+1}, \eta, u)\end{aligned}$$

Since u , A_0 , $\dot{h}(e, \tilde{W}, \tilde{W}_{n+1}, \eta, u)$ are bounded, $\dot{e} \in L_\infty$ and $\dot{\eta} \in L_\infty$. Since $e \in L_2 \bigcap L_\infty, \eta \in L_2 \bigcap L_\infty$, using Barbalat's Lemma we conclude that $\lim_{t\to\infty} e(t) = 0, \lim_{t\to\infty} \eta(t) = 0$. Now using the bouded-ness of $u, S(x), S'(x)$ and the convergence of $e(t)$ to zero, we have that $\dot{\tilde{W}}$, $\dot{\tilde{W}}_{n+1}$ also converges to zero.

□

Remark 5.3 Again from the above analysis we cannot conclude anything about the convergence of the weights to their optimal values. Remark (3.2) provides some details concerning the problem.

□

6 Conclusions

The purpose of this paper is to employ singular perturbation analysis to examine the stability and robustness properties of a dynamical neural network identifier. We have considered many cases that lead to modeling errors, that is parametric, dynamic and dynamic uncertainties. A design methodology is given. To prove convergence results, we have used the well known input-output stability theory, which in turn provides internal stability of the neural network identifier. Finally it is worth noting that only the states of the plant which are related to the reduced order model, are assumed to be available for measurment.

References

[1] K.S. Narendra and A.M. Annaswamy, *Stable Adaptive Systems*, Englewood Cliffs, Prentice Hall, 1989.

[2] K.J. Aström and Wittenmark, *Adaptive Control*, Reading, MA, Addison-Wesley, 1989.

[3] K.S. Narendra, Y.H. Lin and L.S. Valavani, "Stable adaptive controller design, Part II: Proof of stability", *IEEE Trans. Automat. Contr.*, vol. AC-25, pp. 440-448, 1980.

[4] A. Isidori, *Nonlinear Control Systems*, New York, NY, Springer-Verlag, 1989.

[5] H. Nijmeijer and A.J. van der Schaft, *Nonlinear Dynamical Control Systems*, New York, NY, Springer-Verlag, 1989.

[6] D.G. Taylor, P.V. Kokotovic, R. Marino and I. Kanellakopoulos, "Adaptive Regulation of Nonlinear Systems with Unmodeled Dynamics", *IEEE Trans. Automat. Contr.*,vol. 34, no. 4, pp. 405-412, 1989.

[7] I. Kanellakopoulos, P.V. Kokotovic and R. Marino, "An Extended Direct Scheme for Robust Adaptive Nonlinear Control",*Automatica*, vol. 27, no. 2, pp. 247-255, 1991.

[8] S. Sastry and A. Isidori, "Adaptive Control of Linearizable Systems ", *IEEE Trans. Automat. Contr.*, vol. 34, no. 11, pp. 1123-1131, 1989.

[9] I. Kanellakopoulos, P.V. Kokotovic and A.S. Morse "Systematic Design of Adaptive Controllers for Feedback Linearizable Systems", *IEEE Trans. Automat. Contr.*, vol. 36, no. 11, pp. 1241-1253, 1991.

[10] K.S. Narendra, *Stable Identification Schemes in System Identification : Advances and Cases Studies*, New York: Academic, 1976

[11] P.A. Ioannou and P.V. Kokotovic, "An Asymptotic Error Analysis of Identifiers and Adaptive Observers in the Presens of Parasitics", *IEEE Trans. Automat. Contr.*, vol. 27, no. 4, pp. 921-927, 1982.

[12] P.A. Ioannou and P.V. Kokotovic, *Adaptive Systems with Reduced Models*, New York: Springer-Verlag, 1983.

[13] S. Chen, S.A. Billings and P.M. Grant, " Non-linear system identification using neural networks ", *International Journal of Control*, vol. 51, no. 6, pp. 1191-1214, 1990.

[14] Fu-C. Chen, " Back-Propagation neural networks for nonlinear self-tuning adaptive control ", *IEEE Control Systems Magazine*, vol. 10, no. 3, pp. 44-48, 1990.

[15] K.S. Narendra and K. Parthasarathy, " Identification and Control of Dynamical Systems Using Neural Networks," *IEEE Trans. Neural Networks,* vol. 1, no. 1, pp. 4-27, 1990.

[16] P.V. Kokotovic, H.K. Khalil and J. O'Reilly, *Singular Perturbation Methods in Control: Analysis and Design*, Academic Press, New York, 1986.

[17] N. Rouche, P. Habets and M. Laloy, *Stability Theory by Liapunov's Direct Method*, New York, Springer-Verlag, 1977.

The Polynomial Method Augmented by Supervised Training for Hand-Printed Character Recognition

Peter G. Anderson* and Roger S. Gaborski

Imaging Research Laboratories
Eastman Kodak Company
Rochester, New York 14653-5722 USA

Abstract

We present a pattern recognition algorithm for hand-printed and machine-printed characters, based on a combination of the classical least squares method and a neural-network-type supervised training algorithm. Characters are mapped, nonlinearly, to feature vectors using selected quadratic polynomials of the given pixels. We use a method for extracting an equidistributed subsample of all possible quadratic features.

This method creates pattern classifiers with accuracy competitive to feed-forward systems trained using back propagation; however back propagation training takes longer by a factor of ten to fifty. (This makes our system particularly attractive for experimentation with other forms of feature representation, other character sets, etc.)

The resulting classifier runs much faster in use than the back propagation trained systems, because all arithmetic is done using bit and integer operations.

1 Background

The *augmented polynomial method* is a system for classification of alphanumeric characters, such as ZIP codes and the characters written on tax forms. This method extends the system presented in [5] by using iteration to improve classification accuracy. It competes favorably with feed-forward neural network systems trained using back propagation.

1.1 Character Recognition

We only consider digitized characters after a preprocessing step has thinned them to standard width strokes and scaled them into a standard size (we use 30×20 pixels). (We do not address character locating or segmenting issues.)

Broadly speaking, characters of a given classification (i.e., the digit '1,' the character 'A,' etc.) form a cluster in some high-dimensional space. An unknown character is classified by determining which cluster it belongs to, for instance, to which cluster centroid is it nearest.

A straightforward but primitive attempt at clustering is to consider the characters' pixel arrays as points in 30×20 space, and to compute distances between these points using the L^1 or *Hamming* distance. This approach may be satisfactory for matching printed characters of a specified font where error or noise is inverted pixels. Unfortunately, plastic deformations, not inverted pixels, are the deviations from standard for hand-printed characters, so this simple-minded distance calculation totally breaks down. Lines that are only slightly displaced from the standard template produce huge differences in the Hamming distance. (We performed this experiment and achieved under 60% correct classification for digits.)

A more effective approach is to determine a collection of "features" and map the given pixel-coordinate space into a (possibly much higher dimensional) space. This feature mapping is necessarily nonlinear. If the features are well-chosen, the individual character clusters will be much more easily separable; optimistically, the clusters will be linearly separable.

Statistical pattern recognition (see, e.g., [1]) deals with selected features and analytic techniques for determining clusters and cluster membership, whereas neural network pattern recognition seeks to have the features and the discrimination criteria co-evolve through self-organization. The present approach is largely statistical pattern recognition, although it does have an element of iteratively adjusting the cluster boundaries reminiscent of Rosenblatt's *perceptron training algorithm* [2].

Character recognition techniques, whether statistical, neural, or hybrid, work with a data base of labeled characters partitioned into a *training set* and a

*Dr. Anderson is also associated with the Rochester Institute of Technology, Rochester, NY 14623-0887

testing set. The training set is used to construct the recognizer, and the testing set to evaluate its performance. Training tries to get the recognizer to behave as well as it needs to on the training set in order to "generalize" as well as it can on the testing set.

1.2 Shrinivasan's Polynomial Method

Shrinivasan's polynomial classifier ([5]) works as follows. Labeled, hand-printed characters are converted to feature vectors, $\overline{v}$, and are associated with target vectors corresponding to their labels. The components of the feature vectors are F quadratic polynomials formed from the character's pixel array; these features provide evidences of lines throughout the image. The target vector for each character is a standard unit vector $\overline{e}_{k(\overline{v})}$ with the $k(\overline{v})^{th}$ component equal to 1 and all other components equal to zero, where $k(\overline{v})$ is the classification of the character. Standard numerical techniques [3] are used to determine an $F \times K$ matrix, A, to minimize the squared errors, $\sum_{\overline{v}}(A\overline{v} - \overline{e}_{k(\overline{v})})^2$, where the sum runs over all the feature vectors, $\overline{v}$. The *weights matrix*, A, is then used to classify unlabeled characters, by determining the largest component in the product $A\overline{w}$, where $\overline{w}$ is the unknown character's feature vector.

Shrinivasan's method rapidly creates a reasonably good classification system, achieving approximately 93% on digits that have been normalized in a 16×16 array of pixels.

2 Extending the Polynomial Method

We then iterate a process that starts with that weight matrix, A, which was determined using a small number of features and a small sample of the character data base. Subsequent passes ("epochs") identify those training exemplars that are incorrectly classified or are classified with too small a confidence (i.e., the largest component of $A\overline{v}$ fails to exceed the second largest by a sufficient threshold). These poorly classified characters are replicated in the training to strengthen their correct classification, and negative feedback is used to inhibit strong incorrect classifications.

As a result, our system is able to develop a weights matrix with performance approximately 99% on digits, 95.5% on upper case alphabetics only, and 90% on mixed characters. We have also experimented with multiple font, machine-printed alphanumeric characters (i.e., 36 classes), and have achieved 98.7%.

3 Detailed Description

3.1 Feature Sets: Line Evidences

3.1.1 Two-Pixel Products

Shrinivasan's choice of features was logical products of nearby pixels, which can easily be described in terms of the chess moves, king and big knight. A king feature is the product of two pixels that are a king's move apart; that is, any two pixels chosen from the corners of a 2×2 square of pixels. A big knight feature is the product of two pixels that are diagonally opposite in a 2×3 or 3×2 rectangle of pixels. In this design, almost every pixel can participate in eight features (pixels near the edges are used in fewer features), giving nearly 2,000 features.

We use similar features for our characters that are normalized in a 30×20 array. We call our features the *n-king* and *n-knight* features, where the parameter n is a small integer that describes the size in number of pixels in the side of a square. Two pixels on the boundary of such a square, symmetric about the center of the square, forming a stretched version of the chess move, are multipied together (the logical product) to give the feature's value. In the 5×5 square shown below, there are eight features centered around the •: four 5-king features, *ai*, *ck*, *em*, and *go*, and four 5-knight features, *bj*, *dl*, *fn*, and *hp*.

a	*b*	*c*	*d*	*e*
p	.	.	.	*f*
o	.	•	.	*g*
n	.	.	.	*h*
m	*l*	*k*	*j*	*i*

(the parameter n must be even for the n-king *North-South* and *East-West* features to be defined; and n must be of the form $4m+1$ for the n-knight features to be defined.) We have found the following features to be useful: 3-king, 7-king, 5-knight, and 9-knight.

3.1.2 Fuzzy Features

Another type of feature we have experimented with, which seems particularly suitable for this application, we call *fuzzy line features*, and we also label them with chess moves: *fuzzy-n-knight* and *fuzzy-n-king*. As before, these features are extracted from the pixels on the perimeter of an $n \times n$ square, but they are the logical product of the logical sum of pixels. For example, the four fuzzy-3-knight features are built on three-by-three squares of pixels, centered at "•": $(a+b) \times (h+i)$, $(b+c) \times (g+h)$, $(c+f) \times (d+g)$, and $(a+d) \times (f+i)$, as shown below:

a	b	c
d	•	f
g	h	i

And the four fuzzy-5-king features are built around the rim of a five-by-five square of pixels; they are: $(b+c+d) \times (j+k+l)$, $(d+e+f) \times (l+m+n)$, $(f+g+h) \times (n+o+p)$, and $(h+i+j) \times (p+a+b)$:

a	b	c	d	e
p	.	.	.	f
o	.	•	.	g
n	.	.	.	h
m	l	k	j	i

These fuzzy features may be more reliable (lenient) at detecting the presence of lines through the pixel "•" than the first nonfuzzy features.

For hand-printed digit recognition we used 1,500 features using 5-knight and 7-king. For machine printed digits and upper case alphabetics (a combination of three fonts), we used 7-king, 5-knight, and 9-knight, for a total of 600 features. For upper and lower case alphabetic (a total of 40 different characters—the upper and lower case versions of the letters *c, k, o, p, s, t, u, v, w, x, y,* and *z* were treated as the same character), we used 7-king, 5-king, and fuzzy-9-king, and a total of 1,700 features. These choices were determined by trial and error.

3.1.3 Feature Collection Issues

The cost of a large number, F, of features is borne by "training," which involves the creation and inversion of an $F \times F$ matrix (complexity $\mathcal{O}(F^3)$), the maintenance of a weight matrix of size $K \times F$ (where there are K character classes), and the classification-time creation ($\mathcal{O}(F)$) and multiplication ($\mathcal{O}(F^2)$) of the feature vector.

We use feature vectors consisting of between 100 and 1,500 features, which are smoothly distributed over the set of all possible (above-described, quadratic) features (see below for a discussion of the smooth subsampling).

3.2 The Pseudo-Inverse

Our character recognition problem is cast in the following general type of framework: given a matrix, X, and a vector, $\bar{y}$, determine the vector, $\bar{a}$, that minimizes $||\bar{r}|| = ||\bar{a}X - \bar{y}||$. If $\bar{a}X$ is the nearest vector in Range(X) to the given vector, $\bar{y}$, then the difference, $\bar{r}$, must be orthogonal to the subspace, $R(X)$, the range of X; so $\bar{r}X^T = 0$, and $\bar{a}XX^T = \bar{y}X^T$. If XX^T is nonsingular, then $\bar{a} = \bar{y}X^T(XX^T)^{-1}$. $X^T(XX^T)^{-1}$ is called the *Moore-Penrose Pseudo-Inverse of* X [4].

The single-vector problem presented above expands into a multiple-vector problem that we apply to the automatic character classification problem. Suppose that we have a training set consisting of N characters; hence, we have N feature vectors, $\bar{x}_1, \bar{x}_2, \ldots, \bar{x}_N$. Let X be the $F \times N$ matrix, whose i^{th} column is $\bar{x}_i$. Denote the given classification of the ith training vector by k_i, where $1 \leq k_i \leq K$ ($K = 10$ for digit classification problems; $K = 36$ for alphanumeric classification problems). The ideal target for the ith character is $\bar{e}_{k_i}$, the standard unit vector in K-space. Let Y be the $K \times N$ matrix whose columns are $\bar{e}_{k_1}$, $\bar{e}_{k_2}$, …, $\bar{e}_{k_N}$. We seek a $K \times F$ matrix, A, such that $AX = Y$, or rather, that AX approximates Y in the least squares sense. We determine A using the Moore-Penrose pseudo-inverse: $A = YX^T(XX^T)^{-1}$.

A is the weight matrix we seek for character classification. Given an unknown character, we form its feature vector, $\bar{x}$, and evaluate $\bar{y} = A\bar{x}$, which is a list of "evidences" or "classification strengths" for each of the K classes to which our unknown character can belong. Hopefully, one of these K strengths stands out above the others, and we classify the character with high confidence.

3.3 Iterative Training

The training algorithm described above yields a weight matrix, A, that minimizes the sum of the squared errors, $\sum_{i=1}^{N} ||A\bar{x}_i - \bar{y}_i||^2$, where the $\bar{x}_i$ run over the feature vectors of the training data, and $\bar{y}_i = \bar{e}_{k_i}$, the k_i^{th} standard unit vector, denoting that $\bar{x}_i$ is of class k_i. This is approximately what we want. Actually, we want the product vector, $A\bar{x}$, to have its largest component correspond to the correct classification of $\bar{x}$ for as many as possible of the training exemplars, $\bar{x}$. It is not an average error we want to minimize, but the number of incorrectly classified characters.

To apply our least squares technique to a problem involving a different error criterion, we may increase the representation of the incorrectly classified exemplars and exemplars classified with low confidence (called, together *poorly classified* or *hard to learn* exemplars) in our training set.

As described above, the weight matrix, A, is determined by

$$A = YX^T(XX^T)^{-1} \tag{1}$$

Component matrices in equation (1) are maintained as

$$Z = YX^T = \sum_{i=0}^{N} \bar{e}_k \bar{x}_i^T, \tag{2}$$

and

$$W = XX^T = \sum_{i=0}^{N} \overline{x}_i \overline{x}_i^T \qquad (3)$$

where N is the number of training exemplars. If the hard-to-learn $\overline{x}_i$ are repeated with multiplicity m_i,

$$Z = \sum_{i=0}^{N} m_i \overline{e}_k \overline{x}_i^T , \qquad (4)$$

and

$$W = \sum_{i=0}^{N} m_i \overline{x}_i \overline{x}_i^T \qquad (5)$$

If an exemplar, $\overline{x}$, of class k is ill-classified, there is another class, j, such that $(A\overline{x})_j > (A\overline{x})_k - \theta$, where θ is a confidence threshold. We attempt to lower the jth classification strength for $\overline{x}$ by specifying, on at least one epoch, that we want a negative j^{th} classification weight for this exemplar (see the algorithm outlined below).

The confidence threshold, θ, is raised or lowered from one epoch to the next to try to keep the number of ill-classified exemplars near a specified fraction of the training set. We have found that such retraining of 20% of the training exemplars works well. Retraining a larger fraction causes large oscillations. For example, that two character classes 'S' and '5' are initially difficult to separate. When too large a fraction of characters are chosen for retraining, we find that, on one epoch all of these characters will be classified as 'S', and on the next epoch, all will be classified as '5', and so on. With the 20% retraining fraction, a reasonable boundary between the two classes is determined quickly.

We achieve this by the following type of iteration:

initialize the matrices $Z = 0$ and $W = 0$

/* epoch #1 */
for every $\overline{x}_i$
  add $\overline{e}_k \overline{x}_i^T$ to Z
  add $\overline{x}_i \overline{x}_i^T$ to W
compute $A = ZW^{-1}$

for *epoch* $= 2 \ldots$ *epoch_count*
  for every ill-classified $\overline{x}_i$
    /* the correct class k must be strengthened */
    /* the incorrect class j must be inhibited */
    add $(2\overline{e}_k - \overline{e}_j)\overline{x}_i^T$ to Z
    add $\overline{x}_i \overline{x}_i^T$ to W
  compute $A = ZW^{-1}$

During the first epoch we consider every $\overline{x}_i$ to be ill-classified. In subsequent epochs, we use the weight matrix, A, as developed in the previous epoch.

3.4 Subsampling the Training Data

The technique of subsampling the data was originally introduced to speed up the training process. However, this also produced better results, i.e., classifiers with better accuracy.

We use a shuffled data collection, so that in any portion of it we will find approximately equally many exemplars of each character. We start with a small fraction of the training data, say 10%, in epoch one; then we process 20% in epoch two, and so on.

This subsampling gives us the following advantages. Since the classifier functions linearly (in feature space), it forms boundaries consisting of hyperplanes; the character clusters it forms are represented by convex polytopes. If we can sketch out the convex hulls of these polytopes, we may then avoid training (i.e., updating the W and Z matrices) for the majority of the exemplars that are well inside those regions, only focusing attention on exemplars near or on the wrong side of the boundaries.

3.5 Enlarging the Training Set

The training set needs to be big and representative of the characters we want to recognize. However, a lot of the variation in characters can be explained in terms of simple modifications of characters. The technique we use is to translate a given character by eight "king's moves" (i.e., up one pixel, up and right one pixel, right one pixel, etc.), giving nine training exemplars for the price of one.

The effect of this is that the system learns the training data much more slowly, and its performance on the testing data consequently improves greatly. Because of the way we train and retrain, once the system's performance on the training data is very good, then very little additional training can take place.

3.6 Feature Subsampling

We have discovered that the sequence of pixels with coordinates ($11k$ mod 28, $11k$ mod 19) for $k = 1, \ldots, n$ is equidistributed in the 30×20 rectangle of pixels. If we are using an n-king and an m-knight feature, we have approximately eight features centered at every pixel we choose using the above rule. This rule gives us an even covering of the whole image with features, even if the number of features chosen is very small.

4 Implementation Details

4.1 Pseudo-Inverse Calculations

The computation involving the pseudo-inverse, $A = YX^T(XX^T)^{-1}$, seems to require several arrays be built, multiplied, and inverted. Some of these matrices are very large. For example, if we are dealing with N training patterns and F features, and K character classifications, the matrices have the following dimensions:

array	dimensions
A	$K \times F$
X	$F \times N$
Y	$K \times N$

However, because of the simple nature of these matrices and the contexts in which they are used, neither X nor Y needs to be explicitly stored. Recall their definitions: the kth column of Y is a unit vector giving the classification of the kth training pattern; the kth column of X is the binary feature vector of the kth training pattern. Instead of storing X and Y explicitly, we build, on the fly, the matrices $Z = YX^T$ and $W = XX^T$, of dimensions $K \times F$ and $F \times F$, respectively.

We initialize Z to zero; for each training pattern, if its classification is j, we add its feature vector to the jth column of Z.

We initialize W to zero; for each training exemplar's feature vector, $\overline{x}$, we add its outer product square, $\overline{x}\,\overline{x}^T$, to W; that is, add $x_i x_j$ to W_{ij} for every subscript pair, i, j. An outer product square is symmetrical, so W is symmetrical, and we only have to accumulate W_{ij} for $j < i$, i.e., on the lower triangle. Since the feature vector is a sparse binary vector, we only have to add it to selected rows of W.

An important consequence of this elimination of the explicit storage of X and Y is that there is no storage penalty for a large training set.

4.2 Feature Caching

After the above improvement was made to the program, feature extraction—determining an image's feature vector—turned out to be the most time-consuming step in the training system. Consequently, we create files of feature vectors for the training and testing of images. The feature vectors are sparse binary vectors, so we only store the list of indices with nonzero values. The feature vector files are big, and they take a long time to create, but that creation only happens once, and the files can be reused for other experiments.

5 A Growing Machine

Because of our ordering of the pixel centers for feature extractions, a feature vector with F features is a prefix of a feature vector with $F+F'$ features. Consequently, the corresponding matrices, $Z = YX^T$ and $W = XX^T$, for a small feature vector, are submatrices of those for larger feature vectors. The following training algorithm naturally suggests itself.

We determine the largest size feature vector that we believe appropriate (e.g., 1,500 features), and gather feature vectors of that size for every character we process, either in training or testing. However, we will be maintaining classification weight matrices that use only a prefix of that vector, and testing may ignore a feature vector's postfix. For a character used in training, if that character fails to be classified properly by the current weight matrix, its full feature vector will be used to update Z and W. As described above, an epoch consists of a (possibly subsampled or supersampled) pass over the training data. After each epoch, a larger submatrix of Z and W is chosen to compute the new weight matrix.

We obtain several advantages from this system. Since the inversion of W costs time that is proportional to the cube of the number of features actually used, we will be able to rapidly sketch out a classifier that approximates the classifier we will eventually derive. The inexpensive early classifiers allow us to easily filter the training set, avoiding the cost of determining the outer product squares of feature vectors for correctly classified characters. This eliminates another bottleneck.

The form that our training algorithm takes is now:

```
initialize the matrices Z = 0 and W = 0
initialize f, the number of features to be used

for epoch_count iterations
    for every training exemplar's feature vector, x̄
        if x̄ is poorly classified by the current A_f
            add (2ē_k x̄_i − ē_j)^T to Z
            add x̄_i x̄_i^T to W
    increase f
    compute A_f = Z_f W_f^{-1}
```

In this algorithm:

- F features are used to build Z and W, but only f features are used in classification.
- W_f denotes the upper left $f \times f$ submatrix of W.
- Z_f denotes the first f columns of Z.
- A_f denotes the resulting $K \times f$ weights matrix.

6 Two Training Sessions

6.1 Digits

We trained the polynomial classifier to recognize hand-printed digits, using 143,258 training and 15,971 testing exemplars. The training exemplars were shifted to give us four new exemplars for each given one, for a total training set size of 716,290.

The fuzzy-7-knight and the fuzzy-5-king features types were used for 1,500 features. The training started with a feature subset (the f parameter) of 400, and a training exemplar subset of 47,753.

The performance of the training session is shown in Table 1. We found in this example that the artificial enlargement of the training set by shifting kept the training set from ever performing better than the testing set, so that the system never "memorized the noise in the training data" and successfully generalized its learning to the testing data.

The training session took two days on a SUN SPARCstation 2. A feed-forward network trained using back propagation with similar performance took approximately a month to train.

epoch	tss	retrained	ratio	features	testing	training
1	47753	47753	100	400	95.08%	93.67%
2	95507	15714	16	500	97.08%	95.99%
3	143257	16364	11	600	97.31%	96.67%
4	191012	20153	10	700	97.66%	97.16%
5	238762	21273	8	800	97.85%	97.37%
6	286516	25825	9	900	98.11%	97.65%
7	334269	39945	11	1000	98.28%	97.72%
8	382022	31215	8	1100	98.22%	97.93%
9	429775	34003	7	1200	98.26%	97.99%
10	477527	38533	8	1300	98.23%	98.13%
11	525278	40770	7	1400	98.33%	98.17%
12	573032	57852	10	1500	98.38%	98.24%
13	620785	42283	6	1500	98.60%	98.45%
14	668537	43790	6	1500	98.65%	98.53%
15	716290	46752	6	1500	98.65%	98.59%
16	716290	46953	6	1500	98.68%	98.62%
17	716290	47573	6	1500	98.69%	98.64%
18	716290	47900	6	1500	98.68%	98.66%
19	716290	48027	6	1500	98.72%	98.67%
20	716290	48800	6	1500	98.73%	98.68%
21	716290	48062	6	1500	98.71%	98.69%
22	716290	48961	6	1500	98.72%	98.69%
23	716290	48392	6	1500	98.74%	98.70%
24	716290	49373	6	1500	98.74%	98.70%
25	716290	48994	6	1500	98.73%	98.70%
26	716290	49202	6	1500	98.74%	98.71%
27	716290	49245	6	1500	98.75%	98.71%
28	716290	49355	6	1500	98.75%	98.71%
29	716290	49423	6	1500	98.75%	98.71%
30	716290	49754	6	1500	98.74%	

Table 1: Hand-printed digit training session. The second column ("tss") shows a growing subset of the training exemplars. The third column ("retrained") shows how many "ill-classified" exemplars there were, with the "ratio" column giving the ill-classified percentage. The first ten epochs used a subset of the 1,500 features to sketch out a classifier. The last two columns give the classifier's performance on testing and training.

6.2 Machine Print

The second experiment shown here (Table 2) involved machine-printed alphanumeric (36 character classes). The training set consisted of 22,500 exemplars which was created from a set of 2,500 characters by shifting.

The 5-knight, 7-king, and 9-knight features types were used for 600 features. The training started with a feature subset (the f parameter) of 500, and a training exemplar subset of 2,249. This training session took 43 minutes.

epoch	tss	retrained	ratio	features	testing	training
1	2249	2249	100	510	96.73%	90.53%
2	4499	910	20	520	91.33%	87.48%
3	6750	5760	85	530	95.58%	93.19%
4	9000	1712	19	540	98.20%	97.61%
5	11248	6128	54	550	98.04%	97.58%
6	13500	1383	10	560	98.69%	99.07%
7	15750	10846	68	570	98.85%	98.38%
8	18000	2732	15	580	98.69%	99.08%
9	20250	7826	38	590	98.85%	99.11%
10	22500	4548	20	600	98.53%	99.30%
11	22500	6119	27	600	99.02%	99.35%
12	22500	1779	7	600	98.69%	99.40%
13	22500	4723	20	600	99.02%	99.46%
14	22500	2404	10	600	98.69%	99.45%
15	22500	4529	20	600	99.02%	99.51%
16	22500	4081	18	600	98.69%	99.44%
17	22500	4595	20	600	98.69%	99.44%
18	22500	4318	19	600	98.69%	99.44%
19	22500	5545	24	600	99.02%	99.50%
20	22500	3108	13	600	98.69%	

Table 2: Machine-printed alphanumerics training session. (See the caption for Table 1.)

References

[1] R. O. Duda and P. E. Hart, *Pattern Classification and Scene Analysis,* John Wiley & Sons, New York, 1973.

[2] Marvin L. Minsky and Seymore A. Papert, *Perceptrons, Expanded Edition,* The MIT Press, 1988.

[3] William H. Press, Brian P. Flannery, Saul A. Teukolsky, and William T. Vetterling, *Numerical Recipes in C.,* Cambridge University Press, 1988.

[4] G. W. Stewart, *Introduction to Matrix Computation,* Academic Press, 1973.

[5] Uma Shrinivasan, "Polynomial discriminant method for handwritten digit recognition," *SUNY Buffalo Technical Report,* December 14, 1989.

Analysis of Electronic Nose Data Using Logical Neurons

J.D.Mason, E.L.Hines and J.W.Gardner
Department of Engineering
University of Warwick
Coventry CV4 7AL, UK
Email: es238@eng.warwick.ac.uk

Abstract

The object of this study was to evaluate the performance of networks of logical neurons in analysing data from the Warwick Electronic Nose. The results are compared to those previously obtained from a back-propagation network on the same data.

The Warwick Electronic Nose consists of an array of twelve different tin oxide gas sensors, the response to an odour being determined from the conductivity changes of these sensors. Five different alcohol vapours were presented to the nose and the conductivities recorded in each case.

The 'standard' McCulloch and Pitts node used in most topologies - in particular back propagation - has proved successful in many applications. However, implementing nodes of this type in hardware is non-trivial due to the need to store, modify and multiply analogue variables. Logical or Boolean Neurons have their inputs and outputs limited to the set {0,1}. This allows easy implementation in hardware using RAM lookup tables. It is for this reason that Logical Neurons were selected as the basis for this study.

In general, the logical neurons were less successful in classifying the alcohol data than a back-propagation technique. However, a 6 layer ω-state PLN performed almost as well with a 94% success-rate compared to 100% for the MLP. Further work on logical neurons may lead to improvement by fully exploiting their capacity for generalisation.

1 Logical Neurons

A pattern consists of a set of measurements. In order to classify the pattern it can be represented as a point in multi-dimensional space, with one measurement on each axis. If the measurements, and any transformations performed on them, are chosen correctly then similar patterns should be close to one another in this multi-dimensional space. A previously unseen pattern can then be classified according to which cluster of patterns its point lies closest to. Unfortunately, the choice of measurements and transforms to perform is usually a time-consuming and difficult task. The popularity of using neural networks to recognise patterns is due to the fact that they can determine experientially the important features of the data.

If the measurements are limited to the set {0,1} then they can be used to form the address of a lookup table (LUT), the contents of each location in the table indicating which class the pattern belongs to. This method would work for small problems but as the number of measurements is increased then the size of the table soon becomes to large to implement in practice. For instance a 16 by 16 pixel image would need an LUT size of 2^{256}. Also, there is no inherent generalisation. N-tuple sampling [1] provides a method of overcoming these problems. This involves sampling the input N bits at a time, leading to smaller LUTs, the overall output being a combination of these N-tuple outputs. In our previous example, the 16 by 16 pixel image could be sampled by 32 8-Tuples with a total storage requirement of $32 \times 2^8 = 2^5 \times 2^8 = 2^{13}$. N-tuple sampling divides the pattern space into sub-spaces, as with any sampling method this introduces quantisation errors. Any two patterns falling in the same sub-space will give the same N-tuple outputs. Thus N-tuple sampling has some in-built generalisation.

The concept of boolean nodes is as old as the concept of artificial neural networks itself. The very first model proposed by McCulloch and Pitts in 1943 [2] restricted its inputs and outputs to the set {0,1}. Real neurons operate with boolean valued signals, intensity being represented by the repetition rate of the signal. However, the weighted-sum of inputs node became the most widespread, the most common topology being a perceptron combined with the back propagation learning algorithm [3]. Although this topology has proved successful in a variety of applications, the node models and training algorithm are not easily implemented in hardware. To fully exploit the parallel processing capabilities of artificial neural networks, a parallel architecture is better than simulation on a serial computer.

All boolean nodes are based on a lookup table, the first models [4] used fuses which could be blown during training to store the LUT data. The advent of cheap memory in the form of RAM chips allowed boolean nodes to be implemented with off the shelf technology. An N-tuple can be implemented directly in one RAM chip. By connecting the N-tuple's inputs to the address lines of the RAM, an LUT will be formed with each combination of inputs addressing a different location in the RAM. A pattern recognition system, called WISARD, based on this principle was constructed by Igor Aleksander *et al.* [5] and became the first hardware artificial neural network system to be used in commercial applications.

Initially, the RAM locations were only one bit wide. While a '1' in any location indicated a positive response, a '0' could mean either a negative response or represent an unused location. To increase the power of these networks, the nodes were adapted to form Probabilistic Logic Nodes (PLN) [6]. Each location could now store '0','U' or '1', the advantage of this system was that '0' could now be taken to indicate a negative response as untrained locations would contain 'U'. The output of the neuron for a location containing 'U' is '0' or '1' with equal probability. Training is achieved by a global Reward/Punish signal which causes the node to store the last output (Reward) or 'U' (Punish) at the current location.

Spreading algorithms were developed to increase the generalising capacity of the networks. An example of this is the G-RAM [7]. This system adds a phase after training to spread the results over locations still containing 'U's. Also the granularity of the probability states can be increased to try and reduce the amount of knowledge which is erased during a 'Punish' cycle [8].

All of the networks previously described use supervised reinforcement training with a global error signal. Other researchers have implemented different topologies. Self-Organising networks similar to Kohonen's feature map [9] have been produced [10]. Other supervised learning strategies have been developed, for instance the Goal-Seeking Neuron [11] uses a one pass search strategy to train the network.

The common feature of all networks of boolean neurons is their relative ease of implementation in hardware. Also the node models can implement any function of their inputs, whereas the McCulloch and Pitts type node can only perform linearly separable functions. The main disadvantage of boolean nodes is their inability to accept continuous valued inputs[1]. However, this can be overcome by pre-processing the data.

2 The Odour Data

The Electronic Nose consists of an array of twelve tin oxide gas sensors with partially overlapping sensitivities, the response to an odour being determined by the fractional change in conductivity of these sensors [13]. The data to be analysed in this case is the result of an experiment in which five alcohol vapours (methanol, ethanol, butan-1-ol, propan-2-ol, and 2-methyl-1-butanol) are each presented to the sensor array a total of eight times. Thus there are forty sets of results each consisting of twelve conductivity parameters. Details of this experiment can be found elsewhere [14]. Training and test sets are extracted from these readings using the "v-fold" validation technique [15]. One-fold validation was employed, i.e. one example of each alcohol was used in the test set and the remainder in the training set. This provides eight combinations of training/test sets from the one set of data.

3 Pre-Processing

A conductivity signal is analogue in nature, and as such not suitable for direct presentation to the boolean network. They must be processed to convert them to a form acceptable as input to the network. Firstly, the data is normalised over a suitable range. Then it must be coded, there are a number of obvious coding strategies. The simplest is to threshold the data, the problem though is at what level the threshold should be set. It was decided that thresholding would involve the loss of too much information in this application. Each value must therefore be coded into several bits. Straight binary code is unsuitable because values which are similar in magnitude can have vastly different bit patterns, e.g. 7 and 8. Gray Code was therefore chosen as similar magnitudes have similar bit patterns in terms of Hamming distance in this coding method. The analogue to digital conversion process currently has a linear characteristic but in the future a non-linear (e.g. sigmoid) function may be used to suit the distribution of the data. The number of bits used to code each value is selected by the

[1] i-pRAMs [12] are not subject to this limitation.

user. For the purposes of these experiments, 8 bit coding was used.

Gray Code has several problems associated with it:

- A small change in the bit pattern can represent a large change in the magnitude.
- Some of the bits are more significant than others.

These problems can be overcome by using a 'Thermometer' code, where the magnitude is represented by the number of bits set to '1'. However, this coding method leads to unacceptably large training patterns as the number of quantisation levels is increased. An attractive solution is Rank Ordering of the N-tuple inputs [16]. Work is currently in progress to evaluate the performance of the network when the data is coded in this manner.

4 The Simulator

A general purpose simulator has been produced to facilitate testing of networks of logical neurons. As artificial neural networks can be defined concisely in an object orientated style, C++ was chosen as the language for the implementation. The software can accept training/test pattern files in PlaNet or NeuralWorks formats and will simulate topologies defined in a simple text file, an example of which can be seen in Figure 1. Results are produced in a manner which can easily be exported to Matlab for analysis and graphical output. At the moment the user defines the topology to be used, but it is proposed to interface to a Genetic Algorithm front end to simplify optimal topology selection.

In addition to the results of training and testing, the simulator outputs the time taken to train (in both epochs and milliseconds) and also provides a continuous measure of convergence.

5 Network Topologies

Research into the use of artificial neural networks to analyse data from the electronic nose has already been conducted at Warwick [17][18]. Using Multi-Layer Perceptrons (MLPs) and the Back-Propagation training algorithm, 100% accurate classification of previously unseen data can be achieved. The object of this research is to compare the results obtained in the analysis of the same data using various boolean network topologies to those obtained with the MLP.

```
#network to analyse data from the
#electronic nose
retina = 96
outputs = 5
layers = 6
layer = 0
        rams = 48
layer = 1
        rams = 24
layer = 2
        rams = 12
layer = 3
        rams = 6
layer = 4
        rams = 3
layer = 5
        rams = 1
```

Figure 1: A Sample Network Definition File

The different types of boolean network included in these trials were:

- Networks of Probabilistic Logic Nodes (PLNs) [6].
- Networks of ω-state PLNs [8].
- Networks of Goal Seeking Neurons [11].

Several trials were conducted for each type with different combinations of network parameters and their effects on performance observed.

6 Results

Several performance metrics are available. For the purpose of this experiment a simple 'percentage correct' metric was used for easy comparison of the various topologies. Each trial consisted of training the network followed by testing using 100 presentations of the test set to average out any random effects due to the probabilistic nature of the nodes. 224 trials were conducted for each topology; this number resulted from earlier experiments involving several combinations of training parameters but was retained for continuity between sets of results. A summary of the results obtained on unseen data for all topologies is shown in table 1. In the case of the ω-state PLNs (where ω = number of probability states), it was found that networks using nodes with $\omega > 8$ provided little increase in performance in return for much longer training times. For this reason these results are compiled using a network with $\omega = 8$.

The time taken to train each topology is shown in table 2. The times are given in epochs, where one epoch is a complete presentation of the training set.

Topology	Percentage Correct		
	Worst	Average	Best
3-layer PLN	52	71	100
4-layer PLN	64	85	100
6-layer PLN	80	93	100
6-layer ω-state PLN	80	94	100
3-layer GSN	71	81	100
12:7:5 MLP	100	100	100

Table 1: Results

Topology	Epochs		
	Worst	Average	Best
3-layer PLN	706	76	1.6
4-layer PLN	746	118	2.8
6-layer PLN	792	85	6.7
6-layer ω-state PLN	4207	662	71
3-layer GSN	1	1	1
12:7:5 MLP	5186	2375	29

Table 2: Training Time (In Epochs)

7 Conclusions

From the results it is obvious that while boolean networks can sometimes match the performance of MLPs, they cannot do so as consistently. Although the networks have the capacity to generalise with 100% accuracy, the training algorithms do not guarantee convergence to such a solution on every trial. Future work must therefore be centred on the choice of the training algorithm in order to exploit fully the generalising ability of the topologies in use.

In general the boolean networks converged in less epochs than the back-propagation network. The time for one epoch on a Sun IPC workstation was about 0.49 seconds for the boolean networks, no equivalent time is available for the back-propagation. Note that the GSN uses a single pass training and consequently the time for one epoch is much greater.

As the choice of topology affects performance to some degree, a more reliable method than trial and error is needed. It is proposed to employ a Genetic Algorithm to choose the optimal combination of network design and training algorithm. In addition, more algorithms will be made available for simulation (e.g. G-RAM,i-pRAM) - including modifications of existing regimes. The aim is to arrive at a system with capabilities matching those of the MLP, but which can be implemented in hardware with less difficulty.

References

[1] Bledsoe W W and Browning I, *'Pattern Recognition and Reading by Machine'*, Proc. Eastern Joint Computer Conference, Boston, 1959.

[2] McCulloch W S and Pitts W H, *'A Logical Calculus of the Ideas Immanent in Neural Nets'*, Bulletin of Mathematical Biophysics, 5, 1943.

[3] Rumelhart D E, Hinton G E and Williams N, *'Parallel Distributed Processing'*, MIT Press, 1986.

[4] Aleksander I, *'Fused Logic Element Which Learns by Example'*, Electronics Letters, 1, 6, 1965.

[5] Aleksander I, Thomas W V and Bowden P A, *'WISARD - A Radical Step Forward in Image Recognition'*, Sensor Review (GB), 4, 3, 1984.

[6] Kan W K and Aleksander I, *'A Probabilistic Logic Neuron Network for Associative Learning'*, Proc. IEEE 1st Annual Conference on Neural Networks, San Diego, 1987.

[7] Aleksander I, in Eckmiller R et al. (eds.) *'Parallel Processing in Neural Systems and Computers'*, Elsevier Science, Amsterdam, 1990.

[8] Myers C E, *'Output Functions for Probabilistic Logic Nodes'*, Proc. 1st IEE International Conference on Artificial Neural Networks, London, 1989.

[9] Kohonen T, *'Self-Organisation and Associative Memory'*, Springer-Verlag, Berlin, 1984.

[10] Allinson N M, *'Neurons, N-tuples and Faces'*, Computing and Control Engineering Journal, 1, 4, 1990.

[11] Filho E, Bisset D and Fairhurst M, *'A Goal Seeking Neuron for Boolean Neural Networks'*, Proc. International Neural Networks Conference, Paris, 1990.

[12] Taylor J G and Gorse D, *'A Continuous Input RAM-Based Stochastic Neural Model'*, Research Note RN/90/63, University College London, 1990.

[13] Shurmer H V, Gardner J W and Corcoran P, *'Intelligent Vapour Discrimination Using a Composite 12 Element Array'*, Sensors and Actuators B1, 1989.

[14] Gardner J W, Bartlett P N, Dodd G H and Shurmer H V, in Schild D (ed.) *'Chemosensory Information Processing'*,Vol H39, Springer Verlag, Berlin, 1990.

[15] Weiss S M and Kulikowski C A, *'Computer Systems That Learn'* , Morgan Kauffman, San Mateo CA, 1991.

[16] Allinson N M, *'Encoding Non-Binary Data in N-tuple Neural Networks Using Rank Ordering'*, pre-print.

[17] Gardner J W, Hines E L and Wilkinson M, *'Application of Artificial Neural Networks to an Electronic Olfactory System'*, Measurement Science and Technology, 1, 5, 1990.

[18] Hines E L, Gardner J W, Fung W W and Fekadu A A, *'Improved Rate of Convergence in a MLP Based Electronic Nose'*, Second Irish Neural Network Conference, Belfast, Northern Ireland, 1992.

Neural Tree Network Based Electronic Nose

Adhanom A. Fekadu, Evor L. Hines and Julian W. Gardner
Department of Engineering
University of Warwick
Coventry CV4 7AL, U.K.
Email: E.L.Hines@uk.ac.warwick.eng
es791@uk.ac.warwick.eng

Abstract

The training of a multi-layer perceptron using the well known back-propagation algorithm usually requires a priori knowledge of the network architecture. In this paper, results are presented on two practical classification problems which use neural tree classifiers to determine automatically the optimal number of neurons required. The data-set comes from the response of the Warwick Electronic Nose to a set of simple and complex odours.

1. Introduction

The human sense of smell is the primary faculty upon which many industries rely to monitor the flavour of items such as beverages, food and perfumes. The parallel neural architecture existing in the human nose consists of millions of cells in the olfactory epithelium that act as the primary receptors to odorous molecules. These receptors synaptically link into glomeruli nodes and mitral cells which in turn feed into the brain. This architecture suggests an arrangement that could be used in an analogous electronic instrument capable of mimicking the biological system. In the Warwick Electronic Nose, the primary olfactory receptors have been replaced by an array of solid-state sensors that respond differentially to a broad range of chemical vapours or odours. This response may be characterised by a change in the electrical resistance of the sensor array which is processed further in order to identify the vapours/odours [1, 2]. The use of a multi-layer perceptron (MLP) to classify the output has already been studied using the back-propagation technique [3, 4].

The most commonly used neural network is the MLP usually trained by the standard back-propagation algorithm (SBPA) [5]. Training by this method, however, requires a number of user-specified parameters whose values may affect both the training time and the prediction error. Another short-coming associated with SBPA is that the best network architecture to use for a given application is not known a priori. Thus, a considerable time may be spent in finding the appropriate architecture and set of network parameters that produce optimal network, in terms of both the training time and prediction error.

Although a number of methods have been proposed to minimize the training time, the ideal architecture and parameters still need to be found by a set of laborious experiments which may not even give the optimum values.

In this paper, results are presented on simulations carried out on the application of neural tree classifiers to determine automatically an appropriate network architecture for our electronic nose.

A reader who is unfamiliar with decision tree classifiers and neural networks is advised to read [6]-[7].

2. Architecture of neural trees.

Decision trees and neural networks are married together to produce neural tree classifiers. Much research work has been done on this effective marriage [7]-[10]. In this paper, a simulation in the application of neural tree classifiers based on the idea of Sankar and Mammone's Neural Tree Network (NTN) algorithm [10] is presented.

Suppose the training set contains m input feature vectors, x_i, $i=1, 2, ..., m$, each of dimension r. The input feature space, X, is the space containing all the input vectors. Each input feature vector corresponds to one of n classes. A tree structured classifier is constructed by a repeated process of splitting subsets of the input feature space, X, into a set of descendant subsets, beginning with X itself.

A feed-forward neural network with no hidden layer is constructed at each internal (non-leaf) node of the decision tree. Figure 1 shows a neural tree network structure for a 4 input 3 class problem with the neural networks in internal nodes. The neural network at the root of the decision tree is used to divide the input feature space, X, into at most n regions [10]. Each region is then assigned to a child node which is further divided into subregions by the neural network constructed at each child node. This process is repeated in this way until only one class is present in each region (or we are unable to split any further). Each of these single class regions are represented by the leaf nodes which are labelled with the relevant class.

3. Network training

The neural network at internal node t of the tree is constructed with n_t neurons in the output layer, where n_t is the number of classes in the training set contained in feature space, X_t, corresponding to node t. Let the number of input feature vectors contained in X_t be m_t. The size of the input layer in all the networks is the dimension of the input feature vector. In the simulation, the two measures of error defined by (1) and (2) were tested: the L1 norm of errors,

$$E_1 = \sum_{p=1}^{m_t}\sum_{i=1}^{n_t} E_{1pi} = \sum_{p=1}^{m_t}\sum_{i=1}^{n_t} |t_{pi}-y_{pi}| \quad (1)$$

and the L2 norm of errors,

$$E_2 = \sum_{p=1}^{m_t}\sum_{i=1}^{n_t} E_{2pi} = \frac{1}{2}\sum_{p=1}^{m_t}\sum_{i=1}^{n_t} (t_{pi}-y_{pi})^2 \quad (2)$$

where $E_{1pi}=|t_{pi}-y_{pi}|$, $E_{2pi}=0.5(t_{pi}-y_{pi})^2$, t_{pi} and y_{pi} are the target and actual output of neuron i corresponding to the pth input feature vector. The target, t_p, corresponding to a class i input can be constructed by a binary vector that has a 1 at its ith entry, the rest being 0.

Two activation functions given by equations (3) and (4), were tested:

$$f_1(x) = \frac{1}{1+e^{x}} \quad (3)$$

and

$$f_2(x) = \frac{1}{1+e^{-x}} \quad (4)$$

The neuron weight update, equation (5), is found by minimizing the error using the gradient descent technique:

$$\Delta_p w_{ij}(n+1) = \eta \frac{dE_{pj}}{de_{pj}} \frac{df(s_{pj})}{ds_{pj}} x_{pi} + \alpha \Delta_p w_{ij}(n) \quad (5)$$

$$s_{pj} = \sum_{k=1}^{r} w_{kj} x_{pk} + \theta_j , \qquad e_{pj} = t_{pj}-y_{pj} \quad (6)$$

where w_{ij} is the weight from the ith input component to neuron j, η and α are the learning rate and momentum coefficient respectively, θ_j is the bias of the jth neuron. The function E_{pj} in (5) is replaced by E_{1pj} or E_{2pj} from (1) or (2) respectively depending on the error function being used. The function f is the activation function which is replaced by (3) or (4) which ever activation function is used. The derivative of E_1 does not exist when the output of a neuron matches its target output. This does not affect the gradient descent method since no weight update is performed when the output error of a neuron is zero [10].

Two network training algorithms were used: the SBPA and an adaptive back-propagation algorithm (ABPA). In ABPA the learning rate and the momentum coefficient, η, and α are modified dynamically after every training cycle [11].

4. Splitting rule

The neural network at an internal node uses a competition rule to classify an input feature vector x_p. The network at node t classifies x_p as class k if and only if [10],

$$y_{pk} \geq y_{pi}\ ,\ \ \forall\ i \neq k,\ 1 \leq i,k \leq n_t \qquad (7)$$

where y_{pi} is the output of neuron i for input x_p. Feature vectors classified as class k, $1 \leq k \leq n_t$, are assigned a separate region. The neural network splits the region X_t into at most n_t convex regions [12, pp 290-292]. The algorithm grows the number of neurons as it learns until if finds the right number of neurons needed for the classification problem.

5. Simulation results.

The algorithm was tested by applying it to two electronic nose applications: the classification of a set of simple odours (alcohols) and of complex odours (coffees). The alcohol and coffee data-sets were gathered using a version of the Warwick Electronic Nose containing twelve tin oxide gas sensors. The test odours were injected into a glass vessel containing the sensors and the response of each sensor recorded on an IBM PC. Details of the experimental procedure for the alcohols and coffees can be found elsewhere [3,14]. A series of eight tests were carried out on each of the 5 alcohols: methanol, ethanol, butan-1ol, propan-2-ol, and 2-methyl-1-butanol. The coffee data comprises 89 samples from each of the 12 sensors for three different coffee types. 30 samples for each of the first two and 29 samples for the third one.

The alcohol classification problem has been studied previously [4,13]. After repeated experiments (using both the SBPA and ABPA methods of network training), a 12-7-5 fully connected feed-forward network with initial weight range [-1,1], learning rate 1.0 and momentum coefficient of 0.7 was suggested.

The coffee classification problem has also been studied previously [13]. The network studied was a 12-3-3 fully connected feed forward network trained by the method of ABPA.

In order to get the best estimate of classification performance of our neural tree, the "v-fold" validation technique was employed in both problems. In the case of the alcohols, v was set to one, that is one of each alcohol type was used for testing and the rest were used for training. In the case of the coffees, v was set to 3. The network parameters, η and α, were set to 0.7 and 0.9 respectively. The input data was normalised in the range [-1,1]. The objective of the simulation was to find the optimal number of neurons that correctly classify the training data. The initial set of weights affect the rate of convergence and the prediction error of the network when the back-propagation algorithm is used. The algorithm can be trapped by a local minimum. Therefore, whenever a neural tree was unable to split the data at an internal node, the network at that node would be re-trained with a different set of initial weights and with increased number of iterations. The maximum number of re-training was set to 10 after which the algorithm stopped.

The results found are summarised in Table 1. The error functions, E_1 and E_2, and the activation functions, f_1 and f_2, are as defined in equations (1) through (4). The results shown are averages of eight runs each one taking a different set of test data-set. Neural trees with no internal nodes have been found for the alcohols data. This is actually an MLP with no hidden layer. The 100% classification of the test data was achieved when there was no hidden layer. The SBPA failed to split the data when using the L1-norm of errors.

In the case of the coffee classification, the algorithm was not able to split the internal nodes in both the SBPA and ABPA, when used with the L1-norm of errors. The maximum

Figure 1. A neural tree network structure for a 4 input 3 class problem.

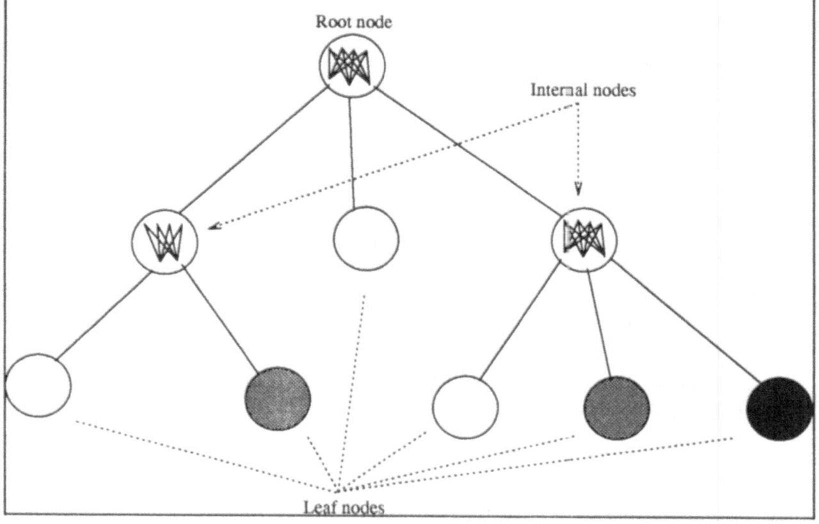

Table 1: Simulation results for the alcohol and coffee classification problem.

Training Algorithm	Error Function	Activation Function	Average % misclassifications		Average Number of internal nodes	
			Alcohol	Coffee	Alcohol	Coffee
ABPA	E_1	f_1	0	n/a	0.0	n/a
		f_2	12.5	n/a	0.125	n/a
	E_2	f_1	25 12.5	10	2.333	1.7
		f_2		10	2.125	1.5
SBPA	E_2	f_1	0	14.4	0.0	1.2
		f_2	0	11.1	0.0	1.1

number of internal nodes found was only 2.

A 100% correct classification found in the case of the alcohols data, however, was not achieved with the coffees data.

Although not all the test data were correctly classified, the algorithm obviously eliminated the choice of the network architecture which would otherwise be required by the back-propagation algorithm.

6. Conclusion

We have applied the neural tree classifier to two practical odour classifications. The results presented here are part of a research programme which is concerned with the application of artificial neural networks in practical instrumentation, such as the Warwick Electronic nose. It has been demonstrated that the neural tree classifier grows the network while learning and constructs a network with the right number of neurons required to classify the sample data.

References.

[1] Shurmer H V, Gardner J W and Chan H T, *'The application of discrimination techniques in alcohols and tobacco using tin oxide sensors'*, Sensors and Actuators, 18, pp 361-371, 1989.

[2] Gardner J W, Bartlett P N, Dodd G H and Shurmer H V, *'Chemosensory information processing'*, ed. D Schild (Berlin: Springer) Vol H39, pp 131, 1990.

[3] Gardner J W, Hines E L and Wilkinson M, *'The application of artificial neural networks to an electronic olfactory system'*, Meas. Sci. Technol. 1, pp 446-451, 1990.

[4] Gardner J W, Hines E L and Tang H C, *'Detection of vapours and odours from a multisensor array using pattern recognition technique. Part 2 : Artificial neural networks'*, Sensors and Actuators B, 15, pp9-15, 1992.

[5] Rumelhart D E and McClelland J L, *'Parallel distributed processing'*, MIT press, chapter 8, 1986.

[6] Lipmann R P, *'An introduction to computing with neural nets'*, IEEE ASSP Magazine, pp 4-22, April 1987.

[7] Safavian S R and Landgrebe D, *'A survey of Decision Tree Classifier Methodology'*, IEEE Trans. Syst., Man., Cybern., vol 21, pp 660-674, 1991.

[8] Brent R P, *'Fast Training Algorithms for Multilayer Neural Nets'*, IEEE Trans. on Neural Networks, pp 346-354, 1991.

[9] Sirat J A and Nadal J-P, *'Neural trees: a new tool for classification,* Network, 1990.

[10] Sankar A and Mammone R J, *'Speaker Independent Vowel Recognition using Neural Tree Networks'.*

[11] Chan L W and Fallside F, *'An adaptive training algorithm for back propagation networks'*, Computer Speech and Language, 2:205-218, 1987.

[12] Mammone R J and Zeevi Y Y, *'Neural Networks: theory and application'*, Academic Press, 1991.

[13] Hines E L, Gardner J W, Fung W and Fekadu A A, *'Improved rate of convergence in a MLP based electronic nose'*, 2nd Irish Neural Networks Conf., Belfast, June 1992.

[14] Leung C W, *'Application of neural networks to the classification of coffee data'*, MSc IT Dissertation, Department of Engineering, University of Warwick, 1991.

Lime Kiln Process Identification and Control: A Neural Network Approach

B Ribeiro *, A Dourado and E Costa
Laboratório de Informática e Sistemas
Universidade de Coimbra
P-3000 Coimbra, Portugal

Abstract

Complex systems exhibiting strong non linearities and time delays such as chemical processes are very demanding in control requirements. In this paper we present a neural network approach for multivariable non-linear kiln process identification and control. Neural networks, in control theory, are attractive because of their powerful capabilities to successfully approximate non-linear functions within a specified approximation error as recent research has proven. They can be used to synthetize non-linear controllers for non-linear processes and it is expected that better results can be obtained as compared to more conventional methods.

The main objective of this work is to train the neural network kiln controller to provide suitable control inputs that produce a desired kiln response. If the neural network plant model is capable of approximating well and with sufficient accuracy the highly non-linear calcination process in the lime kiln, then it may be used within a model based control strategy.

Firstly, the lime kiln process identification is achieved using a feedforward Artificial Neural Network (ANN), namely the plant model. It learns the kiln dynamics through a training process that drives the error between the plant output and network output to a minimum. The non-linear mapping from control inputs to plant outputs is achieved through the use of the backpropagation learning paradigm. The specifications of the neural network to provide the desired system representation are given.

Secondly, a neuralcontroller was designed to adaptively control the non-linear plant. The neural network topology was selected according to common used performance criteria.

Simulation results of non-linear kiln process identification as well as non- linear adaptive control are presented to illustrate the neural network approach. Analysis of the neural networks performance is underlined.

*Supported by Instituto Nacional de Investigação Científica and PORTUCEL

1 Introduction

Neural networks have been used in a wide range of applications In the identification of dynamic systems and synthesis of non-linear controllers in chemical processes, several successfully implementations were reported [1, 2, 3, 4].

The possibility to represent complex non-linear systems results from their well known theoretical ability to approximate arbitrary non-linear mappings as was investigated in [5, 6, 7] for Multilayered Perceptron Networks (MLP). Though not constructive since they do not provide a suitable arquitecture to solve a given problem, they are the basis of current work on the field.

Neural networks can deal with non-linearities and they can learn. The learning is achieved through minimization of the mean square error between required and actual outputs. The back propagation algorithm[8, 9, 10] for training the neural network is often used due to its computational simplicity and parallel structure since learning is distributed to each weight in the network. Neural networks are inherently non-linear and multivariable and are suitable for use in conventional control structures.

In this work, we present an application which aims at studying the feasibility of a neural network approach in system modelling and control of a non-linear multivariable process (two inputs, two outputs) within pulp and paper industry.

The paper is stuctured in six sections. In the following section the control problem is defined and the model used for simulation of the plant is shortly introduced. In the third section the neural network forward modelling for plant identification is described. In the fourth section, the neurocontroller structure is presented as well as its training method. In the fifth section we present the methods for obtaining the training instances, network parameter specifications and the simulation results. Finally, questions of stability and convergence from such approach are addressed and some final conclusions reported.

2 The Lime Kiln Model and Control Problem

Chemical processes need usually to be described by high-order, non-linear distributed parameter equations involving parameters and states that are difficult to determine and /or measure. The processes are usually subjected to external disturbances and transport time delays and an exact mathematical formulation is difficult to obtain. Industrial processes have to operate under changes to environmental conditions and variations in process conditions such as changes in raw materials, fuels, load levels, etc. Therefore a large number and ranges of uncertainties occur in the process parameter values, disturbances and signal inputs. On the other hand, due to industrial costs, it is very difficult to perform all the necessary experimentation to obtain a detailed and reliable mathematical model. These reasons have caused some criticism to many automatic control methods based on analytical methods. However, even for applicability studies some kind of a model is needed, with a minimum degree of exactness.

This work deals with the non-linear control of a rotary lime kiln in the pulp and paper industry. A rotating lime kiln is a multivariable process (Figure 1) which has long time delays and is often under the influence of large, long-term process disturbances. To control it efficiently some type of adaptive controller seems necessary. The choosen controller structure is based on a neural network approach within a model based strategy.

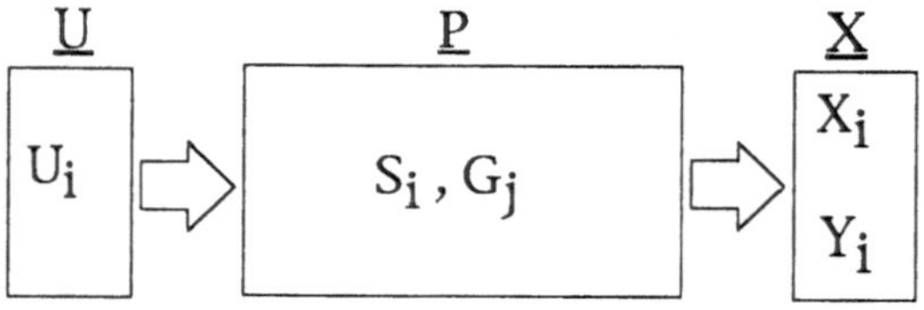

Figure 1: Lime Kiln Plant Model

The dynamic model of a lime kiln is a distributed parameter one which is constituted by a set of partial differential equations with respective boundary and initial conditions given by:

$$\frac{1}{V_s}\frac{\partial X_i}{\partial t} + \frac{\partial X_i}{\partial z} = S_i \qquad i = 1,2,3 \qquad (1)$$

$$\frac{1}{V_g}\frac{\partial Y_j}{\partial t} + \frac{\partial Y_j}{\partial z} = G_j \qquad j = 1,2,3,4 \qquad (2)$$

In these equations, X_i is the solids i-th state variable, V_s is the solids axial velocity and S_i is a generally nonlinear function of space, solids and gases state variables. The symbols Y_j, V_g and G_j have equivalent meaning for the gases variables. The axial length and time coordinates have the notation z and t respectively. The system has split boundary conditions which means that the values of the solids state variables are known at one kiln end (feed end) and that of the gases are known at the other end (burning end).

$$X_i(z=0,t) \qquad Y_j(z=L,t) \qquad (3)$$

The initial conditions are the values of the state variables before some perturbation.

$$X_i(z,t=0) \qquad Y_j(z,t=0) \qquad (4)$$

The gas phase dynamics is neglected due to its small residence times when compared to the solids phase. Thus, the gases PDE's equations are simplified to the form:

$$\frac{\partial Y_j}{\partial z} = -G_j \qquad (5)$$

The task for the multivariable controller is to maintain the temperature profile through the lime kiln at a given value. In complex processes, such as the lime kiln, there is a large number of inputs we can manipulate as well as measurements containing information about the state of the kiln. It is impossible to take into account all these variables to build a model. From the point of view of the control problem we have to find the best control inputs and to select the most representative measurements.

One of the outputs of the model to be identified is the lime quality (defined in terms of calcium oxide available in the final product). However the lime quality measurement is unavailable in real time and, hence, impossible to use it in computer control. The temperature of the product in burning zone is linked closely to the product quality. Too little heat at the burning zone means the material leaves with a too high proportion in calcium carbonate (incomplete calcination). Overheating means inefficiency with excessive burning temperature and a possible kiln wall damage. The manipulated variable for controlling the product temperature in the kiln burning zone is the fuel flow rate. On account of energy savings and environmental pollution, the other choosen controlled variable is the exit gas temperature. The latter is closely related to the excess of O_2 at kiln feed end and can be controlled by the gas flow through the kiln.

Summarizing, one can stabilize kiln temperature profile being the fuel flow one manipulated variable and the gas flow through the kiln the other. The two outputs are the temperatures of the product in the burning zone and the exit gas at the kiln feed end. To achieve an optimal temperature profile and therefore the best kiln control it is important to let the lime quality influence the required product temperature and let the excess of O_2 influence the set point of the exit gas temperature. Note that there is coupled influence between each manipulated variable and each output. Experience has shown, however, that it was the best control strategy.

3 Lime Kiln Identification : Neural Network Forward Modelling

The identification model and the method used to adjust the parameters based on identification error is achieved through a neural network model and a learning rule respectively. The structure for achieving this is indicated schematically in Figure 2.

We assume that the system under consideration belongs to the class of systems that the neural network model can represent. System identifiabilty is widely considered in literature within a classical framework. In his work, Narendra et al. [1] considers a given operator P defining a plant implicitly by the input-output pairs of time functions $u(t)$, $y(t)$, t $\in [0,T]$. For some desired $\epsilon > 0$ and a suitable norm the objective is to determine $\hat{P}$ such that

$$\|\hat{y} - y\| = \|\hat{P}(u) - P(u)\| \leq \epsilon, \qquad u \in U \quad (6)$$

where U is a compact set, $\hat{P} = \hat{y}$ denotes the output of the identification model, and $\hat{y} - y = e$ denotes the error between the output generated by $\hat{P}$ and the observed output y. According to Chen and Billings [11] a wide class of discrete time nonlinear systems can be represented by the nonlinear autoregressive moving average with exogenous inputs model (NARMAX). In the present

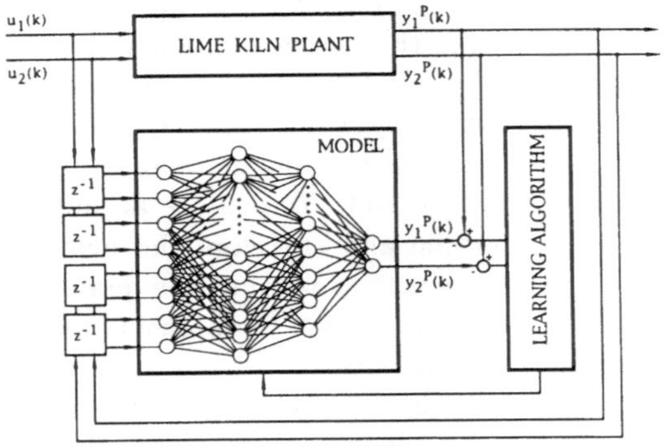

Figure 2: Neural Network Kiln Model

work, a representation of the system in terms of a non-linear function of past inputs, outputs and prediction errors is a simplified form of NARMAX model [12] given by:

$$\mathbf{y}^P(t) = f[\mathbf{y}^P(t-1), ..., \mathbf{y}^P(t-n$$

$$\mathbf{u}(t-1), ..., \mathbf{u}(t-nu)] + \mathbf{e}(t) \quad (7)$$

where

$$\mathbf{y}(k) = [y_1(t), y_2(t)]^T$$
$$\mathbf{u}(k) = [u_1(t), u_2(t)]^T$$
$$\mathbf{e}(k) = [e_1(t), e_2(t)]^T$$

and $ny = nu = 2$ are the corresponding time lags in outputs and inputs. The network response, approximating the underlying system dynamics, is the one step ahead predictor for $\mathbf{y}(t)$ and is described by:

$$\mathbf{y}^M(t) = \hat{f}[\mathbf{y}^P(t-1), ..., \mathbf{y}^P(t-ny),$$

$$\mathbf{u}(t-1), ..., \mathbf{u}(t-nu)] \quad (8)$$

The neural network model, giving the process model, is placed in parallel with the plant and the error between the plant output and network output, i.e., the prediction error is used as the network training signal. The structure above, represents the series-parallel model [1] which means that the network has no feedback. The dependence on the system output is shown in Figure 2.

The learning structure is a supervised one, having as targets the responses generated by the lime kiln simulation model. This is due to the difficulty to obtain reliable industrial data. A class of learning algorithms, known as prediction error algorithms, based on non-linear system identification where derived by Chen et al.[13] for MLP. The objective is to find an optimum set of weigths and thresholds which minimize the error between the system output $\mathbf{y}(t)$ and the network response $\hat{\mathbf{y}}(t,\theta)$:

$$\epsilon(t,\theta) = \mathbf{y}(t) - \hat{\mathbf{y}}(t,\theta) \quad (9)$$

The mechanism of learning is achieved by minimization of $J_N(\theta) : \mathbb{R}^{n_\theta} \to \mathbb{R}$ through a steepest descent algorithm. In case the negative gradient search is used, the back propagation of the prediction error to train the network gives the well kown back propagation algorithm [8, 9]. The optimization criterion or loss function is defined by:

$$J_N(\theta) = \frac{1}{2N}\sum^{N} \epsilon^T(t,\theta)\epsilon(t,\theta) \quad (10)$$

where N is the length of the training data and θ is a n_θ dimensional parameter vector whose elements are the network weights and thresholds.

4 Lime Kiln Controller : Neural Network Inverse Modelling

The proposed controller is a neural network trained in order to minimize the error between the output of the plant model and the reference. For the generation of control inputs $\mathbf{u}(t)$ (as a function of the error signal) to follow a desired trajectory, the existence of some inverse model has to be assumed.

By contrast with direct inverse modelling, also known as generalized learning, in specialized learning, the network inverse model precedes the system and is trained by an input signal representative of the controlled system (the common plant reference). This learning structure as shown in Figure 3 contains also a trained forward model of the system placed in parallel

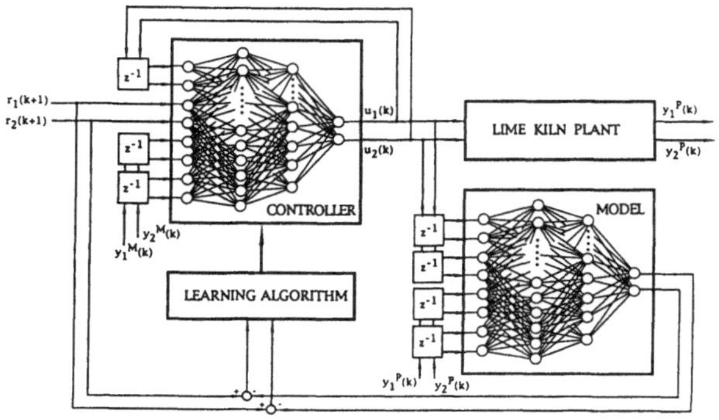

Figure 3: Kiln Controller: Inverse Network Model

with the plant. The error signal for the learning rule is the difference between the training signal reference and the forward model output. The error is propagated back through the forward model (whose weigths are frozen) and the inverse model weigths are then conveniently adjusted. Formally the inverse model training signal should preferably be computed with the process output. To avoid computation of plant jacobian (as in specialized learning), the output of the neural network forward model is used for learning purposes, since it contains the plant jacobian implicitly as was observed by others.

The inverse model network receives as inputs, the current and past system outputs, the reference signal and the past values of the system input. The non- linear input-output mapping given by the network inverse model is:

$$\mathbf{u}(t) = \hat{f^{-1}}[\mathbf{y}^M(t), ..., \mathbf{y}^M(t - ny),$$

$$\mathbf{r}(t+1), \mathbf{u}(t-1)..., \mathbf{u}(t - nu)] \qquad (11)$$

Note that $\mathbf{r}(t+1)$, known usually at time t, replaces $\mathbf{y}(k+1)$ in 7 unknown at time t.

Through the forward and inverse models an identity mapping is forced to be learnt. In this work, we have used an hybrid approach as sugested by Psaltis et al. [14]. Firstly the network inverse model is obtained through generalized learning, following a specialized learning with a training operational signal to fine tune the controller to respond with control inputs that drive the closed loop error to zero.

The non-linear neural networks modelling the plant and its inverse are used in an Internal Model Control (IMC) structure [15]. The validity of this approach is discussed by Hunt et al. [2]. In their paper, a non-linear neurocontroller, within an IMC framework, is implemented for a single input single output system. A similar approach is applied, in this paper, to a multivariable non-linear system (Figure 4).

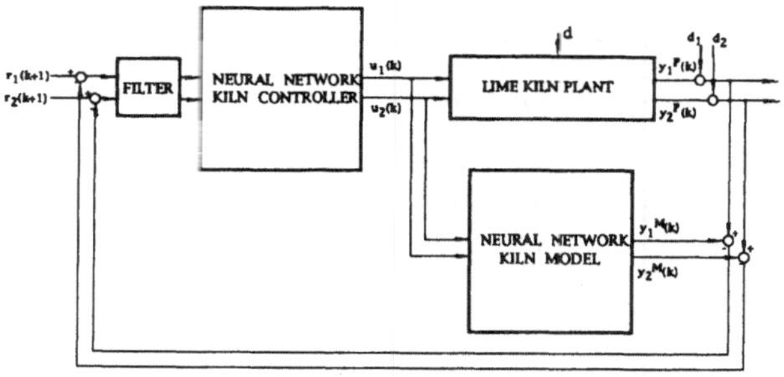

Figure 4: Internal Model Control using Neural Networks

5 Simulation Results

The network training data was obtained by numerical integration of the differential equations describing the physical system. The manipulated inputs (u_1 - fuel flow rate) and (u_2 - gas flow rate) were changed randomly within an uniform distribution. A sampling time of 1 min was achieved. In order the neural network can learn the correct functional system non-linearities, sufficient excitation must be presented in the training data. Therefore, sinusoidal, step and PRBS inputs were also applied to the plant. A data training set of historical data consisting of 4000 input-output pairs was obtained. A testing data set was arranged in the same way. Initialization of the networks weights was done by randomly choosing them from a uniform distribution between -1 and 1 and the input output patterns were also scaled to lie in that range. Using the

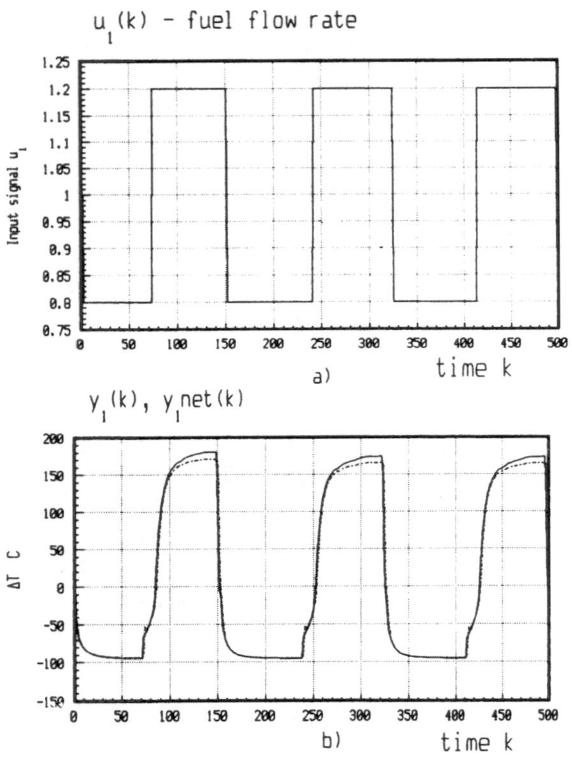

Figure 5: Plant Identification Results a) and b). Process and modelled response to PBRS u_1.

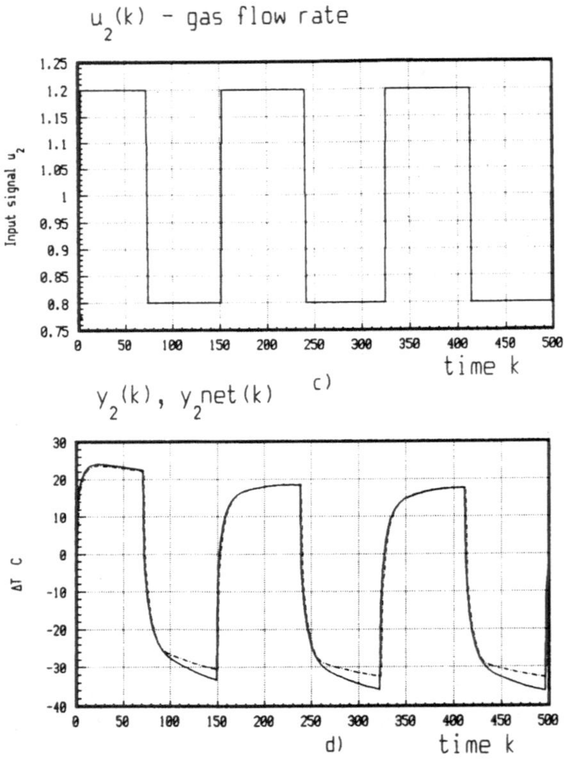

Figure 6: Plant Identification Results a) and b). Process and modelled response to PBRS u_2.

testing data set, the network's prediction performance was periodically observed during the training phase. Figure 5 illustrates the kiln thermal behavior in response to open-loop PRBS test signals. This represents some nine hours of simulation data. The training error (RMS= 0.01125) was $\approx 1\%$ what means the network has implicitly learnt the kiln temperature histories. The neural network IMC approach was tested with randomized step reference values and the tracking behavior of the process is shown in Figure 7 a) b). Simulation results showed a slight offset in the controlled exit gas temperature. Better control was achieved in the product temperature. Figure 7 c) and d) presents the network prediction error for both controlled variables: the product temperature and exit gas temperature. The performance of the neural controller is affected by how well the network can generalize and extrapolate outside the training domain. The neural networks were presented regularly with data provenient from diferent sets of testing data to see how the response fitted on unseen data. When convergence was achieved, (RMS less than 1%), the training was stopped since the networks could generalize well. In this way, some process validation was realized. There is no established technique for evaluating the degree of network performance, i.e., accurate prediction outside the training set. Tables 1 and 2 summarize the topology and specification parameters for both networks. Using conventional notation, the neural network process model as well as the neural controller have the configuration $N^2(8, 20, 10, 2)$, where the superscript indicates the number of hidden layers and 20 and 10 are the number of nodes in each of them. Experimentation was accomplished on several neural network sizes, namely with different number of hidden nodes per layer, and different number of hidden layers. Establishing a compromise between the convergence requirements and the storage capacities, the choosen arquitecture produced satisfactory results.

Layer	Number of Nodes	Learning Rate η	Activation Function
Input	8	-	-
First Hidden	20	0.250	$tanh\gamma = \frac{1-e^{-x}}{1+e^{-x}}$
Second Hidden	10	0.125	$tanh\gamma = \frac{1-e^{-x}}{1+e^{-x}}$
Output	2	0.115	linear

Table 1: Neural Network Configuration

Specifications	Neural Network Process Model & Neuro Controller
Learning Paradigm	Back Propagation
Network Topology	Multilayered, FeedForward
Transfer Non Linear Function	Tanh
Training Data	Historical I/O Data from Simulated Process
Test Data	" "
Input Signal for Training	Step, Sinusoidal and PRBS within 5% 10% 20%

Table 2: Neural Networks Specifications

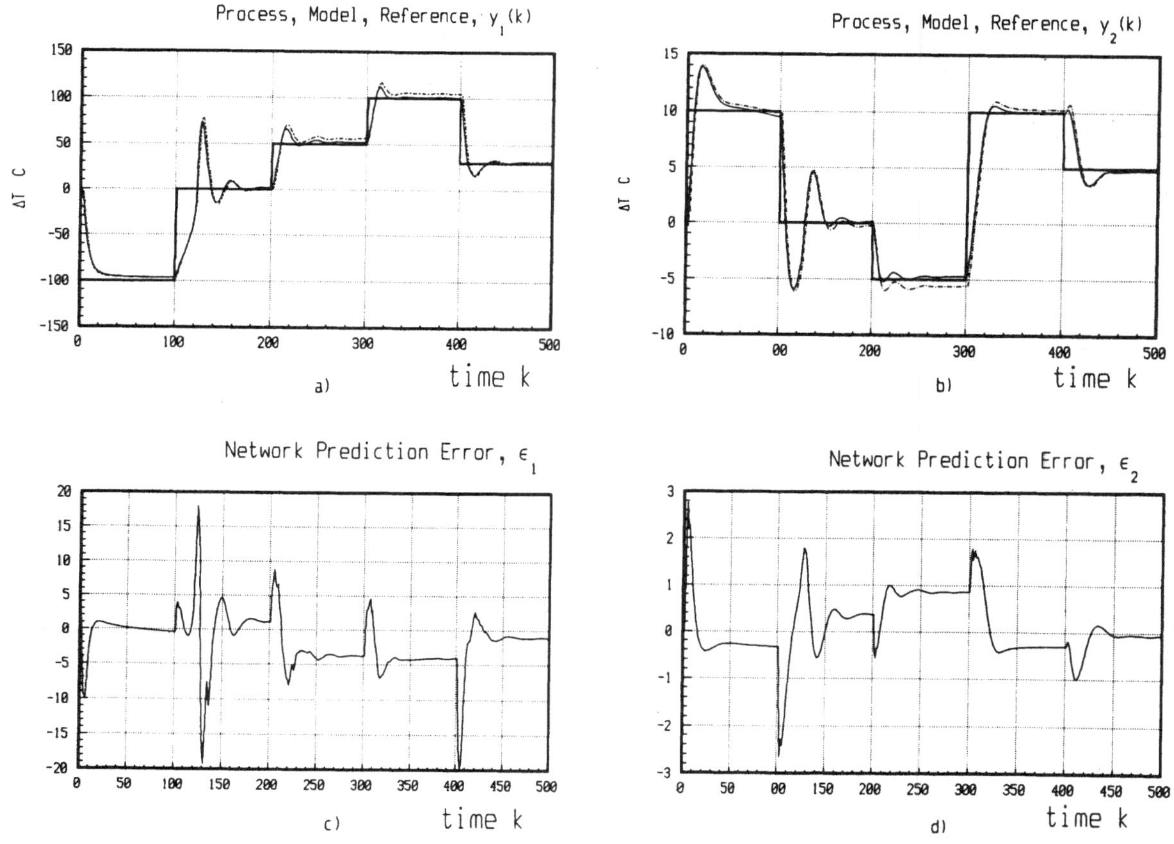

Figure 7: Process Control Simulation Results. Thicker line - reference signal; Thinner line - controlled output; dotted line - modelled output

6 Conclusions

This work reported a neural network approach to the identification and control of an industrial lime kiln which is a non-linear multivariable process. The lack of industrial data, shows that *a priori* knowledge of the system being modelled is needed. The neural network process model was able to identify accurately the underlying system dynamics. It is used in parallel with the plant being the output difference fedback to the system; in this way, it takes in account the model uncertainties. The simulation results show the neuro-controller ability to track arbitrary step changes in reference levels. However, due to inherent modelling errors, the network inverse model does not represent a rigorous inverse of the process model and other methods for determining the inverse, should be worked out. The training of the feedforward non-linear neural networks requires a significant amount of computation. The back propagation algorithm, though slow in convergence, ensures a simple parallel computation and allows satisfactory results. Neural nets constitute a feasible tool for the identification and control of non-linear systems though new engineering techniques are necessary to allow its pratical implementation.

References

[1] K. Narendra and K. Parthasarathy, "Identification and control of dynamic systems using neural networks," *IEEE Transactions on Neural Networks*, vol. 1, pp. 4–27, 1990.

[2] K. J. Hunt and D. Sbarbaro, "Neural networks for non-linear internal model control," *Proc. IEE Pt D.*, vol. 138, pp. 431–438, 1991.

[3] N. Bhat, P. Minderman, and J. T. McAvoy, "Modelling chemical process systems via neural computation," *IEEE Control Systems Magazine*, vol. 10, pp. 24–25, 1990.

[4] D. C. Psichogios and L. H. Ungar, "Direct and indirect model based control using artificial neural networks," *Ind. Eng. Chem. Res.*, vol. 30, pp. 2564–2573, 1991.

[5] G. Cybenko, "Approximation by superpositions of a sigmoidal function," *Math Control Signal Systems*, vol. 3, pp. 303–314, 1989.

[6] K. I. Funahashi, "On the approximate realization of continuous mappings by neural networks," *Neural Networks*, vol. 2, pp. 183–192, 1989.

[7] K. Hornik and M. Stinchcombe, "Multilayer feedforward networks are universal approximators," *Neural Networks*, vol. 2, pp. 359–366, 1989.

[8] D. Rumelhart, J. McClelland, and the PDP Research Group, *Parallel distributed processing: explorations in the microstructure of cognition.* Vol. 1 2, MIT Press, 1986.

[9] P. J. Werbos, "Backpropagation through time: what it does and how to do it?," *Proc. of IEEE*, vol. 78, pp. 1550–1560, 1990.

[10] R. Hecht-Nielsen, "Theory of back propagation neural network," in *Intl. Conf. Neural Networks*, pp. 593–605, IEEE, 1989.

[11] S. Chen and S. A. Billings, "Representation of non-linear systems: the NARMAX model," *Int. J. Control*, vol. 49, pp. 1013–1032, 1989.

[12] S. Chen, S. A. Billings, and P. M. Grant, "Non linear system identification using neural networks," *Int. J. Control*, vol. 51, pp. 1191–1214, 1990b.

[13] S. Chen, C. F. N. Cowan, S. A. Billings, and P. M. Grant, "Parallel recursive prediction error algorithm for training layered neural networks," *Int. J. Control*, vol. 51, pp. 1215–1228, 1990c.

[14] D. Psaltis, A. Sideris, and A. A. Yamamura, "A multilayered neural network controller," *IEEE Control Systems Magazine*, vol. 8, pp. 17–21, 1988.

APPLICATION OF NEURAL NETWORKS TO AUTOMATED BRAIN MATURATION STUDY

L. Moreno, J. D. Piñeiro, J. L. Sánchez, *S. Mañas, J. Merino, L. Acosta, A. Hamilton.
Dept. of Applied Physics. University of La Laguna. Tenerife. Canary Islands. Spain.
*"Nuestra Señora de la Candelaria" Hospital. Dept of Neurophysiology.
La Laguna. Tenerife. Spain

Abstract- An application of Neural Networks (NN) to brain maturation prediction is presented. The problem consists of, given a pattern extracted from electroencephalographic (EEG) signals, state the degree of brain development (low or normal/high). To that end, a population of subjects with their EEG assessed by a neurologist is available. A Backpropagation (BP) neural network is used for this supervised classification task, and a comparison with standard statistical classifiers is made. The effect on performance of several preprocessing techniques such as Principal Components Analysis (PCA), normalization and scaling is investigated. It is found better performance in the NN approach, both in terms of efficiency and consistency.

I. INTRODUCTION

The EEG signals evolve with age. An expert neurologist can determine the maturational level simply by inspecting the multichannel EEG record. The relevance of brain maturation studies lies in the possibility of explaining conduct or schooling problems in children with maturational lag.

Traditionally, quantitative analysis of this problem has been done by means of multivariate analysis: Regression curves, cluster and discriminant analysis, [1], [2]. Many of these methods rely on statistical hypothesis about data. Alternatively, more recent methods based on NN can be used. These methods do not need any *a priori* hypothesis since they are based on learning by example.

For diagnostic purposes, neurologists have considered the EEG signals as composed by different background rhythms (waveforms of approximately constant amplitude, frequency and duration) as well as transients. Usually, background rhythms are classified in different standard frequency bands. The medical diagnosis, carried out by an expert, is obtained through the visual inspection of the multichannel EEG record. In our brain maturation problem, the expert classifies the subjects in two levels (low and normal/high levels) relative to their ages. In order to automate this procedure, we must extract a set of characteristic parameters from the signals. In the next section, we define this problem precisely. In Section III, we analyze in first place the statistical tools used and then go on with the neural network approach. Finally, we compare both methods.

II. PROBLEM DESCRIPTION

Our brain maturational study was made over a population of 130 children without neurophysiological or schooling problems aged 3 to 15 years. A record of 60 seconds of 16 channels was taken with the subject at rest with closed eyes (this is necessary to obtain representative measures of the background brain activity).

Expert diagnosis is based on complex criteria like main background frequencies, topographic differentiation (comparison of the signals in different channels), stability and regularity of the waveforms, and so on. These criteria are heuristic and are acquired mainly with practice. In this work, we have used as characteristic parameters only the relative power in Delta (1.5-3.5 Hz), Theta (3.5-7.5 Hz), Alfa1 (7.5-9.5 Hz) and Alfa2 (9.5-12.5 Hz) bands and the total power, in two occipital and two frontal channels. It is well known that the Alpha band presents a noticeable evolution of its power with age, and that it appears strongly in the occipitals. For completeness, we have also added two other channels in the frontal area since the influence of the occipital sources is much lower and one can expect different trends to appear there.

Due to the non-stationarity of the signals, we took the parameters in the most similar situation for all the subjects (time instants in which power bands presents its absolute maximum). The number of variables varied between 40 and 80.

These parameters were obtained by means of a set of programs [3], for acquisition, real-time display, off-line analysis of 16 channels of EEG and evoked potentials, that we have developed in C and assembly language for a PC-386 with math coprocessor and a data acquisition board with DMA (Direct Memory Access) capabilities.

In addition, we have available the medical maturational diagnostic (Low, Normal/High) made by the expert for each subject of the sample.

III. TOOLS FOR PROBLEM ANALYSIS

In this section, firstly, we will develop the analysis by means of multivariate methods and secondly by means of neural networks. Finally, we realize a comparative study between them. In both cases, we consider total population subdivided in two groups: Training and Test, being the latter roughly a third of the total population and the former the rest of the sample. Moreover, in the training group each class has the same number of members.

In the study, components of both groups are selected at random each trial. This permits us to obtain a correct classification score that not depends on the group selected but on the global population characteristics.

In first place, we must consider if the high number of variables could be reduced without appreciable information loss. This will be carried out by means of Principal Component Analysis (PCA). It is well known that PCA is a linear transformation of a set of correlated variables into a new set of uncorrelated variables (principal components) ordered by its variance. Thus we can reduce the number of variables to study taking into account only those who have the greatest variances. In the Figure 1 appears the percentage of the total variance 'explained' by the reduced set of variables for the occipitals alone and the occipitals and frontals. The first curve rises faster, so it can represent the original trends in data with fewer variables. With the introduction of frontal parameters, new trends appear that need a higher number of variables to account for the same percentage than the occipitals.

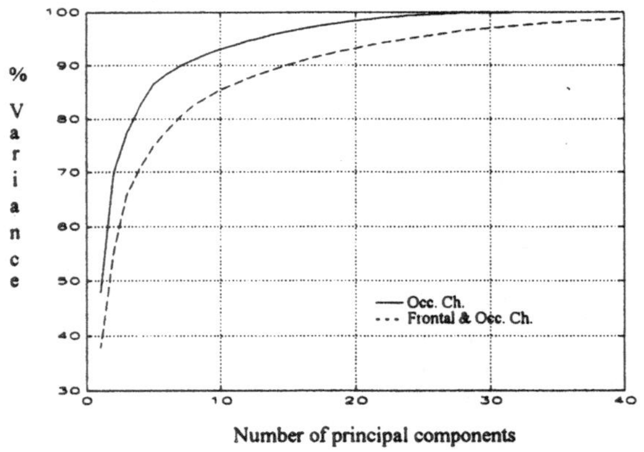

Fig. 1

As is mentioned below, several tests were carried out to check the performance of the classifiers with varying number of input principal components.

Since our variables are highly correlated, we applied PCA and obtained a good compromise with only 5 new variables that account for 87% of the original variance.

Another consideration is the preprocessing of the raw data. The classifiers need data in an homogeneous form, with comparable values and ranges. In the case of NN, it is important to avoid the saturation condition, in which the network is unable to adapt its weights. So it is necessary some sort of normalization or scaling. On the other hand, in some cases, indiscriminate individual scaling of each variable can obscure its relationship with the others, lowering the performance of the classifiers. In this problem our parameters are of similar nature, needing only a global scaling in relative and total powers to the interval [-1,1].

In order to classify subjects, we first use cluster analysis and lineal and quadratic discriminant analysis.

III-A MULTIVARIATE ANALYSIS

Initially, we use cluster analysis to obtain natural groupings based in the concept of distance between subjects without considering the medical diagnosis (non supervised classification), detecting the age evolution of the parameters (absolute maturation). Taking into account that our aim is the maturation level relative to age, it is necessary to apply supervised classification methods [4].

The classification by means of discriminant analysis is a supervised method which depends on statistical hypotheses about the data, such as assuming gaussian distributions, (in the case of linear discriminant analysis, the equality between the covariance matrices of each class). This analysis was carried out using linear and quadratic discriminant functions obtaining best results in the second case, as is detailed in Table 1.

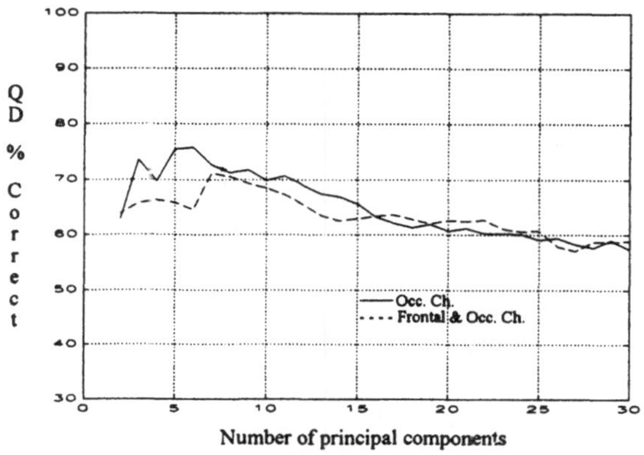

Fig. 2

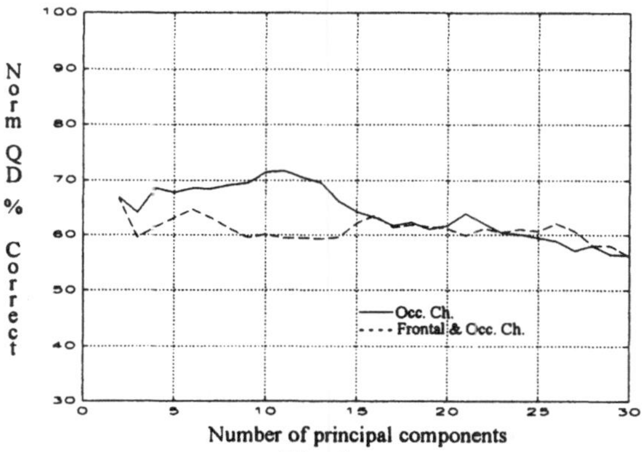

Fig. 3

The Figure 2 shows the performance of the quadratic discriminant function with a variable number of input variables (principal components). The results from the classification using the of occipital and frontal channels are inferior than that of the set of occipitals alone. The latter curve has a maximum in classification success at 5 principal components.

If the individual variables were transformed applying a Z-Score (Subtracting the mean and dividing by the variance of each variable, an usual statistical transformation that scales all variables to the same range), the result curves are in Figure 3. Now the performance has went down and the maximum is attained with around 10 principal components instead of 5. This is an example of the effects of inadequate data scaling referred above.

III-B. NEURAL NETWORK ANALYSIS

In this subsection, we consider the Backpropagation model of NN [5]. To assure that the NN training produces adequate generalization it must be required that:

-Data presented to the network must be representative of the population.

-The network must be able to extract that general characteristics.

These properties must be obtained by the proper selection of the network structure and training parameters. The generalization capabilities are checked classifying another set of data (test group) with the subjects that were not included in the training phase.

The network training is a lengthy iterative process. The training time could be decreased simplifying the network structure, reducing the number of inputs, the number of levels and neurons in each level or using faster convergence algorithms.

In our particular problem, the number of inputs is reduced using PCA as mentioned above.

With regard to convergence algorithms, in addition to the standard method used in Backpropagation (gradient descent with a small fixed step), we have tried other methods from the general theory of optimization [6] like gradient descent with adaptive step and conjugate gradient, however there was no substantial difference between them due mainly to the low number of iterations required in our case.

We had implemented the BP algorithm with a hidden level and with a variable number of neurons in that level. The activation function chosen was the hyperbolic tangent. The inputs to the network were the first 5 principal components. The outputs were the two maturational levels (low and normal/high). Training parameters (α and η), were selected conveniently and the training process was carried out in batch mode. The results are the mean of many independent trials.

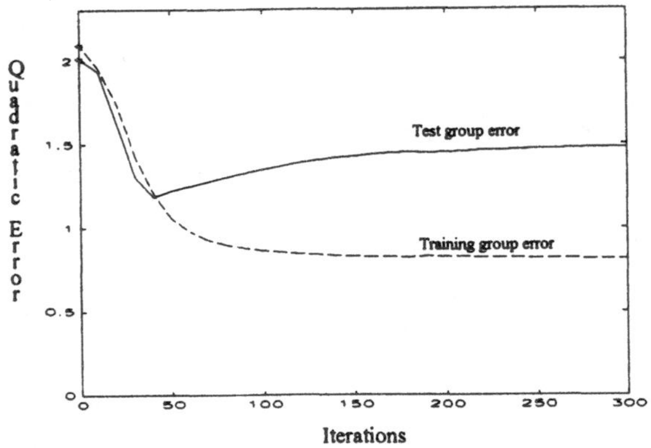

Fig. 4

Figure 4 presents an example of a typical trial, showing the error curves for training and test groups. It shows a minimum in the test group error. This feature is observed in most cases. This minimum marks the point where the network begins to learn particular characteristics from the training group data that are not common to the population. Therefore we can halt the training at this point, ensuring better generalization with less iterations.

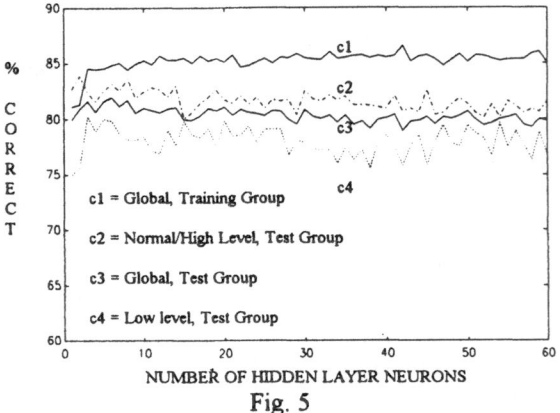

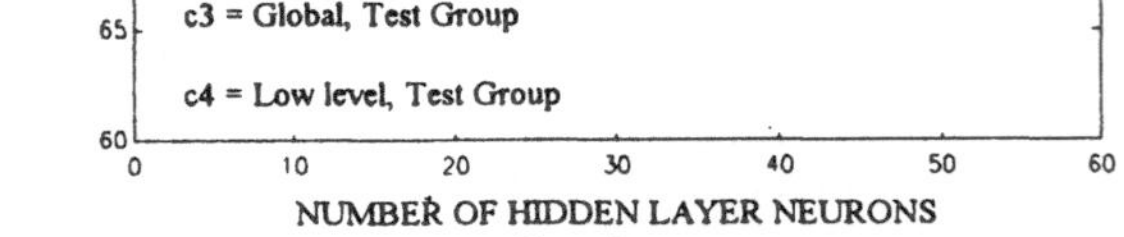

Fig. 5

Figure 5 shows the percentages of subjects correctly classified in the test and training groups versus number of neurons in the hidden layer. Solid lines represent global percentages while dotted and dashed lines represent percentages for both of maturational levels in the test group. It can be observed that these percentages remain roughly constant after an initial increase. This allows us to take a low number of hidden level neurons without adversely affecting performance.

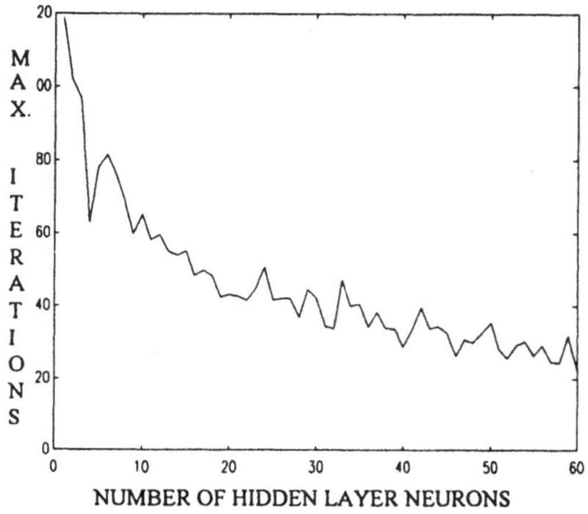

Fig. 6

Figure 6 shows number of needed iterations (with the criteria defined above) versus the number of neurons in the hidden layer. It can be observed that this number decreases as the network becomes more complex.

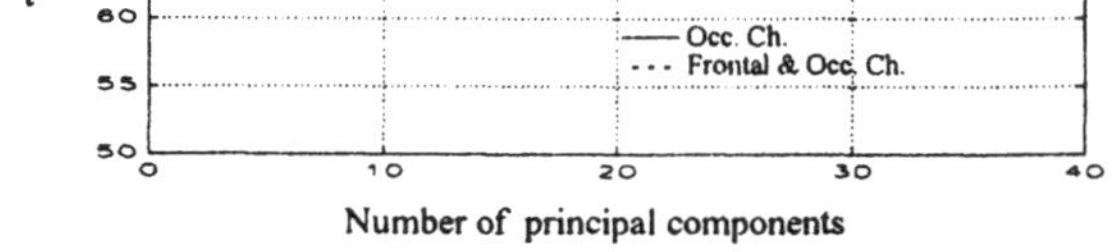

Fig. 7

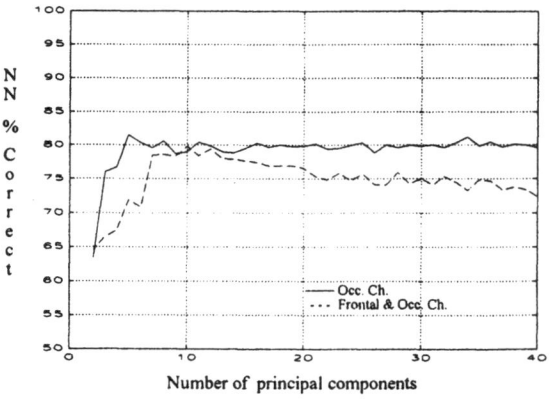

Figures 7 and 8 show the dependence with the number of principal components. As is the case with quadratic discriminant, it is found that the trends added with the inclusion of the frontal channels are not relevant to the problem of maturation. This is shown in Figure 8, where the horizontal axis now represents the percentage of original variance explained by the new variables. For high percentages of variance, the curve obtained from frontal and occipital data even decreases.

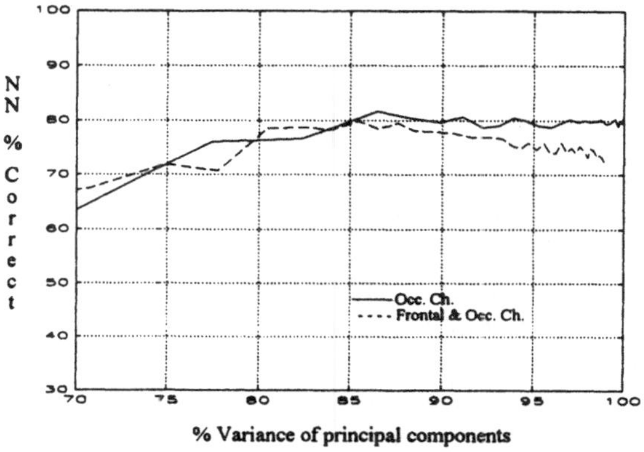

Fig. 8

To summarize this section, it is relatively easy for the network to extract the global properties from the population. This problem is solved by the network with a low number of hidden layer neurons and in a small number of training iterations with adequately preprocessed input data.

IV. COMPARISON BETWEEN CLASSIFIERS

In Table 1 appears the performance of the classifiers in terms of percentage of global correctly classified subjects in test and training groups and percent correct in test group in each maturation level. The variances associated to these figures are also provided.

	Test % correct x±σ	Training % correct x±σ	Low level % correct x±σ	Normal/High levels % correct x±σ
Linear Discr.	75 ± 6	80 ± 4	70 ± 11	78 ± 9
Quadratic Discr.	75 ± 6	80 ± 3	71 ± 11	78 ± 8
Neural Network	81 ± 6	85 ± 2	78 ± 9	82 ± 6

Table 1

We can see from Table 1 that the NN approach do better than the other methods of classification studied here. This good behaviour is revealed not only by global correct prediction percentage, but by a small improvement in the prediction variances.

V. CONCLUSIONS

In this work, we have tried to solve the maturation problem using a relatively limited set of parameters that do not reveal the richness and complexity of brain signals. One obvious extension could be to expand the analysis to include more channels and new variables. However, our main interest lies in elaborating robust procedures for applying and testing NN in brain diagnosis problems. This work constitutes a first step in that direction. Our main goal is the making up of a knowledge-based system that combines rule-based modules with NN to be applied in automated brain signals diagnosis which would exploit the advantages of both approaches.

VI. REFERENCES

[1] E.R. John et al., *'Validity, Utility and Limitations of Neurometric Evaluations in Children'*, in *'Event-Related Potentials in Children'*, A. Rothenberger (Ed.), Elsevier Biomedical Press, 1982.

[2] C. Chatfield, A.J. Collins, *'Introduction to Multivariate Analysis'*, Chapman and Hall, 1980.

[3] L. Moreno, J.L. Sánchez, L. Acosta, J.D. Piñeiro, J.L. Ichaso, S. Mañas, *'Multichannel Digital Processing for EEG Analysis"*, Proc. IASTED Int. Symp. on Circuits and Systems, Zurich 1991.

[4] L. Moreno, J.L. Sánchez, S. Mañas, J. D. Piñeiro, L. Acosta, A. Hamilton, *'Multivariate Analysis and Mapping of EEG'*, 6th European Congress of Clinical Neurophysiology, Lisbon 1992.

[5] D.E. Rumelhart, J.L. McClelland and PDP Research Group, *'Parallel Distributed Processing'*, vol. 1. Cambridge, MA: MIT Press, 1988.

[6] S.S. Rao, *'Optimization Theory and Applications'*, 2nd. Edition, Wiley Eastern Limited, 1984.

A Neuron Model for Centroid Estimation in Clustering Problems.

G. Acciani, E. Chiarantoni

Politecnico di Bari
Dipartimento di Elettrotecnica ed Elettronica
Via Orabona, 4
70125 Bari
Italy

Abstract

In this paper a new model of neuron unit able to search centroid of cluster without competition, as in Unsupervised Competitive Learning (UCL) neural networks, is presented. The drawbacks of classical unsupervised learning laws are investigated analyzing the basic mechanisms of clustering operation and the paradigms of an alternative clustering algorithm are carried out. Starting by this new paradigms, a new neural model where the basic neural element is more complex is obtained. Two simple neural elements are locally interconnected in non-linear feedback and the pair is arranged to form a stand-alone neural unit. The properties of the "neural couple" in combination with the proposed learning law are investigated and it is shown that the proposed neuron unit is able, during learning stage, to perform an automatic switch of learning strategy. A first learning strategy is adopted to find barycentre of the whole input space, then a second strategy allows the centroid of the most populate cluster to be found. Simulation on a test set shows that, given an input space in which different clusters are "well" placed, the synaptic vector of neural unit converges to the centroid of the most populate class at least as fast than other popular learning laws based on competitive learning strategy.

1. Introduction

Many definitions have been used to define a clustering process with regard to an assigned item space [1]. The concept of cluster is strongly connected to the peculiar applications of the partition that can be obtained. In biological system, e.g., clustering is a powerful tool for reduction of data to be stored in a recognition or memory scheme. Given an assigned "item" space, it is possible to obtain dimensional reduction by storing only the primary data (e.g. pattern vectors) regardless of the individual data points. Although the concept of cluster is strongly related to the application peculiarities and it is complex to synthesize a general definition, the clustering problem is normally set up in the following way. Assume that a data set

$$A \equiv \{x_i | x_i \in R^n;\ i = 1,\cdots,h\} \qquad (1)$$

is assigned. This set has to be partitioned into disjoint subsets A_α, $\alpha = 1, \ldots, k$ ($k<h$), defined "clusters" and a set of data prototypes

$$\{y_\alpha | y_\alpha \in R^n;\ \alpha = 1,\cdots,k\} \qquad (2)$$

has to be searched so that, with respect to this division, some functional of error measures , between a data point x_i and a cluster center y_α , describing the quality of grouping, is minimized. The data prototypes y_α are called "centroids" of the clusters, and the assignments

$$\{M_{i\alpha} | M_{i\alpha} \in \{0,1\};\ \alpha = 1,\cdots,k;\ i = 1,\cdots,h\} \qquad (3)$$

of data points x_i to clusters A_α are chosen such that $M_{i\alpha}$=1 denotes that data point x_i is uniquely assigned to the cluster A_α. Generally speaking, the determination of the subsets A_c is a difficult global optimization problem, whereby a set of simultaneous algebraic equations (describing the optimal criterion) are to be solved by direct iterative methods. In literature we have several examples of clustering criterion that utilizes neural networks [3,4,5,6]. Training strategies of artificial neural networks for centroid search belong to two basic categories: supervised and unsupervised laws. In the supervised learning the number of final classes is known and for each input vector the network learns the appropriate output class:

$$x_i \in A_\alpha \;\Rightarrow\; x_i \notin A_k \text{ if } \alpha \neq k \qquad (4)$$

Formally supervised learning depends on class indicator functions that define the clusters:

$$I_{A_\alpha}(x) = \begin{cases} 1 & \text{if } x \in A_\alpha \\ 0 & \text{if } x \notin A_\alpha \end{cases} \qquad (5)$$

$I_{A\alpha}$ indicates if pattern"x" belongs to decision class A_α. Example of this kind of networks can be found in Kohonen Supervised-Competitive-Learning law (SCL) [2]. The unsupervised learning neural networks try to model the distribution of the input space vectors without external indications. The goal of the learning law is essentially to model each neuron element as a "similarity detector" unit. In these networks (typically competitive networks e.g. Kohonen networks) [2,4,8], the input samples are gradually presented to the system according to their random distribution. After sufficient training, the synaptic vectors forces the units' responses in order to become distributed evenly over the input probability distribution. In some cases a topological organization is associated to the input density function: similar vectors tend to excite the same place of network. Self-Organizing Feature Maps is an example of this kind of unsupervised cluster operation. In all unsupervised networks the basic concept is competitive learning mechanism: groups of neurons compete when input sample is presented, to establish winner unit. Only the winner neuron adapts its weight and specialize itself to recognize a specific input class. Many drawbacks of this competitive system have been already denounced [6]: several inhibitive connections are required when the number of elements grows; it is difficult to set the number of starting elements; for particular input space structures and initial distribution of weight vectors, some elements cannot work properly. Several mechanisms have been proposed to overcome the problem of weight vectors distribution [4,5,7]. In all these solutions the competition amongst the neurons of the network is the basic ingredient. In this work it is shown how it is possible to find clusters, removing the requirement of complicated competitive structure, and formulating at the same time a new concept of neuron element.

2. An approach to clustering problem based on density measure

In classical literature [1] cluster searching is carried out either by means of suitable cost function minimizing techniques or via iterative operations of vector space subdivision. A different strategy for cluster searching can be obtained resorting to an approach based on density concept. Let's define hypersphere with radius r the set of all points which lie within a distance r from the origin.[1] Assigned an hypersphere of unitary volume and center in x_i, $S_1(x_i)$, the number of points, n_η, lying within the hypersphere surface depends on the selected point x_i. It is possible to found a cluster by analyzing in every space region the amount of vectors belonging to a conventionally selected volume. To this purpose it is convenient to introduce an indicator of local mean density of elements, say $\rho_\eta(x_i)$. This parameter is defined as the number of vectors, n_η, lying within a unitary hypersphere $S_1(x_i)$ centered in x_i

$$\rho_\eta(x_i) = \frac{n_\eta}{S_1(x_i)} \qquad (6)$$

The whole couple set (ρ_η, x_i) is a peculiarity of the assigned vectorial space. In particular, the presumed existence of clusters involves a not uniform distribution of points and so different values of ρ_η are to be expected varying x_i. It follows that the density ρ_η assumes a maximum local value in correspondence with a cluster, i.e. a space region densely populated. We can therefore define a cluster as a set of vectors belonging to simply connected[2] regions of a space A where the density is greater than a preassigned value ρ_α

$$Cl_\alpha = \left\{ x_i \in A_\alpha \middle| \rho_\eta(x_i) > \rho_\alpha \right\} \qquad (7)$$

As well, the above definition states that the regions characterized by a density value less than an assigned threshold ρ_α are considered as a "noise". With regard to some applications, the above splitting of the input space can be heavily unsatisfactory because the local properties of the space A are not put into evidence. For this reason it is useful to introduce a relative density measure related to a limited region of A. If n_m is the number of vectors contained in a limited area, say S_m, the mean density ρ_m is defined as

$$\rho_m = \frac{n_m}{S_m} \qquad (8)$$

Therefore the relative density γ, for all the regions within S_m, is determined as

$$\gamma(x_i) = \frac{\rho_h(x_i)}{\rho_m} \qquad (9)$$

Unlike ρ_η, the adimensional ratio $\gamma(x_i)$ is independent of the absolute value of space population and it describes local properties with an arbitrary precision depending on the selected limited region. Moreover $\gamma(x_i)$ increases when the volume surrounding x_i is

[1]The n–dimensional volume is a generalization of the three–dimensional one, being an integral of the volume elements expressed in a rectangular coordinate system.

[2] simply connected means a region in which any two points in the region can be joined by a path contained entirely in the region; that is, the region is not composed of disconnected, separated, subregion.

densely populated. For these reasons it is possible to detect clusters looking for simply connected space regions where $\gamma(x_i) > \bar{\gamma}$. It follows that the definition (5) can be stated as

$$Cl_\alpha = \left\{ x_i \in A_\alpha \middle| \gamma(x_i) > \bar{\gamma} \right\} \qquad (10)$$

3. Neural Networks in clustering problems.

The search of dense input space regions can be effectively performed by suitable neural networks. Generally speaking in clustering process by means of neural networks, it is adopted a structure with two layers of neural units: an input layer to normalize the input vectors at unitary length, and a processing layer in which each unit calculates its output value. Depending on the learning law, only some units upgrade their weight according to an appropriate learning law. Consider an elementary neural cell (figure 1) where $f(\bullet)$ is a bounded monotone-nondecreasing signal function, e.g. the logistic signal function $f(z) = (1+e^{-cz})^{-1}$; (c>0), z is the neural unit activation value obtained as a function of x and of the weight vector w. The neural output y is expressed as

$$y = f(z-\sigma) \qquad (11)$$

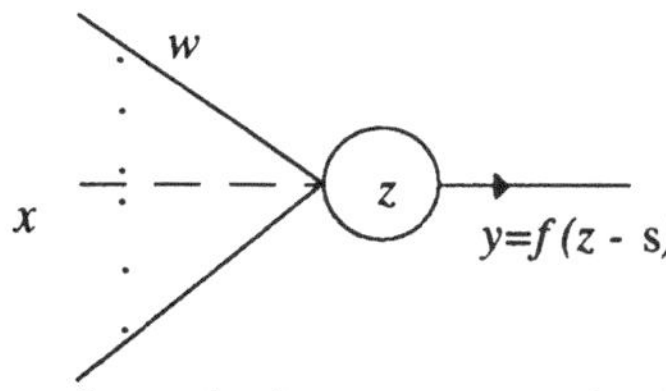

figure 1: elementary neural unit.

In cluster searching, competitive unsupervised learning laws are adopted. The general form of competitive learning equation states

$$\Delta w_j = \zeta(x - w_j)\phi \qquad (12)$$

where Δw_j is the j-th difference of the synaptic weight, ζ is a monotone decreasing learning coefficient, and ϕ is a "selective" term ($\phi \in [0,1]$) which depends on the competitive strategy adopted. In the Kohonen Self-Organizing Map, and Hecht-Nielsen counterpropagation network [1,3], ϕ is related to the winning-lose strategy: after each input sample only the "winner" element changes its weight (ϕ=1) whereas the losing ones are frozen (ϕ=0). The strategy adopted in this neural nets, *substantially* states "learn only if win". With this competitive learning, some neural units are quiescent during the complete learning stage. Under appropriate conditions over the learning coefficient ζ, the Kohonen weight vectors become densest where the input elements are most common, and become less dense (or absent) where the input elements hardly appear. In this way, the Kohonen layer adapts itself so that the probability that an input is closest to w_i (for which the unit i is the winner) is approximately 1/N, for all i, i=1, ..., N. For the simple Kohonen learning law, even on a simply connected region, if the probability density function of the input is far from being constant, also the set of w_i produced is far from being constant as desired. To overcome this problem different variants are possible. De Sieno proposed [7] a mechanism for each neural element that takes in to account the winning information and tends to make ϕ=0 if the unit wins the competition substantially more often than 1/N of the time. In the Differential Competitive Learning (DCL) law, B. Kosko [5], starting by biological consideration, correlated the coefficient ζ (eq. (12)) to the time change of the neuron competition signal $\chi = sgn(f^{k+1} - f^k)$ and hence the (12) becomes

$$\Delta w_j = \zeta \chi (x - w_j)\phi \qquad (13)$$

The DCL improves the performance of the most common Unsupervised learning law in the speed of convergence and accuracy. Again, however, "competitive" winning information ϕ is required to develop the capability of centroids search.

4. The structure of the proposed neural element.

The aim of this paper is to present a "local" learning law based on the strategy " learn only if the density space increases". If the competitive winner signal ϕ is removed in the standard competitive learning law (12), the capability of "selective" learning disappears. All the neural elements in the competitive layer, starting by different initial position, casually fluctuate to go, asymptotically, towards the barycentre of the whole input space. Competitive signal is required to stop an undifferentiated learning process and consequently to select only the regions where a cluster is retained to be present. It has been already observed that the density ρ_η assumes a maximum relative value in correspondence with a cluster. This property can be advantageously used to replace to classical competition a learning law which is sensitive to the density of the space nearby the synaptic vector. If the surrounding space density is growing then the neural cell must learn to avoid the synaptic vector goes away. The density of the regions near to the actual weight vector position can be related to the threshold value σ. Indeed, fixing a threshold value an output value higher than a preset constant quantity $\bar{k}$ (< 1) is obtained when an incoming vector x satisfies the following condition

$$f(\bar{w}\bar{x}-\bar{\sigma}) > \bar{k} \qquad (14)$$

If f is a locally invertible function, setting $\bar{\bar{k}} = f^{-1}(\bar{k})$ we obtain

$$\bar{w}\bar{x} > \bar{\bar{k}} - \bar{\sigma} \qquad (15)$$

It is obvious that varying the threshold value, regions of different width will be able to yield an output neural value higher than the fixed one. When the threshold grows, only nearest vectors will be processed. Supposing that an output value higher than $\bar{k}$ is used for conditioning the learning process (if this is the case, ϕ = 1 if $y > \bar{k}$) it is evident that varying the threshold value it is possible to modify the "capture" zone of vectors and therefore to search for clusters. The above considerations justify the necessity of feedback blocks sensitive to the neural output. Two blocks carrie out a control action over the threshold and the synaptic weight; the proposed new basic neural unit is shown in figure 2.

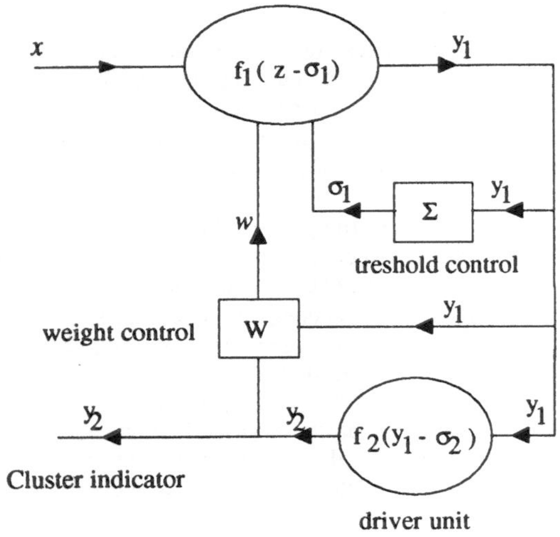

Figure 2: The new basic neural unit

In figure (2) y_1 is the output of the neural unit, while y_2 is the output of the driver unit which supervisions the weight variation law. When y_2=1 then the processed input vector belongs to the detected cluster. The varying threshold of the cell is σ_1, while σ_2 is the constant threshold of the driver unit. $f_1(\bullet)$ and $f_2(\bullet)$ are still two non-linear functions, with generically different slope; further $f_2(\bullet)$ is a nonlinear hard function (e.g. sign function). Σ and W represent the control blocks of threshold value and synaptic weights respectively. In order to obtain the mathematical model of the unit cell, it is necessary to establish relations between σ_1 and y_1 and between w and y_2. The proposed learning law can be stated as:

$$\Delta w^k = (\alpha_1^k + y_2^k \alpha_2^k)(x^k - w^k) \qquad (16)$$

where

$$\alpha_1^k = (1 - \delta\, y_1^k)\, \alpha_1^{k-1} \qquad (17)$$

$$\alpha_2^k = \alpha_2^{k-1} + \delta\, y_1^k\, \alpha_1^k \qquad (18)$$

where the learning coefficient δ, α_1, $\alpha_2 \in [0,1]$ and $\delta \ll 1$. The idea is to interfere with the learning law in order to intercept w when it is going into a cluster orbit and to force it toward the centroid. Note that the above parameter ϕ in relation (12) has been substituted virtually by the output y_2 into (16). As long as $y_2 = 0$ the driver unit is not active and the learning law is controlled by α_1 only. Every time an input vector activating the cell is processed, α_1^k decreases. Therefore during the process the quantity α_1^k move asymptotically towards 0. At the same time, when the neural cell processes input vectors activating the output y_1 only, the weight vector w is still moving towards the entire system barycentre. When y_2=1, which means that input x_k is able to activate the driver unit, we can consider w within an high density zone. The coefficient α_2 (eq. 16) is added to α_1, and the learning capability of the whole neural cell is modified. If the threshold value is adaptively modified by the mean distance of incoming input vectors, the output y_2 will be activated only when the above distance is decreasing. To this purpose the simple variation law for σ_1

$$\Delta\sigma_1^k = \alpha_1^k (y_1^k - \sigma_1^{k-1}) \qquad (19)$$

has been assumed. The above relation holds every time an input vector is presented to the neural cell. σ_1 grows when the output y_1 is active and vice versa. Therefore the combined action of the (16) with (19) allows the vectors belonging to more dense input region (i.e. cluster) to be captured.

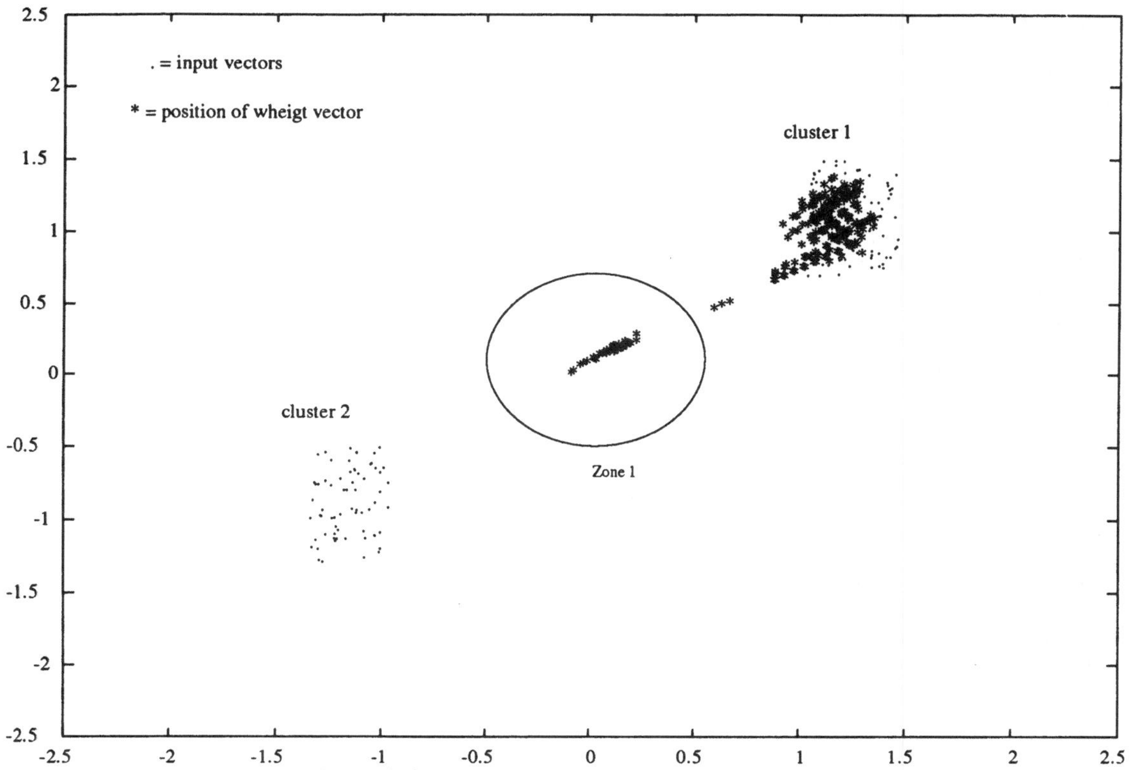

Figure 3: Space configuration at the end of the learn

5. Simulation results

To show the details of the proposed algorithm, a simulation for a single neural unit has been conducted considering an input space bidimensional with 1500 patterns elements randomly positioned and grouped in two clusters, the first with 850 elements and with the centroid at the co-ordinates (1.2, 1.0) and the second with 650 element and with the centroid located at (-1.2, -0.95). The input elements was presented, according to their distribution, to the neural element. The starting values of the learning coefficients and of the algorithm parameters are shown in table I

$$w(0) = 0$$
$$\alpha_1(0) = 0.02$$
$$\alpha_2(0) = 0.28$$
$$\sigma_1(0) = 0.0$$
$$\sigma_2 = 0.9$$
$$\delta = 0.03$$

Table 1: Learning coefficient of the neuron

The simulation results are reported in figure 3 For each step of the learning process, a star character (*) was depicted in the position of the weight of the neuron and a point mark (·) was depicted in correspondence with the position of input vectors. The experimental results confirm the strategy of the searching: at the beginning the tendency is to search the barycentre of input space (zone 1), so, when feedback signal is frequently present, the centroid of the nearest cluster is founded. After only one presentation of the whole input vectors, the algorithm carries the weight vector of the neuron into the position (1.1, 1.1) which is a good approximation of the centroid of the most relevant cluster, located at the position (1.2, 1.0). The rigorous position of the centroid is not obtained because a part of the input vector has been utilized to evaluate the position of the barycentre.

6. Final consideration and future topics

The presented neuron unit is an example of "absence of competition" strategy. An efficient cluster search is performed by cooperation of two simple non linear elements. It is possible to combining "cellular cooperation" with "external competitive strategy": interesting applicable services coming out when more of this neuron units are joined with lateral inibhitory connections to realize a multiple cluster search as in competitive networks. Nevertheless this topics are object of actual study and will be presented in future works.

References

[1] Gaul W. and Schader M. ed., "Classification as a Tool of Research", North-Holland, 1986
[2] Kohonen T., "Self-Organization and Associative Memory" *(2nd Ed.) Berlin, Springer, 1988*
[3] Rumelhart D. E., Zipser D., "Feature discovery by Competitive Learning", *in Parallel Distributed Processing: Exploration in the Microstructure of Cognition, Volume I, (Bradford Books, Cambridge, MA, 1986)*
[4] Hecht-Nielsen R., "Counterpropagation Networks", *Proc. of the Int. Conf. on Neural Networks, II, pp. 19-32, IEEE Press, New York, June 1987.*
[5] Kong S G., Kosko B. , "Differential Competitive Learning for Centroid Estimation and Phoneme Recognition", *IEEE Trans. on Neural Networks, Vol. 2 N.1, January 1991, pp. 118-124.*
[6] Becker S. *"Unsupervised Learning Procedures for Neural Networks", Int. Journal of Neural Systems, Vol. 2 Nos. 1&2, 1991 17-33*
[7] Desieno D., "Adding a conscience to competitive learning", *Proc. Int. Conf. on Neural Networks, Vol. I, pp. 117-124, IEEE Press, New York, July 1988.*
[8] Carpenter G. A., Grossberg S., *"The ART of adaptive pattern recognition by a self-organizing neural networks"*, Computer, Vol. 21, pp. 77-88, Mar. 1988

SYSTOLIC PATTERN RECOGNITION BASED ON NEURAL NETWORK ALGORITHM

D.O. Creteanu[†], V. Beiu[‡,§], J.A. Peperstraete[‡] and R. Lauwereins[‡,①]

[†] National Institute of Hydrology and Meteorology, Str. Ciuce 5, Bucharest 74698, România

[‡] Katholieke Universiteit Leuven, Department of Electrical Engineering, Division ESAT–ACCA Kardinaal Mercierlaan 94, Heverlee, B-3001 Belgium (e-mail: beiu@esat.kuleuven.ac.be)

[§] Bucharest Polytechnic Institute, Department of Computer Science and Engineering Spl. Independentei 313, Bucharest 77206, România

Abstrct. The paper presents a solution for pattern classification, which uses distribute processing both for computing the matching score and selecting the class with the maximum score. The proposed architecture belongs to systolic arrays, being a generalization of the classical priority queue. A detailed description of the elementary processors (EPs) reveals that the algorithm implemented by each EP (which is based on computing the Hamming distance) is common also for neural networks. The overall result is a **O**(M) execution time for M classes (i.e. linear), and **O**(1) execution time with respect to n (the size of the patterns).

For testing the ideas, a simulator has been developed. It has been built starting from a set of C functions for simulating parallel processes. A short description of these functions supports our claim about the improvement of efficiency when developing a simulator starting from these functions. Several results are shortly discussed. Conclusions and further directions of research end the paper.

1. Introduction

Even from the very beginning of artificial intelligence, its models have been studied hoping to achieve human-like performances for digital computing systems. One still fresh and promising direction can be foreseen in the advent of parallel processing arrays like systolic ones [1, 2] or like connectionist models (also known as neural networks) [3, 4]. Their high performances are provided by the massive parallel processing and strong connectivity between the network's elements, in spite of the simplicity and identity of these elements. This identity aspect is quite important when having in mind a hardware implementation, as satisfying the important VLSI criterion of uniformity.

2. Background

A classifier is a system which determines which one out of M classes is the most representative for an unknown input pattern having n elements. The traditional classifier contains two stages (figure 1):

- the first one decodes the input into an internal symbolic representation and an algorithm computes the "matching score" for each of the M possible classes; these results are encoded and sent sequentially to
- the second stage of the classifier stores these scores and selects the class with the maximum matching score; this is sent to the output.

The solution proposed in this paper uses distribute processing both for computing the matching score and selecting the class with the maximum score [5]. The classifier is a linear network with synchronous processors in the nodes. The structure of this network is similar with a "systolic priority queue" (figure 2.a) [1], having one elementary processor (EP) for each

---

① Senior Research Assistant of the Belgian National Fund for Scientific Research.

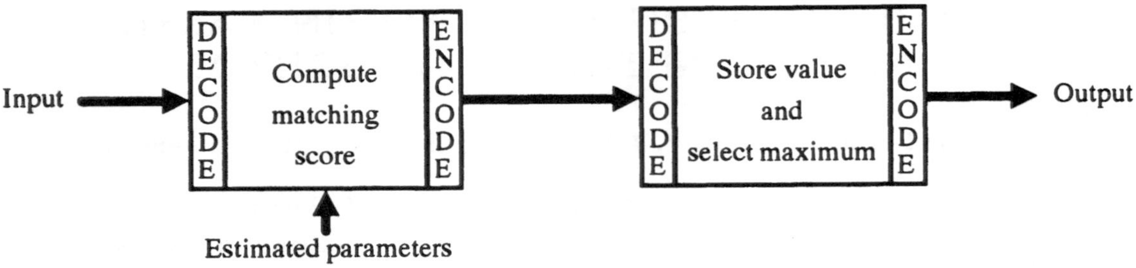

Figure 1. Traditional classifier.

of the M classes. Each EP is connected only with the next and the previous ones (which are called "neighbors"). Exceptions are the first, which is equivalent with a stack-top, and the last, which is equivalent with a stack-base. The first EP connects to a "Network Controller" (figure 2.b). The last EP has the "average" (AVG) output connected to the B input (feedback connection), and the "maximum" (MAX) output indicates an overflow. The network is synchronized from a two-phase non-overlapping clock. All the odd EPs of the network receive the first phase of the clock (φ_1), while the even EPs receive the second (φ_2), thus two neighbors (consecutive EPs), will never be active at the same moment.

The input pattern to be classified is "broadcasted" to all the levels of the network on the X input. As already known [1] this transforms the network from systolic to a semisystolic one (figures 2.b and 3); we can transform it back to a systolic array by properly introducing delay elements between each two EPs [1] (the Mealy machines from figure 3, becoming in fact just one delay element).

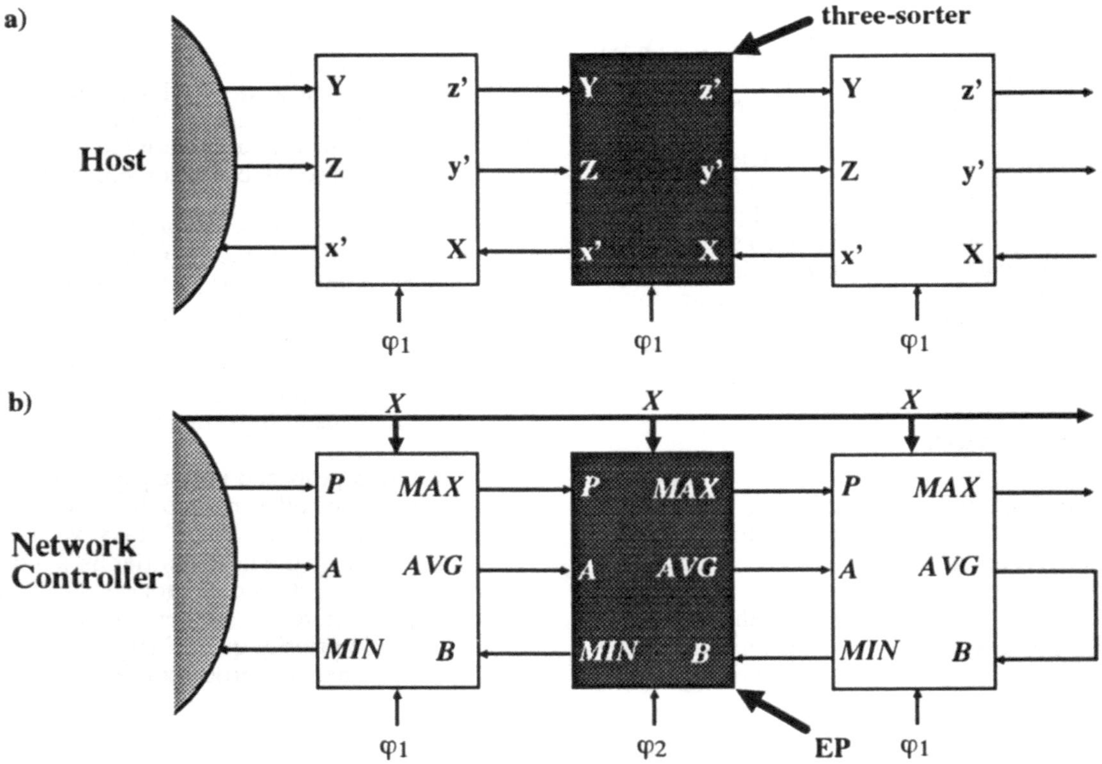

Figure 2. A real time priority queue (a), and the proposed structure (b).

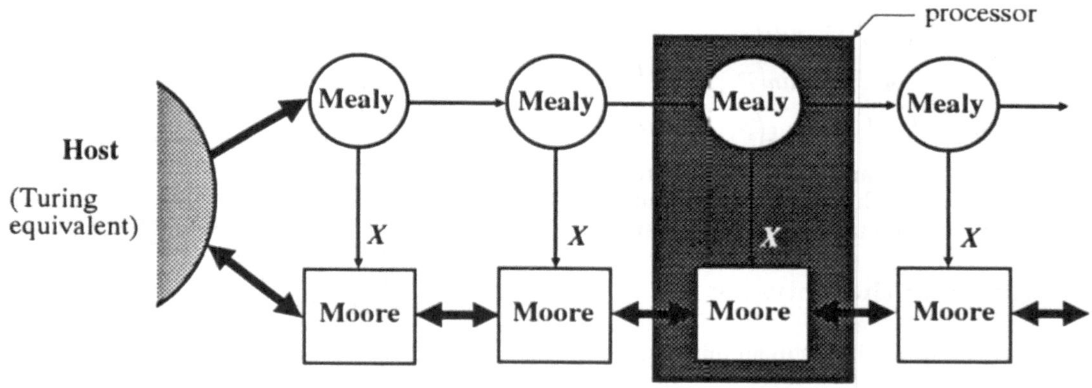

Figure 3. Semisystolic system.

3. The Elementary Processor

Each EP is a synchronous processor (Moore) with a Master-Slave structure (figure 2.b). The EP's inputs are:

- X – the unknown pattern;
- CLK – clock input;
- $P\&A$ – connecting to the previous EP (left neighbor);
- B – connecting to the next EP (right neighbor);

Before defining the *MAX*, *MIN* and *AVG* outputs, we introduce the following notations:

$\mathbf{B} = \{0,1\}$ – the set of binary values;

$\mathbf{I} = \mathbf{B}^n = (i_1, i_2, \ldots, i_n),\ i_k \in \mathbf{B}$

– the set of n bits sequences.

The basic operation used are:

- XOR :: $\mathbf{I}^2 \Rightarrow \mathbf{I}$, where XOR is the bit by bit exclusive OR, which can also be defined as

$$\mathrm{XOR}(a,b) = \left(|a_1 - b_1|, |a_2 - b_2|, \ldots, |a_n - b_n|\right);$$

- $\|\ \|_n :: \mathbf{I} \Rightarrow \mathbf{N_0},\ \|a\|_n = \sum_{i=1}^{n} a_i$ is the number of non-zero bits of a;
- $\delta :: \mathbf{I}^2 \Rightarrow \mathbf{N_0},\ \delta(a,b) = \|\mathrm{XOR}(a,b)\|_n$, being the Hamming distance between the a and b sequences.

We also have to define several simple computations:

- *mux* :: $\mathbf{I}^3 \times \{1,2,3\} \Rightarrow \mathbf{I}$, which is a three inputs multiplexer

$$mux(a,b,c,selector) = \begin{cases} a & \text{if } selector = 1 \\ b & \text{if } selector = 2 \\ c & \text{if } selector = 3 \end{cases}$$

- *max* :: $\mathbf{I}^4 \Rightarrow \{1,2,3\}$, which selects from a, b and p the pattern which has the maximum Hamming distance to x

$$max(p,a,b,x) = \begin{cases} 1 & \text{if } \delta(p,x) \geq \delta(a,x) \text{ and } \delta(p,x) \geq \delta(b,x) \\ 2 & \text{if } \delta(a,x) > \delta(p,x) \text{ and } \delta(a,x) > \delta(b,x) \\ 3 & \text{if } \delta(b,x) \geq \delta(p,x) \text{ and } \delta(b,x) \geq \delta(a,x) \end{cases}$$

- *min* :: $\mathbf{I}^4 \Rightarrow \{1,2,3\}$, which selects from a, b and p the pattern which has the minimum Hamming distance to x

$$min(p,a,b,x) = \begin{cases} 1 & \text{if } \delta(p,x) < \delta(a,x) \text{ and } \delta(p,x) < \delta(b,x) \\ 2 & \text{if } \delta(a,x) < \delta(p,x) \text{ and } \delta(a,x) < \delta(b,x) \\ 3 & \text{if } \delta(b,x) \leq \delta(p,x) \text{ and } \delta(b,x) \leq \delta(a,x) \end{cases}$$

- *avg* :: $\mathbf{I}^4 \Rightarrow \{1,2,3\}$, which selects from a, b and p the pattern which has neither the maximum nor the minimum Hamming distance to x (also sometimes called "average")

$$avg(p,a,b,x) = 6 - max(p,a,b,x) - min(p,a,b,x)$$

Now the functioning of the EP can be concisely described by:

$$MAX^{t+1} = mux\left[P^t, A^t, B^t, max\left(P^t, A^t, B^t, X\right)\right]$$
$$MIN^{t+1} = mux\left[P^t, A^t, B^t, min\left(P^t, A^t, B^t, X\right)\right]$$
$$AVG^{t+1} = mux\left[P^t, A^t, B^t, avg\left(P^t, A^t, B^t, X\right)\right]$$

where the *MAX*, *MIN* and *AVG* outputs are thus equal with the pattern having the maximum, the minimum and the "average" Hamming distance.

4. Operation of the Network

The network has to be set in an initial starting state, where $MAX = MIN = AVG = X$ for all the EPs (figure 4.a).

The commands can now be initiated by the Network Controller on the φ_1 phase clock. As the only connection are through the *A* and *P* inputs of the first EP, the three basic operations (PUSH, POP, NOP) are encoded:

- PUSH– $A \equiv X$, and *P* is set to the pattern to be pushed into the network.
- POP – both *P* and *A* are set to $\overline{X}$ (the logical NOT applied to pattern *X*).
- NOP – is equivalent to a PUSH $\overline{X}$.

The operation of the network can be described by the following algorithm:

- after setting the network in the initial state ($MAX \equiv MIN \equiv AVG \equiv X$), the static input pattern to be classified, is presented to the *X* input;
- one prototype pattern for each class is PUSH-ed into the network (thus *M* PUSH commands have to be performed);
- a POP will now extract the *MIN* from the first EP (which has the minimum Hamming distance to the unknown input pattern); the next *M*-1 POP commands (if performed) will output all the PUSH-ed patterns, in the ascending order of the Hamming distance to the static input pattern *X*.

If the *MAX* output of the last EP has a value different from $\overline{X}$, then too many PUSH operations have been performed, and an overflow has occurred.

A graphic example showing the functioning of the network can be seen in figure 4 (on the next page). In this particular example the network has only 3 EPs, (*M*=3) and the size of input *X* is *n*=4; the sequence of commands performed is: PUSH 1000 (figure 4.b), PUSH 0000 (figure 4.c), PUSH 1001 (figure 4.d), and POP (figure 4.e).

5. Simulation

The network has been simulated by means of a set of C functions, specially developed for simulating parallel processes [6]. Their facilities are not restricted to this particular case, and other applications have proven that they are suitable for a wide class of parallel simulations [6]. In what follows we shall give just a glimpse of the ideas implemented by these functions.

Let's assume that we have to simulate a set of interconnected processes [7]. A process will be fully described by:

- its inputs,
- its internal state, and
- the function which describe the internal state transitions.

The outputs are correlated with the internal state. The simulator takes these principles to the data structure of a process:

```
struct Unit {
        void (*f)(Unit *);
        void * status;
        Queue *in;
        Queue * out;
        int lenin;
        int lenout;
        int wait;};
```

where:

- *f* – is the function that describes both the internal state transitions, and the correlation between this internal state and the outputs; as in normal cases the information used is only from inside the process, the only argument of the function is the address of the process;
- *status* – is a pointer to a user-defined structure containing the information describing the internal state;
- *in* – is a one dimensional array, with the length specified by *lenin*, holding information about the inputs of the process;
- *out* – is an one dimensional array, with the length specified by *lenout*, which describes the outputs of the process;
- *wait* – is a flag which can temporary stop the execution of the process.

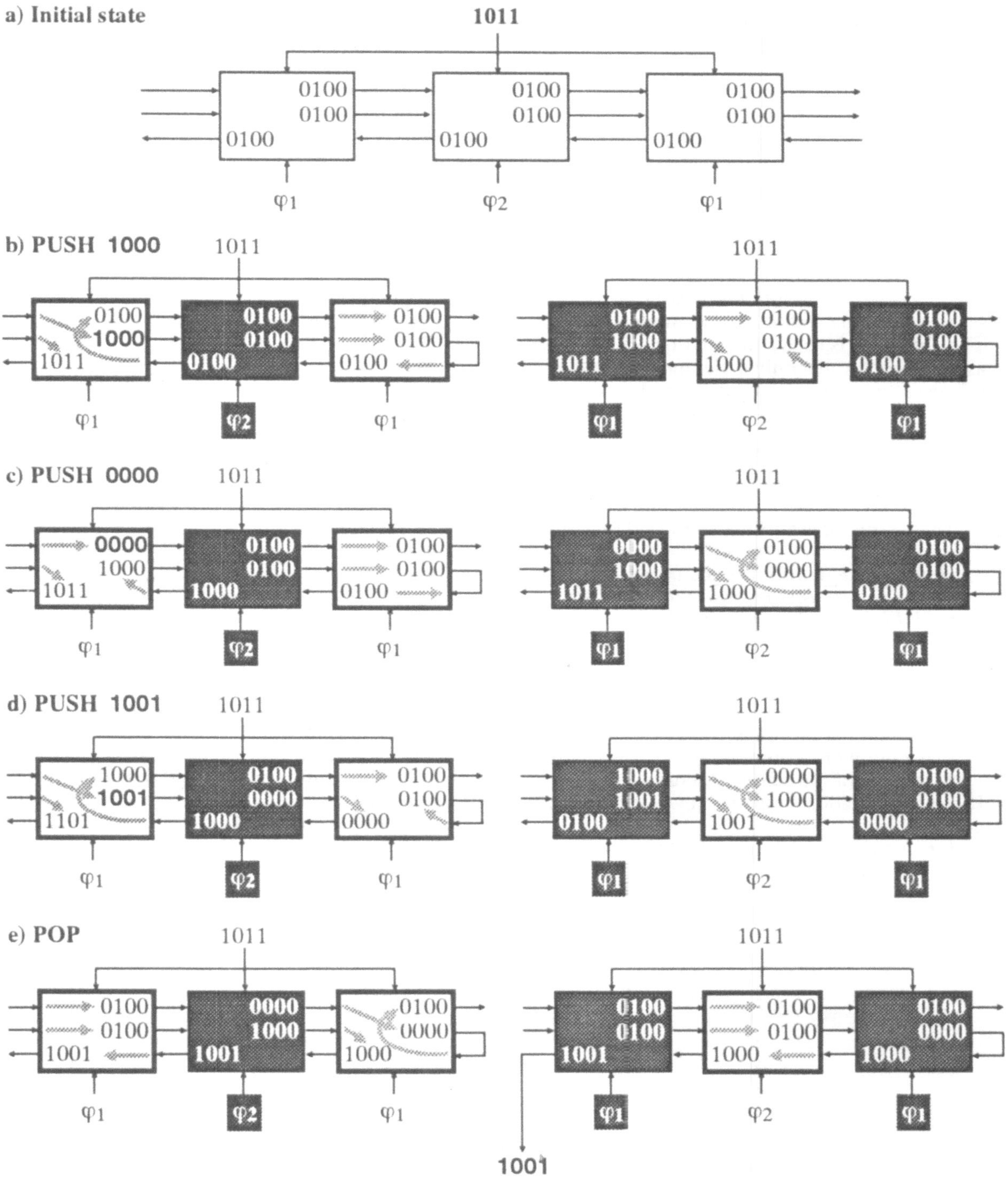

Figure 4. A simple example showing the network at successive time steps: (a) initial state; (b) PUSH **1000**; (c) PUSH **0000**; (d) PUSH **1001**; and (d) POP (which outputs **1001**).

Each element of the *in* vector describes one input, and has the structure:

```
struct cell {
        Unit * next_unit;
        void * mail_addr[2];};
```

where:

- *next_unit* – represents the address of the previous process;
- *mail_addr* – is just a two elements vector, each containing a pointer to a user defined structure (*mail_box*), used for inter-process communications.

The *out* vector is based on the same cell structure as the *in* vector, the only difference being that in this case, *next_unit* is pointing to the next process;

Each link between two processes uses a pair of *mail_boxes* organized as a circular list leading to conflict-free writing and reading of processes (figure 5).

These structures are created by two functions:

```
Unit * Make_Unit(int in_no,int out_no,
                int dim,void (*f)(Unit *));
void Link(Unit * source,int out_no,
                Unit * dest,int in_no,int dim);
```

Make_Unit creates a new process, and has as arguments:

- the number of inputs,
- the number of outputs, and
- the size of the status structure.

It returns the address of the newly created process.

Link creates a connection between two processes; the arguments required are:

- the address of the source process;
- the number of the output to be connected;
- the address of the destination process;
- the number of the input;
- the *mail_box* dimension.

After executing *Link*, the processes will be connected, from the specified output of the source process, to the specified input of the destination process.

The simulation is based on a FILO list structure:

```
Plan {
    Unit * task;
    Plan * next_t;};
```

As one is allowed to make different kind of simulations with respect to timing, three functions take charge of this particular problem. For a synchronous schedule, the *synchron* function is used. It is based on two *Plan* lists: *actualplan* and *newplan*. *Actualplan* stores the set of processes which are in execution, while *newplan* stores the set of successors of the processes from the *actualplan* list. When all the processes from *actualplan* have been executed, by simple pointer operations, *newplan* becomes *actualplan*, and a new *newplan* will be created. For asynchronous simulations there are two possible way of functioning; that is way two functions: *asynchron_one* and *asynchron_two*, have been developed. Both use only *actualplan*. For each executed process its successors are inserted at the end of *actualplan*. *Asynchron_one* executes the processes one by one, starting from the head of *actualplan*. *Asynchron_two* executes them at random (from *actualplan*).

For all these three functions, the simulation

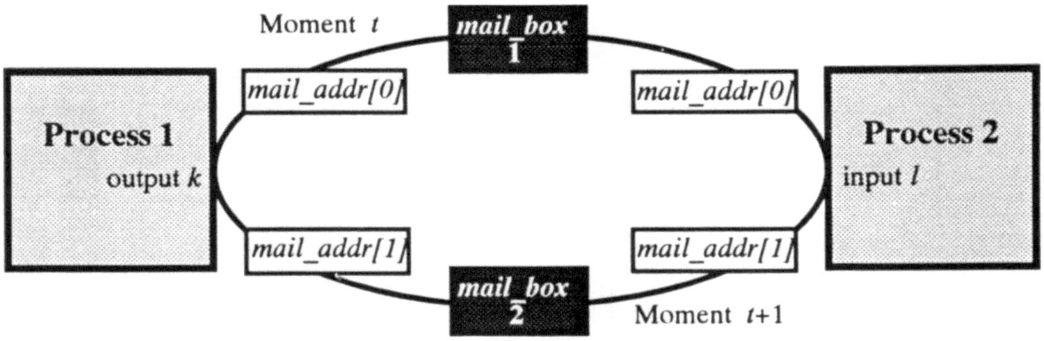

Figure 5. *Mail_boxes* circular list for communication among processes.

is stopped when one of the following situations occurs:

- *actualplan* is empty;
- the *HLT* global flag is set to STOP value;
- all the processes in *actualplan* are in the *wait* state.

To get an idea, for the example presented in figure 4, the code looks like:

Controller – the network controller;
Clock1 – one phase of the clock;
Clock2 – the other phase of the clock;
Stack... – the processing units (EPs) of the network;

```
Controller=Make_Unit(2,2,sizeof(CONTROLLER),controller);
Clock1=Make_Unit(0,3,sizeof(CLOCK),clock1);
Clock2=Make_Unit(0,1,sizeof(CLOCK),clock2);
for(i=0;i<3;i++)
    Stack[i]=Make_Unit(4,3,sizeof(STACK),stack);
Link(Controller,0,Stack[0],0,sizeof(PATTERN));
Link(Controller,1,Stack[0],1,sizeof(PATTERN));
Link(Clock1,2,Controller,1,sizeof(PULSE));
Link(Clock1,0,Stack[0],3,sizeof(PULSE));
Link(Clock1,1,Stack[2],3,sizeof(PULSE));
Link(Clock2,0,Stack[1],3,sizeof(PULSE));
Link(Stack[0],2,Controller,0,sizeof(PATTERN));
Link(Stack[2],1,Stack[2],2,sizeof(PATTERN));
for(i=0;i;i++)
    {
        Link(Stack[i],0,Stack[i+1],0,sizeof(PATTERN));
        Link(Stack[i],1,Stack[i+1],1,sizeof(PATTERN));
        Link(Stack[i+1],2,Stack[i],2,sizeof(PATTERN));
    };
```

6. Results

A more complex example for pattern recognition has also been simulated [3, 5]. The network has M=8 EPs, and the size of the input has been taken n=64. The 8 possible classes were described by the prototypes from figure 6.a. The unknown input has been obtained by randomly changing the bits of one of the prototypes with a given probability (figure 6.b-d). The network was able to recognize input patterns in 100 % of the cases having 10 % distortions, in 98 % of the cases having 20 % distortions, and in 88 % of the cases having 30 % distortions.

7. Conclusions

The paper presents a simple way to enrich the classical systolic architecture of the priority queue for any ordering relation, by including the processing of the order relation in the EPs.

A particular application for pattern recognition is detailed; the ordering relation for this application has been taken as the Hamming distance (used also by neural network algorithms). As the number of operations in recognizing one pattern is M+1 (M being the number of possible classes), the complexity of the recognition algorithm is linear with the number of possible classes.

A set of C functions has been developed, and successfully used for the simulation of neural networks, Petri networks and some simple synchronous structures (systolic arrays), with good speed and size performances. Simulations can thus be developed in a standard, easy-to-write and understandable way. We should also mention the big advantage of the fact that this can be done in a C environment, and also included in other C programs.

The network uniformity offers good features for VLSI implementation [2], showing a possible way to build a high speed pattern recognition coprocessor. In the case when the size of the unknown input is larger than the EP's input size, a bit slice technique can be lightly put into practice.

Other useful applications could be found in associative data base searches.

References

[1] C.E. Leiserson, *"Area Efficient VLSI Computations"*, The MIT Press, 1982.

[2] C.A. Mead, L. Conway, *"Introduction to VLSI Systems"*, Addison-Wesley, 1980.

[3] J.J. Hopfield, *"Neural Networks and Physical Systems with Emergent Collective Computational Abilities"*, Proc. Natl. Acad. USA, **79**, April 1982.

[4] V. Beiu, *"From Systolic Arrays to Neural Networks"*, The Scientific Annals of "Al.I. Cuza" Univ., Iasi (Iassy), Section: Informatics, **35**, 4, 1989.

[5] R.P. Lippmann, *"An Introduction to Computing with Neural Nets"*, IEEE ASSP Magazine, **4**, 2, 1987.

[6] D.O. Creteanu, *"Set of C Functions for Parallel Processes Simulation"*, M.Sc. thesis, Buch. Polytech. Inst., 1991.

[7] V. Aho, E.J. Hopcroft and D.J. Ullman, *"The Design and Analysis of Computer Algorithms"*, Addison-Wesley, 1975.

a)

3C66C3C3C3C3663C 1C3C7C1C1C1C1C1C 7CC3C3061860C0FF 7EC3030E0303C37E

183060C6C6FF0606 FFC0C0FC0603C67C 3F60C0FEC3C3C37E FF03060C18306060

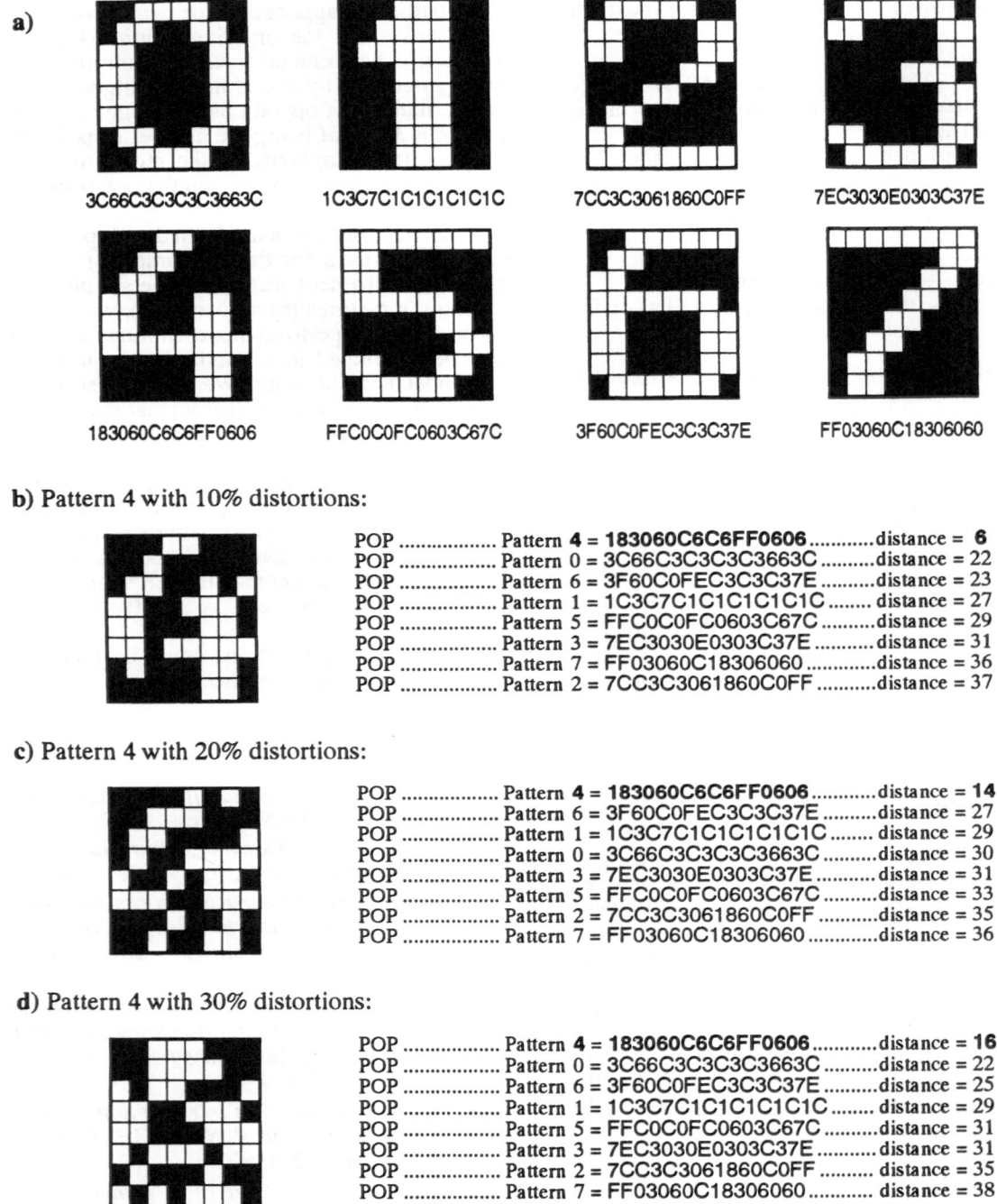

b) Pattern 4 with 10% distortions:

POP .... Pattern **4** = **183060C6C6FF0606** .... distance = **6**
POP .... Pattern 0 = 3C66C3C3C3C3663C .... distance = 22
POP .... Pattern 6 = 3F60C0FEC3C3C37E .... distance = 23
POP .... Pattern 1 = 1C3C7C1C1C1C1C1C .... distance = 27
POP .... Pattern 5 = FFC0C0FC0603C67C .... distance = 29
POP .... Pattern 3 = 7EC3030E0303C37E .... distance = 31
POP .... Pattern 7 = FF03060C18306060 .... distance = 36
POP .... Pattern 2 = 7CC3C3061860C0FF .... distance = 37

c) Pattern 4 with 20% distortions:

POP .... Pattern **4** = **183060C6C6FF0606** .... distance = **14**
POP .... Pattern 6 = 3F60C0FEC3C3C37E .... distance = 27
POP .... Pattern 1 = 1C3C7C1C1C1C1C1C .... distance = 29
POP .... Pattern 0 = 3C66C3C3C3C3663C .... distance = 30
POP .... Pattern 3 = 7EC3030E0303C37E .... distance = 31
POP .... Pattern 5 = FFC0C0FC0603C67C .... distance = 33
POP .... Pattern 2 = 7CC3C3061860C0FF .... distance = 35
POP .... Pattern 7 = FF03060C18306060 .... distance = 36

d) Pattern 4 with 30% distortions:

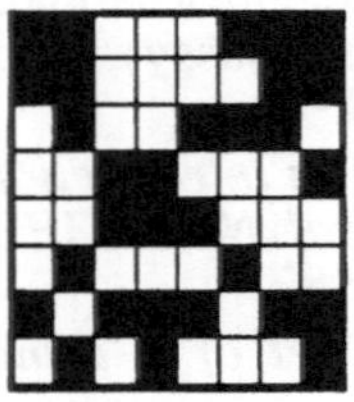

POP .... Pattern **4** = **183060C6C6FF0606** .... distance = **16**
POP .... Pattern 0 = 3C66C3C3C3C3663C .... distance = 22
POP .... Pattern 6 = 3F60C0FEC3C3C37E .... distance = 25
POP .... Pattern 1 = 1C3C7C1C1C1C1C1C .... distance = 29
POP .... Pattern 5 = FFC0C0FC0603C67C .... distance = 31
POP .... Pattern 3 = 7EC3030E0303C37E .... distance = 31
POP .... Pattern 2 = 7CC3C3061860C0FF .... distance = 35
POP .... Pattern 7 = FF03060C18306060 .... distance = 38

Figure 6. The eight digits to be recognized (a); pattent 4 distorted to 10% (b); pattent 4 distorted to 20% (c); and pattent 4 distorted to 30% (d).

Neural Networks versus Image Pyramids

Horst Bischof Walter G. Kropatsch
Inst. for Automation
Dept. for Pattern Recognition and Image Processing
Technical University Vienna
Treitlstr. 3/1832
e-mail: bis@prip.tuwien.ac.at

Abstract

Neural networks and image pyramids are massively parallel processing structures. In this paper we exploit the similarities as well as the differences between these structures. The general goal is to exchange knowledge between these two fields. After introducing the basic concepts of neural networks and image pyramids we give a translation table of the vocabulary used in image pyramids and those used in neural networks. Image pyramids which store and process numerical information (e.g. grey values of pixels) are very similar to neural networks. Therefore we concentrate on "symbolic pyramids". The main idea is to replace a cell of the pyramid by a small neural network, in order to represent and process symbolic information. We will consider local as well as distributed representations for symbolic information. In particular we present a neural implementation of the $2 \times 2/2$ curve pyramid. We derive some general rules for implementing symbolic pyramids by neural networks. Finally we briefly discuss the role of learning in image pyramids.

1 Introduction

Artificial neural networks (henceforth called neural networks) are characterized by massive parallelism and the ability of learning. Their features make neural networks interesting for pattern recognition and computer vision.

Though neural networks have success in many domains, for complex problems, such as vision, the current approach of using fully connected (three-layer) neural networks has severe deficiencies. As Le Cun [25] has stated "Expecting good performance without any a priori knowledge, releying exclusively on learning is wishful thinking". One way to incorporate a priori knowledge is to specify a proper topology of the network. The question we are concerened in this paper, is which neural network topology is suited for computer vision.

Image pyramids have shown to be an efficient data and processing structure for digital images in a variety of vision applications [26, 34]. Therefore we would like to exploit the similarities as well as the differences of neural networks and image pyramids. The general goal of this paper is to exchange knowledge between the fields of neural networks and image pyramids. We would like to emphazise that though both types are massive parallel processing structures, there has been no attempt to view these systems in a common framework.

The structure of this paper is as follows: In sections 2 and 3 we briefly introduce neural networks and image pyramids in order to define the basic concepts. We give a translation table of the vocabulary used in image pyramids and those used in neural networks. In the following sections we compare neural networks and image pyramids in detail. In section 4 we consider the structure of pyramids (regular and irregular). In section 5 we examine the types of information stored in the cells of a pyramid (numeric and symbolic). And in section 6 we examine the processing done by the cells. In particular, networks simulating the curve reduction process of the $2 \times 2/2$ curve pyramid are presented. Finally we give some conclusions and an outlook for further research. Especially the role of learning will be discussed.

2 Neural Networks

The are a variety of different neural network models (e.g. [36, 7]), but there is a common structure to all of them which we describe in this section. A more detailed model can be found in [1]. A neural network consists of a number of highly interconnected processing elements called *units*. The network is characterized by the interconnection scheme, which we call the *topology*, the types of *units*, the processing performed by them, and the *weights* of the connections.

2.1 Network Topology

The topology of the network can be described by a directed Graph $G = \langle U, E\rangle$; U is the set of units and $E \subseteq U \times U$ is the set of arcs between the units. The set of units U is partitioned into three subsets I, H, O (corresponding to input, hidden and output units) such that $I \cup H \cup O = U, (I \cup O) \cap H = \{\}, I, O \neq \{\}$. We call a network *fully connected* if $E = U \times U$. Each arc $\langle i, j\rangle \in E$ has a *weight* $w_{ij} \in \Re$. We call an arc with the associated weight a *connection*.

2.2 Units

Processing of information occurs in the units. A unit i has a *state vector* $s_i \in S \in \Re^N$ which describes the internal state of the unit and an *output value* $o_i \in \Re$ which is sent to other units. The main task of a unit is to compute a new state and a new output value, using the incoming signals (output values of the connected units), the weights of the connections and the own state vector. Formally this process can be stated as the application of an *update function* f_i . Let us call the application of the function f_i, *update* of unit i.

The other task of the unit is to change the weight vector in order to adapt its behavior. Formally this is stated in a learning function l_i. Note that for the update and the learning function only information locally available at the unit is used. The units can therefore operate in parallel and independent of one another.

2.3 Representation

It is important to distinguish between two kinds of representation in a neural network. One is called *local* representation and the other is called *distributed* representation. The information to be represented in a neural network can be manyfold, e.g. characters of an alphabet, greyvalues of pixels, etc.. Let us call the piece of information we are interested in an *item*. We call a representation local if an item is represented by one unit i.e. this unit is activated when the item is present and not activated when the item is not present. In the case of a distributed representation we have a group of units which represent one item, i.e. one item is represented by many units and each unit participates in representing many items.

A local representation is usually much easier to interpret, but a distributed representation is often more economical in terms of units and more robust against noise [16].

3 Image Pyramids

Image pyramids have shown to be efficient data and processing structures for digital images in a variety of vision applications. An image pyramid is a stack of images with exponentially decreasing resolutions [37]. The bottom level of the pyramid is the original image. In the simplest case each successive level of the pyramid is obtained from the previous level by a filtering operation followed by a sampling operator [12]. More general functions can be used to yield the desired reduction. We therefore call them *reduction functions*

Many image processing algorithms run on this hierarchical structure in $O(\log n)$ parallel processing steps (n is the image diameter), whereas they need $O(n^2)$ steps without the use of pyramids.

There are three important properties that characterize a pyramid:

1. Structure: e.g. neighbors, father–son relations between levels
2. Contents of a cell: e.g. pixel, edge, or more
3. Processing performed by the cells: e.g. filtering

3.1 Structure

The structure of a pyramid is determined by the neighbor relations within the levels of the pyramid and by the father–son relations between adjacent levels. We distinguish between

- regular and
- irregular

structures depending on whether the structural relations are the same for all pyramid cells (except on the boundary) or whether they may vary from cell to cell.

3.1.1 Regular Pyramids

Two terms describe the structure of a regular pyramid: *reduction factor* and the *size of the reduction window*. The reduction factor r determines the rate by which the number of cells decrease from level to level. The reduction window (typically a square $n \times n$) associates to every cell in a higher level (called father) a set of cells in the level directly below (called sons). The cells which are neighbors on the same level are called brothers (sisters). The usual notation for describing the structure of a regular pyramid is $n \times n/r$. For example in the classical $2 \times 2/4$ pyramid a window of 2×2 cells forms a new cell of the next higher resolution. Since in this pyramid there is no overlap the number of cells decrease from level to level by a factor of 4.

3.1.2 Irregular pyramids

In irregular pyramids the regularity constraint of regular pyramids is relaxed. These pyramids operate on a general graph structure instead of the regular neighborhood graph as in the case of regular pyramids. There are two ways to construct an irregular pyramid:

1. Parallel graph contraction [35]
2. Decimation of the neighborhood graph [28]

The main purpose for the introduction of irregular pyramids is the rigid behavior (e.g. shift variance) of regular structures [6]. Irregular pyramids offer greater flexiblility [30] for the price of less efficient access.

3.2 Contents of a cell

One can consider the contents of a pyramidal cell as a model of the region which it represents. In the simplest case a cell stores only one (grey) value. We call such pyramids *grey level* pyramids. In more complicated cases several parameters of general models are stored in a cell [13]. But the basic property that numerical values are stored in a cell remains. Subsequently we will call these pyramids *numerical pyramids*.

Besides numerical values it is also possible to store symbolic information in a cell [22]. In this case we have a finite number of symbols, and a cell stores these symbols or relations among them. We call such a pyramid *symbolic pyramid*.

3.3 Processing by a pyramid cell

The main property of processing in a pyramid is that it occurs only local, i.e. every cell computes from the contents of the sons, the brothers, and/or the parents a new value and transmits it to one or more cells of its pyramidal neighborhood. In the bottom–up construction phase input comes from the sons but for some algorithms the flow of information is also in the top–down direction [10].

The type of operations performed by the cells depends of course on the type of the cell's contents. For grey-level pyramids linear filters e.g. Gaussian are commonly used. But also other non–linear filters have some interesting properties, e.g. minimum and maximum filter or filters based on mathematical morphology. In the case of symbolic pyramids other types of reduction functions have to be used. For example [22] introduced curve relations and a reduction algorithm based on the transitive closure of curve relations. In general a finite state machine [18] may be used to perform a symbolic reduction.

Table 1: Translation table

Image Pyramids	Neural Networks
cell	unit
level	layer
structure	topology
contents of cell	activation of unit
bottom-up reduction	activation function
parameters of reduction function	weights

3.4 Translation of terminology

We have introduced the basic concepts of pyramids and neural networks. Since the two research areas have introduced a different terminology we summarize in Table 1 the equivalent notions of the important concepts in these two fields. This should help one being familiar in one of the two fields to translate his knowledge in the other field. In the sequel we will use the vocabulary for image pyramids when talking about image pyramids and that of neural networks when talking about neural networks. But one should keep in mind that the words can be often used interchangeably.

4 Structure

The structure of regular and irregular pyramids can be described by horizontal and vertical graphs. Each level i of a pyramid can be described by a neighborhood graph $G_i = \langle V_i, A_i \rangle$. Where the set of vertices V_i corresponds to the pixels of level i, and $A_i \subseteq V_i \times V_i$ are the neighborhood relations of the pixels. Two vertices $p, q \in V_i$ are connected in G_i if they are neighbors in the structure.

Definition 1 *The neighborhood of vertex $p \in V_i$ is defined by* $\Gamma(p) := \{p\} \cup \{q \in V_i | (p,q) \in A_i\}$

The structure is regular if a well defined neighborhood relation holds for all vertices (except for the boundary).

The vertical structure (i.e. the connectivity between the levels) can also be described by a (bipartite) graph: $R_i = \langle (V_i \cup V_{i+1}), L_i \subseteq (V_i \times V_{i+1}) \rangle$

The receptive field (i.e. the set of all sons) of a cell $q \in V_{i+1}$ is defined as: $RF(q) := \{p | \langle p, q \rangle \in L_i\}$. In a similar manner the projective field of a cell $p \in V_i$ (i.e the set of all fathers) can be defined $PF(p) := \{q | \langle p, q \rangle \in L_i\}$.

Any pyramid with n levels can be described by n neighborhood graphs and n–1 vertical graphs. In the case of regular pyramids we need not store all these graphs because information is given implicitely by the the term $n \times n/r$.

From these considerations it is clear that one can built for any pyramid structure an equivalent neural network topology. Therefore all knowledge about the structure of pyramids can be transfered to neural networks. Also the results of shift variance of regular pyramids [6] hold for equivalent neural networks. Indeed we were able to proof that any rigid locally connected neural network structure has shift variance problems [5].

We can also use neural networks for constructing irregular pyramids. In [4] we have shown how a Hopfield network can be used to decimate a level of an irregular pyramid. The formulation as a Hopfield network is more general than the stochastic decimation, and it naturally includes the concpet of the adaptive pyramid.

5 Contents

We have seen in section 3.2 that we can distinguish pyramids with numeric and symbolic contents in the cells. Whereas neural networks store only numeric information in the units. Therefore cells of a numeric pyramid are from the viewpoint of contents rather similar. The case of symbolic information needs futher investigation.

5.1 Symbolic Content of a cell

Let $\Sigma = \{\sigma_1, \sigma_2, \cdots \sigma_N\}$ be a finite set of symbols, and $R \subseteq \Sigma \times \Sigma$ a binary relation between these symbols. A cell in a symbolic pyramid stores symbols and/or relations between these symbols. Our main concern in this section is on how we can represent this information by a neural network. We are therefore confronted with a representation problem.

Since we have a finite number of symbols (and relations) we can represent a symbol by the activation of a unit e.g. $\sigma_1 = 0, \sigma_2 = 0.1$ etc. This representation is very "unnatural" for neural networks and we would give up some essential characteristics of neural networks. We therefore have to use several units to store symbolic information. The general idea is to replace a cell of the pyramid by a small neural network. We have two possibilities:

1. local representation
2. distributed representation

5.1.1 Local Representation

If we have N symbols we need N units to represent these symbols. Every symbol has a designated unit which becomes active when the symbol is present. If we have to represent relations among the symbols we need $O(N^2)$ units to represent all possible relations. When only one symmetric relation at a time needs to be stored N units are sufficient (i.e. if unit σ_1 and σ_2 are active the relation $\sigma_1 R \sigma_2$ is represented). But if more relations need to be stored we are facing the so called binding problem [8]. We need therefore one designated unit for each relation $\sigma_i R \sigma_j$ which is active when the relation is present. The problem even becomes more severe when several different relations or n-ary relations need to be represented.

5.1.2 Distributed Representation

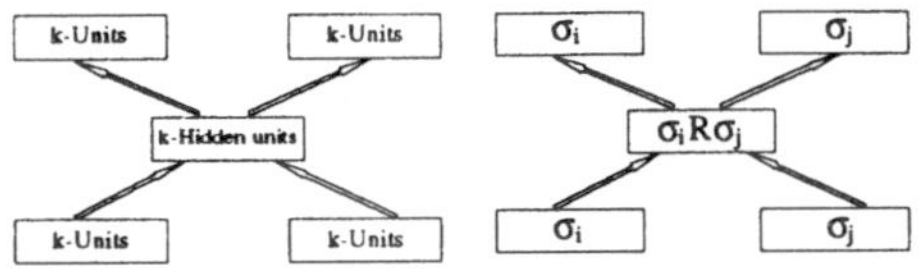

(a) (b)

Figure 1: (a) Autoassociative Network used by Pollack to encode symbolic structures (b) encoding curve relations

Distributed representations are more economical in terms of units and are also more robust than local representations. The general idea is that more than one unit is active when representing one item and a single unit participates in representing more than one item. Many possibilities for distributed representation have been proposed [16, 11, 33, 32]. Many of them are suited for representing numeric information [11],others have been proposed for symbolic information [33, 32].

Recently Pollack [33] has introduced the concept of "Recursive Distributed Representations". He has shown that he can represent binary trees and arbitrary list structures in this framework. The general idea is very simple. An autoassociative network like the one in Fig. 1a is used.

This network is trained to reproduce the input at the output units through a narrow channel of hidden units (half the number of hidden units than input units). The activation of the hidden units is then the distributed representation. This network can now be used iteratively to produce complicated data structures.

For our purpose we can use this scheme to represent relations with only $O(N)$ units (N is the number of symbols to represent). Assume we have a local representation of the individual symbols. We can now represent the relation $\sigma_i R \sigma_j$ with N units like in Fig. 1b.

Since for $N \neq 3$, $2^N \geq N^2$ we can represent all relations with this scheme and have also the possibility to convert it back into a local representation. Moreover we can use constraints on the weights for special relations, e.g. if the weights from the first set of input units are identical to the second set of input units the representation of the relation is symmetric i.e. $\sigma_i R \sigma_j = \sigma_j R \sigma_i$.

6 Processing

The result of a reduction function of a pyramid depends on the contents of the cell, there is no use of performing symbolic reduction on grey-level pyramids and vice versa. Since the processing performed by numeric pyramids is rather similar to that performed by neural networks [4] we concentrate on processing of symbolic information.

6.1 Symbolic Information

As we have seen in section 5.1, in order to represent symbolic information we have to give up the one-to-one correspondence between cells and units; we have to replace a cell of a pyramid by a small neural network. We will now extend this concept to describe processing of symbolic information by neural networks. We will start by designing a neural network for the $2 \times 2/2$ curve pyramid [22]. Finally we present some broader view of neural networks within the concept of symbolic pyramids.

6.1.1 $2 \times 2/2$ Curve pyramid

The $2 \times 2/2$ curve pyramid was introduced in [21]. The basic idea is that linear structures of images are represented by curve relations. A cell of the pyramid is considered as an observation window through which the curve is observed. A single curve intersects this window only twice. Only the intersection sides (N, E, S, W) are stored in the cell (i.e. a curve relation). We denote a curve relation by AB, where $A, B \in \{N, E, S, W, F\}$ (F is the special end code when the curve ends in a cell).

The basic routines of building the next level of the pyramid are (Fig. 2, Fig. 3):

1. Split - subdivision of the cells contents by introducing a diagonal
2. Transitive closure - the curve relations of the four son cells are merged by computing the transitive closure of all relations (i.e. $AB, BC \Rightarrow AC$).
3. Merge - the curve relations of the new cell are selected

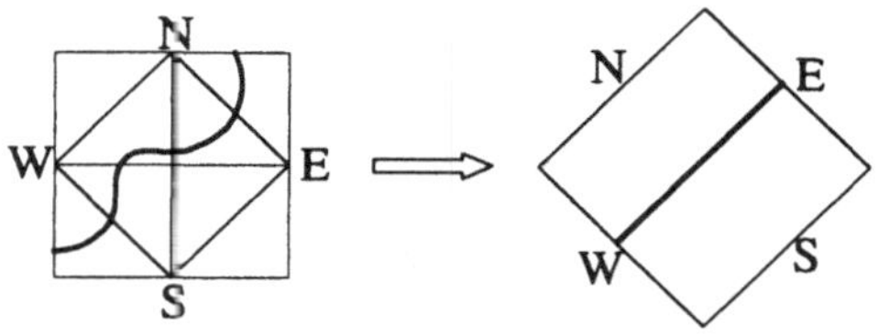

Figure 2: A reduction steps of the $2 \times 2/2$ Curve Pyramid

The $2 \times 2/2$ curve pyramid has several interesting properties like the length reduction property, "structural" noise filtering etc. which are described in [23, 24].

6.2 Neural Network curve pyramid

The $2 \times 2/2$ curve pyramid is an ideal test case to study neural network implementations of symbolic pyramids and to illustrate the concept of replacing a cell of a pyramid by a neural network. In order to simplify the discussion we will not describe end codes and U-turns, but they can be easily included.

The representation of the relation is as follows. Each cell has four output units (N,E,S,W), which have connections to the cells in the next higher level. With this scheme only one relation per cell can be stored unambiguously. In this case exactly two units are activated (i.e. output value = 1) and two units are not activated (i.e. output value = 0). For example if the units N and S are activated a North-South relation is represented. Inside a cell we represent the relations by a local representation as described in section 5.1 (i.e. one cell per possible relation), since we have to represent multiple relations in order to compute the transitive closure.

We have to describe 3 steps in terms of neural networks:

1. Split
2. Transitive Closure
3. Merge

6.2.1 Split

The first operation to perform is to split the curve relations by the diagonal as shown in Fig. 3 This can be done by a neural network like the one in Fig.4

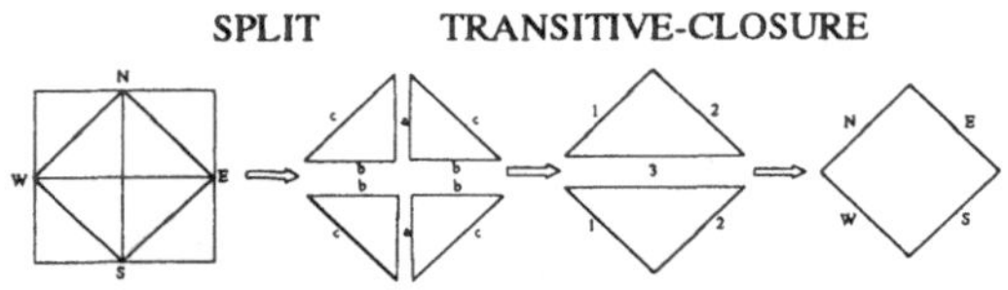

Figure 3: Splitting the Curve Relation by the diagonal

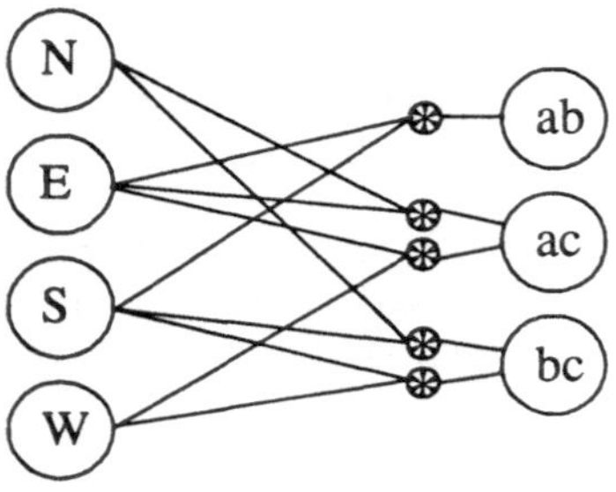

Figure 4: Neural Network Implementation of the splitting operation

The small circles with the star inside indicate multiplicative connections, i.e. only if all inputs are activated a value of one is passed to the units. One should note that three units are sufficient to represent the possible relations in the triangle because the relation is symmetric. The operation depicted in Fig. 4 is done for all four sons of a cell.

6.2.2 Transitive closure

At the next step we have to connect the curve segments in the four triangles by computing the transitive closure of the curve relations. We will first merge the upper and the lower two triangles in parallel, and then merge the the resulting two triangles to the final square, like it is shown in Fig 3

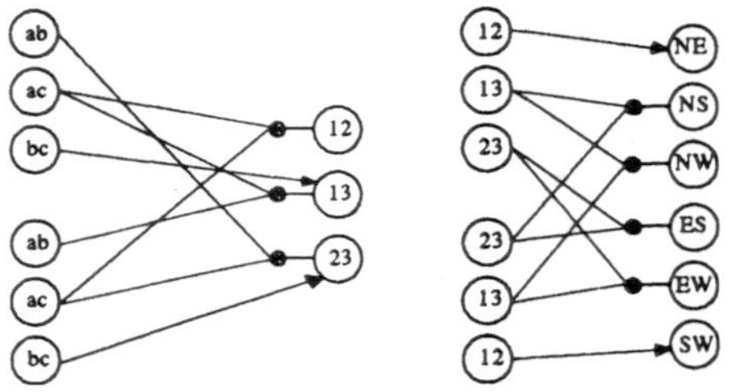

Figure 5: Neural Network Implementation of transitive closure operation. left: for two triangles to another traingle, right: two triangles to a square

The networks for merging two triangles to another traingle and merging two triangles to a square are shown in Fig. 5 (small circles with stars indicate multiplicative connections, small arrows indicate connections with a weight of 1). It is easy to verify that these circuits compute the transitive closure.

6.2.3 Merge

The third operation simply convertes the relation code used for processing the transitive closure to the code used by the output units. This can be done by the network shown in Fig. 6

When putting all the subnetworks together we get a network for performing a reduction step for a single cell (see Fig. 7. One can see that this network is hierarchically structured and consists of many identical subnetworks. In order to handle U-turns and end-codes correctly a few additional units are required. One should note that this network is a designed solution (no learning is required), and performs the same operations as the $2 \times 2/2$ curve pyramid. But it nicely demonstrates that the operations of a symbolic pyramid can be performed by a neural network.

6.3 General neural network scheme for symbolic pyramids

Replacing a cell of a pyramid by a small neural network gives us enough freedom to perform more general computations. In order to process symbolic information we need a network which can act as a finite state machine.

In general these small networks can be considered as grouping small image parts to objects (or parts of objects). Certain constraints have to be satisfied by this grouping process (e.g. the parts meet at a certain angle). Since the grouping is done hierarchically the "combinatorial explosion" of checking all possible combinations can be avoided. Moreover we can use learning algorithms to alter this grouping process. From this considerations we can describe the required properties of the networks.

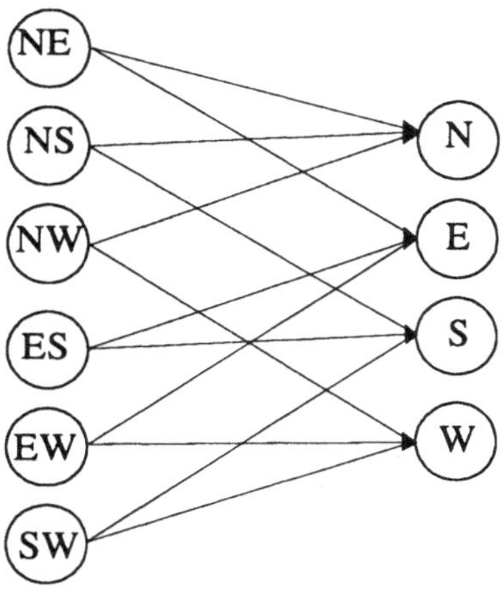

Figure 6: Neural Network for Merging the curve Codes

1. *Recurrent network* - each state of the network can be associated with a symbol
2. *Constraint satisfaction* - the network must converge to a stable state which fullfills as much constraints as possible in order to arrive at a "good" interpretation of the scene.
3. *Distributed representation* - as we have seen in section 5.1 an efficient representation of symbolic structures has to be distributed
4. *Unsupervised Learning* - if learning should be performed at the small networks it has to be unsupervised in order to avoid the problem of teaching each cell of the pyramid separately [9, 38].

There are some network models which fullfill the properties 1–3, e.g. Hopfield network [19, 20] recurrent Jordan network [29] Boltzmann machine [17]. But all these models use a supervised learning scheme. Some research has to be done to find an ideal combination for our purposes. Especially the learning problem is still a major obstacle to perform these tasks.

7 Conclusions and Outlook

In this paper we have rigourosly identified the similarities of image pyramids and neural networks. We have considered the structure of pyramids and the topology of neural networks. Another major point was the contents of a cell. Neural networks store only numerical information whereas in pyramids also symbolic information can be stored in the cells. We have seen that a distributed representation of symbolic information yields an efficient encoding scheme for symbolic information. Also the processing performed by neural networks and pyramids is not very different for the numerical case.

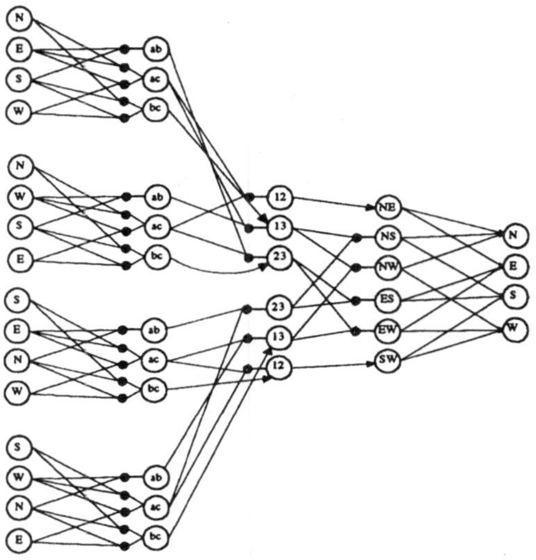

Figure 7: Neural Network for a reduction step in the $2 \times 2/2$ curve pyramid

In the case of symbolic pyramids we have studied the curve pyramid in detail and shown how to implement it by a neural network. Finally we have examined the idea of replacing the cell of a pyramid by a neural network more closely. Several constraints for such a network were derived. We have seen that learning is still an open problem for these architectures.

Learning has to a great extend been excluded from this paper. But one of the main motivations for identifing the similarities between neural networks and image pyramids is to employ learning algorithms on pyramids. An ideal learning algorithm for pyramids has to be unsupervised in order to avoid the problem of long learning times and bad scaling characteristics [15]. A reasonable principle for learning in such pyramidal architectures is the *infomax principle* proposed by Linsker [27]. It states that the cells of each level should preserve as much information as possible. This principle is tightly connected to Hebbian-Learning [14]. We have recently proposed an algorithm [2] based on a modified Hebb-rule of Oja [31], which is able to learn the weights of a numerical pyramid. In a forthcomming paper we will exploit the use of learning algorithms in pyramids in more detail. We would like to emphazise that this extends the capabilities of image pyramids aiming at a common framework for cellular processing structures.

References

[1] Horst Bischof. Modular, hierarchical, and geometrical neural networks. Technical Report PRIP-TR-9, Dept. for Pattern Recognition and Image Processing, TU Wien, 1991.

[2] Horst Bischof. Neural networks and image pyramids. In Bischof and Kropatsch [3], pages 249–260.

[3] Horst Bischof and Walter Kropatsch, editors. *Pattern Recognition 1992*, volume 62 of *OCG-Schriftenreihe*. Oldenbourg, 1992.

[4] Horst Bischof and Walter G. Kropatsch. Neural Networks versus Image Pyramids. Technical Report PRIP-TR-7, Dept. for Pattern Recognition and Image Processing, TU Wien, 1993.

[5] Horst Bischof and Axel Pinz. The invariance problem for hierarchical neural networks. In Chen Su-Shing, editor, *Neural and Stochastic Methods in Image and Signal Processing*, volume SPIE Vol. 1766, pages 118 - 129. SPIE, 1992.

[6] M. Bister, J. Cornelis, and Azriel Rosenfeld. A critical view of pyramid segmentation algorithms. *Pattern Recognition Letters*, Vol. 11(No. 9):pp. 605–617, September 1990.

[7] J.A. Feldman. Connectionist models and their properties. *Cognitive Science*, 6:205–254, 1982.

[8] J.A. Fodor and Z.W. Pylyshyn. Connectionism and cognitive architecture: A critical analysis. *Cognition*, 28:3–71, 1988.

[9] K. Fukushima, S. Miyake, and T. Ito. *Neocognitron*: a neural network model for a mechanism of visual pattern recognition. *IEEE Transactions on Systems, Man, and Cybernetics*, Vol. SMC-13(Nb. 3):pp.826–834, September/October 1983.

[10] William I. Grosky and Ramesh Jain. A pyramid-based approach to segmentation applied to region matching. *IEEE Transactions on Pattern Analysis and Machine Intelligence*, PAMI-8(No.5):pp.639–650, September 1986.

[11] P.J. Hancock. Data representation in neural nets: an empirical study. In *Proc. of the 1988 Connectionist Models Summer School*, pages 11–20. Morgan Kaufmann, 1988.

[12] Robert M. Haralick and Shapiro Linda G. Glossary of computer vision terms. *Pattern Recognition*, 24:pp.69–93, 1991.

[13] R. L. Hartley. *Multi-Scale Models in Image Analysis*. PhD thesis, University of Maryland, Computer Science Center, 1984.

[14] D.O. Hebb. *The organization of behavior*. Wiley, New York, 1949.

[15] G. Hinton. Connectionist Learning Procedures. *Artificial Intelligence*, 40:185–234, 1989.

[16] G.E. Hinton, J.L. McCelland, and D.E. Rumelhart. Distributed Representations. In McCelland Rumelhart, editor, *Parallel Distributed Processing*, volume 1. MIT Press, 1986.

[17] G.E. Hinton and T.J. Sejnowski. Learning and relearning in boltzmann machines. In McCelland Rumelhart, editor, *Parallel Distributed Processing*, volume 1. MIT Press, 1986.

[18] J.E. Hopcroft and Ullman J.D. *Introduction to Automata theory, languages and computations*. Addison Wesley, 1979.

[19] J.J. Hopfield. Neural networks and physical systems with emergent collective computational abilities. *PNAS*, 79:2554–2558, 1982.

[20] J.J. Hopfield. Neurons with graded response have collective computational properties like those of two-state neurons. *PNAS*, 82:3088–3092, 1984.

[21] Walter G. Kropatsch. Hierarchical curve representation in a new pyramid scheme. Technical Report TR-1522, University of Maryland, Computer Science Center, June 1985.

[22] Walter G. Kropatsch. A pyramid that grows by powers of 2. *Pattern Recognition Letters*, Vol. 3:pp.315–322, 1985.

[23] Walter G. Kropatsch. Curve representations in multiple resolutions. In *Proc. Eighth International Conference on Pattern Recognition*, pages 1283–1285. IEEE Comp.Soc., 1986.

[24] Walter G. Kropatsch. Elimination von "kleinen" Kurvenstücken in der $2 \times 2/2$ Kurvenpyramide. In E. Paulus, editor, *Mustererkennung 1987*, Informatik Fachberichte 149, pages 156–160. Springer Verlag, 1987.

[25] Y. Le Cun, O. Matan, Boser B., Denker J.S., Henderson D., Howard R.E., Hubbard W., Jackel L.D., and Baird H.S. Handwritten ZIP Code Recognition with Multilayer Networks. In *Proc. of the 10.ICPR*, pages 35–40. IEEE Computer Society, 1990.

[26] S. Levialdi. Programming image processing machines. In S. Levialdi and V. Cantoni, editors, *Pyramidal Systems for Image Processing and Computer Vision*, volume F25 of *NATO ASI Series*, pages 311–328. Springer-Verlag Berlin, Heidelberg, 1986.

[27] R. Linsker. Self-organization in a perceptual network. *IEEE Computer*, 21:105–117, 1988.

[28] Peter Meer. Stochastic image pyramids. *Computer Vision, Graphics, and Image Processing*, Vol. 45(No. 3):pp.269–294, March 1989.

[29] Jordan M.I. Attractor dynamics and parallelism in a connectionist sequential machine. In *Proc. of 8th Annual Conf. of the Cognitive Science Society*, pages 531–546, 1986.

[30] A. Montanvert and Bertolino P. Irregular pyramids for parallel image segmentation. In Bischof and Kropatsch [3], pages 13–34.

[31] E. Oja. A simplified neuron model as a priniple component analyzer. *J. Mathematical Biology*, 15:267–273, 1982.

[32] Smolensky Paul. Tensor product variable binding and the representation of symbolic structures in connectionist systems. *Artificial Intelligence*, 46:159–216, 1990.

[33] J.B. Pollack. Recursive distributed representations. *Artificial Intelligence*, 46:77–105, 1990.

[34] Azriel Rosenfeld, editor. *Multiresolution Image Processing and Analysis*. Springer, Berlin, 1984.

[35] Azriel Rosenfeld. Arc colorings, partial path groups, and parallel graph contractions. Technical Report TR-1524, University of Maryland, Computer Science Center, July 1985.

[36] David E. Rumelhart and James A. McClelland. *Parallel Distributed Processing*, volume 1. MIT Press, first edition, 1986.

[37] Steven L. Tanimoto. Paradigms for pyramid machine algorithms. In S. Levialdi and V. Cantoni, editors, *Pyramidal Systems for Image Processing and Computer Vision*, volume F25 of *NATO ASI Series*, pages 173–194. Springer-Verlag Berlin, Heidelberg, 1986.

[38] Toru Yamaguchi and Walter G. Kropatsch. A vision by neural network or by pyramid. In *Proceedings of the 6th Scandinavian Conference on Image Analysis*, pages 104–111, Oulu, Finland, June 1989.

APPLICATION OF NEURAL NETWORKS TO GRADIENT SEARCH TECHNIQUES IN CLUSTER ANALYSIS

Stephane DELSERT, Denis HAMAD, Mohamed DAOUDI and Jack-Gérard POSTAIRE

Centre d'Automatique de Lille
Université des Sciences et Technologies de Lille
59655 Villeneuve d'Ascq Cedex, France

Abstract : In this paper, we propose an approach to cluster analysis based on two levels : a neural network and an heuristic clustering algorithm. The neural network provides data compression such as the probability density function (p.d.f.) of the weights must be as near as to the p.d.f. of the data. During the clustering step which is based on the estimation of the gradient of the p.d.f., each available vector weight, or prototype, is moved in the direction of the p.d.f. gradient approximation. The process is iterated until the normalised local gradient is equal to zero, which result near the modes of the p.d.f.

Key words : Clustering, data reduction, Neural networks, Counterpropagation networks, gradient search technique

1. INTRODUCTION

Neural networks are commonly used for pattern recognition. According to the availability of a priori information about the data, there are two approaches : the supervised and the unsupervised pattern recognition techniques. In a supervised context, neural networks learn simultaneously the observations and their associated clusters. They operate as hetero-associative memories. When an observation is presented to the network, it indicates the cluster to which it belongs to. In an unsupervised context, the neural network learns only the observations and memorises their representative prototypes in the connections of the neurones. When an unknown observation is presented to the network, it provides its nearest prototype.

In the context of clustering, i.e. when no a priori information about the data is available, the key problem is to detect the different classes constituting the population of observations. Many classical approaches have been developed, based on the fundamental assumption that the observations are drawn from a multidimensional p.d.f. The p.d.f. is generally estimated by means of one of the well known Parzen's Kernel or K-Nearest Neighbour procedures [1]. Each mode of the p.d.f. can be detected by hill-climbing procedures, using some gradient search technique [2].

However, in many situations it is necessary to reduce the data information in order to reduce storage and computation time. The reduction of information can be of two types : dimension space reduction and number of observations reduction. Classical approaches for reducing the dimension are based on a linear or a non-linear projection of the data on a space of lower dimension [3], [4]. The reduction of the number of observations consists in finding a small number of prototype points such that the p.d.f. estimated from these selected prototypes remains as close as possible to the p.d.f. estimation when all the available observations are used [5], [6].

Neural networks can be used to reduce the data information. The reduction of the dimension can be performed by a multilayer Perceptron [7], [8] while the reduction of size of a sample set can be made by means of the learning vector quantization technique proposed by Kohonen [9].

In this paper, we propose an approach to sample size reduction using counterpropagation networks in autoassociative mode [10]. The counterpropagation network consists of three layers : an input layer that simply multiplex the input signals, an intermediate Kohonen layer, and finally, a Grossberg layer which yields the output signals.

In an unsupervised context, the network has to be trained (section 2). During the training phase, multivariate observations are presented to the input of the network. The Kohonen-Grossberg layers learn these examples and provide data compression such as the p.d.f. of the weights of connections in the Kohonen-Grossberg layers remains as close as possible to the true p.d.f. of the data.

During the clustering step, each available weight vector, or prototype, is moved in the direction of a p.d.f. gradient approximation (section 3). The process is iterated until the normalised local gradient tends to zero, i.e. when the weight vector reaches a mode of the p.d.f.

Some simulations using artificially generated multidimensional data sets are reported in section 4. They show that the performance of the neural network. Neural networks speed up clustering procedures thanks to the parallel implementation of the gradient kernel estimation applied to vector quantization.

2. COUNTERPROPAGATION NETWORK FOR DATA COMPRESSION IN AN UNSUPERVISED LEARNING CONTEXT

Counterpropagation networks, as the multilayer Perceptron, are essentially applied in supervised pattern recognition contexts. The simplified form of the counterpropagation network consists of three feedforward layers: an input layer that simply multiplex the input signals, an intermediate Kohonen layer, and a Grossberg layer which yields the output signals. The Kohonen layer elaborates prototypes of the samples while the Grossberg layer clusters the data. A counterpropagation network operates in two stages : learning and clustering. During the learning stage, the set of pairs constituted by the observations and their associated clusters are sequentially presented to the network which memories these pairs in the weights of its connections. During the clustering operation, the network can be used to find out the class to which belongs any new observation. The network operates in interpolate or extrapolate modes.

In an unsupervised context, the network architecture is modified such that only unlabelled observations are presented to the inputs of the network (Figure 1). During the learning stage, both the Kohonen and the Grossberg layers elaborate prototypes or vector quantization of these observations.

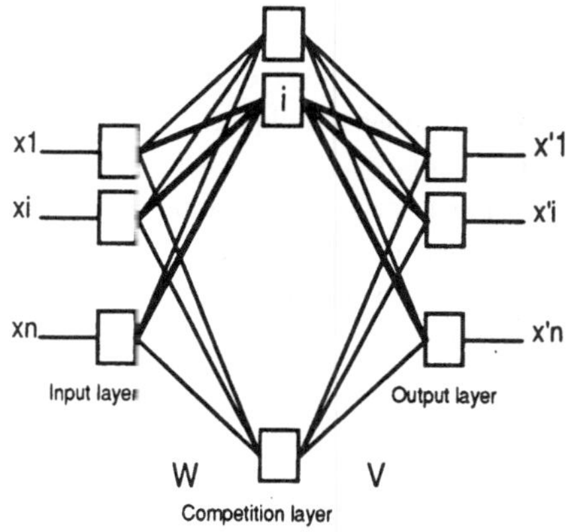

Figure 1. Counterpropagation network during the learning phase.

At each presentation of a sample X, a competition is then held among the units of the Kohonen layer. The winner is the one whose weight vector is the closest to X. The network modifies the Kohonen weight vectors according to equation (1) :

$$W_i(t+1) = W_i(t) + a(t).y_i(t).[X(t) - W_i(t)] \quad (1)$$

where :

- t is the current iteration number.
- $X(t) = (x_1(t), x_2(t), ..., x_n(t))^T$ is the input vector in R^n presented to the network at iteration t.
- $W_i(t) = (w_{i1}(t), w_{i2}(t), ..., w_{in}(t))^T$ is a weight vector, or synaptic vector, in R^n associated to the ith unit.
- a(t) is the learning rate at iteration t, which has to be defined according to the relationships (2) :

$$\sum_t a(t) = \infty \text{ and } \sum_t a^2(t) < \infty \quad (2)$$

- $y_i(t)$ is output of the the winning unit, defined as :

$$y_i(t) = \begin{cases} 1 & \text{if } i = \operatorname{Arg}\left\{\min_k\left[d(X(t), W_k(t))\right]\right\} \\ 0 & \text{else} \end{cases} \quad (3)$$

where $d(X(t),W_k(t))$ is the distance between the observation vector $X(t)$ and the corresponding weight vector $W_k(t)$.

Most of papers dealing with learning vector quantization differ in the choice of the winning unit and in the weight adaptations techniques.

Competitive learning uses the Euclidean distance. It partitions the input space into p domains D_i, each of them being associated to a cluster C_i while the synaptic vector W_i converge towards the centroide of the cluster C_i [11], [12], [13] :

$$R^n = \bigcup_{i=1}^{p} D_i \qquad D_i \cap D_j = \varnothing, \forall i \neq j \quad (4)$$

Competitive learning presents the drawback that the resulting synaptic vectors do not represent correctly the density of observations through the input space.

To overcome this limitation, we use the frequency sensitive competitive learning notion, introduced in [14], [15], [16]. In this approach, each neural unit incorporates a count of the number of times it has been the winner. The distance $d(X(t),W_k(t))$ is modified as follows :

$$d(X(t), W_k(t)) = \|X(t) - W_k(t)\|^2 . f_i(t) \quad (5)$$

where $f_i(t)$ indicates the frequency sensitive coefficient which is the number of times the ith unit wins the competition during the learning stage :

$$f_i(t) = f_i(t-1) + y_i(t) \quad (6)$$

If a given neural unit wins the competition frequently, its count and consequently its distance defined by (5), increase. This reduces the likelihood that the ith unit will be the winner in the next steps. Then the other units with lower count values have more chance to win the competition.

Then we proceed by modifying the synaptic vectors of the winning unit and of its neighbours. When the variance of the frequency sensitive coefficients of units is stabilised we refine the learning by applying a classical competitive learning procedure. This technique allows the synaptic vectors to spread through the input space so that they reflect local density of observations.

The training of the Grossberg layer is similar to the Kohonen layer. Note that, in an unsupervised context, it is easy to demonstrate that the learning phase of the Grossberg layer is unnecessary, so that we can replace the synaptic matrix V (Figure 1) of Grossberg layer by the transposed synaptic matrix of Kohonen layer :

$$V = W^T \quad (7)$$

After completion of the learning phase, the network has to be modified for the clustering stage. The aim of this second stage is to find out the kernels of the modes by means of the estimation of the normalised gradient of the probability density function.

3. THE CLUSTERING STAGE BASED ON A GRADIENT ALGORITHM

The clustering step is based on the estimation of the normalised gradient of the p.d.f. During this step, each synaptic vector is moved in the direction of the normalised gradient approximation. The process is stopped when the gradient reaches a low value which indicates that the weight vector is near a mode of the p.d.f. The number of classes is equal to the number of detected modes. Moreover, the observations which has given rise the same prototype are assigned to the cluster associated to the corresponding mode.

3.1. Nonparametric density gradient estimation

The weight vectors, at the end of the learning procedure, are considered as prototypes of the available observations. They can be represented by points denoted w_i, i=1, ..., p, in the input space.

Let P(w) denotes the underlying p.d.f. of the distribution of the prototypes w_i, i=1, ..., p. The normalised gradient estimation [17] on a weight vector W is given by :

$$\hat{\nabla}_W \operatorname{Ln} P(W) = \frac{\nabla_W (P(W))}{P(W)} = \left(\frac{n+2}{G^2}\right)\left[\frac{1}{m}\left(\sum_{W_i \in S_{G(W)}} W_i\right) - W\right] \qquad (8)$$

- n is the dimension of the input space.
- m is the number of synaptic vectors W_i falling within the domain $S_G(W)$ defined by :

$$S_G(W) \equiv \left\{W_i : (W_i - W)^T (W_i - W) \le G^2\right\} \qquad (9)$$

The term in square brackets in (8) is the mean-shift estimate of the weights in the domain $S_G(W)$ surrounding W. The parameter G depends upon the size of clusters.

3.2. Modification of the neural network to clustering

For the clustering operation, equation (3) is modified such that the winning units of the Kohonen layer are those falling within the domain $S_G(W)$:

$$y_i = \begin{cases} 1 & \text{if } W_i \in S_G(W) \\ 0 & \text{if } W_i \notin S_G(W) \end{cases} \qquad (10)$$

The output vector of the neural network is giving by:

$$X' = \sum_{i=1}^{p} v_i^T y_i \qquad (11)$$

If we replace equations (7) and (10) in (11), we have :

$$X' = \sum_{W_i \in S_G(W)} W_i \qquad (12)$$

where X' is the sum of the weights vectors falling within this domain $S_G(W)$.

The Grossberg vector is modified so as to include a unit which has all its weights fixed to 1 (Figure 2). The output m of this unit provides the number of the units of the Kohonen layer falling within the domain $S_G(W)$. Then the computation of the normalised gradient estimation becomes trivial (see equation 8) :

$$m = \sum_{i=1}^{p} y_i \qquad (13)$$

Thus, the network realises a parallel implementation of the gradient kernel estimation applied to the prototypes.

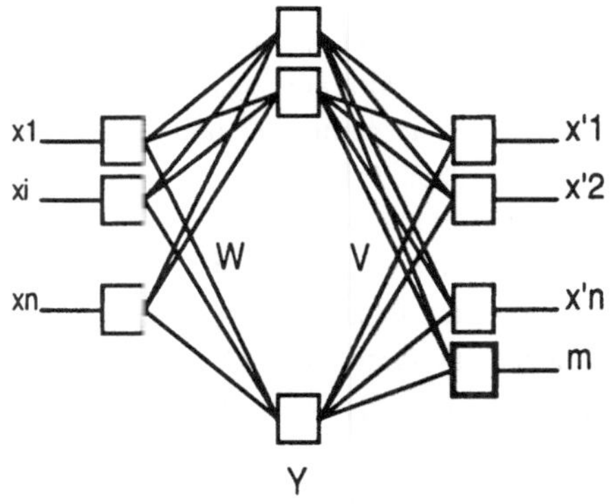

Figure 2. Counterpropagation network during the clustering phase.

The clustering algorithm consists to moving each weight vector, or prototype, in the direction of the gradient computed in w_i. Each weight vector is shifted by some amount proportional to this gradient. This process is iteratively repeated on the transformed weight vector until the gradient tends to zero i.e when the weight vector reaches a mode.

3.3. Clustering algorithm

1. Initialise the radius G of hypersphere $S_G(W)$ to a constant value and let : $W_i(0) \equiv W_i$, $i = 1,2, \ldots, p$

2. For each weight vector :

a- estimate the normalised local gradient defined by (8).

b- move the weight vector recursively, until $W_i(t+1)=W_i(t)$, according to the relationships :

$W_i(t+1) = W_i(t) + b.\hat{\nabla}_W \operatorname{Ln} P(W_i(t))$; b is a positive constant such that $0 < b \leq 1$)

3. Assign the weight vectors which converge towards the same mode to the same cluster.

4. SIMULATION RESULTS

In this section, we present a data set of 1000 observations computer generated. The data set is drawn from 4 Gaussian mixtures density representing 4 clusters of 250 sample points each (see figure 3). Their characteristics are given in Table 1.

The number of units in the input layer and in the output layer of the network are equal to the dimension of the input space. The number of units in the hidden layer is fixed to 225. The parameter G used in the gradient estimation is choose equal to 0.45.

Figure 4 shows the prototypes of the data set. The network preserves the data space repartition such that, either an isolated observation or many observations in a local maximum will be represented by a one prototype. However, the gradient is very sensitive to a local variation of the prototypes so it is necessary to filter them before applying the gradient estimation.

In this paper, we use a simple filter which consists to preserve the prototypes whose observations number is upper to the rate defined by : observations number/prototypes number.

Figure 5 shows the prototypes after filtering and clustering steps. The filtering rate is equal to 4 (4 = 1000/225) and only the prototypes representing more 4 observations are conserved for clustering.

Finally, in figure 6 we present the data clustering results.

	Mean vector	Covariance matrix	Number of samples
cluster 1	1.5 0.0	0.20 0 0 0.2	250
cluster 2	4 0	0.2 0 0 0.2	250
Cluster 3	6.5 1.5	0.25 0 0 0.25	250
cluster 4	6.5 -1.5	0.25 0 0 0.25	250

Table 1 : Characteristics of the data set.

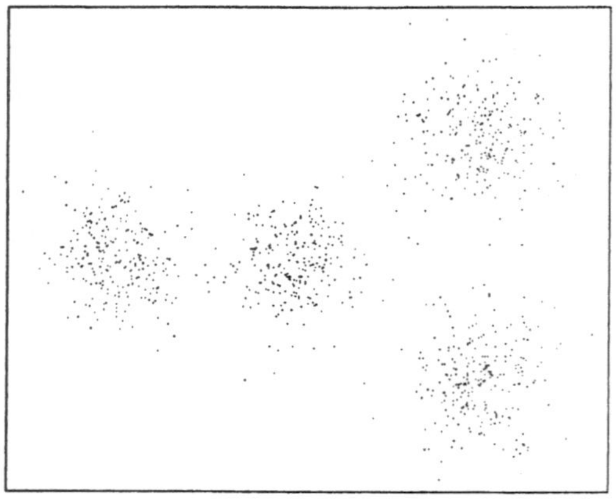

Figure 3 : Original data set.

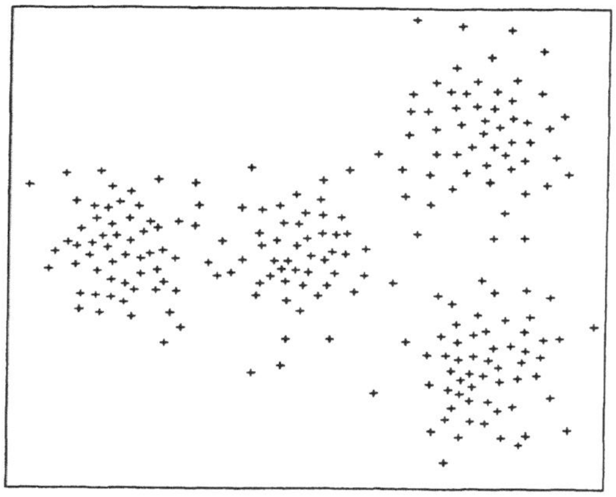

Figure 4 : Prototypes of the data set

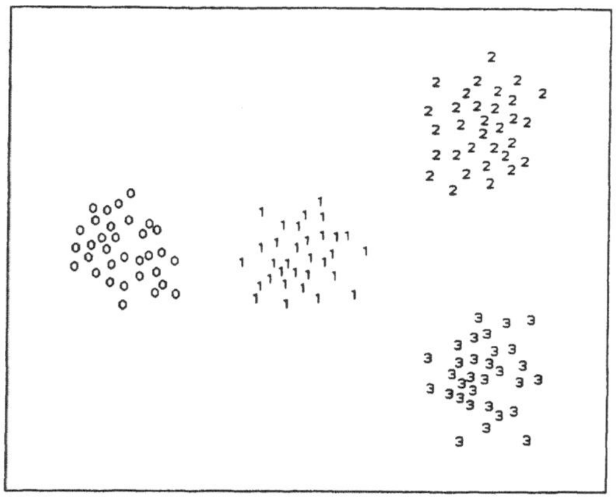

Figure 5. Prototypes filtered and clustered

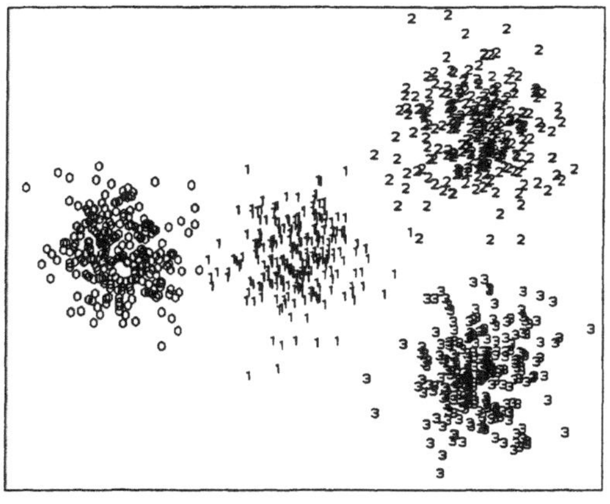

Figure 6. Data set after clustering.

5. CONCLUSION

The proposed counterpropagation network ensures two functions : the data reduction and the pattern recognition of the available observations. The data reduction is ensured in the learning step while the pattern recognition is accomplished in the clustering step. The neural network clustering procedure is accelerated thanks to the parallel implementation of the gradient in the neural network. However, The gradient is very sensitive to the local variation of the prototypes and only the filtered prototypes are used in the clustering step. The number of units in the Kohonen layer has been fixed arbitrarily, it is possible to choice this number according to the minimal information distance between the observations and their prototypes.

6. REFERENCES

[1] Duda R O, and Hart P E, *'Pattern Classification and Scene Analysis'*, Editions J. Wiley, New-York, 1973

[2] Postaire J-G, *'De l'Image à la Décision. Analyse des Images Numériques et Théorie de la Décision'*. Editions Dunod informatique, 1987.

[3] Chien Y. *'Interactive Pattern Recognition'*. New York : Marcel Dekker, 1978.

[4] Sammon J W, *'A Non-Linear Mapping for Data Structure Analysis'*, IEEE Trans. Computer, vol. C-18, pp. 401-409, 1969.

[5] Fukunaga K and Mantock J M, *'Nonparametric Data Reduction'*, IEEE Trans. on Pattern Analysis and Machine Intelligence, vol. 6, N° 1, January, 1984.

[6] Fukunaga K and Hayes R, *'The Reduced Parzen Classifier'*, IEEE Tans. on Pattern Analysis and Machine Intelligence, vol. 11, N° 4, April, 1989.

[7] Baldi P, Hornik K, *'Neural Networks and Principal Component Analysis : Learning from Examples without Local Minima'*, Neural networks, vol. 2, N° 1, pp. 53-58, 1989.

[8] Daoudi M, Hamad D, Postaire J-G, *'Interactive Classification through Neural Networks'*. To be appeared in this conference.

[9] Kohonen T, *'Self-Organisation and Associative Me-mory'*, 2nd Edition New York, Springer-verlag, 1988.

[10] Hecht-Nielsen R, *'Counterpropagation Networks'* Applied Optics 26 (23) : 4979-4984, 1987.

[11] Hecht-Nielsen R, *'Applications of Counter-propagation Networks'*. neural networks Vol.1, pp.131-139, 1988.

[12] Kosko B, *'Neural Networks and Fuzzy Systems, a Dynamical Systems Approach to Machine Intelligence'*, Printice Hall International, 1992.

[13] Rumelhart D E and Zipser D, *'Feature Discovery by Competitive Learning'*, Cognitive science 9, 75- 112, 1985.

[14] DeSieno D, *'Adding a Conscience to Competitive Learning'*. Proceedings to the IEEE International Conference on Neural Networks, pp. 117-124. San Diego, CA : SOS printing, 1988.

[15] Ahalt S C et All, *'Competitive Learning Algorithm for Vector Quantization'*, Neural Networks, vol. 3, pp. 277-290, 1990.

[16] Wai-Chi F et All. *'A VLSI Neural Processor for Image Data Compression Using Self-Organisation Networks'*. IEEE Trans. on Neural Networks, Vol. 3, NO. 3, May, 1992.

[17] Fukunaga K, and Hostetler L D, *'The Estimation of the Gradient of a Density Function, with Applications in Pattern Recognition'*, IEEE Trans. on Information Theory, Vol. IT-21, N°. 1, January, 1975.

LVQ-BASED ON-LINE EEG CLASSIFICATION*

Doris FLOTZINGER, Joachim KALCHER, Gert PFURTSCHELLER
Ludwig-Boltzmann Institute of Medical Informatics and Neuroinformatics, and
Institute of Biomedical Engineering, Department of Medical Informatics,
University of Technology, Brockmanngasse 41, A-8010 Graz, Austria
e-mail: flotzi@fbmtds04.tu-graz.ac.at

Abstract *- In this paper the problems of classifying EEG in both its spatial as well as its temporal aspects are described. The task was to identify the intended side of hand movement (left or right) on the basis of EEG recorded on two channels during one second before movement onset. This is part of a larger project aiming to build a "Brain-Computer Interface" (BCI) which should enable handicapped persons to communicate with their surroundings using their EEG.*

Several solutions based on Learning Vector Quantizers (LVQs) are presented ranging from classification of the whole EEG segment to combinations of classification of several small EEG segments. Two methods of combining these classifications are presented: one using direct classification and majority voting, and the other using a class activation function. The impact of the size of the segments and other parameters on the system performance is discussed. Results of experiments (correct classification rates of between 85-90%) with user-dependent LVQs are reported.

Introduction

The possibility of classifying changes in the EEG with its large inter- and intrasubject variability is a prerequisite of a new communication device, the Brain-Computer Interface (BCI), which uses EEG recorded from the scalp directly for control, e.g., of cursor movement on a monitor. First results of such an EEG-based BCI were reported in [9] where the amplitude of the central mu rhythm was transformed into cursor movement by manually defined and adjusted amplitude ranges for each command. A definition of such ranges demands much experience and skill and influences the performance of the system significantly. To increase performance, biofeedback training was performed over several months, i.e., the 'brain' had to adapt itself to the system. It would be much more preferable to have the system adapt itself to the subject and optimize internal variables using a learning algorithm.

In Graz, a simple Brain-Computer Interface (BCI, for a detailed description see [7]) has been constructed where a subject controls a cursor in one dimension (left/right) on a monitor by means of his brainwaves which are classified by a Learning Vector Quantizer (LVQ), a well-known algorithm which learns from examples. Two experimental paradigms have been set up: the first to create labelled examples for creation of the classifier and the second for on-line cursor control.

Experimental Setup

To be able to control a cursor in one dimension on a monitor, two kinds of patterns have to be distinguished. Because left and right cursor movement can easily be associated with left and right hand movements, the following paradigms were created.

First Paradigm

The subject sat in a comfortable chair in a darkened room and looked at a monitor 100 cm in front of the subject's eyes. The subject was asked to press a microswitch with either the left or the right index finger: the hand to be used (left or right) was indicated by a cue on the monitor one second after an acoustic warning stimulus. The cue was followed one second later by another stimulus in the form of a cross indicating the start of movement. During the second between the cue (left or right) and

* Supported by the "Fonds zur Förderung der wissenschaftlichen Forschung", project P9043.

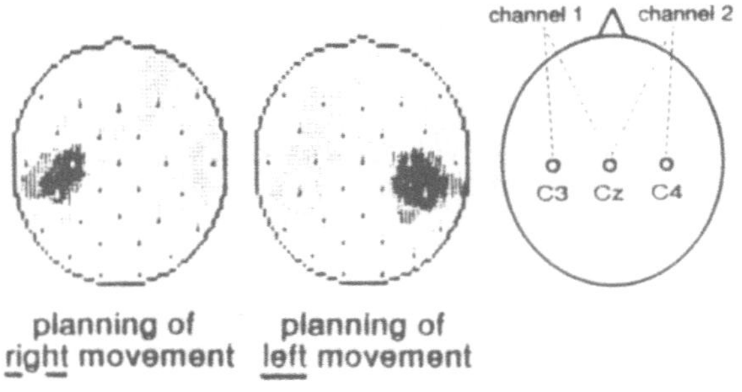

Fig. 1: Event-related desynchronization patterns during planning of left and right hand movements ("black" marks activated cortical regions) and location of electrodes in the experiment.

the presentation of the cross the subject planned to move the indicated index finger and produced an EEG pattern characteristic for the side of movement: it is well known ([1], [3], and [8]) that voluntary one-sided movement results in a clear event-related desynchronization (ERD) of the alpha-band rhythms (8-12 Hz) over the contralateral sensori-motor cortex. Figure 1 shows examples of ERD patterns during planning of one-sided hand movement and the selected location of the electrodes for the experiment. Using the spatiotemporal information of EEG recorded on the 2 electrodes, shown in Fig. 1, over one second, a Learning Vector Quantizer (LVQ) was created to predict which hand will be moved before actual movement onset [6].

Second Paradigm

This LVQ could then be used in an on-line experiment, in which the subject was presented a target (either on the left or the right half of the screen) followed one second later by the presentation of the cursor. This cursor was then moved by the system depending on the classification of the EEG recorded in the second between presentation of the target and the cursor.

In both paradigms the EEG was recorded bipolarly from 2 channels (C3-Cz and C4-Cz, international 10-20 system, see Fig. 1) with a time constant of 0.03 sec, an upper cut-off frequency of 15 Hz and a sampling rate of 64 Hz. From the 64 A/D-converted samples per second and channel the average power was calculated for time segments of 125 msec (8 samples per second) by squaring and then averaging over eight consecutive samples. Nine samples per channel (representing a time segment of 1.125 sec) were available for classification.

Classification of Spatiotemporal Patterns

As a classifier, the Learning Vector Quantizer (LVQ) was chosen, a simple, neurally inspired classifier related to Kohonen's work on self-organizing feature maps [4]. The power, simplicity and speed of LVQ make it a good candidate for on-line classification of EEG.

Classification of time-dependent features has received great attention lately. Most of the developed architectures are extensions of the Multi-Layer Perceptron and incorporate either additional context units or feedback connections. Exploiting the aspect of time within LVQ can be done in two ways (see also Fig. 2):

(1) All timepoints can be concatenated to form one large input vector which represents the whole time interval available. This method provides the classifier with maximum information at one instant, showing the whole development of the signal.
(2) Each timepoint can be classified on its own without further context. In this case, 9 classifications are available which have to be combined to one decision (in our case: left or right). This method has the advantage of being more shift-tolerant than (1) because it does not matter at what timepoints the correct classifications are made as long as they can be made at all.

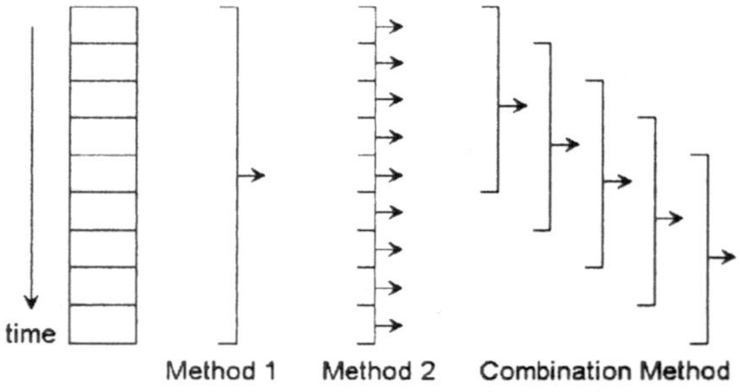

Fig. 2: Possibilities of incorporating time in the classification using LVQ.

Certainly, some sort of combination of these two methods, which takes into account the advantages of both, would be appreciated. This leads to the method of sliding windows (see also Fig. 2): a window of size 1<x<9 is shifted over the data and classifications are again transformed to one decision as in method (2). This method has the advantage of including some time context for each classification and also providing shift-tolerance. Because the exact timepoint where the reaction to the stimulus starts cannot be expected to be constant, the ability of being robust in the case of misaligned patterns is especially important in the case of EEG classification. Furthermore, this method increases the size of the training data set because each trial provides several training vectors.

Using this combined method, the input vectors were formed out of the 9 available timepoints and the 2 channels in the following ways (the spatial information of the 2 electrodes was considered only insofar as both channels were used simultaneously):

(i) Method 1: all timepoints available were used at the same time and the 18-dimensional input vectors (2 channels * 9 timepoints) were classified in one step.

(ii) Method 2: each timepoint was classified individually (2-dimensional input vectors) and these 9 classifications were combined to one decision (left or right).

(iii) Combination method: a sliding window of size 2 or greater was shifted over the 9 timepoints and several 4- or more-dimensional input vectors were formed per trial, the classifications of which had again to be combined to one decision as in (ii).

For combining several classifications to one decision two methods are apparent (see Fig. 3): (a) making a majority vote, i.e., deciding on 'left' if more 'left' classifications were made than 'right' ones. Or (b), if the strength of each classification was stored, then the accumulated stronger side would be the decision [5]. The strength of each classification can easily be calculated by finding the closest reference vector for each class (not just the closest reference vector as in standard LVQ) and defining the activation of class c, A_c, using the distances of the closest reference vector of each class, d_i, to the input as:

$$A_c = 1 - \frac{d_c}{\sum_i d_i}$$

These activations are summed classwise over all classifications and the decision is assigned to the class with the strongest overall activation.

Majority Voting has the disadvantage of depending on an odd number of classifications to be able to define a decision. On the other hand, if the number of classifications is even, a third class of 'no movement' or 'no decision' can be incorporated easily.

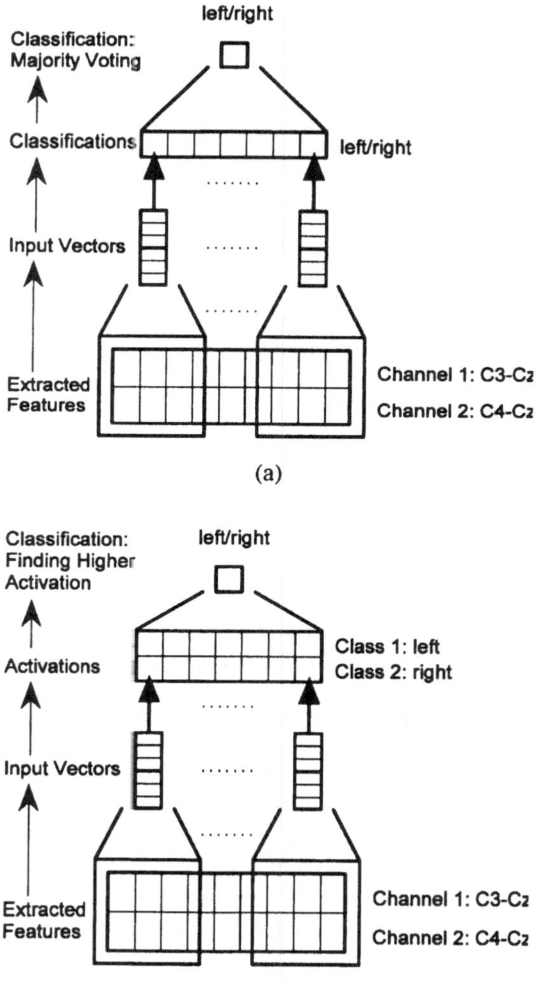

Fig. 3: Combination of several classifications to one decision by (a) Majority Voting and (b) Activation Summation.

Results

One subject participated in five sessions. The first session was recorded using the first paradigm and provided 74 left and 76 right hand movements to create an initial Learning Vector Quantizer. The following four sessions were recorded using the second paradigm and the feedback informed the subject of his success. In the on-line scheme the following setup was used for classification: a sliding window of five timepoints formed 10-dimensional input vectors. Five such vectors were classified seperately and the classifications were combined using majority voting. After each session the LVQ was updated using the new data and the best resulting LVQ was chosen for the next session. The on-line performance of the four on-line cursor control sessions is listed in Table 1.

Session	1st	2nd	3rd	4th
No. left	18 of 40	48 of 53	68 of 74	31 of 35
No. right	40 of 43	36 of 61	30 of 40	36 of 43
Total	70%	73%	86%	86%

Table 1: On-line performance of the four running-phase sessions using a 5-timepoints-window and majority voting. The LVQ3 algorithm was used with a LVQ consisting of 4 reference vectors per class.

To test the dependence of the performance of the classifier on the number of reference vectors, the sliding window size, the LVQ algorithm and the method of combination of multiple classifications, a number of off-line experiments were performed varying the number of reference vectors per class from 2 to 8, the sliding window size from 1 to 9, using both LVQ1 and LVQ3 ([4]), and performing both majority voting and activation summation. For each number of reference vectors per class, 10 LVQs were created from the data recorded using the first paradigm. Each of these 10 LVQs was trained on the data of on-line cursor control sessions 1-3, one at a time, and its performance was tested on the never-before-seen 4th cursor control session using both majority voting and activation summation. Therefore, for each combination of variables an average performance of 10 LVQs was available.

The parameters for LVQ1 and LVQ3 were chosen as follows: the number of iterations was set to 30*(number of trials)*(number of classifications per trial) to give each input vector the same chance to be learned. The size of $\alpha(0)=0.05$ was fixed for both LVQ1 and LVQ3 and decreased to zero during the learning process. The window for LVQ3 shrunk from 80% to 30% of the distance between the two closest reference vectors during the learning process.

Dependence on the number of reference vectors

Fig. 4 shows the performance on the last cursor control session (compare with Table 1, last column) for LVQs trained with LVQ1 on all 9 timepoints with varying sliding window size, the classifications were combined with activation summation. The figure shows an increasing independence of the number of reference vectors as the size of the sliding window increases. This shows the importance of providing context for improved classification.

Fig. 4 also shows that the classification task is not very hard because 2 reference vectors per class already provide excellent results (remember that 1 reference vector per class creates a hyperplane as class border).

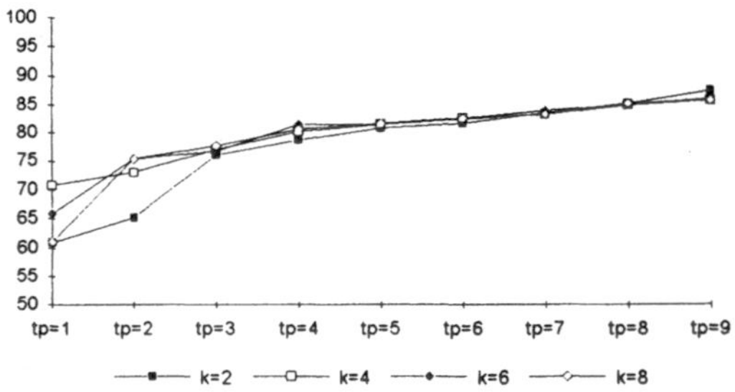

Fig. 4: Averaged performance of LVQs trained with LVQ1 in dependence on the window size (tp) and the number of reference vectors per class (k), combination by activation summation.

Dependence on the sliding window size

Fig. 4 shows the increasing performance with increasing window size. This implies that the LVQ can give a more definite answer if it can see the whole time segment compared to little fractions. This is an astonishing result because (1) the number of training input vectors is lowest if the whole time

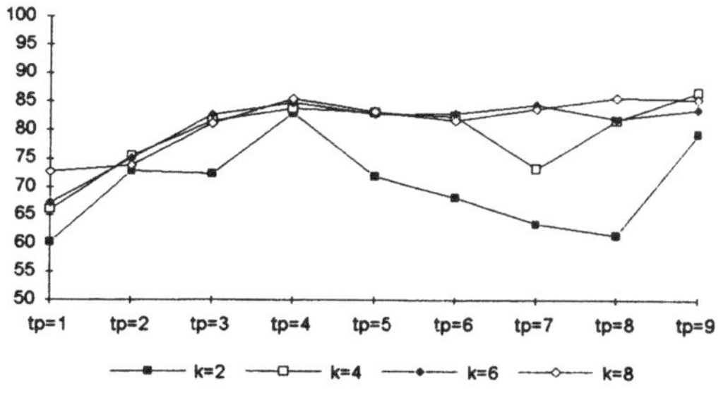

Fig. 5: Same as Fig. 4 but trained with LVQ3.

segment is classified in one step and (2) one wrong classification per trial is fatal whereas with a sliding window wrong classifications can be averaged out by majority voting or activation summation. On the other hand, it stresses the importance of incorporating context information.

The fact that small sliding window sizes give worse performance than bigger ones seems to be due to the differing sizes of the training sets depending on the sliding window size: if the whole time segment is classified in one step then each trial corresponds to exactly one training input vector. If the window size is, e.g., 3 then the number of training input vectors per trial is 7. One might argue that a classification task is easier to learn if more input vectors are available. On the other hand, the more input vectors per trial are available the greater their variance will be and therefore more reference vectors will be necessary to cover the input space.

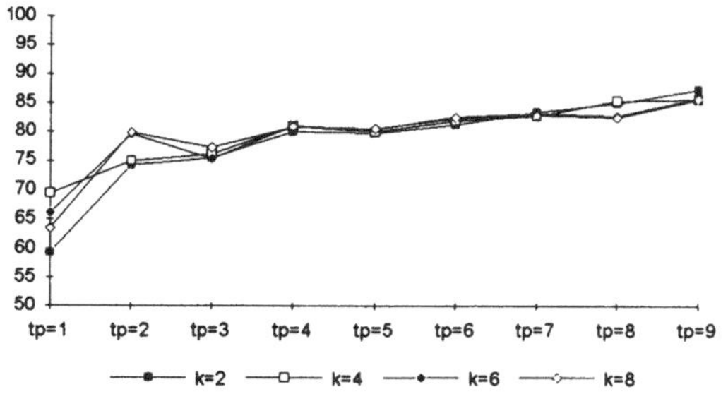

Fig. 6: Same as Fig. 4 but combination by majority voting.

Dependence on the LVQ algorithm

Comparing Figs. 4 and 5, whereby Fig. 5 is the same as Fig. 4 using the LVQ3 algorithm for training, shows a much greater instability of the results of the LVQ3 algorithm. The standard deviations of the results with LVQ3 range between 10 and 30 as opposed to between 1 and 4 with LVQ1 (not shown in the figures). This might be due to the window technique used in LVQ3 which disables some inputs from participating in the update whereas in LVQ1 every input is used. On the other hand the results of LVQ3 are much better in the small sliding window sizes (already 85% for tp=4 opposed to 80% for LVQ1).

Dependence on the classification combination method

No significant difference could be found between the two classification combination methods. Fig. 6 shows the pendant to Fig. 4 with majority voting instead of activation summation. For majority voting the number of classifications to combine must be odd as opposed to the activation summation method which can also cope with even numbers of classifications. The results in Fig. 6 are not as smooth as in Fig. 4 because the number of classifications depending on the sliding window size varies by 2 every two window sizes (9 classifications for window size 1, 7 classifications for window sizes 2 and 3, etc.). Naturally, the results in which the whole time segment is classified in one step (tp=9) are the same for both combination methods because then both can be reduced to standard LVQ.

Conclusion

The Learning Vector Quantizer seems to be a powerful classifier and suited to on-line EEG classification due to a number of facilities:

(1) it is a machine learning method which can learn from very noisy examples;
(2) classification is very quick because only one search through all reference vectors is necessary and their number can be kept very small (4-6 reference

vectors per class already provide satisfactory results);
(3) because the number of values which have to be adjusted is very small even on-line learning would be possible: the LVQ1 algorithm would only introduce the update of the winning reference vector (this possibility has to be investigated more thoroughly for future applications);
(4) LVQ can be step-wise updated with new data without too much disturbing the already learned features.

As for on-line learning, LVQ3 is expected to deliver results which are mostly better than those of LVQ1 but is more sensitive to the order in which the input vectors appear and therefore more unstable. Also, LVQ1 is a quicker method than LVQ3 because it lacks the determination if the input vector falls into a specified window, which suggests better suitability of LVQ1 for on-line learning applications than LVQ3.

Although the sliding window method did not provide much improvement in our application, a comparison between the two methods of classification combination shows preference for the activation summation because of its independence on the number of classifications to combine. Also it offers a very simple way of determining the security with which the decision is formed: the absolute difference between the summed activations of different classes. In our application one decision made the cursor jump right into the target. Using this security measure it would be possible to define a step size for the cursor comparable to that used in [9]: there, certain amplitude ranges were linked to large cursor movements up or down, some to intermediate and small step sizes and another range to 'no movement'. The security measure could be transformed into a step-size by mere multiplication with some yet-to-define constant. Majority Voting only provides the possibility of 'no movement' in the case of an even number of classifications.

References

[1] Chatrian G E, Petersen M C, and Lazarte J A, *'The blocking of the rolandic wicket rhythm and some central changes related to movement'*, Electroenceph. clin. Neurophysiol., 11: 497-510, 1959.

[2] Flotzinger D, *'Neural network-based classification of spatiotemporal EEG-data'*, MSc Thesis, Graz, Technical University 1991.

[3] Jasper H H and Penfield W, *'Electrocorticograms in man: effect of the voluntary movement upon the elctrical activity of the precentral gyrus'*, Arch. Psychiat. Z. Neurol., 183: 163-174, 1949.

[4] Kohonen T, *'Self-organization and associative memory'*, 3rd ed., Springer, Berlin 1988.

[5] McDermott E and Katagiri S, *'LVQ-based shift-tolerant phoneme recognition'*, IEEE Transactions on Signal Processing, 39: 1398-1411, 1991.

[6] Pfurtscheller G, Flotzinger D, Mohl W, and Peltoranta M, *'Prediction of the side of hand movements from single-trial multi-channel EEG-data using neural networks'*, Electroenceph. clin. Neurophysiol., 82: 313-315, 1992.

[7] Pfurtscheller G, Kalcher J, Flotzinger D, *'BCI - a new communication device for handicapped persons'*, Proceedings of the 3rd ICCHP-conference on computer for handicapped persons, 409-415, 1992.

[8] Pfurtscheller G and Berghold A, *'Patterns of cortical activation during planning of voluntary movement'*, Electroenceph. clin. Neurophysiol., 72: 250-258, 1989.

[9] Wolpaw J R, McFarland D, Neat G W, and Forneris C A, *'An EEG-based brain-computer interface for cursor control'*, Electroenceph. clin. Neurophysiol., 78: 252-259, 1991.

A Scalable Neural Architecture Combining Unsupervised and Suggestive Learning

R B Lambert, W P Cockshott and R J Fryer
Dept. Computer Science, University of Strathclyde
Glasgow, G1 1XH. Scotland.
robert@cs.strath.ac.uk

Abstract

Multi-layered perceptrons are, in theory, capable of solving a wide range of problems. However, as the scale of many problems is increased, or requirements change, multi-layered perceptrons fail to learn or become impractical to implement. Self-organizing networks are not so limited by scale, but require a-priori information, typically in the form of preset weights or suitable control parameters to achieve a good categorization of a data set.

Based on research into the behaviour of biological neurons during learning, a new self-organizing neural network has been devised. Moving away from the traditional McCulloch and Pitts model, each neuron stores several independent patterns, each capable of initiating a neuron output. By structuring such neurons into a network, a rapid and equal distribution of data across competitive nodes is possible.

This paper introduces the new network, known as a *Master-Slave* architecture, and learning paradigm. By using competitive and suggestive learning, inputs are distributed across all available classification units, without the need for a-priori knowledge. Two experiments are described, highlighting the potential of the master-slave architecture as a building block for larger networks.

1 Introduction

Of all artificial neural network architectures, the perceptron [1] has proved most popular. With just two hidden layers, designers have a network capable of representing any set of functions [2]. By utilizing the back-propagation training algorithm [3], the network weights necessary to achieve these functions can, in theory, be determined [4]. Yet despite these apparent strengths, the multi-layered perceptron is unsuitable for many practical applications.

Vision is an ill-posed problem and as such has no obvious rule set that can be followed to achieve object recognition. Through a process of learning and constant adaptation, the human brain achieves recognition within a continually changing environment. Given that the brain provides a solution to object recognition, an artificial neural network should, in principal, provide the means to achieving practical machine vision. However, back-propagation has several limitations that restrict its application to small problems with few inputs:

- The back-propagation proof [4] assumes infinitesimally small weight adjustments. This is clearly impractical as it implies an infinite training time. Any practical implementation must use a finite learning rate coefficient, which if too large causes *network paralysis*. As the size of a network or training set increases, the individual adjustments to weights must be reduced to avoid paralysis. Consequently, to achieve convergence, the number of presentations of the training set must be increased, and training time rises.

- By employing gradient descent to reduce a global error, back-propagation can become trapped in poor local minima. As the size of a network and training set increases, the error surface becomes increasingly convoluted, and the probability of becoming trapped in a poor local minima rises.

- The ability to generalize is held up as one of the perceptrons strengths. However, to achieve good generalization, the size of the training set must be significantly greater than the number of network weights [5]. A network with one hidden layer of N nodes applied to an image recognition problem requires approximately pN weights, where p is the number of pixels in the input. Thus, the number of distinct images m in the training set must satisfy $m \gg pN$.

- Generalization cannot (for any realistic problem)

give 100% correct classification for all possible inputs. For instance, a training set may not be representative or complete. In such cases re-training with new inputs may be appropriate. Unlike biological networks, back-propagation requires that training, even with a single new input, involve the re-presentation of the entire training set. Given the storage requirements and training time required, real-time correction of misclassified inputs is impractical.

Many additions to back-propagation have been proposed to reduce training time, reduce the probability of becoming trapped in poor local minima, and to improve generalization. However, in all such cases, the limitations have been shifted and not removed. The recognition of complex objects is still well beyond the capabilities of the perceptron network.

2 Competitive Learning

Given that back-propagation cannot train networks with many inputs, various forms of pre-processing are frequently applied to reduce the information content of a training set. However, pre-processing is typically highly problem dependent.

An alternative, first explored by Hecht-Nielsen [6], is to use a self-organizing network to automatically pre-process and reduce the information content of a training set. Using one or more competitive networks [7], patterns, regularities and correlations within the training set are identified and categorized.

Functionally, a competitive architecture has two layers. An input layer of *matching* nodes each store an adaptive pattern as a weight vector. On presentation of an input to the network, each node calculates the similarity between the input and its weight vector. A second layer of nodes, each receiving an input from every matching node, identifies a single category into which the network input best fits. This network is illustrated in figure 1.

Each matching node has a modifiable weight vector $\mathbf{w}_i$, $1 \leq i \leq n$. To train a competitive network, the weight vector of at least one matching node is updated such that the similarity between the input and weight vectors is increased. The simplest training algorithm, known as *winner-take-all* learning [7], updates the weight vector of the matching node with the best similarity. Thus, a training set $X = \{\mathbf{x}_1, \mathbf{x}_2, \ldots, \mathbf{x}_p\}$, where $\mathbf{x}_q \in X$, is a unit vector, is distributed across n categorization nodes as follows:

1. Randomize all network weights such that $\mathbf{w}_i$ is a unit length vector with random orientation.

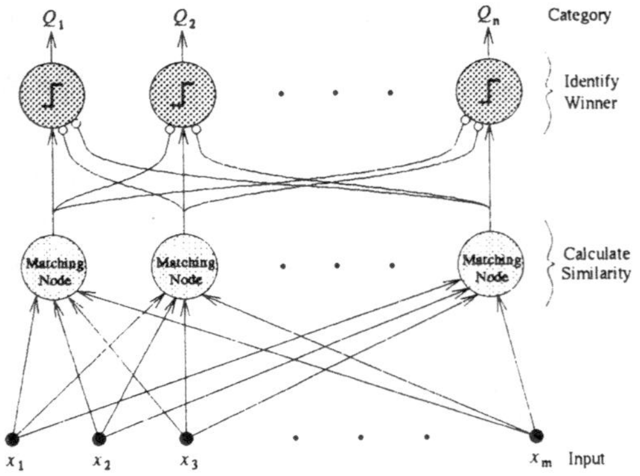

Figure 1: Competitive Network

2. Present a vector $\mathbf{x}_q$ to the network, and calculate similarity S_i for each matching node;

$$S_i = \mathbf{w}_i . \mathbf{x}_q \tag{1}$$

3. Identify a winning category k as the node with $S_k \geq S_i, \forall i$.

4. Update weight vector of winning node such that;

$$\Delta \mathbf{w}_k = \eta[\mathbf{x}_q - \mathbf{w}_k(t)] \tag{2}$$

   where $0 < \eta \leq 1$.

5. Repeat from step 2 until network weights stabilize.

Winner-take-all learning suffers from:

- Instability due to the oscillation of weight vectors between training vectors. Nodes may fail to group a single set of similar inputs, oscillating between two or more unrelated groups. Reducing the learning rate η with time will halt the oscillations, but can result in poor categorization and long training times.

- Unequal distribution of the training set will occur if the initial distribution of the weight vectors does not match the distribution of the training vectors. By setting the initial weights equal to representative members of the training set, improved distribution can be achieved. However, this requires knowledge of the training set.

- Discontinuity between classifications of similar inputs is unavoidable with the winner-take-all learning rule. Small variations to the network input can result in sudden, dramatic changes to the network output. The network is therefore difficult to interpret and susceptible to noise.

Discontinuity between similar inputs can be overcome through feature mapping. By constructing a topological map [8], each categorization node is assigned a fixed position in an h-dimensional geometric space. The difference between any two inputs $\mathbf{x}_a \in X$ and $\mathbf{x}_b \in X$, corresponds to the distance between the nodes used to categorize $\mathbf{x}_a$ and $\mathbf{x}_b$.

Once a winning node k has been identified for an input $\mathbf{x}_q$, *all* network weights are updated such that;

$$\Delta \mathbf{w}_i = \eta(t)\Lambda(d_{k,i}, t)[\mathbf{x}_q - \mathbf{w}_i], \forall i \quad (3)$$

where $d_{k,i}$ is the distance between nodes k and i, and Λ is a scaling function such that nodes close to the winner have their weights modified to a greater degree than those further away. Typically;

$$\Lambda(d_{k,i}, t) = e^{-d_{k,i}/2[\sigma(t)]^2} \quad (4)$$

where $\sigma(t)$ is a monotonically decreasing width function with time. The functions $\eta(t)$ and $\sigma(t)$ are typically started at 1, and fall to zero with time.

While feature mapping virtually eliminates discontinuities, it is at the expense of training time and network size. Many competitive nodes are required, with only a fraction used to categorize a training set. As with winner-take-all learning, a good distribution of the training set is dependent on the initial pre-setting of the network weights.

3 Pre-synaptic Learning

Both the perceptron and competitive network nodes are based on the McCulloch and Pitts model [9] of neuron function. All inputs are assumed to contribute to the activation of the node such that the output is a function of the sum of the weighted inputs.

Studies on the changes during associative learning in the rabbit hippocampus [10, 11] have suggested an alternative neural structure and learning paradigm [12]. Rather than contribute to a global activation, connections from other neurons cluster onto dendritic *patches*, each with its own local activation, and each capable of independently firing the cell.

Each patch has a local activation $\mathcal{A}$ given by the sum of the weighted inputs to that patch. To fire the cell, the local activation must exceed a patch threshold;

$$\xi = (\xi_{\max} - \xi_{\min})e^{-\eta t} + \xi_{\min} \quad (5)$$

where η is a decay constant, $\xi_{\max}$ and $\xi_{\min}$ are the upper and lower limits of the patch threshold, and t is the time since the recovery period following the last initiation of a firing sequence by the patch.

Given that $\xi_{\min} < \mathcal{A} < \xi_{\max}$, the probability of a patch p initiating a firing sequence during a time Δt is:

$$P_p = \frac{\Delta t}{r_p - \frac{1}{\eta}\ln(\mathcal{A}_p - \xi_{\min}) + k} \quad (6)$$

where $r_p \geq \Delta t$ is the patch recovery time, and $k = \frac{1}{\eta}\ln(\xi_{\max} - \xi_{\min})$.

Given m patches, the probability that the neuron body will receive a pulse from any one of its patches during the time Δt is,

$$P = 1 - \prod_{j=1}^{m}(1 - P_j) \quad (7)$$

If the neuron has a recovery time r_c following firing, the firing rate of the cell is,

$$f = \frac{1}{r_c + \Delta t/P} \text{ Hz} \quad (8)$$

Based on Hebbian learning [13], the weights of a perceptron and competitive node are updated according to node state. Given a target output for each node, the difference between that target and actual output is reduced. Note that the identification of a single winning node in competitive learning provides a target output, one or zero, for each network node.

By studying the changes to cells during *Pavlovian conditioning*, learning was found to be independent of cell state. Each patch has one or more unconditioned stimuli (UCS) inputs and several conditioned stimuli (CS) inputs. The weights of the UCS inputs are fixed, while those of the CS inputs are variable. When a UCS input becomes active, all CS inputs simultaneously active *within the patch* are reinforced. In this way unconditioned and conditioned events are associated. Once an association is established, the CS connection *stiffens*, allowing new associations to be made.

4 Network Architecture

Using the idea of dendritic patches, a new neural network architecture has been devised which gives a good distribution of a training set across all available categorization nodes, without discontinuity.

Overview

Designed as a competitive network, each matching node in figure 1 is replaced with a *master* and *slave* node as illustrated in figure 2.

Each master has a single slave, responsible for storing and responding to network inputs. The slave can

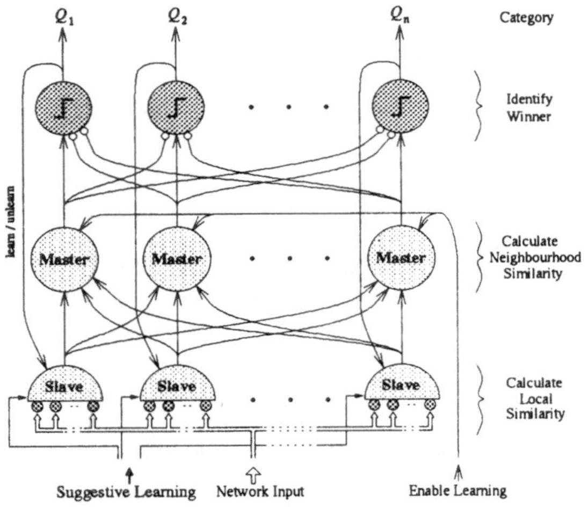

Figure 2: Network Architecture

store a number of distinct patterns, where each pattern corresponds to the weight vector of a dendritic *patch*, fully connected to the network input. On presentation of an input, each patch generates a measure of similarity between the input and its weight vector. The maximum similarity is identified, and becomes the slave output.

To overcome discontinuity between categorizations, a neighbourhood topology is assumed, with each master-slave pair occupying a unique position in geometric space. Each master receives an attenuated similarity measure from every slave, where the attenuation is proportional to the distance between the transmitting and receiving master-slave pairs. These inputs are added to obtain a *neighbourhood* strength for each master. Note that the input from a master's own slave is not attenuated.

From the neighbourhood strength, a master strength is evaluated. If the input is to be categorized, but not learnt, the master strength becomes equal to the neighbourhood strength. However, if the input is to be learnt, the master strength is the sum of the neighbourhood strength and a master activation. The master activation is based on learning history such that each master attempts to learns an equal portion of the training set.

Finally, a single layer of categorization nodes identify the master with the maximum strength. If the input is to be learnt, the winning master's slave will be instructed to learn, while all slaves will be instructed to unlearn.

Slave Node

Given m patches on each slave, the weights on the jth patch of the ith slave, where $1 \leq i \leq n$ and $1 \leq j \leq m$, are given by the unit weight vector $\mathbf{w}_{i,j}$. If a unit vector $\mathbf{x}_q$ from the training set $X = \{\mathbf{x}_1, \mathbf{x}_2, \ldots, \mathbf{x}_p\}$, is presented, each patch calculates a patch similarity,

$$\mathcal{P}_{i,j} = \mathbf{x}_q . \mathbf{w}_{i,j} \tag{9}$$

In addition to the network input, each patch has a suggestive learning input for correction and stabilization of the network categorizations. If the slave's suggestive learning input is asserted high or low, the slave will generate a maximum or minimum output respectively. If the suggestive learning input is unasserted, the slave output will be equal to the maximum patch similarity. The slave output $\mathcal{S}_i$, or local similarity, is summarized in table 1.

Suggestive Learning	Slave Output ($\mathcal{S}_i$)
Asserted High	1
Asserted Low	-1
Un-asserted	$\mathcal{P}_{i,k}$, $\mathcal{P}_{i,k} \geq \mathcal{P}_{i,j}, \forall j$

Table 1: Slave Output

Master Node

Each master-slave pair has a unique position in h-dimensional space such that they are evenly distributed across the surface of an h-dimensional sphere, where $2 \leq h \leq n$.

Every master calculates a neighbourhood similarity $\mathcal{N}_i$ given by:

$$\mathcal{N}_i = \sum_{j=1}^{n} \mathcal{S}_j e^{-d_{i,j}/\rho} \tag{10}$$

where $d_{i,j}$ is the distance across the surface of the sphere between masters i and j, and ρ is a constant controlling the spread of influence.

A master operates in one of two modes according to the *enable learning* input (common to all masters). If asserted low (no learning), the master will generate a master strength $\mathcal{M}_i = \mathcal{N}_i$. However, if asserted high (learn pattern presented to network), the master will add to the neighbourhood strength an internal activation $\mathcal{A}_i$, where $\mathcal{A}_i$ increases if $\mathcal{M}_i$ is less that the maximum master output, and falls to zero if $\mathcal{M}_i$ is the maximum master output. The master output is given in table 2.

Enable Learning	Master Output ($\mathcal{M}_i$)
Asserted High	$\mathcal{A}_i + \mathcal{N}_i$
Asserted Low	$\mathcal{N}_i$

Table 2: Master Output

Finally the winner, and hence the category into which the input best fits is determined by:

$$Q_i = \begin{cases} 1 & \text{if } \mathcal{M}_i \geq \mathcal{M}_j, \forall j \\ 0 & \text{otherwise} \end{cases} \tag{11}$$

Network Learning

A patch can be told to *learn* or *unlearn*. If the network input $\mathbf{x}_q$ is to be learnt, the weight vector is modified such that:

$$\Delta \mathbf{w}_{i,j} = e^{-\omega_{i,j}}[\mathbf{x}_q - \mathbf{w}_{i,j}] \tag{12}$$

where $\omega_{i,j}$ is the patch *stiffness* or resistance to change. The weight is not modified by an unlearn instruction.

The patch stiffness is modified by both the *learn* and *unlearn* instructions. When told to learn, a patch increases its stiffness such that future learning has less influence. If told to unlearn, the stiffness of the patch is reduced to allow greater modification of the patch weight vector. Given that $\omega_{i,j}$ starts at zero;

$$\Delta \omega_{i,j} = \begin{cases} \alpha n & \text{if } \textit{learn} \\ -\alpha & \text{if } \textit{unlearn} \text{ and } \omega_{i,j} > 0 \\ 0 & \text{otherwise} \end{cases} \tag{13}$$

where α controls the rate at which the patch stiffens.

Provided *enable learning* is asserted high, each slave will receive a a *learn* or *unlearn* signal according to the network output as shown in table 3.

Enable Learning	Output (Q_i)	Training
Asserted High	1	Learn
Asserted High	0	Unlearn
Asserted Low	Any	None

Table 3: Training Signal for Slave i

When instructed to *learn*, a slave will instruct the patch with the minimum stiffness to learn the network input. If the slave is told to *unlearn*, the patch which generated the slave output $\mathcal{S}_i$ (the patch with the weight vector most similar to the network input), is instructed to *unlearn*.

Once the slaves have received their training instructions, each master modifies its internal activation. If for a network input $\mathbf{x}_q$, $Q_u = 1$, $1 \leq u \leq n$, then:

$$\Delta \mathcal{A}_i = \begin{cases} -\mathcal{A}_i & \text{if } i = u \\ e^{-\psi_i}[\mathcal{M}_u - \mathcal{M}_i] & \text{otherwise} \end{cases} \tag{14}$$

where ψ_i determines the master's *maturity*. Following the adjustment of $\mathcal{A}_i$, ψ_i is modified such that;

$$\Delta \psi_i = \begin{cases} \beta n & \text{if } i = u \\ -\beta & \text{if } i \neq u \text{ and } \psi_i > 0 \\ 0 & \text{otherwise} \end{cases} \tag{15}$$

where β controls the rate at which a master matures. Every master starts with its maturity equal to zero. This causes each master to learn an equal portion of the training set, regardless of similarities. As learning proceeds and the masters mature, the influence of similarity in determining the winning master-slave pair for each network input increases.

The activation and maturity of the masters are not modified if *enable learning* is asserted low.

Network Operation

Given a training set $X = \{\mathbf{x}_1, \mathbf{x}_2, \ldots, \mathbf{x}_p\}$, the similarity $S^X_{a,b}$ between the unit vectors $\mathbf{x}_a \in X$ and $\mathbf{x}_b \in X$ is, $S^X_{a\,b} = \mathbf{x}_a.\mathbf{x}_b$.

The categorization of X by the network is determined by the similarities between the training vectors and the number of vectors p with respect to the number of available categories n. Given $p \gg n$, each category following training responds to approximately p/n members of X.

If a vector $\mathbf{x}_c \in X$ is chosen at random, an ordered list Y can be derived from X such that $\mathbf{y}_1 = \mathbf{x}_c$, and $S^Y_{1,q} \geq S^Y_{1,q+1}$, $1 < q < p$. If $\mathbf{y}_1$ is assigned to category k, $1 \leq k \leq n$, vectors $\mathbf{y}_2$ to $\mathbf{y}_{p/n}$ have a high probability of being assigned to one of the categories $(k-1) \bmod n$, k, or $(k+1) \bmod n$.

5 Analysis

Two experiments are described in this section, illustrating the categorization and recognition performance of the master-slave architecture.

Image Categorization

The ability to categorize a training set such that common features are identified is one of the key attributes of a self-organizing network. Given no a-priori information (random weights), a master-slave and a winner-take-all network were tested on their ability to categorize a set of noisy images.

A data set consisting of 26 upper-case fixed font roman letters subjected to noise was created, and used to train 26 master-slave pairs and a 26 node winner-take-all network. To identify the categories formed, an output layer of 26 out-star nodes [14] was utilized.

Each network was trained on an equal number of images with a signal to noise ratio of 1:1.0 to 1:1.7 in steps of 0.05. Figure 3 shows some examples of images with a signal to noise ratio of 1:1.0 to 1:1.5.

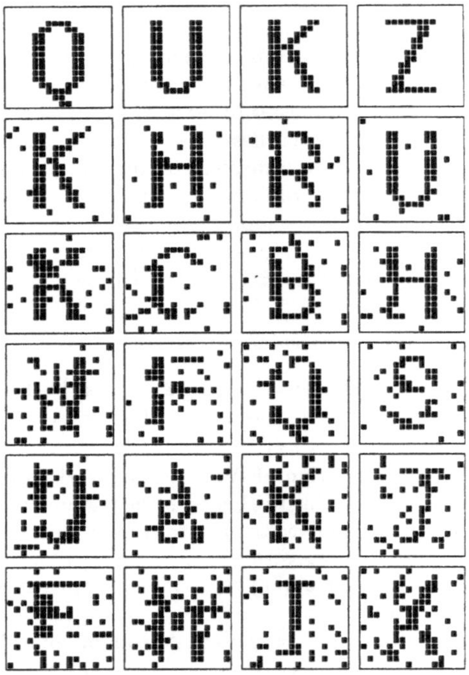

Figure 3: Training data for categorization

To achieve maturity, the master-slave pairs required 10^3 image presentations. The results for both networks are shown in figure 4.

The number of patches required by each slave was also investigated. The categorization results shown in figure 4 are for master-slave pairs each with six patches. Increasing the number of patches per slave beyond six, gave no improvement in categorization, but increased training time. For more general categorization tasks attempted to date, no more than eight patches per slave have been required.

Pattern Recognition

A single layer of master-slave pairs is limited in what it can represent. For complex problems, several layers

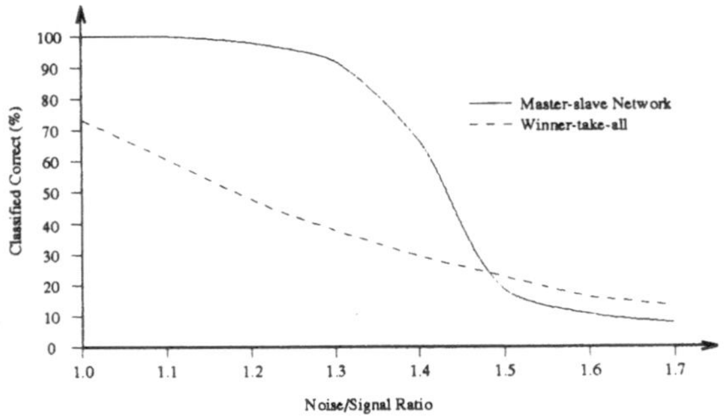

Figure 4: Categorization Performance

are necessary if the number of patches per slave is to be kept low. To recognize upper-case roman letters of various fonts, presented at different positions, and subject to noise, a two layer network was constructed.

An input layer of nine sub-networks of master-slave pairs view overlapping windows, each 8 × 8 pixels, within the input array. The categorizations from each sub-network feed into a second layer with 26 master-slave pairs. See figure 5.

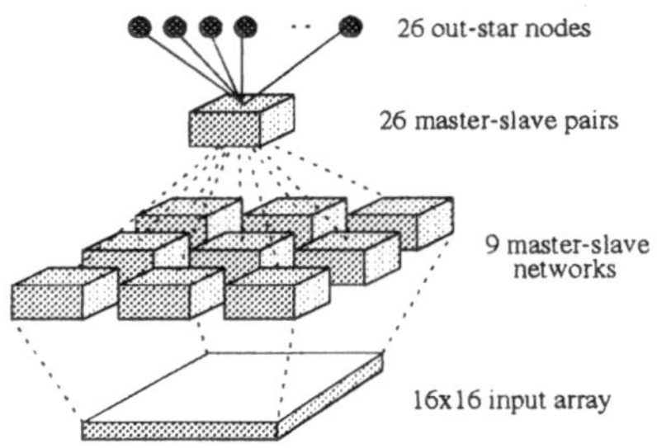

Figure 5: Two layer network

The training data was presented, and the network allowed to self-organize. Once stable (masters are mature), an out-star network was used to identify the distribution of the categories. To correct misclassifications, a teaching signal was introduced by asserting the suggestive learning inputs to the master-slave pairs in the output layer, either high or low as appropriate. Examples of data used to train the network are shown in figure 6.

The number of master-slave pairs used within each first layer sub-network was varied and an acceptable level of recognition achieved with a minimum of 12 master-slave pairs per sub-network.

A multi-layer perceptron with one hidden layer using back-propagation with momentum was trained on the same training set. The results for recognition of the training set are shown in figure 7. Generalization

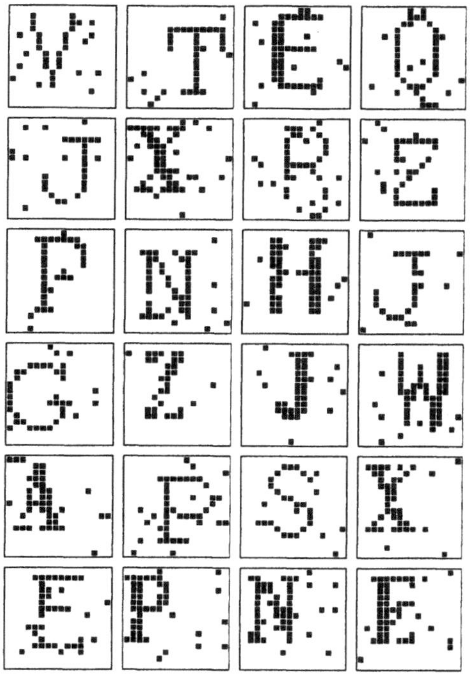

Figure 6: Character recognition training data

results for images not included in the training set are shown in figure 8.

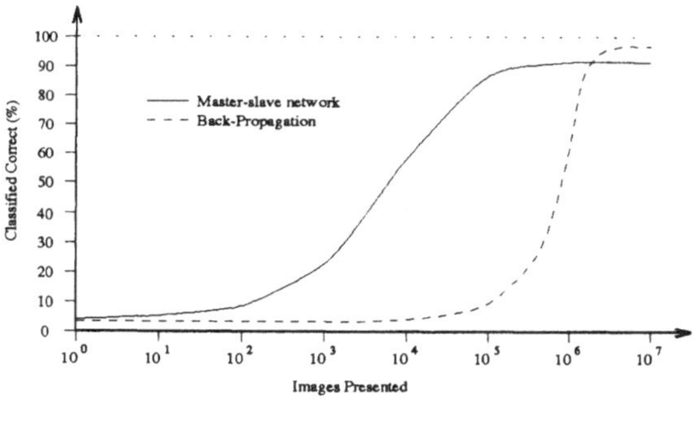

Figure 7: Recognition Performance

6 Conclusion

A self-organizing network capable of rapidly distributing a data set across all categorization units, without the need for a-priori knowledge, has been developed. By allowing each slave to store several patterns, each master-slave pair can learn an equal portion of a training set, ensuring a good utilization

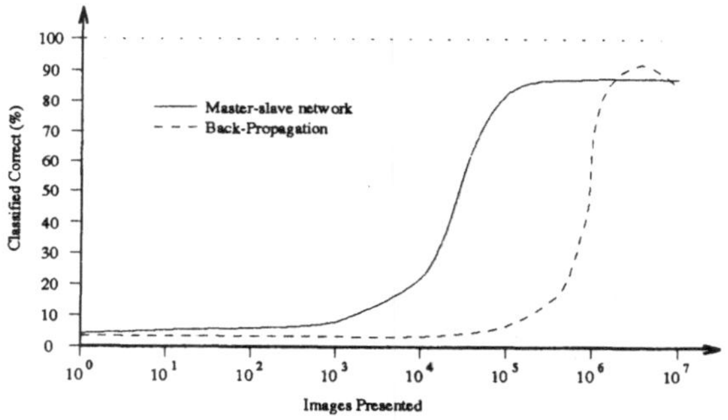

Figure 8: Generalization Performance

of network resources.

With suggestive learning, the response to sub-sets of the training set can be enhanced, and learning quickly stabilized.

While recognition and generalization performance for back-propagation is generally higher than that for a master-slave network, the number of pattern presentation required by the master-slave pairs to achieve a high level of recognition has been found to be at least an order of magnitude less than the number required by back-propagation.

Once the master-slave architecture has self-organized and settled, further training has no effect on generalization performance. This contrasts with the problem of *over-training* in back-propagation. Additionally, the ideal number of hidden units when using back-propagation is highly problem dependent, often requiring long design times while many network variations are analysed. Increasing the number master-slave pairs, and the number of patches per slave beyond the ideal will increase training time and memory requirements, but does not adversely affect the recognition and generalization performance.

Further experimentation and analysis is required to determine the potential and limitations of the master-slave architecture. However the results to date suggest that large networks can be constructed from many sub-networks of master-slave pairs.

References

[1] Rosenblatt F, '*The Perceptron: a probabilistic model for information storage and organization in the brain.*', Psychological Review, 65, 1958.

[2] Cybenko G, '*Approximation by Superposition of a Sigmoidal Function*', Mathematics of Control, Signals, and Systems, 2:303–314,1989.

[3] Rumelhart D E, Hinton G E and Williams R J, '*Learning representations by back-propagating errors*', Nature, 323:533–536, 1986.

[4] Rumelhart D E, Hinton G E and Williams R J, '*Learning internal representations by error propagation*', Parallel Distributed Processing, Vol. 1, Cambridge, MA: MIT Press, 1986.

[5] Baum E B and Haussler, '*What size of net gives valid generalization?*', Neural Computation, 1:151–160, 1989.

[6] Hecht-Nielsen R, '*Counter-propagation Networks*', Applied Optics, 26:4979–4984, 1987.

[7] Grossberg S, '*Adaptive Pattern Classification and Universal Recoding: I. Parallel Development and Coding of Neural Feature Detectors*', Biological Cybernetics, 23:121–134, 1976.

[8] Kohonen T, '*Self-Organized Formation of Topologically Correct Feature Maps*', Biological Cybernetics, 43:59–69, 1982.

[9] McCulloch W S and Pitts W, '*A Logical Calculus of Ideas Immanent in Nervous Activity*', Bulletin of Mathematical Biophysics, 5:115–133, 1943.

[10] Alkon D L, '*Memory Storage and Neural Systems*' Scientific American, July:26–34, 1989.

[11] Olds J L, Anderson M L, McPhie D L, Staten L D and Alkon D L, '*Imaging memory-specific changes in the distribution of protein kinase C within the hippocampus*', Science 245:866–869, 1989.

[12] Alkon D L, Blackwell K T, Barbour G S, Rigler A K and Vogl T P, '*Pattern-Recognition by an Artificial Network Derived from Biologic Neuronal Systems*', Biological Cybernetics, 62:363–376, 1990.

[13] Hebb D O, '*The Organization of Behaviour*', Wiley, 1949.

[14] Grossberg S, '*Some nonlinear networks capable of learning a spatial pattern of arbitrary complexity*', Proceeding of the National Academy of Sciences USA, 59:368–372, 1968.

STABILITY ANALYSIS OF THE SEPARATION OF SOURCES ALGORITHM : APPLICATION TO AN ANALOGUE HARDWARE.

Didier ACHVAR

Laboratoire d'Electronique d'Automatique et d'Informatique
Centre des Systèmes de Production
Ecole Nationale Supérieure des Techniques Industrielles et des Mines d'Alès
6, Avenue de Clavières
30319 Alès FRANCE

Centre d'Electronique de Montpellier
Université des Sciences et Techniques du Languedoc
2, Place Eugène Bataillon
34000 Montpellier FRANCE

Abstract.
We present in our paper a stability analysis of the two-neuron Hérault-Jutten network for separation of two sources from a linear and instantaneous mixture. The non-linear functions used by the learning rule are chosen to be in the form $f(x)=\sinh(x/\alpha)$ and $g(x)=\tanh(x/\beta)$, as we can easily implement them on analogue hardware. Based on simulation results and theoretical considerations, the influence of the non-linear functions and the statistical nature of the source signals on the stability of the algorithm, are pointed out. In order to determine the properties and the limitations of our analogue hardware, the behavior of the algorithm is derived finally for $g(x)=\text{signe}(x)$, as it is actually implemented in our circuit.

Keywords. *Separation of sources, analogue hardware, neural network, non-linear functions, stability, adaptive filter, signal processing.*

Introduction

Consider n independent sources $X_1(t),...,X_n(t)$ observable through n linear mixtures $E_1(t),...,E_n(t)$ as obtained by n linear sensors. $E_i(t)$ are linear combinations of sources through an unknown mixing matrix A. The problem is to restore all sources by only the knowledge of their interference given at the sensor outputs. A solution to this problem was first addressed by Hérault and Jutten [1]. They proposed a completely connected network of n neurons with a permenent unsupervised learning rule. The extensions of their work concentrate mainly on the generalization of the mixing matrix model to convolutive mixtures [2][3][4]. However, for the sake of simplicity, our present study is limited to the separation of two sources from an instantaneous mixture. This paper aims to investigate some convergence properties of the Hérault and Jutten algorithm that we have met when implementing the network on hardware. The first section is devoted to a problem statement. In the second section, building blocks of the circuit realized in our laboratory are discribed. In the final section we present a mathematical analysis of the stability of the learning rule for the particular case of ternary sources. Results are illustrated by means of simulations. We proceede finally to a generalization of our analysis to other kinds of sources by the help of numerical considerations.

1. Problem statement

The most expressive illustration of the source separation problem is the "Cocktail-Party" processing [2][3][4]: Sources are speech signals uttered by n invited persons, and sensors are n microphones. Here, the source separation consists of discriminating each discours from uproars given by microphones (Fig.1). In this context, for two primary sources, and in simplified cases, the convolutive mixture can be modelled as follows:

$$E_i(t) = a_{ii}.X_i(t-\theta_{ii}) + a_{ij}.X_j(t-\theta_{ij}) \qquad \forall\ i\neq j$$

Where the a_{ij}'s are attenuation factors, and the θ_{ij}'s propagation delays of sources.Then we have in vectorial notation :

$$\vec{E}(t) = A(t)*\vec{X}(t) \quad \text{with} \quad A(t) = \begin{pmatrix} A_{11}(t) & A_{12}(t) \\ A_{21}(t) & A_{22}(t) \end{pmatrix}$$

$$\text{and} \quad A_{ij}(t)=a_{ij}.\delta(t-\theta_{ij}) \qquad (1)$$

Where $*$, represents the convolution product, and $\delta(t)$ the Dirac distribution. Then using the Laplace transform equations (1) become :

$\vec{\mathcal{E}}(p) = \mathcal{A}(p).\vec{\mathcal{X}}(p)$ with $\mathcal{A}_{ij}(p) = a_{ij}.\exp(-p.\theta_{ij})$

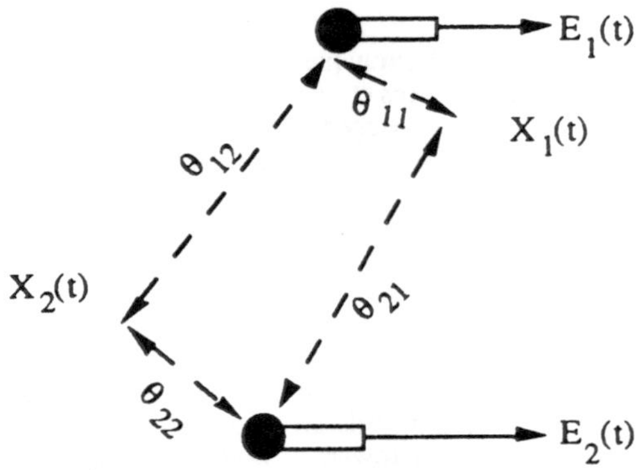

Figure 1 : The "Cocktail-Party" problem.

1.1. Theoretical solution

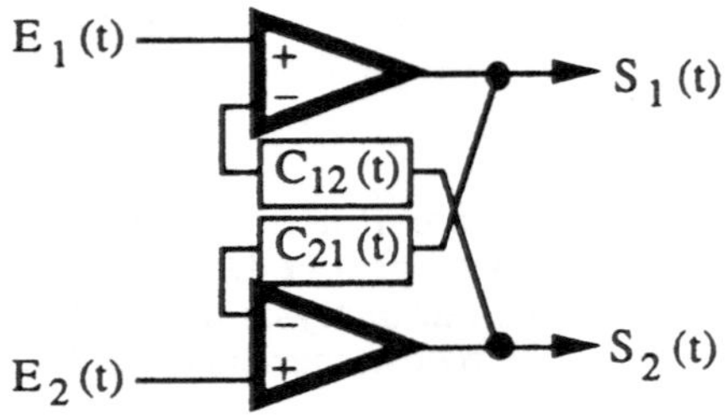

Figure 2 : The two-neurons network .

The network proposed by Hérault & Jutten (denoted H&J) [2][3], presented in Fig. 2, is formed by two operators connected to each other by two synaptic filters with $C_{ij}(t)$ as their impulse response. Hence the systems outputs are given by:

$$S_i(t) = E_i(t) - C_{ij}(t)*S_j(t) \quad \forall\ i \neq j \qquad (2)$$

Then the application of a Laplace transformation on (2) leads to

$$\mathcal{S}_i(p) = \mathcal{E}_i(p) - \mathcal{C}_{ij}(p).\mathcal{S}_j(p)$$

We can write therefore by replacing $\mathcal{E}_i(p)$ from (1) in vectorial form the following relation between sources and the network output :

$$\vec{\mathcal{S}}(p) = \mathcal{F}(p).\vec{\mathcal{X}}(p) \qquad (3)$$

with

$$\mathcal{F}(p) = \frac{1}{1-\mathcal{C}_{12}\mathcal{C}_{21}} \begin{pmatrix} \mathcal{A}_{11} - \mathcal{C}_{12}.\mathcal{A}_{21} & \mathcal{A}_{12} - \mathcal{C}_{12}.\mathcal{A}_{22} \\ \mathcal{A}_{21} - \mathcal{C}_{21}.\mathcal{A}_{11} & \mathcal{A}_{22} - \mathcal{C}_{21}.\mathcal{A}_{12} \end{pmatrix}$$

The synaptic filters $\mathcal{C}_{ij}(p)$ realising the source separation task are those that may cancel one of the two diagonal terms in the matrix $\mathcal{F}(p)$. Two different pairs of filters denoted P and P*, lead consequently toward the two following solutions for the outputs :

if $\theta_{11} \leq \theta_{21}$ and $\theta_{22} \leq \theta_{12}$ (4a) then

$$P = \begin{pmatrix} \mathcal{C}_{12}(p) = \dfrac{a_{12}}{a_{22}}.\exp(-p.(\theta_{12}-\theta_{22})) \\ \mathcal{C}_{21}(p) = \dfrac{a_{21}}{a_{11}}.\exp(-p.(\theta_{21}-\theta_{11})) \end{pmatrix} \qquad (4b)$$

$S_1(t) = a_{11}.X_1(t-\theta_{11})$ and $S_1(t) = a_{22}.X_2(t-\theta_{22})$

if $\theta_{11} \geq \theta_{21}$ and $\theta_{22} \geq \theta_{12}$ (5a) then

$$P^* = \begin{pmatrix} \mathcal{C}_{12}(p) = \dfrac{a_{11}}{a_{21}}.\exp(-p.(\theta_{11}-\theta_{21})) \\ \mathcal{C}_{21}(p) = \dfrac{a_{22}}{a_{12}}.\exp(-p.(\theta_{22}-\theta_{12})) \end{pmatrix} \qquad (5b)$$

$S_1(t) = a_{12}.X_2(t-\theta_{12})$ and $S_1(t) = a_{21}.X_1(t-\theta_{21})$

Hereafter, we use a more compact notation to appoint each couple of solutions P and P* :

$\mathcal{C}_{ij}(p) = c_{ij}\,.\,\exp(-p.\tau_{ij})$ with

$$c_{ij} = \frac{a_{i\sigma(j)}}{a_{j\sigma(j)}} \ ; \quad \tau_{ij} = \theta_{i\sigma(j)} - \theta_{j\sigma(j)} \qquad (6a)$$

Then each output can be written as :

$$S_i(t) = a_{i\sigma(i)}\,.\,X_{\sigma(i)}(t-\theta_{i\sigma(i)}) \qquad (6b)$$

with $\sigma(j)=j$ for P and $\sigma(j)=j$ for P*.

However, for the sake of stability, poles of $\mathcal{F}(p)$ must have negative real parts [4]. Hence, the network is stable if :

$$c_{12}.c_{21} < 1 \qquad (7)$$

It is easy to check that conditions (4a) and (5a), derived from the causality principle [4], and the stability condition (7) cannot be satisfied

simultaneously for P and P*. Therefore, only one of the two previous solutions is possible.
Note that (4a) and (5a) depend on the sources-sensors distances. So we can also verify that the network has to restore the nearest source to the ith sensor on S_i, and inhibit the furthest one in effect.
Theoretically, the architecture allows us therefore to discriminate the unknown sources, and recent works provide the resolution of that simplified case of the "Cocktail-Party" very easily [2][3][4][5]. Nevertheless, for simplicity, we will keep our present study, from now on, limited to instantaneous mixtures (i.e. $\theta_{ij} \approx 0 \ \forall$ i,j). More general discussions are postponed to our future work.

1.2. Adaptive learning rule

According to the inhibition idea related before, the first adaptation rule used by H&J, was based on the minimisation of the variance of the network outputs [1]. Then, the application of a gradient method to the mean square terms $E((S_i^2-E^2(S_i))$ yields the following adaptive rule [1] :

$$\frac{dc_{ij}}{dt} = \mu.f(s_i).g(s_j) \qquad \forall\ i \neq j \qquad (8)$$

where μ is a positive gain less than one and s_i the zero-mean value of S_i. f and g, are two non-linear odd functions.
Let $\phi_{ij}(c_{12},c_{21})$, be the mean value of the gradient, the equilibrium points of (8) are the solution of the following system :

$$\phi_{ij}(c_{12},c_{21}) = E[f(s_i).g(s_j)] = 0 \ \forall\ i \neq j \qquad (9)$$

In fact, high-order moments brought in $\phi_{ij}(c_{12},c_{21})$ by the Taylor expansions of f and g yield ϕ_{ij} to be an independence measurement of the network outputs [1]. These moments are in the form :

$$E(s_i^{2m+1}.s_j^{2n+1}) \text{ with } m,n \in \mathbb{N}. \qquad (10)$$

So, the convergence of (8) is achieved if all the cross-moments introduced in ϕ_{ij} are equal to zero. Then we can easily confirm that the desired filters assigned by P and P* are obvious solutions of (9). But, it has been proven in [6] and [7] that the stability of these equilibrium points depends on statistics of sources, and (8) may converge to a spurious state. Our present study aims to investigate the influence of f and g on the stability of the desired states P and P*. However, no restriction exists for the functions, so we have chosen them to be attenuant and amplifier non-linear functions as they are suggested in [8]. Actually, our analogue hardware is implemented with an hyperbolic sine for f, and a sign function for g. Then we use for our mathematical analysis, and for our simulations:

$$f(x) = \frac{\sinh(x/\alpha)}{\sinh(1/\alpha)} \text{ and } g(x) = \frac{\tanh(x/\beta)}{\tanh(1/\beta)} \qquad (11)$$

as we can implent them on hardware. The performance of the algorithm will be derived at least for strong non-linearities, i.e. for little values of α and β.

2. The analogue hardware

Several version of analogue implementation of the H&J algorithm exist today. The first one, realized with discrete components, was designed by H&J themselves and performs the separation of two sources from linear and instantaneous mixtures [9]. Later, a CMOS integrated circuit of a six-neuron network was also proposed by Arreguit and Vittoz [10]. The one that we have realized in our laboratory is a simplified imitation of the first one [9]. The simplication is in the choice of the amplifier non-linear function. Finally, replacing g(x) by sign(x) underlines the robustness of the H&J algorithm to be wired. So, only building blocks of our circuit are described here.

2. 1. Computing the neurons outputs

Each neuron function (2) is implemented by the help of an analogue multiplier as it is shown in Fig.3 . This circuit realizes the following operation : O = (I1-I2).(I3-I4)/10 + I5. A high-pass filter at its outputs gives the zero-mean value of the neuron outputs.

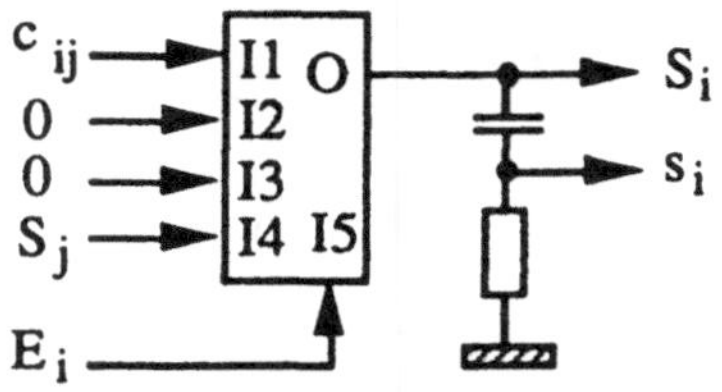

Figure 3 : The neuron circuit.

2.2. Synthesis of the non-linear function f

We have realized this function by improving the natural non-linearity of two diodes wired upside down, shown in Fig.4 . So, for neglected reverse current in each component we have :

$$I(V) = I_S.(\exp(\frac{q.V}{k.T})-1) - I_S.(\exp(-\frac{q.V}{k.T})-1)$$

Where k is the Boltzmann constant, T the absolute temperature, and q the electronic charge. Then the total output current becomes :

$$I(V) = 2.I_S. \sinh(\frac{q.V}{k.T})$$

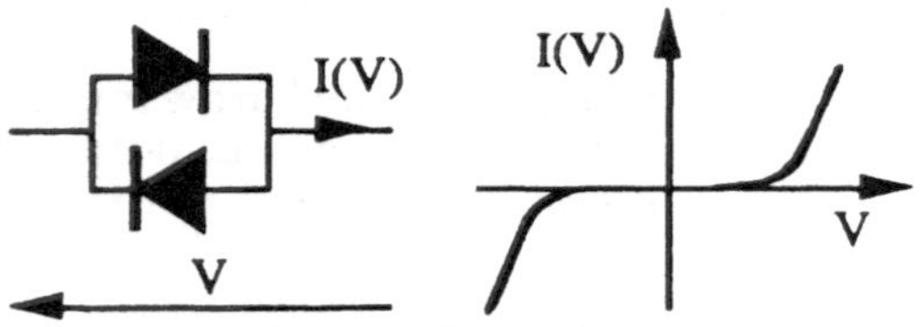

Figure 4 : Synthesis of f

2.3. Synthesis of the non-linear function g

Here, we use the characteristic of a Trigger circuit (Fig.5). Then we have $V_{out}=\text{sign}(V_{in})$.

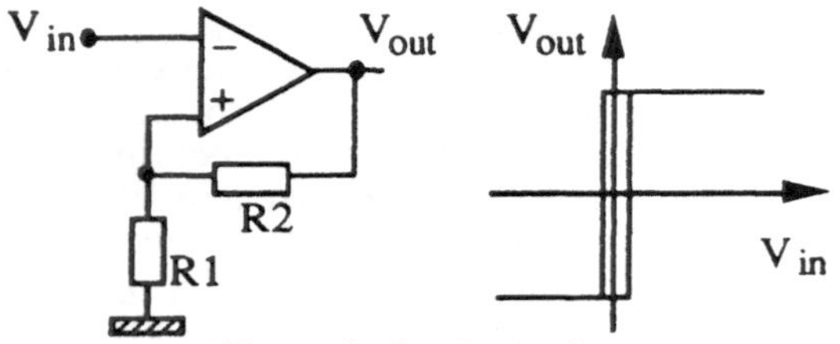

Figure 5 : Synthesis of g

The hysteresis property of that circuit, shaped by the feedback gain is very little, and aims to avoid erratic see-saws that may be produced by the electronic components noise.

2.4. Computing the synaptic weights

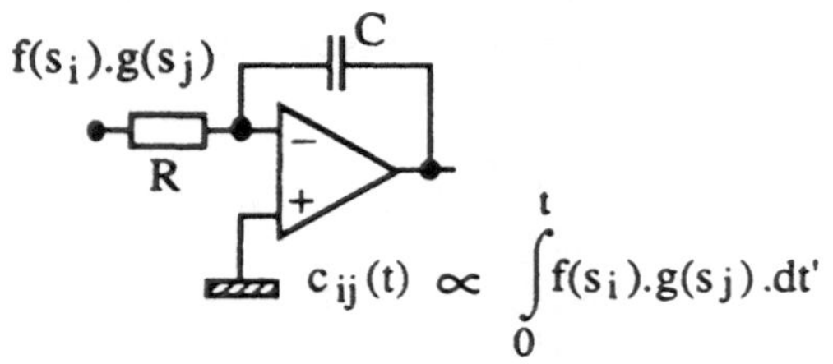

Since g(x) is in binary information, analogue switchs were sufficient to realize the $f(s_i).g(s_j)$ products. Then, according to (8), the synaptic weights are computed by a simple integrator circuit (see figure above).

2.5. Application

The network thus realized in our laboratory can separate sources with frequencies between 2 kHz and 50 kHz, from linear and instantaneous mixtures. The time evolution of the synaptic weights is very fast and the source separation is reached in less then 100 millisecond. Fig.6 shows a real time observation on oscilloscope of the outputs in the case of a mixture of triangle and sine waves.

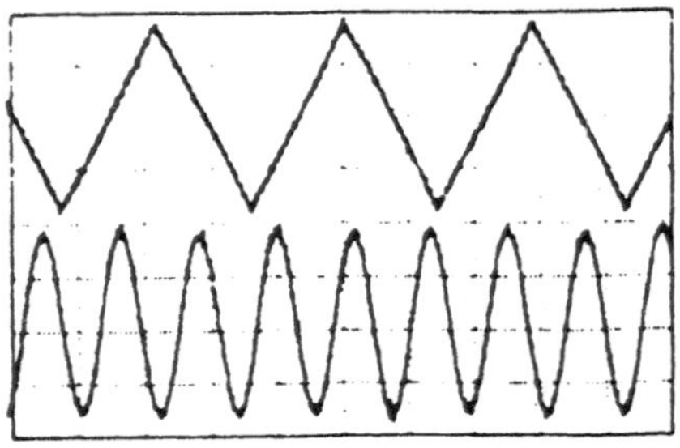
Figure 6 : A real time application on Hardware.

3. Stability analysis

3.1. Mathematical consideration

Consider a small perturbations Δc_{ij} of weights around the immediate neighbourhood of an equilibrium point P or P*. A first order development of (8) leads on average to :

$$\frac{d\,\Delta c_{ij}(t)}{dt} = \frac{\delta \phi_{ij}}{\delta c_{ij}} . \Delta c_{ij} + \frac{\delta \phi_{ij}}{\delta c_{ji}} . \Delta c_{ji} \qquad (12)$$

By differentiating (2) one gets :

$$\frac{\delta s_i}{\delta c_{ij}} = \frac{- s_j}{1-c_{12}.c_{21}} \quad \frac{\delta s_i}{\delta c_{ji}} = \frac{c_{ij}.s_i}{1-c_{12}.c_{21}} \qquad (13)$$

Using (13) and taking into account that around P or P*, all cross-moments like (10) are equal to zero, leads to

$$\frac{\delta \phi_{ij}}{\delta c_{ij}} = \frac{- \mu}{1-c_{12}.c_{21}} . E(s_j.g(s_j).\frac{\delta f(s_i)}{\delta s_i})$$

$$\frac{\delta \phi_{ij}}{\delta c_{ji}} = \frac{- \mu}{1-c_{12}.c_{21}} . E(s_i.f(s_i).\frac{\delta g(s_j)}{\delta s_j})$$

Then (12) can be written in a vectorial form as follows :

$$\frac{d\,\overrightarrow{\Delta c}}{dt} = \frac{-\mu}{1-c_{12}.c_{21}}.J.\overrightarrow{\Delta c} \quad \text{with } \overrightarrow{\Delta c} = \begin{pmatrix} \Delta c_{12} \\ \Delta c_{21} \end{pmatrix},$$

and

$$J_{ii} = E(\, s_j.g(s_j).\frac{\delta\, f(s_i)}{\delta s_i}\,) \,;$$

$$J_{ij} = E(\, s_i.f(s_i).\frac{\delta\, g(s_j)}{\delta s_j}\,) \qquad (14)$$

Thus, the stability of (8) can be examined by considering the sign of the eigenvalues of the matrix **J**. Since $c_{12}.c_{21} < 1$, (condition (7)), and $\mu>0$, the equilibrium point is stable if the eigenvalues of **J** have positive real parts. By noting that J_{ii} and J_{ij} are all positive, the resolution of the characteristic equation of **J**, leads to the following real eigenvalues :

$$\lambda_1 = \frac{J_{11}+J_{22}+\sqrt{(J_{11}-J_{22})^2+4.J_{12}.J_{21}}}{2}$$

$$\lambda_2 = \frac{J_{11}+J_{22}-\sqrt{(J_{11}-J_{22})^2+4.J_{12}.J_{21}}}{2}$$

One can easily verify that λ_1 is always positive, and λ_2 is positive whenever $J_{11}.J_{22}-J_{12}.J_{21} > 0$. The stability condition of the equilibrium points is therefore :

$$\det(\mathbf{J}) > 0 \qquad (15)$$

3.2. Application to ternary sources

Note that near P or P*, i.e. for statistically independent outputs, one can write that :

$$E(\, s_j.g(s_j).\frac{\delta\, f(s_i)}{\delta s_i}\,) = E(\, s_j.g(s_j)\,).E(\frac{\delta\, f(s_i)}{\delta s_i})$$

$$E(\, s_i.f(s_i).\frac{\delta\, g(s_j)}{\delta s_j}\,) = E(\, s_i.f(s_i)\,).E(\frac{\delta\, g(s_j)}{\delta s_j})$$

Now consider two ternary independent sources with probability density functions $f_{X1}(x_1)$ and $f_{X2}(x_2)$, as represented in Fig.7 . By denoting $p_i = f_{Xi}(0)$, replacing outputs from (6b) and non-linear functions from (11) give :

$$E(\, s_j.g(s_j)\,) = \frac{a_{j\sigma(j)}}{g(1)}\,.\,(\,1-p_{\sigma(j)}\,)\,.\,\tanh(\,\frac{a_{j\sigma(j)}}{\beta}\,)$$

$$E(\frac{\delta\, f(s_i)}{\delta s_i}) = \frac{1}{\alpha.f(1)}\,.\,(\,p_{\sigma(i)} + (\,1-p_{\sigma(i)}\,)\,.\,\cosh(\frac{a_{i\sigma(i)}}{\alpha})\,)$$

$$E(\, s_i.f(s_i)\,) = \frac{a_{i\sigma(i)}}{f(1)}\,.\,(\,1-p_{\sigma(i)}\,)\,.\,\sinh(\,\frac{a_{i\sigma(i)}}{\alpha}\,)$$

$$E(\frac{\delta\, g(s_j)}{\delta s_j}) = \frac{1}{\beta.g(1)}\,.\,(\,p_{\sigma(j)} + (\,1-p_{\sigma(j)}\,)\,.\,(1-\tanh^2(\frac{a_{j\sigma(j)}}{\beta}))\,)$$

Then, doing an approximation first for small values of β, and then for small α , the developement of the stability condition (15) shows that the equilibrium point P or P* may be stable if β is superior than a certain threshold value explicited as follows :

$$\beta^2 > \alpha^2\,.\,\frac{p_1.p_2}{(1-p_1).(1-p_2)} = \beta_{th}^2 \qquad (16)$$

Otherwise, the algorithm can possibly converge to a spurious state. This result can be verified by the aid of simulations. For every numerical calculation, the following parameters were kept constant :

- The mixing matrix is $A = \begin{pmatrix} 0,8 & 0,2 \\ 0,4 & 0,6 \end{pmatrix}$

- The non-linear function f is fixed with α=0.2

- The adaptation gain is equal to : μ=0.005 .

According to (6a), the two optimal equilibrium points are therefore :

P (0,33 ; 0,50) and P* (2 ; 3) ,

but P* cannot fulfil the condition (7). So, P only is acceptable. Let, for our first simulations $p_1 = 1/2$ and $p_2 = 1/3$. Then, condition (16) imposes to β to be superior than the following threshold:

$\beta_{th} = 0.14$

We can also determine β_{th} by the estimation of det(J) as a function of β. Then for each value of β to be scanned, elements of the matrix **J** are approximated by :

$$J_{ii} = \frac{1}{N}\,.\,\sum_{t=0}^{N-1} s_j(t).g(s_j(t)).\frac{\delta\, f(s_i)}{\delta s_i}(t)$$

and

$$J_{ij} = \frac{1}{N}\,.\,\sum_{t=0}^{N-1} s_i(t).f(s_i(t)).\frac{\delta\, g(s_j)}{\delta s_j}(t)$$

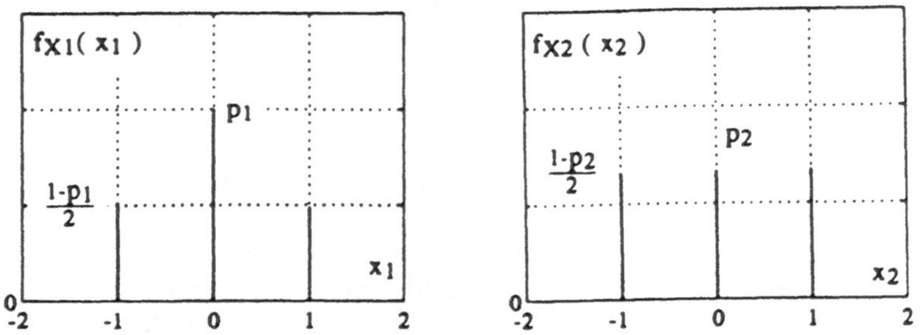

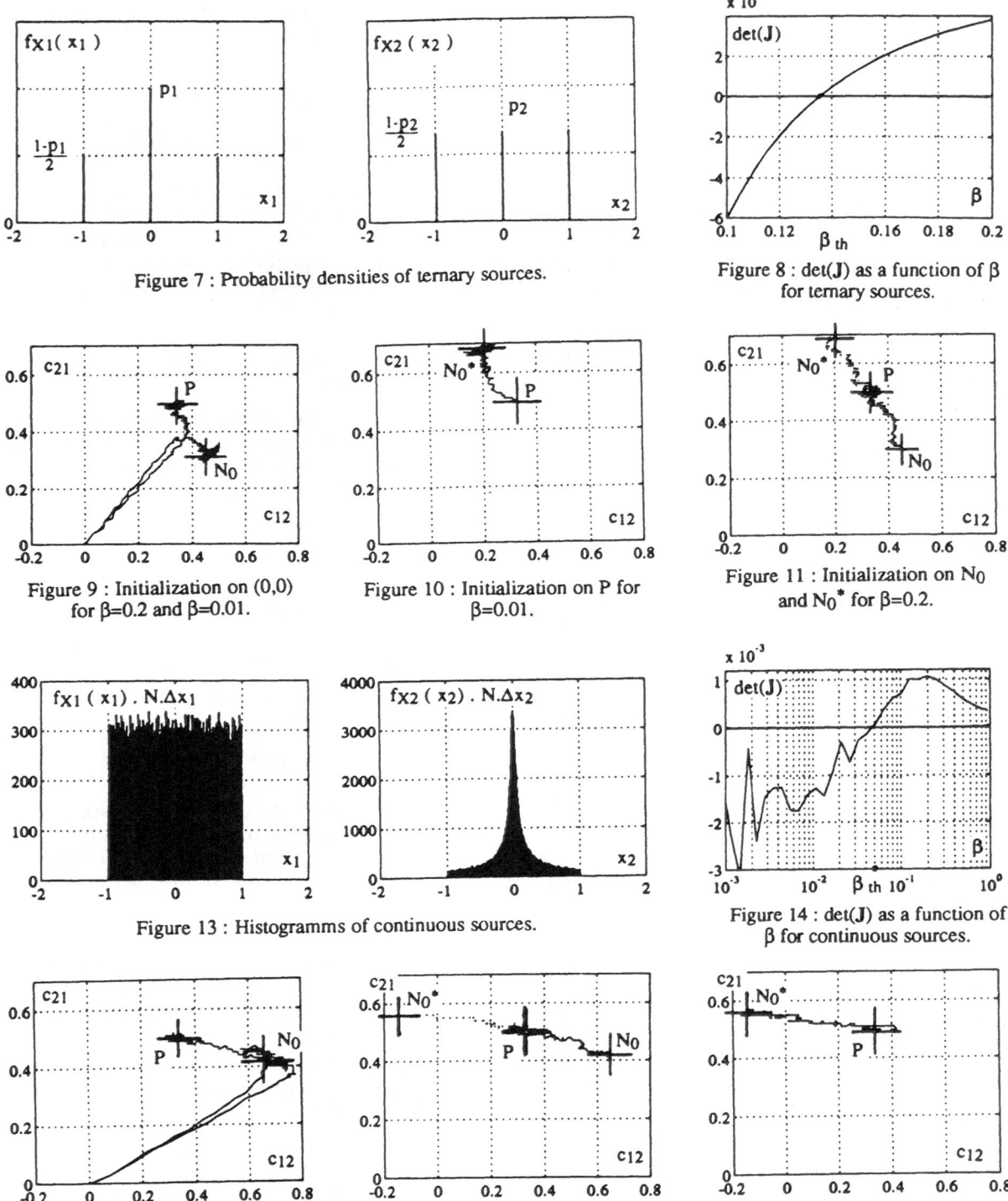

Figure 7 : Probability densities of ternary sources.

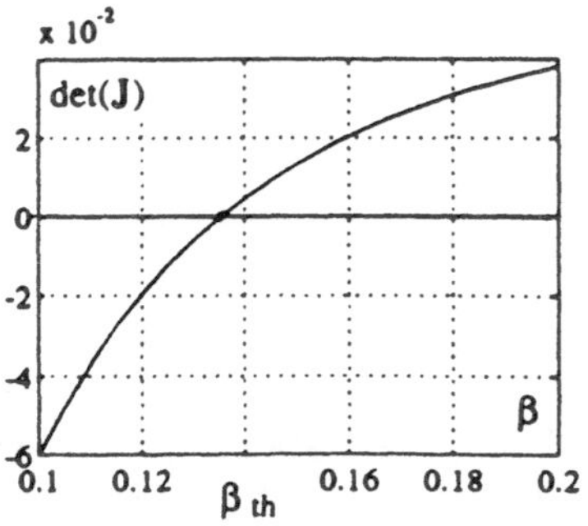

Figure 8 : det(J) as a function of β for ternary sources.

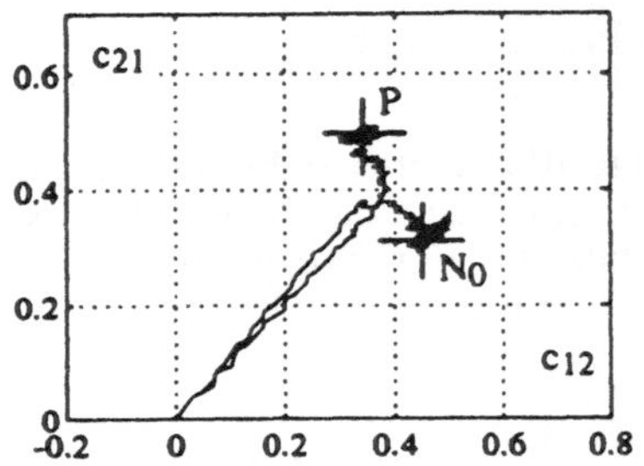

Figure 9 : Initialization on (0,0) for β=0.2 and β=0.01.

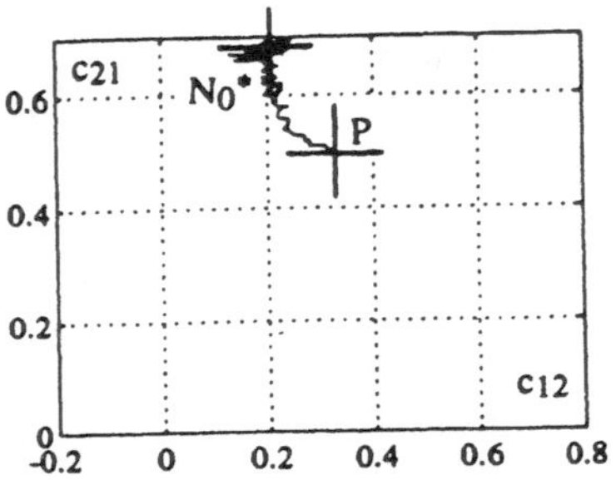

Figure 10 : Initialization on P for β=0.01.

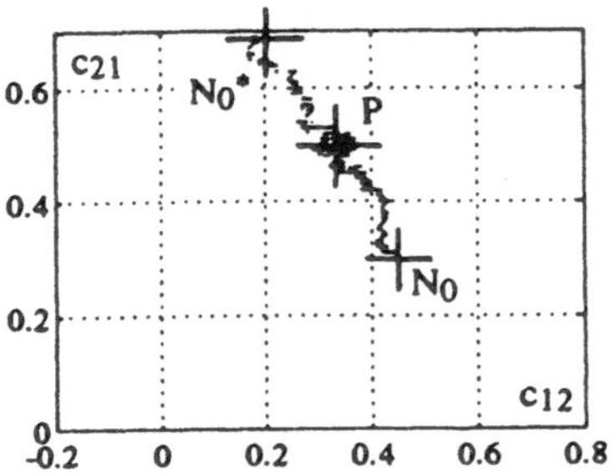

Figure 11 : Initialization on N_0 and N_0^* for β=0.2.

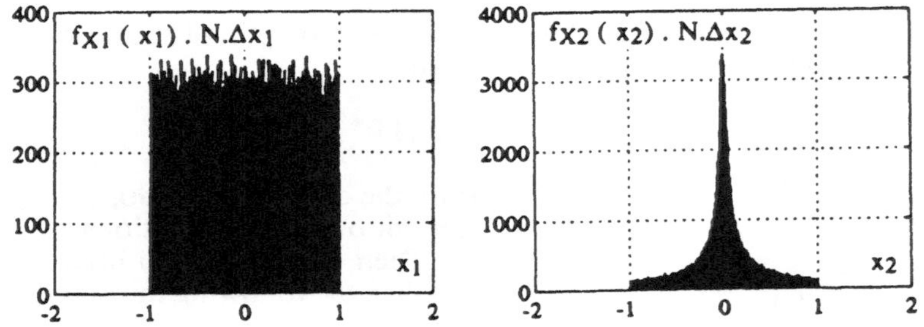

Figure 13 : Histogramms of continuous sources.

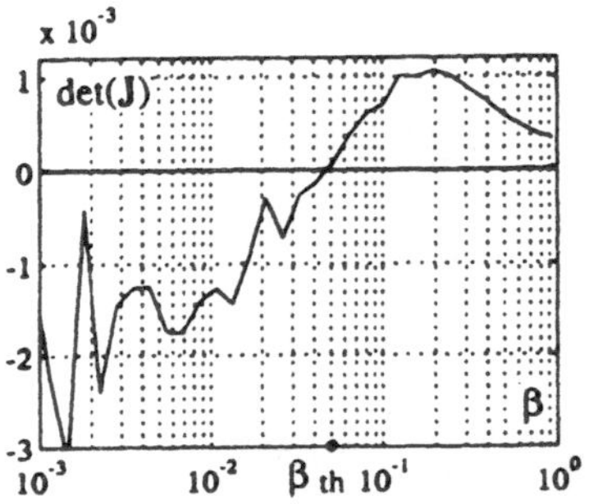

Figure 14 : det(J) as a function of β for continuous sources.

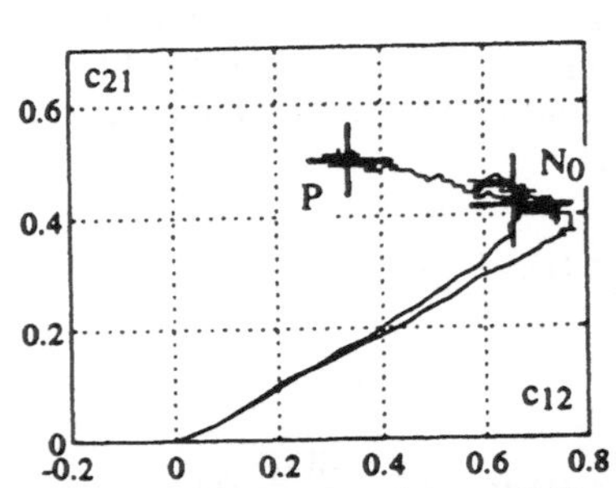

Figure 15 : Initialization on (0,0) for β=0.2 and β=0.01.

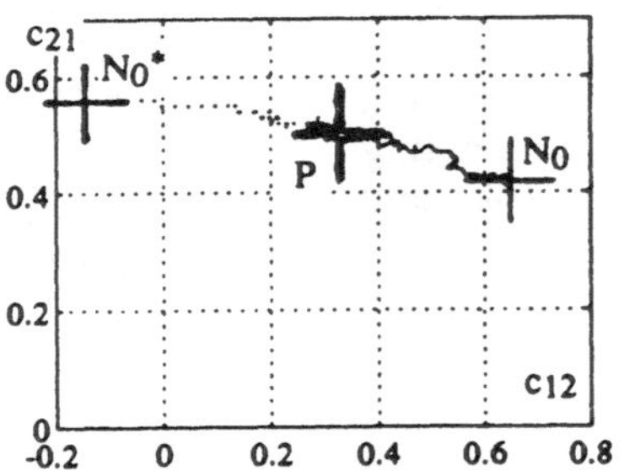

Figure 16 : Initialization on P for β=0.01.

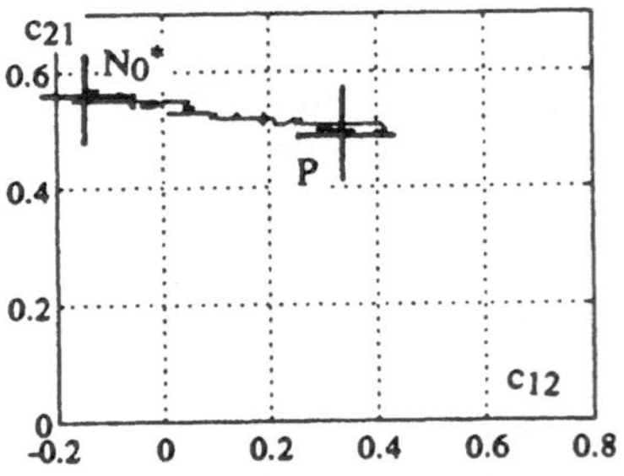

Figure 17 : Initialization on N_0 and N_0^* for β=0.2.

This computation of det(J) obtained for N=20000 , is represented in Fig.8 and appears to give the same value for the threshold.
We have simulated the network for a few values of β and observed the evolution of the networks coefficients for 20000 iterations. Fig.9 represents the evolution of the synaptic weights, initialized to (0,0) in the (c_{12},c_{21}) plane. For β=0.2 (> β_{th}), the algorithm converges toward the desired solution P. But for other values of β less than β_{th} the algorithm converges to other states that fail to achieve the source separation. Let us call those spurious solutions N_0. Then, for small β (=0.01) , that is to say for $g(x) \approx \mathrm{sign}(x)$, we can notice that the algorithm converges to :

N_0 (0.45 , 0.30).

On the other hand, for the same non-linearity (β=0.01), by initializing the weights on P, the simulation of the network (Fig.10) has pointed out an other spurious state. The algorithm seems to converge toward:

N_0^* (0.18 , 0.69).

These examples shed light on that P cannot be a stable equilibrium point for strong non-linearities of the function g. In order to verifiy the existence of N_0 and N_0^* for values of β that satisfy the stability condition, we present in Fig.11 , the evolution of the weights when initialized to a spurious state. The synaptic weights converge to P. So, one can say that N_0 and N_0^* appear only for $g(x) \approx \mathrm{sign}(x)$.

3.3. Generalization for other types of sources

Here we present numerical results and simulations applied to continuous sources as shown in Fig.12. The histogramms $N.\Delta x_i.f_{X_i}(x_i)$ of sources are plotted in Fig.13 for N=100000 samples and a step of Δx_i=0,01. Here, a rigourous stability analysis is very difficult. So, we are contented, with a numerical evaluation of det(J) as a function of β. For each value of β to be scanned, averages that occur in the elements of the matrix **J** are computed with N=100000 samples. The results presented in Fig.14 exhibit the threshold value of β. The stability condition is therefore :

$\beta > \beta_{th}$ and $\beta_{th} = 0.05$

By taking the same steps as before, the influence of g on the stability of the equilibrium state P is illustrated by means of simulations. Fig.15 and Fig.16 , show clearly the existance of two spurious states for β = 0,01 (< β_{th}) .
The first one, revealed by initializing the synaptic weights on zero is :

N_0 (0.65 , 0.42)

and initializing on P reveals the second one :

N_0^* (-0.15 , 0.56).

The equilibrium point P then disappears for $g(x) \approx \mathrm{sign}(x)$.
Finally, for β=0.2 (>β_{th}) and whatsoever the initial state (Fig.15 and Fig.17), simulating the network still shows that these spurious states are due to a strong non-linearity of g.

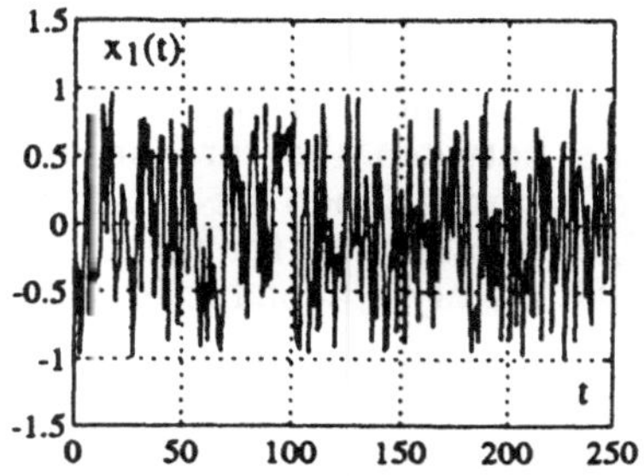

Figure 12a : Time evolution of $x_1(t)$

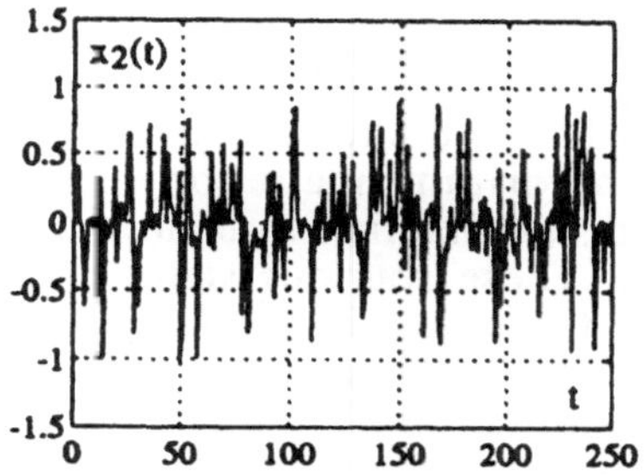

Figure 12a : Time evolution of $x_1(t)$

3.4. Discussion

Whatever be the statistics of sources, our numerical studies for $\beta < \beta_{th}$, encircle the unstability of the optimal equilibrium points and the convergence of the algorithm to an unexpected state. However, the mathematical expression of the threshold (16) is limited to ternary sources. This relation shows that β_{th} depends strongly on the probability densities of sources on zero. Using (16), and for p_1=0.5 , we plot β_{th} for different values of p_2 in Fig.18a

Similarly, for continuous signals, we keep x_1 constant with a uniform probability density and we plot β_{th} for different values of $f_{X2}(0)$ in Fig.18b . These values of the threshold are estimated by searching a root of det(J) as we have done it before.

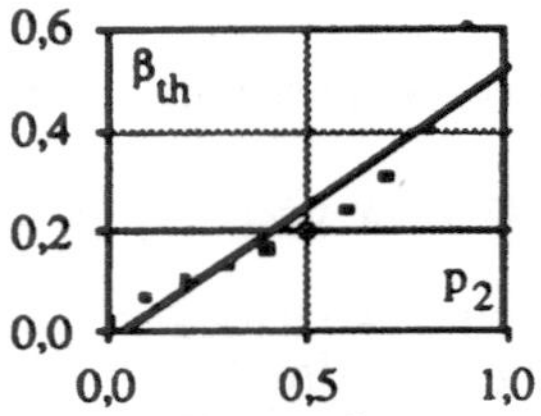

Figure 18.a :

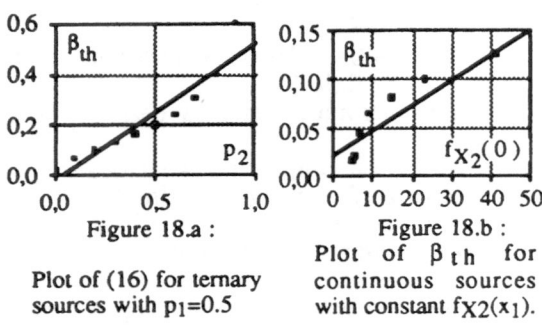

Plot of (16) for ternary sources with p_1=0.5

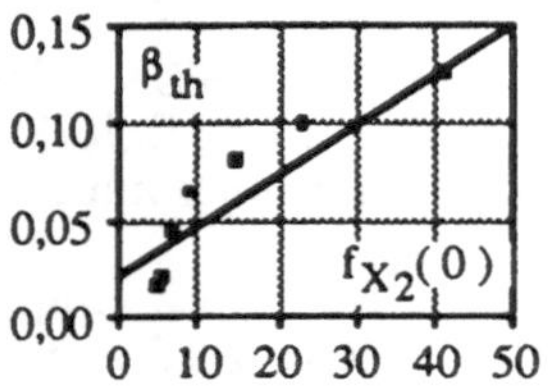

Figure 18.b :

Plot of β_{th} for continuous sources with constant $f_{X2}(x_1)$.

These figures shed light on the same dependance between the threshold and the statistics of sources. Then we can say that the relation (16) established for ternary sources, may be qualitatively acceptable for other kinds of signals.

Conclusion

We have presented in this paper a stability analysis of the H&J algorithm for the separation of two sources from an instantaneous mixture. Using very realistic non-linear functions in the adaptation law, we have pointed out that the learning algorithm may converge to an unexpected state and fails to achieve the source separation. These spurious states depend on non-linear functions. In particular, we have determined that under a value for β, that lets g(x) be similar to sign(x), the desired equilibrium states of the algorithm may become unstable. This threshold value depends on the probability densities of sources in zero. It has been also proved in [7] that the use of high order non-linear functions does not bring an advantage compared to functions like $f(x)=x^3$ and $g(x)=x$. Our results let us confirm that a strong non-linearity for g may damage the convergence of the algorithm. Concretely, for signals as speech or music for instance, where sources contain low energy and very often silent pauses between sounds or words, we are better to take g(x) linear.

Our present study is limited to the the source separation problem from instantaneous mixtures of two sources. More general discussions flowing from the generalization of mixtures model [2][3][4] are confered for our future work.

Acknowledgments

I would like to thank two students who helped during this research, J.M. Olivieri and S. Calpena.

References

[1] C. JUTTEN, J. HÉRAULT. "Blind separation of sources, Part I: An adaptive algorithm based on neuromimetic architecture.". Signal Processing vol. 24 , n°1, pp. 1-10, juil. 91.

[2] C. JUTTEN, L. NGUYEN THI, E. DIJKSTRA, E. VITTOZ, J. CAELEN. "Blind Separation of Sources: an Algorithm for Separation of Convolutive Mixtures". Int. Signal Processing Workshop on High Order Statistics. Chamrousse (France), July 10-12th, 1991.

[3] H.L. NGUYEN THI, Ch. JUTTEN, J. CAELEN, "Séparation aveugle de parole et de bruit dans un mélange convolutif". 13ème colloque GRETSI, Juan-les-pins, 16-20 Sep 91.

[4] S.V. GERVEN, D.V. COMPERNOLLE, "Feedforward and Feedback in a Symmetric Adaptive Noise Canceller : Stability Analysis in a Simplified Case". Signal Processing VI: Theories and Applications. J. Vandewalle, R. Boite, M. Moonen, A. Oosterlinck (eds.). 1992 Elsevier Science Publishers.

[5] D. ACHVAR, A. JOHANNET, J. MAGNIER. "Problème de séparation de signaux comprenant des retards : Illustrations.". Congrès Neuro-Nîmes, Nîmes France, Nov 1992.

[6] P. COMON, "Statistical approach to the Jutten-Hérault algorithm for seperating independent signals", NATO Advanced Research Workshop on Neuro Computing : Algorithms, Architectures and Applications, Les Arcs 1989.

[7] E. SOROUCHYARI, "Blind separation of sources, Part III : Stability analysis". Signal Processing vol. 24 , n°1, pp. 21-29, juil. 91.

[8] L. FETY. "Méthodes de traitement d'antenne adaptées auxradiocommunications." , Thèse, ENST, Paris.

[9] C. JUTTEN, J. HÉRAULT. "Analog implementation of a permanent unsupervised learning algorithm." , NATO Advanced Research Workshop on Neuro Computing : Algorithms, Architectures and Applications, Les Arcs 1989.

[10] E. VITTOZ, X. ARREGUIT. "CMOS Integration of H-J Cells for Separation of Sources." Analog Implementation of Neural Systems, C. Mead and M. Ismail (Eds), Kluwer Academic Publishers, Nowell 89.

Learning with Mappings and Input-Orderings using Random Access Memory - based Neural Networks

Michael H. Gera
Department of Computing
Imperial College,
180, Queens Gate,
London SW7 2BZ,
England.

Abstract
Random Access Memory (RAM)-based systems have been studied for several years. A recent paper by Gera and Sperduti demonstrates how one such RAM-based network can be used to produce a *real-number ordering* for each member of a training set. The coding reflects the relative similarity of input patterns in a highly concise way that is also very economical in its memory requirements. In fact, only one neuron or *discriminator* is required. However, the larger the training set, the closer some codes become. This in turn means that more decimal places are needed for the codes. One possible solution to this problem is described in this paper. A *two-stage* learning method is used. In the first stage, a Kohonen-like RAM-network divides the training set into groupings of similar patterns. Each grouping is then associated with its own Gera/Sperduti discriminator. Since input patterns to each such discriminator are similar, the discriminator itself can be pruned to remove redundant information and maximise output variance.

1. Introduction

Parallel to the mainstream connectionist work on feedforward networks, there has been ongoing work on neural systems based on the use Random Access Memory (RAM) networks as described by Aleksander [1].

RAM-based systems use RAM-like *nodes*. A node has M input lines and 2^M memory locations. The values on the input lines are binary and form the address of a location. Each location stores a real number between 0 and 1. All the experiments discussed in this paper are carried out with M equal to 2 (see Figure 1a). A RAM-based node has 2^M free parameters. A weighted-sum-and-threshold (or feedforward) node has M+1 independent parameters - the weight on each line plus a threshold. Figure 1b shows how such a neuron can be implemented using Ram-nodes. Expressed this way, it is clear that Ram-nodes are a functional superset of feedforward nodes.

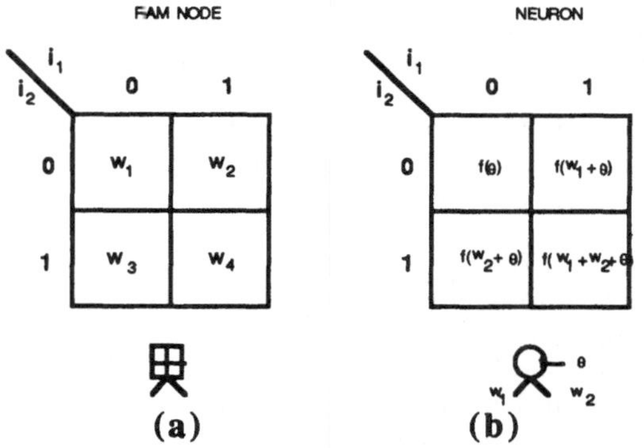

Figure 1. (a) A RAM-based node: i1 and i2 form the binary input lines. Each of the four locations contains an output value (w). (b) Implementation of standard neuron by RAM-based nodes.

A *discriminator* is defined as a bank of RAM-nodes. With input patterns of size L, L/M nodes are used. The node-input-line to input-unit-connections are chosen randomly with the proviso that all input units must be connected to a node. (It is of course possible to increase the number of nodes to beyond L/M. An example occurs in Section 6 below). The pair *(n,l)* refers to the *l*th location in node *n*. The output of a node n, *NOut(n)*, is *v(n,l)* - the value stored in the addressed location *l* in the *n*th node in the discriminator. The discriminator's output is given by equation (1)

$$output = \sum_{n=1}^{K} \frac{NOut(n)}{K} \qquad (1)$$

(The term *output* is reserved for a discriminator's output. The output of some node *n* is always referred to as *NOut(n)*).

Frequency Nodes

Ntourntoufis [2] has used RAM-based nodes to demonstrate Kohonen-like [3] Unsupervised learning. Essentially, he uses a number of discriminators where each node's input lines are randomly connected to one of the inputs. All node memory locations are initialised randomly to real values in [0,1]. The discriminator that responds most strongly to a particular input is rewarded. The other discriminators are punished. His algorithm is especially suited to the formation of clusters of similar inputs.

Ntourntoufis' system does not take frequency effects into account in a satisfactory way. For example, if a pattern appears twice, the RAM locations addressed will have higher values than they would have had if the pattern were only presented once. This, is of course only true for locations that are not already = 1 before either the first or the second training instance. Locations that *are* = 1 lead to non-uniform frequency effects. Thus, C-discriminators are not perfect for frequency-related tasks.

A variant of Ntourntoufis' node called the *Frequency Node* (or *F-node*) is designed to address this problem. F-nodes locations contain integer counts of the number of times a location is addressed during training. The locations are of course initialised to 0. If, during training, the output of a discriminator exceeds the value λ defined in (2),

$$\lambda = thresh * used \qquad (2)$$

where *thresh* is a real constant in [0,1] and *used* is the number of times the discriminator has been accessed, then the discriminator is selected to learn the input pattern. Discriminators constructed out of F-nodes continue to display the same desirable unsupervised learning properties that Ntourntoufis' discriminators display. This is because, λ, which increases in value as a discriminator becomes more used, eliminates frequency effects from the search for a winning discriminator. Thus, grouping of inputs into discriminators is done on *structural* grounds. It is only *within* a group that frequency effects are examined.

GS-Nodes

These nodes are described by Gera and Sperduti [4] and will be referred to as GS-nodes. During learning, the algorithm computes the output for each stimulus. A pair of stimuli is chosen. The algorithm attempts to increase the output variance by increasing the difference in output values between the winner (stimulus yielding the higher output) and the loser (stimulus yielding the lower output value). For each location *l* in node n, the change in the location value is defined by equations 3-5.

$$v(n,l) = v(n,l) + update \qquad (3)$$

$$update = \begin{cases} Beta * v(n,l) * Scaling\_Factor, & \text{(winner)} \\ -Beta * v(n,l) * Scaling\_Factor, & \text{(loser)} \end{cases} \qquad (4)$$

where:

Beta is a real number leaning constant between 0 and 1,

$$ScalingFactor(n,l) = \frac{noOfEpochs}{Times(n,l)Addressed} \qquad (5)$$

The more frequently a particular location *(n,l)* is addressed in the course of an epoch, the greater the number of different stimuli that address it and the less useful it becomes in distinguishing among stimuli. *Scaling_Factor* rewards locations by an amount inversely proportional to the number of times the location is addressed. Thus, less frequently used locations are favoured. A discriminator is termed *teachable* if the locations addressed by at least one of the stimulus pair being taught contain values that allow further learning.

Comparing and Contrasting the two algorithms

The GS-node system possesses the significant advantage of only requiring one discriminator to order the input patterns. However, this can lead to crowding of the output 'space'. For example, if two input patterns that are 512 bits wide differ by just one bit, then only one of the discriminator's 256 nodes (*n* is assumed to be equal to 2) will give a different output for the two patterns. At most, the difference in output will equal $\frac{1}{256}$ or 0.004. In the F-node system, on the other hand, similar inputs cause one set of discriminators to respond most strongly. However, when the system is tested on an input pattern, the output consists only of the number (identifier) given to the discriminator that is responding most strongly. No information about the actual pattern itself is given, unlike the GS-node system where the real-number code positions an input pattern relative to the rest of the training set.

3. New Model

A compromise between the two systems, using a *two-stage* learning process, has been implemented. Figure 2 illustrates this new system. The idea is for the training set to be divided into subsets by the F-node system. Each subset will consist of a set of similar inputs. A GS-node discriminator is associated with each F-node discriminator. If a particular F-node discriminator is trained by some training subset *A*, then the GS-node discriminator that is associated with the F-node is trained on *A*. This does not yet solve the problem of very close output patterns. However, it should now be noted that the input subset to some GS-node now consists of a set of similar patterns. Thus, there is a greater expectation that in many nodes, only *one* location is addressed. These nodes do not distinguish among patterns. The information they hold is thus of no value. They may thus be removed. This *pruning* process is described below.

4. Demonstrating the new system's operation

The model's operation in unsupervised mode is demonstrated by means of a very simple training set that is shown in Table 1.

1 1 1 1 1 1 1 1	Stimulus 0
0 0 0 0 0 0 1 1	Stimulus 1
1 1 1 1 1 1 0 1	Stimulus 2
0 0 0 0 0 0 0 1	Stimulus 3
1 1 1 1 1 1 1 0	Stimulus 4
0 0 0 0 0 0 1 0	Stimulus 5

Table 1. The training set

An F-node discriminator (call it *A*) learned - without supervision - to respond most strongly to Stimuli 1, 3 and 5. Similarly, another F-node discriminator (*B*) learned to respond most strongly to Stimuli 0, 2 and 4.

Having carved up the training set, the GS-node discriminators associated with F-node discriminator *A* were taught on Stimuli 1, 3 and 5. Similarly, the GS-node discriminators associated with F-node discriminator *B* were taught on Stimuli 0, 2 and 4. The results after 17 GS-node learninbg epochs are shown in Table 2. Discriminators 4-7 are the GS-node discriminators. It is clear that the different stimuli have been assigned individual codes. These difference in the codes are due entirely to the rightmost input bits - the only bits that distinguish among the training subset members.

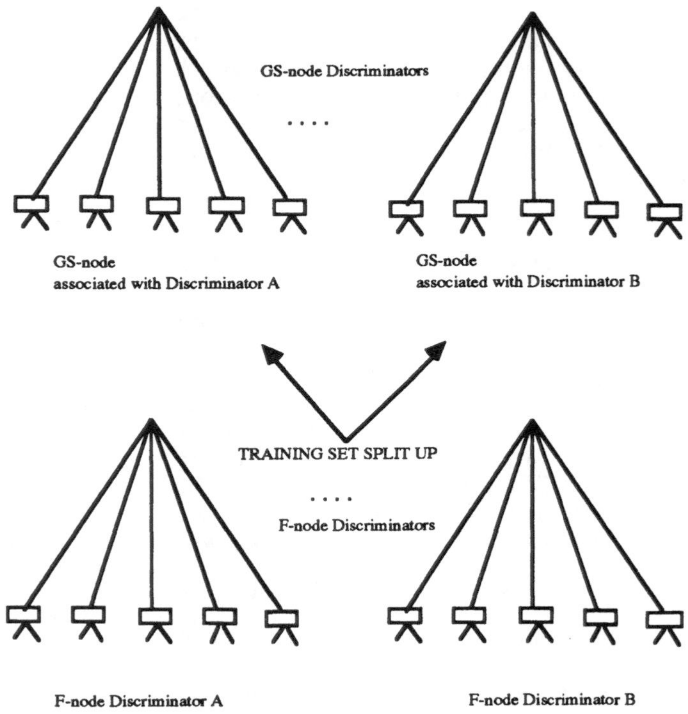

Figure 2. The two stage leaning process. The F-nodes split the training set for the next learning step which is performed by GS-node discriminators.

GS-node Discriminator 4 is associated with F-node discriminator A
Stimulus 1 GS-node discriminator No: 4 Output = 0.34
Stimulus 3 GS-node discriminator No: 4 Output = 0.59
Stimulus 5 GS-node discriminator No: 4 Output = 0.09

GS-node Discriminator 5 is associated with F-node discriminator A
Stimulus 1 GS-node Discriminator No: 5 Output = 0.47
Stimulus 3 GS-node Discriminator No: 5 Output = 0.72
Stimulus 5 GS-node Discriminator No: 5 Output = 0.22

GS-node Discriminator 6 is associated with F-node discriminator B
Stimulus 0 GS-node Discriminator No: 6 Output =0.41
Stimulus 2 GS-node Discriminator No: 6 Output =0.66
Stimulus 4 GS-node Discriminator No: 6 Output =0.16

GS-node Discriminator 7 is associated with F-node discriminator B
Stimulus 0 GS-node Discriminator No: 7 Output = 0.53
Stimulus 2 GS-node Discriminator No: 7 Output = 0.42
Stimulus 4 GS-node Discriminator No: 7 Output = 0.67

Table 2. The results after 17 epochs

Stimulus 1 GS-node Discriminator No: 4 Output = 0.50
Stimulus 3 GS-node Discriminator No: 4 Output = 1.00
Stimulus 5 GS-node Discriminator No: 4 Output = 0.00

Stimulus 1 GS-node Discriminator No: 5 Output = 0.50
Stimulus 3 GS-node Discriminator No: 5 Output = 1.00
Stimulus 5 GS-node Discriminator No: 5 Output = 0.00

Stimulus 0 GS-node Discriminator No: 6 Output = 0.50
Stimulus 2 GS-node Discriminator No: 6 Output = 1.00
Stimulus 4 GS-node Discriminator No: 6 Output = 0.00

Stimulus 0 GS-node Discriminator No: 7 Output = 0.50
Stimulus 2 GS-node Discriminator No: 7 Output = 0.00
Stimulus 4 GS-node Discriminator No: 7 Output = 1.00

Table 3. The pruned output after 17 epochs

Pruning

Pruning the GS-node discriminator is a comparatively simple task. A GS-node discriminator is presented with its training subset and the RAM-locations that are addressed by all the training patterns are identified. Such nodes are identifiable because only one of the memory locations has a count value that is greater than zero. These RAMs are removed. There is no need to re-train the GS-node discriminator. The output is given by equation 6.

$$output = \frac{\sum_{n=1}^{R} NROut(n)}{R} \qquad (6)$$

where $NROut(n)$ is the output of a RAM that is not removed and R is the number of RAMs that are not removed.

Since the variance in the output of the original (unpruned) GS-node discriminator was caused by the RAMs that are not pruned, removing the other RAMs increases the variance. The pruned output at epoch 17 - using the same example as above - is shown in Table 3. Clearly, the variance has now been maximised.

5. Application 2: Prototype Extraction

So far, we have shown how two items of information can be obtained from the new system: an unsupervised classification (or *clustering*) yielded by the F-nodes and a concise coding yielded by the GS-nodes. There is a third information item that can be gleaned from this system: what Robins [5] terms the *domain*.

Robins

Robins finds significant problems with connectionist accounts of cognitive categories. In particular, he finds prevailing connectionist thinking on *prototypes* to be somewhat lacking in psychologically terms. This is because connectionism has treated prototypes as little more than *averages* of a set of inputs. This is not in agreement with Rosch's [6] suggestion that prototypes consist of bundles of *correlated attributes*. Robins attempts to rectify this situation by first noting that within a population with a significant type/category structure, there will be population sub-groups each of which have a degree of 'shared' information. This shared information is termed a *domain* by Robins. Robins then goes on to define a domain-extracting algorithm. This algorithm is not only of cognitive interest, but also of connectionist interest. This is because he defines a bi-modal neural system where nodes - apart from having the usual type of weights - also possess what he terms *co-activation weights*. Co-activation weights are calculated for the input units at a particular layer in a network by using full interconnection among the input units. A co-activation weight is defined by (7)

$$weight_{xy} = \frac{\sum_p x_p y_p}{\sum_p y_p} \quad (7)$$

where $weight_{xy}$ is the coactivation weight from unit x to unit y, x_p is the activation of unit x in pattern p, and y_p is the activation of unit y in pattern p.

Given a certain input, domain extraction proceeds in the following way: For each input unit, the average co-activation with other *active* units is calculated. This value is termed the *centrality* of the unit in question. Now a threshold centrality value - which Robins calls the *criterion* is defined. Having calculated the centrality of each unit, a search is made for a unit that is in *violation* of this criterion. A unit is said to be in violation of the criterion if it is inactive (i.e. equal to '0'), but has a centrality above the criterion, or if it is active (i.e. equal to '1'), but has a centrality value that is below the criterion. In the former case, the unit is activated; in the latter case, it is deactivated. After a unit in violation has been reset, the entire centrality distribution for all units is recomputed. This process is re-iterated until there are *no* units in violation of the criterion. The domain is given by the final activation state on the input units. Robins suggests that this centrality measure is related to the psychological concept of *cue-validity* [6].

6. Domain extraction with F-nodes

Domains can be extracted from F-node discriminators using a technique called *backflow*. Backflow is performed on a trained F-node discriminator to extract its domain. Each node is accessed in turn and the location with the highest value is sought. The binary address of that location is used to *vote* on what the prototypical input for the input units to which that unit is connected should be. Take the F-node illustrated in Figure 3.

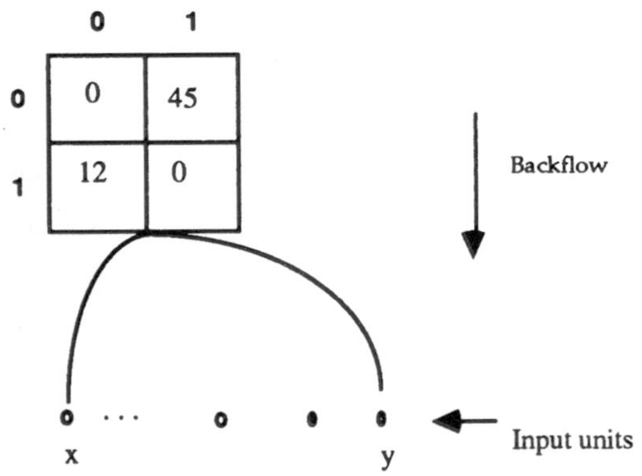

Figure 3. A trained F-node. Location 1 (binary address is [01]) has the highest value and will vote that input unit *x* should be equal to 0 and input unit *y* equal to 1.

The node is connected to input units x and y. The location with the highest stored value is location number 1 (binary address 01). This implies that line x should be equal to 0 and line y equal to 1. Now if each input line is only connected to one node, then in the final domain, unit x and unit y are equal to 0 and 1 respectively. If there is an F-node for all possible input unit pairs, then there will be several votes as to what a unit such as x should be. If the '0' vote is greater than the '1' vote, then in the final domain, unit x is equal to 0. Conversely, if the '1' vote is greater than the '0' vote, then in the final domain, unit x is equal to 1. If the two votes are equal, then the domain value for x is set to either 0 or 1 with equal probability.

Experiments

Two experiments are described, both of which utilise the data set from Robins [7] shown in Table 4.

In the first experiment full-connectivity was used. That it to say, all the 66 possible input pairs for the 12-bit input patterns shown in Table 4 were employed. The value of *thresh* was set to 0.1. One F-node discriminator - without supervision - came to learn Vectors 0 - 4. Another, learned Vectors 5-9. After one epoch, the domains were extracted. They are illustrated in Table 5. These are the same domains that

Robins obtains.

Vector no	Vector
0.	110111001000
1.	111111000000
2.	111010000010
3.	111111000000
4.	111111010000
5.	000100111011
6.	000000101111
7.	000010110011
8.	011000111110
9.	000000111111

Table 4. An data set generated by taking five examples each of two templates and reversing the activation of bits in the ten resulting vectors at 10% probability. From [7]

Domain for input patterns 0-4:	1 1 1 1 1 1 0 0 0 0 0 0
Domain for input patterns 5-9	0 0 0 0 0 0 1 1 1 1 1 1

Table 5. Domains extracted for experiments 1 and 2.

In the second experiment, the number of nodes used was reduced to 6 (i.e. L/M as used above to demonstrate the two-step process) with input lines randomly connected to input units. The same domains were extracted as shown in Table 5 on 17 out of 20 trials where, on each trial, different random input line to input unit connections were used.

Discussion of Experiments

The results of experiment 1 are interesting because they were obtained using an extremely simple equivalent to Robins' co-activation score: the strongest-location-searching in the backflow process. So, even though full pairwise connectivity was used, a co-activation score that is simpler than Robins suffices to obtain the same results as he does. The higher number of parameters afforded by RAM-based nodes (see section 1 above) yield a computationally more powerful network than Robins' weights. Robins' co-activation score for a unit is only calculated with respect to *active* units with a given input. Here, for a given input, the discriminator with the highest output value is ascertained. The domain for the *discriminator* is then extracted using backflow. Information collected from input units that were equal to 0 during training is as useful as that from units which were equal to 1. This is possible due to the nature of RAM-based neurons in which there is no concept of active (i.e. 1) or inactive (i.e. 0).

The power of the RAM-based approach becomes even clearer from the results of experiment 2. Here less than 10% of the co-activations scores that Robins employs were found to be sufficient to obtain the same domains for the absolute majority of trials.

7. General Discussion

We have described a system which yields three different types of information for a given input pattern. The pattern's class is given by simply examining which F-node responds most strongly. A concise coding that indicates the input's overall similarity to other members of its class is also obtainable using the GS-nodes. Finally, the F-nodes also yield the input's domain.

References

[1] Aleksander I, *'The Logic of Connectionist Systems'*. In I. Aleksander (ed.) *'Neural Computing Architectures: The Design of Brain-like Machines'*, London: North Oxford Academic, 1989.

[2] Ntourntoufis P, *'Self-Organization Properties of a Discriminator-Based Network'*, Proceedings of the International Joint Conference on Neural Networks, San Diego, Vol. 2: 319-324, 1990.

[3] Kohonen, T, *'Self Organization and Associative Memory'*. Springer-Verlag, 1984.

[4] Gera M H and Sperduti A, *'Unsupervised and Mixed Learning using Input-Orderings with Weightless Neural Networks'*. Proceedings the International Joint Conference on Neural Networks, Beijing, November, 1992.

[5] Robins AV, *'The Distributed Representation of Type and Category'*. Connection Science, 1: 345-363, 1989.

[6] Rosch E and Mervis C, *'Family Resemblances: Studies in the Internal Structure of Categories'* Cognitive Psychology, 7: 573-605, 1975.

[5] Robins AV, *'Multiple Representations in Connectionist Systems'*, International Journal of Neural Systems, 4: 345-362, 1992.

New Preprocessing Methods for Holographic Neural Networks

Robert Manger

Zagrebačka banka d.d.

Paromlinska 2, 41000 Zagreb, Croatia

Branko Souček

IRIS International Center, Star Service

Via Amendola 168/1, 70126 Bari, Italy

Abstract. We propose two new methods for preprocessing of stimulus data in holographic neural networks. The first method achieves optimal data symmetrization, and it is based on estimating the actual distributions of elements within a stimulus vector. The second method serves for data expansion, and it relies on sine and cosine functions to produce dummy stimulus elements. We describe an implementation of our methods and report on some experiments. Our proposals improve the applicability of holographic networks. Namely, symmetrization assures accuracy in reproducing learned stimulus-response associations, while expansion increases learning capacity.

Key words: holographic neural networks, data preprocessing, stimulus expansion, stimulus symmetrization.

1. Introduction

Holographic networks are a new brand of artificial neural networks, which have recently been proposed by J. Sutherland [4,6]. Although conforming to the general paradigm, this type of networks significantly differs from the conventional "connectionist" type [2,3]. The main difference is that a holographic neuron is much more powerful than a conventional one, so that it is functionally equivalent to a whole conventional network. Another important characteristic is that information is represented by complex numbers operating within two degrees of freedom (phase and magnitude). Holographic networks are available on PC-s through an emulator called HNeT [5].

There is no need to build massive networks of holographic neurons; for most applications one or few neurons are sufficient. In the process of network design, emphasis is shifted from the choice of topology to the choice of adequate *data preprocessing*.

There are two kinds of preprocessing which are of fundamental importance within the holographic neural process. *Stimulus symmetrization* assures reasonable accuracy in reproducing learned stimulus-response associations. *Stimulus expansion* increases the number of stimulus-response associations that can be learned. For both kinds of preprocessing, standard methods are provided in HNeT.

Holographic networks seem to be very suitable for analog problems, where stimuli are given as analog signals, which are in turn represented by very long arrays of real numbers. However, we have tried to apply holographic technology to semi-discrete problems where a stimulus consists of only few values taken from (possibly different) analog or discrete domains. We are motivated by banking applications such as financial ratio analysis, stock forecasting, etc [1]. According to our experiments, holographic networks are also appropriate for semi-discrete problems, provided that some additional preprocessing methods are available.

In this paper we propose two new preprocessing methods, i.e. one for each of the two mentioned types of preprocessing. We also describe the situations where these methods are useful. We believe that our proposals greatly enhance the applicability of holographic networks. The paper is organized as follows. Section 2 reviews the theory of holographic neural networks. Section 3 is concerned with stimulus symmetrization, while Section 4 deals with stimulus expansion. Section 5 briefly describes our implementation of the proposed methods and lists some experimental results. Section 6 gives concluding remarks.

2. Neural Process

A holographic neuron is sketched in Figure 1. There exist only one input channel and one output channel, but they carry whole vectors of complex numbers. An input vector S is called a stimulus and it has the form

$$S = [\lambda_1 e^{i\theta_1}, \lambda_2 e^{i\theta_2}, \ldots, \lambda_n e^{i\theta_n}].$$

An output vector R is called a response and its form is

$$R = [\gamma_1 e^{i\phi_1}, \gamma_2 e^{i\phi_2}, \ldots, \gamma_m e^{i\phi_m}].$$

All complex numbers above are written in polar notation, so that magnitudes are interpreted as confidence levels of data, and phase components serve as actual values of data. Confidence levels for these complex numbers typically extend over a probabilistic scale (0.0 to 1.0).

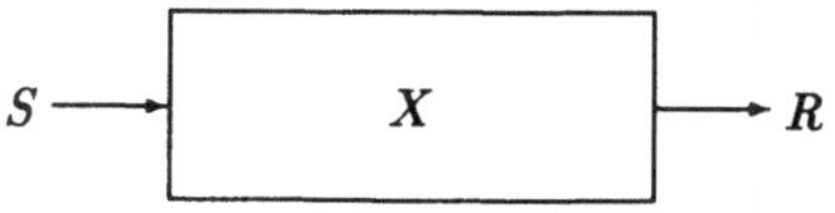

Figure 1. A holographic neuron.

The neuron internally holds a complex $n \times m$ matrix X, which enables memorizing stimulus-response associations. Learning one association between a stimulus S and a desired response R reduces to the (non-iterative) matrix operation:

$$X += S^{\tau} R.$$

Note that all associations are enfolded onto the same matrix X. The response R^* to a stimulus

$$S^* = [\lambda_1^* e^{i\theta_1^*}, \lambda_2^* e^{i\theta_2^*}, \ldots, \lambda_n^* e^{i\theta_n^*}]$$

is computed through the following matrix operation:

$$R^* = \frac{1}{c^*} S^* X.$$

Here c^* denotes a normalization coefficient given by

$$c^* = \sum_{k=1}^{n} \lambda_k^*.$$

The response R^* to a stimulus S^* can be interpreted as a point in the complex plane (i.e. a complex number) composed of many components, as shown in Figure 2.

Each component corresponds to one of the learned responses. If S^* is equal to one of the learned stimuli S, then the corresponding response R occurs in R^* as a component with a great confidence level (≈ 1). The remaining components have small confidence levels ($<< 1$) and they produce a "noise" (error).

Figure 2. Response to an old stimulus.

As we have seen, holographic neurons internally work with complex numbers. Since external data domains are usually real, a suitable data conversion is required. For this purpose the following simple transformation is used: real values from a known range $[a, b)$ are linearly scaled to the range $[0, 2\pi)$ and interpreted as phase orientations of complex values with a unity magnitude. In the rest of this paper we will assume that all data have already been converted to complex.

3. A New Method for Symmetrization

The holographic process requires elements within the stimulus vector to display a highly symmetrical (i.e. uniform) distribution in orientation about the origin of the complex plane. Only symmetry assures that error terms produced in a response will indeed neutralize in a manner analogous to random walk (as shown by Figure 2).

If original stimulus data are not symmetric, some form of preprocessing is needed to assure the confidence of response. Such preprocessing, called ***stimulus symmetrization***, should redistribute phase elements within the stimulus vector to achieve a uniform phase distribution.

The original HNeT system provides a standard method for stimulus symmetrization, which is based on the ***sigmoid*** function [2,3]. According to [5], this method performs a mapping of distributions displaying approximate Gaussian form to a fairly uniform state.

Although satisfactory in many cases, the standard sigmoid transform is not appropriate if the original data distribution differs from Gaussian. The latter can happen quite easily, specially if the stimulus consists of only few elements originating from different domains. Therefore we propose an alternative method for stimulus symmetrization, which takes into account the actual distributions of phase elements within the stimulus vector. The method is derived from the following observations.

Suppose that the starting domain is composed of numbers θ in the range $[0, 2\pi)$ with an arbitrary distribution $f_1(\theta)$. This domain should be converted into another representation, with elements ψ also in the range $[0, 2\pi)$, but with a uniform distribution $f_2(\psi)$. It is necessary to find the proper conversion curve $\psi = g(\theta)$.

Since the starting value θ is a random variable, ψ will also be a random variable. If θ has a distribution function $f_1(\theta)$, then ψ has a distribution function $f_2(\psi)$, which depends on $f_1(\theta)$ and on $g(\theta)$. The easiest way to determine the resulting distribution function $f_2(\psi)$ is as follows: if there is a continuous one-to-one correspondence between θ and ψ, then the probability that the original variable is in the range $(\theta, \theta + d\theta)$ must be equal to the probability that the resulting value is in the range

$(\psi, \psi + d\psi)$. Hence

$$\begin{aligned} f_1(\theta)\, d\theta &= f_2(\psi)\, d\psi, \\ f_2(\psi) &= f_1(\theta) \cdot \frac{1}{d\psi/d\theta} \\ &= f_1(\theta) \cdot \frac{1}{|g'(\theta)|}. \end{aligned}$$

The resulting distribution $f_2(\psi)$ is a function of $f_1(\theta)$ and of the derivative of the transfer function, $g(\theta)$. The absolute value of the derivative is taken because a distribution function cannot have negative values. From the above expression, one can find the transfer function $g(\theta)$, which will transform an original distribution $f_1(\theta)$ into a new desired distribution $f_2(\psi)$.

$$\begin{aligned} g'(\theta) &= \frac{f_1(\theta)}{f_2(\psi)}, \\ g(\theta) &= \int_0^\theta \frac{f_1(\bar{\theta})}{f_2(\psi)}\, d\bar{\theta} + C. \end{aligned}$$

Since we are interested in producing a uniform distribution $f_2(\psi)$, we can write

$$f_2(\psi) = K = \text{const},$$

and consequently

$$\begin{aligned} g(\theta) &= \frac{1}{K} \int_0^\theta f_1(\bar{\theta})\, d\bar{\theta} + C \\ &= \frac{1}{K} F_1(\theta) + C. \end{aligned}$$

Here, $F_1(\theta)$ is the cumulative or integral distribution function of the variable θ. Since the uniform distribution is defined in the interval between 0 and 2π, it follows that $K = 1/(2\pi)$. The constant C can be obtained from the boundary condition for $g(\theta)$:

$$g(2\pi) = 2\pi = 2\pi \cdot 1 + C \implies C = 0.$$

So the required transfer function is given by

$$\psi = g(\theta) = 2\pi \cdot F_1(\theta).$$

Our *optimal* stimulus symmetrization method for holographic networks consists of the following:

- estimate the distribution function $f_1(\theta)$ for a chosen phase element θ within the given stimulus vector;
- compute the corresponding cumulative function $F_1(\theta)$;
- transform the phase element θ into ψ using the above relationship.

Since various phase elements within the same stimulus vector can be drawn from different domains, each element can have its own distribution function. Therefore the process must be repeated for each domain.

4. A New Method for Expansion

As any neural system, a holographic neuron produces only approximate responses to learned stimuli. Even if an ideal symmetry of stimulus data distribution is achieved, the phase response error is still expected to be [5]

$$\phi_{\text{error}} \approx \frac{1}{\pi\sqrt{8}} \tan^{-1}(\sqrt{p/n}).$$

Here n is the length of the stimulus vector (i.e. the number of vector elements), and p is the number of stimulus-response associations learned. By increasing the number of learned associations the noise also increases, until it becomes intolerably high. So the learning capacity is limited, and it depends on the length of the stimulus vector.

For the problems where the stimulus vector length is small, the number of associations that can be accurately encoded is also small. The only way to increase the learning capacity is to artificially increase the stimulus vector length, and to hope that the relation above will still remain valid. Such preprocessing operation is called *stimulus expansion*.

The original HNeT system provides a standard method for stimulus expansion. It is based on switching to *higher order product terms*, generated from elements stored within the original stimulus vector. These terms or "statistics" form a set of unique combinatorial product groups of the specified order. The above error relationship remains valid provided that no two terms have been constructed from the same set of raw input values. Or, differently speaking, only unique terms should be counted when estimating the effect of expansion. Still, the learning capacity can increase considerably in this way, since the number of unique statistics for an initial length n and order of statistic s is

$$\binom{n}{s} = \frac{n!}{s!\,(n-s)!}\,.$$

However, switching to higher order product terms produces an important side effect. As one increases the order of statistic, the neuron's ability to generalize the learned stimulus-response associations becomes smaller and smaller.

The standard stimulus expansion method behaves very well for a moderately short stimulus vector (e.g. between 20 and 100). But for an extremely short vector, switching to higher order statistics does not help too much. There are simply not enough unique terms to produce any substantial improvement. Or one is forced to use a very high order of statistic, and this can cause an unwanted reduction of the neuron's generalization characteristics. For instance, if the vector length is $n = 4$, then the maximum number of unique statistics is obtained with the order $s = 2$, and it ammounts only to 6. If the length is $n = 12$, then we could expand the vector to ≈ 1000 unique terms, but only with a considerably high order of statistic $s = 6$.

Due to the reasons above we propose another method for stimulus expansion, which is suitable for expanding an extremely short vector to a moderate length. Our method adds additional (dummy) elements to the original vector according to the following rules.

1. New elements are uniquely determined by the original elements.
2. Any new element depends on exactly one of the original elements.
3. The dependence is nonlinear.
4. The dependence is continuous.

Rule 1 is necessary in order to produce always the same expanded stimulus for a given original stimulus. Rule 2 tries to neutralize the influence of expansion on the neuron's generalization ability. Rule 3 is needed to avoid replication of the same error patterns in multiple parts of the neuron's memory. Rule 4 assures that similar original stimuli will produce similar expanded stimuli, and this is necessary to enable generalization.

There are many possible variants of the proposed method. We will describe in detail a variant based on *sines* and *cosines*:

(∗ *The original vector elements are* $\lambda_j e^{i\theta_j}$, $j = 1, 2, \dots, n$.
The dummy vector elements are $\lambda_j e^{i\theta_j}$, $j = n+1,\, n+2,\, \dots,\, N$. ∗)
for $j := n+1$ **to** N **do begin**
  $k := \left\lfloor \frac{j+n-1}{2n} \right\rfloor$;
  $r := j + n - 2nk$;
  if $r \le n$ **then begin**
    $\lambda_j := \lambda_r$;
    $\theta_j := \pi(\sin(k\theta_r) + 1)$
  end
  else begin
    $\lambda_j := \lambda_{r-n}$;
    $\theta_j := \pi(\cos(k\theta_{r-n}) + 1)$
  end
end .

Note that the purpose of our sine/cosine method is only to expand an extremely short stimulus vector to a moderate length. Then the standard higher order statistics can further be applied to reach the desired final vector length.

5. Implementation and Experiments

Our preprocessing methods have been implemented as a set of C functions. In this way, a small library of routines has been created, which acts as a supplement to the original HNeT software.

In our library, the sine/cosine expansion method is realized through a single routine. The optimal symmetrization method, on the other hand, is configured as a sequence of three separate routines. The first of them estimates the phase distribution for a list of sample values, the second transforms a distribution into the corresponding cumulative distribution, while the third routine actually transforms phase elements. If more elements are transformed simultaneously, then each element can use its own cumulative distribution function, or alternatively they can share the same distribution. The implemented version of the optimal symmetrization method is slightly more general than the version described is Section 3. Namely, our routines can also produce a uniform distribution over a *contracted* interval $[0, \delta \cdot 2\pi)$, where $0 < \delta \leq 1$ is a chosen contraction factor.

In order to test our preprocessing methods, we used the well-known *iris flowers* classification problem [7], which is supplied in many software packages as a benchmark example. The problem consists of grouping iris flowers into three classes (named Setosa, Versicolor, and Virginica, respectively), according to four attributes (sepal length and width, petal length and width). A training set and a testing set of flowers are given, each comprising 75 examples. A small part of the training set is reproduced in Table 1. In an attempt to solve the problem, we used a single holographic neuron, and we experimented with various combinations of stimulus preprocessing.

#	lngth sepal	wdth sepal	lngth petal	wdth petal	cl-ass
1	0.224	0.624	0.067	0.043	seto
2	0.749	0.502	0.627	0.541	vers
3	0.557	0.541	0.847	1.000	virg
4	0.110	0.502	0.051	0.043	seto
5	0.722	0.459	0.663	0.584	vers
6	0.776	0.416	0.831	0.831	virg

Table 1. Iris flowers classification.

The first group of experiments served for testing our sine/cosine expansion method. The results are presented in Table 2. Each row of Table 2 corresponds to one variant of stimulus expansion. Any variant consists of two phases, as shown by the first two columns of the table. In the first phase our sine/cosine method is used to expand the original stimulus vector to a length manageable by the neuron. The second phase further expands the stimulus to the final length, by applying higher order statistics. For all variants the final vector length is chosen the same, i.e. 400. It means that the neuron always uses the same ammount of physical memory, and this assures a fair comparison among variants.

Columns 3 and 4 of Table 2 describe the network performance, in dependence on the chosen expansion variant. After accomplished training the examples from the training and testing sets, respectively, were presented to the neuron for classification. The percent of correctly classified examples has been recorded in the table. Thus column 3 shows how well the neuron reproduces the learned associations, while column 4 illustrates the quality of generalization.

initial expansion (sin/cos)	final expan (high.ord statistics)	correct (train set)	correct (test set)
omitted	2-nd order 400 terms (6 unique)	60%	56%
from length 4 to 8	2-nd order 400 terms (28 unique)	95%	76%
from length 4 to 12	2-nd order 400 terms (66 unique)	97%	87%
from length 4 to 16	2-nd order 400 terms (120 uniq)	100%	88%
from length 4 to 20	2-nd order 400 terms (190 uniq)	100%	95%
from length 4 to 24	2-nd order 400 terms (276 uniq)	100%	91%
from length 4 to 28	2-nd order 400 terms (378 uniq)	100%	89%

Table 2. Expansion variants.

For the iris flowers problem stimulus expansion is crucial, since the original stimulus vector is extremely short. If no expansion were used, the neuron would have a very limited learning capacity, so that its performance would be completely unreliable. More detailed testing has revealed that without any expansion the neuron is able to memorize only 9 training examples. As we see from row 1 of Table 2, exclusive use of higher order statistics does not help too much, since the number of unique higher order terms is low.

Rows 2 to 4 show how our sine/cosine expansion method increases the learning capacity, so that the neuron eventually becomes able to correctly reproduce all the training examples. The initial expansion to length 20 (row 5) seems to produce the optimal quality of generalization. However, further increase of the initial expansion length (rows 6 and 7) reduces the generalization ability. We explain this phenomenon by the fact that steeper and steeper sine and cosine functions are applied. All experiments presented in Table 2 use the 2-nd order statistics for the final expansion. Choosing an order of statistics higher than 2 would not do any better, since it would also restrict the neuron's generalization capabilities.

The aim of the second group of our experiments was to test the optimal symmetrization method. For these experiments, the whole sample set was initially expanded by sines and cosines in order to increase the stimulus vector length to 20. The produced 20 stimulus elements were treated as original and independent data, with possibly different phase distributions. Then the available symmetrization methods were applied. The results are shown in Table 3. Each row corresponds to one variant of stimulus symmetrization. Again, for all variants the vector length was the same (i.e. it remained 20) so that our comparison is fair.

Columns 2 and 3 of Table 3 have a similar meaning as columns 3 and 4 of Table 2 respectively, i.e. they describe the network performance in dependence on the chosen symmetrization variant. We see that the optimal symmetrization method outperforms the standard sigmoid method.

symmetrization	correct (train.set)	correct (test.set)
sigmoid	83%	81%
optimal, no contraction	99%	91%
optimal, contr.factor 0.9	99%	93%
optimal, contr.factor 0.7	99%	96%
optimal, contr.factor 0.5	99%	93%

Table 3. Symmetrization variants.

But even better results are obtained by the use of contraction combined with optimal symmetrization. Our explanation for this fact is the following. If the full phase range $[0, 2\pi)$ is used, then very small (≈ 0) and very big ($\approx 2\pi$) angles become close. So the neuron can easily regard some quite different stimulus data as being "similar" and produce a wrong generalization. When a contracted phase range is used, big and small angles are more clearly separated and distinguished. However, one must not exaggerate with contraction, as shown by row 5 of Table 3. Namely, choosing a shorter range results in smaller resolution for internally represented phase data.

6. Conclusions

To use holographic networks efficiently, one needs a choice of preprocessing methods. The standard methods of the HNeT system are suitable for problems where the stimulus consists of many values displaying Gaussian distribution. Our methods are useful in situations where the stimulus contains only few values displaying different distributions.

The ideas presented in this paper allow further investigation and experimenting. First, our optimal symmetrization procedure could be simplified for some special cases, e.g. for uniformly distributed discrete domains. Next, our expansion method could be realized with other functions instead of sines and cosines; it would be interesting to find out what is the optimal choice.

References

[1] D.B. GRADDY, A.H. SPENCER (1990) *Managing Commercial Banks*. Prentice Hall, Englewood Cliffs, New Jersey.

[2] R. HECHT-NIELSEN (1990) *Neurocomputing*. Addison-Wesley, Reading, Massachusetts.

[3] B. SOUČEK, M. SOUČEK (1990) *Neural and Masively Parallel Computers*. John Wiley, New York.

[4] J.G. SUTHERLAND (1990) "Holographic Model of Memory, Learning and Expression". *International Journal of Neural Systems*, Vol 1 No 3, pp. 256-267.

[5] J.G. SUTHERLAND (1990) *HNeT Development System, Version 1.0*. AND Corporation, Hamilton, Ontario.

[6] J.G. SUTHERLAND (1992) "The Holographic Neural Method". In *Fuzzy, Holographic and Parallel Intelligence* (B. SOUČEK, Editor). John Wiley, New York.

[7] S.M. WEISS and I. KAPOULEAS (1989) "An Empirical Comparison of Pattern Recognition, Neural Nets, and Machine Learning Classification Methods". *Proceedings of the 11th International Joint Conference on Artificial Intelligence IJCAI-89*, Detroit, Michigan, 20-25 Aug. 1989, Vol 1, pp. 781-787.

A Solution for the Processor Allocation Problem: Topology Conserving Graph Mapping by Self-Organization.

M. Dormanns
H.-U. Heiss

University of Karlsruhe
Department of Informatics
P.O. Box 6980, W-7500 Karlsruhe 1, Germany
E-mail: heiss@ira.uka.de

Abstract

We consider the problem of how to allocate the tasks of a parallel program to the processor elements of a multicomputer system such that the communication overhead of the program is minimized. This problem basically amounts to a graph embedding problem since both the program and the multicomputer can be modeled as graphs. The embedding problem can be characterized as the search for a topology conserving mapping of the source graph to the target graph.

To find such mappings we apply the Kohonen self-organization process. Our main concern in this article is to show how this graph embedding problem can be fitted to the Kohonen technique. To that end, we introduce feature vectors for each node of the source graph to provide topological information exceeding direct neighborhood and considering larger surroundings.

The particular optimization goal of the task mapping problem is being related to known results of the Kohonen process. The behavior and performance of our approach is illustrated by some sample graphs.

1 Introduction

The ever increasing demand for computational power led to the development of large scale parallel computers that already represent the prevailing supercomputer architecture. Their broader acceptance and dissemination, however, is hampered by a serious lack of software support. Architecture dependent programming and low processor utilization are the typical consequences. One of the problems to be solved is the question of how parallel programs should be mapped to a parallel computer to achieve economical machine usage and low response time.

In our paper we consider the mapping problem in *multicomputer systems*, i.e. multiprocessor systems where each processor has its own local memory. They are also known as *MIMD message passing systems*, since due to the absence of shared memory communication between processors has to be done by sending messages over a processor interconnection network. A parallel program is assumed to consist of a set of interacting tasks and the mapping determines which task is assigned to which processor. The goal of such a mapping is to distribute the load (tasks) as evenly as possible over the network while keeping communication costs low, which means that heavily communicating tasks should be placed close together.

Because this problem ist known to be NP-hard, several heuristics like iterative improvement, greedy strategies and problem solving methods from nature like simulated annealing and neural networks have been applied (for a comprehensive classification of problems and algorithms see [1]).

Neuhaus [2] presents a solution using a Genetic Algorithm. Bollinger and Midkiff [3] apply Simulated Annealing not only to the mapping problem, but also to the routing of the messages between the tasks through the communication network of the parallel computer. Other approaches solve the problem by formulating it as a quadratic assignment problem (like in VLSI design) and applying for example evolutionary algorithms [4] to it.

Hemani and Postula [5] present a solution to solve the similar VLSI cell placement problem. They also use a self-organization process based on Kohonen's algorithm to preserve neighborhood relations, but in such a modified way that known analytical results for Kohonen Networks cannot be applied any longer.

In this paper, the problem is formulated as a special instance of the graph mapping or graph embedding problem between the task interaction graph of the program (source) and the processor connection graph of the machine (target). We show how Kohonen's self-organization process can be employed to find a topology conserving mapping. If the mapping preserves topological relations between the nodes of the source graph then - this is the intuition - it is likely that adjacent source nodes will be mapped to adjacent target nodes resulting in minimum or at least low communication costs. Besides the exploitation of the available topological information our approach has the advantage that it can be easily embedded in a distributed load balancing algorithm capable of dynamic assignment decisions at run time.

2 Problem Definition

A multicomputer system can be represented as a so-called Processor Connection Graph (PCG) which serves as the destination graph of our mapping problem :

$$PCG \quad = \quad (P, E^P) \tag{1}$$

Every weighted (undirected) edge $(p,q) \in E^P$ of the PCG indicating the distance between p and q corresponds to the communication cost for the physical (bidirectional) link between two processors p and q. We assume that according to these weights a metric function can be defined to attach a distance to each pair of nodes of the PCG :

$$d : P \times P \rightarrow \mathbf{R}_0^+ \tag{2}$$

In our context, $d(p,q)$ measures the cost to communicate between processor p and q proportional to the length of the shortest weighted path, which is indeed a metric function.

The source graph of our mapping is an abstraction of the parallel program and is called the Task Interaction Graph (TIG) :

$$TIG \quad = \quad (T, E^T) \tag{3}$$

Again, every edge $(x,y) \in E^T$, representing a communication relationship between two tasks x and y, is labeled with a weight according to the distance what in our case is inversely proportional to the amount of communication $c_{x,y}$ between the two tasks.

Now, the aim is to find a mapping:

$$\pi : T \rightarrow P \tag{4}$$

that minimizes the communication costs, defined as:

$$CC \quad = \sum_{(x,y) \in E^T} c_{x,y} \cdot d(\pi(x), \pi(y)) \tag{5}$$

The intuitive hope is that a topology conserving mapping π will also minimize the communication costs. Depending on the particular application there may be further requirements concerning the properties of π, i.e. injectiveness.

3 Applying Kohonen Networks to the Problem

To define the problem in a way that is appropriate to apply Kohonen Networks, we switch from distances to correlations to describe relations between the nodes of the TIG. This is advantageous, because we can define more general, also non-metric relations between nodes without conceptual difficulties, and we avoid the value ∞ for a relation between nodes that we treat as not related at all, what would have led to numerical difficulties and problems during the self-organization process.

So the aim is to map strongly correlated nodes close together, which is equivalent to mapping nodes with small distances in between close together.

3.1 Direct Correlations between Nodes

According to the topological knowledge represented by the graph structure of the TIG, we are only able to relate nodes that are directly connected. To establish an ordered mapping in a reasonable amount of time, this could be not enough (think about the task to order natural numbers if you are only able to relate successive numbers). On the other hand, it depends on the particular problem to what extent it makes sense also to consider not directly connected nodes as correlated (in our problem we are primarily interested in minimizing distances between directly connected nodes). So we seek for a correlation measure between arbitrary non-connected nodes that is tunable with regard to both aspects: the diameter of the surrounding taken into account and how this correlation can be made commensurate with directly connected nodes. We therefore define the *direct correlation* between an arbitrary pair of nodes as:

$$c^*_{x,y} \quad = \quad \begin{cases} \max\limits_{\substack{paths \\ x = k_0 \ldots k_m = y \\ m \le r}} \left[\frac{d^{m-1}}{m} \sum\limits_{i=0}^{m-1} c_{k_i,k_{i+1}} \right] & \\ 1 \quad , \text{if } x = y \text{ (autocorrelation)} & \\ 0 \quad , \text{if no such path exists} & \end{cases} \tag{6}$$

The parameter $r \in \mathbf{N}$ determines the maximal pathlength, up to that nodes are treated as related; $d \in \mathbf{R}$, $0 < d \le 1$ is the decreasing factor that determines the order of magnitude a correlation between two nodes with a pathlength greater than 1 is de-

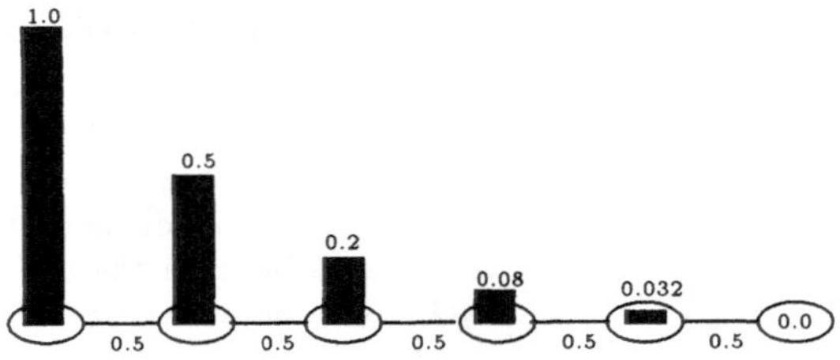

Figure 1: Fading of the direct correlations $c^*_{x,y}$ with increasing distance ($r = 4$, $d = 2/5$, $c_{a,b} = 0.5$)

creasing with growing pathlength[1]. This proposed measure should be general enough to cover a wide class of problems. To illustrate the effect of this calculation, see Figure 1.

3.2 Feature Vectors and Total Correlation

To apply Kohonen Networks we need to construct feature vectors and have to define a 'similarity measure' between those vectors.

A feature vector of a node is constructed out of its direct correlations with other nodes, and is then normalized to unity :

$$x \in T : x \mapsto \vec{x} : x_i = \begin{cases} l\dfrac{c^*_{x,i}}{\alpha_x} & , x \neq i \\ \dfrac{1}{\alpha_x} \quad \left(= \dfrac{c^*_{x,x}}{\alpha_x}\right) & , x = i \end{cases} \tag{7}$$

$$\text{with } \alpha_x = \left(1 + l^2 \sum_{\substack{i=1 \\ i \neq x}}^{|T|} {c^*_{x,i}}^2\right)^{1/2} \quad ; \quad l \in \mathbf{R}, l > 0 \tag{8}$$

We have to normalize the feature vectors, because unlike in other topological structures, in our case the lengths of the vectors say nothing about their relative positions to each other. This indeed has the disadvantage that only the relative distances to all other nodes are taken into account for each node, and not the absolute values. The factor l that is applied to all components except the autocorrelation component enables us to scale the influence of non-direct connections between nodes.

[1] The parameter r is only of pragmatic importance. The exponential decay of $c^*_{x,y}$ with growing pathlength also leads to a vanishing value for large distances between x and y, but with such a fixed diameter r calculation can be restricted to a much smaller set of nodes.

distance	$C^{tot}_{x,y}$	distance	$C^{tot}_{x,y}$
0	1.000000	5	0.038646
1	0.785930	6	0.011046
2	0.485783	7	0.002668
3	0.252270	8	0.000529
4	0.107336	≥ 9	0.000000

Table 1: Values for $C^{tot}_{x,y}$ with increasing distance ($r = 4$, $l = 3.5$, $d = 2/5$, topology of TIG: 2D-mesh with $c_{a,b} = 0.5$)

Now we can define the 'similarity measure' between feature vectors. This corresponds to what we call the *total correlation* and considers not only a path between the two nodes like the direct correlation, but also the paths from both nodes to another, distinct node. It is equal to the scalar product of the two feature vectors :

$$C^{tot}_{x,y} = \vec{x}^T \vec{y} \tag{9}$$

$$= \sum_{i=1}^{|T|} x_i \cdot y_i \tag{10}$$

If $x \neq y$ this is equal to:

$$C^{tot}_{x,y} = \frac{l}{\alpha_x \alpha_y} \Bigg[\underbrace{c^*_{x,y} + c^*_{y,x}}_{\text{direct}} + \underbrace{l \sum_{\substack{i=1 \\ i \neq x,y}}^{|T|} c^*_{x,i} c^*_{y,i}}_{\text{distinct nodes}} \Bigg] \tag{11}$$

Here we can see that the fraction of the total correlation due to correlations of the two nodes to a third, distinct node, can be scaled by the parameter l. If the nodes x and y are directly connected, this simplifies to:

$$C^{tot}_{x,y} = \frac{l}{\alpha_x \alpha_y} \left[2c_{x,y} + l \sum_{\substack{i=1 \\ i \neq x,y}}^{|T|} c^*_{x,i} c^*_{y,i} \right] \tag{12}$$

where the fraction due to direct correlation corresponds directly to the communication cost. To illustrate the effect of calculating $C^{tot}_{x,y}$ in this way, see Table 1.

3.3 Executing the Self-Organization Process

The aim is to map the TIG to the PCG. To do so, we apply the self-organization process, introduced by Kohonen, using the feature vectors as defined above and a network of neurons with the same structure as the destination graph. This enables us to implement and execute it in a very efficient way.

To every node $p \in P$ of the PCG (a processor element) corresponding to a unit of the Kohonen Network, a vector $\vec{p} \in \mathbf{R}^{|T|}$ is assigned that is modified during the self-organization process. For every node $x \in T$ its destination node $\pi(x) \in P$ is defined by:

$$x \in T \mapsto \pi(x) = p^* \in P : C^{tot}_{x,p^*} = \max_{q \in P}\{C^{tot}_{x,q}\} \quad (13)$$

which is equivalent to the condition:

$$||\vec{x} - \vec{p}^*||_2 = \min_{q \in P} ||\vec{x} - \vec{q}||_2 \quad (14)$$

The stimuli-vectors $\vec{x}$ are chosen with uniform probability and the vectors in the PCG are modified according to :

$$\vec{q}^{new} = \frac{\vec{q}^{old} + \epsilon \cdot h_{p^*,q} \cdot \vec{x}}{||\vec{q}^{old} + \epsilon \cdot h_{p^*,q} \cdot \vec{x}||_2} \quad (15)$$

where ϵ is a small value decreasing over time and $h_{p^*,q}$ is the neighborhood function with width σ, defined with respect to the metric function $d(p^*, q)$:

$$h_{p^*,q} = e^{-\frac{d(p^*,q)^2}{2\sigma^2}} \quad (16)$$

3.4 What does Topology Conservation optimize ?

In this section we want to analyze what is optimized by Kohonen's self-organization process regarding our special graph mapping problem and show the relations to the goal of minimizing communication costs in the processor allocation problem.

If there is only a discrete set of feature vectors, we know from analytical results (see for example [6]) that the self-organization process minimizes the potential function:

$$V = \frac{1}{2} \sum_{p,q \in P} h_{p,q} \sum_{x \in F(q)} \frac{1}{|T|} (\vec{x} - \vec{p})^2 \quad (17)$$

where $F(q)$ is the receptive field of node $q \in P$, and $\frac{1}{|T|}$ corresponds to the probability with which $\vec{x}$ is chosen as stimulus vector (uniform distributed in our case).

Expanding $(\vec{x} - \vec{p})^2$ in (17) yields :

$$V = \frac{1}{2} \sum_{p,q \in P} h_{p,q} \sum_{x \in F(q)} \frac{1}{|T|} (||\vec{x}||_2^2 + ||\vec{p}||_2^2)$$
$$-\frac{1}{2} \sum_{p,q \in P} h_{p,q} \sum_{x \in F(q)} \frac{1}{|T|} \sum_{i=1}^{|T|} 2x_i \cdot p_i \quad (18)$$

Taking into account that all vectors are normalized to unity and assuming the cardinalities of the receptive fields to be more or less uniform, the first term is constant. So minimizing (18) is equivalent with maximizing :

$$V' = \sum_{p,q \in P} h_{p,q} \sum_{x \in F(q)} \sum_{i=1}^{|T|} x_i \cdot p_i \quad (19)$$

In the final phase of the self-organization process, and if $|P| \geq |T|$, most receptive fields only consist of one node of the TIG, and the vectors of the nodes in the Kohonen Network are already very similar to those of the nodes in the TIG of their receptive fields, so that the following approximation holds :

$$V' \approx \sum_{x,y \in T} h_{\pi(x),\pi(y)} \sum_{i=1}^{|T|} x_i \cdot y_i \quad (20)$$
$$= \sum_{x,y \in T} h_{\pi(x),\pi(y)} \cdot C^{tot}_{x,y} \quad (21)$$
$$= \underbrace{\sum_{(x,y) \in E^T} h_{\pi(x),\pi(y)} \cdot C^{tot}_{x,y}}_{\text{directly connected nodes}}$$
$$+ \underbrace{\sum_{(x,y) \notin E^T} h_{\pi(x),\pi(y)} \cdot C^{tot}_{x,y}}_{\text{not directly connected nodes}} \quad (22)$$

This can be maximized if those nodes that are directly connected, even with a high value $C^{tot}_{x,y}$, are mapped as close together as possible so that $h_{\pi(x),\pi(y)}$ is as large as possible. The terms in the second sum of (22) are smaller because the values $C^{tot}_{x,y}$ of not directly connected nodes are smaller (and so it is likely that also the values of $h_{\pi(x),\pi(y)}$ are smaller), but serve as important information for establishing a maximum ordered map during the global optimization phase.

Because $C^{tot}_{x,y}$ increases with growing amount of communication between the two tasks if they are directly connected and $h_{\pi(x),\pi(y)}$ decreases with growing distance $d(\pi(x), \pi(y))$, maximizing V' goes in the same direction as minimizing the communication costs CC (5).

4 Simulation Results

We have applied the procedure as described above to several example graphs. In this section some of our results are presented, especially those giving more insight into the procedure. Moreover, the selection of suitable parameters is discussed.

4.1 Injective Mapping

In some situations, it may be necessary to generate an injective mapping. Because the Kohonen process cannot guarantee this in every situation with a reasonable amount of time, we use a very simple after-treatment to enforce an injective mapping: every time more than one node of the TIG is mapped onto the same node of the PCG, all but one are distributed to the nearest nodes with no nodes assigned to.

Clearly, there are better strategies which can also be incorporated into the self-organization process (for example by dynamically scaling the norms of the vectors of the Kohonen Network differently, to make nodes with more than one node mapped onto less attractive for assignment), but because this is not the subject of this paper we restricted ourselves to such a simple strategy.

A non-injective (also called contractive) mapping is not necessarily more topology violating than an injective one, but corrupts our communication cost measure, because this can be trivially smaller with a non-injective mapping (since tasks assigned to the same node are assumed to incur no communication costs at all). So, for non-injective results we mention additionally the degree of parallelism achieved, and indicate the communication costs after applying the after-treatment as an upper bound for what actually can be attained.

But it should be mentioned that there are also circumstances where the way the self-organization process tends to be contractive is highly desirable, because contraction is not arbitrary, but only in those instances that lead to a huge decrease of the communication costs.

4.2 Parameter Dependencies

The interesting parameters are:

- r, the maximum pathlength up to that nodes are treated as related,
- d, the decreasing factor that determines the fading of correlations due to longer paths, and
- l, the scaling parameter to differentiate between direct and non-direct connections.

The other parameters (σ, ϵ and the number of iterations performed) are selected in the same way as in other applications and are not discussed here.

Common to all these parameters is that on the one hand, they have to be selected such that enough, also far reaching topological information is included in the feature vectors, enabling the self-organization process to establish an ordered mapping in a reasonable amount of time, and on the other hand that the optimization goal can be achieved. In our application, this means that direct connections are considered with precedence, so that r must not be too large, and that a maximum degree of parallelism can be achieved.

To make d and l independent of the order of magnitude of the edge weights in the TIG, these should be scaled to a common average (we used 0.5). The parameter r must be selected according to the degree of connectivity of the TIG, so that a reasonable number of nodes can be reached with paths of at most r edges. With a properly adapted r, the parameters d and l do not change very much for different problems. To show the effect of varying these parameters, we experimented with the problem to map an 8×8-mesh onto an 8×8-mesh. It turned out that $r = 4$ is a good choice leading to $d = 2/5$ and $l = 3.5$ as appropriate values (see Table 3 at the end of this paper).

These parameters are not sensitive to the size of the problem, but to the degree of connectivity of the TIG.

4.3 Examples with Irregular Graphs

To evaluate the quality of mappings generated by our procedure more realisticly, we choose some sample graphs as they could occur when solving partial differential equations with the finite element method (see Figure 2), and mapped them to a square mesh. The results are given in Table 2. Because in such irregular examples the optimum communication costs are usually unknown, we also indicate the minimum achieved costs of all trials. Another measure of the degree of topology conservation is the dilation which is the maximum pathlength up to that a singe edge of the source graph is stretched by the mapping. We believe that this is optimum or at least near optimum. One example for the distribution of the resulting communication costs (without after-treatment) is shown in Figure 3.

To illustrate the term 'Topology Conservation', Figure 4 shows the result of a mapping established by the self-organization process for the second graph. The nodes are numbered to show the established mapping. The original graph is mirrored at the diagonal, so that the upper left section is mapped onto the lower right. The right section of the mapped section III is shortened a little bit, to get space for section II that is broader than section I.

Figure 5 shows an example, where a contractive mapping is necessary. The figure shows the achieved partitioning of the TIG. As can be seen, the cardi-

graph	min. cost	if non-opt solution:			after-treatment:	
		avr. largest dilation	avr. cost	avr. parallelism	avr. largest dilation	avr. cost
a	61.5	2.70	55.15	57.23	4.60	69.51
b	67.0	4.04	65.43	57.42	5.84	81.68
c	69.0	6.46	67.63	55.04	6.72	81.90

Table 2: Results for the three irregular graphs. The parameters are: $r = 4$, $d = 2/5$, $l = 3.5$, $\epsilon = 0.6 \cdot 0.2^{t/t_{max}}$, $\sigma = 5 \cdot 0.1^{t/t_{max}}$, number of performed Iterations: $t_{max} = 5000$ (350 trials each).

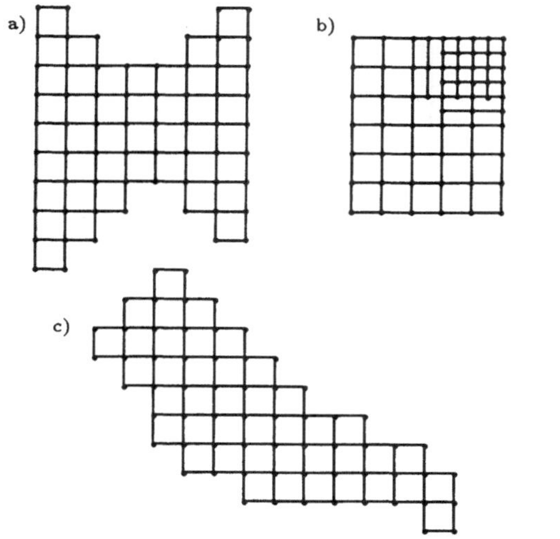

Figure 2: Three example graphs, consisting of 64 nodes each, with a) 106, b) 109 and c) 105 edges.

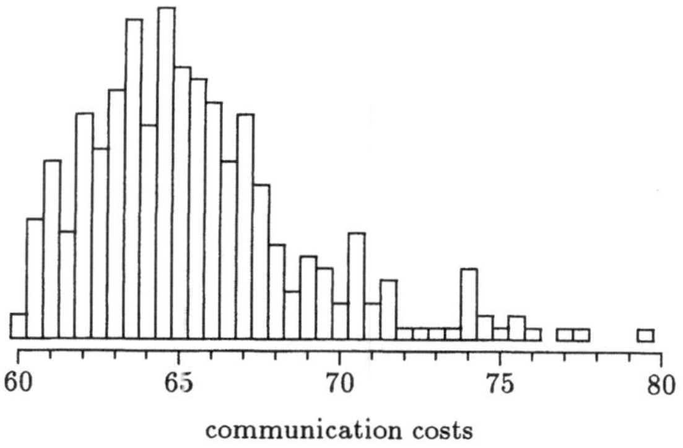

Figure 3: Distribution of the communication costs (before after-treatment) for example graph b (350 trials).

nalities of the partitions are not uniform. We think that the same mechanisms that could be incorporated into the self-organization process to enforce an injective mapping could also be used to establish a more uniform partitioning in the contractive case.

5 Conclusions

The aim was to solve the processor allocation problem in a way that minimizes the communication costs by mapping the Task Interaction Graph to the Processor Connection Graph as topology conserving as possible.

To do so, we developed a procedure to map arbitrary weighted graphs onto each other, using Kohonen's self-organization process. The only assumption is that a metric function on the destination graph can be defined.

Using known analytical results, we showed that this indeed makes the communication costs small. Simulation results for some non-trivial problems show the applicability of our procedure.

We think that the inherent parallelism that also matches the structure of the parallel computer, makes our procedure well suited to apply it to the processor allocation problem. However, some problems still have to be solved.

Because of the normalization of the feature vectors, our solution does not take into account differencies between the total amount of communication of different nodes, but only relative differences among edges of one node. A further step will be to consider this too, by varying the probabilities with which the feature vectors are presented as stimuli vectors, to map nodes and their adjacent nodes with greater importance with a better resolution.

We did also not consider the question how to make some nodes of the destination graph more or less attractive for a node to be assigned to it. This can be used to enforce injectivity, and, in case of the processor allocation problem with dynamic multiprogramming/multitasking, to make processors less attractive for an assignment of another task-node if they already carry heavy load.

We will consider this in our future work.

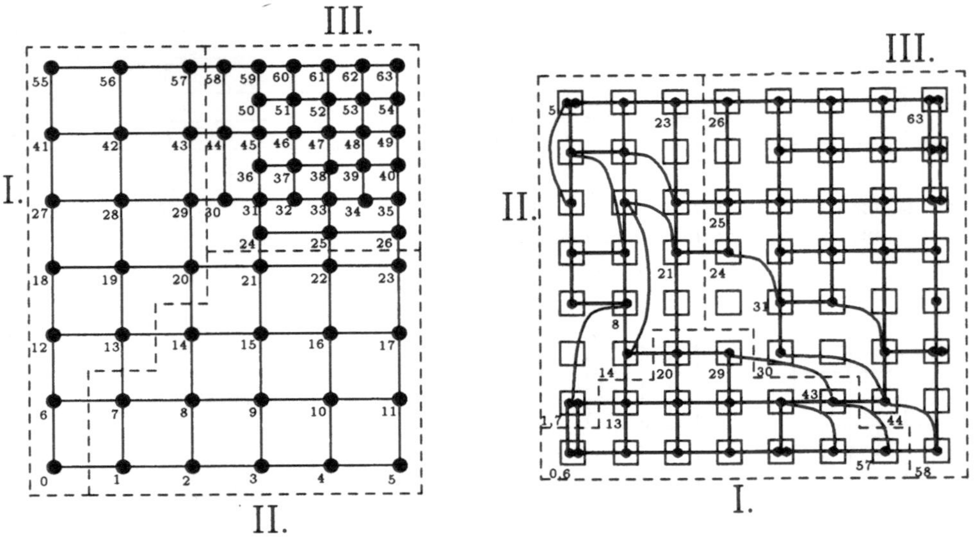

Figure 4: An example of an achieved mapping of graph b). The communication costs in this example are $CC = 63.5$

References

[1] Heiss H-U, *'Classification of Problems and Algorithms for Processor Allocation in Parallel Systems'*, Internal Report 7/91, University of Karlsruhe, 1991.

[2] Neuhaus P, *'Solving the Mapping-Problem - Experiences with a Genetic Algorithm'*, Parallel Problem Solving from Nature 1990, Proceedings, Springer LNCS 496, pp. 170-175.

[3] Bollinger S W and Midkiff S F, *'Heuristic Technique for Processor and Link Assignment in Multicomputers'*, IEEE Trans. on Computers, Vol. 40, No. 3, pp. 325-333, March 1991.

[4] Mühlenbein H , Gorges-Schleuter M and Krämer O, *'New solutions to the mapping problem of parallel systems: The evolution approach'*, Parallel Computing 4, pp. 269-279, 1987.

[5] Hemani A and Postula A, *'Cell Placement by Self-Organization'*, Neural Networks, Vol. 3, pp. 377-383, 1990.

[6] Ritter H, Martinetz T and Schulten K, *'Neural Networks'*, Addison Wesley 1991.

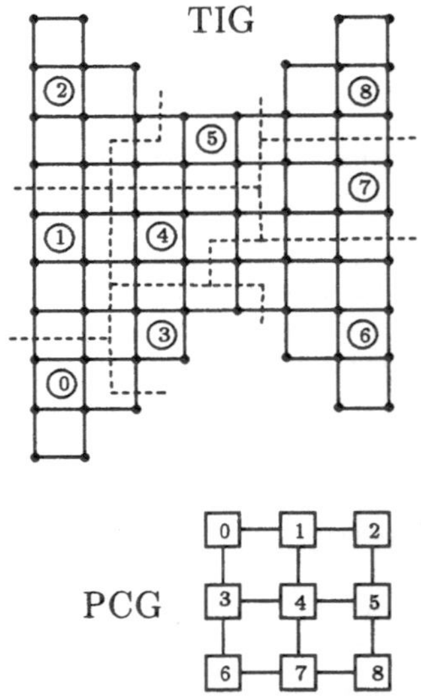

Figure 5: One example for a contractive mapping to a 3×3-mesh. The partitions are labeled according to the processors ($r = 4$, $d = 2/5$, $l = 3.5$, $\epsilon = 0.6 \cdot 0.2^{t/t_{max}}$, $\sigma = 1.7 \cdot 0.3^{t/t_{max}}$, $t_{max} = 1500$).

r	optimal solutions	if non-opt solution:			after-treatment:	
		avr. largest dilation	avr. cost	avr. parallelism	avr. largest dilation	avr. cost
1	3.25 %	6.96	81.81	55.53	7.35	92.33
2	53.8 %	3.01	62.53	60.79	4.04	68.43
3	69.5 %	2.23	58.33	62.04	2.86	61.48
4	68.8 %	2.36	58.42	62.18	3.13	61.47
5	69.3 %	2.20	57.95	62.13	3.05	61.31

d	optimal solutions	if non-opt solution:			after-treatment:	
		avr. largest dilation	avr. cost	avr. parallelism	avr. largest dilation	avr. cost
1/1.5	0.5 %	2.22	59.97	60.83	3.72	65.32
1/2.0	71.8 %	2.16	57.72	62.38	2.96	60.40
1/2.5	71.8 %	2.36	58.73	61.84	3.12	62.55
1/3.0	60.3 %	2.63	59.91	61.39	3.68	65.17
1/3.5	52.8 %	2.78	60.72	60.98	3.77	66.27
1/4.0	43.5 %	3.08	62.46	60.31	4.04	68.56
1/5.0	22.3 %	3.97	66.47	58.90	4.78	74.03
1/6.0	12.8 %	4.32	68.09	58.48	5.18	76.31

l	optimal solutions	if non-opt solution:			after-treatment:	
		avr. largest dilation	avr. cost	avr. parallelism	avr. largest dilation	avr. cost
1.0	6.8 %	2.74	61.37	60.15	3.67	66.86
1.5	31.0 %	2.53	59.41	61.53	3.34	63.31
2.0	52.8 %	2.43	59.18	61.51	3.48	63.63
2.5	69.3 %	2.30	58.64	61.85	3.26	62.31
3.0	71.5 %	2.34	59.12	61.70	3.45	63.53
3.5	70.0 %	2.24	58.53	62.05	3.00	62.01
4.0	66.8 %	2.21	58.02	62.23	2.96	61.04
4.5	58.3 %	2.22	57.98	62.31	2.79	60.47
5.0	43.5 %	2.30	58.00	62.37	2.64	59.98
6.0	21.0 %	2.43	58.51	62.01	2.71	60.29

Table 3: Simulation results for mapping an 8×8-mesh to an 8×8-mesh. The parameters r, d and l are varied. The other parameters are : $r = 4$, $d = 2/5$, $l = 3.5$, $\epsilon = 0.6 \cdot 0.2^{t/t_{max}}$, $\sigma = 5 \cdot 0.1^{t/t_{max}}$, number of performed iterations: $t_{max} = 2400$ (to have a significant amount of non-optimal solutions), 350 trials each. With $t_{max} = 5000$ we achieved a fraction of 98 % optimal solutions. The computation time to do this is 41 sec. on a SUN Sparcstation ELC (including preprocessing of the TIG to generate the feature vectors).

USING A SYNERGETIC COMPUTER IN AN INDUSTRIAL CLASSIFICATION PROBLEM

T. Wagner*, F. G. Boebel*, U. Haßler*, H.Haken†, D. Seitzer*

Abstract

Synergetic Computers (SCs) represent a class of new algorithms which can be used for different pattern recognition tasks. Due to their strong mathematical similarity with self-organized phenomena of physical nature they embody promising candidates for hardware realizations of classification systems. Until now there is still a lack of investigations concerning the importance of synergetic algorithms in the field of pattern recognition as well as concerning their practical performance. One of these synergetic algorithms (SCAP) will be examined in this paper with respect to pattern recognition capabilities. Its capacity of identifying wheels in an industrial environment is discussed. We show that with adequate preprocessing the SCAP reaches recognition rates of 99.3% under variable illumination conditions and even 100% with constant illumination. In addition to this, we try to specify the SCAP with respect to estabished pattern identification algorithms.

1 Introduction

One special pattern recognition task is the problem of identification. There are different attempts to tackle this problem, e.g. parameter estimation, nonparametric approaches, the search for discriminant functions, or neural approaches [6, 8]. But in spite of the theoretical diffences in motivation and deduction of the algorithms, there is also a fundamental equivalence between the different ways of pattern recognition [7].

From the user's point of view the focus is different, since performance, cost and reliability become the most important features. Therefore each algorithm has to be discussed with respect to both aspects, i.e. theoretical background as well as practical results, which is the topic of this paper.

*Fraunhofer-Institute for Integrated Circuits, Wetterkreuz 13, 8520 Erlangen, Germany, Phone: +49-9131-776544, Fax: +49-9131-776599, Email: wag@ep014.pmail.inel.iis.fhg.de

†Lehrstuhl für Theoretische Physik und Synergetik, Pfaffenwaldring 57/IV, 7000 Stuttgart, Germany

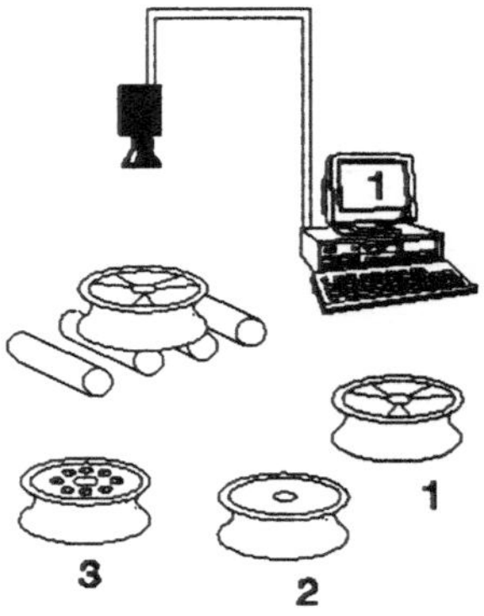

Figure 1: Setup of in situ data acquisition.

The practical problem under investigation was the optical identification of metal car wheels at a foundry (Figure 1), which were offered on a structured background and without information on location and orientation. We discuss the different aspects of identification under industrial conditions for the SCAP algorithm.

Section 2 gives a brief introduction to the theory of SCs in general and the SCAP in particular. Furthermore, we try to specify the SCAP with respect to other classification algorithms. Section 3 describes the experiments performed and serves as a basis for the discussion of the results in section 4. Since there is a close relation between physical synergetic phenomena and SCs, we briefly introduce some ideas for hardware realization of SCs in section 5.

2 Theory

2.1 General remarks

The field of Synergetics deals with self-organizing phenomena in nature [1]. Recently, attempts have been made to describe cognitive phenomena by means of Synergetics [3]. Synergetic Computers (SCs) represent a class of algorithms which make use of these mechanisms of self-organization in order to solve pattern recognition problems [2].

Usually synergetic phenomena are described by sets of differential equations, and therefore they are necessarily associated with the integration of these equations. A potential function which can be chosen arbitrarily describes the goal of the learning and the recognition process respectively.

The specific feature of a synergetic description of a system is that the large number of degrees of freedom in the system is replaced by a few order parameters, which govern the system's evolution in time ("slaving principle"). Therefore the basic features of the system become more obvious, and the number of differential equations under consideration is dramatically reduced.

An example of a synergetic differential equation recognizing the pattern $\mathbf{q}$ is

$$\begin{aligned}\dot{\mathbf{q}} &= \sum_{\mathbf{k}} \lambda_{\mathbf{k}} \mathbf{v}_{\mathbf{k}} (\mathbf{v}_{\mathbf{k}}^{+} \mathbf{q}) - \mathbf{A} \sum_{\mathbf{k} \neq \mathbf{k}'} (\mathbf{v}_{\mathbf{k}'}^{+} \mathbf{q})^{2} (\mathbf{v}_{\mathbf{k}}^{+} \mathbf{q}) \mathbf{v}_{\mathbf{k}} \\ & -\mathbf{B}(\mathbf{q}^{+}\mathbf{q})\mathbf{q}. \end{aligned} \quad (1)$$

(See 2.2 for explanation of the symbols; A and B are constants). Derivation and further explanation of synergetic differential equations can be found in [2].

SCs can perform supervised as well as unsupervised learning processes. Experiments on the practical performance of dynamical unsupervised and supervised learning algorithms for SCs are also the topic of our actual research and will be published soon.

Haken has shown [2] that one special SC algorithm can be reduced to solving a matrix inversion problem instead of integrating a system of differential equations. In order not to mix up the different SC configurations we will call this algorithm SCAP (Synergetic Computer using adjoined prototypes). Only the SCAP will be discussed here; all the other SCs lie beyond the scope of this paper. A comparison of the SCAP's performance with neural nets will be published soon [4].

Originally the SCAP algorithm was designed for autoassociation tasks, which means that there was only one representative for each pattern class. Nevertheless, a slight modification allows to use the SCAP algorithm for identification purposes: We simply average over all the learning samples of each class in order to get just one prototype vector per class.[1]

2.2 Description of the SCAP-Algorithm

In the following, images will be represented as vectors with N elements (N = number of pixels). For C different classes the C prototypes

$$\mathbf{v_i} = \begin{pmatrix} v_{1i} \\ v_{2i} \\ \vdots \\ v_{Ni} \end{pmatrix} \in \mathbf{R^N}, \qquad \mathbf{i} = 1, \ldots, \mathbf{C} \quad (2)$$

were generated by averaging over the P_i learning samples $\mathbf{q_{i,p}}$ of each class i

$$\mathbf{v_i} = \frac{1}{\mathbf{P_i}} \sum_{\mathbf{p}=1}^{\mathbf{P_i}} \mathbf{q_{i,p}} \quad (3)$$

and obey the conditions

$$\sum_{\mathbf{n}=1}^{\mathbf{N}} (\mathbf{v_{ni}}) = 0 \quad (4)$$

$$\sum_{\mathbf{n}=1}^{\mathbf{N}} (\mathbf{v_{ni}^2}) = 1. \quad (5)$$

We are looking for C adjoined prototype vectors $\mathbf{v_i^+}$ which shall fulfill

$$\mathbf{v_i^+ v_j} = \delta_{\mathbf{ij}} \quad (6)$$

in order to identify an arbitrary test pattern $\mathbf{q}$ into the class i with

$$|(\mathbf{v_i^+ q})| = \mathbf{max_j}\{|(\mathbf{v_j^+ q})|\}, \qquad \mathbf{j} = 1, \ldots, \mathbf{C}. \quad (7)$$

The appropriate adjoined prototype vectors are assumed to be a linear superposition of the transposed prototypes $\mathbf{v_k^T}$:

$$\mathbf{v_i^+} = \sum_{\mathbf{k}=1}^{\mathbf{C}} \mathbf{a_{ik}} \mathbf{v_k^T}. \quad (8)$$

[1] Note that "really" synergetic learning algorithms do not average; they can cope with more than one pattern per class, but for the corresponding numerical approach sets of differential equations like Equation (1) have to be integrated.

Multiplying the C Equations (8) with the C $\mathbf{v_j}$'s from the right we get C^2 equations

$$\delta_{ij} = \sum_{k=1}^{C} a_{ik}(\mathbf{v}_k^T \mathbf{v}_j), \qquad i,j = 1,\ldots,C. \tag{9}$$

Defining the matrices W and A

$$\mathbf{W} := \left((\mathbf{v}_k^T \mathbf{v}_j)\right) \tag{10}$$

$$\mathbf{A} := (a_{ik}) \tag{11}$$

and the unit matrix **I**, Equation (9) reads

$$\mathbf{I} = \mathbf{AW}. \tag{12}$$

Equation (12) can be solved by

$$\mathbf{A} = \mathbf{W}^{-1} \tag{13}$$

and from that the adjoined prototypes $\mathbf{v}_i^+$ can be calculated.

Note that the matrix W can only be inverted if the prototypes $\mathbf{v_i}$ are linearily independent. This implies that for the SCAP the number of features always has to be higher than the number of the classes which have to be separated.

In order not to lose information, the square of the length or an arbitrary constant reference level and the offset can be added to the vector as additional components before normalizing it:

$$\mathbf{v} = \begin{pmatrix} v_1 \\ v_2 \\ \vdots \\ v_N \end{pmatrix} \in \mathbf{R}^N \rightarrow \mathbf{v} = \begin{pmatrix} v_1 \\ v_2 \\ \vdots \\ v_N \\ L \\ O \end{pmatrix} \in \mathbf{R}^{N+2} \tag{14}$$

$$\mathbf{L} = \frac{1}{N} \sum_{n=1}^{N} (v_n) \tag{15}$$

$$\mathbf{O} = \frac{1}{N} \sum_{n=1}^{N} (v_n^2) \quad \text{or} \quad \mathbf{O} = \text{const.} \tag{16}$$

This represents the mapping of the N-dimensional feature space to a (N+2)-dimensional hypersphere.

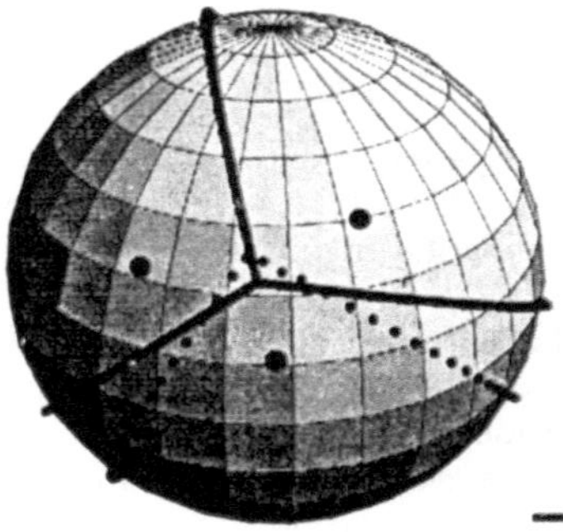

Figure 2: Normalized feature space of a MD- and a SCAP- classsifier: The separating hyperplanes depend on two prototypes each for the MD and all prototypes each for the SCAP.

2.3 Comparison with Minimum Distance Classifier

The conventional classifier which is most similar to the SCAP is the Minimum Distance Classifier using Euclidian measure [8]: It also consists of only one prototype per class and uses the same distance measure.[2].

The differences between these two classifiers lie in the fact that the SCAP algorithm separates regions of the two different classes i and j by a hyperplane **h** constructed from the adjoined prototypes by

$$\mathbf{v}_i^+ \mathbf{h} = \mathbf{v}_j^+ \mathbf{h}, \qquad \mathbf{h} \in \mathbf{R}^N, \tag{17}$$

whereas the minimum distance classifier only uses the prototypes:

$$\mathbf{v}_i^T \mathbf{h} = \mathbf{v}_j^T \mathbf{h}, \qquad \mathbf{h} \in \mathbf{R}^N. \tag{18}$$

A consequence of this is that for the Minimum Distance Classifier (MD) the separating hyperplane between two classes is only a function of these two classes, whereas for the SCAP it is a function of all classes (since the plane depends on the two adjoined prototypes which themselves depend, according to Equation (8) on all prototypes).

Figure 2 shows three prototypes and the regions in the feature space which belong to the different classes.

[2] Due to normalization of the feature vectors, all the vectors lie on the unit hypersphere. Therefore Euclidian measure and dot product are equivalent.)

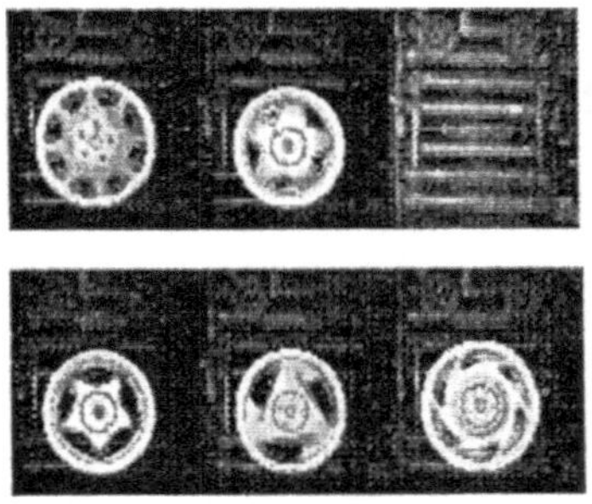

Figure 3: Examples of different metal car wheels to be identified (64*64 pixels).

3 Experiments

3.1 Setup

In our experiments metal car wheels produced in a foundry had to be optically identified (Figure 1).

Due to the founding process, the contours of the wheels could be distorted by rests of aluminum or may simply be dirty. Furthermore, the position and orientation of the wheels was unknown. An image resolution of 64*64 pixels proved to be sufficient (see Section 4.3). Examples of some wheel types are shown in Figure 3. Note the metallic reflections of the background. During a field test at the factory lasting two days, a total of 417 images of 6 wheel types and a class of empty images were aquired. The number of representatives for each class was quite different (Table 1).

Class	1	2	3	4	5	6	7	$\sum$
Samples	156	80	16	46	49	34	36	417

Table 1: Number of representatives per class. Class 3 consists of empty pictures.

The illumination changed abruptly after the first 107 images, because the main illumination of the room was switched off. Nevertheless, one prototype was constructed for images with different illumination belonging to one class. According to Equation (3) averages over the learning sample had to be calculated. The first P_i images of each class represented the learning sample and the rest was used for testing.

3.2 Preprocessing

3.2.1 2D-FFT and Polar Coordinates

A general and robust way of creating a shift- and rotation-invariant pattern in our experiments was the well known combination of twodimensional Fast Fourier Transformation (FFT) and polar coordinates. A description of preprocessing algorithms used here can be found in [5]. The FFT of a 64*64 pixel image was transformed to polar coordinates using a bilinear interpolation. An additional one-dimensional FFT with respect to the angle-coordinate generated a shift- and rotation-invariant pattern. During the rest of this paper, this preprocessing alternative will be referred to as "PP1".

3.2.2 Separation, Polar Coordinates and Correction of Distortion

A different and less general approach for preprocessing was to center the wheel using centroidal coordinates instead of a 2D-FFT. At first, each image was separated from the background using a fixed grey value as threshold. The center of mass was calculated and the separated image of the wheel was centered and transformed to polar coordinates. In order to investigate the effect of simple corrections of distorting effects of the optical lens system the radius of the transformed image was normalized to an average value. Again a one-dimensional FFT with respect to the angle coordinate generated invariance with respect to rotation. This solution for preprocessing is referred to as "PP2".

Due to the separation algorithm it is impossible to classify empty pictures with this preprocessing routine, because the normalization spreads the dark grey levels of the empty picture over the entire grey level scale and thereby causes the threshold operation to fail. As a consequence, the results for this preprocessing routine are calculated neglecting the class of empty pictures.

3.3 Rejection

To further decrease the number of misclassified patterns we made use of a rejection level for the dot product in Equation (7). Only results beyond a certain level were accepted and all others rejected. Though these experiments are not discussed here, it is important for the user to know that it is possible for the algorithm to either increase the recognition rate as much as possible or to reduce the number of misclassified patterns on the cost of the total recognition rate.

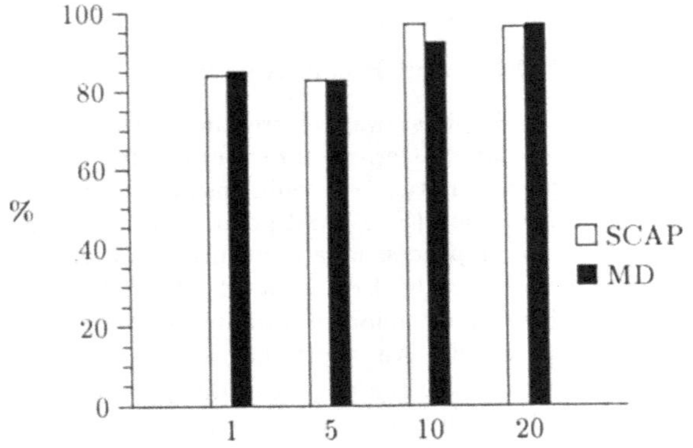

Figure 4: Recognition rate for PP1 as a function of the size of the learning sample for the SCAP and the MDC.

4 Results

4.1 2D-FFT and Polar Coordinates

The results for the images which were preprocessed via PP1 are shown in Figure 4 for the SCAP and the MD classifier. The size of the learning sample used varies from 1 to 10 patterns per class. The recognition rate for the SCAP rises from 84.1% for one representative per class to 96% for 20 representatives per class. The reduction of the recognition rate from 10 to 20 patterns per class is simply a numerical effect: The number of misclassified patterns remains constant (11 errors), but the size of the test sample changes.

The results for the MD classifier are similar to those of the SCAP. If the patterns of each class are distributed gaussian-like with equal variance for each class, the MD classifier with Euclidian distance measure is the best classifier with one prototype per class [8]. The fact that the MD classifier does not always beat the SCAP results from the fact that in practice the distributions within the classes do not necessarily obey normal distributions.

Note that the learning sample consists always of the first P_i patterns of each class, though there is a sudden change in illumination after the 107th image. The results therefore represent the perfomance of a SCAP for wheel recognition with illumination changing during the recognition process.

In an online demonstration this way of performing preprocessing and identification proved to be quite robust with respect to partially covering the wheel with a hand or a sheet of paper and with respect to tilting. Figure 5 shows some examples of properly recogized distorted images. This experiment was performed in the lab with uniform backround and 19 different wheel types.

Figure 5: Examples of distorted images which were recognized properly (64*64 pixels).

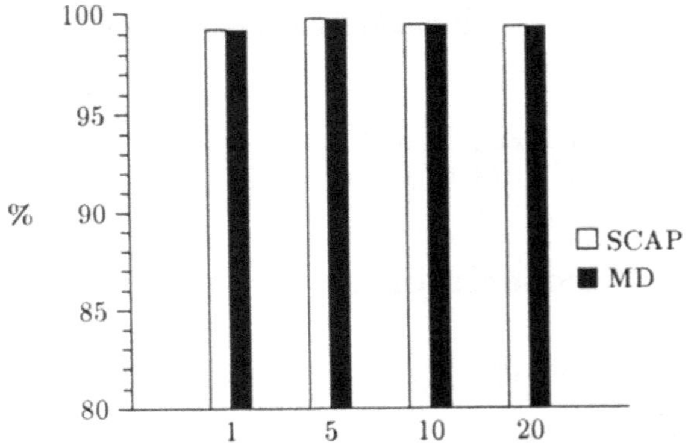

Figure 6: Recognition rate for PP2 as a function of the size of the learning sample for SCAP and MDC. Note the different scale.

4.2 Separation, Polar Coordinates and Correction of Distortion

Figure 6 shows the results for the second set of preprocessing algorithms (PP2). One can clearly see that the preprocessing has become appropriate enough that the difference between MD and SCAP classifier almost vanishes. The recoginition rate for the SCAP rises from 99.2% for one pattern per class to over 99.4% for 10 patterns to 99.3% for 20 patterns (2 errors). Note that these results do not include the classification of the empty pictures, because the separation algorithm does not work on them. We emphasize that a recognition rate of 100% is obtained if the data set is divided into two different series with constant illumination each.

A field test for a longer period of time will be performed during the next months in order to determine the error rate for the SCAP identifier with the improved preprocessing unit PP2.

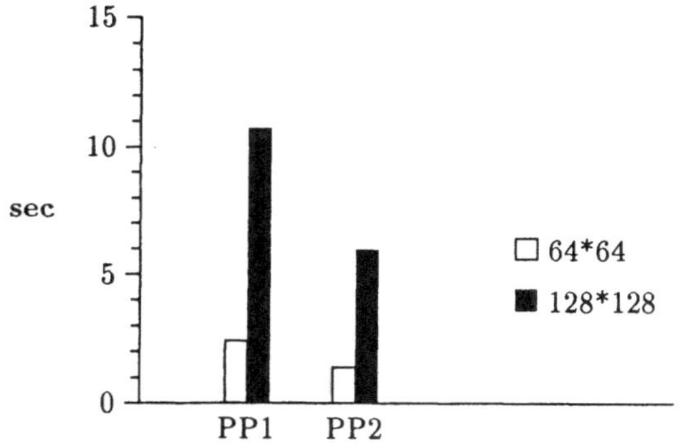

Figure 7: Typical calculation times for different resolutions and preprocessing routines (PC 486).

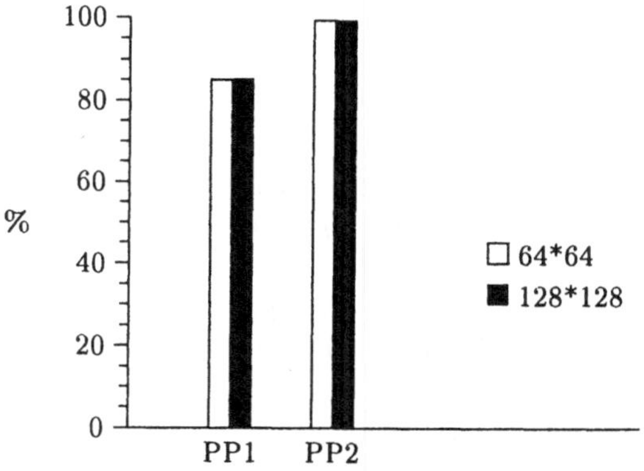

Figure 8: Recognition rate as a function of transformation and resolution.

4.3 Resolution and Time

In Figure 7 one can see typical calculation times for the sets of transformations as a function of image resolution and the sort of preprocessing performed. For our application task, there was no need for further time optimization, but we do not claim that the speed of the routines is already as high as possible.

In our case, a resolution of 64*64 pixels proved to be sufficient. Figure 8 shows the recognition rate as a function of the preprocessing routines for a learning sample size of 10 patterns per class, but the results do not change significantly with higher resolution.

The learning time (assuming that the images are already preprocessed) lies in the range of approximately 5 to 10 seconds for 19 classes and is basically determined by the efforts for the inversion of a C*C matrix.

For recognition the dot products with the prototypes of each class have to be calculated. Therefore the number of multiplications and additions ranges at about C*N.

5 Hardware Realizations

Due to their striking mathematical similarity with self-organized phenomena in physical nature, synergetic algorithms represent promising candidates for hardware realizations.

With the field of Synergetics Haken [1] has proposed a mathematical formalism to cope with self-organized phenomena. Synergetic Computers can be described by the same sort of differential equations which are used to model natural self-organized effects.

Haken discussed a system of coupled lasers which performs a synergetic decision. There are also considerations on the complete setup of an optical synergetic classification system using holographic elements for generating the order parameters.

Investigations of electronic elements based on self-organization (so-called "Synistors") are currently under work. Promising candidates are III-V devices, which employ current filaments in GaAs-structures. They are supposed to show the synergetic winner-takes-all law and could be used as an important element in an integrated SC. Due to their high parallelity the theoretical peak performance is about 10^3 to 10^5 times higher than the newest generation of image processing hardware. A detailed proposal for future hardware realization of synistors will be published soon.

6 Conclusion

For the first time the applicability of the so called SCAP, a special form of a Synergetic Computer, was tested in an industrial identification problem at a factory site. The combination of the SCAP identifier and a quite general preprocessing unit for shift and rotational invariance yielded satisfying results and was surprisingly stable with respect to tilting the object. Furthermore we showed how the recognition rate can be increased up to 99.3% for variable illumination and even up to 100% for constant illumination by improving preprocessing. We discussed the similarity of the SCAP with a minimum distance classifier, though we know that these statements do not hold for arbitrary SCs. We emphasize that having shown the satisfying performance of synergetic algorithms, the huge potential of synergetic hardware is obvious.

Further efforts will aim on industry proof unsupervised learning algorithms for Synergetic Computers and to real world implementation of a Synergetic Computer hardware.

References

[1] Haken H, *'Synergetics. An Introduction'*, Springer, Berlin, 1983.

[2] Haken H, *'Synergetic Computers and Cognition. A Top-Down Approach to Neural Nets'*, Springer, Berlin, 1991.

[3] Haken H and Stadler M (ed.), *'Synergetics of Cognition'*, Springer, Berlin, 1990.

[4] Schramm U et al., *'A Practical Comparison of Synergetic Computer, Restricted Coulomb Energy Networks and Multilayer Perceptron'*, World Conference of Neural Networks, Portland, Oregon, 1993. (submitted)

[5] Jähne B, *'Digitale Bildverarbeitung'*, 2nd ed., Springer, Berlin, 1991

[6] Niemann H, *'Klassifikation von Mustern'*, Springer, Berlin 1983

[7] Schürmann J, *'Neuronale Netze und die klassischen Methoden der Mustererkennung'*, FhG-Berichte 1/91

[8] Duda R and Hart P, *'Pattern Classification and Scene Analysis'*, Wiley & Sons, New York 1973

Connectionist Unifying Prolog

Volker Weber
University of Hamburg, Computer Science Department, Natural Language Systems Division
D-2000 Hamburg 50, Bodenstedtstr. 16, Germany
e-mail: weber@nats4.informatik.uni-hamburg.de

Abstract

We introduce an connectionist approach to unification using a local and a distributed representation. A Prolog-System using these unification-strategies has been build.

Prolog is a Logic Programming Language which utilizes unification. We introduce a uncertainty measurement in unification. This measurement is based on the structure-abilities of the chosen representations. The strategy using a local representation, called ℓ-CUP, utilizes a self-organizing feature-map (FM-net) to determine similarities between terms and induces the representation for a relaxation-network (relax-net). The strategy using a distributed representation, called d-CUP, embeds a similarity measurement by its recurrent representation. It has the advantage that similar terms have a similar representation. The unification itself is done by a back-propagation network (BP-net).

We have proven the systems adequacy for unification, its efficient computation, and the ability to do extended unification.

1 Introduction

Unification is generalized matching. It can be seen as the process of finding unifying substitutions or a unifier: Mappings like $\{X \leftarrow cat, Y \leftarrow cat\}$ are called substitutions. They can be applied to terms resulting in objects that differ only in that each variable appearing to the left of an arrow is replaced by the corresponding term to the right of the arrow. A unifier for two terms is a set of substitutions whose application to either of them produces the same result. Thus each of the substitutions above is a unifier for the terms `hairy(X)` and `hairy(Y)`. For these two terms the effect of any unifier can be obtained by first applying the substitution $\{X \leftarrow Y\}$ and then instantiating further. We call $\{X \leftarrow Y\}$ a most-general-unifier (or mgu for short) for `hairy(X)` and `hairy(Y)`.

Applications for unification There is a wide field of applications for unification like database systems, knowledge acquisition, computer graphics, natural language processing, expert systems, algebra, deduction- and rewrite-systems, as well as logic programming [12].

Unification Rules Classical symbolic unification-algorithms just consist of four rules for unification and two for termination. If t_1 and t_2 are the terms to be unified (**X** variable, t term, **f**, **g** functors, x arbitrary symbol) a system $\mathcal{R} = \{(t_1 = t_2)\}$ is said to be unifyable if after termination of the following algorithm the set $\mathcal{R}'$ equals $\emptyset$ [5]:

1. terminating rules (have to be used, if match):
   $(\mathtt{f}(\cdots)=\mathtt{g}(\cdots)) \in \mathcal{R} \wedge \mathtt{f}\neq\mathtt{g}$ (CLASH)
   $(\mathtt{X}=t) \in \mathcal{R} \wedge \mathtt{X}$ occurs in t (OCCURRENCE)
2. unification rules:
   $(\mathtt{f}(t_{1_1},\cdots,t_{1_n})=\mathtt{f}(t_{2_1},\cdots,t_{2_n}))\in\mathcal{R} \Rightarrow \mathcal{R}' := \mathcal{R}\backslash(\mathtt{f}(t_{1_1},\cdots,t_{1_n}) = \mathtt{f}(t_{2_1},\cdots,t_{2_n})) \cup \{(t_{1_1} = t_{2_1}),\cdots,(t_{1_n} = t_{2_n})\}$ (DECOMPOSITION)
   $(x=x) \in \mathcal{R} \Rightarrow \mathcal{R}':=\mathcal{R}\backslash(x = x)$ (TAUTOLOGY)
   $(\mathtt{X}=t) \in \mathcal{R} \Rightarrow \mathcal{R}':=\{\{\mathtt{X}\leftarrow t\}\sigma|\sigma \in \mathcal{R}\}$ (APPLICATION)
   $(t=\mathtt{X}) \in \mathcal{R}\Rightarrow\mathcal{R}':=\mathcal{R}\backslash(t=\mathtt{X}) \cup \{(\mathtt{X}=t)\}$ (ORIENTATION)

After the application of a unification-rule the set $\mathcal{R}'$ becomes again $\mathcal{R}$ and the algorithm iterates since $\mathcal{R}$ becomes $\emptyset$, or a termination rule has to be used. This rule-system seems to be rather small and to be computed efficient. Automatical theorem provers, like Prolog, however, spend often more than 50% of time for unification. For an efficient computation a skill selection of unification-rules is needed. Also the occur-check is very expensive. Some strategies use infinite unification which does not need to check occurrence.

Strategies Herbrand marks the beginning of logic programming with the first prototype of a unification algorithm [8]. The widest known algorithm is that of Robinson [23, 24]. This algorithm has a exponential time-complexity. Boyer and Moore's 'structure sharing'-algorithm reduced space-complexity but time complexity was still as bad as Robinson's [2]. In 1975 Venturini-Zilli introduced a marker strategy which reduces time-complexity to $O(n^2)$ [34]. Huet gave a almost linear algorithm for a infinite unification strategy (strategy which need not to check occurrence) [10]. Paterson and Wegman introduced the first linear strategy [18]. The disadvantage of their method is a relative complex data-structure which leeds to time consuming computation in the average case. Later approaches often have time-complexities worse to their approach in worst case [15]. There are also some parallel algorithms [7, 35, 39], but the sequential nature of unification has been shown [4, 39]. Some unification strategies use special hardware (transputers, unification-coprocessors, Prolog-machines) [6, 11, 25].

Most connectionist approaches implement only parts of unification. Also, the well known approach of Ballard [1] belongs to this category. Those of Müller, Stolcke and Hölldobler [9, 17, 29] implement unification for different purposes. Their approaches are based on symbolic strategies and implemented with a high degree of parallelism.

Our approach Our approach implements full unification. It is also possible to use infinite unification, and other extensions. Moreover it is possible to set fuzzy-slots which are basis for the uncertainty-management. For this it is in addition possible to fuzzy-unify (fuzzify) terms which are similar. We saw the possibility to build a system with

- efficient computation,
- extended unification,
- fuzzification,
- exact, e.g. proveable and
- adequate strategy and representation.

The next section introduces extensions to unification, like E-unification (unification under a certain equational-theory E) or fuzzification. Section 3 explains the used representations. The basic architecture for the overall system is described in the next section. We will focus on the strategies used by ℓ-CUP and d-CUP in section 5. There we also introduce the networks themselves. The results show some experiments with the system. Before the conclusion we discuss the results and system in comparison to other approaches.

2 Extended Unification

E-Unification There exist some extensions to simple unification like the so called E-unification. E is an equational theory defined by a set of axioms. It induces the reflexive, transitive and antisymetric congruence $=_E$ on the set of terms. If we have a set of E-equations $\{s_1 =_E t_1, \cdots, s_n =_E t_n\}$ than $f(s_1, \cdots, s_n) =_E f(t_1, \cdots, t_n)$ is E-unifyable. In such a theory the terms s, t are called E-equal, iff $s =_E t$. A E-unification with empty equational theory is called syntactic (kind of unification normally used in Prolog). A unification with unrestricted equational theory is called semantic [3]. In Prolog there are only two terms to be unified. It is, however, possible to unify more than two terms. The couple of terms to be unified is called an equational-system. For a E-unification-problem it is necessary to calculate the minimum of solutions. From this it is possible to get all other unifiers of a equational-system. A set of unifiers for an equational-system is called complete, iff every unifier of the equational-system is an instance of one of its elements. This set is called minimal, iff non of its elements is an instance of another. There are four unification-classes: Unitary unifying theories ($\mathcal{U}_u^E$, iff for every unification-problem exists a minimal complete solution-set), finite unifying theories ($\mathcal{U}_f^E$, iff minimal complete solution-sets exist and this sets are finite), infinite unifying theories ($\mathcal{U}_i^E$, iff minimal complete solution-sets exist but there is at least one set which is infinite) and theories of type zero ($\mathcal{U}_0^E$, iff exists a unification-problem in this theory which does not have a minimal complete solution-set). Prolog uses a unitary unification.

Fuzzification The unification strategy described in this paper uses another extension to unification. We call it fuzzification. It was mentioned by Stolcke in 1988 as fuzzy-unification [29, p.88]. It is a technique which allows not only strict unification but also unification of terms which are more or less similar. Therefore a similarity measure has to be introduced. In symbolic systems someone has to define a similarity criterion or if we look at fuzzy logic we have to define fuzzy sets and membership-functions. We make use of the fact that different activations of units may be seen as a similarity measure for terms. In section 3 we see representations which allow to encode similarities of constants or terms (figure 1 and 2).

A Prolog-dialect with Fuzzy-slots In our approach we control the certainty from the symbolic system by fuzzy slots. A fuzzy slot is defined by the predicate:
`cup_fuzzy_slot(`*Fuzzy variable, Fuzzy term, Belief*`)`.
It gives every term compatible with the fuzzy variable a particular belief-value in a context which is compatible with the fuzzy term. For example: By `cup_fuzzy_slot(X, hunt(X, cat(tom)), low)` we set a low belief-value for the first argument of `hunt/2` in the context of `cat(tom)`. If we think of a data-base were `hunt/2` is ment as `hunt(Actor, Recipient)` not only a dog could hunt a cat but in the case of `cat(tom)` also the mouse `jerry` could hunt `tom`. This is possible, due to fuzzification, even if no rule or fact for this is in the data-base.

3 Representations

It is necessary to be able to code terms exact, cause unification is a exact method. Therefore we have to draw special attention to the representation-problem, cause "a poor presentation will often doom the model to failure, and an excessively generous representation may essentially solve the problem in advance" [28].

Matrix-representation for ℓ-CUP For ℓ-CUP we utilize a representation similar to that of Hölldobler [9]. This representation may be called local cause there is one unit for every constant at every (argument-)position of a term. On the other hand there is no local coding of every possible term or every possible substitution (ref. [1]). We represent terms as follows 1: Atoms/functors (constants) are represented by a cube of vectors. Each vector represents one constant. Variables are represented by variable-vectors. All column-vectors (those of the cube and the variable-vectors) have the same number of components. Each component of a vector (or layer of the cube) represents one position of a term. A variable-binding is represented by corresponding

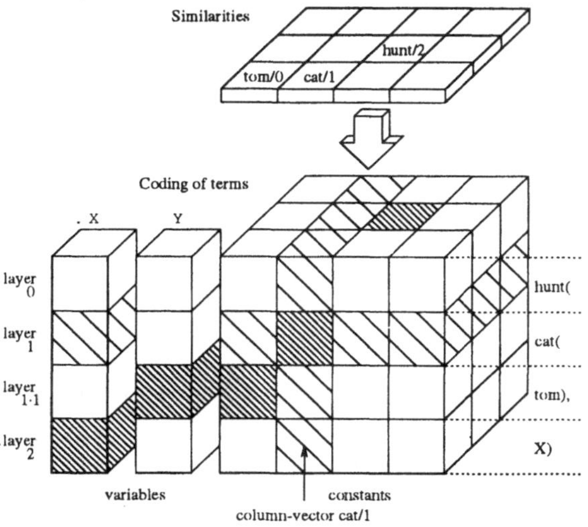

Figure 1: Coding of terms and mapping of similarities

activation-patterns of a variable-vector and a column-vector. For example: In figure 1 variable **Y** is bound to constant **tom/0**. The depicted coding represents the term **hunt(cat(tom),X)**. The units of the representation are activated with values between 0.0 and 1.0 corresponding to their certainty. A FM-net [13] is trained with a symbolic metric to get a order of vectors. This order represents similarities between constants cause of the symbolic metric (see section 5.1.1).

TCO-representation for d-CUP For d-CUP we utilize a representation similar to HRR (Holographic Reduced Representation [20, 21]. Plate introduced this circular convolution based representation for short structures, sequences, and variable-bindings. The convolution-operation given by $\vec{x} \odot \vec{y} = \vec{z}$ with $z_i = \sum_{j=0}^{n-1} x_j \cdot y_{n-j+i}$. There exists now exact inverse of circular convolution. Plate makes use of the circular correlation - an approximative inverse of circular correlation - which may be defined as $\vec{x} \odot \vec{y}^{\star}$ ($\vec{x}^{\star} = \vec{z}$ with $z_i = x_{n-i}$). We construct a new operation called TCO (**T**erm **C**oding **O**peration). It is based on HRR and defined for vectors $\vec{x}, \vec{y} \in [0,1]^n \cup \vec{1}$ by $\vec{x} \otimes \vec{y} = \frac{2}{n} \cdot (\vec{x} \odot (\vec{y} \oplus \vec{1}))$, where $||\vec{x}||_{L_1} = ||\vec{y}||_{L_1} = \frac{n}{2}$ ($||\vec{x}||_{L_n} = \sqrt[n]{\sum |x_i^n|}$, $\vec{1} = (\frac{n}{2}000 \cdots 0)$). The operation $\vec{x} \oplus \vec{y}$ is defined by the vector addition where every component is divided by 2. For the TCO we found the exact inverse every time during experiments. An advantage for representing structured terms is the non-commutativeness of the TCO. This makes it possible to represent order. Other features which are also advantages of the HRR as the associativeness and the great influence of a leftmost member of a sequence for the coding are preserved. Moreover the coding-vectors are all normalized to the sum of absolutes (L_1-norm; $\frac{n}{2}$). The coding of a term with TCO is done in the following way:

> Let $\mathbf{f}(t_1, \cdots, t_m)$ be a term
> (arity m, functor $\mathbf{f}$, arguments $t_1, \cdots t_m$),
> $\vec{t_i}$ the coding of argument t_i, and
> $\vec{f} \in \{0,1\}$, $||\vec{f}||_{L_1} = \frac{n}{2}$ a coding for the functor.
> The TCO-representation of this term is:
> $\vec{f} \otimes \vec{t_1} \otimes \cdots \otimes \vec{t_m}$.

A coding of the term **hunt(cat(tom),X)** will be depicted in figure 2.

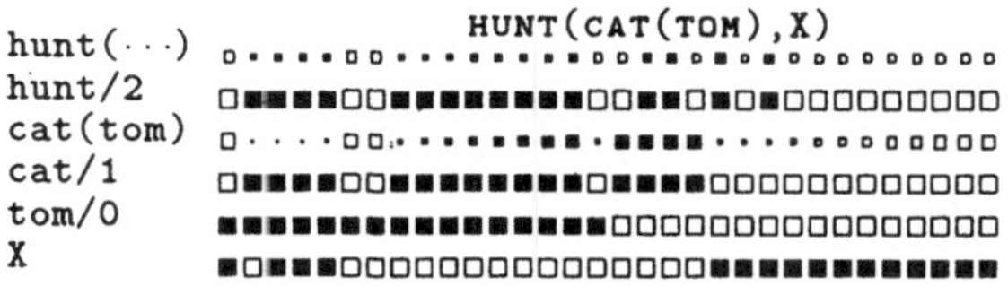

Figure 2: TCO-representation of a term

In our hybrid theorem prover we used vectors $\vec{x} \in \{0,1\}^n$ and $||\vec{x}||_{L_1} = \frac{n}{2}$ for the representation of every constant and variable. With the chosen number of vector-components (32) per term it is possible to distinguish between $\binom{n}{n/2} = 601{,}080{,}300$ constants or variables. The vectors are computed from its internal number by a hashing-function. They are also decoded by a hashing-function. The local coding for constants and distributed representations for terms is depicted in figure 2. Circular convolution is done in the frequence-domain rather than in the time-domain. Therefore we need linear time for convolution and $O(n \log n)$-time for transformation.

4 A hybrid theorem prover

Using the described representations a hybrid theorem prover named CUP (Connectionist Unifying Prolog) was built. The CUP system contains several symbolic algorithms, as well as the local (ℓ-CUP [38]) and the distributed strategy (d-CUP [36]). The theorem-prover itself is a symbolic system. If the system needs to calculate a mgu the currently selected unification-strategy is invoked. The overall-system is depicted in figure 3. The TIV is used to give all constants of a Prolog-program an internal number and determine network-sizes. Also fuzzy slots are extracted and coded as real numbers for fuzzification.

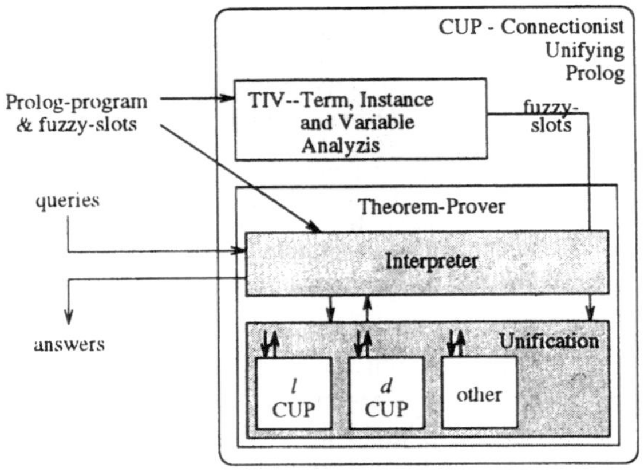

Figure 3: CUP-system

5 Strategies

5.1 ℓ-CUP

The system is divided in two phases: The analysis-phase, where a program has to be analyzed, a coding takes place, and the system is trained. In the unification-phase the network can be used during a symbolic-proof.

5.1.1 Analysis Phase

After the consultation of a Prolog-program the training vectors are encoded and the relax-net (figure 5) is constructed. The map is trained and the order of the constants is mapped to the column-vectors of the relax-net (figure 1).

Training the map The chosen representation of terms allows us to use uncertainty and similarities during processing. Similarities between constants are expressed by neighborhoods in the feature map. Figure 4 shows the representation of such similarities for the following predicates:

```
ped(cat).      ped(dog).      ped(tom).
ped(jerry).    hairy(cat).    hairy(dog).
```

These predicates induce a similarity between `ped/1` and `hairy/1`. For both there are facts $\cdots$`(cat)` and $\cdots$`(dog)`. This is represented by a direct neighborhood (figure 4). The training-vectors for the FM-net are significance-vectors and determined as follows: For every occurring constant in a Prolog-system a vector is generated. This vector contains a component for every occurring constant. These components contain the number of terms, where the constant it encodes and the constant the vector encodes occur in common. The vectors are normalized over all components. Therefore a significance-vector for `ped/1` and our example predicates is:

—	ped/1	cat/0	dog/0	tom/0	jerry/0	hairy/1
ped/1	0.00	0.25	0.25	0.25	0.25	0.00

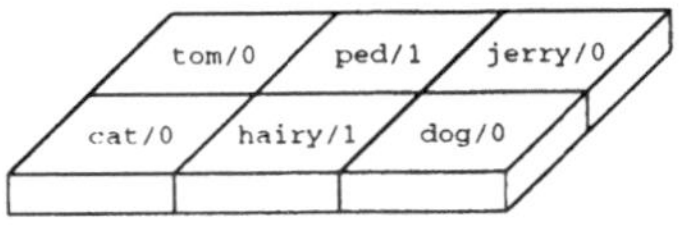

Figure 4: Representation of similarities

Training-vectors are used by the FM-net to find relationships between different terms. Such a coding is adequate since a position in a Prolog-program is often connected to a distinct meaning. If there are different constants at the same position this constants are often used in the same context and therefore similar to some degree. Numbers of similarities between terms can be used to fix similarities between different constants. Similar constants may be members of one class. FM-nets are said to reduce high-dimensional dependencies to a lower-dimensionality conserving many similarities.

5.1.2 Unification Phase

The unification-step in ℓ-CUP is based on the coding described in section 3. Terms to be unified are encoded in the same net and the same variable-vectors by superposition of activations. Superposition is done by activating a unit with the maximum of activations. The relax-net computes the weights of a second FM-net. Activations of units of the stable relax-net represent weights of the map.

Relaxation-Net The coding of two terms is presented to a relax-net (figure 5). After relaxation the variable-vectors are matched into the net to determine the binding. To make use of unification with uncertainty the units of the representation are initialized to values between 0.0 and 1.0. If uncertainty is used also units in the neighborhood are initialized.

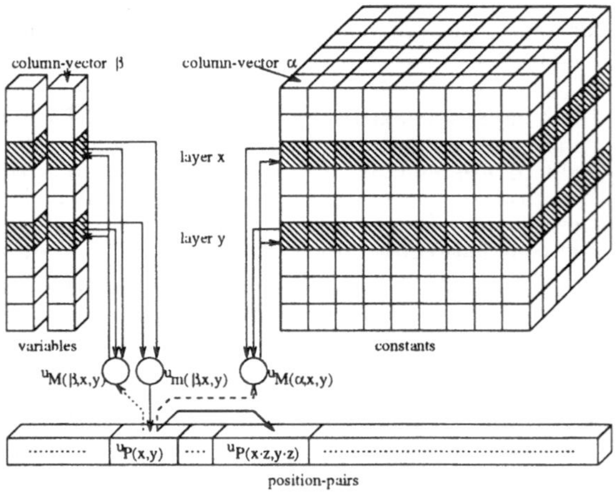

Figure 5: Relaxation network

Units of the relax-net Following we present a description of the unit-types. The connection structure and activation-functions can be found in the appendix and in a paper focusing on ℓ-CUP [38].

Components of the column-vectors $(u_{(x,\beta)})$ are activated if a constant or variable β occurs on position x. Different degrees of activations mean different degrees of certainty.

Minimum units $(u_{m(\beta,x,y)})$ receive activation only through the units of the variable vectors $(\beta \in \mathcal{V})$. They become active if one variable β occurs on two different positions x and y.

The maximum units $(u_{M(\beta,x,y)})$ are the counterpart of this units. If a variable occurs on two different positions x and y it is needed to activate all units representing symbols at position x and y equal. This is done by the maximum units.

The position-pairs $(u_{P(x,y)})$ are needed to control the maximum units. They are also needed to guarantee for the unification of substructures and have therefore connections between each other.

The net utilizes a synchronized update.

Determine stable state The activation of a unit of the position-pairs changes if two units of a variable-vector are active, or an activation through the units of the position-pairs takes place. It follows that the activations of the units of the position-pairs are changed if the stable state is not reached. Therefore only the units of the position-pairs have to be tested for a stable state. The activation of the units of the position-pairs can only decrease or become stable. Their lowest activity is 0. Therefore the net is at least stable if the activity of all position-pair-units is 0.

Determine the mgu After reaching the stable state the variable-vectors are matched through all column-vectors. Two corresponding vectors indicate a variable-binding. All variable-bindings form the mgu.

Determine non unifyability So far it was not mentioned that two terms might not be unifyable. The two terminating-rules CLASH and OCCURRENCE are easy to integrate into the relaxation net. For infinite unification the OCCURRENCE-unit u_O is not needed. A CLASH is indicated by an activation which exceeds 1 as the sum over all units of a layer (figure 6). OCCURRENCE is determined by activating two units of the position-pairs, where one decodes a position which represents a sub-term of the other (figure 6).

How to prove correctness The correctness and completeness of the method can be proven by a mapping to the unification-rules. This is sufficient for a proof cause the correctness and completeness of the unification-rules are proven: The CLASH- and OCCURRENCE-units do in fact the same as the corresponding rules. The system needs no ORIENTATION-rule cause there is no orientation in the representation. The DECOMPOSITION-rule may be found in the

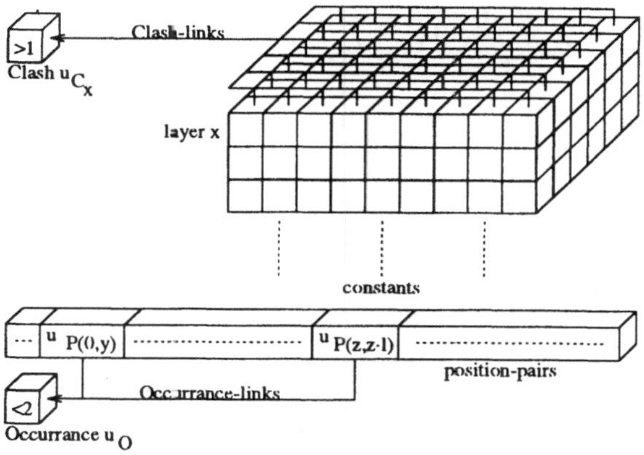

Figure 6: CLASH and OCCURRENCE in ℓ-CUP

representation where the same position in different terms activates units in the same layer. TAUTOLOGY is represented by the activation of the same unit by two different terms and APPLICATION by activation of units of a variable-vector and of a column-vector in the same components. The method terminates always cause the relax-net will reach a stable state. A detailed proof can be found at Weber [37].

5.2 d-CUP

The system is also divided in two phases: The analysis-phase where a program has to be analized, a coding takes place and the system is trained. In the unification-phase the net can be used during a proof.

5.2.1 Analysis Phase

In the analysis phase the TCO-coding of the database takes place (see section 3). We use the PlaNet-simulator [16] for training and graphical presentations of term-encodings. For training the unification network is divided in two parts: the unification problem network (UPN), and the unification decision problem network (UDPN; figure 7). The size of the network depends on the size of the representation for terms.

Setup network UPN's output is a suggested mgu, a term with instantiated variables, and for UDPN a unit for the decision if terms at input are unifyable (figure 7). This unit can be taken as signal for certainty of unifyability. The observation of the certainty-unit of the network allows to introduce uncertainty in unification. This can be controlled from the symbolic system by fuzzy slots which will be compared with the activation of the certainty-unit. Cause of the ability to represent structures with TCO in a similarity conserving way it is also possible to unify similar terms.

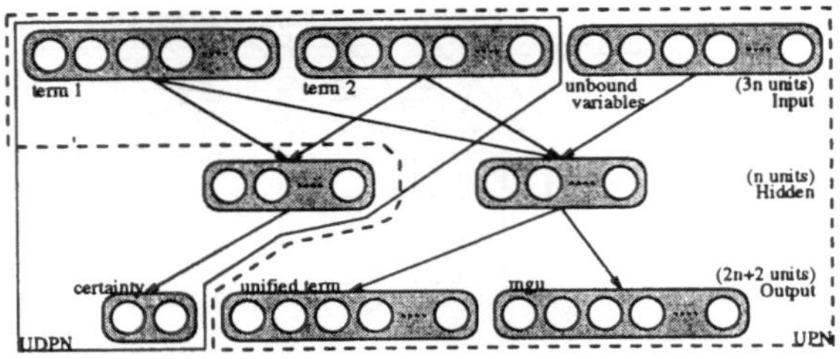

Figure 7: Backprop-architecture for unification

5.2.2 Unification Phase

The coding is the most important part of d-CUP. The unification step itself is quite simple. If we have an encoding of two terms and their unbound variables we can use the trained BP-net to propagate the unifier. After propagation the mentioned comparison between the certainty-unit and the fuzzy slots is done. The representation conserves similarities on the symbols as well as on the connectionist encoding.

It is easy to observe that for the unification rules TAUTOLOGY, ORIENTATION and DECOMPOSITION the representation is as similar as the symbolic terms if the symbols and the numerical distances of representations are compared. An encoding that fulfills this also for the APPLICATION can not be found. The ground for this is: If we find such a representation there would be a numerical encoding of the unification and we could test unifyability of two terms by simple vector-substration in constant time. It is, however, proven that unification belongs to the hardest problems in $\mathcal{P}$. If we had found such a representation we would have also found a highly unlikely result from the complexity view. But we can see that for our purpose the representation works quite well and the network can be trained.

6 Results

Criteria As criteria we want to observe if we are able to fulfill our goals of the introduction.

For sure it is an obvious criteria to look for the efficiency of computation. We want to take the time-consumption in dependence of the length of input-terms. Of course it is necessary to take also the time for (de-)coding. Also an important factor is the training-time of the BP-net. This time is difficult to compare with systems which need not to be trained. Therefore we did not calculate this time. We think our error to this fact is not to big, cause we have to train the system only once.

It is more difficult to look at the extensions of CUP. If we take the E-unification as an extension we can compare it to the theoretical results in unification theory. It is more difficult to rate fuzzification cause it was not implemented before. Therefore we do not have any comparison. It seems, however, to be useful to take the degree we are able to control fuzziness as a criteria for our system. A system that is not able to do strict unification is worthless.

Beside unification we can have a deep view to the representation. As seen in section 3 we can show similarity between representation and symbols with respect to TAUTOLOGY, ORIENTATION and DECOMPOSITION (and in ℓ-CUP also for APPLICATION). The similarity between representations is for d-CUP also independent from the unification method.

Results We first tested CUP's speed. The results for the time consumption are based on five different methods. We examined 64,458 unifications during the proof of queries made with a Prolog-based library-database-system containing about 7,000 clauses. Figure 8 shows the consumed time for different unification algorithms in a logarithmically scaled form. 100% for the square-time method are 26,715,102 unification steps. The result for d-CUP is of course better than ℓ-CUP but we should remark that d-CUP approximates unification. On the other hand d-CUP did right on every unification during tests.

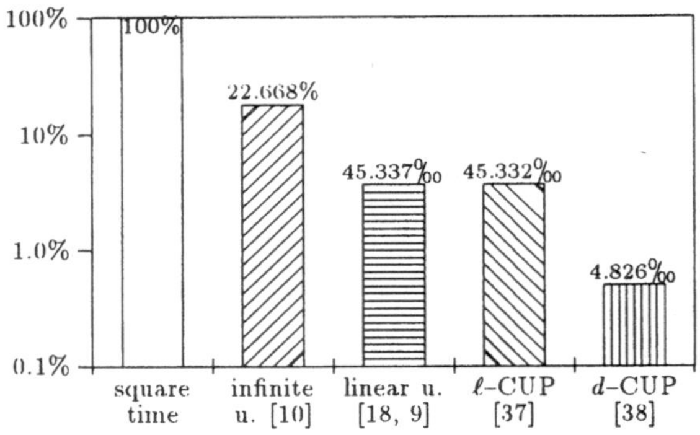

Figure 8: Comparison of time-consumption

We examined several E-unification problems which are based on the axioms in table 1. Table 2 gives an overview over the equational-theories which derive out of this axioms. Some of these theories are decidable some are undecidable. The theories fall in the three classes $\mathcal{U}_u^E$, $\mathcal{U}_i^E$ and $\mathcal{U}_f^E$. If the theory is undecidable there exists no terminating algorithm which decides if at least one admissible unifier exists for this equational-theory. For this theories a heuristic which approximates the decision is useful. A disadvantage is that there are some theories which are not unitary like the Prolog-unification. CUP is designed for unitary unification and will compute for every theory

A		`f(f(x, y), z) = f(x, f(y, z))`
C		`f(x,y) = f(y,x)`
D	D_R	`f(x, g(y, z)) = g(f(x, y), f(x, z))`
	D_L	`f(g(x, y), z) = g(f(x, z), f(y, z))`

Table 1: E-equations for Associativity, Commutativity, and (left & right) Distributivity

only one unifier. This unifier seems to be calculated like a syntactic unifier.

Concluding it can be said that CUP is able to do unification and to represent and interpret similarities in a problem-adequate way. Moreover, as discussed in an earlier paper, the representation is very robust [36]. The system is also able to do extended unifications like E-unification or infinite unification.

axioms	decidable ?	unification-class	ref.
∅	yes	unitary	[8]
A	yes	infinite	[14]
C	yes	finite	[27]
D	?	infinite	[33]
A+C	yes	finite	[22]
A+D	no	infinite	[31]
C+D	?	infinite	[32]
A+C+D	no	infinite	[31]

Table 2: Set of E-unification-problems

7 Discussion

CUP is, as mentioned, able to do (extended) unification. Among the connectionist unification methods only Müller, Stolcke and Hölldobler implement complete unification [9, 17, 29]. Hölldobler's method has also the ability to deal with infinite unification but it is not able to deal with uncertainty.

Stolcke mentioned that activations of a connectionist unification net could be used as signals for uncertainty [29, p.88]. A similar idea also appears in Shastri [26, p.339]. The approach of Stolcke shows the relationship between term-unification, which is done in Prolog, and unification based grammar formalisms. Natural language processing grammar formalisms are often used independently from context. Such a method is syntactically plausible but misses the ability of fault-tolerant analysis. Rules which introduce semantic information promise to improve this. Such extra knowledge is often added to grammars by declarative descriptions [19]. Declarative rules are only capable for cases which are known.

As unification has a sequential nature [4] it is highly unlikely that there is a strict unification method which computes a mgu, like d-CUP, in constant time. ℓ-CUP computes the mgu in linear time and belongs therefore to the fastest – exact – unification methods.

D-CUP opens the possibility of fast fault-tolerant computation which also uses context and similarities of structures.

As mentioned in section 5 there are several hardware implementations of unification strategies. Several of this systems have a simple unification-concept. They try to reduce the unification to other, cheaper methods, like term-matching. Often a type-hierarchy of arguments is introduced during compilation. D-CUP uses a strategy similar to matching. Similarities between term-codings are used to get something like unification. Also Stolcke and Wu developed a distributed approach to do term-matching [30].

Classical symbolic methods distinguish between different unification-rules. Such rule-based methods can only seldom be found in distributed connectionist systems. For ℓ-CUP we compared rules and units while talking about a proof of the correctness (section 5.1.2). The units and network of d-CUP can not be divided in parts which realize this rules. But as we look on the similarity relation and the explanation for its adequacy for this aim we used this rules.

8 Conclusion

Connectionists models for unification has been introduced. The models are able to compute a mgu in linear time/approximate the mgu in constant time and belong therefore to the fastest unification methods. The models are able to do infinite unification. Moreover the ability to deal with uncertainty during unification is given. The use of fuzzification can be controlled from the hybrid theorem prover. For ℓ-CUP the correctness, terminating, and completeness of the strategy doing unification can be proved. Furthermore such a prove can be seen as a good argument for the adequacy of the uncertainty-management. Classical methods do not have a natural equivalence to fuzzification. Such mechanisms could be used in information retrieval, data base systems or natural language systems.

Acknowledgements

This work has been partly supported by the German Research Community ("Deutsche Forschungs-Gemeinschaft; grant DFG-Ha 1026/6-1)". We also would like to thank R. Hannuschka, U. Hartmann, H. Wache, and St. Wermter for their encouragement.

References

[1] D. H. Ballard. Parallel logic inference and energy minimization. TR 142, CS Dep., Univ. of Rochester, NY 14627, (3)86.

[2] R.S. Boyer, J.S. Moore. The sharing of structure in theorem-proving programs. *Mach.Int.*, 101-116, 72.

[3] N. Dershowitz, J.-P. Jouannaud. Rewrite systems. In J. van Leeuven, ed., *Handbook of Theoretical Computer Science B: Formal Methods and Semantics*, 243–320. North-Holland, Amsterdam, 90.

[4] C. Dwork, P.C. Kanellakis, J.C. Mitchell. On the sequential nature of unification. *J.of Logic Programming*, 1:35-50, 84.

[5] N. Eisinger, H. J. Ohlbach. Deduction systems based on resolution. SR-90-12, Univ. K'lautern, 90.

[6] J. Hager, M. Moser. An approach to parallel unification using transputers. In 12th *GWAI*, 83-91, 89.

[7] J. Harland, J. Jaffar. On parallel unification for Prolog. *New Generation Computing*, 5:259–279, 87.

[8] J. Herbrand. Recherches sour la théorie de la démonstration. *Travaux de la Soc. des Science et des Lettres de Varsovie*, 33(128), 30.

[9] S. Hölldobler. A Connectionist Unification Algorithm. TR-90-012, ICSI, Berkeley, CA, (3)90.

[10] G.Huet. *Resolution d'equations dans les langages d'odre 1, 2, ..., ω*. PhD thesis, Univ.de Paris 7, 76.

[11] Y. Kaneda, N. Tamura, K. Wada, H. Matsuda, S. Kuo, S. Maekawa. Sequential Prolog machine PEK. *New Generation Computing*, 4:51–66, 86.

[12] K. Knight. Unification: A Multidisciplinary Survey. *Computing Surveys*, 21(1):93–124, (3)89.

[13] T. Kohonen. *Self-Organization and Associative Memory.* Springer, Berlin, 84.

[14] M. Livesey, J. Siekmann. Termination and decidability results for string unification. CSM-12, Computing Center, Essex University, 75.

[15] A. Martelli, U. Montanari. An efficient unification algorithm. *ACM Trans. on Prg. Lang. and Syst.*, 4(2):258–282, (4)82.

[16] Y. Miyata. *A User's Guide to PlaNet Version 5.6 : A Tool for Constructing, Running, and Looking into a PDP Network.* Dep. of CS, University of Colorado, Boulder, (1)91.

[17] J. Müller. Assoziative Prozessoren und ihre Anwendung für das Theorembeweisen. Master's thesis, Univ. Kaiserslautern, 83.

[18] M. S. Paterson, M. N. Wegman. Linear unification. In *Proc. of the Symp. on the Theory of Comp.*, 76.

[19] F. C. N. Pereira, S. M. Shieber. The semantics of grammar formalisms seen as computer languages. In *Proc. of the 10th COLING*, 123–129. (7)84.

[20] T. Plate. Holographic reduced representations. CRG-TR-91-1, Dep. of CS, University of Toronto, Ontario, Canada, (5)91.

[21] T. A. Plate. Holographic recurrent networks. In C. L. Giles, S. J. Hanson, J. D. Cowan, eds., *NIPS-5*, San Mateo, CA, 92. Morgan Kaufmann.

[22] G. D. Plotkin. Building in equational theories. *Machine Intelligence*, 7:73–90, 72.

[23] J. A. Robinson. A machine-oriented logic based on the resolution principle. *Journal of the ACM*, 12(1):23–41, (1)65.

[24] J. A. Robinson. Computational logic: The unification computation. *Mach.Intell.*, 6:63–72, 71.

[25] P. Robinson. The SUM: An AI coprocessor. *Byte*, 10(6):169–180, (6)85.

[26] L. Shastri. A connectionist approach to knowledge representation and limited inference. *Cog.Sc.*, 12:331–392, 88.

[27] J. Siekmann. Unification of commutative terms. IB 2/76, Institut für Informatik I, Univ. Karlsruhe, 76.

[28] P. Smolensky. On variable binding and the representation of symbolic structures in connectionist systems. CU-CS-355-87, Dep. of CS & Inst. of Cog.Sc., Univ. of Colorado, Boulder (CO), (2)87.

[29] A. Stolcke. Generierung natürlichsprachlicher Sätze in unifikationsbasierten Grammatiken. Master's thesis, Technical Univ. München, 88.

[30] A. Stolcke, D. Wu. Tree matching with recursive distributed representations. TR-92-025, ICSI, Berkeley, CA, (4)92.

[31] P. Szabó. The Undecidability of the d_a-Unification Problem. IB 04/78, Univ. Karlsruhe, 78.

[32] P. Szabó. *Theory of First Order Unification.* PhD thesis, Univ. Karlsruhe, 82.

[33] P.Szabó, E.Unvericht. The unification problem for distributive terms. IB 13/78, Univ.Karlsruhe, 78.

[34] M. Venturini-Zilli. Complexity of the unification algorithm for first-order expressions. Technical report, Consiglio Nazionale Delle Ricerche Instituto per le applicazioni del calcolo, Rome, Italy, 75.

[35] J. S. Vitter, R. A. Simons. Parallel algorithms for unification and other complete problems in $\mathcal{P}$. In *Proc. of the ACM 1984 Annual Conference: The Fifth Generation Challange*, (10)84.

[36] V. Weber. Connectionist unification with a distributed representation. In *IJCNN'92*, III:555–560, (11)92.

[37] V. Weber. Unifikation in Prolog mit konnektionistischen Modellen. Master's thesis, U.Dortmund, (2)92.

[38] V. Weber. Unification in Prolog by connectionist models. In *The 4th Australian Conference on Neural Networks, ACNN'93*, Melbourne, Australia, (2)93.

[39] H. Yasuura. On the parallel computational complexity of unification. TR-027, ICOT, (10)83.

Appendix

Connections of the relax-net

$$
\begin{aligned}
w_{\overleftarrow{u_{m(\beta,x,y)},u_{(z,\beta)}}} &:= 1, \text{iff} \quad (z \in \{x,y\}) \wedge (\beta \in \mathcal{V}) \\
w_{\overleftarrow{u_{F(x,y)},u_{m(\beta,i,j)}}} &:= 1, \text{iff} \quad (x = i) \wedge (y = j) \\
w_{\overleftarrow{u_{F(i,j)},u_{F(x,y)}}} &:= 1, \text{iff} \quad (i = x \cdot z) \wedge (j = y \cdot z) \\
w_{\overleftarrow{u_{M(\beta,x,y)},u_{(i,\alpha)}}} &:= 1, \text{iff} \quad (i \in \{x,y\}) \wedge (\alpha = \beta) \\
w_{\overleftarrow{u_{(j,\alpha)},u_{M(\beta,x,y)}}} &:= 1, \text{iff} \quad (j \in \{x,y\}) \wedge (\alpha = \beta)
\end{aligned}
$$

In all other cases the units are not connected.

Activation-functions of the relax-net

$$
a_{t+1}(u_{(x,\beta)}) := \max(w_{\overleftarrow{u_{(x,\beta)},u_{M(\beta,y,z)}}} \cdot a_t(u_{M(\beta,y,z)}))
$$

$$
a_{t+1}(u_{m(\beta,x,y)}) :=
\begin{cases}
a_t(u_{(x,\beta)}), & \text{iff } 0 < w_{\overleftarrow{u_{m(\beta,x,y)},u_{(x,\beta)}}} \cdot a_t(u_{(x,\beta)}) \\
 & \qquad \le w_{\overleftarrow{u_{m(\beta,x,y)},u_{(y,\beta)}}} \cdot a_t(u_{(y,\beta)}) \\
a_t(u_{(y,\beta)}), & \text{iff } 0 < w_{\overleftarrow{u_{m(\beta,x,y)},u_{(y,\beta)}}} \cdot a_t(u_{(y,\beta)}) \\
 & \qquad < w_{\overleftarrow{u_{m(\beta,x,y)},u_{(x,\beta)}}} \cdot a_t(u_{(x,\beta)}) \\
2, & \text{otherwise}
\end{cases}
$$

$$
\begin{aligned}
a_{t+1}(u_{P_{x,y}}) := \min\big(&a_t(u_{P(x,y)}), \\
&w_{\overleftarrow{u_{P(x,y)},u_{m(\beta,x,y)}}} \cdot a_t(u_{m(\beta,x,y)}), \\
&w_{\overleftarrow{u_{P(x,y)},u_{P(i,j)}}} \cdot a_t(u_{P(i,j)})\big)
\end{aligned}
$$

$$
a_{t+1}(u_{M(\beta,x,y)}) := w_{\overleftarrow{u_{M(\beta,x,y)},u_{j,\beta)}}}
\begin{cases}
a_t(u_{(x,\beta)}), & \text{iff } (j = x) \wedge a_t(u_{P(x,y)}) \\
 & \qquad \le w_{\overleftarrow{u_{M(\beta,x,y)},u_{(x,\beta)}}} \cdot a_t(u_{(x,\beta)}) \\
 & \qquad \ge w_{\overleftarrow{u_{M(\beta,x,y)},u_{(y,\beta)}}} \cdot a_t(u_{(y,\beta)}) \\
a_t(u_{(y,\beta)}), & \text{iff } (j = y) \wedge a_t(u_{P(x,y)}) \\
 & \qquad \le w_{\overleftarrow{u_{M(\beta,x,y)},u_{(y,\beta)}}} \cdot a_t(u_{(y,\beta)}) \\
 & \qquad > w_{\overleftarrow{u_{M(\beta(x,y)},u_{(x,\beta)}}} \cdot a_t(u_{(x,\beta)}) \\
0, & \text{otherwise}
\end{cases}
$$

Plausible Self-Organizing Maps for Speech Recognition

Lionel BEAUGÉ beauge@loria.fr
Stéphane DURAND durand@loria.fr
Frédéric ALEXANDRE falex@loria.fr

CRIN-CNRS / INRIA Lorraine
B.P. 239
54506 Vandoeuvre-les-Nancy Cedex
FRANCE

Abstract

A major problem in connectionist phonetic acoustic decoding is the way to present acoustic signal to the network. Neurobiological data about the inner ear and the primary auditory cortex can be very helpful but are rare. On the other hand, other biological works have shown that structure and functioning of the visual cortex, which have been extensively studied, are very close to the structure and functioning of the auditory cortex.
It has been shown that a simple two-layered network of linear neurons can organize itself to extract the complete information contained in a set of presented patterns. This model has been applied to visual information and has revealed orientation and spatial frequency selective cells.
Such principles have been applied to speech recognition. So we have designed two kinds of maps. The first kind is able to represent frequency characteristics of the signal (e.g.: formantic structure). The second kind takes into account dynamic aspects of the signal (e.g.: formantic transition).
First, these results have been analyzed both from a phonetic and a signal processing point of view and show very interesting representation of the signal. Second, these representations can be used as the input map of a dynamic connectionist network, for speech recognition. The input maps have a selective activity with regard to phonemic structures, and enable dynamic networks to differentiate the phonemes.

Introduction

Connectionism has shown great abilities in classification problems. In order to deal with temporal data such as speech, specific temporal connectionist models have been proposed (e.g: TDNN, [1]). Nevertheless, results obtained with both these techniques are strongly linked to the way information is presented to the network. Concerning static data classification, statistical consideration can be used to design a proper input layer. The same cannot be said for the moment, for temporal input.
Classical input for speech processing is spectrogram, i.e. a frequency/temporal representation. In this case, the temporal aspect is explicitly represented and other specific distributed cues classically used by phoneticians for phonemic structure extraction are poorly accessible to the network.

Our approach is twofold: First, we want the network to dynamically treat time. Instead of an input where time is represented with an axis, we prefer to feed it with a temporal succession of spatial frequency inputs. Second, we want to bring some pieces of phonetic knowledge and design the spatial frequency inputs in such a way that important phonetic cues would be explicitly represented.

From a biological point of view, such an approach has been revealed in cortical maps design. More precisely, visual maps in area 17 have been extensively studied [2]. These studies have shown that a retinotopic organization is preserved in these maps and that neurons are specialized in the detection of important cues for visual treatment (e.g: orientation selectivity). Moreover, the temporal aspect of this treatment (motion) is not represented with an axis but is dynamically treated by the cortex.

From a connectionist point of view, these data about visual cortical map organization have aroused a great amount of modelisation work (Rubner[3], Cooper[4], Marr[5], Alexandre[6]). Out of these, Rubner and Schulten [3] have shown that a simple two-layered network of linear neurons can organize itself to extract the complete information contained in a set of presented patterns. Their model has been applied to visual information and has revealed orientation and spatial frequency selective cells.

In order to develop such an approach for auditory processing, neurobiological data about the inner ear and the primary auditory cortex could be very helpful but are rare [7]. On the other hand, other biological works have shown that structure and functioning of the visual cortex are very close to structure and functioning of the auditory cortex [8]. Consequently our approach will be the following: from an analogical point of view, we will transpose Rubner's model, designed for visual characteristic extraction, to auditory processing and adapt its internal representation to phonetic characteristic.

In this paper we successively present Rubner's model (section I) and the auditory constraints we want to take into account (section II). Then, we detail the two kinds of maps we designed for formantic characteristic extraction (section III) and the results we obtained (section IV). Finally, we discuss this approach and its perspectives (section V).

I The Rubner model and its visual application

In a competitive model, the output units come in competition until a neuron (or possibly several) is active. The adaptation is then concentrated around the winner neuron. Rubner's model is a particular competitive model which extracts characteristics from the input forms.
This model has been applied to visual data in order to obtain visual features detectors by competition.
The system includes:

- an artificial retina which is in fact a 10x10 matrix where each element can be viewed as a simple neuron and with activity in the real domain [-1,+1].

figure 1.1 Example of random retina.

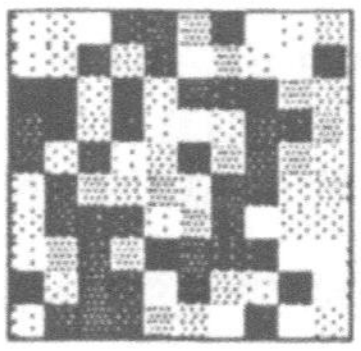

- N output formal neurons. Each neuron in the output layer is connected with all input neurons. Output neurons are connected in cascade series as follows.

figure 1.2 .The Rubner's model

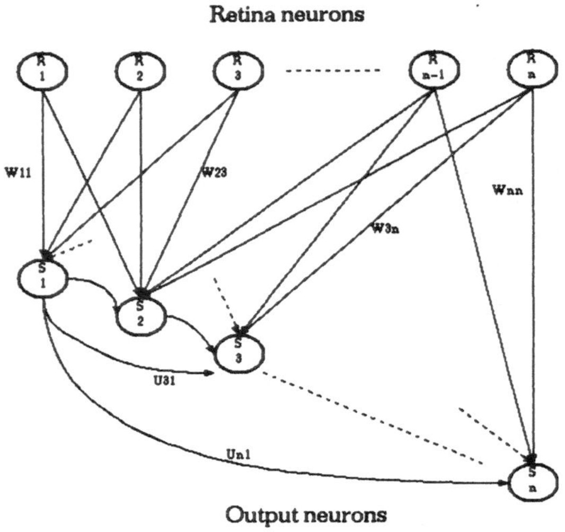

The input for each output unit is computed with the formula

$$E_i = \sum_j W_{ij} \times R_j + \sum_{i<k\leq N} U_{ik} \times S_k$$

where R_j is the activity of the jth retina neuron, S_k is the output value for k^{th} output neuron, W_{ij} is the weight between the input neuron j and the output neuron i and U_{ij} is the weight between output neurons i and j. As there is no threshold function, E_i and S_i are equivalent.

Learning Principle:
A classical Hebbian rule is used between the input and the output layer and to satisfy the competition principle, an anti-Hebbian rule is used between the output units. This can be written as follows:

$$\Delta W_{ij} = \mu_W \times R_j \times S_i \text{ with } 0 < \mu_W < 1$$
$$\Delta U_{ik} = -\mu_U \times S_i \times S_k \text{ with } 0 < \mu_U < 1$$

The W-weights are randomly initialized between -1

and +1 and the U-weights are randomly initialized between -1 and 0 in order to have inhibitory links between output neurons.
The connection's matrix between n output neurons and the retina obtained with a learning corpus created with 2000 random input images is shown above for N=6.

figure 1.3 Connection's matrix for a system with 6 neurons.

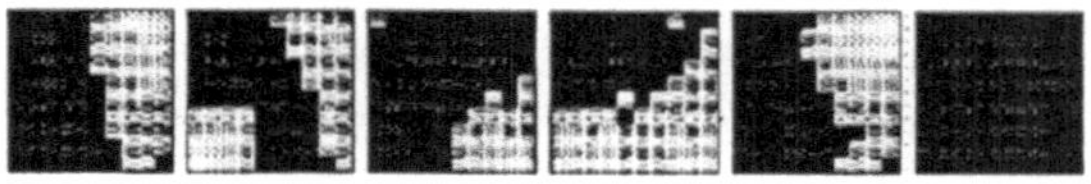

In order to assess the convergence of the system, we compute the quadratic sum of the U-weights; according to Rubner [3], this sum must go towards zero to decrease the correlation between the output neurons. Thus, we assume that each output neuron corresponds to a particular feature in the input map. From a mathematical point of view, Rubner's model computes a principal component analysis where each neuron is linked to a principal axis. More details about the mathematical interpretation can be found in [3].

II Speech signal and auditory constraints:

Speech signal arises from the modulation (by the vocal cords, the mouth and sometimes the nose) of the air blast (the fundamental wave or PITCH) which comes from the lungs. The combination of pitch and modulations gives a wave which is difficult to explain. That is the reason why this wave is analysed by a mathematical method. Generally, one uses a Fourier transform which provides the spectre of the wave in order to have an explicit representation of the signal.
However, for our application, we used a cepstral analysis for which the fundamental wave is ignored and which takes only the modulation part into account. Every two milliseconds, we get a cepstral vector of 128 points corresponding to the frequency scale and for each frequency point the cepstral analysis gives an energy value. The spectrogram is the representation of such an analysis for a span time.

figure 2.1 Example of spectrogram.

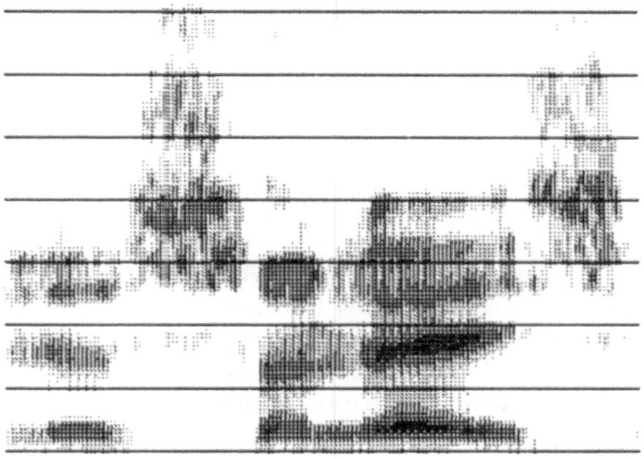

Our study is limited to phonemes which have a formantic structure. A formant is located on the spectrogram by an energy bar as follows.

figure 2.2 A phoneme with a formantic structure.

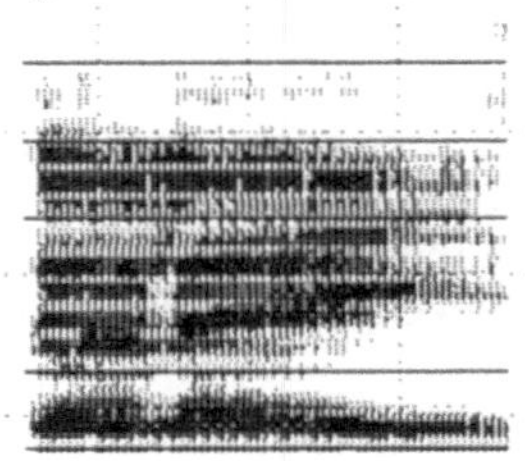

In french, vowels and the consonants 'l', 'm', 'n', 'r' contain formants. Generally, the first formant (~ 900 Hz) and the second (~ 2000 Hz) are used to discriminate the formantic phonemes. Nevertheless, for the same phoneme uttered several times, the formants are never at the same position: formantic location depends on the phonemes before and after. Consequently, a formantic phoneme cannot be learned in a static manner, the temporal aspect being very important.
Below, we develop two kinds of maps to represent the formantic position and the formantic transition (the formant can go up or down).

III Events extract for speech signal

More particularly, we use a combination of the vowels /a/, /i/, /u/ and the consonants /l/, /m/, /n/ /r/. The learning sequence of syllabic words is «ma», «li», etc.

	l	m	n	r
a	la	ma	na	ra
i	li	mi	ni	ri
ou	lou	mou	nou	rou

To characterize a phoneme, we dispose of formants' positions and directions (formantic transitions) to take context into account. We are going to build not one connectionist system but two systems. One is dedicated to the formantic positions (F1xF2 map), the other is dedicated to the formantic transition (FxΔE).

FxΔE map

The speech signal analysis with the cepstral method gives us a vector every 2 ms. To create for an instant the input map, we use a set of 6 vectors, the computation to get the map being the following:
The average between the first three vectors and between the last three ones is carried out. These vectors produce two vectors.
We carried out the difference between both these vectors. We achieve a curve frequency/energy variation.
For each couple (frequency, energy variation) we place the energy of the corresponding frequency of the fourth vector.

figure 3.1 building of FxΔE map.

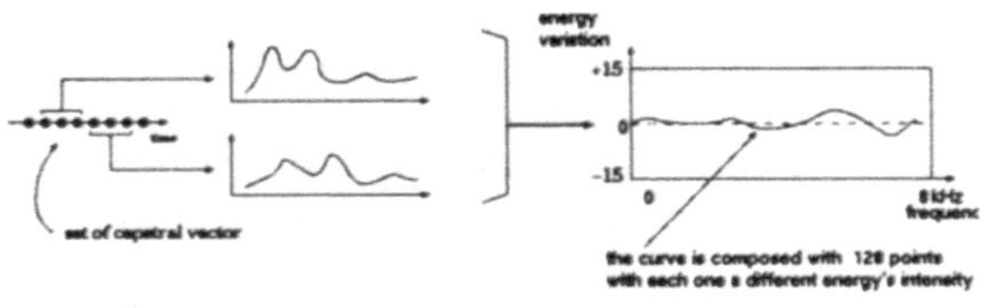

With this processing, the neural net can learn the vectors composing the speech signal knowing a context. The map obtained here can be compared to the artificial retina in the previous paragraph. The FxΔE map has been decomposed into 3 sub-maps; the first map is dedicated to the first formant, the second to the second formant and the third to the other formants. An independent neural net as described in the visual case is connected to each map.
The results achieved for a learning corpus of 70 syllabic words repeated 5 times are the following.

figure 3.2 Connections' matrix for 4 neurons per map (FxΔE map).

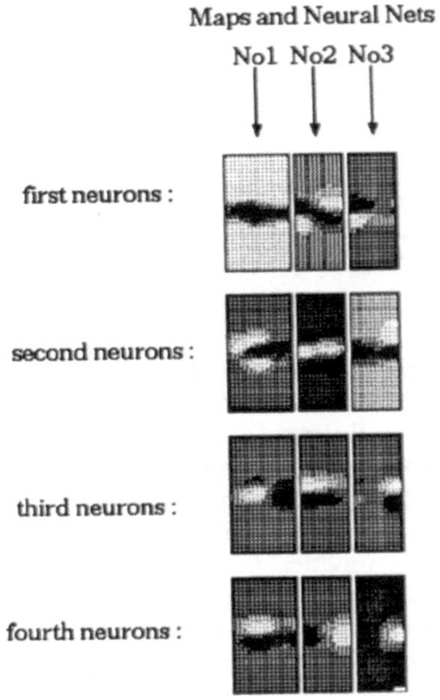

F1xF2 map

This kind of map inspired by phoneticians' studies consists in taking formant position and intensity into account. The map aims to reveal characteristic zones of intensity dispersion on of frequency plane F1xF2.
Abscissa axis F1 corresponds to the frequency domain where the first formant usually appears, ordinate axis F2 corresponds to the frequency domain where the second formant appears and an intensity value at one position on the map results from a continuous linear function of energies E1 and E2, corresponding respectively to a frequency F1 and a frequency F2.

$$k = f((E1,E2))$$

$$f((x,y)) = \sqrt[3]{|x \cdot y^2|}$$

E1 is the energy for the frequency F1 from spectogram and E2 for the frequency F2.
We can note that other maps F2xF3 and F1xF3 could also be constructed to take the third formant F3 into account.

figure 3.3 Construction of the map F1xF2.

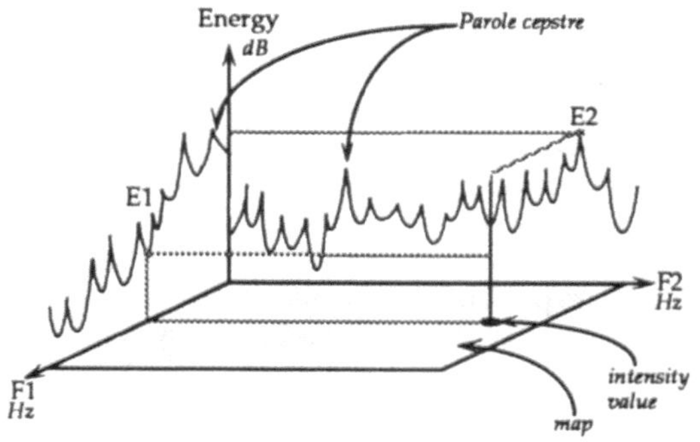

figure 3.4 An example of the map F1xF2 for a vector extract of the syllable m[a].

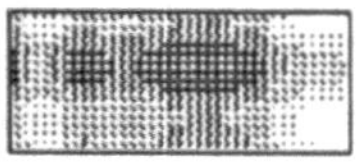

High intensity areas which appear on the map at a presentation of a vector change at each presentation step. This evolution follows formantic transitions. The map obtained can be compared to the artificial retina. In the case of vision, these features tend to look like orientation lines. In the case of speech signals these features can be computed with vocalic triangle, i.e. the maps give prominence to spatial position for different phonemes as well. Moreover the maps give informations from energy.

With this kind of map, the competitive neural networks can learn discriminant acoustical features. In order to improve the precision of the discrimination, the map is split up into sub-maps. This decomposition is in six parts with overlapping zones. For each little map, an independent competitive neural network is connected.

The results achieved for a learning corpus of 70 syllabic words (21000 cepstral vectors or acoustical vectors) repeated five times are illustrated in figure 3.6.

figure 3.5 Connections neurons/map

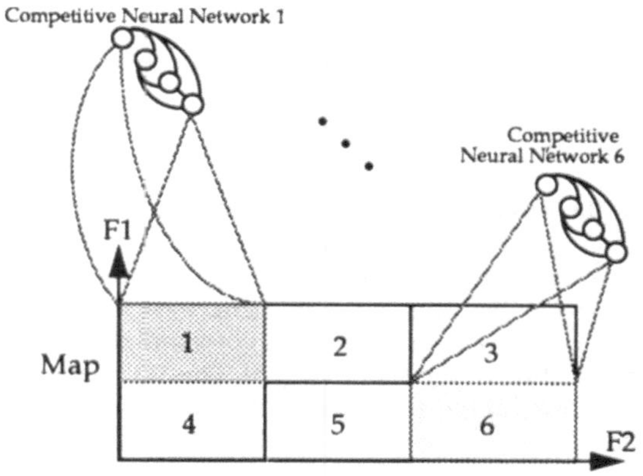

figure 3.6 Masks obtained after a learning of syllabic words.

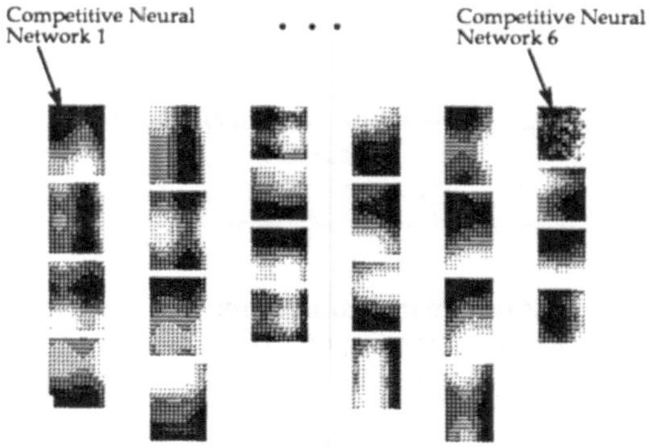

Figure above shows the weight maps of the four output units of the six groups which represent the independent neural networks.

We can note that the configurations of the weight maps of one group testify that each output unit is specialized in the recognition of a specific acoustical feature.

IV Results:

Each neuron from which the connection's matrix is shown in figure 3.2. corresponds to a particular orientation of the curve from the FxΔE map. For instance, the first neuron for the map number 1 activates its output when there is little variation between two cepstral vectors, the third neuron for the second map activates its output when the second formant goes down.

So, for each instant, one neuron has an activity in domain [-1,+1] and we can represent its activity as a time function. Below, we show two activities in time for relevant neurons and for the uttering of the syllable 'ma'.

figure 4.1 Activities of two neurons for the Fx ΔE map.

In the visual case, the competitive model gives neurons specialized in visual feature of the retina. Do auditory features appear from the speech signal?
It seems that the neuromimetic system has converged, i.e. that each neuron corresponds to a special feature of the speech signal. This is shown when one looks at the activities for syllables such as 'la' or 'lou'.

figure 4.2 Activities for the uttering of 'la' (Fx ΔE map).

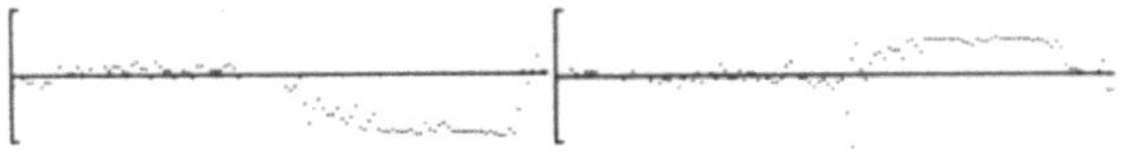

figure 4.3 Activities for the uttering of 'lou' (same neurons)

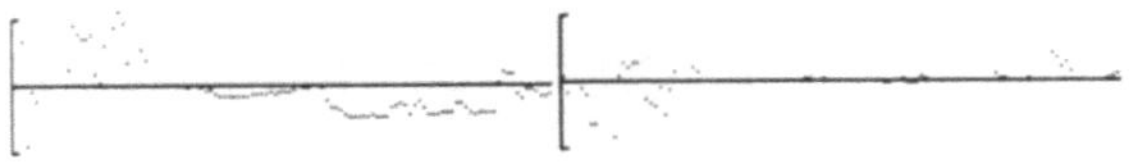

Two kinds of remarks come here:

- first, the crossing of a phoneme ('l' or 'm') to an other ('a' for example) is obvious (the activity changes sign).
- second, it seems that certain neurons correspond to a special phoneme. For example, the first neuron for each figure has an activity close to -1 when 'a' is uttered (this activity is weak when 'ou' is presented); Inversely, the second neuron has a +1 activity when 'a' is uttered while this activity is null when 'ou' is uttered.

V Perspectives and conclusion:

We have detailed a connectionist phonetic acoustic decoding for sounds with formantic structures. The Rubner model we used has shown its capacity to extract features out of inputs. In the visual case we obtain output neurons each one specialized for a particular orientation (vertical, horizontal,...) and this is in agreement with cortical data where, in area 17, we find neurons specialized in orientation selectivity. In the auditory case, we obtain neurons specialized in specific formantic transition and formantic position. The results we have obtained are encouraging but strongly depend on how the input maps are constructed. Our approach to build the inputs is based on the phonetician studies, for phonemes with formantic structure. So, neurons for the F1xF2 map correspond to particular F1xF2 positions and the neurons for Fx ΔE map are specialized for formant direction. Although, the FxΔE map treats variation of formants, the learning of cepstral vectors is static, a phoneme being an ordered sequence of cepstral vectors. Moreover, the acoustic decoding presented here has been trained with a single speaker. In the future, the system will be trained with several speakers and will function for all phonemes that is the reason why we must think to built maps adapted to other types of phonemes such as plosives or fricatives. Phonemes and syllables will be learned with a temporal connectionist system. This could be done with a classical recurrent network [9]. For biological plausibility, we are now testing a temporal connectionist model inspired from cortical data in order to learn temporal sequence of acoustic auditory events.

References

[1] Waibel and al, *'Phoneme recognition using time-delay neural networks'*, IEEE transactions on acoustics, speech, and signal processing, 37, 3, 1989.

[2] Hubel D and Wiesel T, *'Functional architecture of macaque monkey visual cortex'*, Ferrier Lecture Proc. Roy. Soc. Lond.B, (198):1-59, 1977.

[3] Rubner J and Schulten K, *'Development of feature detectors by self-organization'*, biological cybernetic, (62):193-199, 1990.

[4] Bienenstock E L, Cooper L N and Munro P W, *'Theory for the development of neuron selectivity: orientation specificity and binocular interaction in visual cortex'*, The Journal of Neuroscience, 2, (1):32-48, 1982.

[5] Marr D and Hildreth E, 'Theory of edge detection', Proc. R. Soc. Lond. B (207):187-217, 1980.

[6] Alexandre F, *'Une modélisation fonctionnelle du cortex: la colonne corticale. Aspects visuels et moteurs'*, thèse de l'Université de Nancy I, 1990.

[7] Knudsen E, du Lac S and Esterly S, *'Computational maps in the brain'*, American Revue of Neuroscience, (10):41-65, 1987.

[8] Burnod Y, *'An adaptative neural network: The cerebral cortex'*, 2nd edition, Masson, Paris, 1988.

[9] Elman J L and Zipser D, *'Learning the hidden structure of speech'*, J. Acoust. Soc. Am., 1615-1626, 1987.

APPLICATION OF NEURAL NETWORKS TO FAULT DIAGNOSIS FOR HVDC SYSTEMS

K.S. Swarup H.S. Chandrasekharaiah
Dept of High Voltage Engineering
Indian Institute of Science
India

L.L. Lai F. Ndeh-Che
Power & Energy Systems Research Unit
City University
UK

Fax + int + 071 477 8568

Abstract

This paper describes a neural network design and its simulation results for fault diagnosis for thyristor converters and the HVDC power system. Fault diagnosis is carried out by mapping input data pattern, which represent the behaviour of the system, to one or more fault conditions. The behaviour of the converters is described in terms of the time varying patterns of conducting thyristors, pulse zone periods, voltage zone periods and ac & dc fault characteristics.

A three-layer neural network consisting of 24 input nodes, 12 hidden nodes and 13 output nodes are used. 13 different faults were considered, although a lot of research still need to be done, the neural network approach shows a great potential as a more effective strategy for fault diagnosis.

Introduction

Recently there has been growing interest in the application of associative memories and neural networks to problems encountered in expert systems. Most of the earlier works deal with the application of associative memories and neural networks to pattern recognition and mapping (1,2).

Fault detection and diagnosis in real time are areas of research interest in knowledge based expert systems. An approach to fault detection and diagnosis of HVDC systems has been proposed where the conduction patterns of the thyristors of the convertor are used to develop a fault diagnostic expert system (3). The search process involved is exhaustive and hence time consuming, therefore the expert systems are usually too slow to be used in real time environments. Neural networks have considerable advantages over expert systems in terms of knowledge acquisition, additions of new knowledge, performance and speed (4). Reference (5) reported the use of NN to identify faults in ac-dc power systems by the phase voltages and angles. This paper explores the possibility of using neural networks based on conduction pattern of thyristors for fault detection and diagnosis for power convertor used in HVDC power transmission systems. Neural networks are useful in such cases because the nature of the input-output relationship is neither well defined nor easily computable.

Problem Formulation

The convertor is the basic and important component which converts power from ac to dc. References (6,7) summarise the logical behaviour of the convertor under normal and faulted conditions and provides a set of relations and conditions which are used for diagnosis of faults in the convertor. Figure 1 shows the usual configuration of a six-pulse convertor. The switching sequence of the convertor is sequential and periodic under normal conditions (figure 2), it can be observed that the valve conduction pattern is periodic under normal conditions but tends to be unpredictable under abnormal or faulted conditions (figure 3). CF_i is the commutation failure of valve i. The input to the NN is the valve conduction pattern which essentially consists of the voltage available across the valve (termed as voltage zone, VZ), firing pulse available for the valve (termed as pulse zone, PZ), and conducting thyristors (termed as CT). A typical bridge conduction pattern is shown in figure 4.

Figure 5 shows the basic principle of the diagnostic process in mapping the patterns of sensor data to a pattern associated with a fault condition. Each sensor is associated with a classifying system, which determines the fault condition indicated by the sensor. A real number between 0 and 1 is output by each classifier for each possible fault condition. The output of each classifier then is input to an arbitrator which determines the output of the diagnostic system. The arbitrator combines the outputs of the classifiers and outputs a pattern of real numbers corresponding to the output of the system as a whole. The arbitrator may simply average the outputs of the sensors. The algorithm for training the neural networks is written in FORTRAN.

Network Architecture

A three-layer neural network consisting of 24 input nodes, 12 hidden nodes and 13 output nodes are used. Out of the 24 input nodes, 18 nodes are allocated for the thyristor conduction pattern, voltage zone and pulse zone. The remaining six nodes are allocated as follows: two for direct voltage and direct current, three for the ac system bus voltages and the last one for the sampling time. To train neural networks, input data must be presented to the networks. The waveform describing the behaviour of the convertor is represented by a single binary vector 0's and 1's corresponding to the on and off states of the thyristors.

The hidden layer of the network is composed of 12 neurons. This layer extracts input patterns features, i.e. undertakes the nonlinear mapping between the input and output.

Training

The design of a neural network involves two major phases: training and testing. The neural network is trained with the data obtained from digital simulation of a two terminal ac-dc system. The connection weights are determined by using a set of input-output patterns in the training set. A typical different input patterns and output fault conditions derived theoretically for normal and abnormal fault conditions used to train the neural network is given in table 1. Once the connection weights have been figured out, the performance of the neural network is tested using both patterns within and outside the training set. The speed and accuracy of the test results are evaluated in order to decide if modification of the neural structure (number of hidden layers and hidden nodes per layer) or further training of the neural network is necessary.

Results

Simulated sensor data is input to the neural network which has been trained to recognise the difference between convertor faults and faults external to the convertors.

Both the required numbers of iterations and mapping performance were examined for these networks. Because the error criterion was fixed, all networks were expected to perform the mapping with a comparable accuracy. The nodes in the hidden layer were varied from 5 to 25 for every input pattern and the performance of the network in determining the optimum hidden nodes was carried out. Neural network training was started with 5 nodes at the hidden layer, but the network never converged, it did converge when the number of nodes is increased to 10. It was found that 12 nodes in the hidden layer gave the optimum and satisfactory performance in terms of speed of training, fault detection, discrimination and diagnosis.

13 different faults as described in table 2 were considered and the network was able to detect, classify and diagnose the type and duration of the fault. The performance of the neural network for fault detection and diagnosis under the presence of external noise was found to be satisfactory too.

Conclusions

This paper has focussed on the ability of the neural network to correctly diagnose fault conditions of different durations. Input patterns representing either a convertor fault or external fault patterns were presented to the inputs of the top level networks. The system was evaluated on the basis whether the top and lower level networks correctly identify the fault condition and duration of the fault. All real time environments exhibit some level of noise from instrumentation. The effects of noise on the response of the diagnostic system were assessed by randomly perturbing the inputs to the neural networks. The diagnostic systems successfully differentiated between a convertor fault and an external fault.

References

1. L.L. Lai and X.F. Wang, "Application of artificial neural network to power system control', Proceedings of the Ninth Conference on Electric Power Supply Industry, The Association of the Electricity Supply Industry of East Asia and the Western Pacific, Hong Kong, Vol 4, Nov 1992, pp 349-357.

2. D.J. Sobajic and Y.H. Pao, "Artificial neural network based dynamic security assessment for electric power systems", IEEE Trans on Power Systems, Vol 4, No 1, Feb 1989, 220-228.

3. K.S. Swarup and H.S. Chandrasekharaiah, "FDES: fault diagnosis expert systems for HVDC systems", Second Symposium on Expert Systems for HVDC Systems, University of Washington, August 1989, 296-302.

4. R.P. Lippman, "An introduction to computing with neural nets", IEEE ASSP Magazine, April 1987, 4-22.

5. N. Kandil, V.K. Sood, K. Khorasani and R.V. Patel, "Fault identification in an AC-DC transmission system using neural networks", IEEE Trans on Power Systems, Vol 7, No 2, May 1992, 812-819.

6. J. Reeve, "Logic behaviour of h.v.d.c. convertors during normal and abnormal conditions", Proc IEE, Vol 114, No 12, Dec 1967, 1937-1946.

7. J. Reeve, "Direct digital protection of HVDC convertor", Proc IEE, Vol 114, No 12, Dec 1967, 1947-1954.

Acknowledgement

The authors would like to thank the British Council for providing a financial support.

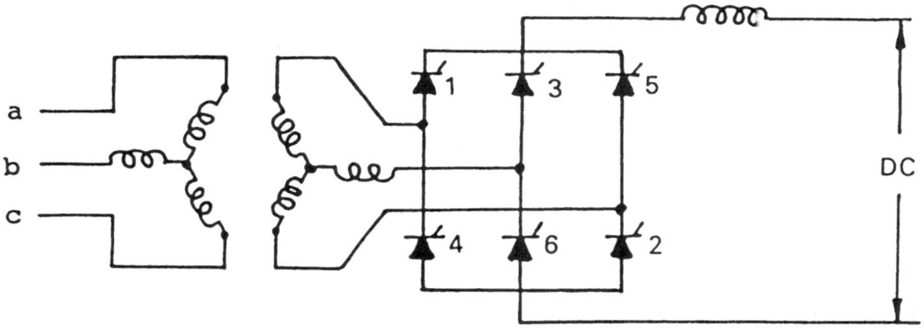

Figure 1 Convertor configuration

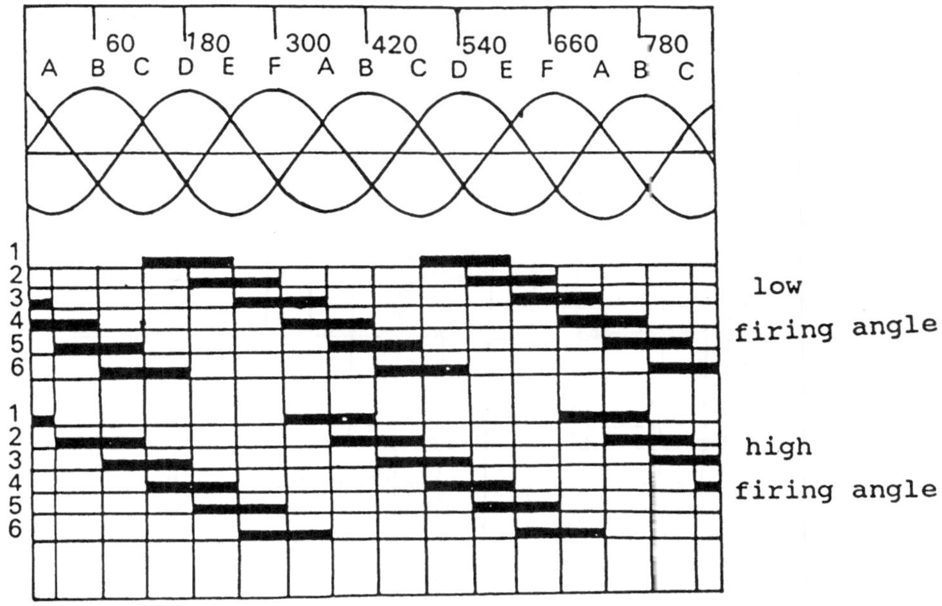

Figure 2 Valve conduction pattern for normal operation

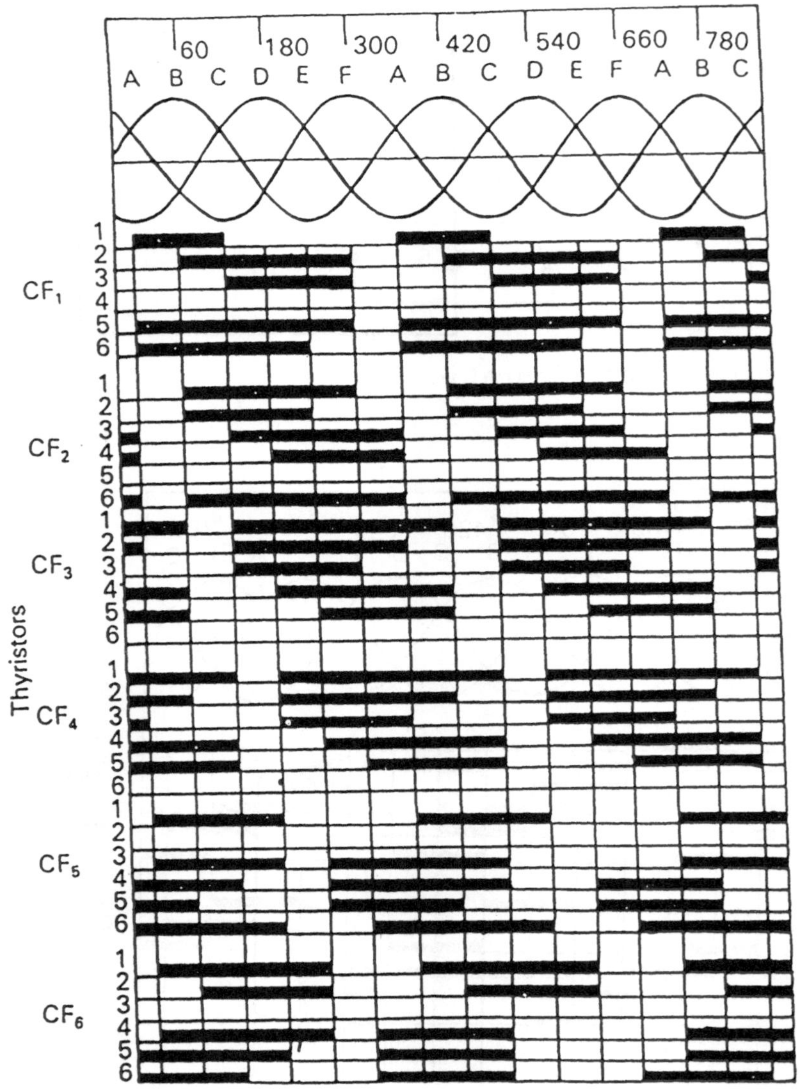

single commutation failure and firing angle less than 120°

Figure 3 Valve conduction pattern for faulted condition

Speaker–Independent Word Recognition with Backpropagation Networks

Bernd Freisleben
Dept. of Computer Science
University of Darmstadt
Alexanderstr. 10
D–6100 Darmstadt, Germany
Email: freisleb@isa.informatik.th-darmstadt.de

Christian–Arved Bohn
Dept. Scientific Visualization of HLRZ
GMD Birlinghoven
P.O. Box 1316
D–5205 Sankt Augustin 1, Germany
Email: bohn@viswiz.gmd.de

Abstract

This paper presents a system that recognizes a limited vocabulary of spoken words in a speaker–independent manner. The system requires only minimal hardware support for acoustic preprocessing. In contrast to other approaches to word–level recognition, it reduces the information content of the speech signals by a compression algorithm before presenting them as inputs to a standard 3–layer backpropagation network. The network learns to recognize the utterances of the speakers in the training set, and the trained network is then used to recognize the spoken words of unknown speakers. Recognition rates of up to 91% were obtained for unknown speakers of the same sex and up to 72% for a mix of both male and female speakers. Since the training times are fast and the system is very cost effective, the approach is practically feasible for a variety of applications.

1 Introduction

The ability to identify spoken words is desirable in a variety of application areas, such as manufacturing, telecommunication and medicine [12], but high–quality speech recognition systems are not easy to built. The challenging computational problems associated with speech recognition and the limited success of the conventional pattern matching techniques proposed to solve them have fostered the development of neural network approaches to speech recognition tasks. The intention is that the generalization properties of neural network learning algorithms are useful to improve the recognition performance.

The proposals made in the literature differ mainly in how the speech signals are converted to a format which can be used as the network input and what the network should learn to recognize. For example, some of the proposals focus on phoneme recognition networks [6, 8, 17], the outputs of which are then processed by further networks or other techniques, such as hidden markov models [3, 14], to achieve word level recognition, while others attempt to recognize words directly [1, 4, 5, 9, 10, 16]. The phoneme recognition systems are conceptually capable of recognizing an unlimited number of words, but they are far more complex, time consuming to develop and difficult to use than word recognition networks, due to the large postprocessing overhead associated with them.

This paper presents an automatic speech recognition system which achieves speaker–independent recognition of a limited vocabulary of spoken words. The system is based on only low cost hardware components (a microphone, a preamplifier and an 8–bit A/D converter) for preprocessing the speech signals, and at the heart of its software design is a backpropagation neural network [15]. The network learns to recognize the spoken words of a set of speakers and is then used to recognize the words of unknown speakers. Several experiments with a speech database containing the recordings of 45 German words from each of 16 different male and female speakers have been conducted to test the performance of the network. The results indicate that the network succeeds in learning all training sets perfectly, but in an experiment with speakers of the same sex it needs to "listen" to 7 speakers before it can recognize the words of unknown speakers with a satisfactory recognition accuracy (91%). For a mix of both male and female speakers recognition rates of up to 72% were obtained. The performance of our system appears to be quite competitive to other results reported in the literature on speaker–independent word recognition, particularly when considering that the number of words included in the speech data used in these approaches was significantly smaller than in our proposal. Apart from the hardware components mentioned above, our system has been fully implemented in *C*. Since the training times of our network are fast and the system is very cost effective, the proposed approach is applicable to various application scenarios.

The paper is organized as follows. Section 2 describes the speech data and the steps taken to preprocess the speech signals in order make them amenable as neural network inputs. In section 3 the network architecture used is presented. The implementation of the proposed speech recognition system and the performance results obtained in the experiments are discussed in section 4. Section 5 concludes the paper and outlines areas for further research.

2 Preprocessing the Speech Data

The speech data used consists of isolated utterances of 45 German words which represent commands for controlling a simple drawing program. A total of 10 male and 6 female speakers were recorded in ambient noise conditions (a normal office environment) with a low cost microphone, preamplified by a 7 kHZ low-pass filter and digitized with an 8-bit analog-to-digital converter sampled at 15625 Hz. Apart from these three components, no other hardware equipment was used to preprocess the speech signals. Thus, our system is very cost effective in comparison to other approaches where higher precision A/D converters or fast signal processors were employed [8].

The individual steps taken to further process the digitized version of each speaker's recording, the sequence of all 45 words with short silences in between them, are graphically illustrated in Figure 1.

In the first step, the preprocessing software of our system automatically extracts the individual words from the speech signal under consideration (1). The extraction algorithm developed succeeds in finding the word boundaries with an error rate of less than 5%, which is quite satisfactory considering that no provisions have been taken to avoid environmental noise.

A 512-point fast Fourier transform analysis, computed every 10.9 milliseconds using a Hamming window [13] (2/3 overlap between successive windows), is then performed to obtain the short-time frequency spectrums of each extracted word (2).

Each spectrum is subsequently transformed into a 15-dimensional speech vector (3) by integrating the (logarithmicly scaled) spectral amplitudes centered at the following frequencies in Hz (taken from [5]; bandwith indicated in parentheses): 130 (30), 164 (38), 206 (48), 260 (60), 327 (76), 412 (96), 520 (121), 655 (152), 828 (192), 1040 (242, 1310 (305), 1650 (384), 2078 (485), 2619 (611), 3300 (770).

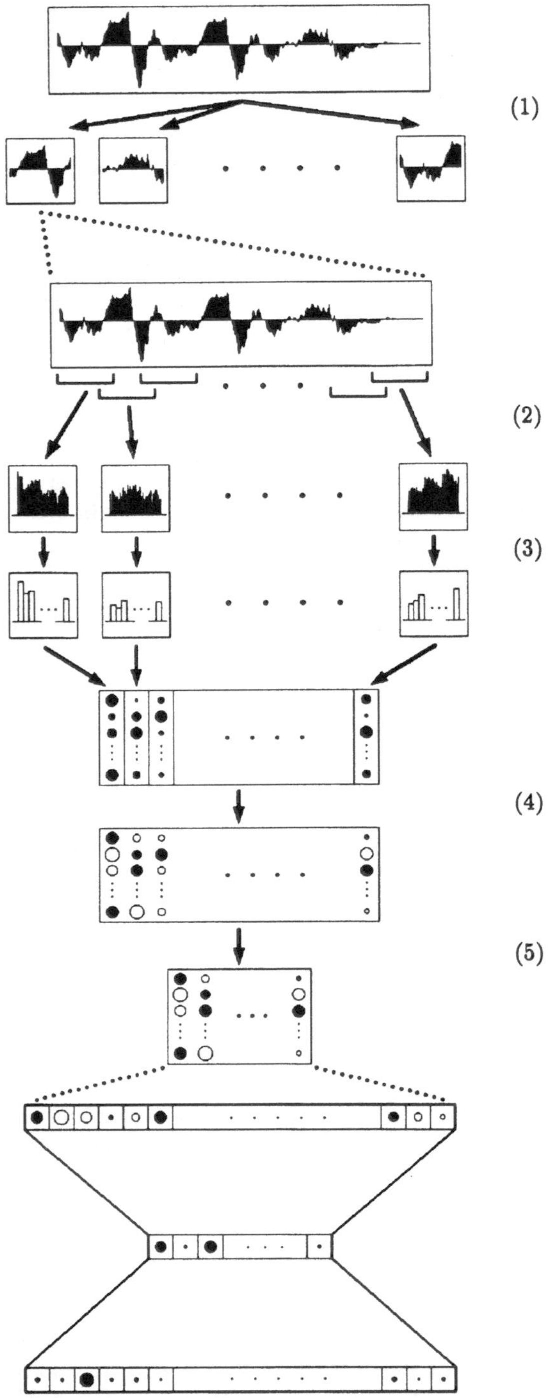

Figure 1: Preprocessing stages

The resulting vectors, between 40 and 80 for each

word, are normalized to the interval [-0.5, 0.5] (4).

In a final step, the sequence of speech vectors of each word is compressed in time by accumulating and averaging them until the sum of their distances exceeds a threshold (5), as proposed in [2]. When this threshold is reached, the sequence of speech vectors is replaced by the average value, leading to 10–17 of such 15–dimensional speech vectors for each word considered. These are used as the neural network input.

The use of a compression algorithm is one of the major differences of our approach to other word recognition endeavours [1, 4, 5, 9, 10, 16] which seem to be based on the assumption that as much as possible of the speech information should be kept to achieve high recognition rates. However, compression does not only reduce the dimensionality of the neural network inputs, which enables the network to learn faster, but also provides a more uniform representation of the utterances, which seems to be beneficial for improving the generalization ability of the network. Figure 2 shows the normalized speech vectors of the German word *"zwei"* before (a) and after compression (b). Large filled circles represent large positive values, and large unfilled circles represent large negative values. The individual speech vectors are displayed vertically, with the lowest frequency at the bottom. The word starts at the left hand side, and the time axis is along the horizontal direction.

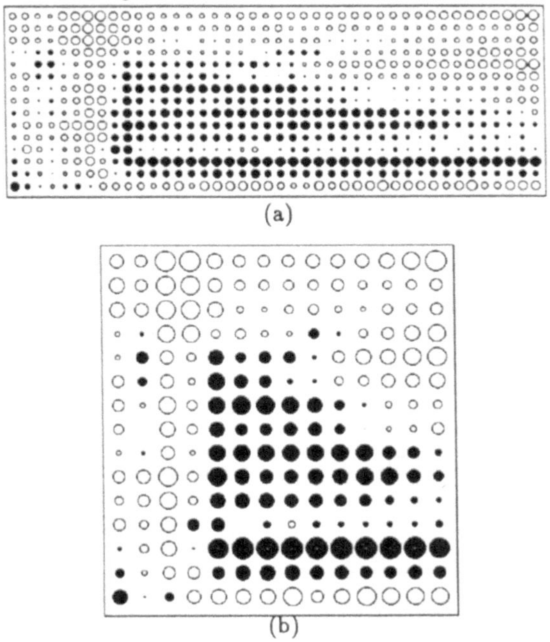

Figure 2: The German word *"zwei"* before (a) and after compression (b)

It is evident that the characteristics of the original word are retained in the compressed version, although the information content is reduced by the compression algorithm. Moreover, small variations in the utterances and the effects of different rates of speech are somewhat eliminated through compression, as shown in Figure 3, where the speech vectors of the German word *"arbeit"*, spoken by the same speaker at two different speeds, are displayed before (a),(b) and after compression (c),(d).

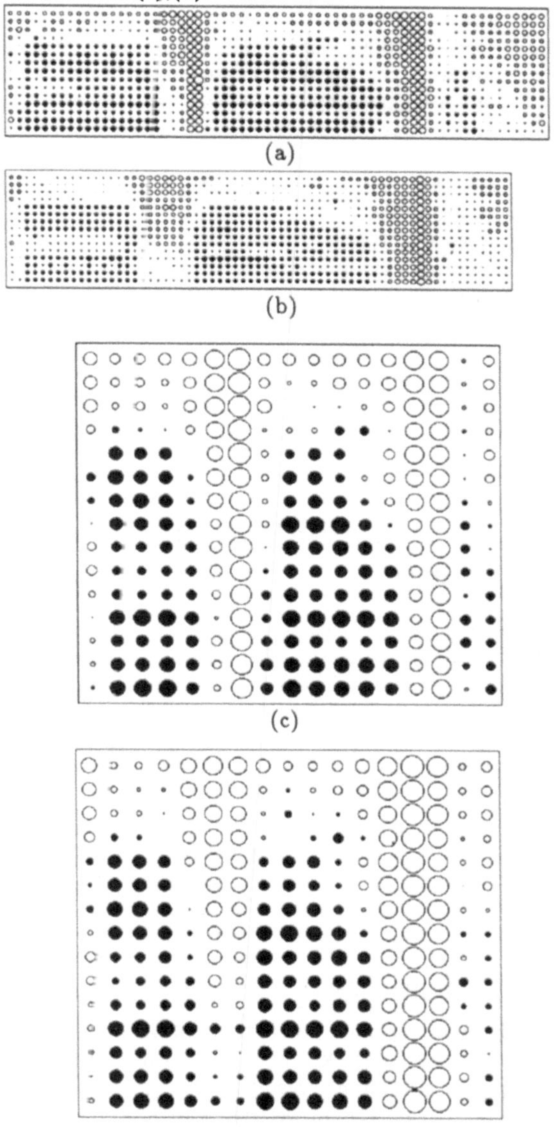

Figure 3: The German word *"arbeit"* before (a),(b) and after compression (c),(d)

3 Network Architecture

We have studied several network architectures with different numbers of hidden units/layers and parameter settings for the single speaker word recognition task and used the net which gave the best performance results in all experiments. The resulting network architecture is a 3-layer feed-forward network with 240 input units, 18 hidden units and 45 output units. The input layer receives the speech vectors of one word ordered into a linear array and the output layer uses a simple 1-out-of-45 coding where the output unit with the highest activation corresponds to the word recognized by the network, as shown at the bottom of Figure 1. If the linearly arranged speech vectors of a word yield a neural network input vector with less than 240 components, the remaining components are set to 0.0.

The learning rule is the standard backpropagation algorithm with the following properties: a) the usual quadratic error function as described in [7] is used; b) errors are accumulated after each input in the training set; c) the learning rate lies between 0.4 and 0.9; d) the momentum term is 0.7; e) the weights are initialized to random values in the range between -0.3–0.3; and f) the input vectors are always presented in the same order.

4 Implementation and Performance

The software for the entire system was written in *C*, and the utterances were recorded and digitized on a Commodore Amiga, equipped with a low cost microphone and an off-the-shelf 8-bit A/D module. A SUN Sparcstation was later used to perform all computations on the digitized speech signals, including preprocessing and neural network training.

In order to test the recognition performance of the network, two different problem areas have been investigated: single-speaker word recognition and multi-speaker word recognition. The results obtained are presented in the following subsections.

4.1 Single-Speaker Word Recognition

In the single-speaker word recognition experiment, all 45 words were spoken by and recorded from a single speaker four times, resulting in four different input sets. Three of them were used as the training set, while the remaining one served as the test set. After only 20–30 presentations of the training set (equivalent to 2–4 minutes computation time on a SUN Sparcstation), the error function was minimized, i.e. the backpropagation algorithm converged pretty fast. This compares favourably to other approaches where several hundreds of iterations were required [4, 16].

The recognition accuracy of the training set was 100%, i.e. the network had properly learned all 45 words. A recall with the test set achieved a recognition rate of 96%, which seems to be slightly superior to the performance results reported in the literature for the single-speaker word recognition problem (see section 4.3).

4.2 Multi-Speaker Word Recognition

The speech data used for the multi-speaker word recognition task consisted of 16 different speakers (10 male and 6 female), each of them recorded once for all 45 words.

In a first experiment, only male speakers were considered. The 10 available input sets were divided into two groups, the first group being the training set and the second group being the test set. Table 1 shows the performance results when the training set contained the speech vectors of 1, 2, 4 and 7 speakers, where in each case the remaining speakers were used as the test set (the percentages represent averages of the speakers in the test set).

#words	#speakers	training set	test set
45	1	100%	37%
45	2	100%	49%
45	4	100%	70%
45	7	100%	91%

Table 1: Recognition rates for male speakers

The results indicate that the network succeeded in learning all training sets perfectly, but it needs to "listen" to 7 speakers before it can recognize the words of unknown speakers with a satisfactory recognition accuracy (91%).

In a second experiment, both male and female speakers were considered. The training and the test sets were assembled in the manner described above. The results are presented in Table 2.

#words	#speakers	training set	test set
45	2	100%	42%
45	4	100%	56%
45	6	100%	59%
45	8	100%	68%
45	10	100%	72%

Table 2: Recognition rates for male and female speakers

The performance results obtained for the training sets are identical to the experiment where only male speakers were investigated. The recognition accuracy for the test sets is not as good, but still acceptable. It should be mentioned that in most cases the "correct" output units had almost equally high activations as the (wrongly) winning unit.

The results suggest that it might be conceivable to use two separate networks, one for male and the other one for female speakers, in order to achieve high recognition accuracies for unknown speakers independent of their sexes.

4.3 Discussion

In this section we discuss some of the observations made during the experiments and compare our results to other results reported in the literature for the word–level recognition.

The number of hidden units seems to have some impact on the generalization ability of the network. Our experiments indicate that a large number of hidden units is usually beneficial to improve the recognition rates of the training set, but leads to worse results for the test set. Since a large number of hidden units also decreases the convergence speed, less hidden units are clearly favourable.

The network behaviour is relatively immune to the settings of the learning and momentum parameters, because the results did not significantly change when the parameters where modified (in a reasonable range). The presentation of the input vectors in a different order, in some applications useful to avoid local minima, did not have an observable effect on the convergence of the network.

Table 3 summarizes the results of other approaches to word–level recognition, in order to allow a comparison to the recognition rates obtained with our proposal.

#words	#speakers	training	test	ref.
146	1	90.6%	58.2%	[16]
17	3	100%	90%	[5]
17	3	100%	94%	[5]
13	1	100 %	92%	[9]
10	1	97.5%	70–100%	[1]
7	1	—	92%	[11]
5	3	84%	62%	[4]

Table 3: Recognition rates of other approaches

Our results are quite competitive to those shown in Table 3. It is worth noting that the speech database in the majority of these approaches (except for the first one) contained fewer words than our speech database, but nevertheless the 96% recognition rate obtained with our network for the test set in the single–speaker case is slightly better, and the 91% recognition rate for multi–speaker word recognition is similar. When our network is trained to recognize up to 30 words, recognition accuracies of 100% for both the training and the test set are obtained. We cannot compare our results for a mix of both male and female speakers to the above approaches, because they do not include information whether similar experiments have been conducted.

The recognition rates alone, however, might not be sufficient to allow a fair comparison, because the recognition rates are clearly dependent on the recording conditions and the articulation of the words. It is always possible to pronounce a word in a manner such that the network is not able to recognize it, and on the other hand, the recognition rate will be improved when particular emphasis is put on pronouncing the same words similarly. A description of the relevant details of the experimental conditions is not provided in the literature on the alternative approaches, but we did not at all try to eliminate environmental noise or take influence on the speakers' pronounciation of the words.

To summarize, our work demonstrates the practical feasibility of building a low cost but high quality speech recognition system which can be used in a flexible manner. The system can be easily trained to recognize the desired words with high accuracy and does not require the assistance of somebody who is deeply familiar with the issues involved in speech processing. It is well suited for applications where a limited vocabulary needs to be recognized, such as in some control applications with voice entry of commands or in tools for supporting the physically handicapped. The system can also be used for *speaker* recognition [12] by simply training the network to learn the mapping between a set of words and a number of speakers, and letting the network recognize who is saying unknown words contained in a test set. Experiments with 5 speakers and a test set of 50 unknown utterances have shown that the network is able to identify the speakers in up to 82% of the cases correctly.

The time for running the preprocessing software and training the network with the backpropagation algorithm is pretty short (about 10 minutes), and it may further be reduced by simply implementing parts of the preprocessing functionality in hardware. An appropriately trained network would then allow word recognition under real–time conditions.

5 Conclusions

In this paper we have presented a speech recognition system that allows to recognize a limited vocabulary of spoken words in a speaker-independent manner. Apart from a few low cost hardware components required for acoustic preprocessing, the system has been implemented in software. A standard 3-layer backpropagation neural network has been used to learn the utterances of the words from a set of speakers, and the trained network was employed to recognize the spoken words of unknown speakers. Before the suitably preprocessed speech signals were presented to the input units of the network, their information content was reduced by a compression algorithm, leading to improvements of the generalization ability and the convergence speed of the network. Experiments have shown that the network performance is quite competitive to other approaches: recognition rates of up to 91% were obtained for unknown speakers of the same sex and up to 72% for a mix of both male and female speakers. Since the system is very cost effective, it is useful in a number of applications.

Among the issues for future research are an evaluation of the network performance when the size of the speech database is increased, the integration of the system in particular application environments, and the study of other learning architectures, such as recurrent networks [7], for word recognition.

References

[1] Behme H, *'A Neural Net for Recognition and Storing of Spoken Words'*, In: Parallel Processing in Neural Systems and Computers, pp. 379–382, Elsevier Science Publishers, 1990.

[2] Bengio Y, Cardin R, and De Mori R, *'Speaker Independent Speech Recognition with Neural Networks and Speech Knowledge'*, In: Advances in Neural Information Processing Systems, Vol. 2, pp. 218–225, Morgan Kaufman Publishers, 1990.

[3] Bourlard H, and Morgan N, *'A Continuous Speech Recognition System Embedding MLP into HMM'*, In: Advances in Neural Information Processing Systems, Vol. 2, pp. 186–193, Morgan Kaufman Publishers, 1990.

[4] Franzini M A, *'Learning to Recognize Spoken Words: A study in Connectionist Speech Recognition'*, In: Proceedings of the 1988 Connectionist Models Summer School, pp. 407–416, Morgan Kaufman Publishers, 1988.

[5] Grajski K A, Witmer D P, and Chen C, *'A Preliminary Note on Static and Recurrent Neural Networks for Word-Level Speech Recognition'*, In: Proceedings of the 1990 International Joint Conference on Neural Networks, Vol. 2, pp. 245–248, Lawrence Erlbaum Publishers, 1990.

[6] Hampshire II J B, and Waibel A, *'Connectionist Architectures for Multi-Speaker Phoneme Recognition'*, In: Advances in Neural Information Processing Systems, Vol. 2, pp. 203–210, Morgan Kaufman Publishers, 1990.

[7] Hertz J A, Krogh A, and Palmer R, *'Introduction to the Theory of Neural Computation'*, Addison–Wesley, Reading, Massachusetts, 1991.

[8] Kohonen T, *'The Neural Phonetic Typewriter'*, IEEE Computer, 3:11–22, 1988.

[9] Kowalewski F, and Strube H, *'Word Recognition with a Recurrent Neural Network'*, In: Parallel Processing in Neural Systems and Computers, pp. 390–394, Elsevier Publishers, 1990.

[10] Lee K, *'Context-Dependent Phonetic Hidden Markov Models for Speaker-Independent Continuous Speech Recognition'*, IEEE Transactions on Acoustics, Speech, and Signal Processing, 38(4), 1990.

[11] Lee Y, and Lippmann R P, *'Practical Characteristics of Neural Network and Conventional Pattern Classifiers on Artificial and Speech Problems'*, In: Advances in Neural Information Processing Systems, Vol. 2, pp. 168–177, Morgan Kaufman Publishers, 1990.

[12] Peacocke R D, and Graf D H, *'An Introduction to Speech and Speaker Recognition'*, IEEE Computer, 8:26–33, 1990.

[13] Rabiner L R, and Gold B, *'Theory and Applications of Digital Signal Processing'*, Prentice–Hall, 1975.

[14] Rigoll G, *'Neural Network Based Continous Speech Recognition by Combining Self Organizing Maps and Hidden Markov Modelling'*, In: Lecture Notes in Computer Science, Vol. 134, pp. 58–65, Springer–Verlag, Berlin, 1990.

[15] Rumelhart, D E, Hinton, G, and Williams, R E, *'Learning Internal Representations by Error Propagation'*, In: Parallel Distributed Processing: Explorations in the Microstructures of Cognition, Vol. 1, 318-362, MIT Press

[16] Sung C, and Jones W C, *'A Speech Recognition System Featuring Neural Network Processing of Global Lexical Features'*, In: Proceedings of the 1990 International Joint Conference on Neural Networks, Vol. 2, pp. 437–440, Lawrence Erlbaum Publishers, 1990.

[17] Waibel A, Hanazawa T, Hinton G, Shikano K, and Lang K, *'Phoneme Recognition Using Time-Delay Neural Networks'*, IEEE Transactions on Acoustics, Speech, and Signal Processing, 37(3):328–339, 1989.

A Neural Learning Framework for Advisory Dialogue Systems

Hans-Günter Lindner and Freimut Bodendorf
Universität Erlangen-Nürnberg, Wirtschaftsinformatik II,
Lange Gasse 20, W-8500 Nürnberg 1
X.400:C=de;A=dbp;P=uni-erlangen;OU1=wiso;OU2=lan;
S=bodendorf; G=freimut

Abstract

A domain independent neural learning framework for advisory dialogue systems (ADS) is suggested. A connectionist view of user and task modeling is introduced that can be implemented in a neural knowledge network. It implicitly interprets man-computer interaction and causes adaptive task support. Adaptive inference is drawn by modifying the causal connections during interaction. The interpretation of the network gives insights into the user's knowledge and preferences. Reasons for misconceptions can be estimated and interpreted by users, designers and rules for network modification.

Neural ADS learn empirically in real-time to raise future system performance but can also be programmed by experts. Additionally, the network can be used for predicting the behaviour of the whole system or its parts.

Advantages are constant retrieval time for associated information, extendability, and variability. Implementing the framework does not require special hardware or neural simulators. To demonstrate the applicability, two prototypical spreadsheet applications are introduced.

1 Introduction

Advisory dialogue systems (ADS) require adaptive behaviour to satisfy user needs in a changing context. Appropriate representations, help, advice and information should be given in accordance with knowledge, facilities, preferences and other individual properties of the user. For efficient use of advisory dialogue systems, a variety of knowledge representing the application's environment has to be acquired and inferred in real-time: knowledge about user, task and interaction features.

To model this complex, changing and interconnected knowledge for user support, that can not be determined in advance, neural networks are a promising approach.

"In summary, neural-based systems represent a new wave of technological potential for advancing systems support.... In order to fully exploit this technology, a fundamentally new orientation for systems support may be useful. Ideally it will be possible to describe the way people represent problem-driven work so that the technology can be integrated to provide support." [11]

The following article shows a step in this direction and proves the applicability for business applications, especially for advisory support.

2 Modeling ADS' environment

2.1 Modeling the user

To behave flexible and adaptive, intelligent ADS have to acquire knowledge about the user in order to adjust itself to the user's requirements. Information about the user is stored in a 'user model' [1]. Background knowledge, experience and individual properties are characteristic long-term parameters. In contrast, reaction time, evaluation of tests, help requests, the path of activated commands and the time spent on usage are short-term parameters. Based on this values, an ADS concludes about appropriate advice.

Information to model users is fuzzy in nature, biased and inconsistent. This problems can be handled by neural networks. For this reason, we used Hecht-Nielsen's backpropagation classifier network in combination with the above mentioned parameters to model users, who use OS/2 [2]. For properly designed networks, simulated users were recognized with a precision of more than 96%. Therefore, neural user modeling seems to be promising for further investigation.

2.2 Modeling the task

The main goal of advice in business applications is to support users while interacting with a computer system. In a problem-driven task, the relations between subtasks are fuzzy but can be deterministic if it is procedure-driven [11]. Connectionist models can handle fuzzy representations of tasks [8].

Lindner developed an atomistic task model that enables adaptive system's advice and behaviour: the user as well as the application cooperatively perform tasks inside an 'action space' [7]. This action space consists of so called *action atoms* which allow task performance. There are three classes of action atoms: *decisions (D), transformations (Tr) and ressources (R)*. Transformation atoms are initiated by decision atoms and are able to change activations while executing a task. Tr and D use R, which are unchangeable functions and data. R can be viewed as hyper-knowledge [3].

The user as well as the application use and modify action atoms for producing service, advice or material objects. An ideal task execution needs situational and individual appropriate atoms. If they are not configured properly, the application has to support users by selecting the correct atoms and by giving advice. A neural framework, which supports task performance is detailed below.

3 A neural framework for adaptive applications

3.1 Connectionist architecture

The adaptive framework is based on a connectionist architecture and enables different learning strategies, e.g. case-based reasoning or rule-inductive learning. It is a two-layered and fully connected network with additional feedback loops from one processing element (PE) to itself. Each PE represents an action atom and is semantically bounded to one of three classes (D, Tr, R). Additional Tr can be added to enhance the complexity of input-output matching or continuous excitation of action atom groups. The neural topology of the framework is shown in figure 1 (a circle is associated with a class of action atoms; Tr_{add} is a class of additional Tr). The input (A) layer contains sensorical atoms receiving messages from the environment whereas atoms of the output layer (B) send messages to it.

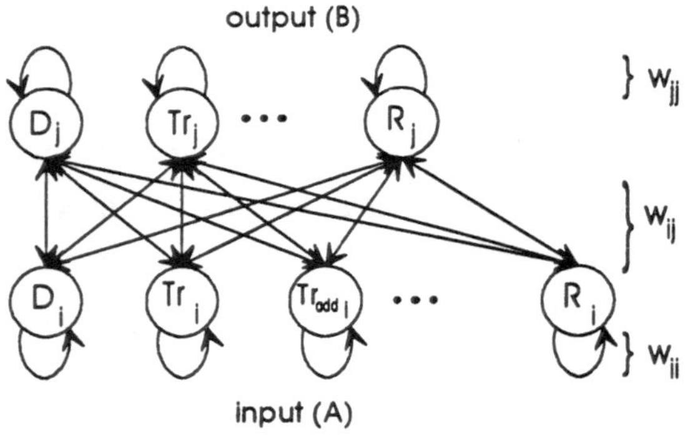

Figure 1: A neural topology for adaptive applications

Connections between all action atoms have synaptic properties. Their weights (w_{ij}, w_{ii}, w_{jj})

take values in [-1,1] and represent their causal relationship. Therefore, the architecture is similar to fuzzy cognitive maps (FCM) which can be reduced to the bidirectional associative memory (BAM) and its derivatives, the temporal associative memory (TAM) [5: 153] and the adaptive BAM (ABAM) [5: 233 ff.].

The transfer function of all action atoms is separated in three parts: propagation, activation and output. Though their structure is similar to conventional PE, their functions can be numeric or symbolic.

With regard to implementation aspects, symbolic rules increase the explicability of the PE's transfer function. For example, determining a sequence of action by rules in a decision atom is easier and needs less calculation time than using an avalanche training process for sequential pattern recognition. Therefore action atoms can be interpreted as active objects [4].

In the simplest case, action atoms contain a weighted summation of input signals and a threshold function. If the activation state exceeds the threshold, the atom fires. That leads to a spreading activation, which influences causal related atoms. Clusters of atoms, semantic fields, asynchronously change their state at a time. They do not affect the plasticity of the network in a short-time period because activations only represent the network's short-term memory.

The synaptic connections encode the long-term memory and can be illustrated by a real-valued matrix with synaptic weights as its elements (figure 2); grey areas represent autoassociative regions. Each weight represents the degree of causality between two action atoms. This enables associated atoms to build a changeable rule, using the atom or the set of atoms at time *t* as the antecedent and the activated atoms at time *t+1* as the consequence.

The meaning of a connection depends on the semantics of paired atoms and its plasticity. When synaptic learning rules are used, their weights are modified during frequent activation changes. All synapses, in combination with activities, model parts of the environment, including user, dialogue, application, and finally the interaction.

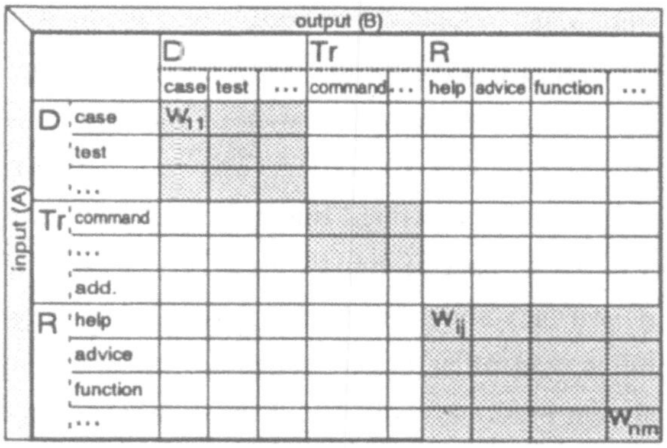

Figure 2: Connection matrix and semantic fields

The following examples show the meaning of semantic fields:

- The *competence of a user* can be deduced from the synaptic strength between commands and R. If the strength is high, it is likely that the use of commands leads to the firing of help or advice atoms and finally active support. Therefore, novices obtain more active support than experts.

- *Preferred command sequences* for a dialogue are modelled by synaptic strengths between command atoms. The actual dialogue history can be obtained from activation states that decay in time, when using self-feedback loops for each atom with a weight in (0,1).
- The *application* is modelled by the action atoms that are implemented. The essential non-adaptive functionality is given by fixed synapses connecting commands and functional R.
- A model of the *interaction* contains all atoms that communicated with the user concerning the task. Revised active advice or interrupted commands are not part of the interaction because they are not causal related to the context.

In accordance with the size of an examinated semantic field, the granularity of the model interpretation can be varied. The user's knowledge can be separated into regions for application use or special domain areas, e.g. an examination may show only an expert level in usage but not in other areas. Semantic separation as shown before is valid for all models mentioned above.

3.2 Learning modes

Three learning modes can be used to construct and varify the connection matrix: deterministic, supervised and unsupervised learning. The learning rules can be assigned to persons who are involved in the development process of an adaptive application: designer, domain experts and users.

A designer uses *deterministic learning* for defining the connection matrix. He fixes the connections between implemented action atoms that are necessary for the essential non-adaptive functionality. His model of the user leads to the design of adaptive properties, semantic areas, their affiliated learning capabilities and modification rules which change matrix states to reach optimal functionality. Deterministic learning can be done by the bipolar Hebbian or outer-product learning, where the connection matrix W is constructed by the product of the designer's k input/output associations with input vector A and output vector B. Each element of a vector A_n, B_n takes values in [-1,1], multiplied by a causality factor cf for each association:

$$W = \sum_{n=1}^{k} cf_n A_n^T B_n \qquad (1).$$

Supervised learning is necessary for users or domain experts who want to adapt the application to their experience and preferences. When the states between single time slices of activation updates have to be adjusted, only two layers of connected PE are necessary for a learning algorithm. Weight modification can be done absolutely by (1) and relatively by a modified form[1] of the differential Hebbian learning law [5: 158], considering existing weights. The derivatives of the neural output signals (S) correlate changes. If the correlation is negative, the weight decreases, and vice versa. Additionally, the modification can be governed by a learning factor LF_{ij} between PE_i and PE_j:

---

[1] In contrast to the original learning law, the passive decay term is neglected. This decay term leads to zero causality between unchanged signals: the predetermined knowledge structure could be eroded.

$$\dot{w}_{ij} = LF_{ij}\dot{S}_i\dot{S}_j \qquad (2).$$

The differential Hebbian learning law leads to receptive semantic fields. When predetermined causalities are enhanced, the affiliated PE can be excited easier than other, e.g. long-term activation of a special domain knowledge (R) causes a higher relevance to the context and makes active support more likely.

If only one output is desired, the competitive version of (2) can be used. When more than one time slice is examined, sequences of actions build networks with hidden layers, and back-propagation learning should be used.

Practical relevance can be clarified when we look at the intentions of domain experts or users who want to individualize an adaptive application. For example, a domain expert wants to associate typical cases in his domain area (activated semantic fields) with methods and information for helping novice users to solve their problems with higher performance; teachers could train such a network for appropriate learner support. Users can adapt the system's behaviour with tools for critique or demonstrational interfaces [9].

Unsupervised learning leads to structure modification during interaction without the direct control of an influencing person; the relevant learning law is the differential Hebbian learning law as mentioned above. Through spreading activation, caused by intentional interaction, clustering of caused PE takes place. Repeated actions enhance the weights of an activation sequence and construct "macros" which minimize necessary commands for performing a task. Frequently used R enhance their relevance and influence active support.

Matrix states allow examination of the user's behaviour. Feedback to the user or designers can be given if semantic fields are defuzzified and dialogue messages are transformed into natural language.

3.3 Recall modes

Interaction between user and application causes activation of action atoms. The set of atoms that are activated in a time slice represent patterns of activation. The associated pattern is stored in the connection matrix. To restore the information, there are three recall modes: a single activation update or BAM and TAM encoding.

The *single activation update* equals a reduced TAM encoding from time t to $t+1$. The input vector is multiplied by W and the resulting output vector provides all activities that should be used for interaction. Perceived actions can be revised or confirmed by the user. Corrections causes learning and change associated patterns. It is normally used during a session because the whole man-computer system is treated as a connection matrix and the final output is a stable pattern corresponding to the initial task (see also section 2.2).

TAM encoding can be used for simulation to find likely interaction results. The output vector at time t is used as the input at time $t+1$. The iteration is done with a copy of the original network until a stable state is reached. If the network's energy increases, an unknown pattern is recalled and the user can be asked actively for avoiding unwanted interaction. A learning phase may start again. Using the transposed weight matrix, the interaction history can be recalled if the connections where fixed.

The *BAM encoding* should also be done with a copy of the original matrix and fixed weights. The activation flows from the input layer to the output layer et vice versa until an equilibrium is reached. Learned patterns are associated to user's inputs to predict active support or failures. Reasons for problems can be traced back in order to avoid a failure before it is done.

Structural stability can be proven for the unsupervised learning network by the ABAM theorem but structurally stable attractors include oscillations or chaotic cyles [5: 242]. Further stability problems arise when using symbolic rules for the PE's transfer function because there exists no mathematical theorem to handle hybrid intelligent systems. Therefore stable behaviour can only be examined by simulation and provided in accordance with the user.

Stability can also be reached when fuzzy modification rules interpret activation and connection states. They feed corrections back to the matrix and govern the system's behaviour by increasing the influence of special domain areas or weakening the impact of active help.

4 Adaptive neural advisory spreadsheets

Individual and adaptive facilities enhance the power of conventional spreadsheet applications. For adaptation, a connectionist knowledge base (CKB) and a macro spreadsheet (MS) have to be combined with the conventional working spreadsheet (WS) as shown in figure 3.

The WS communicates directly with the user. Receptive fields in the WS, which can be treated as the input layer, send signals to the CKB that contains PE and the connection matrix. After the user's action, the CKB is updated. Output signals trigger modifying functions. As a consequence, menues or data of the WS change and appropriate dialogue messages are sent back to the user.

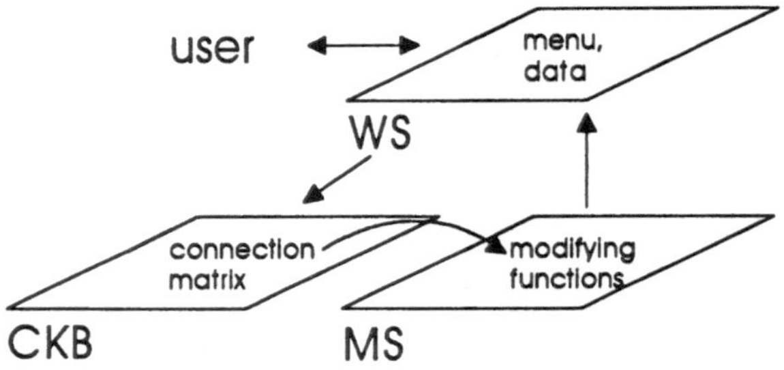

Figure 3: An adaptive spreadsheet architecture

For more detail, figure 4 shows a schematic example of the CKB. It contains a heteroassociative area for connecting external input to output and an autoassociative area, where the input vector at time *t* equals the output vector at time *t-1*. The input can additionally be influenced by the user's critique.

The activation flows from the input to the output, i.e. the input is multiplied by the matrix and results in the output. Fixed or changeable synaptic connections are inside the CKB. Therefore matrix multiplication results in changing connection weights by equation (2) if $LF_{ij} \neq 0$. When an output signal exceeds the threshold value, the activity of the firing neuron is decreased significantly, simulating the neuron's refractory period.

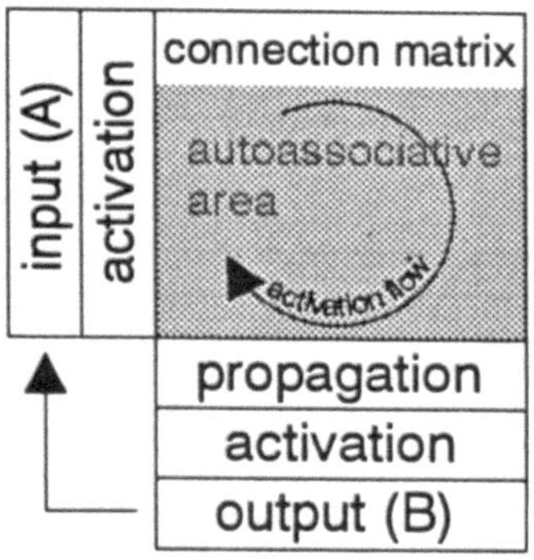

Figure 4: A spreadsheet connection matrix

The spreadsheet architecture was used for a prototypical project information system (ARGUS) and an adaptive support for examining papers at the university (EXAM). The main goal was to prove that adaptation can be done using spreadsheet standard software.

ARGUS individually supports users for enhancing performance. It uses only activation update without modifying weights. The adaptive functionalities are:

- Changing the view according to user and task.
- Visualization of the characteristic calculation in aggregated or detailled form.
- Giving advice for a better use of commands and avoidance of failures through case recognition.
- Changing the menu structure for faster user performance.

ARGUS uses a 100 x 100 connection matrix and needs 6 seconds for a single update on a PC with Intel 80386, 25 MHz, inside. The mechanism works stable and uses fuzzy propagation rules for case recognition as well. The spreadsheet version of ARGUS has demonstrational character but could be implemented using a compiler to reach real-time applicability.

EXAM helps tutors at the university to examine students' tests objectively in accordance with internal correction guidelines [6]. It supports the user by critique and performance feedback. For critique, the system uses R like influencers (positive feedback) and debiasers (negative feedback) [10]. Performance feedback uses defuzzified activation values to show effort, justness, exactness, the mean of awarded points and the variance of granted corrections.

Users can criticize the system as well. With negative critique, firing PE decrease their activations and connection weights change according to (2): When an analogous situation occurs, criticized actions of the system are not activated again. EXAM's behaviour is modifyable through explicit setting of the *LFij* and the self-feedback loops. During Hebbian learning, the usage of messages moves to an individual style of critique.

EXAM uses an 19 x 11 connection matrix and fuzzy rules for input and output interpretation. The response time is 3,6 seconds for a 386-PC with 33 MHz. This time is sufficient for realtime requirements because a user needs more than 4 seconds to read information and enter data into the spreadsheet.

5 Conclusion

We introduced a connectionist framework for advisory dialogue systems. It is based on a neural architecture which models user, task and interaction features. Spreading activation enables user support by giving advice. The network can be determined by designers and is able as well to learn from domain experts or users during interaction. Modification rules which observe and modify the network's behaviour suppress chaotic or faulty attractors.

The applicability can be shown by adaptive spreadsheet applications. The prototype systems work reliable.

Generally, the framework allows fully adaptive support that is only restricted by the number of available 'knowledge nodes'. Inferences are drawn in parallel and change actions according to user-computer interaction. The response time can be calculated in advance.

Future research should be done for improving stability control, enhancing the easiness of design and the power of interpretation facilities.

References

[1] Bodendorf F, '*Benutzermodelle - ein konzeptioneller Überblick*', Wirtschaftsinformatik, 34, 2, 1992: 233-245.

[2] Bodendorf F, Lindner H-G, and Linß H, '*Benutzermodellierung mit Hilfe Neuronaler Netze - Ein Prototyp für OS/2-Nutzer*', Wirtschaftsinformatik-Arbeitspapier 4, Nürnberg, 1990.

[3] Chang A-M, Whinston A B, and Holsapple C W, '*The Hyperknowledge Environment of Model Management Systems*', in Stohr E A and Konsynski B R (eds.), '*Information Systems and Decision Processes*', IEEE Computer Society Press, Los Alamitos et al., 1992: 206-217.

[4] Ellis C A, Gibbs S J, '*Active Objects: Realities and Possibilities*', in Kim W and Lochovsky F H (eds.), '*Object-Oriented Concepts, Databases, and Applications*', Addison-Wesley, Reading, 1992: 561-572.

[5] Kosko B, '*Neural Network and Fuzzy Systems*', Englewood Cliffs, Prentice Hall, 1992.

[6] Lindner H-G, Eggert V and Reinheimer S, '*Computergestützte Individualisierung von Gruppenprozessen am Beispiel einer Examenskorrektur*', Wirtschaftsinformatik-Arbeitspapier 11, Nürnberg, 1992.

[7] Lindner H-G, '*Konzeption eines Modells zur Entwicklung adaptiver und beratender Anwendungssysteme*', Dissertation, Nürnberg, 1993.

[8] Lodewyck R W and Deng P-S, '*Experimentation with a Back-propagation Neural Network - An Application to Planning End User System Development*', Information & Management, 24, 1993: 1-8.

[9] Myers B, '*Demonstrational Interfaces: A Step beyond Direct Manipulation*', IEEE Computer, 25, 8, 1992.

[10] Silverman B G, '*Building a Better Critic: Recent Empirical Results*', IEEE EXPERT, 7, 2, 1992.

[11] Sprague R H Jr. and Hill T R, '*The Nature of Work*', in Stohr E A and Konsynski B R (eds.), '*Information Systems and Decision Processes*', IEEE Computer Society Press, Los Alamitos et al., 1992: 183-190.

Symbolic Learning in Connectionist Production Systems

Wolfgang Eppler

Universität Karlsruhe, FZI
Institut für Rechnerentwurf und Fehlertoleranz
(Prof. Dr.-Ing. D. Schmid)
Haid-und Neustr.10-14, 7500 Karlsruhe, FRG

phone: +49 721 9654 449
telefax: +49 721 9654 459
email: eppler@fzi.de

Abstract. The paper discusses an approach to combine classical artificial intelligence with connectionism. The crucial point with this intention is the implementation of symbols in neural networks. Several proposals are mentioned in [2]. The next important thing to be examined is the increase or modification of knowledge in a connectionist system. In this paper two types of learning are introduced: subsymbolic learning by experience and symbolic learning from linguistic rules. Symbolic learning requires two regions of processing, the language region and the signal region, both of them being coupled with associative links. Symbolic rules are processed sequentially in several cycles, and affect mainly the language region by also influencing the signal region, whereas subsymbolic rules are local to each of the regions and their consequences can be concluded in parallel. In a combined system with both kinds of rules subsymbolic learning and subsymbolic rules are the basis for symbolic learning and the application of linguistic rules. First results with a prototype system are introduced.

Introduction. One of the main problems of the current connectionism is the selection of the topology of the neural network solving a given task. In fact, learning algorithms for self-structuring such networks are known, but the number of units necessary, the number of layers (in multi-layer perceptrons) or the hierarchisation of several networks has to be determined by trial and error. One approach to overcome this lack is *recruitment learning* (e.g. [1]) where new units are created all during the learning process. Of course this is a very time consuming process. Besides this, at the moment there is a trend to process not only totally unstructured problem domains with neural networks but also complex tasks with an at least partially known structure. For these tasks neural networks are chosen because of their massive parallelism and their ability to learn. In this case a combination of conventional methods of computer science and connectionism would provide a good solution. Up to now, rule based systems have been preferred with hardly formalizable problems where they are more suited than procedural programming tools. A combination of these methods will help to cope better with knowledge acquisition and to reduce drastically the response times of interactive systems.

Two types of learning. For a system there are various possibilities to learn (e.g. by example, by doing, from memory, etc.) but two of them are of fundamental importance: (1) learning by experience and (2) learning from linguistic rules. The first one we normally use in neural networks, the last one occurs in conventional expert systems, where an expert has explicitly formulated the rules.

In type-(1) - learning a system is working in a special environment and situation. Learning patterns are presented and the internal structure of the system is modified according to these patterns. This learning process is called subsymbolic and it has, although remotely, similarity with the human learning of skills. If we are talking of rules in this connection only low level rules are meant. We want to understand the processing of information in a network, therefore dependencies of single neurons or neuron assemblies between two layers or two cycles may be expressed by rules. If we have a layered feed-forward network we can map the state of one layer to the state of the next layer and we may describe this mapping with a subsymbolic rule. In a recurrent network rules can describe state transitions from time t to $t+1$. With a general method interpreting state transitions as rule applications also the inverse direction can be gone: with the rules of a production system we may determine state transitions and with them the weights of a neural network ('prestructuring' in [3]). The important thing here is that the name of the symbols is outside the system and cannot be refered in the system itself. When changing the rules we have to put the network offline and prestructure the network by changing the weights.

Although we can deal with symbols in this case, they exist only for an observer (the designer or an interpreter). The system for itself does not cope with symbols. This distinction is fundamental because of the only operational use of symbols in this case. In the rule *"if f(x) then g(x)"* for example the x only indicates the same value in both functions f and g, but the difference between character x and its value $<x>$ is of no importance.

Type-(2) - learning is similar to learning by heart where a teacher tells facts and rules to his scholars. We can speak of symbolic learning if both the name and its meaning are represented as a symbol in the system. If we are working with a neural production system and there is an expert (or a teacher) knowing some new rules the only thing he has to do is to present the new statements as input patterns to the network. In contrast to case (1) the system has to distinguish explicitly between sign and meaning. This is the case if a rule based system accepts facts with variables or a doctor presents his diagnostics to the system, which are related to some signal data from the patient. Best suited for an approach with neural networks seems to be a system with two combined regions - the language and the signal region - being coupled associatively [4].

A combined system. With this sharp distinction of learning types conventional expert systems are only able to learn linguistic knowledge, which can be modified only if words are explicitly changed in definitions of other words. A more powerful system can be achieved by adding other non-linguistic signals (e.g. the ECG or temperature graph in medical expert systems). We then get two regions, the language and the signal region. The great problem here is the coupling of these two regions. Roughly, we can call the language region as the syntactic and the signal region as the semantic part. When we are talking about somebody being feverish, then the word *feverish* will be modelled in the language region, whereas the output of the thermometer, a high temperature, will be processed in the signal region. In conventional expert systems the meaning of words or sentences is defined by other words that can be traced back to basic concepts with a clear meaning (outside the system). In a combined system (or the 'dual coding' approach by Habel/Pribbenow [5] or Kosslyn [6]) to this intentional meaning an extensional one refering to the signal region is added. A first neural approach to a combined system like this can be seen in Figure 1.

The whole system may be viewed from two perspectives, the symbolic and the subsymbolic one. Within the subsymbolic perspective, rules help the designer of the neural network to control specificly the processing from one neural layer to the next and one cycle to the next, respectively. The processing of the rules and the adaptation of the network according to the environment happens massively in parallel. On the other hand in the symbolic perspective the sign of a symbol (located in the language region) is clearly separated from the meaning of the symbol (located in the signal region). Now rules are explicitly occuring as objects in the language region (e.g. (frame, slot, filler)), where they can be represented as tree structures. The implementation in neural networks can be done with the tensor product of Smolensky [7] or similar representations being able to cope with complex data types.

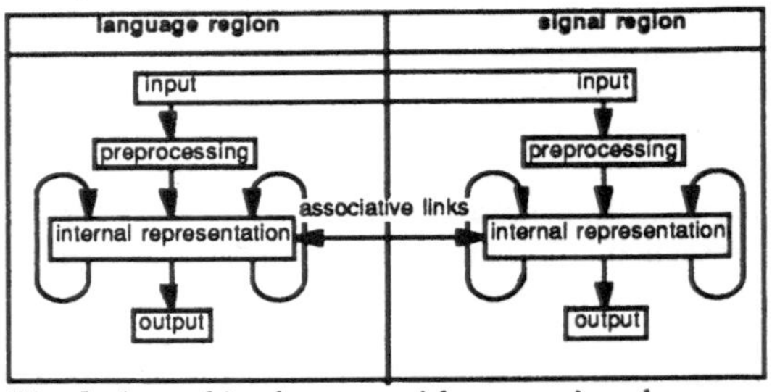

Fig. 1 A combined system with two regions language and signal

Two types of rules. The implemented system can use type-(2) - rules in two different ways. On the one hand rules are stored in the weights of the connections of the language region internal representation. Their antecedents are activated by associative links if the corresponding objects in the internal representation of the signal region are active. In the language region the rule is processed syntactically and logically and new object representations are activated. Because of the associative coupling the objects in the signal region do the same and suitable actions are performed. On the other hand rules can be applied by the explicit input of a rule. After the rule is parsed all objects mentioned are slightly active, they are 'stimulated'. They will be reinforced only if the corresponding objects in the signal space are active and only then objects of the consequent of the rule are activated.

Representation of objects in a Combined System. For both regions, language and signal, object hierarchies can be installed. Objects are represented by frames with slots and values. But not only triples like (frame, slot, value) are allowed. Deeper structures are represented by n-tuples. Each n-tuple is implemented with one or several neurons in one of the two regions of the combined system. These neurons can be distributed over the whole region. Both regions use the same type of representation: in the signal region objects are things or signals with different features, in the language region objects are sentence structures or rules. Rules, for example, are subdivided in antecedent and consequent, these parts are subdivided in object, attribute and value. In a localist representation the leafs of this object hierarchy tree are single neurons. To get a distributed representation only these single neurons have to be multiplied and permutated.

Connections between signal and language. One objective of the system is to couple the linguistic and signal representations in a way that they get the same semantic. If an object is presented to the signal region and the corresponding neurons are activated, the network has to care for the appropriate activation in the language region, and vice versa. Only the conjunction of several language region tripels may affect an activation of a triple in the signal region. Accordingly, the inputs of the signal neurons must be combined with AND-connections, whereas the inputs of the language neurons have to be combined with OR-connections. An example of the connection matrix between these two regions is shown in Figure 2. The matrix shows only the connections from the signal region to the antecedent part of the language region. The connection matrix from the consequent part to the signal region is get if the matrix of Figure 2 is transposed and the OR-connections are replaced by AND-connections. The signal region determines then the rows and the consequent the columns of the matrix.

			signal											
			apple						orange					
			color		taste		ripeness		color		taste		ripeness	
			gr	re	so	sw	tr	fa	gr	re	so	sw	tr	fa
language antecedent	object	ap	V	V	V	V	V	V						
		or							V	V	V	V	V	V
	attribute	co	V	V					V	V				
		ta			V	V					V	V		
		ri					V	V					V	V
	value	gr	V						V					
		re		V						V				
		so			V						V			
		sw				V						V		
		tr					V						V	
		fa						V						V

Fig. 2 Connection matrix from signal to antecedent

In Figure 1 only the basic combined system is shown without special inference rules. No distinction is made between the representation of rules in the connection weights and the representation of rules in the activation values of the neurons. First, the easier case of representing rules in the weights is dealt with. Then a unidirectional connection exists between the antecedent and the consequent of the rule. One further matrix defines these connections between both parts of the language region.

Cross talk. One problem with this representation is the possibility of cross talk. In Figure 3 two different values are bound to two attributes in the signal region. This causes four active tripels in the language region, two tripels that stand for the attributes, and two for the values. With the language region alone it cannot be determined which value is bound to a specific attribute. Therefore, all rules with antecedents that are constituted by arbitrary combinations of the active (attribute, value) pairs are executed. In Figure 5 these are both the rules for x and y. The problem can be solved if the activities of the signal region are considered additionally. In the example the rule "y :- attr1 is val2" would be active only if there is a connection from the signal region (OBJ. ATTR1.VAL2) to the consequent of the language region (rule.cons.y), but this is not the case. The determination of special connections is no cheap trick to solve the cross talk problem because there also exists a learning procedure doing the same. In a first phase a rule may be presented to the network without any activations in the signal space. A single rule does not cause cross talk. Activations in the signal region will emerge in the course of time for the associative connections between the signal and the antecedent region will activate the signal neurons. The task of the learning algorithm is the connection of the installed pattern at the signal region to the consequent part of the language region.

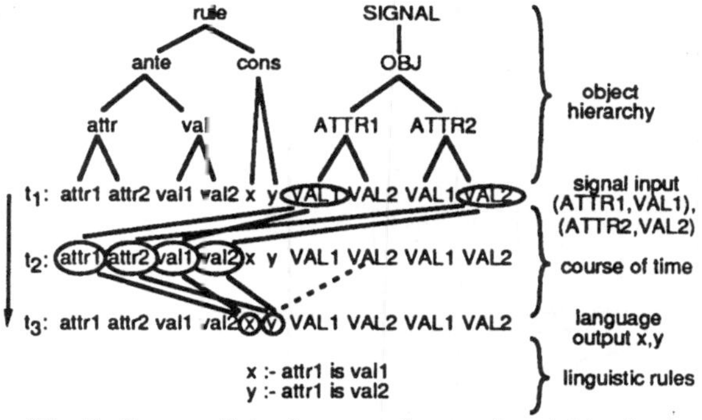

Fig. 3 Cross talk in the case of several activities in the signal region

Long term memory. Even if this method is taken not many rules can be stored at the same time. In the case of storing many rules an intermediate representation is required. This extension concerns only the representation layer of the language region. Rather than a connection matrix between antecedent and consequent we have now connections from the representation layer to the intermediate layer, and vice versa. The main advantage of this additional layer is the much broader coding. In the representation layer features and parts of rules are represented by one or few neurons. Now, in the intermediate layer, whole rules can be represented by single neurons. The broadest possible coding in this manner is the conjunctive coding with one neuron (or one neuron cluster with simultaneous firing neurons) per rule. Another more practical possibility is the coarse coding that needs less neurons for representation because of overlapping activations. The connection matrix between the representation and the intermediate layer contains *k ones* in its columns. The different codings are produced best by a stochastic method. The rows of the matrix determine the activations of the intermediate neurons. Only if complete triples are presented the bias values of these neurons are exceeded. In this intermediate layer (or coarse representation layer) rules are represented by activation values. In one more layer these activations can be stored by an associative learning procedure into the weights between those two layers. Therefore the additional layer is called LTM (long term memory, see Figure 4). The output of the LTM is fed back to the input so that it shows associative behavior. If the antecedent part of a rule is presented the consequent part results after an associative process. Nevertheless, cross talk can occur, and to avoid it input of the signal region is necessary.

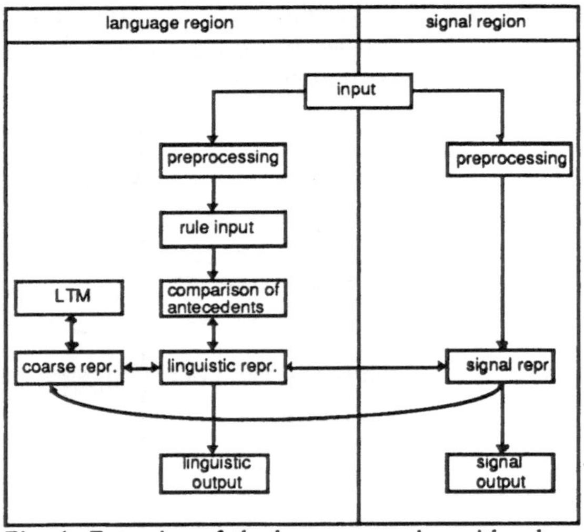

Fig. 4 Extension of the language region with a long term memory (LTM)

Acces of the LTM. An input pattern presented to the signal region stimulates neuron activations in the language region, and it has to be processed according to a stored rule. LTM units only then become active if they are excited both by the neurons of the antecedent part of the language representation layer and the signal representation layer. The LTM completes the consequent part of the rule and activates the corresponding neurons of the language representation layer. These activities cause direct outputs at the language region or indirect outputs at the signal region via the signal representation layer.

Activations representing rules. Another extension concerns the treatment of online rules, i.e. the processing of signal data according to a rule presented at the same time. In contrast to the rules stored in the LTM the preprocessed rule is represented in the rule input layer by activations rather than by weights. The activations of the antecedent part have to be compared with the activations of the antecedent in the linguistic representation. These activities are excited by connections from the signal space. The comparison is a neuron-wise multiplication (or AND-operation) of the antecedents. It is succesful if the result is equal to the antecedent of the rule input. A special neuron (see Figure 5) indicates this fact and passes the consequent of the rule input to the linguistic representation layer (via the comparison layer). The linguistic representation is coupled with the signal space where further actions can be initiated.

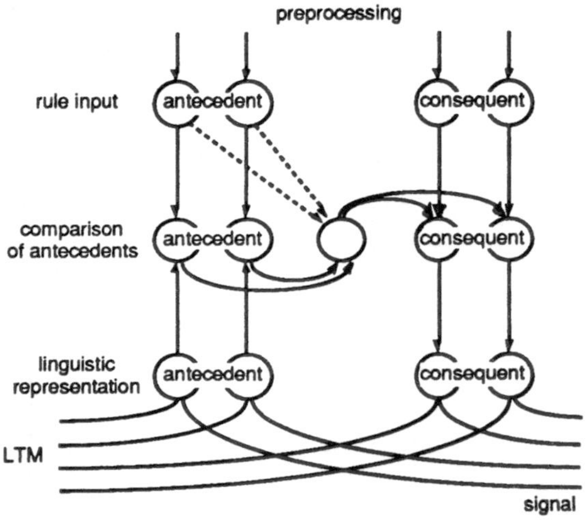

Fig. 5 Closer look at the rule input, comparison and linguistic representation layer

Extended antecedents. Up to now the structure of the rules was very simple for we admit only one condition per antecedent and one consequent. If we look at a consequent neuron we can see that it is activated if a conjunction of three triples is true (e.g. antecedent. object.apple ∧ antecedent.attribute.taste ∧ antecedent. value.sour). We might assume now that we can extend the conjunction to further tripels and get a conjunction with several conditions. But even in this case cross talk occurs so that it is not clear which value is bound to a special attribute or which attribute belongs to a special object. One solution of this problem is to enhance the structure of rule representation by one or several dimensions. If we had triples with the structure (rule part, rule attribute, rule value) we will get now quadruples or n-tuples, e.g. with the structure (rule part, operation, operand, operand attribute, operand value). An example is the quintuple (antecedent, conjunction, operand2, attribute, color).

Different acquisition and application problems. With this basic considerations about a combined system different cases of knowledge acquisition or the application of this knowledge can be described. In the following we will look at this system as a black box and we will take notice of internal processing streams only roughly. In the prestructuring phase subsymbolic rules help to describe the propositional behavior of the system. With these rules a minimal system can be built on the basis of a neural network. Immediately after this phase the processing of simple and ordinary situations is possible (fifth icon in Figure 6). A change of this state always is possible by neural learning in the signal region. In this case (second icon) the language inputs and outputs are not considered. More interesting is the case in which a linguistic rule exists for a special signal input. After one cycle the antecedent of the rule is activated. One cycle later the consequent becomes active and in the last cycle propriate actions of the signal region are initiated (sixth icon). The meaning of the words and rule parts, e.g. the connections from the language to the signal region can be determined at the prestructuring phase. Nevertheless, the system must preserve the capability of learning new concepts and correlations. To achieve this words with the correlated signals have to be presented to the network at the same time (third icon). It depends on the used learning algorithm whether the pattern must be put both at the input and the output of the network (supervised learning) or only at the input (unsupervised learning). A concept relates on a whole class of signal data rather than a single signal pattern. Therefore, the learning patterns must be created over a propriate part of the class. This means that different signal patterns associate the same word in the language region.

Whole rules can be learned by presenting them at the language region without signal data. When using an unsupervised learning procedure the signal neurons are excited by the activities of the language region and they are associated mutually (fourth icon). The two last icons differ mainly in the output (response to a request or initiation of an action) but deal with the same input situation: a linguistic request shall be answered with a special context at the signal region.

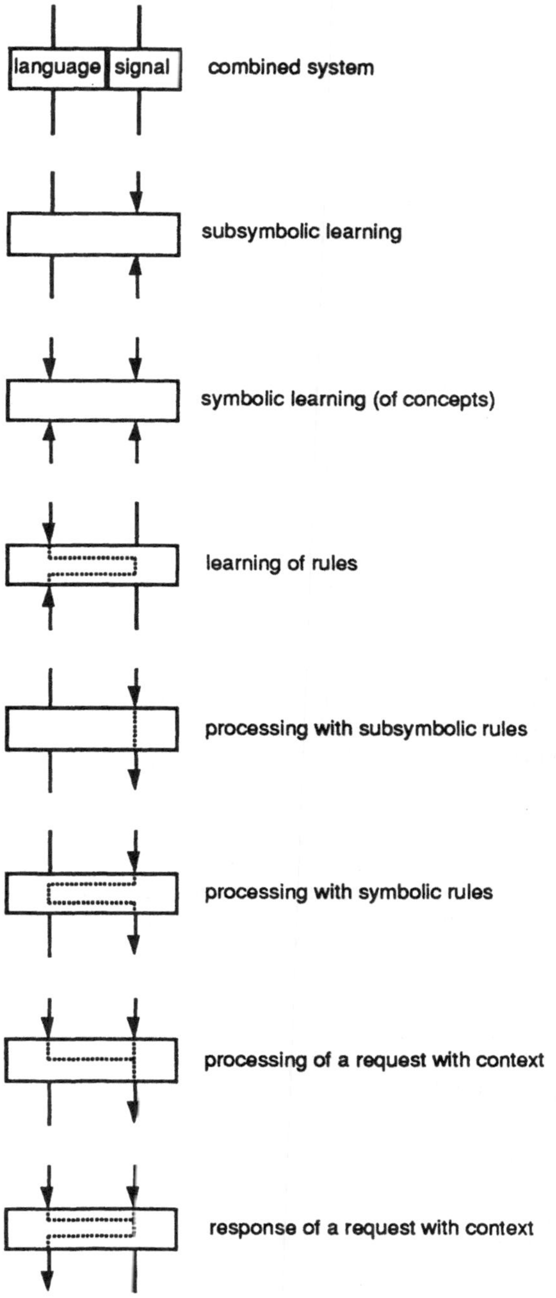

Fig. 6 Knowledge acquisition and application in a combined production system

Application. The following example comes from the signal processing domain. The signals to be recognized look like the signal in Figure 7.

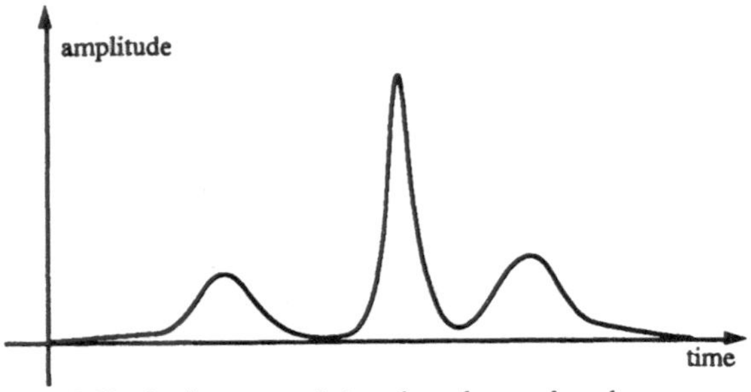

Fig. 7 Typical course of signals to be analyzed

We suppose that several steps of preprocessing were passed, so that noise is eliminated by filters, or values have been normalized. Input values to the combined system are only the amplitude a and the curvature c at time t. The trigger output becomes active if a single peak with a high amplitude and a strong curvature and two smaller peaks with medium amplitude and medium curvature occur. Additionally, it has to be checked whether the correct temporal distances are met.

System design. In the following a design process with several steps is shown. In the first phase a formal rule based description of the system (with crisp and fuzzy rules) is taken to determine the topology and the initializing weights of the network. After this prestructuring phase (see [3]) there remain several faults. Some of them were left deliberately unconsidered because they were tedious to describe, e.g. the fact that the one side peak should not have more than one cycle distance to the main peak if the other side peak shows two cycles distance. The error (total sum of squared output errors over all patterns) after this phase is 12.85. In the second phase 312 test patterns were generated and completed to training patterns with the correct trigger replies. After 14 cycles the error was reduced to 0.087. So the prestructuring phase cared for a very good initial state and an extremly short learning process although the prestructured network contained several faults.

Now we suppose a change of the environment that requires a slight modification of accepted signals. We want to leave the system in the working environment, therefore a new prestructuring phase in a labaratory has to be ruled out. Two different methods dealing with this fact are conceivable. (1) New training patterns must be created. In concrete situations this may be difficult because of the huge number of patterns to be classified. Sometimes it is possible creating them automatically and classifying few peripheral cases by hand. In the current example a further problem occured. Training patterns were created and when testing the network with them the error was 2.078. These patterns were taken to learn the network. After 11 cycles the error still was 2.023. These are very few learning cycles compared to the number of learning cycles used elsewhere. But it shows that the gradient of changing weights is very small and it would take a long time until reaching a state with a reasonable error. Perhaps such a state never will be achieved and the system gets stuck in a local minimum. Therefore, this method will be left unconsidered in the following.

(2) New rules are presented to the network online, and not in a prestructuring phase. For the moment, these rules are represented in a short time memory (STM), i.e. they are coded in the activation patterns of the neurons. In the example the changes of environment can be formulated with an additional rule: "if main peak is in center and side peaks are slightly left or right then activate trigger". The old rule stored in the weights (LTM) went following: "if main peak is in center and side peaks are slightly left or slightly right then activate trigger". When the new rule is presented to the network with the new context of the environment an error of 0.724 occured. This provides a better result than method (1) but is also capable of improvement. Additional 11 learning cycles reduce this error to 0.100.

Linguistic requests. The prestructuring effect is only one advantage of the language region in a combined system. Another advantage is the possibility of asking linguistic requests to the system. An example can be seen in Figure 8, where two alternative states of a combined system are presented.

The only difference in the input of the two states is the request for the location of side peaks (language input matrix, third row, last active neuron in the second state). The other active neurons of this layer stand for the linguistic description of the signal input layer. No current amplitude or curvature inputs are active but the state of previous cycles is preserved. The three active units of the signal input mean: a side peak at time position one, a main peak at position three, and a further side peak at position five (ten old time points are preserved).

The activation of the request neuron 'side peak?' causes stronger activities in the fourth layer of the signal region (main and side peak location). If there are only excitations from the preceding layer of the signal region the predicates corresponding with special neurons are true even if their neurons are active with 0.5 (first, second, fourth and sixth neuron of the location layer with the meaning: main peak between left and center, side peaks left and in center, respectively). Only if there are additional excitations from the language region they may get full active (see second state at location layer). Conversely, activations in this layer do not effect neurons in the request domain of the language region until they exceed 0.5. This is just the case in Figure 8, second state, where the two last active neurons in the language output layer report two side peaks left and in the center, respectively.

Conclusion. Two types of symbols and rules were shown. They motivate a combined system with two regions. Knowledge is represented in both regions. Subsymbolic rules occur in the two regions, symbolic rules only in the language region. Rules are stored statically in the weights of the combined system (LTM) or dynamically as activations (STM). Signals are processed according to these symbolic or subsymbolic rules. Two types of objects are represented in the same way as n-tuples, words in the language and signals in the signal region. The system can be modified offline (by prestructuring) and online (by signal training patterns or by linguistic rules). It was shown that a system like this is able to accelerate learning enormously, to prevent the system falling into a local minimum and to give it an initial structure being open for wide modifications.

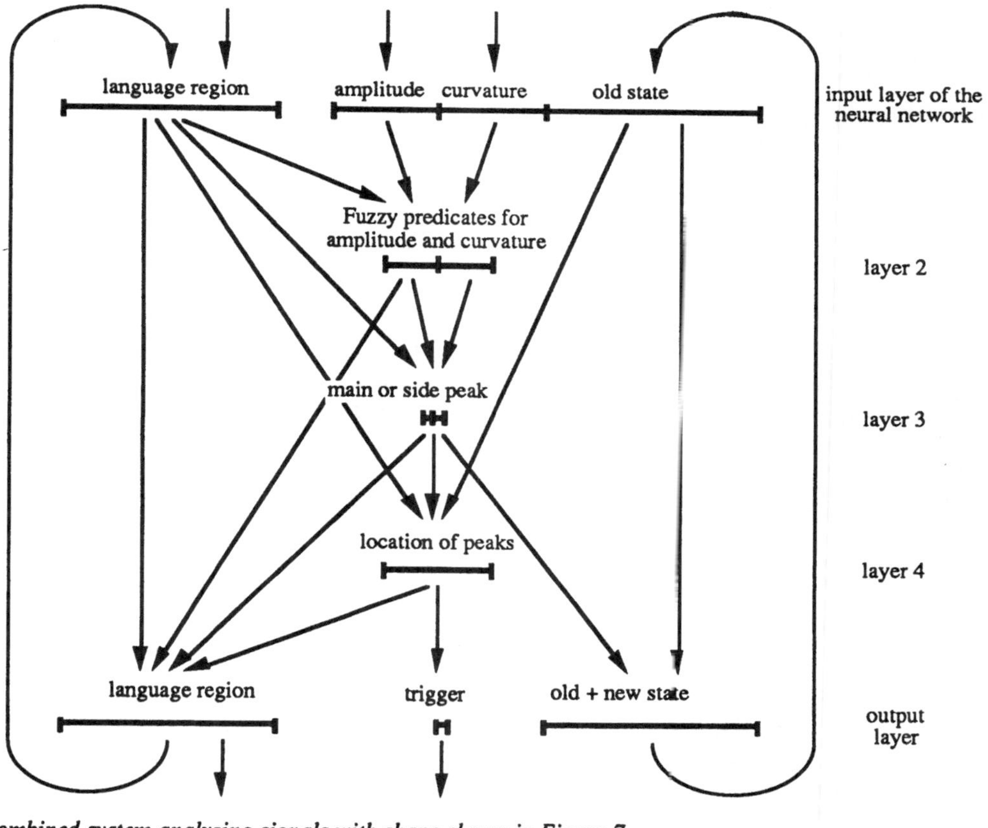

Fig. 8 Combined system analyzing signals with shape shown in Figure 7

References.

[1] Deffuant, G., Dualite local/global & Algorithmes de recrutement, in Proceedings of the Neuro-Nimes '90 (ed. J. Herault), EC2 1990, pp. 147 - 159

[2] Eppler, W., Prestructuring of neural networks with fuzzy rules, in Proceedings of the Neuro-Nimes '90 (ed. J. Herault), EC2 1990, pp. 227 - 241

[3] Eppler, W., Implementation of Fuzzy Production Systems with Neural Networks, in R.Eckmiller et al. (Ed.) Parallel Processing in Neural Systems and Computers, North Holland, 1990, pp. 249 - 252

[4] Eppler, W., Verarbeitung von Symbolen in Neuronalen Netzen, in Güsgen, H.W. et al., Massive Parallelität und Inferenz, Forschungsbericht AIDA-91-16, 1991, pp. 13 - 16

[5] Habel, C., Pribbenov, S., Zum Verstehen räumlicher Ausdrücke des Deutschen - Transitivität räumlicher Relationen, in Brauer/Freksa (ed.): Wissensbasierte Systeme, Informatik-Fachberichte 227, 1989, pp. 139 - 152

[6] Kosslyn, S., Image and Mind. Harvard UP, Cambridge, Mass., 1980

[7] Smolensky, P., Tensor product variable binding and the representation of symbolic structures in connectionist systems, in Artificial Intelligence Vol.46, 1990, pp. 159 - 216

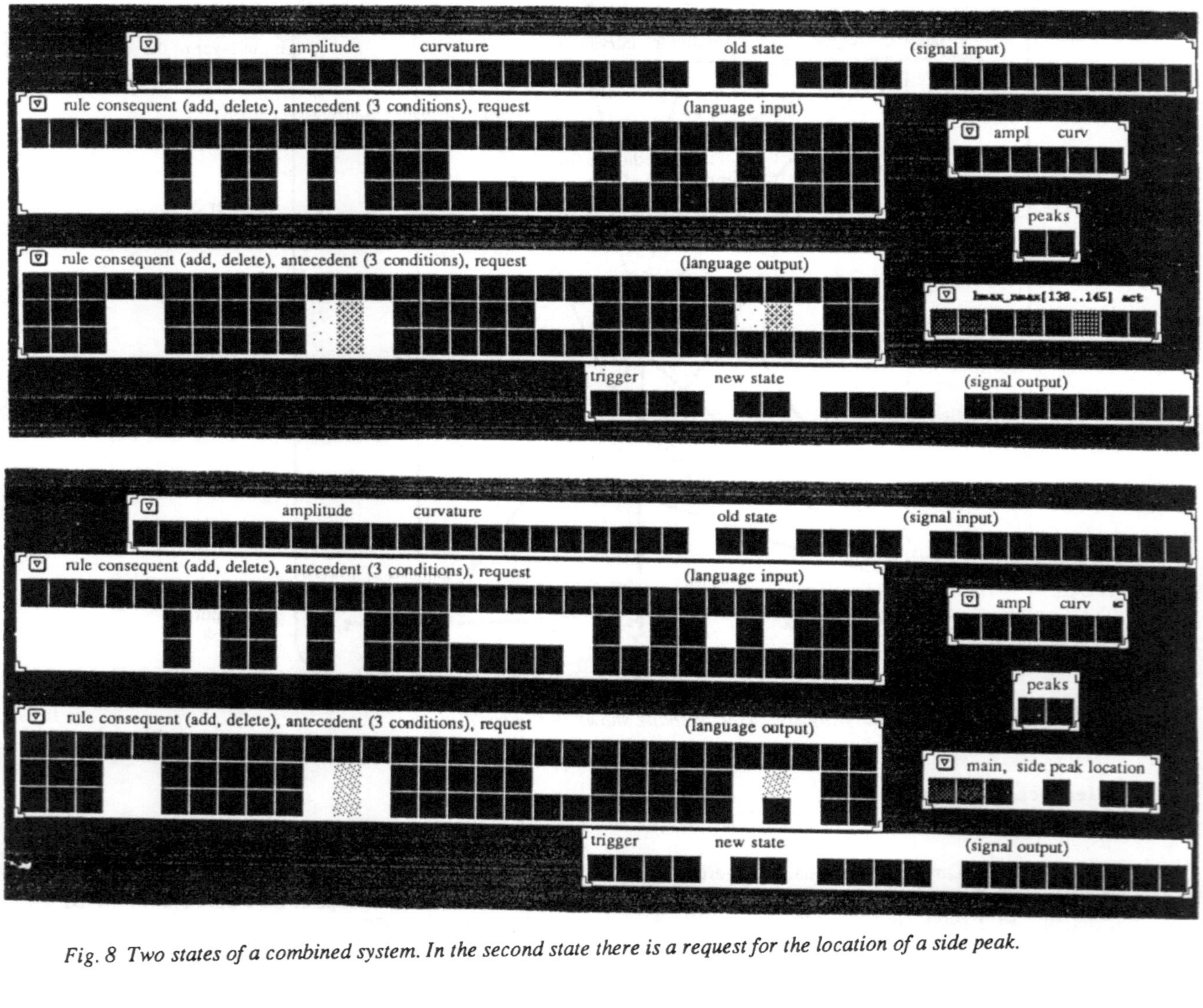

Fig. 8 Two states of a combined system. In the second state there is a request for the location of a side peak.

AN EVALUATION OF DIFFERENT NETWORK MODELS IN MACHINE VISION APPLICATIONS

U. Schramm, K. Spinnler
Fraunhofer-Institute for Integrated Circuits
Wetterkreuz 13, 8520 Erlangen, Germany
Phone: +49-9131-776544, Fax:+49-9131-776599

Abstract

Recently texture segmentation with neural networks has received much interest in fields like remote sensing, medical imaging and autonomous vehicles. We propose to use this approch to improve state-of-the-art machine vision systems. In this paper we present new experimental data to evaluate the performance of different features as well as different neural network models in a segmentation task. We compare features calculated from first-order statistics with features calculated from second-order statistics (cooccurrence features). We investigate different neural network classifers, i.e. multilayered perceptron and restricted coulomb energy model, as well as different conventional classifiers, i.e. minimum distance and nearest neighbour. A real world example of texture segmentation is the detection of defects on industrial treated surfaces.

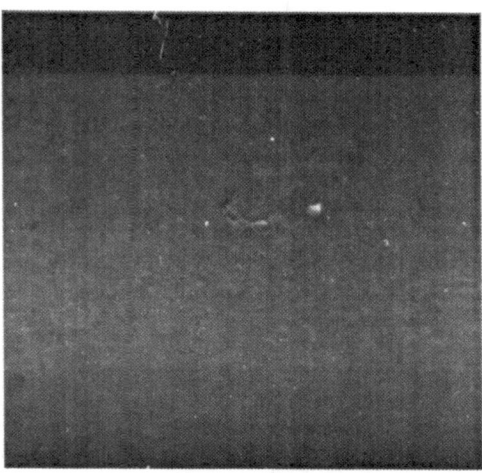

Figure 1: Image of an industrial treated surface with a crack.

1 Introduction

Quality assurance and quality control becomes more and more decisive of competitiveness of companies. At the moment there are great efforts to automate control tasks. The automation of visual inspection is investigated in the field of machine vision or automated visual inspection. Machine vision systems fall short of experts expectations. One reason is that the step of image segmentation has not been solved sufficiently.

State-of-the-art inspection systems use approaches like thresholding or edge detection for segmentation. These approaches are sufficient for simple inspection tasks. In contrast, visual inspection of surfaces is a more difficult task, as the surfaces are not homogenous with respect to their greylevel. The data for our examination are images of industrial treated surfaces. The task is to separate cracks and scratches from normal surfaces (Figure 1).

Therefore we suggest a more powerful segmentation approach, which combines texture analysis and neural networks. We investigate different kinds of neural networks as classifiers, because neural networks are model-free and can be trained simply by presenting examples. Moreover, neural networks offer high throughput and great robustness, features which are imperative to automated visual inspections on the production floor.

Recently texture segmentation with neural networks has recieved much interest. A Markov Random Field Model together with a Hopfield network is applied to Brodatz textures in [1]. Textural features and neural networks are used to classify clouds in LANDSAT images [4]. In order to develop autonomous vehicles, texture segmentation is applied to classify different kinds of terrain [10]. In [8] an application of Kohonen's feature map, the so-called texture map, was introduced. The texture map has been applied

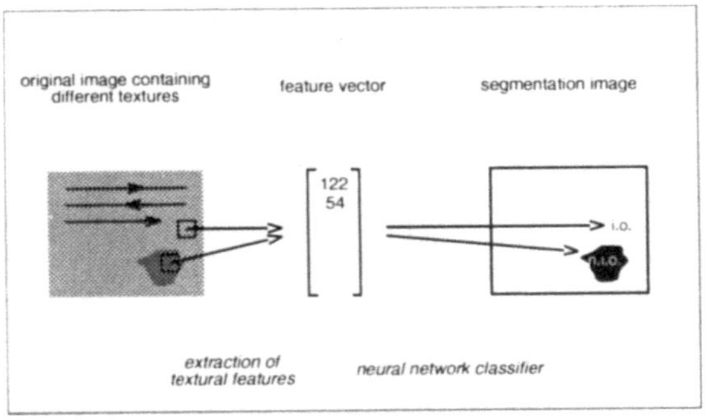

Figure 2: Basic idea of texture segmentation.

to remote sensing data and industrial images.

2 Theory

2.1 Texture Segmentation

Texture segmentation is one problem in the field of texture analysis. In texture segmentation two or more different textures are located on an image. The task is to find the borders between different textures. Figure 2 illustrates the basic idea of texture segmentation. In the left part of Fig.2 an image with two different textures can be seen. This original image is processed by moving small windows across the image. Every window is characterized by textural features. The calculated feature vector has to be classified. The result of texture segmentation is a segmentation image, in which different textures are characterized by different labels.

2.2 Textural Features

We used features calculated from first-order or second-order statistics. In previous studies we presented results using first-order statistics [6], [7]. Examples of that kind of features were the mean greylevel, the median of greylevels and the variance of greylevels calculated in a local window. Additionally the outputs of different gradient operators and directional derivatives were used as features. Overall we used 14 features of this type. The advantage of this feature set is its simplicity concerning computational effort. A database with 9365 feature vectors was generated (ABS_FO_14)[1].

A lot of different textural features were proposed in the field of texture analysis. The cooccurrence features, proposed in [2], are one of the most powerful approaches. In theoretical as well as practical studies best results were reported using these features [5], [9]. Therefore we created a second feature set calculating cooccurrence features on our images of industrial surfaces. We used five features out of the set, i.e. Angular Second Moment, Contrast, Correlation, Sum of Entropy and Difference of Entropy. These features were calculated with two neighborhood relations resulting in a vector with ten features. 9365 feature vectors were calculated at the same image regions as before and were collected in the data base ABS_SO_10.

2.3 Neural Network Classifiers

Neural Networks can be classified into relaxation networks, competitive networks and feedforward networks. With respect to classification problems, it is well-known that feedforward networks can compete with conventional classifiers.

Feedforward networks examined in our investigation are multilayered perceptrons (MLP) and restricted coulomb energy (RCE). A comparison between RCE and MLP concerning error rates was presented in [3]. In contrast to [3] we evaluated the networks not only by the rate of correct/incorrect classifications. We also considered the computational effort for classification, learning and configuration, because these are essential criteria for practical applications.

In both network models the units are arranged in multiple layers. Each unit receives inputs only from units of the previous layer, feedback connections do not exist. If the net is used as a classifier, the input layer is fed with a feature vector and therefore consists of one unit for every feature. The output layer has one unit for every class.

MLP are feedforward networks with a variable number of hidden layers. We used a topology with one hidden layer in our experiments. A unit in a MLP calculates a dot product and afterwards a sigmoid function is used as activation function. Two units build up a hyperplane in the decision space. MLP separate the decision space through complex surfaces by an arrangement of multiple hyperplanes (Figure 3).

A RCE network is a feedforward network with exactly one hidden layer. One unit in the hidden layer can be regarded as one prototype. The hidden layer comprises many prototypes for every class. In the output layer every unit collects the activities of all prototypes per class. A unit in the hidden layer calculates the distance between the input vector and the prototype. Then the distance is compared to a unit specific threshold. If the distance is smaller than the

[1] ABS=workpiece,FO=feature type, 14=number of features

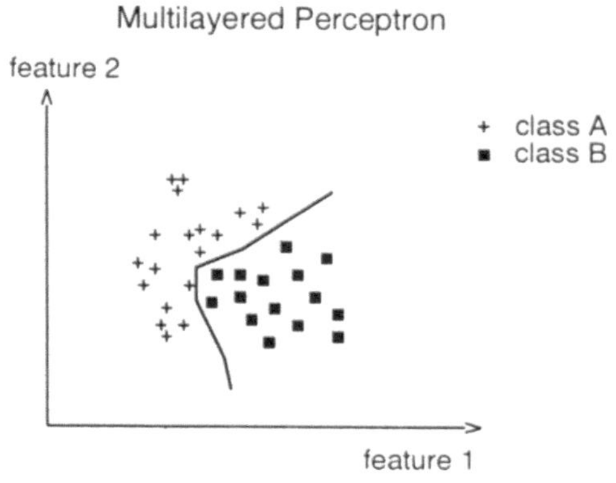

Figure 3: Multilayered Perceptrons separate the decision space by an arrangement of multiple hyperplanes.

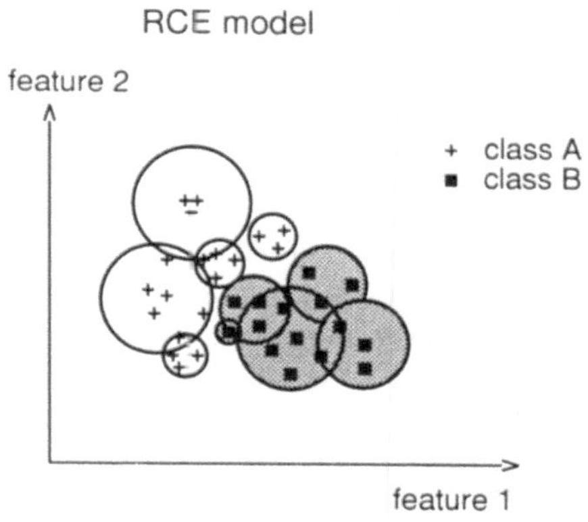

Figure 4: RCE networks model decision regions by an arrangement of multiple hyperspheres.

threshold then the activation of that unit is high. Looking at the decision space a hidden unit corresponds to a hypersphere. The center of the hypersphere is the weight vector, the radius is determined by the threshold. In training mode spheres are generated and the size of the spheres is adjusted. RCE networks build up complex decision surfaces by an arrangement of many hyperspheres (Figure 4).

In addition to neural network classifiers we used the minimum distance classifier (MD) and the nearest neighbor classifier (NN).

3 Experiments

3.1 Setup

The data for our investigation are the data bases with 14 first-order features (ABS_FO_14) and 10 cooccurrence features (ABS_SO_10). Feature vectors had to be classified into one of three classes (background, crack, scratch). The databases were separated into a training sample (6000 vectors) and a test sample (3365 vectors) by random. Three different separations of training and test sample were available. The error rates in the following section are mean error rates of three experiments.

3.2 Results

Table 1 lists the performance of different classifiers on the test sample of ABS_FO_14. The first row shows the error rate, the second the number of prototypes or hidden units, the third the computational effort for classification and the fourth the computational effort for training. MLP and NN yielded about the same error rates. MLP was the best with 1.4%, followed by NN with 1.7%. The error of RCE was slightly higher (3.6%). With an error rate of 21.9% the MD yielded poor results. This was an indication that we had a classification problem with many clusters per class and nonlinear decision regions.

In the second row the numbers of prototypes or hidden units are listed. The smallest number of prototypes was one per class which was used by MD. The number of hidden units increased to 30 for MLP. A strong increase to 460 prototypes was observed with RCE. NN used the whole training sample with 6000 prototypes. There is a strong relation between the number of hidden units and the computational effort in classification mode. In our application the classification of a test vector required the calculation of three distances in the case of MD (40 additions + 80 multiplications $\approx 10^2$ operations). A MLP with 30 hidden units required about 10^3 additions and multiplications. Because the number of hidden units is 460, a RCE network required 10^4 additions and multiplications. NN had to carry out more than 10^5 operations.

With respect to computational effort for learning an opposite tendency was observed. The NN classifier required no training. At the other end the MLP had to carry out about 10^{10} operations to adapt the weights.

Whereas in Table 1 the performance of different classifiers is compared, Table 2 shows the performance on different textural features. Two different types

	Different Classifiers			
	MLP	RCE	NN	MD
Error Rate	1.4%	3.6%	1.7%	21.9%
Prototypes	30	460	6000	3
Classification	10^3	10^4	10^5	10^2
Learning	10^{10}	10^8	0	10^5

Table 1: Comparison of different classifiers. The criteria are the error rate, the number of hidden units, the computional effort for classification and the computational effort for learning.

	Different Feature Sets	
	First Order	Second Order
all	3.6%	0.2%
5 best	2.3%	-
3 best	2.7%	0.3%

Table 2: Comparison of two different feature sets. The error rates on the test samples of both feature sets, first-order and second-order features, are listed in the columns.

of features were examined. The two columns correspond to the two feature sets. The numbers are error rates on the test sample using a RCE network with different feature sets. The first row shows the error rate using all features of one set. Textural features calculated from the second-order statistics outperformed first-order features by far. Using a RCE classifier an error rate of 3-4% was obtained with first-order features, whereas the error rate decreased to 0.2% with second-order features.

Next we reduced the number of features in every feature set (Table 2). Surprisingly the error rate of the five best first-order features was lower than that of the full feature set(2.3% vs. 3.6%). In our oponion, the reason for this behaviour is the kind of modelling used by RCE. Reducing the number of feature to three the error rate of first-order features was about 2.7%. With three cooccurrence features the error rate was still below 0.3%.

4 Conclusion

The classifcation errors of MLP and RCE networks differ only slightly. Also both error rates are comparable with the error rate of the nearest neighbour classifier. A strong advantage of multilayered perceptrons is that only few simple operations are necessary in classification mode. On the other hand multilayered perceptrons demand great effort for configuring and training. Here RCE networks offer advantages, because they are configured automatically in the training phase.

The approach described above is going to be integrated in our machine vision systems. These systems are based on a personal computer. Boards for dedicated image processing algorithms upgrade the computational power of the personal computer. One module extracts textural features. Another module classifies the feature vector. As neurochips for MLP and RCE networks are already availably or are announced for the first half of 1993 it seems possible to integrate MLP as well as RCE networks into our concept in the near future.

References

[1] Chellappa,R. et.al. (1992) *'Texture Segmentation with Neural Networks'*, In B.Kosko (Ed.): Neural Networks for Signal Processing , Prentice-Hall, pp.37-61.

[2] Haralick,R.M. et.al. (1973) *'Textural Features for Image Classification'*, IEEE-T-SMC, vol.3, no.6, pp.610-612.

[3] Hudak,M.J. (1991) *'RCE Networks: An Experimantal Investigation'*, IJCNN'91, vol.1, pp.849-854.

[4] Lee,J. et.al. (1990) *'A Neural Network Approach to Cloud Classification'*, IEEE-T-GR, vol.28, no.5, pp.846-855.

[5] Ohanian,P.; Dubes,R.C. (1992) *'Performance Evaluation for four Classes of Textural Features'*, Pattern Recognition, vol.25, no.8, pp.819-833.

[6] Schramm,U.; Braun,W. (1991) *'Konfigurieren und Trainieren von mehrschichtigen Perzeptron-Netzen'*, In B.Radig (Ed.) Mustererkennung 1991, Informatik Fachberichte 290, S.413-420.

[7] Schramm,U.; Spinnler,K. (1992) *' Neural Network Image Segmentation for Automated Visual Inspection'*, In I.Alexander, J.Taylor (Eds.): Artifical Neural Networks, 2, vol.2, pp.1509-1512.

[8] Simula,O.; Visa,A. (1992) *'Self-Organizing Feature Maps in Texture Classification'*, In I.Alexander, J.Taylor (Eds.): Artifical Neural Networks, 2, vol.2, pp.1621-1628.

[9] Weszka, et.al. (1976) *'A Comparative Study of Texture Measures for Terrain Classification'*, IEEE-T-SMC, vol.6, no.4, pp.269-285.

[10] Wright,W.A. (1992) *'Image Labelling with a Neural Network'*, in P.G.J.Lisboa (Ed.): Neural Networks -Current Applicataions-, Chapman-Hall, pp.149-162.

Combined Application of Neural Network and Artificial Intelligence Methods to Automatic Speech Recognition in a Continuous Utterance

U. Emiliani, P. Podini, F. Sani
Physic Department University of Parma Italy
v.le delle Scienze 43100 Parma
emiliani@vaxpr.fis.unipr.it

Abstract

A very efficient approach to using an artificial supervised neural network in Automatic Speech Recognition in the case of speaker dependent continuous utterance is presented in this paper; it has been tested in the Italian language but in principle not limited to it. An automatic segmentation of the digitized signal and a minimum of human intervention was applied to obtain a phoneme recognition efficiency of 98% on Italian phrases constructed with a limited number of 11 alphabetic classes, defining 20 phonetic subclasses. The efficiency we observed is due to the combined effect of four factors:

- the differential of the detected signal of the utterance was digitized;
- during the parametrization of segments through Fast Fourier Transform (FFT) and critical band definition the effect of a second derivative was simulated: the higher sensitivity in the higher frequency range of ear complex was thus simulated.
- the proper input pattern to be used in the early stages of the training of the neural network was selected by a very sensitive similitude algorithm;
- a dynamic and repetitive training procedure was applied through which the generalization shown by the network after training was used to modify and select the input patterns as well as to control the number of the output nodes used in successive training.

1 Introduction

This paper describes a procedure which, at the end of a proper training period, enables an artificial neural network to analize the constituencies of a new utterance and assign them to the proper phonetic class. The aim is to construct a system able to perform the stated task with a minimum of human intervention: this means that the approach should be the most general possible.

Three main steps can be singled out in the process that a human being carries out to acquire and recognize an utterance: acquisition, sounds identification and comprehension. We shall limit our attention to the first two steps.

1. Utterance Acquisition: a vast literature is available which describes the basic principle of the functioning of the human ear, and although the details are still "fuzzy", many models can be used for computer simulation [1]. Three phenomena are, in our opinion, relevant and are often kept in consideration in developing strategies in Automatic Speech Recognition (ASR) independent from the particular model of the ear:

   - Selective parsing of the utterance does not occur; the ear and the portion of the nervous system directly related to it do not choose the best pattern within a bunch referring to the same phonetic class (a vowel, an explosive or labial for example), but perform automatic and continuous acquisition. [2]
   - The ear system has a non-linear response function that emphasizes the higher frequency portion of the utterance spectra.
   - Critical frequency bands have been identified, through psychoacustic tests, within which perception abruptly changes as a narrowband sound stimulus is modified to have a frequency component beyond the band. When two competing sound signals pass energy through such a critical band the frequency with the higher energy dominates the perception and masks the other frequencies. To a first approximation the amount of

masking is proportional to the total amount of masker energy [3].

2. Sound Identification: this is a more qualitative operation than the previous, where experience and training play a major role. It is clear that the actual processes which take place during this phase are obscure and not well understood at best. We can say that the trained brain associates the incoming sound with a phonetic class. Both in the ear model and in sound analisys a great amount of different approaches are reported in literature.

2 Data Acquisition and Digitization

In designing the equipment for recording and digitizing the utterances, the direct consequence of the observations mentioned in the first paragraph, namely that the ear complex is significantly more sensitive in the higher frequency region of the spectrum, was kept into consideration. After proper filtering the incoming signal was sampled by two sequential Sample and Hold (S/H) driven by two clocks at 11 kHz out of phase of $\pi/2$. The difference of S/H outputs was amplified and sampled again at the same frequency by a third clock with another phase delay of π and converted by a 16 bit AD. The signal was filtered with a 9 pole low pass Butterworth filter with a pass band of 5 KHz to minimize aliasing effects. The acquired signal was then the finite difference over a clock period which emphasizes the high frequency components to simulate the trend of the ear.

3 Segmentation

One of the main problem for ASR in a continuous utterance is to determine the proper length for the time window to be used in segmentation since the utterance itself is a quasi-stationary phenomenon for some sounds while has the characteristic of a transient for others. The time window should be long enough to allow for good frequency resolution but not so long as to average two subsequent phonemes. Generally segmentetion is not an automatic task but is carried out manually. Moreover, in our opinion, the hearing system does not perform such a type of segmentation but uses some kind of persistency (autocorrelation) and reduction (parametrization) criteria. For these reasons an automatic and fixed in time segmentation was chosen, followed by a dynamic weighted average of their FFT as a pre-analysis of the acquired patterns to emulate continuous hearing. The procedure is outlined in the following steps:

1. Segmentation was automatic and overlapping, each segment $f_j(i)$ i=1..128 having a time length of 11.6 msec.
2. Its absolute FFT $F_j(n)$ was then evaluated [4].
3. The subsequent segment was obtained shifting forward by 32 points the 128 points window, and the FFT evaluated again as in point 2.
4. A dynamic average $A_j(n)$ was evaluated by means of the following expression:
   $A_j(n) = [F_j(n)+(m-1)*A_{j-1}(n)]/m$
   where m is a weight depending on the prosody and is evaluated in the following manner: its value is 2 in stationary parts of the signal (the total energy in subsequent F_j is stationary); m decreases, in steps of .25 down to a minimum of 1, if the absolute value of the rate of change of the total energy is greater than a chosen threshold. Moreover the value of m is set to 1 if the total energy is typical of the background noise.
5. With every 4 increments of the j index the corresponding set of values of A_k (k=j/4) was saved, (Fig.-1a) in other words the weighted average of the FFT, as obtained after every 4 shifts of 32 points for a total of 128 points, was saved.

4 Parametrization

The purpose of parametrization is to extract the relevant feature of the spectrum of the phoneme and to form and define what shall be called pattern, eventually allowing an easy synthesis of the signal sound.

As stated, our purpose is ASR, and the model which we assumed is based upon the physiological behavior of the ear complex as reported by [1] and [2] to which a few modifications have been added to make space for the observation concerning the sensitivity shown by the ear in the high frequency range of the spectrum. For each A_k, the following steps are taken:

1. The amplitude of each component of the spectrum is multiplied by its channel number to simulate the effect of a derivative on the acquired signal, which in turn is the differential of the utterance (Fig. - 1b). This has been done because of the empirical observation that by averaging the absolute Fourier spectra over many vocalized utterances the amplitude versus frequency

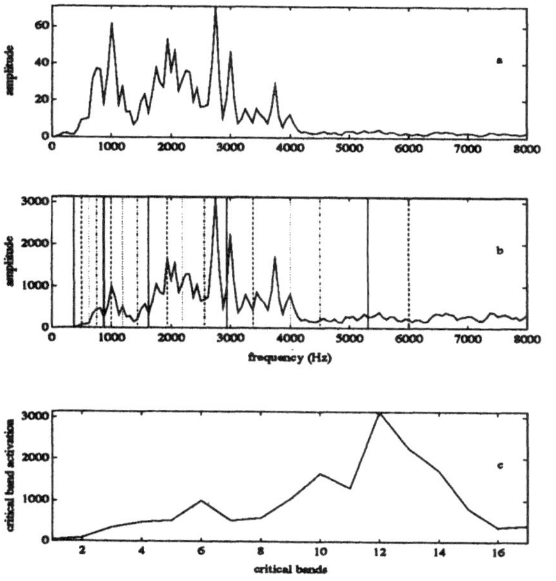

Figure 1: The subsequent steps in the signal elaboration and parametrization are sketched here.
a) - a typical A_k amplitude versus frequency plot obtained with the dynamic average of FFT of subsequent and overlapping segments as described in the text.
b) - an analog plot as in a) where the amplitudes have been multiplied by their respective channel number to simulate the second derivative of the original utterance sound; the 17 critical band ranges used in the parametrization are also sketched on the graph.
c) - the parametric representation of the sound is shown here. The highest value observed within each critical band in b) is selected and assigned as its activation forming a set of values P_k as reduced representation of the original sound.

can be fitted reasonably well by a function of the type Amp/frequency^2

2. The spectrum is divided into sections corresponding to the critical band as reported in Fig.-1.b. In each section the highest amplitude is now determined, to which we refer as the activation of the corresponding critical band.

3. An array of 18 data P_k is created where in the first position the energy associated with A_k have been stored while the activations of the critical bands are inserted in the remaining positions (Fig. - 1c).

This set of parameters is what, for us, represents the sounds, and it is assumed that similar patterns represent similar phonemes. An automatic grouping of the similar patterns is now necessary on which to start the correlation with the relevant classes of phonemes and eventually with the corresponding alphabetical classes.

5 Clustering Algorithm

In this section only the basic principles of the algorithm are described; a complete presentation is given elsewhere [5]. A pattern can be considered as representing a space vector with 17 dimensions, one for each critical band. A similitude type of criteria has been shown to be the most efficient. Let us consider two vectors **A** and **B**
$\mathbf{A} = \{ A_i \}$ $\mathbf{B} = \{ B_i \}$ i = 2..18.
The two vectors are similar if the ensemble of the ratios S=Ai/Bi i=2..18 is constant: this is established by evaluating the standard deviation (SD) of S. If the SD is less or at most equal to a threshold which is established by the algorithm itself, the two patterns are considered similar. After suitable normalization each vector can be represented by one point on the surface in the positive quadrant of a 17 dimensional hypersphere with unit radius. Each phonetic class will generate a cluster of points on the hypersphere surface. In the case of a vowel, because of its stationarity, it will produce a sequence of similar patterns represented by neighboring points in the defined cluster. During a variation from one phoneme to another the representative point will move from the previous cluster toward a new area corresponding to the other phoneme, eventually crossing other cluster surfaces (coarticulation). Our goal is to select, in a continuous utterance, patterns that through their similitude define a cluster. For this purpose the utterance is coded into a matrix where the P_n pattern

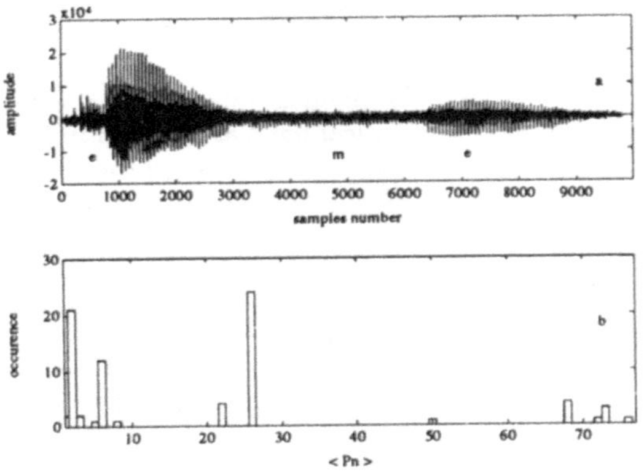

Figure 2: A very simplified sketch of the results given by the clustering algorithm is presented here.
a) - a plot of the acquired signal of an utterance corresponding to the Italian word "emme".
b) - the result of the clustering executed by the algorithm is shown; since in the word there are only two alphabetic classes the most numerous cluster for each alphabetic class is selected for the initial training of the neural network. In the specific case 20 for /e/ and 24 for /m/ (note that the software devised labels the cluster with the position of its earliest pattern). (see also Fig. 3)

is written in the nth row. The search begins by finding the couples of patterns which reciprocally identify each other as the most similar; from the SD values a minimal threshold is evaluated and for each couple their patterns substituted with their average. A new search is started on the new set of patterns looking for couples reciprocally similar within the established threshold. The procedure continues until no other couple is found; the threshold is then upgraded and the search begin again untill the maximum allowed threshold is reached.

In this way the patterns which are clustered together, belong to the same phonetic class. The correlation with the alphabetic class is carried out manually by the operator. During this procedure one observes that often more than one cluster (originating from a different seeding couple) belong to the same alphabetic class. Direct comparison of the spectra, belonging to two different clusters, but assigned to the same alphabetic class, show differences. This observation suggested consideration of each cluster as separate and individual entity, a basic constituent of the speech process, and we thus associate a phonetic class to each cluster.

6 Neural network

The computational method described is quite selective and sensitive and in principle applicable; however the computational load involved is outstanding and time consuming and the capacity to generalize is very limited. Any new phrase is a "new" object and will take the same amount of time to determine its phonetic constituents. On the contrary, a properly trained supervised neural network, is capable of analyzing and discriminating the phonetic classes of an utterance in real time [7]. The time consuming phase is shifted at the training stage, in properly dimensioning the network size and in correctly selecting the significant patterns to be presented to the network input and encoding the desired outputs correctly. Within this outlook, the patterns belonging to the same cluster, as determined by the algorithm described previously, are the inputs to be fed to the network while orthogonal output vectors are selected for the desired outputs. A neural network with hidden units, whose dimensions were optimized on a trial and error basis, was implemented on a Sparc Station 2. A recurrent training and test method has been devised to train the network as described below:

1. An utterance is selected and analyzed through the described algorithm. For each alphabetic class present in the utterance the patterns of the larger cluster are selected as representative of the class itself. An ensemble of orthogonal output vectors is chosen, one for each alphabetic class present.(Fig. 2 a and b)

2. The network is trained using the backpropagation algorithm [6]

3. All patterns of the utterance are now presented to the trained network. Those which are correctly recognized will be grouped with the original set and the network retrained. During this procedure original clusters or groups of patterns belonging to the same alphabetic class may go unrecognized by the network. For these groups a new component for the output vector is created, and the network retrained accordingly. This procedure is repeated until all patterns are recognized correctly apart from a few which are in transition regions of the original utterance.

4. The patterns of a new utterance by the same speaker is presented to the trained network and the steps described in 3 repeated and so on. What has been observed is that at the beginning, the number of the components of the output vector tend to increase to make space for new

phonetic classes assigned to the same alphabetic class. However, after a few repetitions, patterns belonging to different subsets of the same alphabetic class are assigned by the network to both subsets; this suggested merging the two subclass into one during subsequent training. In this way the number of the components of the output vectors of the network does not diverge with the increasing number of utterances during training, but tends to stabilize to a reasonable number.

7 Experimental Results

Phrases were constructed using a reduced alphabet with phonemes belonging to the following (Italian) alphabetic classes: /a/, /e/, /i/, /o/, /l/, /m/, /b/, /r/, /s/, /t/, /sc/. Four phrases were composed in such a way as to carry some logical meaning in the Italian language to allow natural pronunciation. A speaker dependent analysis was chosen. For the elaboration of the data (segmentation, parametrization and clustering) a Matlab Mathwork software package on a Spark-Station was used. The representative patterns obtained were used to train an artificial neural network, with the following architecture: 17 input nodes, 17 hidden nodes and, in the final version, 20 output nodes. Each input node was linked to the activation level of one of the 17 critical bands within the pass band of the filter. The " probability" that the pattern under test belongs to the corresponding phonetic class was associated with each output node. For the present speaker, the eleven alphabetical classes were coded into 20 phonetic classes To each output node is associated. In the training stage three phrases (each with at least 500 patterns) were used while the fourth was used for testing the performance of the network. This procedure was repeated four time from the beginning, interchanging the test phrase each time with one of the set used in the training. In all trials more than 98% of the utterance patterns were reconized and assigned correctly. The patterns not recognized were those between two phonemes or in the trailing of the utterance. (An example of the steps of the procedure is oulined in Fig. 3 in an ultra-simplified example).

8 Conclusions

Although the procedure described above has been tested on a limited number of phonemes, it has shown remarkable efficiency in ASR and in principle can be extended. The procedure described is elaborated and

1	2	3	3	5	6	6
8	6	6	6	6	6	6
6	6	6	6	2	2	2
22	22	22	22	26	26	26
26	26	26	26	26	26	26
26	26	26	26	26	26	26
26	26	26	26	26	26	26
51	52	2	2	2	2	2
2	2	2	2	2	2	2
2	2	2	2	68	68	68
68	72	73	73	73	76	0

E	E	E	E	E	E	E
E	E	E	E	E	E	E
E	E	E	E	E	E	E
*	*	*	M	M	M	M
M	M	M	M	M	M	M
M	M	M	M	M	M	M
M	M	M	M	M	M	M
M	*	E	E	E	E	E
E	E	E	E	E	E	E
E	E	E	E	E	*	*
*	*	*	*	*	*	*

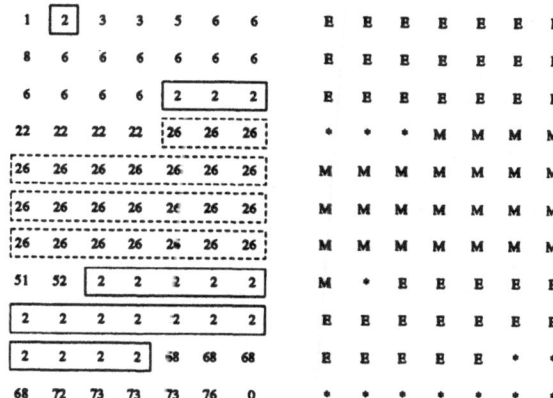

Figure 3: The leftend side illustrates an equivalent way to represent the clustering algorithm results as in Fig. 2b. The positions define the pattern sequence while the numbers represent the cluster label. The clusters labeled 2 and 26 are the most numerous: the first represents the /e/ while the 26 represents the /m/. These patterns are the first to be used in neural network training; after testing the trained network on all the utterance patterns a few patterns belonging to different clusters are recognized as similar to the originals (cluster 2 and 26 in our case) These patterns are added to the new training patterns of the network. Repeating the procedure twice gives the result presented on the right side of the figure where the alphabetic class each pattern is assigned to by the network is reported. The asterisks are patterns that the network does not recognize and are typically transition patterns (exept for the last few which belong to the vanishing part of the recorded utterance). Clarifying the role of the clustering algorithm and that of the network is observing that the algorithm separates the initial and final /e/ in two different clusters (actually there is a slight difference in pronunciation in Italian; the first /e/ is more "open" than the second one and the algorithm distinguishes between them) but the network is able to generalize and assign the two original clusters to the same alphabetic class. (The example reported in the figure is an oversimplified example to outline the procedure used. The results are not significant since the utterance consisted of one word of two phonemes).

time consuming but the results make it worthwhile to use. At the moment a test of the procedure is under way on a much wider set of phonemes of the Italian language in parallel with a test for speaker independent ASR based on the limited original set of phonemes. In both cases preliminary results are comforting. The procedure of treating and representing the sound seems to be efficient and allow good selection of the seeding clusters for the first training stage; the repetitive training procedure adopted amply compensates the algorithm tendency to enlarge the number of the clusters. Therefore it is hoped that when all of the alphabet is used the number of clusters will be limited. A few problems remain to be approached which concern the actual "writing" of the words uttered (orthography):

1. the number of patterns assigned to one phoneme is variable as a result of the automatic segmentation implemented in the procedure (many patterns may be assigned to vowels while only one may be assigned to a plosive phoneme). A problem may be present in correctly recognizing the beginning and the end of a phoneme and identifying transition patterns when they are present.

2. a similar problem is present in separating words in a continuous utterance. A simple pause detection criteria may be not sufficient since some phonemes like /t/ are preceded by a pause (pressure build up pause). For the solution of the above mentioned problems a top -down approach seems to be appropriate.

References

[1] S. Seneff, *A joint synchrony/mean-rate model of auditory speech processing,* Journal of Phonetics, 16 (55) Jan. 1988.

[2] S. Furui, *Digital Speech Processing, Synthesis and Recognition,* Marcel Dekker, Inc. New York, 1989

[3] D. O'Shaughnessy, *Speech Communication,* Addison Wesley, 1987

[4] A. V. Oppenheim, R.W. Schafer, *Digital Signal Processing,* Prentice Hall, 1975

[5] U. Emiliani, C. Oliosi, P. Podini, to be published

[6] D. E. Rumelhart, J. L. McClelland, *Parallel Distributed Processing, Vol. I Exploration In the Microstructure of Cognition,* MIT Press, 1986

[7] P. Cosi, Y. Bengio, R. de Mori, *Phonetically-based Multi-layered Neural Networks for Vowel Classification,* Speech Communication 9 (15) 1990

A Neural Network Based Control of a Simulated Biochemical Process

Abhay B. Bulsari and Henrik Saxén

Heat Engineering Laboratory
Department of Chemical Engineering, Åbo Akademi
Biskopsgatan 8, 20500 Åbo, Finland
Phone: 358-21-654311, Fax: 358-21-654792
E-mail: vt_ai@abo.fi

Abstract

This paper describes an application of feedforward neural networks for inverse plant control of a process with highly non-linear characteristics. A biochemical process was considered where the microorganism, *Saccharomyces cerevisiae*, a yeast, grows in a chemostat on a glucose substrate and produces ethanol as a product of primary energy metabolism. In this process, which is of immense interest to industries worldwide, three state variables were considered: microbial, substrate and product concentrations. The last one is the controlled variable, and the dilution rate is the manipulated variable. In the study, the quality of the control is analyzed for the case where all states are assumed to be measurable and the case where only the product concentration is available.

1 Introduction

Control is an essential part of any plant, and for biochemical processes, where the mathematical models of the process are not accurate and apply to only small operating regions, the conventional methods of control system design are not so reliable. Many attempts have been made to control difficult processes through neural networks [1–6], and several investigators have made use of neural networks for solving other tasks in biochemical processes. This work considers inverse-plant control of a nonlinear biochemical dynamic process [7], which is described by three state variables. The networks, which compute the value of the dilution rate — the controlled variable — based on information about (some of) the state variables, were found to produce a fairly good control of the process.

2 The process and the networks

In the biochemical process studied in this work, a yeast, *Saccharomyces cerevisiae*, is grown on glucose substrate in a chemostat (a biochemical continuous stirred tank reactor) producing ethanol as a product of primary energy metabolism. The process is characterised by three state variables (microbial concentration, X; substrate concentration, S; and product concentration, P) and was simulated by solving a set of differential equations using a program package [8]. The reader is referred to ref. [9] for further details about the process model.

The "process data" consisted of 6 simulations of the system's dynamics. 560 instances were selected from these simulations to train the neural networks described below. The independent variables S_0 (substrate concentration in the inlet) and D (the dilution rate = volume flow rate per chemostat volume) were varied freely in the simulations. All variables are "measured" every 2 hours. For numerical reasons, the variables X, S, P, S_0 and D were reduced to $(X - X^*)/5$, $(S - S^*)/20$, $(P - P^*)/20$, $(S_0 - S_0^*)/50$, and $D - D^*$ respectively. The variables marked with an asterisk are the normal operating values under steady state, as shown in Table 1 of ref. [9]. The normalised variables now stay within the range (–1,1), which permits the use of sigmoidal activation for the output node of the networks. In the figures to be presented, the horizontal axis has time in hours, and the vertical axis has these reduced variables D and P.

Standard multilayer feedforward neural networks were used in this study, where the total input to a node, a_i, is computed as a weighted sum of the outputs, x_j, of the neurons in the layer below. As activation functions, the –1 to 1 sigmoid activation function

$$x_i = -1 + \frac{2}{1 + \exp(-x_i)} \qquad (1)$$

or the symmetric logarithmoid [10]

$$x_i = \frac{a_i}{|a_i|} \log_e(1 + |a_i|) \qquad (2)$$

were used. The latter function is not limited to $(-1, 1)$, and does not saturate for very large inputs.

The Levenberg–Marquardt method [11 – 12] was used to determine the weights in the neural networks by minimizing the square sum of the output errors. The inputs to the networks were information about the states at time t and the (desired) product concentration at time $t + \Delta t$, while the single output was the value of the manipulated variable, $D(t)$, that brings about the desired change in P.

3 The inverse-plant control

This chapter briefly describes the main findings of the work, and reports the performance of the control by different networks.

3.1 Control using knowledge of all variables

In the first phase of this work, neural networks were trained to control the process based on the knowledge of all the variables. The inputs were X, S, P, S_0 and P_{set}, and the output variable was the control variable D.

A linear network resulted in an rms error of 0.01573 on the training set. Using sigmoidal activation functions in a (5,3,1) network (with three hidden neurons), the error reduced to 0.00511. Replacing the sigmoids by logarithmoids did not improve the results (rms=0.00564). A network with one additional hidden node did clearly better: The rms errors for networks with sigmoidal or logarithmoidal activation functions dropped to 0.00375 and 0.00360, respectively. After adding two more nodes, the corresponding rms errors were 0.00260 and 0.00263. This configuration, (5,6,1), was used as a controller for the process.

The controller was tested using an initial condition of $X = 10$, $S = 15$ and $P = 30$. Initially, S_0 was 101.416, and 10 hours later it was reduced to 98.0. The set point for P was 41.232 throughout the duration of the test of the controller. (All variables mentioned above are expressed in the unit g/l.) The controller was stable, and it worked slowly but satisfactorily. The control variable D was limited to between 0.2 and 0.4. Figure 1 shows the network configuration and a typical dynamic behaviour of the process with the controller. In the lower graph, P is presented as a function of time, while the upper graph depicts the dilution rate as a function of time.

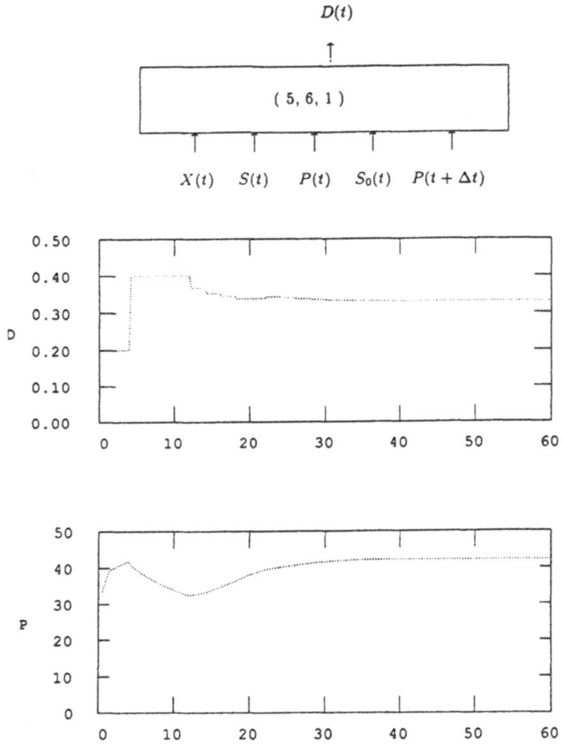

Fig. 1 The (5,6,1) network as a controller of the product concentration. $P(t + \Delta t)$ was used during training, and P_{set} during control.

3.2 Control using only knowledge of P

Usually, not all state variables are measured, for reasons of cost, time, accuracy or practicality. In this second phase, we assume that only the product concentration, the controlled variable, be measurable, besides, of course, the manipulated variable, D. The neural networks can now be trained to control the product concentration based only on the knowledge of its own value, and its set point. Neural networks were trained using the same 560 training instances, as in the first phase.

A linear network gives an rms error of 0.0160, which is a bit more than in the previous stage. The configuration (2,3,1) resulted in rms errors of 0.00819 and 0.00842 using the sigmoids and symmetric logarithmoids as activation functions, respectively. Similarly, the configuration (2,6,1) reduced the errors to about 0.007. This error is considerably larger than the one obtained in the first phase.

The results of using the (2,6,1) network as a controller are shown in Figure 2. The control is slower but better. As before, the upper and lower graphs show the evolutions of D and P, respectively.

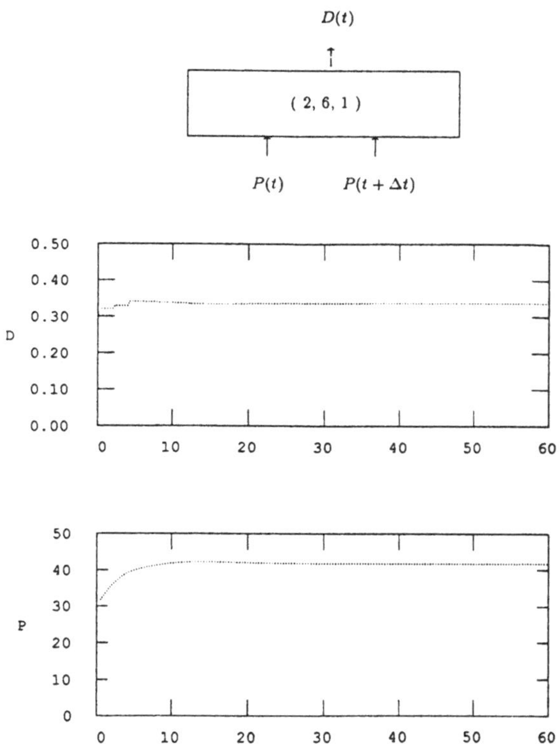

Fig. 2 The (2,6,1) network as a controller of the product concentration. $P(t+\Delta t)$ was used during training, and P_{set} during control.

It may be noted that the disturbance variable S_0 remained in the range of 91 to 111.5 g/l in the training instances. Table 1 shows that the smaller (2,6,1) network has better steady state characteristics, since the point, P_{set}, was 41.232 g/l in all cases. However, the (5,6,1) network generally worked a little faster than (2,6,1), presumably because the former has more knowledge on the state of the process, and on how D affects the state.

4 Conclusions

Earlier work has demonstrated that system identification of this biochemical process can be carried out successfully using feedforward neural networks. Another work [13] showed that state and disturbance estimation for this process can be performed using feedforward neural networks. This investigation illustrates that nonlinear processes of this kind can be controlled using neural networks without prior knowledge of the process. It should be noted that the non-linearities of the process play an important role in this study, because of the long sampling time ($\Delta t = 2$ h vs. 0.25 h in the filtering study of ref. [9]).

References

[1] Bavarian, B., "Introduction to Neural Network for Intelligent Control", *IEEE Control Systems Magazine*, April 1988.

[2] Ydstie, B. E., "Forecasting and control using adaptive connectionist networks", *Comput. chem. Engng.* **14** (1990) p 583–599.

[3] Bhat N. and McAvoy, T. J., "Use of neural nets for dynamic modeling and control of chemical process systems" *Comput. chem. Engng.* **14** (1990) p. 573–582.

[4] Antsaklis, P. J., "Neural networks in control systems", *IEEE Control Systems Magazine* (April 1990) p. 3–5.

[5] Ungar, L.H., B.A. Powell and S.N. Kamens, "Adaptive networks for fault diagnosis and process control", *Comput. chem. Engng.* **14** (1990) p. 561–572.

[6] Miller, W.T., R.S. Sutton and P.J. Werbos, "Neural networks for control", MIT Press, Cambridge, USA, 1990.

[7] Bulsari, A.B. and H. Saxén, "System identification of a biochemical process using feed-forward neural networks", *Neurocomputing* **3** (1991) p. 125–133.

[8] Elmqvist, H. *et al.*, "Simnon user's guide for MS-DOS computers, Version 3", SSPA Systems, Göteborg, Sweden, 1990.

[9] Bulsari, A.B. and H. Saxén, "Application of neural networks for filtering", to be presented at *International Conference on Neural Networks and Genetic Algorithms (ANNGA'93)*, Innsbruck, Austria, April 1993.

[10] Bulsari, A.B. and H. Saxén, "System identification using the symmetric logarithmoid as an activation function in a feed-forward neural

network", *Neural network world* **1** (1991) p. 221–224.

[11] Marquardt, D. W., An algorithm for least-squares estimation of nonlinear parameters", *J. SIAM* **11** (1963) p. 431–441.

[12] A. Bulsari, B. Saxén and H. Saxén, "Programs for feedforward neural networks using the Levenberg–Marquardt method : Documentation and user's manual" Report 90–2, Heat Eng. Lab, Åbo Akademi, Finland, October 1990.

[13] Bulsari, A. and H. Saxén, "Estimation of a disturbance variable using feed-forward neural networks". Proceedings of the *11th IASTED International Conference on Modelling, Identification and Control (MIC'92)*, Innsbruck, Austria, February 1992, p. 248–250.

Table 1 Control using the networks (5,6,1) and (2,6,1).

S_0, 10 hours later	(5,6,1)		(2,6,1)	
	$P(\infty)$	$D(\infty)$	$P(\infty)$	$D(\infty)$
g/l	g/l		g/l	
60	27.74	0.3388	27.97	0.3140
70	31.95	0.3314	32.03	0.3256
80	35.86	0.3323	35.93	0.3290
90	39.65	0.3281	39.04	0.3408
98	42.19	0.3304	41.67	0.3368
110	43.61	0.3383	43.33–43.74	0.3340–0.3415
120	unstable		43.54	0.3415
130	unstable		43.62	0.3423

Applications of Neural Networks for Filtering

Abhay B. Bulsari and Henrik Saxén

Heat Engineering Laboratory
Department of Chemical Engineering, Åbo Akademi
Biskopsgatan 8, 20500 Åbo, Finland
Phone: 358-21-654311, Fax: 358-21-654792
E-mail: vt_ai@abo.fi

Abstract

This paper investigates the applicability of feedforward neural networks for filtering purposes. A biochemical process was simulated to generate the data for training and testing the networks. Delayed measurements of one or more state variables were used as inputs to the networks, which were trained to provide filtered values of the state variables as outputs. The results of filtering were quite accurate. In most cases, linear models *i.e.*, networks with linear activation functions, were found to be adequate. This is due to the short sampling time and the fact that the non-linearities of the process are not very strong in the region of state space considered.

1 Introduction

Artificial neural networks have found several useful applications in process dynamics, *e.g.*, for describing the evolution of the states in biochemical processes [1−4]. Our earlier work demonstrated that system identification [5] and variable estimation [6] can be performed using feedforward neural networks for the biochemical process considered here. Also, disturbance variables can be estimated, even though this is a more challenging task [7]. However, the data was assumed to be free of noise in these previous investigations. The aim of this work is to study filtering and one-step ahead prediction with neural networks. The analysis also clarifies whether partial information about the states can be used for predicting filtered values of all state variables. In the tests of feedforward neural networks for filtering considered here, networks with linear activation functions were, in general, found to suffice, even though the process studied is known to be non-linear. This is explained by the observation that the non-linearity is not very severe for the trajectories studied, and that the sampling time was selected short enough.

2 The process

Biochemical processes are often characterised by highly non-linear behaviour, and, moreover, have operability in limited domains. The process studied in this work is a chemostat (Fig. 1) where a yeast strain, *Saccharomyces cerevisiae*, is produced on glucose substrate. In the metabolism, ethanol is produced. The three state variables considered are thus microbial (yeast) concentration, denoted by X, substrate (glucose) concentration, denoted by S, and ethanol (product) concentration, denoted by P. The feed, F, to the chemostat is sterile, *i.e.*, it contains no microorganisms, and has a substrate concentration denoted by S_0.

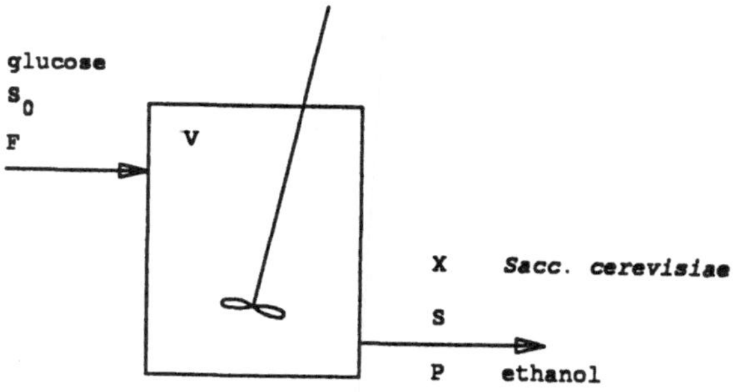

Fig. 1 The biochemical process.

By introducing the dilution ratio, $D = F/V$, where V is the liquid volume in the chemostat, the set of differential equations describing the system can be formulated compactly as

$$\frac{dS}{dt} = D(S_0 - S) - Y_{S/X}\mu X \quad (1)$$

$$\frac{dX}{dt} = (\mu - D)X \quad (2)$$

$$\frac{dP}{dt} = -DP + Y_{P/X}\mu X, \quad (3)$$

where the growth rate, μ, and the so called yield coefficients, $Y_{S/X}$ and $Y_{P/X}$, are given by

$$\mu = \frac{0.427S}{0.245 + S}\left[1 - (\frac{P}{101.6})^{1.95}\right] \quad (4)$$

$$Y_{P/X} = 3.436; \qquad Y_{X/P} = Y_{P/X}^{-1} \quad (5)$$

$$Y_{X/S} = 0.152(1 - \frac{P}{302.3}); \quad Y_{S/X} = Y_{X/S}^{-1} \quad (6)$$

with kinetic and stoichiometric parameters taken from the literature [8]. The upper limit of D is determined by the growth rate, μ, and cannot exceed $0.427\,h^{-1}$. If this value is exceeded, the microbial concentration in the bioreactor decreases towards zero. A typical steady state of the process variables is given in Table 1.

Table 1 Typical steady-state of the bioreactor.

S_0^*	101.4 g/l
D^*	0.345 h^{-1}
S^*	10.0 g/ l
X^*	12.0 g/ l
P^*	41.2 g/ l
μ	0.345 h^{-1}
$Y_{X/S}$	0.131
$Y_{X/P}$	0.291

The model, which consists of equations (1) through (6), is applied to produce both training and testing data for filtering by neural networks.

3 Mathematical treatment

The tests in this work were carried out using standard multilayer feedforward neural networks with at the most one layer of hidden neurons. Since most of the results showed that networks with linear activation functions performed almost as well as, and in some cases even better than, non-linear networks, the results to be presented in this paper are limited to analysis of the linear networks. These networks simply consist of the input and the output layers. The networks were trained by the Levenberg–Marquardt method [9], which is a method originally developed for solving non-linear equations and regression problems. This method, which can be used without inconvenience to solve small or medium-sized training problems, has proved to be much faster than *e.g.* the back-propagation method [10].

The state variables X, S, and P were "normalised" to $(X - X^*))/5$, $(S - S^*)/20$, and $(P - P^*)/20$, respectively, where variables with superscript denote steady-state values (see Table 1); this measure is taken simply for numerical reasons. Eight trajectories, with 97 patters each, were next taken from eight vertices of a cube in state space (see the Appendix). The state variables X, S and P on the 8 trajectories are shown in Figure 2. Noise added is at the most 5 % of the signal strength of the scaled variables. The root mean square noise in the state variables, X, S, and P, is 0.007934, 0.011260, and 0.00500, respectively, which corresponds to an rms noise of 0.00846. The intervals for the scaled variables are $X \in (-0.806, 0.286)$, $S \in (-0.315, 1.027)$, and $P \in (-0.562, 0.188)$.

Further, a test set containing 194 patterns of two trajectories was created. The first trajectory is within the training cube while the second starts outside the range (Sets 8 and 9, respectively, in the Appendix). The corresponding root mean square values of the noise are 0.004693, 0.004153, and 0.001255, or totally 0.006391.

4 Results

Linear networks will in the following be used to tackle the filtering problem, which is formulated in alternative ways. The analysis also considers prediction of state variables based on information about delayed observations. In all the examples, the sampling time has been chosen to equal 0.25 h. The main results will be summarized in Table 4.

4.1 Filtering with noiseless outputs.

In the first study on filtering, the inputs are 5 sets of values of the measured (noisy) state variables at $t-4$, $t-3$, $t-2$, $t-1$, t, while the outputs are noiseless state variables at time t.

Training results in an rms error of 0.00424 on the training set, which corresponds to approximately half of the noise level. This fit must be considered accurate. For the test set, the rms error increases to 0.00733, but the result is still good.

A major limitation of the above study is that the training instances have noiseless outputs,

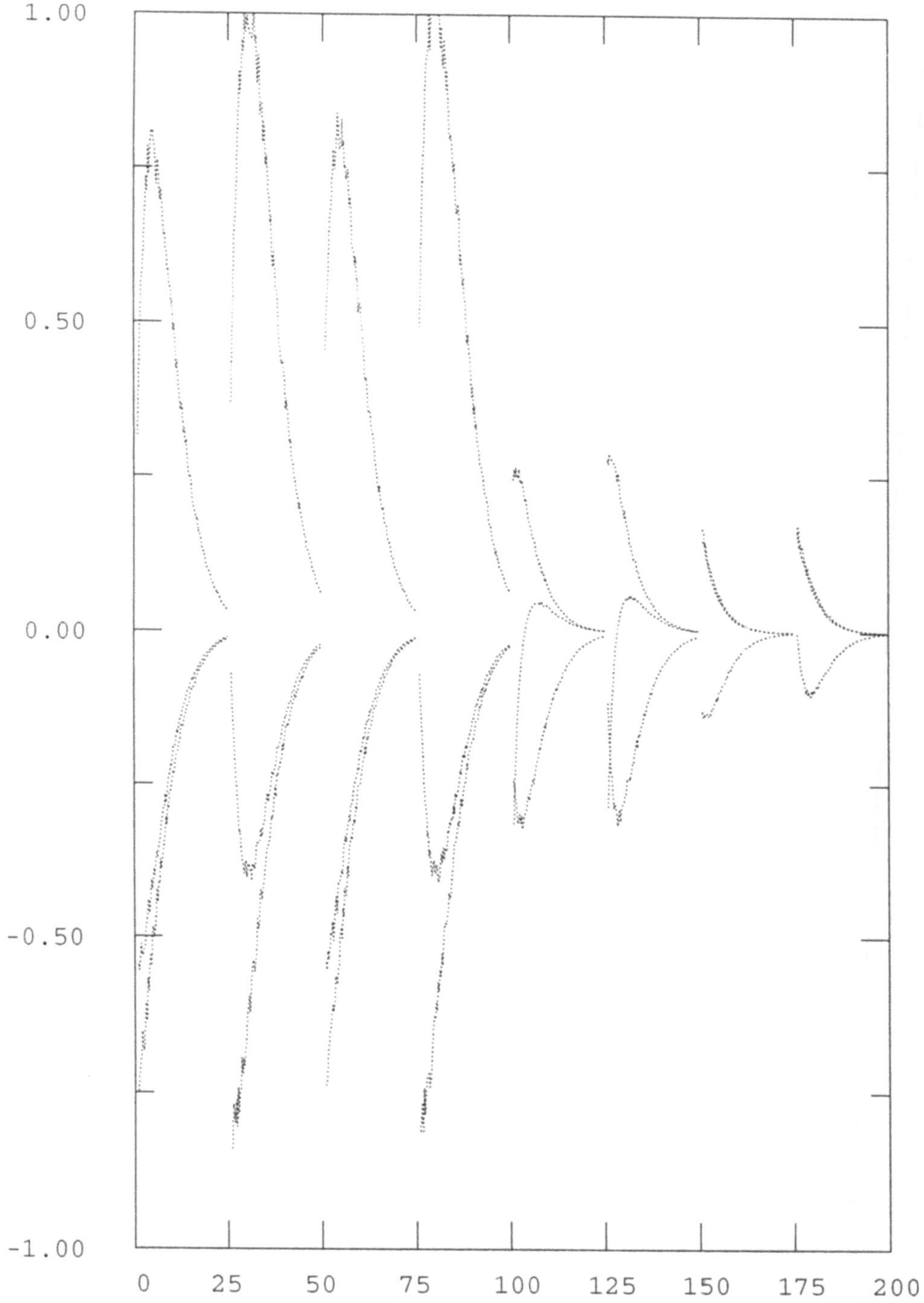

Fig. 2 Noisy observations of X, S and P on the 8 training trajectories.

which cannot be obtained for a real process. If one uses some conventional technique for producing the filtered values, then the neural networks will obviously never learn to do better than the conventional method. These results have, therefore, been included for comparative purposes only. More realistic approaches will be presented in what follows, which leads us into smoothing or prediction problems.

Another severe limitation of the above study is that not all variables are easy to measure in the process, especially not on line. It is therefore not so realistic to assume that measurements of all the state variables are available and can be used as inputs. A more realistic assumption would be to assume one or two variables measurable, and predict the other ones.

4.2 Prediction with noisy outputs for training.

In the second study on filtering, the inputs are sets of lagged values of the measured (noisy) state variables at four times ($t-4$,$t-3$,$t-2$, and $t-1$). The outputs are still the noisy state variables at t, but, since the noise in the signals is white, the network will produce filtered state estimates at t.

After training, the minimum rms error was found to be 0.00985, while the corresponding error on the test set was 0.0126. From this it can be seen that the network performs filtering and prediction quite well from noisy delayed inputs. Figure 3 shows the noisy and filtered state variables on the test trajectories. Still, as was noted in subsection 4.1, the main appreciable limitation of the above study is that not all variables are easy to measure in the real process.

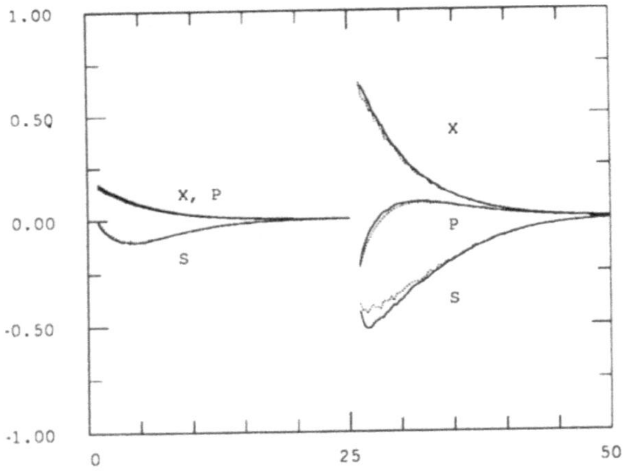

Fig. 3 Noisy and filtered values on test trajectories.

4.3 Estimation with two measured variables.

This study is similar to the one presented in subsection 4.2, but the input information about the microbial concentration, X, is considered not to be available for filtering. This assumption is quite realistic, since the methods for on-line analysis of the microbial concentration are very complicated, slow, or at least expensive. Four sets of values of noisy observations of S and P are still used, and the outputs are the three noisy state variables at time t. After training, the networks should be able to produce the filtered estimates of all state variables at t.

Training results in an rms error of 0.01826, which is lower than the rms error observed for the test set (0.03201). Thus, the prediction errors have grown, but are probably still small enough for control purposes. The state variable X is estimated fairly accurately, but less accurately than the other two variables, which is understandable.

4.4 Estimation with one measured variable.

This section analyzes the possibilities to predict one (P) and estimate two states (S and X) based on lagged observations of only one state (P). The inputs are still 4 previous observations, and the outputs are estimates of all 3 state variables at t.

This analysis results in training set rms of 0.0763, which clearly shows that one cannot estimate the three state variables from delayed values of P. This is in agreement with the findings of earlier studies of the same process [5 – 7]. The largest errors are in S, and there are moderate errors in X. Quite naturally, the errors in P are small.

4.5 One-step prediction of one measurable variable.

The final study clarifies whether a state variable at time t can be predicted using only lagged measurements of the state in question.

In the first analysis, called 5X in Table 2, $X(t)$ is predicted given information about four lagged noisy measurements of the state variable. This gives rise to an rms error of 0.01035 on the training set and 0.00644 on the test set. Obviously, the low error on the testing set indicates that the two trajectories of this set do not correspond to the most difficult (or non-linear) cases. It may be concluded that delayed values of X are sufficient to predict the value of $X(t)$ fairly accurately. With this good accuracy, there is little need to study possible improvements using neural networks with non-linear activation functions.

In a study (called 5S in Table 2), we carry out a similar analysis as above for the substrate concentration. The training and test set rms errors of 0.01620 and 0.00578 show that 4 delayed values of this state variable are enough to predict its next state.

In the final study (5P), $P(t)$ is predicted given information about four lagged noisy measurements of the state variable. The low rms errors on both the training set (0.00768) and the test sets (0.00289) show that the ethanol concentration can be accurately predicted 0.25 h in advance, if lagged measurements are available. Figure 4 shows the results of the analysis.

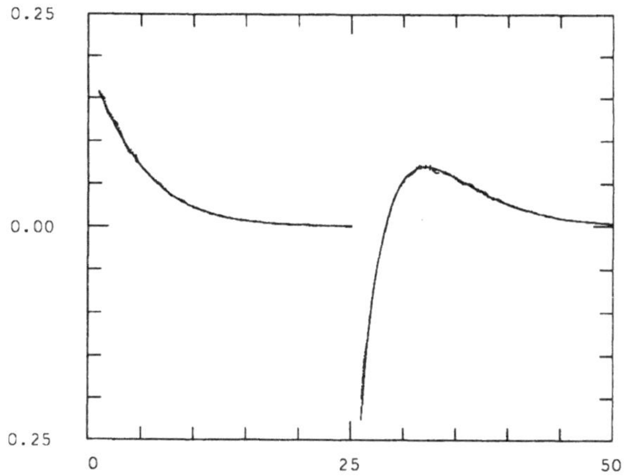

Fig. 4 Noisy simulated and filtered values of P.

Table 2. Rms errors for training and tests

Study	rms train	rms test
1	4.24 E-3	7.33 E-3
2	9.85 E-3	1.26 E-2
3	1.83 E-2	3.20 E-2
4	7.63 E-2	
5X	1.04 E-2	6.44 E-3
5S	1.62 E-2	5.78 E-3
5P	7.68 E-3	2.89 E-3

5 Discussion

Earlier work [5 – 7] has shown that system identification of the biochemical process considered in this work can be performed quite well using feed-forward neural networks. This paper illustrates that filtering and prediction of the states can be carried out by linear models, as long as the sampling time is short, and the behaviour of the process in the region of state space studied is not very non-linear. The first requisite is not considered severe, since a sampling time of 0.25 h in a process with a characteristic time of 3.3 h is not very short. A more severe limitation is the assumption that all state variables be measurable regularly. This and other issues will be studied in forthcoming work.

References

[1] J. Thibault, V Van Breusegem and A. Chéruy, "On-line prediction of fermentation variables using neural networks", *Biotech. Bioeng.* **36** (1990) 1041–1048.

[2] V. Van Breusegem, J. Thibault and A. Chéruy, "Adaptive Neural Model for On-line Prediction in Fermentation", *Canad. J. Chem. Eng.* **69** (1991) 481–487.

[3] D. Psichogios and L.H. Ungar, "Process Modelling Using Structured Neural Networks", Proceedings of *American Control Conference 1992*, Vol. III, p. 1917–1921.

[4] M.N. Karim and S.L. Rivera, "Application of Neural Networks in Bioprocess State Estimation", Proceedings of *American Control Conference 1992*, Vol. I, p. 495–499.

[5] A.B. Bulsari and H. Saxén, "System identification of a biochemical process using feed-forward neural networks", *Neurocomputing* **3** (1991) 125–133.

[6] A.B. Bulsari and H. Saxén, "Estimation of a state variable in a biochemical process using feed-forward neural networks", Proceedings of *Singapore International Conference on Intelligent Control and Instrumentation (SICICI'92)*, Singapore, February 1992, Vol. 1, p. 524–529.

[7] A.B. Bulsari, A.B. and H. Saxén, "Estimation of a disturbance variable using feed-forward neural networks". Proceedings of the *11th IASTED International Conference on Modelling, Identification and Control (MIC'92)*, Innsbruck, Austria, February 1992, p. 248–250.

[8] R.K. Warren, G.A. Hill and D.G. Macdonald, "Improved bioreaction kinetics for the simulation of continuous ethanol fermentation by *Saccharomyces cerevisiae*", *Biotechnology Progress* **6** (1990) 319–325.

[9] D.W. Marquardt, "An algorithm for least-squares estimation of nonlinear parameters", *J.SIAM* **11** (1963) 431–441.

[10] D.E. Rumelhart, J.L. McClelland (eds.), *Parallel Distributed Processing. Explorations in the*

Microstructure of Cognition, MIT Press, Cambridge, Massachusetts, 1986.

Appendix

The input variables S_0 and D are set to the values shown in Table 1. The 9 trajectories start from the following initial conditions (unscaled variables) and approach the operating point (steady-state) given in Table 1.

Trajectory number	X	S	P (g/l)
1	8	8	30
2	8	8	45
3	8	12	30
4	8	12	45
5	13	8	30
6	13	12	30
7	13	8	45
8	13	12	45
9	15	8	30

A Recurrent Neural Network for Time-series Modelling

Abhay B. Bulsari and Henrik Saxén

Heat Engineering Laboratory
Department of Chemical Engineering, Åbo Akademi
Biskopsgatan 8, 20500 Åbo, Finland
Phone: 358-21-654311, Fax: 358-21-654792
E-mail: vt_ai@abo.fi

Abstract

This paper describes an architecture for discrete time feedback neural networks. Some analytical results for networks with linear nodal activation functions are derived, while simulations demonstrate the performance of recurrent networks with nonlinear (sigmoidal) activation functions. The need for models with a capacity to consider correlated noise sequences is pointed out, and it is shown that the recurrent networks can perform state estimation and entertain models with coloured noise.

1 Introduction

Recurrent neural networks have attracted increasing attention during the last few years. The reason for this is that networks with feedback connections can describe the evolution of dynamical systems, such as processes governed by a set of first order coupled differential equations [1]

$$\tau \frac{\partial y_i}{\partial t} = -y_i + f(x_i + I_i) \tag{1}$$

$$x_i = \sum_j w_{ij} y_j . \tag{2}$$

In the networks, y_i is the output of node i, x_i is its internal input, f is the nodal activation function and I_i is an input from the environment. It is easy to show that feedforward neural networks can only describe the fixpoints

$$\frac{\partial y_i}{\partial t} = 0 = -y_i + f(x_i + I_i) \qquad \text{or} \qquad y_i = f(x_i + I_i) \tag{3}$$

of such systems, and only in the case the *weight matrix*, $\mathbf{W} = \{w_{ij}\}$, is lower triangular ($i > j$ in eq. (2)). Recurrent networks, in turn, can describe the *dynamics* of general systems, which implies that they may produce temporal sequences; this is the main difference between a recurrent network and a feedforward network with *tapped delay lines* [2,3].

A partially recurrent neural network in discrete time was introduced by Jordan [4], who made use of feedback connections from the outputs to additional nodes — *state units* — in the input layer. These units were equipped with local self-loops to make it possible for them to evolve smoothly, *e.g.*, follow an exponential decay. The actual inputs entered through so called *plan units*, and the idea was to trigger various time sequences by feeding different plans into the network. Elman [5] introduced feedbacks from (some of) the hidden units in a feedforward network as fictitious inputs to *context units*. This architecture has been used by many investigators for solving tasks in different domains [6 – 8]. Also other architectures have been considered, where intermediate layers of context units [9] or networks with input units equipped with self-loops [10] are used. Networks where all units may use finite impulse response (FIR) and/or infinite impulse response (IIR) synapses have also been developed [11].

In this paper, some properties of a class of partially recurrent neural networks are studied. The architecture used is described briefly in the next section, followed by some analytical results is Section 3. Finally, the model is illustrated by numerical examples and the results are discussed.

2 The network model

We study a class of partially recurrent neural networks, which arises if the following modifications are introduced into standard multilayer feedforward networks: The outputs from some of the units in the hidden and/or output layers are fed back to *fictitious input nodes*. The weights on the feedback connections are fixed to equal B, the backward shift operator defined by $x_{t-1} = Bx_t$. Moreover, *fictitious output nodes* that are fed back may be included, and each fictitious input node has an initial state, which can be estimated by the training method.

Some features of these networks may be pointed out: As a special case, if there are no true inputs and all outputs are fed back, a fully recurrent network results. Moreover, the fact that there are (fictitious) output units without target values during training allows for ***state estimation*** by the networks; the evolution of these nodes is not explicitly governed by the training set, but they can assume the most advantageous states in order to describe the dynamics of the process studied.

The following network notation will be adopted in what follows: A network with N^{in} input nodes N^{hid} hidden nodes and N^{out} output nodes will be denoted by $(N^{\text{in}}, N^{\text{hid}}_L, N^{\text{out}}_M)$, where the subscripts L and M are the (sub)sets of hidden and output nodes that are fed back to the input layer. For purposes of illustration, the $(2, 2_{3,4})$ network is shown below, where the box denotes the bias node.

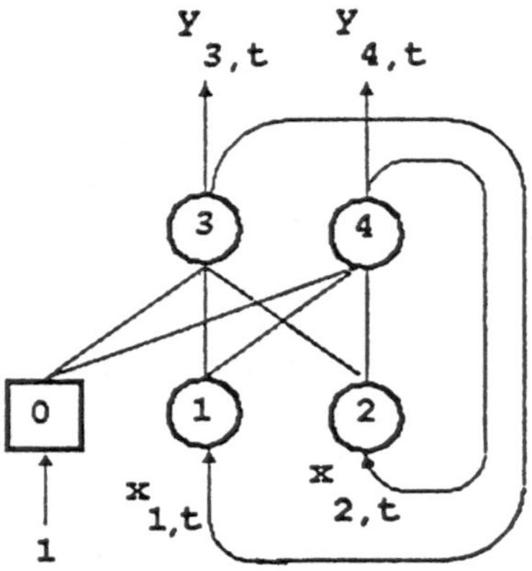

Fig. 1 The $(2, 2_{3,4})$ network.

A recurrent network has an equivalent feedforward representation, which is obtained by ***unfolding*** or ***unrolling*** the network in time [12]. Because of this, any training method developed for feedforward networks can be applied to recurrent networks. In our work, the Levenberg-Marquardt method [13, 14] was used for estimating the weights and initial states of the networks.

3 Some analytical results

In this section we analytically study the ability of networks with linear activation function, $f(x) = x$, to entertain autoregressive moving average models [15] Part of the results presented can be derived through state space representation of time series [16, 17]. Without loss of generality, we consider networks with only one input and study some simple nets. Extension of the results to multiple inputs, which is quite straightforward, is discussed in [18].

3.1 The $(2,1_3)$ network

The simplest recurrent network of practical use, depicted in Figure 2, has an output given by

$$y_{3,t} = w_{30} + w_{31}x_{1,t} + w_{32}x_{2,t}, \tag{4}$$

By this network the parameters of an autoregressive moving average process of the type ARMA(1,1), $x_{1,t+1} = \phi_0 + \phi_1 x_{1,t} + \varepsilon_{1,t+1} + \theta_1 \varepsilon_{1,t}$ (where ε_1 is white noise with zero mean), can be estimated, since

$$\begin{aligned} y_{3,t} &= w_{30} + (w_{31} + w_{32})x_{1,t} - w_{32}(x_{1,t} - x_{2,t}) \\ &= w_{30} + (w_{31} + w_{32})x_{1,t} - w_{32}(x_{1,t} - y_{3,t-1}) \\ &= w_{30} + (w_{31} + w_{32})x_{1,t} - w_{32}(x_{1,t} - \hat{x}_{1,t}) \\ &= w_{30} + (w_{31} + w_{32})x_{1,t} - w_{32}\,\varepsilon_{1,t} \end{aligned} \tag{5}$$

and $y_{3,t} = \hat{x}_{1,t+1}$ give us the conditions

$$w_{30} = \phi_0; \qquad w_{31} = \phi_1 + \theta_1; \qquad w_{32} = -\theta_1 \tag{6}$$

for the weights.

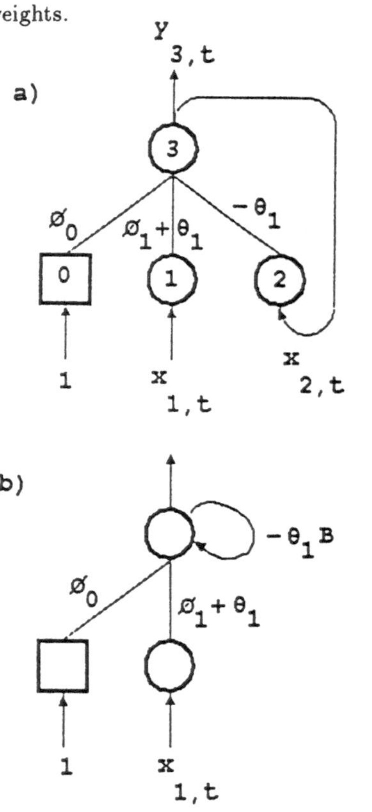

Fig. 2 $(2,1_3)$ network implementing ARMA(1,1).

Note that the weight on the self-loop of the single output (Figure 2b) is $-\theta_1$, because the moving average part has an equivalent infinite autoregressive representation. Naturally, pure AR(1) and MA(1) processes can also be modelled by this network.

3.2 The $(2,2_4)$ network

Another simple network configuration is the $(2,2_4)$ network (Figure 3) where the activation of the first output node is given by eq. (4) and the second output node has been introduced to make state estimation possible.

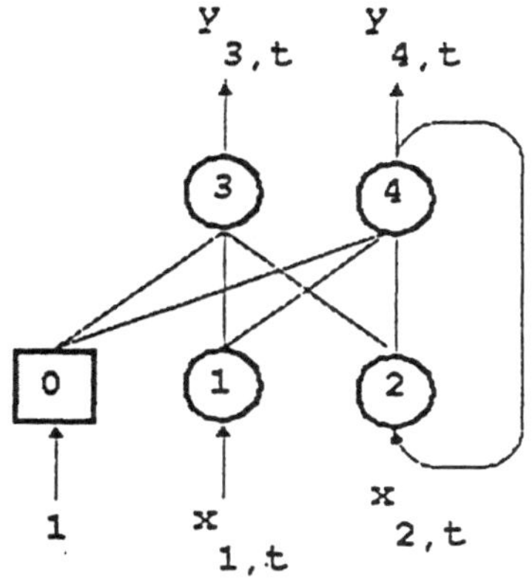

Fig. 3 The $(2,2_4)$ network.

From

$$\begin{aligned} y_{3,t} &= w_{30} + w_{31}x_{1,t} + \\ &\quad w_{32}(w_{40} + w_{41}x_{1,t-1} + w_{42}x_{2,t-1}) \\ &= (w_{30} + w_{32}w_{40}) + w_{31}x_{1,t} + \\ &\quad w_{32}w_{41}x_{1,t-1} + w_{32}w_{42}x_{2,t-1} \end{aligned} \tag{7}$$

it is straightforward to verify that the parameters of an AR(2) process $x_{1,t+1} = \phi_0 + \phi_1 x_{1,t} + \phi_2 x_{1,t-1} + \varepsilon_{t+1}$ can be estimated by training the network with $y_{3,t} = \hat{x}_{1,t+1}$. Here, the fictitious output node acts as a simple lag or delay line.

However, the network can also entertain more complex models. In order to show that it can model an ARMA(2,1) process, $x_{1,t+1} = \phi_0 + \phi_1 x_{1,t} + \phi_2 x_{1,t-1} + \varepsilon_{1,t+1} + \theta_1 \varepsilon_{1,t}$, we expand eq. (7) further

$$\begin{aligned} y_{3,t} &= (w_{30} + w_{32}w_{40}) + w_{31}x_{1,t} + w_{32}w_{41}x_{1,t-1} + \\ &\quad w_{32}w_{42}x_{2,t-1} \\ &= (w_{30} + w_{32}w_{40}) + w_{31}x_{1,t} + w_{32}w_{41}x_{1,t-1} + \\ &\quad w_{32}w_{42}(w_{40} + w_{41}x_{1,t-2} + w_{42}x_{2,t-2}) \\ &= (w_{30} + w_{32}w_{40} + w_{32}w_{40}w_{42}) + w_{31}x_{1,t} + \\ &\quad w_{32}w_{41}x_{1,t-1} + w_{32}w_{41}w_{42}x_{1,t-2} + \\ &\quad w_{32}w_{42}^2 x_{2,t-2} \\ &= \ldots \\ &= w_{30} + w_{32}w_{40} + w_{32}w_{40}w_{42} + w_{32}w_{40}w_{42}^2 + \ldots \\ &\quad + w_{31}x_{1,t} + w_{32}w_{41}x_{1,t-1} + \\ &\quad w_{32}w_{41}w_{42}x_{1,t-2} + w_{32}w_{41}w_{42}^2 x_{1,t-3} + \ldots \end{aligned} \tag{8}$$

and also carry out the corresponding expansion of the ARMA(2,1) model

$$\begin{aligned} \hat{x}_{1,t+1} &= \phi_0 + \phi_1 x_{1,t} + \phi_2 x_{1,t-1} + \theta_1 \varepsilon_{1,t} \\ &= \phi_0 + \phi_1 x_{1,t} + \phi_2 x_{1,t-1} + \theta_1 (x_{1,t} - \hat{x}_{1,t}) \\ &= \phi_0 + \phi_1 x_{1,t} + \phi_2 x_{1,t-1} + \theta_1 [x_{1,t} - \\ &\quad (\phi_0 + \phi_1 x_{1,t-1} + \phi_2 x_{1,t-2} + \theta_1 \varepsilon_{1,t-1})] \\ &= \phi_0 (1 - \theta_1) + (\phi_1 + \theta_1) x_{1,t} + (\phi_2 - \phi_1 \theta_1) x_{1,t-1} \\ &\quad - \phi_2 \theta_1 x_{1,t-2} - \theta_1^2 (x_{1,t-1} - \hat{x}_{1,t-1}) \\ &= \phi_0 (1 - \theta_1) + (\phi_1 + \theta_1) x_{1,t} + (\phi_2 - \phi_1 \theta_1) x_{1,t-1} \\ &\quad - \phi_2 \theta_1 x_{1,t-2} - \theta_1^2 [x_{1,t-1} - (\phi_0 + \phi_1 x_{1,t-2} \\ &\quad + \phi_2 x_{1,t-3} + \theta_1 \varepsilon_{1,t-2})] \\ &= \ldots \\ &= \phi_0 (1 - \theta_1 + \theta_1^2 - \theta_1^3 + \ldots) + (\phi_1 + \theta_1) x_{1,t} + \\ &\quad (-\phi_1 \theta_1 + \phi_2 - \theta_1^2) x_{1,t-1} + (\phi_1 \theta_1^2 - \\ &\quad \phi_2 \theta_1 + \theta_1^3) x_{1,t-2} + \ldots \end{aligned} \tag{9}$$

which can be expressed in a compact form as

$$\hat{x}_{1,t+1} = \varphi_0 + \varphi_1 x_{1,t} + \varphi_2 x_{1,t-1} + \varphi_3 x_{1,t-2} + \ldots \tag{10a}$$

with

$$\begin{aligned} \varphi_0 &= \phi_0 \sum_{j=0}^{\infty} (-\theta_1)^j \\ \varphi_1 &= \phi_1 + \theta_1 \\ \varphi_2 &= -\phi_1 \theta_1 + \phi_2 - \theta_1^2 \\ \varphi_n &= -\theta_1 \varphi_{n-1}; \qquad n \geq 3. \end{aligned} \tag{10b}$$

By equating the coefficients of the two series, we get

the conditions

$$\begin{aligned}
&w_{30} + w_{32}w_{40}(1 + w_{42} + w_{42}^2 + w_{42}^3 + \ldots) = \\
&\quad \phi_0(1 - \theta_1 + \theta_1^2 - \theta_1^3 + \ldots) \\
&w_{31} = \phi_1 + \theta_1 \\
&w_{32}w_{41} = -\phi_1\theta_1 + \phi_2 - \theta_1^2 \\
&w_{32}w_{41}w_{42} = \phi_1\theta_1^2 - \phi_2\theta_1 + \theta_1^3 \\
&w_{32}w_{41}w_{42}^2 = -\phi_1\theta_1^3 + \phi_2\theta_1^2 - \theta_1^4 \\
&\ldots
\end{aligned} \tag{11}$$

The weights can, for instance, be chosen as

$$\begin{aligned}
&w_{30} = 0; \quad && w_{31} = \phi_1 + \theta_1: \quad && w_{32} = 1; \\
&w_{40} = \phi_0; \quad && w_{41} = -\phi_1\theta_1 + \phi_2 - \theta_1^2; \quad && w_{42} = -\theta_1,
\end{aligned} \tag{12}$$

which is represented in Figure 4. The moving average part of the model is again (Figure 4b) realized by a self-loop on the fictitious output unit, with the weight $-\theta_1$.

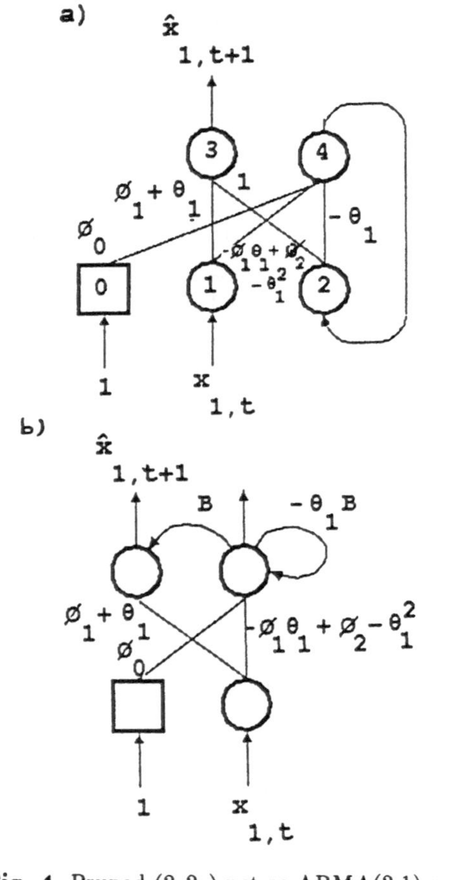

Fig. 4 Pruned $(2, 2_4)$ net as ARMA(2,1) model.

3.3 Extension of the results

General formulae for more complicated network configurations can be derived in an analogous way as in the previous subsections. For networks with one true input, the first output of an $(n,n_{n+2,n+3,\ldots,2n})$ network (with feedbacks from all other output nodes) can model an ARMA$(n,n-1)$ process, whereas the first output from an $((n+1),n_{n+2,n+3,\ldots,2n+1})$ network (with all outputs fed back) can model an ARMA(n,n) process. The former network has some appealing properties in that a variety of actual processes can be described by ARMA$(n,n-1)$ time series models [19]. Pure autoregressive or moving average processes can be entertained by deleting appropriate connections in these networks.

It is also quite straightforward to extend the results to systems with m inputs and m outputs. The reader is referred to ref. [18] for further details.

4 Numerical example

The task of producing points on a circle has been a benchmark problem for recurrent networks [20, 21]. If the training set consists of the K instances

$$\left.\begin{aligned} x_{1,t} &= r\cos(t\Delta\alpha) \\ x_{2,t} &= r\sin(t\Delta\alpha) \end{aligned}\right\} \quad t = 0,\ldots,K-1 \tag{13}$$

it can be shown that a linear fully recurrent $(2,2_{3,4})$ network (Figure 1) can reproduce the sequence exactly [22]. This follows from the fact that the observations, $\mathbf{x} = (x_1, x_2)^T$, can be written as a linear combination of their previous values

$$\mathbf{x}_t = \mathbf{A}\,\mathbf{x}_{t-1} \text{ with } \mathbf{A} = \begin{bmatrix} \cos\Delta\alpha & -\sin\Delta\alpha \\ \sin\Delta\alpha & \cos\Delta\alpha \end{bmatrix}. \tag{14}$$

The weights thus determine the "step-length" on the trajectory, while the initial point sets the radius of the circle.

The advantage of sigmoid nets over linear ones is that the former can learn attractors, whereas the latter ones cannot. Sigmoid networks cannot produce the sequence (13) exactly, but good approximate solutions are found [20] if the amplitude of the trigonometric series — the radius of the circle — does not approach the saturation limits of the sigmoid (as in the example in ref. [23]). It has been demonstrated [22] that the simple $(2,2_{3,4})$ sigmoid network (with $f(x) = \tanh(x/2)$) can be trained to reproduce the (trigonometric) sequence without *teacher forcing*. The underlying reason for the successful training is that the correct phase of the oscillation is detected by estimation of the initial conditions.

As an obvious alternative, it is possible to train a (2,2) feedforward network, with the coordinates of the previous point as inputs, and use the trained weights to hard-wire a $(2,2_{3,4})$ recurrent network. This approach works well if the trajectory is deterministic.

However, consider the case where the sequence is distorted by *e.g.*, white noise, ε,

$$\left.\begin{aligned} x_{1,t} &= r\cos(t\Delta\alpha) + \varepsilon_{1,t} \\ x_{2,t} &= r\sin(t\Delta\alpha) + \varepsilon_{2,t} \end{aligned}\right\} \quad t = 0,\ldots,K-1 \tag{15}$$

This process can be equivalently represented in state-space form, where the outputs, $\mathbf{y} = (y_1, y_2)^T$, can be considered noisy measurements of the states, $\mathbf{x}$,

$$\begin{aligned} \mathbf{x}_t &= \mathbf{A}\,\mathbf{x}_{t-1} \\ \mathbf{y}_t &= \mathbf{x}_t + \boldsymbol{\epsilon}_t, \end{aligned} \tag{16}$$

where $\boldsymbol{\epsilon} = (\varepsilon_1, \varepsilon_2)^T$. The parameters (weights) of this model estimated by a feedforward network with $\mathbf{y}_{t-1}$ as inputs will now be biased, because the noise appears as coloured in the sequence. This is seen if the two equations above are combined to yield

$$\begin{aligned} \mathbf{y}_t &= \mathbf{A}\,\mathbf{x}_{t-1} + \boldsymbol{\epsilon}_t = \mathbf{A}(\mathbf{y}_{t-1} - \boldsymbol{\epsilon}_{t-1}) + \boldsymbol{\epsilon}_t \\ &= \mathbf{A}\mathbf{y}_{t-1} + \boldsymbol{\epsilon}_t - \mathbf{A}\boldsymbol{\epsilon}_{t-1}. \end{aligned} \tag{17}$$

This stresses the need for an appropriate noise model, which can be realized by using recurrence in the network. Theoretically, a good model can also be obtained by feeding several old observations as inputs $(\mathbf{y}_{t-1}, \mathbf{y}_{t-2}, \mathbf{y}_{t-3}, \ldots)$ into a feedforward network, but this approach will be impractical since it leads to large networks with a considerable number of weights to be estimated [24].

Test runs with training sets produced by eq. (15) with $\Delta\alpha = 10°$ and $r = 0.5$ demonstrated that recurrent networks with hard-wired weights, obtained by training feed-forward networks, failed to produce the underlying periodic attractor (Figure 5), because of biased weight estimates. Obviously, increasing the number of training instances does not help. For a (4,2) network with two previous vectors, $\mathbf{y}_{t-1}$ and $\mathbf{y}_{t-2}$, as inputs, the results improved, but the parameter estimates were still strongly biased. Table 1 presents the rms errors, s, for some network configurations tested on data produced by eq. (15) with $\sigma_{\varepsilon_1} = \sigma_{\varepsilon_2} = 0.1$. In the table, M denotes the number of weights, including the initial conditions estimated for recurrent networks.

As for the recurrent network, an attractor was found even though the training set consisted of only one pass ($K = 37$) through the circle and the noise level was considerably increased. This is illustrated in Figure 6, where the training instances (□; $\sigma_\varepsilon = 0.3$) and the trajectory produced by letting the network evolve in time (·····) have been depicted. Table 2 shows that the weight estimates are close to their "optimal" values, *i.e.*, those obtained for noise-free training data [22].

This specific numerical example has not illustrated the utility of using fictitious output nodes. A problem where their use is necessary has been treated in [25].

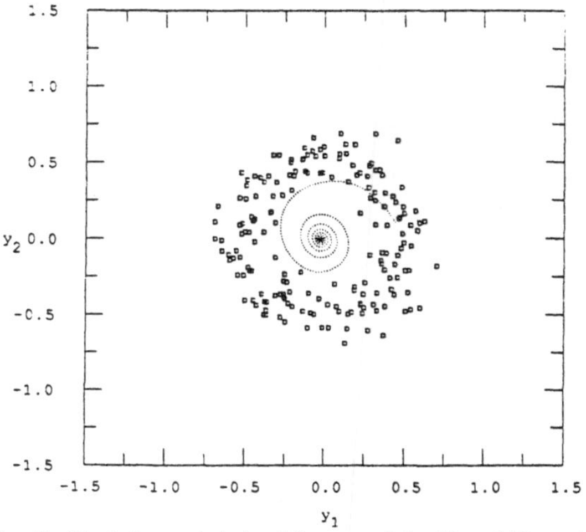

Fig. 5 Training set (□) with $\sigma_\varepsilon = 0.1$, $K = 180$ and trajectory produced by a hard-wired $(2,2_{3,4})$ network, with weights from a (2,2) network.

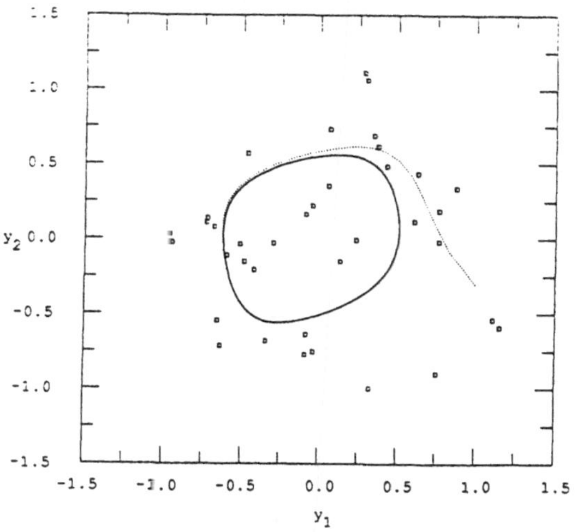

Fig. 6 Trajectory produced by a $(2,2_{3,4})$ network, which was trained on 37 data points (□) with $\sigma_\varepsilon = 0.3$.

Table 1 Number of parameters, M, and prediction errors, s. The training set was produced by eq. (15) with $\sigma_\varepsilon = 0.1$.

Network	M	K	s
$(2,2_{3,4})$	8	37	0.1134
		73	0.1001
		181	0.1046
		361	0.0987
(2,2)	6	36	0.1658
		72	0.1482
		180	0.1494
		360	0.1364
		1800	0.1387
(4,2)	10	35	0.1438
		71	0.1237
		179	0.1283
		359	0.1191
		1799	0.1216

Table 2 "Dominant" weights for $(2,2_{3,4})$ networks. $\sigma_\varepsilon = 0.3$ in all training sets.

K	w_{31}	w_{32}	w_{41}	w_{42}
37	2.252	-0.433	0.436	2.069
73	2.127	-0.369	0.414	2.174
181	2.064	-0.341	0.429	2.172
361	2.083	-0.346	0.417	2.153
Noise-free	2.105	-0.376	0.377	2.106

5. Discussion

This paper has presented a class of recurrent discrete time neural networks, which can emulate several of the partially recurrent network configurations proposed in the literature. A theoretical analysis of networks with linear activation function showed that autoregressive moving average processes can be identified by training the networks. The minimal networks required to entertain ARMA(n,n-1) and ARMA(n,n) models have also been presented. An experimental analysis illustrated that the networks can perform state estimation, *i.e.*, can find an underlying deterministic trend from observing signals corrupted by noise.

References

[1] Pineda, F.J., (1988) "Dynamics and Architecture for Neural Computation", *Journal of Complexity* **4**, 216–245.

[2] Lapedes, A. and R. Farber, (1987) "How Neural Nets Work", *Neural Information Processing Systems* (ed. D.Z. Anderson) 442–456, American Institute of Physics, New York.

[3] Sejnowski, T.J. and C.R. Rosenberg, (1987) "Parallel Network that Learn to Pronounce English Text", *Complex Systems* **1**, 145–168.

[4] Jordan, M.I., (1986) "Attractor dynamics and parallelism in a connectionist sequential machine", Proceedings of the *Eight Annual Conference of the Cognitive Science Society*, Amherst, p. 531–546. Hillsdale: Erlbaum.

[5] Elman, J.L, (1990) "Finding Structure in Time", *Cognitive Science* **14**, 179–211.

[6] Cleeremans, A., D. Servan-Schreiber and J.L. McClelland, (1989) "Finite State Automata and Simple Recurrent Networks", *Neural Computation* **1**, 372–381.

[7] Croft, E.A. and J.P. Huissoon, (1992) "Neural Network Controller for an Autonomous Guided Vehicle", Proceedings of *11th IASTED International Conference: Modelling, Identification and Control* (Ed. M.H. Hamza), p. 46–49, Acta Press, Zurich, 1992.

[8] Pham, D.T. and X. Liu, (1992) "Dynamic System Modelling Using Partially Recurrent Neural Networks", *Journal of System Engineering* **2**, 90–97.

[9] Mozer, M.C., (1989) "A Focused Back-Propagation Algorithm for Temporal Pattern Recognition", *Complex Systems* **3**, 349–381.

[10] Stornetta, W.S, T. Hogg and B.A. Huberman, (1987) "A Dynamic Approach to Temporal Pattern Recognition", *Neural Information Processing Systems* (ed. D.Z. Anderson) 750-759, American Institute of Physics, New York.

[11] Back, A.D. and A.C. Tsoi, (1991) "FIR and IIR Synapses, a new neural network architecture for time series modelling", *Neural Computation* **3**, 375–385.

[12] Rumelhart, D.E., G.E. Hinton and R.J. Williams, (1986) "Learning Internal Representations by Error Propagation", In *Parallel Distributed Processing*, Vol. 1, MIT Press.

[13] Marquardt, D.W., (1963) "An algorithm for least-squares estimation of nonlinear parameters", *J. SIAM* **11**, 431–441.

[14] Bulsari, A.B. and H. Saxén, (1991) "System identification of a biochemical process using feed-forward neural networks", *Neurocomputing* **3**, 125–133.

[15] Box, G.E.P. and G.M. Jenkins, (1976) *Time Series Analysis. Forecasting and Control*, Holden-Day, Oakland, CA.

[16] Aoki, M., (1987) *State Space Modeling of Time Series*, Springer-Verlag, New York.

[17] Bennett, R.J., (1979) *Spatial Time Series*, Pion Limited, London, England.

[18] Bulsari A.B. and H. Saxén, (1991) "A class of partially recurrent neural networks in discrete time", Technical report 91-17, Heat Eng. Lab., Åbo Akademi, Åbo, Finland, 1991.

[19] Pandit, S.M. and S.M. Wu., (1983) *Time Series Analysis with Applications*, John Wiley and Sons, New York.

[20] Pearlmutter, B., (1989) "Learning state space trajectories in recurrent neural networks", *Neural Computation* **1**, 263–269.

[21] Pearlmutter, B., (1990) "Dynamic Recurrent Neural Networks", Technical report CMU-CS-90-196, Carnegie Mellon University, Pittsburgh, USA.

[22] Bulsari, A.B. and H. Saxén, (1992) "A Partially Recurrent Connectionist Model", Proc. of the *10th European Conf. on Artificial Intelligence (ECAI'92)*, (Ed. B. Neumann), Vienna, Austria, 1992, p. 198–202.

[23] Williams, R.J. and D. Zipser, (1989) "Experimental Analysis of the Real-time Recurrent Learning Algorithm", *Connection Science* **1**, 87–111.

[24] Billings, S.A., H.B. Jamaluddin and S. Chen, (1992) "Properties of neural networks with applications to modelling non-linear dynamical systems", *Int. J. Control* **55** 193–224.

[25] Bulsari, A.B. and H. Saxén, (1992) "A Recurrent Neural Network Model", in *Artificial Neural Networks, 2* (Eds. I. Aleksander and J. Taylor) (Proceedings of the *International Conference on Artificial Neural Networks*, Brighton, UK, September 1992), Vol. 2, p. 1091–1094.

The Application of Neural Sensors to Fermentation Processes

S.J. Rawling, D. Peel, S.M.Keith, B. Buxton.
School of Science and Technology,
University of Teesside,
Middlesbrough,
Cleveland,
TS1 3BA
United Kingdom.

Abstract.

This paper describes the training and implementation of Artificial Neural Network models for the real-time, on-line estimation of key biological variables in a fed-batch yeast fermentation process. The neural networks are generic nonlinear models that are configured and trained to act as software sensors for biomass concentration.

In order to successfully customise a neural sensor it is necessary to give considerable attention to the choice of network inputs. The task of choosing these inputs is somewhat eased by knowledge of the yeast's metabolism supplemented where necessary by a statistical analysis of the available inputs and outputs. In this study, the inputs that were chosen reflected both the yeast's metabolic activity and its rate of growth.

The experimental investigation was performed using a Ferranti Process Control Computer connected to a pilot-scale fermentation suite. The neural sensor module was written such that the customised neural network (sensor), configured and trained off-line, could be implemented on-line using the Ferranti PML sequence language. The training and configuration stages of the neural sensor were conducted using software written in-house and designed to be compatible with the Ferranti based neural sensor module.

1. Introduction.

One of the most common constraints concerning the control and monitoring of a fermentation process is the absence of accurate, on-line measurement of key biological variables. This is perhaps most apparent in industrial fermentation control schemes where automated feedback control of biological parameters such as biomass, is not included and therefore control is based on empirical recipes.

Biomass concentration is a commonly measured fermentation process variable, often measured near-to-line by a simple assay procedure that relies on an operator removing a sample of broth from the vessel and filtering a known volume. This is then washed and either weighed to provide a wet weight concentration or dried in an oven to give a dry weight concentration. This procedure takes varying amounts of time with concomitant degrees of accuracy.

Recent advances in impedometric devices have led to the development of a biomass probe [1]. While this probe works well in defined media its accuracy decreases in complex media due to high background capacitance and conductance values.

With this background several algorithmic and heuristic prediction techniques have been developed over recent years for the estimating of biomass in various fermentations. Cooney et al [2] applied a mass balance over a yeast fermentation using microbial state and elemental equations to predict the biomass. However such a technique cannot be applied to a fermentation using complex medium [3]. Lant et al [4] developed an inferential predictor based on a linear parametric model. The inferential mechanism by-passed the need for an absolute measure of the biomass by inferring the biomass concentration from secondary variables such as the carbon dioxide evolution rate. The estimate was then periodically corrected as an actual measure of biomass concentration became available from the sample and assay procedure. Dochain [5], developed a non-linear observer for estimating biomass by reconfiguring a set of differential and algebraic equations that described the fermentation under consideration. Di-Massimo et al [6] described one of the first implementations of artificial neural networks in estimating the biomass concentration in penicillin production. Artificial neural networks are now considered one of the more promising techniques for estimating biomass concentration and other biological parameters.

This paper describes the application of ANN techniques for the on-line estimation of biomass concentration in a fed-batch yeast fermentation. The implementation of the neural network estimators has been performed using a Ferranti industrial process control computer connected to a pilot scale fermentation suite. Consequently the implementations have been performed in a robust industry standard environment.

2. The Biochemical Process.

The term "fed-batch culture" is used to describe batch cultures that are fed continuously, or sequentially with medium without the removal of the fermentation broth [7]. Application of the fed-batch culture technique in industry is directed towards maintaining low residual substrate levels in the medium, in order to prevent oxygen limitation through rapid metabolism of carbon substrate and also to reduce the toxic effects of one or more medium components.

A fed-batch process is generally employed in the production of Bakers yeast (***Saccharomyces cerevisiae***) to minimise the phenomenon known as the Crabtree effect [8]. This regulatory system is described as the repression of respiratory metabolism under an excess of oxygen and glucose, with the subsequent formation of ethanol from pyruvate. Under repressive conditions (i.e., high glucose concentrations) yeast cell growth follows an anaerobic pathway even in an aerobic environment, this is fermentative growth and leads to poor biomass yields. When the glucose falls below a certain critical concentration, cell growth may proceed aerobically provided sufficient oxygen is available leading to high biomass yields.

In this study the following fed-batch control strategies were employed.

2.1 Continuous Feed Feedback Control.

Based on the strategy described by Aiba et al [9] and Woehrer et al [10]. This strategy uses the respiratory quotient (the ratio of carbon dioxide evolved to oxygen consumed), in a PI (proportional and integral) control loop to control the substrate flowrate. Studies on the respiratory quotient have shown that values below 1.0 indicate aerobic growth and therefore high biomass yields and values above 1.0 indicate fermentative growth with poor biomass yields. Optimum performance of the fermenter using this strategy can be attained by maintaining respiratory quotient values of 1.0.

2.2 Pulsed Feed Feedback Control.

To avoid oxygen limitation occurring at high biomass concentrations feed is added in a pulse, such that the time between pulses is constant. When the substrate is added to the fermenter it is rapidly metabolised causing a decrease in the dissolved oxygen concentration. At substrate exhaustion the dissolved oxygen level rises, when this reaches a set level, feed is added to the fermenter. This pulse addition of substrate causes oscillations in the dissolved oxygen and substrate concentration gradients that result in short duration metabolic repression (the Crabtree effect).

3. Fed-Batch Yeast Fermentation.

3.1 Organism.

A commercial strain of *Saccharomyces cerevisiae* (bakers yeast), supplied by Quest (Menstrie, Clackmanannshire, UK) and maintained on malt extract agar slopes, was used throughout this study.

3.2 Media Composition.

The basal medium was composed of the following: potassium dihydrogen phosphate 10g/l, ammonium sulphate 10g/l, magnesium sulphate 3g/l, yeast extract 1g/l and made up with distiled water.

The feed medium was a 50% w/v molasses solution (50% cane, 50% beet).

3.3 Inoculum Preparation.

The inoculum was prepared from a batch fermentation of the bakers yeast on basal medium containing 100 g/l of molasses mix (50% cane, 50% beet) using a Bioengineering 30 litre fermenter for 24 hours at 28°C, pH 4.7, aeration 1 vvm and agitation 350 rpm. After the fermentation was complete, the broth pH was lowered to 3.6 and concentrated via a Westphalia separator (Type KA05-00-105, Olede, Germany), to approximately 1 litre. The concentrate was then stored overnight at 4°C.

3.4 Fed-Batch Operation.

The concentrate was resuspended in 15 litres of basal medium with the following setpoints: 28°C, pH 4.7, aeration 16 l/min. Under a continuous feed the dissolved oxygen tension was controlled to 60% via cascade control on the agitator. With pulse feeding the agitation was maintained at 450 rpm with no dissolved oxygen tension control. pH was maintained with 1M sulphuric acid and 1M ammonium hydroxide.

3.5 Biomass Analysis.

Dry cell weight was evaluated by filtering a known volume of fermentation broth through a Whatman 0.45 micron nitrocellulose filter. The filter paper was dried overnight at 80°C and the biomass calculated in g/l.

3.6 Fermenter-Computer System.

In this study a Bioengineering 30 litre bottom driven fermenter, interfaced through a local controller to a Ferranti PMS 30 Process Control Computer was utilised. The control computer was used in a supervisory mode to monitor and log the operating conditions, and implement the control strategy. The feed pump (a Watson Marlow peristaltic) was controlled directly via sequence control by the PMS 30 computer as dictated by the control strategy.

Analysis of the exit gas was performed on-line via a paramagnetic oxygen analyser, model OA.269 and an infrared carbon dioxide analyser, model PSA401, (Servomex Ltd, Crowbrough, UK), interfaced to the PMS 30 computer. Calculation of the respiratory quotient was performed via a sequence implemented on the PMS 30 that permitted recalibration of the analysers with air.

4. Configuration of Neural Networks.

To successfully configure an artificial neural network (ANN) for estimating biomass in a fed-batch fermentation the following procedure was outlined.

1. Select the network inputs and outputs.
2. Collect and collate the data.
3. Select the network architecture.
4. Select the training algorithm and train.
5. Propagate using the trained network.
6. Evaluate Performance.

4.1 Selection of Inputs and Output.

In this study the choice of output is clearly the biomass concentration and was measured by the assay procedure described above.

The network inputs were;

1. Substrate added.
2. Cumulative alkali pump on-time.
3. Cumulative antifoam pump on-time.
4. Exit carbon dioxide.
5. Initial biomass concentration (from a wet weight estimate)

These inputs were chosen because their theoretical relationship with the cell growth rate, metabolic activity and biomass concentration. Additional input data such as temperature, pH, agitation, and aeration rate was considered unimportant provided these were maintained close to their setpoint values (which were kept constant throughout the fermentation).

4.2 Collection of Data.

Data was collected from four continuous feed fermentations where three respiratory quotient setpoints were used these were 0.8, 1.0 (duplicated) and 1.2. Data was collected from six pulse fed fermentation with a constant pulse duration of 20 minutes (i.e., time between feeds). All data was scaled between 0.1 and 0.8 to improve convergence [11].

4.3 Network Architecture.

In this study a three layer fully connected feed forward neural network architecture was adopted, with one input layer and two neural layers (one hidden and one output). The neurons in the two neural layers used a sigmoidal activation function and each had a trainable bias. Initial weight values were randomly generated to lie between +0.3 and -0.3.

A systematic search for the "best" network was adopted by varying the number of neurons in the hidden layer. A subjective decision based on a performance criterion was then made in order to select the best network. The performance criterion used in this study was the sum of the square of errors made across the training set. This is commonly known as the integral square error or ISE.

4.4 Network Training Algorithm.

The back-error propagation training algorithm popularised by Rumelhart et al [12] was chosen to train the networks due to its simplicity. A learn rate of 0.75 was applied for the first 200 iterations. The best three networks were then selected and trained for a further 400 iterations with the learn rate set to 0.5. From these three networks one was selected and trained for a further 800 iterations with the learn rate set to 0.25.

5. Results and Discussion.

A fed-batch yeast fermentation presents a difficult modeling problem. This is due to switches in metabolism caused principally by the substrate concentration. When in a metabolically repressed state only a small amount of biomass is formed, as glucose is metabolised through to ethanol. However the utilisation rate of glucose is higher than that associated with a non-repressed yeast metabolising glucose. Consequently a higher substrate feed rate is required to maintain the repression. High substrate feed rates can also be demanded by non-repressed cultures with high biomass concentrations. To enable the neural networks to distinguish between the metabolic states an initial estimate of the biomass concentration (based on a wet weight assay) and the current carbon dioxide fraction in the exit gas were included as inputs.

In addition to metabolic influences on biomass concentration, the process of increasing the fermentation broth volume through substrate feeding, pH control and foam control can also effect the biomass concentration. To allow the neural networks to adapt to this problem all the liquid additions to the fermenter were included as inputs.

Further complications are introduced where oscillations in metabolic activity occur due to the feeding strategy, as with pulse feeding. Here the metabolic state of the culture can be said to be chaotic as substrate concentration gradients cause metabolic repression. It is essential that the neural networks adapt to these rapid switches in metabolism, so that the network is not dependant on a particular fed-batch control strategy.

Figure 1 shows the ISE values after 200 iterations for the seven network architectures under investigation. Using the performance criterion described above the networks with 5, 6 and 7 hidden layer neurons (ISE values of 0.064, 0.059 and 0.062 respectively) were selected for additional training. After a further 400 iterations (600 in total) the ISE value was reviewed and the network with 6 hidden layer neurons (ISE = 0.051) selected for further training (Figure 1). Training ceased after 800 iterations (1400 in total) when no significant decrease in the ISE value could be observed (Figure 1).

The performance of this network on a sample of the training data is shown in figure 2a and 2b. This indicates acceptable modelling of the biomass has taken place for each of the fed-batch control strategies. Therefore this configuration of 5 inputs, 6 hidden layer neurons and 1 output neuron, (5-6-1) was considered a reasonable choice for implementation on the process control computer.

The weights, biases and input/output scaling data for this 5-6-1 network, were extracted and transposed to a neural propagation module written in the Ferranti PML language. This module implemented as a sequence on the Ferranti PMS 30 control computer was tested on-line in real time. Figure 3a and 3b compares the estimated values with the measured values for each of the fed-batch feeding strategies employed. The initial estimates provided by the network are inaccurate being up to 15 % in error. This is probably due to an inaccurate initial estimate of the biomass concentration provided by the wet weight assay. The error may also be compounded by the yeast's metabolic activity which can be slow to respond to changes in the fermenter environment in the first hour of feeding. However as the fermentation proceeds the accuracy improves and as is shown in figures 3a and 3b accurate values can be achieved.

6. Conclusion.

The use of feed-forward neural networks to estimate the biomass concentration in a fed-batch yeast fermentation demonstrates that neural computing can be applied to fermentation processes. In the case of a fed-batch yeast fermentation, where two metabolic states can predominate the network must be able to capture the essential process dynamic of switching metabolic state. In order to achieve this it is necessary to choose only the salient network inputs. This selection is made easier with knowledge of the underlying biochemistry and influence of the control strategy on the fermentation. In this case only those inputs effecting the metabolic state, activity and biomass concentration where used.

The procedure followed in selecting the network architecture indicated that a topology of 5 inputs, 6 hidden layer neurons and 1 output neuron, (5-6-1) was a rational choice. Although the accuracy of this network was poor in the initial stages of the fermentation the overall performance was satisfactory, providing encouragement for further use of this technique in biochemical processes.

7. References.

[1] Kell D B and Davey C L, '*Conductimetric and Impedimetric Devices*' in Biosensors, Ed McNeil and Harvey, IRL Press, 1990.

[2] Cooney C L, Wang H Y and Wang D I C, '*Computer-Aided Material Balancing for Prediction of Fermentation Parameters*' Biotechnology and Bioengineering, **19**, 55-67, 1977.

[3] Chattaway T, Demain A L and Stephanopoulos G, '*Use of Various Measurments for Biomass Estimation*', Biotechnology Progress, **8**, 81-84, 1992 .

[4] Lant P A, Willis M J, Montague G A, Tham M T and Morris A J, '*A Comparison of Adaptive Estimation with Neural Based Techniques for Bioprocess Application*', Preprints ACC, San Diego 2173-2178, 1990.

[5] Dochain D, 'On-line Parameter Estimation, adaptive State Estimation and Adaptive Control of Fermentation Processes', PhD thesis, University of Louvain, Belgium, 1988.

[6] Di Massimo C, Willis M J, Montague G A, Tham M T and Morris A J,'*Bioprocess Model Building Using Artificial Neural Networks*', Bioprocess Engineering, **7**, 77-82, 1991.

[7] Yoshida F, Yamane T and Nakamoto K, '*Fed-batch hydrocarbon fermentations with colloidal emulsion feed*' Biotechnology and Bioengineering, **15**, 257-270, 1973.

[8] Crabtree H G, '*Observations on the Carbohydrate Metabolism of Tumours*', Biochem. J., **23**, 536-545, 1929.

[9] Aiba S, Nagi S and Nishizawa Y, '*Fed-batch Culture of Saccharomyces cerevisiae: A Perspective of Computer Control to Enhance the Productivity in Baker's Yeast Cultivation*', Biotechnology and Bioengineering, **18**, 1001-1016, 1976.

[10] Woehrer W, Hampel W and Roehr, M, '*Current Developments in Yeast Biotechnology*', Ed. Moo-Young, 419-424, 1981.

[11] McAvoy T, Te Su H, Wang N S, He M, Horvath J and Semerjian H, '*A Comparison of Neural Networks and Partial Least Squares For Deconvoluting Fluorescence Spectra*' Biotechnology and Bioengineering, **40**, 53-62, 1992.

[12] Rumelhart D E, Hinton G E and Williams R J, '*Learning Representations by Back-propagating Errors*', Nature, **323**, 533-536, 1986.

8. Acknowledgements.

This work was supported by the Science and Engineering Research Council, Swindon, UK and Quest Ltd, Menstrie, Clackmanannshire, UK.

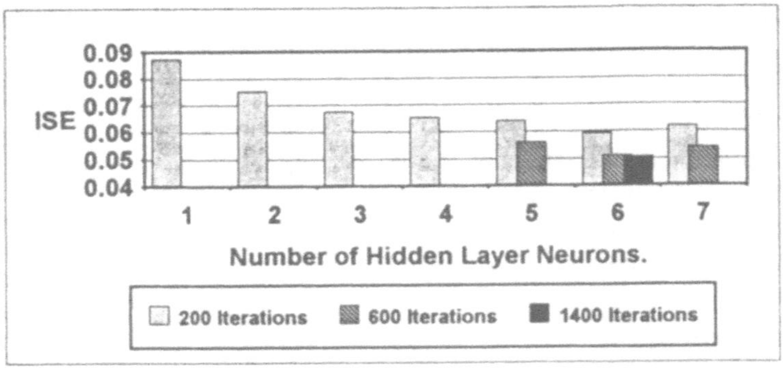

Figure 1. ISE Value After Training.

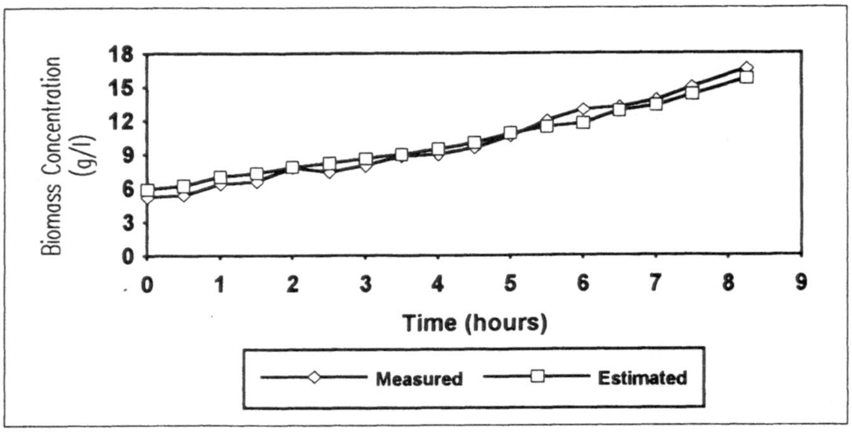

Figure 2a. Continuous Feed Strategy Training Data, Measured vs Estimated Values.

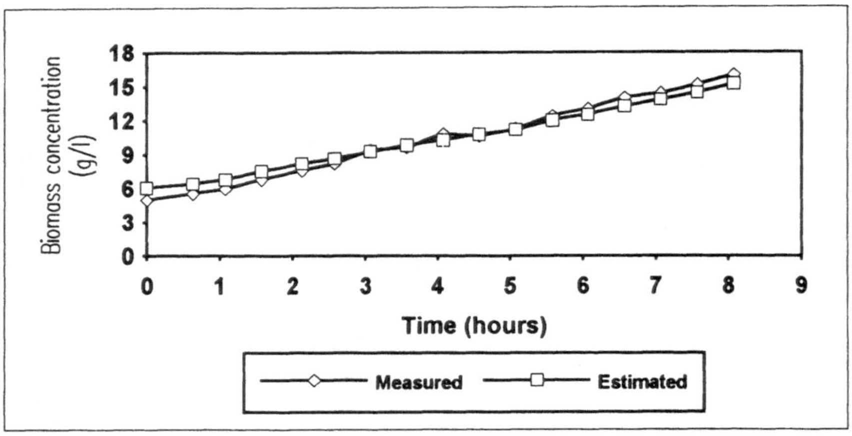

Figure 2b. Pulse Feed Strategy Training Data, Measured vs Estimated Values.

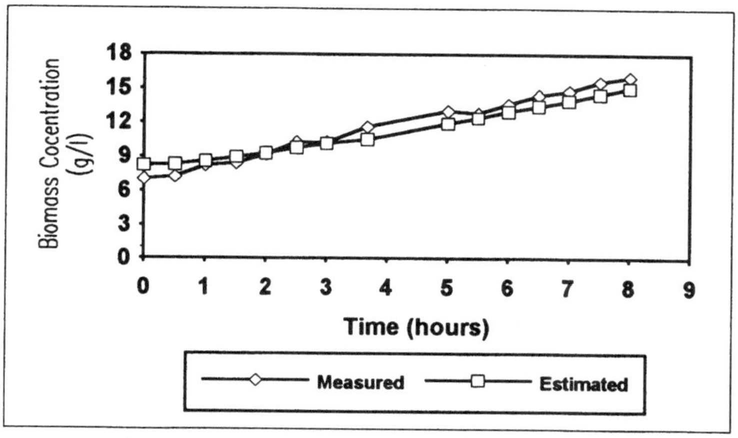

Figure 3a. On-line Testing of 5-6-1 Neural Network Using a Continuous Feed.

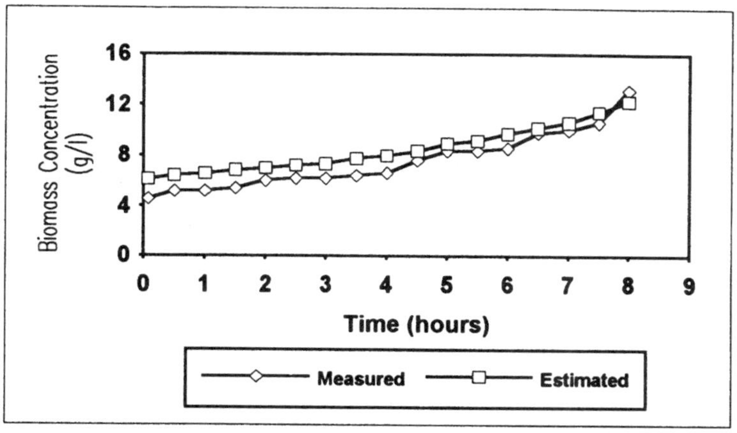

Figure 3b. On-line Testing of 5-6-1 Neural Network Using a Pulsed Feed.

AN APPLICATION OF UNSUPERVISED NEURAL NETWORKS BASED CONDITION MONITORING SYSTEM.

M.A. Javed, A.D. Hope.
Engineering Division
Southampton Institute of Higher Education
Southampton. U.K.

ABSTRACT.

Integrated condition monitoring for fault identification and maintenance planing is increasingly becoming an indispensable activity in today's industrial environment. Expert systems and neural networks are emerging to be the latest tools to be applied for condition monitoring. This paper briefly reviews these techniques and describes applications of artificial neural networks in diagnosing the health of various systems.

The application of neural networks discussed here contemplates to devise an intelligent, self-adaptive monitoring module which can be employed in a wider range of industrial environments. The paper describes a general purpose unsupervised neural networks based monitoring system which categorises the operational routines within the individual application environments of a wide range of industrial machinery. The monitor classifies the sensed data into its respective clusters and demonstrate its potential diagnostic capabilities.

INTRODUCTION.

The emphasis in industry is to prevent unscheduled shut downs due to an unforeseen failure of some critical components by utilising the most effective monitoring techniques available. The widespread approach is to utilise the sensed signals in decision making processes by extracting salient features from these signals which in turn are employed to identify the respective classes of these data.

The recent resurgence of artificial neural networks offer an alternative approach to the conventional condition monitoring methodologies. Neural networks seem an appropriate technique for devising these monitoring systems. Primarily because of their intrinsic capabilities to self-organise and extract salient features from the available data, particularly in noisy environments.

Artificial neural networks models are based on the use of large numbers of elemental processors interconnected in a manner reminiscent of biological neural networks. There is a wide variety of neural network architectures available, each with its own specific learning algorithm[1]. In general these algorithm can be classified as supervised and unsupervised learning. The supervised learning procedure, widely employed by researchers, is the so called error-back-propagation algorithm[2]. These architectures have been shown to exhibit powerful learning and associative recall capabilities for pattern formatted information[3,4,5]. This paper explores the benefits of incorporating neural networks in condition monitoring systems, primarily because of their capability to perform three fundamental tasks. First, they can identify which class best represents an observed pattern, where it may be assumed that the observation has been corrupted by noise. Second, they can be employed as associative memory devices, where an observed pattern can be used to produce a desired class exemplar. A third task these networks can perform is to vector quantise or cluster a set of observations. This is very useful in processing acoustic, vibration and visual data. These data classification, generalisation and association capabilities are all important factor in a condition monitoring and diagnostic systems.

NEURAL NETWORKS BASED APPROACHES.

The capabilities of artificial neural networks, as discussed above, have been employed effectively to extract salient features from the input data, which then forms the basis of a more objective analysis at a subsequent hierarchical level [4,5,6]. At this secondary level the experience gained from processing of the satisfactory machine-behaviour forms the basis of highlighting any presumable reasons for subsequent deviations from it. These deviation are then further classified and associated with their possible causes.

Schram et al[7] have demonstrated the use of these networks to assess the quality of small size motors by inspecting its acoustic impression. A set of ten typical acoustic signatures of good and defective motors were used to train the networks. Defects considered were, faulty bearings, faulty brushes, mass imbalance, etc. The trained network was then employed to recognise the signature and identify the associated defect. Khaparde[8] demonstrated their use in energy management systems to identify variations in demand and to initiate a corresponding remedial response. Javed[9,10,11] and Skitt et al[12] have employed these networks to characterise the operating behaviour of an aircraft engine on the basis of its acoustic noise. The neural networks then monitors the noise from the engine and identifies any deviations from the normal behaviour pattern.

The main reason for employing a neural network in condition monitoring is to take advantage of its capability to self-organise the data providing an objective analysis without any preconceived expectations about the performance of the host machinery. The speed at which a trained neural network operates, certainly provides an opportunity to perform the monitoring and diagnostic functions in near real-time. The technique as proposed here has a broad potential for applications in monitoring the performance of intricate industrial processes without involving costly evaluations by trained experts. Particularly when the nature of these evaluation testing may be intrusive or involve some destructive tests.

NEURAL NETWORKS FOR DISCOVERING CLUSTERS.

Data from multi-sensor systems can be viewed as patterns having points in N-dimensional feature space. It might also be expected that patterns that are similar in some respect, on the basis of class membership or other attribute values, might be close to each other in the N-dimensional pattern space. For example all patterns belonging to class C_i might cluster closer to one another than to any pattern belonging to class C_j.

In supervised learning, discussed above, the neural networks are presented with labelled patterns so that it can learn the mapping between N-dimensional feature space and the interpretation space, the classification space. However, there are circumstances where it might be appropriate to discover how the ensemble of patterns observed in a problem situation is distributed in pattern space. If the mechanism giving rise to the patterns also segregates them into clusters in a meaningful manner, then clearly the procedure that identifies the location and distribution of these clusters is also meaningful and valuable. Such unsupervised learning procedures try to identify several prototypes or exemplars that can serve as cluster centres. A prototype may be one of the actual patterns or a synthesised prototype centrally located in the respective cluster.

The very act of clustering necessitates a choice of metric; that is, it must be decided how distance is to be measured. The choice of Euclidean-Distance leads to the K-means algorithm or to the ISODATA algorithm. However, any other suitable measure of distance, (e.g. Hamming-Distance, or Mahalonobis-Distance), also might be employed as appropriate.

The algorithm for the unsupervised network, which is employed here to discover cluster centres, is described in the following paragraphs. This network is adapted from the Adaptive Resonance Theory (ART) structure of Grossberg and Carpenter[1], but can also accommodate the ideas of KOHONEN[1]. A special case of this network is shown schematically in Fig.(1). It essentially is ART1.

When a (new) pattern $\underline{x} \equiv \{ x_i \}$ is presented to input nodes of the network, a weighted sum y_j can be obtained by bottom up weight processing using the following equation.

$$y_j=\sum b_{ji}\, x_i$$

The Euclidean-Distance can then be calculated to find the upper-level node with the largest y_j value. The pattern $\underline{x}$ is then assumed to be the cluster centred at that node. It is also necessary to verify that $\underline{x}$ truly belongs to the jth cluster by performing top-down processing; that is, from the weighted sum

$$\sum t_{ji}\, x_i$$

Then $\underline{x}$ belongs to jth cluster if

$$\frac{\sum_i t_{ji}\, x_i}{\|x\|} > \rho$$

Where ρ is a vigilance parameter and $\| \underline{x} \|$ is the norm of the $\underline{x}$ vector.

If $\underline{x}$ belongs to the jth cluster then the bottom-up weights, b_{ji}, and top-down weights, t_{ji}, can be updated for that specific j and all i. If $\underline{x}$ does not belong to the one node that was the most likely, that node is deactivated and bottom-up weight processing is continued to obtain a weighted sum y_j as above to start another cluster centre.

The quantities b_{ji} and t_{ji} serve different purposes. For example after all clusters have been formed, the function of b_{ji} is to determine the exemplar that is most like the input $\underline{x}$. Of itself this step does not guarantee that

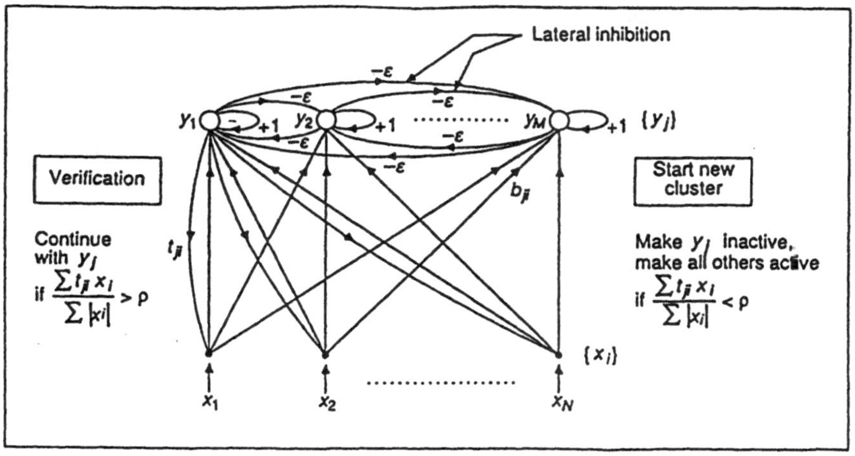

Fig. (1).
Schematic Representation of Network for Unsupervised Learning.

(Adapted from ART1 Architecture)

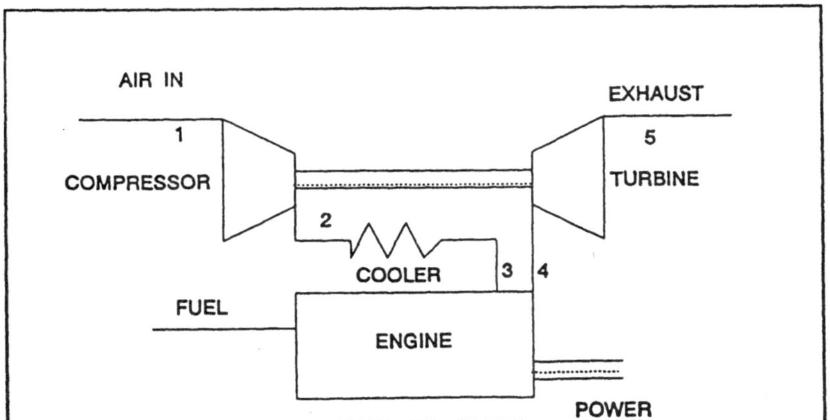

LAYOUT OF PLANT

FIG.. 2

the pattern $\underline{x}$ truly should be considered a member of that cluster. An explicit verification needs to be carried out. This verification is done using the top-down weights t_{ji}.

The quantity $\Sigma t_{ji} x_i$ is essentially a count of the number of coincidences of unit valued features between the jth. pattern and the input pattern. The quantity $\Sigma \| x_i \|$ will always be less than or equal to unity, and it provides a measure of how well the input pattern meets expectations.

If the value of ρ is set high, then conformation requirements are high and a body of patterns is likely to be split up into a large number of separate clusters. On the other hand, if the value of ρ is lowered, then larger departures from expectations are tolerated and the same set of patterns might be organised into a much smaller set of clusters.

DISCUSSION OF MONITORING STRATEGY.

The strategy discussed above for discovering clusters is applied to devise a monitoring system for the turbo-charged, inter-cooled diesel engine based power plant. A schematic diagram of the plant is shown in Fig. (2).

The monitor employs a two tier system comprising of initial categorisation of the sensed data and then subsequent association at the second level. At the initial level the satisfactory operational routines are clustered by processing the more readily available data, e.g. temperatures, pressures, speed. The resulting cluster-centres represent the characteristic behaviour of the plant. A deviation in the characteristics can be identified by monitoring the scatter of these clusters and emergence of new cluster centres. The second level of analysis is invoked in such a case, where a more detailed analysis, incorporating some specific data,e.g. vibration, acoustic signatures etc., is performed to associate the symptom of the deviation to its probable cause.

The satisfactory operational routines of the plant can be classified into a number of clusters, which are centred at specific nodes in the network. This can be seen from Table (1), where the clusters centred around each node epitomise its respective exemplar routine the temperature response of the plant for which is shown in Fig. (3). For example, nodes 0 to 5 are the cluster centres for entire range of satisfactory operational routines. Any malfunction in the plant or subsequent deviations from this satisfactory behaviour is recognised by the emergence of new cluster centres as the data is being processed. This is also presented in Table (1), where cluster centres portrayed by nodes 6 and 7 are due to faults being represented by the sensed data. It is at this stage of the data processing when the second hierarchical stage is invoked to perform the diagnostics in order to associate the symptoms with the causes. This particular aspect is fraught with complications and is being vigorously explored at present. There has been some encouraging progress but a conclusive outcome is yet to be realised.

CONCLUSIONS.

It has been demonstrated that the neural networks can be employed in order to categorise the normal operational behaviour of an industrial plant. The monitoring system can further identify any subsequent deviations in its

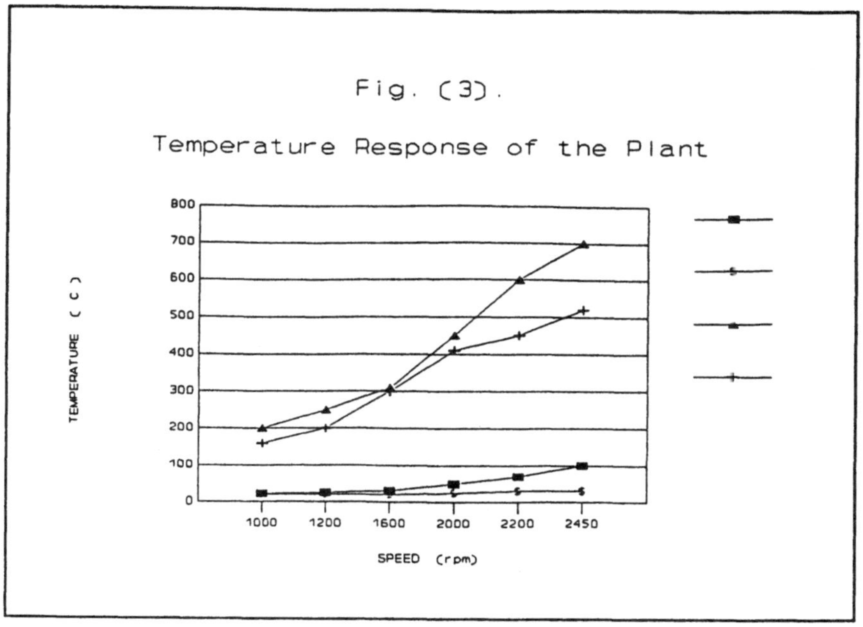

CENTRE No.(NODE)	EXEMPLARS	PATTERN NO.
0	4	0 6 12 18
1	4	1 7 13 19
2	4	2 8 14 20
3	5	3 9 15 21 24
4	4	4 10 16 22
5	4	5 11 17 23
6	1	25
7	1	26

Plnat Response as Cluster Centres.
Table No. (1).

system can further identify any subsequent deviations in its satisfactory operational routines and hence insinuating a faulty condition. The antecedent of the deviation can further be identified by associating symptoms with the causes. This analysis can be carried out by employing another neural network. The work regarding this aspect is still under progress and comprehensive conclusions cannot yet be drawn, however the initial indications are quite encouraging and are going to be discussed.

REFERENCES.

[1]. R.P. Lippman. 'An Introduction to Computing with Neural Networks'. IEEE ASSP Magazine April 1987.

[2]. D.E. Rumelhart, G.E. Hinton, R.J, Williams. 'Parallel Distributed Processing'.MIT Press, Cambridge MA. 1986.

[3]. M.A.Javed, S.A.C.Sanders, M.Kopp. "Numerical Optimisation of the Learning Process in Multi-layer Perceptron Type Neural Networks". IEE Colloquium on Neural Networks: Design Techniques and Tools. IEE Savoy Place. March 1991.

[4]. P.J.C.Skitt, M.A.Javed, S.A.C.Sanders, A.M.Higginson. "Artificial Neural Networks based Quality Monitor for Resistance Welding of Coated Steel". The 3rd. International Conference on Condition Monitoring & Diagnostic Engineering Management. July 1991. Southampton, U.K.

[5]. P.J.C.Skitt, M.A.Javed, S.A.C.Sanders, A.M.Higginson. "Process Monitoring Using Auto-Associative, Feed-Forward Artificial Neural Networks." Journal of Intelligent Manufacturing. Special Issue On 'Intelligent Manufacturing Systems.' Vol. 4, No. 1. 1992. ASME, U.S.A.

[6]. M.A.Javed, S.A.C.Sanders. "An Adaptive Learning Procedure for Neural Networks in Engineering." International Conference on the Application of Neural Networks in Engineering." St. Louis, Missouri, USA. Nov. 91.

[7]. H. Schram, H. Kolb. "Acoustic quality control using a multi-layer neural network". 22nd. Int. Symposium on Automotive Technology and Automation. 1990 Florence, Italy.

[8]. S. Kharpade. "Feasible application of neural networks in energy management system". Int. conference on Automation, Robotics and Computer Vision". 1990 Singapore.

[9]. M.A.Javed, S.A.C.Sanders. "Training Artificial Neural Networks for Applications in Automated Industrial Systems." International Conference on Industrial Electronics, Control and Instrumentation. IECON 91. Kobe, Japan. November 1991.

[10]. M.A.Javed, S.A.Sanders. "Neural Networks Based Learning and Adaptive Control for Manufacturing Systems." IEEE/RSJ International Workshop on Intelligent Robots and Systems." IROS 91.
Osaka, Japan. November 1991.

[11]. M.A.Javed, S.A.Sanders. "Artificial Neural Networks as Intelligent Condition Monitoring Devices". 'Condition Monitoring and Diagnostic Technology', Vol. 2, No. 1, July 1991.

[12]. P. Skitt, R. Witcomb. "The analysis of the acoustic emission of jet engines using neural networks". Condition Monitoring and Diagnostic Technology. Vol. 1. No. 1. June 1990.

A REPORT OF THE PRACTICAL APPLICATION OF A NEURAL NETWORK IN FINANCIAL SERVICE DECISION MAKING

Graham Bazley, MICM

Marks & Spencer Financial Services, Kings Meadow,
Chester X CH99 9QB, England

ABSTRACT

Marks & Spencer Financial Services (MSFS) provides three credit products for its customers; Charge Card, Budget Card and Personal Loans, see Figure 1.

Figure 1

MSFS CREDIT PRODUCTS

	NO. OF ACCOUNTS
CHARGE CARD	2.3M
BUDGET CARD	130K
PERSONAL LOAN	100K

New credit applications are considered using tailor made Credit Scoring Systems as well as independent credit reference bureau data. The Credit Scoring Systems are developed using standard statistical techniques such as multiple regression. Rather than allow the scorecards to be entirely responsible for the credit decision a proportion of applications are "referred" to trained credit appraisal staff for them to make a decision, these staff have the job title, Underwriter, this may entail seeking further information from the applicant. Figure 2 describes the flow of credit applications and the decisions that can be made.

Figure 2

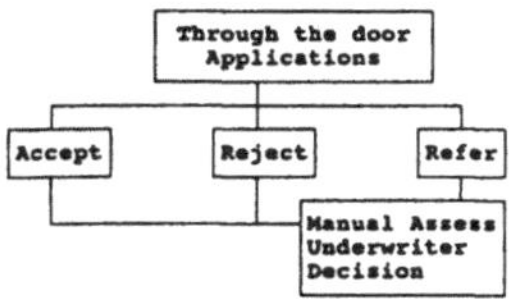

Typically, 20% of applications are referred and the subsequent human intervention is time consuming, decreases customer service and the accuracy of the decision is questionable.

Neural Network software had been acquired by our parent company and they were anxious for this type of technology to be tested within our business.

Automation of the referral process was seen as a suitable area for us to test what benefits might be derived from the use of a Network to make credit decisions.

This paper describes the data upon which a network was developed, the results experienced and a way forward to integrate the Network into our decision support systems using the Champion vs. Challenger concept.

THE PORTFOLIO

Different Credit Scoring Systems and supporting filters are used for different Credit Products. For the Neural Network project, it was decided to focus on the Personal Loan portfolio as this presents the highest risk, both statistically and financially, as well as having one of the highest percentage referral rates.

The Personal Loan Portfolio is recruited from two sources, existing Charge Card Customers (Direct) or new to MSFS attracted from point of sale from one of 300 stores, or through specialised direct mail marketing activities (Indirect). Figure 3 describes the Personal Loan sub populations.

Figure 3

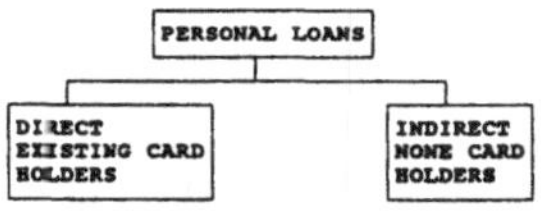

The performance of the two sub populations is very different and for the purpose of the Neural Network test, separate Networks were developed.

DEFINITION OF THE SUCCESS CRITERIA FOR NEURAL NETWORKS

In order to measure the success or otherwise of this project the defined criteria was:

i. Known bad accounts the network would decline.
ii. Previously declined applications

the network would accept.

iii. Known good accounts the network would decline.

The definition of known "good", "bad" and "reject" applications for the project is the same as those used for the credit scoring models, essentially anyone whose payment is overdue 60+ days is classified as delinquent and therefore "bad".

THE NEURAL NETWORK SOFTWARE PACKAGE

The software package used in this project was "Neuralworks Professional II".

Two PC based Neural Networks were developed using data from Direct and Indirect personal loan applications made between 6/88 and 6/91. Figure 4 describes the importance of the "sample window" from which the training data was obtained.

THE TRAINING DATA

MSFS is data rich as a result of retaining information captured at the point of application, at the time a credit reference bureau report was purchased or from the account maintenance system. At the start of the Project 100 characteristics were identified that could be associated with each customer extracted from the sample window see appendix 1.

From this list a subjective assessment was made as to the most likely predictive characteristics in order to reduce the amount of training that would be necessary on the Neural Network. Figures 5 and 6 describes the characteristics selected and their associated attributes.

Figure 4

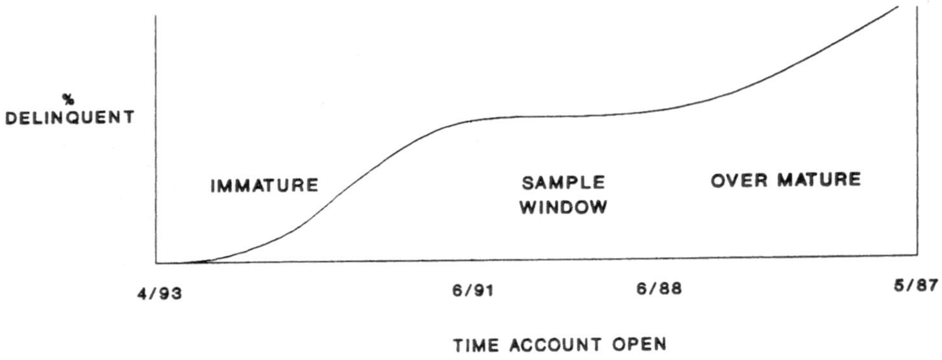

The immature accounts are excluded from the sample because they have not had sufficient time to go bad whilst the overmature accounts are excluded because the very good accounts have been repaid and the accounts closed. The sample window, therefore, reflects a balance of good and bad based upon a mature sample.

Figure 5

DIRECT CHARACTERISTICS/ATTRIBUTES

LOAN AMOUNT

500-2000	2001-5000	5001-7500	7501-10000	Existing customers maximum £10000

APPLICANTS OCCUPATION

BE BO BT,SS	SS	LL PP	NN	RR HH	UU KK	TT FF XX,JJ	CC	GG MM MO	AA AB EE	DD

APPLICANT INCOME

TO 0	TO £6000	TO £12000	TO £20000	TO £25000	£26000+

COMMODITY CODE (LOAN PURPOSE)

01,02 03	07	11 40	50 70	60	63	65	80 90	85	99

REPAYMENT METHOD DIRECT DEBIT/CREDIT TRANSFER

RESIDENT STATUS

OWNER	PARENTS	TENANT	OTHER

CHEQUE CARD YES/NO

TIME WITH BANK

0-2YRS	3-4YRS	5-7YRS	8-12YRS	13PLUS	BLANK

YEARS AT CURRENT ADDRESS

UP TO 1YR	2-4YRS	5-6YRS	7-10YRS	11 PLUS

Figure 6

INDIRECT CHARACTERISES/ATTRIBUTES

LOAN AMOUNT

500-2000	2001-5000	5001-7500	7501-10000	New customers max £7500

APPLICANTS OCCUPATION

BE BO BT,CS	SS	LL PP	NN	RR HH	UU KK	TT FF XX,JJ	CC	GG MM MO	AA AB EE	DD	Decline single Students/Unemployed

MARITAL STATUS

SINGLE	MARRIED	DIVORCED	WIDOWED	BLANK	In conjunction with 7.

% EXPENDITURE TO INCOME

<40%	40-49%	50-59%	60 PLUS	BLANK

COMMODITY CODE

01,02 03	07	11 40	50 70	60	63	65	80 90	85	99

REPAYMENT METHOD DIRECT DEBIT/CREDIT TRANSFER

RESIDENTIAL STATUS

OWNER	PARENTS	TENANT	OTHERS

CHEQUE CARD YES/NO

TIME WITH BANK

0-2YRS	3-4YRS	5-7YRS	8-12YRS	13PLUS	BLANK

YEARS AT CURRENT ADDRESS

UP TO 1YR	2-4YRS	5-6YRS	7-10YRS	11 PLUS

Each attribute value was converted to binary values to avoid scaling problems, however, it has since become apparent that this was not absolutely necessary.

PERFORMANCE DATA

Using the "good", "bad" performance definitions previously referred to accounts were extracted from the Account Management database and assigned an appropriate good or bad flag, only those accounts opened in the mature sample window were selected. The characteristic and attribute data was then added to each record.

For both models three data files were created. A training and a test file containing manual accepts and a further file containing declines that had been manually assessed.

The training and testing files were created with the data randomly spread.

To ensure that the data was not biased towards an event such as a specific marketing mailing campaign, or at a time when it was known that there were inconsistencies in the manual decisions, the training files were created to contain equal numbers of known "good" and "bad" accounts to ensure that the network had sufficient examples of each to be able to distinguish between them accurately.

To achieve this and to leave sufficient "bad" accounts in the test file, the data was split taking 90% of the smallest set (bads) for training and taking an equal number of goods and then leaving the rest for testing, Figure 7 provides an example of this.

Figure 7

INDIRECT LOANS 6/88 - 6/91

	TOTAL	TRAINING FILE	TEST FILE
GOOD	1,546	453	1,093
BAD	504	453	51
	------	---	-----
	2,050	906	1,144

TRAINING THE NETWORK

Using the back propagation technique Neural Network models were built. By an iterative process of changing the parameters which control the Network the best set were established. During prototyping models were developed or discarded based upon the performance of a set of instruments or probes set during training of the Network.

ANALYSIS OF THE RESULTS

For the purpose of this paper only the result of the effect of the Network on the Direct Loan Population will be discussed.

Figure 8 illustrates the sample population that presented itself between 6/88 and 6/91 and the breakdown between accepts/rejects and good/bad customers based upon Underwriter decisions.

Figure 9 illustrates the same sample population as in Figure 8, but with the different decisions made by the Neural Network.

Figure 8

DIRECT LOAN - UNDERWRITER OUTCOMES

DIRECT LOANS

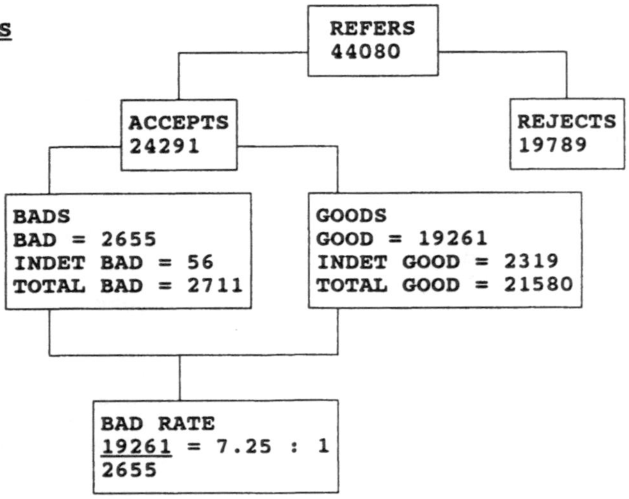

Figure 9

NEURAL NETWORK OUTCOMES

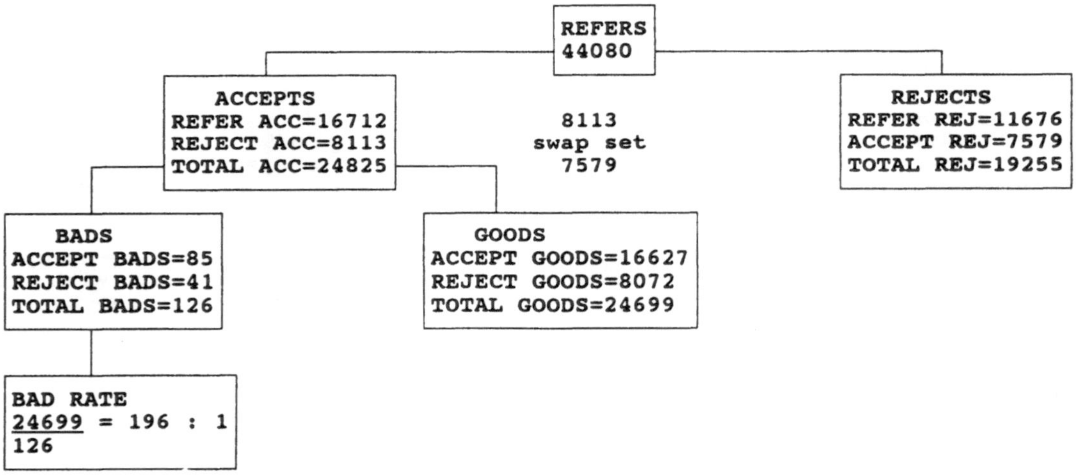

NOTE - We have used the same Bad rate to calculate the Good/Bads from the reject Accepts, as their performance is unknown.

This diagram illustrates the outcome of the same development sample based upon the Neural Network decisions.

The result of the decisions made by the Network were quite remarkable, it had only failed to detect 85 of the known bads that the Underwriters had accepted. However, as with Credit Scoring, the other most significant change was in "reject inference", the Network wanted to accept 8113 of the 19789 the Underwriters had rejected, 41% of the total rejects. Given that the rejects present a high risk, it would be too costly to allow the Network to replace the Underwriters without further controlled testing.

APPLICATION OF THE CHAMPION VS CHALLENGER CONCEPT

The Champion vs. Challenger concept is widely used in MSFS to measure the effect of introducing new working practices within a narrow but controlled environment. New practices are referred to as the Challenger approach whereas the current best practice is referred to as the Champion. The Challenger approach would be applied to, say, 10% of the account base whilst the Champion would continue to be applied to 90%. The effect of the two practices would then be measured over a period of time, usually 3-6 months. If the Challenger improved performance, as illustrated, it would become the new Champion and be applied to 90% of the account base, as illustrated in Figure 10.
It therefore seemed a sensible approach to introduce the Champion vs. Challenger concept to allow the Network to be introduced in a live but controlled environment.

It is proposed to allow the Network to make decisions on 5% of all referrals in future and then measure subsequent performance against Underwriter decisions from the same date of acceptance. Figure 11 describes the population flow that we will expose the Neural Network to.

Figure 10

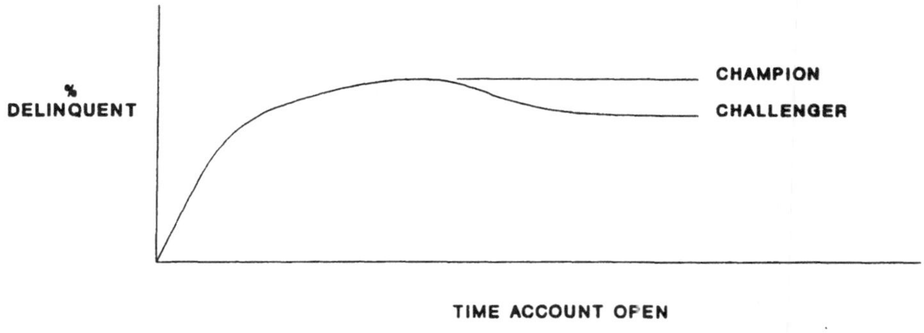

Figure 11

CHAMPION VS CHALLENGER APPLICATION TO A NEURAL NETWORK

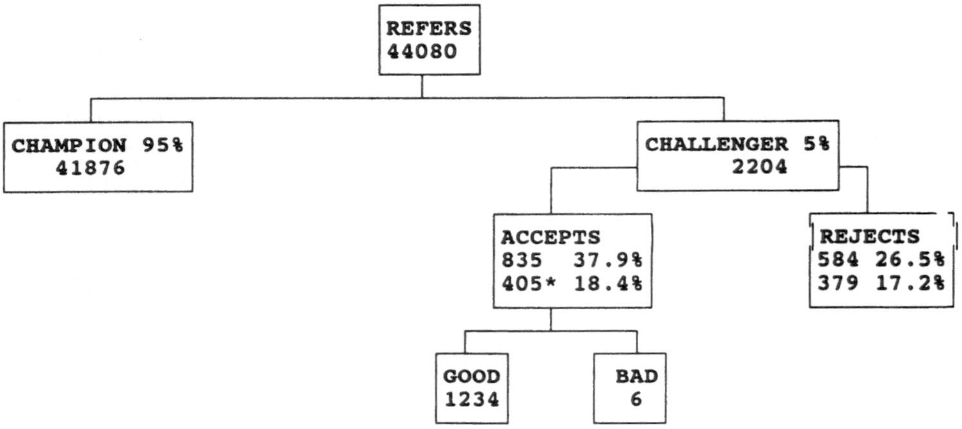

GOOD VS. BAD ON REJECT INFERENCE

BAD RATE	GOOD	BAD	G/B ODDS
5%	385	20	19.2
10%	365	40	10.9
20%	325	80	4.1
30%	284	120	2.4
40%	245	160	1.5
50%	203	202	1.0

In attempting to quantify the break even bad rate to the Underwriter bad rate, we have estimated that we can allow up to 15% of the previously rejected applicants, that the Network wants to accept, to go bad.

CONCLUSION

This project has identified that Neural Networks provide a very significant tool to potentially support many areas of our business that require decision support systems.

In the past, the business has had to rely on expensive external consultants to develop statistical models, Neural Networks provide an opportunity to control our own decision support systems.

Following consultation with Coventry University the results that were obtained were academically supported.

Given the culture of Marks & Spencers we are continually striving to improve the quality of service to our customers, Neural Networks are seen as playing a very important role in the future in contributing to this.

APPENDIX 1

CHARACTERISTICS AVAILABLE FOR ANALYSIS

Bank account number
Application status
Application type
Application date (yymm) year&month
Application date (mmyy)
Application score
Bureau score
Total score
Primary pend reasons
Loan account number
Source code
Branch number
Postcode (1st 4 digits)
Loan amount
Insurance Y/N
Loan amount payable
Monthly payment
Repayment period mth
Deposit/amount payable
A.P.R.
Pause terms mths
Commodity code
Age of application (mths)
Home/work tele number
Residential status
No. of deps.
Applicant occupation
Years in occupation
Applicant income
Total income
Cheque card
No. of credit cards
Access/Visa
Years at bank
Repayment method
Expd. per month
Applicant title
Years at current address
Searches 1st 6 months
Total time at CA + PA

C/A type (current address)
Mosaic
Percentiles cars/house
Percentiles overcrowding
Percentiles fam @ pcode
Searches 1st 2 days
Own cais at CA/PA
Worst active SN cais L6 C&P add
Tot Bal SN cais L3 CP add
Non MO cais 1 SN CP add
O/S Bal cais 2 CP add
Active cais 2 CP add
Overlap on ccj CA
Overlap on cais 8/9 CA
SN CA/PA mths ccjs
SN CA/PA # ccjs
SN CA/PA val ccjs
SN CA/PA cais 8/9
SN CA/PA mths cais 8/9
SN CA/PA val cais 8/9
SN CA/PA cais 4
SN CA/PA cais 3
Initial decision
Final decision
Date final decision
Decline review
Account type
Wst SN cais L6m
SNSI tot val cais 8/9
SNSI tot o/s bal
SNSI cais 8/9 time
SNSI wst status L6m
SNSI wst status L6m>12
SNSI wst status L12>12
SNSI wst status L6m M&S
SNOI cais 8/9 time
SNOI tot val cais 8/9
SNOI wst status L6m
SNOI wst status L6m>12
SNOI tot o/s bal
SNOI tot o/s bal M&S
Source indicator

O/S= Outstanding Balance; Bal=Balance; Val=Value; MO=Mail Order;
CA=Current Address; PA=Previous Address; WST=Worst; L=Last;
SN=Same Name; SNSI=Same Surname Same Initial; SNOI= Same Surname Other Initial

Performance Evaluation of Neural Networks Applied to Queueing Allocation Problem

Junichi Takinami, Yutaka Matsumoto and Norio Okino
Division of Applied Systems Science
Faculty of Engineering, Kyoto University
Kyoto 606-01, Japan
Email: matumoto@kuamp.kyoto-u.ac.jp

Abstract

In this paper we consider the dynamic allocation of customers to queues and study the performance of neural networks applied to the problem. The queueing system consists of N parallel distinct servers, each of which has its own queue with infinite capacity. A controller allocates each arriving customer to one of the servers at arrival epoch, who maximizes the probability of starting service for the customer in the earliest time. A neural network is incorporated into the controller, so that the neural controller can make an allocation decision adaptively to changing situations. We present a simple training method for the neural controller. We consider two types of neural networks (BP and LVQ3) and compare their performance in numerical examples.

1 Introduction

The queueing allocation problem is the problem which treats how to allocate customers to one of parallel servers upon arrival so as to maximize or minimize a certain objective function. We assume no jockeying among queues is permitted. The allocation can be *static* if it is predetermined by a schedule or *dynamic* if it depends on the information about the system state such as the number of waiting customers, the service process at each queue, and the arrival process of customers. The objective function to be maximized can be the discounted number of customers to complete their service in any time t, or the one to be minimized can be the mean delay of customers. The solutions for these objective functions are called *social optimal* policies. On the other hand, if the objective function to be minimized is, for example, the expected delay of an individual customer, the solution is called an *individual optimal* policy.

Under some assumptions, the dynamic allocation problem has been formulated and solved in Markov decision theory [9]. Given that all servers are stochastically identical and the service time is i.i.d. (independent and identically distributed) exponential, and the arrival process is Poisson, Winston [11] proved that the SSQ-Policy (Send-to-the Shortest Queue Policy) maximizes, with respect to stochastic order, the discounted number of customers to complete their service in any time t. Weber [10] extended Winston's result to the case where the service time distribution has an IFR (Increasing Failure Rate) and the arrival process is a generic point process.

The queueing allocation problem has important practical applications in routing for computer communication networks [4] or in load balancing of distributed processors [1]. In these applications, however, the servers cannot be identical, nor can the service time distribution have an IFR, so that the theoretical result is not directly applicable.

Recently, the application of neural networks to classification problems has been paid much attention [2],[8]. Especially, the BP (Back Propagation) algorithm [6] and the LVQ (Learning Vector Quantization) algorithms [3],[5] have been successfully applied to those problems.

The prime purpose of this paper is to examine the performance of neural networks (BP and LVQ3) when they are applied to the dynamic allocation problem, where the servers can be stochastically distinct and the service time distributions must not have an IFR. By incorporating a neural network, it is expected that the controller can make an allocation decision adaptively to changing situations. Based on the information only about the arrival time and the service starting time of customers, we present a simple training method for the neural controller, which seeks an individual optimal policy.

This paper is organized as follows. Section 2 reviews the mathematical framework of the dynamic allocation problem in Markov decision theory. In Section 3 we summarize the BP and LVQ3 algorithms and present when and how to update the weights when they are applied to the dynamic allocation problem. Simulation results are shown in Section 4, where we compare and discuss the following: the mean waiting time, the percentage of correct allocations by the neural networks, the effect of elapsed service time on allocation decisions, and the adaptation ability to changing situations. Finally we conclude this paper in Section 5.

2 Dynamic Allocation Problem

We consider a queueing system with N parallel servers, where each server has in general a distinct service time distribution and its own queue with infinite capacity. Customers arrive at the system according to a given stochastic process and upon arrival the controller chooses an action from a set of options in terms of the assignment of the customer to some server's queue. Once a customer is allocated to a queue, no jockeying among queues is allowed. Each server serves its customers according to the first-in first-out discipline.

If the service time distributions of all servers are mutually independent and the service time of each server is i.i.d. exponential, then the system state can be defined as the number of customers at each queue because of the memoryless property of the exponential distribution. Then the system state can be labeled by non-negative integers, $i = 0, 1, 2, \ldots$. Upon arrival of a customer, the controller has N possible options, that is, to assign the customer to server l $(l = 1, 2, \ldots, N)$. Let X_k and A_k denote the system state and the action chosen at the arrival epoch of the k-th customer $(k = 1, 2, \ldots)$, respectively. Then the next state of the system is determined by transition probabilities $P_{ij}(l)$,

$$P_{ij}(l) = \Pr\{X_{k+1} = j | X_1, A_1, X_2, A_2, \ldots, X_k = i, A_k = l\}. \quad (1)$$

If action l is chosen according to a function f which maps the state space into the action space, the sequence of states $\{X_k : k = 1, 2, \ldots\}$ forms a Markov chain with transition probabilities $P_{ij} = P_{ij}(f(i))$. A set of actions defined over the state space is called a *policy*.

The goal of the dynamic allocation problem is to obtain such a policy that maximizes or minimizes a certain objective function E. Usually

$$E = \int_0^T e^{-\alpha t} dD(t), \qquad \forall T > 0, \quad (2)$$

is employed as an objective function to be maximized, where α is a discount factor $(0 < \alpha \leq 1)$ and $D(t)$ represents the probability distribution function for the number of customers who receive service by time t.

We note that the above Markov chain formulation is possible only for the case that the service time of each server is i.i.d. exponential. In our neural network approach, there is no restriction on the service time distribution. We attempt to train the neural controller so that upon arrival, it allocates the customer to the server who maximizes the probability of starting service for the customer in the earliest time. This policy is individually optimal and it is obvious from [10] that this individual optimal policy agrees with the social optimal policy satisfying Eq.(2) in case of the i.i.d. service time with an IFR at all identical servers. In general, however, the mean waiting time under any individual optimal policy becomes longer than the one under the social optimal policy satisfying Eq.(2).

3 Application of Neural Networks

3.1 BP and LVQ

The BP is a learning algorithm which is used for multi-layered feed-forward neural networks [6]. In the BP, interconnecting weights $\{w_{ij}\}$ are updated so as to minimize a certain energy function E. In this paper the following Kullback divergence is employed for E to reduce the learning time [7].

$$\begin{aligned} E &= \sum_p E_p \qquad (3) \\ &= \sum_{p,i}\{t_i^p \log \frac{t_i^p}{o_i^p} + (1-t_i^p)\log\frac{(1-t_i^p)}{(1-o_i^p)}\}, \end{aligned}$$

where o_i^p and t_i^p represent the output of the i-th unit for the p-th pattern and its corresponding target output, respectively. Whenever an input pattern is presented, weights are modified by

$$w_{ij}(t+1) = w_{ij}(t) - \eta\frac{\partial E_p}{\partial w_{ij}(t)}, \qquad (4)$$

where $\eta > 0$ is the learning rate.

For the output function at each neuron, we use such a sigmoid function as

$$f(x) = \frac{1}{1-e^{-x/\tau}}, \qquad (5)$$

where τ is called a temperature.

The LVQ is a nearest neighbor classifier which is implemented by neural networks with two layers [3]. Let m_i represent the codebook vector (weight vector) for the i-th output neuron. Usually more than one output neurons are assigned to each class and we denote a function which maps each neuron into a class by $g(i)$. Given an input vector x, the output neuron with the closest codebook vector to x is selected by

$$c = \arg\min_i\{\| x - m_i \|\}, \qquad (6)$$

and the closest codebook vector to x is denoted by m_c. Then x is classified into class $g(c)$. To update codebook vectors, there are basically three variants in the LVQ family (i.e., LVQ1, LVQ2 and LVQ3) [3]. In this paper we adopt the LVQ3 for comparison with the BP because it performed best among the three in a series of our numerical experiments.

The LVQ3 algorithm is summarized as follows [3]. Let m_i and m_j be the two closest codebook vectors to x, and d_i and d_j be the Euclidean distances of x from m_i and m_j, respectively. If d_i and d_j satisfy

$$\min(\frac{d_i}{d_j}, \frac{d_j}{d_i}) > s, \quad \text{where} \quad s = \frac{1-w}{1+w}, \qquad (7)$$

then x is referred to as *falling into a window of relative width* w. The codebook vectors are updated by the following equations:

$$\begin{aligned} m_i(t+1) &= m_i(t) - \alpha(t)[x(t) - m_i(t)], \\ m_j(t+1) &= m_j(t) + \alpha(t)[x(t) - m_j(t)], \end{aligned}$$

where m_i and m_j are the two closest codebook vectors to x, whereby x and m_j belong to the same class, while x and m_i belong to different classes, respectively; furthermore x must fall into the window;

$$m_k(t+1) = m_k(t) + \epsilon\alpha(t)[x(t) - m_k(t)],$$

for $k \in \{i,j\}$, if x, m_i, and m_j belong to the same class;

where $0 < \alpha(t) < 1$ and $0 < \epsilon < 1$.

3.2 Training Method

The *waiting time* of a customer is defined as the time interval between his arrival epoch and the start of service, while the *delay* is defined as the time elapsed from the arrival epoch to the end of service. We attempt to train the neural controller, so that it allocates an arriving customer to the server who maximizes the probability of starting service for the customer in the earliest time. Thus we are basically focusing on the waiting time. This policy might be useful in routing for computer communication networks to reduce unnecessary sorting of delivered packets.

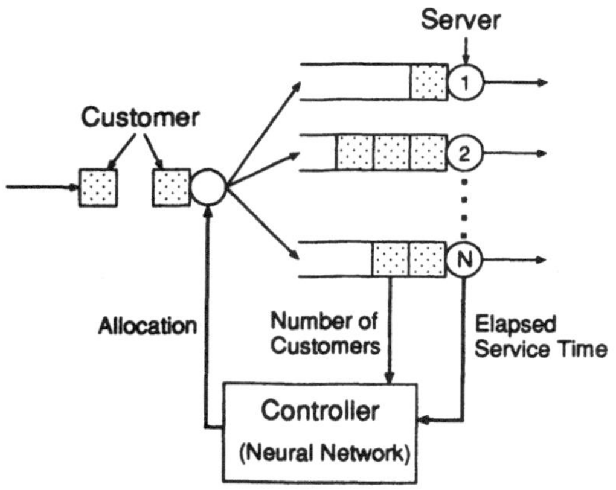

Figure 1: Queueing system and neural network.

We assume that the information available at the arrival epoch of a customer is 1) the number of customers and 2) the elapsed service time (if any customer is being served) at each server (see Figure 1). In the following, the former for server l ($l = 1, 2, \cdots, N$) at time t is denoted by $n_l^{(t)}$, while the latter by $e_l^{(t)}$. Then the information $S^{(t)}$ used to make an allocation decision is given by

$$S^{(t)} = (n_1^{(t)}, n_2^{(t)}, \ldots, n_N^{(t)}, e_1^{(t)}, e_2^{(t)}, \ldots, e_N^{(t)}). \quad (8)$$

We note that in case of the exponential service time, $e_l^{(t)}$ ($l = 1, 2, \cdots, N$) have no effect because of the memoryless property.

The problem is how to produce target data through the evolution of the queueing system. In practical situations, available statistics on each customer are usually the arrival time and the starting time of service, and sometimes the completion time of service as well. This paper considers a simple training method by making use of the arrival time and the start time of service. We produce target data based on whether there is any customer, who arrived later than a tagged customer, was assigned to a different server from the tagged customer, and started receiving service earlier than the tagged customer.

Formally, the neural controller is consulted at the arrival time of a customer and updated at the beginning of service in the following manner. Suppose that a customer has arrived at the queueing system at time t, where this customer is referred to as a *tagged* customer below. Then the tagged customer is assigned to the server, whose image neuron has the biggest output when $S^{(t)}$ is presented as an input to the neural network. Suppose that the tagged customer is assigned to server i and the service to the tagged customer starts at time t'. Let $a_j^{(t')}$ denote the arrival time of the customer who is being served (if any) by server j at time t'. At time t', we train the BP by input data $I = S^{(t)}$ and the following target data $T = (t_1, t_2, \ldots, t_N)$:

1. If $^\exists j$ such that $n_j^{(t)} = 0$, then select one of j's (say, k) with equal probability, and set $t_l = 1$ for $l = k$ and $t_l = 0$ for $l \neq k$.

2. Otherwise,

   (a) if $^\exists j$ such that $a_j^{(t')} > t$ or $n_j^{(t')} = 0$, then select one of j's (say, k) with equal probability, and set $t_l = 1$ for $l = k$ and $t_l = 0$ for $l \neq k$,

   (b) otherwise, set $t_l = 1$ for $l = i$ and $t_l = 0$ for $l \neq i$.

In case of the LVQ3, the neural controller is taught to classify the input data I into class k with $t_k = 1$.

We note that for given I, the target data T change stochastically, so that a small learning rate is suitable to prevent oscillation. Moreover, training data sets obtained in sequence are correlated. To decrease the influence of correlation, we teach the neural networks at time t' by using, in addition to (I, T), a past training data set which is randomly chosen from last 100 training data sets. In consequence, the neural networks are trained twice as much as the number of customers who started receiving service.

4 Simulation Results

Simulation Models

Throughout our simulation experiments, the arrival process of customers is assumed to be Poisson with arrival rate λ. We denote the service rate of server i by μ_i. The waiting room at each queue has a capacity of 1,000 customers and in case customers overflow this capacity, the simulation program is stopped and the data are not counted. The neural network models used in our simulation are as follows:

BP: A three-layered feedforward network is employed. The number of hidden units is twice as much as that of servers. The learning rate is $\eta = 0.01$ before 1,000 customers start receiving service and $\eta = 0.001$ after that. The temperature of the sigmoid function is $\tau = 3.0$. The weights $\{w_{ij}(t)\}$ are initialized by random numbers between $-0.3 < w_{ij}(0) < 0.3$.

LVQ3: The number of codebook vectors is five times as much as that of servers. $\epsilon = 0.1$ and $\alpha = 0.05$ before 1,000 customers start receiving service, and $\alpha = 0.001$ after that. The window width is $w = 0.5$. Because a proper initialization of codebook vectors is indispensable for the LVQ algorithms [3], we initialize them by the SSQ-Policy for the first N input neurons with random numbers for the last N input neurons.

Mean Waiting Time

First of all, we evaluate the performance of the neural controller with respect to the mean waiting time of customers. We consider the case that the number of servers is $N = 4$ and the service processes are mutually independent among the servers and the service time is i.i.d. exponential at each server with $\{\mu_i\} = \{0.4, 0.5, 0.8, 1.0\}$. Under these assumptions, we can easily calculate the individual optimal policy that allocates each arriving customer to the server who maximizes the probability of starting service for the customer in the earliest time.

1. If $^\exists j$ such that $n_j^{(t)} = 0$, then assign the customer to one of the servers with $n_j^{(t)} = 0$ with equal probability.

2. Otherwise, assign the customer to server s such that

$$s = \arg\max_j\{\left(\frac{\mu_j}{\mu}\right)^{n_j^{(t)}} \prod_{i=1,\neq j}^{N} \sum_{k_i=0}^{n_i^{(t)}-1} \frac{(n_j^{(t)} - 1 + \sum_{l=1,\neq j}^{N} k_l)!}{(n_j^{(t)} - 1)! \prod_{m=1,\neq j}^{N} k_m!} \prod_{n=1,\neq j}^{N} \left(\frac{\mu_n}{\mu}\right)^{k_n}\},$$

where μ is defined as $\mu = \sum_{i=1}^{N} \mu_i$. The mean waiting time is evaluated by the average of 5 runs with different initial values, where in each run the statistic is collected for 100,000 customers after 100,000 customers start receiving service, at which the training of the neural networks is stopped.

λ	Indi. Opt.	BP	LVQ3
1.0	0.089	0.098	0.106
1.5	0.308	0.316	0.364
1.8	0.553	0.586	0.666
2.1	1.078	1.133	1.223
2.4	2.374	2.454	2.626

Table 1: Mean waiting time.

Table 1 compares the mean waiting time for the individual optimal policy, the BP and the LVQ3. As we can see, there is no great difference between the individual optimal policy and the BP. In comparison between the BP and the LVQ3, the BP outperforms the LVQ3 for all the arrival rates.

Percentage of Correct Allocations

If all servers are stochastically identical and the service time is i.i.d. exponential at each server, then the SSQ-Policy is socially as well as individually optimal. For this case we examine the percentage of correct allocations by the neural networks. The arrival rate is now $\lambda = 1.0$, the number of servers is $N = 5$, and the service rates are $\{\mu_i\} = \{4.0, 4.0, 4.0, 4.0, 4.0\}$. The neural networks are trained until 100,000 customers start receiving service. Then 10^5 cases from $\{n_i^{(t)}\} = \{0,0,0,0,0\}$ to $\{9,9,9,9,9\}$ are presented to the neural networks to check whether the output agrees with the SSQ-Policy, where the elapsed service times are fixed at $\{e_i^{(t)}\} = \{0.0, 0.0, 0.0, 0.0, 0.0\}$.

Table 2 shows the percentage of correct allocations, where frequent situations are defined as the situations which occurred more than 100

	Frequent Situations	Rare Situations
BP	97%	82%
LVQ3	91%	96%

Table 2: Percentage of correct allocations.

times until 100,000 customers started receiving service, while the other situations are classified into rare situations. We can see that the BP learns frequent situations better than rare situations. On the other hand, we notice that the LVQ3 has a higher percentage of correct allocations in rare situations than in frequent situations. This is an inheritance of the correct initialization. Namely, given an input vector x, the LVQ3 updates only two nearest codebook vectors, so that the codebook vectors close to the input vectors corresponding to rare situations are seldom updated. Thus the LVQ3 tries to approximate only frequent situations by using a limited number of codebook vectors, while the correct initial codebook vectors for rare situations remain unchanged.

Effect of Elapsed Service Time

If the service time is exponential, the allocation is not affected by the elapsed service time because of the memoryless property of the exponential distribution. Otherwise, the elapsed service time is important information for the allocation decision. Next we examine the effect of using the elapsed service time in the cases of the exponential service time and the deterministic service time. The arrival rate is now $\lambda = 0.13$, the number of servers is $N = 2$, and the average service times are $\{1/\mu_i\} = \{10.0, 15.0\}$.

In Figures 2 to 5, the number of customers and the elapsed service time for server 2 are fixed at $n_2^{(t)} = 1$ and $e_2^{(t)} = 0.0$, while those for server 1 are varied from $n_1^{(t)} = 1$ to 3 and $e_1^{(t)} = 0.0$ to 10.0. In this case, if the output for server 1 in the BP network exceeds 0.5, the customer is to be allocated to server 1, and otherwise, to server 2.

From Figure 2, we can see that the allocation decision of the BP is not affected by the elapsed service time in case of the exponential service time at the two servers. In other words, the neural network learned the memoryless property of the exponential distribution. In case of the LVQ3, however, we have incorrect decisions between $e_1^{(t)} = 1.8$ and $e_1^{(t)} = 4.0$ for $n_1^{(t)} = 1$ in Figure 3. In case of the deterministic service time, the allocation decision changes at $n_1^{(t)} = 2$ and $e_1^{(t)} = 5.0$ in Figures 4 and 5, which is a theoretically correct decision boundary. Thus we can conclude that the information about the elapsed service time is correctly used in the decision-making of the neural networks.

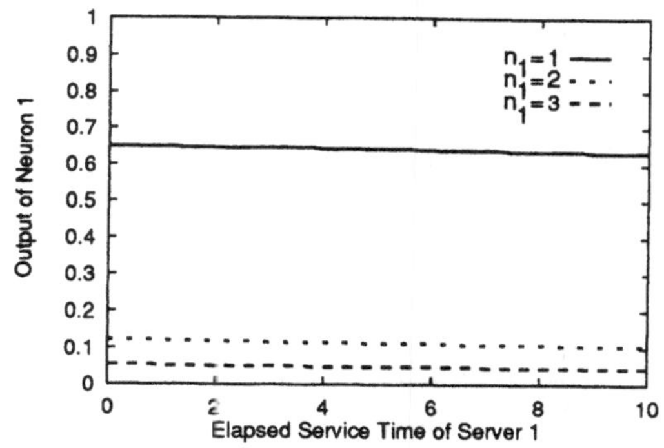

Figure 2: Effect of elapsed service time for BP and exponential service time.

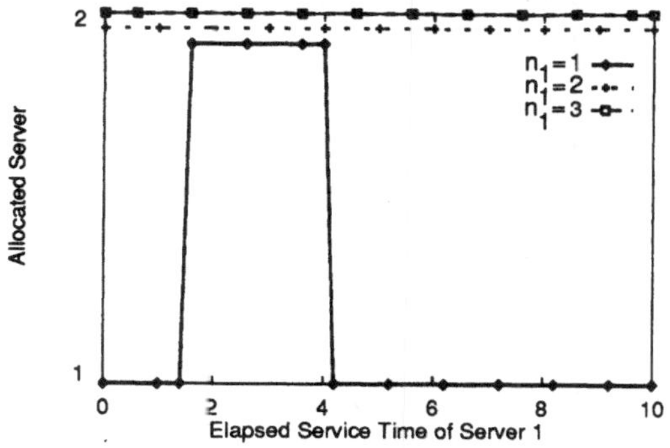

Figure 3: Effect of elapsed service time for LVQ3 and exponential service time.

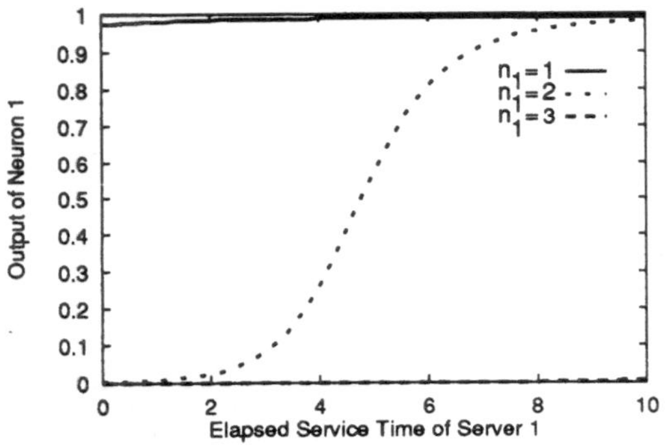

Figure 4: Effect of elapsed service time for BP and deterministic service time.

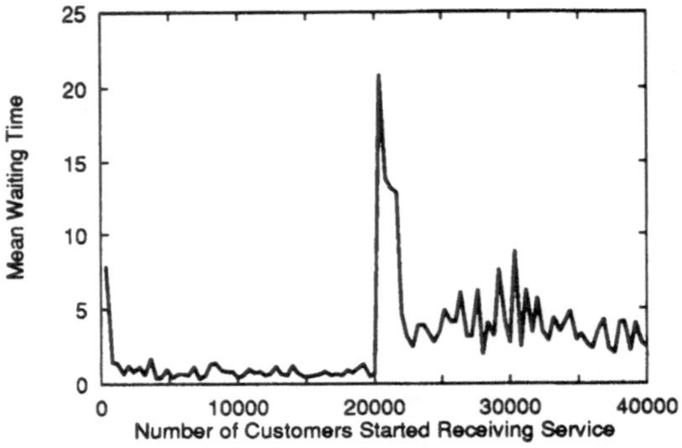

Figure 6: Adaptation of BP to changing situations.

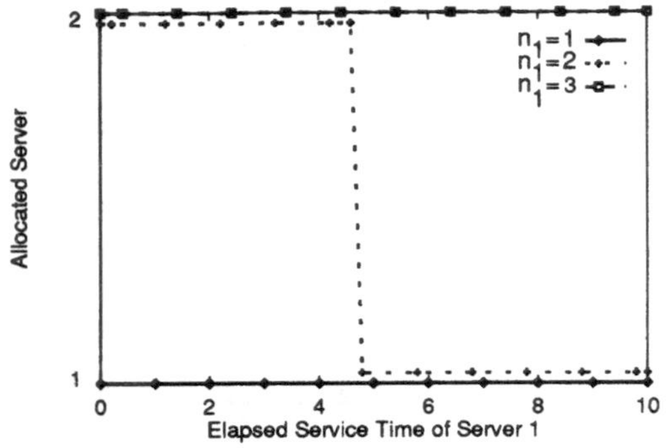

Figure 5: Effect of elapsed service time for LVQ3 and deterministic service time.

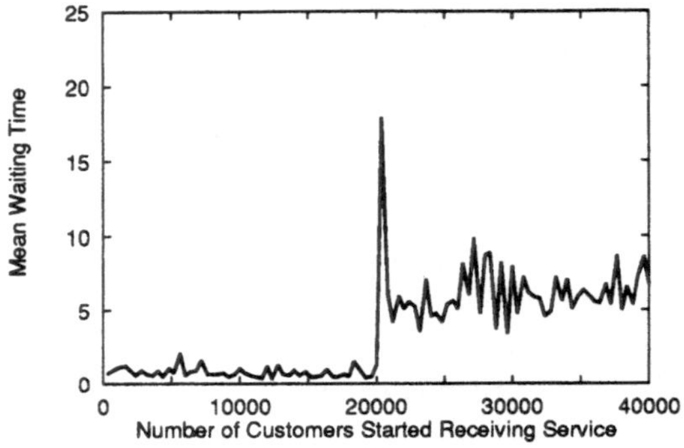

Figure 7: Adaptation of LVQ3 to changing situations.

Adaptation to Changing Situations

Finally, we examine the adaptation ability of the neural networks to changing traffic situations. In this experiment we assume that there are $N = 4$ servers in the system and the service time is deterministic. We begin with $\lambda = 2.3$ and the average service times $\{1/\mu_i\} = \{1.0, 1.25, 2.0, 2.5\}$. When 20,000 customers have started receiving service, the arrival rate λ is suddenly changed to 0.25, while the average service times are changed to $\{1/\mu_i\} = \{15.0, 15.0, 10.0, 10.0\}$.

The convergence of the mean waiting time after the sudden change of traffic situations is plotted in Figures 6 and 7, where the mean waiting time is calculated by the average of the preceding 400 customers. From Figures 6 and 7, we can see that the neural controllers adaptively learn the change of traffic situations.

5 Concluding Remarks

In this paper we applied the BP and LVQ3 neural networks to the dynamic allocation problem and compared their performance. Based on the information only about the arrival time and the service starting time of customers, we presented a simple training method for the neural networks to assign an arriving customer to the server who maximizes the probability of starting service for the customer in the earliest time.

From the numerical examples, it turned out that the dynamic allocation by the neural networks exhibits satisfactory performance in terms of the mean waiting time, the percentage of correct allocations, and the adaptation to changing situations. It was also shown that the neural networks can make correct use of the elapsed service time for the allocation decision. In comparison between the BP and the LVQ3, the BP showed overall better performance than the LVQ3, for which we sometimes experienced overflow at some queues when the arrival rate is high. The reason for the overflow is explained as follows. In the LVQ3, the codebook vectors are not updated in such cases as 1) x and m_j belong to the same class, while x and m_i belong to different classes, but x does not fall into the window, and 2) x belongs to a different class from m_i and m_j. Therefore, once all codebook vectors are located so that 1) or 2) occurs for most input vectors, the LVQ3 hardly learns correct allocations and some queues overflow. In a series of experiments, we observed that 2) happened for most input vectors before some queues overflowed. In this sense, the BP is more robust than the LVQ3 if the neural network is used for online control. Another disadvantage of the LVQ3 is that as pointed out in [3], it requires a proper initialization, which may be very difficult in some practical cases. Further improvement of the training method and application to the bulk-arrival case remain for future research.

References

[1] L. Kleinrock, *Queueing Systems, Vol. II: Computer Applications*, John Wiley & Sons, NY, 1976.

[2] T. Kohonen, G. Barna, and R. Chirsley, "Statistical Pattern Recognition with Neural Networks : Benchmarking Studies," *Proc. IEEE Int. Conf. Neural Networks*, vol. I (San Diego, CA), pp. I-61-68, July 24-27, 1988.

[3] T. Kohonen, "The Self-Organizing Map," *Proceedings of the IEEE*, vol. 78, no. 9, pp. 1464-1480, September 1990.

[4] K. R. Krishnan, "Markov Decision Algorithms for Dynamic Routing," *IEEE Communications Magazine*, pp. 66-69, October 1990.

[5] Z. P. Lo, Y. Yu, and B. Bavarian, "Derivation of Learning Vector Quantization Algorithms," *Proc. IJCNN'92*, vol. III (Baltimore, MD), pp. III-561-566, June 7-11, 1992.

[6] D. E. Rumelhart, J. L. McClelland, and the PDP Research Group, *Parallel Distributed Processing Volume 1: Foundations,* Cambridge, MIT Press, 1988.

[7] S. A. Solla, E. Levin, and M. Fleisher, "Accelerated Learning in Layerd Neural Networks," *Complex Systems 2*, pp. 625-640, 1988.

[8] A. C. Tsoi and R. A. Pearson, "Comparison of Three Classification Techniques, CART, C4.5 and Multi-Layer Perceptrons," *Neural Information Processing Systems 3*, pp. 963-969, Morgan Kaufmann Publishers, CA, 1991.

[9] J. Walrand, *An Introduction to Queueing Networks,* Prentice Hall, NJ, 1988.

[10] R. R. Weber, "On the Optimal Assignment of Customers to Parallel Servers," *J. Appl. Prob.*, vol. 15, pp. 406-413, 1978.

[11] W. Winston, "Optimality of the Shortest Line Discipline," *J. Appl. Prob.*, vol. 14, pp. 181-189, 1977.

Real-data-based Car-following with Adaptive Neural Control

Walter Weber, Jost Bernasch

Bavarian Research Center of Knowledge Based Systems (FORWISS)
Orleanstr. 34, D-8000 Munich 40, Germany
weber,bernasch@forwiss.tu-muenchen.de

February 25, 1993

Abstract

Our goal is to train a neural network to control the longitudinal dynamics of a real car. Input signals to the neural controller are its own *speed*, its own *acceleration*, *distance* and *change of distance* to the car ahead. Net output are control signals for *throttle* and *brake*. The acceleration of our own car and the distance to the car ahead are computed in discrete time-steps. Therefore all networks were trained with discrete-time data.

We performed experiments with several architectures including Jordan net, Elman net, a fully recurrent network and a recurrent network with a slightly different architecture (using 'normal' hidden units without recurrent connections as well) to analyze different behaviour and the efficiency of these neural network types. We used standard backpropagation to train the Jordan- and Elman-style networks and the backpropagation through time algorithm to train the recurrent nets.

Currently, we are working on a two-level predictor-hierarchy, where a higher-level network is intended to solve problems the lower-level net can not. This is achieved by restricting the high-level network's input to those data items in time in the case of which the low-level net had difficulties to produce a matching output.

The most promising results have been achieved using an algorithm which is a kind of backpropagation through time, in the case of which the time to look back is restricted to a specific amount of time steps. We call this backpropagation through truncated time.

Throughout our experiments we have been observing that the way of encoding data has an important influence on the performance of a network. Several runnings confirmed that this has a stronger impact than all of the well-known parameters (e.g. learning rate, momentum) of backpropagation. It could be shown that backpropagation through truncated time and our way of encoding data both lead to far better results than conventional approaches.

1 Introduction

Up to now in the development of intelligent cruise control a lot of aproaches have been made to control the longitudinal dynamics of a car in order to achieve an appropriate driver assistance which can not only maintain constant speed but also includes a strategy to keep the right distance. PID-controllers can do this quite well, but they have the disadvantages of:

1. being difficult to design. One needs a complete algorithmic description of the problem that is to be controlled.
2. being non-adaptive.

In the case the state of the system changes in the course of time, it would be of advantage to have an adaptive controller. The system to be controlled is a car, its state is suceptible to slight changes in the course of time. A new car will have other dynamics than the same car that was already driven for 50.000 km or even more. A controller that takes these changes into account and adapts itself in an appropriate way is therefore desirable. There is hardly any PID-controller the performance of which will not decline after a change in the system's state. Thus a new algorithmic description of the changed problem would have to be designed.

2 Problem Description

We made use of data which were produced by the simulation tool PELOPS. PELOPS is used at BMW in the PROMETHEUS project to examine different strategies for the control of the distance to a car ahead. The utterly accurate description of real cars used in this simulator makes the training of neural networks with quasi-real data possible. All nonlinearities that occur in the mapping from the position of the driving pedals to the resulting dynamics of a real car are accuratly reproduced in PELOPS [LDL92].

The following interface between the neural controller and the environment was provided. The net gets the following input:

- its own speed $v(t)$
- its own acceleration $a(t)$
- distance to the car ahead $d(t)$
- change of distance to the car ahead $\dot{d}(t)$

The network is supplied with control signals in discrete time steps every $2/10\ s$. From these input signals it produces control signals for throttle and brake. We combine these signals in one and call it $ped(t)$. A signal $ped(t) < 0$ defines the control signal for the brake, $ped(t) > 0$ refers to the desired throttle-position. If $ped(t) = 0$ neither throttle nor brake are activated.

We limited our attention to the longitudinal dynamics of a car, input like lateral acceleration or steering angle were not taken into account. After some experiments we could show that other possible input like relative speed to the car ahead or speed of pedal movements were not needed. On the contrary, these additional data had a negative influence on the training procedure. However less input than the four mentioned would not suffice to solve the problem.

The longitudinal dynamics of a real car are non-linear to a great extent:

1. The relation of rpm to engine output is non-linear. Thus the mapping from the position of the throttle to the power at the propeller shaft is non-linear.
2. Friction influences the braking efficiency. Thus the mapping from the

JORDAN	
epochs	error
5	17.75
10	17.15
20	11.88
30	6.01
50	2.86
100	1.85
500	0.26

Table 1: Total output error of a Jordan-style network.

ELMAN	
epochs	error
5	10.24
10	7.57
20	1.34
30	1.05
50	0.90
100	0.26
500	0.14

Table 2: Total output error for an Elman-style network.

position of the brake-pedal to resulting braking-power is non-linear.

3. There are several types of friction that influence the power transmission from the propeller shaft to the road.

This proves that the appropriate method of training a neural controller has to be a dynamic one. One needs cyclic or recurrent networks for dealing with time varying sequences of data.

3 Jordan-style Network

In our first experiments we used a Jordan-style network with recurrent links from the output units back to a recurrent layer that provides the hidden layer with additional input [Jor86]. Table 1 shows the output error of a Jordan-style network. We trained our networks on a trajectory with a length of $100s$. We were forced to choose such a short and relative simple training sequence. The Jordan- and Elman-style networks could not learn on more difficult trajectories. The error reported is the total error over the whole length of the training-sequence. We chose a learning rate of 0.2 and a momentum term of 0.3.

4 Elman-style Network

The next experiments were carried out on an Elman-style network [Elm88]. In this type of network the complete hidden layer is copied 1:1 to a so-called context layer at every time-step. The context layer is connected back to the hidden layer. Thus at every time-step the hidden layer not only receives new external input but can also look back to its own previous activation. Training results for that type of network are listed in Tab. 2. The Elman-network performed best on a learning rate of 0.2 and a momentum term of 0.3.

5 Recurrent Networks

Further work was done on different recurrent architectures which were trained by employing backpropagation through time (BTT). We used three-layered networks with one input, one output and one hidden layer. The hidden layer did not consist only of recurrent units, we also used 'normal' hidden units. Several runnnings using these different types of hidden layers confirmed that the additional use of hidden units increases performance. The effect is similar to connecting the input directly to the output, but it seems more convenient to let hid-

BTT	
epochs	error
5	0.88
10	0.49
20	0.27
30	0.15
50	0.06
100	0.02
500	0.01

Table 3: Total output error of a recurrent network that was trained with BTT.

den units find representations of a direct input-output mapping. The best configuration of learning rate and momentum was 0.5 and 0.9 respectively.

Experimentally, we could show that truncating the input history, — by using only a restricted amount of time-steps to perform backpropagation through time —, led to better results. The assigning of the credit of errors which occured late in time to actions which were taken many time-steps before might confuse the learning process.
Another advantage of our modification is that it allows training on a sequence the exact length of which is not known beforehand.

One could compare this method of putting a finite duration input history into the network with using a tabbed delay line on the input units. However, we use this 'time-window' for training dynamic recurrent networks with backpropagation through time. You can visualize this time-window as a sliding window that is pushed along the training-sequence. We considered a window length of 20 time-steps to be the best for our task.

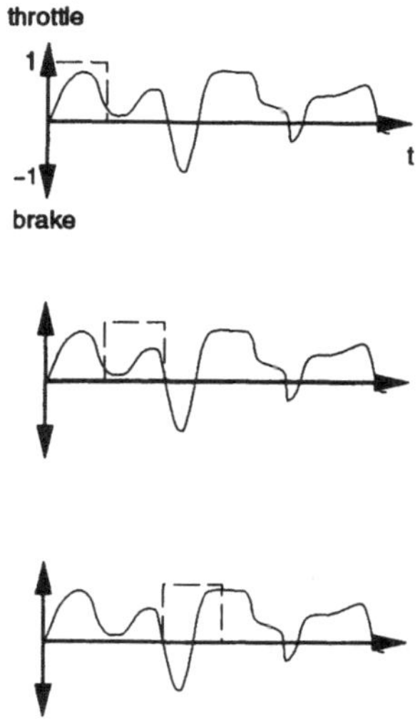

Figure 1: Pushing a sliding window along the training-sequence.

6 History Compression and Chunking Sequences

The underlying idea of the principle of history compression is to order a temporal sequence hierarchically [Sch91]. A lower-level network is trying to find the matching outputs and is passing its input to a higher-level network whenever it is not able to find a solution that is good enough. The higher-level network, called *chunker*, not only receives this external input but also the internal state of the lower-level net, the *automatizer*, at the time-steps in question. Thus the chunker is intended to recognize a higher-level temporal structure of a sequence. The automatizer tries to adapt the chunker's internal states as well as the targets. Thus the automatizer learns the meaningful internal states of the chunker. After some time of training the reduced description of the input sequence that is passed to the chunker will decrease more and more until the automatizer has learned to work on the whole training sequence successfully.

This method is based on 1. the automatizer's ability to learn the chunker's internal state and 2. on the chunker's ability to learn the automatizer's state. This however, also implies a problem. A representation of the state of a network (i.e. the activations of its units) is too complex to serve as an additional target for another network. All state-representations which we considered to be valid have been showing the tendency to rather disturb and slow down the learning process than support it.

7 Topological Encoding of Input-Data

A straightforward way of encoding the sensor data which serve as an input to our network is to normalize every data item in [0;1]. This ,however, only provides a rough encoding. A much better approach is to topologically encode numerical input values onto several input nodes [GB92]. This can be done either in a linear way or the activation of every input node is computed with the gaussian function. A simple example may illustrate this: The input range of the speed is between 0 and $50m/s$. We used 5 input nodes to encode the speed. The first node encodes a speed of $10m/s$, the next $20m/s$ and so on. If we use linear encoding and the current input value is $25m/s$, input nodes 2 and 3 will have an activation of 0.5 each. The other input nodes will not be activated at all. This encoding of input data provides several advantages:

- The numerical accuracy of the input values can be used efficiently. The accuracy is increasing exponentially with the number of nodes which are used to encode one input value.
- The network can extract more information from the input data if input values are encoded topologically. If one numerical value is encoded in the activations of several input units the item 'similarity' includes also 'numerical similarity'. This can be very important for the network's ability to generalize.
- Ranges of input data can be encoded with different granularity. Supposing you want to encode the speed-range between 20 and $30m/s$ with smaller steps than other ranges you only have to use 4 nodes to encode this specific range.

Figure 2 shows how a numerical value of $18m/s$ for the speed input is topologically encoded in the input nodes.

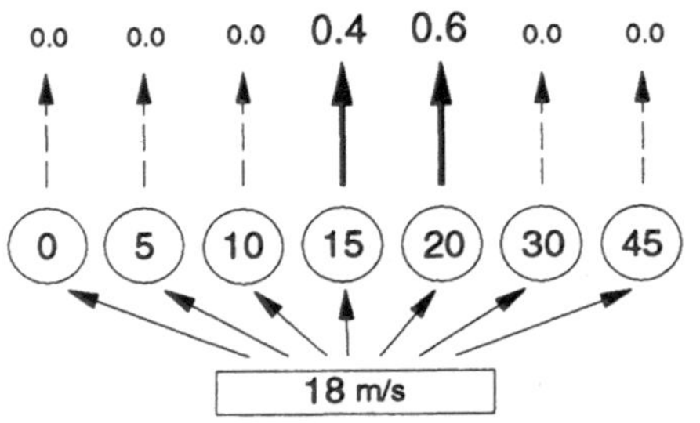

Figure 2: Topological encoding (linear) of the input value *speed*

In our experiments we used 9 units to encode a speed-range from 0 to $50m/s$, 12 units for a range of acceleration values from -10 to 5 m/s^2, 8 units for a distance from 0 to 120 m and 13 units to encode the change of distance that can occur in one maximum time step of $1s.$, -15 to 20 m. Tab. 4 shows the output error of a BTT-network that was trained with topological encoded data as well as

epochs	error	
	encoded	normalized
5	22.61	18.58
10	11.60	16.60
20	6.57	14.20
30	4.93	12.05
50	3.34	7.86
100	0.2	1.87

Table 4: Output error of an optimal configured BTT-net with normalizing or topological encoding input data.

with normalized input data. This experiment was done using a $300s$. duration input sequence.

On account of the increased number of weights as a result of the numerous connections, the encoded model gets off to a worse start. That is a drawback of topological encoding. Due to the many weights that are to be adapted, networks that use encoded data are slower.

8 Problems and Future Work

Future work has to be done on the 2-level chunking system. The internal state of each net that is an additional training target for the other one is too complex. We represent an internal state of a network with the activations of its hidden and output units. Of course this is the best representation of an internal state of a network but it is too complex to serve as an additional target for another network. However any other representation would not suffice.

One important topic of future interest is the use of on-line adapting neural controllers which are able to adapt different ways of driving starting from a stable initial position. One can reset the network at any time and thus effect robust control of the longitudinal dynamics of the car; additional learning will cause the network to act like the driver in question.

One would, however, need a 100% robust algorithm for the on-line training of recurrent networks to be successful in this respect. Besides, one would be faced with another problem if one wants to copy a 'human controller': Man does not act on the basis of exact sensor-information like a neural network. He is able to take key information that is not available for the network (e.g. sentiment of acceleration, peripheral vision) into account.

9 Summary

We showed that dynamic neural networks are able to learn the control signals for throttle and brake of a real car. Provided with the input vector $[v(t)\ a(t)\ d(t)\ \dot{d}(t)]^T$ at every time step t, recurrent networks can learn to produce the matching control signal $ped(t)$. We pointed out that a good selection of the network-topology is a very important supposition to succeed.

The encoding of input data — above all if it is taken from a physical plant — has also a great influence on learning-success. Our way of encoding numerical values topologically in several input units has proved better than conventional normalized input values.

10 Acknowledgements

Simulation data for this work was provided by the simulation tool PELOPS which is installed at the Forschungs- und Ingenieurszentrum of BMW in Munich. We would like to thank Michael K. Arras from the Bavarian Research Center for Knowledge Based Systems (Forwiss) in Erlangen to make his Forwiss Artificial Neural Network Simulation Toolbox available.

References

[Arr92] Michael Karl Arras. *Forwiss Artificial Neural Network Simulation Toolbox*, version 0.9.8 beta edition, December 1992.

[Bar90] Andrew G. Barto. Connectionist learning for control: An overview. In W. Thomas Miller III, Richard S. Sutton, and Paul J. Werbos, editors, *Neural Networks for Control*, chapter I, pages 5–59. MIT Press, 1990.

[Elm88] Jeffrey L. Elman. Finding structure in time. Technical Report CRL 8801, Center for Research in Language, University of California, San Diego, 1988.

[Fah88] Scott E. Fahlman. A empirical study of learning speed in back-propagation networks. Technical Report cmu-cs-88-162, 1988.

[GB92] Dr. Hans Geiger and Dr. Lilian Bungarten. Netz-werkzeugkasten. *Elektronik*, 2:87–91, 1992.

[GSR91] Aaron Gordon, John P. H. Steele, and Kathleen Rossmiller. Predicting trajectories using recurrent neural networks. In Cihan H. Dagli, Soundar R. T. Kumara, and Yung C. Shin, editors, *Intelligent Engineering Systems Trough Artificial Neural Networks*, pages 365–370. ASME Press, New York, 1991.

[Jor86] M. I. Jordan. Serial order: a parallel distributed processing approach. Technical Report ICS Report 8604, Institute for Cognitive Science, University of California, 1986.

[LDL92] Jens Ludmann, Rainer Diekamp, and Georg Lerner. Pelops - ein programmsystem zur untersuchung neuer laengsdynamikkonzepte im verkehrsfluss. In *VDI-Tagung: Berechnung im Automobilbau*, pages 191–207, 1992.

[Pea90] Barak A. Pearlmutter. Dynamic recurrent neural networks. Technical Report cmu-cs-88-196, School of Computer Science, Carnegie Mellon University, 1990.

[Sch91] Juergen Schmidhuber. Neural sequence chunkers. Technical Report FKI-148-91, Institut fuer Informatik, Technische Universitaet Muenchen, 1991.

[SSR92] Gary M. Scott, Jude W. Shavlik, and W. Harmon Ray. Refining pid controllers using neural networks. In John E. Moody, Steve J. Hanson, and Richard P. Lippmann, editors, *Advances in Neural Information Processing Systems 4*, pages 555–562. Morgan Kaufmann Publishers, San Mateo, 1992.

[Wil90] Ronald J. Williams. Adaptive state representation and estimation using recurrent connectionist networks. In W. Thomas Miller III, Richard S. Sutton, and Paul J. Werbos, editors, *Neural Networks for Control*, chapter I, pages 97–114. MIT Press, 1990.

[WZ88] R. J. Williams and D. Zipser. A learning algorithm for continually running fully recurrent networks. Technical Report ICS Report 8805, University of California, San Diego, La Jolla, 1988.

[WZ89] Ronald J. Williams and David Zipser. Experimental analysis of the real-time recurrent learning algorithm. *Connection Science*, 1:87–111, 1989.

An Adaptive Plan.

R. Garigliano, A. Purvis, P. A. Giles, D. J. Nettleton ‡

Artificial Intelligence Systems Research Group,
School of Engineering and Computer Science,
University of Durham, DH1 3LE, UK.

Abstract

This paper introduces a novel form of adaptive plan that is significantly different from current ones. The plan presented does not rely on the generation of expected numbers of solutions and as such can use a new means of sampling the solutions which are used as parents in the crossover process.

A mathematical analysis of some parts of the adaptive plan is presented followed by a discussion of some of the plan's main advantages. Finally, the successful use of the adaptive plan in a genetic algorithm applied to a problem in shape representation is discussed.

1 Introduction

Genetic algorithms have recently been successfully applied to a wide range of problems. Holland [1] identifies the following four components of such an algorithm: an environment of the system; a set of structures; an adaptive plan and a function for measuring the performance of each structure. The adaptive plan is crucial to the success of a genetic algorithm, since it is this that controls mutation and crossover, and hence the production of new (child) solutions from old (parent) ones. Many currently available adaptive plans have problems in that, to ensure a constant population size, there is often an arbitrary assignment of the number of times any particular solution can be used as a parent. In this paper a novel adaptive plan is presented which has several advantages over current ones.

Section 2 begins by introducing some of the notation and terms connected with genetic algorithms that are relevant to the rest of this paper. Adaptive plans are then discussed and an example given together with a discussion of some of its drawbacks. Section 3 introduces a novel adaptive plan together with a mathematical analysis of some of its properties. A discussion of some of the merits and drawbacks of this new adaptive plan is then given. In section 4 a discussion of the implementation of the adaptive plan to a problem in shape representation is given and section 5 provides a conclusion to the paper.

‡e-mail: D.J.Nettleton@durham.ac.uk or Roberto.Garigliano@durham.ac.uk

2 Preliminaries

Holland [1] identifies the following four components as requirements of an adaptive system:

1. an environment of the system, E;
2. a set of structures, $\mathcal{K}$;
3. an adaptive plan, τ, which modifies the system structures;
4. a measure, μ, of the performance of each structure.

The purpose of an adaptive system is to update iteratively a subset of structures, $K \subseteq \mathcal{K}$, based on the information it receives from its environment, so that the average performance of individual structures $k \in K$ improves. In general, the form of the performance measure will depend on the environment, and so it would be usual to adopt Holland's notation of writing $\mu_E(k)$ to represent the performance of a structure in the environment, $E \in \mathcal{E}$, where $\mathcal{E}$ is the set of all possible environments. However, this notation becomes cumbersome when describing the n'th structure in a set at a time t, and since normally the environment is fixed, explicit reference to it shall be missed out. μ_{nt} shall therefore be written as shorthand for $\mu_E(k_n(t))$, but it is worth bearing in mind the proper meaning.

2.1 Schemata

In the analysis of Holland, solutions are represented as binary strings. Since then several other, usually problem specific, solution representations have been used. However, binary strings are often still used in examining the theory of genetic algorithms, and since any set of solutions which can be represented by a string of numbers can obviously be represented as a binary string they are still widely used in solution representation.

Borrowing the terminology from genetics, the use of a binary representation means that each point (or locus) on the string can be occupied by one of only two 'alleles'. This simply means that there can be either a '1' or '0' at each point of the string. A section of the string can then be represented as follows:

$$\cdots - 1 - 0 - 0 - 1 - 1 - \star - \star - \star - \star - 0 - 1 - \cdots$$

where the $\star$ stands for a locus at which the allele value is of no importance to the current discussion. Holland gives the following definitions:

Definition A schema is an n-tuple of defining positions along a binary string.

That is to say each solution string can be viewed as a compound entity consisting of a combination of different loci groupings. The groupings are allowed to overlap and a single locus is permitted to be a member of more than one distinct schema. Thus a string of length l loci contains 2^l different possible schemata. A specific association of allele values to the defining positions of a schema is called an instance or realisation of that schema.

Definition Let a schema, ϵ, have n defining positions $i_1, i_2, \ldots, i_n$ along a binary string. The length of the schema is defined to be:

$$l(\epsilon) = i_n - i_1$$

Hence the example schema has a length of ten units.

Schemata are treated as the random variables in a population and as the real entities being evaluated when the fitness of a solution is calculated. An analysis of the change in the relative proportions of schema instances leads to an explanation of the power of genetic algorithms.

2.2 Adaptive Plans

Baker [2] considers the adaptive plan to be comprised of two distinct parts:

1. determination of the individual's expected values;
2. conversion of the expected values to discrete numbers of offspring.

The algorithm used to convert the real expected values to integer numbers of offspring is called the sampling algorithm and must ensure a constant population size, whilst providing an accurate, consistent and efficient means of sampling.

The following technique is the basis for the most commonly used sampling algorithms, and is repeated until the required number of samples have been selected.

- Calculate **R**, the sum of all the expected values of the individual solutions of a generation,
- 1-1 map the individual solutions to a section of the real line [0,**R**) such that each individuals section is equal to its expected value,
- generate a random number in the range [0,**R**),
- select the individual whose section of the real line spans the random number.

This is commonly referred to as the "spinning wheel" method, since it can easily be compared to a wheel in which each slice is proportional to its expected value.

Clearly the means by which the expected values are determined is very important since this directly affects the number of times a solution can be expected to be selected for use as a parent solution. In determining the expected values it is important to ensure that solutions with a high fitness level are more likely to be selected than those with a low fitness level.

One popular means of determining the expected value of a solution $k_n(t)$ is to use $\mu_{nt}/\bar{\mu}_t$ where μ_{nt} is the fitness of the n'th solution in the t'th generation and $\bar{\mu}_t$ is the mean fitness of all the solutions in the t'th generation. There are several disadvantages with this. For example, if in an early generation a solution emerged that was much fitter than the other solutions in that generation, then that solution would be used in the production of a large number of child solutions. However, if that solution was far from being globally optimum then convergence to a local optimum is likely — this is commonly called exploitation.

There are many other possible ways to determine expected values and of sampling them so that the population size remains constant from generation to generation. In the next section a new approach to the selection of solutions is presented which does not rely on calculating expected values and trying to sample them as effectively as possible; as such it is fundamentally different to other adaptive plans.

3 Plan P1

The crossover operator combines the material of two parent solutions to form a child solution. Since the amount of material donated by each of the parents is chosen at random (the crossover point), it is equally probable that the majority of material in the child solution is contributed by the less fit parent. Hence, the only consistent way of classifying the child is as the offspring of both parents.

Each parent can therefore be considered as having one of two shares in each child it is involved in producing, and so the child generation contains a total of $2N$ shares where N is the population size. Clearly now each solution in a generation can be allocated at least one share in the next generation, whilst there being adequate capacity for allocating extra shares on the basis of relative fitness. Further, this allocation can be achieved in a natural way by use of the adaptive plan P1 :

Adaptive Plan P1

1. Set $t = 0$ and randomly select N possible solutions.

2. For each solution, $k_n(t)$, select at random a set of r solutions without replacement, $K_r(t)$, from the current population, $K(t)$, where $1 < r \leq N$.

3. Select the fittest solution, $k^*(t) \in K_r(t)$.

4. Apply the crossover operator to $k_n(t)$ and $k^*(t)$, and store the result in the next generation as $k_n(t+1)$.

5. Apply the mutation operator to $k_n(t+1)$ with probability P_m.

6. Increment t by one.

7. Repeat steps two to six until $t = T$, where T is the number of generations to be run.

3.1 Analysis

As stated earlier an analysis of the change in the relative proportions of schema instances leads to an explanation of the success of genetic algorithms. Since the adaptive plan controls the selection and recombination of solutions it is vital to the success of a genetic algorithm. This section examines several mathematical properties of the adaptive plan P1.

To analyse the performance of P1 the following definition is needed:

Definition 1 — Let the number of solutions in a population at a time t with fitness value between μ and $\mu + \delta\mu$ be $\mathcal{N}_t(\mu)$. Then define:

$$\nu_{nt} = \frac{1}{N} \int_0^{\mu_{nt}} \mathcal{N}_t(\mu) d\mu$$

where N is the population size. (It has been assumed that the μ_{nt} are positive, real numbers, but in general the lower limit of the integration is the lower limit of the range of fitness values.)

From its definition ν_{nt} is clearly the probability that solution $k_n(t)$ is at least as fit as any other single solution chosen at random from the current population.

The following Lemma concerns the number of shares each solution can expect to contribute to the next generation.

Lemma 1 — For an adaptive plan of type P1, the expected number of population shares allocated to each solution is bounded by:

$$N(k_n(t), r) = 1 + r(\nu_{nt})^{r-1}.$$

Proof — The probability of a given solution $k_n(t)$ not being selected at all in a random sample of r solutions chosen without replacement from a population of size N is:

$$\frac{N-1}{N}\frac{N-2}{N-1}\cdots\frac{N-r}{N-(r-1)} = \frac{N-r}{N}$$

and so the probability of it being chosen is $1-(N-r)/N = r/N$. There are then $r-1$ other distinct solutions in the random sample and so the probability of $k_n(t)$ being fittest is $(\nu_{nt})^{r-1}$. Therefore, the expected number of shares allocated to $k_n(t)$ by random selection without replacement over N trials is:

$$\begin{aligned} N(k_n(t), r) &= 1 + \frac{r}{N} N(\nu_{nt})^{r-1} \\ &= 1 + r(\nu_{nt})^{r-1}. \end{aligned}$$

□

Definition 2 — $\langle(\nu_{\epsilon t})^{r-1}\rangle$ is the average value of $(\nu_{nt})^{r-1}$ for all $k_n(t)$ that are instances of ϵ at time t.

Theorem 1 — Let $P(\epsilon, t)$ be the probability of a given solution $k_n(t)$ being an instance of ϵ at time t. For an adaptive plan of type P1 using both crossover and mutation operators, the expected change in $P(\epsilon, t)$ over one generation is bounded by:

$$P(\epsilon, t+1) \geq (1 - \tfrac{[1-P(\epsilon,t)]l(\epsilon)}{l-1}) P(\epsilon, t) r\langle(\nu_{\epsilon t})^{r-1}\rangle (1-P_m)^L$$

where P_m is the probability of a mutation occurring at any given locus and L is the number of defining positions of the schema ϵ.

Proof — In addition to the one guaranteed share in the next generation, each solution is expected to be allocated $r(\nu_{nt})^{r-1}$ more — Lemma 1. Let $B_\epsilon(t)$ be the set of individuals that are instances of the schema ϵ. The total number of offspring expected for the set $B_\epsilon(t)$ is:

$$\begin{aligned} N_\epsilon(t+1) &= \sum_{\{n: k_n(t) \in B_\epsilon(t)\}} r(\nu_{nt})^{r-1} \\ &= N_\epsilon(t) r\langle(\nu_{\epsilon t})^{r-1}\rangle \end{aligned}$$

where $N_\epsilon(t)$ is the number of instances of schema ϵ at time t. Since the probability of crossover selection in P1 is 1 and if $l(\epsilon)$ is the length of ϵ, then a proportion $l(\epsilon)/(l-1)$ of the schema offspring will have a crossover falling within its defining positions i.e. the schema is split. When an instance of ϵ is crossed with another instance of ϵ the result is also an instance of ϵ, otherwise the result may or may not be an instance of ϵ. Let $P(\epsilon, t)$ be the probability of a given solution $k_n(t)$ being an instance of schema ϵ at time t. If, having already selected an instance of ϵ, the probability of crossing it with another instance of ϵ is just $P(\epsilon, t)$, so no more than a proportion $\frac{[1-P(\epsilon,t)]l(\epsilon)}{l-1}$ of the modified offspring of ϵ can be expected to be instances of schemata other than ϵ, the remainder $\left(1 - \frac{[1-P(\epsilon,t)]l(\epsilon)}{l-1}\right)$ will be instances of ϵ. Therefore if N is the number of solutions in the population and $N'_\epsilon(t+1)$ is the number of instances of ϵ that survive into the next generation:

$$\begin{aligned} P(\epsilon, t+1) &= \frac{N'_\epsilon(t+1)}{N} \\ &\geq \left(1 - \frac{[1-P(\epsilon,t)]l(\epsilon)}{l-1}\right) \frac{N_\epsilon(t+1)}{N} \\ &= \left(1 - \frac{[1-P(\epsilon,t)]l(\epsilon)}{l-1}\right) \frac{N_\epsilon(t)}{N} r\langle(\nu_{\epsilon t})^{r-1}\rangle \\ &= \left(1 - \frac{[1-P(\epsilon,t)]l(\epsilon)}{l-1}\right) P(\epsilon, t) r\langle(\nu_{\epsilon t})^{r-1}\rangle \end{aligned}$$

If the probability of one of the defining loci being mutated is P_m, then the probability of each defining position being unchanged is $(1-P_m)$, and the probability of all L loci being unchanged is $(1-P_m)^L$. Then:

$$P(\epsilon, t+1) \geq (1 - \tfrac{[1-P(\epsilon,t)]l(\epsilon)}{l-1}) P(\epsilon, t) r\langle(\nu_{\epsilon t})^{r-1}\rangle (1-P_m)^L$$

□

Corollary 1 — The condition necessary for a schema to increase its probability of occurrence from one generation to the next under the adaptive plan P1 is:

$$r\langle(\nu_{\epsilon t})^{r-1}\rangle \geq \left(1 + \frac{l(\epsilon)}{l-1}\right).$$

Proof — From Theorem 1 instances of schema will increase so long as the factor of $P(\epsilon, t)$ is greater than 1 i.e.

$$\left(1 - \frac{[1-P(\epsilon,t)]l(\epsilon)}{l-1}\right) r\langle(\nu_{\epsilon t})^{r-1}\rangle (1-P_m)^L \geq 1$$

$$r\langle(\nu_{\epsilon t})^{r-1}\rangle(1-P_m)^L \geq \frac{1}{\left(1-\frac{[1-P(\epsilon,t)]l(\epsilon)}{l-1}\right)}$$

Taking $c = \frac{[1-P(\epsilon,t)]l(\epsilon)}{l-1}$ then since c is the product of two probabilities $c \leq 1$. Also, $1/(1-c) \geq (1+c)$ since $(1-c)(1+c) = (1-c)^2 \leq 1$ and so:

$$r\langle(\nu_{\epsilon t})^{r-1}\rangle(1-P_m)^L \geq \left(1+\frac{[1-P(\epsilon,t)]l(\epsilon)}{l-1}\right)$$

$P(\epsilon,t)$ will be very small for any given ϵ since the total number of schemata is in general very large, and assuming that P_m is small enough such that $(1-P_m)^L \approx 1$. Then, to ensure an increase it is required that:

$$r\langle(\nu_{\epsilon t})^{r-1}\rangle \geq \left(1+\frac{l(\epsilon)}{l-1}\right). \quad \square$$

Theorem 2 — In a population of size N where for each solution, $k_n(t)$, a random set of r solutions, $K_r(t)$ is selected, the expected number of solutions in one generation that will need to be evaluated using the evaluation function is:

$$E(N,r) = N\left(1-(\tfrac{N-r}{N})^N\right).$$

Proof — The probability of a given solution $k_n(t)$ not being selected at all in a random sample of r solutions chosen without replacement from a population of size N is, $\frac{N-r}{N}$. Since in one generation there are N selections of $K_r(t)$ the probability of a $k_n(t)$ not being chosen to be in any of them is $(\frac{N-r}{N})^N$. The probability of a solution being selected to be in at least one of the N, $K_r(t)$ is, therefore, $1-(\frac{N-r}{N})^N$. Hence the expected number of solutions that will be selected in a generation of size N (and so require evaluating) is $N\left(1-(\frac{N-r}{N})^N\right)$. $\square$

3.2 Discussion

Some of the advantages of the adaptive plan P1 are:

1. There is no need to make an arbitrary choice of the exact numbers of offspring to allocate to each solution since this is handled automatically.
2. Each solution is guaranteed at least one share in the next generation regardless of how low its fitness value may be — this ensures that no solution is completely neglected during a single generation.
3. Each solution receives a total number of shares in the next generation dependent on its relative fitness value ensuring that, as in other adaptive plans, the better solutions donate more genetic material than do poorer ones.
4. By choosing r to be small the advantage of good solutions over poor ones can be reduced so encouraging exploration over exploitation and lessening the risk of premature convergence. For example, with $r=2$ the expected number of shares for the solution $k_n(t)$ is:

   $$1+2\nu_{nt}$$

   Clearly no matter how much better a particular solution is than the rest it can have $\nu_{nt}=1$ at best and so contribute on average only three shares to the next generation. An adaptive plan in which offspring numbers are allocated proportional to $\mu_{nt}/\bar{\mu}_t$ could have the best structure contributing nearly all of the offspring.
5. Only the fitnesses of the solutions selected ($K_r(t)$) need to be evaluated. If the fitness evaluation function is computationally expensive any reduction in the number of times the function is used is advantageous.

   The following table shows the expected number of calls to the fitness evaluation function for various values of r and with $N=100$ (Theorem 2):

N=100	r		
	2	3	4
E(N,r)	86.738	95.245	98.313

6. Since it is the best solution from the random subset of the whole population that is required, it may not always be necessary to completely evaluate the fitness of each solution. Instead it may (depending on the problem) be possible to determine which is the best solution in a given set by using approximate or partial evaluation of the solutions.

Of course there are also certain disadvantages associated with an adaptive plan of type P1, namely:

1. The use of very small values of r makes it possible that a good solution could be allocated fewer shares than it warrants or that it could be allocated just its single share. This is necessary however if the effects of exploitation are to be reduced.
2. Since all trial solutions may not be evaluated the best-so-far performance may be degraded. This is especially likely to occur with large populations and small sampling sizes.

4 Implementation

In implementing the adaptive plan P1 the only variable that needs to be set is that of the random sample size r. Lemma 1 gives the number of shares in the next generation that can be expected for each solution under such a plan and this is used in helping to determine r for practical purposes.

It is important to consider the number of shares that the fittest solution can expect to receive in the next generation as this has relevance in determining whether or not premature convergence to a suboptimal may occur. If $\hat{k}$ is the fittest solution in a generation then it's corresponding value of ν_{nt} will be 1 and so from Lemma 1:

$$N(\hat{k}, r) = 1 + r$$

This clearly demonstrates that as the sample size is increased the best solution can expect to receive a greater proportion of the available shares. In the limit it is allocated them all and the other solutions receive only their one share. The limit case corresponds to the fittest solution always being chosen as the mate and will clearly result in that solutions descendants dominating the population. With $r = 2$, however, the average number of extra shares that the best solution can expect to receive is two, hence restricting its ability to become dominant too quickly. This is just the effect required to prevent premature convergence. By combining such a small sample with the ability to use approximate evaluation, and only evaluating solutions as and when necessary, the implementation of P1 is made highly efficient.

The decision to take $r = 2$ is further supported by noticing that Corollary 1 now simplifies to

$$2\langle \nu_{\epsilon t} \rangle \geq 1 + \frac{l(\epsilon)}{l - 1}$$

The interpretation of this is that for even the shortest schemata to increase their population share, it is necessary that, on average, solutions containing their instances be more likely to be chosen in preference to any other randomly selected solution. It should be noted at this point that although setting $r = 2$ gives a theoretical minimum advantage to above average solutions, it is possible that this goes too far and results in adverse effects on the algorithms performance. An extensive investigation of the effects of sample size is beyond the scope of this paper. However, a genetic algorithm with $r = 2$ has been successfully implemented in an attempt to solve a problem in shape representation [3] [4] — this is now discussed.

4.1 P1 and Shape Representation

The problem to be discussed is that of two dimensional shape representation. An input shape is to be represented in terms of an iterated function system (a set of contraction mappings), which itself can be represented as a set of numbers. The theory of iterated function systems (IFSs) has been well developed by Barnsley [5] [6] [7] and only a brief summary is given here — a more detailed discussion being beyond the scope of this paper [3] [4].

The IFS of a shape is a collage of smaller, continuously altered copies (affine transformations) of that shape exactly covering the original [6]. This means of representation is particularly advantageous since the primitive shapes (those used to build up the original) will automatically have the correct morphology and there is no arbitrary assignment of primitives for the shape to be encoded.

The space of IFS encodings in which the shape representations exist is very complex, with many suboptimal and optimal solutions. The remainder of this section discusses the application of a genetic algorithm (GA), using the adaptive plan P1, to search this space.

The environment of the system is an image file created by a simple image processing program applied to the shape to be encoded and contains information such as pixel position. Since the environment is time independent, the information is invariant and need only be read in once at the beginning of a run of the GA.

Since an IFS can be represented as a set of numbers, the structures used in the GA are binary strings. The adaptive plan used is P1 and the crossover operator is a two point crossover. The mutation operator simply changes the value at a randomly chosen point of the binary string.

The evaluation function must in some way determine how alike two shapes are. There are many possible such measures available each with their own advantages/disadvantages — a combination of point coverage and the Hausdorff distance is used. The cost of evaluating solutions is by far the most computationally intensive part of the GA. However, in using the adaptive plan P1 with $r = 2$ it is not necessary to evaluate all solutions and since, only the best solution from a set need be determined, it is possible to use approximate evaluation techniques in order to speed up the GA runtime.

The GA applied as discussed above, with $r = 2$, was run with a population of 100 individuals over 100 generations. The crossover operator was allowed to operate over the full length of solutions and the probability of mutation was kept low at 0.01. Although successfully encoding test shapes the runtime was in some cases several hours — this being due to the evaluation function being computationally expensive. However, the implementation does show that a genetic algorithm using P1 can be successful.

5 Conclusion

In this paper a novel form of adaptive plan P1 has been introduced. Unlike many other adaptive plans P1 does not rely on the generation of expected numbers of offspring and as such a new means of sampling parent solutions can be used. Some theoretical aspects of P1 have been discussed, together with the plan's main advantages. The selection of the sampling constant r for the purpose of a practical implementation was discussed. Finally the successful implementation of P1 to a problem in shape representation was presented.

References

[1] Holland J H, '*Adaption in Natural and Artificial Systems*', University of Michigan Press, 1975.

[2] Baker J E, '*Reducing Bias and Inefficiency in the Selection Algorithm*', Genetic Algorithms and their Applications: Proc. Second Int. Conf. Genetic Algorithms, ed J.J.Grefenstette, Lawrence Erlbaum Associates, 14–21, 1987.

[3] Giles P A, '*Iterated Function Systems and Shape Representation*', PhD Thesis, University of Durham, 1990.

[4] Garigliano R, Purvis A, Giles P A and Nettleton D J, '*Genetic Algorithms and Shape Representation*', Proc. Second Ann. Conf. Evolutionary Programming, San Diego, USA, February 1993.

[5] Barnsley M F and Demko S, '*Iterated Function Systems and the Global Construction of Fractals*', Proc. Royal Society London A399, pp 243–275, 1985.

[6] Barnsley M F, Ervin V, Hardin D and Lancaster J, '*Solution of an Inverse Problem for Fractals and other Sets*', Proc. Nat. Acad. Science USA, vol 83, pp 1975–1977, Apr. 1986.

[7] Barnsley M F, '*Fractals Everywhere*', Academic Press, 1988.

[8] Booker L, '*Improving Search in Genetic Algorithms*', in Genetic Algorithms and Simulated Annealing, ed Davis L., Pitman 1987.

[9] Davis L and Steenstrup M, '*Genetic Algorithms and Simulated Annealing: An Overview*', in Genetic Algorithms and Simulated Annealing, ed Davis L., Pitman, 1987.

MAPPING PARALLEL GENETIC ALGORITHMS ON WK-RECURSIVE TOPOLOGIES

I. De Falco, R. Del Balio, E. Tarantino, R. Vaccaro

Istituto per la Ricerca sui Sistemi Informatici Paralleli (IRSIP)-CNR
Via P.Castellino, 111, 80131 Naples (ITALY)
Email address: ivan@irsip.na.cnr.it

Abstract

In this paper a parallel simulator of Genetic Algorithms is described. The target machine is a parallel distributed-memory system whose processors have been configured in a WK-Recursive topology. A diffusion mechanism of useful local information among processors has been carried out. Specifically, simulations of genetic processes have been conducted using the Travelling Salesman Problem as an artificial environment. The experimental results are presented and discussed. Furthermore, performance with respect to well-known problems taken from literature is shown.

1. INTRODUCTION

Nowadays, the process known as adaptation is becoming important in Artificial Intelligence (AI). In Robotics, for instance, it is not possible to program a robot which is able to handle every situation it will encounter. It is clear that it has to adapt itself to the changes of the environment it lives in. More generally, many applications in Machine Learning can be viewed as adaptation problems. While these problems are not too simple to solve for the AI researchers, the adaptation is successfully faced by natural evolution. Current research is demonstrating not only its applicability to a wide variety of problems, including image processing, stock market forecasting, neural networks and large data-base searching, but also that its search efficiency may overcome that provided by more traditional approaches [1 - 4]. This formulation of problems whose solution is deduced by the mechanisms of natural and biological evolution has inspired the class of algorithms known as Genetic Algorithms (GAs) introduced by Holland [5].

With the availability of parallel computers, the remarkable power of this novel search technique has shown up and, consequently, several attempts to design and implement parallel versions of GAs have been made [6, 7, 8].

The paper is organized as follows: the next section provides a brief review of GAs and parallel approaches to them. In the third, a description of the Travelling Salesman Problem (TSP) and its coding are reported, while, in the fourth, experimental results for this problem are presented and discussed. The last section is dedicated to the final observations and to a brief outline of future works.

2. GENETIC ALGORITHMS

2.1 Traditional Genetic Algorithms: a review

GAs are a robust adaptive optimization technique which allows an efficient probabilistic search in high dimensional space [9]. In order to apply genetic evolutionary concepts to a specific problem, two issues are to be faced: a) the encoding of potential solutions and b) the objective function (*fitness function*) to be optimized that plays the role of the environment. The genetic representation is just a vector of components referred to as *chromosome*. Each element of the vector is called *gene*.

An initial population of chromosomes is processed by means of various biologically inspired operations reproducing, and simultaneously altering, their genetic composition. The most important operations involved are random mutation and/or inversion of data in individual chromosomes, crossover and recombination of genetic material contained in chromosomes of different parents. Optimization is made by (a) selecting pairs of individuals with a probability proportional to their fitness and (b) mating them to create new offspring. The replacement of bad individuals with the new ones is effected on the basis of some fixed strategies. The processes of reproduction, evaluation of fitness and replacement of individuals are repeated until a fixed stopping condition becomes satisfied.

2.2 Parallel approaches to Genetic Algorithms

The classical GAs are sequential algorithms which pose some problems since their execution may require many generations and a large number of individuals in the population. Time and memory limitations make these algorithms ineffective. The availability of parallel machines and, thus, the introduction of Parallel Genetic Algorithms (PGAs) [6, 7, 8], at the end of the eighties, have been very useful to overcome these limitations and to efficiently exploit the capability of this optimization technique.

There are two possible alternative parallel approaches to be considered: the first is the parallelization of the control structure of the GA itself, namely the selection process and the genetic operators, that leads to a fine-grained PGA [6]; the second consists of a number of classical GAs running in parallel and interacting from time to time by exchanging individuals. This is a coarse-grained PGA [7, 8]. Several versions of the latter differ in the spatial structure proposed for the subpopulations and in the linkage of the runs. With reference to the first point, three different mathematical models for spatial population structure have been defined: the *island model*, the *stepping stone model* and the *isolating by distance model*; details on them may be found in [10]. As regards the linkage of the runs, for example Tanese introduces two parameters: the *migration interval* representing the number of generations between one migration and the next, and the *migration rate* denoting the percentage of individuals to be taken into account for each migration [7].

The approach we follow is the coarse-grained one below described: the PGA consists of a set of classical GAs, each assigned to a different processing element, running in parallel on different subpopulations. The model chosen for the spatial structure of the set of subpopulations is the stepping stone in which an exchange of useful information takes place among neighbours only. More specifically, we carry out a *diffusion process* in which only the current local best values are exchanged among the neighbouring subpopulations. Therefore, differently from other strategies [7], the number of individuals sent is fixed to one. The number of generations between successive diffusions is called *diffusion interval*. In the approach proposed, each GA has both a cooperative and a competitive nature: the former is represented by the reproduction and diffusion processes while the latter is exemplified by those of the selection and replacement. The main idea of this "cooperative-competitive" approach for searching in a large configuration space is to create a dynamic system that develops a global optimization from local optimized searches. This seems sound as it has been proved that N linked searches perform better than N independent ones [8].

Schematically, each GA has to:
1) generate randomly an initial subpopulation;
2) evaluate the fitness function for each individual;
3) communicate the current local optimum to a collector process;
4) select parents and generate new offspring by making use of the genetic operators chosen;
5) evaluate the fitness for the descendants and, if better, replace the parents with the new offspring;
6) communicate with the neighbouring processing elements;
7) select among the local best and the received ones and replace bad individuals in the subpopulation;
8) go to 3 to perform the next epoch until a prefixed stopping condition becomes true.

It should be noted that the step 3 is not strictly necessary, but its only function consists in providing the user with an interactive control of the evolution process.

The PGA uses different smaller subpopulations of "active" individuals (with reference to their possible diffusion) against the large population of "passive" individuals typical of the classical GA. During the first communication phase, each process sends its current local result to a collector process which finds and saves the best among all the optima received. Then, each GA operates independently improving the average subpopulation fitness during its lifetime by means of the steps 4 and 5. With the second communication phase, an exchange of current local optima among neighbouring subpopulations is performed, so as not to have simply independent searches but, more appropriately, linked searches in parallel. Then, a selection among the local best and the received ones takes place and the winner becomes the new local optimum. However, the other data replace, if better, bad local elements randomly selected so that the subpopulation size remains unchanged. A significant advantage of the exploration of many search subspaces in parallel with information exchange among neighbouring subpopulations is the introduction of new diversities so that the possibility of a premature convergence to a suboptimal solution is reduced [10].

3. THE TRAVELLING SALESMAN PROBLEM AS COMPUTATIONAL ENVIRONMENT

Several strategies, such as the classical Simulated Annealing, have been proposed to solve optimization problems [11, 12], but they generally lead to unsatisfactory solutions, so a wide variety of problems has been solved with GAs. Among these there is the Travelling Salesman

Problem (TSP). The TSP is one of the most typical test beds for combinatorial optimization problems. Given n cities, the task is to find a tour sequence with a minimum "cost" starting from a city, visiting each other city once and only once and returning to the starting point. The final aim is the minimization of the objective or cost function here represented by the euclidean length of the tour. The problem requires compiling a list of $(n-1)!/2$ alternative tours, a number that grows faster than any finite power of n. Generally, it is possible to find the exact shortest path only for small-sized problems since the TSP lies in the class of NP-complete problems. In many applications it is often desirable to reach suboptimal solutions in a reasonable time, so that the quality of a solution has to be estimated not only on the basis of the value found, but also on the time spent to achieve it.

The first efforts to find solutions to the TSP by using genetic strategies are those reported in [13, 14]. The operators they used did not always guarantee the admissibility of the offspring tours: this means that tours undergoing crossover might produce illegal offspring tours which should be corrected. Goldberg and Lingle overcame this problem suggesting a new type of crossover operator, the Partially Mapped Crossover (PMX) [15].

In many applications, for reasons of simplicity of analysis, elegance of available operators (in terms of their implementation easeness and of their resemblance to biological mechanisms) and requirements for speed, the chromosomes are simply strings of 0s and 1s.
Vice versa, in this work, each individual is designed to be specifically adapted to the features of the computational environment. Therefore, a tour (chromosome) is represented by a string, namely by an array of n distinct elements of the set $\{1, 2, ..., n\}$, where the order indicates the sequence in which the cities are visited. As a matter of fact, this complicates the genetic operators but makes the solutions more comprehensible. In our algorithm, three standard genetic operators are employed to manipulate the individuals in order to generate a better tour: the mutation (MUT) which, for the TSP, consists of a random swap between two cities; the inversion (INV) in which, chosen randomly two points, the section between them is reinserted in the reverse order and, at last, the PMX crossover (PMX) which uses two crossover points (the section between these points defines an interchange mapping) [16]. Thus, these components may be viewed as performing a pseudo-darwinian evolution of the trial solutions in which the genetic material of the individuals survives through mating and reproduction.

4. EXPERIMENTAL RESULTS

Since our PGA consists of a set of communicating sequential GAs, it runs with maximal efficiency on an MIMD distributed memory machine: in fact, there is only local communication to be done and, as all the subpopulations are treated in the same way, the algorithm is simply scalable to any number of processing elements. Because of this, the target machine is a distributed memory MIMD system, the Meiko Computing Surface. This machine is a microcomputer array based on T800 INMOS transputers, each containing four bidirectional 20 Mbits/s communication links and capable of 1.5 Mflops (20 MHz). The OCCAM2 programming language has been utilized to implement the simulator.

For our experiments, a WK-recursive topology with 16 nodes has been configured [17]. The PGA consists of genetic processes (each of which allocated onto a node of the topology) organized in clusters running in parallel. A generic cluster, constituted by a subpopulation and its neighbours, is depicted in grey in Figure 1.
It is clear that individuals going from a cluster to another and surviving the selection processes spread within the topology until they reach and influence the evolution of each subpopulation. The rapidity with which such a diffusion process takes place is related to the topology parameters, in particular to its diameter [17]. For our topology and for the fixed number of processors, the diffusion process is carried out in three steps, which is the minimum possible for 16 processors.

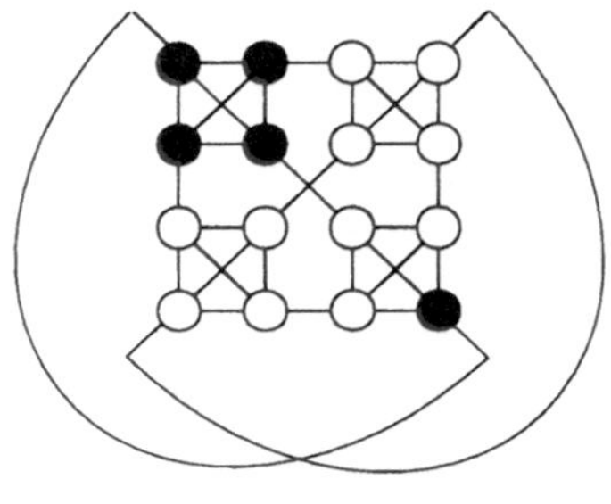

Figure 1. A WK-recursive topology with an example of a cluster.

The evolutionary dynamics of the subpopulations is influenced by the reproduction mode, the reproduction probability, the population size, the number of genes, the diffusion interval and obviously by the environmental-fixed constraints (fitness function).

Preliminary tests have been performed on tours with *a priori* known solutions (namely their euclidean lengths) using different genetic operators, in order to find the best combination for the problem at hand.

The probabilities of applying the operators have been fixed as follows: P(PMX)=0.7, P(MUT)=0.008 and

P(INV)=0.008 when not used on its own and 0.7 otherwise. The PGA runs until the minimum or the prefixed maximum number of epochs are not reached. The trials have been repeated over 5 runs. The average results are reported in Table 1. This table reports also the values of the parameters involved: *sub_size* and *diff_int* which denote the subpopulation size and the diffusion interval, respectively.

Cities	Convergence times			*sub_size*	*diff_int*
	PMX and MUT	PMX and INV	INV		
30	1' 30"	14.8"	4.47"	30	70
50	2' 42"	2' 17"	23.27"	50	100
100	22' 44"	18' 17"	6' 44"	50	100

Table 1. The convergence times using different genetic operators.

As it may be seen from Table 1, the inversion operator is the most suitable to solve the TSP. An intuitive explanation is that, every time a possible offspring tour contains an intersection between two edges, it can be furtherly optimized, so that the intersection is removed [18]. Probabilistically, this can be very easily achieved through the above mentioned operator in one single step, whereas an operator like the mutation would need several steps to remove it [19]. A more detailed explanation of the superiority of the inversion operator with respect to the mutation has been made elsewhere [20]; suffice it to say now that, in order to remove a loop containing m cities in an N-city tour, the former has a probability $1/N^2$ to do it in one step, whereas the latter has a probability $1/N^m$ to perform it in m/2 steps.

Figure 2 shows the number of epochs (*n_epochs*) needed for the convergence to the optimum.

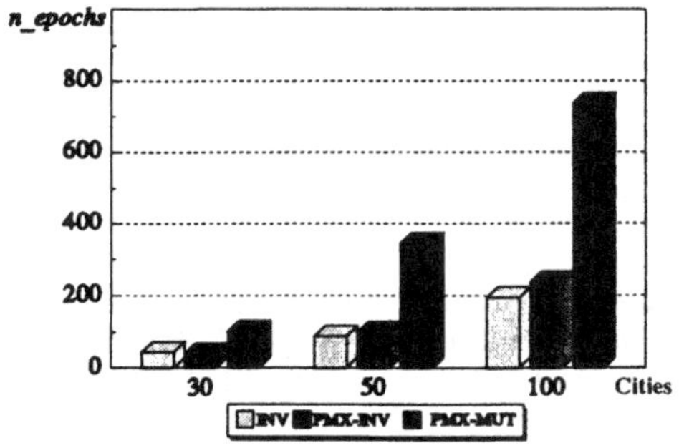

Figure 2. The number of epochs for different operator combinations as a function of the number of the cities.

It appears evident that the inversion outperforms the approaches with other operators as regards not only the convergence time, but also in terms of the evolution process.

The tests above have been performed for several values of the subpopulation size in a wide range and the conclusions we have reached about the inversion remain the same. An interesting point to evidence is that, for each fixed length of individuals (i.e. for each given number of cities), it has been verified that there is an optimal subpopulation size and that it is not useful to work with larger subpopulations.

It is worth explaining the historical reasons why we made this set of experiments: actually, we started performing an usual GA as described in literature, i.e. by using its classical operators PMX crossover and mutation. In a second phase, we wondered whether another asexual operator like inversion could adequately substitute mutation and we found out it actually can: the results were quite encouraging, as the new operator combination outperformed the old one. Finally, we noticed that the behaviour of the PMX and of the inversion is such that, in a sense, they are "mutually exclusive", so we tried to use them both separately: as a result, inversion on its own was excellent and PMX was not comparable with it (the results of this latter configuration are not reported in this paper).

Another set of experiments by using only the inversion has been performed to ascertain the effectiveness of the diffusion process. For a fixed problem, namely an *a priori* known 100-city tour with a *sub_size*=70, we have started from the limit condition in which no diffusion (*no_diff*) takes place (i.e. we have in this case performed 16 independent searches) and then we have decreased the diffusion interval. The convergence time (*conv_time*) as a function of the *diff_int* parameter is reported in Table 2.

diff_int	*no_diff*	400	300	200	150	100	75	50	25
conv_time	16' 53"	3' 18"	3' 3"	2' 44"	2' 41"	2' 38"	2' 28"	2' 7"	2' 2"

Table 2. The convergence time as a function of the diffusion interval.

The experiments reveal that, as the diffusion interval decreases, the convergence time diminishes, so that the validity of the diffusion process has been proved.

It is to point out that the combined effects of the simultaneous alteration of the parameters (*sub_size* and *migr_int*), as a function of the probability of applying the operators, have not been investigated in this paper.

Further tests have been performed for the solution of classical TSPs, namely Oliver's 30-city, Eilon's 50-city and 75-city problems [21]. The best values reached (*fin_value*), the convergence times, the number of epochs and the control parameters involved are shown in Table 3.

Problem	Cities	*fin_value*	*conv_time*	*sub_size*	*n_epochs*	*diff_int*
Oliver	30	*423.949*	*9.36"*	*30*	*84*	*100*
Eilon	50	*438.132*	*1' 23"*	*50*	*166*	*200*
Eilon	75	*557.685*	*3' 28"*	*70*	*183*	*300*

Table 3. The values obtained and the convergence times for classical TSPs.

With reference to the results reported in literature [22], it should be noted that high quality solutions have been found in very low convergence times.

In the following the paths reached for the problems above mentioned are outlined.

Oliver's 30-city problem

(22,60) (25,62) (7,64) (2,99) (41,94) (37,84) (54,67) (54,62) (58,69) (71,71) (74,78) (87,76) (83,69) (64,60) (68,58) (71,44) (83,46) (91,38) (82,7) (62,32) (58,35) (44,35) (45,21) (41,26) (25,38) (24,42) (18,40) (13,40) (4,50) (18,54)

Eilon's 50-city problem

(61,33) (58,27) (51,21) (59,15) (46,10) (39,10) (36,16) (30,15) (32,22) (27,23) (25,32) (17,33) (20,26) (21,10) (13,13) (5,6) (10,17) (5,25) (7,38) (12,42) (8,52) (5,64) (17,63) (16,57) (25,55) (21,47) (30,48) (32,39) (31,32) (40,30) (48,28) (45,35) (42,41) (38,46) (37,52) (42,57) (31,62) (27,68) (37,69) (43,67) (52,64) (63,69) (62,63) (57,58) (49,49) (58,48) (62,42) (52,41) (52,33) (56,37)

Eilon's 75-city problem

(67,41) (62,48) (62,57) (70,64) (57,72) (50,70) (47,66) (55,65) (55,57) (50,50) (55,50) (55,45) (51,42) (50,40) (54,38) (55,34) (50,30) (45,35) (45,42) (41,46) (40,37) (38,33) (33,34) (29,39) (33,44) (35,51) (30,50) (21,45) (21,48) (22,53) (26,59) (30,60) (35,60) (40,60) (40,66) (31,76) (10,70) (17,64) (15,56) (9,56) (7,43) (12,38) (21,36) (20,30) (11,28) (6,25) (12,17) (16,19) (15,14) (15,5) (26,13) (30,20) (27,24) (22,22) (26,29) (36,26) (35,16) (36,6) (44,13) (40,20) (43,26) (52,26) (48,21) (50,15) (54,10) (50,4) (59,5) (64,4) (66,8) (66,14) (60,15) (55,20) (62,24) (65,27) (62,35).

5. CONCLUSIONS

In the present paper the increasing importance of PGAs has been pointed out and a parallel simulator of GAs has been presented and discussed. The implementation has been made on a coarse grain multicomputer whose processors have been configured in a WK-recursive topology. The TSP has been investigated as test bed.

The analysis of the results related to various combinations of genetic operators has led to prove that the inversion operator is more suitable than the PMX crossover usually utilized. Another set of experiments has been carried out to establish the effectiveness of the diffusion process. The results obtained have demonstrated that the smaller is the diffusion interval the less is the convergence time. Furthermore, with the tests performed on classical TSPs taken from the literature, high quality solutions in low convergence times have been reached.

Since the results achieved seem to be sound, our future goal is the application of the simulator to practical problems in both scientific (neural networks and image understanding) and financial (classifier systems for stock market forecasting) fields.

REFERENCES

[1] Bhann B, Lee S and Ming J, '*Self-Optimizing Image Segmentation System Using a Genetic Algorithm*', Proc. of the Fourth Int. Conf. on Genetic Algorithms, Belew R K and Booker L B eds., Morgan Kaufmann Pub., 362-369, 1991.

[2] Treleaven P and Goonatilake S, '*Intelligent Financial Techniques*', Proc. of the Workshop on Parallel Problem Solving from Nature: Applications in Statistics and Economics, EUROSTAT, 1992.

[3] Reeves C and Steele N, '*Problem-solving by simulated genetic processes: a review and applications to Neural Networks*', Tenth IASTED Int. Symposium on Applied Informatics, Hamza M H ed., 269-272, Innsbruck, Austria, February 10-12, 1992.

[4] Cui J, Fogarty T C and Gammack J G, '*Searching databases using Parallel Genetic Algorithms on a Transputer Computing Surface*', Proc. of the Third Annual Conference of the Meiko User Society, Manchester Business School, University of Manchester, April 9-10, 1992.

[5] Holland J H, '*Adaptation in Natural and Artificial Systems*', University of Michigan Press, Ann Arbor, 1975.

[6] Manderick B and Spiessens P, '*Fine-grained Parallel Genetic Algorithms*', Proc. of the Third Int. Conf. on Genetic Algorithms, Schaffer J D ed., Morgan Kaufmann Pub., 428-433, 1989.

[7] Tanese R, '*Distributed Genetic Algorithms*', Proc. of the Third Int. Conf. on Genetic Algorithms, Schaffer J D ed., Morgan Kaufmann Pub., 434-439, 1989.

[8] Mühlenbein H, Schomisch M and Born J, '*The parallel genetic algorithm as function optimizer*', Parallel Computing, 17, 619-632, North-Holland, 1991.

[9] Goldberg D E, '*Genetic Algorithms in Search, Optimisation, and Machine Learning*', Addison-Wesley, Reading, Mass., 1989.

[10] Mühlenbein H, '*Evolution in Time and Space - The Parallel Genetic Algorithm*', Foundations of Genetic Algorithms, Rawlins G ed., 316-337, Morgan Kaufmann Pub., 1992.

[11] Kirkpatrick S, Gelatt C D and Vecchi M P, '*Optimization by Simulated Annealing*', Science, 220, 671-680, 1983.

[12] Beightler C S, Phillips D T and Wilde D J, 'Foundations of optimization', Englewood Cliffs, NJ: Prentice-Hall, 1979.

[13] Grefenstette J, Gopal R, Rosmaita B and Van Gucht D, '*Genetic Algorithms for the Traveling Salesman Problem*', Proc. of the Int. Conf. on Genetic Algorithms and Their Applications, Grefenstette J ed., 160-165, 1985.

[14] Davis L, '*Job shop scheduling with Genetic Algorithms*', Proc. of the Int. Conf. on Genetic Algorithms and Their Applications, Grefenstette J ed., 136-140, (Erlbaum L, 1988, original proceedings 1985).

[15] Goldberg D and Lingle R, '*Alleles, loci and the Traveling Salesman Problem*', Proc. of the Int. Conf. on Genetic Algorithms and Their Applications, Grefenstette J ed., 154-159, (Erlbaum L, 1988, original proceedings 1985).

[16] Reeves C, '*An Introduction to Genetic Algorithms*', 33rd OR Society Conference Tutorial Papers, Birmingham, 1991.

[17] Della Vecchia G and Sanges C, '*A Recursively Scalable Network VLSI Implementation*', Future Generation Computer Systems, 4, 3, 235-243, North Holland, Amsterdam, 1988.

[18] Ambati B K, Ambati J and Mokhtar M M, '*Heuristic combinatorial optimization by simulated Darwinian evolution: a polynomial time algorithm for the Traveling Salesman Problem*', Biological Cybernetics, 65, 31-35, Springer-Verlag, Berlin, 1991.

[19] De Falco I, Del Balio R, Tarantino E and Vaccaro R, '*Simulation of Genetic Algorithms on MIMD Multicomputers*', to appear in Parallel Processing Letters, 1993.

[20] De Falco I, Del Balio R, Tarantino E and Vaccaro R, '*A Self-replication-based Parallel Adaptive Strategy for the Travelling Salesman Problem*', submitted to the Fifth Int. Conf. on Genetic Algorithms, to be held at the University of Illinois, Urbana-Champaign, Illinois, USA, July 17-22, 1993.

[21] Eilon, Watson-Gandy and Christofides, '*Distribution Management: mathematical modeling and practical analysis*', Operational Research Quarterly, 20, 309, 1969.

[22] Whitley D, Starkweather T, Fuquay D A, '*Scheduling Problems and Traveling Salesmen: The Genetic Edge Recombination*', Proc. of the Third Int. Conf. on Genetic Algorithms, Schaffer J D ed., Morgan Kaufmann Pub., 133-140, 1989.

DIVERSITY AND DIVERSIFICATION IN GENETIC ALGORITHMS: SOME CONNECTIONS WITH TABU SEARCH

Colin Reeves
School of Mathematical and Information Sciences
Coventry University
UK
Email: CRReeves@uk.ac.cov.cck

Abstract

Genetic Algorithms (GAs) have been used very successfully to solve a variety of optimisation problems, but despite their successes, there are a number of outstanding problems in their implementation. One of the most pervasive problems is that of premature convergence of the process, usually associated with a loss of diversity in the population of chromosomes.

In this paper, we will first review some of the existing solutions to the problem of preserving diversity. These all use the basic GA framework; there are also extreme solutions such as the invariant GA which dispenses with the fundamental selection process altogether.

We argue that an underlying (and largely unaddressed) problem is the GA's lack of memory: it is here that some connections with the concept of Tabu Search (TS) may prove fruitful. In TS the search is characterised in terms of the twin concepts of *intensification* and *diversification*. Intensification relates to the ability of the search strategy to focus on a particular area (or particular areas) of the search space in order to find improved solutions. In this sense, a GA as customarily conceived is clearly an intensifying process. Diversification is achieved by the incorporation of memory into its basic structures—that is, structures are devised which can record the history of the search.

We will describe these mechanisms in more detail, with particular reference in the context of this paper to their diversifying effects. From this standpoint we will then suggest some ways in which these mechanisms can be adapted so as to offer a systematic and coherent framework for diversification within the Genetic Algorithm paradigm.

1 Introduction

Genetic Algorithms (GAs) have been developed from original work by Holland [1] on adaptive systems. They have been used very successfully to solve a variety of optimisation problems: some of these applications are summarised in books by Goldberg [2], Davis [3] and Reeves [4]. GAs work by employing populations of structures (called *chromosomes*) which encode candidate solutions to a problem. These structures are allowed to undergo changes by means of so-called *genetic operators*—chiefly *crossover* in which pairs of chromosomes exchange material, and *mutation* in which changes occur to the individual *genes* which make up the chromosome. The choice of the chromosomes which undergo re-combination by means of crossover is controlled by a selection process which encourages reproduction of those chromosomes which are relatively 'fitter'.

Tabu search (TS) has also received recent attention as a general methodology for solving optimisation problems. The most comprehensive treatment to the subject is given by Glover and Laguna [5]. It can be viewed as a development of existing neighbourhood search techniques which can cross boundaries of feasibility or local optimality normally treated as barriers. This is achieved by systematically imposing and releasing constraints to allow exploration of these otherwise forbidden regions. (It is the 'forbidden' nature of these search areas that explains the use of the adjective 'tabu'.) At the simplest level, this enables the search process to recognise regions which it has already explored, and thus provides a basis for focusing the search.

The search may be further characterised in terms of the concepts of *intensification* and *diversification*. These ideas will be explained further in section 3, where we will also suggest some ways in which they can be interpreted in the GA context.

2 Implementation Problems in GAs

Despite the very many real successes of GAs, there are a number of outstanding problems in their implementation. One of the most pervasive reported problems is that of premature convergence of the process, almost invariably associated with a loss of diversity in the population of chromosomes. That this is a problem is attested by the number of ways which have been proposed to alleviate the problem. We will first review some of these solutions, before considering how some TS concepts might fruitfully be used. In a paper such as this it is not possible to do more than sketch the outlines of the topics covered; many are discussed in greater detail in [4].

2.1 Population sizing and selection

One of the most obvious questions relating to the implementation of a GA is what size population to use. In principle, it is clear that small populations run the risk of seriously undercovering the solution space, while for a given amount of computational effort, larger populations allow the examination of fewer generations, and therefore may preclude convergence to the optimum in a reasonable amount of time. Some research has been reported on this problem by Goldberg [8]. Using the criterion that we should try to maximise the expected number of new schemata per individual, this work indicates that the optimal size for binary-coded strings grows exponentially with the length of the string n. Some refinements of this work are reported in [9], but they do not change the overall conclusions significantly.

However, this would imply extremely large populations in most real-world problems, and the *practical* performance of the GA would be quite uncompetitive with other optimization methods such as simulated annealing and tabu search. Fortunately, empirical results from many authors (see e.g. Grefenstette [6] or Schaffer *et al.* [7]) suggest that population sizes as small as 30 are quite adequate in many cases, while some experimental work by Alander [10] suggests that a value between n and $2n$ is optimal for the problem type considered. Some later work by Goldberg [11], based on a different argument, goes some way to supporting the use of populations rather smaller than his earlier work suggested. Nevertheless, little has been published that is relevant to *really* small populations. Some recent work by Reeves [12] has considered the question of a minimal population size for q-ary alphabets. This shows that using the conventional binary coding, relatively small populations can ensure reasonable diversity in the initial population, although for higher-cardinality alphabets, considerably larger populations are required.

2.2 Modifications to reproductive selection

Many variations on the basic theme of parent selection in the reproductive phase have been proposed. Fitness scaling [2] has a twofold intention: on the one hand to maintain a reasonable differential between relative fitness ratings of individuals (particularly as the population becomes more homogeneous), while on the other hand to prevent a too-rapid takeover by a 'super-individual'. From the latter point of view, scaling clearly acts to prevent loss of diversity. Fitness *ranking* [13] has a similar effect, and avoids the need for extra scaling parameters.

De Jong [14] introduced the concept of *crowding* to the selection process. With this approach, using a steady-state GA where population replacement is incremental rather than *en bloc*, we choose c individuals at random as candidates for deletion from the population and that individual which is closest to the the new chromosome is selected. In this way the rate of loss of diversity can be reduced.

Another related idea is that of *incest prevention*, introduced by Eshelman and Schaffer [15], where the selection strategy rather than the replacement strategy takes on the rôle of preserving diversity. Here, mating of chromosomes is only allowed if their Hamming distance exceeds a threshold value which is gradually reduced as the population converges.

2.3 Fitness evaluation

Real populations do not favour indiscriminate mating, for the result (even if it is biologically possible) is usually sterile. In a similar way, if we try to optimize a multi-modal function by means of a GA, the result from crossover of two chromosomes which are close to *different* optima may be much worse than either. The answer in nature is the existence of *species*. Goldberg and Richardson [16] were able to induce this behaviour by defining a *sharing function* over the population which is used to modify the fitness of each chromosome. The function could take many forms, but a simple linear function $h(d)$ is effective, where d is the *distance* between two chromosomes. The sharing function is evaluated for each pair of chromosomes in the population, and then the sum $\sigma_j = \sum_{i \neq j} h(d_i)$ is computed for each chromosome j. Finally the fitness of chromosome j is adjusted by dividing by σ_j; the adjusted fitness values are then used in place of the original values. The effect of this is that chromosomes which are close will have their fitness devalued relative to those which are fairly isolated, thus creating *niches* in the overall population. This is important in the context of multi-modal functions, but it also has the effect of inculcating useful diversity.

2.4 Chromosomal representation

Another way of tackling the diversity problem is to focus on the coding used to represent the parameters of the problem. One possibility is to modify the coding as the GA proceeds, using a form of dynamic parameter adjustment such as that described by Shaefer [17] and Schraudolph [18]. This can be used to get an initially 'coarse-grained' estimate of potentially good areas of the parameter space, which can then be searched more intensively. A similar idea has been introduced by Whitley [19] under the name of *delta coding.* In this approach, whenever a significant amount of diversity is lost, a new population is introduced which is however related to the best solution of the previous phase.

While having an obvious intensifying effect, these procedures do also manage to maintain some diversity by virtue of their continual reorientation of the search.

2.5 Modified operators

In many applications, the simple (one-point) crossover operator has proved extremely effective, but it is clear that it can lead to severe loss of diversityin some cases. Eshelman *et al.* [20] discuss this in the context of crossover 'bias': simple crossover has no distributional bias, which limits the exchange of information between parent chromosomes. The addition of mutation helps, but it may need to be set fairly high in order to preserve a reasonable level of population diversity, and this in turn may slow up the algorithm.

Booker [21] and Fairley [22] have independently suggested that before applying crossover, we should examine the selected parents to find the crossover points which would produce offspring which differ from the parents. Fairley implemented this idea by his 'string-of-change' crossover (Booker used the term 'reduced surrogates'), which entails computing an 'exclusive-OR' (XOR) between the parents. Only positions between the outermost 1s of the XOR string will be considered as crossover points. This again helps to encourage diversity. Fairley investigated this operator in the context of solving knapsack problems, and found significantly improved performance compared with the standard crossover operator.

There seems no reason why the choice of crossover point should be restricted to a single position, and Booker [21] for example, has empirically observed that increasing the number of crossover points to two has improved the performance of a GA. The most thorough investigation of multi-point crossovers is that by Eshelman *et al.* [20], whose arguments and empirical evidence strongly supported the contention that one-point crossover is not the best option.

Several reports have suggested ways of generalizing the simple cross-over operator. The latter can itself be represented as a binary string: for example, the operator representation `1 1 1 0 0 0 0` may be taken to mean that the first 3 elements of a 7-bit chromosome are taken from the first parent, and the last 4 from the second. This can clearly be generalized very easily by allowing the pattern of 0s and 1s to be generated stochastically using a Bernoulli distribution. Syswerda

[23] investigated this 'uniform crossover' and discussed its advantages and disadvantages over simple crossover. The main advantage is that 'building-blocks' (in Goldberg's phrase) no longer have to be encoded compactly (i.e. as short schemata) in order to survive. Uniform crossover is obviously completely indifferent to the length of a schema: all schemata of a given order have the same chance of being disrupted. Of course in some problems this may not be an advantage, but it certainly helps to prevent loss of diversity.

Sirag and Weisser [24] take yet another route, by modifying the basic genetic operators in the spirit of simulated annealing. At high temperatures, their operator can be expected to behave rather like the generalized uniform operator; as temperatures moderate the number of 'switches' between parents decreases and it becomes more like the standard simple crossover, while at very low temperatures it just copies one of the parents.

Recently, many of these extensions have received a theoretical treatment by De Jong and Spears. In their paper [25] they are able to characterize exactly the amount of disruption introduced by a given crossover operator. In the course of this work, they further generalize the uniform crossover operator: in Syswerda's original description it was implicitly assumed that the Bernoulli parameter was 0.5, but De Jong and Spears show that (as is intuitively expected) the amount of disruption can be reduced by choosing different values.

2.6 Summary

We have briefly reviewed some of the many ways in which loss of diversity can be alleviated, but all of these rely on the basic GA framework. One novel approach is that adopted by Culberson [26] with his *invariant GA*. This may be thought of as an extreme solution to the diversity problem (although that is not its original motivation). Its most striking departure from 'normal' GA practice is to insist that offspring always replace their parents; mutation is not used, so that none of the alleles present in the original population are ever lost, only re-combined. One source of diversity loss is thus eliminated straightaway. Current experience with this model is limited, but it seems worth investigation.

However, an underlying problem that none of these approaches addresses is the GA's lack of *memory*. A GA has no explicit way of knowing when it is merely repeating the sampling of regions of the parameter space (or its encoding) which it has already searched before. In some ways, of course, this is deliberate: a GA is designed to be a local algorithm which as Holland [1] has shown, nevertheless cleverly obtains non-local information by the implicit processing of similarity subsets or *schemata*.

Incorporating memory explicitly into a GA would mean keeping all sorts of statistics which would probably make the algorithm completely unworkable. However, the development of tabu search has shown that it is possible to *simulate* both short-term and long-term memory structures in a compact way, so that memory can be relatively easily exploited in search procedures.

3 Tabu Search

TS may be conveniently characterized as a form of neighbourhood search (NS) over a solution space $\mathbf{X}$, provided we define neighbourhood search in a less restricted fashion than usual. Frequently, for example, constructive and destructive procedures are excluded, whereas such procedures and their combinations are routinely subjected to the guidance of TS. In neighbourhood search, each solution $x \in \mathbf{X}$ has an associated set of neighbours, $N(x) \subset \mathbf{X}$, called the neighbourhood of x. Each solution $x' \in N(x)$ can be reached directly from x by an operation called a *move*, and x is said to move to x' when such an operation is performed. Normally neighbourhoods are assumed to be symmetric.

The failing of NS is well-known: its propensity to deliver solutions which are only *local* optima. Various strategies based on some form of randomization have been suggested for alleviating this problem. Tabu search, in contrast to these methods, employs a somewhat different philosophy for going beyond the criterion of terminating at a local optimum. Randomization is de-emphasized, and generally is employed only in a highly constrained way, on the assumption that intelligent search should be based on more systematic forms of guidance.

The notion of exploiting certain forms of flexible memory to control the search process is the central theme underlying tabu search. The effect of such memory may be envisaged by stipulating that TS maintains a selective history H of the states encountered during the search, and replaces $N(x^{now})$ by a modified neighbourhood which may be denoted $N(H, x^{now})$. History therefore determines which solutions may be reached by a move from the current solution, selecting x^{next} from $N(H, x^{now})$.

In TS strategies based on short term considerations, $N(H, x^{now})$ is typically a subset of $N(x^{now})$, and the tabu classification serves to identify elements of $N(x^{now})$ excluded from $N(H, x^{now})$.

TS also uses history to create a modified evaluation of currently accessible solutions. This may be expressed formally by saying that TS replaces the objective function $c(x)$ by a function $c(H, x)$, which has the purpose of evaluating the relative quality of currently accessible solutions. The relevance of this modified function occurs because TS uses aggressive choice criteria that seek a best x^{next}, i.e. one that yields a best value of $c(H, x^{next})$, over a candidate set drawn from $N(H, x^{now})$. Moreover, modified evaluations are often accompanied by systematic alteration of $N(H, x^{now})$, to include neighbouring solutions that do not satisfy customary feasibility conditions (i.e. that strictly speaking do not yield $x \in \mathbf{X}$). Reference to $c(x)$ is retained for determining whether a move is improving or leads to a new best solution.

The ways in which TS accomplishes these aims are described in detail in Glover and Laguna [5]. We summarise some of these in section 3 below. In a GA context, we have no direct analogue of a TS move, but the steady-state version of a GA in particular does suggest some parallels. In a GA there are two fundamental types of 'components': the chromosomes, and the operators. It is at least possible that by implementing some sort of 'history' for these components, we can encourage diversification. In the next section, we also suggest some ways in which this might be implemented.

3.1 Diversification versus randomization

As Glover and Laguna observe, seeking a diversified collection of solutions is very different from seeking a randomized collection of solutions. They provide a set of conditions for generating a diversified sequence of points, but point out that such generation is computationally very demanding, which would seem to rule out any attempt to incorporate them into a GA. Even a relaxed version of these conditions described in [27] would seem to pose considerable computational demands.

There are important exceptions, however. The initial population of strings for a GA is conventionally chosen randomly, and Reeves [12] has shown that while this is reasonable for binary strings and large populations, in cases where we wish to use q-ary strings ($q > 2$) and/or small populations, this is inferior to using a more diverse initial population. When we consider only the initial population, the computational burden is negligible, since we can access diversified collections of strings generated by *error-detecting codes* and/or *experimental design* techniques. Further details may be found in [12]. We suspect that it may always be worth considering starting with a diversified population in any GA implementation.

3.2 Recency

One objective in TS is to encourage exploration of parts of the solution space that have not been visited previously. This can be achieved in practice by prohibiting the reversal of previous moves—that is, these (reverse) moves become 'tabu'. However, to make this prohibition absolute would confine the search to a straitjacket, so we prohibit the reversal of the most *recent* moves only. Recency may be construed as a fixed parameter defined as the *tabu tenure* of a move, or it may be allowed to vary dynamically during the search. It is also often beneficial to focus on some component or *attribute* of a move rather than on the complete move itself. Finally, in many implementations tabu moves are allowed to be overridden under certain conditions specified by an *aspiration* criterion.

In a GA context, several ways suggest themselves in which this type of approach could be adopted, at a variety of levels. For example,

- in one-point or multi-point crossover, we could treat a *crossover point* as an attribute so that the GA is encouraged to explore the effect (within the short-term) of crossing over at different places;

- in a steady-state GA, we could make the newly created chromosome itself tabu, effectively forbidding it to reproduce until its tabu tenure had expired. (In this case, we would incidentally be imitating the naturally observed process of *maturation* of an organism before reproduction can occur. As seen above in the case of speciation, the simulation of macro-processes observed in real populations has been useful in previous GA implementations.)

3.3 Frequency

Recency simulates *short-term* memory; Glover and Laguna [5] suggest that a form of *long-term* memory can be implemented by the use of a variety of frequency measures. Two types of frequency measure are identified: residence measures and transition measures. The former relate to the number of times a particular attribute is observed, the latter to the number of times an attribute changes from one value to another. In either case, the frequency measures are usually used to generate penalties which modify the objective function $c(x)$ to $c(H, x)$ as discussed above. Diversification is thereby encouraged by the generation of solutions that embody combinations of attributes significantly different from those previously encountered.

Again, to apply this in a GA context we have the choice of interpreting attributes in the context of the chromosome or of the operators. For example,

- we could use frequency measures to promote the exploration of crossover points;

- similarly, we could use frequency measures to guide the mutation or otherwise of particular bits of a chromosome;

- we could institute penalties for chromosomes on the basis of frequency measures of their bit-values, thus affecting their fitness values and hence their chance of reproduction.

3.4 Quality and Influence

The last suggestion links in to TS concepts of *quality* and *influence*, which also relate rather more directly to diversification strategies previously suggested for GAs. Quality in TS usually refers simply to those solutions with good objective function values; a collection of such solutions may be used to stimulate a more intensive search in the general area of these *élite* solutions. This clearly has something in common with ideas of dynamic parameter encoding and delta-coding, as discussed above. For diversification purposes, in the GA context, this could usefully extended to encompass not only solutions whose objective function value is above average, but also those whose quality is high in terms of their diversity relative to the population as a whole. The frequency-based penalty measures discussed above are one possible means of accomplishing this goal.

Influence is roughly a measure of the degree of change induced in solution structure—commonly expressed in terms of the distance of a move from one solution to the next. In these terms, in a GA context, it also connects to direct measures for ensuring a diverse population. From this perspective, for example, we can see that some of the ideas discussed earlier, such as those relating to crowding and the string-of-change crossover, are measures to encourage high-influence 'moves'.

3.5 Some preliminary experimental results

We carried out some tests on two specific ideas mentioned above. The object was to maximise a cubic function defined by a 5-bit binary integer string. We ran 30 trials using a steady-state type of GA which had been modified so that newly created chromosomes had to wait for T iterations before being allowed to reproduce ($T = 0$ is the standard case). We also examined the possibility of using a frequency-modified mutation rate, where the probability of mutation was allowed to vary between two values (α, β) according to the number of alleles present at each locus. A full report is in preparation, but we may summarise our findings as follows:

- using the recency-based selection mechanism generally led to a more diversified population for values of T around $M/2$ (where M is the population size);
- online performance suffers slightly using recency-based selection, but in terms of finding the optimum, performance was not degraded;
- using global information to modify mutation probabilities also increased diversity significantly, and reduced online performance;
- frequency-based mutation improved the chances of finding the optimum about as much as an ordinary fixed mutation rate, but diversity was maintained at a much higher level.

We emphasize that these findings are tentative, and relate to one series of experiments on one simple function; however, they accord fairly well with what we might intuitively suspect, and certainly do not suggest that the exploration of these TS-related ideas is of nugatory importance.

4 Conclusions

This papers has focused on some possible links between GAs and tabu search in one particular area. Many other suggestions and connections have been made: several possibilities are raised by Glover and Laguna [5], while some more recent papers [27, 28] lay the foundations for search strategies that combine elements of genetic algorithms with other approaches in a yet more fundamental way. To those who take a purist view of artificial intelligence in general, and the genetic algorithm in particular, the implicit use of non-local and problem-specific information may be a stumbling-block. However, as Glover [28] has observed, there is no reason whatsoever for restricting ourselves to 'primitive' processes for solving problems, when the addition of more sophisticated tools may enable us to find worthwhile improvements. To put it another way, if we want to solve complex problems, we should feel free to use a little real intelligence to supplement the artificial variety!

It is our belief that TS deserves to be much more widely known, and we hope that this introduction to the area, and exploration of connections with GAs, will stimulate the GA community to consider how the two methodologies together might generate yet better ways of solving difficult optimization problems.

References

[1] J.H.Holland (1975) *Adaptation in Natural and Artificial Systems.* University of Michigan Press, Ann Arbor.

[2] D.E.Goldberg (1989) *Genetic Algorithms in Search, Optimization, and Machine Learning.* Addison-Wesley, Reading, Mass.

[3] L.Davis (Ed.) (1991) *Handbook of Genetic Algorithms.* Van Nostrand Reinhold, New York.

[4] C.R.Reeves (Ed.) (1993) *Modern Heuristic Techniques for Combinatorial Problems.* Blackwell Scientific Publications, Oxford.

[5] F.Glover and M.Laguna (1993). Tabu Search. In [4].

[6] J.J.Grefenstette (1986) Optimization of control parameters for genetic algorithms. *IEEE-SMC*, **SMC-16**, 122-128.

[7] J.D.Schaffer, R.A.Caruana, L.J.Eshelman and R.Das (1989) A study of control parameters affecting online performance of genetic algorithms for function optimization. *In* [29], 51-60.

[8] D.E.Goldberg (1985) *Optimal initial population size for binary-coded genetic algorithms.* TCGA Report 85001, University of Alabama, Tuscaloosa.

[9] D.E.Goldberg (1989) Sizing populations for serial and parallel genetic algorithms. *In* [29], 70-79.

[10] J.T.Alander (1992) On optimal population size of genetic algorithms. *Proc. CompEuro 92*, 65-70. IEEE Computer Society Press.

[11] D.E.Goldberg and M.Rudnick (1991) Genetic algorithms and the variance of fitness. *Complex Systems*, **5**, 265-278.

[12] C.R.Reeves (1993) Using genetic algorithms with small populations. (Submitted to ICGA5).

[13] D.Whitley (1989) The GENITOR algorithm and selection pressure: why rank-based allocation of reproductive trials is best. *In* [29], 116-121.

[14] K.A.De Jong (1975) *An analysis of the behavior of a class of genetic adaptive systems.* Doctoral dissertation, University of Michigan.

[15] L.J.Eshelman and J.D.Schaffer (1991) Preventing premature convergence in genetic algorithms by preventing incest. *In* [32], 115-122.

[16] D.E.Goldberg and J.Richardson (1987) Genetic algorithms with sharing for multimodal function optimization. *In* [30], 41-49.

[17] C.G.Shaefer (1987) The ARGOT strategy: adaptive representation genetic optimizer technique. *In* [30], 50-58.

[18] N.Schraudolph and R.Belew (1990) *Dynamic parameter encoding for genetic algorithms.* CSE Technical Report No. CS90-175.

[19] D.Whitley, K.Mathias and P.Fitzhorn (1991) Delta coding: an iterative search strategy for genetic algorithms. *In* [32], 77-84.

[20] L.J.Eshelman, R.A.Caruana and J.D.Schaffer (1989) Biases in the crossover landscape. *In* [29], 10-19.

[21] L.B.Booker (1987) Improving search in genetic algorithms. *In* [31], 61-73.

[22] A.Fairley (1991) *Comparison of methods of choosing the crossover point in the genetic crossover operation.* Dept. of Computer Science, University of Liverpool.

[23] G.Syswerda (1989) Uniform crossover in genetic algorithms. *In* [29], 2-9.

[24] D.J.Sirag and P.T.Weisser (1987) Towards a unified thermodynamic genetic operator. *In* [30], 116-122.

[25] K.A.De Jong and W.M.Spears (1992) A formal analysis of the role of multi-point crossover in genetic algorithms. *Annals of Maths. and AI*, **5**, 1-26.

[26] M.Lewchuk (1992) *Genetic invariance: a new type of genetic algorithm.* Technical report TR 92-05, Dept. of Computing Science, University of Alberta.

[27] F.Glover (1992) Tabu search for nonlinear and parametric optimization (with links to genetic algorithms). *Discrete Applied Mathematics.*

[28] F.Glover (1992) Optimization by ghost image processes in neural networks. *Computers & Ops.Res.*, (to appear).

[29] J.D.Schaffer (Ed.) (1989) *Proceedings of the 3rd International Conference on Genetic Algorithms.* Morgan Kaufmann, Los Altos, CA.

[30] J.J.Grefenstette(Ed.) (1987) *Proceedings of the 2nd International Conference on Genetic Algorithms.* Lawrence Erlbaum Associates, Hillsdale, NJ.

[31] L.Davis (Ed.) (1987) *Genetic Algorithms and Simulated Annealing.* Morgan Kauffmann, Los Altos, CA

[32] R.K.Belew and L.B.Booker (Eds.) (1991) *Proceedings of the 4th International Conference on Genetic Algorithms.* Morgan Kaufmann, San Mateo, CA.

Clique Partitioning Problem and Genetic Algorithms

Dominique Snyers *

Abstract

In this paper we show how critical coding can be for the Clique Partitioning Problem with Genetic Algorithm (GA). Three chromosomic coding techniques are compared. We improve the classical linear coding with a new label renumbering technique and propose a new tree structured coding. We also show the limitation of an hybrid approach that combines GA with dynamic programming. Experimental results are presented for 30 and 150 vertex graphs.

1 Introduction

Clique Partitioning Problem (CPP) is a classical NP-hard optimization problem. de Amorim et. al [1] use simulated annealing and tabu search. We report here some new results using genetic algorithms on the same data sets.

Section two introduces the Clique Partitioning Problem and more specifically its use in a classification problem referred to in the litterature as Règnier Problem (see [8]).

Section three presents the Genetic Algorithm paradigm. Three coding techniques are introduced: linear coding, permutation coding and decision tree coding. Linear coding is associated with the classical GA. Permutation coding is used in an hybrid approach mixing GA with dynamic programming. And finally, decision tree coding is introduced to improve efficiency and accuracy on large graphs. We compare indeed these three approaches on two data sets: the classification of 30 Cetacea and the classification of 158 United Nation countries.

*Télécom Bretagne, B.P. 832, 29285 Brest Cedex, FRANCE. e-mail: snyers@enstb.enst-bretagne.fr

2 Clique Partitioning

Given :

- an undirected complete graph $G = (V, E)$ where V is a n vertex set and E is the edge set,
- a set C of costs c_{ij} on the edges between vertices i and j,

the Clique Partitioning Problem (CPP) consists in finding an equivalence relation on V minimizing the sum of the edge costs between vertices in the same equivalence class (or clique). With this equivalence relation $\mathcal{P}$ represented by a characteristic vector p_{ij} ($p_{ij} = 1$ if vertices i and j are in the same equivalence class and 0 otherwise) this partitioning problem becomes (Règnier's formulation [8]):

$$\text{Min} \quad \sum_{1 \leq i,j \leq n} c_{ij} \cdot p_{ij};$$

$$\begin{aligned} \text{with} \quad & p_{ij} = p_{ji}, && \forall i,j \in V; && (1)\\ & p_{ii} = 1, && \forall i \in V; && (2)\\ & p_{ij} + p_{jk} - p_{ik} \leq 1, && \forall i,j \in V; && (3)\\ & p_{ij} \in \{0,1\}, && \forall i,j \in V. && (4) \end{aligned}$$

Constraints (1), (2) and (3) respectively denote the symmetry, reflexivity and transitivity of the equivalence relation $\mathcal{P}$ represented by the characteristic vector p_{ij} (constraint (4)).

The clique partitioning problem has been used in a wide range of applications from VLSI circuit design ([9]), mapping parallel programs on parallel architectures [11], to image segmentation ([3]). In the sequel, we shall only focus on a classification problem called the *Règnier's Problem* in which objects have to be classified into equivalence classes based on their feature values (see [8], [1]).

Règnier's Problem

Assume we have a set V of n objects described by q variables as for example the eight individuals described by some physical characteristics shown in Table 1 (example taken from [6]). We can associate an equivalence relation $\mathcal{R}_k$ with each of these q variables in such a way that objects i and j belong to the same equivalence class when they agree upon the value taken by the given variable k. Objects 1 and 3 from Table 1 for example, are in the same equivalence class for relation R_{Height} since they both agree on variable *Height*. This leads us to the characteristic vector r_{ij}^{height} shown in Table 2.

Indiv.	Height	Hair	Eyes
1	tall	dark	blue
2	short	dark	blue
3	tall	blond	blue
4	tall	red	blue
5	tall	blond	brown
6	short	blond	blue
7	short	blond	brown
8	tall	dark	brown

Table 1: Individual physical characteristics

r_{ij}^{height}	1	2	3	4	5	6	7	8
1	1	0	1	1	1	0	0	1
2	0	1	0	0	0	1	1	0
3	1	0	1	1	1	0	0	1
4	1	0	1	1	1	0	0	1
5	1	0	1	1	1	0	0	1
6	0	1	0	0	0	1	1	0
7	0	1	0	0	0	1	1	0
8	1	0	1	1	1	0	0	1

Table 2: *Height* variable characteristic vector

Règnier's problem consists in classifying these objects in an optimal way, i.e. partitioning these objects into equivalence classes that minimize the number of disagreements between this partitioning equivalence relation, called $\mathcal{P}$, and the other equivalence relations $\mathcal{R}_1, \cdots \mathcal{R}_q$ associated with each variables. Règnier's problem can therefore be expressed as follows:

$$Min \sum_{k=1}^{q} \delta(\mathcal{P}, \mathcal{R}_k)$$

where $\delta(\mathcal{P}, \mathcal{R}_k)$ defines a distance in terms of the number of disagreements between equivalence relations $\mathcal{P}$ and $\mathcal{R}_k$, respectively represented by the characteristic vector p_{ij} and r_{ij}^k:

$$\delta(\mathcal{P}, \mathcal{R}_k) = \sum_{i \leq i,j \leq n} (p_{ij} - r_{ij}^k)^2.$$

Règnier's problem thus becomes:

$$Min \sum_{k=1}^{q} \sum_{i \leq i,j \leq n} (p_{ij} - r_{ij}^k)^2 \qquad (5)$$

subject to (1), (2), (3) and (4) since p_{ij}^k and r_{ij}^k are the characteristic vectors of equivalence relations.

Equation (5) successively simplifies into

$$Min \quad Cst + \sum_{k=1}^{q} \sum_{i \leq i,j \leq n} p_{ij}.(1 - 2.r_{ij}^k)$$

and

$$Min \quad Cst + \sum_{i \leq i,j \leq n} p_{ij}.(q - \sum_{k=1}^{q} 2.r_{ij}^k). \qquad (6)$$

Therefore, we can associate every object to be classified with a vertex of a graph whose cost values c_{ij} are given by :

$$\begin{cases} c_{ii} = 0, \\ c_{ij} = q - 2.\sum_{k=1}^{q} r_{ij}^k \quad \forall i,j : i \neq j, \end{cases}$$

With these values, Règnier's problem of (6) becomes a special instance of the Clique Partitioning Problem:

$$Min \sum_{i \leq i,j \leq n} c_{ij}.p_{ij},$$

subject to (1), (2), (3) and (4).

Table 3 gives the $c_{i,j}$ coefficients computed from Table 1. Figure 1 shows us its associated graph with the dotted (resp. plain and dashed) edges corresponding to the -1 (resp. 1 and 3) values of the $c_{i,j}$ costs.

A possible clique partioning of this graph is shown on Figure 2.

3 Genetic Algorithm

CPP is a well known NP-hard problem [12], therefore requiring some good heuristics to explore its potentially huge search space. de Amorim et. al [1] use

c_{ij}	1	2	3	4	5	6	7	8
1	0	-1	-1	-1	1	1	3	-1
2	-1	0	1	1	3	-1	1	1
3	-1	1	0	-1	-1	-1	1	1
4	-1	1	1	0	1	1	3	1
5	1	0	-1	1	0	1	-1	-1
6	1	-1	-1	1	1	0	-1	3
7	0	1	1	0	-1	-1	0	1
8	-1	1	1	1	-1	0	1	0

Table 3: c_{ij} costs

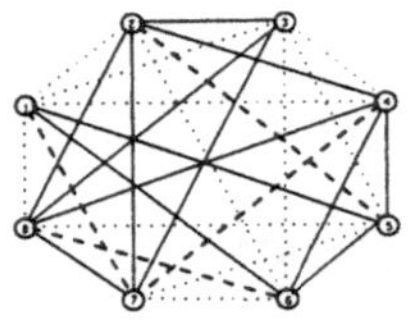

Figure 1: Complete weighted graph

simulated annealing and tabu search approaches. We report here some new results using the genetic algorithm approach on the same data sets.

Genetic algorithm is an optimization algorithm loosely inspired by the evolution theory. A candidate solution is encoded in a data structure called a *chromosome* (usually a bit string), whose fitness can be evaluated by the function to be optimized: the *fitness function.* A population of such candidate solutions is first randomly chosen; this population is then successively modified by mating the fittests amongst individuals, and this in turn, leads to a new generation of individuals mixing positive dominant characteristics of the two well-fitted parents, hopefully for the better. In other words, the mating strategy acts as a heuristic to guide the search into solution space regions extrapolated from the original position of the

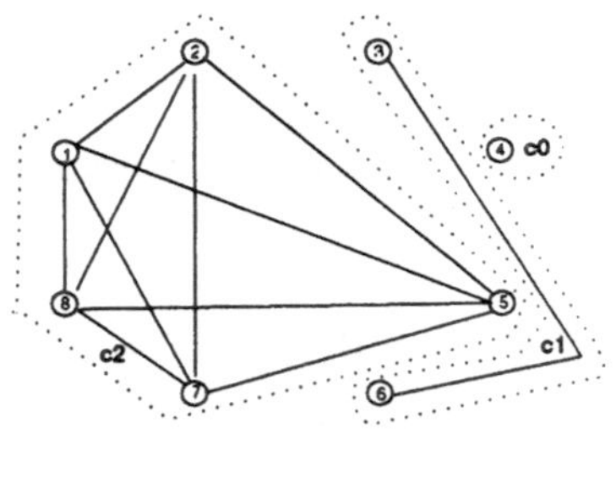

Figure 2: Three clique partitions

fittest parents. The genetic algorithm runs in four steps:

1. Random **initialization** of the population;
2. **Selection** of the fittest parents;
3. Creation of a new generation by chromosome **crossover** between parent pairs;
4. **Mutation** to avoid early convergence.

In this paper we use a generational genetic algorithm (similar to Grefenstette's Genesis program) in which new offsprings replace their parents in the population. We also use an elitist strategy ensuring the best individual to be kept from one generation to another.

Different crossover and mutation operators may be defined, mostly depending on the kind of coding techniques chosen for the problem at hand. In the sequel we shall present three different coding techniques for the CPP and their associated genetic algorithms.

3.1 Explicit Clique Label Coding

Let us start with the natural way to encode a clique partitioning problem in which the n vertices of the graph $G = (V, E)$ have to be partitioned in cliques (i.e. vertex set of a complete subgraphs) in order to minimize the criteria detailed in Section 2. Every such a clique corresponds to an equivalence class of the partitioning relation $\mathcal{P}$ and can therefore easily be encoded as a list of vertex labels.

3.1.1 Linear coding

We impose an upperbound k to the number of cliques. Each candidate clique partition therefore, can be represented by a list of labels encoded as a "chromosome" in the following way:

- the chromosome is split up into n equal length subparts called *alleles* (one allele for each vertex);
- each allele encodes an integer;
- this integer is taken between 0 and k and corresponds to the clique label associated with the corresponding graph vertex.

For example, the chromosome corresponding to the clique partitioning of Figure 2 is illustrated on Figure 3.

With this coding the search space has a $O(l^n)$ complexity.

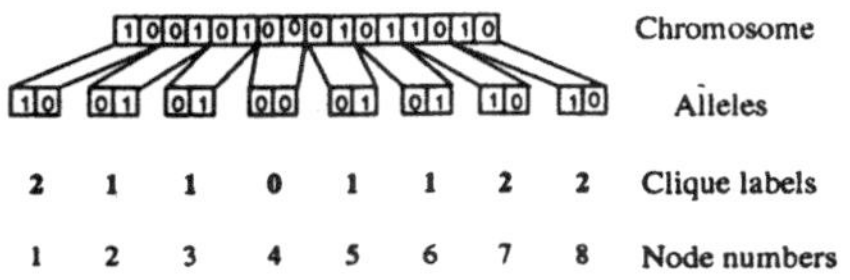

Figure 3: Linear Coding

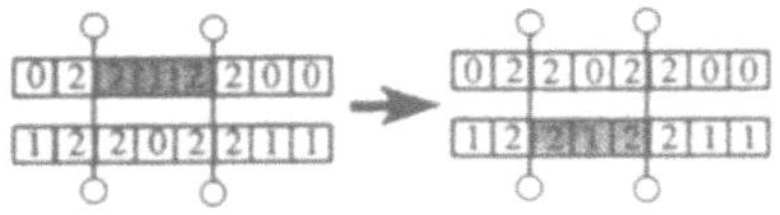

Figure 4: Double Point Crossover

Crossover is then applied on the fittest among these chromosomes in order to create a new generation of potential partitions. Figure 4 depicts at the allele level the classical "two point crossover" in which substrings between two randomly selected points are swapped.

With this partition coding however, the crossover operation may often combine alleles with different semantics, i.e. label definitions. The two chromosomes from Figure 4, for example, are isomorphic (permutation of clique labels) therefore corresponding to the same partitioning. By crossing over these two chromosomes however, alleles with different label definitions are mixed and this annihilates the crossover drive to combine new positive features while keeping common features.

3.1.2 Sequential ordering

In a first attempt to solve this problem, Jones and Beltramo ([4]) imposed an arbitrary ordering to these allele labels: by renumbering them, they force the clique to appear in increasing order within the chromosome. Chromosomes on Figure 4, for example, are renamed (see Figure 5) and the allele labels can now be put in correspondence with each other, leading to a meaningfuler crossover heuristic. With this ordering restriction, all the isomorphic chromosomes are eliminated from the search space, hence also reducing the complexity by a factor $k!$.

We tested this renumbering techniques on two real data sets (taken from [1]): the classification of 30 Cetacea described by 15 feature values and the classification of 158 countries based on their votes on 14 United Nation resolutions. Figures 6 and 7 show us the mean result of 20 GA runs on these two data sets. The x-axis represents the trial number (the number of

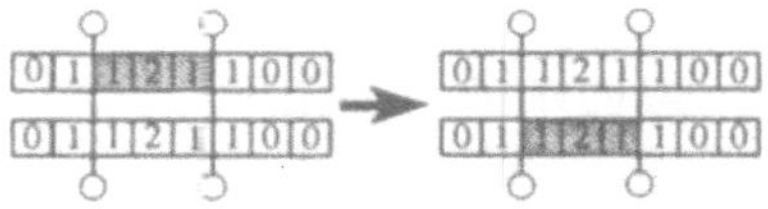

Figure 5: Sequential Renumbering

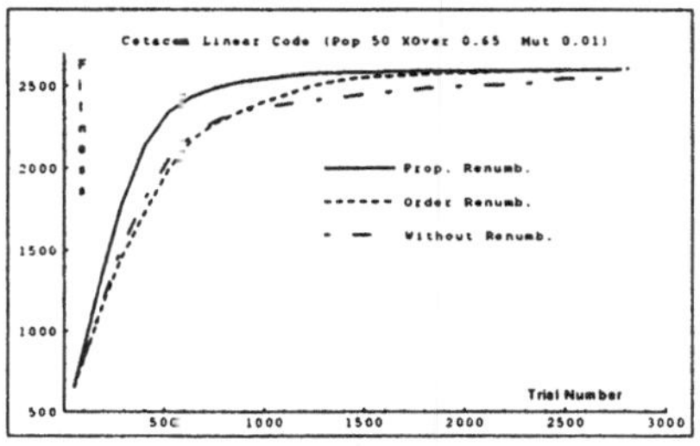

Figure 6: Cetacea classification by linear code

time the fitness function is activated) and the y-axis the fitness value

Combined with mutation however, this renumbering technique can also have some rather disruptive effects. Let us consider, for example, the allele string: 0 0 1 1 2 1 2 0. If the first allele is mutated to 2 we get after renumbering: 0 1 2 2 0 2 0 1. It differs completely from the chromosome before mutation. This is contrary to the normal mutation behavior in the genetic algorithm context where only local changes are performed on the population. Sequential renumbering combined with mutation may thus lead to some totally random behaviors.

3.1.3 Proportional ordering

In a new attempt to solve this crossover problem, we propose a new renumbering strategy, robuster for large chromosomes and less sensitive to these disruptive mutations. Instead of imposing the order of ap-

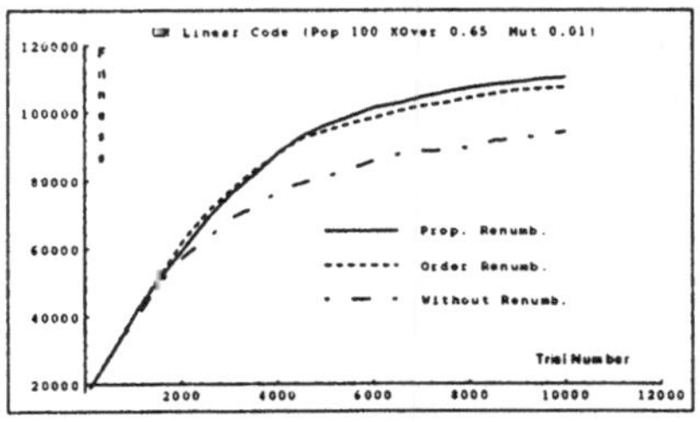

Figure 7: linear coding on UN resolutions

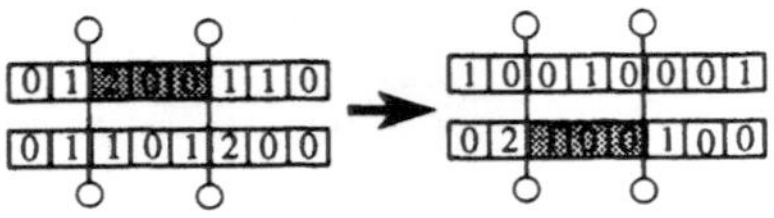

Figure 8: Proportional Renumbering

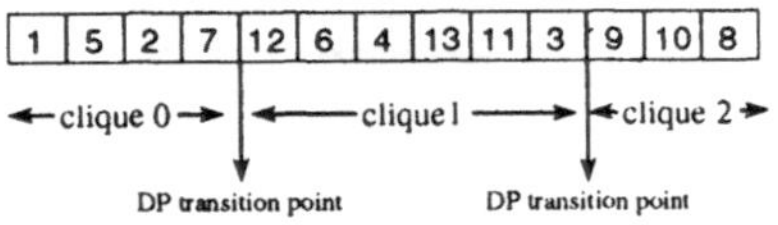

Figure 9: DP chromosome partitioning

pearance as clique labels, we order them according to the clique cardinality, the largest clique receiving label 0, the second largest clique label 1, etc. (see Figure 8 for an example).

Figure 6 and 7 show us how this proportional ordering renumbering technique slightly outperforms the other two approaches.

3.2 Hybrid approach

Jones and Beltramo ([4]) proposed another approach to avoid running into these renumbering problems. They use permutations of graph vertex numbers as an indirect way to encode clique partition in chromosomes. In their hybrid approach indeed, the dynamic programming algorithm (DP) slices the chromosome sequence by grouping in the same clique the vertices appearing between the optimal transition points found by DP as shown in Figure 9. Each permutation can be sliced in many different ways but only the optimal partition (the one found by DP) is kept, and its associated cost returned as fitness value.

In this hybrid approach thus, the genetic algorithm acts as a heuristic in isolating regions of the solution space to be searched with a polynomial complexity by a dynamic programming search for locally optimal solutions. We use Holland's permutation crossover operator, also called the *PMX crossover* (see [2]) to generate the new regions of this solution space. This PMX crossover is similar to the regular two point crossover with an additional step forcing the resulting chromosome to remain a vertex permutation. It swaps groups of vertices from the selected parents, therefore combining promizing clique assignments hopefully for the better.

Permutation coding does not suffer from any kind of renumbering problems since PMX crossover directly deals with vertices instead of clique labels as

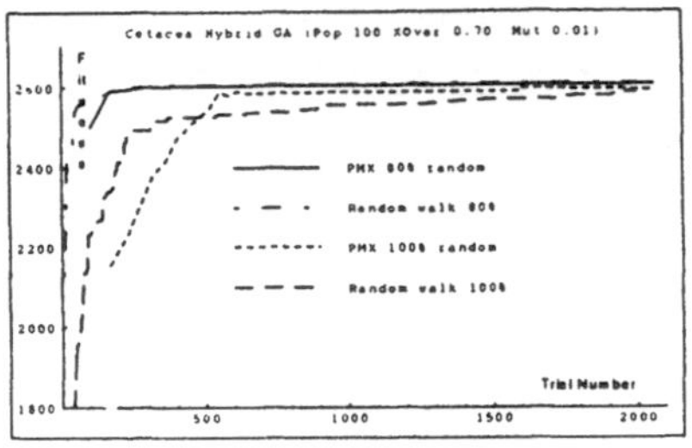

Figure 10: Hybrid GA for Cetacea classification

in the previous cases.

Figure 10 compares the average results on 20 runs of the hybrid algorithm with a 100% random initial population and with a 80% random initial population. The remaining 20% initial chromosomes were chosen as follows: randomly chose the first one, select its closest neighbor next, and so on. Figure 10 also presents the results of a random walk on permutations. On this 30 vertex cetacea graph, this smart initialization makes indeed most of the difference, leading the random walk to perform nearly as well as the GA. On the UN 158 country classification problem, fitness computation increases dramatically due to the intrinsic $O(n^4)$ complexity (see Appendix A) and experimentation becomes really time and resources consuming. We have not been able to improve much from the 80% random initialization as shown on figure 16.

3.3 Decision tree coding

We propose here a new coding technique to the Règnier's problem in terms of decision trees. A *decision tree* has its vertices labeled with a variable (associated with a feature of the object to be classified), its arcs labeled with the legal values of the corresponding feature variable, and its leaf vertices associated with the clique labels. Figure 11 depicts such a decision tree.

Chromosomes here, encode decision trees as simple classification programs to assign vertices to cliques. For example, Figure 11 decision tree applied to the objects of of Table 1 leads us to the clique partition shown on Figure 2.

With linear coding, genetic algorithms searched a partition space of complexity $O(k^n)$. With decision trees, clique coding becomes implicit and the genetic algorithm searches another solution space: the decision tree space which has a complexity[1] $O(k^{f^q})$ where

[1] Indeed, there is a maximum of f^q leaves in a decision tree with q features variable accepting at most f values as can be

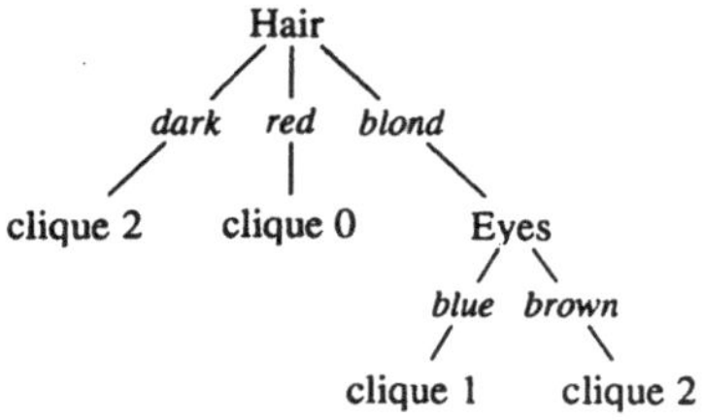

Figure 11: Decision Tree

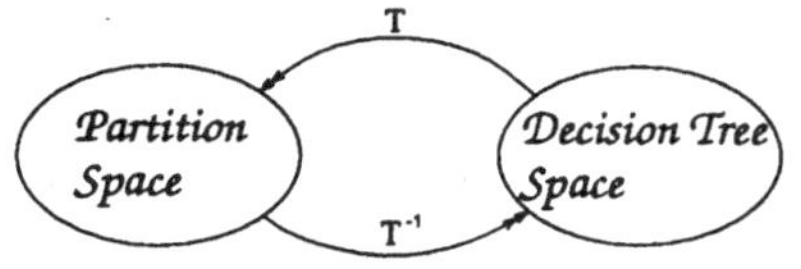

Figure 12: Two search spaces

k is the number of classes, q is the number of features (or equivalence relations $\mathcal{R}_k$ in Règnier's formulation) and f is the maximal number of values accepted by a feature variable. The solutions found in these two search spaces are equivalent as can be proved by showing how to go from one space into the other one (see Figure 12). The labeling algorithm of the decision tree can be used as a transformation T to go from the tree space into the partition space. In order to find the inverse transformation T^{-1}, we must prove that: for all objects X with a set of partition $\{\mathcal{R}_k\}$ leading to a partition $\mathcal{P}$ we can find a tree T such as $T(X) = \mathcal{P}$. And this can always be proved by constructing such a tree, layer by layer, with each layer representing a partition $\mathcal{R}_k$ on a variable k as shown on Figure 13 and by labeling the leaves accordingly to the clique partion $\mathcal{P}$.

In most cases, complexity of the decision tree space is larger than the partition space, many trees leading to the same clique partitions. With real data however, some feature variables take more importance than others and leads to solutions represented by small trees. The genetic algorithm can thus benefit from this increased redundancy of the decision tree space while restricting his search to small tree only. Figures 15 and 16 clearly shows us the power of this decision tree coding on our data sets.

seen on Figure 13). One of the k class should then be associated with every one of these leaves, hence leading to the given complexity.

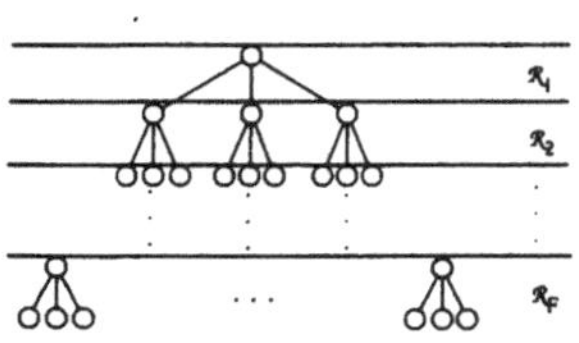

Figure 13: Proof tree

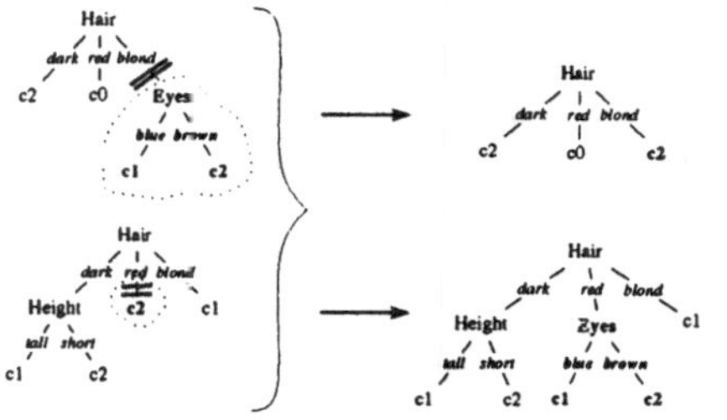

Figure 14: Tree CrossOver

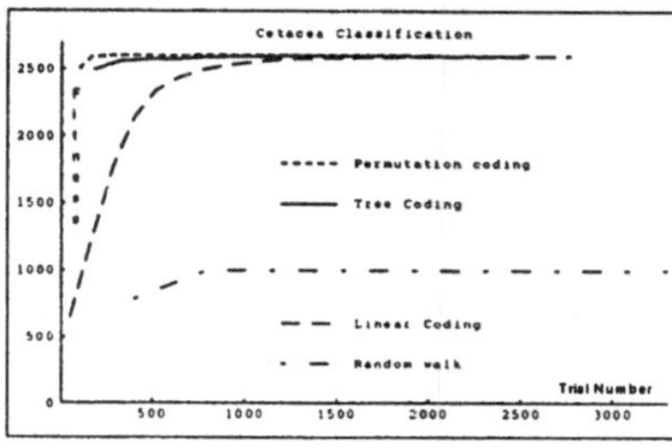

Figure 15: Cetacea classification

Furthermore, we also hope to improve the efficiency of the crossover operator. Indeed, when we dealt with explicit coding only clique labels were available, but with decision trees, more information is taken into account and feature value combinations are now manipulated by crossover instead of just clique labels.

3.3.1 Tree Crossover

Crossover on trees was introduced by John Koza for his research on evolution of Lisp programs by genetic algorithms (see [5] for example). Tree crossover consists in exchanging subtrees between chromosomes as depicted on Figure 14. These subtrees may be regular subtrees or leaves (terminal vertices) depending on a leaf/subtree ratio given as a parameter to the genetic algorithm. Other parameters can also be added to impose a maximal depth to the trees and/or to characterize the initial population of trees.

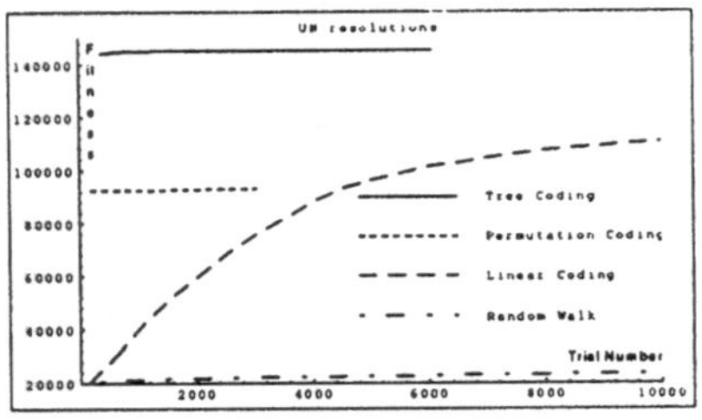

Figure 16: UN country classification

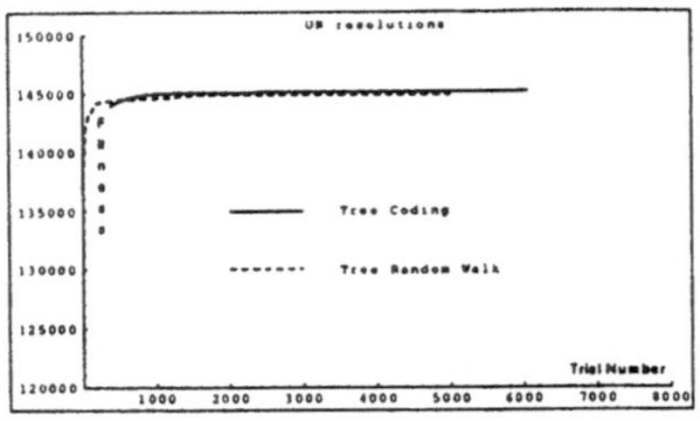

Figure 17: Random Walk on Trees

3.4 Experimental results

Figure 15 and 16 compare the average results of 20 runs of this decision tree based GA with the two previously presented approaches (plus a random walk on linear codes). We compare these three approaches on the two data sets of the Cetacea and UN resolutions. Crossover on trees were performed with a 0.5 probability of cutting at the leaf level. Initial trees on the other hand had a maximal depth of 4 with a 0.5 leaf/subtree ratio.

Hybrid GA with permutation coding slightly outperforms tree coding on Cetacea, but as it was pointed out in Section 3.2 this is mostly due to the smart initialization of 20% of the initial population. This results does not seem to hold for larger graphs and on UN country classification, permutation encoded GA does not improve much the smart initialization. However, due to the cpu time required by the DP on a 158 vertex graph (see Table 4), we stopped investigate any further this approach.

On the UN country classification many runs of the regular GA do not converge to the optimal partition, even with our new renumbering proportional to the clique cardinality. This is due to the fact that GA cannot take advantage of any kind of locality in the clique partitioning problem. Moving a vertex from a clique to another one indeed, can drasticly change the fitness value. Crossing over two promizing chromosomes therefore, does not give us any more chances to get good children. The GA shemata model [2] does not seem to hold for this clique partioning problem. Furthermore, the redundancy rate of optimal solutions is rather low in comparison with other problems for which GA have been proved to work well and this further handicaps the GA search.

Only the decision tree based GA converges to the optimal solution in a reasonable number of iterations. This fast convergence however, might also be due to the data themself. Some UN resolutions indeed have more discriminant power than others in classifying countries from their votes. Small and simple trees are therefore often sufficient to reach the optimal solution and random walks on such trees might find them as well. Figure 17 shows us how tree encoded GA outperforms only slightly pure random walks on trees with a maximal depth of 4 (4 on the 14 potential UN resolutions). This shows us the power of decision tree coding in comparison with linear or permutation coding on real life data.

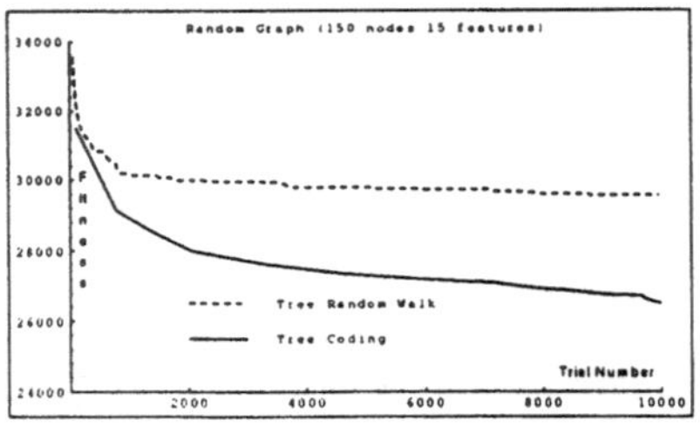

Figure 18: Random Graph

In order to show the positive contribution of GA, we have generated a random graph from a random description of 150 objects with 15 variables potentially taking 6 different values each and 8 different leaf values (only 8 cliques allowed). Results are plotted on Figure 18 which indicates the contribution of GA's in solving this random graph.

Table 4 presents the average cpu time of one fitness value computation and the average cpu time required to reach convergence in the three approaches. These experiments were ran on a NeXtStation and code optimization was far from being our main concern. Just as a comparison the cpu time required to solve the UN country classification with simulated annealing and tabu search on a SUN SPARC workstation were 7.51 sec and 14.95 sec respectively (see [1] for more details).

Coding	cpu/trial	Conv. cpu
Cetacea		
Linear	2.6 msec	7.8 sec
Permut.	116 msec	23 sec
Tree	11 msec	13 sec
UN		
Linear	25 msec	300 sec
Permut.	23 sec	9200 sec
Tree	34 msec	130 sec

Table 4: CPU time

4 Conclusion

In this paper we used Genetic Algorithms for a special case of the Clique Partitioning Problem applied to the classification of objects described by a set of feature values (Règnier's problem). We show how critical coding can be. We proposed a new coding technique and we compared it with the other approaches.

Linear code is the most widely used GA coding technique for the CPP ([4],[10],[11]). However, we faced rather poor results on the partitioning of large graphs. We somewhat improved these results with a new label renumbering technique but too many runs did not get to the optimal solution. We discussed why the Clique Partitioning Problem does not seem to fit GA paradigm well. Further studies should be carried out though to confirm this point, for example applying Manderick's statistical fitness landscape analysis tools [7].

Anyway, we tried Jones and Beltramo's hybrid approach mixing dynamic programming with GA. It worked rather well on small graphs but as soon as the graph dimension grows, the $O(n^4)$ DP complexity increases the fitness computation time too much to be useful in the GA context. We also observed the important impact of a smart initialization

We finally proposed the use of decision tree structured chromosomes to solve Règnier problem with GA. We showed how this coding technique efficiently lead the GA to optimal solutions on our data sets. We also stressed the power of this coding technique on real life data by comparing tree encoded GA with a pure random walk in the small decision tree solution space. We then tested the robustness of this decision tree coding technique on a purely random tree. We think that this coding technique could also be used with success within the framework of other optimization paradigms as Simulated Annealing or Tabu search.

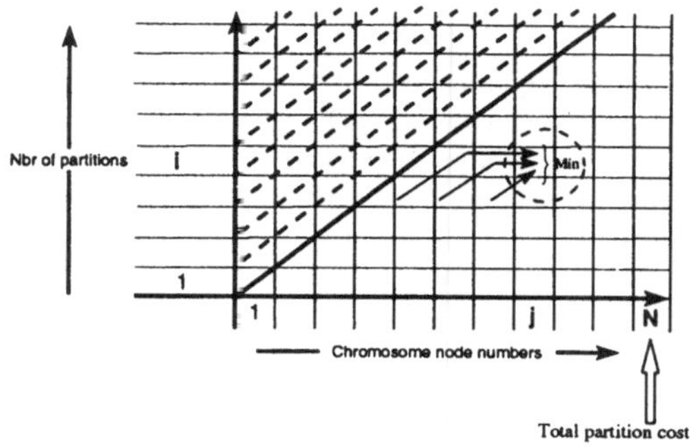

Figure 19: DP cost computation

A Appendix: Dynamic Programming

Dynamic programming is a very efficient technique to find optimal solutions without any redundant computation when the Bellman criteria applies, i.e. when an optimal solution can be constructed from optimal partial solutions. In our case, the optimal solution to be found is the optimal partition of the n vertex sequence and the partial solutions are the partitions of j vertex subsequences ($j < n$).

Figure 19 depicts the dynamic programming partial partition cost table. The horizontal axis represents the chromosome vertex number sequence and the vertical axis the number of partitions. An element $c(i,j)$ of this table represents the cost of optimally partitioning the first j chromosome vertices in i cliques. The minimum value of the last column gives us the optimal partition cost and its associated clique cardinality.

This table can be filled line by line: $c(i,j)$ is computed from the elements of the previous line. Indeed, if the last clique of this optimal partitioning has k elements the optimal cost will be the optimal cost partitioning the first $j-k$ chromosome vertices in $i-1$ clique plus the cost contribution of the last clique:

$$c(i,j) = c(i-1, j-k) + last(j,k).$$

To satisfy the Bellman criteria, every admissible k value must be checked and only the optimal one kept. Here follows the dynamic programming algorithm :

```
FOR every line i DO
   c(i,i) = 0
   c(i, i+1) = c(i-1, i)
   FOR j = i+2 UNTIL j < n
      c(i,j) = c(i-1, j-1)
      FOR k=2 UNTIL k ≤ j-i+1
```

$$c(i,j) = max_k[c(i\text{-}1, j\text{-}k) + last(j,k)]$$

One can easily notice that the last clique contributions are identical for each element of the same column. These contributions have to be computed in advance thus and this is the critical part of the DP algorithm leading to its $O(n^4)$ complexity on the number of vertices. Here follows the algorithm to fill in the *last* table:

```
last(0,1) = 0
FOR L=1 UNTIL L < n
  last(L,1) =0
  FOR i=2 UNTIL i ≤ L+1
    FOR j=0 UNTIL j < i-1
      FOR k=1 UNTIL k < i-j
        last(L,i) = last(L,i) +
                    w(chr(L-j),chr(L-k-j))
```

where $w(i,j)$ is the weight associated with the edge between vertices i and j and $chr(i)$ is the i-th vertex in the chromosome vertex sequence.

Acknowledgement:

The author would like to thank Saul G. de Amorim for sending him the data sets, Jean-Pierre Barthélemy for his suggestion to try tackling Règnier's Problem, and Eric Rose for his help in programming NeXtstations.

References

[1] S.G. de Amorin, J.-P. Barthélemy, C.C. Ribeiro, "Clustering and Clique Partitioning: Simulated Annealing and Tabu Search Approaches", *Journal of Classification*, **9**:17-41, 1992.

[2] D. Golberg, *Genetic Algorithm in Search, Optimization and Machine Learning*, Addinson Wesley, Reading, 1989.

[3] L. Hérault, J-J. Niez, "How neural networks can solve hard graph problems: A performance study on the graph K-partitioning", *Neuro-Nimes'89 Int. workshop on Neural Networks & their applications*, Nimes, France, pp.237-255, Nov 1989

[4] D. Jones, M. Beltramo, "Solving Partitioning Problems with GA", *4-th conf. on Genetic Algorithms*, 1991.

[5] J. Koza, *Genetic Evolution and Co-Evolution of Computer Programs*, Proc. of Artificial Life II, Santa Fee, 1990.

[6] Y.-H. Pao, *Adaptive Pattern Recognition and Neural Networks*, Addison Wesley, Reading, 1989.

[7] B. Manderick, M. de Weger and P. Spiessens, "The Genetic Algorithm and the Structure of the Fitness Landscape", *4-th conf. on Genetic Algorithms*, 1991.

[8] S.Règnier, "Sur quelques aspects mathématiques des problèmes de classification automatique", *I.C.C Bulletin*, **4**: 175-191, 1965.

[9] R.L.Russo, P.H.Oden, P.K.Wolff, "A heuristic procedure for the partioning and mapping of computer logic graphs", *IEEE Trans on Comp.*, Vol.C-20, **12**: 1455-1462, Dec 1971.

[10] E-G. Talbi, P. Bessiere, "*A Parallel Genetic Algorithm for the graph partitioning problem*", ACM Int Conf on Supercomputing, Cologne, Germany, June 1991.

[11] E-G.Talbi, T. Muntean, "*Static allocation of communicating process on a parallel architecture*", Int Conf. on High Speed Computation II, Montpelier, M.Durand and F.Dabagh (Editors), Elsevier Science Pub, North Holland, pp. 71-82, Oct 1991.

[12] Y. Wakabayashi "Aggregation of Binary Relations: Algorithmic and Polyhedral Investigations", Doctoral Thesis, Universität Augsburg, 1986.

Self-Organization of Communication in Distributed Learning Classifier Systems

Norihiko Ono and Adel T.Rahmani
Email: ono@is.tokushima-u.ac.jp
Department of Information Science and Intelligent Systems
Faculty of Engineering, University of Tokushima
Tokushima 770, JAPAN

Abstract

In this paper, an application of learning classifier systems is presented. An artificial multi-agent environment has been designed. Mate finding problem, a learning task inspired by nature, is considered which needs cooperation by two distinct agents to achieve the goal. The main feature of our system is existence of two parallel learning subsystems which have to agree on a common communication protocol to succeed in accomplishing the task. Apart from standard learning algorithms, a unification mechanism has been introduced to encourage coordinated behavior among the agents belonging to the same class. Experimental results are presented which demonstrate the effectiveness of this mechanism and the learning capabilities of classifier systems.

1 Introduction

Machine learning paradigms requiring a supervisor are not appropriate when faced with learning tasks for which no such supervision may be provided. For example, autonomous agents, such as robots cannot be given explicit information about the adaptive behavior corresponding to different situations they may encounter. Required knowledge must be inferred through interaction with the environment. Furthermore, since this knowledge is not available all at once, they must learn it incrementally. These two characteristics are the two most important distinguishing features of a class of learning methods called incremental reinforcement learning.

In a reinforcement learning problem, a learning agent receives sensory information from its environment and choose an action to send to the environment, hence changing the environmental state. In addition, a reward is assigned with some of these states. The goal of learning is to maximize the cumulative reward received over time.

Classifier systems (CSs), proposed by John Holland, are promising learning paradigm to fulfill these needs. They have been used to study the behavior of adaptive organisms and autonomous agents.

2 The Problem Domain

The primary task domain we have used is mate finding in a population of artificial animals. This problem has been also studied by G. M. Werner and M. G. Dyer [8] using a recurrent neural network approach. We have tried to adhere to most of their assumptions and constraints so that the results obtained by the two approaches may be compared. In their implementation, they have used genetic operators to combine the animals' genome which encode the weights and biases of their neural networks, and no learning takes place in those networks. This almost corresponds to the strength adjustment of classifiers by Bucket Brigade Algorithm (BBA).

The environment is defined as a 50 by 50 grid where edges are continued toroidally. Each cell is either empty or occupied by an animal. We randomly place 50 females and 50 males into the environment, so the population density is 4%. A typical environment is shown in Figure 1. Females are represented by "F" and males are represented by characters indicating their orientations. Empty positions are shown by blanks.

Our female animals have the capability to look at their surroundings and when detecting a male in a nearby position (within her visual field) then producing a sound which is represented as a signal pattern. They have a repertoire of eight such signals. When they do not detect any male, they just keep silent or the emitted signals do not have any significance since no male can hear them.

On the other hand, male animals can only hear the sound which has been emitted by the nearest female in their "auditory field". Upon receiving a sig-

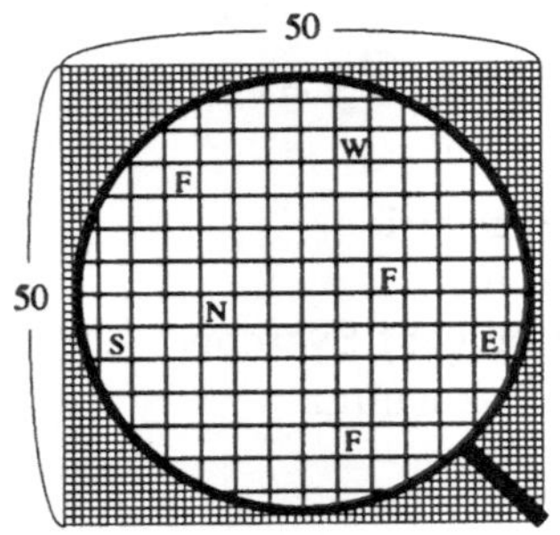

Figure 1: Environment for Artificial Organisms.

nal they interpret it as one of four possible actions they can take (MOVE_FORWARD, TURN_RIGHT, TURN_LEFT, STAND_STILL). If no signal is heard then they just take an action as dictated by their rule set.

As there are two types of animal in our simulated environment, we consider two sets of classifier systems to represent each class. Each animal is associated with a classifier system of its own class which models the animal's behavior. For both systems we use classifiers with just one condition and one action. Our classifiers for the female system can be represented by the following IF-THEN rule:

- *if* (*a male is detected*) & (*its position is P*) & (*its direction is D*) *then emit signal S*.
- *if* (*no male is detected*) *then emit some specific signal s*.

And for the male system as follows:

- *if* (*signal S is detected*) *then take action A*.
- *if* (*no signal is detected*) *then take action A*.

At every iteration, at first all female animals scan their nearby positions in a specified order, and accordingly generate their signal patterns, each represented as three-bits. Then male animals detecting these signals make a move and if a mating happens between a male and a female (i.e. the male gets to the female's position) then they both receive a payoff from environment (as an indication for their need satisfaction). After the mating happens, they are placed into new random positions. Our mating scheme seems a little odd since no offspring is generated as a result of it.[1]

This problem, as just was described, falls within the category of *animat problem* suggested by Wilson [10] and may be considered as: incremental learning of multiple disjunctive concepts (though limited) under delayed payoff.

3 The Basic System Description

In standard classifier systems, there is only a single set of rules (classifiers) which interacts with the environment. Also in the parallel or distributed versions of classifier systems, generally a number of similar classifier systems are considered. Here we deal with a system having multiple copies of two different sets of classifier systems (i.e. sets of males and females).

Basically, there are two learning mechanisms in classifier systems. The first is accomplished by the credit assignment algorithms by adjusting the associated strengths of a fixed set of rules. In Holland's model of classifier systems (Michigan approach) [3], BBA, a model based on an analogy with a service economy, is responsible for this task. The second mechanism is creation of new rules by the rule discovery algorithms. The new rules are inserted into the rule set to compete with other rules and to be evaluated under BBA. Again, in standard classifier systems Genetic Algorithms (GA) have been the main rule discovery algorithm. Since our main concern has been to observe self-organization of a kind of communication protocol among a population of artificial animals, to make things simpler we have preferred not to use GA as rule discovery mechanism. Instead all the CSs are initialized with a fixed set of rules and only their strengths are adjusted mainly under BBA. Here we apply GA to encourage these CSs to organize coordinated behavioral strategies and eventually to agree on a common communication protocol to succeed in accomplishing their tasks. Individual CSs are treated as individuals in GA; their strength vectors are regarded as their chromosomes, and a newly-proposed genetic operator, called *unification* operator, is applied to them. This experiment shows how effectively BBA in conjunction with unification operator can assign strength to classifiers under a delayed reward scheme and even form short classifier chains without classifiers being coupled to each other explicitly.

The basic execution cycle also differs from the standard one as shown in Figure 2. The action of winner classifier from a female classifier system, after being

[1] Werner and Dyer [8] treated males and females as the same kind of animals (i.e. they share common chromosome structures), while we treated them as totally different ones from each other. In this sense, we attempt to let males and females co-evolve a common communication protocol.

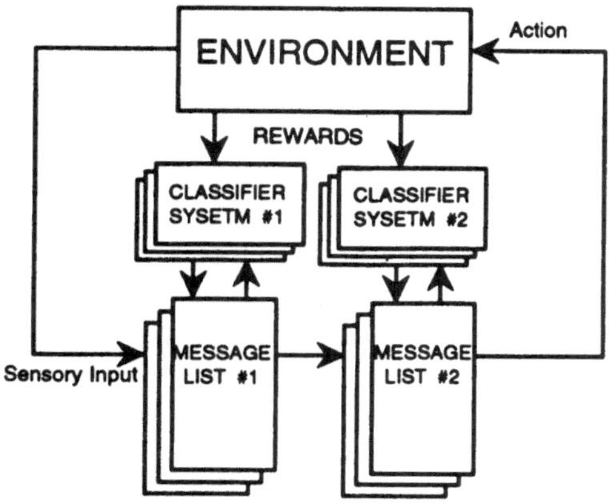

Figure 2: Structure of Distributed Learning Classifier Systems.

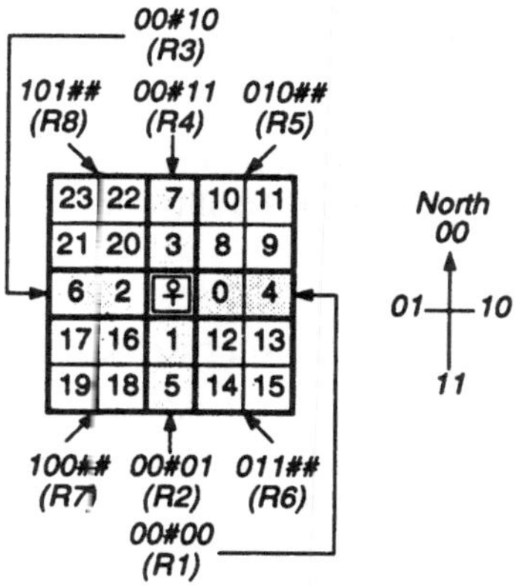

Figure 3: Representation of Male's Positions and Orientations. These 24 positions are classified into 8 regions, $R1$-$R8$. By using don't case symbols, each of these regions can be represented as a single pattern of symbols.

placed into message list, instead of effecting the environment directly, is moved into the message list of a male classifier system in a nearby position. Then the action of winner classifier in this classifier system will indicate how the environment will change.

4 Implementation

4.1 Representation

One of the most important steps in implementing a classifier system is deciding how to represent or encode the environmental states. This problem has been addressed by some researchers [1, 7], but probably there is no unique solution to handle this problem in all domains although there might be some general guidelines. To this end, devising a good representation remains as an art. One characteristic of a good representation is to reduce the number of rules which has to be learned by the system(solution set size) by allowing some generalization to be possible.

To make this argument clear, let us consider an example from our problem domain. There are 24 positions around a female which constitute her visual field. We have to consider an ordering for these positions to be scanned by the females. If we randomly or sequentially number those positions, then for the 24 positions and 4 directions, the classifier system has to learn about 96 rules to deal with all possible situations. Instead if we adopt the ordering shown in Figure 3 with an appropriate encoding of the four directions, we can reduce the number of rules which must be learned to 24 (reducing the solution set size by 75%!).

4.2 Bucket Brigade Algorithm

We use a standard Bucket Brigade Algorithm similar to the one described in [6]. We select the winner classifier by a probabilistic selection algorithm based on the effective bids of the competing classifiers. The effective bid is calculated as a power of each classifier's bid. We used a power of 2 to bias the selection toward classifiers with higher strength. We impose a biding tax of 5%. No head (life) tax is necessary since we use a fixed set of rules.

We reduce the taxes from classifiers' strengths only if their strengths are more than the initial strength. This means that they have to pay these taxes from what they have gained, something like an *income tax.* This puts more pressure on the winning classifiers and if they cannot compensate for this tax then soon their strengths are reduced to the level of other competing classifiers. This simple strategy turned out to be very effective, specially in preventing looping. For example, when the males are trapped in turning left and right successively or just moving forward when they are out of visual field of any female.

In standard Bucket Brigade Algorithm, a reward is paid to the current winner classifier(s) when the system gets into a state with such an associated reward. Since as part of BBA we keep track of previous winner, we think it will be useful if we give a portion of the reward to the last winner which is readily available. This simple extension will speed up the reinforcement and also helps the formation of short classifier chains.

Also to speed up reinforcement, we used a low initial strength (10) for all classifiers and a high amount of fixed reward (1000).

4.3 Unification Mechanism

Apart from the bucket brigade algorithm, we use a unification mechanism to encourage coordinated behavior of these animals.

During the initial stages of our experiments, we noticed that if we let these animals behave freely, then almost always all the possible signals would be interpreted as MOVE_FORWARD by the male animals. This was not surprising for a number of reasons: first the rules with a MOVE_FORWARD action are the only possible candidates for receiving external payoff from the environment, second from the females' point of view, there are four situations from which a mating will happen by a male's MOVE_FORWARD. So even in the best case if we assume that a female will use only one signal in all those four cases, then without any coordination among females, we need about 50 (number of females) different signal patterns just to cover those situations.

One way to overcome this difficulty was to make our females more intelligent and let them know that when they use a signal in a situation to mean MOVE_FORWARD then in all other cases which they want to encourage males to moving forward, they can use the same signal. Moreover, the females had to share this knowledge globally among themselves so that they all use only a few signals to be interpreted as MOVE_FORWARD by the male animals. Since communication capability of these animals is fairly limited, this approach may not be feasible.

Instead we decided to let them share their experiences locally without making our animals superficially intelligent. At some fixed intervals (say every 100 steps) a female is selected randomly and if she can find another female in a nearby position, then the strengths of all classifiers in their CSs are averaged pairwise. The same procedure is also performed for a pair of male animals. We call this a *unification* operator, because it will allow our animals to unify and coordinate their behavior with each other.

Here we regard strength vector of each animal as its chromosome, and unification operator corresponds to a crossover operator in GA; males and females attempt to learn their behavioral strategies both empirically and genetically, by using BBA and the unification mechanism, respectively.

5 Results

5.1 Performance of Organized Communication

To obtain a performance reference for comparing our results with, we performed a number of experiments with both CSs for males and females initialized with a perfect set of rules. That is we assigned one signal for each of the three useful moves (excluding STAND_STILL) and then initialized the females' set of rules with 25 classifiers; 24 classifiers for the 24 possible situations plus one for no male detection. Similarly, we initialized males' set of rules with 7 classifiers; 3 for when a signal is received from a female and 4 classifies for no signal case. So to see how our learning system works, we compare our results with the results we get from the above simulations.

Figure 4 shows the performance when we use a moving average of the number of steps to mating as the performance measure. We let all of animals have at least 200 number of matings and record an average over every 10 matings.

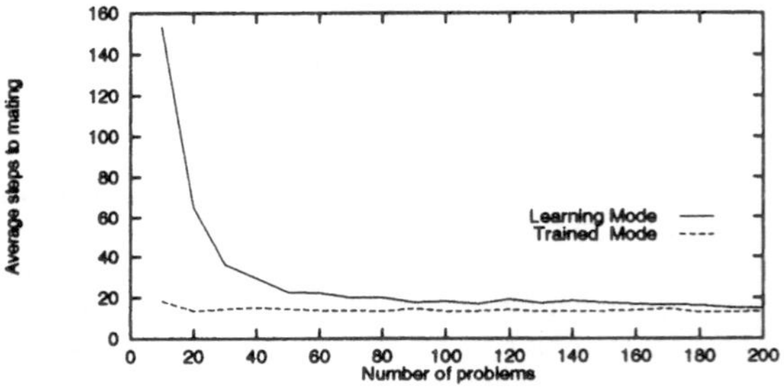

Figure 4: Male animals learn how to reduce average steps to mating.

One point to mention about our reported results is that usually in experiments with genetic algorithms and classifier systems, an average result over a number of simulations (usually a minimum of 5 runs) is reported. Because of stochastic nature of these algorithms, the results may differ significantly from run to run. But we got a fairly uniform results in many experiments we performed due to the fact that our system internally operates with a number of agents and hence the effect of stochastic errors are eliminated or reduced.

Another performance measure we considered was the percentage of number of correct steps taken by

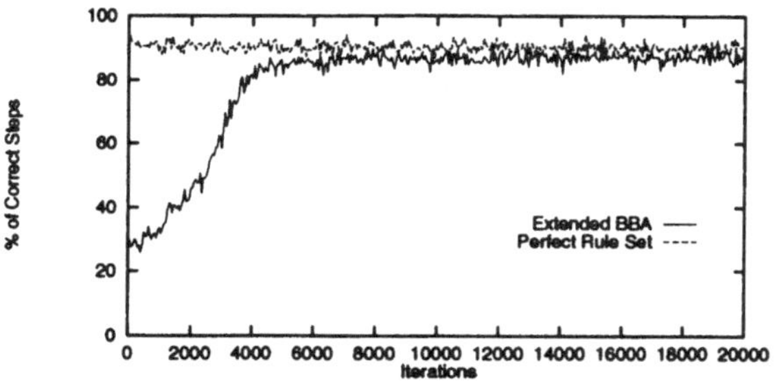

Figure 5: Performance of System.

males. We count the number of correct steps to mating when a male is within the visual field of a female and report the result every 50 iterations. Figure 5 shows the result.

Again this is compared with the same measure obtained from experiments with perfect set of rules. The reason that even under a perfect set of rules the system does not reach a 100% correct performance is that there are many misleading interactions among these animals. For example, it is possible that more than one male be within the visual field of a female, but the female can only detect the nearest one and send signal appropriate for that one's position and orientation. But the same signal is received by the other male which may not be a good one in its case. This shows how noisy our environment may be. This result is also better than a previous one we had reported earlier in [5]. The improvement is due to the two-step reward scheme which was explained before.

5.2 Organization Process of Communication

During runs of our simulation, collective behavior of males and females changes as follows:

1. At initial iterations, both males and females behave randomly. Males have not established a common understanding of input signals (Figure 6 (a)).

2. At iteration 3000, those males that stand still have been extinguished. By this time no male interprets any signal as standing still (Figure 6 (b)). At the same time, males and females have totally agreed upon some signals ("001" and "111").

3. After 10000 iterations, males have already evolved a unified interpretation for all signal patterns (Figure 6 (c)). Since those tasks performed by females are much more complicated than those by males, their behavioral patterns have not been converged even in this stage. However, once males have evolved to interpret female signals in such a uniform manner, they start to converge, and a communication protocol as shown in Figure 7 is organized gradually.

This evolutionary process of communication is different from that of Werner and Dyer's results [9]. According to their results, males that usually go straight take over the population at early stage of the evolution; then males appear that turn appropriately when in the same row or column as a nearby female. On the other hand, we have never observed such phenomena in any runs of our simulation; there have been no stages that most of males simply keep moving forward irrespective of the signals they receive.

Table 1 shows the final result. By observing females' responses and males' interpretations, we can see a communication protocol has been organized among them. All the male animals interpret signals "001", "010", "100", and "111" as MOVE_FORWARD, signals "000" and "110" as TURN_LEFT, signals "011" and "101" as TURN_RIGHT and no signal as STAND_STILL. Then when the females detect a male in region 1 facing West, they send signal "111" which is interpreted as MOVE_FORWARD by males. Or when they detect a male in region 8 facing North, majority of them send signals "011" or "101" which means to turn right. Only in a few cases there remains some misunderstandings, but the overall result is quite satisfactory.

6 Concluding Remarks

This paper has presented a learning classifier system. Two subtasks have been learned in order to perform a goal task. The results indicate that the Classifier Systems are promising learning paradigm in studying adaptive behavior of autonomous agents and artificial organisms.

We have also seen that how effectively Bucket Brigade Algorithm can apportion the credit which is essential to any learning system and that it can work in conjunction with our unification operator even in a multi-agent environment.

There are many interesting ways we can add to the complexity of our problem domain. All these remain for the future research.

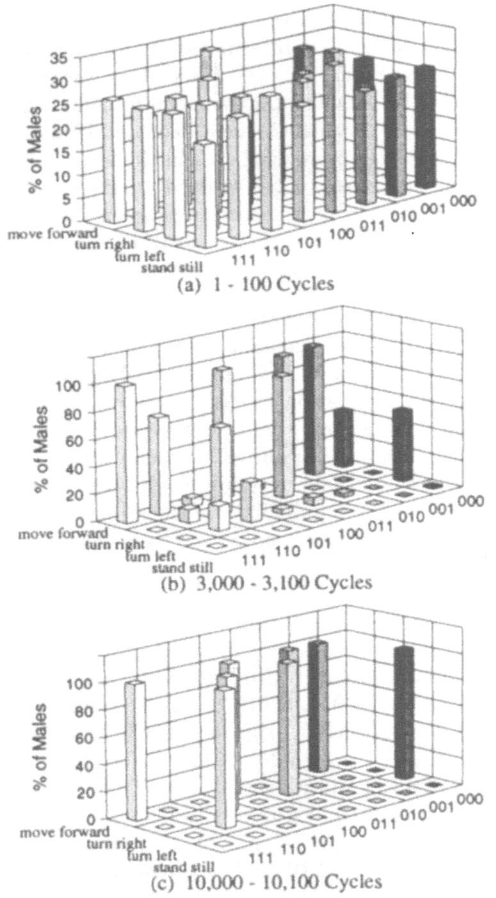

Figure 6: Typical Transition of Responses of Males. In this case, males evolve that interpret signals "000" and "110" as TURN_LEFT, signals "011" and "101" as TURN_RIGHT, and the remaining ones as MOVE_FORWARD.

REGION	DIR	000	001	010	011	100	101	110	111
R1	N	50	0	0	0	0	0	0	0
R1	W	0	0	0	0	0	0	0	50
R1	E	0	0	0	50	0	0	0	0
R1	S	0	0	0	50	0	0	0	0
R2	N	0	0	0	0	0	0	0	50
R2	W	0	0	0	50	0	0	0	0
R2	E	50	0	0	0	0	0	0	0
R2	S	2	0	0	48	0	0	0	0
R3	N	0	0	0	50	0	0	0	0
R3	W	0	0	0	49	0	1	0	0
R3	E	0	50	0	0	0	0	0	0
R3	S	50	0	0	0	0	0	0	0
R4	N	0	0	0	0	0	0	50	0
R4	W	50	0	0	0	0	0	0	0
R4	E	0	0	0	50	0	0	0	0
R4	S	0	50	0	0	0	0	0	0
R5	N	16	0	0	1	5	0	28	0
R5	W	0	50	0	0	0	0	0	0
R5	E	1	0	3	23	3	16	0	4
R5	S	0	0	50	0	0	0	0	0
R6	N	0	0	0	0	0	0	0	50
R6	W	0	0	0	0	50	0	0	0
R6	E	17	3	3	1	0	8	15	3
R6	S	0	1	2	12	5	16	9	5
R7	N	0	0	50	0	0	0	0	0
R7	W	0	0	0	7	1	42	0	0
R7	E	0	50	0	0	0	0	0	0
R7	S	11	2	1	3	2	0	28	3
R8	N	0	0	0	16	1	28	3	2
R8	W	36	0	1	0	1	2	7	3
R8	E	0	0	50	0	0	0	0	0
R8	S	0	0	0	0	50	0	0	0
MOVE_FORWARD		0	50	50	0	50	0	0	50
TURN_RIGHT		0	0	0	50	0	50	0	0
TURN_LEFT		50	0	0	0	0	0	50	0
STANDSTILL		0	0	0	0	0	0	0	0

Table 1: Typical Communication Protocol.

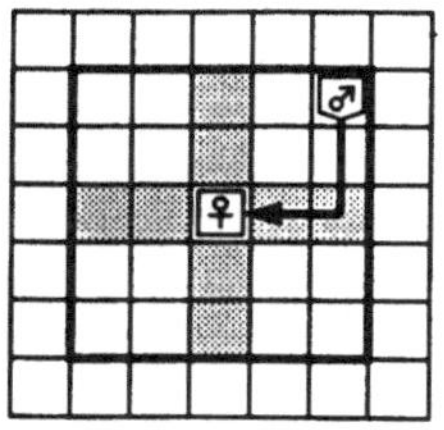

Figure 7: Organized Communication Protocol. When finding a male in cross-shaped (shaded) regions, a female tells the male to keep turning until he is facing her, by sending appropriate signals. Then she changes her signal to let him to move straight to her. Males entering her visual field but outside of these regions are told to simply move forward.

References

[1] Booker, L. B., *'Representing Attribute-based Concepts in a Classifier System'*, Foundations of Genetic Algorithms, 1991.

[2] Holland, J. H., *'Properties of the Bucket Brigade'*, Proceedings of the First International Conference on Genetic Algorithms and Their Applications, 1985.

[3] Holland, J. H., *'Escaping Brittleness: the possibilities of general-purpose learning algorithms applied to parallel rule-based systems'*, Machine Learning: an Artificial Intelligence Approach, Vol. 2, 1986.

[4] Goldberg, D. E., *'Genetic Algorithms in Search, Optimization, and Machine Learning'*, Addison Wesley, 1989.

[5] Rahmani, A. T., Ono, N. (1992), *'Genetic Evolution of Communication in Distributed Classifier Systems'*, Proceedings of the 45th National Conference of Information Processing Society of Japan, 1992.

[6] Riolo, R. L., *'Bucket brigade performance: I. Long sequences of the classifier'*, Proceedings of the Second ICGA, 1987.

[7] Schuurmans, D., Scaeffer, J., *'Representational Difficulties with Classifier Systems'*, Proceedings of the Third ICGA, 1989.

[8] Werner, G. M., Dyer M. G., *'Evolution of Communication in Artificial Organisms'*, Artificial Life II, Addison-Wesley, 1991.

[9] Wilson, S. W., *'Knowledge growth in an artificial animal'*, Proceedings of the First ICGA, 1985.

[10] Wilson, S. W., *'Classifier systems and the animat problem'*, Machine Learning, **2**, 1987.

Design of Digital Filters with Evolutionary Algorithms

Thomas Görne and Martin Schneider*
Technical University Berlin

Abstract

A recursive digital Filter with infinite impulse response (IIR) is characterized by its recursive and non-recursive filter coefficients and the corresponding filter structure. In the well-known filter design algorithms the structure has generally to be chosen beforehand together with the maximum allowed filter length.

Speed of computation can be an important aspect in determining the maximum permissible filter length and therewith the filter structure in an application. Specialized 'slim' IIR filter (SIIR) structures are proposed as a generalized class of recursive filters including the classical forms. In SIIR structures a large number of coefficients is set to zero, thereby reducing the total number of non-zero coefficients and thus the computational cost.

1 Introduction

Filter designers are used to handle a multitude of different filter structures. Distinctions run between e. g. finite impulse response (FIR) and infinite impulse response (IIR) filters, direct and casçaded forms, lattice structures etc. Common to most filter design procedures is the fact that the designer has to specify a filter structure beforehand, compute the filter coefficients and then decide if the chosen filter fulfills the requirements.

This procedure can generally lead to satisfying results if used diligently. The chosen filter structure may nonetheless prove not to be the optimal one, as most design algorithms concentrate on less complicated structures. Furthermore there is the difficulty of "nulling out" negligeable coefficients, increasing the efficiency, for a defined allowed error in the design. We therefore decided to investigate an approach with more degrees of freedom, leaving especially the number of filter coefficients and the filter structure to be determined by the algorithm [1].

For the design of recursive (i. e. IIR) filters there exist two different commonly used methods. The more traditional approach is based on the well-defined design rules for active filters, which are translated to the digital domain by the bilinear transformation. The computer aided design (CAD) approach relies on iterative algorithms for the minimization of an approximation error which is given as a mean squared sum of distances between the aspired filter behaviour and the temporary filter [2] [3]. The approach to be described in the following concentrates on the aim to design filters with a non-predetermined structure and greatest efficiency.

*This paper is a result of an informal project at the Institutes of Communication Science and Bionics & Evolution Technique of the TU Berlin. Mailing address is now: Görne Akustik, Gustav-Müller-Str. 34 a, D-1000 Berlin 62.

2 'Slim' IIR Filters

Of the well-known filter structures, the 'direct' one shown in Fig. 1 is one of the simplest. The filter is characterized completely

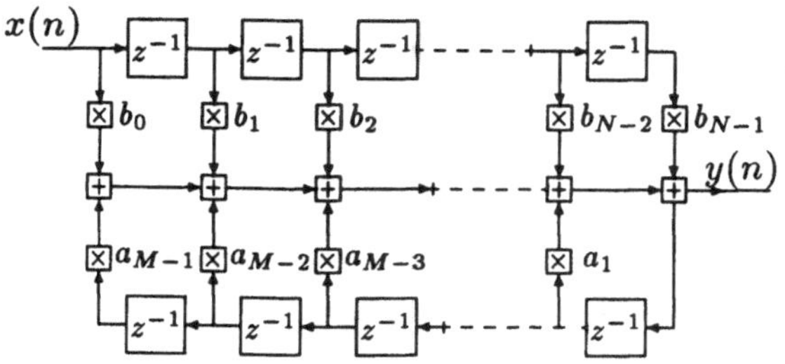

Figure 1: Direct form of an IIR filter

by its difference equation

$$y(n) = \sum_{i=1}^{M-1} a_i y(n-i) + \sum_{j=0}^{N-1} b_j x(n-j). \quad (1)$$

The impulse response of the filter can be easily determined from (1) and will be the design criterion used in the following examples. In general $M + N - 1$ multiplications and $M + N - 2$ additions per input sample have to be computed when filtering.

The basic idea behind SIIR filters is to work with a non-predetermined filter structure in such a way that the filter length is not restricted, but with a maximum of coefficients set to zero. As a multiplication with zero needs not to be performed, increasing the efficiency, flag values α_i, β_j can be introduced for each of the coefficients, indicating coefficients with values $a_i, b_j \neq 0$

$$a_i = \alpha_i c_i \ , \ b_j = \beta_j d_j \text{ with } \alpha_i, \beta_j \in \{0, 1\}. \quad (2)$$

In the case of a slim filter in direct form one then obtains a 'slim' (or 'thinned out') filter structure as shown in Figs. 2 and 3.

A multiplication (and following addition) then has just to be performed if the corresponding flag is set. The impulse response

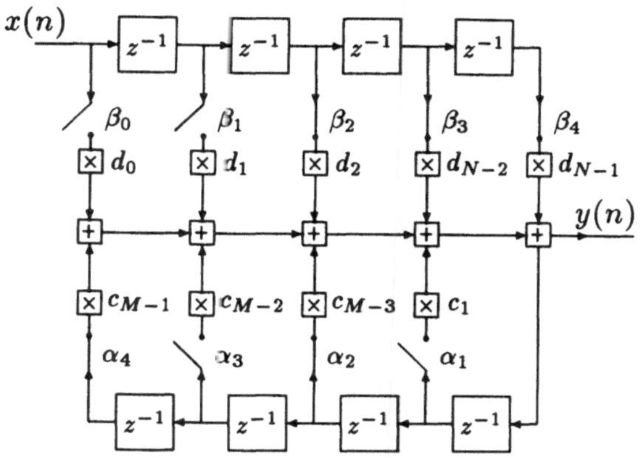

Figure 2: 'Slim' IIR filter in direct form. Flags are symbolized by switches α, β indicating coefficients $a, b \neq 0$. Arrows indicate the direction of the signal flow.

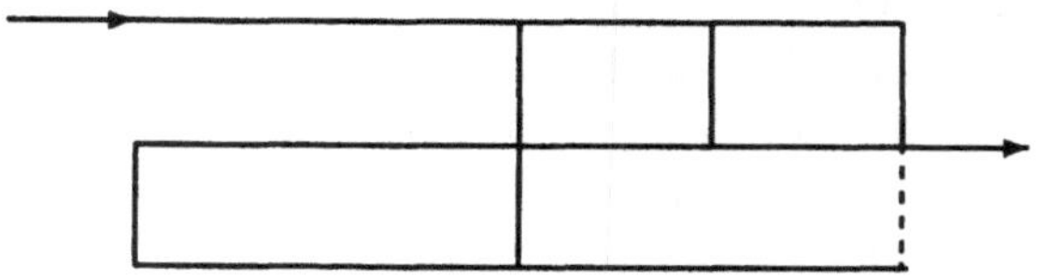

Figure 3: SIIR filter, short-hand notation (open switches α, β are displayed as missing connections).

$g(n)$ of a SIIR filter is thus determined by (1) and (2) as

$$g(n) = \sum_{i=1}^{M-1} \alpha_i c_i g(n-i) + \sum_{j=0}^{N-1} \beta_j d_j \delta(n-j) \quad (3)$$

where $\delta(n)$ is the delta function

$$\delta(n) = \begin{cases} 1 & \text{for } n = 0 \\ 0 & \text{for } n \neq 0. \end{cases} \quad (4)$$

As the computational cost of a flag check is much lower than that of any mathematical operation, the existence of a maximum number of zeroes is favourable (The cost of a floating point multiplication on e.g. a standard IBM-compatible PC with mathematical coprocessor is $\approx 25 \ldots 100$ times that of a flag check operation). The maximum cycle length realizable in such a SIIR structure is thus basically independent of the computational cost, disregarding the cost of flag checking.

The latter is an important aspect, as the quality of the approximation of a finite impulse response by an IIR filter can depend on the number of coefficients as well as on the maximum cycle length.

3 Design Algorithm

As every Evolutionary Algorithm acts as a random–controlled hill–climb algorithm, it should be no problem to equal the CAD filter design approach mentioned earlier in the determination of the filter *coefficients*. This kind of filter design, however, requires a certain amount of heuristics in getting a proper filter *structure* as well as filter order (i.e. number of coefficients).

As in SIIR design the filter structure shall be subject to the optimization, the algorithm is meant to deal not only with the coefficient optimization, but also with a high–dimensional discrete optimization problem (allowing the use of $M-1$ recursive and N non–recursive coefficients, the possible SIIR structures in direct form span a $M+N-1$–dimensional discrete search space with $2^{M+N-1} \approx 10^{0.3(M+N-1)}$ possible states).

A possible solution for both problems is offered with the Evolution Strategy and the close related Structure Evolution algorithm.

Evolution Strategy — introduced by Rechenberg in 1964 and modified by Rechenberg, Schwefel and collaborators since then [4] [5] [6] — is mainly based on the biological principles of mutation, selection and inheritance. The objects for optimization are to be seen as a population of individuals. In the evolution process the object is described by a real–valued vector $\vec{o}$ in the search space. A set of μ randomly initialized parent objects $\vec{o_p}$ create λ offspring by means of gaussian distributed mutation:

$$\vec{o} = \vec{o_p} + \sigma \cdot \vec{r} \tag{5}$$

where σ determines the mutation stepsize and $\vec{r}$ is a vector of gaussian distributed random numbers. A population in generation γ consists of parents and offspring. A fitness function $Q(\vec{o})$ is used to select the μ parent objects for the generation $\gamma+1$ out of the population of generation γ. The stepsize σ is adaptively controlled by the same mechanisms. The possibility for recombination (i.e. crossing over) of several individuals is provided, but in the examples shown here it is not used, as the algorithm works well without.

A modified Evolution Strategy for the optimization of structured objects is employed here in order to determine real–valued coefficients and the corresponding filter structure at the same time. This "Structure Evolution" [7] [8] is a modified Evolution Strategy for the optimization of structured objects. Structure Evolution works with multiple populations, where all individuals in a single population are of the same structure type. The set of populations is treated as one meta–population, and the competition of populations is performed after a certain isolation time τ. The generalized object vector $\vec{o_s}$ in structure optimization consists of real–valued parameters and the discrete structure code[1]. During the optimization structure mutation is performed by a mutation operator S which is dependent on the structure coding:

$$\vec{o_s} = S(\vec{o_{sp}}, \sigma_s). \tag{6}$$

In order to perform a local search in the structure space it is necessary to define a proper stepsize σ_s. Since the hamming distance d

[1]In the following, the index s shall denote objects and strategy parameters in the meta–population.

gives a measure for the distance between binary coded structures [9], we decided to design mutation procedures that perform random steps in the search space with a distinct "hamming stepsize" $\sigma_s = d = const.$ σ_s is given by the probability of inverting a structure flag p_{mut} :

$$\sigma_s = p_{mut} \cdot (M + N - 1). \quad (7)$$

After τ generations of parameter optimization the parent structures for the next iteration of structure optimization are selected (Fig. 4).

In SIIR design, valuation of the filter structures relies on an error criterion (here denoted as Q_1) and a discrete structure fitness function Q_2. Q_1 is given by the mean squared distance between a target impulse response and the SIIR response

$$Q_1 = \frac{1}{L} \sum_{n=0}^{\infty} (g(n) - f(n))^2 \quad (8)$$

calculated for a finite number of samples, where L denotes the length of the target impulse response. Q_2 is the total number of non–zero coefficients in the temporary structure

$$Q_2 = \sum_{i=1}^{M-1} \alpha_i + \sum_{j=0}^{N-1} \beta_j. \quad (9)$$

Because we must not loose sight of our original filter design problem, we have to make a decision if the temporary filters (the fittest of each population) fulfill the design requirements. Therefore we introduced a maximum allowed approximation error ϵ to distinguish between "good" and "bad" structures, independent of the filter size. Progressive diminuition of ϵ in the evolutionary process leads to the smallest possible structures matching the design requirements.

Calculation of a highscore list for structure valuation with respect to ϵ is schematically shown in Figure 5. Other optimization problems, however, may lead to different methods of structure evaluation.

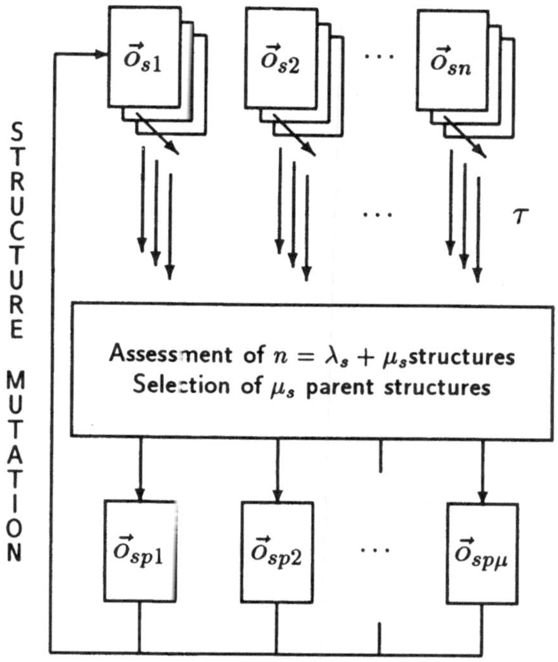

Figure 4: One iteration in Structure Evolution

4 Experimental Results

In the following, SIIR design shall be illustrated with a 22–coefficient perfect reconstruction FIR solution from [10]. SIIR filters have been designed to replace a set of FIR filters within a perfect reconstruction filter bank [11].

For the SIIR design a Structure Evolution algorithm was implemented starting with 20 randomly initialized structures. The maximum number of recursive coefficients $M - 1$ was set to 31, the maximum number of non–recursive coefficients N to 32. The structure flags were initially set with a probability $p_{start} = 0.8$ so that the (rather 'dense') filters of the generation $\gamma = 0$ had a mean of

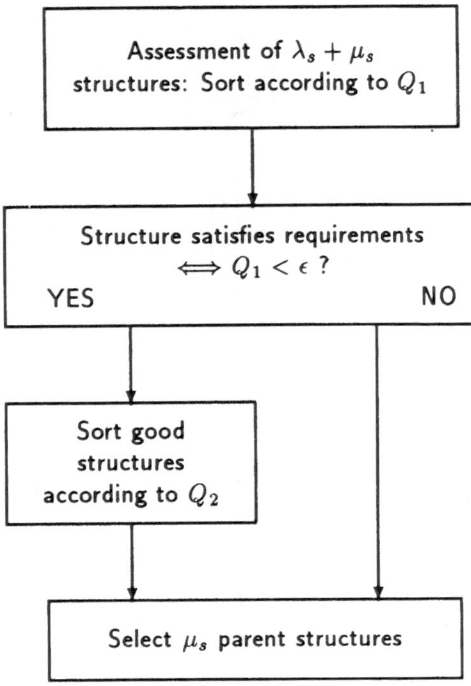

Figure 5: Structure selection algorithm

48 non–zero coefficients. In the optimization process a SIIR filter was represented by an object vector $\vec{o_s}$ consisting of the coefficients c_i, d_j and the structure variables α_i, β_j:

$$\begin{aligned}\vec{o_s} &= (c_1, \ldots, c_{M-1}, \\ &\quad d_0, \ldots, d_{N-1}, \\ &\quad \alpha_1, \ldots, \alpha_{M-1}, \\ &\quad \beta_0, \ldots, \beta_{N-1})\end{aligned} \tag{10}$$

The approximation error $Q_1(\vec{o_s})$ as in (8) was used to adjust the filter coefficients. The taget imulse response is determined by the $L = 22$ coefficients of the FIR. Approximation of (8) with a finite length of K samples of the impulse response leads to

$$Q_1(\vec{o_s}) = \frac{1}{L} \sum_{n=0}^{K-1} \left(g(n) - \sum_{l=0}^{L-1} f_l \cdot \delta(n-l) \right)^2 \tag{11}$$

where f_l are the FIR solution coefficients and $g(n)$ denotes the impulse response of the SIIR filter as in (3). Calculation of $K = 96$ samples turned out to be sufficient in the design process.

It turned out that the remaining mean squared error after 200 iterations is close enough to the absolute remaining error to allow a valuation of the filter, as depicted in Fig. 6. Isolation time τ was thus set to 200.

Mutation of the filter structure was accomplished with a constant mean hamming stepsize $\sigma_s = 2$ within the given filter order. Moreover the filter order, determined by M and N, was varied by using discrete Gaussian distributed random numbers.

The development of a 12–coefficient SIIR approximating the 22–coefficient FIR took 50 iterations in the meta–population. Experiments with initially 'sparse' filters ($p_{start} = 0.2$) led to similar SIIR filters within even less iterations (Fig. 7, 8, 9).

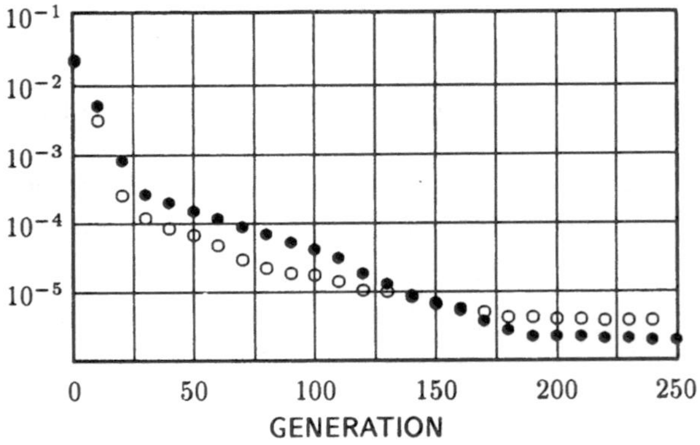

Figure 6: Progress of the mean squared error Q_1 during the coefficient optimization; two different SIIR structures: • $Q_2 = 12$, ∘ $Q_2 = 10$. Every 10th generation displayed.

During the optimization a total of $50 \cdot 20 = 10^3$ SIIR filters out of approximately 10^{19} possible structures were evaluated. The maximum allowed error ϵ was progressively diminished down to $3 \cdot 10^{-6}$. The designed filters fulfilled the requirements.

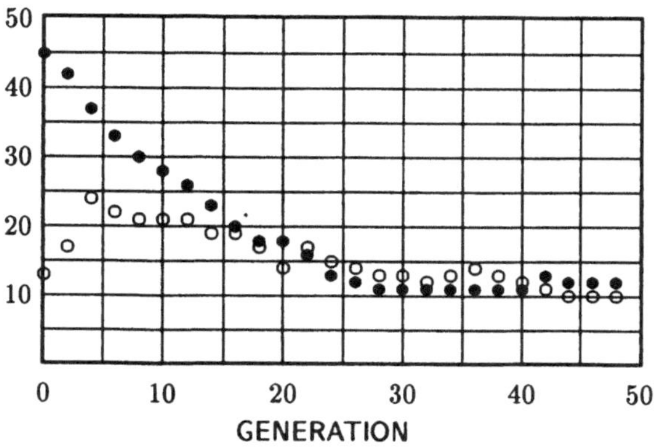

Figure 7: Progress of the structure size Q_2 (number of non–zero coefficients) during the structure optimization; two different initial random populations: • $p_{start} = 0.8$, ∘ $p_{start} = 0.2$. Every 2nd generation displayed.

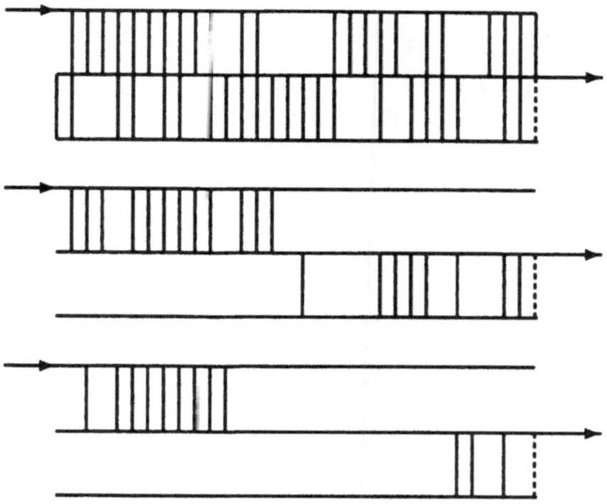

Figure 8: SIIR structure samples out of the design process with $p_{start} = 0.8$. Generation a) 0, b) 19, c) 49.

5 Conclusion

Evolution Strategy has been introduced as a tool for the design of recursive digital filters. The proposed approach has proven to be efficient in determining filters with generalized structures that cannot be handled by traditional filter design algorithms. Evolutionary parameter optimization converges to stable results (a fact that is not self–evident in traditional design algorithms). Structure optimization has fulfilled the expectancy in providing 'slim' IIR filters within relatively few iterations.

The experiments described exemplify the strategy within a limited class of filter structures in the direct form. Further studies aim towards even less restricted structures.

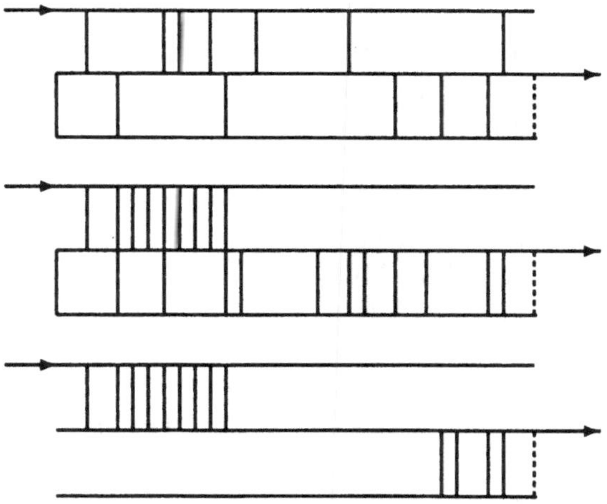

Figure 9: SIIR structure samples out of the design process with $p_{start} = 0.2$. Generation a) 0, b) 15, c) 38.

References

[1] Lohmann R, Görne T, Schneider M and Warstat M, *'Digital Systems and Structure Evolution'*, Prepr. 94th Conv. Audio Eng. Soc., Berlin, 1993.

[2] Azizi S A, *'Entwurf und Realisierung digitaler Filter'*, Oldenbourg, München, 1990 (in German).

[3] Oppenheim A V and Schafer R W, *'Digital Signal Processing'*, Prentice-Hall International, London, 1975.

[4] Rechenberg I, *'Cybernetic Solution Path of an Experimental Problem'*, Royal Aircraft Establishments, Library Translation 1122, Farnborough, 1964.

[5] Rechenberg I, *'Evolutionsstrategie – Optimierung technischer Systeme nach Prinzipien der biologischen Evolution'*, Frommann–Holzboog, Stuttgart, 1974 (in German).

[6] Schwefel H P, *'Numerical Optimization of Computer Models'*, Wiley, Chichester, 1981.

[7] Lohmann R, *'Selforganization by Evolution Strategy in Visual Systems'*, in: Voigt H M, Mühlenbein H, Schwefel H P (eds), Evolution and Optimization '89, Akademie–Verlag, Berlin, 1990.

[8] Lohmann R, *'Structure Evolution in Neural Systems'*, in: Souček B and the Iris Group, Dynamic, Genetic and Chaotic Programming, Wiley, New York, 1992.

[9] Hamming R W, *'Coding and Information Theory'*, Prentice–Hall, Englewood Cliffs, 1980.

[10] Dorize C and Villemoes L F, *'Optimizing Time–Frequency Resolution of Orthonormal Wavelets'*, Proc. IEEE Int. Conf. Acoustics Speech & Signal Proc., 1991.

[11] Rioul O and Vetterli M, *'Wavelets and Signal Processing'*, IEEE Sig. Proc. Magazine, October 1991.

An Empirical Study of Population and Non-population Based Search Strategies for Optimizing a Combinatorical Problem

Antti Autere
Helsinki University of Technology
Department of Computer Science
Otakaari 1 A
SF-02150 Espoo, Finland
email: aau@cs.hut.fi

Abstract

The performance of several algorithms for optimizing a combinatorical problem is compared empirically. The following topics are studied:

- Can population based search strategies find good solutions faster than non-population ones ?
- Are strategies that recombine two individuals to form new solutions better than those who use one individual only ?
- Does the size of the population affect on the performance ?
- Does the way how individuals are selected from the population affect on the performance ?

The test problem used in these experiments is *knapsack*. 5000 variations of the test problem were solved. The number of the test function evaluations was recorded in the simulations. Statistical data was obtained in the form of cumulative frequencies and average values.

Based on these simulations the short answers to the four questions above are: yes; *not necessarily*; yes, but it depends on the selection strategy; and yes.

1 Test Problem

The test problem is to maximize a function $f(\mathbf{x})$:

$$f(\mathbf{x}) = \sum_{j=1}^{M} c_j x_j, \qquad (1)$$

subject to $f(\mathbf{x}) \leq L$, where $x_j \in \{0,1\}$ and the coefficients c_j are integers. If the constraint is violated, i.e. if $f(\mathbf{x}) > L$, then $f(\mathbf{x})$ is set to zero. In these experiments $M = 20$, and $L = 1000$.

Let the set of the coefficients be $C = \{c_1, ..., c_M\}$. Finding the global optimum f^* means that all the subsets of C must be generated and sums of their elements calculated. The number of the subsets is $2^{20} = 1048575$ that is the size of the search space.

The subsets of C are "generated" by the variables $x_j = 1, s_j \in S$, $x_j = 0, s_j \notin S$, $(j = 1, ..., 20)$. Each subset of C is thus represented as a binary vector $\mathbf{x}_i = (x_1, ..., x_{20})(i = 1, ..., 2^{20})$ Each binary vector corresponds to one point in the search space.

As an example, consider
$C = \{32, 31, 45, 12, 22, 28, 21\}$. A 7-bit vector $\mathbf{x} = (0,0,1,1,1,0,1)$ generates a subset $\{45, 12, 22, 21\}$. The value of the function is $f(\mathbf{x}) = 45 + 12 + 22 + 21 = 100$.

1.1 Problem Selection

The 20 integers of the set C were drawn from the rectangular distribution [0..400]:
$C = \{346, 130. 182, 290, 56, 317, 395, 15, 148, 326, 204, 158, 371, 79, 92, 160, 212, 321, 63, 247\}$.

The global optimum is $f^* = 1000$. Figure 1 shows the first 1000 values of $f(\mathbf{x})$. X-axis contains the binary vectors $\mathbf{x}_i, i = 1, ..., 1000$ in the gray code order, [RND77].

The purpose of these experiments is not to find the global optimum of the test problem. Five goals of varying complexity were solved instead. These goals are to find values of $f(\mathbf{x})$ that are

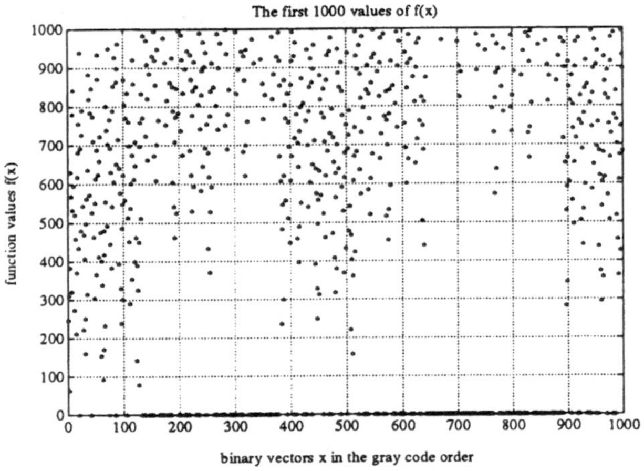

Figure 1: The first 1000 values of f(x)

bigger than some reference values, e.g. 90 % of the global optimum.

The five goals, G1, G2, G3, G4 and G5 are to reach 90 %, 95 %, 97.5 %, 99 % and 99.5 % of the global optimum respectively. The percentages are not important, however. The important thing is the complexity of the goals.

To measure the complexities of the goals, all the 1048575 test function values were evaluated. Table 1 shows the relative frequencies of the function values bigger than the five goals, $f(\mathbf{x}) \geq G_i, (i = 1,2,3,4,5)$. These frequencies are identified to the corresponding probabilities $P(X \geq G_i)$, where $f(\mathbf{x})$s are now considered as values of the random variable X.

Let us calculate the size n of the random sample needed to achieve the goals "almost surely", i.e. with high probability, say 0.9. This is the same as calculating one minus the probability that a particular goal has not been achieved by the sample:

$$1 - [P(X < G_i)]^n = 0.9. \qquad (2)$$

Solving this equation for the sample size n gives

$$n = \frac{log(0.1)}{log(P(X < G_i))} \qquad (3)$$

The last row of table 1 shows these sample

Goals	G1	G2	G3	G4	G5
Rel. freq.	0.0087	0.0049	0.0026	0.0011	0.00061
Samp.size	263	470	885	2038	3771

Table 1: The complexities of the five goals

sizes: the goals G1, and G2 are fairly easy to achieve whereas G4, and G5 are more complex.

2 Testing Method

There are 20! permutations of the coefficients c_i in the set C, which corresponds to the different codings of the test problem. 5000 permutations of the c_is were randomly generated as the test data used at the experiments. These are represented as *ordered sets* or vectors $C_1, ..., C_{5000}$.

All 5000 test problem variations and the five goals, G1,G2,G3,G4, and G5 were solved by the algorithms tested. The number of the test function evaluations for each algorithm and for each goal were recorded.

Mean values were calculated from the 5000 simulation runs. To see more details cumulative frequencies were formed from the simulation data. These cumulative frequencies are identified with the corresponding cumulative distribution functions, i.e.

$$F(x) = P(X \leq x), \qquad (4)$$

where x now denotes the number of the test function evaluations.

Table 2 shows the mean values for random search. Figure 2 presents the resulting cumulative distribution functions. Y-axis denotes the number of the test runs, and X-axis the number of the test function evaluations.

Figure 2 shows that 4500 runs out of 5000 reached the goal G4 by evaluating the test function about 2000 times. It means that the goal G4 is achieved by the probability $P = 4500/5000 = 0.9$ by making 2000 trials (function evaluations). Remember, the corresponding value was calculated beforehand in the previous section and presented at the table 1. The two values agree well. See also the similarities between the correspond-

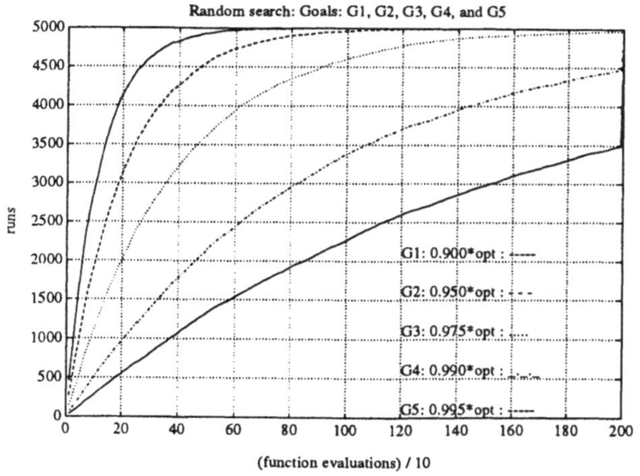

Figure 2: Random search

method	*mean values*				
Goals	G1	G2	G3	G4	G5
Random search	113	204	388	897	1638

Table 2: Random search

ing calculated and simulated values for the goals G1,G2 and G3 [1].

3 The Algorithms Tested

No specialized algorithms developed for knapsack problems were tested because the purpose is not to focus attention to the test problem itself. The following search strategies have been studied.

3.1 "Simulated Annealing": SA

This algorithm is a simplified version of the *simulated annealing* algorithm, see [KGV83]. A new solution is found either in the neighborhood of the old one or at random. The probability of that decision is not changed during the simulation, which is the difference between this and the simulated annealing.

The following pseudo code illustrates the algorithm SA:

[1] The value for G5 does not fit into the figure

```
initialize(s);
WHILE (NOT(GoalsReached))
      IF (random(0,1) < pm)
      THEN mutate(s1)
      ELSE neighbor(s,s1);
      s := s1;
END
```

The initialize operation generates one solution s at random.

The mutation operation generates a new solution at random. Mutation is applied with a probability p_m

The neighbor operation changes randomly some bits of the solution. It first determines the amount n of random bits. This is a random variable reaching from 1 to $MaxStep$ where $MaxStep$ is, e.g. half the length of the binary strings. $MaxStep$ is called *the maximum step size.* Next, n random positions of the bits to be changed are determined.

For example, suppose the old solution is:

0000000000

Next, suppose n is four, and the first, fifth, seventh, and tenth bits are generated at random. Then the result may be:

1000100001

3.2 Greedy:G

The algorithm SA does not keep the best solution in memory. This may be a drawback. *The greedy* algorithm maintains the best solution found so far. A new solution is then sought either in the neighborhood of the best one or at random. The probability of the decision is p_m. If the new solution is better than the old one then it is remembered instead.

The following pseudo code illustrates the algorithm G:

```
initialize(s);
WHILE (NOT(GoalsReached))
      IF (random(0,1) < pm)
      THEN mutate(s1)
      ELSE neighbor(s,s1);
      IF (f(s1) > f(s))
      THEN s := s1;
END
```

The initialize, *the neighbor*, and *the mutation* operations are the same that SA has.

One can see the greedy algorithm as a special case of the following multigreedy algorithm with the population size *one*.

3.3 Multigreedy: MG

Multigreedy maintains a set or a population of potentially good solutions. At each step it selects one solution and then finds another from its neighborhood. If the new solution is good enough then it is included in the population.

The following pseudo code illustrates the algorithm MG:

```
initialize(S);
WHILE (NOT(GoalsReached))
      select(S,s);
      IF (random(0,1) < pm)
      THEN mutate(s1)
      ELSE neighbor(s,s1);
      update(s1,S);
END
```

The initialize operation generates randomly a starting population S.

The selection of a solution from the population is either random or is based on the *fitness values* of the solutions. Different selection strategies are explained in more detail in the context of the genetic algorithm.

The neighbor operation is the same as before.

A solution candidate may also be generated at random with the mutation operation and with the mutation probability p_m, as before.

The update operation first adds the new solution $s1$ to the population S and then removes the worst individual from S.

The name multigreedy has also been used for the algorithms that every time explore the whole neighborhood of the selected solution, see e.g. [Man91]

This multigreedy algorithm differs from the genetic algorithm only in a sense that it selects *one* instead of two parent solutions. This would correspond to parthenogenetic reproduction found among some insects and plants. Thus one could also call the algorithm *the parthenogenetic algorithm*.

3.4 Genetic Algorithm: GA

Genetic algorithm maintains a population of potentially good solutions. At each step, it selects two individuals from the population as parents and then breeds them to produce a child. The reproduction is done by the *crossover* operation. If the new solution is good enough then it is included in the population.

The following pseudo code illustrates the algorithm GA:

```
initialize(S);
WHILE (NOT(GoalsReached))
      select(S,s1,s2);
      IF (random(0,1) < pm)
      THEN mutate(s3)
      ELSE crossover(s1,s2,s3);
      update(s3,S);
END
```

The mutation and *the update* operations are the same that the multigreedy algorithm has.

3.4.1 Crossover Operators

Four crossover operators were studied: *one-point*, *two-point*, *three-point* and *uniform crossover*.

One-point crossover chooses a *crossover point* at random for the two parent binary strings. This point occurs between two bits and divides each individual into left and right sections. Crossover then swaps the left (or the right) section of the two parents to form their children. In this study only one child is produced, however.

As an example, consider the two parents:

Parent 1: **1010101010**
Parent 2: 1000010000

Suppose the crossover point randomly occurs after the fifth bit. Then one of the children is

Child: **10101**10000

Similarly, two-point crossover would produce a child:

Child: **10**0000**1010**,

where the two crossover points occurred after the second and the fifth bit. Three point crossover functions analogously to the two-point one.

Uniform crossover does not select a set of crossover points. It considers each bit position of the two parents, and copies the bit from either the first or the second parent with a probability of 0.5 [2].

Suppose the first, third, fourth and ninth bits are copied from the first parent. Then the child is

Child: 1010010010

3.4.2 Selection Schemes

Three strategies for selecting parents from the population were studied: *uniform, proportional* and *ranking* selection.

Let S^t denote the population of the individuals a_i^t at the generation t: $S^t = \{a_1^t, a_2^t, ..., a_n^t\}$, where n is the size of S^t. Selection schemes assign *selection probabilities* $p_s(a_i^t)$ for the individuals in the population. Parents are then selected according to these probabilities.

Uniform selection selects two parents from S^t with a probability

$$p_s(a_i^t) = 1/n. \tag{5}$$

Thus every individual have the same chance to be selected.

Proportional selection, [Gol89a], selects two parents from S^t with the probabilities proportional to their fitness values $f(a_i^t)$:

$$p_s(a_i^t) = f(a_i^t)/\sum_{j=1}^{n} f(a_j^t) \tag{6}$$

Ranking selection, see e.g. [Whi89], [BH91], requires the individuals of the population S^t to be sorted in an ascending order of their fitness. Let S^t now be represented as *an ordered set* or *a vector*: $S^t = (a_1^t, a_2^t, ..., a_i^t, a_{i+1}^t, ..., a_n^t)$. Two parents are selected from S^t with probabilities proportional to their *rankings* i:

[2]The reference [SJ91] considers uniform crossovers with other probabilies, too

method	*mean values*				
Goals	G1	G2	G3	G4	G5
Greedy	117	134	177	316	582

Table 3: Greedy

$$p_s(a_i^t) = i/\sum_{j=1}^{n} j \tag{7}$$

All the selection strategies studied here are *preservative*, [BH91], since every individual has a non-zero selection probability.

4 Experimental Results

Remember, 5000 test function variations and five goals, G1, G2, G3, G4, and G5 were solved. The number of criteria function evaluations were reccrded and cumulative frequencies [3] were formed.

4.1 Non-Population Algorithms: SA and G

Figure 3 shows the simulation results for the algorithm SA. The simulations were done with the maximum step sizes: $MaxStep = 10$, and 5, and the probabilities: $p_m = 0.05$, and 0.3.

The results indicate that SA does not perform any better than random search, see also figure 2.

Figure 4 and table 3 show the results for the greedy method: G. The simulations were done with the maximum step size: $MaxStep = 10$, and the probablity $p_m = 0.3$.

The greedy method performed better than random search and SA in spite of the easiest goal G1, see also figure 2, table 2, and figure 3.

4.2 Multigreedy: MG

Figure 5 shows the results for the multigreedy algorithm MG. The data for the goals G3, and G5 is presented. The parameters used at the simulations were the population size: 20, $MaxStep =$

[3]or distribution functions

$5, 10, 20$, and $p_m = 0.05$. The "parent" was selected at random from the population.

$MaxStep = 10$, and 20 were a little better than $MaxStep5$. MG with $MaxStep = 10$, or 20, performed better than G when the goal G5 were solved. The goal G3 were solver equally well by MG and G.

One may say that the population based multi-greedy performes better than the non-population greedy, at least when the more complex goals were involved.

4.3 Crossover Operations

The goal for these simulations is to study the effect of different crossover operations, one-point, two-point, three-point, and uniform, for the genetic algorithm GA. The corresponding results for MG with $MaxStep = 10$, are included as a reference.

Two population sizes, 20 and 200, and two selection strategies, random and ranking, were tested. Figures 6,7,8, and 9 show the simulation results for the two goals G3 and G5.

Uniform crossover performed best among the crossover operations. However, the differencies were small with the population size 200, and random selection. These results agree with the results of the reference [SJ91] where different versions of the uniform crossover operations were studied.

A somewhat surprising result is that MG, with the maximum step size 10, performed *better* or at least as well as GA with uniform crossover. It means that only one parent and "parthenogenetic" reproduction seems to be enough for solving the problems of this paper. This is, of course, simpler than selecting two parents every time.

4.4 Population Sizes

The algorithm MG with $MaxStep = 10$ was tested with the population sizes: 5, 20, 50, 100, and 200 to see if some sizes were better than others. The data for the goals G3, adn G5 is presented.

Figure 10 shows the results when the parents were selected at random. Note, that GA with the uniform crossover is included in the figure, too. The population size 20 is the best. Larger and smaller sizes need more trials to solve the goals, e.g. with the probability $4500/5000 = 0.9$.

Again, GA with the uniform crossover and MG performed equally well.

Figure 11 shows the results when the selection was based on the fitness values. The population sizes 50, and 100 are the best ones. The size 5 is the worst. The size 200 is not good when the goals were solved with small probabilities. However, when the solving probability is 0.9 it is as good as 50, or 100.

Figure 12 shows the results when the selection was based on the rankings. The population size 50 was good when the goal G5 is solved. However, the size 5 was the best when the goal G3 was considered.

The sensibility of MG to the population sizes was small when proportional and ranking selection was used. Bigger differencies were found when the selection was done at random.

The effects of the population sizes has been studied in the reference [Gol89b]. There, the optimum sizes depend on the assumed convergence time: whether it is a constant or a logarithmic function of the population size. When constant convergence time is supposed the optimum population size is about 30 for binary strings of length 20. The results of this paper roughly agrees to those ones, except in figure 12, the goal G3.

4.5 Selection Schemes

The algorithm MG with $MaxStep = 10$ was tested with the three selection schemes: random, proportional, and ranking. The population sizes 20, and 200. Figures 13, and 14 show the results when the goals G3, and G5 were solved.

The ranking and the proportional selection were better than the random one. However, the performance of the proportional and the ranking selections did not differ very much from each other, as opposed to the reference [Whi89].

5 Conclusions

The performancies of four optimization algorithms were compared empirically. 5000 variations of a knapsack function and five sub-optimal goals of varying complexities were solved. The goals were 90 %, 95 %, 97.5 %, 99 % and 99.5 % of the global optimum respectively.

The optimization algorithms fall into two categories: non-population and population based ones.

Two kinds of non-population algorithms were studied: the one that does not keep the best solution in memory, and the other that does. The former performed as badly as pure random search. The latter was better when more complex goals were involved.

The population-based algorithms studied were the multigreedy and the genetic algorithm. The multigreedy uses only one old solution to form new ones in contrast to the genetic algorithm that recombine two parent solutions.

The sensibility of the algorithms to population sizes were small. Only when random selection was used were the differencies bigger. The "optimum" population sizes ranged from 20 to 100.

Uniform crossover was the best of the crossover operations tested. A somewhat surprising result, however, was that multigreedy performed *better* or as well as the genetic algorithm with uniform crossover. This clearly deserves a closer look.

Finally, ranking, and proportional selection performed better that random selection. However, the differencies between ranking and proportional selections were small.

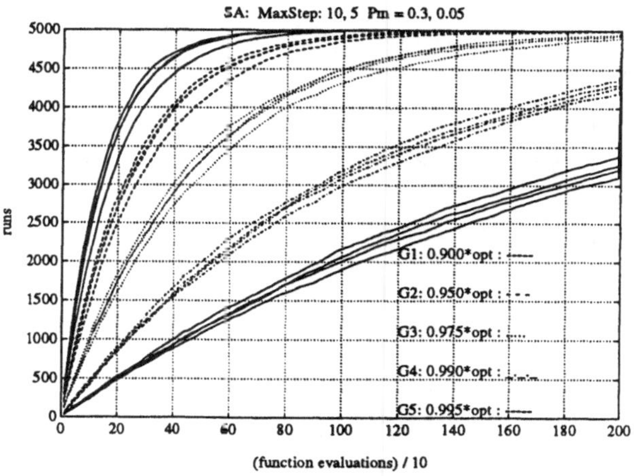

Figure 3: SA

References

[BH91] Thomas Back and Frank Hoffmeister. Extended selection mechanisms in genetic algorithms. In Richard K. Belew and Lashon B. Booker, editors, *Fourth International Conference on Genetic Algorithms*, San Diego, July 1991.

[Gol89a] David E. Goldberg. *Genetic Algorithms in Search Optimization and Machine Learning*. Addison-Wesley, 1989.

[Gol89b] David E. Goldberg. Sizing populations for serial and parallel genetic algorithms. In J. David Schaffer, editor, *Third International Conference on Genetic Algorithms*, George Mason University, June 1989.

[KGV83] S. Kirkpatric, C. D. Gelatt, and M. P. Vecchi. Optimization by simulated annealing. *Science*, 220(4598), 1983.

[Man91] Vittorio Maniezzo. The rudes and the shrewds: an experimental comparison of several evolutionary heutistic applied to the qad problem. Technical Report 91-042, Politecnico di Milano, 1991.

[RND77] Edward M. Reingold, Jurg Nievergelt, and Narsing Deo. *Combinatorical Algorithms: Theory and Practice*. Prentice-Hall, 1977.

[SJ91] William M. Spears and Kenneth A. De Jong. On the virtues of parameterized uniform crossover. In Richard K. Belew and Lashon B. Booker, editors, *Fourth International Conference on Genetic Algorithms*, San Diego, July 1991.

[Whi89] Darrel Whitley. The genitor algorithm and selection pressure: Why rank-based allocation of reproductive trials is best. In J. David Schaffer, editor, *Third International Conference on Genetic Algorithms*, George Mason University, June 1989.

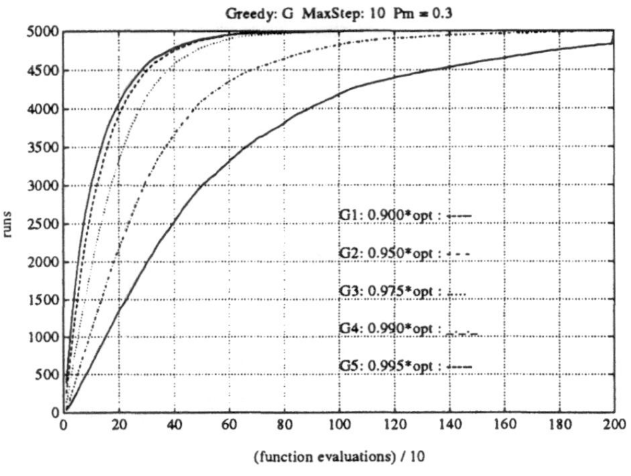

Figure 4: Greedy

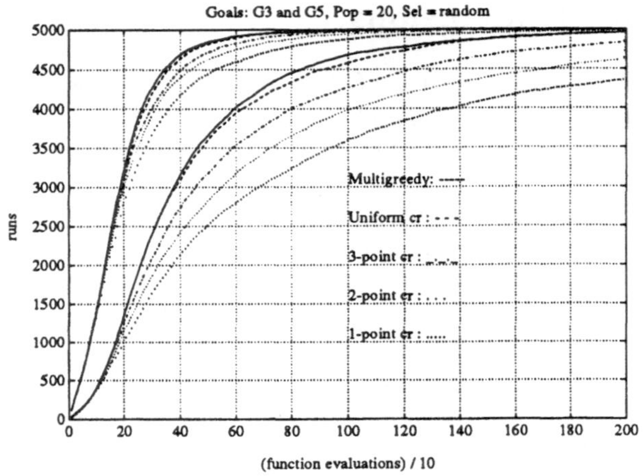

Figure 6: Different crossover operations. Goals: G3, G5. Pop: 20. Selection: random

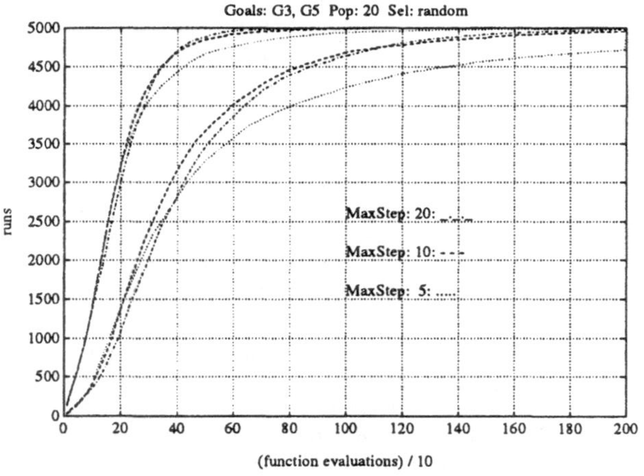

Figure 5: Multigreedy: MG

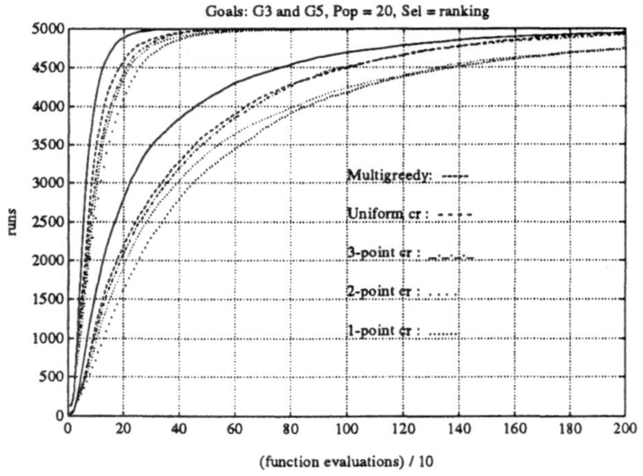

Figure 7: Different crossover operations. Goals: G3, G5. Pop: 20. Selection: ranking

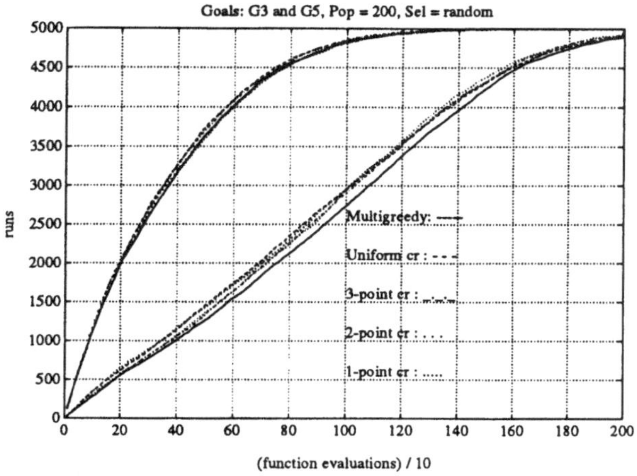

Figure 8: Different crossover operations. Goals: G3, G5. Pop: 200. Selection: random

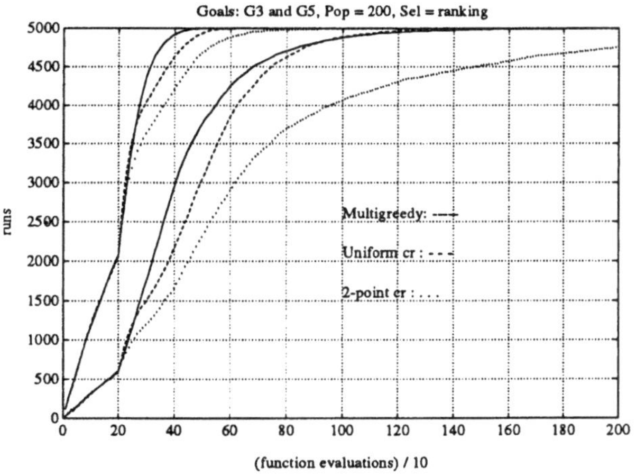

Figure 9: Different crossover operations. Goals: G3, G5. Pop: 200. Selection: ranking

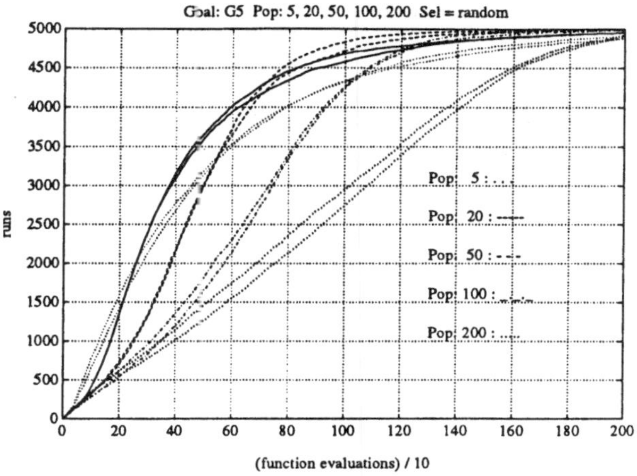

Figure 10: Different population sizes for MG, and GA (uniform crossover). Goal: G5. Selection: random

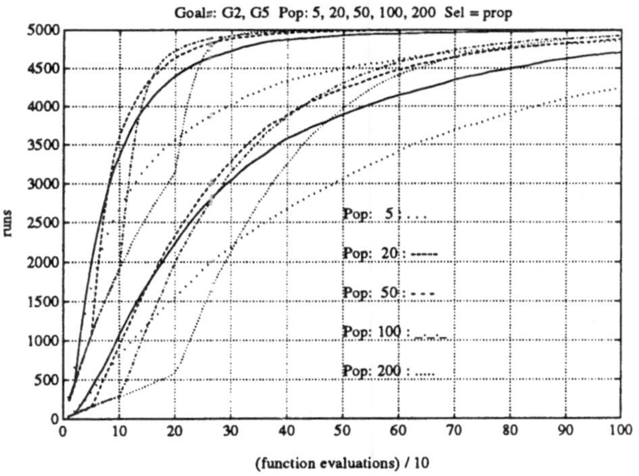

Figure 11: Different population sizes for MG. Goals: G3, G5. Selection: proportional

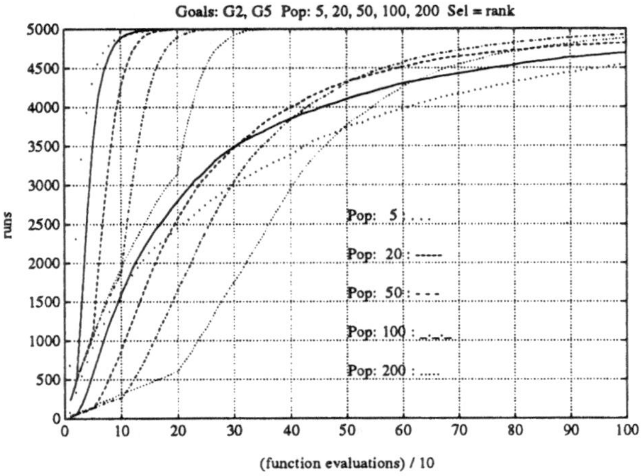

Figure 12: Different population sizes for MG. Goals: G3, G5. Selection: ranking

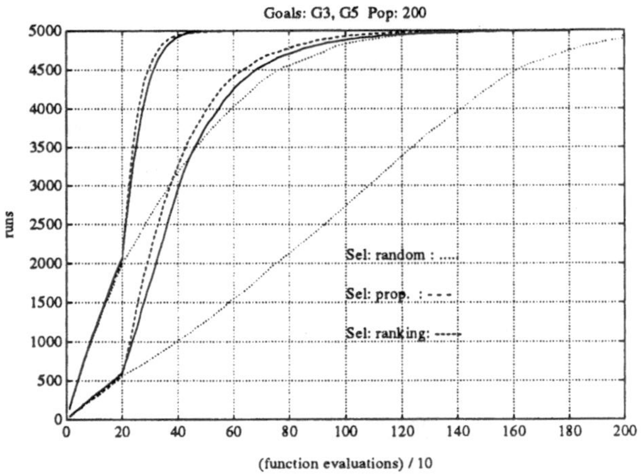

Figure 14: Different selection schemes for MG. Goals: G3, G5

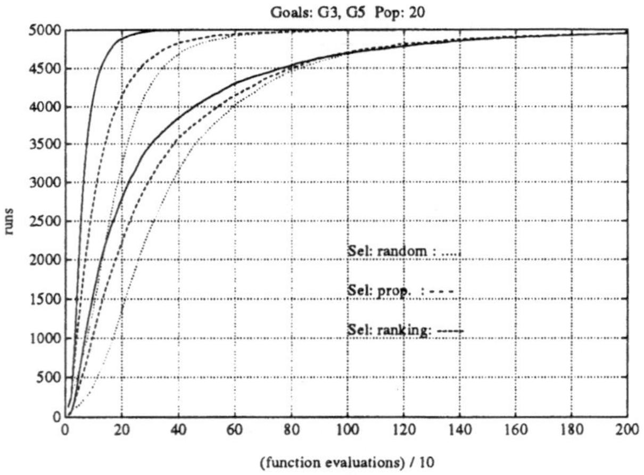

Figure 13: Different selection schemes for MG. Goals: G3, G5

The Parallel Genetic Cellular Automata: Application to Global Function Optimization

Marco Tomassini
Centro Svizzero di Calcolo Scientifico
Via Cantonale, CH-6928 Manno, Switzerland
e-mail: mtomassi@cscs.ch

Abstract

This work describes massively parallel genetic algorithms inspired by cellular automata models and by large, spatially distributed populations of individuals, as suggested by biological analogies. Models with strict locality and a variety of pseudo-diffusion models are presented. The models are applied to the the global optimization problem of multiextremal multimodal functions. They are tested on a suite of hard standard test functions. Results are then discussed for the various models taking into account the unusual population sizes, their diversity and the role of individual's migration.

1 Introduction

Genetic algorithms have been successful as robust algorithms for optimization, searching, classification and other difficult tasks. Surprisingly enough, most genetic algorithms studies to date, both practical and thoretical, have been done within the constraints of sequential computing in spite of the fact that a great deal of parallelism exists in GA's. Now that parallel machines are becoming commonplace the study of genetic algorithms in a more natural setting becomes possible. Here I will present highly parallel GA's based on the concept of Cellular Automata, a computational model that has been widely used in modeling physical phenomena. The behaviour of these parallel algorithms will then be empirically studied in a well-known application domain in which established test cases exist: the global function optimization problem. Possible reasons for the effectiveness of the parallel algorithms will then be hinted at and future developments discussed.

2 Cellular Automata and Genetic Algorithms

Cellular Automata (CA) are discrete dynamical systems whose evolution is completely determined by local rules. More formally, a cellular automaton is a K-dimensional lattice in which each lattice site is occupied by a finite-state automaton. Let S be the (finite) set of states in which each automaton can be. The state s of a given automaton i at step $n+1$ will depend on the state of the automata belonging to some neighborhood N_i of i at step n, where, by convention, i also belongs to N:

$$s_i^{n+1} = F(s(N_i)^n),$$

where $s_i^{n+1}, s(N_i)^n \in S$.

The transition function F can be deterministic or stochastic and is a mapping from the set of neighbor states to the set of automaton states. All lattice sites can, in general, be updated simultaneously, which together with the locality of the transition rules, make a CA system highly parallel. An introduction to CA's is given in [1].

Genetic algorithms are based on the co-evolution of a population of individuals where each individual represents a feasible solution in some problem space. The evolution is adaptive in the sense that the population evolves in discrete steps towards good global solutions by favoring "fit" individuals and by exchanging information among individuals by means of combination operators. Instead of being "blind" as pure random search, the process is thus biased towards promising regions of the search space. A standard reference on GA's is Goldberg's book[2].

Let us now suppose that a large population of individuals lies on a 2-dimensional grid, one individual per grid point, and that there exist a transition rule for computing the next state of any individual from the current states of the cells in a given neighborhood. If the updating rules are similar to typical genetic operators, then we can call this CA a Genetic Cellular Automata (GCA). GCA's as defined here are general systems. In the next section I will consider their specialization to the global function optimization problem.

3 Data-parallel Function Optimization

The global optimization (minimization) problem can be stated as follows. Given a function f(x):

$$f : R^n \mapsto R$$

and a set $D \in R^n$, find x^* such that:

$$f(x^*) = \min\{f(x) \mid \forall x \in D\}$$

where $x = (x_1, x_2, \ldots, x_n)^T$.

We may assume that the objective function f is always ≥ 0 in D.

Genetic algorithms have been quite successful on this kind of problems, especially when standard mathematical methods are hard to apply [2]. GA's approach the problem by taking a covering discrete subset S of points x in D:

$$S = \{x \mid x \in D, x_i \in [a_i, b_i]\}$$

where $a_i < b_i$ and $i = 1, 2, \ldots, n$.

In this work each of the n coordinates x_i is represented by a binary string. The string is decoded as an unsigned integer and mapped into the real interval $[a_i, b_i]$. The concatenation of the n coordinate substrings form a binary string that is the representation of a particular point $x \in D$. The length of the bit-string depends on the required precision i.e.,the number of points in the intervals $[a_i, b_i]$. This is a standard encoding for real quantities in GA's [2]. Other encodings are also possible. For example, straight floating-point representations have been successfully used [3]. Here I prefer to stick to the simplest encodings that allow the use of straigtforward crossover and mutation operators.

The problem now can be recast as a GCA in the following way. Consider for example a 2-D toroidal grid with the initial population of trial points $x \in D$ randomly distributed, one point per cell. The fitness function corresponds to the value of $f(x)$ and can be computed at any grid site independently. Crossover is applied in the following arbitrary way: a given cell selects the fittest among its four nearest neighbours and single point crossover is performed among the bit-string representations of the given individual and of that neighbour, the crossover point being chosen at random [2]. The fittest of the two offspring and the original point replaces the original individual. Here "fitter" means the point with the lower function value. Other definitions of crossover might be used. For example, one might discard one of the offspring at random and keep as the new individual the better of the remaining offspring and the original individual. Since a rigorous analysis is lacking, there is no reason for preferring one particular rule. Mutation can be applied with a given probability to any cell by randomly flipping a single bit in the trial point bit-string representation. Note that "point", "cell" and "individual" are used here with the same meaning, although the individual is actually represented by a record containing the coordinates $x_1, \ldots, x_n$ of point x in floating-point format, their binary representation and the $f(x)$ value.

The crossover operator as defined here is a local operator: selection and recombination take place within a small neighborhood. This is different from the way in which crossover is traditionally used: that is, some form of biased stochastic selection of individuals from the entire population, followed by random mating. I choose the rule for crossover for reasons that are empirical at this stage. One is simplicity: ranking all the individuals in the grid according to fitness and then doing crossover among arbitrarily distant pairs is possible but lengthy. The second is to take advantage of the easy parallelism and straightforward communication features of the local grid model since it has a direct counterpart in hardware. The intrinsic diversity of the large population (on the order of tens of thousands of individuals instead of hundreds) in the GCA somehow compensates for the local character of the recombination, giving rise to many possible combinations between "good" individuals. In successive models I will reintroduce some degree of longer range information transmission by means of a pseudo-diffusion device and I will compare the various models.

The evolution of the simple GCA can now be described by the following pseudo-code:

for each cell i **do in parallel**
  `generate a random value` $x_i \in D$
end parallel do
while `not done` **do**
  for each cell i **do in parallel**
    `evaluate` $f(x_i)$
    `get` f^N, f^S, f^E, f^W
    `get` x^N, x^S, x^E, x^W
    $f_i^m \leftarrow \min\{f^N, f^S, f^E, f^W\}$
    $(x_i', x_i'') \leftarrow x_i \otimes x_i^m$

```
        evaluate f(x_i') and f(x_i'')
        f_i^m ← min{f(x_i), f(x_i'), f(x_i'')}
        x_i ← x_i^m
        mutate x_i with probability p_m
    end parallel do
end while
```

where N, S, W and E stand for the north, south, west and east neighbor respectively and ⊗ represents the crossover operator. Note that x_i represents here a point in R^n, not a single coordinate value, and x_i^m is the point that has the least function value f_i^m. The algorithm is synchronous and fully parallel and it will henceforth be called a Parallel Genetic Cellular Automata (PGCA0). Only recently I become aware that similar models have been proposed by other researchers [4].

Cellular automata have been very successful in modeling diffusion phenomena [5,6]. In order to partially compensate for locality, I will introduce a diffusion-like process in which individuals take short random walks around their grid positions to meet other individuals that do not belong to their immediate neighborhood. This process only resembles genuine diffusion since there is no "empty space" in the grid, all sites being occupied. An individual takes its random walk to a nearby grid point in one of the N, S, E or W directions and compares itself with the individual at that grid point. If the arriving individual is better then it replaces the original one. If more than one individual (four at most) compete for the same grid point with the original cell, the fittest survives. The process is one of "jumping with replacement" and tends to replicate good individuals. After a while, this replication might give rise to a certain uniformity of the population with the consequent risk of premature convergence. It will be shown in the next section that this is not the case and that the pseudo-diffusion model is better on the average than the diffusionless one. Once again the unusual size of the population seems to play an important role in maintaining diversity while evolving "islands" of fit individuals. If one takes a biologically-oriented view, the present model roughly simulates a large population having spatial extension in which breeding activity between individuals takes place on a local scale. Although it is not implied that we have to literally follow natural models, they can provide us with useful analogies and ideas.

The following pseudo-code describes the new model (PGCA1), which adds a *migration* phase to PGCA0:

```
for each cell i do in parallel
    generate a random value x_i ∈ D
end parallel do
while not done do
    for each cell i do in parallel
    {selection and crossover step}
        same as PGCA0
    {migration step}
        Choose a random direction
        N,S,E or W;
        Jump a distance L in the
        chosen direction from
        site j to site i
        f_i^m ← min{f_i, f_j}
        x_i ← x_i^m
    end parallel do
end while
```

Where f_j is the fitness of the individual(s) coming to site i from site(s) j. In PGCA1 the individual's movements are limited to a small region (L=1 or 2) around the original position. It would be interesting to consider a model with unrestricted jumps i.e., any individual can migrate to any other site in the grid. Although this model seems attractive, it implies a time tradeoff since unrestricted movement is a costly operation because of the routing and collision handling required. For this reason, the individual's diffusion process is only done at every "frequency" generations. The algorithm, called PGCA2, is the same as PGCA1 except for the *migration step*:

```
for each cell i do in parallel
    {migration step}
    if generation mod frequency = 0 then
        choose site j at random
        and get the individual j;
        f_i^m ← min{f_i, f_j}
        x_i ← x_i^m
    end if
end parallel do
```

Where `generation` is the generation number and `frequency`=10 in this work.

Note that the "jumping and replacing" policy is only one of several possibilities. If we select a "jumping and mate" rule instead, we obtain a whole new set of PGCA's. Here is one exponent of this family derived from PGCA1:

```
for each cell i do in parallel
    generate a random value x_i ∈ D
end parallel do
while not done do
    for each cell i do in parallel
        evaluate f(x_i)
        get f^N, f^S, f^E, f^W
        get x^N, x^S, x^E, x^W
        if fitness_i > Threshold then
            choose a random direction;
            jump by distance L ≥ 2
            in that direction
        end if
        f_i^m ← min{f_i^N, f_i^S, f_i^E, f_i^W, f_i^j}
        (x_i', x_i'') ← x_i ⊗ x_i^m
        evaluate f(x_i') and f(x_i'')
        f_i^m ← min{f(x_i), f(x_i'), f(x_i'')}
        x_i ← x_i^m
        mutate x_i with probability p_m
    end parallel do
end while
```

Here **Threshold** is some fixed fitness value; one possible choice for it is the current average population fitness. Individuals that have an above-average fitness migrate to other cells at a distance L in a random direction. F_i^j represents the function value of the individual(s) coming to site i from site(s) j. I have had little experience with this kind of automaton and it will not be discussed further here.

All the models have been implemented and tested on the Connection Machine CM-2. This architecture is particularly suited to cellular automata studies since it can be configured as k-dimensional (k=1 to 31) rectangular grid of processors, with fast synchronous nearest neighbor communications and the possibility of communicating between arbitrarily distant processors through the hypercube network. Each cell of the CA is assigned to a single CM-2 processor and N CM-2 physical processors can transparently simulate $2^k.N$ virtual processors, with k a small integer. This makes it possible to study very large systems. For more details on the CM-2 see [7].

4 Test Cases and Results

To test the previous massively parallel algorithms it is necessary to define some suitable test problem. De Jong's functions have been very popular as test cases in the literature on function optimization with GA's [2]. However, these functions are by no means "hard" by modern standards and the problems are too small to be suitable for a massively parallel computing setting. I only used three of the De Jong functions to validate the algorithms. All of the functions always quickly converged to the expected global minimum in a series of 30 runs for each function. To actually test the PGCA models I employed functions f_6 from [8] and f_7 and f_8 from [3]. These are difficult multimodal test functions used in mainstream mathematical optimization research and are defined as follows:

$$f_6(x) = \sum_{i=1}^{6}[x_i^2 - \cos(18x_i)],$$

$$f_7(x) = \sum_{i=1}^{10} -x_i \sin(\sqrt{|\, x_i \,|}),$$

$$f_8(x) = \sum_{i=1}^{10} x_i^2/4000 - \prod \cos(x_i/\sqrt{i}) + 1,$$

With $x_i \in [-5.12, 5.12]$ for f_6, $x_i \in [-500, 500]$ for f_7 and $x_i \in [-600, 600]$ for f_8. Function f_8 is particularly hard. The global minimum $f = 0$ is at $x_i = 0, i = 1, \ldots, n$. In the ten-dimensional case there are four suboptimal minima with $f(x) \approx 0.0074$ at $x \approx (\pm\pi, \pm\pi\sqrt{2}, 0, \ldots, 0)$. These functions are described more fully in [3,8].

Genetic algorithms, as other search methods with a stochastic component, do not offer convergence guarantees for arbitrary problems. If the process is ergodic, then as $i \rightarrow \infty$,

$$\min f(x_i) \rightarrow \min f(x),$$

$i = 1, 2, \ldots, n$ and $x \in D$. However, nothing can be said *a priori* on the rate of convergence. On arbitrary problems it is difficult to establish precise stopping conditions. Should the search stop after a given number of steps? Or should it stop when the result is within a given percent of the sought (and generally unknown) value? Or when the variation in the result remains smaller than a fixed amount during a given number of steps? Since in our case both the location and the value of the global minimum is known a priori, one could use a stopping criterion of the type:

$$||f_{best} - f_{min}|| < \epsilon$$

where f_{best} is the best value found so far, f_{min} is the true minimum value and the value of ϵ would depend on the resolution i.e., on the length of our binary strings and the width of the searching interval. However, since the location and value of the global minimum is unknown in real-life problems, I decided to use the following, more experimental methodology. A standard number of steps (generations) to convergence of the PGCA is empirically determined on several runs of a typical problem. I used f_8 as the standard problem since it is the hardest one and therefore constitutes a sort of "upper bound" on the number of steps. This number was thus determined to be 50. The algorithms are then run 50 times, 50 generations per run for each test case. Quoting the number of function evaluations is standard practice in optimization work. In our case, however, since the parallel algorithms are synchronous, $3 \times N$ evaluations are done simultaneously at each generation (one for each cell plus two for the offspring), where N is the number of grid points i.e., the population size. Therefore, the total number of function evaluations is simply $3 \times N \times$ *generations.* The results of running PGCA0, PGCA1 and PGCA2 50 times each on the test cases f_6, f_7 and f_8 is shown in Table 1, where the percentage of success for each model is quoted. The population size is 8K=8192 for the runs reported on the Tables.

In spite of the questionable statistical validity of the results, some general conclusions may still be drawn. From Table 1 one sees that PGA0 is the least efficient model, failing to converge to the global minimum in 85% of the runs. PGCA1 was run with jump distance L=1 and L=2. PGCA1 constitutes an improvement over PGCA0, even when L=1. In this case many neighboring sites end up containing the same individuals, which slows down further evolution if crossover happens repeatedly between them. With L=2, the spreading of good individuals is better, as they may replace points that are not in their immediate neighborhood. Indeed, PGCA1 with L=2 gives consistently better results. PGCA2 gives the best results. It always converges to the global minimum in each run for all the test cases and, because the general "reshuffling" of cells is only allowed every `frequency` generations (see section 3), it is also faster than PGCA1 and nearly as fast as the simple PGCA0. Average cpu times, measured on an 8K CM-2 for a 50 generations run of each model for function f_8 are given in Table 2. The cost of evaluating the function is high. For f_8 it amounts to about 35% of the total time for model PGCA2. For more difficult, real-life functions, this cost will tend to swamp the cost of the rest of the algorithm, which makes PGCA2 even more interesting.

It is interesting to note that for all methods the global minimum is nearly always found in several regions of the population grid at the same time. This "spatial redundancy", due to the relative segregation of the regions, improves the robustness of the algorithms since it helps avoid being trapped into suboptimal minima.

From Table 1 it is not possible to discriminate further among the different methods, since 50 generations are too many for f_6 and f_7. It is thus interesting to observe in more detail the convergence rate using the various PGCA's. This is given in Tables 3 and 4. These results merely represent trends, since they are an average over twenty successful runs only for each problem.

It is clear from Tables 3 and 4 that not only the models in which migration is allowed are more successful on the average (see Table 1), they also show a faster converge rate. And their average rates of convergence are in the same order than the percent success rate.

Overall, PGCA2 gives the best results in terms of success percentage, convergence rate and running time. However, in an experiment that I performed, PGCA1 with L=2 also had a success rate of 100% on problem f_8 over 50 runs of 70 generations each. Note also that an hypothetical hardware implementation for PGCA1 would be simpler and cheaper than that for PGCA2.

Another point of interest is the behaviour of the PGCA's as a function of the population size. Using PGCA1 with L=2 as an example, I found that doubling or quadrupling the number of grid points i.e., the population size, from 8K to 16K and 32K improves both the success percentage and the convergence rate. This was expected, since it greatly enhances population diversity. The computing time also increases but less than linearly. The limitation to powers of two numbers is due to the CM-2 structure.

It might well be that, in spite of the good results, too much selection pressure is imposed by the "jump and replace" policy. It is possible to diminish this effect by allowing individuals to jump only when they have a fitness value that exceeds a given threshold. Preliminary experimentation showed that there is no difference in practice with the straight method.

Finally, the role of the mutation operator should be discussed. With simple crossover mutation is theoretically needed for the process to be ergodic i.e., to make the exploration of the whole search space possible in the long run. However, for PGCA's the role of mutation is only marginal. Diversity comes already from the large size of the population and there is a low probability that some space regions are impossible to get to. Running the algorithms without mutation showed very similar results on the average.

I have run Grefenstette's GENESIS package, a standard public-domain genetic algorithm, on the same test problems. The PGCA and the standard algorithms are very different and cannot meaningfully be compared. However, it is interesting to note that

Function	PGCA0	PGCA1 (L=1)	PGCA1 (L=2)	PGCA2
f_6	100	100	100	100
f_7	100	100	100	100
f_8	15	60	80	100

Table 1: Success percentage of models PGCA0, PGCA1 (with L=1 and L=2) and PGCA2 on test functions f_6, f_7 and f_8.

Algorithm	CPU time
PGCA0	3.92
PGCA1 (L=1)	4.58
PGCA1 (L=2)	5.08
PGCA2	4.00

Table 2: CPU times (in seconds) for a 50 generations run of function f_8. Times are averages over 20 runs for each algorithm.

GENESIS always failed to find the global minimum in 50 runs on f_8, with a population size=50 or 100, up to about 10000 generations per run, with hundreds of thousands of function evaluations. The algorithm always found itself trapped in one or the other nearby local minima. On f_6, the easiest problem, GENESIS is effective converging to the global minimum in nine runs out of ten, with a population size of hundred and 100000 trials. Problem f_7, Schwefel's function, was correctly solved in 30% of the cases, the remaining 70% converged to the second best minimum. Although I did not try to tune the algorithm and I cannot offer conclusive evidence, it seems that there is something fundamentally different in the PGCA with migration capabilities that makes it more suitable for difficult, large-scale problems. At least as far as the type of test problems considered here is concerned.

5 Conclusions and Future Work

The PGCA, a massively parallel genetic evolution model, has shown its effectiveness on difficult multimodal multiextremal global function optimization problems. One of the advantages of the algorithm is simplicity and generality. Although I used it for the classical problem of function optimization in this preliminary investigation, work is in progress with the PGCA in the field of combinatorial optimization. I kept the algorithms purposely unsophisticated and general. Although it is well-known that genetic algorithms are more effective if problem-specific knowledge and hybrid methods are used, only the simplest genetic representations and operators were employed here. It has been shown that the diffusion-like processes play an important role in the PGCA by making the combination of good individuals more likely and by largely avoiding early population uniformity. It does not seem that mutation plays an important role in the PGCA. The binary encoding used in this work has the drawback of being too cumbersome when high precision is required in multidimensional problems. It would therefore be useful to study the behaviour of the PGCA using floating-point representation and to compare it with the straight binary encoding. It would also be interesting, and straightforward, to extend the present models to higher dimensions. This is very easy to do for the CM-2 and the effect of the neighborhood size on the average convergence rate could then be studied. It is also possible to study larger neighborhoods in the 2-D case, as in [4]. It is clear that applying the PGCA's to some difficult real-life problem where very little is known about the function would be a more stringent test for the algorithms. Although they are intuitively appealing, at this point

Generations	PGCA0	PGCA1 (L=1)	PGCA1 (L=2)	PGCA2
10	321.85	213.8	188.90	326.5
20	31.02	16.54	14.23	12.87
30	4.61	1.91	1.54	0.82
40	0.87	0.28	0.21	0.002
50	0.14	0.002	0.002	0.002

Table 3: Convergence behavior for the test function f_7. The columns give the best function value at the given generation number, averaged over twenty runs, for each PGCA model.

Generations	PGCA0	PGCA1 (L=1)	PGCA1 (L=2)	PGCA2
10	0.624	0.528	0.557	0.619
20	0.242	0.206	0.183	0.165
30	0.117	0.079	0.069	0.059
40	0.061	0.029	0.029	0.015
50	0.031	0.010	0.009	0.003

Table 4: Convergence behavior for the test function f_8. Same conditions as for f_7 in Table 3.

in time I can only offer empirical evidence for the validity of the models. A careful mathematical analysis would be needed to demonstrate the soundness of this approach. Finally, I remark that the CM-2 is a nearly ideal machine for implementing large-scale, highly parallel synchronous cellular automata models.

Acknowledgement

I want to thank my colleagues C. Mazza for useful discussions and B. Chopard for kindly making available the CM-2 environment of the University of Geneva.

References

[1] T. Toffoli and N. Margolous, *'Cellular Automata Machines'*, MIT Press, Cambridge MA, 1989.

[2] G. Goldberg, *'Genetic Algorithms in Search, Optimization and Machine Learning'*, Addison Wesley, Reading, MA, 1989.

[3] H. Muhlenbein, M. Schomish and J. Born, *'The Parallel Genetic Algorithm as Function Optimizer'*, Parallel Comput. **17**, 619, 1991.

[4] P. Spiessens and B. Manderick, *'A Genetic Algorithm for Massively Parallel Computers'*, in *'Parallel Processing in Neural Systems and Computers*, North-Holland, 1990.

[5] L. Brieger and E. Bonomi, *'A Stochastic Cellular Automaton Simulation of the Non-linear Diffusion Equation'*, Physica D. **47**, 159, 1991.

[6] B. Chopard and M. Droz, *'Cellular Automata Models for the Diffusion Equation'*, J. Stat. Phys. **64**, 859, 1991.

[7] *Connection Machine CM-200 Series Technical Summary'*, Thinking Machine Corporation, Cambridge, MA, 1991.

[8] A. Torn and A. Zilinskas, *'Global Optimization'*, Springer-Verlag, 1989.

Optimization of Genetic Algorithms by Genetic Algorithms

Bernd Freisleben and Michael Härtfelder
Department of Computer Science (FB 20)
University of Darmstadt
Alexanderstr. 10
D–6100 Darmstadt
Germany
Email: freisleb@isa.informatik.th-darmstadt.de

Abstract

This paper presents an approach to determine the optimal Genetic Algorithm (GA), i.e. the most preferable type of genetic operators and their parameter settings, for a given problem. The basic idea is to consider the search for the best GA as an optimization problem and use another GA to solve it. As a consequence, a primary GA operates on a population of secondary GAs which in turn solve the problem in discussion. The paper describes how to encode the relevant information about GAs in gene strings and analyzes the impact of the individual genes on the results produced. The feasibility of the approach is demonstrated by presenting a parallel implementation on a multi–transputer system. Performance results for finding the best GA for the problem of optimal weight assignment in feedforward neural networks are presented.

1 Introduction

The performance of a Genetic Algorithm (GA) applied to an optimization problem is intimately related to a set of control parameters which must be appropriately determined by the designer of the GA in order to obtain high quality solutions in an efficient manner. Several proposals have been made in the literature to find optimal settings for such parameters like the mutation probability, the crossover probability and the population size [1, 3, 8, 9, 15] for a particular type of GA. However, the designer is not only faced with the problem of determining the *values* of the parameters for particular genetic operators, but also *what* parameters he or she should consider, i.e. what instances of the genetic operators are best suited to solve the problem at hand. For example, the decisions to be made may include choices among the different variants of the selection, crossover and mutation operators which have been suggested in recent years [2]. Approaches in that direction have already been proposed [6, 11], but these were somewhat limited in scope. For instance, Grefenstette [6] included choices between two different selection strategies in his optimization technique, and Mercer and Sampson [11] studied the effects of two crossover and three mutation operators on a particular optimization problem.

In this paper we present an approach that is based on an extended range of features defining a GA in order to decide what type of GA is likely to produce the best results for a given problem class. Similar to the approaches suggested in [6, 11], the basic idea of our proposal is to consider the search for the "best" GA as an optimization problem which may then be solved by another GA. In other words, a *primary* GA operates on a population of *secondary* GAs which in turn work on the search space of the original problem. The paper describes how to encode the relevant information about GAs in gene strings and analyzes the impact of the individual genes on the results produced. In contrast to the proposals cited above, our approach tries to determine the best variants of a wide range of genetic operators. This leads to a much larger number of GA features considered than previously involved, and also enables us to investigate the significance of a decision for or against an operator variant versus the significance of determining the optimal parameter value for a particular given operator variant. It will be shown that making the right choice for some operator variants during particular phases of the search process is more crucial than for others, and thus some light is shed on how the various GA features interact with each other. The feasibility of the approach is demonstrated by presenting a parallel implementation on a multi–transputer system. Performance results for finding the best GA for the problem class of optimal weight assignment in feedforward neural networks are presented.

2 General Approach

The basic idea of treating the search for the best GA for a given problem class as an optimization problem leads to a two-level optimization approach. On the bottom level, the secondary GA operates on a population of gene strings which represent possible solutions of the problem to be solved. On the top level, the primary GA works on a population of secondary GAs, each of which is represented as a separate gene string. Each of the secondary GAs runs independently to produce a solution of the problem considered, and the fitness of the solution influences the operation of the primary GA. The number of generations created on the two levels is independent of each other. The string with the highest fitness in the last primary generation is expected to be the best GA for the original problem. The two-level optimization architecture developed for implementing our approach is shown in Figure 1.

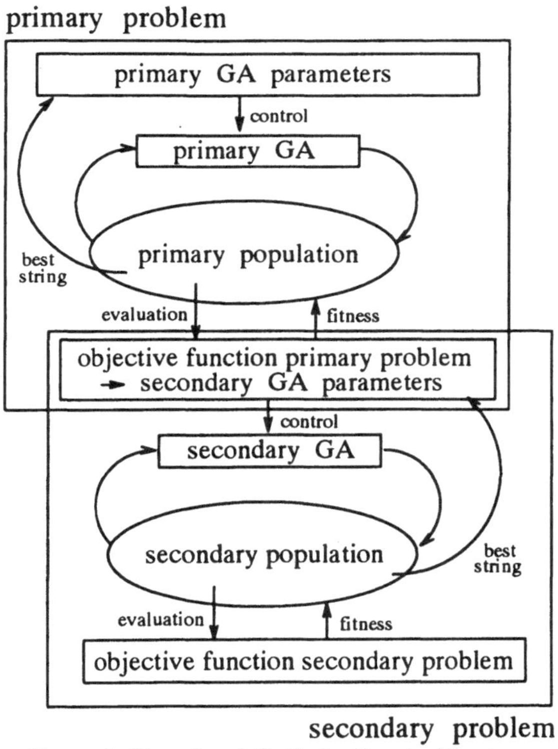

Figure 1: Two-Level Optimization Architecture

Provided that it is possible to appropriately encode the relevant information of GAs as gene strings in the primary population, there is one problem with the two-level approach that still needs to be discussed, namely that of determining the kinds of genetic operators and their parameter settings for the primary GA. In the worst case, these must be specified by the designer, which is by no means different from the usual method of genetic algorithm development pursued at present. In Grefenstette's proposal, the control parameters of the primary GA were simply set to those identified by De Jong [9] in a number of experiments, while Mercer and Sampson defined special types of meta-operators which were different from the genetic operators used in the secondary GAs. However, two further approaches are conceivable:

1. The genetic operators and parameter values are initially determined by the designer and the primary GA is applied to find the best GA for a secondary problem which closely resembles the properties of the primary problem. Such a problem would require, for example, real-valued genes, the possibility to treat logical subgroups of genes as an atomic unit and a sufficiently complex search space with multiple suboptimal peaks. The values obtained for the best secondary GA in *this* scenario are then copied and used as the parameter settings of the primary GA for optimizing the secondary GAs for the particular problem to be solved.
2. The type of primary GA and the parameter values are initially determined by the designer and then copied from the best secondary GA for the problem in discussion (possibly after every primary generation). Thus, the control parameters of the primary GA are allowed to change during the computation. This approach, however, is only possible if the structural properties of both are identical. For example, it does not work if the secondary GA operates on strings for reordering problems, such as in combinatorial optimization problems like the *traveling salesman problem* [4, 13, 14].

The second approach is the one that was used in our experiments. Although we are aware that we cannot expect to obtain the optimal primary GA for our purpose with this method, it proved to be quite successful in the sense that the quality of the secondary GAs found steadily increased with increasing primary generations.

3 Representation of GA Populations

In this section we discuss how to represent the information associated with a GA in gene strings of the primary GA population. A GA is characterized by its genetic operators and the parameters to control them. The basic operators are *selection*, *crossover* and *mutation*, but for each of them several variants with different semantics exist. Thus, there are two kinds of information that need to represented: first, the particular variant of a genetic operator and second, the parameter values required for the selected variant.

The coding scheme we propose takes these two kinds of information into account. The genes in the string representing a GA are either *decisions* or *parameters*. A decision is numerically represented by the probability that a particular variant of a genetic operator is selected among a limited number of variants of that operator. A parameter specifies a real-coded value associated with the selected variant.

Let us now consider the information represented by the individual genes in the strings of the primary population. The criteria selected should be regarded as one possibility to characterize a GA, and the strings can be easily extended to include other information if desirable. In our current proposal, each GA is defined by 19 components, where 10 of them are decisions and the remaining 9 are parameters. The meaning of the components, each of which may occupy several genes in the gene string representing a GA (in total: 139 genes), is summarized in Figure 2.

No.	Type	Information Represented
1	parameter	population size
2	parameter	sharing function exponent
3	decision	selection method
4	decision	elitist model
5	decision	double crossover
6	parameter	number of crossover points
7	parameter	crossover probability
8	decision	crossover units
9	decision	crossover points
10	decision	permutation crossover
11	parameter	distance of crossover points
12	decision	crowding method
13	parameter	crowding factor
14	decision	mutation units
15	parameter	mutation probability
16	decision	mutation function
17	parameter	exponential distribution
18	parameter	normal distribution
19	decision	mutation value replacement

Figure 2: Components of Strings for GA Representation

The individual components are briefly explained in the following:

1. The meaning of this parameter is obvious. The number of genes in the population plays an important role in any genetic algorithm approach.
2. This parameter determines the shape of the sharing function introduced in [5] in order to allow the formation of stable parts in the population (species) serving different domains of a function (niches). It essentially measures the distance between two strings and modifies their fitness in the sense that strings in the same neighborhood are forced to share their fitness among another, which effectively limits the uncontrolled growth of particular species within a population.
3. The decision how to select two strings from the population in order to produce offsprings is reflected in this component. Four variants of the selection operator are provided:
   - stochastic sampling with replacement [9]
   - remainder stochastic sampling without replacement [1]
   - stochastic tournament [2]
   - ranking number selection [2]
4. This component is used to decide if the best string generated up to time t should be included in the population of generation $t+1$ (variant 1) or not (variant 2). This so called *elitist* model was proposed in [9].
5. The crossover operator potentially produces two complementary offsprings, and it must be decided if only one (variant 1) or both of them (variant 2) should be included in the new generation.
6. This parameter determines the number of crossover points for the crossover operator. In our approach, it is computed as a percentage of the number of genes in a string.
7. This parameter contains the crossover probability for the method of determining a variable number of crossover points (see item 9).
8. This component reflects the decision if the crossover operator should, as usual, consider genes as the smallest atomic entity (variant 1) or if it should be applied such that logical subgroups of genes stay together as a structural unit (variant 2) [12].
9. This decision refers to two variants for determining the crossover points: in variant 1 the

crossover probability is used to determine a variable number of crossover points, while variant 2 determines the location of a fixed number of crossover points according to some distribution function.

10. This component requires the decision among three different crossover operators for reordering problems:
    - partially matched crossover [4]
    - order crossover [14]
    - cycle crossover [14]
11. This parameter determines the maximal distance between two crossover points in the crossover operator for reordering problems.
12. This component is used to determine if strings should be replaced according to the *crowding* technique proposed in [9] (variant 1) or not (variant 2). The crowding technique has been introduced to induce nichelike behaviour in GA search in order to maintain diversity in the population.
13. This parameter determines the crowding factor needed for variant 1 of the previous item.
14. Similar to its analogue for the crossover operator, this component refers to the decision if only one gene is mutated (variant 1) or units consisting of logical subgroups are mutated together (variant 2).
15. This parameter determines the probability of applying the mutation operator to a gene or unit.
16. This component represents the decision if the mutation amount of real-valued genes is determined according to a normal distribution (variant 0) or an exponential distribution (variant 1).
17. This parameter determines the mean value of the exponential distribution of the mutation amount (see item 16).
18. This parameter determines the density function of the normal distribution of the mutation amount (see item 16).
19. This component reflects the decision if the mutation operator should overwrite the old gene (variant 1) or if it should add the new value to the old value (variant 2).

Although our list of features for representing GAs is certainly not exhaustive, it is at least sufficient to serve our purposes. The two-level GA-based optimization approach enables us to investigate all 19 features *simultaneously*. We explicitly distinguish between decisions and parameter values and provide several alternatives for the decision components. In contrast, the proposal made in [6] was based on 6 control parameters (population size, crossover rate, mutation rate, generation gap, scaling window and selection strategy), while the approach described in [11] focused on 5 different GA features (2 crossover operators and three mutation operators). To the best of our knowledge, our proposal is the first approach where such a relatively large number of GA parameters are considered together.

3.1 Indicators

The probability of selecting a particular variant does not say anything about the impact of that variant on the solution process. For example, the decision to either include only one or both offsprings produced by the crossover operator in the next generation may be less significant for the solution quality than the decision about the method of determing the crossover points. In order to be able to evaluate the impact of a particular decision, we introduce a special construct, called an *indicator*. An indicator is computed for each variant of an operator by extracting information from all strings of the primary population after each primary generation:

Let W_i^G be the indicator for the operator variant V_i after primary generation G, *popsize* the number of strings in the population of generation G, $\#V_i^n$ the number of decisions made for variant V_i in the string n and $\#var$ the number of variants, then

$$W_i^G(V_i) = \frac{1}{popsize} \cdot \sum_{n=1}^{popsize} \frac{\#V_i^n}{\sum_{k=1}^{\#var} \#V_k^n} \quad (1)$$

The range of values for $W_i^G(V_i)$ is the interval [0..1]. At the start of the optimization process, an indicator has the value $1/\#var$. The faster an indicator approaches one of the interval boundaries, the larger is the impact of the corresponding decision on the solution process, either in a positive sense (values near 1) or in a negative sense (values near 0). An operator variant whose indicator is more or less stable around the starting value or oscillates within the interval without an observable trend is relatively unimportant for the solution of the secondary problem considered. In order to additionally avoid decisions which are based on short-term fluctuations of the corresponding indicators, the actual indicators used in our implementation are calculated as moving averages over the last 5 generations. Both Grefenstette [6] and Mercer and Sampson [11] do not provide means to measure the impact of a control parameter on the solution process, and thus our approach promises to be well suited to enable investigations about the relationships between the control parameters defining a GA.

3.2 Fitness

In the absence of an a priori known best or worst fitness of a string, our method of fitness calculation is based on the idea of measuring the *increase* of fitness created by the secondary GA. Let S_n^m be the best fitness of a string in the secondary population for test problem instance m after generation n and $\#prob$ the number of test problems, then the fitness of a secondary genetic algorithm S_{GA}^n after generation n is given by

$$Fitness(S_{GA}^n) = \frac{1}{\#prob} \cdot \sum_{m=1}^{\#prob} \frac{max(0, S_n^m - S_0^m)}{1 - S_0^m} \tag{2}$$

An additional motivation for this method is to avoid distorted fitness values arising from possibly different degrees of complexity of the test problem instances. For example, when the test problems are *traveling salesman problems* with different numbers of cities but the same tour length, then the average tour lengths increase with increasing numbers of cities.

3.3 External Values

The possibility to use test problem instances with different complexities leads to a further problem which must be addressed. In this case it does not make sense to assume that a particular parameter value in the GA strings will be optimal for all problem instances in a class of problems, because for different complexities of the problem instances different values for a particular parameter will certainly be required. The solution we propose to cope with this problem is based on providing an external value which specifies the complexity of each problem instance. An example of such an external, problem instance specific value is the number of genes.

The impact of the external value x on a GA parameter v_i is expressed by a function $f(x)$ which includes constant, linear, polynomial and exponential terms to generally represent the possible relationships between a parameter and the problem complexity:

$$f(x) = v_1 \cdot x^{v_2} \cdot e^{(x \cdot v_3)} + v_4 \tag{3}$$

Since the different genes in the GA strings may take significantly different values, the function $f(x)$ may also produce different values. In order to restrict the values produced by $f(x)$, we embed it into another function $g(x)$ which garantuees that all values produced are in the interval [0..1], no matter what values for the $v_i's$ and x are used:

$$g(x) = \frac{1}{\pi} \arctan(f(x)) + \frac{1}{2} \tag{4}$$

This mechanism, however, does not work for decision components with more than two alternatives, since the probabilities are mutually dependent on each other. In order to allow the use of the function also for the decision components, the decision probabilities are determined in the following manner:

1. Compute $f_i(x)$ for variant V_i with equation (3)
2. Normalize $f_i(x)$ according to

$$f_i^*(x) = \frac{|f_i(x)|}{\sum_{k=1}^{\#var} |f_k(x)|} \tag{5}$$

where $\#var$ is the number of variants for the decision considered.

The introduction of the external values allows to adapt the parameter values of the secondary GAs to the complexity of the problem instances, a property which is not present in Grefenstette's proposal [6]. Mercer and Sampson [11] provide such a facility, though in a completely different manner, by a parameter adjustment scheme which is based on the involvement of a parameter in creating new population members.

4 Implementation and Performance

We implemented the proposed approach on a multi-transputer system, the MEIKO Computing Surface [10], consisting of 72 transputers T800. A SUN 4/390, equipped with 4 transputers which are connected to the remaining 68 transputers installed in a separate cabinet, is used as the host computer. The implementation was written in the *C* programming language on a SUN Sparcstation, employing the MEIKO cross compiler and the communication features offered by the *MEIKO CS Tools* programming environment. A master/slave organisation, in which a unique master task handled the I/O functions, collected statistics and performed the computations associated with the primary GA and several slave tasks were responsible for solving the test problem instances by executing the secondary GAs, was chosen as the fundamental communication structure. After having worked on a problem instance for some number of generations, the slaves returned the obtained fitness values, together with information required for evaluating the solution process, to the master. The master computed the indicators and determined the parameter values of the currently best secondary GA, the string with the highest fitness in the primary population.

In order to investigate the performance of our system, we have selected a class of test problems, namely the optimal determination of connection weights for a feedforward neural network. This class of problems is well suited as a test case, because a) performance results are available in the literature [7, 12]; b) it requires the more complex real-valued gene coding; c) there are examples where the global optimum is known and therefore can easily be used for fitness calculations; and d) it allows operators for units of logical subgroups of genes to be explored.

4.1 Indicators

As already explained in the previous section, the importance of particular decisions and their variants for the solution process is reflected in the corresponding indicators. Figure 3 shows the indicators which evolved as the ones with the largest impact for the problem class of neural network weight assignment. The most significant indicators, computed as averages over 5 runs, are those of the components 19, 8, 3, 16 and 4 of Figure 2 (ordered by priority).

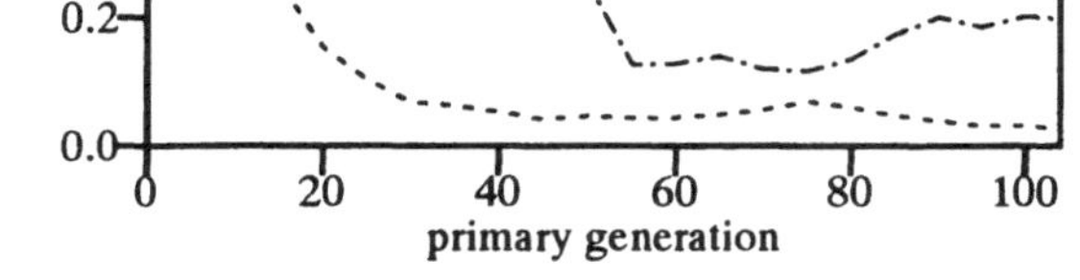

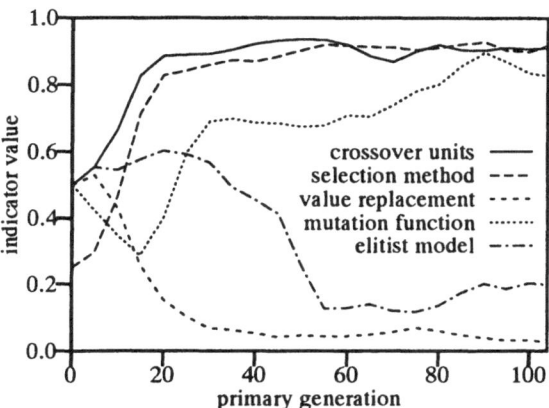

Figure 3: Indicators of the Most Significant Decisions

Indicators with a delayed and weaker reaction are those of the components 12 and 14. They become important after the 60th primary generation, i.e. when other decisions and parameters have already been appropriately determined. The indicators with the least significant decisions within the range of the 104 primary generations investigated are the components 5 and 10.

4.2 Parameters and Decisions

In this subsection we present the results for the development of the decisions and parameters during the execution of the primary GA. The results are based on the secondary GA with the highest fitness after 104 primary generations. They are summarized in Figure 4.

In the following we explain some of the results shown in Figure 4 in more detail.

The optimal population size (component 1) as determined by the primary GA consists of 100 strings. This is the upper limit allowed in our implementation in order to limit the computation time, indicating that large populations are preferable.

Among the alternatives provided for the selection method (component 3), stochastic tournament clearly dominates the other variants. This is graphically illustrated in Figure 5.

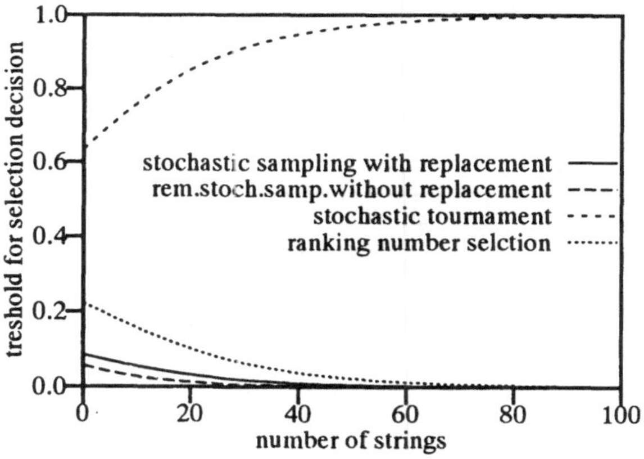

Figure 5: Decision About the Selection Method

The optimal crossover probability (component 7) evolved as 0.49. This value remained nearly constant for all string lengths investigated.

The optimal mutation probability (component 15) is illustrated in Figure 6. It increases with increasing numbers of genes in a string.

Number	Component	Best Variant/Value of Parameter
1	population size	100 (maximal)
2	exponent of the sharing function	1.8
3	selection method	stochastic tournament
4	elitist model	yes
5	double crossover	no
6	number of crossover points	48 % of crossover units
7	crossover probability	0.49
8	crossover units	unit: weights of one neuron
9	crossover points	variable number
10	crossover method for reordering problems	not applicable
11	distance of crossover points	not applicable
12	crowding method	yes
13	crowding factor	$2.4 \equiv 2$ strings
14	mutation units	units: weights of one neuron
15	mutation probability	0.8–0.93
16	mutation function	normal distribution
17	exponential distribution of mutation	not relevant
18	normal distribution of mutation	standard deviation: 0.5–2.0
19	mutation value replacement	use old values

Figure 4: Results for Neural Network Weight Optimization

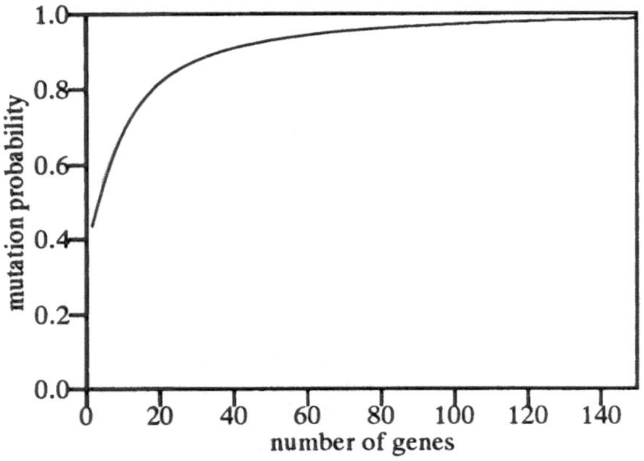

Figure 6: Optimal Mutation Probability

4.3 Performance Profile

In this subsection we intend to demonstrate that the fitness of the best secondary GA increases with increasing primary generations. In order to do so, we selected the best secondary GA after every 5th primary generation, applied it to 50 randomly generated test problems with different numbers of weights to optimize (for 100 secondary generations) and measured its performance. In the implementation, the work was distributed to 50 transputers, each of them being responsible for one test problem. The total computation time required for performing the whole cycle (104 primary generations and a total of 105000 secondary generations for the 50 test problems every 5th primary generation) was 45 hours. This whole procedure was repeated four times. Figure 7 shows the results, where the final values are averages over the four experiments.

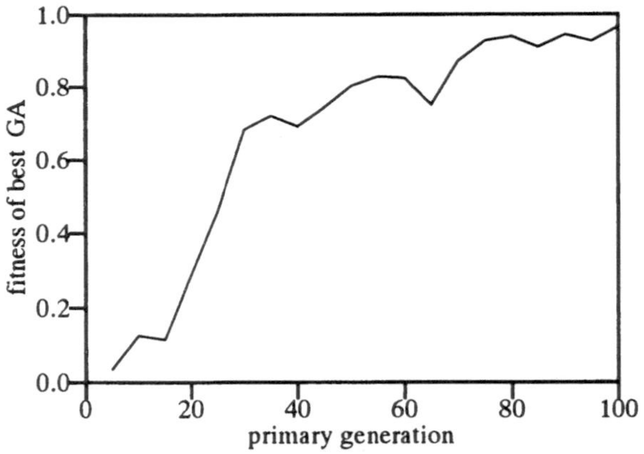

Figure 7: Performance Profile

Up to primary generation 20, the increase in quality is pretty slow, but then it significantly gets larger during a few primary generations. In the third phase, the increase again slows down. The reason is that the quality of a GA seems to be dependent on a few important parameters and operator variants. Once, these have been roughly determined, the GA will produce acceptable results. In subsequent primary generations, the constellation is then refined.

5 Conclusions

In this paper we have presented an approach to determine the optimal genetic algorithm, i.e. the most preferable type of operators and their parameter settings, for a given problem class. The basic idea is to consider the search for the best GA as an optimization problem and use another GA to solve it. A primary GA operates on a population of secondary GAs which in turn are responsible for solving the problem instance in discussion. We have presented the issues associated with representing the relevant information about GAs in strings of genes and have shown that the approach is feasible by implementing it in a multi–transputer environment. Performance results for finding the best GA for the problem of optimal weight assignment in feedforward neural networks have demonstrated the quality of our implementation.

There are a number of issues for future research, such as improving the runtime performance of our implementation, including other genetic operators in the architecture and investigating the results of further test problems in more detail.

References

[1] Booker L, *'Improving Search in Genetic Algorithms'*, In: L. Davis, editor, *Genetic Algorithms and Simulated Annealing.* Pitman, London, 1987.

[2] Goldberg D E, *'Genetic Algorithms in Search, Optimization and Machine Learning'*, Addison-Wesley, Reading, Massachusetts, 1989.

[3] Goldberg D E, *'Sizing Populations for Serial and Parallel Genetic Algorithms'*, Proc. of the 3rd Int. Conference on Genetic Algorithms and their Applications, pp. 70–79, 1989.

[4] Goldberg D E and Lingle R H, *'Alleles, Loci and the Travelling Salesman Problem'*, Proc. of the 1st Int. Conference on Genetic Algorithms and their Applications, pp. 154–159, 1985.

[5] Goldberg D E and Richardson J, *'Genetic Algorithms with Sharing for Multimodal Function Optimization'*, Proc. of the 2nd Int. Conference on Genetic Algorithms and their Applications, pp. 41–49, 1987.

[6] Grefenstette J J, *'Optimisation of Control Parameters for Genetic Algorithms'*, IEEE Transactions on Systems, Man and Cybernetics, 16(1):122–128, 1986.

[7] Heistermann J, *'The Application of a Genetic Approach as an Algorithm for Neural Networks'*, Proc. of the 1st Int. Workshop on Parallel Problem Solving from Nature, pp. 297–301, Lecture Notes in Computer Science 496, Springer–Verlag, 1990.

[8] Hesser J and Männer R, *'Towards an Optimal Mutation Probability for Genetic Algorithms'*, Proc. of the 1st Int. Workshop on Parallel Problem Solving from Nature, pp. 23–32, Lecture Notes in Computer Science 496, Springer–Verlag, 1990.

[9] De Jong K A, *'An Analysis of the Behaviour of a Class of Genetic Adaptive Systems'*, PhD Thesis, University of Michigan, 1975.

[10] MEIKO, *'Computing Surface Documentation Vol. 1/2'*, MEIKO, 1989.

[11] Mercer R E and Sampson J R, *'Adaptive Search Using a Reproductive Meta–Plan'*, Kybernetes, 7:215–228, 1978.

[12] Montana D J and Davis L, *'Training Feedforward Neural Networks Using Genetic Algorithms'*, Proc. of the 11th Int. Joint Conference on Artificial Intelligence, pp. 762–767, 1989.

[13] Mühlenbein H, *'Parallel Genetic Algorithm, Population Dynamics and Combinatorial Optimization'*, Proc. of the 3rd Int. Conference on Genetic Algorithms and their Applications, pp. 416–422, 1989.

[14] Oliver I M, Smith D J and Holland J R C, *'A Study of Permutation Crossover Operators on the Travelling Salesman Problem'*, Proc. of the 2nd Int. Conference on Genetic Algorithms and their Applications, pp. 224–230, 1987.

[15] Schaffer J D, Caruna R A, Eshelman L J and Das R, *'A Study of Control Parameters Affecting Online Performance of Genetic Algorithms for Function Optimization'*, Proc. of the 3rd Int. Conference on Genetic Algorithms and their Applications, pp. 51–60, 1989.

Achieving Self-Stabilization in a Distributed System Using Evolutionary Strategies

Dwight Deugo and Franz Oppacher
Intelligent Systems Research Group
School of Computer Science, Carleton University
Ottawa, Canada, K1S 5B6
E-mail: dwightdeugo@scs.carleton.ca, oppacher@scs.carleton.ca

Abstract

In this paper we present a genetic self-stabilization protocol for the canonical distributed problem of leader election. A self-stabilizing distributed system is one that can be started in any global state, and, during its execution, will eventually reach a legitimate global state(s) and henceforth remain there, maintaining its integrity without any kind of outside intervention. Current self-stabilizing systems either program the stabilizing feature into their protocols, or they use randomized protocols and special processors to stabilize the system. We believe that self-stabilization should be an emergent property of a distributed system, and, by transforming a distributed problem to a model of evolution, which is inherently self-stabilizing, we demonstrate how the emergence of self-stabilization can be achieved. We attempt to achieve more than just solving a problem with a distributed genetic algorithm: we take a distributed problem and show how analogies from evolution can be used to solve it.

1 Introduction

A distributed system contains a set of sequential processors, represented by state machines. A processor communicates with its neighboring processors by exchanging messages. The graph formed by representing each processor as a node and connecting all neighboring nodes by an edge is called the system's *communication graph* or network. Each processor can receive messages, perform internal transitions and send messages, and can, therefore, be viewed as a RAM program. The collection of the system's processor programs is called the *protocol* of the system. In general, each processor independently follows the same protocol in order to solve the problem, and there is no central synchronizing processor or central storage [1].

A distributed system's primary goal is to solve its given problem. Although not often explicitly mentioned, an equally important goal is for its behavior to be, and remain, stable. The stability of the global state, or configuration, of a distributed system can be thought of as a function of the states of its processors. Although every processor in the system may be stable, the union of their 'local' states can give rise to an unstable, illegitimate global state. Ideally, a distributed system's global state should be legitimate - a known and accepted configuration.

Can one develop distributed systems that always remain in legitimate configurations? The answer is obviously yes. However, these systems often assume an ideal world, where faults and inconsistencies never occur [2]. We do not, however, live in such a world; therefore, one should think about how to detect and handle the situation of a distributed system entering an unstable global state.

Instead of just pre-programming the stability of a distributed system's behavior, it seems more desirable to endow it with a capacity for **self-stabilization**. As shown in figure 1, after detecting its global state to be 'unstable', a self-stabilizing system should force itself back to, and keep itself in, a legitimate, stable, global state.

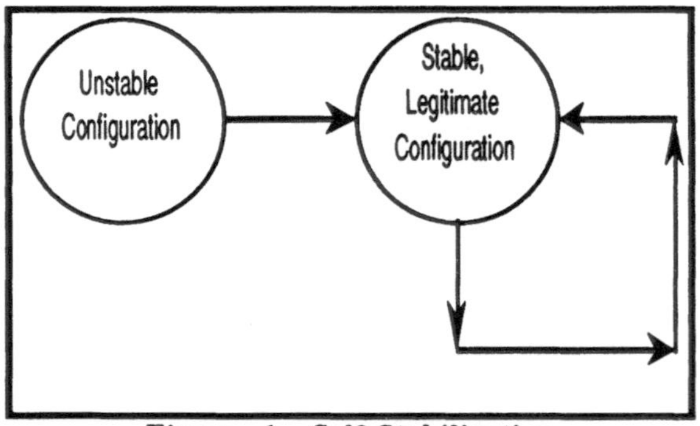

Figure 1. Self-Stabilization

Formally, a self-stabilizing distributed system is one that can be started in any global state, and, during its execution, will eventually reach a legitimate global state(s) and henceforth remain there, maintaining its integrity without any kind of outside intervention.

The idea of self-stabilization was first introduced by Dijkstra [3]:

> 'We call the system *self-stabilizing* if and only if, regardless of the initial state and regardless of the privilege selected each time for the next move, at least one privilege will always be present and the system is guaranteed to find itself in a legitimate state after a finite number of moves.'

To demonstrate the notion of self-stabilization, Dijkstra considered three protocols for the mutual-exclusion problem in a ring using state machines with k-states, 4-states, and 3-states. Dijkstra's seminal paper makes some specific restrictions. First, he assumes that there is a global criterion identifying when a system is in a legitimate global configuration. Consider, as Dijkstra did, a system with n nodes, each node having a state machine of k-states, $k > n$. If the global configuration of the system is an n-tuple, c = $(q_1, q_2, \ldots, q_n)$ where q_i is the state of processor i, $1 <= q_i <= n < k$, then there are at least n^n different global configurations. It seems, then, because of the enormous size of the global configuration space, that this criterion can only be met in the simplest of problems.

Other restrictions of Dijkstra's include: at least one processor is active in a legitimate global configuration; every move from a legitimate global configuration will produce another legitimate global configuration; and there is always a sequence of moves between two legitimate global configurations.

It seems impossible to meet these restrictions in very large configuration spaces. In particular, it seems hopeless to try to account for all possible legal or illegal configurations and their transitions to other configurations.

Dijkstra noted that, due to the symmetry condition, there is no deterministic, uniform, self-stabilizing, mutual exclusion protocol for rings of composite size. Therefore, he and many since him [4, 5, 6] have used a semi-uniform model. In such a model, one processor is running a different protocol from all others, and is used to maintain the global integrity of the system. This model, which deviates from the general view of a distributed system, is one of **enforced stabilization** rather than self-stabilization - one processor manages and enforces the stabilization of the network. This model is also less fault-tolerant: if the 'special' processor dies, so, too, does the stabilization feature.

A self-stabilizing system should be less concerned about mapping out the configuration space for detection of illegal configurations, and it should be more concerned with directing itself towards the final goal configuration(s), regardless of whether it is in a currently legal or illegal configuration. A self-stabilizing protocol should not restrict itself to a predefined set of legal and illegal situations, but, rather, should try to handle all situations, even unexpected ones.

The use of randomized self-stabilizing protocols [7] is a step towards solving some of the above problems. Randomized self-stabilizing protocols continuously 'guess' at the answer to the distributed system's problem until the correct one is found. They are only concerned with producing a final answer and do not rely on execution sequences or illegal configuration detection for their operation. Since they are randomized protocols, they are not directed towards producing better guesses; they simply wander about the solution space until the correct configuration is found. Can we somehow direct the randomized protocols towards the final goal, making better guesses, and not just random ones? Below we show that this is indeed possible with a model of self-stabilization we call *genetic self-stabilization*.

Genetic self-stabilization uses analogies from evolution to produce a randomized, distributed self-stabilization model. The model, by using old guesses as the bases for new ones, has a strong inductive bias that helps to direct the distributed system towards its goal configuration. This feature of directed convergence is achieved by transforming a distributed system's protocol and problem to a model of evolution: fitness proportional reproduction and genetic operators, such as crossover and mutation, are applied to a population of entities. Messages from neighboring processors and uniform random events are used to change the state of a local processor, but only those states that are better (fitter) than the previous ones are retained and used by a processor in the future. The analogy from evolution is that good past solutions are used to guide future solutions. The benefit of the model is that the stabilization of the distributed system is achieved using only uniform, local processor actions, with little or no knowledge of the global task, and no concern of the current global configuration.

A self-stabilizing system should, after any fault, regain its goal stable configuration in a finite amount of time, or at least approach it. Since it is practically impossible to anticipate all possible faults, self-stabilization should not be achieved by explicit programming. Under our model, self-stabilization is an emergent property of the system.

In this paper we present a genetic self-stabilization protocol for the canonical distributed problem of leader election. Leader election is a fundamental problem of distributed computing and involves electing one and only one processor as the leader of the network. There are many well known deterministic id-based protocols [8, 9] for this task. This paper describes the first genetic-based approach for leader election in a ring, in a complete network, and in a complete network in which messages are not sent on all edges. We show that self-stabilization can be an emergent property of the system. We also show that adding communication links to processors does not necessarily lead to a faster, more efficient model.

We attempt to achieve more than just solving a problem with a distributed genetic algorithm [10]: we take a distributed problem and show how analogies from evolution can be used to solve it.

2 Model

The computation model used in this paper is an *asynchronous message passing system*. A message passing system is a collection of processors that exchange information only by transmitting messages using FIFO communications channels between neighboring (connected) processors. Processors are asynchronous, i.e., there is no bound on their processing speeds. Messages are transmitted asynchronously, i.e., there is no bound on either the delivery time or message channel capacity. It is assumed that every message sent is eventually received.

The system used here is a uniform distributed system consisting of n processors, denoted $P_1, P_2, P_3, \ldots, P_n$. Each processor contains the same protocol, each has a unique identity, 1 <= identity <= n; and each has one state variable: **Leaders**. The Leaders state variable contains a bit string of length n and an integer reference point that is initialized with the processor's id. Since we are concerned with self-stabilization, the initial bit values of Leaders are not stated - they are randomly initialized. Each processor can be regarded as a RAM whose program is composed of atomic steps. An atomic step consists of the reception of messages, an internal computation, and the optional sending of messages to other processors.

Messages are sent between neighboring processors, i.e., processors that are connected in the system's communication network. In this paper, we consider three different communication networks: a ring, a complete network, and complete networks where messages are not sent on all links. In a ring network, each processor is connected to only two other processors. In a complete network, each processor is connected to all other processors, although it may not choose to send messages on certain links.

3 Protocol

In this section, we present a genetic self-stabilization protocol for election in a ring and in a complete network, both with and without full messaging capability[1]. The goal of the protocol is for each processor to agree on a common Leaders string, i.e., a string with a single bit set. Once a processor finds that it and its neighbors agree on the same leader - they have the same Leaders string - it assumes that its Leaders string indicates the network's leader. The network will eventually stabilize itself, with all processors agreeing on the same leader. However, processors may come to again disagree on the network's leader, as a result of a fault, but, because of the protocol's self-stabilization feature, the network will stabilize itself again.

The genetic election protocol presented in section 3.2 relies on analogies from evolution; therefore, before describing it, we examine how evolution can be used as a self-stabilization strategy for distributed systems.

3.1 Genetic Strategy

The Darwinian process of evolution involves repeated cycles of selection (biased by fitness) and reproduction applied to a population of structures. This model of survival of the fittest selects the best structures from the current population for reproduction into the next one. Adaptation occurs during reproduction when two strong structures are crossed with one another. Besides crossover, small mutations also occur in the new structures, and necessitate further adaptations. The crossover and mutation of the two selected structures results in two new structures that differ slightly from the original ones.

Genetic Algorithms (GAs) [11] have successfully applied the model of evolution to difficult computation problems. For example, good results have been achieved using GAs on various NP-Complete problems [12, 13, 14].

GAs have two features that make them potentially useful for a distributed system. First, a GA can be viewed as a self-stabilizing algorithm. GAs are search algorithms that use a population of 'guesses' to conduct the search. As the GA is searching, the population of guesses tends to converge to one guess - the answer to the search problem. Once the population has converged, it tends to remain in its stable configuration, unless it is reset. If a new guess is introduced into the population, it is overwhelmed by the majority consensus of the other guesses, and within a few generations it is replaced by the existing majority guess. It is an important property of a GA that its population usually converges regardless of the initial population, although the time it takes to converge may vary. Therefore, the application of the evolutionary model to distributed computing enables the construction of a general model of self-stabilization.

Second, GAs often differ only in their fitness functions - the function that evaluates the fitness of each individual in the population - because fitness is dependent on the problem domain. Everything else in the system remains unchanged across different problems. GAs use the same basic algorithms for reproduction, selection, mutation, and crossover. Therefore, when an

1 The ability or lack of ability of a processor to send messages to all of its neighboring processors.

evolutionary model is applied, only one general protocol is required for many different problems.

To solve a distributed problem in an evolutionary framework, one must be able to specify the following:

1) An encoding of potential answers to the problem as a genetic string.
2) A population of strings.
3) Methods of selection and reproduction for replacing the population.
4) A fitness function to compare potential answers with one another.

The operations of selection and reproduction are surprisingly simple, but they require strings, not necessarily bit strings [15], for their operation. Therefore, a distributed system must be able to encode possible answers to its given problem as strings. For example, a bit string of size *n*, the number of processors, can be used as an encoding for the election problem. The *i'th* bit position represents whether the *i'th* processor is a possible leader. A string that represents a possible final answer to the leadership question has every bit, except one, set to 0. There are 2^n strings altogether, but only *n* strings represent possible legal answers to leader election, i.e., those with only one bit set. How do we get agreement from *n* processors on which of the 2^n strings to use? This question is answered in the next section.

The second requirement of the evolutionary model is that we have a population of strings in the distributed system. One solution is for every processor to maintain a population of strings as part of its state [16]. For this paper, we prefer another population model, according to which each processor contains, as part of its state, one Leaders string [17, 18].

The third requirement of the evolutionary model is for the distributed system to be able to select and reproduce strings. This is not a difficult requirement for a distributed system. For example, a processor can select a mate from one of its neighboring processors. One question that arises is how big does that neighborhood have to be for the algorithm to be effective? In a ring, each processor has only two neighbors, but in a complete network a processor can select any other processor as a mate. As our results show, a larger population from which to select mates does not necessarily make the system converge faster to a final answer. The added cost of messages has to be considered as well.

The fourth requirement of the evolutionary model is that there exist a fitness function that determines the strength of any string. Ideally the fitness function has one global optimum and very few local optima. The ruggedness of the fitness function is usually a good indicator of the speed of convergence. A simple fitness function for the previous encoding of leader election could return *n* minus the number of zeros in a string, penalizing a string that has all zeros - possibly making it equivalent to a string with two ones. Therefore, the best string would have only one bit set.

In the next section, we show how the population of processors in a distributed system makes the task of solving a distributed problem in an evolutionary framework relatively easy.

3.2 Genetic Leader Election Strategy

In this section, we look at an evolutionary model for leader election. Before describing the protocol, we look at the encoding of the problem, how the population is assembled, and the problem's fitness function.

3.2.1 Encoding

We encode a solution to the leader election problem as described in the last section: each processor has a string (called **Leaders**) consisting of *n* bits and one integer reference point. Each bit position represents whether a processor is, or is not, a leader. A string with one bit set means that the corresponding processor at that position is a leader. A system is stable when all processors have the same Leaders string with 1 bit set.

3.2.2 Population

Since every processor has a single Leaders string, our algorithm considers its population the set of Leaders strings from the processors. We wish to point out here, although further details are given in the Selection and Reproduction section, that the network configuration determines which Leaders are selected to reproduce with one another; mates are not always selected, as in the case of a GA, from the entire population.

3.2.3 Fitness

In our model, a Leaders' fitness value is calculated as follows:

```
FitnessOf: Leaders
  | zeroCount oneCount value locationOfOne
    offset secondOffset |

  zeroCount := number of zeros in Leaders.
  oneCount := n - zeroCount.
  locationOfOne := last location of a 1 in
                                    Leaders

  "First fitness range: If Leaders has only one
   bit set, return a value between 0.5 and 1.0"
  (oneCount = 1)
  ifTrue: [
```

```
offset := reference point of Leaders -
                              locationOfOne.
(offset positive)
ifTrue: [
  value := 0.75 - (offset / (16 * (n - 1)))]
ifFalse: [
  value := 0.75 + (offset / (16 * (n - 1)))].

secondOffset := reference point of Leaders -
                                    processor id.
(secondOffset positive)
ifTrue: [
  value := value - (secondOffset / (16*(n - 1)))]
ifFalse: [
  value := value + (secondOffset / (16*(n - 1)))].

value := value + (reference point of Leaders /
                                    (4 * n)).
^ value].

"Second fitness range: If Leaders has no bit
 set, return a value just less than 0.5"
(oneCount = 0) ifTrue: [^ (n - 1) / (2.0 * n)].

"Third fitness range: If Leaders has more
 than one bit set, return a value between 0 and
 the second fitness range value"
^ |zeroCount - 1| / (2.0 * n)
```

This fitness function has three levels of fitness. The first level is between 0.5 and 1.0. This level is reserved for strings with one bit set. The next level is just below 0.5, and is reserved for a string with all zeros. The final level of fitness is reserved for strings with more than one bit set. The fitness of a string in this range is decreased proportionally to the number of 1's in it.

The first level is used to break the fitness symmetry between n different strings with only one bit set. However, we do not want to bias the fitness function so strongly that the same string always wins the leader election. To help break symmetry, we use a reference point which is located at the *n'th* plus one position in the Leaders string. When a processor initializes, besides generating random bits for its Leaders string, it also writes its id into the reference point. This is done only once. From then on, Leaders strings are generated only by the crossover and mutation operators. Crossover ensures that the reference point is also crossed, and the mutation operator has no effect on it.

The fitness of a string at the first level starts at 0.75 and is decreased proportionally to the distance of its set bit's position from the reference point. This ensures that a string whose set bit's position is close to the reference point is stronger than one whose set bit's position is further away from it.

The fitness of a string is decreased further proportionally to the distance of its processor id from its reference point. This ensures that a processor whose id is close to the reference point is stronger than one whose id is further away from it.

Finally, the fitness of a string is increased proportionally to its reference point.

The intent of this fitness function is to characterize as the strongest a string whose set bit's position, reference point, and processor id correspond to one another. Therefore, the same string will have different fitness values on different processors. It is only when strings are equally distant from the reference point that the processor id is needed to break the symmetry.

3.2.4 Selection and Reproduction Protocol

In all network configurations, a processor receives messages[2] from its neighbors. The processor then uses its and its neighbor's Leaders strings as the population from which to select mates for the production of its new Leaders string. In general, the reproduction protocol is as follows:

```
"Collect one Leaders message from all neighboring
 incoming message links"
messages := self gatherIncomingLeadersMessages.
messages add: (self leaders).

" Select two leaders to mate"
mate1 := self selectLeadersFrom: messages.
mate2 := self selectLeadersFrom: messages.

"Next, probabilistically mutate the Leaders bits
   of the two selected mates"
mate1 mutate.
mate2 mutate.

"Crossover mates"
self timeToCrossover
ifTrue: [
   child1 := mate1 crossoverWith: mate2.
   child2 := mate2 crossoverWith: mate1.
ifFalse: [
   child1 := mate1.
   child2 := mate2].

"Chose best child to be new Leaders of the node"
newLeaders := child1 bestOf: child2.

"Update processor Leaders"
self leaders: newLeaders.

"Broadcast the new Leaders to neighbors"
self sendLeadersMessage].
```

After receiving Leaders messages from its neighbors, a processor selects two Leaders to mate with one another

---

[2] A message contains a Leaders string.

by applying roulette wheel selection [11] to a population consisting of its neighbor's Leaders strings and its own Leaders string. Then, the mutation operator is applied to both mates. Next, a new Leaders string is produced by applying two-point crossover [19] to the two selected mates and then selecting the best of the two resulting Leaders strings. The new Leaders string replaces the processor's old one, and then it is sent to all of the processor's neighbors.

In a ring, a processor waits until it receives Leaders messages from its two neighbors. In a complete network, a processor waits until it receives Leaders messages from all other processors. In a complete network in which processors do not send messages to all neighbors, a processor waits for a fixed time period and then receives Leaders messages from any incoming link that has one.

4 Results

In this section, we examine the results of the application of the genetic self-stabilizing model to the leader election problem using different distributed network configurations. All of the following experiments were repeated 11 times on each network, and the results taken from the median run of each particular experiment. A run is the time (in generations) taken for the network to converge to a common Leaders string. A generation is considered as a round of *each* processor receiving its neighbor's messages, reproducing, and informing its neighbors of its new Leaders string.

4.1 Rings and Complete Networks

In the experiments on rings and complete networks, with full messaging capability, we worked with four different network sizes: 5, 10, 20, and 40 processors. The first result, shown in figure 2, provides a comparison of the number of generations required for each network to converge to a common Leaders string.

As one would expect, the number of generations required for a complete network to stabilize to a common Leaders string is less than the number of generations required for a ring. There is, after all, more information passed in a complete network; accordingly, each processor selects a new Leaders string from a larger, more representative population.

An interesting aspect of figure 2 is that it appears to take $O(n)$ generations for the network to stabilize. However, to compare with existing self-stabilization protocols, we must look at the number of messages sent between processors.

Figure 3 displays the total number of messages sent by each of the four sizes of networks in order to stabilize. There are two points to note about this figure. The first point is that the number of messages required for stabilizing the networks is of $O(n^2)$. This result is achieved in many existing self-stabilizing designs [20], but they often include more design requirements, such as synchronous processors with many states.

The second point is that, although the number of generations required to stabilize a complete network is less than the number required to stabilize a ring, the cost of stabilizing a complete network, expressed as the number of messages sent, is considerably more than the cost of stabilizing a ring.

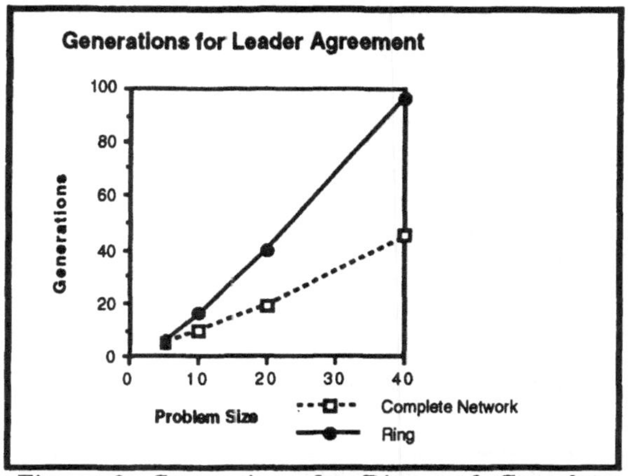

Figure 2. Generations for Ring and Complete Network Agreement

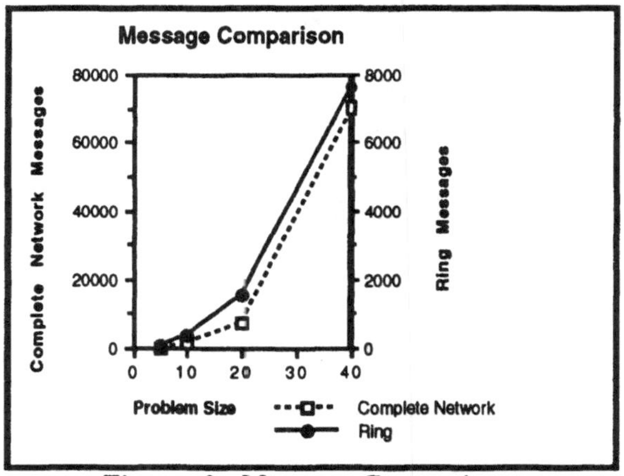

Figure 3. Message Comparisons

This raises a question: how many messages does a processor in a complete network need to send to the other processors to maintain the self-stabilization rate of complete message passing? Stated in another way, how large does a processor's population of Leaders messages need to be in order to maintain the self-stabilization rate of complete message passing? Another question is what is the trade-off between messages sent, or similarly

Leaders message population sizes, and self-stabilization rates? The answers to these questions are the subject of section 4.2.

4.2 Complete Networks with Incomplete Message Passing

In this experiment, we used a fully connected 10 processor network, but varied the number of Leaders messages a processor sends, each generation, from 1 to 9. Rather than sending Leaders messages to all its neighbors, a processor randomly selects, without replacement, the neighbors it will send messages to. The results of this experiment, shown in figure 4, show that there is a definite trade-off between messages sent and generations needed to stabilize. It appears that permitting processors in a 10 processor network to send more than 4 Leaders messages provides little increase in the stabilization rate, and thus constitutes a waste of messages. In fact, for 5, 20, and 40 processor networks, the results are the same: having processors send just less than $\frac{n}{2}$ messages in each generation maintains the same stabilization rate as sending n messages.

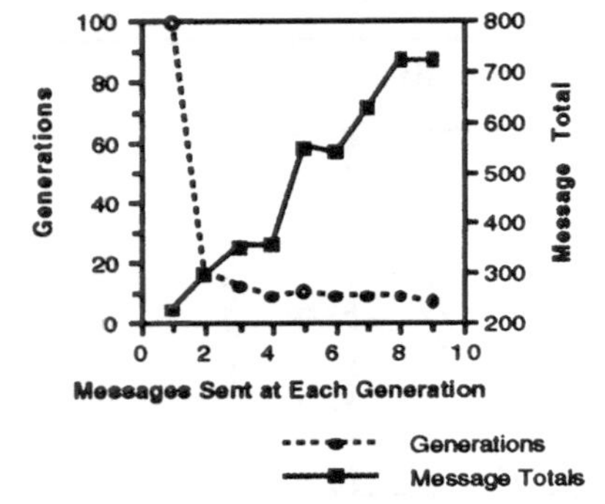

Figure 4. 10 Processor Problem with Variable Message Sending

5 Conclusion

From the results, we can conclude that the genetic self-stabilizing model achieves it goals. First, its general model of computation forms the basis from which many distributed problems, with differing network configurations, can be solved, provided an encoding, a population, a fitness function, and a selection and reproduction schema can be identified. The major work involved in transforming a new distributed problem to the evolutionary model is providing an encoding and fitness function for it.

Second, the results show that the genetic self-stabilizing model is self-stabilizing. From a random configuration, the system always manages to find its way back to the goal configuration without any outside intervention. What is even more impressive is that the system does not rely on the underlying distributed problem for its stabilization. In an open system, evolution never stops. However, in a closed system, evolution stabilizes to a final configuration. The fact that we are modeling a closed system, using the evolutionary process, means that it will converge to a stable configuration - stabilization is an emergent property of the system.

Finally, the results show that the genetic self-stabilizing model directs itself to better and better answers. Stabilization is not something that should happen at once, but, rather, something that should evolve. The stabilization of the network is achieved by first stabilizing a small part of the system, and then stabilizing more and more of it until the entire system is stabilized, which is in contrast to the randomized self-stabilizing systems which never partially stabilized.

This work has raised several issues that should be considered. The first issue deals with symmetry breaking. Processor ids are required to break symmetry. A question for future research is: is there a way to break symmetry without using processor ids? A second issue deals with population sizes and communications. Can one propose general theories on population sizes for genetic self-stabilization of distributed systems? Finally, what other distributed problems can benefit from an evolutionary model?

We believe that our results demonstrate the potential of genetic self-stabilization, and that our model should be explored further.

6 References

[1] Lamport L., *'Time, Clocks and the Ordering of Events in a Distributed System'*, Communications of the ACM, V:21, N:7, 558-565, 1978.

[2] Gallager R.G., Humblet P.A., and Spira P.M., *'A Distributed Algorithm for Minimum-Weight Spanning Trees'*, ACM Transactions on Programming Languages and Systems, 5, 2, 66-77, 1983.

[3] Dijkstra E.W., *'Self Stabilizing Systems in Spite of Distributed Control'*, Communications of the ACM, 17, 11, 643-644, 1974.

[4] Brown G.M., Gouda M.G., and Wu C.L., *'A Self-Stabilizing Token System'*, Proceedings of the Twentieth Annual Hawaii International Conference on System Sciences, 218-223, 1987.

[5] Burns J.E., *'Self-Stabilizing Rings Without Daemons'*, Technical Report GIT-ICS-87/36, Georgia Institute of Technology, 1987.

[6] Dolev S., Israeli A., and Moran S., *'Self Stabilization of Dynamic Systems Assuming Only Read/Write Atomicity'*, Proceedings of the Ninth Annual ACM Symposium on Principles of Distributed Computation, 103-118, 1990.

[7] Israeli A., and Jalfon M., 'Self Stabilizing Ring *Orientation'*, Proceedings of the 4'th International Workshop on Distributed Algorithms, 1990.

[8] Korach E., Moran, S. and Zaks S., *'Tight Lower and Upper Bounds for Some Distributed Algorithms for Complete Network of Processors'*, Proceedings of the 3rd Annual ACM Symposium of Principles of Distributed Computing, 199-207, 1984.

[9] Korach E., Kutten S., and Moran S., *'A Modular Technique for the Design of Efficient Distributed Leader Finding Algorithms'*, ACM Trans. Program. Lang. Syst. 12, 1, 84-101, 1990.

[10] Tanese R., *'Distributed Genetic Algorithms'*, Proceedings of the Third International Conference on Genetic Algorithms, Morgan Kaufmann, 434-439, 1989.

[11] Goldberg D.E., *'Genetic Algorithms in Search, Optimization, and Machine Learning'*, Addison-Wesley, 1989.

[12] Deugo D.L. and Oppacher F., *'Explicitly Schema-Based Genetic Algorithms'*, Proceedings of the Ninth Biennial Conference of the Canadian Society for Computational Studies of Intelligence, 46-53, 1992.

[13] Whitley D., Starkweather T., and Fuquay D., *'Scheduling Problems and Traveling Salesmen: The Genetic Edge Recombination Operator'*, Proceedings of the Third International Conference on Genetic Algorithms, Morgan Kaufmann, 133-140, 1989.

[14] De Jong K. and Spears W., *'Using Genetic Algorithms to Solve NP-Complete Problems'*, Proceedings of the Third International Conference on Genetic Algorithms, Morgan Kaufmann, 124-132, 1989.

[15] Deugo D.L. and Oppacher F., *'Improving the Quality of Case Memory Using Genetic Techniques'*, Proceedings of the Eight Biennial Conference of the Canadian Society for Computational Studies of Intelligence, Morgan-Kaufmann, 161-168, 1990.

[16] Cohoon J.P., Martin W.N., and Richards D.S., *'A Multi-Population Genetic Algorithm for Solving the K-Partition Problem on Hyper-Cubes'*, Proceedings of the Fourth International Conference on Genetic Algorithms, Morgan Kaufmann, 244-249, 1991.

[17] Gorges-Schleuter M., *'ASPARAGOS: An Asynchronous Parallel Genetic Optimization Strategy'*, Proceedings of the Third International Conference on Genetic Algorithms, Morgan Kaufmann, 422-427, 1989.

[18] Manderick B. and Spiessens P., *'Fine-Grained Parallel Genetic Algorithms'*, Proceedings of the Third International Conference on Genetic Algorithms, Morgan Kaufmann, 428-433, 1989.

[19] Syswerda, G., *'Uniform Crossover in Genetic Algorithms'*, Proceedings of the Third International Conference on Genetic Algorithms, Morgan Kaufmann, 2-9, 1989.

[20] Chang J., Gonnet G., and Rotem D., *'On the Costs of Self-Stabilization'*, Information Processing Letters, Vol. 24, 311-316, 1987.

Improving Simple Classifier Systems to alleviate the problems of Duplication, Subsumption and Equivalence of Rules

A. Fairley and D.F. Yates,
Department of Computer Science
University of Liverpool, U.K.

Abstract

For new, potentially improved rules that is, the search performed by a classifier system's genetic algorithm is guided by the relative strength of the rules in the extant rule base. This paper identifies three general types of rule whose presence in a plan can affect the relative strength of rules in a rule base and thereby provide the potential to compromise the effectiveness of the genetic algorithm. The nature and extent of relative strength distortion is investigated and a method to combat the distortion which involves adaptation of the standard bucket brigade algorithm, is proposed.

1.0 Introduction

A *Machine Learning System (MLS)* is a computer based system that is capable of 'learning' an appropriate set of rules for solving problems in a specified domain. Essential to any successful MLS are: a means for distinguishing the utility, in respect of problem solving, of the individual rules in the available rule set (*rule base* or *plan*), and a mechanism for discovering new improved rules which are used subsequently to supplement the rule base. A *Classifier System* is a particular type of MLS.

In such a system, each rule (or *classifier*) is represented by a character string which is divided into 2 parts: a *condition* part and an *action* part. The condition part specifies those conditions which must be satisfied if the action specified in the corresponding action part is to be considered. The utility of a classifier in the rule base is represented by a value called its *strength*. Each time a classifier is found to be effective in a problem solving episode, the system rewards it by increasing its strength via, and according to, an agent referred to as a *credit assignment algorithm*. To effect the discovery of new rules, a classifier system employs a *genetic algorithm (GA)*. Inasmuch as the search performed by a GA in a classifier system is intimately linked to relative rule strength, the effectiveness of the search in discovering improved rules may be impaired if the relative strengths of classifiers in the extant plan do not accurately reflect their relative utility. The work reported in this paper focuses on the set of circumstances in which such distortion of classifier strength can occur.

The very nature of a genetic algorithm is such that when a new plan for a classifier system is generated as a result of the algorithm being applied, the new plan may contain rules with identical actions, call them A and B, but which are such that either:

(1) A and B are duplicates;

(2) A subsumes B (B is subsumed by A);

(3) A is equivalent to B.

Here, the terms duplicate, subsume and equivalent are defined as follows.

A is said to be a *duplicate* of B if the condition parts of A and B are identical.

A is said to *subsume* B if A and B are not duplicates and B is active $\Rightarrow$ A is active but A is active $\not\Rightarrow$ B is active. If A subsumes B then, A is referred to as a *subsuming* rule and B as a *subsumed* rule.

A and B are said to be *equivalent* if A and B are not duplicates and A is active $\Leftrightarrow$ B is active.

This paper reports the results of an investigation of the effects on relative rule strength in a classifier system's plan as a result of the presence of duplicated, subsuming and equivalent rules.

Further, it establishes the nature and effects of such rules in respect of relative rule strength, proposes approaches for circumventing the effects, and in respect of equivalent rules, demonstrates the efficacy of the suggested approach.

2.0 The Test Model

The problem used to investigate the effects of duplicate, subsuming and equivalent rules on the performance of various plans in SCS1 was a simplified version of the *Two Tanks Problem (TTP)*. In its most general form, the problem, set in three dimensions, requires a tank, T1, which possesses a gun as armourment, to destroy an identical tank T2. In the simplified version of this problem adopted here, the setting was restricted to a 'one-dimensional board' composed of an infinite strip of identical contiguous squares. Also tank movement was restricted to one square backwards or forwards per game-move and gun movement to increasing/decreasing gun elevation by 'one notch'- equivalent to increasing/decreasing gun range by one square.

Initially, the two tanks are placed randomly at a distance of $10 \leq x \leq 20$ units apart, with the range of both T1 and T2 set to $x \pm i$, where the integer i is selected at random as either 1, 2, or 3. An *episode* (the term is borrowed from Grefenstette, [2]) is a complete play of the game, starting with the two tanks in their initial positions, each tank taking alternate moves (T1 takes the first), and ending when either a tank hits its opponent (achieved by firing when the gap is equal to the firing range), or neither tank has won after t moves, where t is the maximum length of an episode.

On any move, each tank has a non-null set of actions, namely:

- move forward 1 square;
- move backward 1 square;
- fire;
- do nothing;
- increase elevation by 1 notch;
- decrease elevation by 1 notch.

All other complicating features, of the more general problem, such as availability of ammunition and the weather, are assumed to be irrelevant.

This problem was chosen because its simplicty readily facilitated the interpretation of the results of the experiments performed in respect of duplicate, equivalent and subsuming rules.

3.0 The Classifier System, SCS1

The structure of SCS1 is depicted in figure 1. The *World Model* represents the current state of the game, and each cycle of the system determines the next move to be made by tank T1.

The knowledge in SCS1 is held in the form of a tactical plan used to control T1. Classifiers in the plan communicate with the environment (world model) and other classifiers via *messages*. There are two types of message:

(i) Environmental (Detector) Messages

An environmental message, E, is a six bit binary string which is used to represent the state of the world model at a given time.

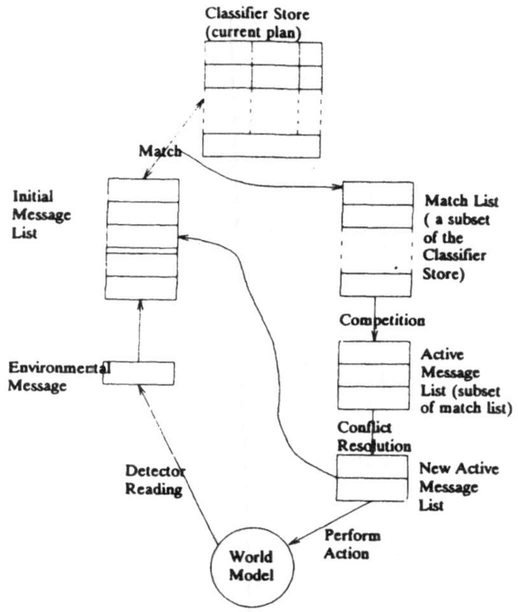

Figure 1: The classifier system SCS1 without rule discovery

Each bit, E_i, $1 \leq i \leq 6$, represents a property within the world model. All environmental messages have bits 1 and 2 (the leftmost bits) set to 00, with the remaining 4 bits set as follows:

Bit 3 - 1 if opponent altered elevation,
0 otherwise;
Bit 4 - 1 if opponent moved, 0 otherwise;
Bit 5 - 1 if opponent did nothing, 0 otherwise;
Bit 6 - 1 if opponent fired, 0 otherwise.

(ii) Internal Messages

Internal messages are used to enable classifiers to communicate with one another and, as such, provide a one move memory for the system. They specify the next action to be performed by the system and as with environmental messages, they are six bits long. All internal messages have the first two bits set to 01. Bits 4 to 6 which represent the six possible actions, are set as follows:

001 - move forward 1 square;
010 - move back 1 square;
011 - fire;
101 - increase elevation by 1 position;
110 - decrease elevation by 1 position;
000 - do nothing.

Bit 3 in internal messages is redundant and is always set to zero.

Each classifier is represented in the IF..THEN form with a conditional part and an action/message part. The conditional part consists of 1 or 2 conditions relating to either or both types of message. Conditions are strings of length six over the alphabet {0, 1, #}, where # is a wildcard ('don't care') symbol. The action/message part is an internal message which specifies the action to be performed by that classifier. Examples of rules of this form are shown below:

000001 : 010110
- IF T2 fired THEN decrease elevation

000100, 010001 : 010011
- IF T2 moved AND T1 advanced THEN fire

All rules in this paper are assigned an initial strength of 10.0.

At the start of a cycle, the detector reads the current state of the world model into the environmental message, E, which is subsequently added to the message list. On the first cycle of an episode, the message list will be empty but on later cycles, it will contain internal messages from previous cycles.

The next step in the cycle is the *Matching* phase. Here, the condition part of each classifier in the plan is compared with the messages on the message list. A match occurs if, for every condition, C, in the classifier, there exists a message, M, on the message list, for which every bit in the condition either matches the equivalent bit in the message ($C_i = M_i$) or the condition bit is set to the wildcard character ($C_i = \#$). A list is kept of all matching classifiers and upon completion of the match phase, a competition is held to reduce the number of matches to the size of the message list. This is performed by calculating a *bid* for each competing classifier, j, as follows:

$$bid_j = b \cdot specificity_j \cdot strength_j$$

Here b is a constant called the *bid co-efficient* and *specificity* is a value relating to how specific or general the classifier is, and is calculated as the ratio of the number of non-wildcard conditions to the total number of conditions. The n classifiers (where n is the size of the message list) with the largest bids are then selected for the message list. At this stage, all classifiers pay a *life tax*, which is fixed at 1% of its current strength, to the system.

Some of the messages posted may suggest alternative actions and thus *Conflict Resolution* must be performed so that the messages sent to the output interface are consistent. Conflict Resolution involves calculating the cumulative bid of each action specified by messages on the message list and adopting the action with the largest total. All messages not consistent with the selected action are then deleted from the message list, and in respect of those remaining, the classifiers which sent them repay their suppliers - those classifiers which were responsible for activating them. In the special case where a classifier has been activated by the environmental message, the payment is made to the environment instead of another classifier. This method of paying previously active classifiers is called the *Bucket Brigade Algorithm*.

After the action has been selected, it is passed via an effector message to the world model, which is subsequently updated. An action then needs to be selected for the opponent, T2. For the purposes of this paper, T2 was assigned the plan of firing on every move.

If an episode is successful (T1 is victorious) then a pay-off of P, is assigned to each active classifier, a defeat results in zero pay-off while a draw results in a percentage, Q, of P being awarded. For all experiments reported in this paper P is set to 10.0 and Q to 50%.

4.0 The Tests and Their Results

This section contains a description, and discussion of, the results deriving from a series of simple tests aimed at investigating the distortion of rule strength induced by the presence of duplicate, subsuming and equivalent rules in a classifier system's plan.

4.1 Tests involving Duplicated Rules

To serve as a control experiment, SCS1 was invested with a simple plan, P_1, containing but the single rule:

A: 00#### : 010011

and applied to a series S of 60 instances of the test problem. The resulting value of A's strength was found to be 0.6247.

P_1 was then replaced, in turn, by plans $P_2, P_3, \ldots P_8$, and the experiment repeated under conditions of no competition (that is, using a message list sufficiently large to hold all simultaneously active classifiers). Here, P_i denotes a plan containing only i copies of A. Each experiment gave analogous results, namely and perhaps not surprisingly, that using plan P_j, each of the j copies of A acquired a strength of $0.6247/j$, the combined strength of the copies equating with the strength of A in the control experiment.

Although indicating that the presence of duplicate rules can significantly distort rule strength, the above experiment is somewhat unrealistic in that no competition was involved either in terms of message list size or from other active classifiers. Correspondingly, to investigate the influence of message list size, SCS1 was invested with plan P_6, a message list of size k and, for $k=2,...,5$, applied to the problem set S (the cases $k=1$ and $k=6$ replicate the control and previous experiment respectively).

It had been anticipated that, with a message list size of k, $(6-k)$ of the classifiers would achieve a strength of zero and each of the remaining k classifiers a strength of $0.6247/k$. Such proved to be the case. Thus, it can be said that although reducing the size of the message list does reduce the distortion in rule strength caused by duplicated rules, it is unlikely that the distortion will be completely eliminated unless a message list of size one is adopted - and this is too limiting for many applications.

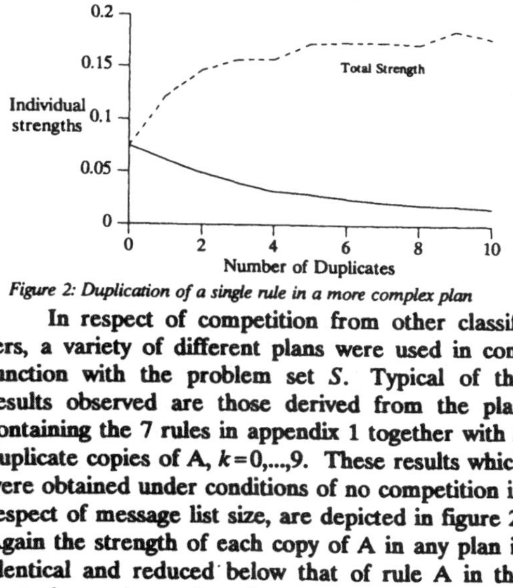

Figure 2: Duplication of a single rule in a more complex plan

In respect of competition from other classifiers, a variety of different plans were used in conjunction with the problem set S. Typical of the results observed are those derived from the plan containing the 7 rules in appendix 1 together with k duplicate copies of A, $k=0,...,9$. These results which were obtained under conditions of no competition in respect of message list size, are depicted in figure 2. Again the strength of each copy of A in any plan is identical and reduced below that of rule A in the control experiment, but the extent of the reduction is not as great as that experienced when the plan contains *only* copies of a single rule. Correspondingly, the sum total of the individual strengths of the copies of A in a plan falls below the strength of A derived from the control experiment, but this value increases as the number of duplicates is increased. The reason for this is as follows. When only A and its k duplicates are active on a given cycle, each of these rules receives $1/(k+1)$th of the pay-off. However, rules other than A and it's duplicates will be active on some some cycles and hence the pay-off will be divided equally between A, its duplicates, and the other active rules. On such cycles, the rules differing from A are, in terms of the effects of pay-off, acting as if they were duplicates of A!

The distortion of rule strength deriving from the prescence of duplicated rules within a plan is undesirable. Fortunately, however, the problem can easily be circumvented by checking any newly generated rules against the extant rule-set prior to that rule's assumption into the set. More problematic, however, is dealing with new rules which turn out to subsume, be subsumed by or be equivalent to, rules already in a plan.

4.2 Tests Involving Equivalent Rules

Consider two equivalent rules, R_1 and R_2 in some plan. Let S_i, $spec_i$, and $payment_i$ denote respectively the strength and specificity of rule i, and the payment made to i after a successful bid, $i=1, 2$. During an episode which lacks competition, R_1 is active $\Leftrightarrow R_2$ is active and in the steady state, $payment_1 = payment_2$, hence:

$$b * spec_1 * S_1 = b * spec_2 * S_2$$

and thus
$$S_1/S_2 = spec_2/spec_1 \qquad (1)$$

This simple analysis suggests that the ratio of the strength of two equivalent rules might be expected to be the inverse of the ratio of their specificities. To investigate this, a simple plan, denoted P_E, was applied to the problems in test set S but with the length of each episode constrained to L cycles, L=1, 2,...,200. P_E contained the three rules

X:- 01#### : 010011
Y:- 010### : 010011
000010 : 010011

In effect, rules X and Y are equivalent since the 3rd bit (lexicographically) is redundant and always set to zero. The third rule is only active on the first cycle of an episode. The results derived with P_E are depicted in figure 3. The shape of the graph does not accord with the above analysis. However its shape can be understood if it is compared with the graph of figure 4 which was obtained

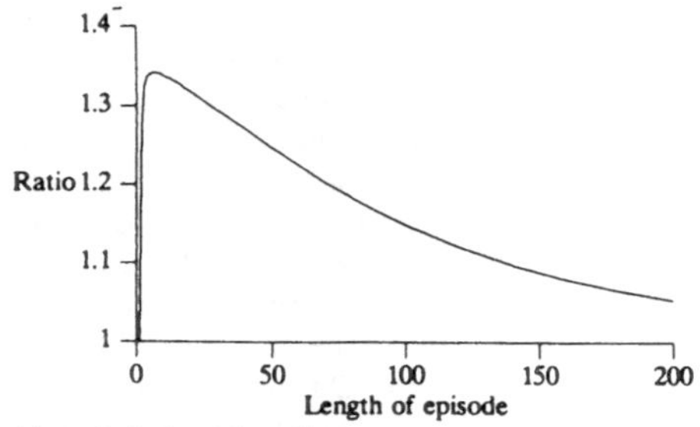

Figure 3: Ratio of S_X to S_Y

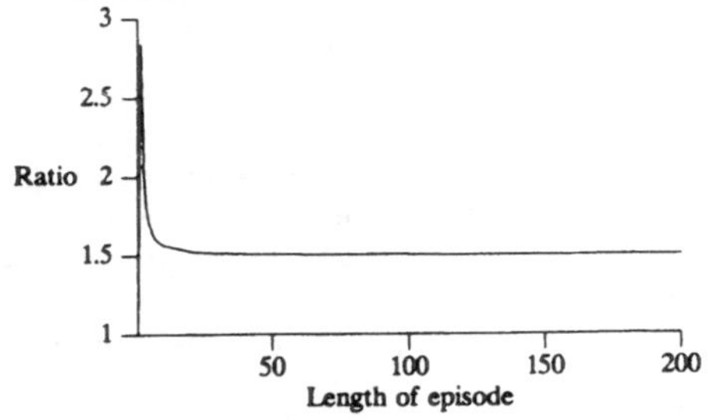

Figure 4: Ratio of S_X to S_Y when payoff is zero

by repeating the experiment but allocating no pay-off at the end of any episode. From the graphs it may be readily concluded that it is system pay-off which most radically influences the ratio of the rule strengths and that, as a result of the effects of taxation, its influence increases with the length of episode such that $S_X/S_Y \rightarrow 1$.

Although equivalent rules tend to have equal strength, the above experiment gives no indication of how the introduction of an equivalent rule into a plan will distort the strength of a rule, already in the plan, to which it is equivalent. Correspondingly, the above experiment (with pay-off restored) was again repeated with first X and then Y removed from P_E. The strengths of X and Y derived using P_E are depicted in figure 5, and those with X and then Y removed from P_E in figure 6. These graphs suggest that the introduction of an equivalent rule tends to halve the strength of the rule to which it is equivalent.

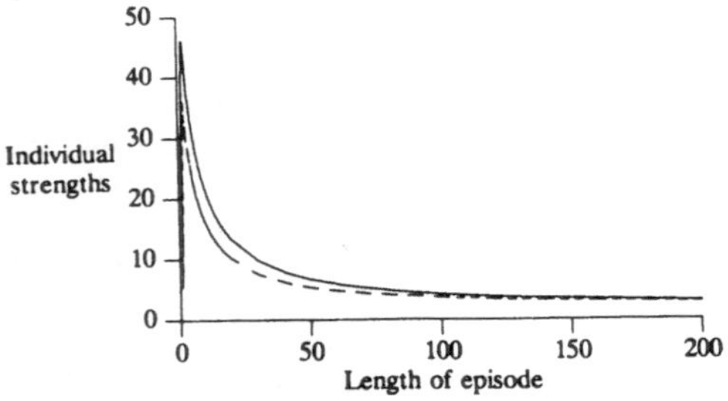

Figure 5: S_X and S_Y in P_E

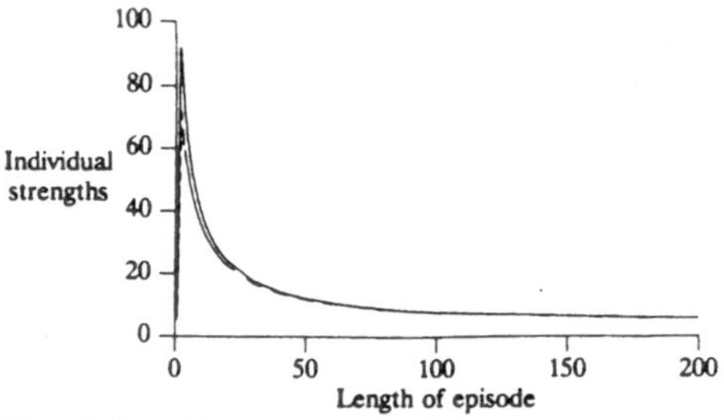

Figure 6: S_X and S_Y when used alone

Other experiments (not reported in detail here) involving the presence of $k > 2$ equivalent rules show that this effect generalises in such a way that the strength of each equivalent rule is *strength/k*. That is, equivalent rules tend to share 'true strength' in much the same way as do duplicate rules.

4.3 Tests Involving Subsuming Rules

Defining the nature of the influence on 'true' rule strength resulting from the co-existence of subsuming and subsumed rules on a plan is a far more complex task than those involving duplicate or equivalent rules. The reason for this is the potential for complex relationships between several rules, each pair of which is linked by subsumption. It is not possible for the authors to make completely general statements concerning strength distortion in respect of the subsumption relationship for, as yet, the investigations are incomplete. However, for the less complex situations, the effects of co-existing subsuming/subsumed rules was found to depend upon the relative utility of the rules. Results for such situations are reported below.

When a subsumed/subsuming rule is introduced into a plan containing corresponding subsuming/subsumed rule and the individual rules are of approximately equal utility (irrespective of whether the rules are good or poor), experiments analogous to those in section 4.2 revealed that the subsumed and subsuming rules acted as if they were equivalent. That is, they tended to share strength. (The results are not reported in detail here as in essence, they are very similar to those reported in section 4.2. Specific details can nonetheless be found in Fairley and Yates [ ]). When, however, the subsumed and subsuming rules have different utility, the results obtained can reflect a potentially undesirable outcome such as that reported in tables 1 and 2. These two sets of results were obtained by applying the plans specified in the tables to the test set S.

Rule 4 (table 2) is a subsuming rule with poorer utility than the rule, already present in the plan, which it subsumes (rule 3). Notwithstanding, the subsuming rule assumes all the strength when it is introduced.

Table 1: Results without subsuming rule

No	Classifier	Strength	Relative St'gth
1	000010 : 010011	0.9085	0.20
2	01#011 : 010110	7.7936	1.71
3	01###0 : 010011	5.0012	1.09

Table 2: Results with subsuming rule

No	Classifier	Strength	Relative St'gth
1	000010 : 010011	0.3814	0.14
2	01#011 : 010110	0.0000	0.00
3	01###0 : 010011	0.0000	0.00
4	01#### : 010011	10.6139	3.86

4.4 Discussion of Results

The results of sections 4.1-4.3 have demonstrated that when a rule, R_A say, in a plan is supplemented with either duplicate, equivalent or approximately equally effective subsuming rules (call these *analogues* of R_A for ease of exposition), the 'true' strength of R_A is distorted to the extent that it is shared with its analogues. Such distortion may well lead to a reduction in a classifier system's capability to produce an effective plan. However, it is not only a distortion of absolute strength that is observed, the strength of R_A relative to that of other rules in the plan may also be affected. For example, consider a plan P_1 containing no rule analogues and assume that when used on a given problem episode, two of its rules, R_A and R_B, are active on a given cycle of the episode. In this case, R_A and R_B would each receive the same payment q/2 say. If however P_1 is supplemented with two analogues of A and this new plan applied to the same episode, then on the given cycle, R_B, R_A and its two analogues could be active. Consequently all four rules would receive a payment of q/4, but the strength of R_B relative to the combined strength of R_A and its analogues in this plan will not in general be the same as the relative strength of R_A and R_B in the plan P_1. Unfortunately, such distortion of relative rule strength can impair the effectiveness of a classifier systems genetic algorithm and thereby again reduce the systems ability to produce an effective plan.

It would appear that distortions in rule strength, and relative rule strength are linked to the fact that the relative strength of a rule and any of its analogues is unity or approximately so, and, as shown in section 4.2 for equivalent rules, this is linked in turn to the nature of the credit assignment scheme. Correspondingly, a number of alternative credit assignment schemes were investigated to determine whether one which induced a relative strength between a rule and its analogue that differs from unity, might reduce distortion of rule strength.

5.0 Alternative Credit Assignment Schemes

In the past, alternatives to the 'standard' credit assignment scheme have been investigated by Wilson [3,4]. Whilst maintaining the standard scheme for inter-rule payment, Wilson focused attention on environmental pay-off. Unfortunately, the informal experiments he performed on the alternative schemes considered, revealed none as being any more effective than the standard.

As a consequence of the results presented in section 4, four credit assignment schemes which differ from the standard in respect of either inter-rule payment (I), environmental pay-off (E), or both, suggest themselves. These are:

(a) I divided equally and E divided in proportion to the bids made;

(b) I and E both divided in proportion to the bids made;

(c) I divided in proportion to the bids made, E divided equally;

(d) I and E both divided in proportion to specificity.

Here, I and E respectively denote inter-rule payment and environment pay-off, and (of course), it is assumed that payment is made only to active classifiers.

5.1 Investigation of Suggested Schemes

To investigate the four schemes, each was implemented in SCS1 in turn, and, using plan P_E of section 4, applied to the set S of test problems. The results from the four experiments were derived in the form of the relative strength (denoted S_1/S_2) of the two principal rules in P_E for lengths of episode in the range [1, 200] and are depicted in figures 7→10 respectively.

Scheme (a)

Figure 7 differs little from figure 3 (pay-off divided equally amongst the active rules) because, during an episode, S_1/S_2 again converges to the ratio of rule specificities (see equation(1)) whence, the bids made by the rules are identical, they receive the same pay-off, and therefore, as the length of episode increases, $S_1/S_2 \to 1$.

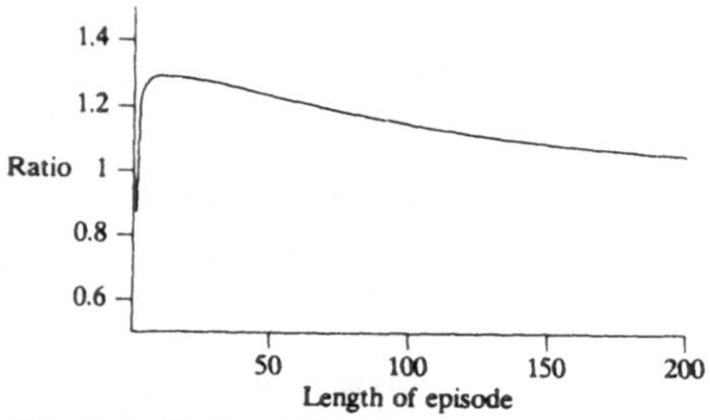

Figure 7: S_1/S_2 for scheme (a)

Scheme (b)

Figure 8 clearly shows that, even for episodes of small length, only the more specific rule gains strength. Why should this be when it appears that, during and episode, both classifiers are paid the same amount as they bid? This view is however not quite correct because if a rule were to be satisfied on x, say, moves of an episode including the last, it would receive (x-1) payments from other classifiers. Therefore, the rule which pays the most on the first play will lose out by the corresponding amount over the period of the episode. Thus, as the number of episodes increases, this 'deficit' is compounded multiplicatively, and hence $S_1/S_2 \to 0$.

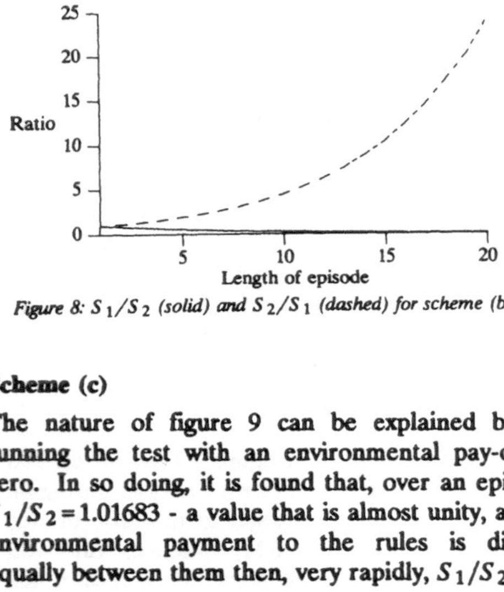

Figure 8: S_1/S_2 (solid) and S_2/S_1 (dashed) for scheme (b)

Scheme (c)

The nature of figure 9 can be explained by re-running the test with an environmental pay-off of zero. In so doing, it is found that, over an episode, $S_1/S_2 = 1.01683$ - a value that is almost unity, and as environmental payment to the rules is divided equally between them then, very rapidly, $S_1/S_2 \to 1$.

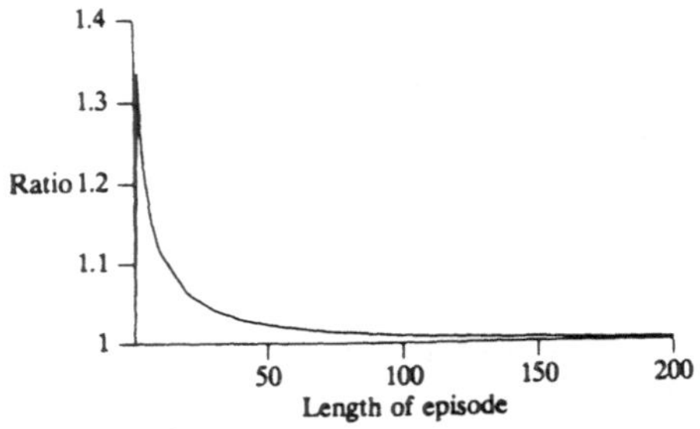

Figure 9: S_1/S_2 for scheme (c)

Scheme (d)

The shape of the graph in figure 10 conforms entirely to intuition and simple analysis; the relative strength of the rules converges with increasing episode length to the ratio of their specificities.

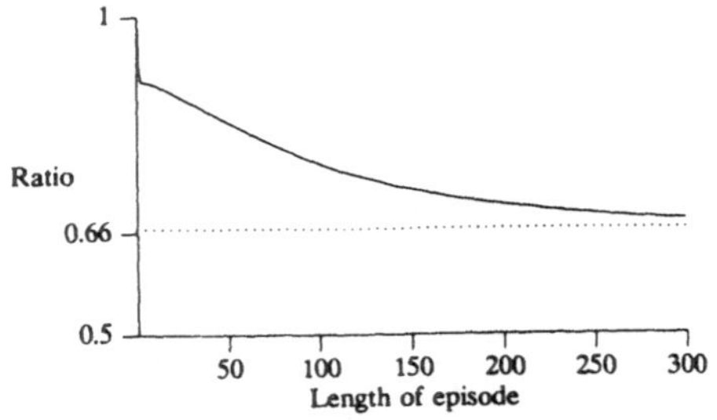

Figure 10: S_1/S_2 for scheme (d)

5.2 Comparison of Schemes

As suggested in section 4.4, convergence of S_1/S_2 to unity may well be undesirable, and correspondingly, the results of section 5.1 suggest that credit assignment schemes (a) and (c) should be eschewed. To investigate whether the two remaining schemes could improve on the standard, each of (b), (d) and the standard were used, in conjunction with the rule set given in appendix 1 and a more extensive test set $S\prime$ containing 300 problems.

In respect of the rule-set, rules 1 and 2 represent the overall defaults, whilst the bulk of the solution strategy is effected by either the rule-pair 3 and 6 or the rule-pair 4 and 5. The rule set, in effect, possesses a memory of only the previous move, and therefore the optimal strategy to win against an opponent who must necessarily be either outside or inside the range of the gun, is to select one of these rule-pairs and not deviate from it. Rule 7 is only active on the first move of an episode.

The three plans which were developed are as follows.

Plan 1: using the standard scheme

The plan developed here adopted the strategy corresponding to rule-pair 3 and 6 as its main-stay. However, the strength of rule 3 is sufficiently small in comparison with those of rules 1 and 2 that, on most occasions, the strategy adopted will merely be to fire on every move. As predicted by the earlier experiments, the 2 default rules, which are very similar, have approximately equal strength. The strength of rules 4 and 5, as well as that of rule 7 (outbid by default rule 1) approach zero as a result of taxation.

Table 3: Plan 1

No	Classifier	Strength	Relative St'gth
1	00#### : 010011	3.3399	2.32
2	01#### : 010011	3.3393	2.32
3	00#0##, 01#011 : 010101	0.0333	0.02
4	00#0##, 01#011 : 010110	0.0000	0.00
5	01###0 : 010011	0.0000	0.00
6	01##0# : 010011	3.3734	2.34
7	0000010 : 010011	0.0000	0.00

Plan 2: I and E divided in proportion to the bid

Again the rule pair 3 and 6 (alternate lowering and firing of the gun) has been selected as the principle strategy. The rule-pair will be active on each move of open play except the first when default rule 1 will be active. Here too, the predictions of earlier experiments are in evidence - the strength accorded to the two default rules is lodged in only one of them, namely rule 1.

Table 4: Plan 2 proportional to bid

No	Classifier	Strength	Relative St'gth
1	00#### : 010011	0.3390	1.28
2	01#### : 010011	0.0000	0.00
3	00#0##, 01#011 : 010101	0.7517	2.84
4	00#0##, 01#011 : 010110	0.0000	0.00
5	01###0 : 010011	0.0000	0.00
6	01##0# : 010011	0.7639	2.88
7	0000010 : 010011	0.0000	0.00

Plan 3: I and E divided in proportion to specificity

This plan is similar to plan 1 both in its election of rule pair 3 and 6 and the relatively reduced strength of rule 3 when compared with those of the default rules. As with plan 1, this will often lead to a policy of constant firing of the gun. Again the strength of the default rules accord with the results of previous experiments - the ratio of their strength converges to the ratio of their specificities, and as they have equal specificity, this converges to unity.

Table 5: Plan 3 proportional to specificity

No	Classifier	Strength	Relative St'gth
1	00#### : 010011	0.0045	0.08
2	01#### : 010011	0.0045	0.08
3	00#0##, 01#011 : 010101	0.0022	0.04
4	00#0##, 01#011 : 010110	0.0000	0.00
5	01###0 : 010011	0.0000	0.00
6	01##0# : 010011	0.3860	6.67
7	000010 : 010011	0.0078	0.13

The success of the three plans as measured by the number of successful episodes achieved by each when applied to the test problem set S/ is reported in figure 11. It is readily observed from the graph that plan 1, based on the standard credit assignment scheme, was the least successful. In fact, plan 2 performed over three times more successfully than plan 1 and was well over twice as successful as plan 3. Thus, it is concluded that adopting credit assignment scheme (b) can have beneficial effects in respect of reducing the distortion caused by duplicate, equivalent and subsuming rules.

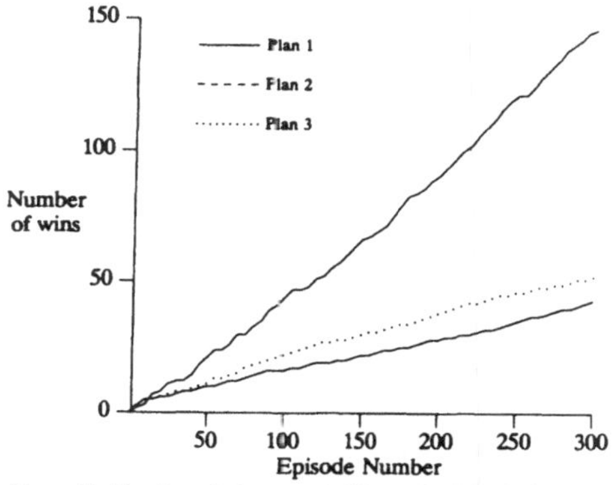

Figure 11: Number of wins recorded by each of the 3 plans

6.0 Conclusions

The results presented in this paper have shown that the existence of duplicate, equivalent or subsumed/subsuming rules in the plan of a classifier system can distort the relative strengths of the rules within the plan. An undesirable consequence of this is the potential misfunctioning of the system's associated genetic algorithm. Fortunately, the introduction of duplicate rules can be prevented by performing tests at the stage of rule generation. Detecting and removing equivalent and subsumed/subsuming rules is however more problematical.

When equivalent rules are present in a plan, they effectively share (equally) what should be a rule's true strength (although for short episodes the ratio of any two may be closer to the inverse of their specificities). The use of competition represents one means of reducing the effects arising from the presence of such rules but, if the competition is too severe, the equivalent rules may dominate other useful rules and consequently distort further their strengths (and relative strengths). An alternative method of approaching this problem, namely that of adopting a credit assignment scheme that differs from the standard, has been investigated. Four different schemes have been examined. One of these, that of dividing both inter-rule payment and environmental pay-off in proportion to the bids made, possesses characteristics which facilitate implicit discovery and the (effective) removal of equivalent rules. For, when such a rule is introduced into a plan, the effect of this credit assignment scheme is to reduce the strength of the less specific of the two equivalent rules to zero. The effectiveness of this scheme relative to the standard scheme has also been established in respect of plan performance on the Two Tanks Problem. Whether the scheme does in fact facilitate the improved functioning of the genetic algorithm has yet to be investigated, however, the authors hope to report on this in the near future.

References

[1] Goldberg D E, *'Genetic Algorithms in search, optimization and machine learning'*, Addison-Wesley, Reading, MA, 1989.

[2] Grefenstette J J, Ramsay C L and Schultz A C, *'Learning Sequential Decision Rules Using Simulation Models and Competition'*, Machine Learning, Vol 5, 1990.

[3] Holland J H, Holyoak K J , Nisbett R E and Thagard P R, *'Induction: processes of inference, learning and discovery'*, MIT Press, Cambridge, MA, 1986.

[4] Wilson S W, *'Bid Competition and Specificity reconsidered'*, Complex Systems, 2, 6, 1988.

[5] Wilson S W and Goldberg D E, *'A critical review of classifier systems'* in Proceedings of the Third International Conference on Genetic Algorithms, Morgan Kaufmann, Fairfax, VA, 1989.

[6] Fairley A and Yates D F, *'The effects of subsuming and subsumed rules upon rule strength in simple classifier systems'*, Internal Report, The University of Liverpool, 1993.

Appendix 1:

The 7 rules used in many of the experiments in this paper are:

1. 0 0 # # # # : 0 1 0 0 1 1
   IF whatever T2 did THEN fire
2. 0 1 # # # # : 0 1 0 0 1 1
   IF whatever T1's last move THEN fire
3. 0 0 # 0 # # , 0 1 # 0 1 1 : 0 1 0 1 0 1
   IF T2 didn't move AND T1 fired on last move THEN increase elevation
4. 0 0 # 0 # # , 0 1 # 0 1 1 : 0 1 0 1 1 0
   IF T2 didn't move AND T1 fired on last move THEN reduce elevation
5. 0 1 # # # 0 : 0 1 0 0 1 1
   IF T1 didn't fire on last move THEN fire
6. 0 1 # # 0 # : 0 1 0 0 1 1
   IF T1 didn't fire on last move THEN fire
7. 0 0 0 0 1 0 : 0 1 0 0 1 1
   IF T2 moved THEN fire

Genetic Algorithm Selection of Features for Hand-printed Character Identification

Roger S. Gaborski Peter G. Anderson* David G. Tilley
Christopher T. Asbury†

Imaging Research Laboratories
Eastman Kodak Company
Rochester, New York 14653-5722 USA

Abstract

We have constructed a linear discriminator for hand-printed character recognition that uses a (binary) vector of 1,500 features based on an equidistributed collection of products of pixel pairs. This classifier is competitive with other techniques, but faster to train and to run for classification.

However, the 1,500-member feature set clearly contains many redundant (overlapping or useless) members, and a significantly smaller set would be very desirable (e.g., for faster training, a faster and smaller application program, and a smaller system suitable for hardware implementation). A system using the small set of features should also be better at generalization, since fewer features are less likely to allow a system to "memorize noise in the training data."

We tried several genetic algorithm approaches to search for effective small subsets of features, and we have successfully found a 300-element set of features and built a classifier whose performance is as good on our testing set as the system using the full feature set.

1 Introduction and Summary

Starting with a hand-printed digit recognition algorithm, with a 97–99% correct recognition rate, which functions as a linear discriminator based on a collection of 1,500 binary features extracted from a 30×20-pixel array, we determine some 300-element subsets of those 1,500 features and build a new classifier that achieves the same recognition rate. (Related experiments are reported by Chang [3], who studied *all* subsets of a set of 15 features, and by Siedlecki and Sklansky [8], who searched for an optimal subset of 30 features. Both of these papers used the size of the chosen features subset to penalize the genetic algorithm's fitness function. We froze the size of our feature subset at 300. Brill [2] coupled a similar genetic search with an approximate, rapid fitness evaluator for searching for good feature subsets, which would later be applied to a counterpropagation network.)

Two separate experiments were performed. One involved using the features subset (i.e., the "individual" of the genetic algorithm) to train and test a classifier (the result of this testing is the genetic algorithm's "fitness" value). The other involved using the $10 \times 1,500$-element discrimination matrix we had for the good classifier and extracting a 10×300-element submatrix according to the GA individual. This second experiment short-circuits any training; it goes directly to testing. The first experiment has been eminently successful, creating small classifiers with an error rate (i.e., wrong classification) of just under 1%. The second experiment was never able to achieve better than a 3% error rate; we will not pursue this topic further here.

The smaller feature set is desirable for run-time space and speed of the algorithm as well as for potential hardware implementation [6]. Additionally, a smaller feature set would not so easily over-fit the training exemplars ("memorize the noise in the training data"), thus providing expectation of better generalization.

Another disadvantage of large feature sets is that an F-element feature set implies that the training algorithm must invert an $F \times F$ array, a process whose time complexity is $\mathcal{O}(F^3)$.

We had tried to reduce the feature set manually during development of the original recognizer, but that attempt was unsuccessful. The search space is massive, and it was difficult to effectively represent the sets of features. A rule for features evenly spaced over a character's image was easy to derive and state. Cleverly constructed subsets, perhaps

*Dr. Anderson is also associated with the Rochester Institute of Technology, Rochester, NY 14623-0887

†Mr. Asbury is a cooperative education student from the Rochester Institute of Technology

problem-dependent ones, were not.

Our search for good feature subsets was conducted using several versions of a genetic algorithm. We varied the strategies for population management, child creation, and parent selection, as well as the parameters for population size and mutation rate. The results were, probably, what one should have expected (except for the success of locating feature subsets whose performance matched that of the full 1,500 set); i.e., larger population sizes combined with more conservative parent selection rules resulted in higher quality results that took longer to achieve. Those exercises clearly illustrated such principles (to the extent that a number of experiments could be run, each of which took the full power of a SUN SPARCstation 2 for several days).

2 The Polynomial Algorithm

We have been working with a learning system applied to the classification of hand-printed alphanumeric characters known as the *polynomial method* . Character classification is done as follows: first, extract a binary feature vector from a normalized character; second, multiply the feature vector by a matrix, A; finally, determine the largest component in the product vector, whose index is the classification of the unknown character.

In [7], Uma Shrinivasan described a learning algorithm to determine the matrix, A, based on classical least-squares-error minimization, using a data base of correctly labeled training exemplars and the associated target vectors. These target vectors are simply standard unit vectors with the single nonzero entry in the labeling position.

The features used are, generally, the logical products ("and") of two nearby pixels (hence "polynomial"). We have been able to create accurate classifiers using 1,500 features resembling dilations of the king and knight chess moves in a 30 × 20 array of pixels. Each pixel could be the center of up to eight such chess moves, which gives nearly 4,800 possible features. The feature count gives the size of a matrix that must be inverted, so we chose a relatively equidistributed set of these features to hold the total to a manageable yet effective number.

Specifically, suppose we have N labeled character training exemplars for K character classes and F features (here, $N = 100,000$, $K = 10$, and $F = 1,500$.) Construct an $F \times N$ matrix, X, whose N columns are the F-element feature vectors of the training exemplars, and a $K \times N$ matrix, Y, whose N columns are the K-element "target vectors" corresponding to the correct classifications for the respective training exemplar. A target vector for a character of classification k has value 1 at subscript position k and 0's elsewhere. We determine a "classification matrix," A, which satisfies, in the least-squares-error sense,

$$AX = Y \tag{1}$$

This is achieved using the Moore-Penrose pseudo inverse,

$$A = YX^T(XX^T)^{-1} \tag{2}$$

Character recognition is achieved using this classification matrix A by extracting a feature vector $\overline{x}$ from an unknown character, calculating

$$\overline{y} = A\overline{x} \tag{3}$$

and assigning classification k to the unknown character, where

$$\overline{y}_k > \overline{y}_i, \forall i \neq k \tag{4}$$

We augmented the polynomial method with an iterative technique inspired by perceptron and adaline training [1]. Simply stated, our method strives to determine the training exemplars that are near the boundaries for their particular classification and builds the matrices X and Y shown in (1) and (2) with those boundary cases over-represented. Although the resulting classifier is constructed using a least-squares-error rule, it functions more like one whose goal is to achieve more correct classifications than one whose goal is to identify and separate classification clusters based on centers of mass. This approach improved the performance of the 1,500-feature classifier from 97.06% correct on the testing data to 98.71%. Or, in other words, the incorrect classification rate is reduced from 2.94% to 1.29%.

3 GA Hill-Climbing

We have been experimenting with genetic search algorithms to locate suitable classifiers using F in the range 100–500, concentrating on 300 features. The search space is gigantic (e.g., there are $\binom{1500}{300} \approx 5 \times 10^{144}$ possibilities using 300 features), so genetic search seemed particularly appropriate. An "individual" in the gene pool is a subset of F of the 1,500 features used in the original, working classifier. The "fitness function" is the classification accuracy (percent correct) of a classifier built using Shrinivasan's one-shot learning technique, where the system is built using a training set consisting of 30,000 hand-printed digits and a testing set of 20,000.

Fitness evaluation is the principal timing bottleneck in this process. We tried various approaches to this problem; these are described below.

(See [4] and [5] for detailed treatments of the theory of genetic algorithms.)

4 Creating Children

In our GA experiments we represent individuals as a sequence of 1,500 1's and 0's in which the number of 1's is exactly F. Two individuals are chosen to become parents according to their fitnesses (competition techniques for choosing individuals for parenthood are described below). We use a form of uniform crossover [5] to combine genetic material from the chosen parents to create the children. In this process, if both parents agree (0 or 1) at some position, the two children will both inherit the value that the parents agree on. If the parents disagree at a position, the two different values are assigned to the two children randomly. However, we do ensure that each child is created with exactly F 1's.

5 Selecting Parents

Parents are selected from the current population of individuals according to their fitness. We choose individuals to be parents with the probability of being chosen a monotonically increasing function of their fitness.

The first interpretation of "monotonically increasing function of their fitness" is that individuals may be chosen to be parents *proportional to their rank* when rank-ordered by fitness. This could be implemented by sorting the population according to fitness, and then choosing individuals' ranks by a simulated, biased roulette wheel [5]. A simpler method, which achieves the same result, is to pick two individuals from the population (picking with uniform distribution), and then choose the one with greater fitness.

In our initial experiments, this rule took the following form. We used a random shuffling procedure to select four individuals. The fittest of the first pair and the fittest of the second pair are chosen as the two parents.

Later experiments were generalizations of this first one. We shuffle and select $2N$ individuals; the fittest of the first N and the fittest of the second N are chosen as the two parents. (From probability theory we have the following. Suppose that a random number from the interval $[0, 1]$ has uniform distribution; i.e., $f(x) \equiv 1$. Let $x = max(x_1, x_2)$, where x_1 and x_2 are chosen independently from the uniform distribution; then the probability distribution for x is $f(x) = 2x$. Generally, if $x = max(x_1, x_2, \ldots, x_N)$, then $f(x) = Nx^{N-1}$.) For very large N, this will degenerate to the choice of the two fittest individuals to serve as parents. This would cause a rapid loss of "genetic material" with a consequent failure to search large portions of the space of all individuals. As we expected, the quality of the solutions deteriorates with larger N, but, as a trade-off, the speed of reaching good solutions increases. Whitley [9] discusses a similar phenomenon, calling the increased tendency toward selecting the high-fitness individuals as parents "selective pressure." Tables 1 and 2 show experimental results.

6 A Sequence of Populations

We experimented with several different methods for maintaining populations. We used the notion of distinct generations as well as that of a single, evolving population. GA folklore seems to indicate that these algorithms are sufficiently robust, that almost any approach will give solutions to the optimization problems of equivalent quality, but that some methods may locate the champions much faster than others. Some of our experiments can take several days, so we wanted to search out the rapid techniques. The first four population methods we tried were:

- *Steady state.* This was the first technique that we put together. When the two children are created, they replace the two losers of the two two-way competitions that selected the two parents.

- *Replace two worst.* This method is similar to the first, in that there is a single evolving population. However, here we replace the two least fit individuals of the population with the two new children. (We do assure that neither of the parents are one of the two worst performers.)

  Both the first and the second method assure that the current fitness champion stays in the population pool.

- *Simple generational.* In this method, we introduce the notion of "generation," where the children of the current population are used to create an entire new population. Here, the current champion can be lost.

- *Keep best half.* This fourth method is a variation on the third. The children of parents in generation n are used to build generation $n + 1$. Then, the union of these two generations is sorted by fitness, and the top (fittest) half becomes generation $n + 1$. Again, we assure that the best performers are not lost from the population pool.

7 Fitness Testing

In this section, we describe the sequence of experimental arrangements we used, and how we addressed problems of efficiency.

In order to save training and testing time in the iterated polynomial algorithm, feature vectors (1,500 features per) were extracted from all the training and testing exemplars and cached in disk files. So, our genetic search program used these "features files" instead of actual characters. For every genetic individual to be fitness tested, we read 30,000 feature vectors from the training set, extracted the indicated features to form a training vector of the desired length (generally 300), and we formed the classification weights matrix $A = YX^T(XX^T)^{-1}$, then processed 20,000 characters of the testing set, by similarly extracting the subset of the features and multiplying the smaller vector by A. The fitness for the individual in question was the fraction of correctly classified characters. We took that fraction, multiplied by 1,000, and converted to an integer, to represent the fitness. Thus, we express fitness values in the range 0–10,000.

It occurred to us that we could achieve the same testing if we were to pre-evaluate the two matrices, YX^T and XX^T, for the full set of 1,500 features, which are then used in the evaluation of A. The components for evaluation of the A-matrix corresponding to a selected subset of the full feature set are simply the obvious submatrices of these large YX^T and XX^T matrices. This pre-evaluation takes one or two hours, depending on the computer chosen and its current load, but it would double the speed for individual fitness evaluation. When the development of this technique stabilized, we were able to cache these two arrays on disk, with only the I/O cost as overhead.

The one-shot training (Shrinivasan's algorithm), using 30,000 training exemplars and 1,500 features, resulted in a classifier with 97% accuracy on the 20,000-character testing set. Our genetic search was able to find feature subsets of size 300 that performed slightly better on this testing set. (Caveat: genetic hill-climbing uses a fitness measure that is derived using what we had called the "testing set." This set is now intimately involved with (genetic) training, although the individual elements do not directly affect the classifier we are building. To be totally fair, a third set of characters is needed for final testing.)

We had hoped that, using the good, small subset of features iteratively in the polynomial training algorithm, we would be able to achieve an even better classifier. That hope did not materialize. The one-shot trained A-matrix was as good as we could get with that technique.

What had started out as simply a time-saving measure—caching the large matrices—suggested a new approach that has led us to a better result. The iterative polynomial training technique develops the two matrices YX^T and XX^T, but had no plans to save them. Now they were valuable: we re-ran the training, which gave us 98.71%, saved the two large matrices, and used them with our genetic search. This approach was successful. We were able to discover a set of 300 of the 1,500 features that yielded 99.01% (same caveat as above).

8 Summary of Results

After preliminary experiments had been tried, we settled on the use of the data mentioned above, namely the arrays YX^T and XX^T that were a by-product of the training session of the polynomial algorithm that produced the digits classifier whose performance was 98.71%. We determined that a continuous population evolution rather than explicitly separate generations was desirable (separate generations made children of good performers wait too long to have good children of their own), and that the replace-two-worst strategy worked best.

Our goal was a classifier that used only 300 of the 1,500 features. The available parameters were the population size, the parent-choosing method, and the mutation rate.

We tried population sizes of 100, 200, and 300. A population of 100 converged rapidly, but we did not achieve the best we could; 200 was better than 100; and 300 worked marginally better than 200, but took a long time to find the performers we were seeking.

We selected parents using the "best of N." Large values of N heavily skew the parent selection distribution, which causes rapid convergence but degrades the fitness of the champions it discovers. $N = 3$ seems to be an optimal trade-off point.

Figure 1 shows the hill-climbing achieved by the genetic algorithm as we evolved a population of 300 individuals. Parent selection was performed by locating the most fit in two disjoint four-member competitions ("best of four").

Tables 1 and 2 show the results of several experiments which were run with no mutation and were terminated when the population converged to copies of almost equal individuals. They do not show monotonic results as we have described above. The specific results depend, to some extent, on the outcome of the pseudo-random number sequence (we used *drand*48()), and several runs with different random seeds for each parameter would presumably smooth out these artifacts. Our approach was to sample the parameter space as widely as we could rather than to repeat the experiments for fixed parameters.

	Population size		
parent competition	100	200	300
best of 2	3,700	16,000	25,000
best of 3	3,300	9,100	18,000
best of 4	1,450	4,850	10,900
best of 5	1,400	4,650	9,200

Table 1: Number of individuals evaluated until convergence.

	Population size		
parent competition	100	200	300
best of 2	9837	9893	9901
best of 3	9841	9884	9894
best of 4	9793	9845	9872
best of 5	9802	9856	9876

Table 2: Fitness of the best individual discovered.

Figure 2 shows the feature subset in the best we have found. Each feature is the logical product of two factors, where each factor is the logical sum of three pixels. The three pixels' centers are a scaled-up version of a chess king's or knight's move. The figure shows the chosen "chess moves" as line segments connecting the two centers.

We implemented mutation as follows. When two new children were created, we would *mutate* a specified number of times, which we denote by *MUTATE_COUNT*. A single *mutate* step consists of randomly choosing two bits in the string of length 1,500 and interchanging their values. So, the probability that a single *mutate* will actually modify an individual is 0.32.

With a *MUTATE_COUNT* of three to nine, we were able to discover slightly better individuals in approximately the same time as it took to *converge* to the best without mutation; see table 3. (Without mutation, searches converge, so that all individuals in the population are nearly identical and have the same fitness. With mutation, searches never "converge.")

	MUTATE_COUNT			
parent competition	0	3	6	9
best of 4	9793	9809	9802	9800
best of 5	9802	9802	9827	9804

Table 3: Fitness of the best individual after 1,500 evaluations using various levels of mutation. The entries in the first column are *converged*.

Higher *MUTATE_COUNT* rates spoil the hill-climbing. Any successes with high rates of mutation seem to be attributable to nothing more than luck.

References

[1] Peter G. Anderson and Roger S. Gaborski, "The polynomial method augmented by supervised training for hand printed character recognition," *Proceedings of the International Conference on Neural Networks and Genetic Algorithms,* 1993.

[2] Frank Z. Brill, Donald E. Brown, and Worthy N. Martin, "Fast genetic selection of features for neural network classifiers," *IEEE Transactions on Neural Networks*, Vol. 3, No. 2, March 1992.

[3] Eric I. Chang, Richard P. Lippmann, and David W. Tong, "Using genetic algorithms to select and create features for pattern classification," *Proceedings of the International Joint Conference on Neural Networks*, 1990.

[4] Lawrence Davis (ed.), *Handbook of Genetic Algorithms*, Van Nostrand Reinhold, New York, 1991.

[5] D. E. Goldberg, *Genetic Algorithms in Search, Optimization, and Machine Learning*, Addison-Wesley, New York, 1989.

[6] A. Rao, P. G. Anderson, R. S. Gaborski, and K. S. Jaiswal, "A hardware polynomial feature net for handprinted digit recognition," *Proceedings of the Third IEE International Conference on Artificial Neural Networks,* 1993.

[7] Uma Shrinivasan, "Polynomial discriminant method for handwritten digit recognition," *SUNY Buffalo Technical Report*, December 14, 1989.

[8] W. Siedlecki and J. Sklansky, "Constrained genetic optimization via reward-penalty balancing and its use in pattern recognition," *Proceedings of the Third International Conference on Genetic Algorithms*, 1989.

[9] Darrell Whitley, "The *GENITOR* algorithm and selective pressure: why rank-based allocation of reproductive trials is best," *Proceedings of the International Conference on Genetic Algorithms*, 1989.

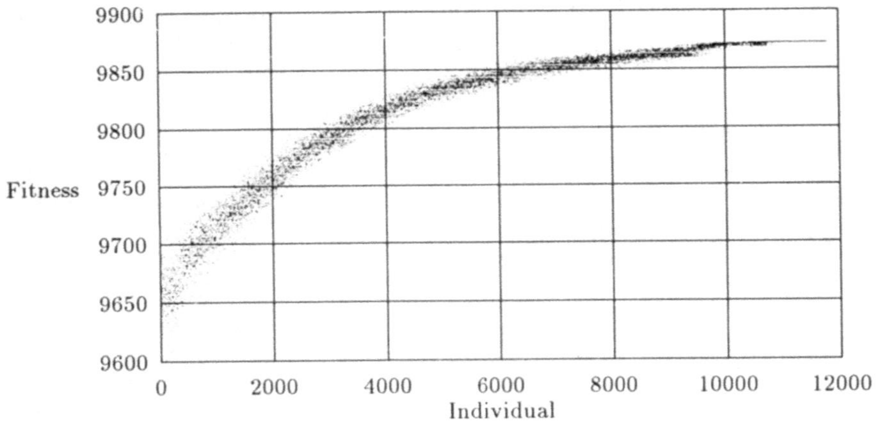

Figure 1: The fitness of each of the individuals encountered in our search for the best 300-element subset of features. This search converged with the eventual fitness of 98.72%. The population size was 300; the parents were chosen using the *best-of-four* competition.

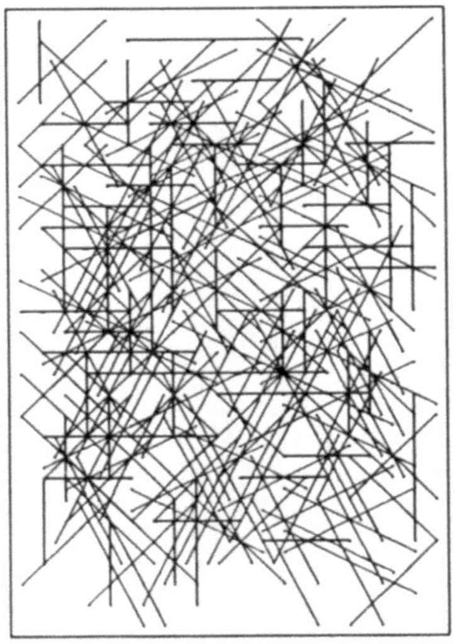

Figure 2: The 300 features of the best feature subset we have located.

Analysis and Comparison of different Genetic Models for the Clustering problem in Image Analysis

Rita Cucchiara

Istituto di Ingegneria
Universita' degli Studi di Ferrara
via Scandiana, 21 Ferrara, ITALY
Tel 39-532-65121 Fax 39-532-740983
E-Mail Rita@Deis36.cineca.it

Abstract

This paper presents several genetic approaches to the clustering problem of N elements in an n-dimensional Feature Space. This process has been applied in an Image Analysis context in order to divide a set of objects into a fixed number of groups, dependending on their characteristics. The partitioning models are based on very general issues so they can be used in many different clustering applications, as well as real objects grouping. The genetic paradigm has been choisen because the cluster Solution Space has to be explored without any 'a priori' or heuristic knowledge and also because the performed parallel search can elude the relevant number of local minima in the solution optimisation. Different cluster models and genetic operators have been analysed in order to exploit the genetic algorithm power in an Image Analysis environment. A performance comparison between solutions is shown, using several chromosome codes and genetic operators.

1.Clustering In Image Analysis

The goal of cluster analysis is to find a convenient, valid data organisation without any 'a priori' knowledge of category labels but based only on inter-data relationships. The absence of identifiers distinguishes cluster analysis from discriminant analysis such as classification, pattern matching and decision analysis [1] In general terms, element partitioning is guided by a concept of object 'similarity', which can be suggested by the kind of application; thus a cluster could be defined as a set of elements which are alike, while entities from different clusters are not.

Cluster analysis has been adopted in many application fields such as operating research, social science, economics, engineering and artificial vision too. Different goals may be achieved with clustering, depending on the data and the considered environment: entity partitioning in real typologies, knowledge modelling, behaviour prediction or pre-assumption for classification.

In a computer vision environment cluster partitioning could be required at different levels of image processing, image analysis and image understanding. During low-level vision [2] cluster analysis is used in segmentation processes in order to divide the pixels of an image into contiguous regions or curves. After segmentation, a separation between objects and background is performed and an image analysis phase begins by feature extracting for each object. The result of this task is the data structure transformation from the real Image Space to an n-dimensional Feature Space, where every object is described by a vector of n feature values. In this work geometrical and topological features in 2D images have been considered such as area, perimeter, connectivity with Euler Number or symmetrical axis. In this context the clustering can be applied to image analyse thanks to the objects features examination.

Clustering can be considered as the final goal in several automatic inspection applications to divide objects into tolerance classes or to find similarity aspects between elements, while in image understanding the clustering could be used to create a training set in classification phase. An example is shown in figure 1:

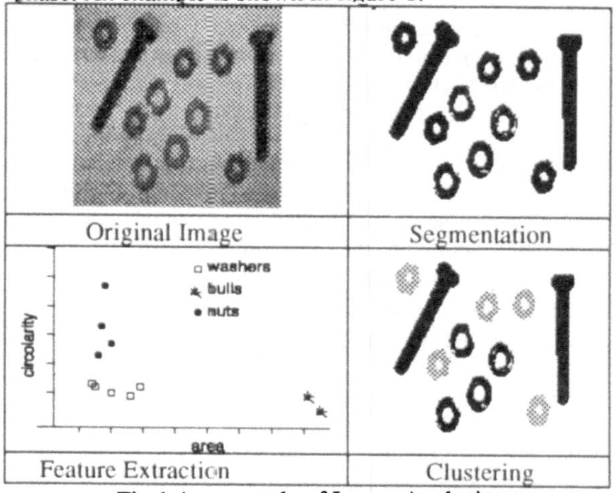

Fig.1 An example of Image Analysis

The complexity of the clustering problem is given by different factors:

1) The clustering is an NP-HARD problem, therefore an exhaustive approach is not practicable due to the exponential dependency of the potential partitions from the input data. In fact it is possible to demonstrate[1] that the number of partition of N elements in m cluster is given by

$$C(m,N)=\frac{1}{m!}\sum_{j=1}^{m}(-1)^{m-j}\binom{m}{j}(j)^{m} \qquad (1)$$

For instance, the partitioning of 22 objects into 5 clusters has 1.91 E^{13} possible solutions.

2) The clustering complexity grows if the number of groups is unknown. In such a case the number of solutions becomes

$$C(m,N)=\sum_{k=1}^{m}\frac{1}{k!}\sum_{j=1}^{k}(-1)^{k-j}\binom{k}{j}(j)^{k} \qquad (2)$$

because the number of clusters can go from one to N (object number). In the previous example increases to about E^{18}. Generally, the cluster number is given, especially in quality inspections, or can vary between a minimum and a maximum value. In this work, m is fixed but the model can be extended to the case with m unknown.

3) It is very difficult to translate the concept of 'similarity' in a unique mathematical model, but this depends on the clustering goal as will be shown in the next sections.

For these reasons many techniques have been proposed, especially based on heuristic methods, that can be very efficient but strongly dependent on applications, data and initial conditions. In this paper a genetic approach is considered which has been successfully employed in several NP-HARD problems[3] and also presents good tradeoff between robustness and efficiency.

2.Clustering with Genetic Algorithms

This section provides a general overview of Genetic Algorithm structure and explores various operators and techniques applied to the clustering problem. A GA is a stochastic computational model that seeks the optimal solution to an objective function.The search is performed through an iterating procedure applied to a "population of individuals", i.e. a set of feasible solutions. The main feature of GA's is that the searching strategy is similar to biological evolution: better solutions are reproduced, whereas worse solutions are discarded. Thus, the search strategy is based on the possibility to discriminate between elements in order to resolve which is a good solution of the fitness function and therefore has a good chance of reproducing and generating new elements with its genetic inheritance.A GA scheme is sumarised in fig.2

```
Genetic_Algorithm ()
{    <Initialise population>
     while <Not (Stop condition) > do {
              <Fitness Evaluation>
              <Selection>
              <Reproduction and Mutation>
              }
     <Choose final solution>
}
```

Fig. 2 Scheme of a Genetic Algorithm

A peculiar characteristic of GAs is the functional separation between the problem-dependent data structure representation and cost function (fitness) evaluation and the robust reproduction phase that may be common to each application. For this reason, this work concerns the advisability of creating a good genetic model for clustering and adopting efficient operators for the optimization. The clustering problem may be formulated as follows:

Let Φ be the n-dimensional Feature Space and S a subset of Φ with N elements Xj represented with n coordinates in Φ:

$$S \subseteq \Phi : S = \{Xj \mid Xj \in \Phi, j=1..N\} \qquad (3)$$

The problem searches the best Ψ partition of S in m clusters , such as $\Psi = \{Ci, i=1..m\}$ where $Ci = \{Xj, Xj \in S\ j=1..Ni\}$

$$\cap Ci=\emptyset \qquad (4)$$

$$\cup Ci=S \text{ or } \sum_{i=}^{m} N^{i} = N \qquad (5)$$

and the "best" idea is correlated to the concept of "similarity" between objects Xj.

Coding

The first step is efficient data representation with a binary code, considered [3] the most appropriate for a genetic process. Each Ψ solution is coded as a chromosome, composed of a set of variables or genes.

A first formulation may consider N genes (one for each object) which can assume a discrete value between 1 and m to indicate the belonging cluster; m could be fixed or indicate the maximum group number. Therefore, the binary code needs N x Int(log2 (m)) bits. For instance, dividing 30 objects into 5 classes, each chromosome has a 60-bit length. Referring to fig. 4b the chromosome is

0	1	1	1	1	0	1	0	1	1	0	1	1	0	1	1
X1		X2		X3		X4		X5		X6		X7		X8	

A second possibility considers a gene as a cluster and indicates which elements belong to it with a Boolean matching code [4] using (m x N) bits

1	0	0	0	0	1	0	0	0	0	1	1	0	0	1	0	0	1	0	0	1	0	0	1

C1 C2 C3

Both solutions present several differences: the 1st is shorter and thus faster, the other redundant and memory-consuming; the 2nd, again, due to the redundancy, requires in homologous bits of each cluster only one value '1', the others '0', because each object can belong only to one cluster. In terms of schemata theory [5][6] this methods presents many correlated schemata in the different genes, and needs to translate the idea of schema and its evolution into a gene instead of a chromosome. For instance, if a gene contains the schema **10 (where * is the indifferent condition), all the others must contain the schema **0*. Thus, the code call to modify the reproduction task in order to generate only consistent solutions; on the other hand, experimental tests show that it requires more or less the same population size as the first coding technique, although the chromosome is longer, because it has fewer independent schemata, and this quantity can influence the more efficient size of initial population [7].

As to the initial population size and the average number of generations for the algorithm, several comparison have been performed: these results concern average values of 60 tests to divide 24 objects into 5 clusters and 22 objects into 4. The results are shown in Table 1a, 1b where Bfit is the best normalized fitness found, Varf is the variance during the different tests, Bgen is the average generation number which finds the best fitness and Varg is the Bgen's variance.

Population size	Bfit	Varf	Bgen	Varg
50	2.12	0.412	196	34
100	1.88	0.296	180	28
150	1.34	0.188	164	21
200	1.0	0.0	144	16
250	1.0	0.0	138	22
300	1.0	0.0	135	31

Table 1a GA with boolean matching code

Population size	Bfit	Varf	Bgen	Varg
100	1.482	0.212	156	29
150	1.0	0.0	112	22
200	1.0	0.0	109	20

Table 1b GA with compressed binary code

The tables show the convenience of the compressed binary code, from the time efficiency point of view, but it requires a more complex crossover reproduction operator: the relevant chromosome length with independent schemata calls for careful estimation of the crossover probability in order not to excessively differentiate the generated sons from the parents, i.e. not to excessively change the exploration hyper-planes in the solution space. On the contrary, with the first technique the crossover operator is applied gene by gene and this effectively corresponds only to shifting few objects from one cluster to another using simple crossover operators.

The further advantage of the boolean matching code is the direct explication of the presence or the absence of an object in a cluster, that can be applied exploited in the fitness evaluation parallel model, while the compressed code is more useful in a sequential implementation. For all these reasons the first method has been preferred. Looking at the table, the average population size of 200 elements has been settled on for a problem with twenty or thirty objects, because this dimension offers the best performance with a lower computational delay.

Selection

In this phase the reproducible solutions are selected; it is convenient to search for a good compromise between selecting only the best elements to increase the algorithm convergency and avoid a block around local minima of the cost function. Different methods have been proposed in literature [3]. The classical Montecarlo statistic selection has been adopted. In addicton, a sort of 'elitist selection' [8] has been investigated to conserve the best solutions achieved in each generation , by introducing a given number of 'clones' in the new generation. The clonation percentage has to be carefully detected in order not to limit the genetic model and concentrate the research only in one direction.

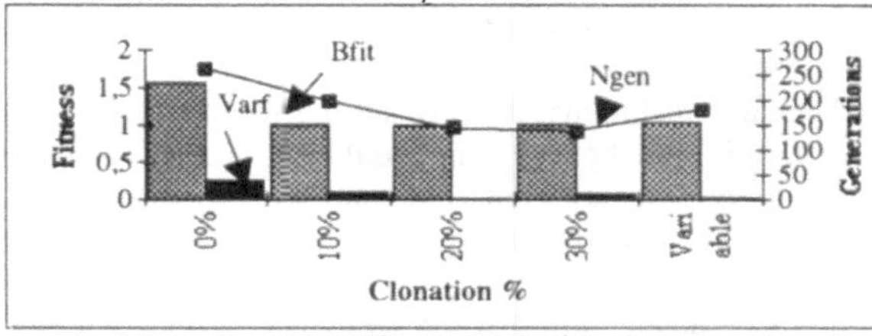

Fig.3 Results with different clonation percentages

The graph shows several tests with 0%, 10%, 20% and 30% clones and a clonation percentuage linearey variable from 10% to 50%. This proof is based on the idea allowing an inital, large Solution Space exploration and concentrating the research around the best directions over the last generations. The best percentage seems to be 20% maximum because a higher value demonstrates an average fitness of more than one, suggesting that several tests do not find the best solution and are trapped in local optima. Variable clonation could be a good proposal but depends on the application domain.

Reproduction methods

The crossover is the basic operator to reproduce new solutions from those selected in a given generation. Using the reduntant Boolean code, the GA applies the crossover not over all chromosomes but only between two random genes of two selected chromosomes, then adjusts the other genes to create an admissible string. For instance, observe the figure 4a and 4b, before crossover and after:

1	0	0	0	0	0	0	1	0	0	1	1	1	1	1	0	0	1	0	0	0	0	0	0
1	1	0	0	0	1	0	0	0	0	0	0	0	0	1	0	0	0	1	1	1	0	0	1

after crossover:

1	0	0	0	0	1	0	1	0	0	1	1	0	0	1	0	0	1	0	0	1	0	0	0
1	1	0	0	0	0	0	0	0	0	0	0	1	1	1	0	0	0	1	1	0	0	0	1

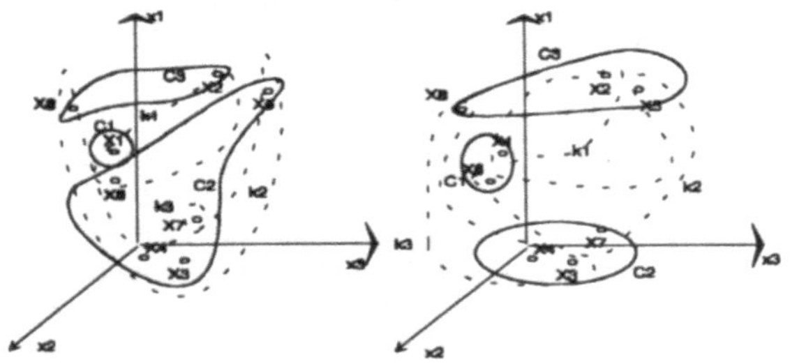

Fig. 4a, 4b cluster evolutions

Even the crossover is separately applied on the genes, nevertheless the schema theory is still valid; for example in the figure the schema *10 is conserved after the reproduction (to maintain the points X3 and X4 in the same cluster). In table 2 a comparison between one-point, two-point and uniform crossover, is presented (using the same data stucture as the previous tables). The three techniques achieve similar results the uniform methodis the better. This could be justified by the intrinsec symmetry of the uniform crossover with respect to the unbalancing of the others which privilege the moving of only a given object subset or, according to the [9] considerations, if the schemata are long enough, they have a higher recombination probability with uniform crossover.

Another recombination method has been investigated, using only a chromosome (and not a pair) to create a new individual, in a sort of "asexual" reproduction.[4] The crossover operators work over two genes of the same chromosome with the physical result to move a number of objects from one cluster to another. As is obvious, this algorithm does not converge to a unique solution because all elements of the initial population separately explore the Solution Space in different directions. This type of procedure may seem very closed to heuristic methods and may hurt the nature of genetic algorithms because it controverts the conservation schema theory (two genes with boolean code are necessarily different in homologous bits), but offer high performance results in ga-deceptive[3] problems with many local minima. A proposal could be to mix both approaches, using the standard coupled reproduction with clones and a concurrent "asexual" generation. In all reproduction phases the mutation operator has been applied with a 0.01.. 0.001 variable percentage [10] to permit an alternative manifold hyperplane exploration and avoid premature convergence.

The Fitness model

As was explained in precedenting sections, the definition of Cluster and Clustering Problem is difficult and related to the application field. For instance, if the goal is to group objects with very dissimilar features, clustering could be defined as the search for regions with a relatively high density of points separated by other regions with low density.

In such a case, the model must exalt the difference between good and less good solutions in order to converge to the optimum solution, accepting also unbalanced clusters or clusters with a single element. On the contrary, in applications such as quality inspections, where a set of similar objects has to be separated into groups, there are often many acceptable solutions, and the goal could be to promote a balanced distribution of objects in a given number of clusters. In this context the "fuzzy theory" may be combined with the genetic model, for instance by putting a value between 0 and 1 in the boolean cluster code to act as object belonging probability.

Given the variety of goals, many models have been examined and several of them allow exploitation of the genetic power. In this work two models are presented: The first adapts the nearest neighbor paradigm to the clustering without a given pattern set by dividing the points in Feature Space regions, calculating their centroids and evaluating the average distance between objects and their centroids.

The second is oriented to balance points into clusters, searching for the partition that minimizes the distance between each object and the others of the same group. Both methods can use the same concept of distance but the first, the "centroid based" clustering search to minimize an n-spherical density volume in the n_dimensional space, while the other, the "inter-objects distance based", explore the similarity between neighbour elements. Figure 5 shows the different results achieved using the two fitness finction models.

The metric used is the classical Euclidean distance between points, but other metrics can be devised dependending on the speed requirements for the process.

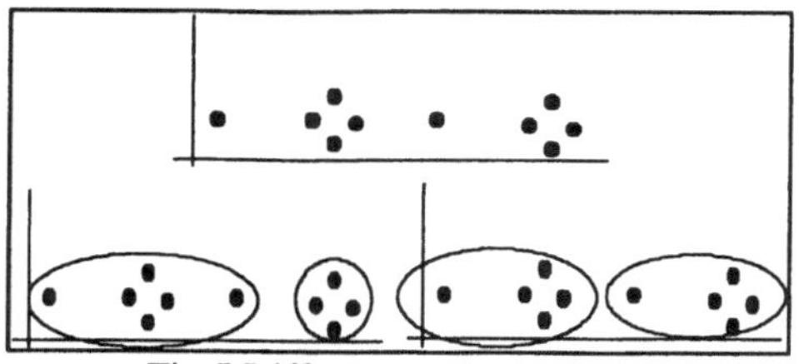

Fig.5 Different cluster methods

To evaluate the fitness of a feasible solution Ψ, as previously defined, with the centroid based method, the centroid for each cluster has to be calculated. The centroid is given by:

$$\overline{X}^i = \{\overline{x}_1^i, \overline{x}_2^i .. \overline{x}_n^i\} \text{ and } \overline{x}_k^i = \frac{\sum_{j=1}^{N^i} x_{jk}}{N^i} \quad k=1..n \tag{7}$$

The Cost Function to be minimzed is given by

$$f(\Psi) = \sum_{i=1}^{m} \frac{1}{N^i} \sum_{j=1}^{N^i} d(X_j, \overline{X^i}) \tag{8}$$

$$\text{or } f(\Psi) = \sum_{i=1}^{m} \frac{1}{N^i} \sum_{j=1}^{N^i} (\sum_{k=1}^{n} (x_{jk} - \overline{x}_k)^2)^{1/2} \tag{9}$$

In the second clustering model the Cost Function becomes

$$f(\Psi) = \sum_{i=1}^{m} \frac{1}{\frac{N^i(N^i-1)}{2}} \sum_{j=1}^{N^i} \sum_{t=j+1}^{N^i} (\sum_{k=1}^{n} (x_{jk} - x_{tk})^2)^{1/2} \tag{10}$$

This model is more time-consuming but generally more efficient as is testified by a comparative test over the same data stucture, reported in table 2 (last row).

Crossover	Bfit	Varf	Bgen	Varg
one-point (centroid-based)	1.0	0.0	107	24
two-point (centroid-based)	1.0	0.0	118	23
uniform (centroid-based)	1.0	0.0	105	19
uniform (inter-object)	1.0	0.0	28	9

Table 2 Crossover operators comparison

The effective best performance of the second method as is testified in the Ngen value can be caused by the nature of the data. In the observed test, in fact, all the objects are relatively similar and the element movement from one cluster to another does not excessively modify the centroids positions but can generate a considerable variation in the intra-object distance.

3.Conclusions

This work aims exploit the effectiveness of the genetic algorithms in clustering problems, especially in a context where the data dimensions have not been supported by exhaustive solutions and the problem generality does not always find a good solution with heuristic tecniques. The compared results could be used to create a robust model for clustering in image analysis related to the applications goals.

Acknowledgements

The author would like to thank Professor Salmon and his research group at the University of Bologna for many valuable discussions. The author is also grateful to Luigi Di Stefano and Massimo Piccardi for their help in preparing this presentation.

Bibliography

[1] J.M. Jolion, P Meer, S.Batouche '*Robust Clustering with Applications in Computer Vision*, IEEE Trans. on P.A.M.I., vol. 13, no. 8, Aug. 1991

[2] D.Vernon '*Machine Vision: Automated Visual Inspection and Robot Vision*' Prentice Hall 1991

[3]D.E. Goldberg, '*Genetic Algorithms, in search Optimization & Machine Learning* ', Addison-Wesely 1989

[4]C. Alippi, R. Cucchiara, '*Cluster Partitioning in Image Analysis Classification: a Genetic Algorithm Approach*,' Proc. of the IEEE Int. Conf. COMPEURO 92, 1992The Hague.

[5]K.A. De Jong '*Adaptive System Design: a Genetic Approach*', IEEE Trans. on Systems, Man & Cybernetics SMC-10,9, pp.566-574

[6] M.D. Vose '*Generalizing the notion of schema in genertic algorithms*' Artificial Intelligence 1991 n. 50 pp 385-396

[7] D.E. Goldberg, '*Sizing Population for Serial and Parallel Genetic Algorithms*', Proc. of the Third Int. Conf. on G.A. and their Applications, pp. 70-79, Morgan Kaufmann 1989

[8] J.J. Grefenstette, '*Optimization of Control Parameters for Genetic Algorithms*', IEEE Trans. on Systems, Man & Cybernetics SMC-16, 1, 1986

[9]G.Syswerda '*Uniform Crossover in Genetic Algorithms*', Proc. on the 3 Int. Conf. on G.A. and their Appl. pp. 2-9,1989

[10]T.C. Fogarty '*Varying the Probability of Mutation in the Genetic Algorithm*, Proc. of the Third Int. Conf. on G.A. and their Applications, pp. 104-109, 1989

Evolving Recurrent Dynamical Networks for Robot Control

D. T. Cliff, P. Husbands and I. Harvey
School of Cognitive and Computing Sciences
University of Sussex
Brighton BN1 9QH, England
email: davec philh inmanh @ cogs.susx.ac.uk

Abstract

This paper describes aspects of our research into the development of artificial evolution techniques for the creation of control systems for autonomous mobile robots operating in complex, noisy and generally hostile environments. At the heart of our method is an extended genetic algorithm which allows the open-ended formation of control architectures based on artificial neural networks. After outlining the rationale for our work, and giving the background to our techniques and experimental method, results are presented from experiments in which we contrast the behaviours of robots evolved under the same evaluation function but with different sensory capabilities. One set of robots have tactile sensing only, whereas the other have both tactile and primitive visual sensors. Although the evaluation task does not explicitly rely on vision, results show conclusively that evolution is able to exploit visual input to produce very successful controllers, far better than those for robots without vision.

1 Rationale

In oder to establish sufficient context for a clear explanation of our motivations a brief detour into the recent history of AI is called for.

Over the past decade there have been a number of increasingly urgent calls for a serious re-evaluation of many of the central assumptions of Artificial Intelligence.[1] The most influential of these, backed up with highly suggestive practical research results, have come from Rod Brooks and colleagues at MIT [3, 2]. Although some of the old guard remain unimpressed, and continue to claim that traditional formal logic-based AI techniques are about to come of age [4], there is evidence of a gradual shift of sympathy within the field towards the New AI [5]. Chief among those assumptions of traditional AI attacked by the new movement are:

- The dogma of functional decomposition: that an intelligent system's architecture should be organised around well defined functional modules with simple clean communication links. This goes hand-in-hand with assumptions about top-down hierarchical design methodologies.
- That both naturally and artificially intelligent systems should necessarily be dependent on internal representations to enable reasoning within and communication between modules.
- That the role of an intelligent system's coupling with the environment is either of no importance at all or a minor design detail to be handled at the last minute.

Instead, new approaches attach central significance to:

- Embodiment. It is claimed that interactions with the environment are of extreme importance in understanding natural intelligence and in developing artificial forms.
- The generation of adaptive behaviour. This is seen as the prime role of animal nervous systems and of great importance in useful artificial systems.
- The bottom-up development of entire artificial 'creatures'. The previous two items point to the study of complete behaviour generating systems for autonomous agents acting in realistically complex and uncertain environments. Because entire systems are involved, a gradual movement from the simple to the more sophisticated is deemed pragmatic. Many studies with

[1] Only the briefest of summaries is given here, see [3, 2, 9, 15, 1] for details.

these general characteristics have asked awkward questions about the fundamental need for reasoning and representation within intelligent systems.

Whilst aligning ourselves with the New AI, we believe there are serious questions to be asked about how such an endeavour is approached. How should we go about developing complete autonomous agents? One possible answer to this question is the subject matter of this paper.

We have argued elsewhere [10] that there is growing evidence that control systems for autonomous mobile robots (artificial creatures) acting in complex uncertain dynamic environments will involve many *emergent* interactions between many constituent parts, with the number of interactions growing exponentially with the addition of new parts. Because of this, the hand design of such systems will become intractable as new parts are added to generate more sophisticated behaviours. We advocate artificial evolution as the most promising way to tackle this problem. In this approach behaviour-oriented evaluation schemes are used to measure the fitness of each member of a population of genotypes coding for control systems. These fitness measures can then be used to allow selective breeding in the ways familiar from standard genetic algorithm (GA) practices. However, the demands of this sort of task are quite different from those of the usual function optimisation applications of GAs. As will be shown later, these lead to the need for an extended form of GA, and for significant modifications to usual practices.

2 The Sussex Evolutionary Robotics Project

More detailed discussions of the foundations of our work can be found in [10, 7, 13, 6], this section will summarise our approach to date.

The general aim of our research is to investigate the use of artificial evolution for the development of control systems for autonomous mobile robots engaged in visually guided navigation-based tasks. Mimicing nature, we feel that the best way to achieve this is incrementally: new capabilities are built on top of old as the evaluation tasks are made gradually more difficult. Instead of searching through the entire space of robot control architectures for each new task, the GA is used to search through the space of possible adaptations to the existing architectures, using operators which allow increases in the sophistication of the control systems. The population being evolved is always a genetically converged *species*, and increases in genotypic expressive power (e.g. through an increase in length) can only happen gradually enough to maintain a correlated fitness landscape. The basis for extending GAs to work within this open-ended framework is given in [12], which describes the species adapted genetic algorithm (SAGA) used here and developed to cope with dynamic length geneotypes. Genetic convergence signals the end of the road in standard fixed dimension GA applications, in [11] it is shown how our extended GA allows significant improvements in fitness *after* convergence.

What should we evolve? There are good grounds [10] for thinking that generalised connectionist networks are the most appropriate form for the control systems. We advocate arbitrarily recurrent dynamical networks with time delays on connections, continuous real valued signals, and non-uniform neuron dynamics. The following sub-section gives brief details of the particular neuron model and genetic encoding scheme we used in the experiments described later.

2.1 Neuron Model and Genetic Encoding

The neuron model employed to date has been designed for its applicability to low-level control, we intend to investigate various other models in the near future. In its operation there are separate channels for excitation and inhibition. Real values in the range [0,1] propagate along excitatory links subject to delays associated with the links. The inhibitory (or veto) channel mechanism works as follows. If the sum of excitatory inputs exceeds a threshold, T_v, the value 1.0 is propagated along any inhibitory output links the unit may have, otherwise a value of 0.0 is propagated. Veto links also have associated delays. Any unit that receives a non zero inhibitory input has its excitatory output reduced to zero (i.e. is vetoed). In the absence of inhibitory input, excitatory outputs are produced by summing all excitatory inputs, adding a quantity of noise, and passing the resulting sum through a simple linear threshold function, $F(x)$, given below. Noise was added to provide further potentially interesting and useful dynamics and to give an indication of the properties of a physical implementation of a unit which would be likely to include naturally occurring noise. The noise was uniformly distributed in the real range [-N,+N].

$$F(x) = \begin{cases} 0, & \text{if } x \leq T_1 \\ \frac{x-T_1}{T_2-T_1}, & \text{if } T_1 < x < T_2 \\ 1, & \text{if } x \geq T_2. \end{cases} \qquad (1)$$

In all our work to date the networks have been simulated in software. Their continuous nature is modelled by using very fine time slice techniques, as discussed later. In the experiments described in this paper the following neuron parameter setting were used: N=0.1, T_v=0.75, T_1=0.0 and T_2=2.0. The last two threshold values give a slope of 0.5 to the linear section of the excitatory transfer function. The networks are hardwired in the sense that they do not undergo any lifetime adaptation. However, as will be seen, their dynamic recurrent nature allows a single net of this kind to generate a range of adaptive behaviours, even under very noisy conditions.

The genotypes encoding networks of such neurons are character strings that are interpreted sequentially. Each string is divided into three distinct areas for input units (attached to sensors), hidden units and output units (attached to actuators) respectively. Firstly the input units are coded for, each preceded by a marker. The first part of a unit's gene encodes node properties such as threshold values; there then follows a variable number of sections each representing a connection from that node. Each section specifies whether it is an excitatory or inhibitory connection along with other properties such as delays, and then the target unit to connect to is indicated by jump-type and jump-size. The jump-type allows for both relative and absolute addressing. Relative addressing is provided by jumps forwards or backwards along the genotype order; absolute addressing is relative to the start or end of the genotype. These modes of addressing mean that offspring produced by crossover will always be legal. The internal nodes and output nodes are handled similarly with their own identifying genetic markers. Clearly this scheme allows for any number of internal nodes. This type of genetic encoding is relatively straightforward. Its extension to allow more complex developmental schemes including reuse of string sections (subroutining), is likely to greatly increase the power of the technique and is something we are actively pursuing.

2.2 Experimental Setup

All experiments to date have been based on careful simulations of a physical robot constructed at Sussex. The robot is cylindrical in shape with two wheels towards the front and a trailing rear castor. It has front and back bumpers and four whiskers — two front and two back, each at 45° to the central axis of the robot (see Figure 1), these tactile sensors give binary on/off signals. Later we describe how primitive visual sensors were added for some experiments. The wheels have independent drives allowing turning on the spot

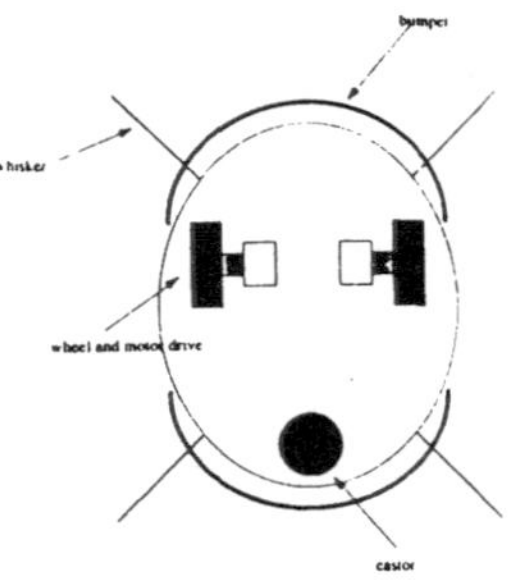

Figure 1: Plan view of simple robot.

and fairly unrestricted movement across a flat floor. The signals to the motor can be represented as a real value in the range $[-1.0, 1.0]$. This range is divided up into five equal portions with fuzzy boundaries, and depending on which portion the signal falls into, the wheel will either remain stationary or rotate half/full speed forwards/backwards. The motor signal range is achieved by using two output neurons per motor, and subtracting one's output from the other's. An input unit is assigned to each sensor, but the number of hidden units is not specified. Starting from randomly wired networks, the evolutionary algorithm is used to produce robust noise tolerant controllers for various robot tasks.

The continuous robot kinematics were modelled by using fine time slice simulation methods. At each time step the motor outputs were sampled and used to calculate the new position of the robot with appropriate kinematic equations. Noisy actuators were modelled by introducing stochastic elements into the equations. The simulation of collisions with walls was based on observation of the physical system. The continuous nature of the networks was simulated by using a very fine time slice approach. At the beginning of each robot step sensor reading were taken and fed into the network, unit inputs and outputs were synchronously updated over N cycles where N was randomly chosen on each robot step from the uniform range [40,60], and then motor signals were sampled. The important point to note is that unit time in the network simulations is much smaller than in the kinematics, this has important consequences for the properties of delays on network connections.

In early experiments networks capable of generating simple robust obstacle avoidance behaviours in cluttered environments were evolved using tactile sensing [10], but recently we have concentrated on uncluttered minimalist environments to allow close analysis of results. However, these environments are still noisy as are the (realistically) simulated robot kinematics. The results included in this paper are derived from experiments in which the robot moves in a cylindrical arena (see Figure 2). In each of the experiments populations of size 60 were evolved for 100 generations under the following evaluation function:

$$\mathcal{E} = \sum_{\forall t} \exp(-s|\mathbf{r}(t)|^2) \qquad (2)$$

Where $\mathbf{r}(t)$ is the 2D vector from the robot's position at time t to the centre of the arena, and $\forall t$ denotes the duration of the evaluation run (in all experiments discussed in this paper this was set at 100 time steps). $\mathcal{E}$ drops off very rapidly the more time the robots spend near the walls. This implicit evaluation function was chosen to study, with different sensing capabilities, how well artificial evolution produced networks to keep the robots as far away from the walls as possible for as long as possible. The walls of the arena are black while the floor and ceiling are white, this gives a potentially useful source of visual information.

In order to encourage robust controllers, each member of each generation was evaluated 8 times using $\mathcal{E}$ with each run starting from a different randomly generated position and orientation near the walls. The *worst* of these evaluation scores was then taken as the fitness of the individual. After measuring each member's fitness the population was ranked and parents were chosen by a quadratic ranked-based selection function. Offspring were produced via crossover and mutation, the entire population being replaced on each generation, and with the best of the previous generation being carried over unchanged. Variable length genotypes were allowed by using a crossover operator that exchanged homologous sections between parents, sometimes resulting in small increases or decreases of length. The initial population was randomly generated with small variations in the number of hidden units and the number of links from each unit.

3 Results: Blind v Sighted

The following results compare the behaviours generated by networks evolved for different sensory capabilities. The implicit task defined by $\mathcal{E}$ can theoretically be achieved by both blind robots (tactile sensing only) and sighted robots. The blind robots might conceivably develop timing circuitry to move the correct distance from the walls to the centre, while the sighted robots may develop networks to correlate visual input with position. In both cases this would have to be achieved in the face of noise at all levels. It was thought interesting to compare results from evolution under the two varieties of sensing.

In these experiments the crossover rate was set at 1.0, the mutation rate per genotype at 0.7, and the neuron parameters as stated in Section 2.1.

3.1 Blind Robots

The blind robots had access to the four whiskers and two bumpers shown in Figure 1. In the first set of experiments each evaluation run was for 100 robot time steps and all network connections were given fixed time delays of 1 network cycle and fixed unit weights. Ten different runs of 100 generations were made. In each case the fitness of all members of the initial random population was very close to zero (less than 0.05). By the end of each run significant improvement had been achieved. In 8 of the runs the best member of the final population achieved average scores of 15 or greater, worst scores of 10 or greater and best scores of 25 or greater. The theoretical optimum score, for a robot that headed straight for the absolute centre and then stayed there, is about 80. All of these fittest individuals displayed behaviours very similar to that shown in Figure 2. The outer circle is the boundary of the arena and the interior of the inner circle roughly equates to the high scoring area under $\mathcal{E}$.

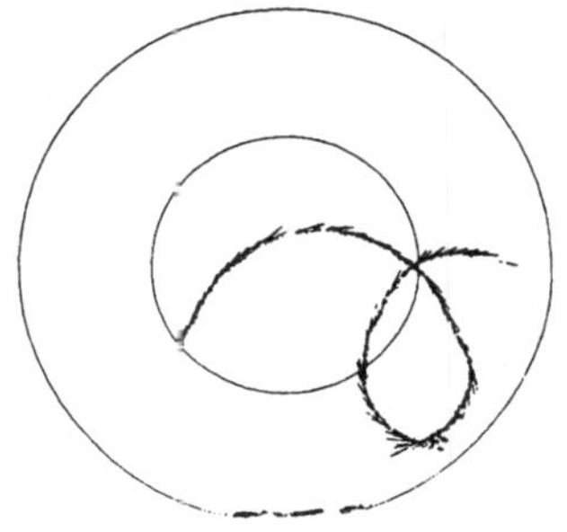

Figure 2: Fit behaviour for blind robots. The arrows indicate the orientation of the robot at each of 100 time steps.

The network responsible for generating this behaviour is shown in Figure 3. A detailed qualitative and quantitative analysis of this network is given in [14], showing how sensory inputs are used to produce tight turns away from the walls and, significantly, how internal neuron noise has been used to produce a right

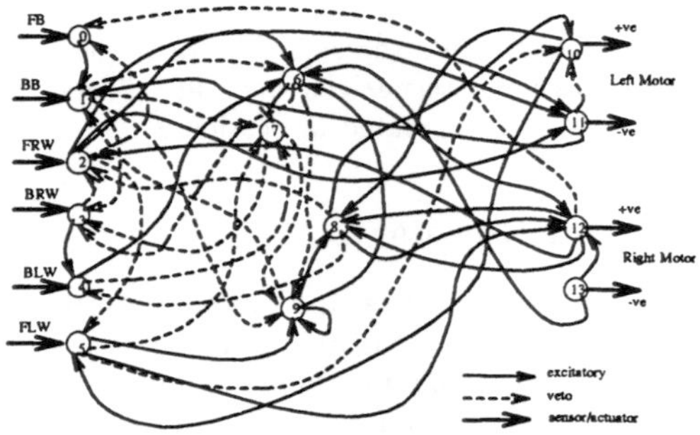

Figure 3: Network responsible for behaviour shown in figure 2. BB=back bumper, FRW=front right whisker and similar for other sensory inputs.

motor output that oscillates steadily between full forwards and half forwards At the same time the left wheel is jammed on half forwards by virtue of being fed by a *generator* unit. A generator unit is multiply connected to itself via various feedback loops and, as proved mathematically in [14], has its excitatory outputs clamped on 1 through an amplificatory build up of internal noise. The combination of these motor signals is an oscillation between straight line and circling movements giving the very wide sweeping paths shown. This is a robust strategy, combined with the tight turns away from the wall, to score moderately well under $\mathcal{E}$.

Central to this behaviour is the use of the noise-based oscillator. Interestingly, all the other equally fit networks from other runs were based on the same principle although they all had different architectures and the details of how the oscillators worked varied.

3.1.1 Variable Delays

A second set of experiments investigated the use of genetically specified time delays on network connections. The sections of the genotypes coding for connection properties were extended to indicate whether or not a connection had a delay greater than the default (1 network cycle) and if so what that value was. Non-default delay values could be one of eight discrete values, {10,20, ... 70,80}, where the delay unit is a network cycle. Other aspects of the experiments were the same as before. Note that the longer delay values are of durations greater than a robot time step.

In each of five runs all members of the initial population again scored close to zero. In each run significant improvement was made faster than in the fixed delay runs. The fittest members of the final populations all had worst scores of greater than 12, average scores of greater than 17, and best scores of greater than 28. In other words the variable delay networks were able to provide slightly more robust controllers. Figure 4 shows the behaviour of a robot controlled by one of these fittest networks and run for 200 time steps. It can be seen that the behaviour is similar to those produced in the previous experiments and scores reasonably well. Figure 5 shows the network responsible for the behaviour while Figure 6 gives time plots for sensor readings, neuron outputs and motor activities.

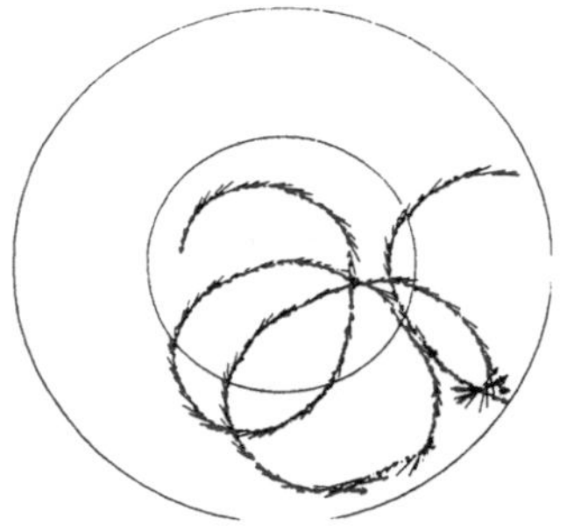

Figure 4: Fit behaviour for blind robots with genetically specified delays on network connections. The arrows indicate the orientation of the robot at each of 200 time steps.

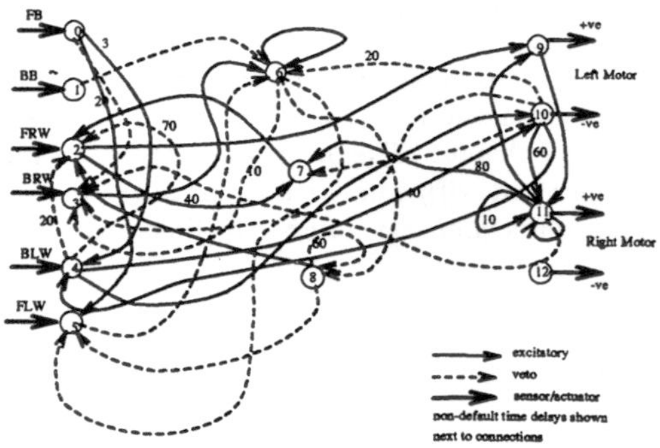

Figure 5: Network responsible for behaviour shown in Figure 4.

Study of the network and its time plots over numerous runs of the robot readily reveals that units 12, 3, 6 and 8 play no part in any aspect of the control process. By further eliminating all units whose

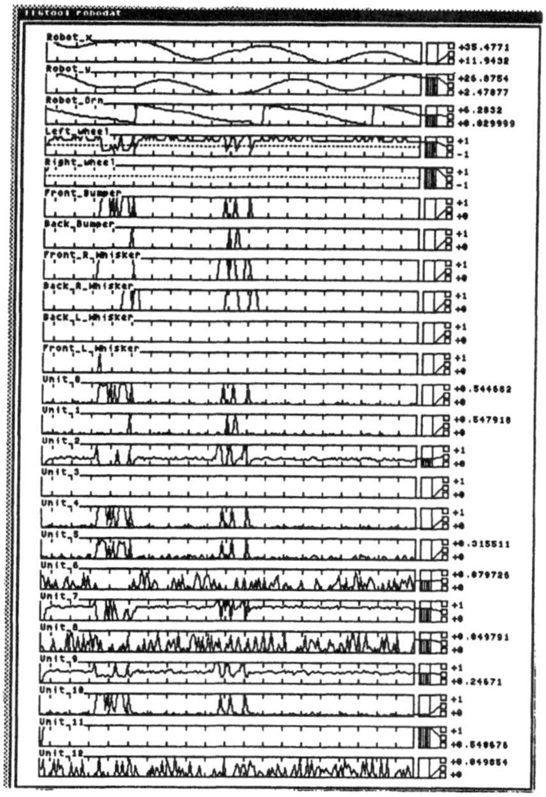

Figure 6: Time plots for sensor readings, neuron outputs and motor activities corresponding to behaviour in Figure 4.

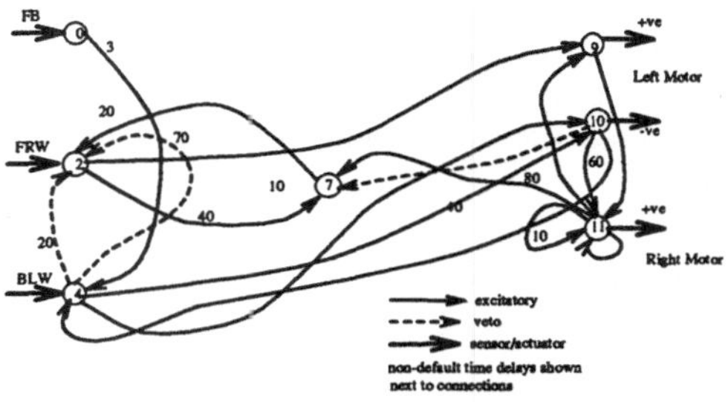

Figure 7: Reduced network with all redundant units and links from Figure 5 removed.

connections were only to/from these redundant ones, it is possible to draw the greatly simplified network shown in Figure 7. The default wide path behaviour is produced by a noise-based oscillator network similar to those found before. Unit 11 is a generator unit which keeps the right wheel clamped on fast forwards. Unit 9 has excitatory inputs from 11 and 2. Input for 2 comes from 11 via 7 and from the mutual feedback loop between 2 and 7. Remembering that a signal is reduced by a factor of 0.5 as it passes through a unit, the input to 9 (ignoring noise for the moment) will be 1.38, giving an output of 0.69. Following the mathematical analysis in [14], none of the feedback loops in the 11-7-2-9 circuit are sufficient to allow build-up of noise. Hence the output of 9 performs a random walk close to 0.69. Given that the fuzzy boundary between the full and half forwards motor signals is centred on 0.67, this is enough to produce an oscillating left wheel speed. The delays on some of the links in the circuit tend to act to smooth the noise, resulting in more tightly tuned, and slightly more reliable behaviour than in the previous experiments. In earlier generations networks with oscillators based on delayed veto feedback loops were predominant but they died out as the noise-based variety took over. The operation of the network in collisions with the walls is also easily understood. Only the front bumper, front right whisker and back left whisker are used; this is reasonable given the simplicity of the environment and the fact that the robot always moves forwards. Taking one of the possible cases, if the front bumper is hit, input to 10 via 4 goes high (courtesy of the double feedback loop between 4 and 10) causing 7 to be vetoed which results in a drop in the signal to 9. The combination of these changes is to throw the left wheel speed to half backwards while the right wheel speed remains full forwards. This results in the tight turn away from the wall. The long delay on one of the connections from 4 to 10 prolongs the turn after sensory input has ceased. This results in more reliable turning behaviour than possible with fixed unit delay nets.

3.2 Sighted Robots

A further set of experiments was performed where two very simple visual sensors were added to the robots. The orientation and acceptance-angle of these photoreceptors were placed under evolutionary control by having a separate 'vision chromosome' coding for these parameters. In this way the visual sensor morphology was evolved concurrently with the control network. Ray-tracing techniques were used to accurately simulate the physics of vision. For full details see [7].

For eight separate runs using the evaluation function $\mathcal{E}$ and with fixed unit delays on all network connections, each member of each initial population scored close to zero, as with the blind robots. After 100 generations the fittest members of each population scored significantly higher than this, with about half the runs producing networks capable of generating near-optimal behaviours. These consistently achieved average scores of between 60 and 70, markedly better than the blind robots. The next section discusses the evolutionary trajectory of one of the runs.

3.3 Evolutionary Trajectory

Recording the evolutionary trajectory of a network may appear to be a daunting task: each individual in every evaluated population is a potential candidate for consideration. However, the task is made simpler by use of summary statistics concerning the evolutionary history of the population within which the network evolved. Some of the most useful statistics are those connected with the evaluation function. As was stated above, each individual in the population is evaluated eight times, and the lowest evaluation score from the eight trials is taken as the measure of the individual's fitness; so that the individual with the highest worst score is viewed as the most fit in the population. At the end of each generation, we recorded three values: the fittest individual's worst score; the fittest individual's average score over its eight trials; and the highest evaluation score on a single trial recorded by any member of the population (referred to as the *best* score) – note that the best score is not necessarily a score recorded by the fittest individual; it could have been a score recorded for an individual whose highest worst score was much lower than that of the fittest's. A graph of these three values over 100 generations is shown in Figure 8.

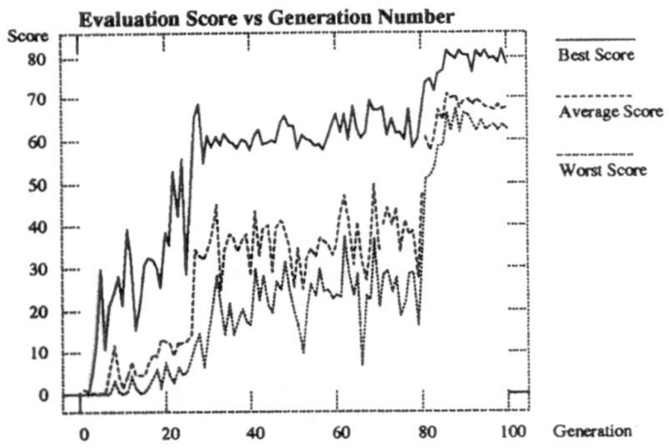

Figure 8: Evolution scores over 100 generations for visually-guided robots. See text for further details.

The data shown in Figure 8 can be characterised as follows: from the first generation onwards, all three score statistics show a fairly steady climb up to around generation 30; from generation 30 to generation 80, the scores are roughly constant (i.e. the best scores are almost all in the range 55-65, and the average scores are in the range 25-45); then, after generation 80, the scores show a sudden sizeable increase, followed by a continued more gradual climb until generation 90, after which they maintain their high levels but are again roughly constant. To give an impression of the evolutionary trajectory of the behaviours exhibited by the robot, Figure 9 shows typical behaviours of the best individual in the population at generations 40, 75, 85, and 100. As can be seen, by generation 40 the visually guided robots are exhibiting behaviour similar to that shown by the better blind robots. At generation 75, the robots can move to the centre and then perform circling movements to stay there. By generation 80, the robot spins "on the spot" very near to the centre: the change from 'circling' to 'spinning' seems to account for the sizeable increase in scores noted at generation 80 in Figure 8). The increase in scores between generation 80 and generation 100 appears, from the behaviour plot for generation 100, to be due to an improvement in the robot's ability to take a direct path to the centre.

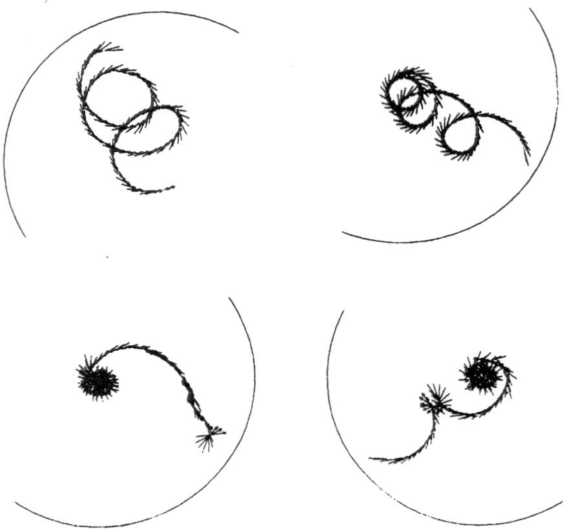

Figure 9: Typical behaviours of the best individual in the visually-guided population at (clockwise from top left) generations 40, 75, 85, and 100.

Using techniques similar to those outlined in Section 3.1.1, analysis of the network controllers for the sighted robots showed that they relied almost exclusively on visual input to produce the very fit behaviours observed (see [8] for details). To confirm this we performed experiments with two of the fittest networks, here referred to as C1 and C2, in which visual input was completely disabled. Keeping all other conditions identical to those used in their evolution, and taking the average score over 80 evaluations, C1 scores 65.104 with visual input, and 0.014 without, and C2 scores 60.259 with visual input, and 0.323 without. Clearly without visual input the controllers are useless; they have evolved, concurrently with the visual morphology, to make very good use of visual information in the environment.

4 Conclusions

After presenting the rationale for our work in evolutionary robotics and giving the technical background to our research, we showed results from experiments in which behaviours of sighted and blind robots, evolved under the same evaluation function, were compared. Although the implicit evaluation task does not explicitly rely on any particular form of sensory input, evolution has been able to exploit vision to produce very fit controllers, far fitter than the best of the blind controllers. The results also suggest that the use of genetically specified time delays on network connections, and the concurrent evolution of control network and sensory morphologies, are powerful weapons in the evolutionary arsenal. A major conclusion is that robots capable of sophisticated autonomous behaviour need vision. Realistic simulations of environments more complex than those used here become very expensive; this has led us to develop specialised visuo-robot equipment which allows us to conduct evolutionary experiments, with their very heavy evaluation demands, while abandoning simulation all together [7].

References

[1] R. D. Beer. *Intelligence as Adaptive Behaviour: An Experiment in Computational Neuroethology.* Academic Press, 1990.

[2] R.A. Brooks. Intelligence without reason. In *Proceedings IJCAI-91.* Morgan Kaufmann, 1991.

[3] R.A. Brooks. Intelligence without representation. *Artificial Intelligence*, 47:139–159, 1991.

[4] A. Bundy. Clear thinking on artificial intelligence. *The Guardian*, 5th Dec., 1991.

[5] W.J. Clancey. The frame of reference problem in the design of intelligent machines. In K. vanLehn, editor, *Architectures for Intelligence: The 22nd Carnegie Symposium on Cognition.* Lawrence Erlbaum Associates, 1991.

[6] D. Cliff, I. Harvey, and P. Husbands. Incremental evolution of neural network architectures for adaptive behaviour. In *Proc. 1st European Symposium on Artificial Neural Networks*, Brussels, 1993.

[7] D. Cliff, P. Husbands, and I. Harvey. Evolving visually guided robots. In H. Roitblat J. Meyer and S. Wilson, editors, *Animals to Animats: Proceedings of the 2nd International conference on the Simulation of Adaptive Behaviour.* MIT Press/Bradford Books, 1993.

[8] D. Cliff, P.Husbands, and I. Harvey. Analysis of evolved sensory-motor controllers. Technical Report CSRP-264, COGS, University of Sussex [submitted to ECAL 93], 1993.

[9] D. T. Cliff. Computational neuroethology: A provisional manifesto. In J.-A. Meyer and S.W . Wilson, editors, *From Animals to Animats: Proceedings of The First International Conference on Simulation of Adaptive Behavior*, pages 29–39. MIT Press/Bradford Books, Cambridge, MA, 1991.

[10] I. Harvey, P. Husbands, and D. Cliff. Issues in evolutionary robotics. In H. Roitblat J. Meyer and S. Wilson, editors, *Animals to Animats: Proceedings of the 2nd International conference on the Simulation of Adaptive Behaviour.* MIT Press/Bradford Books, 1993.

[11] I. Harvey, P. Husbands, and D. Cliff. Genetic convergence in a species of evolved robot control architectures. In *(submitted to) Proc. 5th Int. Conf. on GAs.* Urbana-Champaign, July 1993.

[12] Inman Harvey. Species adaptation genetic algorithms: The basis for a continuing SAGA. In *Proceedings of the First European Conference on Artificial Life*, pages 346–354. MIT Press/Bradford Books, Cambridge, MA, 1992.

[13] P. Husbands, I. Harvey, and D. Cliff. An evolutionary approach to situated AI. In *Proc. 9th bi-annual conference of the Society for the Study of Artificial Intelligence and the Simulation of Behaviour (AISB 93).* IOS Press, 1993.

[14] P.Husbands, I. Harvey, and D. Cliff. Analysing recurrent dynamical networks evolved for robot control. In *Proc. 3rd IEE Int. Conf. on ANNs.* IEE Press, 1993.

[15] F. Varela, E. Thompson, and E. Rosch. *The Embodied Mind.* MIT Press, 1991.

GENETIC ALGORITHMS FOR ON-LINE SYSTEM IDENTIFICATION

Kevin Warwick and Yong Ho Kang

Department of Cybernetics
University of Reading
Whiteknights
Reading, UK.

ABSTRACT

When required for batch processing, system identification can be used in a number of ways to find a good fit in terms of modelling the system structure, time delay and characteristic parameters of a plant. The best structure and delay selection can often be found in terms of a computationally simple model order testing procedure, with a range of different candidate models being considered.

Due to the need for computational efficiency for on-line identification, this is often restricted to a more straightforward parameter estimation exercise, the structure being selected as a fixed term, usually of low order. Any structural tuning is then carried out in terms of sampling period variation, which can mean that vital plant information does not appear in the identified model.

This paper presents an on-line technique for system identification, making use of genetic algorithms for structure and time delay estimation. The identification scheme is multi-level, the bottom level consisting of the more usual parameter estimation exercise, with the upper level carrying out structure identification. The first of these can update in real-time, whilst the second operates in its own time. At any instant, one model is selected as "best" in terms of both structure and parameters, and this is the one employed as on-line identification.

INTRODUCTION

On-line system identification or parameter estimation can be looked at in terms of finding the best structural model and subsequently to find the best fitting parameters within that model. Often with recursive estimators in adaptive control the structural aspect is either deemed to be known or a simple, lst or 2nd order, model is specified and employed even if this might not be the best one to use. Some work has employed such as Bayesian estimation or Maximum Likelihood techniques to provide a concurrent on-line estimation of system structure [1,2], but in most cases these techniques are fairly cumbersome and certainly computationally expensive.

For off-line, batch system identification, an approach to deal with the structural problem is that of Model Order Testing, i.e. to consider a number of potential structural solutions and to quantify these in terms of a fairly simple measure, most likely by assessing the sum of squares error for each model structure.

A new approach, that is relatively light on computational load, and is a potential solution for on-line structural selection is to employ Genetic Algorithms. Essentially an initial population of models is specified, with the "best" of these being employed for on-line processing. The population is enhanced by means of genetic algorithms, new structures being genetically introduced in a number of ways, e.g.mutation or crossover. Periodically the weakest member of the population is removed and a new member introduced, by means of measurement against overall defined goodness objectives. By this means if a newly introduced structure proves to be of little value it will soon be allowed to die, whereas if it is seen to be of importance it can gain in strength and subsequently be employed on-line.

Genetic algorithms are search and ordering techniques which have their foundations in natural selection and genetics [3]. In principle they are very simple and can be very efficient and relatively computationally light. An approach to system identification in which the Genetic Algorithm is used directly as part of the recursive estimator, has already been reported [9]. The approach taken here differs in that the parameter estimation procedure is considered to be of the more usual Least Squares form, with the Genetic Algorithms operating on a population of usual, low-level estimators. At the low-level estimator stage, it is assumed that the single-input, single-output plant to be identified can be described by the discrete-time difference operator equation:

$$A(z^{-1})y(t) = z^{-k}B(z^{-1})u(t) + c(z^{-1})e(t) \qquad (1)$$

where $\{y(t) : t\varepsilon\tau\}$ and $\{u(t) : t\varepsilon\tau\}$ are sequences of system output and input signals respectively. Also $\{e(t) : t = 0, \pm 1, \pm 2, ...\}$ is a white noise sequence with zero mean and finite variance. The integer part of the system time delay is $k \geq 1$ such that the b_o term must be no-zero in the following polynomial definitions:

$$\begin{aligned} A(z^{-1}) &= 1 + a_1 z^{-1} + \ldots\ldots a_{n_a} z^{-n_a} \\ B(z^{-1}) &= b_o + b_1 z^{-1} + b_{n_b} z^{-n_b} \qquad (2) \\ C(z^{-1}) &= 1 + c_1 z^{-1} + \ldots\ldots + c_{n_c} z^{-n_c} \end{aligned}$$

in which the delay operator has effect such that $z^{-i}y(t) = y(t - i)$.

In the following section the identifier, which is aiming to primarily estimate the coefficient of the A and B polynomials along with k, on the assumption that C is unity, i.e. only white noise is present.

HIERARCHICAL IDENTIFICATION

The overall system identifier is split into two sections, see Figure 1, the low level element being a standard identification procedure [4,8], such as recursive least squares, which is operating on-line, the upshot of this being that the parameter updates must be obtained within the sample period specified. This low level identifier, with its particular model structure, transport delay and forgetting factor [5] will continue to operate, in its own way, unless the upper level identifier decides that an alternative is necessary, at which point the new identifier takes over.

The upper level of the identifier consists of a small population of alternative identifiers, generally of the same type, but of different model structure, transport delay and/or forgetting factor. The size of the population, should be relatively small, possibly four or five and is directly dependent on the amount of computing power and memory available and the sampling period specified.

A further user specified element is N, the window of samples over which a comparison is made. At the end of a particular period of N samples a count is made, for each of the identifiers, of the sum of squares of the modelling errors, over N, in a similar way as for model order testing in standard system identification [6]. The sum of squares is then modified simply by multiplying the sum by the number of parameters estimated in that model, to give Σ.

This modified sum of squares is the critic with which the different members of the population are placed in a league table. The top identifier in the population, after a particular period N, is the one with the lowest Σ, and this identifier is selected as the contender. The bottom identifier in the population is the one with the highest Σ over the period N, and this identifier is a potentially replaceable (relegatable) set-up.

Once a contender identifier has been selected after a particular window of N samples, although it remains within the population, it takes on a further special position, as follows. At the end of the next N samples its Σ over that N is compared firstly with that of the on-line, low-level identifier over the same period. If the contender is found to have a lower Σ, its parameter set, structure, time delay, forgetting factor etc. all take over the on-line role, i.e. the contender switches places with the low-level identifier.

Whichever loses out of the contender and the original on-line identifier, i.e. whichever has the highest Σ over the most recent N, is compared with the Σ's of the remainder of the population, and takes its appropriate place, i.e. it could become (or remain) the contender over the next window N, it could fill the relegation spot or it could take up a mid-league position.

The identifier in the relegation spot is not automatically ousted from the population, and in general it is expected that an identifier must fill the relegation slot on m instances in the last (m + p) opportunities, where $m \geq 1$ and $p \geq 0$ are integers. On initial start-up of the identifiers, it is useful for m and p to be small, possibly both unity, whereas for process control in the steady-state, $m = 20$, $p = 0$ could be acceptable. Particular care must be taken not to automatically relegate an on-line identifier or contender on fall from grace, unless they remain in the relegation spot for a reasonable period. Also a newly entered identifier must be given time to settle in and indeed may fill the relegation spot at first when its parameter set is tuning in. One problem with actually relegating any identifier from the population is that it might be quite some time before it finds its way back into the population, i.e. it is best to be sure that an identifier has not performed well for quite a period before ousting it.

But conservatism over actually relegating a poorly performing identifier must be balanced with the desire to introduce new, genetically inspired identifiers. Indeed it is, in some instances, best if an operator interrupt is allowed in order to automatically bring in a new genetically inspired identifier at the expense of the identifier in the relegation spot. On the other hand if a change occurs in the actual plant whose characteristics are being identified, the Σ of one or more of the identifiers in the present population will most likely soon become much larger. Relegation can then be enforced on an identifier whose Σ increases above an upper band.

The nature of the genetic algorithms employed to select new identifiers for introduction to the population, and their relationship with the basic identifier structure is given in the next section. This section concludes with a number of practical problems to be overcome in the use of such identifiers.

Organisation of the relegation spot must necessarily be dependent on the nature and type of plant being identified. Plant whose characteristics remain essentially constant or vary slowly with respect to time, only really need a relegated identifier with corresponding new genetic input very infrequently, i.e. m is very large. Indeed it may well be useful not to automatically relegate an identifier, but to keep it as an external member of the population, to compare its Σ with the newly entered identifier over a pre-defined number of N windows, and to make a subsequent relegation decision.

Another type of plant has characteristics which vary more often and tend to vary in one direction, often through ageing or because of plant modification. This type of plant does not really require any special arrangements.

A third type of plant has characteristics which vary quite often, but are, in most cases, cyclic, such as those encountered in the motors of a robot manipulator, as the manipulator moves through positional changes. In this case it is possible that an identifier which has, for the moment, gone out of favour and is filling the relegation spot, may well have the potential to return to a challenging position at a later, and possibly not too distant time. In this case it is possible to extend to a second division in which a small number of identifiers exist, all of whom can quickly be promoted to the main league, should their Σ value become small. The argument for a second division, which necessarily means much more computational effort, is, in the case of a robot manipulator, severely set back by the fact that the relatively short sampling period necessary for motor control severely restricts the number of identifiers that can be considered.

One problem can occur with cycling of the bottom two or three identifiers, each one spending a short time in the relegation spot, before handing over to one of the others, only to return in the near future. Although this can sometimes be picked up by actually relegating in response to m out of m+p instances, it may be necessary to include a double check based on the above, along with a count of the most frequent relegation spot holder over a longer period, say 100 instances.

The inclusion of a new identifier raises a number of problems. Firstly, should present counts such as m, continue or be restarted. It is felt best, but not necessarily so, that all counts are restarted when a new identifier is introduced, such that all identifiers are competing on the same terms. A bigger problem, however, is how to provide the new identifier with a seeded parameter set, and possibly corresponding seeded forgetting factor. If the plant input and output are reasonably exciting this is not so much of a problem in that the new identifier can be given an initially fairly low forgetting factor, possibly exponential, thereby allowing the identifier to tune in quickly. Even then the first two or three N windows may need to be ignored where signal excitation is less, then parameter seeding is best done by means of the new identifier taking on board the latest parameter estimates of the identifier with the most similar structure, following the prioritised schema (i) time delay, (ii) denominator parameters, (iii) numerator parameters (iv) forgetting factor.

GENETIC ALGORITHM FOR STRUCTURE IDENTIFICATION

The genetic algorithm employed is based on the supposition that, at any particular time instant, the on-line identifier is the best fit at that moment, and that any modification should take that as a starting point, i.e. it is taken as a modification on the fittest at the time.

The most frequently employed approach is to match the on-line identifier and the contender, e.g. selecting time delay and denominator of the identifier with the numerator and variable forgetting factor of the contender, i.e. crossover [7]. Other variations in the string of four selections can also be invoked, a check being made that the identifier thereby initiated does not already exist and has not existed for a sufficient period.

A less frequent crossover invoked is that of matching the on-line identifier with the most successful genetically induced identifier, where one exists, on the provision that the identifier was able to gain a place in the population.

The final genetic element introduced is that of mutation, which merely involves mutational actions on the on-line identifier. This can be a purely pseudo- random operation on the four selections of the identifier, within realistic bounds, or can be in terms of structured mutations involving only slight variations from the on-line identifier, and only one feature at a time.

CONCLUSIONS

A Genetic Algorithm structure for on-line system identification has been proposed. The approach is particularly useful for adaptive systems where an adaptive model structure is also required. The method of estimator structure selection has distinct advantages over more computationally heavy algorithms, not only because of the actual computing requirements, but also because of its comprehensibility.

Trials are now being carried out to further investigate the potential of the genetic algorithms approach to system identification, as described.

REFERENCES

[1] Karny, M. *'Algorithms for Determining the Model Structure of a Controlled System'* Kybernetica, 19, 2, 164-178, 1983.

[2] Karny, M. and Kulhavy, R., *'Structure Determination of Regression-type Models for Adaptive Prediction and Control'*, in Spall J.C. (ed.), Bayesian Analysis of Time Series and Dynamic Models, Marcel Dekker, 1988.

[3] Goldberg, D., *'Genetic Algorithms in Search, Optimization and Machine Learning'*, Reading MA, Addison-Wesley, 1989.

[4] Soderstrom, T and Stoica, P., *'System Identification'* Prentice-Hall Inc., 1989.

[5] Fortescue, T.R., Kershenbaum, L.S. and Ydstie, B.E., *'Implementation of Self-tuning Regulators with Variable Forgetting Factors'* Automatica, 17, 931, 1981.

[6] Warwick, K., *'System Identification'*, chapter in "Industrial Digital Control Systems", rev. 2nd ed., K. Warwick and D. Rees (eds.), Peter Peregrinus Ltd., 1988.

[7] Kiernan, L.A. and Warwick, K., *'Developing a Learning System Capable of Hypothesis Justification'*, Proc. Int. Conference Control 91, Edinburgh, 272-276, 1991.

[8] Leith, D.J., Murray-Smith, D.J. and Bradley, R., *'Combination of Data Sets for System Identification'* Proc. IEE, Part D, 140, 1, 11-18, 1993.

[9] Kristinsson, K. and Dumont, G.A., *'System Identification and Control using Genetic Algorithms'*, IEEE Trans. on Systems, Man and Cybernetics, 22, 5, 1033-1046, 1992.

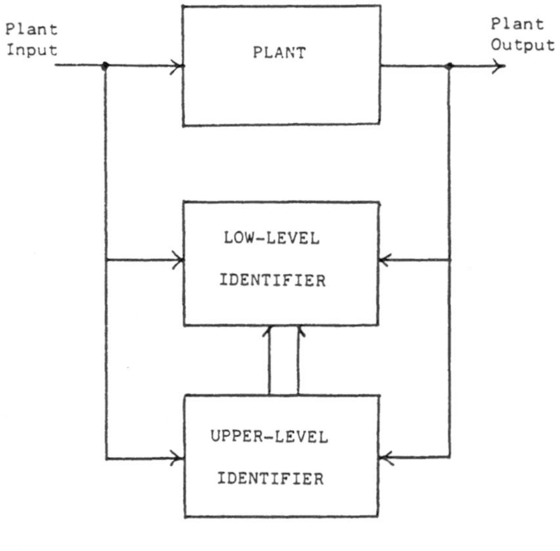

FIGURE 1

EVOLVABLE HARDWARE
Genetic Programming of a Darwin Machine

Hugo de Garis

Brain Builder Group,
Evolutionary Systems Department,
ATR Human Information Processing
Research Laboratories,
2-2 Hikari-dai, Seiko-cho, Soraku-gun,
Kyoto, 619-02, Japan.
tel : + 81 7749 5 1440,
fax : + 81 7749 5 1408,
email : degaris@hip.atr.co.jp

Keywords : Darwin Machines, Evolvable Hardware, Software Configurable Hardware, FPGAs (Field Programmable Gate Arrays), HDPLDs (High Density Programmable Logic Devices), Hardware Accelerators, GP (Genetic Programming), GAs (Genetic Algorithms), Complexity Independence of GAs, GenNets (Genetically Programmed Neural Network Modules), ALife, Artificial Nervous Systems, Biots (Biological Robots), 1000-GenNet Biots, Darwinian Robotics, GenNet Accelerators, Molecular Scale Technologies, Nanotechnology, Complexity Problem, Avogadro Machines, "Embryonics" (Embryological Electronics), "Embryofacture" (Embryological Manufacture).

Abstract : For the past three years, the author has been dreaming of the possibility of building machines which are capable of evolution, called "Darwin Machines". As a result of several brain storming sessions with some colleagues in electrical engineering, the author now realizes that hardware devices are on the market today, which use "software configurable hardware" technologies that the author believes can be used to build Darwin Machines within a year or two. This paper suggests there are at least two approaches to be taken. The first approach uses "software configurable hardware" chips, e.g. FPGAs (Field Programmable Gate Arrays), HDPLDs (High Density Programmable Logic Devices), or possibly a new generation of chips based on the ideas that FPGAs etc embody. The second approach uses a special hardware device called a "hardware accelerator" which accelerates the simulation in software of digital hardware devices containing up to several hundred thousand gates. Darwin Machines will be essential if artificial nervous systems are to be evolved for biots (i.e. biological robots) which consist of thousands of evolved neural network modules (called GenNets). The evolution time of 1000-GenNet biots will need to be reduced by many orders of magnitude if they are to be built at all. It is for this reason that Darwin Machines may prove to be a breakthrough in biotic design. When molecular scale technologies come on line in the late 1990s, the Darwin Machine approach will probably be the only way to build self assembling, self testing molecular scale devices.

1. Introduction

This paper presents some ideas which the author hopes will create a whole new subfield in such specialities as Genetic Algorithms, Artificial Life, Neural Networks, and Electronics in general, namely the concept of "evolvable hardware", or "Darwin Machines". Such devices will be essential if artificial nervous systems for biots (biological robots, artificial creatures) are to be evolved in an acceptably short time. There is a growing recognition that biots will need to be evolved rather than be handcrafted, because the complexity of behavioral units and their interconnections has reached breaking point. However, this complexity problem can be more easily overcome if one uses an evolutionary approach to biot design.

The author and John Koza have been writing a lot of papers over the last few years on the concept of Genetic Programming. This term was introduced independently by both authors in 1990 [de Garis 1990a], [Koza 1990] with rather different definitions, but the essential message was the same, i.e. that one could evolve solutions to complex problems. Koza's definition of GP is to evolve LISP programs [Koza 1990, 1992] to solve a variety of problems. The author's definition, which is more general and includes Koza's as a special case, is using GAs (Genetic Algorithms [Goldberg 1989]) to build/evolve complex systems, e.g. neural network dynamics, artificial embryos etc [de Garis 1990, 1991, 1992]. The author's definition of GP changes the traditional focus of GAs away from being seen essentially as optimizers, towards being seen as builders, and builders of systems which are too complex in their dynamics or structures to be predictable or even analyzable.

This shift in emphasis was a result of perceiving what the author called the "complexity independence" of GAs. To clarify what is meant by this concept, consider the evolution of the weights of fully connected neural networks, so that the time-dependent outputs of these networks control leg component orientations, so that a simulated quadruped is able to walk, to turn left, turn right, etc. These neural nets (called GenNets) are complex non linear devices whose dynamics are extremely complex. However, GAs do not care about the complexity of the systems they are evolving, so long as their fitness measures keep increasing, i.e. so long as the system being evolved is evolvable. This then shifts emphasis onto the question of what are the criteria for good evolvability.

This GA indifference to the complexity of the systems it evolves allows one to evolve very complex yet functional systems, i.e. systems more complex than those which can be produced by any other technique. We know that the evolutionary construction (GP) technique is effective, because we have the existence proof of ourselves, of our brains, or embryos etc. Nature uses GP.

GP then has proved itself capable of building black box systems which function the way one wants, but suffers the disadvantage that the inherent complexity of the black box systems is probably too great for analysis. GP therefore tends to counteract the tendency of 300 years of modern science, which has largely taken an analytic approach to understanding, i.e. the traditional method of breaking down a complex system into simpler components and then studying them. GP on the other hand is inherently synthetic. It assembles components in complex ways and then tests whether the ensemble works. Those that (by chance combination) do work, have more copies made of themselves, to be mutated and crossed over, to generate a further set of trial combinations. Nature stumbled upon this evolutionary synthetic approach and it has been spectacularly successful. It produced the most complex structure in the known universe, the human brain.

As a result of several recent ALife conferences [e.g. Ecal91, ALife III] a general awareness has arisen that the "way to go" in building biots is to take the evolutionary approach, as mentioned above. Recent notable converts to this "Darwinian Biotics" have been Beer (who handcrafted neural circuits for his artificial cockroach [Beer 1990, 1991]), and Brooks (who build Genghis and Attila, the hexapod insect robots [Brooks 1990, 1992]).

Recently, the author used GenNet modules to build an artificial nervous system for a simulated quadruped biot [de Garis 1993]. The outputs of each GenNet controlled the time dependent orientations of leg components, such that the biot walked, turned, etc. For each different behavior, there was a separate GenNet. Behaviors were switched simply by taking the outputs of one GenNet (i.e. one behavior) and inputting them into the next GenNet (which generated the next behavior). No matter at what moment the switching occured, the second (qualitative) behavior evolved for, was generated.

Five such behaviors were evolved, namely :-walking straight ahead (where the fitness definition was simply the distance covered by the biot in a given time), turning clockwise, turning anticlockwise (where the fitness definitions were simply the appropriate angles rotated by the biot in a given time), pecking, and mating. A series of detector GenNets was also evolved, such as a frequency detector, a signal strength detector, a signal strength difference detector etc. These detectors were used to send inputs to decision GenNets, which behaved like production rules (e.g. IF (input1 > 0.4) & (input 3 < 0.5) THEN action5). The outputs of these decision GenNets triggered one of the behavior GenNets.

This simulation was quite successful as far as it went. It showed that the GenNet approach to artificial nervous system design works. The author would now very much like to scale up to a 100 or 1000 GenNet biot. But there's a catch! It took a lot of time to evolve each GenNet. In practice (on a Mac II family computer) it would take a day of computer time. If ever a 1000 GenNet biot is to be built, GenNet accelerators will be essential. Somehow, the whole GP process needs to be accelerated. Hence the thoughts of the author have been directed lately towards the idea of Darwin Machines, i.e. taking advantage of hardware (as against software simulation) speeds by evolving neural circuits (GenNets) directly in hardware.

2. Software Configurable Hardware

The basic idea behind a Darwin Machine, is to use a type of hardware device which accepts some bit string software instruction and uses it to configure (i.e. to "wire up") a hardware circuit and not just once, but repeatedly. The outputs of this newly configured hardware circuit are then tested with another piece of software configurable hardware to determine the quality of its performance, i.e. its fitness. Those bit string instructions which configure high fitness circuits, survive (a la GA), to reproduce in the next generation. Thus every time a new bit string instruction is input, a new circuit is generated or configured. Modern "software configurable hardware" technologies [Broesch 1991] (e.g. using EEPROM (electrically erasable PROMs) or SRAM (static RAMS)) allow this possibility, so the idea of "evolvable hardware" can become a reality. It is a most exciting prospect, and may have a major impact not only upon biotics, the immediate concern of the author, but on the whole of electronics. As electronic and computer circuits reach complexity levels which are beyond human capacities to comprehend, the Darwin Machine may end up being the only effective circuit design tool.

FIG. 1 shows one suggestion for a Darwin Machine architecture. It contains a single master circuit and a population of slave circuits which function in parallel. The master circuit sends each slave a "GA chromosome" (i.e. a bit string instruction which is used to configure a hardware circuit), the population fitness definition (i.e. a second bit string instruction which is used to configure the hardware circuit which measures the fitness of the outputs from the first circuit), and initial input values. Each slave takes its chromosome, configures a circuit according to the instructions contained on the chromosome, measures the circuit's fitness according to the fitness definition, and reports this value back to the master, which then calculates the next generation of chromosomes in a GA like way, to complete the cycle.

Each slave therefore consists essentially of two circuits, one which evolves, and the other which does not, but is used simply to measure the fitness of the circuit which does evolve. However, this raises a new problem. The whole point of a Darwin Machine, besides accelerating the evolution time, is to enable the evolution of complex circuits, in typical GP fashion.

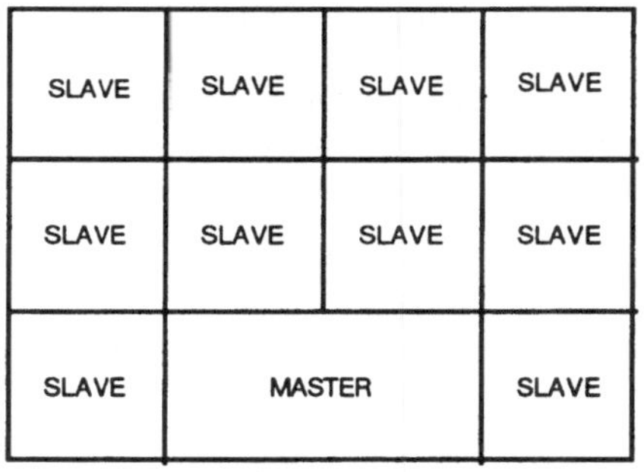

FIG. 1 A Darwin Machine

That is, the circuits which evolve, may be too complex for human comprehension. However, the fitness measuring circuit must (at some level) be humanly comprehensible, because it has to be humanly specifiable. At some point in the whole procedure, a human genetic programmer must specify the fitness definition, and provide the corresponding bitstring instruction which configures the fitness measuring circuit.

One can imagine however that the fitness measuring circuit is itself evolved using a simpler fitness definition. Thus a whole chain of fitness circuit evolutions becomes possible. At the end of the chain is a humanly comprehensible fitness definition and corresponding configuration instruction.

Once the above problem is solved, there remains another problem, and that is how does one evolve motions of real world biots? For example, imagine that one wishes to evolve the circuits of a neural controller which sends instructions to the legs of a real world (physical) biot to make it walk. Measuring the fitness of such a creature takes a long time, namely minutes, rather than milliseconds as in a computer simulation. Thus evolving such a controller may be a painfully slow process. The size of the chromosome population would need to be very small, and techniques would need to be devised to shorten the mechanical fitness measuring time.

One way round this problem might be to simulate as closely as possible the real world biot's mechanics and evolve a "ball park" set of chromosomes for a particular motion. These chromosomes could then be ported to the real world biot (one at a time to a single biot, or in parallel to a set of identical biots) for some fitness polishing. If the simulation is reasonably "real world accurate" then this should accelerate the evolution.

Using these techniques, one can probably build up a library of motion GenNets, which could be stored in ROMs and linked to a common bus to the effectors (legs, arms etc). The real challenge in building biots, is evolving the middle decision layer. In the human brain, only a small minority of neurons are concerned with detector input and motion output. Most neurons are concerned with internal chores, such as building associations between multisensor inputs, or for storing memories, or for constructing speech, etc etc. The evolution of these internal circuits would be ideal for Darwin Machines, because it would not require the real world mechanical speeds to measure the fitnesses of motions. One could then evolve increasingly complex neural circuits to perform increasingly complex functions, but at computer hardware speeds.

3. FPGAs, HDPLDs, and Hardware Accelerators

This section presents the above ideas in a little more detail, although not a lot. This paper is not a presentation of completed work. Research projects based on the evolvable hardware and Darwin Machine concepts are presently underway at the author's lab, (ATR in Japan).

One of the aims of this paper is to persuade more GA, ALife, and Neural Network people to get into electronics, with the intention of filling in the details, which at the present time are lacking in this paper. It is hoped that the basic notion of "evolvable hardware", or "software configurable hardware", and hence the possibility of building genuine Darwin Machines within a year or so, will be sufficiently inspiring to motivate readers to start talking with colleagues in electronics about how to build Darwin Machines in itty-gritty detail.

There appear to be two main approaches to Darwin Machine design. One can either evolve the hardware directly, using technologies which underly such devices as FPGAs or HDPLDs (whose characteristics will be described shortly), or one can use special hardware accelerators, which are special hardware boxes designed to accelerate the software simulation of hardware designs containing up to several hundred thousand logic gates. To gather information on FPGAs, one might try contacting Xilinx Corporation (San Jose, CA). For HDPLDs (High Density Programmable Logic Devices), try Lattice Corporation (Hillsboro, Oregon), and for hardware accelerators, try Zycad Corporation (Menlo Park, CA). For a general overview of such devices, see [Broesch 1991].

If one uses a hardware accelerator, it might be possible to evolve at a higher level of abstraction than at individual gate level, as would be the case with FPGAs and HDPLDs.

There are arguments both ways. FPGA and HDPLD chips are rather cheap now, around several tens of dollars, whereas a hardware accelerator costs tens of thousands of dollars. What one might gain in terms of flexibility of evolutionary level with a hardware accelerator, might be more than offset by its price, compared to FPGA or HDPLD gate level evolution.

We present now a more detailed description of one of the software configurable hardware devices, namely FPGAs (Field Programmable Gate Arrays). A similar story could also be given for HDPLDs (High Density Programmable Logic Devices), but will be omitted for reasons of space. This description of FPGAs is provided to give the reader unfamiliar with such devices, a feel for what software configurable hardware can do. As will be shown in the next section, these devices in their present form, are probably unsuitable to serve as a basis for evolvable hardware.

FIG. 2 shows a fairly typical FPGA chip architecture. It consists of three basic element types :- CLBs (Configurable Logic Blocks), IOBs (Input Output Blocks), and Interconnects. Quoting from Xilinx's technical literature [Xilinx 1991], "Like a microprocessor, the ... device is a program-driven logic device. The functions of the ... configurable logic blocks and I/O blocks, and their interconnections, are controlled by a configuration program stored in an on-chip memory. The configuration program is loaded automatically from an external memory on power-up or on command, or is programmed by a microprocessor as a part of system initialization". "Since ... FPGAs can be re-programmed an unlimited number of times, they can be used in innovative designs where hardware is changed dynamically ...". The CLBs (Configurable Logic Blocks) consist essentially of "programmable combinatorial logic and storage registers. The combinatorial logic section of the block is capable of implementing any Boolean function of its input variables ... " (usually between 4 and 9 input lines). The I/O Blocks can be programmed to connect CLBs to the input/output pins of the FPGA chip. The interconnects (which run vertically and horizontally between the columns and rows of the CLBs) are used to connect the CLBs to other CLBs or to IOBs.

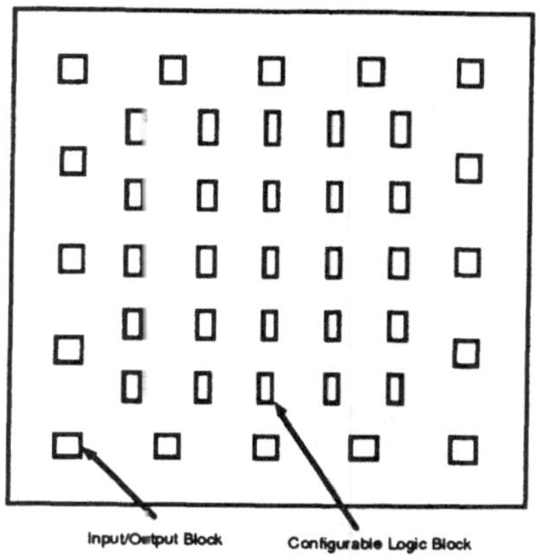

FIG. 2 FPGA ARCHITECTURE

The number of CLBs in a (Xilinx) FPGA ranges from 64 up to 900. The number of IOBs ranges from 58 to 240. The number of equivalent gates ranges from 1200 to 20,000. In the Xilinx 4000 series, the internal RAM configuration memory can range in length from 2000 to 28,800 bits. (Note, it is this configuration memory, which may serve as the chromosome for the circuit evolution).

In the case of a hardware accelerator, its input is usually in the form of VHDL (i.e. VHSIC (i.e. very high speed integrated circuit) Hardware Description Language). The accelerator then uses this description to simulate the hardware device so described.

4. Evolvable Chip Research

The section of the lab in which the author works, recently purchased a software configurable hardware development system to help explore the concept of evolvable hardware, and for other purposes. For those interested, it was a Lattice Corporation ispLSI 1032 system. The author is now convinced that there are too many problems with such systems for them to be immediately suitable as a means to evolve hardware. The main obstacle is speed. For the evolvable hardware concept to be useful, it needs to satisfy at least two requirements. One is that the circuits evolved be both functional and too complex for human understanding, otherwise they could be humanly designed in the traditional way. The other requirement is that the circuits be evolved quickly, i.e. a lot more quickly than simulating their evolution in software using a general purpose computer, otherwise it would be easier and cheaper just to use computer simulation.

The designers of FPGAs and HDPLDs did not have the concept of evolvable hardware in mind when they conceived their products. What they did have in mind was the idea that a generic hardware device could be wired up by software, to meet the requirements of individual users, and that the configuration could be changed easily and quickly (e.g. in less than an hour) by those users, when design modifications in their hardware became necessary. These generic devices became very popular amongst electronic engineers. They enabled the number of chips on a board to be reduced, and they increased the speed with which new hardware designs could be built and tested.

However, when one begins thinking of using such devices as a means to evolve hardware, there are several problems. Consider the following :-

a) The configuration bitstring (i.e. the software instruction which is used to configure or wire up the generic device) is inputted serially, i.e. one bit at a time, for thousands of bits. Even if one "mutates" only one bit, the whole bitstring has to be re-inputted (serially). This "downloading" typically takes about 30 seconds.

b) The routing, i.e. the choice of connections between the logic blocks and the I/O blocks is done by software and usually takes several minutes, depending upon the complexity of the circuit.

c) Generating the fuse map (i.e. mapping all the routes between logic and I/O blocks to "fusible" gates, to generate the so-called "JEDEC file") takes about a minute.

d) Most of the details as to how the circuit is wired up are company secrets, so one would be unable to know which bits in the bit string corresponded to which fusible gates.

But still, this process is far quicker than the usual time taken to change a hardware design (i.e. days to weeks), but for hardware evolution, it is unacceptable. It would be quicker to simulate in software. But, the technologies underlying these software configurable hardware devices can be used to design special evolvable chips, which could be used in special architectures to build Darwin Machines.

What is needed is an "evolvable chip" which can be sent a configuring bit string (partially) in parallel, e.g. 64 bits at a time. The chip would store each 64 bit "slice" of the configuring bit string (which may be thousands of bits long) into its "chromosome register". This (partially) parallel downloading need only be done once. The downloaded chromosome then remains in the register, to be subjected to mutation. To mutate one bit of the chromosome, the chip could be sent a "mutation address" (e.g. 16 bits), which causes a single bit in the chromosome to flip. Occasionally, portions of chromosomes could be swapped between two devices.

Further work then needs to be undertaken to design a whole system of such evolvable chips. A lot of work recently has been devoted to the parallelization of Genetic Algorithms [e.g see papers in GA 1989, GA 1991, PPSN91]. The ideas contained in these papers will very probably be needed (perhaps in modified form) to build parallelized evolvable chip systems. For example, how frequently should one crossover chromosomes (a time costly operation), or should one rely largely on mutation etc. There are many possible parallel GA algorithms.

However, the real message of this section is that the technologies to build such evolvable chips exist today, e.g. EEPROM and SRAM. What is needed is a research effort to design evolvable chips and systems, using these existing technologies. The author would like to make an appeal that such research be initiated in many research labs around the world.

5. The Remaining Challenges

Many challenges remain, if one wishes to use Darwin Machines to evolve electronic (neural) circuits and artificial nervous systems, e.g.

a) How are evolvable chips systems to be configured so as to behave like neurons?

b) How are fitness circuits to be designed?

c) How are GenNet modules to be combined into higher level circuits?

Research work is also currently underway to software configure analog (as distinct from digital) hardware [e.g. Ning et al 1991]. This work would obviously be useful for the evolution of the author's (analog) GenNets. In time, if the concept of the Darwin Machine proves useful, software configurable hardware devices might be designed with Darwin Machine applications specifically in mind. The author's lab (ATR in Japan) is currently working on setting up a Darwin Machine research project. We hope to be the first in the world to simulate and build a true Darwin Machine. Results will be reported in later papers.

To close this paper, it might be interesting to mention the link between Darwin Machines and nanotechnology. Nanotechnology [Carter et al 1988, Schneiker 1989, Drexler 1992] is molecular scale engineering. The intention is to build nano scale robots (nanots) which pick up atoms here and put them there to build any substance, including copies of the same nanots. Our lab is heavily involved in nanotech, as it is considered a critical technology that will dominate 21st century economics. However, building such "Avogadro Machines" (with a trillion trillion components) with demand self assembly techniques to be practical. But the complexities will also be enormous. Therefore it will be virtually impossible to predict function from structure with these self assembled Avogadro Machines. Therefore GP techniques will be needed to build/evolve them, and hence the need for molecular scale Darwin Machines. Molecular scale self assembling manufacture has been called "embryofacture" by the author. It looks as though the Darwin Machine concept will have a critically important future.

References :

[ALife III] *"Proceedings of the 3rd International Artificial Life Conference"*, Santa Fe, New Mexico, 15-19 June, 1992. To appear.

[Beer 1990] *"Intelligence as Adaptive Behavior : An Experiment in Computational Neuroethology"*, Beer R.D., Academic Press, 1990.

[Beer & Gallagher 1991] *"Evolving Dynamical Neural Networks for Adapative Behavior"*, Beer R.D. & Gallagher J.C., Technical Report 17, 1991, Dept. of Computer & Engineering Science, Case Western Reserve University, Cleveland, Ohio.

[Broesch 1991] *"Practical Programmable Circuits : A Guide to PLDs, State Machines, and Microcontrollers"*, J.D. Broesch, Academic Press, 1991.

[Brooks 1990] *"Elephants Don't Play Chess"*, Brooks R.A., in Designing Autonomous Agents : Theory and Practice from Biology to Engineering and Back, ed. Maes P., MIT Press, 1990.

[Brooks 1992] *"Artificial Life and Real Robots"*, Brooks R.A., in "Toward a Practice of Autonomous Systems : Proceedings of the First European Conference on Artificial Life", eds. F.J. Varela and P. Bourgine, MIT Press, 1992.

[Carter et al 1988] *"Molecular Electronic Devices"*, F.L. Carter, R.E. Siatkowski, H. Wohltjen eds. North Holland, 1988.

[de Garis 1990a] *"Genetic Programming : Modular Evolution for Darwin Machines"*, Hugo de Garis, IJCNN-90-WASH-DC, (Int. Joint Conf. on Neural Networks), January 1990, Washington DC, USA.

[de Garis 1990b] *"Genetic Programming : Building Nanobrains with Genetically Programmed Neural Network Modules"*, Hugo de Garis, IJCNN-90 SANDIEGO (Int.Joint Conf. on Neural Networks), June 1990, San Diego, California, USA.

[de Garis 1990c] *"Genetic Programming : Building Artificial Nervous Systems Using Genetically Programmed Neural Network*

Modules", Hugo de Garis, in Porter B.W. & Mooney R.J. eds., Proc. 7th. Int. Conf. on Machine Learning, pp 132-139, Morgan Kaufmann, 1990.

[de Garis 1990d] "*Genetic Programming : Evolution of a Time Dependent Neural Network Module Which Teaches a Pair of Stick Legs to Walk*", Hugo de Garis, ECAI-90, (9th. European Conf. on Artificial Intelligence), August 1990, Stockholm, Sweden.

[de Garis 1991a] "*Genetic Programming : Artificial Nervous Systems, Artificial Embryos and Embryological Electronics*", Hugo de Garis, in "Parallel Problem Solving from Nature", Lecture Notes in Computer Science 496, Springer Verlag, 1991.

[de Garis 1991b] "*LIZZY : The Genetic Programming of an Artificial Nervous System*", Hugo de Garis, ICANN91, Int.Conf. on Artificial Neural Networks, June 1991, Espoo, Finland.

[de Garis 1991c] "*GenNETS : Genetically Programmed Neural Nets : Using the Genetic Algorithm to Train Neural Nets Whose Inputs and/or Outputs Vary in Time*", Hugo de Garis, IJCNN91 Singapore, Int.Joint Conf.on Neural Networks, November 1991, Singapore.

[de Garis 1991d] "*Genetic Programming*", Hugo de Garis, Ch. 8 in book "Neural and Intelligent Systems Integration", ed. Branko Soucek, WILEY, 1991.

[de Garis 1992a] "*Steerable GenNETS : The Genetic Programming of Controllable Behaviors in GenNets*", Hugo de Garis, ECAL91 Paris, Proceedings of the 1st. European Conference on Artificial Life, MIT Press.

[de Garis 1992b] "*Artificial Embryology : The Genetic Programming of an Artificial Embryo*", Hugo de Garis, Ch. 14 in book "Dynamic, Genetic, and Chaotic Programming", ed. Branko Soucek and the IRIS Group, WILEY, 1992.

[de Garis 1992c] "*Genetic Programming : Evolutionary Approaches to Multistrategy Learning*", Hugo de Garis, chapter 21 in "Machine Learning : A Multistrategy Approach, Vol. 4", R.S. Michalski & G.Tecuci (Eds.), Morgan Kauffman, 1992.

[de Garis 1993] "*Genetic Programming : GenNets, Artificial Nervous Systems, Artificial Embryos*", Hugo de Garis, WILEY manuscript.

[Drexler 1992] "*Nanosystems : Molecular Machinery, Manufacturing and Computation*", Drexler K.E., Wiley, 1992.

[Ecal91] "*Toward a Practice of Autonomous Systems : Proceedings of the First European Conference on Artificial Life*", eds. F.J. Varela and P. Bourgine, MIT Press, 1992.

[GA 1989] "*Proceedings of the Third International Conference on Genetic Algorithms*", J.D. Schaffer ed., Morgan Kaufmann, 1989.

[GA 1991] "*Proceedings of the Fourth International Conference on Genetic Algorithms*", R.K. Belew and L.B. Booker eds., Morgan Kaufmann, 1991.

[Goldberg 1989] "*Genetic Algorithms in Search, Optimization, and Machine Learning*", D.E. Goldberg, Addison-Wesley, 1989.

[Higuchi et al 1993] "*Evolving Hardware with Genetic Learning : A First Step Towards Building a Darwin Machine*", T. Higuchi, T. Niwa, T. Tanaka, H. Iba, H. de Garis, T. Furuya, in "Proceedings of the 2nd Int. Conf. on the Simulation of Adaptive Behavior (SAB92), MIT Press, 1993.

[Koza 1990] "*Genetic Programming : A Paradigm for Genetically Breeding Populations of Computer Programs to Solve Problems*", Koza J. R., Stanford University Comp. Sci. Dept. Technical Report, STAN-CS-90-1314, June 1990.

[Koza 1992a] "*Genetic Programming Paradigm : Genetically Breeding Populations of Computer Programs to Solve Problems*", Koza J. R., Ch. 10 in "Dynamic, Genetic and Chaotic Programming", ed. Branko Soucek and the IRIS Group, WILEY 1992.

[Koza 1992b] "*Genetic Programming" : On the Programming of Computers by Means of Natural Selection*", Koza J. R., MIT Press.

[Lattice 1992] "*pLSI and ispLSI Data Book and Handbook*", Lattice Corporation, Hillsboro, Oregon, 1992.

[Lattice 1992] "*GAL Data Book*", Lattice Corporation, Hillsboro, Oregon, 1992.

[Ning et al 1991] "*SEAS : A Simulated Evolution Approach for Analog Circuit Synthesis*", Zhen-Qiu Ning, Ton Mouthaan, and Hans Wallinga, in Proceedings of CICC, IEEE, 1991.

[PPSN91] "*Parallel Problem Solving from*

Nature", Lecture Notes in Computer Science, No. 496, Schwefel H.-P. and Manner R. eds., Springer Verlag, 1991.

[Schneiker 1989] "*Nano Technology with Feynman Machines : Scanning Tunneling Engineering and Artificial Life*", Schneiker C., in "Artificial Life", Langton C.G. ed., Addison Wesley, 1989.

[Xilinx 1991] "*The Programmable Gate Array Data Book*", Xilinx Corporation, San Jose, CA, 1991.

A Fast Genetic Algorithm with Sharing Scheme Using Cluster Analysis Methods in Multimodal Function Optimization

Xiaodong YIN† and Noël GERMAY
Laboratoire d'Electrotechnique et d'Instrumentation
Catholic University of Louvain, Place du Levant, 3
1348 Louvain-la-Neuve (Belgium)
†E-mail : yin@fort.ucl.ac.be

Abstract

Genetic algorithms with sharing are well known for tackling multimodal function optimization problems. In this paper, a sharing scheme using a clustering methodology is introduced and compared with the classical sharing scheme. It is shown from the simulation on test functions and on a practical problem that the proposed scheme proceeds faster than the classical scheme with a performance remaining as good as the classical one. In addition, the proposed scheme reveals unknown multimodal function structure when *a priori* knowledge about the function is poor. Finally, introduction of a mating restriction inside the proposed scheme is investigated and shown to increase the optimization quality without requiring additional computation efforts.

1. Introduction

Genetic algorithms (GAs) have been successfully applied in various domains of search and optimization. They are highly effective and robust over a broad spectrum of problems. The multimodal function optimization problem was tackled through the use of some methods inspired from natural notions of niche and species. Cavicchio[1] introduced a *preselection* scheme in which a child replaces the worse parent if the child's fitness is higher than that of the worse parent. De Jong[2] suggested a similar replacement scheme (*crowding* scheme) in which the new individual is compared not to his parent, but to the nearest individual in the population on the basis of their string similarity. These two schemes were reviewed in [3] and were shown not quite effective for maintaining diversity due to large replacement errors, i.e. replacement of a member of some niche by a member of other one. Goldberg and Richardson[4] suggested a *sharing* scheme where idea is to enforce the individuals of the population to share available resources by dividing the population into different subpopulations according to the similarity of individuals. To implement the sharing scheme, a sharing function $\mathrm{Sh}(d_{ij})$ as a function of the distance d_{ij} between two individuals i and j is defined as : if d_{ij} is smaller than a threshold value σ_{share}, then $\mathrm{Sh}(d_{ij})=1-(d_{ij}/\sigma_{share})^{\alpha}$; otherwise, $\mathrm{Sh}(d_{ij})=0$. Where σ_{share} is the maximum distance between strings required to form as many niches as there are peaks in the solution space. The niche count m_i for individual i is calculated by summing sharing function values contributed by all N individuals of the population : $m_i=\sum_{j=1}^{N} sh(d_{ij})$. The original fitness f_i is derated due to the presence of other strings in the same niche and the shared fitness f_i' is given as : $f_i'=f_i \ / \ m_i$. This permits the formation of stable subpopulations around the different optima in the multimodal space. The sharing scheme was shown in [5] be better able to preserve diversity than the crowding scheme and has been successfully applied to solve a variety of multimodal functions[5,6]. However, since the computation of the niche count m_i is of complexity $O(N^2)$, this sharing scheme is computationally expensive. In addition, the estimation of the parameter σ_{share} requires prior knowledge about the number of peaks in solution space. In a situation where the number of peaks is *a priori* unknown, the sharing method will face with the difficulty of choosing an adequate value of σ_{share}. It would then be appealing to call for some methods to alleviate the process of calculation and to make the sharing method more performing.

Cluster analysis methods which will be presented here are hoped to be suitable for this purpose. Their objective is to group either the data units or the variables into clusters such that the elements within a cluster have a high degree of "natural association" among themselves while the clusters are "relatively distinct" from one another [7]. They have been applied to various domains such as biology, botany, psychiatry, pattern recognition, artificial intelligence, civil engineering, etc. To link with genetic algorithms, the *data units* in cluster analysis methods correspond to the *individuals* in GAs and the *clusters* are analogous to the *niches* in GAs. Many methods are available for clustering data units into different clusters. Some methods can not only assign the data units to a given known structure but also reveal *a priori* unknown structure which actually exist among them. In the remainder of this paper, a brief review of the cluster analysis methods is made, then a clustering algorithm is proposed and implemented in GA with sharing. The proposed sharing scheme is compared with the classical sharing scheme through a test on two multimodal functions and an application to a concrete electrical problem. Finally, the implementation of mating restriction developed in [5] is also tried by using the proposed sharing scheme.

2. Review of Cluster Analysis Methods

Cluster analysis methods are normally categorized into two families of algorithms : *hierarchical algorithms* which aim to build a tree from branches toward the root in each step by fusioning the two nearest classes; and *nonhierarchical algorithms* whose objective is to classify the data units into a single partition of k clusters, where k is the number of clusters which is either specified *a priori* or determined by the algorithm. Here, the number of clusters is ideally associated with the number of peaks in multimodal function. Since our problem is to search a partition of individuals in the population and the number of individuals in each cluster but not a hierarchy of all individuals, the attention is then specially focused on the nonhierarchical algorithms which are generally more economical than the hierarchical algorithms.

We present here one kind of the nonhierarchical algorithms called centroid methods. Generally speaking, centroid methods consist of two processes :

1.Process of representation : a cluster C_c is represented by its centroid $G(C_c)$:

$$G(C_c) = (\bar{x}_{c1}, \bar{x}_{c2}, \cdots, \bar{x}_{cj}, \cdots, \bar{x}_{cp}) \quad c = 1 \cdots k \quad (1)$$

$$\text{with } \bar{x}_{cj} = \frac{1}{n_c} \sum_{i=1}^{n_c} x_{ij}$$

where p : number of variables in a data unit
k : number of clusters
n_c : number of data units in cth cluster
$\bar{x}_{cj}$: mean on the jth variable in the cth cluster
x_{ij} : value on the jth variable for the ith data unit
x_i : vector for the ith data unit

2. Process of assignment : each data unit is assigned to the cluster with the nearest centroid :

$$x_i \in C_c, \quad \text{if } d(x_i, G(C_c)) = \min_{l=1,k} d(x_i, G(C_l))$$

The centroid methods alternate iteratively these two processes until they converge to a stable configuration. For starting the process, an initialisation process is needed in order to get a first set of centroids. Two representative centroid methods will be introduced in the following sections [7] : an algorithm with k fixed called *MacQueen's KMEAN algorithm* and another algorithm with variable k named *Adaptive MacQueen's KMEAN algorithm*.

2.1. MacQueen's KMEAN Algorithm

MacQueen's KMEAN algorithm proceeds to sort N data units into k clusters in the following steps :

1. Take the first k data units in the data set as clusters of one member each.
2. Assign each of the remaining N-k data units to the cluster with the nearest centroid. After each assignment, recompute the centroid of the gaining clusters.
3. After all data units have been assigned, take the existing cluster centroids as fixed seed points and make one more passing through the data set assigning each data unit to the nearest seed point.

The algorithm minimizes implicitly the partition error E, i.e. the total within group error sum of squares :

$$E = \sum_{c=1}^{k} \sum_{i=1}^{n_c} \sum_{j=1}^{p} (x_{ij} - \bar{x}_{cj})^2 \quad (2)$$

where $\sum_{j=1}^{p} (x_{ij} - \bar{x}_{cj})^2$ is the squared Euclidean distance between the centroid in the cth cluster and the ith data unit in that cluster. A local convergence is then attain rapidly in just one iteration. In the algorithm, k*(2N-k) squared distance calculation and (k-1)*(2N-k) squared distance comparison are carried out. The algorithm is of complexity $O(2Nk)$.

The MacQueen's KMEAN algorithm with k fixed *a priori*, calls for the user's prior knowledge for the fixation of the number of clusters. However, in most multimodal function optimization problems, the number of peaks in the solution space is often unknown *a priori*. Furthermore, in the process of GA with sharing, stable subpopulations are not formed at the begin of generations, they are formed after several generations through competition and sharing within and across niches. If k is fixed during all the process, it arrives that the algorithm is forced to always produce k clusters which may not correspond the real structure of data. A variation of the MacQueen's KMEAN algorithm presented in the next section does not fix k *a priori*. It can find a suitable number of clusters which fits the real data structure.

2.2. Adaptive MacQueen's KMEAN Algorithm

Adaptive MacQueen's KMEAN algorithm allows the number of clusters to vary during the initial assignment of the data units. On the other hand, it needs two supplementary parameters : minimal distance (coarsening parameter) d_{min} and maximal distance (refining parameter) d_{max}. The algorithm proceeds in the following steps :

1. Choose an initial value of k, take the first k data units as initial clusters of one member each. Compute all pairwise distance among these first k data units. If the smallest distance is less than d_{min}, then merge the two associated clusters and compute the distance between the centroid of the new cluster and all remaining clusters. Continue merging nearest clusters as necessary until all centroids are separated by a distance at least as large as d_{min}.
2. Assign each of the remaining N-k data units to the cluster with the nearest centroid. After each assignment, update the centroid of the gaining cluster and compute the distance to the centroids of the other clusters. Merge the new cluster with the cluster having the nearest centroid if the distance between centroids is less than d_{min} and continue merging as necessary until all centroids are at least d_{min} distant apart. If the distance to the nearest centroid is greater than d_{max}, then take the data unit as a new cluster of one member.
3. After data units $k+1$ through N have been assigned, take the existing cluster centroids as fixed seed points and reallocate each data unit to its nearest seed points.

Like the MacQueen's KMEAN algorithm with k fixed, the algorithm stops after the first reallocation without attempting to realize further convergence. The computation time is then very economical. In addition, by fusioning the clusters with nearby centroids, the algorithm avoids creating fine distinctions which artificially divide natural clusters. By creating new clusters when a data unit is distant from all existing centroids, the algorithm permits the existence of outliers. The choice of the clustering parameters d_{min} and d_{max} requires some experience. Normally, d_{max} is taken quite large and d_{min} quite small. Since on the one hand, GA works in conjugation with exploitation and exploration and on the other hand, the clustering algorithm is not expensive, GA with proposed sharing can be repeated several times with different values of d_{min} and d_{max}, in an effort to reveal the intrinsic structure of the multimodal function.

3. Implementation of Clustering Algorithm in GA with Sharing

After the reproduction and the crossover processes in each generation, the clustering algorithm is applied to cluster all the individuals in population into different niches. The number of niches and the number of individuals in each niche are determined. Then, for individual i, its potential fitness f_i is divided by the approximated number of individuals in the niche to which it belongs m_i':

$$f_i' = \frac{f_i}{m_i'} \tag{3}$$

$$\text{with } m_i' = n_c - n_c * (d_{ic} / 2d_{max})^{\alpha} \quad x_i \in C_c \tag{4}$$

where α is a constant, d_{ic} is the distance between individual i and its niche's centroid $G(C_c)$.

Apparently, the proposed sharing scheme seems to be coarse with respect to the classical sharing scheme. In fact, the classical niche count formula can be rewritten as :

$$m_i = \sum_{x_j \in C_c} (1 - (d_{ij} / \sigma_{sh})^{\alpha}) = n_c - \sum_{x_j \in C_c} (d_{ij} / \sigma_{sh})^{\alpha} \quad x_i \in C_c$$

The first term of above expression is normally preponderant with respect to the second term. Difference between two schemes is then not significant.

It is known that in the initialization process of clustering algorithms, a good choice of seed points and reordering of population often improve the efficiency of algorithm. In GA, this can be realized by using the information of fitness in each individual. In our work, before the application of clustering, the population is firstly sorted in descending order according to the individual's fitness height. By this reordering, the individuals having the highest fitness are put in front of the population and taken as the initial seed points.

4. Testing Results

Both the MacQueen's KMEAN algorithm and the adaptive MacQueen's KMEAN algorithm have been implemented in simple genetic algorithm with stochastic remainder selection[8]. We present here the results of the sharing scheme using the adaptive MacQueen's KMEAN algorithm. The Euclidean distance in the phenotypic space is chosen as distance - metric. The program is run on a IBM PC machine with INTEL 386, 33 MHz processor. For comparing the performances of classical sharing and proposed sharing, the tests have been carried out on many functions. In this paper, two multimodal functions used in [5] and the load flow function of an electrical power system in [9] are taken again.

4.1. Two Multimodal Functions

The first function F1 has 5 peaks of equal height in the interval $0 \leq x \leq 1$:

$$F1 \;:\; f_1(x) = \sin^6(5\pi x)$$

The second function F2 has 5 peaks of nonequal height in the interval $0 \leq x \leq 1$:

$$F2 : f_2(x) = \exp^{-2\ln 2\left(\frac{x-0.1}{0.8}\right)^2} \sin^6(5\pi x)$$

Besides the fact that on-line performance and off-line performance measures are used for judging the performance of different sharing schemes, chi-squared like distribution error measure developed in [5] is also used for comparing the actual distribution of individuals in generations to an ideal distribution. In the measure, we consider that individuals having fitness values higher than ϵ = 80 % of the representative peak fitness values are associated with that peak. Five runs are made with different initial populations generated at random for each of two sharing schemes, and an average statistics is calculated for the performance measures. For a fair comparison of the two schemes, the same initial populations and genetic parameters are used. The parameter x is coded as a 30-bit string. GA is run with population size N=100, crossover rate P_c=0.9 and mutation rate P_m=0.0. The sharing parameters are α=1 and phenotypic σ_{share} = 0.1, the following clustering parameters are selected :

initial number of clusters k = 10
minimal distance d_{min} = 0.05
maximal distance d_{max} = 0.1

Scheme	Function F1	Function F2
Classical	14 m. 54 s.	15 m. 14 s.
Proposed	01 m. 54 s.	02 m. 06 s.

Table 1 Comparison of the average computation times by the two schemes on functions F1 and F2

Table 1 lists the average computation time of 5 runs using the two schemes for 200 generations on functions F1 and F2 respectively. It is noted that the proposed scheme proceeds 7 - 8 times faster than the classical one. The on-line performance measure on function F1 after 200 generations with the classical scheme and the proposed scheme is 0.864 and 0.883 respectively. Figures 1 and 2 show the chi-square like deviations with the two schemes in generations on functions F1 and F2. The distribution errors with the two schemes are comparable and are both very small what means that the proposed scheme is also capable to maintain the distribution of individuals quasi - proportional to the fitness heights of each peak.

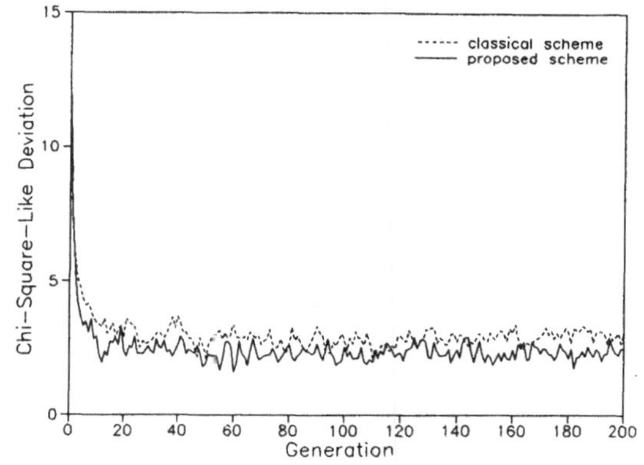

Figure 1 Comparison of the chi-square-like deviation of the two schemes on the function F1

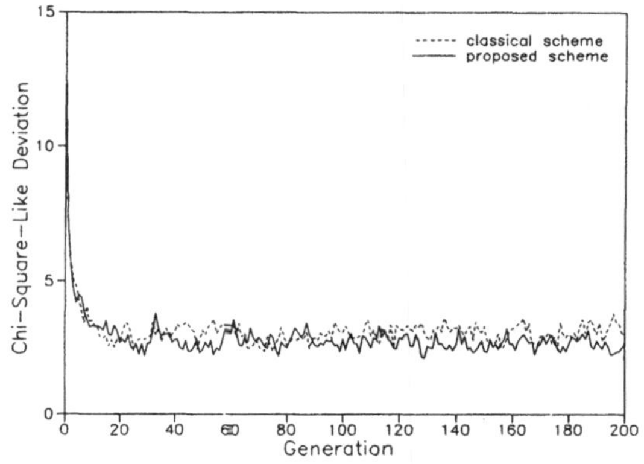

Figure 2 Comparison of the chi-square-like deviation of the two schemes on the function F2

All individuals in the population at the 200th generation of a run with the proposed scheme can be visualised in Figure 3 for function F1 and in Figure 4 for function F2, and their corresponding clusters and centroids are listed in Tables 2 and 3 respectively. It is

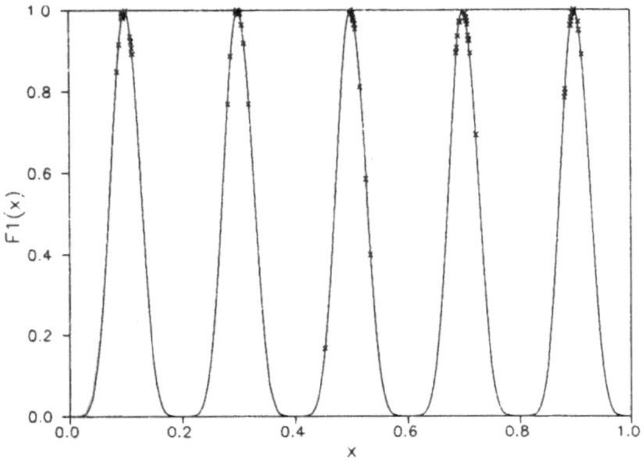

Figure 3 Situation of population at generation 200 with the proposed scheme on the function F1

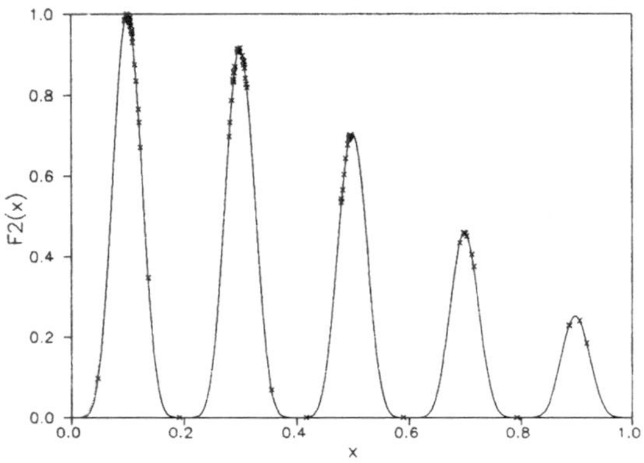

Figure 4 Situation of population at generation 200 with the proposed scheme on the function F2

shown from Tables 2 and 3 that the number of individuals of each main peak cluster is very close to the ideal number of individuals distributed in each peak, and their centroids approximately correspond to the peak locations (0.1,0.3,0.5,0.7,0.9). The 6th cluster in Table 3 contains only one individual and its centroid is at least 0.1 (d_{max}) distant from the main clusters' centroids. Such "lethal strings" which do not belong to any niches are detected by the algorithm and are formed as new clusters.

Figures 5 and 6 plot the number of clusters in generations of the run with the proposed scheme on functions F1 and F2 respectively. In each generation, the clustering algorithm starts with the initial number of clusters, and finishes by finding a final number of clusters suitable for the situation of population in generation. Except one or two outlier cluster(s) in certain generations, the number of clusters is found to be 5 which fits the real structure of function.

Cluster	n_c	Centroid
1	20	0.0984
2	18	0.3006
3	23	0.5040
4	19	0.7031
5	20	0.8973

Table 2 Clustering result with the proposed scheme at generation 200 on the function F1

Cluster	n_c	Centroid
1	32	0.1069
2	25	0.2996
3	23	0.4882
4	14	0.7068
5	5	0.8966
6	1	0.5889

Table 3 Clustering result with the proposed scheme at generation 200 on the function F2

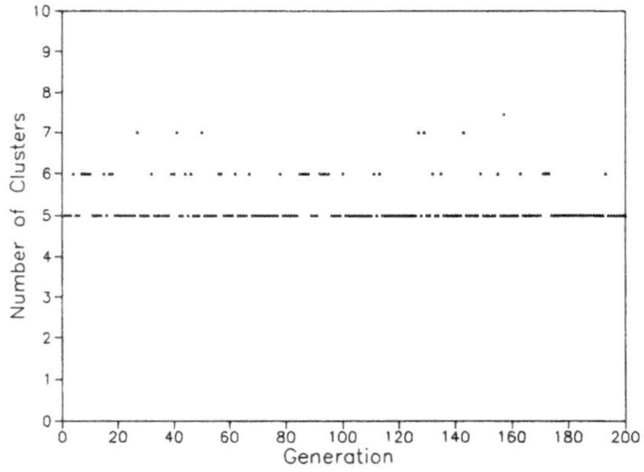

Figure 5 Variation of number of clusters with the proposed scheme on F1

It is obvious that the obtained result depends on the

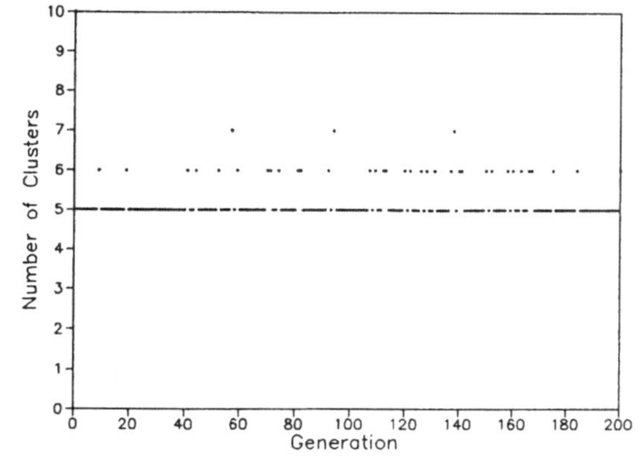

Figure 6 Variation of number of clusters with the proposed scheme on F2

choice of minimal distance d_{min} and maximal distance d_{max}. They play an important role in the processing of clustering. However, the tolerance of parameters is relatively large. It was found by empirical runs that the proposed scheme remains efficient if d_{min} is chosen between 0.02 - 0.1 and d_{max} between 0.06 - 0.18 with d_{max} 2 - 3 times greater than d_{min}. The initial number of cluster can be chosen at random.

We have attempted to apply the two sharing schemes to a more complicated function. Figure 7 shows the result obtained with the proposed scheme for such a function. The algorithm is able to detect all 11 peaks but a correct choice of parameters d_{min} and d_{max} is more delicate. If they are too small, some peaks especially the flat or dissymmetric ones carry more than one cluster so that the total number of niches is abounding. On the contrary, if d_{min} and d_{max} are too large, the number of clusters becomes too small and some very fine peaks are lost.

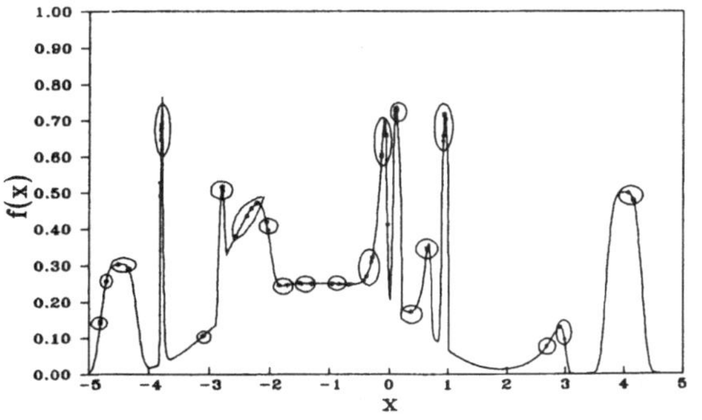

Figure 7 Distribution of 100 points at generation 100 by using the proposed scheme with d_{min} = 0.1 and d_{max} = 0.2. The obtained 20 clusters are circled.

4.2. Load Flow Function

In [9], the classical sharing scheme was applied to solve a practical multimodal function optimization problem - finding simultaneously multiple load flow solutions in 3-bus electrical power system. The detailed function formulation and the system data are referred in [9]. There are 4 variables named E_2, E_3, F_2 and F_3 and each of them was coded by a 9-bit substring in the interval of [-1.5, 1.5]. GA was run with N=200, P_c=0.9 and P_m=0.01. Phenotypic σ_{share}=0.25 was chosen. The computation time was about 39 minutes for 100 generations and 4 load flow solutions were found simultaneously by GA with sharing function. In the application of the proposed scheme to the same problem, the same genetic parameters are used, and the following clustering parameters are selected :

initial number of clusters k	= 20
minimal distance d_{min}	= 0.25
maximal distance d_{max}	= 0.50

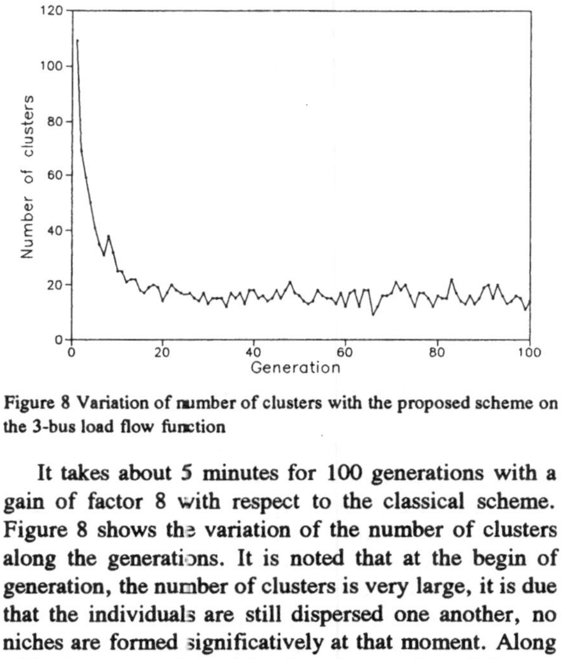

Figure 8 Variation of number of clusters with the proposed scheme on the 3-bus load flow function

It takes about 5 minutes for 100 generations with a gain of factor 8 with respect to the classical scheme. Figure 8 shows the variation of the number of clusters along the generations. It is noted that at the begin of generation, the number of clusters is very large, it is due that the individuals are still dispersed one another, no niches are formed significatively at that moment. Along with the intervention of sharing, the number of clusters diminishes very rapidly, then become stable. Table 4 lists the obtained clusters, their memberships and the 4 variables' centroids. We can easily distinguish that there are four clusters having many individuals (marked with *) which represents the solution clusters. The clustering algorithm finds thus by itself four load flow solutions. In Table 4, individuals in the main peak clusters occupy

No.	n_c	Centroids			
		E_2	E_3	F_2	F_3
1*	58	0.0428	0.0217	-0.0237	0.0100
2*	52	0.0461	0.6941	-0.0203	0.0133
3*	40	0.4578	0.0323	-0.0230	0.0087
4*	34	1.0303	0.9787	-0.0250	0.0183
5	4	1.0656	0.0954	-0.0572	0.0088
6	4	0.4212	0.0044	-0.7720	0.0044
7	1	0.0000	1.0303	-0.0029	0.0029
8	1	0.1027	0.1145	-0.0205	0.7661
9	1	0.0147	0.6840	-0.0147	0.7720
10	1	1.0538	0.0000	-0.0147	0.0029
11	1	0.0382	-0.8190	-0.0147	0.0147
12	1	0.0029	0.0088	1.4824	0.0029
13	1	0.9540	1.0303	1.5000	0.0088
14	1	0.5078	0.0029	-0.0029	-1.5000

Table 4 Clustering result at generation 100 with the proposed scheme on the 3-bus load flow function

more than 90% of the total number of individuals in the population The remaining clusters have few or only one individual(s) which are considered as lethal strings. Generally, these lethal strings have quite low fitness values. We can exclude these individuals by putting them into a unclassified cluster according to certain criterions, the fitness height, for example. Thus the efficiency of algorithm is expected to be further improved.

5. Implementation of Mating Restriction Using Clustering

For improving the on-line performance of classical sharing, a phenotypic mating restriction scheme was developed in [5]. The scheme permits to reduce the crossover between individuals of different niches by restricting individuals be mated within a distance σ_{mating}. In [5], the parameter σ_{mating} was set be equal to σ_{share}. Recall that in the clustering process, once individuals have been classified into different niches, the membership of each individual is then identified. If we similarly set the clustering parameters equal to the phenotypic mating restriction parameters, the mating restriction scheme can be implemented in a straight-forward way by using the result of clustering:

> To find a mating companion of individual i, if an individual having the same membership with that of individual i is found, mating is carried out, otherwise another individual is tried; if no such individual is found in the population, a random individual is chosen.

With this scheme, the mating takes place directly between individuals within the same niche, with no need to recalculate the distance between each individual.

The above mating restriction scheme has been added in the proposed sharing scheme and tested also on the early used test functions. The average computation time of 5 runs with the same previously used initial population and parameters is 1 min 53 sec and 2 min 6 sec on functions F1 and F2 respectively. The computation time by the proposed sharing scheme with mating restriction is thus nearly the same as without mating restriction. From another point of view, the use of mating restriction which has reduced the presence of lethal strings in population, improves the clustering process. The chi-square-like deviation value after 200 generations has been lowed to 0.2414 and 0.7258 on function F1 and F2 instead of 0.2 and 0.25 without mating restriction. Figures 9 shows the comparison of performances with proposed sharing alone and with proposed sharing with mating restriction on function F1.

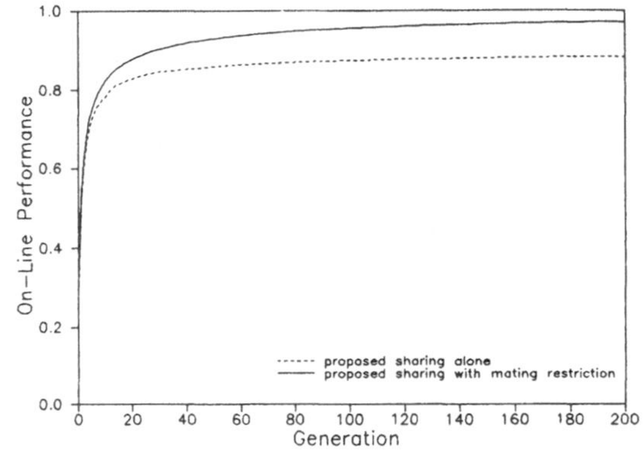

Figure 9 Comparison of the on-line performance with proposed sharing and with proposed sharing with mating restriction on function F1

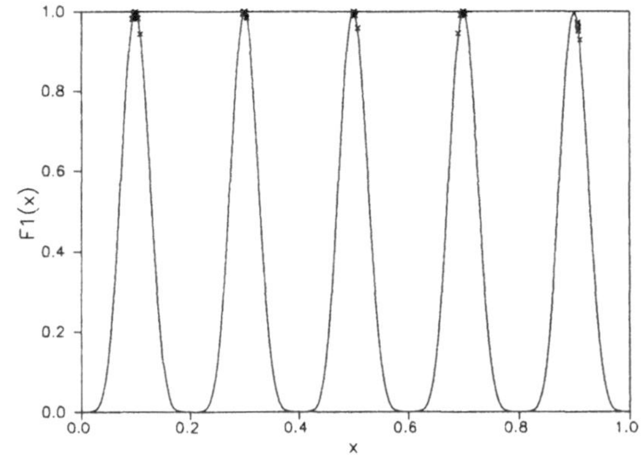

Figure 10 Situation of population at generation 200 with the proposed scheme with mating restriction on function F1

Cluster	n_c	Centroid
1	21	0.0995
2	19	0.2995
3	21	0.5016
4	20	0.6992
5	19	0.9073

Table 5 Clustering result at generation 200 with proposed sharing with mating restriction on function F1

It is shown that the on-line performance is increased significantly and by the mean time, the chi-square-like error is much reduced by using the mating restriction. Figure 10 plots all individuals at generation 200 of a run, and their corresponding clustering result is tabulated in Table 5. The individuals are shown more concentrated around the peaks and the obtained centroids with mating restriction are much closer to the peak locations than those in Table 2 without mating restriction. Since the performance of the algorithm is improved without additional computational efforts, the mating restriction scheme using clustering is recommended in practical application.

Conclusions

In this paper, a fast sharing scheme based on the adaptive MacQueen's KMEANS algorithm has been presented.

It is shown in the experiment that the proposed scheme performs as efficiently as the classical scheme, but faster than the classical one, due to the fact that the complexity of the proposed scheme is reduced. In fact, it does not need to calculate the niche count as precisely as the classical scheme did, the proposed sharing scheme is economical but remains effective.

Furthermore, the proposed scheme breakthroughs the limitation of *a priori* requirement on the number of niches, and is shown to be able to reveal the real structure of the function. The proposed scheme provides an exploratory way to discover the function structure in presence of uncertainty.

Mating restriction has been implemented using directly the result of clustering. The performance of the algorithm is improved without requiring any additional computation efforts.

References

[1]. D.J.Cavicchio, '*Adaptive search using simulated evolution*', Ph.D. Thesis, Univ. Michigan,1970

[2]. K.A.De Jong,'*Analysis of the behavior of a class of genetic adaptive systems*', Ph.D. Thesis, Univ. Michigan, Univ.Abst.Inter.36(10),5140B, 1975.

[3]. S.W.Mahfoud,'*Crowding and preselection revisited*', Proc. 2nd Conf. Parallel Problem Solving from Nature, pp.27-36, 1992.

[4]. David E.Goldberg and Joh Richardson, '*Genetic algorithms with sharing for multimodal function optimization*', Proc. 2nd Inter. Conf. Genetic Algorithms, pp.41-49., 1987

[5]. Kalyanmoy Deb and David E.Goldberg, '*An Investigation of Niche and Species Formation in Genetic Function Optimization*', Proc. 3rd Inter. Conf. Genetic Algorithms, pp.42-50, 1989

[6]. D.E.Goldberg, K.Deb and J.Horn, '*Massive multimodality, deception, and genetic algorithms*', Proc. 2nd Conf. Parallel Problem Solving from Nature, pp 37-46, 1992.

[7]. Michael R.Anderberg, '*Cluster Analysis for Applications*', Academic press, 1975.

[8]. Goldberg, '*Genetic Algorithms in Search, Optimization, and Machine Learning*', New York : Addison - Wesley, 1989.

[9]. Xiaodong YIN and Noël GERMAY, '*Investigations on Solving Load Flow Problems by Genetic Algorithms*', Electric Power System Research, Vol.22, pp.151 - 163, 1991.

An Interactive Genetic Algorithm for Controller Parameter Optimization

Bogdan Filipič, Đani Juričić

Jožef Stefan Institute
Jamova 39, 61000 Ljubljana, Slovenia
E-mail: *Bogdan.Filipic@ijs.si*

Abstract

Genetic algorithms are stochastic search algorithms inspired by biological phenomena of genetic recombination and natural selection. They simulate the evolution of string individuals encoding candidate solutions to a given problem. Genetic algorithms proved robust and efficient in finding near-optimal solutions in complex problem spaces. They are usually exploited as an optimization method, suitable for both continuous and discrete optimization tasks.

In this paper, genetic algorithms are investigated as an engineering optimization tool. The work focuses on tuning parameters in control system design. This domain has already been approached with genetic algorithms, but most of the experiments have been done using computer simulations of the devices to be controlled. We present the Interactive Genetic Algorithm (IGA) for controller parameter optimization. IGA carries out the evolution of parameter values in a traditional manner, but differs from a conventional genetic algorithm in that it allows interaction with real-world environment. The algorithm suggests the trials to be performed to explore the parameter space, and accepts results of the trials from the environment. The paper describes IGA and its application in tuning the parameters of a PID regulator operating on a laboratory device.

1 Introduction

Genetic algorithms are search algorithms imitating principles of biological evolution, such as reproduction, genetic recombination and selection. They were devised to study evolutionary phenomena in biological systems and to apply these principles in software engineering. Due to their generality and computational efficiency, genetic algorithms are nowadays exploited as an optimization and learning method in a number of domains, like robotics, network routing, control systems design, image interpretation, production scheduling, etc. The pioneering work in the field was done by Holland [1]. Recent reviews of genetic algorithms include [2, 3], while some typical applications are presented in [4].

A genetic algorithm operates on a set of candidate solutions, called a population. Candidate solutions are usually encoded as fixed-length binary strings and evaluated according to a problem-dependent fitness function. The population is created randomly and evolved in generations. In each generation, strings are reproduced selectively with respect to their fitness values, and modified by genetic operators, such as crossover, inversion and mutation. The selection phase ensures the survival of the fittest population members, while the genetic recombination provides the information flow among the string individuals.

By executing simple syntactic operations on strings, genetic algorithms process the so called schemata, i.e. the similarity templates corresponding to numerous individuals not actually present in the current population. This feature, known as *implicit parallelism,* distinguishes genetic algorithms from other search and optimization methods. Moreover, the *building block hypothesis* states that progressively better strings are constructed from the best substrings from past generations. As a consequence, the algorithm usually converges to highly fit population members representing high-quality solutions to the given problem. Detailed examinations of this properties can be found in [2, 3].

The genetic algorithm approach fits well within the scope of control systems design. Designing control involves two fundamental steps: defining controller structure and tuning controller parameters. Genetic algorithms have already been applied in both tasks. Examples from the literature include the application of reproduction plans based on population genetics to dynamic systems control and parameter optimization [5], and genetic-algorithm-based learning of rules

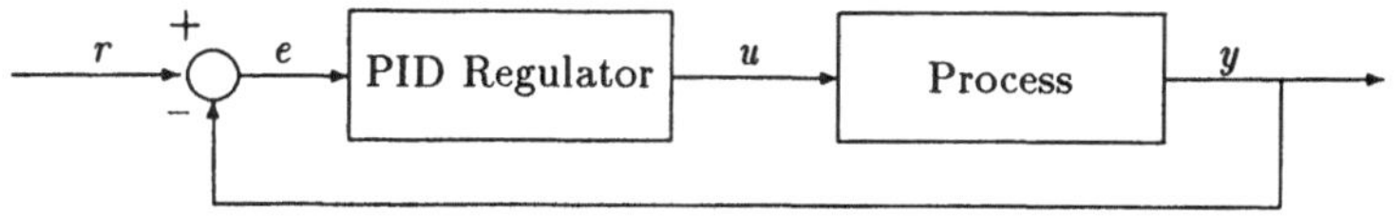

Figure 1: Controlling a process with a PID regulator.

to control a simple inertial object and a natural gas pipeline [6]. Moreover, a genetic algorithm was employed to induce control rules for a simulated pole-cart system [7]. The same control task was considered in [8], where control rules were synthesized without prior knowledge by a genetic algorithm, refined by an inductive learning algorithm, and finally fine-tuned, applying a genetic algorithm again. In [9], genetic optimization of controllers with a fixed structure was performed in frequency domain. In addition, a series of papers on genetic algorithms and learning techniques in control can be found in [10].

We have used genetic algorithms to perform time-domain optimization of a controller with a predefined structure. The classical proportional-integral-derivative (PID) regulator has been considered and its parameters tuned to satisfy optimality criteria preferred by human operators. Experiments were carried out on a testing laboratory device, representing an unstable and nonlinear dynamic system. To carry out the project, the Interactive Genetic Algorithm (IGA) has been devised, enabling real-world evaluation of parameter values.

In the next section, the PID regulator and the optimization task are discussed. Section 3 presents the Interactive Genetic Algorithm and Section 4 the experimental setup. Experiments and results are described in Section 5. We conclude with a summary of results and observations on IGA as an engineering optimization tool.

2 PID Regulator and Parameter Tuning

The PID regulator operates in a feedback loop with the process to be controlled, as shown in Figure 1. The regulator affects the process through the control input u, determined by the difference between the actual process output y and the reference value r. The value of the control input u is obtained in the following way:

$$u(t) = k_P\, e(t) + k_I \int_0^t e(\tau)d\tau + k_D\, \dot{e}(t) \qquad (1)$$

The control input and the process output are determined by the values of parameters k_P, k_I and k_D. Knowing either the mathematical model of the process or its impulse response, the process output can be computed, and the parameter values set by means of optimization. Traditional mathematical optimization involves some error measure, such as quadratic error integral or integral time absolute error (ITAE), that is to be minimized. Unfortunately, mathematically optimal regulators may produce responses not acceptable in practice. In addition, mathematical criteria may not be suitable for human assessment. Regulator design specifications can also include conflicting requirements. Consequently, optimizing regulator parameters is time consuming and frequently done on a trial and error basis.

Nevertheless, PID regulators are widely used in industrial process control due to their simplicity and generality. In industrial applications, optimization approaches, such as the Ziegler-Nichols method [11], have been traditionally used to tune regulator parameters. The Ziegler-Nichols approach consists of formulas that give parameter values ensuring predefined proportions among the peaks in the transient response of a controlled process. The results are satisfactory in the sense of settling the process rapidly, but may produce unacceptably high overshoots [12].

It is known that human operators usually perform parameter tuning intuitively, without exact theoretical knowledge. They have their own preferences for the optimality of a transient response. These typically include time-domain features of the response, such as settling time and overshoot (see Figure 2). This observation motivated knowledge engineers to develop expert systems for controller parameter tuning, based on response attributes. The expert system presented in [12], for example, is based on a set of rules about how the changes in parameter values affect the response characteristics.

The objective of our work was to tune the regulator parameters with respect to the criteria preferred by human operators, but hard for traditional mathematical approach. The considered problems included finding parameter values yielding minimum settling time, minimum overshoot, and certain combinations of the two criteria, such as finding the minimum settling time given maximum allowed overshoot, and vice versa.

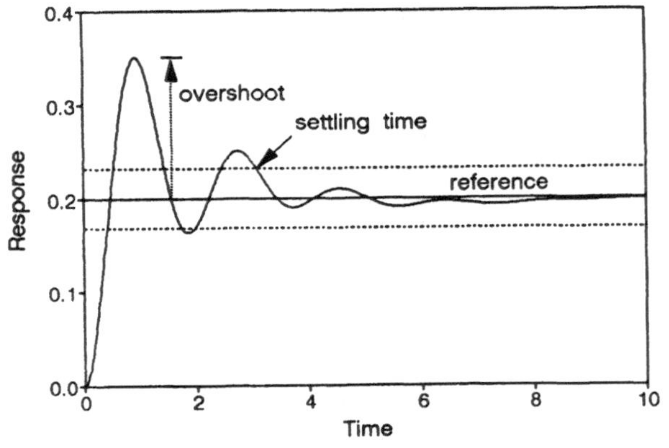

Figure 2: An example of a process transient response and its time-domain characteristics: overshoot and settling time.

3 Interactive Genetic Algorithm

To design a genetic algorithm for a given problem, certain problem-dependent algorithm elements need to be defined. They influence both the efficiency of the algorithm and the quality of its results. These elements include:

- representation (encoding) of candidate solutions,
- fitness function to evaluate candidate solutions,
- genetic operators,
- algorithm parameters, such as population size, number of generations and operator probabilities.

In IGA, the traditional binary string representation of solutions is employed. Each of the parameters $k_i, i = P, I, D$, is subject to interval constraints $a_i \leq k_i \leq b_i$, and Δk_i is the corresponding discretization step. Values of k_i from the interval $[a_i, b_i]$ are represented as binary strings of $\lceil \log_2(\frac{b_i - a_i}{\Delta k_i}) \rceil$ bits. Three such strings are concatenated into a binary "chromosome", representing a point in the parameter space to be searched by the algorithm.

Candidate solutions are evaluated through a fitness function. The fitness function embodies a quality measure defined over time-domain attributes of a process transient response. The evaluation is performed in interaction with the real-world environment and proceeds as follows. After a solution is generated, the algorithm invokes a trial on a device controlled by the PID regulator. A trial consists of settling the process at a reference point, changing the reference value, and monitoring the step response of the observed process. Attributes of interest are extracted from the response and returned to IGA to calculate the fitness of the candidate solution.

IGA incorporates two additional mechanisms to utilize the evaluation procedure. First, trials giving unacceptable responses (e.g. extremely high overshoots) are interrupted immediately after failure detection. In such cases an error status is returned instead of the attribute values. Second, a database of trials is kept during the execution, containing the regulator parameters and their fitness for each trial. Prior to each evaluation the database is checked to avoid unnecessary trial repetitions. The described mechanisms reduce the amount of time spent on evaluating the solutions.

To present an example of the fitness function, let us assume the optimization task is to find the parameter values minimizing settling time, given maximum allowed overshoot $\Delta y_{\max}$. Let k denote a triple of parameter values, $t_{\text{set}}(k)$ the settling time and $\Delta y(k)$ the overshoot of the step response determined by k. In addition, let t_{obs} be the observation time for which the process is monitored. Then the fitness of k is obtained as follows:

$$f(k) = \begin{cases} f_0, & \text{status} \neq \text{OK} \\ f_0 + (f_1 - f_0)\frac{\Delta y_{\max}}{\Delta y(k)}, & \Delta y(k) \geq \Delta y_{\max} \\ f_1 + (f_2 - f_1)\frac{t_{\text{obs}} - t_{\text{set}}(k)}{t_{\text{obs}}}, & \Delta y(k) < \Delta y_{\max} \end{cases} \quad (2)$$

where f_0 and f_2 are fitness function upper and lower bounds used to normalize fitness values, and f_1 is a threshold fitness value discriminating parameter settings satisfying the maximum overshoot constraint from the ones violating this requirement. The above defined fitness function ranks candidate solutions with respect to the degree of satisfying the optimality criterion, thus producing the so called selective pressure during the simulated evolution.

The genetic operators applied in IGA are multiple-point crossover and guaranteed mutation. The multiple-point crossover generates two offspring by exchanging a predefined number of alternate subsections between two parent strings. Crossing sites are selected independently from the PID parameter sections in a string. Mutation is implemented as altering bit values at randomly selected string positions. IGA is of the elitist type, preserving the best individual from previous generation in the current population.

Default parameter settings are as follows: crossover probability 0.7 and mutation probability 0.05. Population size, the number of generations and the number of crossing sites are to be adjusted depending on the given problem. In general, our experience from tuning the PID regulator parameters is that a rather low number of trials are sufficient to obtain satisfactory results (see the example in Section 5).

4 Experimental Setup

The environment used in our investigation consists of four elements:

- IGA,
- the SIMCOS software package [13], activating trials on request received from IGA, monitoring the controlled process, and returning response attributes to IGA,
- a laboratory device constructed for the purpose of control studies,
- hardware including the regulator and the interfaces to the device and to the computer.

The four elements are integrated in the sense that no user assistance is required during the execution of the algorithm. The environment therefore operates as a self-tuning regulator.

The testing laboratory device consists of an object ("rocket") floating on an air bubble in a water container. The vertical position of the object depends on the size of the bubble, which depends on the water pressure, and the water pressure is influenced by the pump voltage. The device is shown in Figure 3.

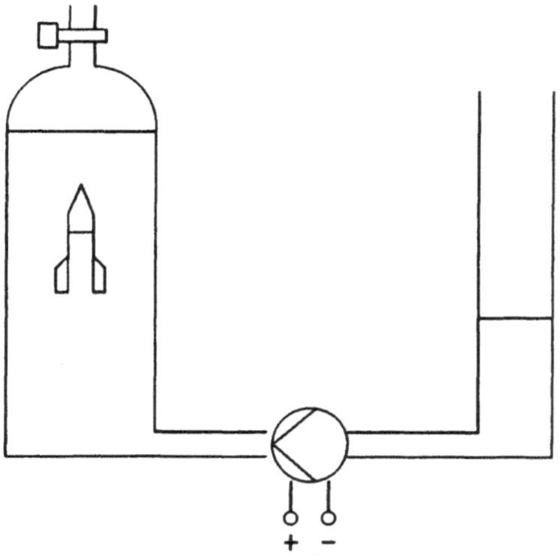

Figure 3: Laboratory device used in experiments.

The control task is to keep the object at a prescribed position by adjusting the voltage as the process input. A more detailed description of the device is given in [14]. The corresponding mathematical model is nonlinear, of the third order.

5 Experiments and Results

A series of parameter tuning experiments were performed with IGA for various problem types and different parameter spaces. Initially, some attempts were needed to reveal and eliminate the difficulties due to unstable operating conditions and inaccurate sensoring of the floating object position.

The following is a presentation of a typical experiment, illustrating the environment setting and the algorithm performance. Given the step change of the reference position $\Delta ref = 0.2$ m, the maximum allowed overshoot of 20% Δref, and assuming the transient is settled when oscillation remains within $\pm 5\% \, \Delta ref$, the PID regulator parameters were to be found that would minimize the settling time. The observation time was 35 seconds.

Candidate solutions were encoded as 21-bit strings, consisting of three 7-bit sections representing three PID parameter values. For each parameter, the search interval lower bound was 0.0 and the discretization step 0.2, giving the search interval $[0, 25.4]$. PID parameter settings were evaluated according to the fitness function from Equation (2). Fitness function constants f_0, f_1 and f_2 were 0.01, 0.5 and 1.0, respectively. The IGA parameters were as follows: population size 10, number of generations 30, crossover probability 0.7, number of crossing sites 3, and mutation rate 0.05. However, we believe that due to the known robustness of genetic algorithms reasonable results would also be produced under different parameter settings.

Best-of-generation performance in terms of settling time is shown in Figure 4 for three runs under the above conditions and compared with the best result found in a series of 300 random trials.

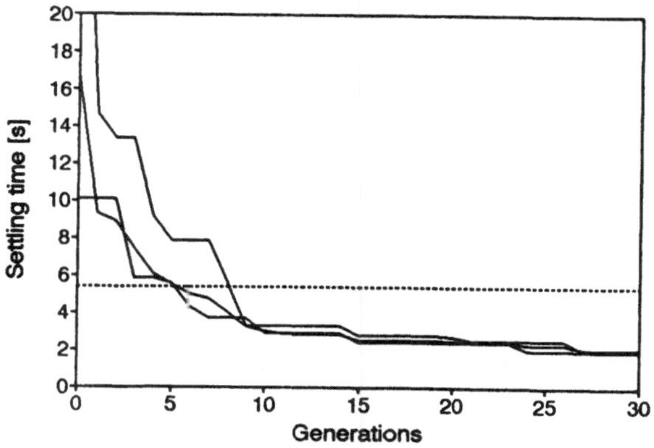

Figure 4: Best of generation performance of three IGA runs. Dotted line represents the best result found in 300 random trials.

Besides the obvious difference in final results, significant difference between the adaptive search performed by the genetic algorithm and the random search was confirmed, i.e. only about 20% of IGA trials were unsuccessful (returning error), whereas 84% failed during the random search. Furthermore, as first comparisons show, IGA also outperforms the traditional pole-placement method applied to the same control task [15].

6 Conclusion

IGA is an implementation of the adaptive search principle, supporting interaction with the environment. The approach was applied in tuning the parameters of a PID regulator operating on a laboratory device. Populations of binary strings representing points in the parameter space were evolved by the algorithm. During the simulated evolution, candidate solutions were evaluated with respect to the time-domain attributes of the process step response. As a result, high-quality solutions were found after examining a small portion of the search space.

Genetic optimization requires no *a priori* knowledge about the given problem. Instead, it exploits adaptivity, recombination and selection mechanisms while searching for good solutions. By the PID regulator case-study we have also demonstrated the possibility of considering the optimality criteria understandable to human operators.

IGA operates as a shell that efficiently guides the search through parameter space by manipulating the encoded solutions and suggesting trials to be performed. In our view, the approach can serve as a general-purpose engineering optimization tool.

Acknowledgement

The work has been supported by the Slovenian Ministry of Science and Technology. The authors are grateful to Tanja Urbančič for her cooperation in this study, and to Sašo Džeroski and Aram Karalič for reading the manuscript.

References

[1] Holland J. H., *'Adaptation in Natural and Artificial Systems'*, University of Michigan Press, Ann Arbor, 1975.

[2] Goldberg D. E., *'Genetic Algorithms in Search, Optimization and Machine Learning'*, Addison-Wesley, Reading, 1989.

[3] Rawlins G. J. E., *'Foundations of Genetic Algorithms'*, Morgan Kaufmann, San Mateo, 1991.

[4] Davis L. (ed.), *'Handbook of Genetic Algorithms'*, Van Nostrand Reinhold, New York, 1991.

[5] De Jong K., *'Adaptive system design: A genetic approach'*, IEEE Trans. Systems, Man, and Cybernetics, Vol. 10, No. 9, September 1980, pp. 566–574.

[6] Goldberg D. E., *'Dynamic system control using rule learning and genetic algorithms'*, Proc. 9th Int. Joint Conf. on Artificial Intelligence, Los Angeles, 1985, pp. 588–592.

[7] Odetayo M. O. and McGregor D. R., *'Genetic algorithm for inducing control rules for a dynamic system'*, Proc. 3rd Int. Conf. on Genetic Algorithms, Morgan Kaufmann, San Mateo, 1989, pp. 177–182.

[8] Varšek A., Urbančič T. and Filipič B., *'Genetic algorithms in controller design and tuning'*, To appear in IEEE Trans. Syst., Man and Cybernetics, 1993.

[9] Hunt K. J., *'Optimal control system synthesis with genetic algorithms'*, Männer R. and Manderick B. (eds.), Parallel Problem Solving from Nature 2, Elsevier, Amsterdam, 1992, pp. 381–389.

[10] ____, *'Preprints IFAC/IFIP/IMACS Int. Symp. on Artificial Intelligence in Real-Time Control'*, Delft, The Netherlands, 1992, Session 31: Genetic algorithms and learning, pp. 579–610.

[11] Ziegler J. G. and Nichols N. B., *'Optimum settings for automatic controllers'*, Trans. A.S.M.E., Vol. 64, 1942, pp. 759–768.

[12] Litt J., *'An expert system to perform online controller tuning'*, IEEE Control Systems, Vol. 11, April 1991, pp. 18–23.

[13] Zupančič B., *'SIMCOS Language for simulation of continuous and discrete dynamic systems'*, Faculty of Electrical Engineering and Computer Science, University of Ljubljana, 1991.

[14] Urbančič T., Juričić Đ., Filipič B. and Bratko I., *'Automated synthesis of control for nonlinear dynamic systems'*, Preprints IFAC/IFIP/IMACS Int. Symp. on Artificial Intelligence in Real-Time Control, Delft, The Netherlands, pp. 605–610.

[15] Pavlinič A., *'Modelling and computer control of an unstable laboratory process'*, B.Sc. Thesis, Faculty of Electrical Engineering and Computer Science, University of Ljubljana, 1991 (in Slovenian).

ROBUSTNESS AND EVOLUTION IN AN ADAPTIVE SYSTEM APPLICATION ON CLASSIFICATION TASK

J. BIONDI
UNIVERSITY OF NICE-SOPHIA ANTIPOLIS
I3S-CNRS
bat. 4, rue A. Einstein 06560 Valbonne, France
Phone : (33) 92-94-26-55 FAX : (33) 92-94-28-98
E-mail : jb@mimosa.unice.fr

Abstract

In this paper, we proposed an approach to a single-step Classifier System, in which the useful population is built by progressively specializing classifiers. It has been applied to a classification task in a medical domain. To permit the system to explore alternatives without making decisions earlier in learning stages, all the classifiers that might be selected are triggered and receive the resulting reward corresponding to their action. The payoff function involves the classifier's performance, its specificity and the system's performance (its robustness). Genetic operators are activated with a probability which depends on the system's robustness. During the test stages, no further learning takes place and the system's performance is measured by the percentage of correct classification made on the second set of examples. When the measure of performance is the highest, the population is stabilized and contains the correct classifiers (the payoff function and genetic operators have no more effect on classifiers). This approach achieves convergency more quickly and makes it possible to have a final accurate population without over-specializing.

1 - Introduction

The medical domain we worked on concerns blood-flow diseases. The surgeon concerned by our study used to evaluate the operation after three months, in two ways. An objective evaluation (it is a percentage) takes into account very precise and evaluable features : rules have been given by the surgeon and an expert system has been developed. The value of the subjective evaluation reflects what the surgeon thinks of the satisfaction of the patient. This kind of appreciation is very difficult to express. It depends on a lot of different features related to the personality of the patient, his habits, his stress with regard to diseases, pain, his activities etc. We are interested in knowing what features discriminate the satisfaction value of the patient concerning his operation and in the possibility of predicting the satisfaction value of a new patient. The data we gathered consist of responses to a questionnaire filled up by patients before (for features) and after (for satisfaction) the operation.

Several methods can be used for learning classification tasks. Neural networks using Back Propagation is one way of performing supervised learning and they are very efficient for learning from preclassified patterns. We investigated the Genetics Based Machine Learning approach, to implement a learning system which captures the main relationship between the features and the categories and makes a prediction when a new example is given to it. Experimental comparaisons [4] have been made in the context of preclassified examples and they have shown induction tasks are very accurately performed. A better quality of the Classifier Systems is their ability to extract knowledge in the form of comprehensible rules which are useful for a better understanding of the relations between features and categories. Nevertheless, CS's performances depend closely on several parameters: the message list size, the initial population, the rule strength updating function. In this paper, we are mainly interested in the choice of the initial population, the dynamical control of genetic operators and in the system's robustness.

Before giving further detail about the way we did it, we will give in section 2, a short overview on Classifier Systems and on problems related to them. In section 3, we will describe our approach. Section 4 presents the first results obtained, their implications; we will conclude with the future work.

2 - An overview of Classifier Systems and some difficulties related to.

2.1 Multiple-step Classifier

Classifier Systems [9] are GBLM which learn incrementally by interacting with their environment to adapt themselves to it. They are a class of adaptative systems used in learning situations where the only information available for the learning system is a payoff saying how useful a given action was with respect to the system goal. CS are based on production rules or classifiers. A measure is assigned to each rule, called its strength and it indicates the utility of the rule to the system's goal which is to be in adequation with its environment. The strengths of the rules are updated by schemes such as Bucket Brigade Algorithm [10]. Learning in classifier systems is driven by genetic algorithms that operate by repeatedly selecting rules on the basis of their strengths and applying genetic operators to these rules to produce new rules that then enter into competition with existing rules. The Classifier System adaptation may be viewed as a composition of a learning process and an evolution process. The evolution process modifies the structure of the population (rules) and the learning process, strength of the rules.
Recent results indicate that Classifier Systems are highly sensitive to particular encodings and parameter choices. Particularly, they are some difficulties with the size of the population (number of rules), the choice of initial rules, the loss of good rules and the representational capabilities of the rules. There are many proposals for improving the performance of CS's with regard to these problems.

One of the problems in CS's is the forgetful behaviour in which the good rules which were apparently well learned can be lost, either by crossing-over (with replacement) or because the rule has not been applicable for long periods of time (Bucket Brigade Algorithm). Several studies have been made to provide an alternative to the problem of forgetfulness in Classifier Systems. Wilson's approach in the "animat" study [18] was to increase the population size in the hope of providing enough storage to retain knowledge that might not be applicable for long periods of time. Grefenstette and Zhou [19] proposed the use of a long-term memory which retains knowledge during the life of the system and makes it possible for the system to transfer a problem-solving expertise to a solution of a similar new task. This extension to the CS's has been implemented in CSM (Classifier System with Memory).

The second problem we are concerned with is the representational difficulties. Historically, CS's employ a simple, sub-symbolic representation for rules, based on Holland's Schema Theory. The usual knowledge representation consists of fixed length strings of alleles over the simple {0, 1, #} alphabet. Schuurmans and Schaeffer [15] discussed some of the representational problems which would appear to hinder genetic search processes. Shu and Schaeffer proposed to introduce variables in Classifier Systems [13]: the variables are used to describe abstract relations in a succint manner, reducing the size of the solution set for many problems. In SAMUEL system [7], Grefenstette used a restricted high-level rule language and genetic operators suitable for the language.

A third problem deals with the choice of initial rules and the size of the population. In most of the CS's, a random initialization process is involved to generate initial rules. Some studies have investigated the seeding of the initial population with available knowledge. So, Wilson used a create operator: when Animat is confronted with an environmental message with no matching classifier, the create operator is involved. This operator simply takes an imprint of the input message and, with specified probability, generalizes each position of the imprint (replacing 1 or 0 by #). A random action is then selected and appended to the create taxon. In SAMUEL, Grefenstette used an adaptative initialization. In the initial population, rules are completely general rules and rules are specialized according to their early experiences: the Specialize operator creates a new rule with the same right-hand side as the maximally general rule but with a more specialized left-hand side. The system GABL, which has been applied in a classification task [6], creates a 100% correct rule set for a first single exemple. This rule set is used to predict the classification of the next exemple.

If the prediction is incorrect, GABL is invoked to generate a new rule set using the two exemples.

2.2 One-step Classifier

In a one-step classifier, each classifier is immediately rewarded (or not) as a result of its current action. The Bucket Brigade Algorithm is not used. However, these three difficulties may occur even in a single-step CS. BOOLE [18] is a single-step CS that learns increasingly difficult multiplexer problems. In this system, the probability of crossover is dicreased or increased with regard to the change in CS entropy. Wilson uses an adaptative crossover control to overcome the first problem. Newboole [4] is a modified version of BOOLE which learns from preclassified examples; the authors chose to use a natural representation for instances and rules rather than code them in the form of a binary string. They want to privilege a more powerful representation to make future enhancements and the exploitation of results easier (comprehensible solutions). For the third problem, in Newboole, the initial population is built by using the first P instances (P is the size of the population) to create P initial rules; the condition part of the new rule is a copy of the instance in which 'don't care' symbols are inserted with a probability of 0.5. The conclusion part is chosen randomly among the possible conclusions.

3 - Our approach

As in Newboole, we separated the examples in two parts : one used during the learning stage, the other for testing; we also chose a natural representation for instances and rules. However, the mechanism we used differs somewhat from the one-step systems from which we have taken our inspiration. First, the useful population is built by progressively specializing classifiers. Initially, the population contains the most general classifiers: it is the smallest complete population, that is, it may generate all permissible actions (classes). A creating process takes place at each message generation to create the optimal rule by partially specializing previous classifiers. The classifiers's specificity is increased all along the learning stages. We don't limit the population size but the rules are not duplicated and this process may not be involved at each stage (it is a parameter of the algorithm). Second, we explore simultaneously the search space from a set of points. To permit the system to explore alternatives without making decisions earlier in the learning phases, all the classifiers that have their conditions satisfied are triggered and receive the resulting reward corresponding to their action. There is no competition between classifiers during the learning stage. This corresponds to the full exploration of the different possibilities. During the test stage, the CS exploits knowledge obtained during the learning stage; the decision is determinist; it is based directly on the highest strength. This corresponds to the exploitation phase. The system's peformance is evaluated by the success rate during the test stage; it is the percentage of correct classifications made by the system on the second set of exemples during which no further learning takes place. Last, a crossover operator is involved with a probability inversely proportional to the system's performance.

The third point concerns the choice of a good payoff function. Since the behavior of CS's is driven by reinforcement received from the environment, the mechanisms for distributing or assigning (for a single-step) this reward are primary concerned in CS's. As, in the learning stage, the population is built by progressively specializing classifiers and all the matching classifiers are activated, so we selected a payoff function involving the rules's specificity and the system's performance: the more specific a rule is, the more involved in the decision it is. This allows to privilege the good specific rules during the test stage (their strength grows up faster than the strength of general rules). Because the payoff function depends on the success rate, the environment has no more effect on the system which has learnt. So, the learning system becomes robust: the payoff function and the genetic operators do not operate any more on classifiers.

3.1 Knowledge Representation

We chose a natural representation for coding the examples (questionnaire's answers): each example is completely described by 19 attributes representing the patient's characteristics and one of the 6 distinct categories of satisfaction. Each attribute and class can take a set of discrete mutually exclusive values. The example base consists of 110 examples. One part of the base is used during the learning stage, the other part for the testing. We will see the influence of the ratio learning set size/testing set size in section 4.

3.2 Initial Population

The search space is very large and we have a very small base of examples, so there is little chance to get a learning system in choosing a randomly initial population. Rather, we initialize the population with the six most general rules: each rule left-hand side contains only the "don't care" symbol and the right-hand side is one of the 6 permissible classes. Each rule's strength value is initially 0; this value is not modified by the payoff function (cf. 3.6).

3.3 Adaptation Cycles

The adaptation system is achieved through a succession of cycles (figure 1). Each cycle is composed of three stages:

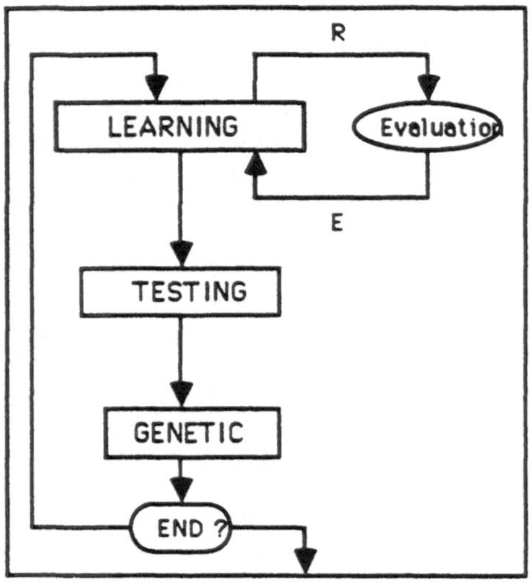

Figure 1

- the learning stage: each instance of the Learning base is presented as an input to the system. A matching set is formed (it is never empty); its rules are evaluated (cf. 3.4). Create/specialize operator may be or not involved (cf. 3.5) and the strength of each rule is updated (cf. 3.6).
- the test stage : the rule population is tested on the entire Testing base ; the System success rate is the percentage of correct decisions.
- the genetic evolution stage : only the crossover operator (cf. 3.7) is used to discover new rules from rules which are selected probabilistically according to strength (roulette wheel selection).

3.4 Rules Evaluation

We evaluate each rule of the matching set (all rules whose conditions match the input example description) by computing the difference between the correct answer (Ac) and the matching rule consequent (actual answer: Aa). The reward is simply a decreasing function of the error:

$$R = (1 - | Ac - Aa |) / Emax$$

Emax is the maximum error. The reward is always between 0 and 1; R = 1 when the answer is correct and R = 0 when Ac-Aa = Emax (the worst answer).

3.5 Create/Specialize Operator

As in SAMUEL, we used an adaptative creation of rules by specializing the best rule of the matching set. When an example from the learning set is proposed as a system input, each rule of the matching set is evaluated (cf. 3.4). The create-specialize operator creates a new rule with the best evaluated rule of the matching set by randomly specializing some attributes (with example attribute values) and by keeping the same right-hand side. In case several rules are candidates, the most specific one is used. If the conflict set still contains more than one classifier, the rule's strength is taken into account and, at the end, a random choice will be applied. In this way, a set of trees is created, in which each node is a rule, the leaves being the most specific rules and the roots the most general ones. Each time the create/specialize operator is applied, the number of attributes to be specialized is given by the specialization rate which is a system's parameter. This rate has a great influence on the system performance as we will see in section 4. By applying the crossover operator (cf. 3.7), two nodes of different trees may be crossed to create two new rules. System adaptation is reached when, in the testing stage, the used rules are enough specialized to be meaningful and able to classify the new examples correctly.

3.6 Reinforcement Algorithm

The initial base contains the six more general rules (which match all examples!). The reinforcement algorithm has to allow the emergence of useful and meaningful rules which classify unseen examples. The more specific a rule is, the more important the implication of this rule is in the decision. The updating of the strength of rule i at time t+1 depends on the rule evaluation (R_i), the rule's specificity (sp_i) and the rule's strength at time t:

$$S_i(t+1) = S_i(t) * (1- spi) + R_i * sp_i$$

The rule's specificity is the ratio of the number of instanciated attributes on the total number of attributes. If $sp_i = 0$ then $S_i(t+1) = S_i(t)$: the six most general rules have a constant strength value equal to 0 and if $sp_i = 1$ then $S_i(t+1) = R_i$: the entirely specified rule (with none #) received the full payoff.

Considering the strength variation between two strength updates:

$$\delta Si = S_i(t+1) - S_i(t) = sp_i * (Ri - S_i(t))$$

we may observe what we called [3] "the surprise and inurement effect"; in natural learning this effect has been pointed out by psychologists [12] : the higher the contradiction between a reward and the expected reward, the more effective the reward; and the more the reward is expected, the less effect it has. This implies a versatile system and in order to control the variability of strength, we introduced in MAGE system [3] the concept of robustness. The surprise effect has to be less important as the systeme is improving its performances. So, we introduce a factor (1- ρ) in the above formula in which ρ is the system's robustness which is, in our case, the measure of system success rate:

$$\delta Si = sp_i * (Ri - S_i(t)) * (1 - \rho)$$

So, the strength updating function has no more effect when the system's robustness is maximum :

$$S_i(t+1) = S_i(t) * (1- spi* (1 - \rho)) + R_i * spi* (1 - \rho)$$

3.7 Genetic Operators

After the learning and testing stages, n rules are selected probabilistically according to their strength. The Crossover operator generates two children from each pair of selected parents (one point crossover). The resulting offsprings are added to the population in case they were not. The "parent" rules are not deleted (reproduction operator). The Mutation operator is not used because of the way we create classifiers (imprints from examples). The crossover probability P is given by :

$$P = (1- \rho)* p_c$$

p_c is a system's parameter ; in experiments, it has been set to 0.8. So, the crossover's probability decreases as the success rate increases. To determine the strength of the new rules, we bear in mind the strength and the specificity of the parents. The children's strength is obtained as follows :

- let be P1 and P2, the two parent rules,
- let's consider a one point crossover randomly set in condition part of parent rules,
- let's consider, for each parent rule, the two parts l and r (left and right) on both sides of the crossover point (random mating) : $sp_l(P1)$ (resp. $sp_l(P2)$) is the specificity of left side of P1 (resp. P2), $sp_r(P1)$ ($sp_r(P2)$) is the specificity of right side of P1 (resp. P2),
- S(P1) (resp. S(P2)) is strength of P1 (resp. P2),
- let be C1 and C2 the two offspring rules; the following formulas allow a child rule to inherit the parent rule s strength proportionally to the specificity of the inherited part (number of instancied attributes/number of attributes of left or right part). So, the new rule's strength is all the higher as the rule inherits of a high specificity.

$$S(C1)=(sp_l(P1)*S(P1)+sp_r(P2)*S(P2))/ (sp_l(P1) + sp_r(P2))$$

$$S(C2)=(sp_r(P1)*S(P1)+sp_l(P2)*S(P2))/ (sp_r(P1) + sp_l(P2))$$

4 - Experiments and Performance

In this section, we report on an experimental evaluation of the performance and behaviour of our method applied to the classification task. First, we describe the setting of the parameters' values, then we will present and discuss the results obtained in our experiments.

4.1 Parameter values

Our objectives in these first experiments are to set the system's parameter values : the specificity rate (used by create/specialize operator) and the learning set size with regard to the testing set size. We carried out several series of tests to study the best choices. For each test, the genetic operators are turned off. A test is evaluated on a single cycle (one learning and testing stage) by examining the final population size, the success rate, the partial failure rate (partial failure on a testing set example is when the best rule in the matching set doesn't have the correct consequent), the total failure rate (total failure is when the matching set is empty), the maximal strength and maximal specificity of the rules, the average specificity of the rules.

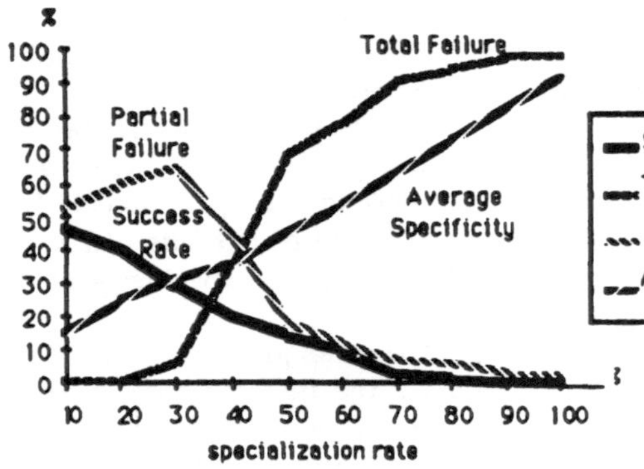

Figure 2

Figure 2 plots the success rate, the partial failure rate, the total failure rate and the average specificity with learning set size equals to 70% of examples base (30% for testing set) and different specialization rate. The results represent the average over 10 experiments for each specialization rate. Examining the rule base, we can observe that the number of rules which have been created from the six more general rules is larger when the specificity rate is high and the success rate decreases when the specialization rate increases. So, we set this parameter to 30%. The second experiment concerns the learning set size.

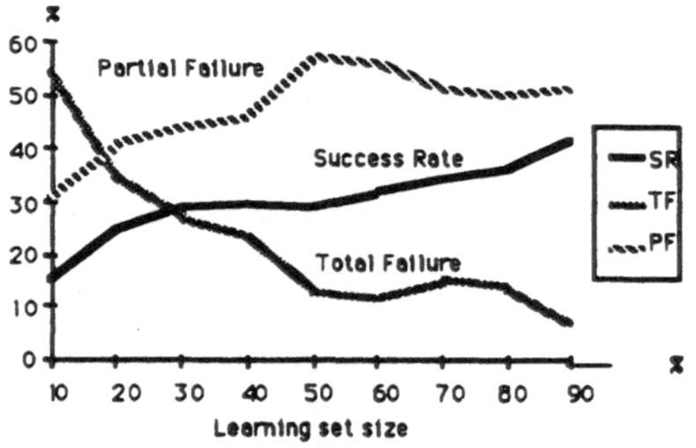

Figure 3

Figure 3 plots performances (average over 10 experiments) by increasing the learning set size (percentage of examples base). Good exploration of search space is obtained from 70% of the examples chosen for learning. In the further tests, the values of this two parameters will not change.

4.2 Experimental conditions

A learning stage consists in presenting all the learning set examples (77 examples); testing is run on 33 examples. Each experimental run is composed of cycles. A system's cycle corresponds to these two stages and possibly to a genetic stage. They are many ways to set each cycle running: by using or not the create/ specialize operator, by using or not the crossover operator and setting different values to crossover probability (p_c). When the two operators are turned off, only the reinforcement algorithm is running (update of classifiers strength). The only way to create new rules and to make their specificity change is by involving the two operators. Several tests were performed to compare the effectiveness of these two operators. We present three of them. To ensure a fair comparaison, the couple Learning set/ Test set was the same. The create/specialize operator is always invoked during the first two cycles (creation of a new rule from every example of the learning set). Test T1: crossover is turned off; create/specialize operator is invoked at each cycle. Test T2: every ten cycles, the crossover is involved (pc=0.8). Test T3: every ten cycles, the crossover (pc=0.8) and create/ specialize operators are used. For each cycle, the system outputs are : the total number of rules (NbR), the success rate (SR), the average specificity over all the rules (ASp), the number of rules used in testing stage (NRU), the average specificity of the rules used (ASRU), the minimum specificity (Smin) and the maximum specificity (Smax) of the rules used. For each rule, we can also know: the creation date, the number of uses in the learning and in the testing stages, the number of offsprings, the parent rules, the number of its total failures, partial failures and successes, and of course, for every cycle, the value of its specificity and its strength.

4.3 Results and analysis

The table 1 shortly sums up the results obtained in these three tests. For each test, we stopped the

	NbC	NbR	SR	ASp	NRU	ASRU	Smin	Smax
T1	20	430	43%	0.63	14	0.36	0.26	0.63
T2	40	490	34%	0.36	9	0.36	0.26	0.47
T3	50	1124	34%	0.59	11	0.43	0.26	0.74

Table 1

system run when no more change occurred in the success rate or in the specificity of the rules used in the testing stage: the first column value is the number of cycles which the system performed during every test.
The figure 4 plots the system evolution over 20 cycles during test T1. All 11 rules used in the testing stage are created during the first 4 cycles. The success rate decreases as the specificity of the rules used increases. From cycle 6, the create operator is ineffective (new rules cannot be created), the population size remains equal to 430; the average specificity and the success rate do not improve.

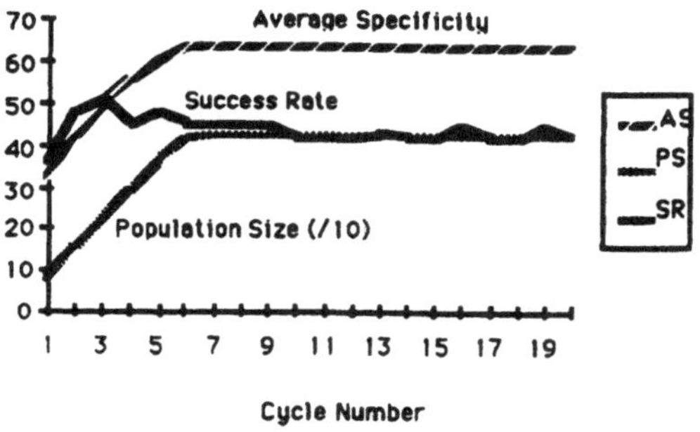

Figure 4 (test T1)

In the test T2 (crossover operator with p_C=0.8), more rules are created (490 rules) but the average specificity and the success rate have lower values than those obtained in the previous test. During the last testing stage, 9 rules were used: two were created by the create/specialize operator (first two cycles) and 7 by crossover. The figure 5 plots the results obtained in the test T3. Of the 11 used rules during the last cycle, 5 were created with create operator and 6 by crossover.

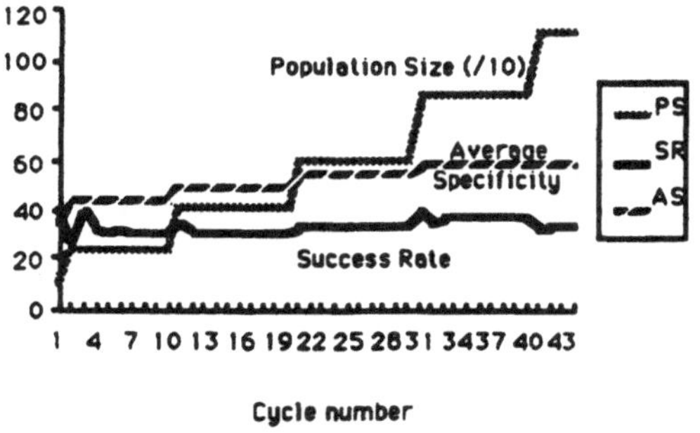

Figure 5 (test T3)

We can observe the shape of the curves which is varying by degrees every time the two operators are involved: the average specificity and the success rate increase every ten cycles. The strength of the new created rules is updated during the following cycles (reinforcement algorithm): the reinforcement process is effective only during 2 or 3 cycles.

By alternating reinforcement cycles with one creation/specialization cycle followed by one crossover stage, we get the best compromise between the success rate and the average specificity of the useful rules. A close examination of the last useful rules (fired in last testing stage), reveals that most of them are leaves of the trees. Their specificity depends on the depth of the branch which they belong to. We have been able to provide the surgeon with some interesting and meaningful rules concerning the classes 2 and 6. Nevertheless, the system cannot extract useful rules from the data set for the other 4 classes. The two main reasons are the bad repartition of the examples over the 6 classes (the classes C2 and C6 are prevalant in the base as they correspond to 70% of the examples), and the fact that some attributes are not only not discriminant but even have contradictory values in the same class. As the domain is difficult, the size of the data set is small, and all the examples are not pertinent, so we think that's the reason why the performance of the learning system is limited.

5 - Conclusion

This paper presented a single-step Classifier System in which the useful population is built by progressively specializing classifiers and by invoking the crossover operator. The objective of the approach which we put forward is to privilege the most specialized rules and to reach a balance between the success rate of the system and the average specificity of useful rules. We proposed strength updating function and crossover probability which depend on the system performance. The performance of the system has been studied on a complex and uncertain environment. To better know the learning data used, we will make use of statistical classification techniques over these data and we will compare the results obtained in each methodology. We will also examine the effectiveness of the proposed adaptative mechanisms on other data sets which may be considered more reliable. The system we developed gives us a lot of information, not only on the system's behaviour, but also on the comportment of each rule. We have not exploited all these data yet. So, many issues remain open to improve the system's performance. How can we use such item of information as the number of children, the rule activity level, the rule location in the hierarchy of each class ... to select and crossover the rules. What will the effect of the Merge and the Delete operators [8] be on the system's performance? We are also studying the progressive specialization from general rules of motion in the Bucket Brigade algorithm which is applied to the adaptative control of a moving robot which must reach a target in an unknown environment with obstacles to overcome.

Acknowledgements
This work is a follow-up to DEA project : I thank C. Delplanque for her contribution to this work especially in programming the learning system. I regret she was not able to collaborate to this paper. I am very grateful to Dr. Lepivert for allowing us to gather the experimental data used in this work (200 questionnaires have been sent to his patients!). I thank Pr. J.L. Cavarero, without whom I never would have thought doing research on that application; thanks also for many helping discussions and for having read the paper and made some useful remarks.

References

(a) Book

[1] Goldberg D E, '*Genetic Algorithms in search, Optimisation, and Machine Learning*', Addison Wesley, Reading, MA, 1989.

[2] Davis L, '*Handbook of Genetic Algorithms*', Van Nostrand Reinhold, New York, 1991.

(b) Article

[3] Biondi J, '*MAGE, un Modèle d'Apprentissage GEnéral*', thèse d'Etat, Univerité de Nice-Sophia Antipolis, 1990.

[4] Bonelli P and Parodi A, '*An efficient Classifier System and its Experimental Comparison with two Representative learning methods on three medical domains.*', Proceedings of the 4th ICGA, Morgan Kaufmann Publishers, San Mateo, California, 1991.

[5] Booker L B, '*Improving the performance of genetic algorithms in classifier systems.*',Proceedings of the first ICGA, Hillsdale,New jersey, Lawrence Erlbaum Associates, 1985.

[6] De Jong K A and Spears W M, '*Learning Concept Classification Rules Using Genetic Algorithms*', Proceedings of 12th IJCAI, Sydney, Australia, 1991.

[7] Grefenstette J J, '*A system for Learning Control Strategies with Genetic Algorithms*', Proceedings of the 3rd ICGA, Morgan Kaufmann Publishers, Inc., San Mateo, California, 1989.

[8] Grefenstette J J, '*Lamarkian Learning in Multi-agent Environments*', Proceedings of the 4th ICGA, Morgan Kaufmann Publishers, California 1991.

[9] Holland J H, '*Adaptation in Natural and Artificial Sytem*', The University of Michigan Press, Ann Arbour, 1975.

[10] Holland J H, '*Escaping Brittleness: The Possibilities of General-Purpose Learning Algorithms Applied to Parallel Rule-Based Systems*', in Machine Learning II, R.S. Michalski, J. G. Carbonell and T.M. Mitchell (ed.), 1986.

[11] Kelly J D and Davis L, '*A Hybrid Genetic Algorithm for Classification*', Proceeding of 12th IJCAI91, Sydney, Australia, 1991.

[12] Rescorla R A and Wagner A R, '*Variations in the effectiveness of reinforcement ans nonreinforcement*', in Classical Conditioning II : Current Research and Theory, Ed. A.H. Black and W.F. Prokasy, pp. 64-99, New York, Appleton-century-Crofts, 1972.

[13] Shu L and Schaeffer J, '*VCS : Variable Classifier Systems*', Proceedings of the 3rd ICGA, Morgan Kaufmann Publishers, 1989.

[14] Shu L and Schaeffer J, '*HCS : Adding Hierarchies to Classifier Systems*', Proceedings of the 4th ICGA, 1991

[15] Schuurmans D and Schaeffer J, '*Representational difficulties with Classifier Systems*', Proceedings of the 3rd ICGA, Morgan Kaufmann Publishers, 1989.

[16] Sedbrook T A, Wright H and Wright R, '*Application of a Genetic Classifier for Patient Triage*', Proceedings of the 4th ICGA, 1991.

[17] Valenzuela Rendon M, '*The Fuzzy Classifier System : A Classifier System for Continuously Varying Variables*', 4th ICGA, 1991.

[18] Wilson S W, '*Classifier Systems and the Animat Problem*', in Machine Learning Journal 2 , Kluwer Academic Publishers, Boston, 1987.

[19] Zhou H H and Grefenstette J J,.'*Learning by Analogy in Genetic Classifier Systems*", Proceedings of the 3rd ICGA, Morgan Kaufmann Publishers, Inc., San Mateo, California, 1989.

On Robot Navigation Using a Genetic Algorithm

Jarmo T. Alander
Helsinki University of Technology,
Department of Computer Science, SF-02150 Espoo, Finland
email: ja@cs.hut.fi

Abstract

In this work we have studied the possibility of using genetic algorithms in mobile robot navigation. The autonomous robot has a map of the room it moves within and some simulated sensors including range sensors to measure the distance between the robot and the other objects in the room. The location estimation method is based on minimizing the fitness function that depends on the measured data and the environment model by a genetic algorithm. The potential benefits of the genetic algorithms in this application area include robustness, parallel nature, generality, flexibility, incrementality, and simplicity. The obvious drawbacks include slow and stochastic processing. This work is a preliminary one in examining the applicability of genetic algorithms to solve computational problems of industrial and autonomous robots. In general the proposed method of finding the vector that gives the optimal fitting to the model used can be applied in many other calibration type problems as well in robotics as many other fields, too.

Keywords: autonomous mobile robots, calibration, computational geometry, genetic algorithms, modelling, navigation, optimization.

1 Introduction

An autonomous robot is a wheeled or legged robot that can move from one place to another in a given environment. The environment can be a simple one easily described by a geometric model, e.g. a building, or a more irregular outdoor terrain.

When an autonomous robot moves from place to place it is important to know its location and orientation as precisely as possible. In practise, however, various sources of noise and cumulation of error cause both random and systematic error in location estimates. Sooner or later the location estimate is usually so erroneous that it is useless and the robot must calibrate its location before continuing its activities.

The fast-rate control loop of a mobile robot usually use dead-reckoning techniques to estimate the current robot location. This technique ultimately yields large cumulative uncertainty. That is why often a higher-level, slower-rate control loop utilizing environment sensors is used to reduce this location uncertainty. An estimate of the robot's location is computed by matching the sensory data against the model of the environment.

In this work we examine the possibility to use genetic algorithm optimization methods for continuous enchancement of location estimate. We are dealing with a simulated robot that moves along prescribed paths in a closed room containing the robot and some other objects or obstacles, that do not move around. The robot has a set of optical or sonar range sensors that can measure the distance between the robot and the nearest obstacle along the line of sight (see fig. 3). In addition the robot has sensors to measure the distance traveled by the robot, and a sensor to measure the direction of movement. All sensor signals consists of the exact simulated values plus simulated noise.

1.1 Related work

The application of genetic algorithms in industrial and autonomous robotics has been dealt with at least in the following papers: [9, 8, 7, 10, 12, 17, 21, 19, 20, 26, 6, 14, 15, 27]. Most of the papers describe work on simulated robots, path planning and classifier systems.

Mobile robot navigation and geometric modelling has been studied much (see e.g. [13, 3, 16, 5]). Ayache and Faugeras has used Kalman filtering to estimate the environment geometry and further the location of the mobile robot [3]. Latombe et al have studied e.g. the navigation of mobile robots and sensory uncertainties [25]. They have analyzed the location precision as the function of the environment geometry. The resulting sensory uncertainty field is further used in robot path planning in order to create paths that avoid points where the location estimation evaluated from the environment sensor data is especially error prone.

2 The genetic algorithm

2.1 The problem

The problem is to estimate the location and orientation of a robot from the measured sensor values and the given geometric model of a simulated environment, which consists of two dimensional polygons (fig. 3).

2.2 Crossover operator

The parameters of the problem, the robot coordinates x and y, the orientation angle ϕ, and the values of the range sensors S_i, form the chromosome of the problem. The chromosome is represented by a vector of floating point numbers p. The elements of the parent chromosomes i.e. the parameter vectors $p_i^{(1)}$ and $p_i^{(2)}$ are combined by using the well known basic analytic crossover:

$$b_i = (1 - t_1)p_i^{(1)} + t_1 p_i^{(2)}, \qquad (1)$$

where t_1 is a uniform random variable belonging e.g. to the interval $[-.25, 1.25]$.

2.2.1 Adaptive crossover

The obvious drawback of the basic analytic crossover operator is that it does not care about the relative fitness of the parents. A faster convergence of the method is expected if the data contained in the fitness values of the parents more or less guides the crossover process.

B. Roysam et al have used a crossover operator that gives more weight to the parent gene that is more fit [4].

In this work we have tested the following two adaptive crossover operators. The easiest way to adapt the above basic analytic crossover method to parent fitness values is to transform the random parameter t in eq. (1) in such a way that the area around the better parent or the most probable solution area gets higher probability than the rest of the interval. This can be done e.g. by shifting linearly the search area towards the better parent value (linear adaptive crossover):

$$t_l = t_1 \pm \Delta$$

or by the following quadratic transformation (quadratic crossover), which gives more probability to the potentially better end of the search interval:

$$t_2 = \begin{cases} t_1^2 & \text{, if } f(p^{(1)}) < f(p^{(2)}) \\ (t_1 - 1)^2 & \text{, otherwise} \end{cases}$$

where f is the fitness function giving smaller values for the better estimates.

The third crossover operator tested was the following one, which we call Gaussian crossover:

$$t_G = \begin{cases} N(0,1)/4 & \text{, if } f(p^{(1)}) < f(p^{(2)}) \\ N(0,1)/4 + 1 & \text{, otherwise} \end{cases}$$

where $N(0,1)$ is a Gaussian random variable with $\mu = 0$ and $\sigma = 1$.

2.3 Mutations

The above crossover operators are all such that they generate new alleles so that a separate mutation operator is not inevitable. However, we have used a mutation operator with mutation rate 0.01. The mutated value is a uniform random variable chosen from the interval of average gene values of the parents added to the unmutated gene value.

2.4 Fitness function

The fitness function gives a positive value which tells us approximately how far the proposed solution is from the situation where the data fits perfectly the model. The minimum values of the fitness function is intended to give the best possible estimate of the true location of the robot when both measurement and model data of the environment is used. The fitness function used is:

$$f(M,S) = \sum_{i\in I}(S_i - M_i)^2,$$

where S_i is the ith simulated measured value containing an unknown error E_i, $M_i = M_i(S_i)$ is the corresponding function value evaluated from the model, and I is the index set contain all the indexes of the sensors the values of which are accepted.

An example of the fitness function of the test arrangement of the figure 3 is shown in figures 1 and 2. As can be seen from these figures the fitness function in multimodal and discontinuous at several points. As a whole, however, the fitness function seems reasonably well behaving and suitable for a genetic algorithm fitness function.

2.5 Selection

In every generation the n best items are selected. The tests and theory show that the genetic algorithm works best when n has values ranging from 10% to 50% of the total population size N [1, 18] In this work we have used the ratio $n/N = 20/50$.

3 The test arrangement

The simulated system is shown in figure 3. The system is simulated exactly and the simulated noise is added to the exact sensor values which are further input to the genetic algorithm optimizer. The noise is a random variable with Gaussian distribution.

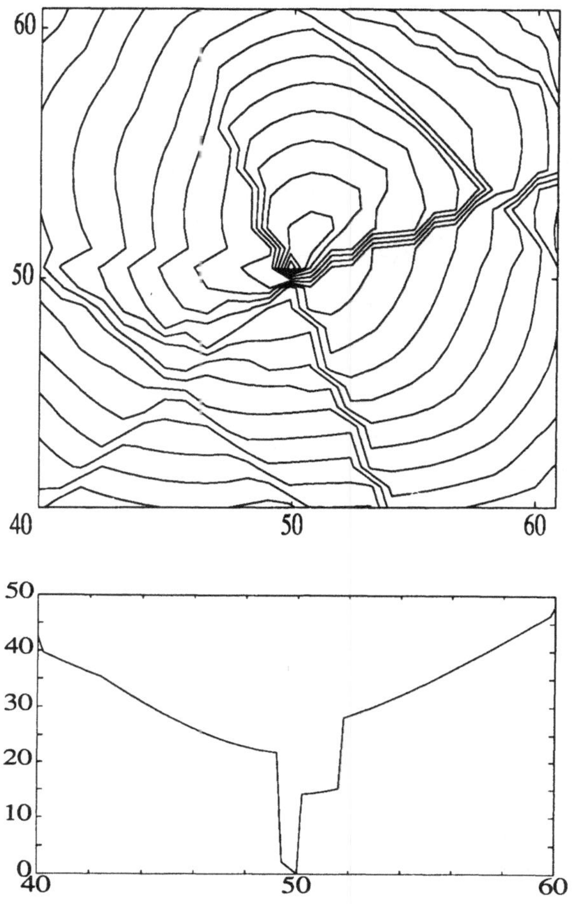

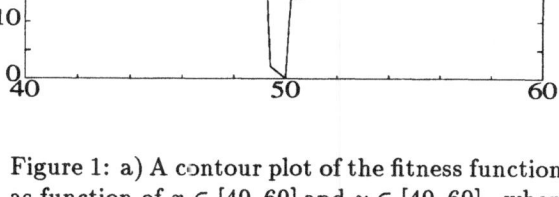

Figure 1: a) A contour plot of the fitness function as function of $x \in [40,60]$ and $y \in [40,60]$, when $\phi = 0$. b) The same fitness function when $y = 50$.

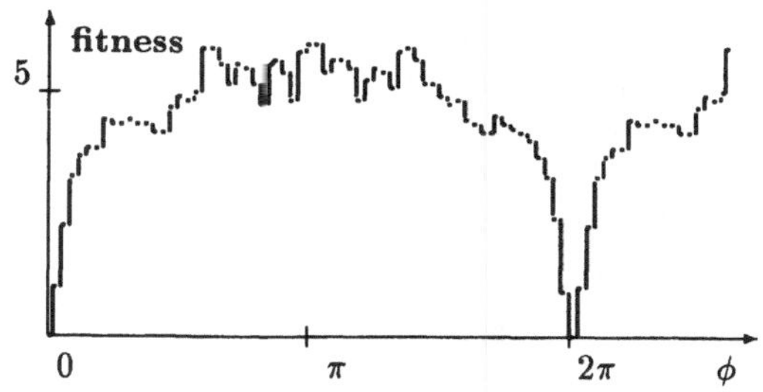

Figure 2: The fitness as the function of the directing of the movement ϕ at point (50,65).

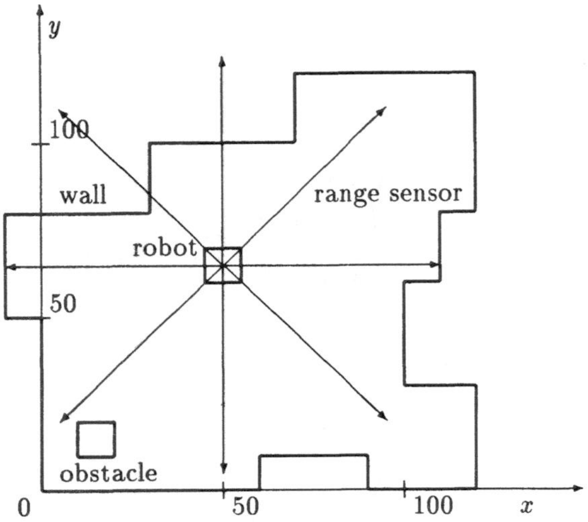

Figure 3: The simulated room and a robot having 8 range sensors.

In the test program procedures **ran1** and **gasdev** from [22] were used to generate uniform and Gaussian random numbers respectively.

The robot is moved by the simulator and the corresponding sensor values are evaluated. This data is input to the genetic algorithm that tries to find the exact values of the error terms using the geometric model of the environment via the fitness function.

3.1 Program size and speed

The test program was coded in C++ [24] and it consists of approximately 1300 lines. Most of the code deals with primitive geometry and test sampling, while the GA part of the code is only approximately 400 lines. The class structure of the test program is shown in figure 4.

No particular execution time optimization has been done. The average execution speed is approximately 10 points / CPU min on a SPARC-station 2, when the population size is 50, out of which 20 best items are selected, and when 10 generations are evaluated.

```
class point {           // (x,y) coordinate
      real  X,Y;        // coordinates
      ...
      real  dist2(point&);    // Eucl. distance
};

class line {            // planar line
      point *A,*B;      // end points
      ...
      point* crossPoint(line&);
                        // cross point of 2 lines
};

class room {            // 2-D room model
      line** walls;     // walls and obstacles
      int    NofWalls, // number lines in model
             MaxWalls; // maximum allowed lines
      ...
};

class robot {           // robot model
      real  X,Y,Fii;    // coordinates
      line  *Rays;      // range sensor "rays"
      int   NRays;      // number of range sensors
      char  *Name;      // the name of the robot
      ...
      void  locate(real,real,real);
                        // set robot coordinates
      void  simulate(room&,real*,int);
                        // simulate sensors
};

class specimen {        // one parameter set
      real *Genes;      // parameter array
      int  n;           // number of parameters
      real Fitness;     // fitness value
      ...
      specimen* breed(specimen*);
                        // breed two specimens
};

class population {      // population of specimens
      specimen **Unsorted; // all specimens
      specimen **Selected; // selected specimens
      int       n;         // population size
      int       nSelected; // number of selected
      ...
      specimen* best();    // the best of generation
      void      selection();// select the best specimens
      void      generation(); // one generation step
};
```

Figure 4: The class structure of the test program. Only the classes and their variables are shown. In addition each class contains 10 - 20 methods, of which only the most important are shown here.

4 Results

In figure 9 there are shown two points to be estimated and the corresponding solution candidates generated by a GA. As can be seen the convergence rate towards the known points is reasonable.

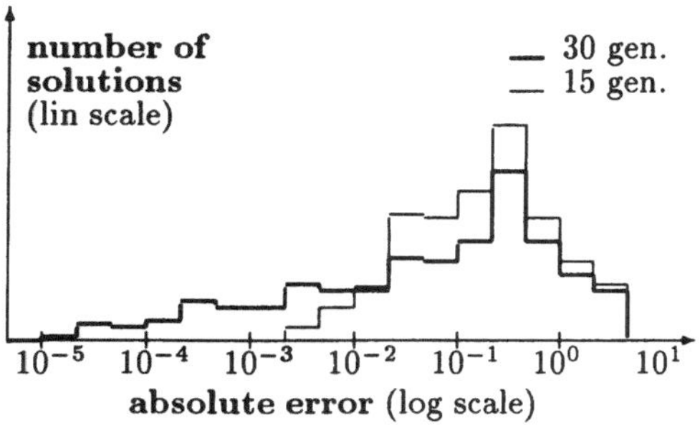

Figure 5: The distribution of the absolute estimation error, when the 16 range sensors are errorless, 1000 samples along a circular path of radius 30, 15 and 30 generations, population size 50 of which 20 best items are selected to the next generation, and linear adaptive crossover is used.

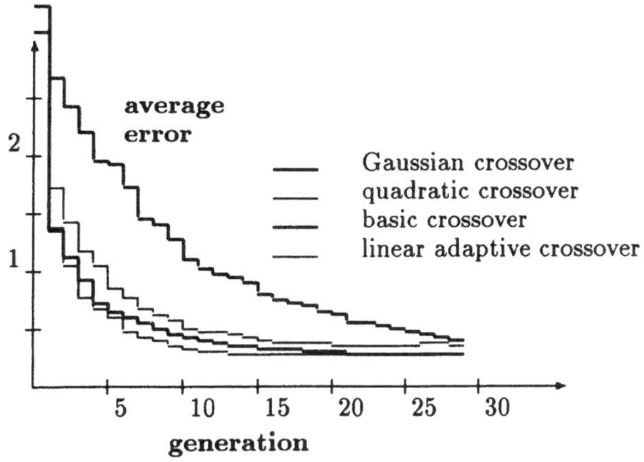

Figure 6: The average absolute estimation error as function of the number of generations and crossover method. 100 points estimated per each case. The same test case as in the previous figure.

4.1 Noiseless range sensors

The first test of the method was to estimate the values of x, y, and ϕ when the simulated 16 range sensors were noiseless. The resulting location estimation distributions are shown in figure 5. The test path was a circle of radius 30 and center point (50,50) shown in figure 3.

In figure 6 is shown the average estimation error as function of the number of generations. Three different crossover operations were tested: the **basic analytic crossover**, the **linear adaptive crossover**, which means that the search area is shifted 0.2 units (/1.5 units) towards the better parent value, and the **quadratic crossover**. As can be seen the effect of the tested crossover operators on the average estimation error is not great. At first sight this seems to be somewhat surprising and disappointing. However, a closer look at the properties of genetic algorithms reveals that they are rather insensitive to this kind of parameters [2].

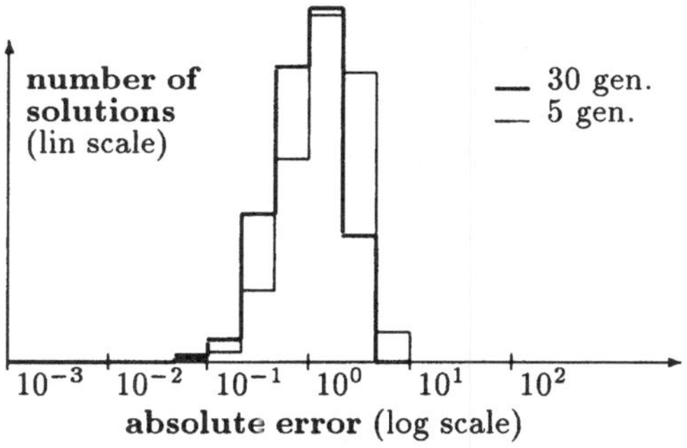

Figure 7: The distribution of the absolute estimation error, when the range sensors contain Gaussian noise ($\sigma_{R_i} = 3$). 1000 points were estimated and linear adaptive crossover was used. Other parameters are the same as in the previous figure.

4.2 Noisy range sensors

In figure 8 is shown the average estimation errors when the range sensors contain Gaussian noise having $\sigma_R = 0, 1, 2$, and 3 units, while the x and y sensors have Gaussian noise for which $\sigma_{xy} = 5$, and the orientation sensor has Gaussian noise with $\sigma_\phi = 1$. In figure 7 is shown the distribution

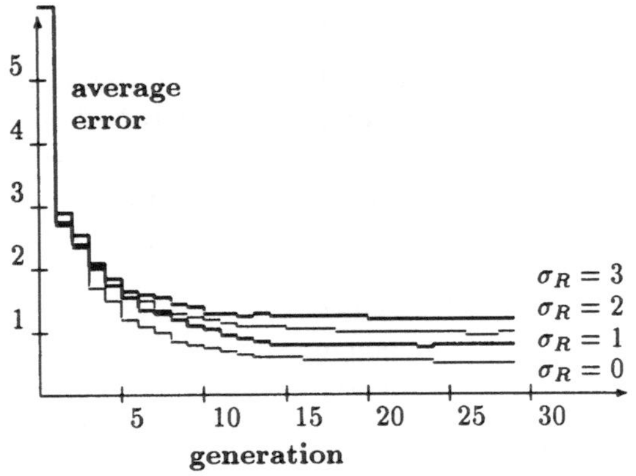

Figure 8: The average absolute estimation error as function of the number of generations and range sensor noise ($\sigma_R = 0, 1, 2$, and 3). 100 points per each case was estimated and linear adaptive crossover was used. Otherwise the same test parameters as in the previous figure.

of the estimation error in case of $\sigma_R = 1$ unit Gaussian noise at generations 5 and 30.

Acknowledgements

The author wants to thank Mr. Sam Sandqvist for his kind help with the proofreading of the manuscript.

5 Conclusions and future

In this work we have represented a computational method to aid robot navigation. The estimate of the location and orientation of an autonomous mobile robot is evaluated using a genetic algorithm to fit the values from different location sensors to a simple geometric model of the robot's environment. The proposed method can be applied to many other similar calibration and estimation type problems, in which we have in addition to measured data some extra knowledge or model of the object system, as well as in robotics and many other fields of technology. The genetic algorithm and simulation of the object system allow us to benefit from much more complicated models than classical estimation methods.

It seems that about ten generations is enough at population size 50 in practise to produce good estimates. The processing speed at this number of generations is about 10 points per CPU minute, which is reasonable. By using a multiprocessor system we should be able to speed up processing easily to about one point per second. The fitness function seems rather well behaving and this should give us more means to further speed up processing by concentrating search mainly to the best area.

Three different real-valued crossover methods were tested, but it seems that the crossover method does not have much influence on the convergence speed of the genetic algorithm.

For practical purposes it would be beneficial to develop the estimation method so that it also would handle more rigorously the noise and noise statistics inherent in physical sensors. We could e.g. utilize already done estimations by using their statistics in generation of the starting populations when estimating new path points.

References

[1] Jarmo T. Alander. On finding the optimal genetic algorithms for robot control problems. In *Proceedings IROS '91 IEEE/RSJ International Workshop on Intelligent Robots and Systems '91*, volume 3, pages 1313–1318, Osaka, 3.-5. November 1991. IEEE Cat. No. 91TH0375-6.

[2] Jarmo T. Alander. On optimal population size of genetic algorithms. In Patrick Dewilde and Joos Vandewalle, editors, *CompEuro 1992 Proceedings, Computer Systems and Software Engineering, 6th Annual European Computer Conference*, pages 65–70, The Hague, 4.-8. May 1992. IEEE Computer Society, IEEE Computer Society Press.

[3] Nicholas Ayache and Olivier D. Faugeras. Maintaining representations of the environment of a mobile robot. *IEEE Transactions on Robotics and Automation*, 5(6):804–819, 1989.

[4] Anoop K. Bhattacharjya, Douglas E. Becker, and Badrinath Roysam. A genetic algorithm for intelligent imaging from quantum-limited data. *Signal Processing*, 28(3):335–348, 1992.

[5] Dong Woo Cho. Certainty grid representation for robot navigation by a Bayesian method. *Robotica*, 8(2):159–165, 1990.

[6] Alberto Colorni, Marco Dorigo, and Vittorio Maniezzo. Genetic algorithms and highly constrained problems: The time-table case. In Schwefel and Männer [23], pages 55–59. (Proceedings of the 1st Workshop on Parallel Problem Solving from Nature (PPSN1), Dortmund, 1.-3. October 1990).

[7] Yuval Davidor. *Genetic Algorithms and Robotics: A heuristic strategy for optimization.* World Scientific Publishing, Singapore, 1990.

[8] Yuval Davidor. Lamarckian sub-goal reward in genetic algorithm. In Luigia Carlucci Aiello, editor, *9th European Conference on Artificial Intelligence*, pages 189–194, Stockholm, 6.-10. Aug. 1990. Pitman Publishing.

[9] Yuval Davidor. Robot programming with a genetic algorithm. In *1990 IEEE International Conference on Computer Systems and Software Engineering*, pages 186–191, Tel-Aviv, 8.-10. May 1990. IEEE Computer Society Press.

[10] Yuval Davidor. *A Genetic Algorithm Applied to Robot Trajectory Generation*, chapter 12, pages 144–165. In [11], 1991.

[11] Lawrence Davis. *Handbook of Genetic Algorithms.* Van Nostrand Reinhold, New York, 1991.

[12] Marco Dorigo and Uwe Schnepf. A bootstrapping approach to robot intelligence: First results. Technical Report 90-068, Politecnico di Milano, Dipartimento di Elettronica, 1990.

[13] Alberto Elfes and Larry Matthies. Sensor integration for robot navigation: Combining sonar and stereo range data in a grid-based representation. In *Proceedings of the 26th Conference on Decision and Control*, volume 3, pages 1802–1807, Los Angeles, CA, December 1987. IEEE.

[14] John R. Koza. *Genetic Programming: On Programming Computers by Means of Natural Selection and Genetics.* The MIT Press, Cambridge, MA, 1992. (in press).

[15] John R. Koza and James P. Rice. Automatic programming of robots using genetic programming. In *AAAI-92 Proceedings Tenth National Conference on Artificial Intelligence*, pages 194–201, Jan Jose, California, 12. - 16. July 1992. AAAI Press/ The MIT Press.

[16] David J. Kriegman, Ernst Triendl, and Thomas O. Binford. Stereo vision and navigation in buildings for mobile robot. *IEEE Transactions on Robotics and Automation*, 5(6):792–803, 1989.

[17] J. R. McDonnell and W. C. Page. Mobile robot path planning using evolutionary programming. In Ray R. Chen, editor, *Proceedings of the Twenty-fourth Asilomar Conference on Signals, Systems & Computers*, volume 2, pages 1025–1029, Pacific Grove, California, 5.-7. Nov. 1990. The Computer Society of IEEE/Maple Press.

[18] Heinz Mühlenbein and Dirk Schlierkamp-Voosen. The distributed breeder genetic algorithm i. continuous parameter optimization. Technical Report 92-121 (?), GMD, 1992.

[19] René Natowicz and Gilles Venturini. Learning the behaviour of a simulated moving robot using genetic algorithms. In M. H. Hamza, editor, *Artificial Intelligence Application & Neural Networks (AINN'90)*, pages 49–52, Zürich, 25. - 27. June 1990. The International Association of Science and Technology for Development - IASTED.

[20] René Natowicz and Gilles Venturini. Learning the behaviour of a simulated moving robot using genetic algorithms. In Teuvo Kohonen and Françoise Fogelman-Soulie, editors, *Cognitiva 90 At the Crossroads of Artificial Intelligence, Cognitive Science, and Neuroscience, Proceedings of the Third COGNITIVA Symposium*, pages 645–654, Madrid, 20.-23. November 1990. North-Holland.

[21] Joey K. Parker, Ahmad R. Khoogar, and David E. Goldberg. Inverse kinematics of redundant robots using genetic algorithms. In *Proceedings of the 1989 IEEE international Conference on Robotics and Automation*, volume 1, pages 271–276, Washington, D.C., 14.-19. May 1989. IEEE Computer Society Press.

[22] William H. Press, Brian P. Flannery, Saul A. Teukolsky, and William T. Vetterling. *Numerical Recipes in C.* Cambridge University Press, Cambridge, 1988.

[23] H.-P. Schwefel and R. Männer, editors. *Parallel Problem Solving from Nature*, volume 496 of *Lecture Notes in Computer Science.* Springer-Verlag, Berlin, 1991. (Proceedings of the 1st Workshop on Parallel Problem Solving from Nature (PPSN1), Dortmund, 1.-3. October 1990).

[24] Bjarne Stroustrup. *The C++ Programming Language.* Addison-Wesley, Reading, 1985.

[25] Haruo Takeda and Jean-Claude Latombe. Sensory uncertainty field for mobile robot navigation. In *Proceedings of the 1992 IEEE International Conference on Robotics and Automation*, pages 2465–2472, Nice, France, 12. - 14. May 1992. IEEE Computer Society Press, Los Alamitos, California.

[26] Gilles Venturini. Characterizing the adaptation abilities of a class of genetic based machine learning algorithms. In Francisco J. Varela and Paul Bourgine, editors, *Towards a Practice of Autonomous systems, Proceedings of the First European Conference on Artificial Life (ECAL91)*, Paris, 11.-13. December 1991. The MIT Press.

[27] Min Zhao, Nirwan Ansari, and Edwin S. H. Hou. Mobile manipulator path planning by a genetic algorithm. In *Proceedings of the IROS'92*, 1992.

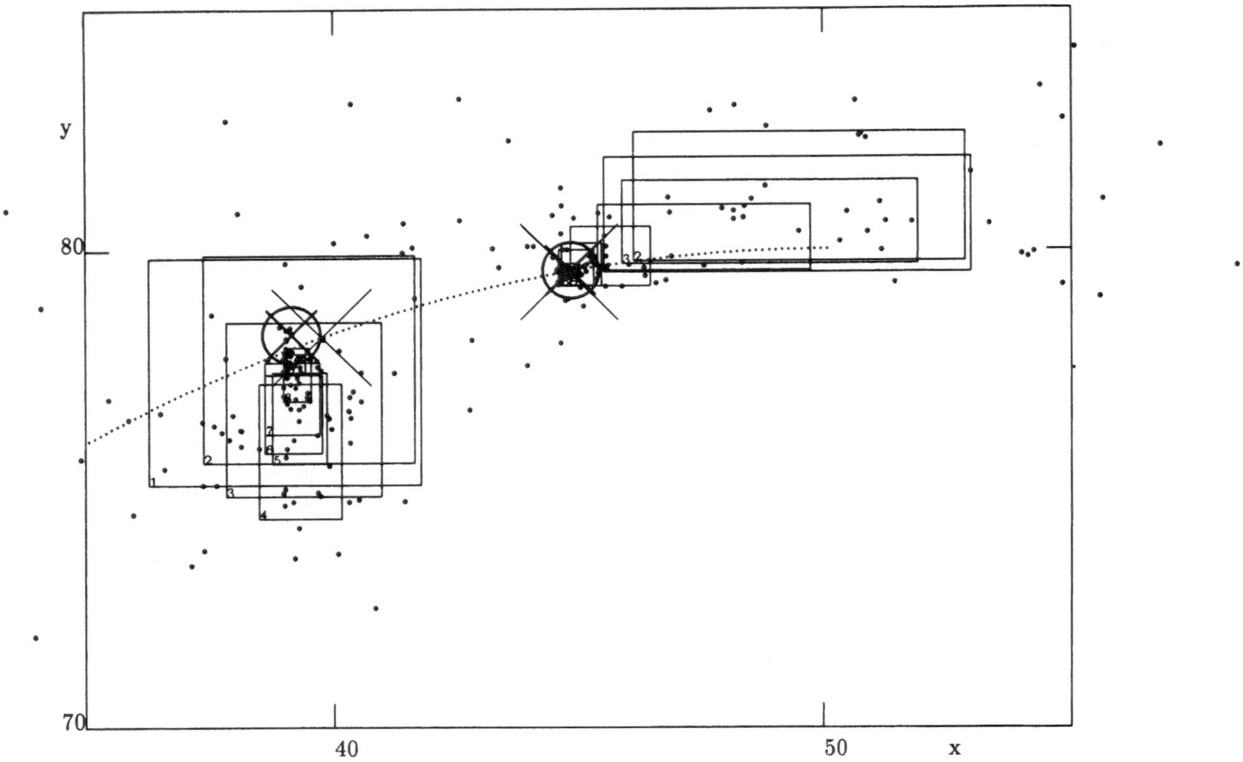

Figure 9: The x and y coordinates of solutions ($\bullet$) of 10 generations at two points ($\times$) of the circular test trajectory ($\cdots$), when the range sensors are errorless, population size 50 of which 20 best items are selected to the next generation. Rectangles show the average intervals of x and y coordinates in different generations. The best point estimated by the genetic algorithm is shown by $\otimes$.

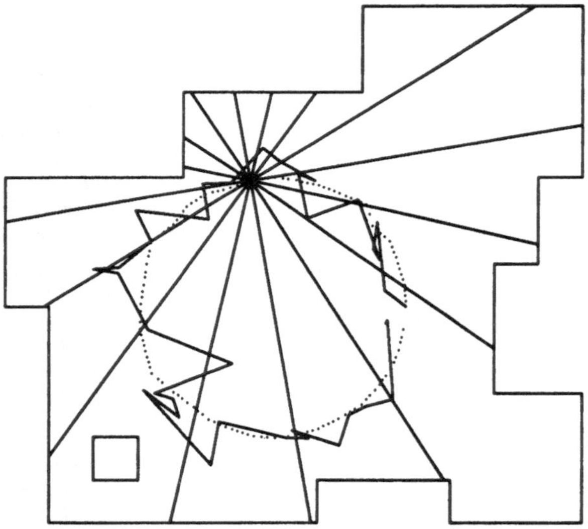

Figure 10: The measured (—) and the 10th generation location estimates ($\cdots$) at 36 points along the test circle. The range sensor values are shown at the 10th path point. $\sigma_R = 5$ and $\sigma_\phi = 1$.

Genetic Algorithms and Classifier Systems in simulating a Cooperative behavior

Antonella Carbonaro, Giorgio Casadei, Aldopaolo Palareti
Dipartimento di Statistica "P.Fortunati," Università di Bologna

Abstract

Genetic Algorithms and Classifier Systems are often used in biologic-like and evolutionary behaviors' simulations. The basic example is Wood7 Wilson's world. In this environment it is interesting to study some problems: Can evolve the cooperative behaviors of organisms present in the world? How and when do the behaviors evolve? Some preliminary results show the conditions under that cooperative behavior rules are developing rapidly.
Particularly we have pointed out the likely of following observations:

a. The cooperative behavior develops more easily if the initial population start from the same point.
b. It exists some thresholds under that the cooperative behavior can't evolve; these thresholds depend to the population size.

1. Introduction

The goal of this research was the construction of an artificial system with a some type of intelligent behavior.
The psychological literature offers a number of useful definition of intelligence; we have used the following, from the physicist Van Heerden:
intelligent behavior is to be repeatedly successful in satisfying one's psychological needs in diverse observably different situations on the basis of past experience.

Therefore, if we want to respond the formulated tasks, we must consider an approach characterizing the interaction between the model and the environment in which it is in.
Briefly, the most interaction between a system and his environment need ability to store and manipulate the information. From this point of view, the way to construct an intelligent system is to study an efficient strategy to manipulate, store a and load symbolic expression in the lesser time and in the best possible way.
The basic problem to realize intelligent behavior is not only relative to the identification of the powerful technique to realize some type of procedure but, especially, is relative to flexibly represent and efficient utilize repertoire of knowledge.
In other words the most important needs are to access to the most possible relevant information, to automatic acquiring knowledge, to dynamic modifying knowledge and to suggests new ideas when the system reach the limit of its capacity.
The traditional knowledge based approach to artificial intelligence shows some fundamental deficiencies in the generation of the powerful reasoning techniques. Explaining the cognitive ability of the brain purely in terms of symbol manipulation as in current AI implementation seems to lack the flexibility and expressiveness of natural cognitive system (1986 D. E. Rumelhart, J. L. McClelland).

An autonomous agent must posses an adaptive power itself in order to adapt its behavior to any change in the environment. In fact, the adaptability can not be demand to some coordination techniques for the selection of the different possible action. Even if the coordination techniques are powerful it is inappropriate to lead complex behavioral sequences characteristics of the flexible and robustness systems (1991 IEEE, M. Dorigo, U. Schepf).
For this reasons we had want to develop a genetic system able to absolve one task: increasing own survival chance.
In the GA's story, even if they represent a recent resolution methodology, exist a large number of different techniques utilized. In the CS's story the different implementations are so many too. Therefor it is difficult to compare results and to determinate why they reach so different performances.
The reproduction of previously obtained results is furthermore difficult because of the lack of "robustness" intrinsic to this kind of systems. Small variations of parameter values can in fact lead to substantial performance differences.
For this reasons we have thought to realize an adaptive learning system, based on Classifier System, like Animat. It was developed in 1985 from Wilson. Roberts developed this kind of problem too, but the two application's performances was substantially different. The base goal of this research was the reproduction of a model with the learning and survival property like that obtained in the previously researches. We want in fact analyze the results and justify the differences.
In order to realize the most general possible system, we had not utilize some auxiliary knowledge that had been used in Wilson's research. We know that this choice will be influent in the performance's system but we prefer to build a model that use only the small information quantity necessary to the learning.
Then we have thought to develop the system for a more complex learning; we have so modify the environment and the tasks that the model must learning.

2. Folder results

The original problem was proposed and analyzed by Wilson in 1985; we describe the results given by his model. The only documentation about his work regards data computed in 8000 problems (iterations) executed on a word called Wood7. Wilson analyze two different kind of parameters: the performance, that is a mobile average on the number of steps done to reach food, and the generality, that is the utility level of the classifier produced in problems that are similar to the original one. The performance reported is about 5 steps at the 8000th. problem. The conclusions he takes are this: "In its simply way, Animat meets the definition of intelligence stated at the beginning. Animat becomes good at satisfying its need for food in a Woods of diverse object configuration on the basis of experience. Though not yet tested, Animat's rule generalization over time suggests that performance would be maintained in a somewhat different Woods, or if the Woods slowly changed."

3. Some type of trail

The best trail. In this simulation the organism has a total cognition of the word in which it is in; so it needs not to have a direct contact with an object to perceive it but immediately knows where to go to reach a needs' satisfaction font. This kind of trail represents the best performance that is possible in a specific environment for an animal with human capacity. The average, on all the iterations, for this distances gives the value of about 2.2 steps (for Wood7) and this results represents the best for omniscient automata.

Random trail. This kind of simulation is conceptually different from the folder. In fact we suppose that the animal is driven, in its choice, by completely random actions. The system does not perceive any object, neither far nor in contact, and it decides in a random way where to go. This kind of trail represents the worst performance that is possible in a certain environment for an animal with no sensorial capacity. The average, on all the iterations, for this distances gives the value of about 41 steps and this results represents the worst case.

Pseudo-random trail. In this kind of simulation the system has got the same kind of knowledge that the Animat has. So we could compare most properly behaviors of the Classifier System with the ones of a traditional algorithm. In this case we give to the animal the capacity to avoid trees and go directly to the food when it could see it and also the capacity to move in random way when it do not perceives any signal. The average, on all the iterations, for this distances computed in this way, gives, for Wood7, the value of about 8-9 steps.

Rules-driven random trail. The last traditional algorithm implemented gives an idea on the computational capacity of a ruled based system that does not depend from a payoff distributed by the environment to the best messages.

4 Results and behavior of basis system

Our target is to investigate and compare the merits of Wilson and Roberts, the two researchers that worked on ANIMAT problem with different results. The results obtained with our system are more similar to Roberts' one, as well it's very hard to check likeness with work almost unknown.

However in the three systems (Our, Wilson and Roberts) the behavior of the animal in from of different situation is the same.

The performance we will describe are obtained as the average results of subsequent independent elaboration. This to decrease the consequence of random choices that drive the system.

The results obtained on WOOD7 environment (the same used by Wilson and Roberts) are the followings:

Death-rate - Initially the percentage of death paths is 70-80%; then this value decreases very quickly (20% at 100th iteration) to values of 2-3%.

Path length to reach food - Initially the path length is about 20-25; this value, especially at the beginning of the elaboration, is inversely proportional to death-rate. Afterwards the path length decrease to 10-12 at 2000th iteration and to 7-8 at 8000th iteration.

On the contrary Wilson's path length at 8000th iteration is 4-5 steps and Roberts' one
is 6.5 steps.

Our results are obtained with the evolution parameters:

- decreasing rules payments
- message covering technique
- initial rules set representative of WOOD7 environment;

- genetic algorithms runs every 400 iterations;
- variable number of rule coupling to reproduce.

Result are shown in the following graphs:

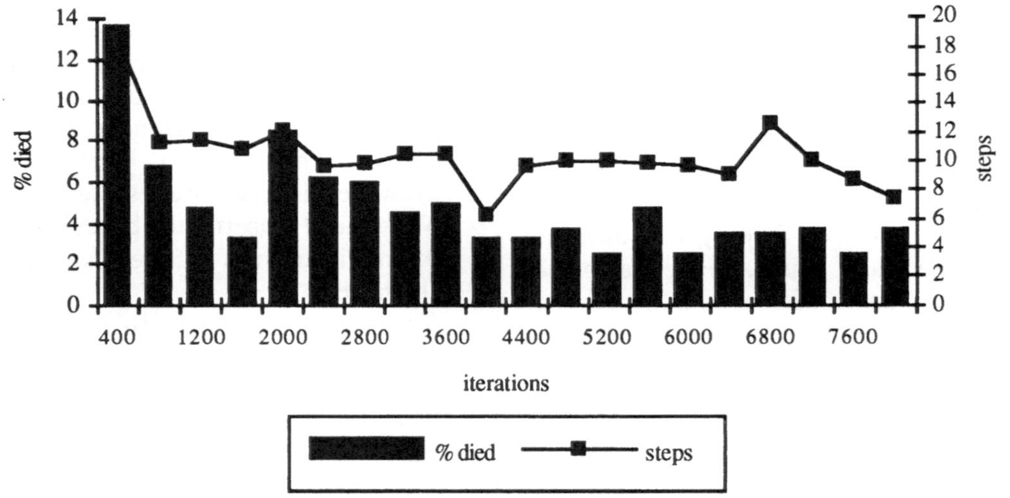

Figure 1; Performance graphs.

The system respect the definition of intelligence proposed previously; in fact the model learns to satisfy our needs in a particular environment.

5. Implemented development for a cooperative system

We expend Animat problem with introduction of other animals of the same species in the environment. The present of few animals living in the same environments increases the difficult of learning for every one.

So, if we supposed that animals must cooperate in searching and capture food, it is very important to set correctly the number of animals.

A presence to poor of organism will not allow a sufficient degree of cooperation, so that animals will not survive.

6. Result

In this way we do different independent trials with a variable number of animals and consider the average that rate. Obtained the result show that with a single animal the death rate is 100%; increasing the number of animals, that rate decrease more and more and assumes values as following:

1 animal	100%
3 animals	79%
5 animals	70%
7 animals	64%
10 animals	56%

This behavior show that setting appropriately animals number setup a cooperation mechanism between animals.

This result is not a banal consequence of increased animal number.

When the system work with only one organism, a 400 rules population allows learning of intelligent behavior. With more organisms the possible configurations of sensorial inputs grow considerably.

So we can predict the insufficient dimension of population so that the system is not able to create a population set containing the best rules to face all possible situation. In fact we found that with only one animals, the system has no useful rule in 0.6% of trial. This percentage increases with the number of animals:

1 animal	0.6%
3 animals	12%
5 animals	25%
7 animals	36%
10 animals	49%

7. Conclusions

This behaviors do not allow the system to stabilize the population set. When the system is not in possession of an applicable rule, it randomly generate another rule. So, with values so high, the system cannot work correctly.

We give the basis to extend the following concepts:

- the cooperation between animals of the same species;
- the concurrency of animals on the same resource kind;
- the coesistence of more needs to satisfy.

Bibliography

D. E. Rumelhart, McClelland J. L. (1986). Parallel Distributed Processing Exploration in the Microstructure of Cognition, MIT Press.

L. B. Booker (1982). Intelligent Behavior as an adaptation to the task environment. Doctoral dissertation, Departement of Computer and Communication Sciences, University of Michigan, Ann Arbor.

L. B. Booker (1988). Classifier System that Learn Internal World Models. Machine Learning, 3: 161-192.

M. Dorigo (1989). Genetics Algorithms: the State of the Art and some Research Proposal, Report n. 89-058. Dipartimento di Elettronica, Politecnico di Milano.

M. Dorigo, U. Schnepf (1991). Genetics_Based Machine Learning and Behavior_Based Robotics: a new syntesis. Report n. 91-044. Dipartimento di Elettronica, Politecnico di Milano.

M. Dorigo, U. Schnepf (1991). Organization of Robot Behavior Trough Genetic Learning Processes. To appear in the Proc. of the Fifth IEEE International Conference on Advanced Robotics.

D. E. Goldberg (1989). Genetics Algorithms in Search, Optimization and Machine Learning. Addison-Wesley.

North, Holland. Future Directions in artificial intelligence, P. Flatch, A. Meersman Editors.

Roberts, Gary, Departement of Artificial Intelligence, University of Edimburgh, Scotland. A Rational Reconstruction of Wilson's Animat and Holland's CS-1.

S. W. Wilson (1985). Knowledge Growth in an Artificial Animals. Proceedings of the First International Conference on Genetic Algorithms and their Applications (pp.16-23). Pittsburg, PA: Lawrence Erlbaum.

S. W. Wilson (1987). Classifier System and the Animat Problem. Machine Learning, 2:199-228.

Dynamic Management of the Specificity in Classifier Systems

Cathy Escazut **Philippe Collard** **Jean-Louis Cavarero**
University of Nice-Sophia Antipolis, I3S-CNRS,
250, av. A. Einstein 06560 Valbonne, France
Tel : (33) 92 94 26 17 - Fax : (33) 92 94 28 98
e-mail : escazut@mimosa.unice.fr, pc@mimosa.unice.fr, jlcava@mimosa.unice.fr

Abstract

The estimation of the rule usefulness in a classifier system is faced to the credit-apportionment problem. Usually, the apportionning of payoffs process is performed by the bucket brigade algorithm. However, some works have shown that this algorithm presents some difficulties.

Generally, the condition part of a rule is defined on an alphabet containing a "don't care" symbol. That is why a same rule can be fired in different contexts. In such conditions, it is impossible to use too generalized classifiers because of the incoherence of the strength management.

The solution we propose here, can solve the problem : general classifiers belonging to a success-ending sequence are dynamically specialized. In order not to store all the sequence actions, the Bucket Brigade algorithm is applied to the new-created rule specificity. So, the closer a classifier is from the end of the solution sequence, the more specific it is. This new algorithm is presented here and applied to an automous moving robot which must learn how to move in an environment with obstacles.

Keywords : Classifier Systems - Bucket Brigade Algorithm - Specificity - Back Specialization - Autonomous Moving Robots.

1 Introduction.

In this paper, we are interested in classifier systems. Such systems learn how to solve problems which solution is a multi-step task. Learning is performed thanks to payoffs sent by the environment. They translate the exactness of the answer proposed by the system. Then whether they are positive or negative, these payoffs are back-propagated through the whole sequence of actions using the Bucket Brigade algorithm. With this mechanism each rule, also called classifier, is evaluated in accordance with its role played in obtaining the searched solution. This measure is computed by updating a coefficient named strength for each classifier. This value is used to determine the rule to be activated.

The presence of a "don't care" symbol (#) in the classifier representations is important because it allows us to generalize classifiers by expressing disjunction in the condition parts of classifiers. So, a classifier with "don't care" symbols can belong to two different sequences of actions one ending in failure, and the other one ending in succes. In the first case, the feedback mechanism will reduce the classifier's strength, and in the second one it will increase this same coefficient. The measure of the rule exactness becomes unreliable : the strength value is corresponding to which context ?

Different solutions to solve this problem had been proposed. We expose some of them in section 3. Our solution is described in the section 4. Finally, the last section presents an implementation of the new algorithm with the test results.

2 The credit assignment problem.

A classifier system is a machine learning system that learns syntactically string rules called classifiers [1]. All the classifiers form the *population*. Each of them is specified by a condition part defined over the alphabet {0, 1, #} where the "don't care" symbol # represents either a 0 or a 1, an action part defined over the same three-letters alphabet {0, 1, #} but the # symbol is playing the role of a "pass through" in the sense that wherever the # symbol occurs in the action part, the corresponding bit in a message satisfying the condition part is passed through into the outgoing message. Moreover, a strength is corresponding to the reliability rate of each rule. The updating of this value is necessary to estimate the classifiers in accordance with the role they played in the obtention of the solution. This ajustement follows the 7 steps of the bucket brigade algorithm :

1. All the messages sent by the environment are posted in a message list.
2. All the messages in the message list are compared to each classifier of the population. The set M composed with the matching classifiers is formed.
3. Each classifier C_i in M makes a bid B_i proportional to its strength S_i. In this way rules that are highly fit are given preference over other rules.
4. A new set, M' with all the classifiers making an important bid, is created.
5. The strength S_i of each classifier C_i in M' is reduced by the amount of its bid B_i. This same amount is shared out among all the classifiers wich sent the message which C_i matched.
6. A new message list is formed from the action parts of the classifiers in the set M'.
7. If a payoff, P, is received by the environment, it is divided in equal share among all the classifiers in M', increasing each classifier's strength by the share amount.
8. Return to step 2 until the termination condition is satisfied.

In fact, the classifier strength is updating following the expression :

$$S_i(t+1) = S_i(t) - B_i(t) + R_i(t)$$

where $S_i(t)$ is the strength of the classifier C_i at time t, $B_i(t)$ represents the bid offers by C_i, and R_i is the C_i's receipts from its successors in the solution sequence or from the environment reward.

We have seen the bid offers by a classifier is proportional to the strength, so we can quantify B_i as the product of the strength by a coefficient bid named C_{bid}. Moreover, the receipts, R_i, are in fact the bids offer by C_i's successors, noted C_{i+1}. Then we can write $R_i = C_{i+1} \cdot C_{bid}$. So the classifier strength can be expressed by :

$$S_i(t+1) = S_i(t) \cdot (1 - C_bid) + S_{i+1} \cdot C_{bid}$$

The bucket brigade algorithm has advantages of simplicity. However some researchers have shown that this algorithm presents some drawbacks. The two main ones are that it may not adequaly reinforce long action sequences [9], and it is inappropriate in approximating the rule usefulness functions under multiple contexts.

In order to explain clearly this last apportionment of credit problem, let us take an example (Figure 1). Suppose the message sending by the environement is 11000, two classifiers can be fired. If the classifier, C_1, 1####:11100 is applied, then the activation of the rule C_2, #1###:10#10 gives a success, making the C_2's strength higher. In the next cycle, faced to the same environmental message, the classifier C_2 is fired because of its high performance. Unfortunately this directly leads to a failure.

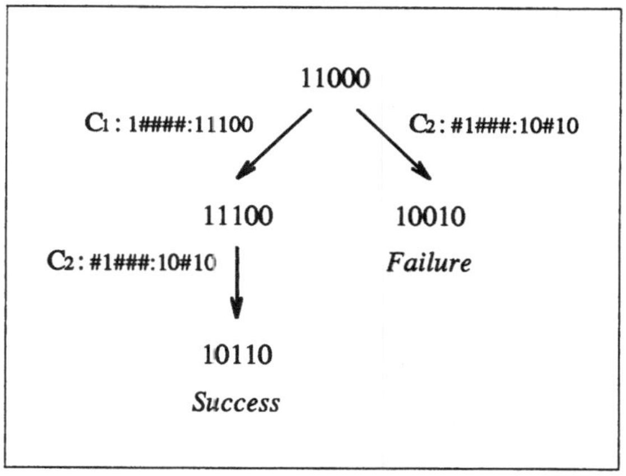

Figure 1: The credit apportionment problem.

This example permits to point out the problem caused by generalized classifiers : the exactness of a multi-context classifier cannot be evaluated by a single strength.

3 Related works.

There are some alternative approaches for solving the problem caused by the use of general classifiers.

The naive one consists in using for each context the more specific rule that matches the context. In this case the drawback is that it leads to use a large number of rules : one for each context possible, and the "don't care" symbol becomes useless.

Nevertheless the presence of # symbols in the part condition is absolutely necessary. They permit the classifiers to organize themselves into a structure called ***default hierarchy*** by Holland [3]. Default hierarchies are overlapping sets of rules where general classifiers cover general conditions and more specific ones cover the exceptions. In [4], Goldberg uses in his tests an initial population organized in a default hierarchy. Such populations have two advantages over nonoverlapping rule sets : the population is smaller, and the solution space is enlarge.

Exception rules (more specific classifiers) must be fired in context where default rules (general rules that are correct in some situations, but incorrect in others) would conduct to a failure. So the conflit resolution mechanism must choose the classifier with high specificity. This suggests that the bid value has to be proportional to the product of strength and some linear function of specificity.

This solution allows us to work with a population formed with generalised classifiers (the default rules). But this also supposes that all the exceptions possibles are known (the exception rules). Moreover, these schemes cannot be expected to perform adequately in arbitrary classifier system environment. Interested readers will refer to [5] for more details.

An other solution consists in associating to each classifier not a single strength but an array of strengths : one for each activation context [6]. In that way the algorithm proposed separates the contexts into context subsets. Then it employs array-valued bids to estimate rule usefulness at the context subset level. The strength updating is performed by the bucket brigade algorithm.

Even though it is interesting, this solution presents an important weakness : it assumes that we know how many contexts will be activated in order to define the array size. Moreover, it needs a large memory space fore storing all the strengths arrays : one per classifier.

In the next section, we propose a solution in order to overpassing these problems.

4 The back specialization algorithm.

The solution we propose can solve the problem without any knowledge on the number of activation contexts [7]. Our suggestion is to apply the bucket brigade algorithm not only for the rule strength management, but also for the new created rule mechanism. The idea of the back-propagation is here applied to the rule specificity. The specificity of the classifier C_i, noted Spe^{C_i}, is a measure of the generalization degree of the rule. We define this value by the rate : $Spe^{C_i} = p/l$ where p is the number of instanciated positions of the condition part, and l represents the length of the condition part. So if $Spe^{C_i} = 0$, the condition part classifier is composed only with "don't care" symbols. And if $Spe^{C_i} = 1$ the condition part of C_i contains no # symbol.

The back specialization algorithm principle is simple : a new classifier more specific is created from a generic rule only if this rule belongs to a success ending sequence. (See figure 2). The back propagation of the specificity is used in order not to store all the classifiers fired in an answer. In this way a classifier will be specialized if its successor in the solution has been previously specialised.

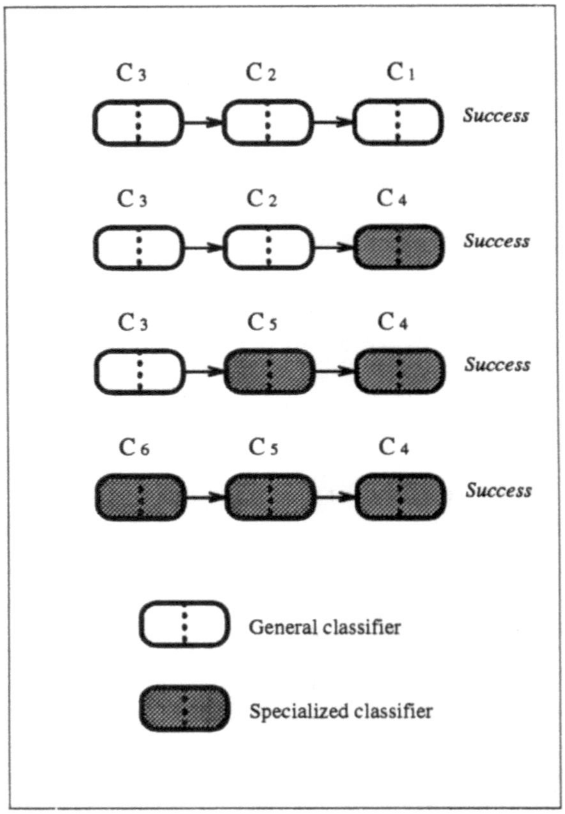

Figure 2: The back specialization mechanism.

With this specialization mechanism the population evolves in size : initially, all the classifiers are totally general ($Spe^{C_i} = 0$), and new more specific rules are progressively inserted. But what about the specificity of the new created ruled ? Two mechanisms have been used. They can be either completely specialized or progressively specialized.

4.1 The complete back specialization.

In this first method the specialization permits the creation of totally specialized classifiers. If we take again the example described by the figure 1, the initial population is composed with the two general classifiers :

C_1 : 1####:11100
C_2 : #1###:10#10

After the first success the classifier C_2 has to be specialized : it is the last rule fired in the success ending sequence. The totally specialized classifier C_3 is so added to the population :

C_1 : 1####:11100
C_2 : #1###:10#10
C_3 : 11100:10110 (from C_2)

The second time the system applies this success ending sequence, the population will be increased by the new classifier C_4, created from C_1, because its successor,C_3, is an already specialized rule. The population becomes :

C_1 : 1####:11100
C_2 : #1###:10#10
C_3 : 11100:10110 (from C_2)
C_4 : 11000:11100 (from C_1)

So, the final population is composed with the initial classifiers, and with all the totally specialized rules belonging a success ending sequence. Like in the solution presented earlier, it is necessary to privilege the activation of specialized classifiers. Tests have proved that this solution is not as good as it seems [8]. That is why a variante of this mechanism has been envisaged : the partially back specialization process.

4.2 The partially back specialization.

This method is inspired by the previous one. But now classifiers are progressively specialized. This means that the specificity evolves as the system learns. With this mechanism, the population became rapidely important but also more diversified. That is why a success ending sequence can be found more quickly.

The new classifier specificity is compute using an equation similar to the one used to update the classifier strength by the bucket brigade algorithm :

$$Spe^{C_i} = Spe^{C_i} \cdot (1 - C_{Spe}) + Spe^{C_{i+1}} \cdot C_{Spe}$$

where C_{Spe} is the specialization coefficient.

This formula is not directly applicable to the last classifier of the solution sequence : it has no successor. In this case, the new specificity is 1.

In order to illustrate this new mechanism let us take the figure 1 example. The initial population is the same as in the totally back specialization :

C_1 : 1####:11100
C_2 : #1###:10#10

When the system takes for the first time the succes sequence, the last classifier, C_2 is totally specialized creating C_3 :

C_1 : 1####:11100
C_2 : #1###:10#10
C_3 : 11100:10110 (from C_2)

The second time the system applies the success ending sequence, the last but one classifier, C_1 is in its turn partially specialized. C_4's specificity is compute from these of its successor : the classifier C_3, and from its parent : the rule C_1 :

C_1 : 1####:11100
C_2 : #1###:10#10
C_3 : 11100:10110 (from C_2)
C_4 : 1#00#:11100 (from C_1)

On the next cycle, classifiers C_4 is one more time specialized, giving a new population :

C_1 : 1####:11100
C_2 : #1###:10#10
C_3 : 11100:10110 (from C_2)
C_4 : 1#00#:11100 (from C_1)
C_5 : 1100#:11100 (from C_4)

In this way the closer a classifier is from the end of a solution sequence, the more specific it is. We can say that the specificity is back propagated from the last to the former classifier of the sequence.

The test results of this technique are presented in the next section.

5 Experimental results.

The mechanism of the partially back specialization had been used for the simulation of an autonomous moving robot. This classifier system application is inspired from numerous works [9, 10].

The robot must learn how to navigate in a completely unexplored maze-like environment. The system has to find and learn obstacle-avoiding paths. To do this, our robot must have a good description of the area where it is moving. The environment perception can be considered to be the robot's field of vision. This last notion will be used as messages in order to fired classifier.

The structures used here refers to the one employs by Wilson in [11]. The messages are not the usual one-dimensional string pattern. Instead, a 5×5 array of binary characters are representing the field of view (See figure 3). In order to anticipate moves, the field of view is wider than the allowed moves (represented by a dot on the robot's field of vision).

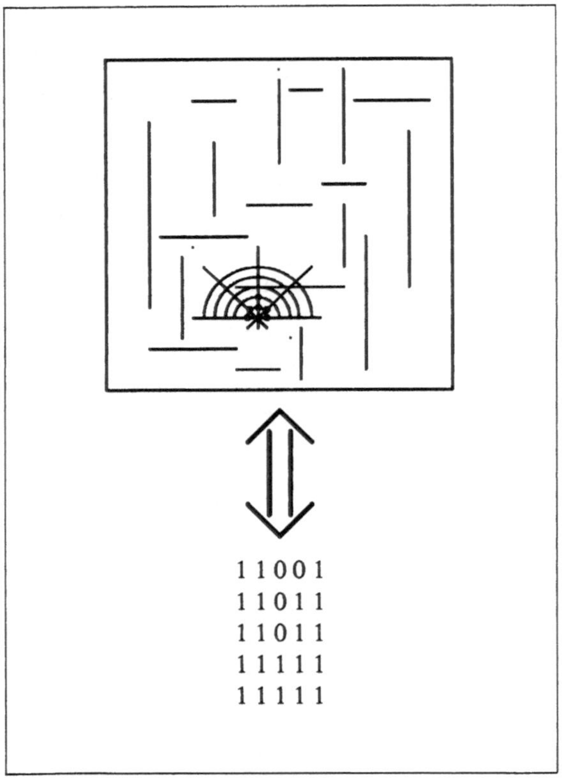

Figure 3: The robot's field of vision.

Since the messages are responsible for the classifier activation, the condition part is a 5×5 array of ternary characters. The action part allows to compute the new position of the robot from the distance covered and the rotation executed. That is to say that the action of a classifier is composed by the polar coordinates of the robot's new position comparatively to the old one.

For instance the message of the figure 3 can fired the following classifier :

$$\left|\begin{array}{ccccc} 1 & \# & 0 & 0 & \# \\ 1 & \# & \# & 1 & 1 \\ 1 & 1 & 0 & 1 & 1 \\ 1 & \# & \# & \# & 1 \\ \# & 1 & \# & 1 & \# \end{array}\right| : 1\ 000$$

Activing the action part "1 000", the robot will move to two steps (the first part of the condition : 1) straight away (the second part of the condition : 000).

The initial population we used in our experiments, contains six completly general classifiers. Their condition part is entirely filled with "don't care" symbols, and their action part represents a legal move. All the initial strengths are set to zero. In order to validate our solution we used a standard classifier system, and compared the results with one using our solution.

During the first cycles, the standard classifier system and the one doted with the back specialization have a similar behavior (Figure 4). In the case of the back specialization mechanism, this phase is a learning one with specialized rules creation.. After this, it uses only the specialized classifiers making the success rate growing up. Even though, the system working only with general classifiers can't learn anything : its success rate decreases.

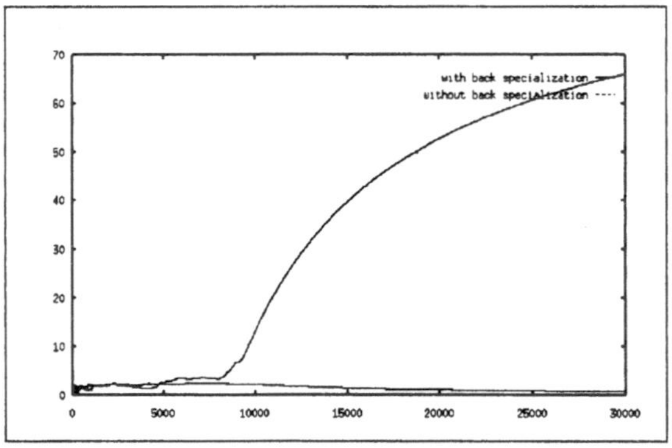

Figure 4: Evolution of the success rate with time.

During the "learning phase" (aproximatively 9500 cycles), the back specialization mechanism creates all the necessary partially specialized rules in order the robot can move in the environment avoiding obstacles. After this phase, all the robot's moves are correct (Figure 5).

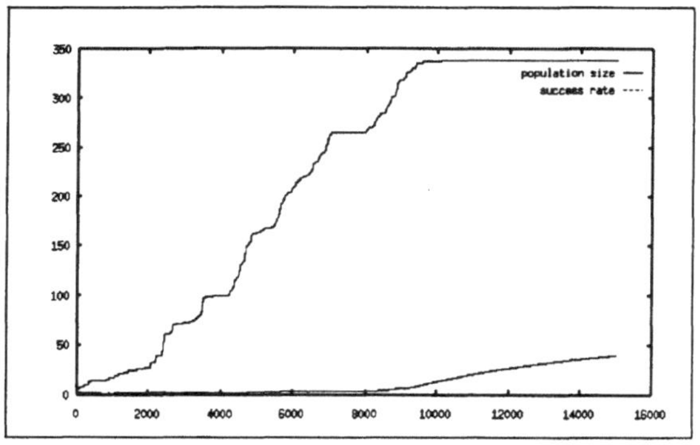

Figure 5: Evolution of the population size with time.

The figure 6 shows the evolution of the population average specificity. We can notice that the maximal specificity is reached about the cycle 2500. Then the average decreases and finally tends to stabilize.

The initial phase of growth is the result of the total specialization of the latest rules of ending sequences. The phase of decrease corresponds to the creation of partially specialized classifiers, due to the utilization of the totally specialized rules.

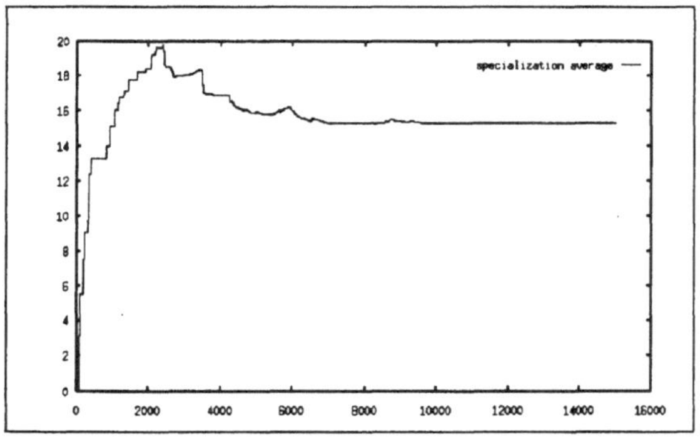

Figure 6: Evolution of average specificity with time.

The final population is composed with the six general classifiers having a low strength, and with more or less specialized ones which strength is proportional with their specificity : the more a classifier is specialized, and therefore reliable, the more its strength is important.

Conclusions and future perspectives.

In this paper we presented a solution which can allow the use of an initial population containing only general classifiers. We pointed out the interest of a dynamic growing population in classifier systems performed by a back specialization mechanism. This solution presents the advantage of having pertinent classifiers created as the system learns.

Moreover, this solution constitute a coherent model since the same algorithm : the bucket brigade, is used for the strength and specificity rules management.

As the back specialization create new classifiers, this mechanism can be assimilated to a genetic operator. So it may be interesting to compare our solution with a genetic based one.

Further works will conduct us to consider the adaptability degree of the robot : after a first learning phase, we would like to change the robot's environment in order to make the system independant and autonomous.

References

[1] Holland J.H., *'Escaping brittleness : The possibilities of general-purpose learning algorithms applied to parallel rule-based systems'*. In R.S. Michalski, J.G. Carbonell and T.M. Mitchell (Eds.), Machine Learning II (pp. 593-623), 1986.

[2] Wilson S.W., *'Hierarchical credit allocation in a classifier system'*. In L. Davis (Ed.) Genetic algorithms and simulated annealing (pp. 104-115), 1986.

[3] Holland J.H., *'Adaptation in natural and artificial systems'*. Ann Arbor : The university of Michigan Press, 1975.

[4] Goldberg D.E., *'Genetic algorithms in search, optimization, and machine learning'*. Addison-Wesley, 1989.

[5] Smith R.E. and Goldberg D.E., *'Variable default hierarchy separation in a classifier system'*. In G.J.E. Rawlings (Ed.) Foundations of genetic algorithms (pp. 148-167), 1991.

[6] Huang D., *'The context-array bucket-brigade algorithm : an enhanced approach to credit-apportionment in classifier systems'*. Proceedings of the third international conference on genetic algorithms and their applications (pp. 311-314), 1989.

[7] Escazut C., *'L'attribution des récompenses au sein d'un système apprenti'*. Rapport de DEA, University of Nice, 1991.

[8] Collard P., *'Back specialization with the bucket brigade algorithm'*. Proceedings of the tenth IASTED international conference (pp. 130-133), 1992.

[9] Wilson S.W., *'Classifier systems and the Animat problem'*. In Machine learning, Vol (2), (pp 199-228), 1987.

[10] Millán J. del R. and Torras C., *'Learning to avoid obstacles through reinforcement'*. In Machine Learning, proceedings of the eigthth international workshop, (pp. 298-302), 1991.

[11] Wilson S.W., *'Adaptative "cortical" pattern recognition'*. Proceedings of an international conference on genetic algorithms and their applications, (pp.188-196), 1985.

DYNAMIC SEQUENCING OF A MULTI-PROCESSOR SYSTEM: A GENETIC ALGORITHM APPROACH

Colin Reeves
School of Mathematical and Information Sciences
Coventry University
UK
Email: CRReeves@uk.ac.cov.cck
Helen Karatza*
Aristotle University of Thessaloniki
Thessaloniki
Greece

Abstract

The dynamic sequencing of jobs through a multi-processor system is one to which little attention has been paid, in contrast to the extensive literature on static sequencing problems. Yet in many real problems, to assume, as the static model does, that we know about all jobs that will arrive in the course of a processing cycle is hardly realistic.

Existing solutions usually assume a queueing-theoretic orientation, rather than an optimization one, in which the decision as to which job should be processed is made on the basis of some simple selection criteria, such as First-Come First-Served, or Shortest Processing Time.

Here we investigate the use of a Genetic Algorithm (GA) to solve the successive sequencing problems generated by finding a near-optimal sequence for those jobs available just before successive *event times*—the times at which the job being processed on the first processor completes its processing. Some comparisons are made between using the GA approach versus some simple rules.

1 Introduction

Genetic algorithms have been shown [1, 2, 3] to be robust methods of solving static sequencing problems in which processing times are fixed and known in advance. Recently, they have been extended to cope with problems involving stochastic processing times [4], in which a simple GA outperformed naive search methods, as well as a more sophisticated simulated annealing procedure. We believe this is because the GA works with a population of solutions, so that particular subsequences receive multiple trials, thus overcoming the inherent weakness of 'one-shot' methods for sequencing in a noisy environment.

In this paper, we turn to the consideration of another weakness of conventional sequencing models: the assumption that the jobs are all present at the outset. In reality, this is only approximately true; in problems involving the scheduling of jobs through a computer, for instance, we have a queue of jobs which is being continually augmented at one end even as jobs leave the queue at the other. In a survey made over 10 years ago, Graves [5] observed:

> It may be quite easy to construct a schedule; what is difficult is the constant schedule revision required by the dynamic environment.

More recently, Dudek *et al.* [6] have suggested several reasons why so much flowshop scheduling research has found so little application, among which was the fact that

> Real flowshop situations usually are dynamic rather than static.

Here we consider a problem of this type, where jobs arrive at a job pool before passing through m processors arranged in series. The time t_{ij} required for processing job i on processor j is known. There is infinite buffer storage between consecutive processors, and set-up times on each processor are assumed to

*On sabbatical leave in Dept. of Computer Science, City University, London, for the period of this research.

be sequence-independent. No job pre-emption is allowed. Initially there are n' jobs in the pool, but further jobs arrive as time passes.

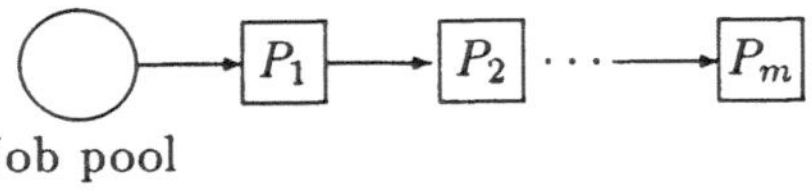

The problem is at any stage to determine the sequence in which the jobs in the pool should be processed in order to optimise some measure of performance. This is clearly a dynamic problem, since as more jobs arrive the current 'best' sequence may have to change. Even this is of course an approximation to what really happens—in real problems jobs may not call on all processors in the same order, they may need to visit a subset of processors more than once, and so on. However, our purpose in studying this simplified version of the problem, as outlined above, was to test the hypothesis that a GA would be a suitable means of approaching such a problem.

There are a number of ways of assessing the performance of a system like that described. It was decided that the most natural performance measures would be the mean *response time* (in a machine shop context, this is often called the mean *flow time*),

$$R(n) = \sum_{j=1}^{n}(C_j - A_j)/n$$

where n jobs have been processed, and job j arrived at time A_j, and was completed at time C_j; and the *throughput rate*,

$$T(n) = n/(C_n - A_1).$$

These performance measures are of course correlated to some extent—normally, the higher the throughput rate the smaller the mean response time—but while response time refers to the system performance from the viewpoint of the jobs, throughput rate measures performance from the server's perspective.

2 Implementation

A simulation model of the system described above was programmed, as shown in the box below:

- Initialise job pool;
- Compute job sequence;
- Schedule 1st job;
- Compute 1st event time T_E.
- Repeat
  If no arrivals before T_E then
  1. schedule next job;
  2. compute next event time T_E;
- else
  1. add additional job(s) to current job pool;
  2. re-compute job sequence for current pool;
  3. schedule next job;
  4. compute next event time T_E;
- Until simulation time exceeds a specified limit.

Clearly, the GA enters at the points where a re-computation of the 'best' sequence of the current job pool is required. The simple scheduling rules would also be implemented at this stage.

It is important to realize that by re-computing the best sequence from the current job pool, we make the assumption that a good overall solution will be obtained if we try at any stage to sequence the currently available jobs as if no more jobs will arrive. It is this hypothesis that we shall evaluate by comparing with the more traditional job selection criteria.

2.1 The genetic algorithm

The GA used to solve the sequencing problem was adapted from that described in [3], whose characteristics can be summarised as follows:

- an initial population of 30 chromosomes using a sequence representation;
- parent selection using ranking;
- incremental population replacement (also known as a steady-state GA);
- replacement of a randomly chosen string of below-median fitness;
- a sequence-based crossover (see [3] for details);
- an adaptive mutation rate;
- a termination condition of $\mathcal{O}(nm\log[m+n])$ objective function evaluations.

At the first stage, the initial chromosomes were chosen at random, and this could also be done at each subsequent application of the GA. However, by basing, at each stage, the initial population on the population of solutions obtained at the previous stage, we found that good solutions to the current problem were determined more rapidly, which may be an advantage when a decision on the next job to be sequenced is needed in real-time.

2.2 Other selection criteria

There were 3 obvious candidates for simple selection criteria instead of the GA: we could use

- job arrival order (FCFS);
- shortest (first machine) processing-time order (SPT(1)).
- shortest (total) processing time order (SPT(all));

The first of these corresponds to doing nothing, simply scheduling on a First-Come First-Served basis; the other two attempt to take into account the likely delay to other jobs that could be incurred by scheduling a specified job now. Clearly, by scheduling a job with a large processing-time requirement when other (shorter) jobs are available, the response-time for those other jobs is likely to be increased. The first machine is of course the most important in this simplified model, since once the current job completes processing on the first machine, we are free to schedule another. These types of criteria have been studied for some special cases of single-processor scheduling problems using a queueing-theoretic framework, and Conway *et al.* [7] have an interesting discussion which shows that, under certain conditions, the Shortest-Processing Time criterion is optimal for single-machine problems. This has not been shown to be true for multi-processor problems—in fact, as we have remarked above, by proposing a re-scheduling of the current job pool, we are assuming that it is not.

3 Test problems

Several sets of test problems were generated. In each case, the arrival rate and service (processing) rate of jobs were assumed to be the same: clearly, if the arrival rate is greater than the service rate, the size of the job pool will increase without bound, which would not be tolerated in a real system. Job arrivals were assumed to occur according to a Poisson process, but job-processing times were generated from 3 different distributions—an exponential distribution (corresponding to a Poisson process), an Erlang-k (hypo-exponential) distribution with $k = 4$, and a branching-Erlang distribution. These correspond to the cases where the coefficient of variation of processing-times is 1, less than 1, and greater than 1 respectively. Further details of these distributions and their characteristics can be found in [8].

In each case, 30 jobs were assumed to be in the pool initially, and the simulation was continued until a further 200 jobs had entered the system. The values of $T(n)$ and $R(n)$ were measured at $n = 50, 100, 150$ and 200. Below we tabulate the results for a typical data set—although we have only reported this one example in detail, they represent a pattern that was consistently observed.

Erlang-4				
n	$R(n)$			
	GA	FCFS	SPT(1)	SPT(all)
50	16.73	34.20	32.21	29.56
100	23.11	37.90	35.60	32.68
150	26.16	40.60	37.83	33.09
200	28.28	45.00	42.52	36.57
	$T(n)$			
	GA	FCFS	SPT(1)	SPT(all)
50	0.970	0.869	0.911	0.857
100	0.956	0.891	0.889	0.902
150	0.940	0.911	0.907	0.917
200	0.942	0.919	0.913	0.915

Exponential				
n	$R(n)$			
	GA	FCFS	SPT(1)	SPT(all)
50	21.18	42.67	37.72	30.94
100	27.82	47.68	40.09	34.92
150	28.78	50.11	41.47	37.85
200	32.68	55.42	46.72	43.19
	$T(n)$			
	GA	FCFS	SPT(1)	SPT(all)
50	0.904	0.814	0.844	0.792
100	0.935	0.857	0.893	0.836
150	0.952	0.870	0.896	0.838
200	0.932	0.857	0.857	0.839

	Branching-Erlang			
n	$R(n)$			
	GA	FCFS	SPT(1)	SPT(all)
50	22.66	60.70	40.00	23.06
100	28.86	75.20	43.24	26.26
150	34.53	88.00	48.90	30.93
200	39.01	99.90	57.61	37.58
	$T(n)$			
	GA	FCFS	SPT(1)	SPT(all)
50	0.750	0.656	0.697	0.667
100	0.947	0.589	0.775	0.707
150	0.868	0.663	0.812	0.772
200	0.916	0.694	0.821	0.808

It can be seen that in every case dynamically scheduling jobs using the genetic algorithm approach has increased the throughput rate relative to the simple selection criteria, and in most cases by a substantial margin. Except in the case of the SPT(all) criterion for branching-Erlang processing times, the mean response time was significantly reduced as well. This has occurred because the SPT(all) selection criterion introduces a lot of processor idle time into the system, and leaves the few very long jobs that do arise with this statistical distribution to be done only when no short jobs are left.[1]

It should also be pointed out that the actual computing time needed (basing each new initial population on the previous final one, as discussed above) for each re-scheduling was relatively insignificant: no more than 40 seconds on a Sequent S82 computer. This of course depends on the number of jobs to be re-scheduled at each stage, but given that the job arrival rate is limited to be no greater than the service rate per machine, the only factor that would affect this substantially is the size of the *initial* pool.

4 Conclusions

This research has demonstrated that using a genetic algorithm is a viable and successful strategy for dynamically scheduling jobs in a multi-processor flowshop system. From our initial simulations we found that the improvement in terms of mean throughput rate and response time was substantial, at a computational cost which would mean the procedure could be implemented in real time. However, more work needs to be carried out to investigate how performance varies with the size of the pool, and with different job processing time distributions.

It is also clear, as pointed out in the introduction, that in real problems processing times are not actually known in advance, although we may be able to predict them with a fair degree of accuracy. Genetic algorithms have been found effective for stochastic flowshop sequencing, and future work will also investigate the potential for using GAs in multiprocessor systems which are both dynamic and stochastic.

Finally, GAs have also been used for more

[1]This result actually points up a slight failing in our choice of performance measure. Our $R(n)$ measure has only taken account of *completed* jobs. Although we do not expect it to affect the general conclusions from these comparisons, we intend to rectify this in future work.

general *job-shop* scheduling problems in the static, deterministic case. It would clearly be worth examining their application to stochastic and/or dynamic versions of this problem.

References

[1] G.A.Cleveland and S.F.Smith (1989) Using genetic algorithms to schedule flow shop releases. *In* J.D.Schaffer (Ed.) (1989) *Proceedings of the 3rd International Conference on Genetic Algorithms.* Morgan Kaufmann, Los Altos, CA.

[2] H.M.Cartwright and G.F.Mott (1991) Looking around: using clues from the data space to guide genetic algorithm searches. *In* R.K.Belew and L.B.Booker (Eds.) (1991) *Proceedings of the 4th International Conference on Genetic Algorithms.* Morgan Kaufmann, San Mateo, CA.

[3] C.R.Reeves (1993) A genetic algorithm for flowshop sequencing. *Computers & Ops.Res.*, (in review).

[4] C.R.Reeves (1992) A genetic algorithm approach to stochastic flowshop sequencing. *Proc. IEE Colloquium on Genetic Algorithms for Control and Systems Engineering.* Digest No.1992/106, IEE, London.

[5] S.C.Graves (1981) A review of production scheduling. *Operations Research*, **29**, 646-675.

[6] R.A.Dudek, S.S.Panwalkar and M.L.Smith (1992) The lessons of flowshop scheduling research. *Operations Research*, **40**, 7-13.

[7] R.W.Conway, W.L.Maxwell and L.W.Miller (1967) *Theory of Scheduling.* Addison-Wesley, Reading, Mass.

[8] C.H.Sauer and K.M.Chandy (1981) *Computer Systems Performance Modelling.* Prentice-Hall, New Jersey.

Genetic Algorithms versus Tabu Search for Instruction Scheduling

Steven J. Beaty *
NCR Microelectronics
2057 Vermont
Fort Collins, Colorado 80525
Steve.Beaty@ftcollins.ncr.com
beaty@longs.lance.colostate.edu

Abstract

Most scheduling problems require either exponential time or space to generate an optimal answer [7]. Instruction scheduling is an instance of a general scheduling problem and Dewitt [8] uses this fact to show instruction scheduling is a NP-complete problem. This paper applies Genetic Algorithms, Tabu Search, and list scheduling to the instruction scheduling problem and compares the results obtained by each.

1 Introduction

Sequencing is defined by Ashour [1] as being "concerned with the arrangements and permutations in which a set of jobs under consideration are performed on all machines." That is, what is the order the jobs will be performed; what is the priority of each job? Sequencing thereby ranks the jobs to be executed. Baker [2] states "scheduling is the allocation of resources over time to perform a collection of tasks." Scheduling usually places already ordered jobs into slots, often accounting for conflicts in resource usage. The combined sequencing/scheduling (order/place) process produces the desired outcome: jobs placed on machines capable of performing the desired tasks in the correct order at a correct time.

Instruction scheduling (IS) involves the placement of atomic machine operations into machine instructions. A data dependence DAG (DDD) is often used to describe the necessary operations and their order. The nodes in a DDD contain the operations, and the edges denote a partial order on the nodes. This partial order is used to guarantee both program dataflow and machine resource requirements. The edges of a DDD do not constrain the order nodes are scheduled, only the order they appear in the final schedule. The solution space may be viewed as an incomplete n-dimensional hypercube, where n is the number of operations to be performed. Each operation might be executed at a variety of locations in the code, and each dimension represents the range of instructions that operation can be placed. IS is complicated by both the inherent dataflow ordering between operations in the source code and the complexities of the architecture of the target machine. The architecture may have complex timings between operations, a number of different field encodings, and a limited number of resources that can perform any given operation.

Most existing IS methods rely on heuristics to remove the examination of parts of the search space that appear fruitless. Using heuristics can be difficult when attempting to arrive at an efficient yet efficacious scheduler. This difficulty is compounded by several factors.

- The heuristics generally must be regenerated for each machine targeted.
- The heuristics themselves are not in a form easily understood by humans, thus making it difficult for humans to correctly choose and modify a scheduler's behavior.
- It is possible that the heuristics do not address an issue that has great influence on the final code.
- Heuristics that work well for one ordering of operations may not work well for another.
- Heuristics are also picked before the execution of the instruction scheduling routine and remain static throughout. They have no ability to learn from previous runs or to take advantage of anomalous situations existing in specific situations that lead to shorter code sequences.

*the author is an affiliate faculty member at Colorado State University

With these difficulties in generating schedulers, several stochastic methods have been attempted to solve the IS problem. In Jacobs et al. [16], the Metropolis Monte Carlo technique was used. De Gloria and Faraboschi [11, 12] used a Boltzmann Machine approach to good effect. In [4, 3], Genetic Algorithms were shown to produce good results.

List scheduling (LS) is a general [7] scheduling method often used for instruction scheduling [10]. LS builds a ready set that contains all jobs that are not waiting on the results of another job. In a DDD, this is represented as nodes with no unscheduled predecessors. In finding the ready set, LS performs a topological sort of a DDD, thereby reducing the search space of the scheduling problem and increasing the chances of finding a valid schedule. List scheduling has an implicit heuristic: scheduling nodes with no predecessors results in valid orderings more often than scheduling nodes with predecessors. As with all heuristics, there are instances where this assumption does not hold.

A difficulty with using LS in combination with stochastic methods is it requires time $O(n^2)$ [19, 10], where n is the number of nodes in the graph. Most stochastic methods require the evaluation of numerous of node orderings, creating a desire for a scheduling method with less time complexity. It is unnecessary to use the topological ordering of list scheduling if it can be replaced by a strong method of sequencing. Lookahead scheduling [5] addresses these concerns by providing a fast method to scheduling operations, without incurring the overhead of list scheduling.

2 Genetic Algorithms

Genetic Algorithms have been used successfully to perform TSP [23, 24], job shop [24], and flow shop [6, 21] optimization problems. Encouraging results from these problems drove the use of GAs for instruction scheduling.

The GENITOR GA program, developed by Whitley [22, 25], was used for these studies. It has some differences with "standard" GAs that appear to increase performance. It does not replace the entire population with each. generation. Instead it probabilistically chooses two parents to reform into two offspring. Recombination and mutation occur, then one of the offspring is discarded randomly. The remaining offspring is placed in the population according to its fitness in relation to the rest of the strings. The lowest-valued string is discarded. This keeps high-valued strings within the population, directly accumulating high-performance hyperplanes. It also bases the reproductive opportunity upon rank with the population, not upon a string's fitness value in comparison with the average of the population, reducing the impact of selective pressure fluctuation. It also reduces the importance of choosing a proper evaluation function for fitness in that the difference in the fitness function between two adjacent strings is irrelevant.

An evaluation function that ranks the fitness of a string in the population must be produced. Choosing a proper function, i.e., one that represents a string's relative worth in the population without inordinate bias, is important. For instruction scheduling, a minimization problem, the result of the evaluation function must reflect the length of the final schedule that a member of the population generates. A difficulty encountered is that not all members will produce valid final schedules. Failures will occur when a conflict arises (e.g. timing, resource, or field) due to the order of scheduling the operations. It is not surprising that certain orders will fail to produce valid schedules for a given DDD; the impact of ordering on the production of valid schedules is emphasised in all previous instruction scheduling methods.

After consideration, the evaluation function selected performs a "worst-case" evaluation when a string fails to produce a valid schedule. This evaluation is produced by assuming all unscheduled operations have no parallelism available in them, necessitating their serial placement. The calculation of the evaluation function is then trivial; it is the number of instructions that contain operations so far, plus the length of the path containing the serial ordering of all the unscheduled operations. This produces a good estimate in the event of schedule failure; those schedules with more operations placed will receive a better evaluation. It also produces an exact evaluation in the presence of a valid schedule.

Six different recombination operators were studied. These are described in Starkweather et al. [21] and include two order crossovers, partially mapped crossover, cycle crossover, position-based crossover, and edge recombination. Starkweather et al. demonstrate that each operator will perform differently for each problem domain. The performance difference can be measured in the speed of convergence to a good solution. For example, edge recombination finds good solutions more rapidly on the TSP while performing more poorly than the others on scheduling problems.

The number of generations should be related to the relative difficulty of producing an optimal schedule for a given DDD. DDDs with a few simple operations do not require as many generations to find good schedules as do those with many complex operations. For these experiments, the number of generations is n^2, where n is the number of operations. The size of the genetic population is n. The strings in the population are of strings of non-repeating integers. This representation is consistent with those used in the TSP and shop scheduling problems previously mentioned. All strings are randomly initialized. No effort is made to optimize GA parameters for IS in this study. The selection bias is 1.5. There is no mutation, adaptive or otherwise.

3 Tabu Search

Tabu search (TS) is an optimization method that uses a form of short-term memory used to keep a search from becoming trapped in a local minima. A tabu list is formed that keeps track of recent solutions. At each iteration in the optimization process, solutions are checked against the tabu list. A solution that is on the list will not be chosen for the next iteration (unless it overrules its tabu condition by what is called an aspiration condition.) The tabu list forms the core of tabu search and keeps the process from cycling in one neighborhood of the solution space.

At each iteration, a steepest-descent solution that does not violate the tabu condition is chosen. If no non-tabu improving solution exists, the best non-improving solution is taken. The combination of memory and gradient descent allows for diversification and intensification of the search. Local minima in the search space are avoided while good areas are well explored.

Two bits of pseudo-code will show the basic of the TS method used here. The first, in Figure 1, is the overhead procedure. It controls the number of iterations, updating of the best solution so far, and controls the tabu list. The second, in Figure 2, finds the next move in the search by a swapping procedure. All the possible swaps in the sequence are tried, and the best non-tabu swap is chosen. The routine shown finds the best move from the current location in the search space to a neighboring position. An alternative is to find the first location in the neighborhood that is an improvement over the current one. These two possibilities are termed *best improving* and *first improving* respectively.

TS has been effectively used for a number of problems related to the IS problem. Glover and McMillan [13] used it for employee scheduling, Eck [9] studied Job shop scheduling, and Laguna et al. [18] applied TS to machine scheduling. The success in these areas helped motivate this study of instruction scheduling.

```
tabu_search ()
{
    for (i = 1; i <= # iterations; i++)
    {
        value = best_move ();
        make best_move;
        make best_move tabu;
        if (value < global_best)
        {
            global_best = value
        }
    }
}
```

Figure 1: Tabu Search

```
best_move ()
{
    for (i = 1; i < n; i++)
    {
        for (j = i + 1; j <= n; j++)
        {
            swap (sequence[i,j]);
            value = evaluate (sequence);
            if (tabu[i][j] &&
                value > global_best)
            {
                continue;
            }
            if (value < best_so_far)
            {
                best_so_far = value;
                best_move = [i,j];
            }
        }
    }
    return best_so_far;
}
```

Figure 2: Best Move

For this study, two different resequencing operators were applied. The first swapped two machine operations in the sequence and evaluated the resulting sequence. The second performed an insertion procedure by removing one operation from the sequence, shifting the remaining elements to fill in the open spot up a certain point and then placing the removed operation into that spot. In some cases, the insertion procedure may prove more suited for a sequencing task if the sequence is almost completely optimized. For example, if an optimal sequence is 1 2 3 4 5 6 7 8 9, and the current solution is 1 9 2 3 4 5 6 7 8, the swap procedure would take more iterations to arrive at the optimal solution. This can occur in IS when the last node is a branch instruction and must be placed last in the block.

There are a number of different types of information that can be kept on the tabu list. For example, when a operation is moved from one position in the sequence to another, one could make moving that operation back to its original position tabu. One could keep the relative position information for each operation. The total sequence could be saved. In these experiments, the contents of the tabu list for the two procedures was different. For the swap procedure, the list contained pairs of operations that had been swapped recently. This kept operations from reversing their current relative positions. Insertion is more difficult to express as a relative condition as each insertion changes the position of many different operations. For this case, the tabu list contained the actual permutations from the recently effective evaluations.

The same evaluation function (lookahead scheduling) used with GAs, was used for TS. The first improving move scheme was employed. The tabu list size was of length seven for both the swap and insert procedures. Various other lengths were studied (e.g. length n, where n was the number of MOs) and found to produce very similar results. This suggests that the local minima neighborhoods are fairly small for this instance of the IS problem. The number of evaluations was limited to n^2 as in the GA approach. A running average was kept in order to compare directly with the GA results.

4 Comparisons and Conclusions

A number of different programs were run through the compiler in order to compare the effectiveness of GAs, TS, and list scheduling. The compiler was targeted to produce code for the IBM RS/6000 architecture [15]. A representative example is shown in Figure 3. This graph represents the major block found in the forth Lawrence Livermore kernel [20]. The `EDGE_RECOMB`, `ORDER1`, `ORDER2`, `PMX`, `CYCLE`, and `POSITION` lines are the six genetic operators. The `LIST` line represents the list scheduling result. Note it is drawn to give a reference, it requires only one evaluation to compute. The `INSERT` and `SWAP` lines are the two tabu operators. Both GAs and tabu search worked well for finding good solutions to IS problems.

Genetic operators emphasizing order converged faster than those emphasizing adjacency. This comes as no surprise; all previously effective methods for IS also emphasize order. This evidence does however shed additional light on the nature of of the instruction scheduling process by providing more controlled, empirical evidence. The ordering of the placement of nodes by the genetic algorithm mirrors the approach used by human coders. The nodes with the greatest impact on final schedule length are placed first, with those having lesser impact placed later. The order of placement that ensures validity is also reflected.

In this study, both the swap and insert operators demonstrated very similar behavior, pointing to the fact that absolute order is not of ultimate importance for this particular IS problem. Both were able to avoid local minima and find competitive solutions. It took tabu search longer to find the better solutions than the best genetic operators. This may be a reflection of the fact that GAs are more suited to the IS problem. It could also be that the genetic technique used is more highly "evolved" having been used for a number of sequencing problems before.

References

[1] S. Ashour. *Sequencing Theory*. Springer-Verlag, New York, 1972.

[2] K. R. Baker. *Introduction to Sequencing and Scheduling*. John Wiley and Sons, Inc., New York, 1974.

[3] S. Beaty. Genetic algorithms and instruction scheduling. In *Proceedings of the 24th Microprogramming Workshop (MICRO-24)*, Albuquerque, NM, November 1991.

[4] S.J. Beaty. *Instruction Scheduling Using Genetic Algorithms*. PhD thesis, Mechanical Engineering Department, Colorado State University, Fort Collins, Colorado, 1991.

[5] Steven J. Beaty. Lookahead scheduling. In *Proceedings of the 25th Annual International Sym-*

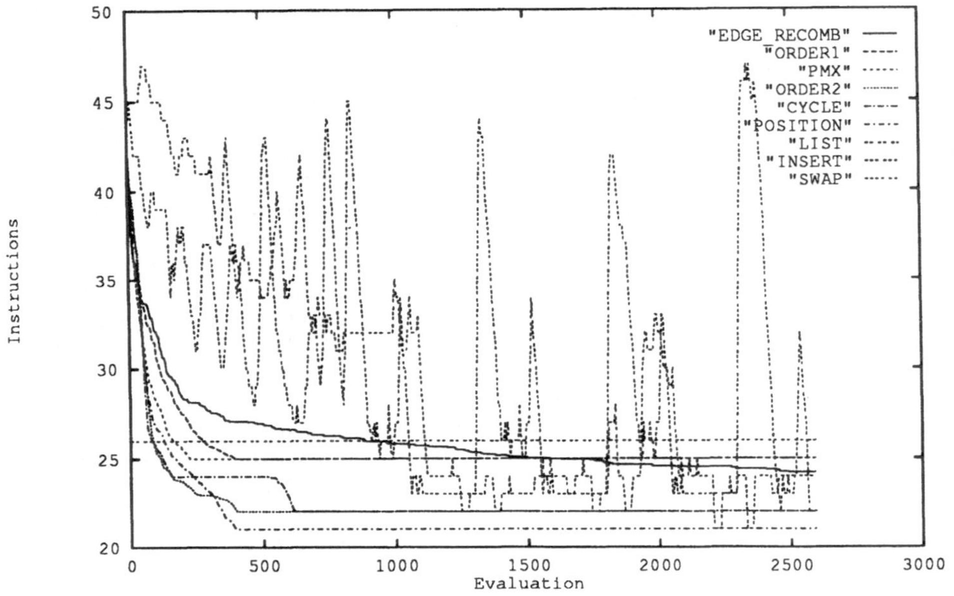

Figure 3: Comparative Results

posium on Microarchitecture (Micro-25), pages 256–259, Portland, Oregon, December 1992.

[6] Gary A. Cleveland and Stephen F. Smith. Using genetic algorithms to schedule flow shop releases. In *Proceedings of the Third International Conference on Genetic Algorithms*. Morgan Kaufmann, 1989.

[7] E.G Coffman. *Computer and Job-Shop Scheduling Theory*. Jon Wiley & Sons, New York, 1976.

[8] D.J. DeWitt. *A Machine-Independent Approach to the Production of Optimal Horizontal Microcode*. PhD thesis, Department of Computer and Communication Sciences, University of Michigan, Ann Arbor, MI, 1976.

[9] B.T. Eck. Good solutions to job shop scheduling problems via tabu search. Technical report, Department of Industrial Engineering and Operations Research, Columbia University, New York, May 1989.

[10] F. Gasperoni. Compilation techniques for vliw architectures. Technical report, Courant Institute of Mathematical Sciences, New York University, March 1989.

[11] A. De Gloria and P. Faraboschi. A boltzmann machine approach to code optimization. *Parallel Computing*, 17:969–982, December 1991.

[12] A. De Gloria, P. Faraboschi, and M. Olivieri. A non-deterministic scheduler for a software pipelining compiler. In *Proceedings of the 25th Annual International Symposium on Microarchitecture (Micro-25)*, pages 41–44, Portland, Oregon, December 1992.

[13] F. Glover and C. McMillan. The general employee scheduling problem: An integration of management science and artificial intelligence. *Computers and Operations Research*, 13(5):563–593, 1986.

[14] Fred Glover and John Knox. Application of tabu search to the placement problem in vlsi design. Technical report, Colorado Institute for Artificial Intelligence, 1990.

[15] IBM. *IBM Journal of Research and Development*, January 1990.

[16] Dean Jacobs, Jan Prins, Peter Siegel, and Kenneth Wilson. Monte carlo techniques in code optimization. In *Proceedings of the 15th Annual Workshop on Microprogramming (Micro-*

15), pages 143–148, Palo Alto, California, December 1982.

[17] John Knox and Fred Glover. Tabu search: An effective heuristic for combinatorial optimization problems. In *Proceedings of the Third Annual Rocky Mountain Conference on Artificial Intelligence*, pages 306–321, June 1985.

[18] M Laguna, J.W. Barnes, and F. Glovér. A tabu search method for scheduling jobs on parallel processors. Technical report, Department of Mechanical Engineering, University of Texas-Austin, November 1989.

[19] D. Landskov, S. Davidson, B.D. Shriver, and P.W. Mallett. Local microcode compaction techniques. *ACM Computing Surveys*, 12(3):261–294, September 1980.

[20] F.H. McMahon. The livermore fortran kernels: A computer test of numerical performance range. Technical report, Lawrence Livermore National Laboratory, December 1986.

[21] T. Starkweather, S. McDaniel, K. Mathias, C. Whitley, and D. Whitley. A comparison of genetic sequencing operators. In *Proceedings of the Fifth International Conference on Genetic Algorithms*. Morgan Kaufmann, 1991.

[22] D. Whitley and J. Kauth. Genitor: a different genetic algorithm. In *Proceeding of the Rocky Mountain Conference on Artificial Intelligence, Denver, Co.*, pages 118–130, 1988.

[23] D. Whitley, T. Starkweather, and D. Fuquay. Scheduling problems and traveling salesmen: The genetic edge recombination operator. In *Proceedings of the Third International Conference on Genetic Algorithms*. Morgan Kaufmann, 1989.

[24] D. Whitley, T. Starkweather, and D. Shaner. The traveling salesman and sequence scheduling quality solution using genetic edge recombination. In L. Davis, editor, *The Genetic Algorithms Handbook*. 1990.

[25] Darrell Whitley. The GENITOR algorithm and selective pressure: Why rank - based allocation of reproductive trials is best. In *Proceeding of the 3rd International Conference on Genetic Algorithms*. Morgan Kaufmann, 1989.

A Massively Parallel Genetic Algorithm on the MasPar MP-1

Markus Schwehm
Institut für Mathematische Maschinen und Datenverarbeitung VII,
Universität Erlangen-Nürnberg, Martensstr. 3, D-8520 Erlangen, Germany
schwehm@immd7.informatik.uni-erlangen.de

Abstract

This contribution describes the implementation of a fine-grained parallel genetic algorithm 'MPGA' on the MasPar MP-1, a massively parallel mesh connected array processor with global router and 1024 (up to 16384) 4-bit processing elements. The implementation uses object oriented methods to provide a large set of standard strategies which can be adapted for a given application. Report modules support the investigation of the performance of the GA. The Implementation shows a good performance compared to other implementations on parallel hardware.

1 Introduction

A basic property of genetic algorithms is the use of a population of solutions (compared to the single solution that is handled with, using simulated annealing). This population is modified simultaneously by genetic operators. This property allows a straightforward implementation of genetic algorithms on massively parallel SIMD architectures. Each individual of the population is manipulated by the same genetic operators synchronously.

The implementation presented in this paper is designed as an experimental system. So rapid prototyping is an issue. The user should be able to re-use as many (problem independent) modules as possible. In many cases the user only has to provide a fitness function and to select a set of strategy modules. Tuning of parameters can be done automatically.

The next section describes the target machine, a massively parallel array processor. The third section discusses population structures and their impact on selection pressure. The implementation itself is described in more detail in the fourth section. Section five presents some successful applications of the environment and compares them with other implementations.

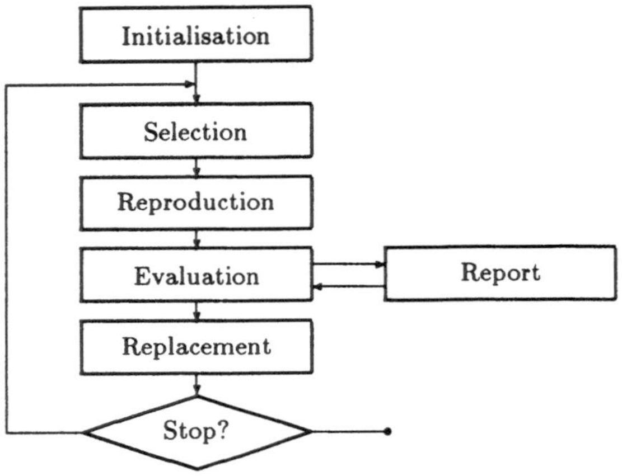

Figure 1: Program overview

2 Target Machine

The MP-1 from MasPar Computer Corporation is a fine-grained, massively parallel SIMD computer. It consists of a UNIX subsystem, an array control unit, the processor element array and some fast I/O facilities. The UNIX subsystem (in our case a VAX 3520) provides the interface to the outer world. Here runs the compiler, debugger and job manager. The UNIX subsystem transmits programs and data to the array control unit, where scalar code is run and parallel instructions are decoded and broadcast to the processor array. The processor array consists of 1024 (up to 16384) 4-bit processing elements (PEs) with 16 k-bytes of local memory.

The PEs are arranged in a 2-dimensional cyclic mesh (torus). Local communication is provided via X-net, which connects each PE to its eight nearest neighbours. The mesh is divided into 8×8 clusters of 4×4 PEs. Each cluster has a bidirectional link to the global router, which can realize any communication pattern but with lower bandwidth. The router is also used as link to high speed I/O devices.

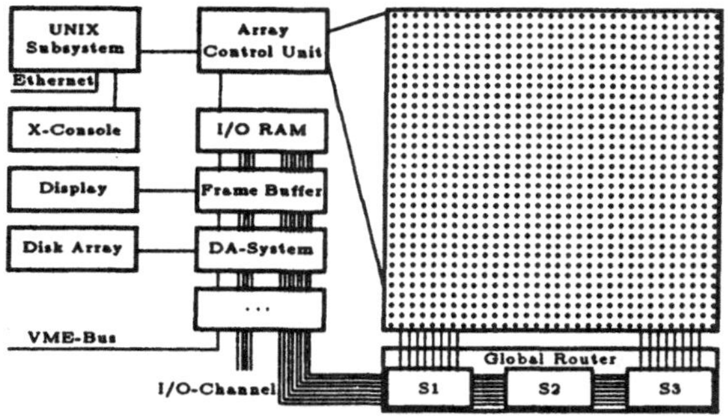

Figure 2: The MP-1 array processor

The MasPar is provided with MPL, a data parallel extension of C. The main conceptual difference between MPL and C are parallel objects. Variables that are declared plural are physically located in the local memory of each processing element and can have different values on each PE. C operators are naturally extended to parallel objects and expressions for the use of communication by X-net and router are added.

3 Population structure

The classical genetic algorithm (GA) uses a single population of individuals. Selection and replacement of individuals is based on global criteria, so the selection pressure per individual is proportional to its relative fitness in the whole population. Using this method, premature convergence can occur, if a relatively good individual spreads too fast over the whole population due to the global selection pressure and thus decreases diversity.

Parallel genetic algorithms (PGA) divide the populations into subpopulations [1]. On each subpopulation runs the same genetic algorithm, additionally from time to time some individuals are distributed to other subpopulations. Since the exchange of information between individuals is faster inside a subpopulation than between them, most of the selection pressure is local and several regions of the solution space can be searched simultaneously. This algorithm is well suited for a multiprocessor architecture, where each subpopulation is assigned to a processor and the intra subpopulation exchange of individuals is realized by low bandwidth message passing. A drawback of this approach is the fixed number of subpopulations since the optimal number of subpopulations is dependent of the structure of the fitness function and thus of the problem itself.

In massively parallel genetic algorithms (MPGA) each processing element processes a single individual [2][7]. Since each individual only interacts with individuals of a local neighborhood, selection pressure is also local. Locality is defined by the communication topology of the processing elements, in our case the X-net of the MP-1, but experiments to emulate different communication topologies have shown, that other topologies like deBruijn or hypercube are also useful. Even a randomized topology with four neighbors per PE showed good performance [4]. It can be observed, that regions of similar individuals grow, and depending on the problem and its fitness function these regions remain small or collapse into one global region.

The MPGA also allows to assign a subpopulation to each processing element. Due to the limited memory size of each PE, this is feasible only for small problems where phenotype and multiple copies of the genotype fit into the local memory of each PE. For these problems, the increase of population size did not yield to better results than the single individual per PE assignment.

4 Implementation

A genetic algorithm consists of a set of operators (methods) that manipulate a given population. For each operator exists a set of strategies to choose from. Some of the operators are problem independent, they can be re-used for different applications. Problem dependent modules are to be supplied by the user and should easily be linked to the system. To prevent inadmissible configurations, restrictions must be checked. All these demands for an easy to use genetic algorithm environment would be supported by an object oriented language like C++. When this project started, there was no object oriented language available for the MasPar MP-1.

But it is also possible to use object oriented methods with a standard programming language like MPL, if the organisation of classes and methods is coded explicitly. Tools are available to support this tedious task. The structure of the object oriented code is described in a syntax similar to C++. This definition file is converted automaticaly into a code template. Now MPL code can be filled into the body of the functions. This works, but nevertheless a data parallel extension of C++ for the MasPar is highly recommended.

Without going further into details how to do object oriented programming with MPL, the basic structure of the MPGA is as follows:

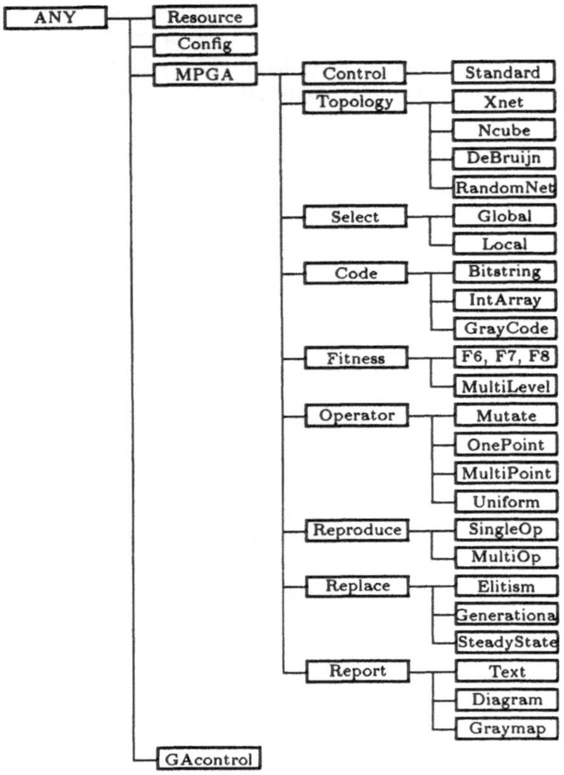

Figure 3: The class hierarchy of MPGA

```
Config.Control:                 Standard
Config.Code:                    IntArray
Config.Select:                  Tournament
Config.Operator:                FastMutate
Config.Operator:                MultiPoint
Config.Fitness:                 PackFitness
...
*Mutate*mutate_rate:            0.005
*MultiPoint*crossover_rate:     0.05
*IntArray*bits:                 14
*IntArray*dimension:            34
```

Figure 4: Example ressource file

4.1 Initialisation

First the resource files are read. These files do not only contain parameter settings like the crossover rate. They also contain information about the desired configuration of the algorithm, e.g. which one of the available crossover operators is to be used. See figure 4 for a selection of configuration parameters as they are used in the example of section 5. For each selected method a startup procedure is initiated, which execute tasks like the generation, decoding and evaluation of an initial population.

4.2 MPGA Control

The MPGA module itself consists of a set of methods, which together perform the succession of generations. They are controlled by a module which consists so far of only one standard strategy. This module would have to be executed if for example local hillclimbing should be added or if the paradigm of memetic algorithms should be implemented [5].

4.3 Selection Process

Selection in the MPGA means that for each individual (each located on different processing elements) a mating partner must be found in the rest of the population. This task has much influence on the speed of the information spreading over the whole population and thus controls the selection pressure. A good choice here determines if premature convergence will occur or not. The selection process is divided into three parts:

- Definition of an underlying topology. This defines the set of neighboring individuals if a local selection strategy is used.
- Definition of a deme, i.e. the subset of the population from which the mating partner is to be selected. If the deme equals the whole population, we have a global (panmictic) selection strategy and the MPGA should behave like the standard GA described in section 3. A local deme definition can be a subset of the local neighborhood (tournament) or the elements passed during a random walk. Both strategies can be performed for each generation or once during the initialisation phase.
- Selection of the mating partner from this deme. One could choose the best individual or any one with probability proportional to its fitness or its rank relative to the deme. Alternatively *both* parents can be selected from the deme.

4.4 Reproduction and Evaluation

In this implementation, reproduction is regarded as problem independent. Each mutation or crossover applied to the bitstring of an individual must result in a valid solution of the seach space.

The bitstring is evaluated by a user defined fitness function. There exist several modules to interpret the bitstring as a vector of (grey coded) integer or real values. Other interpretation methods—as for example as a graph—can be supplied by the user. Standard test functions provided by the system are F6, F7, F8 and the multilevel graph partitioning problem.

4.5 Replacement

Selection pressure is already introduced during the selection process. To raise selection pressure would increase convergence speed together with the chance of premature convergence. Best results were archieved using generational replacement or even a randomized strategy, instead of using a strategy based on the fitness values of parents and offspring, like steady state or elitism replacement. Elitism had a chance against generational replacement only if the underlying topology had a very short diameter.

4.6 Report Modules

Report modules don't operate on the population and thus don't contribute to the optimisation process. But they are necessary during the development of an algorithm, since they allow to observe what happens inside the population. MPGA provides modules for collection and display of data about:

- Tracking a single individual: fitness, bitstring, phenotype.
- Local environment of the individuals: fitnes maps, hamming distance maps.
- Observing the whole population: average fitness, max-min-reports, number of replaced individuals, diversity, inbreeding, fixation, number of substitutions or classes.
- Comparing the performance of several runs of the same or different configurations of the MPGA: Convergence speed and reliability, robustness.

The MPGA environment supports the output of data as plain ASCII text to the terminal as standard output. Moreover graphical output is possible as Encapsulated Postscript code, which can be piped into a file, a previewer or directly to a laser printer. Postscript was choosen since this is a well established standard and available on most platforms. Phenotypes can be displayed if a corresponding output routine is provided by the user. The user can also overwrite these report modules with his preferred output language, for example the phenotypes displayed in figure 5 were directly output by the MPGA as LaTeX code.

4.7 Meta-GA

Due to the modularity of the MPGA, a large variety of strategies and parameters have to be choosen. It turns out, that for each application, a different setting of the parameters seems to be optimal. Finding the optimal setting of the parameters is itself an optimisation problem. MPGA provides methods (GAcontrol) to find an optimal configuration by applying a genetic algorithm itself. This is realized by using the performance of a MPGA run as fitness value for the meta-GA. The search space is the set of possible configuration and parameter settings.

5 Experiments and Results

Figures 4–9 belong to a sample run of the two dimensional packing problem. Kröger et al [6] have solved this problem on a transputer network. The bitstring there was interpreted as a binary tree, which described a packing scheme. Since the decoding function is not total, specialized mutation and crossing over operators had to be introduced to exclude impossible packing schemes. The implementation on the MPGA uses a total decoding function; essentialy the bitstring only determines a permutation of the list of rectangles and decides which rectangle is rotated, before it is placed by bottom- or bottom-left strategy. The decoding function is total, so any standard mutation and crossover operator can be used.

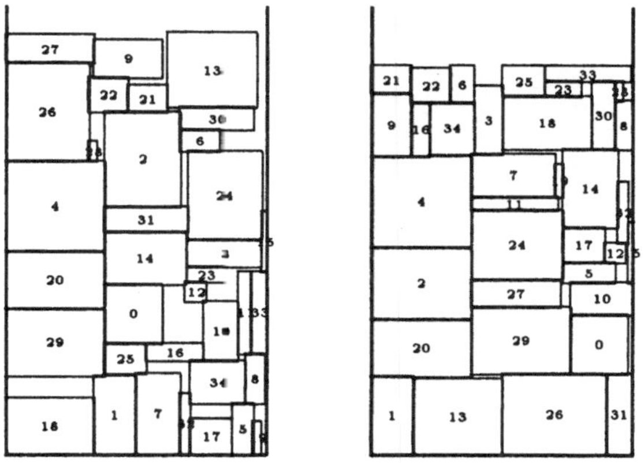

Generation 0 Generation 100

Figure 5: Two-dimensional packing

A comparision of the MPGA on the MasPar MP-1 with the implementation of an MPGA on the Connec-

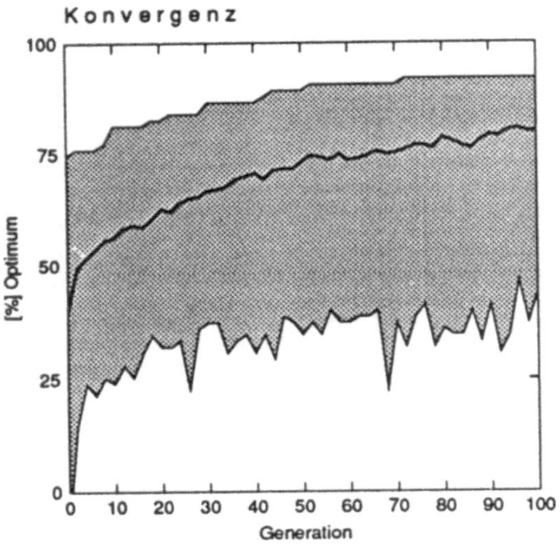

Figure 6: Max-Min report

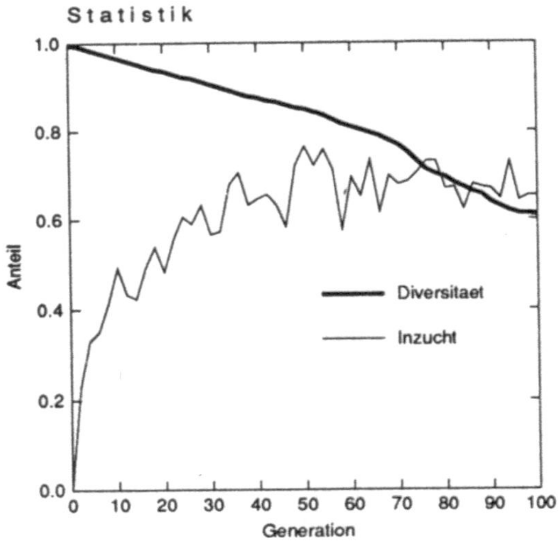

Figure 8: Statistics

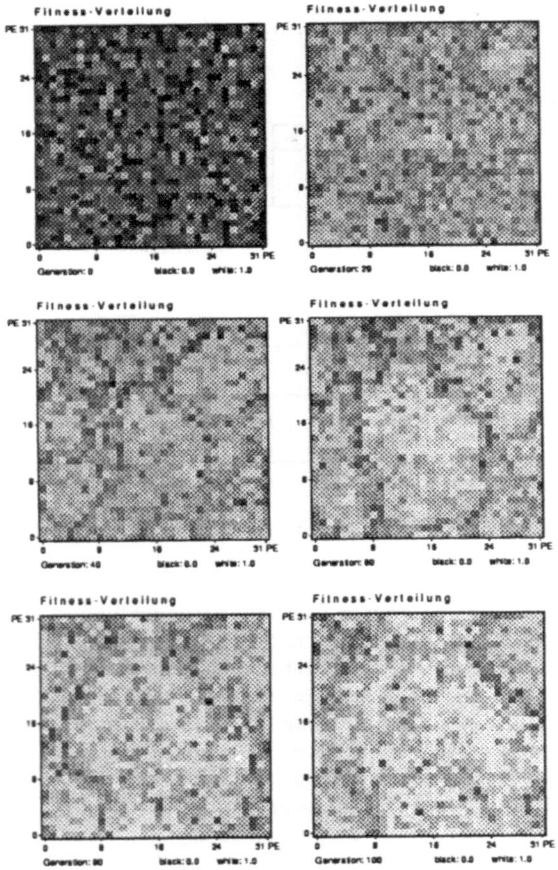

Figure 7: Fitness map

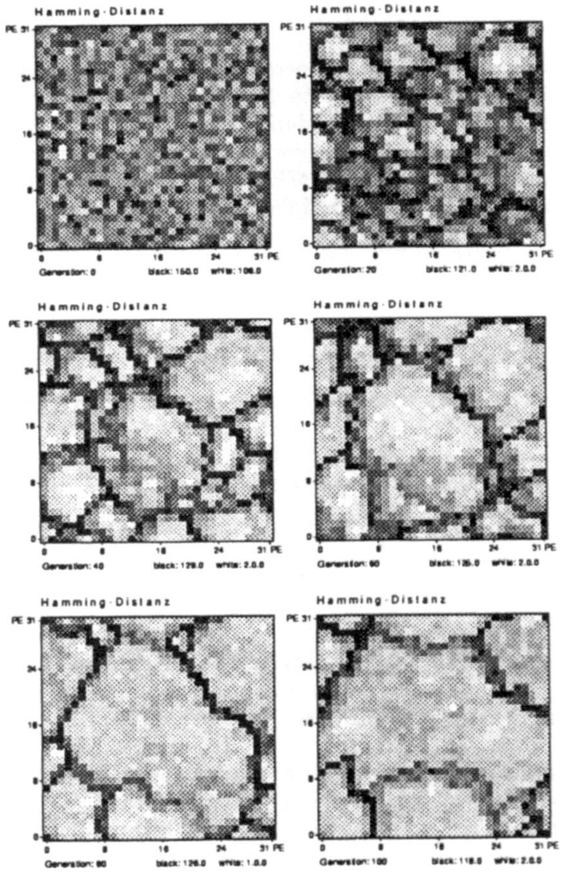

Figure 9: Hamming distance map

tion Machine CM-2 by John Collins [7] is shown in figure 10. Both algorithms solve the multilevel graph partitioning problem on a 2-dimensional torus with random walk selection.

Computer	CM-2(16k PE)		MP-1(1k PE)	
Path length	1	10	1	10
# Generations	43	13	58	17
Elapsed time/s	7	4	3	1
Time/generation	0.16	0.26	0.05	0.07

Figure 10: Comparision of CM-2 and MP-1

Figure 11 shows the performance of the MPGA compared to a PGA implemented on a 16-transputer system by Mühlenbein et al [1]. The optimisation problem is F8 in ten dimensions. The MPGA needed much more evaluations since it did not use local hill-climbing.

Computer	16 Transputer	MP-1 (1k PE)
# Evaluations	59520	ca. 200000
Elapsed time/s	16.84	ca. 10
Time/evaluation	0.00282	ca. 0.000048

Figure 11: Comparision of 16 transputer to MP-1

The MPGA has also been used to solve a taskgraph mapping and scheduling problem. The solution even optimized the topology of the target transputer network. A comparision with an algorithm for the exact solution on a workstation showed that the optimal solution was found reliable. Moreover the genetic algorithm was able to solve problem sizes which were out of scope for the exact algorithm because of memory limitations.

6 Conclusion

A massively parallel genetic algorithm environment has been presented. Altough there is no object oriented language available yet on the MasPar, object oriented methods have been applied sucessfully to allow re-use of code and to construct a new application program by selection of methods from a growing collection. Several configurations of the algorithm and parameter settings can be tested without recompilation. It is also possible to find an optimal configuration automatically by the use of a meta-GA. A set of report modules helps to collect statistics about the performance of the MPGA. Visualisation of this data helps to understand what is going on inside the population. Moreover the MPGA has a good performance compared to other implementations of genetic algorithms on parallel hardware. Genetic algorithms are well suited for massively and data parallel architectures.

It is planned to port the code to a data parallel extension of C++ as soon as such a language is available for the MasPar. A scalar version of the code is in preparation, too.

Acknowledgements

I would like to thank Frederik Siegmund for implementing the genetic algorithm and Thomas Walter, Sönke-Sonnich Gold and Christoph Schäftner who were responsible for a very high load of the MasPar during the last year while using the MPGA.

References

[1] Mühlenbein, H. and Schomisch, M. and Born, J.: *"The Parallel Genetic Algorithm as Function Optimizer"*, Parallel Computing 17 (1991), pp. 619–632

[2] Manderick, B. and Spiessens, P.: *"Fine-grained Parallel Genetic Algorithms"* in Proc. Third Int. Conf. on Genetic Algorithms", ed. J.D. Schaffer, Morgan Kaufmanmn, San Mateo, CA, 1989

[3] Gorges-Schleuter, M.: *" Explicit Parallelism of Genetic Algorithms through Population Structures"* in Schwefel, H.P.: Parallel Problem Solving from Nature, Springer, Berlin, 1990

[4] Schwehm, M.: *"Implementation of Genetic Algorithms on various Interconnection Networks"* in: Valero, M. et al (Eds.) Parallel Computing and Transputer Applications, Part I, pp.195-203, IOS Press, 1992

[5] Moscato, P.: *"On Evolution, Search, Optimisation, Genetic Algorithms and Martial Arts: Towards Memetic Algorithms"* Caltech Concurrent Computation Program Report 826, CalTech, Pasadena CA, (1989)

[6] Kröger, B. and Schwenderling, P. and Vornberger, O.: *"Genetic Packing of Rectangles on Thansputers"* in: Schwefel, H.P.: Parallel Problem Solving from Nature, Springer, Berlin, 1990

[7] Collins, R. J. and Jefferson, D. R.: *"Selection in Massively Parallel Genetic Algorithms"* in: Proc. Fourth Int. Conf. on Genetic Algorithms", ed. R.K. Belew, Morgan Kaufmanmn, San Mateo, CA, 1991

Standard Cell Routing Optimization Using A Genetic Algorithm

Steve Bassett, Michael Winchell

NCR Corporation, Ft. Collins Colorado.

March 1, 1992

A typical approach to cell-based analog design involving resistors is to maintain the standard cell height by using an array of standard resistor legs. As the array becomes larger and more taps are placed on the resistor, random orientation of the resistor legs can lead to routing problems as signals cross back and forth across the cell routing channel. Heuristic optimization of the resistor leg orientation to reduce signal crossing becomes a complex search through a 2^{n-1} search space, where n is the number of legs to be oriented.

An alternate approach is to use a general optimization algorithm that is good at searching large spaces. Genetic algorithms are of this class. In the current application, the genetic algorithm approach was able to gain a 50% routing improvement over manual techniques on large resistor arrays.

1.0 The Resistor Routing Problem

1.1 From Circuit Design to Physical Design

In a circuit requiring a network of resistors, the circuit designer typically creates a circuit-level schematic such as the one shown in Figure 1. To the designer, the important characteristics of the resistor network are the ratios of the resistors comprising the network and in some cases the precise absolute resistor values.

FIGURE 1. An Example Circuit Designer's Requirement: A Simple Resistor Divider Network.

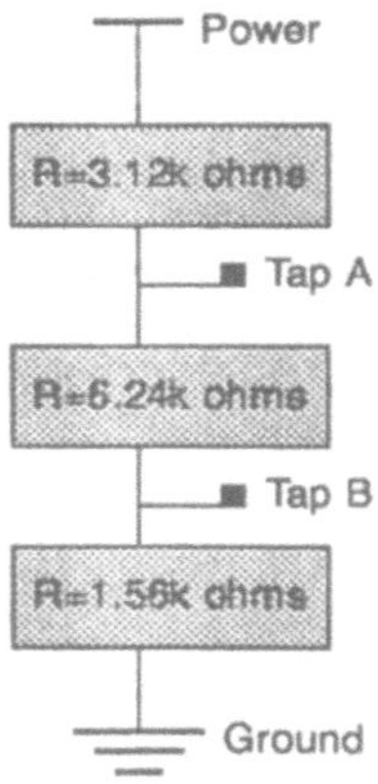

Many issues add to the manufacturing complexity of a resistor network on an integrated circuit (IC). Although Figure 1 shows a simple example, it may be used to illustrate some of these issues.

In cell-based design of an integrated circuit, circuit components, or cells, of standard height are used to allow for efficient placement of cells into rows with routing channels between each cell row [1].

Resistor networks tend to be different in each circuit application. They are therefore treated as custom cells, requiring a new physical design each time. In cell-based integrated circuit design, custom cells must also adhere to the standard cell height constraint. For resistor networks, this has led to the use of standard fixed-height resistive elements, or legs [2]. To reduce the physical design burden, the legs are also of fixed width and, depending on which resistive material is used, of fixed resistive value.

FIGURE 2. The Physical Schematic Representation of Figure 1

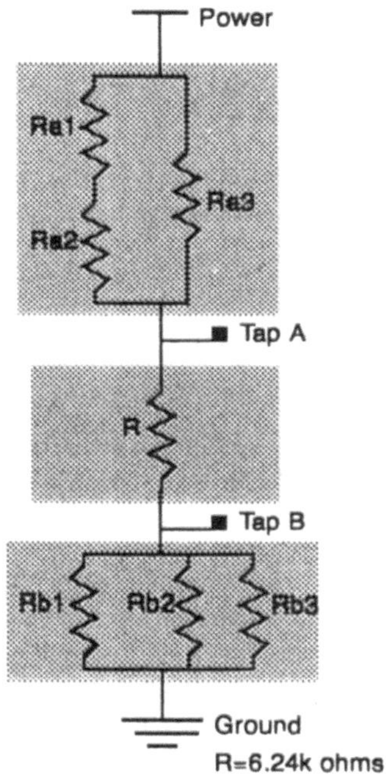

The first step in reducing the resistor network to a physical realization is to generate a physical-level schematic that uses the standard cell legs but which is electrically equivalent to the circuit-level resistor network that the designer specified. Figure 2 shows a possible physical schematic representation of the network shown in Figure 1 using standard resistor legs.

The next step in the creation of the custom resistor network cell is to map the physical-level schematic into the silicon-level description that will drive the manufacture of the IC. The rest of this paper deals with the representation of the resistor network at the silicon level.

1.2 The Physical Design

In the silicon-level design of a resistor network cell, the standard resistor legs are oriented vertically and placed side by side, forming a one-dimensional array of resistor legs with connections on the top and bottom of the array. This basic arrangement is illustrated in Figure 3.

FIGURE 3. Basic Resistor Network Array

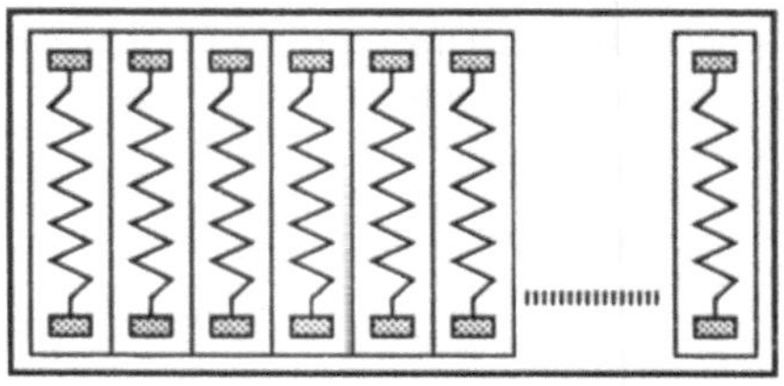

The array of legs forms two rows of terminals representing the top and bottom physical connections of the resistor legs.

Variations in electrical characteristics exist in all integrated circuits, and typically consist of silicon processing variations, temperature range effects, and power supply tolerances.

To minimize the effects of variations in processing across a silicon wafer, the legs of each of the resistors that make up the network are interdigitated, or alternately placed, so that any local variations in the electrical characteristics of the integrated circuit will tend to affect all of the resistors in the network equally. Figure 4 shows the resistor legs of Figure 2 as they might be interdigitated and how a local process variation might affect each resistor [3].

FIGURE 4. Resistor Element Interdigitation

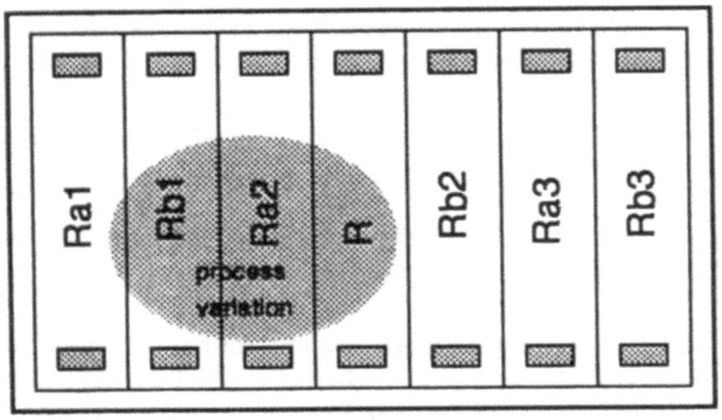

Interdigitation is important to maintain tight resistor ratio matching, usually the most important characteristic of a resistor network. As another example, Figure 5 illustrates the physical connections that would be needed in an interdigitation of two unconnected resistors each with an equal number of legs.

FIGURE 5. Interdigitated Resistor Network

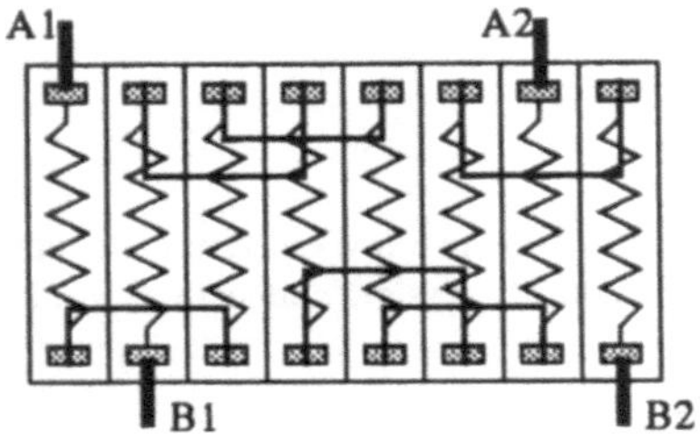

Due to the standard width of the fundamental resistor leg, the metal wires routed in the vertical direction for leg interconnection can be no wider than the standard leg width and in fact must be narrower to allow for typical metal spacing design rules. For this reason, if the resistor is generated with vertical wires routed with the maximum wire width, then the number of vertical routing tracks is equal to the number of resistor legs forming the network. Similarly, the number of horizontal routing tracks contained in the channel between the resistor leg ends is a function of the standard length of the resistor leg and the desired width of the horizontal routing layer wires. Figure 6 illustrates the horizontal and vertical routing tracks available.

FIGURE 6. Physical Network Routing Tracks

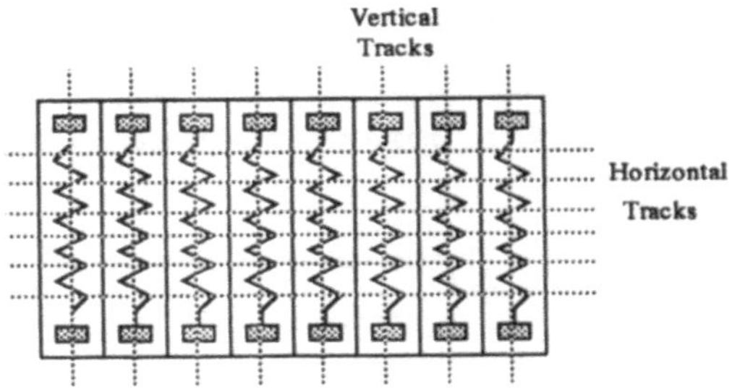

The space in which to interconnect the legs of the network is thus bounded by the standard leg length and width. This limited space in which to connect the resistor ends to form the network may not provide enough room to complete all interconnections without wires overlapping, thus causing electrical shorts between wires, an unacceptable situation.

Within the resistor array, it is possible to invert the orientation of any resistor leg while retaining the connectivity of the leg's respective ends. By executing such a "flip" of a given leg, the leg may now be oriented in the array such that the connections to its respective ends no longer must cross from the array top to the bottom to connect to other legs but may now have connections only along the array top and array bottom, respectively. This concept can be illustrated with a resistor network created from resistor legs connected in series, known as a serpentine resistor due to the appearance of its physical layout. In the worst interconnect case, the bottom connection of the first resistor would connect to the top of the next resistor in series, as illustrated in Figure 7.

FIGURE 7. Worst Case Serpentine Interconnection

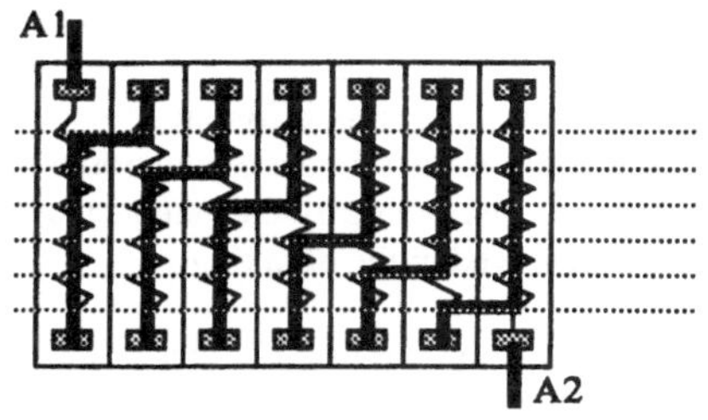

If the second, fourth, and sixth legs are flipped, then the connections are simplified and reduced to adjacent terminals on like sides. The greatly reduced interconnect congestion in Figure 8 is evident by the unused horizontal routing tracks. Although this example is greatly simplified, applying the same procedure to a typical resistor network of fifty legs typically results in dramatically reduced interconnect congestion.

FIGURE 8. Best Case Serpentine Interconnection

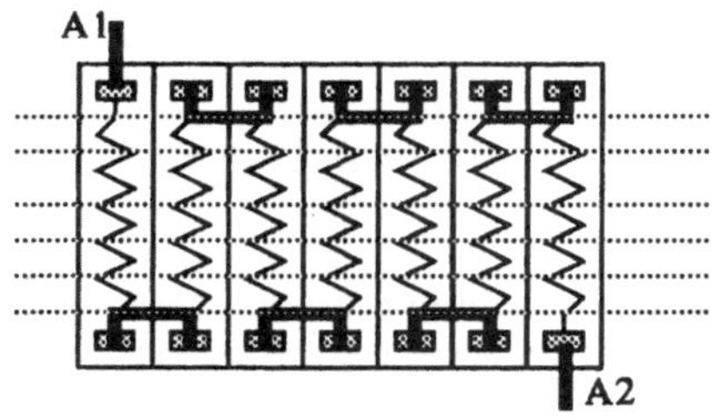

One strategy is to orient the resistor legs to reduce the number of signal interconnect wires which must cross the array in the vertical direction. The serpentine example suggests that this strategy will reduce channel congestion.

Routing congestion within the leg array tends to be proportional to the number of legs in the array. With more legs, generally more connections are required and yet no additional horizontal routing tracks are allowed due to the fixed height of the resistor leg. A badly oriented resistor array may not have a routing solution that adheres to the constraint posed by the fixed cell height.

As shown in Figure 7, wires that must cross the routing channel consume a portion of a horizontal routing track and contribute to interconnect congestion. By orienting the resistor legs to minimize the number of wires crossing the channel, the congestion can be minimized and the chances for completing the network interconnection are greatly improved.

2.0 Application of the Genetic Algorithm to the Resistor Leg Orientation Optimization Problem

Several possible approaches exist to minimize the number of routing channel wire crossings. An exhaustive approach would require evaluation of every possible resistor leg array orientation to look for the one that results in the minimum number of channel crossings. To see that this is not feasible, consider the following.

There are 2**n possible orientations of n resistor legs. Since half of the orientations are mirror images of the other half, the number of combinations that needs to be considered is 2**(n-1). To exhaustively search for the optimum in a one hundred leg resistor network, approximately 6 E29 evaluations are required. To put this into perspective, on a 1000 MIP machine, the optimum solution would be guaranteed to be found in somewhere around 19 trillion years!

An alternative approach is to use heuristics to limit the exhaustive search. Many such schemes may be thought up, each with advantages and disadvantages. One common disadvantage is that it is difficult to guarantee that the heuristic algorithm will find an optimum or near optimum solution. Careful analysis is required to determine just how well a heuristic might do.

Instead of using heuristics to reduce the search space, general optimization algorithms may be applied. In the present problem, heuristics are difficult to evaluate and their programming is ad hoc. General optimization algorithms, such as genetic algorithms, are well known and easy to program. Although the general optimization algorithm doesn't guarantee that the optimum will be found, it will at least find a local minimum and can often do better than a heuristic approach.

2.1 Why is the Genetic Algorithm Suitable?

Genetic algorithms comprise a set of powerful search algorithms that do not use domain knowledge to guide their search. Instead, genetic algorithms search through the solution space by trying to combine positive aspects of previous solution attempts. A challenges is to create a suitable solution representation with associated genetic operators such that a simple (i.e. fast) evaluation of solutions is possible.

A suitable solution representation, operator set and evaluation function must allow for an operator set that preserves favorable aspects of previous solution attempts.

A good discussion of the representation and suitable operator challenge is contained in [4]

The algorithm used in this application is a variation of the GENITOR algorithm created by Whitley et al at Colorado State University. [5]

In this variation of the genetic algorithm, a ranked population of solutions is maintained. The initial population is seeded with random solutions. Parent solutions are randomly chosen from the population and a genetic crossover operator is applied. One resulting child solution is then ranked and bubble sorted into the appropriate slot in the population. The worst member of the population is discarded.

The larger the population, the more potential diversity exists in the initial population. However, even with large populations, the diversity rapidly diminishes. The mutation operator can be used to reintroduce diversity, while the crossover operator is used to do the genetic search. Due to the expected rapid convergence of the population, and the expected need to use mutation, a small population was used initially and found to produce adequate results even without using mutation.

The crossover operator used two randomly chosen crossover points in order to avoid any bias in the solution search. To see that this is necessary, note that with one crossover point, the first and last resistor leg in an offspring would always come from different parents. Thus the favorable aspects of the relative orientation of the first and last legs as discovered by the parents would not necessarily be preserved in the offspring.

Using the binary encoding of the solutions as described below, the crossover operator does not disrupt what was good in the parent solutions. To see that this is so, note that if an array orientation allows for efficient routing, then subsections of the resistor array also exhibit good leg orientation for efficient routing.

2.2 Problem Encoding

The representation of the leg orientation in the resistor array is a natural one. A one dimensional bit array is established with length equal to the number of standard resistor legs in the network. The binary string representing the orientation as input to the application is termed the baseline orientation. Each bit represents an individual standard resistor leg. A zero in a particular leg (bit) position represents a leg orientation as read from the Spice CAD file while a one represents the same leg flipped vertically about its mid-point relative to the baseline orientation. The array is referred to as the orientation pattern.

Each standard leg of a resistor network connects to two electrical nodes or signals, one on each leg end. For a leg with a flipped orientation, the electrical nodes connected to the leg ends would be exchanged.

The population of solutions is seeded with random binary numbers.

2.3 The Evaluation Function

The ideal evaluation of a specific leg orientation pattern would be to route the wires of the network and compare the congestion to that of other orientations. This is not feasible as the routing algorithm is time consuming.

Instead, a faster evaluation is applied. The congestion of the interconnect channel running between the ends of the resistor leg array is proportional to the number of signals which must be connected to both top and bottom leg ends within the array. Such a connection represents a channel crossing from top to bottom.

Those signals which have connections only on the leg array top or only on the bottom will consume only one horizontal routing track. Signals which have connections on both the top and bottom of the leg array consume at least one extra horizontal routing track which is needed to cross the channel.

During the evaluation of an orientation, two lists are maintained as images of the resistor leg array top and array bottom. The top list identifies the signal connected to the top end of each leg in positional order. The bottom list likewise identifies the signal connected to the bottom end of each leg in order. Signal names appearing in both lists represent a channel crosing. The total number of channel crossings required for an orientation is used as a relative measure of the routing congestion which can be expected from the orientation of the resistor legs.

If a leg is flipped in an orientation relative to the baseline orientation, the top and bottom signals are swapped in the top and bottom signal lists.

An additional design constraint that must be taken into account is that it is desirable to have tapped signals available on both the top and bottom of the cell to allow for more efficient routing between cells during the chip-level place and route. Therefore, where it is desirable for non-tap signals to connect only to the top or only to the bottom of the resistor leg array to minimize cell routing congestion as described above, a tap signal must cross the cell routing channel at least once. The evaluation rule counts the channel crossings and the number of tap signals that do not cross the channel. Optimization then becomes the minimization of the evaluation function.

2.4 The Optimization and Genetic Application

The initial population of orientations is created by generating nine random binary numbers with one bit per standard leg. The baseline orientation is added to the population for a total of ten individuals. Each orientation is evaluated for channel crossings and non-accessible tap signals and stored in ranked order.

During the generation of a new individual from the existing population, parent solutions are randomly chosen without any bias.

New crossover points are randomly chosen for each application of the crossover operator to the parent orientations. To generate the child orientation, the leg orientation of the first parent is copied bitwise up to the first crossover point. The bits are then copied from the second parent until the second crossover point is reached. When the second crossover point is reached, the rest of the bits are taken from the first parent orientation. The result is a child built from two parent orientations sampling different portions of the solution hyperspace.

The child orientation is then evaluated and bubbled into the appropriate position in the population according to its ranking. The worst member of the population is discarded.

End conditions are established to halt optimization when a better solution is unlikely to be found. One empirical end condition is that a solution is found that has no more than twenty five percent of the signals crossing the channel. A limit is also set on the number of consecutive generations which yield no improvement in the best orientation.

The best orientation is then used by the automatic cell routing program that is also part of this application.

3.0 Results

Application of the genetic algorithm to the routing congestion problem was found to add approximately twenty-five percent to the runtime for generation of a physical resistor network cell. Total run times to automatically route the resistor network cell for networks of ten to fifty standard legs were on the order of one to five minutes which compares favorably to the hours or days required to layout the cell manually.

The use of the genetic algorithm is user selectable. To save time, some users at first turned the genetic optimization off. If a resistor network could not be routed successfully in the standard cell height, the user then would rerun the resistor layout generator with the genetic optimization enabled. However, after initial usage, it was found that the run time with the genetic optimization enabled in the first place was found to be minimal compared to a trial and error approach involving possible reruns. Today the genetic algorithm is enabled as the default for all resistor networks regardless of leg count.

Many networks were evaluated ranging from ten to three hundred standard legs, with the typical requirement of the tool being ten to fifty legs. The use of the genetic algorithm was found to reduce the channel crossings measure by approximately fifty percent in nearly all cases. The improvement of the routing congestion for networks exceeding one hundred legs was for most cases absolutely necessary to avoid unroutable orientations. A three-hundred leg example, contrived to break the tool, was successfully routed only when the genetic optimization was used.

4.0 Conclusions

The resistor layout generator has resulted in an approximate forty to one hundred times productivity improvement for the resistor layout portion of a cell-based design. In some recent cases, manual resistor layout would not have been possible. While genetic optimization is not required for all resistor networks, it results in better routes, using less interconnect, and extends the range of resistor networks that may be successfully routed.

5.0 Bibliography

[1] Weste N and Eshraghian K, 'Principles of CMOS VLSI Design', Addison-Wesley Publishing Company, 1985, pp236-248.

[2] Gray P R and Meyer R G, 'Analysis and Design of Analog Integrated Circuits', John Wiley & Sons, 1977, p114.

[3] Gray P R and Meyer R G, 'Analysis and Design of Analog Integrated Circuits', John Wiley & Sons, 1977, pp389-394.

[4] Grefenstette J, 'Incorporating Problem Specific Knowledge in Genetic Algorithms', Genetic Algorithms and Simulated Annealing, Davis L, Morgan Kaufmann, 1987.

[5] Whitley D and Kauth J, 'GENITOR: a Different Genetic Algorithm', Proceedings of the Rocky Mountain Conference on Artificial Intelligence, 1988, pp118-130.

6.0 Authors

Stephen J. Bassett
NCR Corporation, Microelectronic Products Division
2001 Danfield Ct.
Fort Collins, Colorado 80525
email: Steve.Bassett@FtCollinsCO.NCR.com

Michael A. Winchell
NCR Corporation, Microelectronic Products Division
2001 Danfield Ct.
Fort Collins, Colorado 80525
email: Mike.Winchell@FtCollinsCO.NCR.com

Efficient Parallel Learning in Classifier Systems

U. Hartmann

Department of Informatics, University of Dortmund
Marthastr. 2, D-4300 Essen 1, Germany
e-mail: hartm@marlies.informatik.uni-dortmund.de

Abstract

Classifier systems are simple production systems working on binary messages of a fixed length. Genetic algorithms are employed in classifier systems in order to discover new classifiers. We use methods of the computational complexity theory in order to analyse the inherent difficulty of learning in classifier systems. Hence our results do not depend on special (possibly genetic) learning algorithms. The paper formalises this rule discovery or learning problem for classifier systems which has been proved to be hard in general. It will be proved that restrictions on two distinct learning problems lead to problems in *NC*, i.e. problems which are efficiently solvable in parallel.

1 Introduction

Classifier systems are simple rule based systems which employ genetic algorithms to discover new rules. In order to analyse classifier systems it is useful to distinguish three main components of these systems [4, 2]:

1. performance system,
2. credit assignment system, and
3. rule discovery system.

The *performance system* represents the lowest level of a classifier system. It is a rule-based system like most expert systems, but classifiers are special rules which are message-passing, highly standardised and applied in parallel [2]. The performance system of a classifier system is a (simple) production system. It receives *input* from the environment which describes special environmental situations and it produces *output* in order to react to these situations. Appropriate reactions of the system are rewarded through *payoff* by the environment.

The working memory or central data structure of a classifier system is the *message list*. This message list contains binary messages of a fixed length. The *input interface* encodes the information received by the so-called *detectors* into a set of messages. These messages are posted to the message list **M**. *Effectors* represent the output of a classifier system. An effector is activated by a special message in the message list. The *output interface* checks whether any message in the message list activates an effector.

A *classifier* is a production rule containing a finite number (≥ 1) of ternary conditions and one ternary message specification. In a classifier's condition the special symbol '#' is a wild-card, i.e. it matches arbitrary values at the message position. A classifier is allowed to post its message to the message list if each condition is matched by at least one message in the message list. Some systems also permit negated classifier conditions as well as 'positive' conditions. Such a negated condition, labeled '-', is assumed to match a message list whenever there is no matching message in the list. We also permit '#' symbols in the message of a classifier; here the symbol denotes a distinct operation, namely 'pass through' of the value at same position of the message which activated the classifier.

Messages can be generated by two sources: either by the environment using the detectors or by a classifier which posted its message. On the other hand messages can be used in two ways: either in order to activate classifiers or to stimulate effectors which generate output of the system. In practice classifier systems make use of a message prefix, a so-called *tag*, which indicates both the origin and the purpose of the message. This tag usually consists of the first two bits of a classifier. However, we do not use any tags since this does not affect our results.

In general the classifiers in the classifier system may be of different usefulness and there may also be completely useless classifiers. An individual classifier should be rated according to its role in achieving reward or payoff from the environment. The *credit assignment system* keeps track of the usefulness of the classifiers in the performance system. There are various ways of doing this. However, most systems use a method John Holland called the *bucket brigade* algorithm [7].

The *bucket brigade* algorithm is intended to solve the credit assignment problem for classifier systems. In a classifier system the problem is to decide

1. which of the matching rules shall be allowed to post their message, and

2. which of the classifiers shall be used by the rule discovery system to build new classifiers.

This task is especially difficult in environments where rewards are rare.

The classifier system only contains a small portion of the possible rules (i.e. classifiers) and even the best of the rules in the system might not achieve much payoff from the environment. The *rule discovery system* uses the existing classifiers in the system to generate new classifiers which are supposed to achieve payoff. These generated rules replace the least useful classifiers. The rule discovery system often employs so called *genetic algorithms*. Genetic algorithms are heuristics which are intended to generate new, useful rules [6, 2, 4].

In this paper we use methods of the complexity theory to analyse the inherent difficulty of learning in classifier systems. Our results do not depend on special (possibly genetic) learning algorithms. The computational complexity of learning within resource bounds has been studied first by Leslie Valiant who introduced the model of *probably approximately correct (PAC) learning* [10]. Stephen Judd analysed the so-called *loading* problem in neural networks [8]. We develop a formalisation of the learning problem in classifier systems which is quite close to the loading problem in neural networks.

2 Parallel Random Access Machine

The *parallel random access machine* (P-RAM) is one of the models more commonly used to design parallel algorithms [3]. This machine model avoids constraints which might derive from the use of special hardware. Hence the P-RAM provides a model to design parallel algorithms in a straightforward way.

The P-RAM does not set any constraints on the number of possible links between processors and memory locations. This is certainly not realisable on available hardware but it has been shown that P-RAMs can be efficiently simulated by processors with a constant number of links [3]. Moreover the costs of the simulation are only polylogarithmic in the size of the input n, i.e. $O\left(\log_2^k(n)\right)$ for any constant k.

The processors of the P-RAM use synchronously a common global memory and communicate through this common global memory. Each processor is a uniform cost random access machine (RAM) and is equipped with the usual operations and instructions. The processors of the P-RAM execute simultaneously the same programme (main control programme) but may work on different data (distinct memory locations). The P-RAM model is also called single-instruction, multiple-data-stream (SIMD) model.

P-RAM models often differ in the way simultaneous access of the same memory location is permitted. We use the following conventions [3]:

1. Simultaneously *reading* from the same memory location is permitted for any number of processors, and

2. Simultaneously *writing* into the same memory location is not allowed.

We will design parallel algorithms which can be implemented on the P-RAM. However, we will use a high-level language to specify our algorithms. The basis of this high-level language will consist of usual instructions of a programming language (as Pascal) which is to be extended by a parallel statement:

for all $x \in \mathcal{X}$ **in parallel do** *instruction(x)*

where x is an element of the set $\mathcal{X}$ and the execution of the statement consists of the following steps:

1. A processor is assigned to each element $x \in \mathcal{X}$. We require that the elements of $\mathcal{X}$ can be encoded by distinct positive integers. Processor $P_{code(x)}$ will be assigned to element x in constant time where *code(x)* is an encoding of x as a positive integer.

2. The assigned processors execute in parallel all the operations specified by command or procedure *instruction(x)*. When p is the maximum number of processors used in the computation $\log_2 p$ is the initial threshold time required to activate each processor.

In order to keep the P-RAM model realistic there are no links between the processors themselves. Therefore the initial threshold time of the P-RAM is necessary to activate all the processors involved in the computation of the programme: Starting with one active processor, the number of active processors is doubled in each time step since each active processor can activate another through the common global memory.

The execution terminates when all p involved processors have completed their computations.

We assume that complexity classes *NC*, *P*, *R*, and *NP* as well as the concept of *NP-completeness* are known [1, 3].

3 A Formal Notion of Learning

Throughout this paper we are only interested in analysing the difficulty of learning in classifier systems. Therefore we do not discuss the performance of the rule-based production system of the classifier system.

A *message* m_i in a classifier system is a binary string which is of a fixed length $\ell : m_i \in \mathbb{B}^\ell$ with $\mathbb{B} = \{0, 1\}$. We will denote the j-th letter of a message m by $m_{(j)}$ while m_i denotes the i-th message. These messages are collected in a *message list* $\mathbf{M} = \langle m_1, \ldots, m_p \rangle$ which is a finite list of p messages $m_i = (m_{i(1)}, \ldots, m_{i(\ell)})$, with $m_{i(j)} \in \mathbb{B}$. A classifier *condition* d_i is a string of finite length $|d_i| = \ell$ over the set $\mathbb{T} = \{0,1,\#\}$: $d_i \in \mathbb{T}^\ell$.

A *classifier* c_i is a production rule containing a finite number ($n \geq 1$) of conditions d_l ($|d_l| = \ell$) and one message specification m_c: $c_i = (d_1, \ldots, d_n, m_c) = (c_{i(1)}, \ldots, c_{i((n+1)\cdot\ell)})$ where $c_{i(j)} \in \mathbb{T}$. The *message* of the classifier c_i is specified by its message specification $m_c \in \mathbb{T}^\ell$.

An arbitrary classifier condition $d_i = (d_{i(1)}, \ldots, d_{i(\ell)})$ *matches* a message $m_j = (m_{j(1)}, \ldots, m_{j(\ell)})$ iff for each $q, 1 \leq q \leq \ell : d_{i(q)} = m_{j(q)}$ or $d_{i(q)} = \#$.

Let $\mathbf{M} = \langle m_1, \ldots, m_p \rangle$ be a message list and $c_i = (d_1, \ldots, d_n, m_c)$ be a classifier where $(d_1, \ldots, d_n)$ are the conditions and m_c specifies the message of the classifier. A classifier c_i *matches* a message list $\mathbf{M}$ iff each condition d_j ($1 \leq j \leq n$) matches at least one of the messages m_q ($1 \leq q \leq p$) in the message list.

The main learning task to be solved in a classifier system is to find a set of classifier conditions which agree with a set of message lists, so-called tasks, to be learned. We distinguish positive and negative learning tasks. The question whether a classifier agrees with a message list can be specified in two ways:

1. The *strong interpretation* is to say that the set of classifier conditions has to be consistent with all the tasks. According to the strong interpretation the conditions are required to match exactly the positive training examples; given negative examples none of the conditions is allowed to match.
2. The *weaker interpretation* requires the classifier conditions to match all of the positive tasks without considering any negative ones.

The weaker interpretation turns out to be too weak to characterise the learning problem adequately. The reason is that the problem is trivial because the classifier condition which entirely consists of '#'-symbols matches every message of length ℓ. So we have to add some constraints to make the problem more interesting:

Classifiers are required to be as specific as possible.

In practice the credit assignment system of a classifier system employs two distinct mechanisms in order to keep the rules as simple as possible: the *specificity* of the classifier and the *bid tax*. Hence we require the rule discovery system to look for rules which are as specific as possible. In complexity theory such optimisation problems are normally formalised as the problem to decide whether there are conditions of a certain length k since an efficient solution for this decision problem also implies an efficient solution for the optimisation problem.

Classifier systems are intended to be able to adapt their behaviour in changing environment. In addition such systems have to deal with the difficulty that they have only got an unspecific performance parameter (pay-off) and no knowledge whether certain classifiers might possibly be improved.

Like Stephen Judd [8] we will consider fully specified learning tasks. One might object against this strong restriction of the learning task since it does not represent the learning task of classifier systems with all its difficulties. On the contrary we argue that such a restriction is sensible since even this *easier* task turned out to be hard [5]. Therefore such a complex learning task would be even harder to solve or require some restrictions on the complexity of the environment.

However, we have not yet discussed why to consider only classifier conditions. In general classifiers of a classifier system possess n conditions. So we do not only consider classifiers which possess one single condition. Classifiers are allowed to send messages which might be used as a memory of the system. When considering a fully specified learning task such a memory is no help at all since the task will only be extended to finding classifiers (or classifier conditions) matching each message list of the task *and* such 'internal' messages.

In practice such classifiers which implement a memory might be useful in environments with time dependent behaviour. As we argued before this would increase the complexity of the learning problem.

4 Loading in Classifier Systems

At first we will examine a general loading problem for classifier systems. The objective is to find classifier conditions which match each of a set of positive tasks but which do not match any of the given negative tasks. Such a problem is easily solvable since we can easily test whether one message in each positive task is consistent with all the negative examples. Therefore the trivial solution of this problem consists of t_+ matching classifier conditions where t_+ is the number of positive tasks.

A classifier system is normally required to find a set of at most k classifiers which match each of the positive tasks and none of the negative, i.e. to distinguish positive from negative tasks.

Definition 1: k-DISTINGUISHINGCLASSIFIERSET (k-DCS)
Let $\mathcal{T}^+ = \{\mathbf{M}_0^+, \ldots, \mathbf{M}_{t_+}^+\}$ be a set of positive tasks, $\mathcal{T}^- = \{\mathbf{M}_0^-, \ldots, \mathbf{M}_{t_-}^-\}$ be a set of negative tasks where each task $\mathbf{M}_i^{+/-} = \{m_1, .., m_p\}$ consists of a set of at most p messages $m_j \in \{0,1\}^\ell$, k a positive number, and $\mathcal{C} = \{c_1, .., c_k\}$ be a set of k classifier conditions.
k-DCS $= \{\langle \mathcal{T}^+, \mathcal{T}^- \rangle | \exists \mathcal{C} = \{c_1, \ldots, c_k\}$: at least one message m of each task $\mathbf{M}_i^+$ in $\mathcal{T}^+$ is matched by a classifier condition $\mathcal{C}$ and no classifier condition matches a task $\mathbf{M}_i^-$ in $\mathcal{T}^-\}$

It turns out that this language k-DISTINGUISHINGCLASSIFIERSET is *NP-complete* for $k \geq 2$, i.e. there is no effective way to decide if a given instance is in DISTINGUISHINGCLASSIFIERSET unless $NP = R$, $NP = P$ respectively [9, 5]. Restricting the number of classifiers to be found leads to a problem which is tractable in parallel.

Theorem 1:
1-DCS $\in \mathcal{NC}$ if $|\mathbf{M}| = 1$.

Proof:
First, we define a procedure *melt*(m_1, m_2) which combines two messages (two classifier conditions respectively) to a condition which matches both messages (classifier conditions respectively).

```
procedure melt(m_1, m_2) :
begin
    for all j ∈ {1, .., ℓ} in parallel do
        if m_1(j) = m_2(j)
            then c_(j) ← m_1(j)
            else c_(j) ←'#';
    return c;
end
```

Clearly, this can be done using polynomial (ℓ) processors in constant time (not considering the time required to activate ℓ processors).

Second, we use the procedure *melt*(m_1, m_2) to compute a classifier which matches each task in the procedure *matching classifier ()*. Let $t = 2^m$ be the number of tasks and the array $d = d(2^m), .., d(2^{m+1})$ contain the tasks $d_1, .., d_t$. If t is not a power of two, then a (minimum) number of dummy elements can always be added to ensure this condition.

Figure 1 shows how the algorithm computes the classifier condition $d = d(1)$. Clearly, this can be done in logarithmic time ($log_2(t)$ times sequentially calling $O(t)$ parallel calls of *melt*(m_1, m_2)) using polynomial $(t \cdot \ell)$ processors.

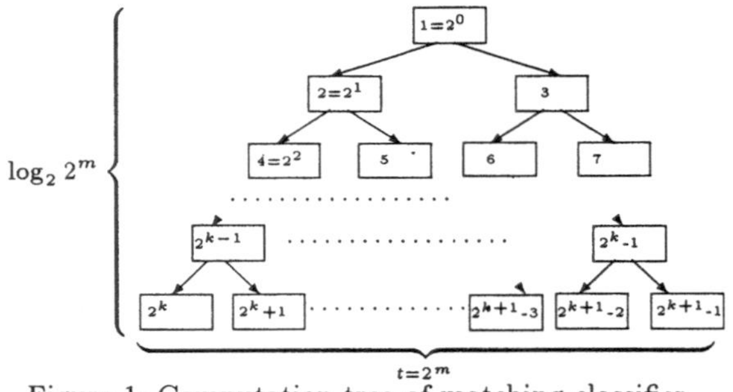

Figure 1: Computation tree of matching classifier

```
procedure matching classifier (d):
begin
    for k ← (m − 1) step -1 to 0 do
        for all j ∈ {2^k, .., 2^(k+1) − 1} in parallel do
            d(j) ← melt(d(2j), d(2j + 1));
    return d(1);
end
```

We decide whether it is possible to distinguish the negative examples from the positive ones using the matching classifier. Therefore we have to check whether the computed classifier condition matches any of the negative examples in $\mathcal{T}^-$.

The procedure *check(c,m)* tests whether the classifier condition c matches the message m.

```
procedure check (c,m):
begin
    matching ← match;
    for all i ∈ {1,..,ℓ} in parallel do
        if c_(i) ≠ '#'
            then if c_(i) ≠ m_(i)
                then matching ← mismatch;
    return matching;
end
```

Now we use the procedure *check(c,m)* to test whether the classifier condition mismatches each negative task (procedure *check classifier()*). Clearly, this procedure can be computed in polylogarithmic time using ℓ processors.

Let $t = 2^m$ be the number of negative tasks in $\mathcal{T}^-$ and $ms_{(1)}$, .., $ms_{(t)}$ contain these tasks. (As mentioned before, we can always add a minimum of dummy elements in constant parallel time to ensure that t is a power of two.)

The procedure *check classifier(c)* computes the matches, mismatches respectively, of classifier condition c and each negative task $m_{a(i)}$ with $(t \cdot \ell)$ processors using $\lceil log_2(t) \rceil$ calls of procedure $check(m_1, m_2)$. It continues to compare the resulting matches of each negative example and the classifier condition with each other.

```
procedure check classifier (d,m_a):
begin
    for all i ∈ {1,..,t} in parallel do
        m(2^m + i − 1) ← check(d,m_a(j));
    for k ← (m − 1) step -1 to 0 do
        for all j ∈ {2^k,..,2^(k+1) − 1} in parallel do
            if m(2j) = m(2j+1)
                then
                    m(j) ← m(2j)
                else
                    m(j) = 'mismatch';
    return m(1);
end
```

Figure 1 shows how the algorithm computes the match *m(1)*. Clearly, this can be done in polylogarithmic time $(O(log_2(t)))$ using polynomial $(t = 2^m)$ processors.

□

Now we will relax the requirements of our loading task in order to discuss whether this 'easier' task is also easier in the sense of complexity theory. We only request the system to find classifier conditions which match each positive task.

Once again there is a trivial solution because the classifier condition which entirely entirely of '#' matches every message of length ℓ. However, we are not only interested in an arbitrary matching condition; we are interested in finding a specific matching condition, i.e. a condition which contains as little 'wild card' symbols ('#') as possible. We define the language MATCHINGCLASSIFIER to formalise our notion of specific matching classifier conditions.

The reason that we require specific classifier conditions is that such a condition is more likely to distinguish a positive example, i.e. a detector assignment that should be matched, from a negative example, i.e. a detector assignment that must not be matched. Notice, however, that we do not require the system to guarantee that it only matches positive examples.

Definition 2: k-MATCHINGCLASSIFIER (MC-k) Let $\mathcal{T} = \{\mathbf{M}_0, .., \mathbf{M}_t\}$ be a set of tasks where each task $\mathbf{M}_i = \{m_1, .., m_p\}$ consists of a set of at most p messages $m_j \in \{0,1\}^\ell$, $c = c_{(1)}, .., c_{(\ell)}$ be a classifier condition and k an integer.

MC $= \{\langle \mathcal{T} \rangle | \exists$ a classifier condition c: c contains at least k $c_{(j)} \neq$ '#' and matches each task $\mathbf{M}_i$ in $\mathcal{T}$ (at least once) }

The language k-MATCHINGCLASSIFIER turns out to be *NP-complete* for $k \geq 3$, i.e. untractable unless *NP* = *R*, *NP* = *P* respectively [5]. Now we consider more restricted learning tasks where we limit the length of the message list.

However, if we consider a more restricted learning task where we only allow message lists of length one ($p = 1$) it turns out that MATCHINGCLASSIFIER becomes tractable in a very strong sense.

Theorem 2:
MC-1 $\in$ *NC*.

Proof:
Firstly, we use the procedure *matching classifier()* which is defined in the proof of theorem 1. This procedure computes a maximal matching classifier for all the positive examples in $\mathcal{T}^+$ (in polylogarithmic time).

Secondly, we use the scheme shown in figure 1 in order to see that the number of symbols $\neq$ '#' can be computed in $O(log_2(\ell))$ time using ℓ processors. This number can be compared with k in constant time.

□

In general, learning in classifier systems proved to be hard. Even when we restrict our task to positive examples only it remains *NP-complete*. There are special learning problems which are efficiently parallelisable; however, these problems are quite restricted.

Efficiently parallelisable	Intractable
NC	*NP-complete*
1-DISTINGUISHING CLASSIFIERSET 1-DCS	k-DISTINGUISHING CLASSIFIERSET k-DCS
1-MATCHINGCLASSIFIER MC-1	k-MATCHINGCLASSIFIER MC-k

Table 1: Loading in classifier systems

5 Related Work

Leslie Valiant developed the theory of *probably approximately correct (PAC) learning* which provides a formal framework to distinguish tractable learning tasks from intractable tasks [10]. Within his model he employs Boolean formulae in order to characterise families of learning tasks. The PAC-learning model requires the learning programme to classify correctly a complete domain of Boolean concepts, given a polynomial number of examples. These examples are presented according to a fixed, but unknown probability distribution. However, there might be some bizarre examples which occur with low probability; therefore the learned programme has to agree on most of the distribution but not on all the examples. A sequence of examples which is presented to the learning algorithm might be highly unrepresentative which introduces another source of error. This case is also required to occur with a certain probability. Furthermore the learning programme has to run in polynomial time.

Normal forms of Boolean formulae are characterised in the following way: A *monomial* is a conjunction of literals. A formula is in *disjunctive normal form (DNF)* when it is a disjunct of terms and each term is a monomial. A formula is *monotone* whenever it contains positive literals only. A formula is in *monotone disjunctive normal form (monotone DNF)* whenever it is a DNF-formula and each term consists of positive literals only. A formula is in *ℓ-disjunctive normal form (ℓ-DNF)* whenever it is a DNF-formula and each term consists of at most ℓ literals.

We can interpret a classifier system as a certain normal form of Boolean formulae, if we regard a single string position in a message as a Boolean variable. All the bits of a message must be matched by one of the classifier conditions, i.e. these Boolean variables are connected by a conjunction. Each message in a message list might be matched by each classifier, i.e. messages are matched by a disjunction of classifier conditions. As Leslie Valiant pointed out such monotone DNF formulae are PAC-learnable [10].

Since messages as well as conditions of a classifier are of a fixed length ℓ we consider Boolean terms of fixed length ℓ. Formulea in ℓ-DNF are PAC-learnable [10].

Notice that these PAC-learning results deal with message lists containing one single message too. Moreover they investigate the tractability of the learning problems rather than their parallelisability.

6 Conclusion

We presented our own results on the computational complexity of learning in classifier systems. We restricted the general loading problem to find only one consistent classifier condition out of a set of positive and negative examples which contain one message each. This restricted problem turned out to be parallelisable efficiently.

Relaxing the requirement that the classifier condition has to distinguish positive from negative examples we looked for classifiers of a certain specificity which match each of a set of positive tasks. If these tasks only contained one single message each, the problem were efficiently solvable in parallel. However, permitting these positive tasks to consist of sets of messages again turns out to be an intractable problem.

Acknowledgements

I would like to acknowledge the proof-reading of the manuscript by Ruth Hannuschka, Volker Weber, and Georg Pietrek. I am also grateful for the final language proof-reading by Eva Dreyer; owing to her assistance the style of the paper was considerably improved. Prof. Norbert Fuhr and Ulrich Pfeifer kindly provided the facilities to actually write this paper at the University of Dortmund.

References

[1] J. L. Balcázar, J. Díaz, and J. Gabarró. *Structural Complexity I.* EATCS, Monographs on Theoretical Computer Science. Springer-Verlag, Berlin, 1 edition, 1988.

[2] L. B. Booker, D. E. Goldberg, and J. H. Holland. Classifier systems and genetic algorithms. *Artificial Intelligence*, 40:235 – 282, 1989.

[3] A. Gibbons and W. Rytter. *Efficient Parallel Algorithms.* Cambridge University Press, Cambridge, UK, 1988.

[4] D. E. Goldberg. *Genetic Algorithms in Search, Optimization & Machine Learning.* Addison-Wesley Publishing Company, Reading, Massachusetts, 1989.

[5] Uwe Hartmann. Computational complexity of neural networks and classifier systems. Diplomarbeit, University of Dortmund, PO.Box 50 05 00, D-4600 Dortmund 50, Germany, July 1992.

[6] J. H. Holland. Adaptation. In R. Rosen and F. M. Snell, editors, *Progress in Theoretical Biology IV*, pages 263 – 293. Academic Press, New York, 1976.

[7] John H. Holland. Properties of the bucket brigade algorithm. In John J. Grefenstette, editor, *Proceedings of an International Conference on Genetic Algorithms and Their Applications*, pages 1 – 7, Pittsburgh, PA, 1985.

[8] Stephen Judd. *Neural Network Design and the Complexity of Learning.* Neural Network Modeling and Connectionism. The MIT Press, Cambridge, MA, 1990.

[9] Leonard Pitt and Leslie G. Valiant. Computational limitaions on learning from examples. *Journal of the Association of Computing Machinery*, 35(4):965 – 984, October 1988.

[10] Leslie G. Valiant. A theory of the learnable. *Communications of the ACM*, 27(11):1134 – 1142, November 1984.

CO-EVOLVING COMMUNICATING CLASSIFIER SYSTEMS FOR TRACKING

Lawrence Bull
Faculty of Engineering
email l_bull@csd.uwe.ac.uk

Terence C Fogarty
Faculty of Computer Studies and Mathematics
email tc_fogar@csd.uwe.ac.uk

University of the West of England
Bristol, BS16 1QY, England.

Abstract

In this paper we suggest a general approach to using the genetic algorithm (GA)[1] to evolve complex control systems. It has been shown [2] that although the GA may be used to evolve simple controllers, it is not able to cope with the evolution of controllers for more complex problems. We present an architecture of co-evolving communicating classifier systems [3] as a general solution to this, where the only restriction is that each classifier system is responsible for one simple behaviour. Thus the ecology of sub-problems evolves its own organisational structure at the same time its constituents evolve their solutions. Whether this structure ends up as a democratic soup, a hierarchy, or something in between, is determined by co-evolution rather than prescribed a priori by a creator. We use the trail following "tracker task" to compare the performance of a single classifier, responsible for the control of the whole system, evolved for this task with the performance of a co-evolved controller using our approach. The resulting interactions of the classifier systems are also examined.

The Tracker Task

Jefferson et al [4] used a genetic algorithm to solve the problem of generating an "ant" that could perform the tracker task, ie to follow a winding broken trail, the "John Muir trail", across a grid environment. They successfully evolved two types of equivalent controller for the ant; a three layer recurrent neural network and a finite state automaton. Each of these controllers had effectively a five bit memory and was represented by a genome of about 450 bits long providing a very large search space for the GA. An ant's lifetime is two hundred steps, where at each discrete step it is given an indication of whether it faces a part of the trail and returns one of four possible actions. The ant can go forward one square, turn left or right on the spot, or do nothing and is scored according to how much of the trail it followed (max 89). The genetic algorithm then takes a percentage of the top scorers and uses them to create the next generation - using multiple point crossover and mutation for creating one child from two parents. A population size of 64K individuals was used on a connection machine [5].

Evolving a Tracker

To make the single classifier system controller equivalent to Jefferson et als representation requires a five bit memory and a genome of around 450 bits. To achieve a five bit memory we give the classifier a string length of five and allow the highest bidder's action string (effector) from the previous step to be posted onto the message list along with the detector string. Because the message list contains two strings at each time step the classifiers are made bi-conditional. Each classifier's strength is given in four bits (binary), resulting in a gene length of nineteen bits, allowing twenty four of them to a system giving a genome of 456 bits. All learning is done solely by the GA, with the Pittsburgh approach adopted.

Figure 1 shows the typical evolutionary progress of 5K classifier ants at the Tracker task, with the crossover rate at 0.01, bit mutation at 0.001 and the selection fraction at the top 5%, over fifty generations. The parameters have been altered from those used by Jefferson et al to make the runs more manageable; it is not our intention here to make a direct comparison between the classifier system representations and the ANNs of FSAs [6].

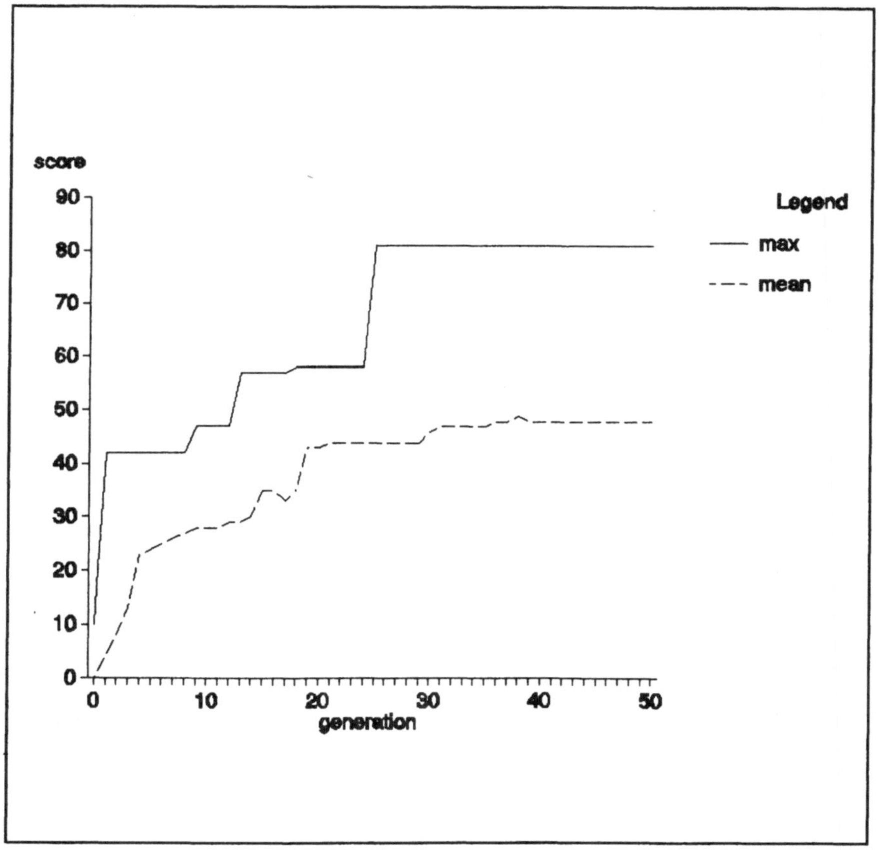

Figure 1

Co-evolving a Tracker

To demonstrate the concept of co-evolving communicating classifier systems for control the problem of guiding a two tracked vehicle along a broken road is now presented. There is a motor for each of the tracks which can be on (forwards or backwards) or off at each particular time step. If either motor is off then nothing happens. If they are both on (forward) then a forward movement results. If either of the motors is on (forward) and the other is on (backward) a right or left turn on the spot results. Reversing is not allowed. The only information available to the controllers of the motors about the road is whether or not there is a piece of it immediately ahead. Each controller also gets the effector of the other side, along with its own effector, from the previous time step. The problem is equivalent to the Tracker task, however rather than evolve a single classifier system to control the two motors each motor is assigned its own classifier system of half the size; each contains twelve classifiers. The genome of each controller is now 228 bits, with them stored in two arrays where they are paired by their opposite number (left[0] with right[0], etc) and where both receive the score that vehicle achieved.

Figure 2 shows the typical co-evolutionary progress of a population of 5K vehicles at the Tracker task, using the same GA parameters as in Figure 1 for the single system model, over fifty generations.

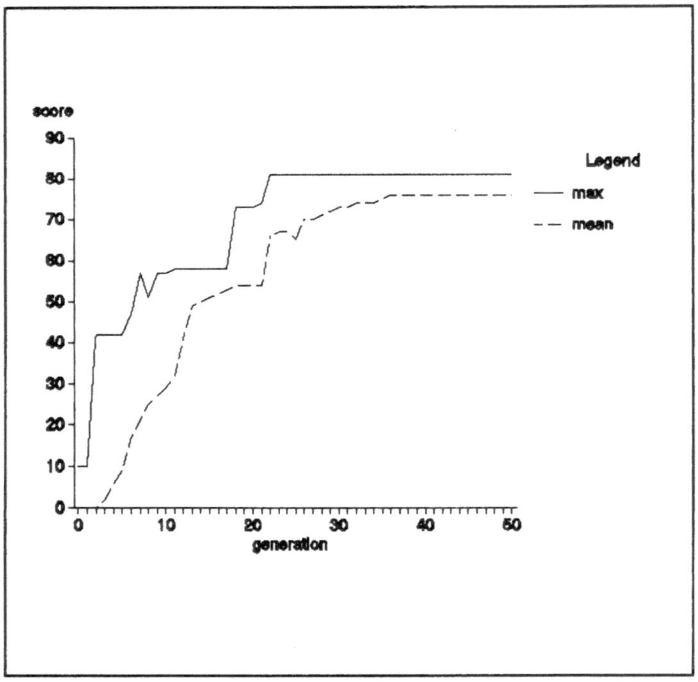

Figure 2

Results

Figures 1 and 2 show that, as Jefferson et al state, a basic logic capable of scoring 81 within the 200 time steps is not particulary difficult to evolve, but after this specific features of the trail must be exploited which consistently proves more difficult in runs with these settings. It should be noted that we have been able to evolve and co-evolve logics capable of scoring over 81 at the John Muir trail though. The rate of progress can be seen to be about the same, with the co-evolved systems on average being perhaps slightly faster to develop the 81 logic. The noticeable difference between the two approaches is that the co-evolved systems are much more converged than those in Figure 1. The single systems mean score is around forty points below the highest scorers, in contrast to the five or six points below we see in Figure 2. We are only able to get similarly converged populations in the single systems when we increase the size up to around 10K ants, but these systems then appear to find it more difficult to do better than a high score of 81.

As stated at the beginning of the paper, we allow any form of structure to develop along with the "species" themselves. In this system the approach yields two types of relationship between the motors; a democracy or a hierarchy where one motor is more developed to its partner. In these hierarchy/oligarchy structures one motor develops rules matching the trail infront detector and the no trail infront detector strings, where in both cases it will give an effector that turns it on (forward). The opposite controller develops six rules which activate on combinations of effectors (its own and the other sides) and the detectors. The emergence of this type of relationship was not unexpected; in most naturally occurring symbioses the partners are usually different sizes, with one or more of the symbionts confined to specific regions of a host [7]. In the democratic soup both controllers evolve six rules that match the detector string and each others effector alternately. For example, the strategy that evolved in most runs was for the ant/vehicle, on loosing the trail (due to a break or corner), to turn right up to four times and then move forward a step until the trail was seen again, as Jefferson at al show, ie :-

Forward A : Both match solely on the detector indicating trail and turn on (f).
1st Right : Left matches detector and own output, turns on (f). Right matches detector and lefts output, turns on (b).
2nd Right : Left matches detector and right output, turns on (f). Right matches detector and own output, turns on (b).
3rd Right : Left matches detector and own output, turns on (f). Right matches detector and lefts output, turns on (b).
4th Right : Left matches detector and rights output, turns on (f). Right matches detector and own output, turns on (b).
Forward B : Left matches on its own output, turns on (f). Right matches on lefts output, turns on (f).

We have noticed that these democratic relationships are able to co-evolve to the 81 optima slightly quicker than the hierarchy. Kauffman and Johnsens [8] investigations into the effects of interdependence on co-evolution reported a similar phenomena finding that the more tightly coupled the partners fitness landscapes, the quicker the system as a whole reaches equilibrium - explained as due to an increase in ruggedness of the landscapes causing more local optima. Both types of co-evolved systems have been able to develop a vehicle capable of scoring 83 on the trail, matching the single systems best.

Conclusion

We have presented the initial results from a general approach to using the GA to co-evolve controllers for complex problems. The outcome from using communicating classifier systems would seem to indicate that for quite a large search space an increase in comparative mean performance can be seen, along with a slight decrease in time taken for convergence. This is perhaps as expected; the continuous adaption to each others (incremental) changing fitness landscapes - a constant moving of the target - generates a more efficient search method. We have shown that evolutionary pressure does produce relationships between the controllers during their adaption and have subsequently found (results not shown) that when the ability to communicate is removed performance drops significantly. Dorigo and Schnepfs [9] work on robot controllers, using a hierarchy of learning classifier systems, reports improvements over a single system when the learning of the classifiers is manually controlled. Their approach is based on Tinbergens [10] model of animal behaviour, a development of Bookers [11] GOFER system, in which the lower level behaviour classifier systems develop with respect to each other and those controlling the interactions between them higher up. We are now building a model to see how our more general architecture scales to problems with many communicating classifier systems.

References

[1] Holland J H, 'Adaption in Natural and Artificial Systems', Univ. of Michigan Press, Ann Arbor, 1975.

[2] Fogarty T C, 'Evolving Controllers', IEE Digest, 106, 1992.

[3] Holland J H, 'Escaping brittleness : the possibilities of a general purpose machine learning algorithm applied to parallel rule-based systems', Machine Learning II, Los Altos, 1986.

[4] Jefferson D, Collins R, Cooper C, Dyer M, Flowers M, Korf R, Taylor C and Wang A, 'Evolution as Theme in Artificial Life : The Genesys/Tracker System', Artificial Life II, Addison-Wesley, 1989.

[5] Hillis W D, 'The Connection Machine', MIT Press, 1985.

[6] Bull L, 'Classifiers and the Tracker Task', Internal Report, Bristol Polytechnic, 1992.

[7] Smith D C, 'The Biology of Symbiosis', Arnold Edward, 1987.

[8] Kauffman S A and Johnsen S, 'Co-Evolution to the Edge of Chaos: Coupled Fitness Landscapes, Poised States, and Co-Evolutionary Avalanches', Artificial Life II, Addison-Wesley, 1989.

[9] Dorigo M and Schnepf U, 'Genetics-based Machine Learning and Behaviour Based Robotics : A New Synthesis', IEEE Trans. on Sys., 22, 6, 1992.

[10] Tinbergen N, 'The Study of Instincts', Oxford University Press, 1951.

[11] Booker L, 'Classifier Systems that Learn Internal World Models', Machine Learning, 3, 3, 1988.

On Finding Optimal Potential Customers from a Large Marketing Database - a Genetic Algorithm Approach

Sam Sandqvist
Oy MGlobal Finland Ab,
Sirrikuja 4C13,
SF-00940 Helsinki, Finland

Abstract

In this work we have studied the problem of finding potential customers from large marketing databases using a genetic algorithm. The problem is that it is far from clear, in advance, what constitutes a good potential customer in the database in question, even if criteria for how a customer should be graded can be formulated. The genetic algorithm approach uses this grading as the basis for a fitness function, crucial to the genetic algorithm, and effectively applies the genetic algorithm to classify the database accordingly. As a consequence, the result directly tells us both how well the grading succeeds in finding good customers and gives us a set of found optimal customers.

Keywords: Genetic algorithm, search, optimisation, large database, marketing

1 Preface

Modern marketing relies more and more on very large databases containing potential customers and prospects. The task of keeping these databases both up-to-date and readily usable for marketing is very expensive. This paper describes a novel approach to utilise the information in such a database. Because marketing is an ongoing, dynamic, very expensive process it is crucial that information culled from it is both quickly produced and qualitatively good. Using the approach described the search time is virtually independent of the size of the database (i.e. in constant time), and even so produces potential customers that are as good as those found using more traditional methods.

2 Introduction

2.1 Marketing Database

Marketing databases always contain information on either companies or people, or both, information that may describe both existing customers and potential customers. There are two kinds of marketing databases: business-to-business for inter company marketing, and consumer databases for company to private consumer marketing. For the purpose of this study these are the same.

A database used for marketing is characterised by having, in addition to the common information held in company and contact databases, several very volatile items, such as consumer behaviour, ordering history, paying behaviour, as well as campaign information describing actions (and reactions) that have been targeted at the contact in question. It is this information that forms the basis for grading potential customers.

2.2 Optimal Potential Customer

Given a certain product (or product mix) to market, it is certain that some people react more favourably to an offer for it that some other people. An optimal potential customer is one that is more likely than others to respond to an offer. Optimality can be simply based on address: a contact on the coast is certainly more likely to buy a boat than someone in the mountains. It can be based on ordering history: someone who bought a book on natural history may respond favourably to an offer for a world atlas, for instance.

2.3 Genetic Algorithm

For a thorough discussion of genetic algorithms, see. e.g. [Go89], [Da91].

2.4 The Problem

The basic problem is thus to find the optimal customers from the marketing database. However, we usually do not need all of them; frequently only a few hundred to a few thousand suffice. Given the high cost of direct marketing, it is necessary and desirable to address as small a number as possible, but as good a potential set as possible as well.

Since we do not know the distribution of potential customers in the database, given the criteria we are interested in, which of course vary from offered product to offered product, it is impossible to ascertain the effectiveness of a direct marketing campaign in advance if our campaign is based on a random sample.

Because a fully indexed (i.e. indexed on all fields) database is not feasible, the problem is to find the optimal potential customers as well as the optimal criteria for both the actual database in question and the grading criteria we are using.

Below, the problem has been simplified to finding a small number of potential customers, e.g. 100, as well as the optimised selection criteria for the search. The latter result may be used to characterise, or classify, the database in the light of the grading specifications given in advance.

One may also view the problem as one of pattern recognition: the task is to find the optimal vector in n-space, n being the number of fields we are interested in in the database.

2.5 Related Work

The application of genetic algorithms to database access and selection is novel; it was first described in [Sa92]. Specifically for marketing the only reference found is [GS87], but this is not in a database environment. Genetic algorithms have been used in database query optimisation, see e.g. [BFI91]. They have also been investigated for the problem of finding optimal indexes for databases, e.g. [GA90]. A work describing genetic algorithms for distributed databases is [SF92].

The task of finding optimal criteria for marketing purposes has also been addressed in other, non-computing fields; e.g. [Pe92].

3 Alternative Approaches

Since the problem is trivial in a fully indexed (i.e. all fields have an index) database, but this is not a realistic approach because the number of fields may be huge, numbering hundreds, several other approaches may be used.

3.1 Random Search

It is certainly possible to search randomly through the database in order to obtain potential customers. There are several problems with this approach. One, it is not guaranteed that we will ever obtain good potentials. Two, with a large database, the time required to obtain enough samples to have a statistically decent chance to obtain a good sample is of the order of $O(n)$, n being the number of records in the database, which in most cases is too large. Three, this approach does not enable us to draw any further conclusions regarding possible patterns in the database.

3.2 Linear Search

A linear search may be a feasible approach for small databases; search time is still $O(n)$. However, since we do not know where in the database the good potentials are we may have to resort to searching the whole database. Even if existing indexes may be used, they distort the obtained sample by providing the absolutely best potentials only for the indexed information ignoring any other information that might give much better samples.

3.3 Cumulative Search (Greedy Algorithm)

The greedy algorithm remembers previously obtained best samples and will combine these with newly obtained samples, using any of the algorithms above. It is clear that any algorithm will use cumulative results since the technique offers much better actual performance; combined with any of the previous algorithms, however, will not result in qualitatively better results unless the number of sought after records is both small and they occur early on in the search.

4 The Genetic Approach

For an overview of genetic algorithms in general, see e.g. [Go89]. A preliminary description of the approach used in this work appeared in [Sa92].

The practical work has been undertaken using the following software.

Genesis TSP Software, Inc.

The basic genetic algorithm package that provides the foundation for this work. Written in C. See [GDC91].

SoftC SoftC, Ltd.

Database access function library used in the modified version of Genesis. Written in C. See [So90].

GA Database Processor

User interface for Genesis and query editor / processor. Written by the author in Smalltalk/V for MS Windows.

The packages interact in the following way.

1. First, the user selects, using the GA Database Processor, an algorithm file. This file provides the input to Genesis, and may be freely edited using the Algorithm Editor, which is part of the Processor.

2. The user also selects a database query file, which, in the same way as the algorithm, may be freely edited using the Query Editor supplied with the Processor.

3. Lastly, the user also selects a database to process. This provides the Query Editor with all field characteristics; names of fields, types, lengths etc.

4. The user may store all file definitions in a Workspace file, which may be retrieved separately to fill in all the file definitions above.

5. When the user is satisfied with his query and saves it, a C subroutine *eval.c* will be generated corresponding to the query. This file contains the fitness function and other query/database-specific functions used by the modified version of Genesis to actually run the genetic algorithm on the database.

6. When the user presses the *Run*-button on the main screen, the GA Database Processor will spawn a process to initiate the following sequence of events:
   - copy the files to their proper locations
   - *make* a new *ga.exe* program corresponding to the query and algorithm
   - run *ga.exe*
   - print the results on the printer using the Genesis *report* program

The software package is intended to provide quick results for evaluation purposes. It has at this time not been optimised, nor have the reporting functions been geared towards casual, non-specialist use.

For screen shots of the three main screens of the package, see Appendix I.

4.1 Description of Genetic Algorithm Used

For a preliminary description of the algorithms, see [Sa92]. Please note that that system was written in the database language M (formerly MUMPS), which limited its usefulness in a commercial environment. As previously indicated, the present system is written in a combination of C and Smalltalk.

The population consists of chromosomes with genes based on the actual query in the following way. Every rule in the query corresponds to one gene. The chromosome thus corresponds to a whole query. The purpose is to maximise the query, i.e. to obtain a query with selection criteria optimised for the fields and database in question.

The basic idea is to match each rule in the query to a randomly selected record in the database. If the rule is true, i.e. the record field (or fields, as the rule may be a complex one with several conditions and fields) matches, the corresponding gene's value (a real number) will be multiplied be a weight assigned to that rule in the query.

Population size	100	smaller values do not offer enough variation to obtain stable populations.
Crossover Rate	0.6	a good, standard value.
Mutation Rate	0.001	standard value; higher values fragments population.
Generation Gap	1.0	indicates that whole population is replaced each generation. Not clear how smaller values would affect performance.
Scaling Window	5	adjust maximum fitness values every 5 generations. Used to prevent fitness value stagnation to cause fitness evaluation to fail to distinguish between very similar values. For details, see [GDC91].
elitist strategy	yes	in actual runs, elitist strategy was used. However, testing wihout did not change results substantially.
Trials	2000	take 2000 samples from database.

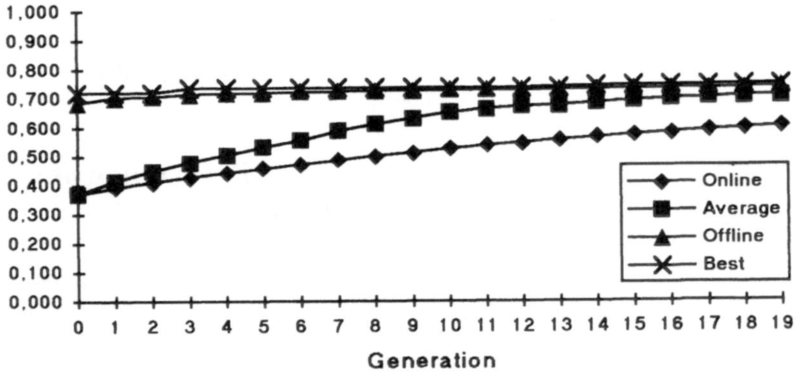

The fitness function, discussed below, will combine the gene values to get the fitness for that chromosome, and this for that query. Natural selection will then optimise the population of chromosomes according to the genetic algorithm used.

Several variants of genetic algorithms [GDC91] offered by the Genesis package were tried, and the best results so far were obtained using the parameters on the previous page.

In actual use a query could have been similar to the following.

```
TL1 == "90-" weight: 0.9
US1 != " " weight: 0.2
(US7 == "MONT") ||
 (US6 contains: "B") weight: 1.0
TL1 != " " weight: 0.5
```

This query tests fields `TL1` (a telephone number), `US1`, `US6` and `US7` (codes with different meanings) assigning separate weights to all rules.

The syntax specified using the Query Editor clearly shows the C-like conditions. Note that the weight, although normally in the interval (0 .. 1], actually may be any real value. Negative values would then indicate a negative influence of the corresponding rule on the query as a whole.

4.2 The Fitness Function

The fitness function is based on the query and the weights of its rules, as indicated above.

Given a chromosome with *n* genes we treat it as a *n*-vector; likewise with the rule weights.

If each rule gets the value 0 when false, 1 when true, we obtain, for any rule *i*

$$f_i = w_i * r_i$$

where f_i is the fitness for rule *i*, w_i its weight and r_i its truth value.

The fitness for the whole chromosome fit_{chr} is taken to be the weighted mean of the fitnesses of its rules:

$$fit_{chr} = \frac{\Sigma f_i}{\Sigma w_i}$$

This is passed back as the value of the *eval* function.

5 Results and Comparisons

For early results, see [Sa92]. For a run using four rules, all with equal weights, on a moderately sized database with approximately 66000 records, with 70 fields each, we obtain the following results indicated in the figure.

The curves depict the following performance characteristics:

- *Online* the mean of all evaluations so far
- *Offline* the mean of the current best evaluations

- *Average* the average performance of current generation
- *Best* the current best value

As shown, a steady improvement is indicated, c.f. the average curve. More important, though, is the fact that the mean of all generations (as indicated by the offline curve) is settling into a recognizable pattern.

6 Conclusions

See also [Sa92]. Preliminary results indicate that the approach produces at least as good potential customers as any other approach. The crucial difference is that using the genetic algorithm we obtain these in essentially constant time.

The time can easily be calculated as

$$O(trials * population * sample_{trial})$$

In other words, constant time dependent only on the number of trials, whether we sample several records per trial (which evens out wildly different chromosomes; see [Sa92]) and the populations size.

For the test results above, running time was about 30 seconds.

More important, we obtain a pattern, a profile of the database for the particular selection criteria we are interested in. This pattern recognition capability is extremely interesting, and potentially very useful in the system the author has investigated.

References

Go89 Goldberg, David E.
Genetic Algorithms in Search, Optimization and Machine Learning
Addison-Wesley, Reading 1989

Sa92 Sandqvist, Sam
On Finding Optimal Customers from a Large Marketing Database
STEP-92 New Directions in Artificial Intelligence, Proc. of Finnish Int. AI Conf., Vol 3, p. 36ff

SF92 Siegelmann, Hava T. and Frieder, Ophir
Document Allocation in Multiprocessor Information Retrieval Systems
Tech. Report IA-92-1, George Mason University, Fairfax, 1992

Ga90 Galarce, Carlos E.
Adaptive Systems and the Search for Optimal Index Selection
M.Sc. Thesis, Wayne State University, Detroit, 1990

IK90 Ioannides, Y.E. and Chang, Y. C.
Randomized Algorithms for Optimizing Large Join Queries
Proc. 1990 ACM-SIGMOD Conf. on Management of Data, ACM Press, 1990

BFI91 Bennett, Kristin; Ferris, Michael C. and Ioannides, Yannis E.
A Genetic Algorithm for Database Query Optimization
Proc. Fourth Int. Conf. on Genetic Algorithms, Morgan Kaufmann Pub, 1991 p. 400-407

Da91 Davis, Lawrence (ed)
Handbook of Genetic Algorithms
Van Nostrand Reinhold, New York, 1991

GS87 Greene, David Perry and Smith, Stephen F.
GBML System Learns Rules Describing Customer Preferences

Proc. Second Int. Conf. on Genetic Algorithms, Morgan Kaufman Pub., 1987 p. 217-223

So90 SoftC Ltd.
SoftC Database Access Library documentation
Electronic media, with the software. 1990

Pe92 de Pelsmacher, Patrick
What are your customers looking for? - An illustration of conjoint analysis shows how to elicit customer preferences and price perceptions for different product characteristics
Operational Research Insight, Vol. 5, issue 2 (March - June 1992), p. 16-21

GDC91 Grefenstette, John J; Davis, Lawrence, and Cerys, Daniel
Genesis & OOGA: Two Genetic Algorithm Systems
TSP, P.O.Box 991, Melrose, MA 02176, 1991

Appendix I.

Main screen of GA Database Processor.

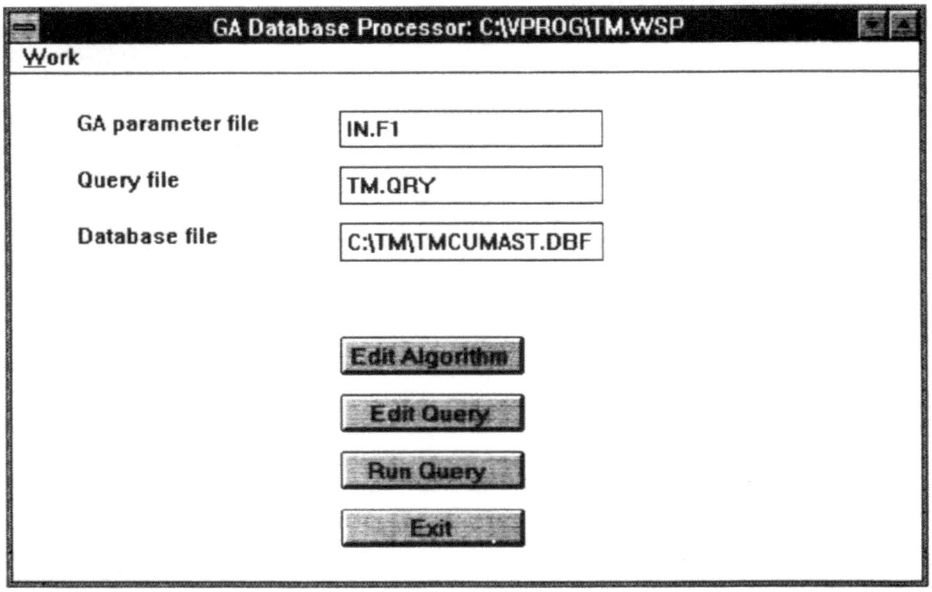

The Algorithm Editor

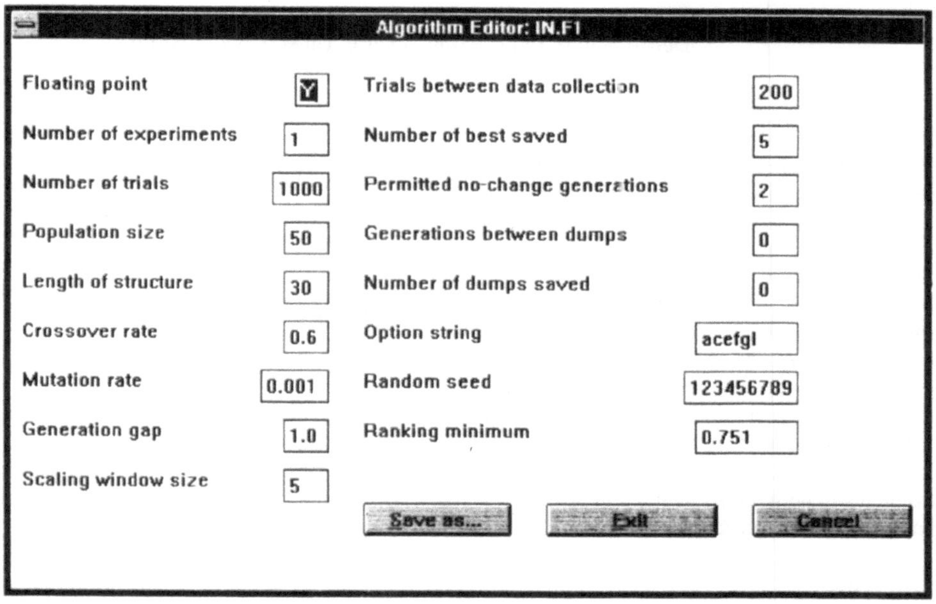

The Query Editor:

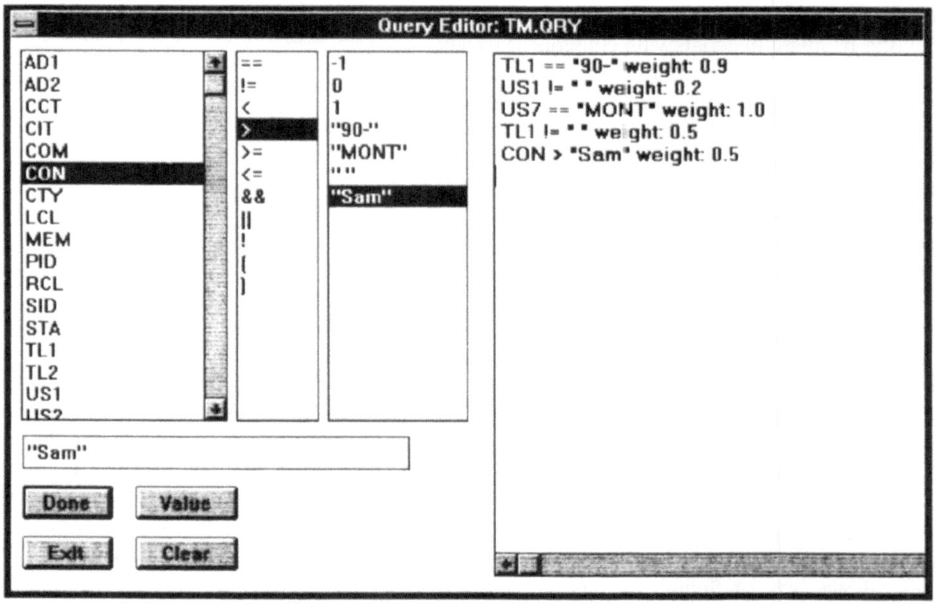

STRUCTURAL DESIGN FOR ENHANCED NOISE PERFORMANCE USING GENETIC ALGORITHM AND OTHER OPTIMIZATION TECHNIQUES

Dr. A.J. Keane andy.keane@uk.ac.ox.eng
Department of Engineering Science, University of Oxford,
Parks Road, Oxford, OX1 3PJ, U.K.

Abstract

The control of structural vibration in aeroplanes and ships is of great importance in achieving low noise targets. Currently, such control is effected using viscoelastic coating materials although much current research is concerned with active, anti-noise based control measures. Recent studies using Genetic Algorithm (GA) optimization methods in the field of Statistical Energy Analysis (SEA) suggest that it may be possible to design passive noise filtration characteristics into such structures. This paper reports initial work in this field: it compares GA's with more classical optimization methods and shows how improvements in noise performance can be obtained for the simple structures considered.

1. Introduction

Many engineering structures suffer from exposure to noise and vibration sources. These sources often excite unwanted structural vibrations which can cause damage or the transport of vibrational energy to distant parts of the structure where they cannot be tolerated. The most common treatment for such problems is to coat the structural elements with heavy viscoelastic damping materials with consequent weight and cost penalties. Clearly, if the vibrational energy could be contained near to the points of excitation there would be a reduced need for damping treatments and, additionally, they could be concentrated in regions where they were most effective. The upshot of this problem is the desire for some kind of structural filter design that can be built into the structure and that is capable of carrying static loads but blocks higher frequency motions.

During part of a recent programme of work looking at improved noise flow prediction methods Statistical Energy Analysis (SEA) was being applied to simplified aerospace structural models. During this work the behaviour of anomalous structural configurations became of interest[1]. The primary reason for this was the desire to quantify the tails in the statistical descriptions of the responses of structures with random physical properties. It became clear that the most direct way of finding configurations where energy flowed well or badly was to employ optimization techniques to maximize or minimize the flows. This proved difficult using classical approaches and a survey of methods, including the Genetic Algorithm (GA) was carried out. The results of this survey are given here and these will form the basis of a more detailed investigation into whether some of the structural configurations produced can be adopted to yield the passive filtration characteristics that designers desire.

2. A Simplified Model

Perhaps the simplest system that can be formulated when considering the flow of vibrational energies between connected structures consists of two, axially vibrating rods which are mutually coupled at a point through a spring. To simulate the noise control design problem the mass properties of the individual rods are taken as unknown, but bounded variables that vary as functions of position. Interest then focuses on the coupling power receptance function for these two subsystems, i.e., it is this quantity that must be minimized if energy is to be prevented from flowing from one structure to the other. Thus, when two rods are coupled at $x_i=a_i$ and excited by statistically independent point forces $F_i(t)$ acting at $x_i=b_i$, $i=1,2$, the coupling power receptance for flows from the first to the second is given by[2, 3, 4]

$$H_{12}(\omega)=\frac{\omega^2 k_c^2 c_2}{m_1^2 m_2 |\Delta|^2}\sum_{r=1}^{\infty}(\psi_r^2(a_2)/|\phi_r|^2) \qquad (1)$$

$$\times\left|\sum_{i=1}^{\infty}(\psi_i(b_1)\psi_i(a_1)/\phi_i)\right|^2$$

with

$$\Delta = 1+(k_c/m_1)\sum_{i=1}^{\infty}(\psi_i^2(a_1)/\phi_i)+(k_c/m_2)\sum_{r=1}^{\infty}(\psi_r^2(a_2)/\phi_r)\ .$$

Here ϕ_i is $\omega_i^2-\omega^2+ic_1\omega$, similarly for ϕ_r The summations over the indices i and r respectively denote summations over the modes of the first and second rods. $\omega_{i,r}$ and $\psi_{i,r}$ are the natural frequencies and mode shapes of the rods. The quantities $c_{1,2}$,$m_{1,2}$ and k_c denote the coefficients of viscous damping, total masses and coupling spring constant.

When the masses of the individual rods are modelled as varying quantities, the natural frequencies and mode shapes vary in a very complex fashion. This results in the receptance function having properties that are extremely difficult to predict. As has been noted, the aim of the present investigation is to obtain a minimum value of $H_{12}(\omega)$ as a function of the mass profiles of the two rods. A general analytical solution to this problem is currently not possible. However, for every realization of the pairs of rods, the natural frequencies and mode shapes can be calculated for the individual rods and this information incorporated into equation (1) to generate the receptance function. A crucial step in this calculation involves obtaining the natural frequencies and mode shapes of the rods. When the mass properties vary along the lengths of the rod special means are needed to derive these frequencies and shapes[1].

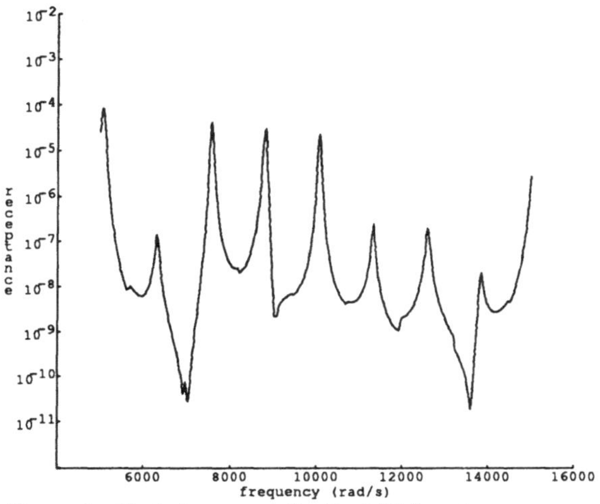

Figure 1 - $H_{12}(\omega)$ versus ω for the initial system.

3. Initial System

All design problems start from some simple initial system which is refined as the design process progresses. Here the initial values of the system properties are taken to be m=21.54kg, L=5.182m, AE=0.1785MN and c =80s^{-1}. The coupling spring constant is taken to be k_c=0.5x10^{7}N/m and the points of coupling fixed at a_1=2.3m (0.44L) and a_2=3.3m (0.64L). The forces are assumed to act on only the first subsystem with the second subsystem remaining externally unforced. The point of forcing is taken to be b_1=1.192 m (0.23L) and the driving frequency range of interest is taken to be 5,000 rad/s to 25,000 rad/s. This represents a very simple configuration with two similar systems having uniform properties and arbitrarily chosen drive and coupling points.

To evaluate this configuration in terms of energy flows $H_{12}(\omega)$ must be calculated and here the infinite summations appearing in equation (1) have been carried out over modes occurring in a frequency band of width 8,796 rad/s centred at the driving frequency ω. Figure 1 shows the resulting power cross receptance function and it can be seen to be dominated by the natural frequencies of the two systems Notice the logarithmic scales used to plot the function; it shows massive variations in behaviour from frequency to frequency. The behaviour of systems with non-uniform properties is similarly complex and even harder to predict *a priori*. It is the goal of optimization methods to be able to rapidly explore such problems without prior knowledge, finding useful configurations as they go.

4. Optimization

When formulating this problem as an optimization study, a decision must be made on how many sections to discretize the rods into, i.e., on how many variables will be used to describe the problem. Clearly, if few are used the optimization problem will be simplified but the range of options studied will be restricted and good, novel configurations may not be found. Conversely, too fine a discretization may waste computing time leading to the search being stopped too early and again the possibility of missing good designs. As in all things a compromise must be made and in order to see the effect of this decision two levels of discretization have been studied, with the rods being broken into 2 or 20 piecewise uniform elements of equal length, such that $\rho_{im}, i=1,2, m=1,2(20)$, is the mass per unit length of the ith rod in the mth section. This variation also allows study of the ability of the optimizers to handle different orders of problem, i.e., four or forty variables. The optimizers are then required to determine values of $\rho_{im}, i=1,2, m=1,2(20)$, which lead to the minimum of the function $H_{12}(\omega)$ at ω=10,000 rad/s.

The decision to minimize energy flow at a single frequency considerably reduces the length of time taken to study this problem, as in reality the response would be integrated over a frequency range and the resulting integral would be minimized. Such integration requires many more function evaluations but it does smooth the function being examined and this somewhat simplifies the problem, i.e., here a quicker but harder problem has been chosen in the interests of research into the applicability of 'modern' optimization methods in this area. The results of applying the best method found here to the integral of $H_{12}(\omega)$ are given towards the end of this work to demonstrate the great potential benefits optimization may bring in producing enhanced noise performance.

All real design problems are bounded in some sense; in structural dynamics such bounds often reflect the need for structures to carry static or low frequency loads, also real designs may need to use stock material sizes and so on. It is quite common for optima to be created by the intersections of these constraints in some way or other. Such constraints are often difficult for optimizers to handle since they represent massive local distortions in the optimization surface. Here the design is arbitrarily constrained by ensuring that the total mass of the individual rods take values in the range $3<m_i<50$ kg ; $i=1,2$ along with the restriction on each individual subsection that $0.5<\rho_{im}<10.0$ kg/m.

4.1. Optimization Methods

Having decided on the function to be optimized, the number of variables involved and the constraints to be applied, attention must then be focussed on the method to be used. Optimization has been studied for a great many years, both as an abstract mathematical problem and also from the viewpoint of engineering design. A great many methods have evolved which are detailed in a sizable literature with each method having its champions and detractors. A good survey of what may be called 'classical' and 'heuristic' methods, presented in an engineering context, is given by Siddall[5]. A common feature of these methods is that they date from before the advent of really cheap powerful computers. They have thus tended to be developed with low order problems in mind, since the computational power to test them on high dimensional problems was not available at their inception. They commonly make use of derivatives and hill climbing methods. Optimization methods can also be found in most libraries of commercially produced numerical software and that produced by NAg Ltd. is typical[6]. Finally, in recent years two competing, modern methods have received much publicity, i.e., the Genetic Algorithm (GA) and Simulated Annealing (SA, sometimes known as stochastic relaxation). Both these methods are acquiring a significant literature detailing their performance and variants on the basic themes[7, 8].

Since the problem being studied here is known to be complex and in one form, of high order, it was decided to take a representative selection across these methods and study them comparatively. The methods tried were i) Genetic Algorithm, ii) Simulated Annealing, iii) Linear Approximation with Linear Programming, iv) Hooke and Jeeves Direct Search and v) NAg routine E04UCF. In all these cases the underlying methods are detailed in the references with the following variations/options being adopted

(1) the GA uses binary encoding, a sixteen bit word length, survival proportional to fitness, elitism, single point crossover, inversion and low level mutation;

(2) the SA uses an annealing schedule controlled by three parameters with one governing the number of temperatures in the schedule, one their spread and the third their absolute values with the temperatures being spread logarithmically within the desired range;

(3) the linear method is used as given except that the variable bounds are strictly enforced by using them as constraints;

(4) the Hooke and Jeeves method is used as given except that the variable bounds are strictly enforced by using them as constraints and

(5) the NAg routine is used as given except that no derivative information is supplied to the routine.

The first four of these are unconstrained methods and when using these optimizers the constraints have been enforced by applying a simple one pass external penalty function of the type described by Siddall[5].

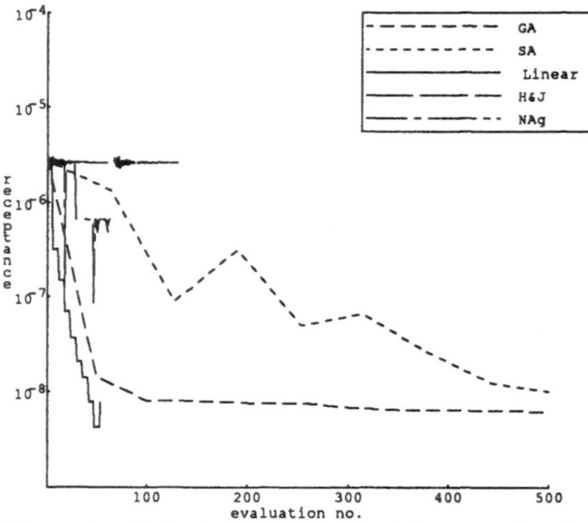

Figure 2 - Objective function $H_{12}(\omega)$ versus function evaluation number for four variables; all methods.

4.2. Four Variable Problem

To begin with, each method was applied to the problem with an allocation of up to 500 evaluations of the objective function, using four variables to describe the rods. The first part of table 1 summarizes the results of these calculations and it may be seen that the best result is obtained with the GA although the linear approximation reaches nearly the same result in around 12% of the time. The Hooke and Jeeves search did not make significant changes to the start point while the NAg routine made only modest improvements. The SA method performs nearly as well as the GA although showing fluctuations around its steady downward progress. These features may be clearly seen in figure 2 where the curves of $H_{12}(\omega)$ versus evaluation number for all the methods used are plotted on a common base. Notice that in this and subsequent figures illustrating the optimization progress, only the best result of each generation or annealing temperature are plotted for the GA and SA methods since their other values show considerable scatter and obscure the graphs.

method	no. vars.	range consts.	min.	feasible	no. trials
GA	4	yes	6.02E-9	yes	500
SA	4	yes	9.81E-9	yes	500
Linear	4	yes	7.79E-9	yes	54
H&J	4	yes	2.62E-6	yes	132
E04UCF	4	yes	6.54E-7	yes	65
Linear	4	no	1.68E-11	ranges	259
H&J	4	no	1.61E-11	ranges	988
GA	40	yes	6.89E-9	yes	500
SA	40	yes	4.12E-8	yes	500
Linear	40	yes	7.83E-7	yes	336
H&J	40	yes	8.64E-7	yes	3895
E04UCF	40	yes	2.62E-6	yes	228
Linear	40	no	1.73E-11	ranges	3403
H&J	40	no	2.19E-9	yes	7426

Table 1 - Comparison of various optimizers.

Two features common to both the linear and Hooke and Jeeves methods, as implemented by Siddall[5], are that, by default, they both take bounds on the design variables as being advisory and they often use more function evaluations than desired if significant improvements are being achieved. To demonstrate these features the previous calculations were repeated for these two optimizers but with the explicit variable bounds constraints removed. In both cases significant improvements in the objective function are obtained although the bounds constraints are violated, with the Hooke and Jeeves search using nearly 1,000 steps. These two additional results are summarized in the second part of table 1. The equivalent graph showing rates of convergence is given as figure 3. This shows that despite the additional freedom made available to the optimizers to drive the notch in the curve to the frequency chosen for optimization, the rates of convergence are no faster, c.f., figure 2.

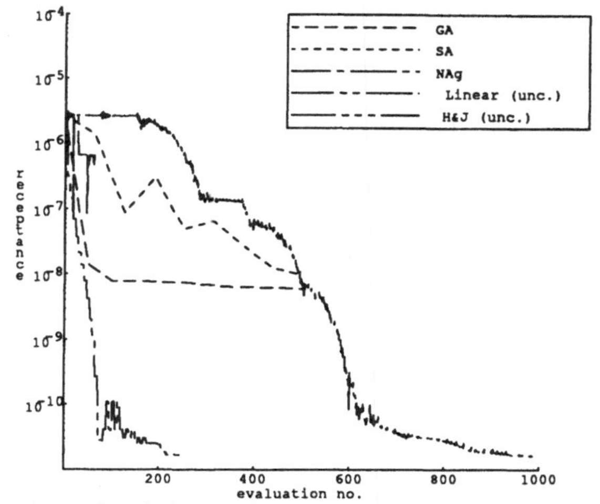

Figure 3 - Objective function $H_{12}(\omega)$ versus function evaluation number for four variables; all methods, Linear Approximation and Hooke and Jeeves Search without variable bound constraints.

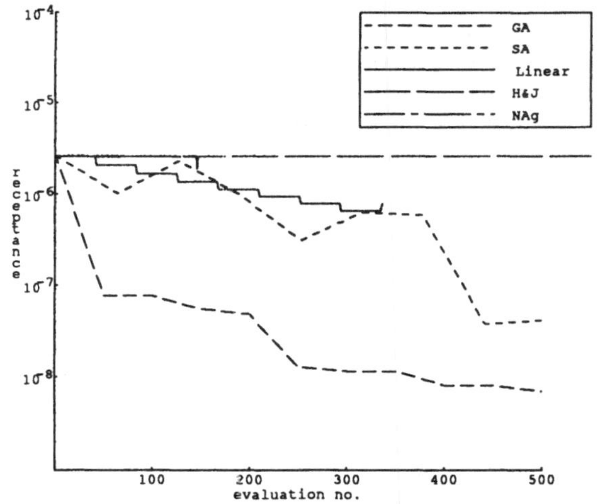

Figure 4 - Objective function $H_{12}(\omega)$ versus function evaluation number for forty variables; all methods.

4.3. Forty Variable Problem

Turning to the case of forty variables, the third part of table 1 shows the results of these calculations. Again some of the methods aborted the search before using the entire allocation of 500 trials, while the Hooke and Jeeves search continues until nearly 4,000 evaluations have been used. It may be seen that the best result is once more obtained with the GA, although it is slightly worse than for the four variable case using the same number of function evaluations. Here, the linear approximation does significantly worse than before, with the Hooke and Jeeves search coming second, albeit at the cost of around eight times as many evaluations as the GA. The NAg routine this time fails to achieved significant improvements on the start point. The SA method performs quite well but with less rapid convergence than the GA. These features may be seen in figure 4 where the curves of $H_{12}(\omega)$ versus evaluation number for all the methods used are plotted, c.f., figure 2.

Relaxing the variable bound constraints again gives the linear and Hooke and Jeeves methods more room for manoeuvre, see the final section of table 1. Notice that in this case the Hooke and Jeeves search produces a feasible result despite the relaxed constraints. Moreover, the final result represents the best achieved throughout this initial survey, being some 64% less than that achieved by the GA with four variables. It does however require nearly 7,500 evaluations to achieve this. Figure 5 shows the equivalent rates of convergence with that for the GA allowed to run for 5,000 evaluations, c.f., figures 2-4. Figures 2 and 5 allow an important conclusion to be drawn; viz., the rates of convergence of all the methods except the GA are affected, to a greater or lesser extent, by the number of variables used in the problem. In particular, the rapid convergence of the linear method when dealing with low order problems is lost when tackling high order cases.

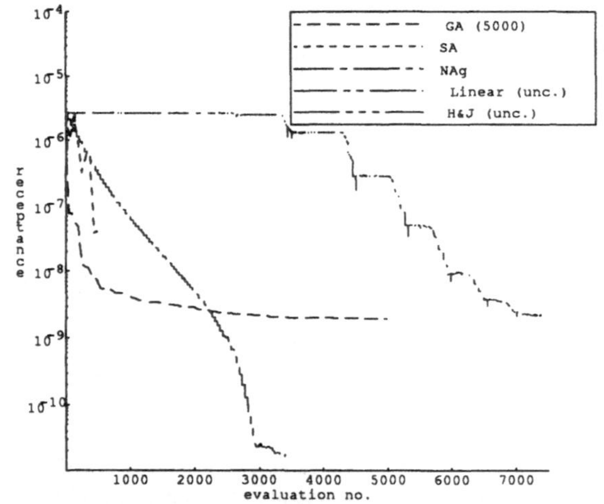

Figure 5 - Objective function $H_{12}(\omega)$ versus function evaluation number for forty variables; all methods, Linear Approximation and Hooke and Jeeves Search without variable bound constraints. Shows $H_{12}(\omega)$ for the Genetic Algorithm with 5,000 function evaluations.

As has just been noted figure 5 shows the effect of allowing the GA to proceed for 5,000 evaluations, demonstrating that it can better the Hooke and Jeeves method when given a similar number of evaluations. This result forms the starting point for a survey of GA control parameters described next, the numerical results are thus given as line one of table 2, where it is seen that a 68% improvement is achieved over the result for 500 evaluations giving an 11% improvement on the longest Hooke and Jeeves search carried out.

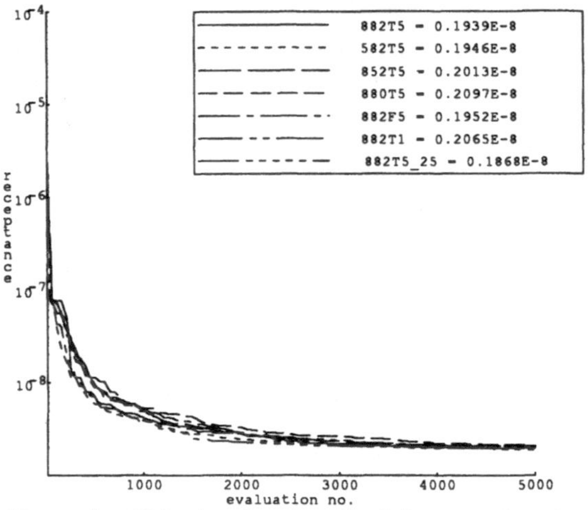

Figure 6 - Objective function $H_{12}(\omega)$ versus function evaluation number for forty variables; Genetic Algorithm with various parameter variations and 5,000 function evaluations, see text for explanation of key.

P[best]	P[cr.]	P[inv.]	P[mut.]	prop.	Npop	min.
0.8	0.8	0.2	0.005	yes	50	1.94E-9
0.5	0.8	0.2	0.005	yes	50	1.95E-8
0.8	0.5	0.2	0.005	yes	50	2.01E-9
0.8	0.8	0.0	0.005	yes	50	2.10E-9
0.8	0.8	0.2	0.01	yes	50	1.95E-9
0.8	0.8	0.2	0.005	no	50	2.07E-9
0.8	0.8	0.2	0.005	yes	25	1.87E-9

Table 2 - Survey of Genetic Algorithm Parameters

5. Survey of Genetic Algorithm Parameters

Having applied a number of optimizers to the structural dynamics problem in hand without attempting to specifically tune them to this problem area, the final survey described here considers the effects of varying some of the parameters used by the GA. Of course, it is to be expected that fine tuning would allow any method to handle specific problems with improve results or speed of convergence. In fact, for general purpose use it is desirable for such tuning to have *little* effect since a method that achieves reasonable results without fine tuning is clearly preferable. This is one of the known drawbacks of the SA; the precise annealing schedule adopted can very seriously alter the method's performance. Conversely, one of the strengths of the GA is its great robustness in this respect.

To investigate this aspect of the GA, six specific parameters were varied

P[best]
: the proportion of the population that survive to the next generation (default 0.8);

P[cross]
: the proportion of the surviving population that are allowed to breed (default 0.8);

P[invert]
: the proportion of the surviving population that have their genetic material re-ordered using the inversion operator (default 0.2);

P[mutation]
: the proportion of the new generation's genetic material that is randomly changed (default 0.005);

Proportionality Flag
: decides whether the new generations are biased in favour of the most successful members of the previous generation or alternatively if all P[best] survivors are propagated equally (default TRUE) and

Population Size
: the number of trials used per generation which is therefore inversely related to the number of generations, given a fixed number of trials in total (default 50).

These six parameters have been varied in turn, one at a time, for the forty variable problem using 5,000 trials, i.e., in comparison with the last result of the previous section. The results achieved are compared in figure 6 which shows the changes arising in the objective function and in table 2 which summarizes the results (in the figure the codes in the key relate to the parameters values detailed in the table in the order given there, i.e. 8825T represents the default; the numbers are the final minima achieved). As can be seen from these results the GA is remarkably stable for significant changes to the parameters used, with the second best result being gained with the default set. Using more generations of fewer members gave the best results of all but even then only a few percent better than for the standard case, the entire range of variations giving only a 12% spread in the minima. The variation of mass profiles and $H_{12}(\omega)$ produced using the reduced population size are given as figures 7 and 8. As can be seen a rather complex profile has been arrived at for this design and it yields a curve with a significantly reduced response at the frequency of interest.

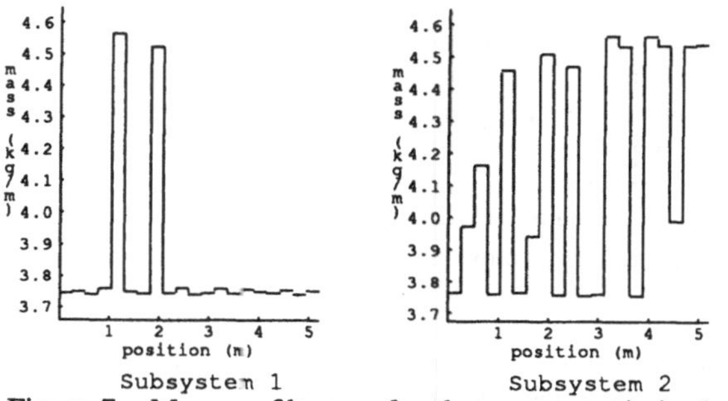

Figure 7 - Mass profiles ρ_{im} for the system optimized using the Genetic Algorithm with forty variables, 5,000 function evaluations and reduced population size.

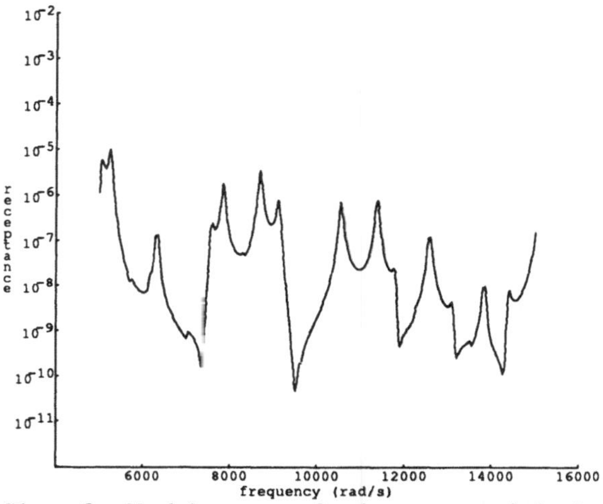

Figure 8 - $H_{12}(\omega)$ versus ω for the system optimized using the Genetic Algorithm with forty variables, 5,000 function evaluations and reduced population size.

6. Minimization of Energy Flow over a Range of Frequencies

As has previously been mentioned, to be useful in producing enhanced noise performance the integral of the energy flow with respect to drive frequency must be minimized. This is a very time consuming calculation and so has not been used as the basis of the previous rather wide ranging surveys. Nonetheless, to illustrate the power of the GA with default parameters it has been applied to this problem with an integration width equal to 8,796 rad/s centred at the previous frequency of interest of 10,000 rad/s. The resulting mass profiles and curve of $H_{12}(\omega)$ are shown as figures 9 and 10, c.f., figures 1, 7 and 8. Figure 10 also shows the curve for the original system; it is plotted on linear scales and the areas under these curves then represent the total energy transfers. It is clear from this last figure that the optimizer has been able to identify a configuration that reduces the energy flows over a range of frequencies. Moreover, although the resulting mass profile shows some tendencies towards geometric repetitions it is shows no precise patterns. As such, it can confidently be expected to give similar reductions in noise transmission when slightly modified, either deliberately or as a result of manufacturing inaccuracies. This rather random nature of the resulting mass profile is, of course, inherent in the GA approach to design since each new variant is the result of essentially random changes from a previous design as it is the selection process that the GA governs rather than the synthesis of new designs.

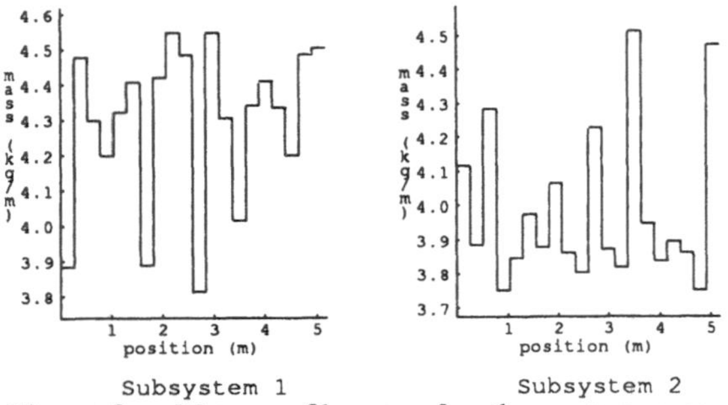

Figure 9 - Mass profiles ρ_{im} for the system range optimized using the Genetic Algorithm with forty variables.

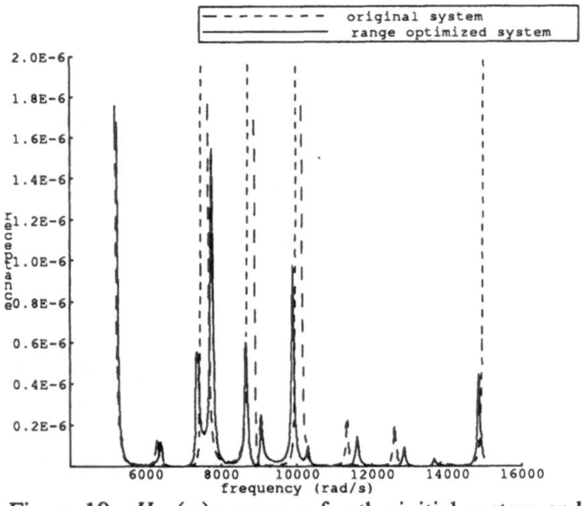

Figure 10 - $H_{12}(\omega)$ versus ω for the initial system and that range optimized using the standard Genetic Algorithm with forty variables.

7. Conclusions

The present study has considered the application of various optimizers to a highly idealised structural noise control problem. These surveys have shown that the Genetic Algorithm (GA) performs very well on such tasks when compared to a range of other methods if large numbers of variables are used to describe the designs. It allows dramatic reductions in energy flows between coupled structures to be achieved, and moreover such reductions can be made to hold over significant frequency bandwidths while at the same time generating designs that can be expected to remain impervious to minor structural modifications. It is also not sensitive to the exact choice of control parameters used when running the optimizer. This arises because the flow of vibrational energy around a complex structure is dominated by the many resonances exhibited by such structures and also the large number of physical parameters needed to specify typical structural designs (in the case studied here up to forty, many hundreds in a full ship or aircraft structural model); both features which the GA is seen to deal with well. In summary it may be said that

(1) the highly complex nature of the eigenproblems found in structural dynamics leads to difficult optimization domains;

(2) when few variables are used to describe the problem classical, slope driven or linearized methods are very rapid and powerful despite the complex relationships;

(3) when many variables are used to describe the problem the classical techniques may still work but they become very expensive to use;

(4) Genetic Algorithms and Simulated Annealing methods are then more useful and

(5) unlike other methods, Genetic Algorithms do not appear to require careful tuning to gain the best results.

References

1. A. J. Keane and C. S. Manohar, "Power Flow Variability in a Pair of Coupled Stochastic Rods," *J. Sound Vib.* **164**(2)(1993).

2. H. G. Davies, "Random vibration of distributed systems strongly coupled at discrete points," *J. Acoust. Soc. Am.* **54**(2) pp. 507-515 (1973).

3. P. J. Remington and J. E. Manning, "Comparison of Statistical Energy Analysis power flow predictions with an 'exact' calculation," *J. Acoust. Soc. Am.* **57**(2) pp. 374-379 (1975).

4. A. J. Keane and W. G. Price, "A Note on the Power Flowing between Two Conservatively Coupled Multi-Modal Sub-systems," *J. Sound Vib.* **144**(2) pp. 185-196 (1991).

5. J. N. Siddall, *Optimal Engineering Design: Principles and Applications*, Marcel Dekker, Inc., New York (1982).

6. NAg E04UCF, *NAg Mark 12 Reference Manual*, Numerical Algorithms Group Ltd., Oxford (1987).

7. D. E. Goldberg, *Genetic Algorithms in Search, Optimization and Machine Learning*, Addison-Wesley, Reading (1989).

8. S. Kirkpatrick, C.D. Gelatt, Jr., and M.P. Vecchi, "Optimization by simulated annealing," *Science* **220**(4598) pp. 671-680 (May 1983).

THE CONCRETE ARCH DAM

AN EVOLUTIONARY MODEL OF THE DESIGN PROCESS

I. C. Parmee, Plymouth Engineering Design Centre, University of Plymouth, Plymouth, UK. ian@uk.ac.plym.cx

Abstract

The Plymouth Engineering Design Centre (PEDC) is one of the seven Science and Engineering Research Council funded Design Centres that have been established at various academic institutions within the UK. The initial aim of the PEDC is to carry out fundamental research into the application of adaptive search techniques to engineering design. The following paper describes the application of the adaptive search technique known as the Genetic Algorithm (GA) to the multi-variable problems associated with the design of a double curvature Concrete Arch Dam.

The evolutionary characteristics of the Algorithm and their relevance to the designer are briefly described and the main features of the Arch Dam design geometry are defined. The development of an interactive designer / GA design tool and problems concerning the integration of related design aspects are discussed.

The current status of the project is considered and further work at the Centre involving the application of the Genetic Algorithm to other engineering design domains is outlined.

The Genetic Algorithm

The Genetic Algorithm offers a robust, non-linear search technique that is particularly suited to design problems involving large numbers of variable, interactive parameters. Unlike other optimization techniques the GA has the ability to avoid convergence upon local optima and can successfully negotiate search spaces that are discontinuous in nature to finally identify the global optimal solution without the need for gradient information. The algorithm achieves this by the random exchange of information between increasingly fit parameter combinations and the introduction of independent random change.

The designer must first select those parameters that appear to affect the overall design of an element / system and determine an appropriate objective function. At this stage little, if any, knowledge of the interactive relationships of the chosen parameters is required. Binary coding is introduced to represent the values of the selected parameters.

This binary representation provides the basis for the biological analogy of the evolutionary strategy. The binary string can be considered to represent a chromosome and the individual digits of the string represent genes. Alleles, which in a natural system, cause variations in the characteristics of a gene, are represented by the value of each digit.

Each parameter string in turn is processed by a computer simulation of the system under design and the relative fitness assessed from its performance. Those strings of low fitness are then rejected whilst those of medium fitness are reproduced singly. High performance examples may be reproduced twice. This procedure is known as Fitness Proportionate Reproduction [1,2] and is analogous to the selection procedures of natural systems.

A closer analogy is then achieved by introducing Crossover and Random Mutation Operators. Crossover allows a random interchange of information between the reproduced parameter strings / chromosomes whilst the Random Mutation Operator causes randomly selected digits / genes to change value according to a preset mutation probability.

It is a combination of Fitness Proportionate Reproduction and Crossover which ensures that each successive generation consists mainly of increasingly fit parameter strings / chromosomes whilst Random Mutation ensures that widely varying areas of the design space are constantly being sampled thereby preventing the Algorithm from converging upon local optima.

The Concrete Arch Dam

Unlike other dams which rely upon self-weight to resist the hydrostatic forces generated by the impounded reservoir the Concrete Arch relies upon its shape alone. The designer must therefore implement horizontal and vertical curvatures that will transmit thrust to the dam's foundation and abutments in a manner which ensures that uniform compressive stresses and limited tensile stresses are generated within the structure under all loadings.

The constituent curves may take elliptical, circular or polynomial forms. Alternatively they may be based upon a logarithmic spiral. It is probable that an overall dam geometry will incorporate some if not all of these forms. In addition, a number of horizontal curves of differing radii may be combined at any particular elevation and the radii will vary with depth to better accommodate increasing hydrostatic force. During this initial stage of the project, work is being concentrated upon one particular geometrical configuration. This is a three-centred configuration which utilizes three separate plan curves at each elevation to increase the angle of incidence at the valley sides. The increased angle results in a better transmission of forces into the main body of rock.

The designer must first establish vertical curvatures into which the horizontal, three-centred curves can be integrated to form the smooth internal and external faces of the dam (known respectively as the intrados and extrados). In order to achieve this, a reference plane is established and a reference cylinder defined as shown in Figure 1. The intrados and extrados curves can then be established from secondary and tertiary coordinate systems as illustrated. The origins of these planes are positioned to provide a vertical section which, when combined with the horizontal curvatures, will perform within safe limits under a large number of load conditions.

To ensure that the horizontal curvatures vary in a satisfactory manner with depth their radii are defined by the lines of centres shown. The slope or curvatures of these lines of centres can be varied in order that the increasing hydrostatic forces are better transmitted via the dam curvature into the abutments.

A plan of the resulting geometry is shown in Figure 2. Eleven main parameters control the flexibility of this geometry and, assuming that each parameter varies between set constraints at a reasonably low resolution, a design space consisting of around 1×10^{15} different geometries can be assumed. The addition of elliptical, polynomial, and logarithmic spiral curves to the geometric model will increase the size of this search space by several orders of magnitude.

The addition of these other curves will also necessitate the development of a more sophisticated GA. Each parameter string of the modified GA will contain all the information required to construct a number of geometries based upon either a three-centred, elliptical, polynomial or logarithmic spiral arch. Alternatively, this information could be used to construct a hybrid geometry incorporating some or all of the four curves. It will therefore be necessary to introduce controlling genes into each parameter string / chromosome to define geometry type and to then activate the relevant parameters within the string. This approach will allow unrestricted crossover of genetic information whilst ensuring that potentially useful information can remain dormant within each string until activated at the appropriate time [3].

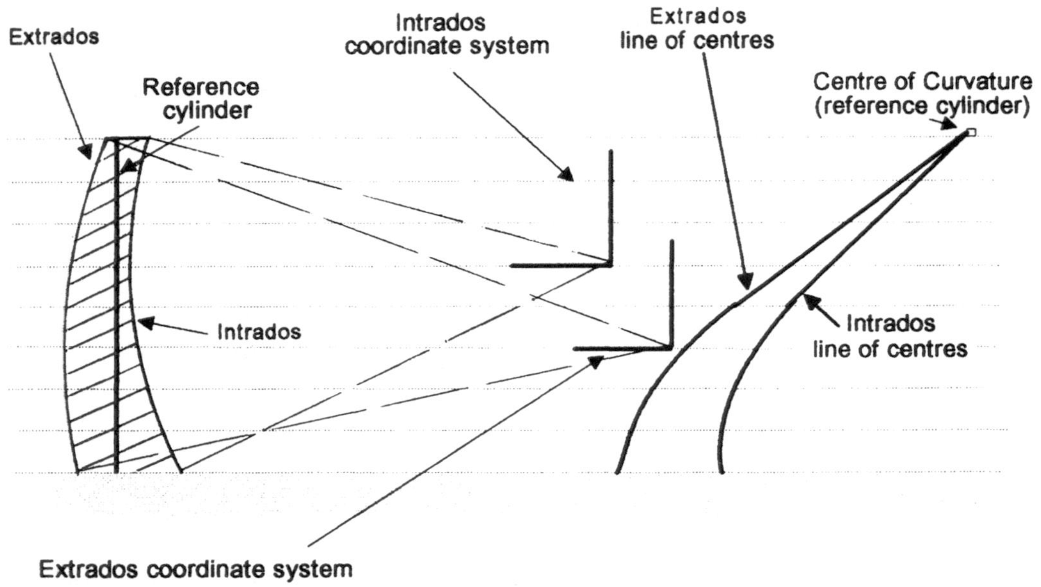

Figure 1: Elevation of Vertical Curve Coordinate Systems and Lines of Centres for Horizontal Curve Radii.

The fitness of each parameter combination is assessed by a structural analysis of the dam using one of a number of suitable methods. The level of sophistication of the

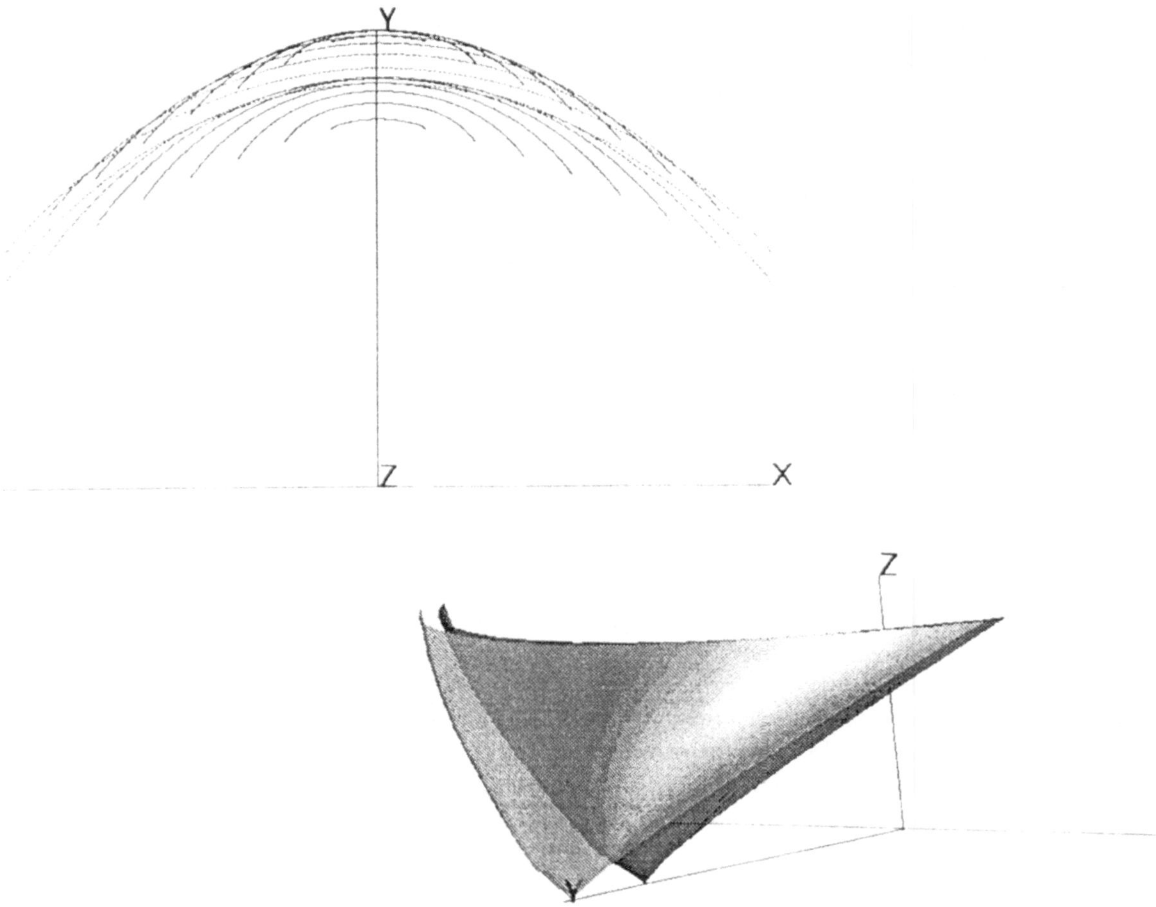

Figure 2: Plan and Elevation of Concrete Shell Geometry

method of analysis in use will depend upon the design detail required. For instance, at pre-feasibility stage it will be necessary to rapidly assess a number of 'good' designs in order to achieve a competitive outline costing. At this stage a basic analysis of the structure would be undertaken in order to minimize computer time. However, a sophisticated finite element analysis of the dam shell, the abbutments, the foundation and the reservoir itself will be required at the detailed design stage. The Centre is currently utilizing the US Bureau of Reclamation's finite element model, EADAP, which performs both a static and dynamic analysis of the combined aforementioned elements [4,5].

The Design Process

The Arch Dam designer, having been provided with the main physical characteristics of the proposed site of the dam, is faced with a formidable choice of possible design directions. Those already discussed have concerned the concrete shell geometry alone and represent a multi-dimensional design space that could only be superficially sampled by the designer given the likely time constraints within which he will be working.

The designer must therefore rely upon previous personal experience and known examples to enable him to reduce the design space to manageable proportions. He will, by necessity, be restricted to a few isolated regions that are already familiar to him and he will attempt to improve the design within the apparent constraints of those regions. The benefits of this approach are twofold. In the first instance an acceptable design can be achieved within time and related budget constraints and secondly, previous experience concerning the performance of similarly designed dams will provide a degree of confidence. This design process is therefore, evolutionary in nature. The 'better' dam designs tend to survive to seed the initial, tentative designs of a new project. The relative infrequency of large-scale Arch Dam design and construction however, ensures a particularly lengthy development path unless breakthroughs can be achieved via related research and / or the introduction of computer-assisted design techniques.

It is within these preliminary stages of the design process that the immediate benefits of the Genetic Algorithm become apparent. Although the achievement of a global optimum solution via the GA may utilize a significant amount of computing resource, the initial identification of better design regions can be achieved in a relatively short period of time. The designer's efforts therefore, need not be concentrated in those few areas with which he is familiar. He can utilize the GA in an investigatory manner to rapidly negotiate the design space and identify those better areas that appear worthy of further analysis. Within each generation the GA will concentrate its search in a number of fit regions of the design space. The designer will be able to access the geometric data defining the best example of each region. This approach will provide potentially 'better' initial designs that are directly related to the main physical characteristics of the project.

Detailed Design

It is envisaged that during the detailed design stage the evolution process will be allowed to proceed until convergence upon an optimal design is achieved. However, a great deal of designer / GA interaction will be a feature of this process and will directly contribute to the development of an overall optimal design. Visualization techniques are currently under development at the Centre which will allow the designer to access and manipulate three-dimensional images of the 'better ' designs of each generation. The designer will utilize his intuition and experience relating to both Arch Dam design and the specific site requirements to identify those interim geometries that potentially suit the current situation. He can then draw off the relevant data and further develop these designs off-line to suit his particular requirements. It is possible that the results from such off-line developments will be used to seed the initial chromosomes of further GA runs. In this manner, experience and specific knowledge relating to the site would be implied within the genetic code.

In addition to the three-dimensional images, research is also being undertaken to establish abstract visualization techniques that will enable the designer to develop a better understanding of the multi-dimensional aspects of the Arch Dam problem. The objective here is to facilitate understanding of the interactive relationships between the many parameters and provide the designer with information concerning aspects of the search space and the behaviour of the search process within it.

Current Status of the Project

Preliminary research has concentrated upon three-centred concrete arch configurations. Flexible geometrical models which characterize acceptable dam arrangements have been developed in-house. A basic GA manipulates these models and a simple structural analysis technique has provided the fitness function required by the Algorithm. The objective function has been to minimize dam volume whilst maintaining uniformity of compressive stress and minimal tensile stresses in the lower portions of the structure. The computational simplicity of the fitness function has ensured rapid improvement in overall dam configuration.

A structured Genetic Algorithm [3] has recently been introduced in order to allow the simultaneous processing of data relating to three potential dam sites. The Algorithm's structure allows conflicting geometric characteristics peculiar to each site to be successfully manipulated during crossover and mutation. The sophistication of the structural analysis has also been increased by introducing the finite element model of the concrete shell and foundation rock. The aim is to allow the designer to rapidly assess the relative merits of three potential sites using the less rigourous analyses before utilizing the GA-driven finite element model to achieve a more detailed single optimal design.
The initial structural analysis has concentrated upon one particular load case involving the deadweight of the structure combined with the hydrostatic thrust associated with the impounded reservoir. Other load cases that must be taken into account include uplift pressure, shrinkage, temperature effects and seismic disturbance. Although it will not be necessary to analyse each load case at every call to the model it will be necessary to introduce some form of structured approach that ensures that 'good' design solutions meeting the major load criteria also perform adequately under all other load conditions.

Computational expense represents a major drawback when considering the GA processing of complex finite element / CFD design problems. This problem is being addressed by fundamental research at the Centre involving the analysis of GA behaviour. The overall objective of this work is to develop a GA variant that represents a robust adaptive search technique. This technique will manipulate and process design information in an optimal manner thereby minimising calls to the design model in order to reduce computational cost. Another approach to this problem is an intended investigation of the parallelisation of the processing techniques utilised by the Algorithm.

Other PEDC Projects

Initial research concerning the Genetic Algorithm at the University involved the generation of an optimal chamber design for a novel, low-head hydropower device

[6,7] and the application of the technique to the design of FIR digital filters [8].

Since the establishment of the Engineering Design Centre, in addition to the Concrete Arch Dam work, research has concentrated upon the integration of the Genetic Algorithm within the following engineering design domains:

The Design of Gas Turbine Components with Rolls Royce, plc.

The GA has been applied to two areas of gas turbine design namely the design of blade cooling hole geometries and the design of the turbine annulus. Application has been at the preliminary design stage in both cases. The cooling hole geometry problem has involved the development of a simple, one-dimensional, steady-state model of a cooling passage situated at the leading edge of a high pressure gas turbine blade. This preliminary simulation models the characteristics of a number of cooling methods such as the inclusion of ribs and pedestals within the passage and the introduction of surface cooling. The objective function is to minimize cooling flow requirements by achieving an optimal cooling passage configuration. Results have been sufficiently encouraging to promote further development of both the turbine blade model and the GA methodology. It is expected that this increased sophistication will lead to the development of a valid preliminary design tool.

The Design of Computationally Efficient Finite Impulse Response Digital Filters with GEC Plessey Semiconductors.

A digital filter frequency response can be realized by cascading primitive sections together. These have the advantage that when the multipliers are reduced power of two coefficients, the multiplication process can be simplified to a simple bit shift and add operation. This is especially useful for high sample rate applications such as television encoding (ie CCIR 601-1 recommendations), since multipliers have a large propagation delay time-relative to delays and adders and their relative chip area is large.

The problem is to decide how many sections of which particular type, with certain delays and coefficients, to cascade together in order to produce the desired frequency response. The Genetic Algorithm is utilized to select and alter these parameters, minimizing the frequency error to zero, whilst optimizing the number and computational complexity of the sections.

Summary

The double curvature Concrete Arch Dam offers an extremely rich design environment which, although amenable to the application of conventional optimization techniques, presents such techniques with severe difficulties. The multi-dimensional aspects of the dam geometry combined with the many load conditions and physical constraints create a complex design space which no doubt contains numerous local optima. The introduction of different types of curvature plus the ability to assess a number of dam sites each possessing differing characteristics is likely to introduce a degree of discontinuity.

It is probable that the powerful, non-linear search techniques of the Genetic Algorithm will negotiate the design space in an optimum manner thereby providing the designer with a highly interactive, design tool that is both powerful and robust.

Fundamental research at the Centre is addressing the development of visualisation techniques that will increase the transparency of the GA processing technique. Analysis of the Algorithm's behaviour within the engineering design domain is also contributing to the overall knowledge base at the Centre. This analysis will provide a basis for the development of high-performance Algorithms which will reduce computational requirements.

Acknowledgements

The work described within the paper represents one of the three core projects funded by the UK Science and Engineering Research Council and currently being undertaken at the Plymouth Engineering Design Centre. The industrial collaborators for this project are Knight Piesold and Partners, a civil and environmental engineering consultancy. Knight Piesold have supplied the necessary Arch Dam design software and are continuing to support the project with the provision of expertise.

References

[1] Holland, J. H. ' *Adaptation in Natural and Artificial Systems.'* Ann Arbor, The University of Michigan Press, ISBN 0-472-08460-7, 1975.

[2] Goldberg, D. E. *'Genetic Algorithms in Optimization and Machine Learning'.* Addison-Wesley Publishing Co., ISBN 0-201-15767-5, (1990).

[3] Dasgupta, D. and Mcgregor, D. R. *'A Structured Genetic Algorithm'.* Unversity of Strathclyde, 1991.

[4] Ghanat, Y. and Clough, R. W. *'EADAP - Enhanced Arch Dam Analysis Program.'* University of California at Berkley, 1989.

[5] Serafim, J. L. and Clough, R. W. *'Arch Dams - International Workshop on Arch Dams, Coimbra, 5-9 April, 1987.'* A. A. Balkema, ISBN 90-6191-8650, 1990 .

[6] Parmee, I. C. *'Pneumatic Hydropower Systems.'* PhD Thesis, Polytechnic South West, UK, 1990.

[7] Parmee, I. C. *'The Operational Optimization of a Pneumatic, Low-head Hydropower System using Evolutionary Design Techniques.'* Proceedings of the Ises Solar World Congress, Denver, USA, 1991.

[8] Suckley, D. *'Genetic Algorithm in the Design of FIR Filters.'* IEE Proceedings, Vol 138, No. 2, April, 1991.

THE GENIE PROJECT - A Genetic Algorithm Application to a Sequencing Problem in the Biological Domain.

J.D.Walker, P.E.File, C.J.Miller, W.B.Samson
Department of Mathematical and Computer Sciences
Dundee Institute of Technology
Bell Street
DUNDEE DD1 1HG
UNITED KINGDOM

Abstract

This paper describes the current development and implementation of a form of genetic algorithm (GA) suitable for tackling a complex sequencing problem in the biological domain - the building of restriction "maps" from the results of partial digest experiments. Building restriction maps is a time-consuming and lengthy activity which relies on human judgement of inexact data.

The paper is organised into the following sections. The procedure for building restriction enzyme maps is described in section 2. There are several aspects of map assembly which make it a relevant problem to study from a GA point of view and these are outlined in section 3. The GENIE project is an ongoing project and the way in which the problem is being tackled by developing a GA and the implementation issues are discussed in section 4. Preliminary results are shown in section 5, the paper is summarised in section 6 and section 7 highlights future developments.

1 Introduction

GAs are search procedures based on the mechanics and analogy of natural selection. They were introduced by [1] - for an introduction to the subject refer to [2]. Map assembly is an example of a difficult sequencing problem which requires some form of search to find a good solution from a large problem space of feasible solutions. Traditional GAs are not effective for sequencing problems as illegal solutions can be generated.

Two of the goals of the GENIE project are to develop some form of GA which can successfully tackle sequencing problems and to incorporate an evaluation function based on human assessment of subjective and error prone data. In order to meet these goals, the traditional GA has been modified in several ways to produce a hybrid GA.

The concept of a hybrid GA has been suggested by a number of authors [3, 4, 5, 6]. A hybrid GA is a GA which incorporates problem specific information or various search techniques. Research conducted in this area has shown a hybrid GA approach to be promising for combinatorial optimisation problems such as the Travelling Salesman Problem [7, 8, 9, 10, 11] and scheduling problems [12, 13].

In the GENIE project, a hybrid GA approach is being applied to find a good solution to the map assembly problem.

2 Map building procedures

Geneticists worldwide are attempting to identify and isolate genes and to generate a "map" showing the position and function of each gene on the human DNA (known as genetic mapping).

A current technique which has had a significant impact on genetic mapping is the use of restriction enzymes (REs). REs are used to cut up DNA into fragments. There is a wide range of these enzymes which have different properties and will cut at particular sites. By observing the fragments produced and using already known probes (genetic markers), geneticists can produce restriction maps showing the sites where the enzymes cut and the position of known probes. Using this technique allows regions of DNA to be characterised.

An example of a RE map is shown in Figure 1 and the data from which it was generated is shown in Figure 2 [14]. The numbers in Figure 2 represent the lengths of the DNA fragments. Each fragment contains a probe and has been cut at either end by the RE shown.

When one RE is used, the fragments obtained are called "single digests" and each end of the fragment has been cut by that RE. When two REs are used, the fragments obtained are called "double digests".

Figure 1: RE Map assembled from the data in Figure 2 (not to scale) showing the ordering of the RE cut sites and probes.

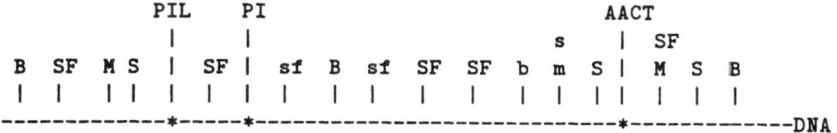

```
            PIL     PI                           AACT
              |     |                        s    | SF
 B  SF   M S  |  SF |  sf  B  sf  SF  SF  b  m  S |  M  S  B
 |  |    | |  |  |  |  |   |  |   |   |   |  |  | |  |  |  |
--------------*-----*-----------------------------*-------------DNA

NOTES

1. PIL, PI, AACT - probes
2.  B,  SF, M, S   - REs
3.  REs shown in capital letters represent complete cut sites,
lower case indicates partial cut sites.
```

Here, each end of the fragment has been cut by one or other of the REs.

RE cut sites are said to be either "complete" or "partial". A complete cut site is a site on the DNA which is always cut by the RE. A partial cut site is a site which is cut on some pieces of DNA, and left intact on others. The occurence of partial cut sites leads to the presence of long fragments. Determining the position of the probes in relation to one another relies on the presence of such fragments. Long fragments are likely to contain two or more probes if the probes are adjacent.

When all fragments present in Figure 2 are combined, taking into account the error present in the lengths, the choice of RE cutting in the double digests and the nature of the cut sites, a RE map is generated which fits the data best. The map reveals the ordering of the probes, the cut sites of the restriction enzymes, and whether or not the cut sites are partial or complete. The map assembled using the data in Figure 2 is shown in Figure 1.

3 Map building characteristics

Map building is essentially an ordering problem - the map is linear and after the number and types of cut sites have been determined, they must then be placed in the correct sequence. There are several aspects of map building which make it a difficult activity.

3.1 Maps do not fit perfectly.

No map ever fits perfectly as there are errors present in the amount and quality of the data used to build a map which are due to experimental limitations. "Weak" fragments are sometimes obtained. Most weak fragments should fit into a map but as some are due to error, they tend to be ignored when a map is assembled but should in theory fit into the solution. A "good" map is one in which the data, including most of the weak fragments, fits well.

There has been a lack of criteria by which the correctness of maps can be assessed. As part of the GENIE project, appropriate criteria have been identified and incorporated into an evaluation function. This is discussed further in section 4.1.3.

3.2 Fragments overlap.

All fragments obtained for a particular probe must overlap the region of the probe. Many orderings of fragments are possible and it is difficult to determine where the cut sites are.

3.3 More than one solution may be possible.

Due to the overlapping nature of the fragments and the error in the fragment lengths, it is possible that many feasible map solutions may be generated from the same data.

4 A genetic algorithm approach to solving the problem

The traditional GA has been modified to handle the map assembly problem. The representation and optimisation techniques employed by the expert are used to ensure that the domain knowledge embodied in

SINGLE DIGEST DATA

B

PIL	PI	AACT
250	250	355
225	225	275
	[65]	230

M

PIL	PI	AACT
350	350	350
260	260	80
	[180]	

S

PIL	PI	AACT
255	255	70

SF

PIL	PI	AACT
165	[190]	[195]
[10]	[135]	[175]
	80	135
65		
10		

DOUBLE DIGEST DATA

B+M

PIL	PI	AACT
250	250	230
225	225	105
130	130	
	[65]	

B+S

PIL	PI	AACT
125	125	70
	[65]	

B+SF

PIL	PI	AACT
165	[190]	135
[10]	50	105
	10	

M+S

PIL	PI	AACT
255	255	70
	[180]	30

M+SF

PIL	PI	AACT
85	80	135
	65	85
[10]	10	

S+SF

PIL	PI	AACT
	[130]	30
70	80	
(10)	65	
	10	

NOTES

1. PIL, PI, AACT - probes
2. B, SF, M, S - REs
3. Fragments contained in brackets indicate "weak" fragments.

Figure 2: Table containing Experimental Data used to generate the map shown in Figure 1.

the encoding is preserved. The GA operators are tailored to apply to the new representation and include domain-based heuristics, and an evaluation function which is based on subjective assessment has been developed. Other problem specific knowledge has guided the development of the overall algorithm.

The modifications made to the traditional GA may appear to be at odds with "pure" GA research, as one of the main aims of GA research has been to develop an algorithm that is robust across a variety of problem domains and operates without problem specific information. It is the intention here to see if the power of the traditional GA can be harnessed and used as the basis of an effective algorithm for a real- world complex problem. Although the traditional GA is robust across a wide range of problems, it is unlikely to be the best algorithm to use for any specific application.

4.1 Implementation issues

A "front-end" menuing system has been developed for the GA which allows for various features to be selected. This has facilitated the implementation of the modified GA and allowed the effect of changing particular features to be observed.

There are several areas which have been addressed during the development of the modified GA and these are discussed in the following sections.

4.1.1 Representing the problem

Traditional GAs generally represent chromosomes as binary vectors. However, having considered several options, it was decided that in the modified GA, the chromosome syntax should be changed to reflect the problem as shown in Figure 1.

4.1.2 Developing a set of genetic operators

Applying the traditional genetic operators to map assembly has similar difficulties with applying them to other sequencing problems such as the Travelling Salesman Problem (TSP). In the case of the TSP, there are constraints placed on the symbol string that represents a tour of the cities, in that no city can appear more than once. The traditional recombination operators rearrange symbols on a chromosome independently of each other. When solutions are coded as sequences and the traditional operators applied, cities can appear more than once or not at all.

When developing operators for map assembly, taking into account previous work [2, 15], a way of breaking up the chromosomes that was natural for the problem was sought. The traditional operators have been developed to meet the constraints on the chromosomes while preserving the motivating principles behind these operators. Crossover provides an opportunity for the best attributes of both parent strings to be incorporated into the offspring. Mutation is a mechanism for introducing necessary attributes into an individual when those attributes do not already exist within the current population.

4.1.3 Developing the evaluation function

The way in which chromosomes are assessed is critical to the success of any GA as this has a direct influence on the parents of the next generation. It is essential that an evaluation function captures the essence of a good or bad map.

Developing such an evaluation function for the GA is complicated. It is not easy to arrive at a mathematical expression which indicates the correctness of a map.

Historically, there has been a lack of objective criteria by which the quality of existing maps can be assessed or the quality of potential maps can be predicted. Such a facility would allow geneticists to make an objective appraisal of new and old maps.

4.1.4 Setting parameter values

There are several parameters in a GA that require to be set to appropriate values — population size, number of trials, operator probabilities and evaluation normalisation techniques.

There are established parameter settings described in the literature for GAs using binary representation, binary crossover and mutation [16]. Finding good settings for non-binary representations is not a trivial task [17] as the techniques available can take a great deal of time. Davis [17] has devised a system for parameterising operator probabilities for GAs that differ from the traditional type. His technique is being used to measure how effective each of the operators devised are and to obtain appropriate settings.

4.1.5 Generating the initial population

Traditional GAs generate the initial population at random. However, it has been recognised [18] that if domain specific knowledge is available, it can be usefully exploited in the GA.

The modified GA will use the domain specific information in the form of the experimental data in order to generate maps.

4.1.6 Selecting a reproduction technique

Three types of reproduction technique are being investigated generational replacement, generational replacement with elitism and steady-state replacement without duplicates. (These are reviewed in [6].)

5 Results

5.1 Representation

Chromosomes have been represented as probe maps which denote a simplified version of the complete problem. The choice of representation exemplifies a natural way of splitting up the problem as it is a strategy that the expert may adopt when building maps. A chromosome (probe map) consists of a number of cut sites and the probe, in a particular order as shown below.

eg   B b M m S F * S M F B

The position of the probe is indicated by an asterisk. The length of the chromosome depends on the amount of probe data.

5.2 Genetic operators

Two reproduction operators have been developed "side swap", which is a modified type of crossover (crossover occurring at the position of the probe), and "order swap".

Side swap swaps the LHS of one parent with the RHS of the other parent, as shown in Figure 3 overleaf.

Order swap, as shown in Figure 4, swaps the order of 2 different REs in the child chromosome as long as the swap is legal. A complete cut site cannot be moved closer to the probe than any of the partial cut sites. A partial cut site cannot be moved further away from the probe than its complete cut site.

5.3 Evaluation function

An evaluation function has been developed which generates a value indicating the goodness of fit between the proposed probe map and the experimental data.

From discussions with geneticists on the features of good and bad maps and through the use of a detailed questionnaire, the important characteristics of maps have been identified. A marking system has been developed which provides a means for scoring characteristics of maps including subjective assessments made by the geneticist. The scoring of the various characteristics can be thought of as sub- evaluations. Three types of sub-evaluations are carried out based on the fit of the single digest results, the fit of the double digest results and the fit of the weak fragments. While each sub-evaluation is not considered appropriate as the sole means of evaluating maps, each has a significant contribution to make to the overall evaluation function.

5.4 Initial population

A "probe map builder" has been developed which takes as input items of probe data (as shown in Figure 2) selected at random and creates legal probe maps for the initial population.

6 Summary

This paper has outlined the current development and implementation of a modified GA for a difficult sequencing problem.

7 Future developments

The evaluation function developed will be used to validate existing published maps. Once all the results from the GENIE project are collected, the maps generated by the modified GA will be assessed both by geneticists and by comparison with current validated maps which will enable the success of the GA to produce "best fit" maps to be determined. The performance and limitations of the modified GA will be analysed and the implications of the results for other sequencing problems and for problems which require a more complex evaluation function will be established.

References

[1] Holland J H, "Adaptation in natural and artificial systems", Ann Arbor: The University of Michigan Press, 1975.

[2] Goldberg D E, "Genetic Algorithms in Search, Optimisation and Machine Learning", Addison Wesley, 1989.

[3] Bethke A D "Genetic Algorithms as Function Optimisers", (Doctoral dissertation, University of Michigan), Dissertation Abstracts International 41(9), 3503B (University Microfilms No. 8106101), 1981.

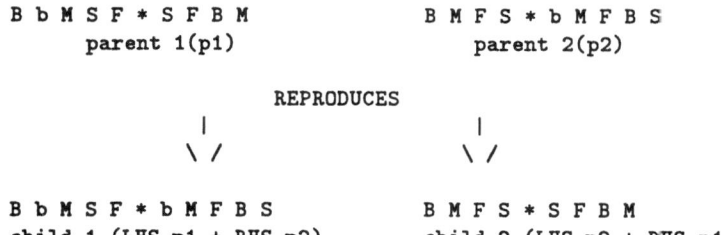

```
B b M S F * S F B M                 B M F S * b M F B S
      parent 1(p1)                       parent 2(p2)

                          REPRODUCES
                  |                       |
                 \ /                     \ /

B b M S F * b M F B S               B M F S * S F B M
child 1 (LHS p1 + RHS p2)           child 2 (LHS p2 + RHS p1)
```

Figure 3: Side Swap

```
            B b M S F * b M F B S
position    0 1 2 3 4 5 6 7 8 9 10


order swap positions 2 and 4 -> B b F S M * b M F B S


order swap positions 0 and 3 - ILLEGAL -> B b M S F * b M F B S
                                              map unchanged
```

Figure 4: Order Swap

[4] Bosworth J, Foo N, Zeigler B P, "Comparison of Genetic Algorithms with Conjugate Gradient Methods", (CR-2093) Washington, DC: National Aeronautics and Space Administration, 1972.

[5] Goldberg D E, "Computer aided Gas Pipeline Operation using Genetic Algorithms and Rule Learning" (Doctoral Dissertation, University of Michigan), Dissertation Abstracts International, 44(10), 3174B (University Microfilms No. 8402282), 1983.

[6] Davis L, "Handbook of Genetic Algorithms", Van Nostrand Reinhold Publishers, 1991.

[7] Goldberg D E, Lingle R, "Alleles, Loci, and the Travelling Salesman Problem", Proceedings of an International Conference on GAs and their Applications, 154-159, 1985.

[8] Grefenstette J J, Gopal R, Rosmaita B J, Van Gucht D, "Genetic Algorithms for the Travelling Salesman Problems" Proceedings of an International Conference on GAs and their Applications, 160-168, 1985.

[9] Oliver I M, Smith D J, Holland J R C, "A Study of Permutation Crossover Operators on the Travelling Salesman Problem", Genetic Algorithms and Their Applications: Proceedings of the Second International Conference on GAs, 224-230, 1987.

[10] Whitley D, Starkweather T, Fuquay D "Scheduling Problems and Travelling Salesman: the Genetic Edge Recombination Operator", Proceedings of the Third International Conference on GAs, San Mateo,CA: Morgan Kaufmann Publishers. pp133-140, 1989.

[11] Whitley D, Starkweather T, Shaner D, "The Travelling Salesman and Sequence Scheduling: Quality Solutions Using Genetic Edge Recombination", Handbook of Genetic Algorithms, ed L. Davis, Van Nostrand Reinhold Publishers, New York, 1991.

[12] Davis L, "Job Shop Scheduling with Genetic Algorithms", Proceedings of an International Conference on GAs and their Applications, 136-140, 1985.

[13] Syswerda G, "Schedule Optimisation Using Genetic Algorithms", Handbook of Genetic Algorithms, ed L.Davis, Van Nostrand Reinhold Publishers, New York, 1991.

[14] Sefton L, Kelsey G, Kearney P, Povey S, Wolfe J, "A Physical Map of the Human PI and AACT Genes", Genomics 7, 382-388, 1990.

[15] Fox B R, McMahon M B, "Genetic Operators for Sequencing Problems", Foundations of GAs, Rawlins G (ed), Morgan Kaufmann Publishers, San Mateo, California, 1991.

[16] Schaffer J D, Caruana L J, Eshelman L J, Das R, "A Study of Control Parameters Affecting Online Performance of GAs for Function Optimisation", Proceedings of the Third International Conference on GAs pp 51-60. San Mateo, CA:Morgan Kaufmann Publishers, 1989.

[17] Davis L, "Adapting Operator Probabilities in Genetic Algorithms", Proceedings of the Third International Conference on GAs pp 61-69. San Mateo, CA:Morgan Kaufmann Publishers, 1989.

[18] De Jong K A, "Learning with GAs: an overview" Machine Learning, 3(2), 121-138, 1988.

Hybrid Genetic Algorithms for the Traveling Salesman Problem

P. Prinetto M. Rebaudengo M. Sonza Reorda

Politecnico di Torino

Dipartimento di Automatica e Informatica

Corso Duca degli Abruzzi 24

I-10129 Torino - ITALY

E-mail: {prinetto,reba,sonza}@polito.it

Abstract

A comparative analysis is performed on an experimental basis among four different cross-over operators. In order to exploit the benefits of the different operators, a new one (called Mixed Cross-over*) is introduced, trading-off the CPU time requirements and the obtained results. A new operator is then proposed, whose goal is to include in the genetic mechanism some heuristic knowledge drawn from the already proposed local-optimization techniques. The performance of the new operator is discussed.*

1 Introduction

Genetic Algorithms (GAs) [Holl75] have been deeply investigated in the last decade as a possible method for solving function optimization and combinatorial problems.

Many works have been published, concerning the use of GAs for solving classical optimization problems, and different solutions have been proposed, concerning the representation of each individual, the size of the population, the cross-over and mutation operators, the initialization strategy.

On the other hand, a large number of successful deterministic algorithms and heuristics have been proposed for many optimization problems, and even the best evolutionary technique is not likely to effectively compare with them. Nevertheless, by including the existing knowledge into the genetic mechanism one could hope to improve the performance of both the deterministic and the genetic approach.

The goal of this paper is first to analyze the effectiveness of some proposed techniques from an experimental point of view, with particular emphasys on cross-over operators; a hybrid approach is then presented, aiming at including heuristic techniques into the pure genetic algorithm schematon.

The experiments have been performed focusing on the Traveling Salesmen Problem using the classical benchmark set provided in the literature [Rein90].

In the next section the principal parameters characterizing a GA solution for the TSP are presented; Section 3 concerns the comparative analysis of 4 cross-over operators and introduces the new one; Section 4 presents an heuristic approach to the TSP; Section 5 introduces a new operator based on the hybrid approach, and Section 6 draws some conclusions.

2 TSP and GAs

The TSP has been deeply investigated by researchers involved in GAs ([GGRV85] [OSHo87] [SuVG87]).

The implementation of a GA is characterized by a number of parameters. The most significant ones are:

1. coding strategy;
2. parents selection;
3. *cross-over* mechanism;
4. mutation;
5. initialization mechanism;
6. population size;
7. population renewal rate;
8. strategy for sons inclusion.

In the following we will discuss each of them in some detail.

Coding Strategy It is the way according to which the single individual is coded. The following solutions have been proposed:

- *Path Representation*: the tour is described by a vector of N integers, whose i-th element holds the value j iff the city j is reached at the i-th step;
- *Ordinal Representation* [GGRV85]: the tour is described by a vector of N integers, where the i-th element value can range from 1 to $N-i+1$ and represents the position of the city reached at the i-th step in the *Free List* composed by all the cities not yet reached at that step;
- *Adjacency Representation* [GGRV85]: the tour is described by a vector of N integers, whose i-th element holds the value j iff the tour includes the edge form city i to city j.

Despite the fact that a tour is not uniquely represented using the first representation, this seems to be the more effective from a computational point of view, and we will refer to it for the following.

Parents Selection The mechanism according to which individuals are chosen for reproduction must emulate the mechanism of the natural selection in which the most suitable individuals will survive. So, the *selection operator* must work in such a way that the individuals with best fitness will reproduce more frequently. The chosen algorithm is the well-known *roulette wheel* one.

Cross-Over Mechanism Four cross-over mechanisms have been considered in the literature:

- *Order Cross-over* (OC): two cut points are randomly chosen and the portion of the first parent between them is copied to the offspring; the rest of the offspring is taken from the second parent;
- *PMX Cross-over* (PC): two cut points are chosen: the portion in the two parents they select is used to select several swap operations between cities;
- *Cycle Cross-over* (CC) [OSHo87]: portions of the tour called *cycles* are chosen from each parent in a random fashion and transferred to the offspring;
- *Heuristic Cross-over* (HC) [GGRV85]: a starting city for the tour is randomly chosen; then the shortest between the edges leaving the city in the two parents is taken, and the process is repeated on the new city, and so on.

Mutation Some experiments on different forms of mutation have been performed. Some researchers [OSHo87] defined a *mutation operator* SWAP, but the results we obtained using it have been quite poor, so we will not consider any mutation operator in the following.

Initial Population Selection of *initial population* should be made in a careful way, aiming at guarantying that the starting average fitness is not too bad, while randomness is still present. To combine these two goals we used a mixed approach, based on a very draft heuristic. the elements of the initial population are chosen by means of the *nearest neighbor* technique:

- starting from an initial city randomly selected, we proceed by choosing as the next element in the tour the nearest city from the current one.

Population Size Experimental results showed that as much as the population grows results improve. On the other side, the growth of the population causes a significant increase in the elaboration time and in the memory used. As a result of this, a good trade-off has been found with a population having a number of elements equal to the number of cities.

Population Renewal Rate The number of sons per generation is the parameter which determine the renewal rate inside the population. It must be neither too low, to avoid an excessive slackness of the population nor too high, to avoid a possible worsening of the average fitness. The number of sons has been set to the 40% of the population size.

Strategy for sons inclusion After every reproduction, sons replace the worst elements of the population according to the *elitism* strategy.

3 Evaluation of Cross-over mechanisms

In order to evaluate the different cross-over operators proposed in the literature, we wrote a C program

Problem	NG	PS	OC	HC	CC	PC
eil51	100	100	436.0	434.7	442.6	438.9
st70	100	100	822.8	819.7	832.0	831.7
pr76	100	100	668.4	666.0	682.4	680.3
kroA100	100	100	22720	22038	24155	22911
lin318	1000	100	83482	83463	83452	83498

Table 1: Experimental Results for the 4 Cross-Overs: Average Final Fitness

implementing the genetic algorithm for the TSP. The tool has then been run by activating in turn one of the four operators and performed several experiments in order to verify their effectiveness using a subset of the classical benchmark set of TSP provided in the literature [Rein90]. The program runs on a Sun4/300 with 8 Mbyte under the SunOS Operating System and amounts to about 1,000 code lines.

Several runs have been performed of each experiment and the average among the obtained results has been computed.

The results have been analyzed using two main parameters:

- the *average final fitness* produced by each cross-over after a constant number of generations: results are reported in Tab. 1 for 4 small problems, and show that HC is the best mechanism. In the table, NG represents the number of generations and PS the population size;
- the *speed* of each cross-over in terms of CPU time required to compute a fixed number of generations; Tab. 2 reports the percent ratio of the CPU time required by each operator with respect to the slowest one. Results show that HC is the more expensive technique from the CPU time point of view, while the time required by the other three operators is about of the same order of magnitude.

Problem	NG	PS	OC	HC	CC	PC
eil51	100	100	10	100	25	9
st70	100	100	12	100	23	11
pr76	100	100	11	100	26	11
kroA100	100	100	10	100	23	10
lin318	1000	100	12	100	27	11

Table 2: Experimental Results for the 4 Cross-Overs: CPU time

As a conclusion, results prove that HC is the best cross-over mechanism as far as the average final fitness is considered, but also the slowest one.

Making a step by step analysis of the best individuals produced by the HC operator, it is possible to see a good behavior in the first generations while in the latest ones the best fitness tends to become stable and not to improve any more. This seems to be due to a concentration occurs around local minima.

Best results have been achieved by using a *mixed cross-over (MC)* allowing to exploit the advantages in terms of CPU time effectiveness and average fitness of all the considered operators. According to the new operator, in every reproduction sons are generated by randomly choosing one of the four cross-over mech-

Problem	NG	PS	MC	
			Fitness	CPU time
eil51	100	100	434.0	33
st70	100	100	811.2	35
pr76	100	100	662.3	33
kroA100	100	100	21986.1	32
lin318	1000	100	81366	34

Table 3: Experimental results for the mixed cross-over

anisms and applying it to the construction of a new individual.

Using the mixed cross-over, the elaboration time decreases and the average final fitness improves, as demonstrated in Tab. 3.

Moreover, by mixing the different cross-over operators together, the risk that the population falls into a local minimum decreases, thus improving the effectiveness of the resulting algorithm.

4 Heuristics

In order to include some hill-climbing mechanism able to improve the population created through the standard genetic approach, we considered the following three local-optimization techniques:

- *2-opt* [LiKe73]: an individual is taken, and all the possible edge couples $e_{a,b}$, $e_{c,d}$ are considered; if the substitution with the couple of edges $e_{a,c}$, $e_{b,d}$ reduces the tour length, it is done;
- *Or-opt* [Or76] [LLRS85]: for each connected string of 3 cities in the individual, a test is performed, to check whether the string can be relocated between two other cities in the tour at a reduced cost; if this is the case, the relocation is done;
- *Group Optimization*: a group of n adjacent cities is randomly chosen, all the permutations of the n cities are considered, and the best one is taken; for our experiments n is set to 5.

To compare the effectiveness of the three heuristics to the one of the pure GA the following mechanism has been implemented:

1. an initial population is randomly generated;
2. each of the three heuristics is applied to each individual, thus possibly generating new individuals;
3. the elitism strategy is applied;
4. steps 2 and 3 are repeated for a given number of generations.

Tab. 4 reports the final average fitness and the CPU time obtained using this mechanism. Provided that the same population size and the same number of generations is considered, the elaboration time required by heuristic approach is approximately 500 times the one required by the approach based on the MC operator, while the results are by far better than the ones of the pure GA.

Problem	NG	PS	Heuristics	
			Fitness	CPU time
eil51	100	100	432.0	510
st70	100	100	680.0	498
pr76	100	100	549.0	520
kroA100	100	100	21350	502
lin318	1000	100	42978	491

Table 4: Experimental results for the heuristic approach

On the basis of this consideration, a new operator is introduced in the following Section in order to include in the GA some effectiveness drawn from the heuristic approach at the cost of an acceptable overhead.

5 Hybrid algorithm

Some authors [MSBo91] already underlined the importance of hill-climbing as a way to significantly improve the efficiency of a GA; the results presented in the previous Section confirm the validity of the heuristic approach.

We then experimented a hybrid approach in which a new operator, called *Heuristic Operator*, is activated at each generation on a subset of the individuals composing the current population. The operator applies one of the three described heuristics on the individual it is activated on: each heuristic is assigned an activation probability inversely proportional to the computational cost of its activation. The sum of the three probabilities is equal to 1.

Problem	NG	PS	OC	HC	CC	PC
eil51	100	100	428.4	427.4	429.5	429.1
st70	100	100	678.8	678.8	680.0	679.7
pr76	100	100	544.4	542.7	547.4	547.8
kroA100	100	100	21285	21284	21330	21309
lin318	1000	100	42844	42830	42978	42972

Table 5: Experimental results for the hybrid algorithm

Results show that the combination of the three heuristic mechanisms with the genetic approach significantly improves the final performance, especially in the first phase of the search. Tab. 5 reports the average final fitness produced by the experiments, where the *Heuristic Operator* is used, together with one of the four cross-over. From this table it is clear that the new operator can significantly improve the results. The activation probability is 50% for *Or-opt*, 30% for *2-opt* and 20% for *Group Optimization*.

The cost in term of CPU time is however high, as the required time once the operator is inserted is increased up to 300 times, changing for the pr76 problem from less than 1 sec for 200 generations to about 300 sec. The size of the population is 100 individuals, and 40 new elements are computed at each generation. New individuals are inserted in the population, which is then sorted and reduced to the original size by deleting the worst individuals.

The comparison between the results of Tab. 1 and Tab. 5 shows that an improvement has been reached for all the considered problems.

The final proposed algorithm is a mixed approach

Problem	NG	PS	4x3eur	
			Fitness	CPU time
eil51	100	100	427.2	92
st70	100	100	678.8	87
pr76	100	100	541.8	98
kroA100	100	100	21279	91
lin318	1000	100	42780	86

Table 6: Experimental results for the 4x3eur algorithm

among the four cross-over mechanisms and the three heuristics techniques (*4x3eur*), exploiting both the MC operator and the Hybrid one.

Tab. 6 reports the final average fitness and the CPU time for the 4x3eur algorithm.

6 Conclusions

An experimental evaluation of the effectiveness of some proposed Cross-over mechanisms has been performed: the results show that the Heuristic Cross-over provides the best results when the standard library of Traveling Salesman Problems is considered, although the CPU time it requires it is larger than the one required by any other operator. To trade-off effectiveness and speed, a Mixed Cross-over has been introduced, and its behaviour experimentally verified.

Moreover, some heuristic knowledge has been incorporated into the genetic mechanism. A new operator is defined, called ***Heuristic Operator***, aiming at applying some local-optimization technique to a given subset of the population at each generation. Its main effect is a speed-up in searching the optimum solution at the cost of an increased CPU time. An activation probability has been introduced to trade-off results with required CPU time. Experimental results show the effectiveness of the proposed approach.

References

[GGRV85] J. Grefenstette, R. Gopal, B. Rosmaita, D. Van Gucht: "Genetic Algorithms for the Traveling Salesman Problem," Proc. First Int. Conference On Genetic Algorithms, Pittsburgh, PA (USA), 1985, pp. 160-168

[Holl75] J.H. Holland: "Adaption in Natural and Artificial Systems," University of Michigan Press, Ann Arbor, MC (USA), 1975

[LLRS85] E.L. Lawler, J.K Lenstra, A.H.G. Rinnooy Kan, D.B. Shmoys: "The Traveling Salesman Problem," John Wiley & sons, New York, USA, 1985

[LiKe73] S. Lin, B.W. Kernighan: "An Effective Heuristic Algorithm for the Traveling Salesman Problem," Operations Research, 1973, pp. 498-516

[MSBo91] H. Muhlenbein, M. Schomisch, J. Born: "The Parallel Genetic Algorithm as Function Optimizer," Parallel Computing, Vol. 17 , 1991, pp. 619-632

[Or76] I. Or: "Traveling Salesman-Type Combinatorial Problems and their Relation to the Logistics of Regional Blood Banking," Ph. D. Thesis, Northwestern University, Evanstone, IL (USA), 1976

[OSHo87] J.M. Oliver, D.J. Smith, J.R.C. Holland: "A Study of Permutation Operators on the Traveling Salesman Problem," Proc. Second Int. Conference On Genetic Algorithms, Cambridge, MA (USA), 1987, pp. 224-230

[Rein90] G. Reinelt:
TSPLIB 1.1
Institut fuer Mathematik, Universitaet Augsburg, 10 Dec. 1990

[SuVG87] J.Y. Suh, D. Van Gucht: "Incorporating Heuristic Information into Genetic Search," Proc. Second Int. Conference On Genetic Algorithms, Cambridge, MA (USA), 1987, pp. 100-107

Using a Genetic Algorithm to Investigate Taxation Induced Interactions in Capital Budgeting

R. H. Berry G. D. Smith
University of East Anglia, Norwich NR4 7TJ, UK.
gds@sys.uea.ac.uk
rhb@sys.uea.ac.uk

Abstract

The capital budgeting problem, as analysed here, involves selecting a combination of projects from the set of all possible combinations of projects to maximise the net present value of cash flows. The traditional investment appraisal approach of carrying out a project by project analysis and selecting accordingly is invalid because of features of the many current taxation system, which introduce non-linear interdependences between the projects.

Using a serial implementation of the genetic algorithm toolkit GAmeter, we investigate this effect using aspects of the UK taxation system on a set of standard capital budgeting problems and compare the results with those obtained using a more traditional approach and a mixed integer programming approach. The capital budgeting model developed includes features such as "carry forward" of capital allowances, multiple rates of corporation tax and capital rationing constraints. The model can be extended to include other tax features found in many economies.

1 Introduction

Let there be n economically independent projects in which a firm can invest. The capital investment for each project, as well as the net cash flow from the project over its life, is given for each project in the set. The capital budgeting problem is to choose the combination of projects which yields the best possible discounted present value. The number of possible combinations is $2^n - 1$. With as few as 20 possible projects therefore, there are over 1 million possible combinations. A good search technique is required to find the optimal combination.

It is assumed, for the purposes of comparison, that the investment decisions are based on ***discounted net present values***, henceforth referred to as NPVs, and that a uniform discount rate is applied to all cash flows. Throughout the paper, we use i, where i = 1, 2, ..., n, as the subscript which refers to a particular project and j, where j = 1, 2, ..., m, as the subscript to specify the year since the start of the planning period.

Let d_i be a binary variable which equals 1 if project i is selected and is equal to 0 if it is not. Let $(NPV)_i$ be the pre-tax NPV of project i. The objective of the capital budgeting problem is therefore to maximise

$$\sum_{i=1}^{n} d_i(NPV)_i - TAX \qquad (1)$$

where TAX is the aggregate corporation tax paid by the firm over the period being considered.

On a pre-tax basis, i.e. maximising only the first term of equation 1, the problem reduces to a straightforward optimisation problem in which each project may be analysed individually and if it delivers a positive pre-tax NPV over the period of interest, it is accepted while if NPV is negative, it is not accepted. However, it has long been accepted that capital investment decisions should be based on after-tax cash flows. A standard approach, see [1], to this after-tax case is to continue to consider each project individually, and to calculate the post-tax NPV of each project as if it were the only project accepted, making strong, and often implicit, assumptions about the tax situation of the company considering the project. If this post-tax NPV is positive, accept the project; if not, then reject it. The objective may then be realised by summing the NPVs of all those projects accepted.

However, Buckley [2] and Berry and Dyson [3] have shown that interactions may occur between project cash flows and the ongoing cash flow of the firm, and also between the cash flows of projects which are economically independent on a pre-tax basis. This leads to a non- linear system and the problem becomes one of selecting the optimal ***combination*** of projects, rather than a series of individual project decisions. Berry and Dyson [3] have developed a mixed integer programming (MIP) model which calculates the total tax payable and thereby selects the optimal combination using a branch and bound technique. A development of that MIP system is included in [4].

Despite the fact that an optimal solution to this problem can be found using a mixed integer programming approach, there are strong arguments for reanalysing the problem using a genetic algorithm. Mathematical programming approaches to management problems have low acceptability. Production and operations applications exist, but marketing applications are fewer, and financial applications fewer still. There are many possible reasons for this, but one appears to be the restrictions on problem representation which use of a mathematical programming approach involves.

The tax interactions problem involves consideration of $2^n - 1$ project combinations, where n is the number of economically independent projects. This number rapidly exceeds that which could be feasibly dealt with without some kind of structured search for attractive combinations. Mathematical programming provides an efficient, combined search and evaluation algorithm, but at a cost; the problem representation is tightly controlled. A simple example demonstrates this.

In the UK, the function relating taxable profit to tax paid is piecewise linear. There is a small company rate, a large company rate, and a shifting average rate in between. If the small company rate, payable on total profits when they are less than £100,000 (L_1) is 0.29 (t_1), and the large company rate, payable on total profits when they are over £500,000 (L_2) is 0.35 (t_3), then tax payable on intermediate profits, say Y, is given by the simple function:

$$Tax = Yt_3 - \frac{L_2 - Y}{F} \qquad (2)$$

where F is the *formula fraction* determined from the given taxation parameters. The implied intermediate tax rate is 0.365 (t_2).

A spreadsheet formulation, capable of evaluating a combination of projects, can be achieved by a simple IF...THEN...ELSE... structure. However, the mixed integer programme defined in Ashford, Berry, & Dyson [4], capable of both selecting and evaluating combinations of projects, requires five constraints, and four variables *for each year of the planning horizon* to provide a link between taxable income and tax payable. In addition, other integer variables must be introduced to cope with the dichotomy of whether or not tax is due at all for each year. This form of representation is not accessible to a typical manager. It is a barrier to model implementation and use.

The genetic algorithm approach separates the model representation from the search procedure. Project combinations can be generated by the genetic algorithm and submitted to an evaluation function which is readily understood, and therefore has the ability to be experimented with, by most managers. Ideally, any genetic algorithm used in financial applications should be front-ended with a spreadsheet model, allowing a level of interaction which most managers can apparently accept and use.

1.1 Characteristics of Taxation Systems

Features of the U.K taxation system will be used in order to demonstrate the impact of taxation on the project selection process. Two major sources of project interactions in taxation systems are the system of capital allowances, and the multiplicity of tax rates. Allowances are determined by the purchase cost of capital assets. Capital allowances replace depreciation in the calculation of taxable profit. In simple terms, assuming that both turnover, and all non-depreciation costs are cash items,

$$AccountingProfit = Turnover - Costs - Depreciation.$$

If a company carries out an investment, the cost of that investment is not an allowable expense in the calculation of taxable profit. In calculating taxable profit, depreciation is added back to accounting profit, and capital allowances deducted. Thus

$$TaxableProfit = Turnover - Costs - Allowances$$

and in turn, the net cash flow (NCF) is defined as

$$NCF = Turnover - Costs - Tax - Investments.$$

This NCF, discounted using a standard discount rate, is the basis of our objective function.

Changes in tax regimes occur frequently, and the availability of allowances can also change. In the UK, pre 1984, 100% of the cost of plant and machinery was available as an allowance against tax in the year the plant and equipment was acquired. Over a period of years, this system was replaced by a 25% reducing balance system. The reducing balance system works as follows: If an asset costing C is purchased in year 1 of a planning horizon, an allowance of

$$\alpha(1-\alpha)^{j-1}C \qquad (3)$$

will become available in year j, where α equals 25%. At the end of year j, after the available allowance has been claimed, the amount of unused allowance is

$$(1-\alpha)^j C. \qquad (4)$$

By setting α to be 1 (equal to 100%) in equation 3 and equation 4, we are considering 100% first year allowance, while any other value, say 25%, refers to the reducing balance method. We shall consider both these cases to allow comparison with the results of [4].

Thus, generalising the above description, let C_j represent any capital purchases in year j and let P_j be the pool of available allowances in year j generated by this set of capital investments. Then

$$P_j = (1-\alpha)P_{j-1} + C_j\ , \quad P_1 = C_1. \tag{5}$$

The capital allowance A_j which may be claimed against the corporation tax for year j is then given by

$$A_j = \alpha P_j. \tag{6}$$

Let R_j be the net cash flow from all the operations for year j, i.e. the cash flow from the projects plus the firm's ongoing cash flow. If $(R_j - A_j)$ is non-negative, i.e. income exceeds available allowance, then the taxable income for the year j is based on $(R_j - A_j)$. If, however, $(R_j - A_j)$ is negative, then the available allowances exceed the income. In this case, no tax is paid in this year and an amount of unused allowance is generated. If we let U_j represent this unabsorbed allowance for year j, then

$$U_j = A_j - R_j, \quad R_j > 0, \tag{7}$$

$$U_j = A_j, \quad R_j < 0. \tag{8}$$

At present in the UK, there are two mechanisms of making use of this unused allowance. These are:

(a) - The quantity of unused allowance may be *carried forward* to succeeding years until it is used up. Under UK tax law, this may be done over an indefinite period of time. As a result, the capital allowance available in year j is given by

$$A_j = \alpha P_j + U_{j-1} \tag{9}$$

instead of equation 6.

(b) - Another possibility is that the unused allowance U_j may be *carried back* to the previous year to absorb any surplus profit which resulted then. This *carry back* facility was limited to one year under the 1984 Finance Act, but has subsequently been extended to three years. If this unused allowance is more than enough to cover the profit in the previous year(s), any that is still remaining is carried forward, as above.

This paper considers *carry forward* only. We acknowledge that *carry back* weakens the role of allowances in generating project interactions. In any given tax system, the balance between rate of allowances and the length of *carry back* available determines the extent of interactions.

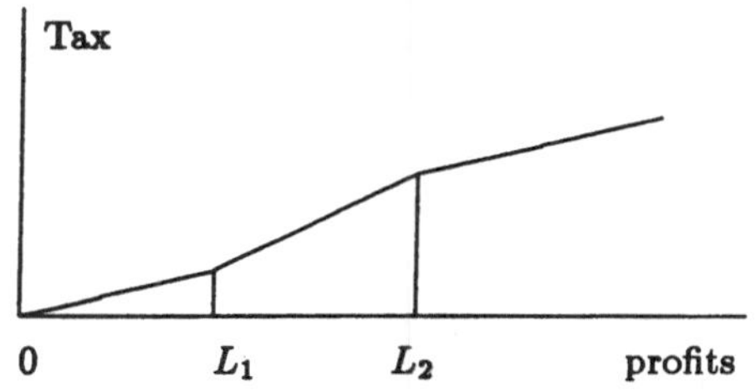

Figure 1: The Multiple Tax-Rate system in the UK

In 1992, the depressed state of the UK economy led the Chancellor to introduce a 40% first year allowance, with 25% of the residual as a writing down allowance, in each subsequent year. The reducing balance method of allowance calculation was maintained, but with a slightly higher first year availability. This situation can be modelled with a simple modification to the above equations.

Corporation Tax is payable on any profit remaining after the available allowances have been used. In order to test our model, we shall initially adopt the two different tax models used in [4], but then extend these to include some new features. These systems are:

Single tax rate - In this model, a single tax rate t (=35%) is used and the tax paid will be t times the taxable income, $(R_j - A_j)$, if this is non-negative.

Multiple tax rate - In the current UK tax system, there are three levels of tax rate: t_1, t_2 and t_3. Below the limit L_1 of taxable profit, the tax rate is t_1. Above the limit L_2 of taxable profit, where $L_2 > L_1$, the tax rate is t_3, while between the two limits, the tax rate is t_2. See Figure 1.

In practice, the values t_1, t_3, L_1 and L_2 are given and t_2 is derived. The values we shall use are $t_1 = 0.29$, $t_3 = 0.35$, $L_1 = 100000$ and $L_2 = 500000$.

2 A GA Approach

The problem is very amenable to solution by a genetic algorithm (GA). A GA is a search paradigm which is based on the mechanics of natural population genetics. Details of the general technique can be found in [5, 6, 7]. The particular GA software we use is *GAmeter*, a toolkit developed by the genetics group at the University of East Anglia, and which has the following features:

- Implementations available on Macintosh MACII, Unix and the Meiko transputer systems.
- Full graphics user interface, including windows displaying
  1. the current population, chromosomes plus fitness values, updated by double-clicking on the window,
  2. a line graph of solution v/s time or generations, updated by double-clicking on the window,
  3. the current values of the parameters, e.g. crossover rate, mutation rate, size of population, etc. These are controlled by slider buttons and can be altered at any time.
- The ability to stop the run and edit individual chromosomes which are subsequently evaluated and included in the current population.
- The facility to add new operators to suit the application, and many other features.

See [8, 9, 10] for more details.

In this particular instantiation of GAmeter, the chromosome is simply the vector of binary variables d_i, for i = 1,2, ... n. The basic fitness value is given by equation 1.

The optimal solutions achieved with this objective function are not subject to any capital rationing. It is assumed in these cases that there are sufficient funds available to invest in the projects deemed to be acceptable. In Weingartners original application of mathematical programming to the capital budgeting problem [11], there was no mention of tax induced interactions. His concern was with the interactions generated by a scarcity of capital. However, finance theorists have argued that the presence of capital rationing constraints is incompatible with the application of an objective function based on net present value, with discount rates taken from the capital market. The discount rate represents an opportunity cost of funds. If a capital rationing constraint is binding, this should affect the opportunity cost of funds, so the discount rate becomes endogenous to the model.

Capital rationing can however be found in the real world. Managers do find the scale of their capital budgets set by their superiors. The rationale of these constraints may have nothing to do with inability to raise funds for additional projects. It may be the case that the constraint is a simple numerical method to ensure that a manager does not take on more projects than he/she can cope with given other demands on time. A capital rationing constraint could also reflect a strategic decision within a corporation, to ensure a balanced growth among business units for example.

Therefore it is not unreasonable to have apparent capital rationing constraints in a capital budgeting model, but the rationale should have nothing to do with ability to raise funds. In the presence of capital rationing, a standard technique for dealing with constraints, that of ***pricing out***, is used. Thus the fitness value as specified by equation 1 is subject to a penalty, which may either be linearly related to the amount of overspending on capital investment, or dependent on the number of projects over a specified maximum. Thus

$$Fitness = \sum_{i=1}^{n} d_i(NPV)_i - TAX - \beta(penalty), \quad (10)$$

where β is a parameter which allows us to alter the weighting of the penalty.

3 Results

In order to allow comparison to be made with the results of Ashford, Berry and Dyson [4], the same data set is used and the same assumptions are made. This data set is taken from Weingartner [11], tables 9A.1 and 9A.2 and consists of net cash flows for each of 14 projects and for the firm over ten years. See Table 1. In this table, a positive cash flow represents net revenue while a negative cash flow represents a capital expenditure.

Additional assumptions have to be made in order to apply the model. These are:

- The uniform discount rate is assumed to be 5%.
- The ongoing cash flow of the firm is reduced by £300,000 in each year to ensure that an unabsorbed allowance is generated in the initial and final years of the 10 year period.
- The initial capital allowance pool is assumed to be nil.

As previously stated, the standard textbook approach assumes that sufficient taxable profit is available in the company to make use of all allowances generated by a new project, as soon as those allowances are legally available. This is an optimistic treatment. Buckley [2] argued that such an assumption was unnecessary, and that the actual profit of the investing

project	year									
	1	2	3	4	5	6	7	8	9	10
1	-100	20	20	20	19	19	18	16	14	11
2	-100	20	18	18	18	18	14	14	14	14
3	-100	15	15	15	15	15	13	13	13	13
4	-100	20	6	11	7	16	5	14	18	3
5	-100	-60	-60	80	74	66	56	44	30	14
9			-120	25	25	30	35	30	25	20
10		-100	18	17	15	12	8	-10	18	17
11	-150	20	20	20	20	20	20	20	20	20
12	-100	20	18	16	14	12	10	4	-20	20
14	-50	-100	-175	50	55	60	65	60	50	40
19	-75	-75	-40	40	40	40	35	35	30	25
22		-85	20	20	16	15	13	10	7	3
23		-270	-100	125	115	105	80	60	35	25
28		-355	60	70	80	70	55	40	25	15
ongoing	100	60	20	-20	-60	-100	-140	-180	-220	-260

Table 1: Weingartner's data set

company should be used during the process of investment appraisal. Given that a company may have low profits, allowances may not be used when they are generated and may have to be carried forward. This reduces their value. Thus some projects with positive NPVs under the optimistic assumption may become negative when Buckleys method is applied. Berry and Dyson [3] recognised that an extension of Buckleys approach was necessary, with combinations of projects, rather than individual projects being evaluated.

Under these three approaches, project after-tax NPVs can obviously differ. More importantly, project choice can differ. Projects with positive NPVs under the optimistic appraisal may become negative NPV projects under Berry & Dysons more realistic assumptions, while projects with negative NPVs under Buckleys approach may become positive NPV projects. Ashford, Berry & Dyson [4] have demonstrated the existence of project interactions for a variety of different tax systems.

The software GAmeter was run on the set of data in Table 1, using different settings of the parameters specifying the allowance and tax systems. For some settings, the optimal solution could be read from [4] thus allowing the runs to stop when this optimal solution was attained (experimental mode). For each setting, GAmeter was run a number of times in order to ascertain the average behaviour, but it quickly became apparent that the optimal solution was being found every time and that there was very little variation in the times to do so. Where the optimal solution was not known in advance, GAmeter converged quite rapidly to a solution which was subsequently checked for optimality using an exact technique, the MIP model. Thus GAmeter found the optimal solution in every case, with times varying from a few seconds (for the simple cases) to approximately 50 seconds(for the capital rationing cases). In each case, the optimal solution was fed into a specially designed spreadsheet model which in turn gave the breakdown of the solution in terms of what each of the accepted projects would contribute to the optimal solution.

The different sets of conditions under which the system was run, and the corresponding results are as follows:

1. *(Pre-tax values.)* Table 2 shows the pre-tax NPVs of the individual projects, with the resulting aggregate NPV if only those projects with a positive pre-tax NPV were to be accepted, i.e a 1 in the d_i column.

2. *(Simple tax-rate of 35%; 100% first year allowance.)* If we consider for the moment the case when $\alpha = 1.0$, equivalent to 100% first year allowances, we find that the aggregate NPV and the set of accepted projects is identical to the pre-tax case. This is a consequence of the particular pattern of cash flows in the example. Substantial capital allowances are generated by the investments. There is no profit against which they can be offset, and they are therefore carried forward, sheltering future project cash flow from tax.

Project	Pre-Tax NPV	d_i
1	£26353	1
2	£18732	1
3	£1061	1
4	-£19915	0
5	£72523	1
9	£34077	1
10	-£20479	0
11	-£7844	0
12	-£20224	0
14	-£17745	0
19	£3505	1
22	£2860	1
23	£79158	1
28	-£6483	0
aggregate	£238267	

Table 2: The Pre-Tax NPVs of the projects. The 1s in the column headed d_i define the set of accepted projects.

Project	Buckley NPV	Buckley d_i	Berry & Dyson NPV	Berry & Dyson d_i
1	£32801	1	£18832	1
2	£25816	1	£13970	1
3	£10763	1	£1413	1
4	-£9022	0	-£13794	0
5	£96014	1	£48307	1
9	£40426	1	£28238	1
10	-£11908	0	-£16929	0
11	£8337	1	-£4757	0
12	-£13141	0	-£20956	0
14	£4438	1	-£14687	0
19	£27354	1	£5115	1
22	£8654	1	-£881	0
23	£105507	1	£49214	1
28	£19867	1	-£18538	0
aggregate	£144252		£193960	

Table 3: Post-tax NPVs of each project assuming a simple tax-rate of 35%, and a 25% reducing balance method.

3. *(Simple tax-rate of 35%; 25% reducing balance.)* In Table 3, the columns headed "Buckley" show the NPV contributions of the individual projects, and the resulting set of accepted projects, under Buckley's approach. As can be seen, a different pattern of projects from the pre-tax set is accepted. The simple tax system of 35% tax rate and 25% capital allowances on a reducing balance basis has distorted the business decision. On this basis, projects 4, 10 and 12 result in a negative NPV after tax and are hence rejected. Also, with this choice of projects, the aggregate NPV is £144252.

   However, using GAmeter to find the optimal combination of projects under these conditions, we find that the optimal aggregate NPV is £193961, with a set of accepted projects which is different to both the "Buckley" and the pre-tax set. Each entry in this column headed "Berry & Dyson" is the contribution made by each project when all other accepted projects are taken into consideration. It is readily seen from this simple example that the tax/allowance system invalidates the traditional approach of dealing with investment on a project-by-project basis.

4. *(Simple tax-rate of 35%; 25% reducing balance; capital rationing.)* In this case, we impose a simple constraint, namely that a fixed amount of capital, K, is available to be spent in the first year of the planning horizon. Provided $K <$ £475000, the first year investment costs of the previous "Berry & Dyson" optimal solution, GAmeter will be forced to find an alternative optimal solution. The value of β used in the run was 1.0, since the penalty adopted, namely the amount of overspending, is of the same order of magnitude as the other terms in the objective equation 10. The results are shown in Table 4:

K	$NPV(opt)$	d_i
£400000	£192546	11001100001010
£300000	£186269	11001100000010
£200000	£170709	10001100000010

Table 4: Optimal aggregate NPVs assuming a simple tax-rate of 35%, and a 25% reducing balance method, with first year capital spending limited to K.

The results are as expected, and in fact could have been easily derived from the "Berry & Dyson" optimal solution, using the spreadsheet model to flip certain accepted projects to "rejected" until the capital rationing constraint was met. However, it is encouraging to know that GAmeter finds the new optimal solution in less than a minute in all cases. Given a single constraint, the model is effectively ranking project NPV against unit of scarce resource consumed.

Project	Buckley		Berry & Dyson	
	NPV	d_i	NPV	d_i
1	£31696	1	£20121	1
2	£24601	1	£14786	1
3	£9100	1	£1353	1
4	-£10890	0	-£14843	0
5	£91987	1	£52458	1
9	£39338	1	£29239	1
10	-£13379	0	-£17538	0
11	£5563	1	-£5286	0
12	-£14355	0	-£21131	0
14	£635	1	-£15211	0
19	£23265	1	£4839	1
22	£7661	1	-£669	0
23	£100990	1	£54347	1
28	£153506	1	-£17714	0
aggregate	152704		201065	

Table 5: Post-tax NPVs of each project assuming a multiple tax-rate, and a 25% reducing balance method.

What is more important is that the changes necessary to implement this into the fitness function took very little time at all, compared to the time taken to deal with such a constraint in the mathematical programming model.

Alternatively, we could impose a constraint that no more than k projects can be managed at any given time. If the number of accepted projects exceeds this limit, a penalty should be incurred. In this case β will have to to be increased to something of the order of the other terms in the objective equation in order for that penalty to be felt, but the result will be the same. This type of rationing is used for the multiple tax case.

5. *(Multiple tax-rate; 25% reducing balance; no capital rationing.)* The multiple tax system described earlier is now applied and the results are given in Table 5. It is noted that the genetic algorithm is generating the same results as the mixed integer programming model in [4].

   We also note that this tax regime gives rise to the same set of optimal projects as the simple tax system, and thus continues to vary from that derived from the project-by-project approach.

6. *(Multiple tax-rate; 25% reducing balance; capital rationing.)* Finally, we consider the same tax regime as in the previous case, but now including capital rationing. However, in this case we take the constraint to be that no more than k projects may be taken on at any time. The results are shown in Table 6. For this set of runs, the value of β is 30000, a value which is significantly larger than the average pre-tax contribution of each project and thus is able to be effective in penalising any solution which violates the constraint. Thus, in this simple case, GAmeter is seen to be able to deal with capital rationing constraints by a differential *pricing out* mechanism.

k	NPV(opt)	d_i
7	£201065	11101100001010
6	£199713	11001100001010
5	£193910	11001100000010
4	£177807	10001100000010

Table 6: Optimal Net Cash Flows assuming a multiple tax-rate and a 25% reducing balance method, with no more than k projects being undertaken at any time.

4 Conclusions

The results in this paper confirm that under many tax systems, the text book treatment of investment appraisal is inappropriate. Project by project analysis must give way to a search among project combinations. The mixed integer programming approach in [4], while capable of solving the problem, is not management-friendly. A description of tax rules in a mathematical programming framework is difficult. The genetic algorithm approach allows the search for a good combination to be separated from the evaluation of that combination. The evaluation function can be implemented in a variety of forms. The use of a spreadsheet can only enhance management acceptance, and hence improve the quality of investment appraisal processes.

References

[1] Franks J.R. and Broyles J.E. *Modern Managerial Finance.* Wiley and Sons, London, 1979.

[2] Buckley A. The distorting effects of surplus advance corporation tax. *Accounting and Business Research*, 1975. summer.

[3] Berry R.H. and Dyson R.G. A mathematical programming approach to taxation induced in-

terdependencies in investment appraisal. *J. Bus. Fin and Acc.*, 6(4):425–441, 1979.

[4] Ashford R., Berry R.H., and Dyson R. Taxation induced interaction in capital budgeting. Working Paper.

[5] Holland J.H. *Adaptation in Natural and Artificial Systems*. University of Michigan Press, Ann Arbor, MI, 1975.

[6] Goldberg D.E. *Genetic Algorithms in Search, Optimization, and Machine Learning*. Addison-Wesley, Reading, Mass., 1989.

[7] Reeves C.R., editor. *Modern Heuristic Techniques*. Blackwell Scientific, 1993.

[8] Kapsalis A. and Smith G. D. The gameter toolkit manual. School of information systems, computing science technical report, University of East Anglia, 1992.

[9] Kapsalis A., Rayward-Smith V.J., and Smith G.D. Solving the graphical steiner tree problem using genetic algorithms. *J. Oper. Res. Soc.*, 1993. (to appear).

[10] Kapsalis A., Rayward-Smith V. J., and Smith G. D. Fast sequential and parallel implementation of genetic algorithms using the GAmeter toolkit. Submitted for publication, 1993.

[11] Weingartner H.M. *Mathematical Programming and the Analysis of Capital Budgeting Problems*. Kershaw Publishing Company, London, 1974.

Fast Sequential and Parallel Implementation of Genetic Algorithms using the GAmeter Toolkit

A. Kapsalis V. J. Rayward-Smith
G. D. Smith
University of East Anglia, Norwich NR4 7TJ, UK.
email: gds@sys.uea.ac.uk

Abstract

A General Search Paradigm is formulated using a higher order function and, in this context, we discuss the properties which characterize genetic algorithms, tabu search, simulated annealing, etc.

From the specification of this general search algorithm, we develop a formal specification of a class of genetic algorithms. By suitable settings of input parameters, we show that a wide variety of (genetic) algorithms can then be instantiated.We have developed kernel software to implement this specification on different architectures. In particular, we have versions running sequentially on Macintosh computers and under Unix and another, parallel version on a Meiko transputer rack. The user interface is identical and the user of the system need have no architecture-dependent knowledge to use this software. We briefly describe these implementations and show how simple it is to port genetic algorithms from one achitecture to the other.

The toolkit implementing the kernel together with its associated interface is called *GAmeter* and we report on the use of this software on a number of case studies in combinatorial optimisation. We have encouraging results on a range of examples including the undirected and directed Steiner tree problems.

The main strengths of our software is its excellent graphics interface and its wide applicability. However, it also has a number of novel features including dynamic control of parameters such as population size and crossover/mutation probabilities, thus giving additional control to the user.

1 Formal Specification of a Search Algorithm

Let us assume that we have a universe, U, of (potential) solutions to a problem. We seek solutions that meet certain constraints, or, put another way, that lie in some subset, S, of U. Assuming each problem can be assigned a value in some totally ordered set, W, an optimisation problem will seek one or more elements of S of maximum value. Thus, given a set $S \subset U$, a totally ordered set, W (often the reals, $\mathcal{R}$) and a function, $value : U \rightarrow W$, an optimisation problem is of the form

$$\text{maximise } value(p)$$
$$\text{such that } p \in S \subset U.$$

A common technique used in optimisation is to "price out" the constraints. This is done by defining *value* on all elements of U in such a way that any $p \in U \setminus S$ has such a small value that it is guaranteed not to provide the maximum value. In the case where $W = \mathcal{R}$, this is usually achieved by ensuring $value(p)$ is a large negative number for any $p \in U \setminus S$. The optimisation can then be expressed simply as

$$\text{maximise } value(p)$$
$$\text{such that } p \in U.$$

The last decade has seen a considerable increase in interest in optimisation techniques based on directed or random search. Hill climbing, random search, genetic algorithms, tabu search, simulated annealing and a range of hybrid algorithms all fall into this category. The essential idea is to start with an initial pool, P, of zero or more potential solutions to the problem. If these solutions are not satisfactory then some subset, Q, is selected and used to create a new set of solutions, R. The pool, P, is then reconstructed depending on the original set of solutions, the selected subset and the newly generated set. This process is repeated until the pool P is deemed satisfactory. We define this paradigm more formally in Figure 1.

In a sequential, purely random search, P is initially empty but, thereafter, contains the single solution of greatest value found to date. The function *create* makes no use of its argument, Q, but instead randomly selects a number of solutions from U. Then,

```
Search
{objective is to maximise value(p) such
that p ∈ S ⊂ U }
P, Q, R : set of solutions ⊂ U;
initialise(P);
while not finish(P) do
     begin
     Q := select(P);
     R := create(Q);
     P := merge(P, Q, R)
     end
end Search
```

Figure 1: The General Search Paradigm

the function *merge* merely delivers a singleton set containing the solution of greatest value from $P \cup R$.

In a simple, hill climbing search, *initialise* selects some starting set of solutions, $U_0 \in U$ and assigns it to P. In the sequential case, this set P is usually constrained to have cardinality one. In the **while** loop, *select* is the identity function, *create* generates all the neighbours of the elements in Q and *merge* selects a subset of the solutions in $P \cup R$ which are of maximum *value* subject to some limitation on cardinality. The *finish* function can ensure termination either after some given number of iterations or when the value of the best solution in P is not being (significantly) improved.

Tabu search [1, 2] and simulated annealing [3, 4] are both variants of simple hill climbing. In tabu search, the function, *create*, selects the neighbours of the elements in Q but omits those specified to be "tabu". The "tabu list" is a dynamic set of user-defined rules which defines those neighbours which are forbidden. This dynamic list is often implemented as a circular list of forbidden moves, in which the last moves are added to the list and are forbidden until a fixed number of iterations, usually the length of the list, has passed; see [5] for example.

Simulated annealing varies from hill climbing in the function, *merge*. This function selects a subset of the solutions in $P \cup R$ favouring those with highest value. Initially, there may be no strong preference for the higher valued solutions, but as the temperature parameter reduces, so this favouring becomes stronger. Nevertheless, there is always some chance that a solution with a less good value will be kept in preference to a better solution. Considerable effort is expended in simulated annealing algorithms to get the cooling rate correct so that this degree of randomness introduced into hill climbing is beneficial. See, for example, [6].

Both tabu search and simulated annealing are generally viewed as sequential algorithms where P has cardinality one but it is obvious from the preceding discussion that they can be viewed equally easily as parallel algorithms. Moreover, there is obvious scope for a hybrid algorithm which uses the tabu list to limit R together with simulated annealing techniques within *merge*.

The general search paradigm becomes a genetic algorithm in the case where:-

1. P contains more than one element
2. The function *create* uses genetic operators to generate new solutions.
3. Both *select* and *merge* depend on the evaluation of solutions and favour the preservation of solutions with higher *value*.

It is worth noting at this stage that the general genetic algorithm, as specified by Figure 1 and the above conditions, allows for the possibility of replacing only a portion of the population with new solutions as well as the option of replacing the entire population with the newly generated set. Many specifications, see [7] for example, limit the user to a default *merge* of the entire population, whereas studies have shown that incremental replacement is often superior. We have used these observations in the specification of our toolkit, GAmeter, which is discussed in the next section.

2 The toolkit GAmeter

The toolkit provides a friendly and easy to use environment for the development and experimentation of optimisation problems using Genetic Algorithms (GAs). Its open architecture and design based on the general search paradigm of Figure 1 makes it a GA system that can be easily transformed and expanded in accordance to the findings of current and future research in the field. The toolkit can even be adapted to implement any of the search techniques briefly described in the previous section, and mixtures of them. Work on these hybrids is in progress.

Before launching into a description of the toolkit interface, a description of the options available in the functions *initialise*, *select*, *create* and *merge* is appropriate.

Firstly, an initial population, P, of chromosomes must be created and evaluated. This is done in the *initialise* function. In the basic version of GAmeter, this initial population is created randomly. However, in specific instantiations of GAmeter, other options have been included, equivalent to seeding this initial population with a solution derived from some or other heuristic, see [8, 9].

Once the population, P, has been generated, a selection process is used to generate the "gene pool", Q. The optional mechanisms available in GAmeter for this process are (see also Figure 2),

- Random : *select* using random numbers drawn from a uniform distribution.
- Roulette Wheel : *select* with probability determined from the relative fitness of each solution (see, for example,[10]).
- Fibonacci : *select* according to a fixed sequence of rank in the pool. In this case a Fibonacci sequence is used, thus favouring the better solutions.
- Exponential : *select* according to rank, but using random numbers drawn from an exponential distribution.
- Tendency : *select* randomly, but in such a way as to favour selection from the best sub-population% of solutions in P, where sub-population% must be entered.
- Exclusive : *select* solutions randomly from the sub-population

The choice of which mechanism is used is achieved using the **Settings** window, described below. See also Figure 2. The choice of the *select* mechanism may be changed during the execution of GAmeter for purposes of experimentation.

Once the "gene pool", Q, has been produced in this way, genetic operators are applied to members of Q to *create* the "offspring pool", R. The operators available in the basic version of GAmeter are the standard crossover and mutation operators, see for example [10]. Once again, however, other genetic operators have been included in some specific implementations of the toolkit. For example, a *trim* operator was introduced in [11, 8] to remove vertices of degree one from the subgraph induced by the chromosome in the Steiner tree problem in graphs.

The final stage of each generation is to *merge* the three sets P, Q and R in some way to arrive at a new population, P, from which to create future generations. The options available in GAmeter for this process are currently

- BestFit : With this mechanism, the best solution in R replaces the worst solution in P until all solutions in P are better than the solutions in R or $R = \emptyset$. Random noise is introduced in the form of an *Inaccuracy% parameter*, with 0% value meaning no noise.
- New Solutions : With this mechanism, only the solutions in R that do not already exist in P and are better than the current members of P are accepted. These accepted solutions replace solutions in P that are worse than them.
- Tendency : Uses the Tendency mechanism to select a number, $n = |Q|$, of solutions in R. The n worst solutions are removed from P and the newly selected ones are inserted in their place.
- Replace : Given an *AcceptPressure* parameter, all solutions in R with fitness value better than the product of this parameter and the fitness value of the worst solution in P, replace the solutions that already exist in P.
- Random : *merge* P and R and then randomly remove solutions from the new set P so as to maintain a fixed population size. The solutions that are allowed to be removed should not belong to Q.

Once again, these are chosen using the **Settings** window; see Figure 2.

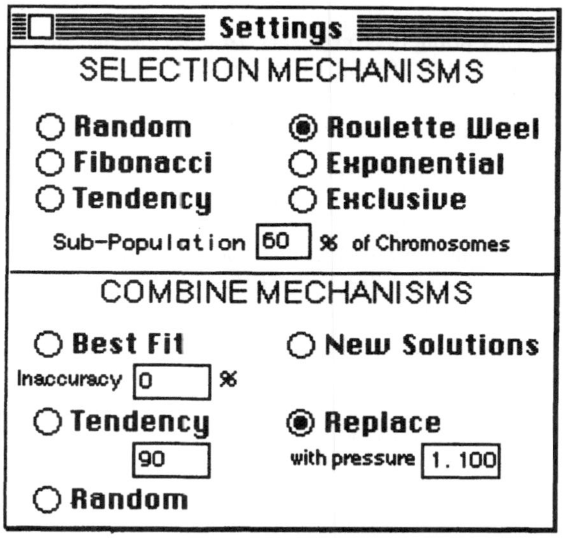

Figure 2: The **Settings** window

2.1 The User's View

At this point it is necessary to describe the area of responsibility of the user in applying GAmeter to the solution of a particular problem. There are three levels of implementing the toolkit. These are as follows:

- The shallow level is a specific instantiation for a particular problem. In this case, the source code is not required and the user need only be given the user manual [12] and details of the structure of the data files in order that he or she can apply the software to his or her own data set.
- The intermediate level is the basic version which allows the user to code an evaluation function pertaining to a problem, compile this and integrate it into the toolkit. The user is not given sufficient information to alter the options available in the kernel software.
- The deepest level is to give the user sufficient information to edit the toolkit in such a way that the alphabet of the chromosome and the functions *initialise*, *select*, *create* and *merge* can be altered to include other options, some of which may be problem specific.

The following description is aimed at the first two levels.

Let *Gene* be the set defining the alphabet for a particular problem. This set is generally user defined but in the basic version of the software is set at {0,1} by default. Let *Chromosome* be a sequence of the elements in *Gene*. The elements of the solution space, S, are of type *Chromosome*. The GA operates within S and therefore the way S corresponds to solutions will necessarily affect the GA's performance. For an alphabet of two elements and a chromosome size n, the size of S is 2^n. The size of the bit string and the details of the representation of solutions by binary strings is necessarily a user decision.

Apart from this chromosomal representation of a solution to a problem, a fitness value for that chromosome is also needed by the GA. This value, which is called *FitnessValue*, indicates how well a particular chromosome describes a solution to the problem at hand. The type of the *FitnessValue* should be capable of representing the entire range of the fitness values of the solution space. In practice

$$FitnessValue = LongInteger$$

is sufficient for most practical purposes and this is the default type. Given a chromosome and its size, the *EvaluateChromosome* function returns the fitness value of the chromosome. This function is problem dependent and must therefore be user-defined. For the top two levels of interaction with GAmeter, these are all the user essentially requires, apart from minor administrative procedures to initialise data structures and read data files, for example.

2.2 The User Interface

The interface is based on windows. Most of the operations are performed, with the help of a pointing device (for example a mouse), by selecting items in pull-down menus or clicking into parts of windows. The toolkit works under the windowing environment of the system for which it is built and does not try to impose its own. Thus, in the Macintosh implementation, the Macintosh window system is used and in the Unix implementation, the X-Windows system is used. What will be described here are the operations common to all implementations.

Once the GAmeter application is loaded, a small **report** window appears near the top of the screen displaying the following information (see also Figure 4):

- the number of the current generation,
- the average fitness value of all solutions in the population pool,
- the change in the average value compared to the previous generation,
- the minimum and maximum fitness values in the pool,
- the number of seconds the optimisation has been running,
- the number of function evaluations,
- the number of evaluations after which the best solution was found
- the number of evaluations performed per second.

All operations and some additional windows are selected/displayed with the use of the following pull-down menus (see Figure 4):

The File Menu contains operations for loading the data for a problem and saving the solution found by the GA, loading and saving a special "state" file that enables the user to stop the optimisation at any point and continue it at some later time, saving a text file describing the progress of the optimisation, and finally quitting the application. The operations to load and save data files are problem dependent and must be defined by the user if required. See [12] for details.

The Optimise Menu is used to start, stop or freeze and restart the optimisation from the beginning. It also contains an operation that copies the contents of the **population** and **graph** windows, described below, to a clipboard so that they can be pasted and transferred to other applications.

The Utilities Menu is used to select the information that will be used by the toolkit to graphically display the progress of the optimisation, manually edit any chromosome in the population pool and set/reset the toolkit to/from experiment mode. When in experiment mode a number of optimisations, specified by the user, are executed automatically. In this mode an optimisation concludes either after a specified amount of time has passed or the specified solution (if known) is found. The ability to edit any chosen chromosome from the current population, P, and replace the new chromosome, complete with fitness value, back into the population, is an ideal tool to help monitor the operation of the application during development for the solution of a new problem.

The Windows Menu is used to display or hide the following windows :

1. The **Population** window (see Figure 4) displays a list of all the solutions in the population pool that exist in a particular generation. The first number in the solution line(s) shows the ranking of the solution in the pool, with 1 meaning the best solution in the pool. This number is followed by the fitness value for the solution and then the bit-string representation of the solution. To make the identification of a particular bit in the bit-string easier the | character is inserted at the beginning and the ending of every 16 bit block. In addition to this solution list, information about the best solution found so far is displayed. This information consists of the bit-string representation and the fitness value of the best solution, the generation that it was found, the number of evaluations and finally the time (in seconds) after which this solution was found.

   The problem-developer has the option of adding as many lines of text as he wishes at the top of this window. This means that problem dependent information can also be included. For the example implementation of the Steiner Problem in Graphs (SPG), one line is added showing the number of vertices, edges and special vertices for the problem being solved (See Figure 4).

   This window is not refreshed every generation but by a process of double clicking inside the window

2. The **Parameters** window is used to display and change the current GA parameters; namely the crossover rate, the mutation rate, the size of the population pool and the minimum and maximum number of solutions allowed to participate in the creation of new solutions. The number of solutions that participate in the creation of an equal number of new solutions is a random value between minimum and maximum, obtained for every generation.

   An added feature of GAmeter is the facility to alter these during an execution, either by use of sliders or by entering the new values in the appropriate box in the window. See Figure 3.

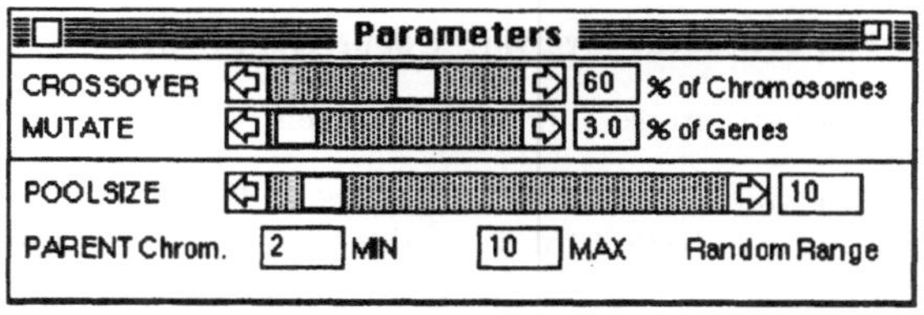

Figure 3: The **Parameters** window

3. As previously discussed, the **Settings** window is used to choose the mechanism that is used to *select* solutions to join the gene pool, Q, and to choose the mechanism which will be used to *merge* the new solutions created, R, with the solutions in P and Q to create the new population, P. See Figure 2. At a more sophisticated level, the user can add other selection and merging options.

4. The **Graph** window (ee Figure 4).is used to display a line-graph with x-axis and y-axis values selected by the user from the **Utilities** menu. Double clicking in the contents of this window updates the graph so that it reflects the current state of optimisation.

The User Menu consists of operations that the user has added to the toolkit. Most of the operations under this menu are problem dependent.

An example of this would be an option to seed the initial population.

All operations described above can be performed while an optimisation is running. As a result, new values of GA related parameters will be applied in the next generation. Thus the toolkit allows dynamic control of the optimisation.

Current implementations of the toolkit include a Pascal version for the Apple Macintosh, a C version for the Unix system and a parallel version running on a Meiko transputer system.

3 Instantiations of GAmeter

GAmeter is now in use at a number of sites in Europe, and is being tailored for a wide variety of problems. Restricting attention to the work at the University of East Anglia, GAmeter has been, or is being, applied to the following problems:

- the Steiner Tree Problem in Graphs
- the Directed Steiner Tree Problem in Graphs
- a non-linear investment appraisal problem
- a non-intersecting paths problem

The Steiner problem in graphs (SPG) is that of finding a subgraph of a weighted graph which contains at least a given set of special nodes and is of minimum edge weight. This is a classic NP-complete problem on which a considerable amount of research effort has been expended (see, for example, [13]).

A GA algorithm to solve this problem has been developed on each of our implementations of the GAmeter toolkit. Results for the Apple MACIIFx implementation are reported in [8]; these results were for the testbed problem set, B , as used by Beasley [14, 15, 16]. Both the Unix and the Transputer implementations solved all these problems to optimality. As might be expected, the Unix implementation when used on a Sun Sparc Station gave an increase in speed of nearly 9 times that of a MACIIFx. Similarly, excellent results have been obtained for the directed Steiner tree problem in graphs, see [9], and for the tax-induced investment appraisal problem, see [17].

GAs yield very naturally to parallelism, see, for example, [18, 19]. The obvious scope for parallelism is that the *EvaluateChromosome* function applied to newly generated solutions can be computed on separate processors. Thus, in the master-slave approach, the slave processors are assigned the task of applying *EvaluateChromosome* to chromosomes distributed by the master processor. The remaining tasks of *initialise*, *select*, *create*, *merge* and *finish* are performed sequentially by the master processor. This approach may result in under-utilisation of the slave processors and is only advisable for problems requiring a considerable amount of processing time to evaluate chromosomes. This is indeed the case for SPG, where an evaluation involves computing the cost of a minimum spanning tree of an induced subgraph.

An alternative paradigm for parallelism uses a multi-GA approach. In this case, every processor runs a complete GA program for the same problem and solutions from neighbouring processors are exchanged on a regular and controlled basis. This approach is likely to be more suitable when evaluation of chromosomes takes relatively little time. We are currently researching to determine precisely when this approach is to be preferred.

In both these approaches, the interface is controlled by another processor which is usually that of the host computer.

Currently, our parallel GAmeter implementation uses the master-slave approach and our experiments have only concerned the SPG problem. Because of the random nature of GAs, the number of evaluations per second has been used to measure the speedups obtained. Speedup critically depends not only upon the number of processors and the size of the "gene pool", Q, but also upon the amount of time required to perform an evaluation. Initial results for the SPG problem are encouraging. For example, a parallel implementation on a Meiko transputer rack produced a speedup of nearly 5 when 6 processors were used and $|Q| = 6$. The speed was similar to that of a Macintosh MACIIFx when 1 processor was used.

4 Conclusions

We would like to summarize our paper with the following points:

- By applying a more abstract approach to genetic algorithms, we see that they are merely a special case of a more general search strategy.
- This abstract approach provides a natural framework for the development of hybrid algorithms, for which toolkits similar to GAmeter can be developed.
- The toolkit, GAmeter, provides a friendly user-interface for genetic algorithms which enables fast and efficient implementations.

- Multi-platform implementations of GAmeter enable simple porting of GA software from one machine to another.
- In particular, parallel versions of the toolkit provide easy access to users with little or no experience of parallel architectures.
- Implementations of a range of problems using GAmeter have proved successful and have been reported

References

[1] Glover F. Tabu search, part 1. *ORSA J. Comp.*, 1:190–206, 1989.

[2] Reeves C.R., editor. *Modern Heuristic Techniques.* Blackwell Scientific, 1993.

[3] Kirkpatrick S., Gelatt C.D., and Vecchi M.P. Optimization by simulated annealing. *Science*, 220(4598):671–680, 1983.

[4] Dowsland K.A. Simulated annealing. In Reeves C.R., editor, *Modern Heuristic Techniques*, chapter 2. Blackwell Scientific, 1993.

[5] Adenso-Diaz B. Restricted neighborhood in the tabu search for the flowshop problem. *EJOR*, 62(1):27–37, 1992.

[6] Osborne L.J. and Gillett B.E. A comparison of two simulated annealing algorithms applied to the directed Steiner problem on networks. *ORSA Journal on Computing*, 3(3):213–225, 1991.

[7] Tam K.W. Genetic algorithms, function optimization, and facility layout design. *EJOR*, 63(2):322–346, 1992.

[8] Kapsalis A., Rayward-Smith V.J., and Smith G.D. Solving the graphical Steiner tree problem using genetic algorithms. *J. Oper. Res. Soc.*, 1993. (to appear).

[9] Kanatis S. The directed Steiner tree problem: An application of genetic algorithms. Master's thesis, University of East Anglia, 1992.

[10] Goldberg D.E. *Genetic Algorithms in Search, Optimization, and Machine Learning.* Addison-Wesley, Reading, Mass., 1989.

[11] Kapsalis A. Steiner trees. Master's thesis, University of East Anglia, 1991.

[12] Kapsalis A. and Smith G. D. The GAmeter toolkit manual. School of information systems, computing science technical report, University of East Anglia, 1992.

[13] Winter P. The Steiner problem in networks: A survey. *Networks*, 17:129–167, 1987.

[14] Beasley J.E. An algorithm for the Steiner problem in graphs. *Networks*, 14:147–159, 1984.

[15] Beasley J.E. An SST-based algorithm for the Steiner problem in graphs. *Networks*, 19:1–16, 1989.

[16] Beasley J.E. Or-library: distributing test problems by electronic mail. Technical report, The management school, Imperial College, 1990.

[17] Berry R.H. and Smith G.D. Using a genetic algorithm to investigate taxation-induced interactions in capital budgeting. Submitted for publication, 1993

[18] Macfarlane D. and East I. An investigation of several parallel genetic algorithms. In *Proc. 12th Occam User group, Technical Meeting*, pages 60–67, 1990.

[19] Petty C.B., Leuze M.R., and Grefenstette J.J. A parallel genetic algorithm. In *Genetic algorithms and their applications: Proc 2nd Int. Conf. Genetic Algorithms*, 1987.

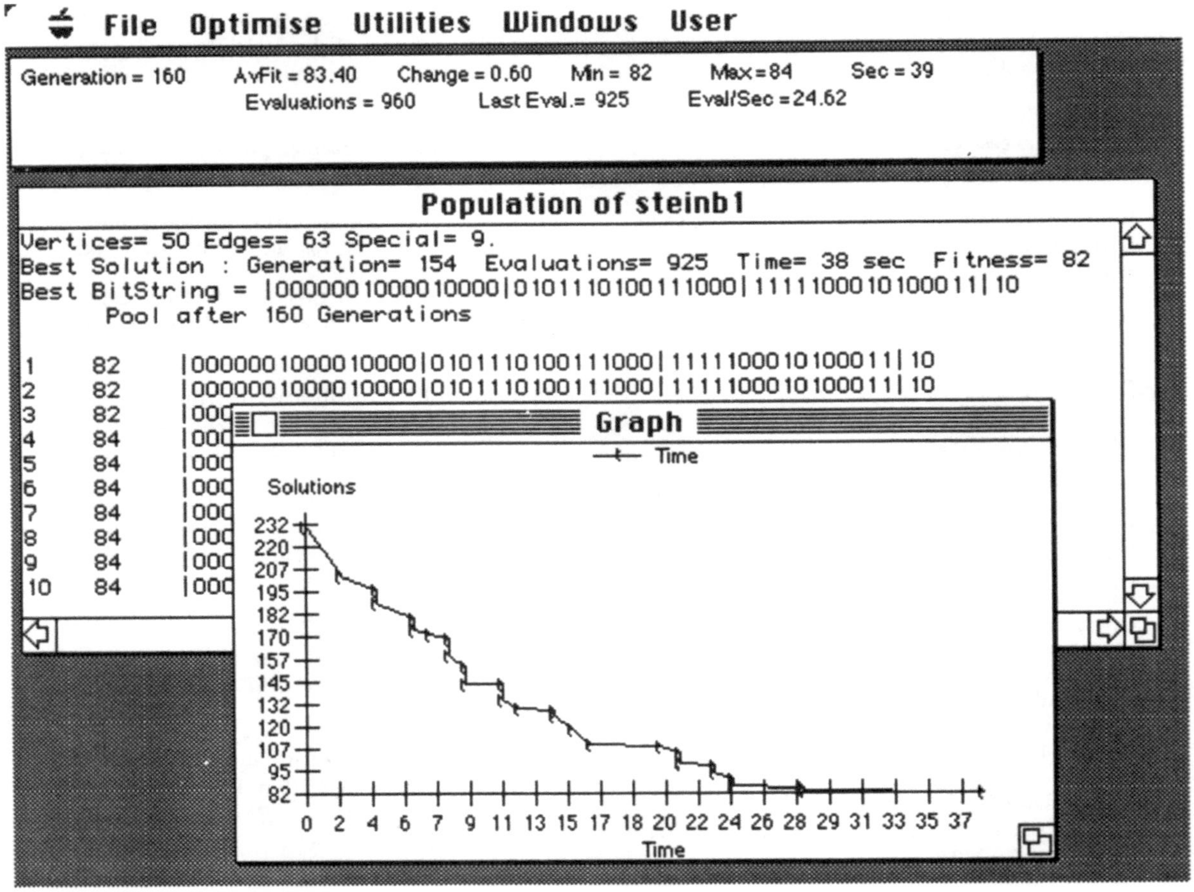

Figure 4: An example screen showing the **report** window at the top of the screen, the **population** window and the **graph** window at the end of a typical run.

Locating Pressure Control Elements for Leakage Minimisation in Water Supply Networks by Genetic Algorithms

K. S. Hindi *   Y. M. Hamam †

1 Abstract

The complex problem of choosing the types of pressure-control elements, locating them and determining their settings in order to minimise leakage in water supply networks is addressed. The difficulty stems from the discrete nature of the choice/location variables, as well as from the fact that the problem is invariably large-scale, due to the size of the networks involved, in addition to being non-linear and non-convex due to the nature of the head-flow relationships. The paper describes the design of a genetic algorithm to solve the problem and compares the results obtained with those obtained by a mixed-integer programming model developed earlier by the authors.

2 Overview of Problem

Losses through leakage from water supply networks can be significant. They are estimated to vary in Britain between 10 % and 50% of the total supply of various water authorities, with an average of the order of 25% [1]. Several factors affect leakage such as the state and quality of pipes and fittings and the characteristics of the soil in which they are laid. In particular, the higher the pressure is, the higher is the rate of leakage [1, 3, 4]. Hence, it is advantageous to reduce pipe pressures as far as possible, particularly that pressure is the only factor among those affecting leakage that can be easily controlled in an existing network. On the other hand, it is necessary to maintain throughout the network pressures sufficient to meet consumer demands.

There is, thus, a need for striking a compromise between these two conflicting objectives. To this end, use is made of two types of control elements: pressure-reducing valves (PRVs) and closed gate valves. The problem of determining the settings of PRVs in order to minimise leakage has been addressed in [12, 5, 11] and by the present authors [8, 9]. The objective of the work reported is to develop a computational technique for solving the associated planning problem described below.

Given a water supply network and a set of representative demands, the planning engineer starts by nominating a set of possible locations (sites) for locating the control elements. At each site, it is possible to install either a PRV, at a certain given cost, or a closed gate valve, also at a given lower cost, or not install either, naturally at no cost. It is, then, required to choose one of these options for each site and determine the appropriate

*Department of Computation, University of Manchester Institute of Science and Technology (UMIST), P.O.Box 88, Manchester M60 1QD, Britain

†Ecole Superieure d'Ingenieurs en Electrotechnique et Electronique, 2 Boulevard Blaise Pascal, B.P.99, 93160 Noisy-Le-Grand, France

settings of the PRVs, in order to minimise a composite cost function consisting of the weighted sum of the capital costs of the valves and the cost of leakage, subject to a large set of non-linear, non-convex constraints. These can be expressed as:

$$C_d Q = -D \qquad (2.1)$$

$$p_a^s - p_a^t = f(q_a) \quad \forall a \in \mathcal{A} \qquad (2.2)$$

$$p_i \geq p_i^{\min} \quad \forall i \in \mathcal{D} \qquad (2.3)$$

$$q_a \text{ unrestricted} \quad \forall a \in \mathcal{A} \qquad (2.4)$$

$$q_l \text{ unrestricted} \quad \forall l \in \mathcal{L} \qquad (2.5)$$

where

$\mathcal{A}$ = set of pipe arcs

$\mathcal{L}$ = set of arcs representing nominated locations (sites) for control elements

$\mathcal{D}$ = set of all nodes excluding source nodes

C_d = arc-node incidence matrix for all nodes $\in \mathcal{D}$ with

$$c_{ij} = \begin{cases} +1 & \text{if flow in arc } j \text{ leaves node } i \\ -1 & \text{if flow in arc } j \text{ enters node } i \\ 0 & \text{otherwise} \end{cases}$$

p_i = head at node i

q_a = flow in arc a

Q = vector of arc flows

D = vector of demands

p_a^s = head at sending (initial) endpoint of arc a

p_a^t = head at receiving (terminal) endpoint of arc a

p_i^{min} = minimum head required at node i

Constraints (2.1) represent the nodal balance equations. Constraints (2.2) represent the head-flow relationships, which are usually taken to be of the form:

$$f(q_a) = kq_a \mid q_a \mid^{\alpha-1} q_a$$

where $1 \leq \alpha \leq 2$ and is usually taken to be 1.85. Constraints (2.3) ensure that the head at each demand node is greater than a specified minimum to ensure security of supply.

Knowing the heads at the sources of supply, the settings of the PRVs and the loads, the equations representing constraints (2.1) and (2.2) can be solved by steady-state, hydraulic network simulation to find the pressures and flows throughout the network. It is then possible to ascertain whether constraints (2.3) are satisfied.

3 Solution by a Genetic Algorithm (GA)

Although a mixed-integer non-linear programming model developed earlier by the authors [10] reliably produces good solutions, attaining global optima is not guaranteed due to lack of convexity; hence the justification for studying the application of genetic algorithms to this problem. This section presents how the issues involved in designing a GA for this purpose were resolved, assuming that the reader is familiar with general outline of GAs.

Overall Procedure Two sets of decisions are involved. The first pertains to the choice of control element to install at each site, if any; and the second relates to the settings of the chosen PRVs, so that leakage losses are minimised. clearly the two sets of decisions are hierarchically related, with the first at the upper level. The solution process mirrors this hierarchy, with the first set of decisions represented by an evolving population of strings. Determining the fitness value of each string is carried out in its turn by a genetic search which strings represent the settings of the PRVs.

Determining the Solution Space As far as the choice of control element is concerned, the solution space is evident; at each site, we have one of three choices: either install a PRV, or install a closed gate valve or install nothing.

The solution space for the settings of the

PRVs is determined by the combined feasible range of settings. The upper limit of the range can be set equal to the value of the highest pressure in the network before any pressure reduction, which can be found by steady-state simulation of the network. The lower limit can be set to the value of the minimum permissible pressure, since, obviously, the pressure setting of any valve cannot, in any feasible solution, be lower than that value.

String Lengths and Population Sizes The length of each string representing a choice of control elements was set such that each site was represented by a 2 bit substring. This would for a rather large problem instance involving 10 sites give a string of 20 bits. The 2 bits cover 4 values, whereas the choices involved are 3. However since in the optimal solution, a larger proportion of the sites will have nothing installed in them, it was decided to consider that a substring valve of 0 or 1 signifies installing nothing, of 2 signifies installing a closed gate valve and of 3 signifies installing a PRV.

The length of each substring representing a PRV was set to 7. This, for an instance involving 5 such valves would involve 35 bits. Since the maximum range over which a valve setting could vary is in the region of 30 m, a 7 bit substring gives a resolution slightly better than 0.25m in the worst case and substantially better in most cases.

The size of the population of the strings at the upper level was set to 40 for problem instances involving 5 sites or less (string length $\leq$ 10 bits) and to 80 for problem instances involving between 6 and 10 sites (12 $\leq$ string length $\leq$ 20). Population size at the lower level was set to 80 for problem instances involving 3 PRVs or less (string length $\leq$ 21) and to 160 for instances involving between 4 and 6 PRVs (28 $\leq$ string length $\leq$ 42).

Fitness Function the objective of the optimisation is to minimise a composite objective function with two weighed parts; the first representing capital cost (i, e., cost of installing gate-valves and PRVs) and the second representing the cost of leakage losses. The leakage losses are usually considered to vary linearly with the pressures in the network. This assumption is normally justified on the basis that the exact relationship between leakage losses and pressures is not known with any degree of certainty. Moreover, the existing optimised planning procedure requires the linearity assumptions, since it employs a linear programming relaxation. Although the GA employed here would work with any computable objective function and hence accept any pressure-leakage loss relationship, linearity was assumed, for the sake of comparing results with those of earlier work.

Since the fitness function needs to be maximised, it was considered to be $M - \Gamma$, where M is a suitably large number and Γ is the value of the objective function. M was chosen to be an upper bound on the value of the objective function and can be calculated simply by assuming the capital cost to be that of installing a PRV at each site and the leakage losses to be those corresponding to no pressure reduction.

In calculating the leakage loss part of the objective function value, if a minimum pressure constraint was violated, then a penalty was added. The penalty was calculated to be

$$0.2M(p_i^{\min} - p_i)^2.$$

Reproduction, Crossover and Mutation The three customary operators of reproduction, crossover and mutation were used to generate successive populations. Based on a normalised fitness $F_n(i)$, the number of offsprings for each individual was calculated. First, the fitness was normalised with the average value of the fitness, $F_n(i) = F(i)/\bar{F}$. The strings with higher-than-average fitness received more than one offspring, and those with below-average fitness less than one offspring on the average. To achieve this, the strings were selected according to what has become known as the stochastic remainder selection without replacement [6]. Accordingly, the strings receive a number of offsprings at least equal to the integer value of their normalised fitness. Then, the population is filled up by choosing another offspring for each of the strings with probability equal to the fractional part of the normalised fitness until the total number of offsprings equals the population size N.

The best string in the population was always identified and if it was not in the new population it was put in, replacing another string chosen at random.

Also to preserve diversity, ranking was introduced [2]. Whenever a certain ratio of the normalised fitness was receiving no offspring, the strings are sorted according to their fitness value. Then, instead of normalising the fitness with the average, the normalised fitness is computed according to

$$F_n(i) = \frac{2(\max -1)}{N-1}\text{rank}(i)+1-(\max -1)\frac{N+1}{N-1}$$

with max user defined and $1 \leq \max \leq 2$. The range of normalised fitness is then $[2 - \max, \max]$.

Crossover and mutation were carried out in the normal way [6], with a crossover frequency $p_c = 0.8$ and a mutation frequency $p_m = 0.01$.

Termination The algorithm was terminated after examining 100 generations at the upper level. Also, each objective function value evaluation, which is equivalent to optimising the settings of the PRVs, was terminated after 100 generations.

Network Simulation Clearly, network simulation has to be carried out very many times. It is thus essential that it be effected efficiently. In the present work, a highly efficient simulator was used [7]. Also the burden is alleviated somewhat by the fact that in many instances, network conditions change very little from one simulation to the next, so that comparatively little computation is needed to update the solution. Nevertheless, a high degree of simulation accuracy is not needed in the present context. It is, therefore, sufficient to employ a larger convergence tolerance than is usual, thereby reducing computation time substantially.

4 Results

A substantial number of problem instances were studied. In each case, solution was carried out by the GA and by the mixed integer programming model developed earlier [10]. Both algorithms gave comparable results. However, the valve settings given by the two procedures were different, though leading to approximately the same leakage loss values. This would seem to indicate that the leakage loss minimisation problem either has many minima or is in fact convex, despite the nonconvexity of the individual head-flow relationships, and is flat bottomed. This is an issue being investigated by the authors.

5 Conclusion

The GA, as described here, does not seem to offer any advantage over the mixed integer programming model developed earlier [10]. However, the following points are noteworthy:

- The possibility that the choice of parameters of the GA could be improved upon can not be discounted at this stage.

- Although the GA is inferior to the mixed integer programming model in terms of computational efficiency on a serial computer, its inherently parallel nature would probably make it very efficient on a parallel computer.
- Unlike the mixed integer programming model which is based on the assumption that leakage losses vary linearly with the weighted sum of pressures, the GA is capable of dealing with any computable leakage loss-pressures relationship.

References

[1] Water Authorities Association (1985) *Leakage control policy and practice.*

[2] Baker, J. E., "Adaptive selection methods for genetic algorithms, " Proc. Int. Conf. Genetic Algorithms and Their Applications, 1985, pp. 101–111.

[3] Bessey, S. G. (1985) Progress in pressure control, Aqua, No. 6, pp. 325–3330

[4] Bessey, S. G. (1985) Some developments in pressure reduction, Journal of the Institution of water Engineers and Scientists, Vol. 39, No. 6, pp. 501–505 (1985).

[5] Germanopoulos and Jowitt, P. J. (1989) Leakage reduction by excess pressure minimization in water supply networks, Proceedings of the Institution of Civil Engineers, Vol. 87, pp. 195–214.

[6] Goldberg, D. E., Genetic algorithms in Search, Optimization and Machine Learning. Reading, MA: Addison-Wesley, 1989.

[7] Hamam, Y. M. and Hindi, K. S. (1989) Steady-state solution of pipe networks: a new efficient optimisation-based algorithm, *DTG Report, Computation Department, UMIST.*

[8] K. S. Hindi and Y. Hamam (1991), 'Pressure Control for Leakage Minimisation in Water Supply Networks, Part 1: Single-Period Models', *International Journal of Systems Science*, Vol. 22, no. 9, pp. 1573–1585.

[9] K. S. Hindi and Y. Hamam (1991), 'Pressure Control for Leakage Minimisation in Water Supply Networks, Part 2: Multi-Period Models', *International Journal of Systems Science*, Vol. 22, no. 9, pp. 1587–1598.

[10] K. S. Hindi and Y. Hamam (1991), 'Locating Pressure Control Elements for Leakage Minimisation in Water Supply Networks: an Optimisation Model', *Engineering Optimisation*, Vol. 17, pp. 281–291, 1991.

[11] Jowitt, P. J. and Xu C. (1990) Optimal valve control in water distribution networks, ASCE Journal of water Resources Planning and Management, Vol. 116, No. 4, pp. 445–473.

[12] Sterling, M. J. H. and Bargiela, A. 'Leakage reduction by optimised control of valves in water networks', Trans. Inst. of Measurement and Control, Vol. 6, No. 6, 1984, pp. 293- 298.

A "Noise Gene" for Econets

Orazio Miglino* Roberto Pedone** Domenico Parisi**

*Department of Psychology, University of Palermo
**Institute of Psychology, National Research Council
e-mail:Roberto@irmkant.bitnet

Abstract

Genetically controlled noise is applied to the weights of neural networks trained with a genetic algorithm. Networks simulate simple organisms living in an environment. Reproduction is based on the ability of each network, during its life, to respond to sensory information from the environment with appropriate motor action. Each network has an amount of noise which is genetically inherited (in the 'noise gene') with mutations and it varies interindividually. Noise modifies the value of a weight differently for each spreading of the activation through the network. Such noise has a positive effect on the evolutionary increase in fitness and it makes fitness less dependent on the initial choice of a random population. Evolutionarily, whatever its initial amount, noise reaches an intermediate amount during the first third of the evolutionary process and then it goes near zero.

1. Introduction

Biological systems contain much noise and their functioning tends to be helped rather than hindered by noise. This is why noise is a useful addition to models of behavioural capacities which are inspired by biological systems while it is usually a problem for models such as symbolic models which are not so inspired. Hanson (1991) [1] has shown that neural networks can learn more easily a particular task using backpropagation if the value of a connection weight is not fixed but is randomly selected from a distribution of such values each time the connection is used to compute the network's output. Nolfi, Parisi, and Pedone [2] present simulations showing that backpropagation learning results in better generalization if some amount of noise is added to the network's input patterns during learning. In a different context, Hinton and Nowlan (1987) [3] have argued that the performance of a genetic algorithm based on selective reproduction and crossover is improved if some noise is introduced in the phenotypes.

In the present paper we apply the idea of genetically controlled noise to ecological neural networks, that is, networks which simulate simple organisms living in an environment (Parisi, Cecconi, and Nolfi, 1990) [4]. A population of such networks reproduces differentially based on the ability of each network during its life to respond to sensory information from the environment with appropriate motor actions. While there is no learning (i.e. directed change in connection weights during life) in such networks, some amount of noise is added to the inherited connection weights each time they are used to determine the network's output (motor action) on the basis of input (sensory information). The amount of noise varies from one individual network to another and it is encoded as a 'noise gene' in the inherited genotype of the individual together with the matrix of connection weights of the individual. Therefore, evolution controls the amount of noise present in the population at any given instant (generation). The aim of the research was to examine how noise changes evolutionarily as a function of various amounts of initial noise and how noise affects the evolution of the capacity to respond with appropriate output to the network's input.

2. Evolution with genetically controlled noise

The ecosystem used in the simulations has been described in Nolfi, Elman, and Parisi (1990) [4],[5],[6]. Individual neural networks live alone in a bidimensional environment for a fixed lifetime (number of activation spreadings). The environment contains randomly distributed food elements. At the of life each network is assigned a fitness value which is the total number of food elements reached (eaten). There are two input units which encode direction and distance with respect to the organism's facing direction of the nearest food element. The two output units encode in a binary fashion one of four possible motor actions: move one step forward, turn 90 degrees right or left, or do nothing. The intermediate layer includes 7 hidden units.

An initial generation of 100 individuals is created all with the same network architecture but randomly generated connection weights. The weights are selected from a flat distribution in the interval between +1 and -1. At the end of their life (5000 cycles) the 20 best networks generate 5 copies each of their weight matrix. This is repeated for 100 generations. Five of the 28 total connection weights are mutated at birth by adding a quantity randomly selected between +1 and -1 to their current value.

In a control simulation no noise is added to a network's weights during the network's life. In the simulations with noise each time a particular weight is used to compute the network's output a quantity is added to the weight's value which is based on two factors. One factor is the weight's value itself. The other factor is the noise gene of the particular individual. The noise gene specifies the interval within which the noise to be added to the weight's value is randomly selected. Noise genes vary from one individual to another and they are inherited from one's parent together the set of connection weights. The noise gene is a parameter which is subject to random mutation like the weights themselves. In other words, the offspring of a particular reproducing individual will have an amount of noise added to the their weights which depends on the noise range specified by the noise gene of their parent - except for mutation of the inherited parameter.

Technically, the noise value is randomly chosen with a flat distribution within the following interval:

$$+ (NG * W_i) <= DW <= - (NG * W_i)$$

where NG is the noise gene, W_i is the network's ith weight, and DW the quantity to be added to the W weight.

The noise gene is a value included between 0 and 1. A different quantity of noise is added to each weight at each spreading. For example, given a noise gene of .4 for a particular network and a weight value of .9 for a particular connection of the network, in a particular spreading a quantity of noise randomly chosen in the interval between +.36 and -.36 is added to the weight's value. Each reproducing parent transmits to offspring its NG value with a mutation randomly selected between +.1 and -.1.

We have run three sets of simulations with noise by varying the noise gene's initial value. In one set of simulations the noise gene is initially at zero value for all individuals but mutations can change this value so that noise can be added to networks with an inherited noise gene's value greater than zero. In a second set of simulations the initial value of the noise gene is maximum, i.e. 1, for all networks. In a third set it is intermediate, that is, .5. For each of the four conditions (no noise and three levels of initial noise) we have run four simulations with different initial populations of weight matrices (different random seeds).

Figure 1 shows how fitness values change across the 100 generations for the simulation without noise and for those with three different initial noise values (.0, .5, 1). The results are presented separately for each of the four different initial populations. Each graph presents data concerning the average fitness and the fitness of the best individual for each generation. The same data are summarized in Figures 2 and 3 averaging on the four different initial populations.

Two results emerge from these data. The first result is that noise tends to have a beneficial effect on fitness, that is, on the evolutionary increase in the capacity to approach food based on information on food location. This effect concerns both average and peak fitness and it tends to be restricted to the initial evolutionary period of fitness increase whereas it almost disappears in the final stages when fitness values become more stable (Figures 2 and 3).

The second result is that noise tends to decrease the dependency on initial conditions, i.e. on the initial population of weight matrices, and to reduce the variability from one generation to the following one of the fitness values of the best individuals (Figure 1). In fact, the fitness curves for the four different populations tend to be more similar in the simulations with noise than in those without noise, and the similarity increases with increasing initial noise. Furthermore, the shape of the curves for the best individuals tends to be more rugged in the simulations without noise and to become more smooth with increasing initial noise.

In the simulations which include a noise gene the initial value of the noise gene varies in the three sets of simulations but it then is left free to vary. It is therefore possible to determine the evolutionary course of the value of the noise gene in the three different cases. Figure 4 shows how the average value of the noise gene changes across the 100 generations for the simulations where the initial value of the gene is zero. (This curve is based on the average of the four simulations with different initial populations.) This figure clearly shows that the quantity of noise (range within which noise values are randomly selected) increases during the first third of evolutionary time (up to around generation 30) and then it decreases until it stabilizes at values near to zero in the last third of evolution. (Zero is never reached presumably because of mutations.)

Figures 5 and 6 present the same data for the simulations with intermediate and maximum initial value of noise. It appears that, whatever the initial value, the average value of the noise gene tends to stay or to reach an intermediate level for the first third of evolutionary time and then to decrease to almost zero, a value which remains constant until the end of evolution.

3. Discussion

Adding genetically modulated noise to the functioning of neural networks has a positive effect on the evolutionary increase in performance of a population of networks which reproduce selectively and with random mutations. Furthermore, noise appears to make the evolutionary changes in fitness less dependent on initial conditions, i.e. on the particular random number generator which is used to generate the initial population of weight matrices, and to reduce variability in maximum performance in successive generations. Finally, since noise level is itself a parameter which varies inter-individually and is subject to evolution, we observe a tendency for evolution to generate an intermediate level of noise in the first third of evolutionary time, then a reduction of noise level until a quasi-zero level of noise stabilizes in the last third of evolution. These results emerge whatever the initial noise level in the population (zero, intermediate, maximum).

Many of these results can be explained if we represent individual networks (weight matrices) as points in a multidimensional space of weight values. The space represents all possible weight matrices. Each point (weight matrix) corresponds to a given height (fitness value) on the fitness surface. However, while networks without noise (our first, baseline, simulation) are represented by a single point in weight space which corresponds to a single fitness value on the fitness surface, networks with noise added to their functioning during life are represented by a cloud of points around the inherited point (weight matrix) in weight space. Consequently, their fitness value will be the average of the fitness values of the different points in the cloud. This allows networks with noise to explore the portion of the fitness surface surrounding their inherited fitness value (surrounding region). More specifically, networks with better surrounding regions (i.e. regions containing high fitness values) will be more likely to leave offspring than networks with less good surrounding regions. This has a beneficial effect on evolutionary increase in fitness because, as a result of mutations, networks with better surrounding regions are also more likely to have good offspring than networks with less good surrounding regions (Parisi, Nolfi, and Cecconi, 1992) [6].

The exploration of surrounding regions can also explain why noise can make evolution less dependent on initial conditions. Different initial populations correspond to different distributions in weight space of the initial generation of individuals and therefore to different distributions on the fitness surface. However, by allowing an exploration of larger portions of the fitness surface (surrounding regions) and by making computed fitness values an average of many different values, noise tends to make fitness surfaces more smooth and therefore more similar indipendently of initial distribution.

Finally, the evolutionary process appears to be able to regulate noise level in the population so that, whatever the initial level, an intermediate level of noise is present in the first stages of evolution and then noise is almost non-existent in the later stages. We conclude that noise tends to be useful during the stages in which evolutionary learning occurs but then it becomes a liability when there is not much more to be learned and it disappears. This final result can also be explained by using the notions of weight space, fitness surface, and evolutionary movement in weight space and, therefore, also on the fitness surface. The phase in which evolutionary learning does occur (the first third of evolution) is that during which individual points (networks) tend to move, generation after generation, to higher levels on the fitness surface. Hence, groups of networks tend to concentrate on elevated regions of the fitness surface (hilltops) and to abandon lower regions. Fitness ceases to increase when this process has reached a point when further movement cannot but bring networks to lower fitness values. No further improvement is possible. There may be higher regions on the fitness surface which are unexplored and uninhabitated but they are too distant to be reached with available mutations.

It is clear that noise is useful to explore higher regions of the fitness surface when such regions are reachable. This is what happens during the first third of evolution in our simulations, when most of the evolutionary increase in fitness occurs. However, when no more higher regions exist which can be explored, noise can only damage performance by pushing networks towards lower regions. Therefore, selective reproduction and mutation acting on noise genes operate together to reduce noise level in the population until a stable level of almost zero noise is reached.

References

[1] Hanson, S.J. *A stochastic version of the delta rule.* In S. Forrest (ed.) Emergent Computation. Cambridge, Mass., MIT Press, 1991.

[2] Nolfi, S., Parisi, D., and Pedone, R., *How noise helps generalization in feed-forward networks.* In E. Caianiello (ed.) Parallel Architectures and Neural Networks. Singapore, World Scientific, 1992.

[3] Hinton, G.E. and Nowlan, S.J. *How learning can guide evolution.* Complex Systems, **1**, pp. 495-502., 1987.

[4] Parisi, D., Cecconi, F., and Nolfi, S. *ECONETS: neural networks that learn in an environment.* Network, **2**, pp. 149-168, 1990.

[5] Nolfi, S., Elman, J.L., and Parisi, D. *Learning and Evolution in Neural Networks.* CRL Technical Report 9019, University of California, San Diego., 1990.

[6] Parisi, D., Nolfi, S., and Cecconi, D. *Learning, behavior, and evolution.* In P. Bourgine and F. Varela (eds.) Towards a Practice of Autonomous Systems. Cambridge, Mass., MIT Press, 1992.

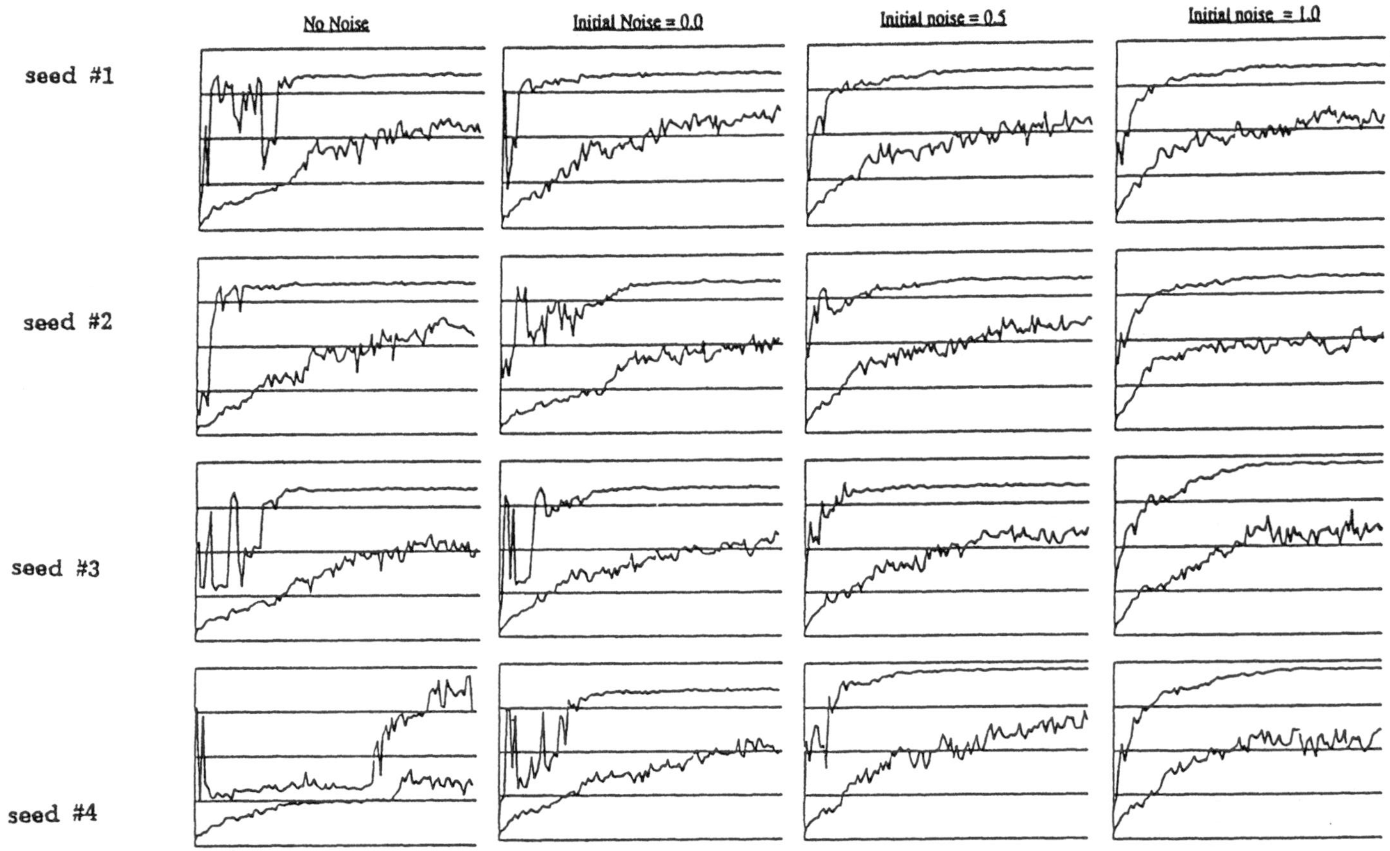

Fig. 1 : Changes in fitness values (y) across evolutionary time (x) for average and best individuals, four different initial populations (seeds): no noise population and three populations with different initial noise levels (0.0; 0.5; 1.0).

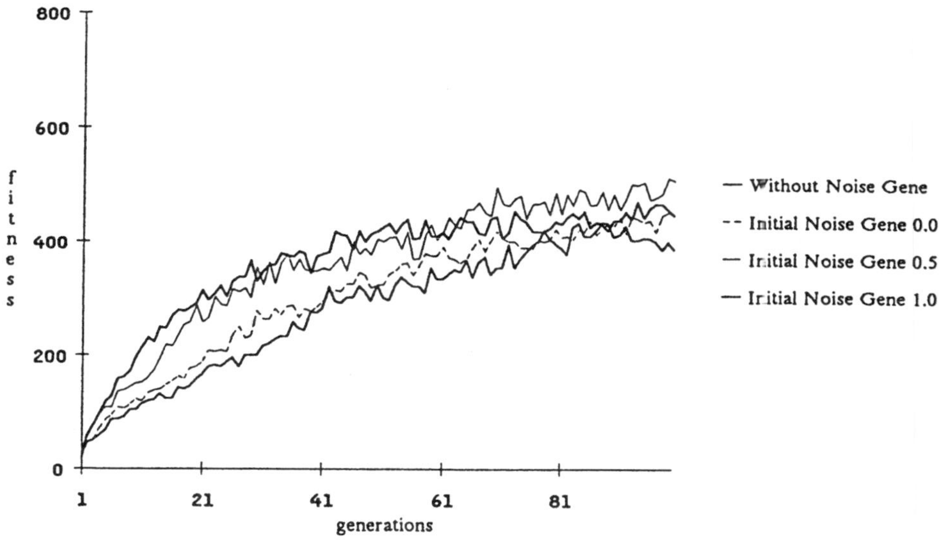

Figure 2: Changes in fitness values for the average individual across 100 generations for the no-noise populations and the noise gene populations. Each curve is the average of 4 different simulations with different initial seed.

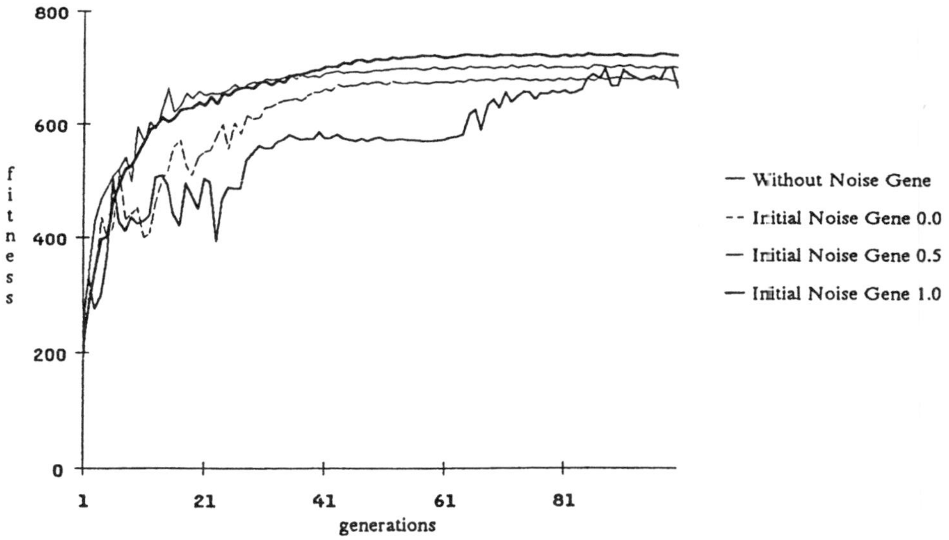

Figure 3: Changes in fitness values for the best individual across 100 generations for the no-noise population and the noise gene populations. Each curve is the average of 4 different simulations with differentt initial seed.

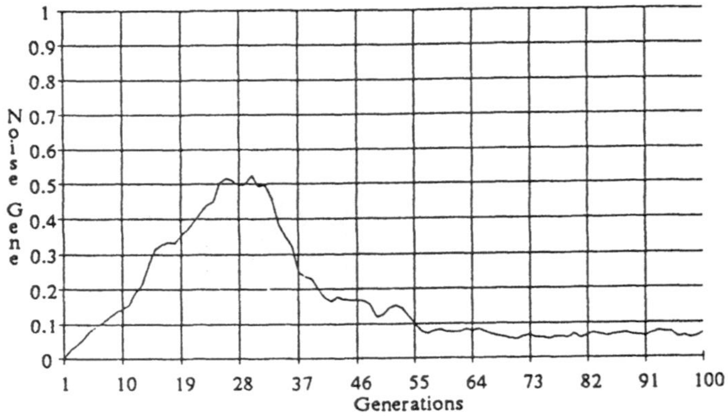

Figure 4: Changes in noise level for the population with zero initial level of noise (average of 4 different simulations).

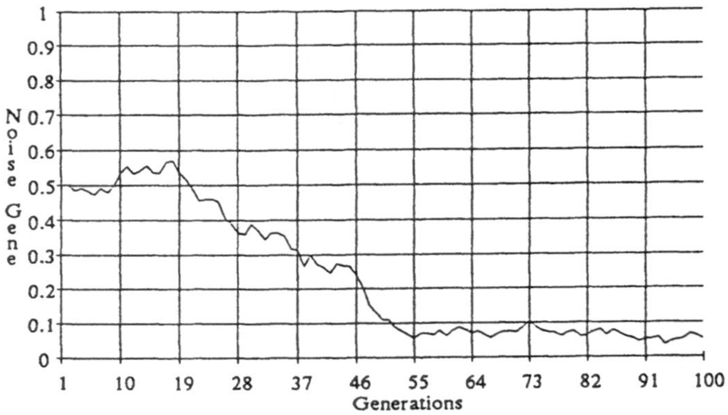

Figure 5: Changes in noise level for the population with 0.5 initial level of noise (average of 4 different simulations).

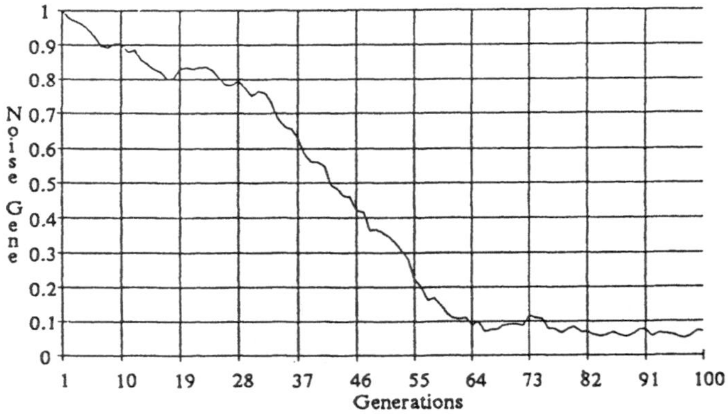

Figure 6: Changes in noise level for the population with 1.0 initial level of noise (average of 4 different simulations).

A Solution to Global Illumination by Genetic Algorithms

Brigitta Lange, Christoph Hornung

Fraunhofer-Institut für Graphische Datenverarbeitung
Wilhelminenstr. 7
D-6100 Darmstadt
F.R.G.
E-Mail: lange@igd.fhg.de

Abstract

A new approach to optimize the computer simulation of radiant light transfer by means of evolutionary techniques for the generation of photorealistic images is introduced.
The formulation of radiant light transfer in a model leads to a system of complex integral equations, which currently have been solved by Monte Carlo Methods. One of the major problems in Monte Carlo sampling is to determine the location and density of sample points in order to reduce the variance of the estimates.
Here a solution is provided by applying evolution strategies to calculate the global illumination. Thus exploiting the search space, i.e. the hemisphere of incident radiation to a point on a surface in a very efficient way through maintaining populations of rays and applying selfadaptive genetic recombination operators.
The simulation process now becomes selforganizing and the transition of one state into another is no longer independent of previous states which allows the system to adjust optimally to a particular lighting situation.

1 Introduction

The production of realistic images requires the ability to simulate the propagation of light in an environment, i.e. the ability to completely account for the global illumination arising from complex interreflections within the environment.
In principal it should be possible to trace the progress of radiant energy through any system by applying *Maxwell's Equations* and the associated boundary conditions. In practice, however, this is often an impractical if not impossible task. Each of the incoming directional intensities is a reflected intensity leaving another surface in the surrounding. This leads to a complex expression since the outgoing directional intensities must be solved for all surfaces simultaneously.
An alternative approximation to the Maxwell Equations for the simulation of radiant light transfer is referred to as *Rendering Equation* [7]. This equation is helpful in providing a physical picture of the light transport. By invoking a probabilistic model of the radiation exchange process and applying *Monte Carlo* sampling techniques it is possible to approximate the Rendering equation.
The fundamental problem encountered in the simulation of radiant light transfer is *sampling*. That is, where should the light intensity function be sampled to guarantee some bound on the variance with the fewest number of sample points? A related problem is the determination of the solution error at any point.
The techniques presented here contribute towards solutions to these problems by enabling "*evolution*" of radiant light transfer models. In contrast to previous models they exploit in a very efficient way information about local and global intensity extrema during the simulation process thus allowing an optimal adaptation of the propagation model to the specific lighting situation without having to make in advance any assumptions about the light distribution and without any expensive book-keeping of important lighting directions. Thus a better convergence towards the solution of the integral equation is achieved which in turn results in an improvement of image quality.
In the following a survey of light transfer models and

related simulation techniques is given and a new approach of simulating the global illumination by means of *evolutionary algorithms* is presented. This approach evaluates the extremely noisy and discontinuous objective function recursively by using random choice and *stochastic ray tracing* as a tool to guide a highly exploitative search through the parameter space, i.e. the hemisphere of incident ray directions.

2 Radiant Light Transfer Models

Rendering is the process of creating an image from a model. In computer graphics, an image is rendered by a computer program which simulates the radiant light transfer in a mathematical model and records the results digitally. The light transfer model and the simulation algorithm used to solve it are the most important characteristics of a renderer. The light transfer model governs the quality of the simulation results. The simulation algorithm affects the simulation speed and the accuracy of the results. Radiant light transfer between surfaces can be characterized by an integral equation. In classical radiation theory, the equation models the intensity of light leaving a point on a surface in a particular direction [12].
The directional intensity of a surface is the sum of the emitted intensity and the reflected intensity in that direction. To obtain the reflected intensity I of an elemental surface area dA in direction (θ_r,φ_r) (see Figure 2-1) from energy incident from all directions the reflected intensity has to be integrated over the total hemisphere Ω of incoming intensity from all directions $I(\theta,\varphi)$, which leads to the following integral equation:

$$I(\theta_r,\varphi_r) = \varepsilon(\theta_r,\varphi_r)+\int_{\Omega}\rho\,(\theta_r,\varphi_r,\theta,\varphi)\,I(\theta,\varphi)\cos\theta\,d\omega; \tag{2.0}$$

where $\varepsilon(\theta_r,\varphi_r)$ is the emitted intensity in direction (θ_r,φ_r) and $\rho(\theta_r,\varphi_r,\theta,\varphi)$ is the bidirectional reflectance of a surface.

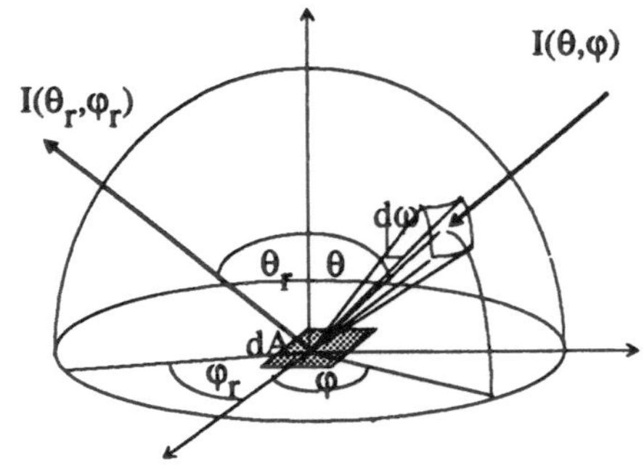

Figure 2-1 Geometry for radiant light transfer calculations

The Rendering Equation, is an alternative form of the integral equation presented above [7] which subsumes nearly all existing illumination models used in image synthesis. Algorithms such as classical *ray tracing* [15], and *radiosity* [4] can be viewed as approximations to global illumination in which specific simplifying assumptions lead to tractable methods of solution.
Since it is not possible to find a closed form analytical solution, there are two general methods for solving the integral equation governing global illumination: finite element methods and Monte Carlo methods. The former approach yields radiosity algorithms where the light leaving the light source is followed in a forward direction and the latter approach yields ray tracing algorithms which start from the eye and follow light paths in reverse direction.
Finite element methods are most appropriate for scenes containing only diffuse surfaces because then the radiance is a function of surface position only.
Monte Carlo Methods are usually used for general scenes, where the radiance is a function that depends not only on surface position but also on view-direction. Thus stochastic ray tracing procedures can be applied that will be discussed in the following section.

2.1 Monte Carlo Simulation

For a point on a reflecting surface the total hemisphere of incoming radiation has to be accounted for in order to calculate the emitted energy. But it is impossible to follow all of the rays needed to simulate the radiative exchange process. Thus we use Monte Carlo methods and *path-tracing* [7] to sample the hemisphere of incoming directions for a point on a surface by a finite set of randomly selected rays. Although it is possible to approximate a solution to the Integral Equation using uniform stochastic sampling and sample-mean Monte Carlo integration [11] the convergence under most conditions is so slow that such a solution is impractical.

The key to fast convergence is in deciding what are the best sample locations, that is in the context of the rendering equation altering the probability density in such a way, that the samples are distributed in directions in which the incident illuminance is large. But such a process is infeasible, since it would require previous knowledge about the illuminance distribution.

There have been made several attempts to increase the statistical efficiency of the Monte Carlo solution through well known techniques like *importance sampling* or *stratification* ([14] [8] [1]). The former technique alters the sampling distribution in order to invest most of the computation effort in collecting samples that contribute significantly to the solution. Usually this is achieved by distributing the samples according to the surfaces reflection properties. The latter is very effective if the domain of integration can be partitioned into strata within which the variance is smaller than the difference between their means. In such a case estimating the integral of each stratum independently is superior to integrating over the entire domain. In the context of rendering stratification is preformed by estimating the integral of direct and indirect illuminance independently. Although these techniques reduce the number of samples they heavily rely on a priory information, i.e. using known properties of a given scene and structuring the sampling and reconstruction process and even by combining stratification techniques and importance sampling it requires tracing hundred's of rays to reduce the variance of the indirect illuminance to tolerable levels.

Since the intensity distribution of indirect illuminance over the hemisphere above a point on a surface is in general a noisy discontinuous objective function, having many local peaks of different magnitude and both, location and magnitude of these local intensity maxima are unknown in advance we have to apply a global search technique which is superior to Monte Carlo sampling, in that the majority of samples can be located around all of these local peaks. Due to path-tracing, Monte Carlo methods are prone to errors in that concrete conclusions about illuminance on higher levels in the path are impossible on a single trial of a stochastic process.

In global illumination simulation we have to find a procedure enabling the illuminance calculation for every point to adjust efficiently to the local environment. Such a procedure is presented in the following sections.

3 Genetic Algorithms

In nature, evolution, which is the process of adaptation of living organisms, can be regarded as a very powerful optimization method. Thus developing nature analog problem solving strategies seems to be promising. Genetic algorithms are search algorithms with a great versatility, that mimic the effects of evolution and natural selection. They are based on *genetic operators* and differ from normal optimization and search procedures in that they work with the coding of a parameter set and that they start their search from a whole population of points, not a single point. Furthermore they do not rely on any auxiliary information or any deterministic rules.

Genetic algorithms have shown to be a useful method of searching large spaces using simulated systems of variation and selection. Thus they seem to be quite suitable in the process of sampling the space of incoming light to a point on a surface and have potential to improve the simulation process towards optimal convergence.

Genetic algorithms achieve much of their breadth by ignoring information except that concerning payoff and it is proven that they can tolerate extremely noisy function evaluation, if this in turn permits resources to be used for exploring other points in the search space [3]. Furthermore they find near optimal results quickly, after searching only small portions of the search space. Due to these properties they are perfectly suited to optimize the simulation of global illumination in order to approximate the Rendering Equation.

First we will examine how genetic algorithms apply to the calculation of the indirect illuminance and then an implementation concept is presented.

3.1 Evolving Incident Radiation

A genetic algorithm has to be designed that produces successively better approximations to the integral equation by exploring the total hemisphere of incident radiation searching for those solid angles or ray direc-

tions that contribute significantly to the illuminance, i.e. light rays with maximum reflected intensity. Furthermore the genetic process has to adapt itself to the different local illuminance situations (for instance convergence towards local optima, leaving of local optima). Thus the genetic operators have to be changed during the runtime of the process.
The intensity equation represents the *phenotype* of the global illumination problem whereas the *genotype* is the coding of the different parameters of the system. The free parameters or variables of the global illumination system are light rays that can be represented by ray directions (i.e. vector quantities) which are defined over the space of the total hemisphere of incoming directions. A floating point representation of the ray directions is more appropriate than a binary representation, since it allows a problem specific design of the genetic operators. Furthermore rays that are close to each other in the representation space are also close in the problem space [10].
An initial population of individuals representing ray directions can be produced by generating a fixed number of random samples equally distributed over the hemisphere where the number of rays should vary according to the particular lighting situation.
A new generation is produced by recombination of individuals (i.e. by mating only within certain 'solid angles'). In order to increase the overall fitness of the next population the process of selection has to be such that the next generation rays are chosen in those directions where they reduce the variance of the total sample set.
The fitness or quality of each individual is obtained by substituting the set of light ray intensities into the objective function of the problem and calculating the difference between the population mean and the individuals intensity, where the fitness is in inversely proprtional to the difference.
The accurate evaluation of the objective function in the simulation of global illumination in order to produce a meaningful fitness value is probably the most challenging part in the design of the genetic process. Thus we will proceed with an examination of the objective function.

3.2 Recursive Objective Function Evaluation

The *indirect illumination* of a point visible to a pixel can be calculated by approximating the integral over the hemisphere of incoming directions (θ_j, φ_j), $j = 1,\ldots,N$ with a discrete set of N samples that do not intersect the light sources. (See Figure 3-1)

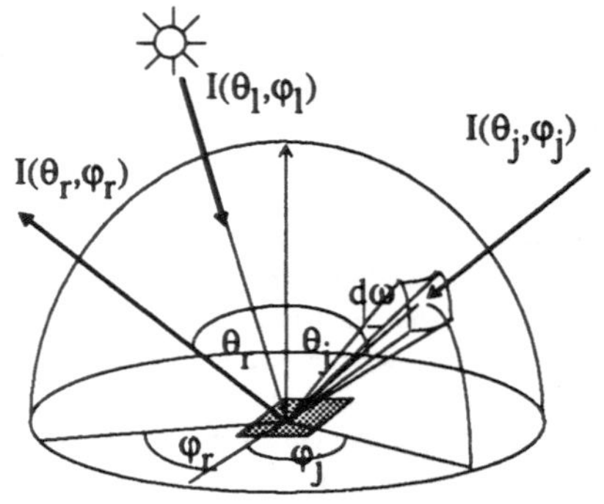

Figure 3-1 Discrete set of sample rays over the total hemisphere of incoming directions

The objective function thus can be expressed as:

$$I(\theta_r,\varphi_r) = (1-w) \cdot \sum_{j=1}^{N} \frac{I(\theta_j,\varphi_j) \cdot \rho(\theta_r, \varphi_r, \theta_j, \varphi_j)}{\sum_{j=1}^{N} \rho(\theta_r, \varphi_r, \theta_j, \varphi_j)}$$

$$+ w \cdot \sum_{l=1}^{L} I(\theta_l,\varphi_l) \cdot \rho(\theta_r, \varphi_r, \theta_l, \varphi_l);$$

(3.0)

where $I(\theta_r,\varphi_r)$ is the intensity reflected from the point on a surface in the direction of the incident ray; $I(\theta_j,\varphi_j)$ is the intensity of a sample ray obtained by recursively evaluating equation (3.0), since it is the reflected intensity of another point in the environment.

$\rho(\theta_r,\varphi_r,\theta_j,\varphi_j)$ is the bidirectional reflectance function of the surface. $I(\theta_l,\varphi_l)$ is the intensity from the light source direction, that varies also with the bidirectional reflectance function of its incident angles, to keep the variance low the light sources are calculated in an extra component, where L is the number of light sources. In order to conserve energy, the contributions of indirect and direct illuminance have to be weighted by w, which is a fator in accordance to their solid angles.
Separating the direct lighting calculations from the stochastic process will reduce the number of samples and increase accuracy. Thus we obtain a constrained optimization problem with the constraint that no ray may intersect a light source. By applying penalty methods [3] this problem can be transformed into an unconstrained search technique.
The genetic algorithm is now used to calculate the indirect illuminance (i.e. the first term of equation 3.0), where the discrete parameter set of ray directions has to be optimized such that for a fixed number of sample rays equation (3.0) becomes the best approximation to the integral equation (2.0).
The determination of the intensity of a sample ray is again an optimization problem where indirect illuminance has to be approximated by sample rays on higher levels, i.e. other points in the environment.
A first approach to calculate the fitness value of these secondary rays in the path is by applying the same stochastic ray tracing procedure as used in the Monte Carlo process [9]. This would result in rapid calculation but coarse approximations of higher level ray intensities due to taking only one random sample at each level. Consequently we have to apply the genetic algorithm recursively in order to obtain reasonable fitness values. Thus at each level of the path a genetic algorithm is used to evaluate the indirect illuminance. With the implication that the fitness value used for the selection process can only be determined if each of the corresponding higher level genetic algorithms has already converged.
Since higher level fitness values contribute proportionally less to the final fitness the demands on convergence behavior of the genetic algorithms on these levels can be rated low. Therefore genetic algorithms on higher levels require much less generations and the computational cost genetic algorithm declines from level to level. In order to stop the simulation process the path length has to be finite thus the path depth is controlled by a fixed maximum level of recursion.

3.3 Implementation Concept

Since the evaluation of ray directions by ray-tracing is computationally very expensive, the goal in global illumination calculations is to achieve a greater confidence about the approximation results by finding an optimal sample distribution for a fixed number of sample rays instead of gaining confidence by increasing the number of rays (i. e. supersampling).
To find the optimal distribution of N rays for a point of a surface by means of evolutionary techniques two different strategies have been followed.
In a first approach it is assumed, that the initial population Ω is represented by a set of random ray directions equally distributed over the total hemisphere of a surface point visible to the eye:

$$\Omega = \{(\theta_1,\varphi_1),(\theta_2,\varphi_2),\ldots,(\theta_N,\varphi_N)\},$$

with

$$\theta_i \in [0,\frac{\pi}{2}] \wedge \varphi_i \in [0,2\pi], \forall i \in \{1,\ldots,N\}. \tag{3.1}$$

Each individual ray is then evaluated by the ray-tracing procedure, resulting in an intensity value that is weighted according to the bidirectional reflectance function. The fitness of a ray is determined by the ratio of the rays reflected intensity to an assumed background intensity. This is due to the fact, that rays hitting no objects return a constant background color. Therefore the overall contribution of the background intensity to the hemisphere can be calculated in advance and the efforts of the genetic search procedure can be spend on those directions where the intensity differs significantly from the background intensity, i.e. directions where the gain in information for the evaluation of the objective function (reflected intensity) is high.
To produce successive generations a modified genetic algorithm [10] with the three operators *reproduction* (roulette wheel selection), *arithmetical crossover* with constrained random mating, and *mutation* is applied to the stochastic ray tracing program.
A new population is formed from the old one by selecting independently $n < N$ rays via roulette wheel for recombination and n distinct rays to die. This selection procedure has the advantage that now only the n new

individuals that have been generated through recombionation operators (crossover and mutation) have to be evaluated by the ray-tracing procedure, since the remaining (N-n) rays are just copied into the new population. Thus the computational cost decreases.

Arithmetical crossover is performed by weighted linear combination of two ray-directions (i.e. vector addition with scaling of the ray directions by a factor proportional to their fitness) and results in one intermediate offspring being closer to its fitter parent. Since crossing of two rays with a far distance to each other gives no rise to hope that the new offspring delivers a better result, crossover is only allowed for rays lying within the same solid angle.

The mutation of ray-directions is performed randomly within a predefined maximum mutation range (i.e. a 3D-box) in order to guarantee that the mutated ray lies in the convex hull of the hemisphere.

All recombination operators are applied to successive populations of rays and are self-adaptive in that the probabilities for mutation and crossover and also the mutation range and the size of the solid angle within a crossover is performed adjust appropriately to the population state.

Since the illumination is a multiple-peak function with many local maxima the simple GA will distribute almost all of its points around the highest peak [3]. But in global illumination we are interested to find all intensity extrema in order to keep variance low. By theory of niche and species we apply advanced operators like crowding [6] and sharing [2] so that subpopulations can be allocated to those solid angles with peak variances in intensity in proportion to the magnitude of the solid angle.

Based on the observation that the indirect illuminance tends to change slowly over a surface ([13], [14]) because the direct component has already been accounted for; we initialize our start populations for every pixel with the fittest "sample locations" of its neighboring pixels thus starting with an usable estimate from previous calculations and avoiding aliasing artefacts. The size of the start population is varied by the mean number of high quality rays of neighboring pixels.

The above described optimization technique can also be interpreted as a *(1,1)-evolution strategy* [10], where the population of rays is seen as a single individual which undergoes mutation. since not a single ray but only the total set of rays form a potential solution to the objective function.

In a second approach to global illumination calculation the initial population consists of a set of individuals Ω_i with i = 1,..., N, where each individual is defined as a set of random rays distributed over the hemisphere following equation (3.1). Thus each individual is a potential solution to the objective function (3.0), and its fitness is the inverse ratio to the deviation from the population means. This multimembered evolution strategy is performed with genetic operators that compare with those of the first approach.

4 Conclusion and Future Directions

In global illumination the objective is to calculate the indirect illuminance by approximating the integral equation with a discrete set of sample rays that do not intersect light sources. In order to minimize the variance we have to concentrate the distribution of the sample rays in those parts of the hemisphere that are of major importance in terms of contribution to the reflected intensity and relatively few rays in directions where the incoming light is sparse. By recursively applying genetic algorithms a very efficient and adaptive simulation of global illumination can be achieved.

In our future research in genetic algorithms we will concentrate on developing advanced genetic operators to improve the performance and apply genetic algorithms also in the process of sampling the direct light sources. Because genetic algorithms are highly suitable for parallelization an implementation of the genetic algorithm on a system of parallel processors will follow.

References

[1] Drettakis; G.; Fiume, E., *'Structure-Directed Sampling, Reconstruction and Data Representation for Global Illumination'*, Proceedings of the Second Eurographics Workshop on Rendering, 1991

[2] Goldberg, D.E.; Richardson, J.Holland, J.H., *'Genetic algorithms with sharing for multimodal function optimization'*, Genetic Algorithms and their applications: Proceedings of the Second International Conference on Genetic Algorithms, 1987, pp. 41-49

[3] Goldberg, D.E., *'Genetic Algorithms in Search, Optimization and Machine Learning'*, Addison-Wesley Publishing Company, Inc., 1989

[4] Goral, C.M.; Torrance, K.E.; Greenberg, D.P.; Battaile, B., *'Modeling the Interaction of Light Between Diffuse Surfaces'*, Proceedings of SIGGRAPH'84, In Computer Graphics, Vol. 18, No. 3, July 1984, pp. 213-222

[5] Heistermann, J., *'Zur Theorie Genetischer Algorithmen'*, Interne Berichte am Fachbereich Informatik der Univ. Frankfurt, 6/1991

[6] De Jong, K.A., *'An Analysis of the behaviour of a class of genetic adaptive systems'*,Dissertation Abstracts International 36(10), 5140B; Doctoral Dissertation, University of Michigan, 1975.

[7] Kajiya, J.T., *'The Rendering Equation'*, Proceedings of SIGGRAPH'86, In Computer Graphics, Vol. 20, No. 4, August 1986, pp. 143-150

[8] Kirk, D.; Arvo, J., *'Unbiased Variance Reduction for Global Illumination'*, Proceedings of the Second Eurographics Workshop on Rendering, 1991

[9] Lange, B., *'The Simulation of Radiant Light Transfer with Stochastic Ray-Tracing'*, Proceedings of the Second Eurographics Workshop on Rendering, 1991

[10] Michalewicz, Z., *'Genetic Algorithms + Data Structures = Evolution Programs'*, Springer-Verlag, 1992

[11] Rubinstein, R.Y., *'Simulation and the Monte Carlo Method'*, John Wiley & Sons, 1981

[12] Siegel, R.; Howell, J.R., *'Thermal Radiation Heat Transfer'*, Hemisphere Publishing Corporation, Washington DC., 1981

[13] Ward, G.J.; Rubinstein, F.M.; Clear, R.D., *'A ray tracing solution for diffuse interreflection'*, Proceedings of SIGGRAPH'88, In Computer Graphics, Vol. 22, No.4, August 1988, pp. 85-92

[14] Ward, G.J., *'The RADIANCE Lighting Simulation System'*, Global Illumination, ACM SIGGRAPH'92, Course Notes of the 19. Annual Conference&Exhibition on Computer Graphics and Interactive Techniques, July 1992

[15] Whitted, T., *'An Improved Illumination Model for Shaded Display'*, Communications of the ACM, Vol. 23, No. 6, June 1980, pp. 343-349

NeXTGene: A graphical user-interface for GENESIS under NeXTStep

Christian Jacob and Axel Burghof
Lehrstuhl für Programmiersprachen
Universität Erlangen-Nürnberg
Martensstraße 3, D-8520 Erlangen, Germany
Email: *jacob@informatik.uni-erlangen.de*

Abstract

Experimenting with genetic algorithm (GA) systems can be substantially supported by graphical user interface (GUI) frontends to GA-kernels. We briefly describe our ideas about what a GUI for evolutional systems might look like. Our first results in designing an interface with the help of NeXTStep's Interface Builder [1] are presented.

1 Introduction

GENESIS is a well-known software system for experimenting with genetic algorithms. Originally developed by J.J. Grefenstette [1], GENESIS has been enhanced by N.N. Schraudolph [2] under the name of GAucsd [2], which is publicly available (version 1.4) and belongs to the state-of-the-art programs in the field of genetic algorithm research.

Due to its proper modularization we chose this system as our basic kernel platform in order to develop a graphic-oriented GA-system which we are using for the evolution of neural networks. GENESIS is interactively controlled by shell-scripts and puts its results (reports about GA-parameters, evolved strings, population descriptions etc.) into files; thus it is easy to adjust parameters to problem specific values and the files provide an easy interface to other programs (as e.g. a neural network simulation system).

However, GENESIS still has some disadvantages or let's call them "inconveniences": It is not as easy to automatically keep track of a large number of GA-experiments which all have different parameter settings for the same problem domain. A kind of project manager is missing which might be controlled with the help of a graphical interface. Furthermore, GENESIS does not say much about its performance during simulation runs; only at the end of these runs report files will give information about what has happend.

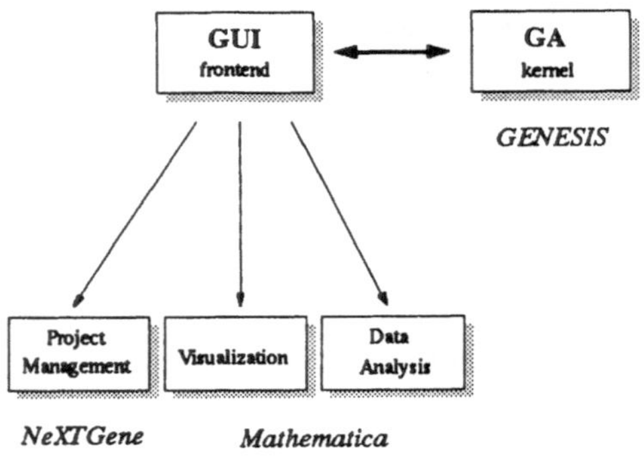

Figure 1: Graphical user interfaces for genetic algorithm systems

We consider the following ideas to be important for successful GA research and simulation (fig. 1):

- Experimenting with genetic algorithms is tedious when there is no proper user-interface available; a graphical interface connected to a kernel GA-system like GENESIS being available would help to carry out GA research more effectively.

- As long as much simulation has to be done in order to get some insight into the performance of genetic operators and codings, visualization tools are very important and help in getting a better (intuitive) understanding of what is going on during simulation runs.

- Data analysis tools should be available for post-processing GA-protocol data.

[1]NeXT, NeXTStep and NeXTStep Interface Builder are trademarks of NeXT Computer, Inc.

[2]With 'GENESIS' we will both refer to GENESIS and GAucsd.

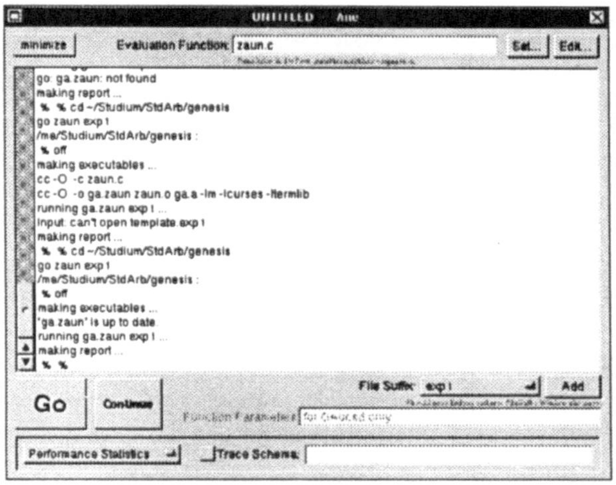

Figure 2: A simulation control window

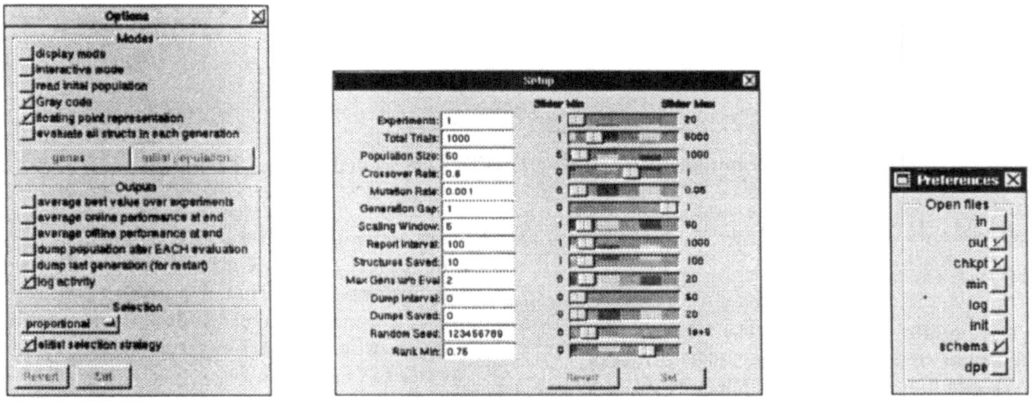

a. b. c.

Figure 3: Examples of setup windows for GENESIS options and parameter settings

2 Graphical user interfaces

With "Genetic Algorithm Workbench" [3] and "The Evolution Machine" [4] two GA-systems have been presented which support experimentation with a graphical user inferface. Although both systems are very easy to use and partly visualize GA dynamics in an intuitive way, program sources are not publicly available restricting the usefulness of these systems for research purposes. For our GA research system we chose GENESIS as the basic platform providing a library of GA-related routines, and developed a graphical interface to its kernel (fig. 1).

We chose NeXT machines as our hardware platform in order to learn in how much NeXT's Interface Builder supports the development process for graphical user interfaces which is mostly a difficult and tedious job (as we know from our experience with OSF/Motif interface design systems). The first observations we made using NeXT's tools were the possibility of rapid (in its true sense) prototyping and the ease of coping with the problem of creating graphical interfaces for systems like GENESIS. We will show this by presenting some of our first results; it might be interesting to note that we started with no detailed knowledge about how to use the NeXTStep Interface Builder and the object-oriented Objective-C programming language used by NeXT.

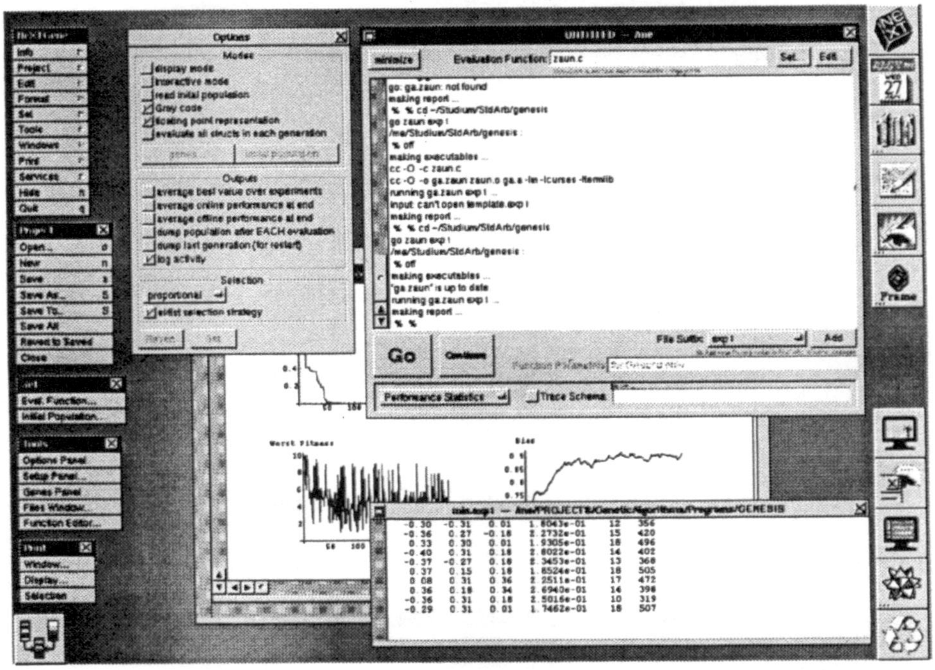

Figure 4: An example session with GENESIS controlled by NeXTGene

3 The NeXTGene user interface in detail

One idea to structure GA-simulation data is that several GA-runs can be put into socalled 'projects'. Each project can be viewed as a collection of GA-experiments for a certain problem domain, where each GA-run might have slightly different GA control parameters (mutation rate, crossover rate, dump interval etc.).

For each project the following window classes are available:

- Option windows control mode settings like coding (Gray or other), parameter representation (bit strings or floating point vectors), average performance statistics etc. (see figure 3a.).

- Setup windows define parameter settings which are mostly floating point values out of a predefined interval (number of experiments, population size, report interval etc., see figure 3b.).

- Special setup windows are used to edit more complex parameter settings like e.g. lists of gene specifications (see figure 5).

- Control windows interactively control simulation runs together with a number of parameters like the evaluation function or performance statistics used. For each control window there is a kind of console window where system messages are directed to (see figure 2).

- Display windows serve as special visualization tools for different GA-parameters (see figure 6). In the simplest case this might be a functional plot of the (average, best, worst) fitness values or a pixel-oriented display of selected bit-strings. Option panels (like in figure 3c.) control which display windows should be visible during simulation runs.

Figure 4 gives an overview of the menu structure and functions offered by the user interface and shows an example session with diverse control panels. Specialized scripts are available via the graphical interface to call filter programs which first extract selected data from the files (see e.g. the contents of the *min.exp1*-file [3] in the lower right window) produced by GENESIS and then post-process and visualize these data with the help of Mathematica.

[3] which is automatically opened if the *min*-button is checked in the *Open files* preferences panel (fig. 3c.)

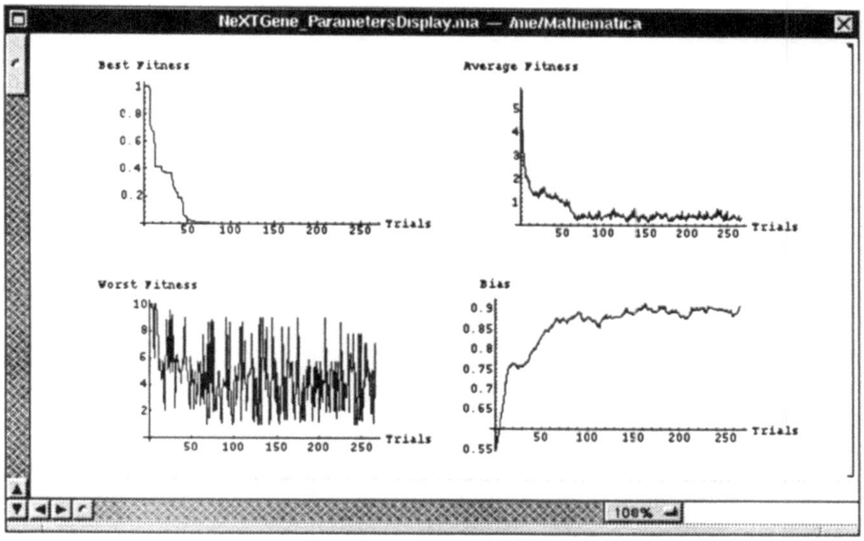

Figure 6: Example of a graphics window for dynamic GA parameters

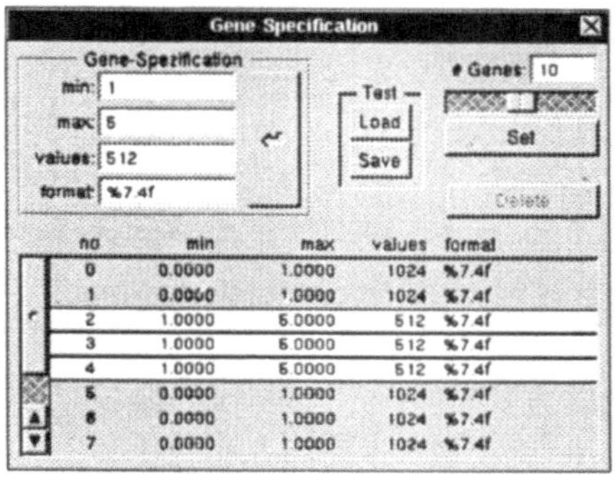

Figure 5: Setup window for gene specification

4 Interfaces to other applications

In order to use other software, especially for off-line data analysis, we are developing interfaces to other UNIX [4] or NeXTStep applications (e.g. StatLab, a simple statistical analysis tool) or to commercially available software like Mathematica [5] which provides a huge variety of data visualization and analysis libraries (see e.g. figure 7). One first idea concerning Mathematica is to create animated graphics that give insight into GA-performance by depicting two-dimensional projections of the problem domain.

5 Conclusion and further research

Our main goal with the NeXTGene project is to develop an easy to use, modularized and object-oriented software environment for interactive experimentation with genetic algorithms. The graphical user interface makes experimenting much more easy and efficient, if the whole system is flexible enough to be quickly enhanced with new features (e.g. new genetic operators, new evaluation functions, new analysis tools), and if control of these features can be smoothly integrated into the current user interface. We found that NeXT's software environment meets these requirements. This does not mean, however, that these concepts are not available on or could not be ported to other machines as well. [6]

The developed NeXTGene-GENESIS system will serve as our basic control interface for an evolution system of neural net topologies, connection weights and neuron functionality. We intend to enhance the system by special modules to support genetic programming techniques as well(see e.g. [5]).

[4] UNIX is a registered trademark of AT&T

[5] Mathematica is a registered trademark of Wolfram Research, Inc.

[6] We are indeed trying to apply our experience with NeXT and the NeXTStep user interface to other UNIX platforms with OSF/Motif.

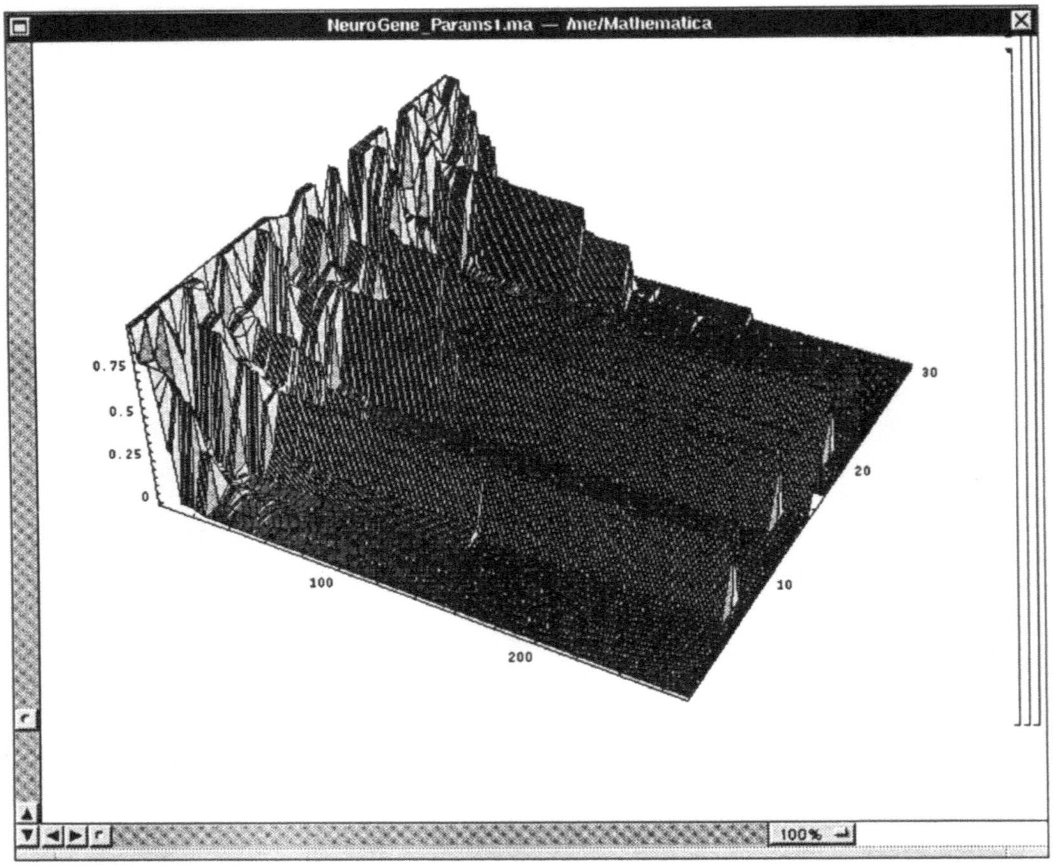

Figure 7: Example of a comparative plot of best-fitness values for 30 GA runs with 250 generations each

References

[1] Grefenstette J J et al., *GENESIS and OOGA: Two Genetic Algorithm Systems*, Melrose, 1991.

[2] Schraudolph N N et al, *A User's Guide to GAucsd 1.2*, University of California, 1991.

[3] Hughes M, *Genetic Algorithm Workbench Documentation*, Cambridge, 1989.

[4] Voigt H M and Born J, *The Evolution Machine, Manual Version 2.1*, Berlin, 1991.

[5] de Garis H, *Genetic Programming: Building Artificial Nervous Systems with Genetically Programmed Neural Network Modules*, in: Soucek B (ed.), *Neural and Intelligent Systems Integration*, New York, 1991.

Using Genetic Algorithms in Economic Modelling: The Many-Agents Approach

Joachim Stender#, Kifah Tout#, Peter Stender*

#Brainware GmbH, Gustav-Meyer-Allee 25, W-1000 Berlin 65, Germany
*Gesellschaft für Wirtschaftsforschung und Informationsmanagement(GWI) mbH, Marburger Str. 16, W-1000 Berlin 30, Germany

ABSTRACT

The theory of economies with many agents was developed in the sixties, stimulated by two papers of Aumann (Aumann 1964, Aumann 1966). The mathematical properties of this approach have been analysed by Dierker (Dierker 1974). According to Dierker the theory of economies with many agents allows removing both the convexity and the rationality hypothesis in economics. The idea of replacing the rationality hypothesis by the theorem of "bounded rationality", has been the center of more recent work in economics (Schelling 1978, Arthur 1991).
As it is possible to map the framework of the theory of economies with many agents directly onto PBE-automata, Principle-based engineering offers interesting routes to explore in the area of economics; especially in Economic Modelling. Principle-Based Engineering (PBE) is the process of designing a system through the definition of interacting elements where each element is an example of a principle (Addis 1991). Higher order systems can be constructed from principles that have, in their turn, been engineered as principle-based sub-systems. The behaviour of a principle-based system (PBS) is "emergent" through the resultant interaction of the elements and with an 'environment'.
Using the PBE approach, economic theorist's might directly model the behavior of many economic agents with partially conflicting interests and therefore gain insights in the results of the interdependencies of these agents for complete markets or market segments etc. This type of work is currently being actively researched at Brainware in the framework of the ESPRIT project PAPAGENA.

1. Introduction

Macroeconomic models have traditionally been constructed from solely aggregating demand and supply relationships. These macroeconomic models in turn have to be justified by "micro-foundations" describing people or companies (agents) maximizing their utilities and would on average behave according to these relationships. To obtain average aggregates from individual behaviours numerous assumptions have been made about the distribution of preferences and the used technology. Important in this respect are economic models which can be used to evaluate decisive effects of changes like a change of policy regime or other exogenous variable processes which avoid all the drawbacks of traditional economic models. For this reason a 'deep structure' micro approach is necessary; the one followed here is based on a theory of economies with many agents.

The theory of economies with many agents was developed in the sixties, stimulated by two papers of Aumann (Aumann 1964, Aumann 1966). A systematic presentation is given in Hildenbrand´s book (Hildenbrand 1972). The mathematical properties of this approach have been analysed by Dierker (Dierker 1974). An interesting aspect indicated by Dierker is the fact that the theory of economies with many agents allows removing both the convexity and the rationality hypothesis in economics. Especially the last aspect - replacing the rationality hypothesis by the theorem of "bounded rationality" - has been the center of more recent work in economics (Schelling 1978, Arthur 1991).

The development of highly advanced information technology, including transputer-based parallel architectures, has enabled more powerful approaches and instruments to be used in economic modelling, such as **Genetic Algorithms (GA´s)** and **Principle Based Engineering (PBE)**. As it is possible to map the framework of the theory of economies with many agents directly onto so-called PBE-automata, Principle-Based Engineering offers interesting routes to explore in the area of economics; especially in Economic Modelling. Using the PBE approach, economic theorist´s might directly model the behavior of many economic agents with partially conflicting interests and therefore gain insights in the results of the interdependencies of these agents for complete markets or market segments etc. Thus the power of prediction, today, is beyond the pure mathematical approach.

2. Principle Based Engineering

Principle Based Engineering is the process of designing a system through interacting principles (Addis 1991). In one interpretation this approach is just a different perspective on conventional computing since any system could be described in terms of interacting elements which are principles. In another interpretation the PBE approach challenges the Church-Turing thesis and suggests a radical conclusion that there is a more general model of computation than the Turing machine.

This controversial idea is suggested because the PBE method of computing does not have the same notion of algorithm that a Turing Machine does. The Turing Machine provides a way of specifying effective procedures which leads to a notion of algorithm where every step in a computation is well defined. The PBE notion of algorithm is in terms of interacting principles which do not specify what to compute or how to compute it. The general principles underlying a system are modelled and the behaviour of the system emergences and evolves, computing items without precisely defining how these are obtained. The Church-Turing approach focuses on the results and designs the system to compute them, while the PBE approach focuses on the system and allows the consequences of that system to emerge without precisely defining what they are.

Principles are simple laws that are true of a system. They abstract the most general properties of the system and would guide but not determine the behaviour of the system. To some extent they are rules but only in the capacity of guiding the behaviour of the system; not in determining the behaviour of the system as an effective procedure does nor the consequences of that behaviour (the items computed) as an effective procedure would.

It is possible to identify general and specific principles. Specific principles are those that are true about a particular domain such as language. It is possible to extract principles that underlie rules in some specific domain previously described by rules and explain the data (and perhaps even more) but with greater simplicity. Each principle is itself quite simple and it is the interaction that results in an item that would require a complex rule to describe.

General principles are those that are not domain specific but are applicable and true in a wider class of systems. The principle of survival of the fittest in the theory of evolution is such a principle. Though this theory began with the evolution of the species it can be applied to other systems (such as economies and social structures) where the system evolves through the principle of the survival of the fittest. These general principles might be used to explain the interaction of specific principles and in that capacity a PBE system could be regarded as the interaction of many principles with the environment.

The general idea of such principles evolving a system towards an ordered complexity goes against entropy which would make such a system progress towards chaos, but this is something which is true of natural systems which should tend towards chaos but which are evolving. Though principles do evolve the system in no way do they specify exactly what will be created. It is the interaction of principles among themselves and with the environment that do this.

The fact that the PBE automata does not use effective procedures to compute items means that it need not suffer the limits of formalised systems.

The PBE automata is 'defined' by the interaction of modules. It is important to understand what exactly the interaction of modules means. The strong modularity of Fodor (Fodor 1983) would imply that a system was strongly decomposable because each module would be computationally autonomous sharing neither data nor memory with other modules. In contrast modularity where there was such interaction between components would be only weakly decomposable. The PBE automata would be modular in the strong sense with each module being completely computationally autonomous.

The fact that the PBE automata utilises strong modularity means that it is distinct from other models of computation which could also be described as "components interacting". For example connectionism might be an example of components interacting. In this case the module degenerates into a unit in a network, where each 'module' is dependent on the other modules, in at least the data that they communicate between each other.

3. Mapping the framework of the theory of economies with many agents onto PBE-automata

When mapping the framework of the theory of economies with many agents onto PBE-automata, three main principles are to be followed. Firstly **system organisation** which consists of system activity relying on many homogenous or hetrogeneous elementary agents. Within the system synchronised execution is not a vital issue and no globally coordinated control is enabled.

The system is controlled solely by meaningful stimuli input through external (border) links which are the only communication paths for the system to the environment. Secondly the **communications issue** which states that the agents interconnection structure is simple and regular and kept as stable as possible. Communication is only allowed through interconnection structure. Two agents may communicate to each other only if they are directly linked and the communication may result both in data and control information exchange, however communication is as reduced as possible.

Figure 1: Key Issues for Economic Modelling

- Modelling the bevaviour of many independent economic agents (´micro-foundations´) vs. macroeconomical modelling
- ´Bounded Rationality´ vs. ´Homo oeconomicus´
- Nash Equilibria
- Modelling the non-linear behaviour of the global system through the many-agents approach
- Learning: agents adapt the strategies of more successful agents
- Practicability now enabled by modern computing architectures (Parallel Processing)

The final principle, **agent capabilities**, concerns each agent acting on its own account, although with no global representation of the system. The agent activity is directed only by the internal "program" under the external conditions created by the information obtained from the immediate neighbours, resulting in the possiblity of the program of an agent to change dynamically.

4. The Genetic Algorithm Manipulation Environment (GAME)

4.1. Introduction

GAME is a general-purpose programming and simulation environment for Genetic Algorithms (GAs) and Parallel Genetic Algorithms (PGAs). It is intended that GAME will provide a multi-utility toolkit for the programming and simulation of the majority of algorithms ranging from multi-start hillclimbing, through parallel simulated annealing, through standard GAs to artificial life. In addition to the parametrised applications and algorithms GAME should provide extensive tools for building new applications and thus a high degree of modularity is needed. GAME is being developed with the aim to use it later as a European standard for the design of genetic adaptive systems. To achieve that goal, GAME has to make use of the state-of-the-art programming techniques. Therefore, the object-orientated and API (Application Programming Interface) approach were adopted. The programming language used to implement GAME is *parallel C++*.

There are three levels of interaction with the GAME environment namely: the user level, the application programmer level, and the system programmer level.

The GAME environment consists of six major components.These are: the *Graphical Interface* (using XWindows), the *User Monitor*, the High Level Language(HLL), the *Algorithm and Application Libraries*, the *Compilers*, and finally the *Virtual Machine(VM)*.

4.2. The Virtual Machine (VM)

This object is the GAME´s machine independent low level code, responsible for the management and execution of actual *genetic manipulations*. The VM consists of five principal objects:

- *Population Manager (PM):* is responsible for coordinating operations over several different *pool* objects, over *individuals*, *chromosome*, and *gene* objects.

- *Pool:* is a structure which holds definitions and parameters for creating a population.

- *Fitness:* is user´s problem dependent. This object defines the way the fitness of an *individual* in the *pool* is calculated.

- *Parallel Support (PS):* is responsible for the management of the low level parallelism offered by VM.

- *Graphic Monitor Support (GMS):* is the object that offer runtime support or application specific data monitors. It works in conjunction with GAME´s Graphical Interface.

5. An Example: Locational Modelling

The Locational Modelling Application is consisting of four basic components:

1. *The Labour-Market Model*
2. *The Enterprise Model*
3. *The Location Model*
4. *The Global Economic Model*

The models 1-3 are constructed on the basis of the

theory of many agents [Aumann 64, Aumann 66, Hildenbrand 73] by using the GAME environment in the way of a Principle-Based Automoton. Model 4, the *Global Economic Model*, allows the integration of all components into a total system as well as the calculation of the fitness values for each economic agent.

Models 1 and 2, namely the *Labour-Market Model* and the *Enterprise Model*, are considered as demand-side models, model 3 as the supply-side. We assume that the enterprises will be modelled as the demanders for locations, while the employees are demanders for both locations (3), and jobs (2).

For both components of the demand side the fundamental equation of the Hotelling-Theorem [Hotelling 1929, Sutton 1991] have been chosen: *U (p,t,d) = U*-p-dt, with:*

U = Utility
U = Maximum Utility*
P = Price
t = Structure of Preferences

d = Distance

The Preference Structure for the economic agents (labour and enterprises) and the decision factors for locations have been used for a similar model developed by Beta Research, Syosett, N.Y. and Fast Forward, Portland, Ore. for the American magazine MONEY [Smith/Englander 1992]. They have carried out six consecutive annual studies with the aim to identify the most attractive locations in the USA.

6. Conclusion

Macroeconomic models in general have microeconomic foundations and the Many Agent theory follows the same principles. However, using expectations as a grounding the Many Agent theory goes beyond the ability of the traditional expectation and game theorys that state what is rational for individual agents is necessarily rational for a whole economic model.

The Many Agent theory is a dynamic model in which many individual agents obey the rules of rational expectations by maximising their utility given certain constraints. Each agent has parameters which are defined by assumptions specific to the agent. Using computers enables the behaviour of an individual to be integrated into a whole economic model consisting of many individual agents acting under various parameters. Although the individuals may act with rational expectations the aggregate of the economic model as a whole is not necessarily rational.

7. Acknowledgements

The work described in this paper is supported by the European Community ESPRIT III-Project No. 6857 PAPAGENA (Programming Environment for Applications of Parallel Genetic Algorithms).

References

[Addis 1991] Addis, "Principle-Based Engineering as applied to subpopulation modelling, PAMINA Project Proposal, 1991".[Addis 1991]

[Arthur 1991] B. Arthur, "Designing Economic Agents that act like Human Agents: A Behavioral Approach to Bounded Rationality, Papers and Proceedings of the American Economic Association," American Economic Review, 5/1991.

[Aumann 1964] R. Aumann, "Markets with a continuum of traders", Econometrica, 32(1964), 39-50.

[Aumann 1966] R. Aumann, "Existence of competitive equilibria in markets with a continuum of traders", Econometrica, 34(1966), 1-17.

[Dierker 1974] E. Dierker, "Topological Methods in Walrasian Economics", 1974.

[Fodor 1983] Fodor, "Modularity of Mind," 1983.

[Hildenbrand 1972] W. Hildenbrand, "On Economies with many agents", Journal of Economic Theory, 5(1972a), 152-162.

[Schelling 1978] T. Schelling, "Micromotives and Macrobehavior", New York 1978.

[Smith/Englander 1992] M. T. Smith and D.W. Englander,"The Best Places to live in America, in: Money" 9/1992, pp. 110-124.

[van Damme 1987] E. van Damme, "Stability and Perfection of Nash Equilibria", Springer, Berlin 1987.

[Hotelling 1929] H. Hotelling "Stability in Competition",Economic Journal, Vol. 39, pp 41-57

[Sutton 1991] J.Sutton," Sunk Costs and Market Structure", Cambridge/London 1991

GIRS: A genetic approach to Information Retrieval

Max Höfferer
Institute of Statistics and Computer Science
Department of Information Systems
Liebiggasse 4/3, A-1010 Vienna, Austria
email: mh@ifs.univie.ac.at

Abstract

The fundamental problem in information retrieval (IR) is to identify the relevant documents from nonrelevant ones according to a particular user's request. This paper describes a genetic approach to probabilistic information retrieval for improving recall and precision of an information retrieval system. The presented prototype model - GIRS: Genetic Information retrieval System - consists of the following parts:

1. a Genetic Algorithm (GA) is used for (a) the redescription of documents (indexing component) and (b) the clustering of co-relevant documents depending on user relevance judgements.
2. Classifier System (CS): the system evolves sets of rules containing clustered descriptions of similar documents to improve the recall and presicion of the system.

1. Introduction

Information retrieval (IR) is concerned with the representation, storage, organisation, and accessing of information items like documents [1]. There are two fundamental problems in IR:

(1) the representation of documents by storing some form of description in a database and the representation of a user's information needs by queries, and

(2) to identify the relevant documents from nonrelevant ones according to a particular user's request.

Traditional IR methods suffer from a lack of adaptive behavior aimed toward promoting learning.

This paper outlines the design and ongoing research in implementing GIRS, a Genetic Information Retrieval System based on adaptive search to redascribe documents and a Classifier System to learn from examples made by the user to seperate relevant from non-relevant documents.

The paper is structured in the following way. Section two gives a short introduction to the information retrieval domain. In the third section a GA for document redescription is considered. The fourth section suggests how a GA can induce descriptively similar clusters of co-relevant documents Finaly section five describes an approach to an adaptive information retrieval system.

2. Problem domain: Information retrieval

The purpose of an information retrieval system (IRS) is to provide information in response to user queries. An IRS consists of the following components and operations:

- *Textual database*: Original documents or surrogats of original documents (abstracts, lists of assigned or derived keywords and phrases).
- *Indexing*: maps the documents' contents and the user's request to an

indexing language. Indexes to index terms, phrases, and references within the original document or the surrogate.

- *Terminological control*: thesauri, stop word lists, citations and other devices.
- *User interface* - Queries of various types:
  - sets of index terms or phrases, some of which may be fixed attribute values, boolean combinations of elements in these sets,
  - weighted sets of words, phrases or attribute values.
- *Operations*:
  - *selection*: identifying a subset of documents or surrogates or parts of these based on whether or not they satisfy a query,
  - *ranking*: ordering of documents or subsets of documents with respect to the values of a matching function between documents and queries.

The earliest IRSs were based on the boolean model [2] where stored documents are identified by sets of index terms or phrases - with the help of an inverted file structure - and queries are expressed by using boolean combinations - with the boolean operators: AND, OR, NOT - of index terms. Only selection operations are possible and documents are selected if they satisfy the boolean query. To overcome the various disadvantages of the boolean model (e.g. difficulties in controlling output, no document ranking, no weigths attached either to the document or to the queries) the vector space model [3] was proposed which permitted ranking operations. The document space is seen as a matrix in which row vectors (*i*) represent documents and column vectors (*j*) index terms. A particular document is identified by a collection of terms or descriptors $term_1$, $term_2$, ... , $term_t$, where $term_{ji}$, is assumed to represent the weigth, or importance, of term *j* assigned to document *i*.
Queries are also represented as vectors of term weigths. Retrieval involves a ranking of the documents with respect to the query on the basis of some document-query matching function such as the *Cosine, Taniamoto, Dice,* or *Jacquard* measure [4]. In IR it is customary to measure the effectiveness of retrieval by using two parameters:
Recall = proportion of relevant terms actually obtained divided by the total number of relevant terms contained in the collection (that is, the proportion of relevant documents retrieved) and
Precision = proportion of relevant terms actually obtained divided by the total number of retrieved terms contained in the collection (that is, the proportion of retrieved documents that are relevant).

Probabilistic theory has also been used as a means for modeling the retrieval process. In the probabilistic IR model [5], [6] the probability of relevance is computed for each document with respect to a query where each document is described by a binary vector. A decision rule can be used to assign a document to a relevant set if the probability of the document being relevant - given the evidence in the document representative - is greater than the probability of the document being nonrelevant, that is, if

$$P(Rel|Doc) > P(Nonrel|Doc).$$

As it is too difficult to estimate these probabilities directly the Bayes' theorem is used to get an equivalent decision rule:

$$P(Doc|Rel)P(Rel) > P(Doc|Nonrel)P(Nonrel)$$

where P(Rel) and P(Nonrel) are the a prior probabilities.
This model is based on the probabilistic ranking principle (PRP) [7] which states that optimum retrieval is achieved when documents are ranked according to decreasing values of their probability of relevance with respect to the current query. As [8] pointed out, PRP's assumption that index terms are independently distributed in both relevant and nonrelevant documents is not valid leading to suboptimal ranking.

Using a probabilistic retrieval model to provide satisfactory results the model should rely on independence assumptions and feedbach data should not be based on a small set of users.

3. Genetic redescription of documents

One way to improve document description is to perform the description process repeatedly. The goal is to determine from past users how a document should have been described so that its description can be modified and made more satisfactory for future users. In this approach a Genetic Algorithm (GA) [9] is applied to the task of document indexing to change the way a document is represented [10].

```
procedure ga_redescription
begin
        t = 0
        Initialize P(t)
        Evaluate P(t)
        while (not termination-condition)
        do
        begin
                t = t + 1
                Select P(t) from P(t-1)
                Recombine P(t)
                Evaluate P(t)
        end
end
```

In each generation g for document x there will be a set of descriptions for that document. Each document is a binary vector:

$$des\_x\_gen_i = \langle \overset{T_1}{1} \quad \overset{T_2}{1} \quad .. \quad \overset{T_k}{0} \rangle$$

where each T_i $(1 \leq i \leq k)$ is an index term that is either being employed in describing a document (1) or is not (0).

The initialization step of procedure *ga_description* represents the indexing component of GIRS. Each document - GIRS's document base consists of decisions of the Austrian Supreme Court in civil law - is described by an initial set of index terms. Students and experts made relevant judgements about the legal documents and provided a set of relevant queries for these documents. The initial set of document descriptions and the set of relevant queries for that document are identical considering the fact that the query a user uses to find a relevant document and the description he provides for that document should be the same.

The Evaluation procedure measures the performance of competing document description. Each description is matched - using the Jaccard similarity function [4] - with a set of relevant queries and an average recall matching score is calculated with respect to the relevant query set (see Figure 1, appendix). For each description $des\_x\_gen_i$ a relative fitness is measured by $rel\_fit\,(des\_x\_gen_i) = (des\_x\_gen_i)/F$, where

$F = (1/N) * \Sigma fit(des\_x\_gen_m)$ with N descriptions of document x and m relevant queries.

The purpose of the Select operator is to cause the best performing descriptions in the current generation to be disproportional represented in the next generation. This is achieved by simple roulette wheel selection [9] which creates $rel\_fit\,(des\_x\_gen_m)$ copies of $des\_x\_gen_m$ $(1 \leq m \leq N)$.

The Recombine operator replaces the set of descriptions by composing new descriptions from parts of the best-performing descriptions in the current set. Cross-over randomly divides the newly created set of N descriptions into floor $(N/2)$ pairs (and an additional remaining description if N is odd).
A cross-over point p_j, $1 \leq p_j \leq k-1$ with k = length of the vector is selected randomly for each pair j.
The generation $g+1$ set of document descriptions is created as follows:

$$\text{init}(\text{des-pair}_{j1}) + \text{final}(\text{des-pair}_{j2})$$

init(des-pair$_{j2}$) + final(des-pair$_{j1}$)
where
des-pair$_{j1}$ and des-pair$_{j2}$ are the pair of document descriptions in the jth pair, init(des-pair$_t$) gives the first p_j positions in vector des-pair jt (t = 1,2) and final(des-pair$_{jt}$) are the last (k - p_j) positions in vector des-pair$_{jt}$ (t = 1,2).

The following two descriptions comprise the jth pair and p_i is randomly selected to be 3:

before crossover

T1	*T2*	*T3*	*T4*	...	*Tk*
<1	1	1	0	...	1 >
<0	1	1	0	...	0 >

after crossover

T1	*T2*	*T3*	*T4*	...	*Tk*
<1	1	1	0	...	0 >
<0	1	1	0	...	1 >

The new set of document x descriptions will replace those in figure 1 (appendix) and the entire adaptive process will be repeated.
The GA operating on document descriptions causes the set of document descriptions associated with a document to move over time toward those queries to which they are relevant. Documents which tend to be relevant to similar queries will have descriptions that move closer to each other. Therefore clustering documents based on their descriptions will aid retrieval speed and effectiveness.

4. User-based Clustering of co-relevant documents

In IR the cluster hypothesis states that 'closely associated documents tend to be relevant to the same queries [3]. If two documents have similar descriptions they are called co-relevant and as [11] pointed out the cluster hypothesis also holds for co-relevant documents. Clusters are user-based if their information results from users' relevance judgements [12]. In the experiments clusters conceive a set of k documents relevant to a given set of queries. Each of the documents in the clusters receive an initial set of descriptions which are adapted over time. To support genetic adaption the following procedure generates description sets for in our case one potential cluster [13].

```
procedure cluster_description
begin
      Initial_description
    { Start with a document with n
    descriptions, des_m, 1 ≤ m ≤n}

      Create_description_set
    { Create empty descriptions for each
    of k documents in the cluster }

      Create_k_sets, S_i, 1≤ i≤k
      For i:= 1 to k Do S_i= [].

      Fill_description_sets
      Set num_des < n
      { number of descriptions per set S_i}
      For j = 1 to num_des * k  Do
      begin
      Set x = j mod n (if x=0, set x=n)
      Select a random set, S_i, with less
      than num_des descriptions
      Add des_m to S_i.
      end
end
```

The procedure randomly distributes copies of the descriptions of the topic among the documents in the cluster where each document receives *num_des* descriptions.

5. A classifier system for information retrieval

A Classifier System (CS) is a parallel, message-pasing, rule-based system with the ability to learn [14]. A simple CS consists of the following parts:
* Detectors and effectors,
* Message System (with input, output, and internal message lists),
* Rule system (population of classifiers),
* apportionment of credit system (bucket brigade system),

* Genetic alogorithm (reproduction of classifiers).
The environment sends a message which is accepted by the CS's detectors and placed on the input message list. The detectors decode the message into one or more decoded messages and place them on the internal message list. The messages activate classifiers, where the strongest ones place messages on the message list. These new classifiers may activate other classifiers or send some messages to the output message list. The effectors code these messages into an output message, which is returned to the environment. The environment evaluates the action of the system and the bucket brigade algorithm updates the strenghts of the classifiers.
The evolution of individual rules is known as the Michigan approach and is based on Holland's CS-1 system [15]. The counterpart of the michigan approach is the Pitt approach which is a blend of a production system [16] and a CS and evolves sets of rules instead of individual rules [17]. These rule sets are evaluated, crosses , mutated, or otherwise genetically altered to create new possibly better rule sets for evaluation in future generations.
The presented genetic approach to IR can be viewed as a special solution to the problem of concept learning (or inductive learning) in artificial intelligence. The concept to be learned in IR is the user concept of relevant documents based on examples made by the user.
The presented genetic information retrieval system applies sets of rules instead of single rules. GIRS consists of the following parts (see Figure 2 - appendix).
* Detector modul:
Decode user queries (sets of descriptors) into messages and place them on the input message list.
* Knowledge base modul:
The document space can be viewed as the basic component of a simple knowledge representation system where the knowledge about documents is expressed through assignment of terms and weigths to the documents.
GIRS = $< DS, S_i, Rel, V_t, V_{Rel}, f >$
where DS are clusters of similar documents, S_i is the set of document descriptions, the attribute *Rel* describes the user judgement about the relevance status of a document and the knowledge function
$f: DS \times S_i \rightarrow V_t$, and
$f: DS \times \{Rel\} \rightarrow V_{Rel}$.

If all rules sets in the knowledge base have been evaluated the GA is invoked to construct new document descriptions.

* Genetic Algorithm modul:
GA's work is to improve document desciption repeatedly over time. The genetic operators cross over and selection are applied to rule set structures.

* Clustering modul:
The clustering modul forms cluster of co-relevant documents with the help of the GA. As a result the modul forms subsets of documents into rulesets of similar documents which are used by the problem solving component.

* Problem solving component:
The problem solving component consists of rule sets describing clusters of similar documents. The working memory (WM) consists of fixed length binary strings, where each WM element consists of a signal and a data part. The production memory consists of sets of rules where each rule is a fixed length string. The condition of a rule consists of k fixed patterns; the first i attend to i detectors, and the remaining $k-i$ attend to signals contained in working memory.
Detector: e.g. 01101 01010
Rule: e.g. 1##0 10#0## 0100# 0011 $\rightarrow$ *Retrieve* (des_m)
Working memory: signal data
010011 10011

GIRS's inference engine scans down the working memory for matching grouping patterns. If a rule is completely matched on the left-hand side it posts its signal to the WM and the corresponding action is executed.

* Effectors:
The effectors code the messages into an output message, which is returned to the environment. The user is provided with refined document descriptions and enters his feedback data via the detectors to the system.
The output of GIRS is a ranking of relevant document descriptions.

6. Conclusion and further research

In this paper an adaptive information retrieval system called GIRS has been suggested, where the major concept of this newly approach is the adaptive behaviour of the system.

The test implementation of GIRS in C is running on a personal computer. At the moment we are implementing the problem solving component, which is the main part of the system. The database contains more than 500 decisions of the Austrian Supreme Court in Civil Law where each document contains approximately 3000 words. To start with an initial population students and experts judged the relevance of these decisions. Based on these results we will start with a database of more than 8000 decisions by randomly generating starting descriptions for the GA to begin with.

References

[1] G. Salton and M.J. McGill, '*Introduction to Modern Information Retrieval*', McGraw Hill, New York, 1983.

[2], G. Salton, '*Automatic Information Organization and Retrieval*', McGraw Hill, New York, 1968.

[3] G. Salton (Ed.), '*The Smart Retrieval System - Experiments in Automatic Text Processing*', Prentice Hall Inc., Englewood Cliffs, New Jersey, 1971.

[4] T. Noreault, M. McGill and M.B. Koll, '*A performance evaluation of similarity measures, document term weighting schemas and representations in a boolean environment*', Information Retrieval Research, R.N. Oddy et. al. (Eds.), Butterworths, London, 1981.

[5] M.E. Maron and J.L. Kuhns, '*On relevance, probabilistic indexing, and information retrieval*', Journal of the ACM 7, 3, 1960.

[6] S.E. Robertson and S. Jones, '*Relevance weigthing of search terms*', JASIS, 27, 1976.

[7] S.E. Robertson, '*The probability ranking principle in IR*', Journal of Documentation, 33, 1977.

[8] C.J. Van Rijsbergen, '*A theoretical basis for the use of co-occurence data in information retrieval*', Journal of Documentation, 33, 1977.

[9] D.E. Goldberg, '*Genetic Algorithms in Search, Optimization and Machine Learning*', Addison-Wesley, Reading, Mass., 1989.

[10] M. Gordon, '*Adaptive subject description in document clustering*' Doctorial Dissertation, University of Michigan, 1984.

[11], C.J. Van Rijsbergen, '*Information Retrieval*', Second Edition, Butterworths, London, 1979.

[12] V. Raghavan, and J.S. Degoun, '*User-oriented document clustering: A framework for learning in information retrieval*', International Conference on research and development in information retrieval, Pisa, Italy, 1986.

[13] M. Gordon, '*User-based document clustering by redescribing subject descriptions with a genetic algorithm*', Journal of the American Society for Information Science, 42, 5, 1991.

[14] L.B. Booker, D.E. Goldberg and J.H. Holland, '*Classifier Systems and Genetic Algorithms*', Artificial Intelligence, 40, 1989.

[15] J.H. Holland and J.S. Reitman, '*Cognitive Systems based on adaptive algorithms*', R.S. Michalski et al. (Eds.), Machine Learning: An Artificial Intellligence Approach, Morgan Kaufmann Publishers, Los Altos, CA, 1983.

[16] N.J. Nilsson, '*Problem Solving Methods in Artificial Intelligence*', McGraw-Hill, New York, 1971.

[17] S.F. Smith, *'Flexible learning of problem solving heuristics through adaptive search'*, Proc. 8th Int. Conference on Artificial Intelligence, Morgan Kaufmann Publishers, Los Altos, CA, 1983.

Appendix

Figure 1: Matching of descriptions with relevant queries. Each description is matched with M queries.

	$rel\_x\_g_1$	...	$rel\_x\_g_M$
$des\_x\_g_1$	$J(g_1,q_1)$		$J(g_1,q_M)$
...			
$des\_x\_g_N$	$J(g_N,q_1)$		$J(g_N,q_M)$

The column indicates N descriptions of document x in generation g. For each row the average matching score is calculated: $1/M \, \Sigma J(g_i,q_i)$. Summing up the row and column matching scores gives the overall matching score for the document descriptions in force in the current generation g.

Figure 2: GIRS's module structure

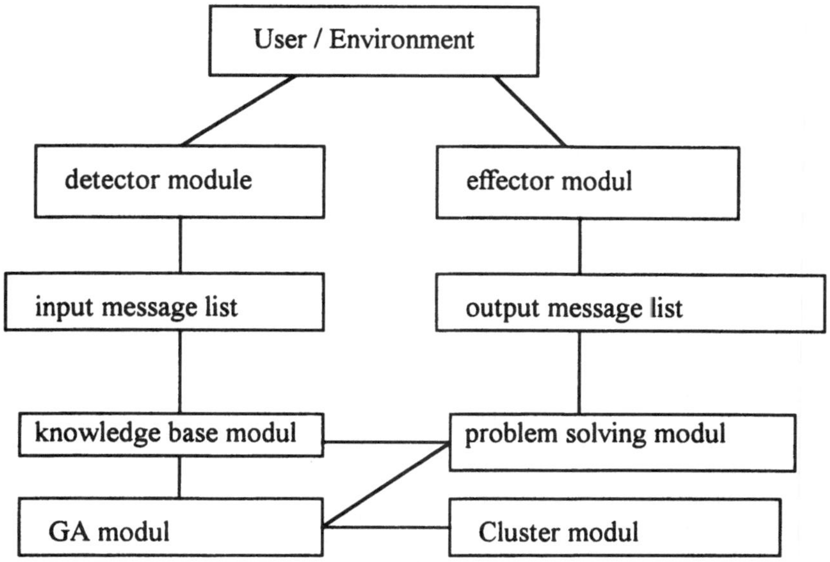

Classifier System in Traffic Management

Giorgio Casadei, Aldopaolo Palareti, Gianluca Proli
Dipartimento di statistica "P.Fortunati", Università di Bologna

Abstract

The systems of controlling and improving traffic movement have been studied for several years now. The usefulness of these systems is that they can modify and change the lights signals of traffic lights. It is not enough to intervene when the situation has reached a critical point such as a traffic jam. The system has to work out how the traffic will flow. The ideal solution would be a system that works out and foresees the situation on the roads based on a model of motorists' behaviour. This research shows how to best utilise the classifier systems so that it would be possible to create a model that is similar to that of the real world.

Introduction

In recent years the traffic problem, particularly in urban areas, has become more acute and urgent, both with the increase in the number of vehicles and the unsuitability of the road network.
The study of the road network has been under examination for several years now. Several nations have already put some of the computer based tools into practice out in the field. In these systems the information of the traffic conditions comes from the traffic monitors placed at important road intersections. As a result of the information obtained from the monitors, these systems can modify the frequency of the light signals changes of traffic lights, and can synchronise them with one another. In this way, it is possible to improve, by means of reducing the amount of time, the motorist uses the road network, and as result reducing the number of traffic jams and enabling public transport to have top priority.
The difficulty of putting these systems into practice lies in the fact that it is not enough to intervene only when the situation has become critical (typically seen by means of a traffic jam). In these cases the flow and quantity of traffic was not foreseen before happened. Thus it is essential to know how to foresee a critical situation arising and to be able to prevent it from happening.
Most of the systems used today (at Nice and Hamburg for example) use forecasts based on experience. We can assume that a flow model that incorporate the routes that satisfy the needs of motorists travelling from one place to another; would minimise the amount of time road users are on the road network. What is more, we must take into account that many cars on the road can cause incovenience for each other.
The ability of drawing up a model of motorists' behaviour, in response to the different needs of getting from on place to another can emerge together, hence allowing them to be calculated which the minimum quantity of data available the flow of traffic using the different roads. This system could be used to plan the road network.
This would enable us to:

- locate the critical junctions that are more subject to traffic jams;
- understand and to isolate the factors that affect city traffic, as well as a better understanding of the knowledge that road users have of the road network;
- formulate reliable preventive measures regarding the redistribution of traffic, overcoming shortcomings in the road planning policy that could give rise to a critical situation; in that these measures would modify one or more routes (introducing one way systems, prohibiting turnings and parking) and the result of which, would change the road network.

This could prove to be a useful tool in advising small to medium local authorities interested in investing a small amount of money in collecting traffic data, and to analyse the movement of traffic in their area and to develop strategies to resolve any problems in them.
Furthermore, the local authorities would be able to develop systems that could give up to minute information to the public on current traffic situation.

Formalisation of the problem

The data is the topography of the road network together with a set of connection requirements. The problem consists of finding a model that represents the situation in true life; the routes that the motorists take. That is, a model that fits as closely as possible to the routes that satisfy the

needs of motorists while, at the same time, trying to minimise the time the motorists are on the road network.
A direct graph, in which the nodes meet at the intersections or cross-roads and the arcs meet at the roads, represents the topography of the road network.

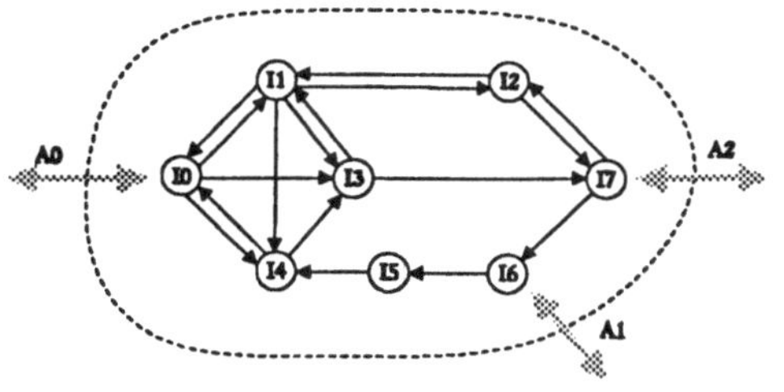

figure 1; an example of the topography of a road network

The set of connection requirements that the model has to satisfy is written inside the Origin/Destination matrix. This matrix is derived from the departure and arrival intersection of every route request.

to / from	Cross 10	11	12	13	14	15	16	17	Total	
10		5	40				50	20	115	27,6%
11	31		10						41	9,8%
12	5				5				10	2,4%
13										
14										
15										
16	60	10		1	50			55	176	42,2%
17	15	5	5		50				75	18,0%
Tot.	111	20	55	1	105		50	75	417	100,0%
	26,6%	4,8%	13,2%	0,2%	25,2%		12,0%	18,0%	100,0%	

figure 2; an example of Origin/Destination matrix, based on the topography of figure 1

We would like to bring the rush hour from 7 to 8 in the morning, when the road network is under its greatest strain, to your attention.
The Origin/Destination matrix is calculated from the General Population Census data, because the costs would be too much to carry out a survey to collect data of this kind. This choice has its drawbacks in that certain types of transport movements can not be noted, due to the limitations of the Census data. Some types of transport movements that the Census does not include are transport of goods, public transport and other occasional transport.
The length of every route is measured in seconds, necessary for motorists to travel along them. This amount of time is also dependent on the kind of traffic found travelling on the arcs at the same time.

The length (in seconds) of every route is calculated on the sum of the lengths (in seconds) of every arc; in this way the time taken to cross every intersection was considered constant, in this first release of the system.
The length (in seconds) of every arc is determined by the length (in meters) of the same arc and by the velocity of vehicle movement travelling along that arc. The flow function, used by the latter, depends on the characteristics of the roads and also on the number of cars that are found on those very same roads. These characteristics are the width, the numbers of lanes, and so on, and are known as "road capacity",

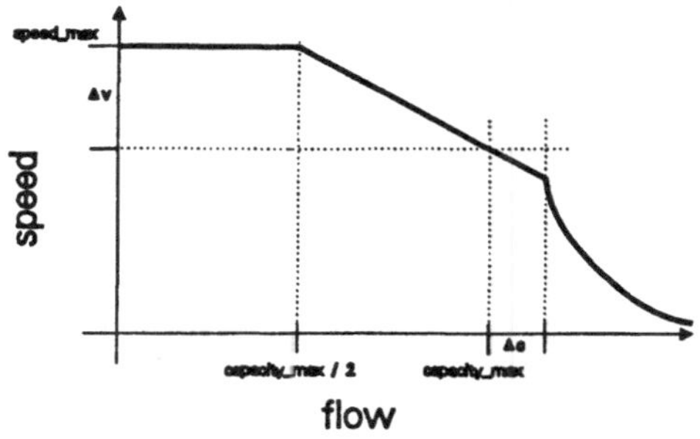

Figure 3; the flow function used in our study.

We defined our flow function in agreement with civil engineers and based it on the standard flow function [MSH61] [Orl74]. As it can be seen in figure 3, when the flow of the road is less then half of maximum capacity (the number of cars) the cars can travel at a maximum velocity of 50 km/h. However, if the flow increases, this results in a decrease of velocity. This decrease is initially linear and then, when it has reached a critical level, it becomes cubic.
Even today most civil engineers still have to confront this problem and they have to overcome it in an unsophisticated way using the "consecutive approximation" method.
The first step of the approximation method is to find the "all or nothing" distribution. That is, to find a particular distribution of traffic flow of motorists travelling, to their destinations, using the shortest route available, calculated in meters. This hypothetical situation is not realistic in real life in that it is not feasible that motorists can use every road at a maximum velocity.
The second step of the approximation method consists of modifying the "all or nothing" distribution, considering that cars are a nuisance to each other on the road. The second step is repeated until a more acceptable solution arises. The solution arising out of this long process certainly can not be fully satisfactory, in that it is impossible to take into account all the factors and how they correlate to each other.

The method described above is the only one that is used in real applications up to now. However, we propose a system that goes beyond this, in that it can be used to obtain an automatic solution of the problem.

How the system works

The strategy we proposed here is based on the use of a Classifier System. This strategy allows us to obtain an acceptable solution, that comes pretty close to the optimum level in reasonable time. What is more, this strategy allows us to construct a model based on reality.
Since the Classifier System is a learning system that carries out tasks by means of interaction with the environment; it creates a model, that behaves according to the rules, within itself.
The Origin/Destination matrix solely determines the initial positions of the cars on the routes that are shown on the graph (figure 1). The cars are ready to depart from their node-origin and eventually to travel to their destination.
Therefore, the Classifier System plans the routes that the cars will have to follow and it moves the cars from one place to another, one after the other, based on the rules of its system. This procedure finishes when every car has reached its destination or is "suspended" due to lack of rules applicable, in the given situation. At this point, the rules receive their rewards either as a prize in the case of successful rules or as punishment in the case of suspensions.
This process ("iteration") is repeated a number of times. It respects the rules and only modifies their strengths, according to the rewards received.
The set of iteration is called "generation"; when it terminates, the Genetic Algorithm creates a new set of rules at the same time. This set will then form the basis of future generations.
In this way, generation after generation, it is possible to work out consecutive models that have a high probability that would bring us closer to the solution of the problem.

```
Load the road network topography
Load the Origin/Destination matrix
Randomly formulate an initial set of rules
for generation=1,M
        for iteration=1,N
                Put cars in their initial node
                while there remain cars to move
                        Move the cars using the
rules
                end while
                Give rewards to rules
        end for
        Formulate a new set of rules
end for
end.
```

Figure 4; schemata of procedure

Some modifications were needed for the standard Classifier Systems with regard to the structure and the way they work, considering the particular problems that needed to be resolved.
So, we looked at and implemented in our system, a particular type of rules that we then coded, in such a way as to make the rules consistent with the topography under examination.
Furthermore we altered the way the message system works, when there is more than one rule that could be used.
Finally, we introduced a new attribute for every classifier, called "copies", and introduced into the system a message covering mechanism that kept the semantics of the problem in consideration.

The prototype of traffic management system presently in use

Firstly, we put the prototype of traffic management system to the test using a simplistic road network and the Origin/Destination matrix of that network.
The results obtained showed that a stabilising system occurred after 10 to 50 generations. They also demonstrated good use in correspondence to the capacity of the each road on the network.
We carried out research on a number of case studies of which a few had deterministic solution for the "all or nothing" distribution. In these last few cases we are able to reach nearly optimal solutions. For example, in a typical case that was repeated ten times, the rate between the mean of solutions found and the deterministic solution was 1.21 with a confidence interval at 95% of 1.14-1.28. In this same case the random solution was worst, at a rate of 2.55, with a confidence interval at 95% of 1.54-3.58. The variance analysis demonstrates a significant difference of $P<0.01$ between our system solution and the random one.

Application of this system in a real situation

We tested our system in a real situation. The situation chosen was the traffic circulation in the Forlí area (north-east Italy). The first step was to find the topography of the route network for the Forlí area, as well as the values for its Origin/Destination matrix.

We only choose the more important roads used by motorists, and constructed a graph on results obtained. The graph has 205 arcs and 62 nodes.

The information of every road represented on the graph is related to the corresponding arc of the road network. These pieces of information are: the direction of the vehicles (one way system or not), the length of road measured in meters, the capacity of the roads in terms of vehicles per hour, calculated using the charts in the Highway Capacity Manual 1985 [NRC85].

The Origin/Destination matrix was work out using the data from 1991 Italian General Population Census, that was given to us by the local authority for that area. We found 26837 individual needs for getting from one place to another; when similar needs are grouped together, this figure can be reduced to 670 different kinds of requirements for routes.

Secondly we considered only one hundred main kinds of requirements for routes, out of the total of 18705 motorists. Great emphasis only on the major requirements because when we tried to use all the motorists' requirements this did not improve the results but only brought about a reduction in the efficiency of the system.

The values of the parameters we used are as follows:

number of independent trial runs	30
number of generations for every run	200
number of iterations for every generation	10
number of classifiers (rules)	600
percentage of wild cards at the start of every run	5%
percentage of modifiers changed at the end of every generation	10%
percentage of modifiers made, using the covering method	20%
crossover probability	50%
mutation probability	0.01%
use of consistency factor (On or Off values)	On
the factor used to penalise classifiers with routes that have not been finish	1000
the factor used to penalise classifiers with routes that can not be finish	10000
the factor used to penalise unused classifiers	14
the factor used to reduce the age of useful classifiers	1000
the factor used to increase the age of dangerous classifiers	100

In our opinion, the results obtained have been found to be completely positive.

The system showed great efficiency: on average it obtained a representative model of the real traffic distribution in only 80 minutes of running the program. A personal computer 486 with 33 MHz clock was used.

An important indicator of system effectiveness is the valuation of Performance. This is defined as the rate between the length in time of routes obtained from the systems and the length of optimal routes (without traffic problem). This ratio would then be normalised dividing it by the number of cars in the graph (figure 1). We formulised it as follows:

$$Performance = \frac{\sum_{routes} \frac{cars_p \times time_p}{better\_time_p}}{total\_cars}$$

with the following criteria:

- *routes* – a set of routes required (every route has a starting point a_p and a finishing point b_p);
- $cars_p$ – a number of cars that require route p;
- $time_p$ – the sum of time needed for each car to travel along route p; in the case of suspended routes, we used the average time for the completion of the route in a random fashion;
- $better\_time_p$ – the theoretical minimum time needed for every car to reach their final destination in the absence of other motorists and going at maximum speed of 50 km/h;
- *total_cars* – total number of cars in the trial run;

The following is always true:

A performance equal to 1 shows that the model developed suggests motorists use only the shortest

(in meters) routes. Our aim was that, by incrementing the number of generations, the Performance measure would become closer to the minimum value.
We must take in consideration that is a theoretical minimum. However, it can not be achieved in real situation. The particular structure of the road network and the quantity and destination of the vehicles on that very same road network determines whether it is possible to get close to the minimum theoretical performance possible. Therefore, it is not possible to propose an *a priori* threshold for the system taking into account the above factors.
However, the results can be valued using two special performances:

- Random Performance, that is leaving out motorists that move randomly on the road network;
- "All or nothing" Performance [Orl74], that is leaving out every motorist that uses the shortest route defined in terms of meters.

Here is an outline of the results obtained in our trail test. At the beginning, the Performance measure was 41. Already at the 11th generation it decreased to the Random Performance value (30.98). The Performance value became equal to the "all or nothing" Performance value (5.62) at the 82nd generation. The Performance value decreased further to the value of 4.20 at 187th generation. The following Performance values after this, remained more or less stable, until they reached a value of 4.51 at the 200th generation.
In Figure 5, for every consecutive generation of the ten independent trial tests, we show the standard deviation values from the average Performance values of the tests.
The graph in figure 5 shows the "robustness" of the system. The results of the independent trial tests show a more stable pattern when the generations are consecutive.

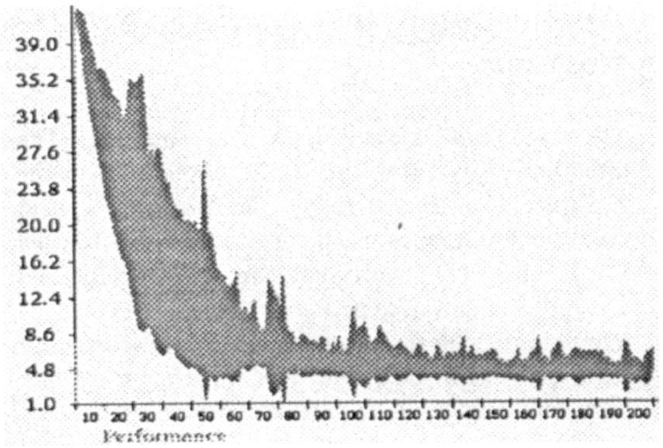

Figure 5; the graph of Performances of ten independent trials.

Furthermore, we looked at the average flow values calculated at the end of the trial tests with those that were measured on real roads carried out by the local authority for the Forlí area [Fil85].
The main difference in percentage was found in the heart of the city, within the old city walls. Figure 6 shows some of the data, with regard to the flow of cars on the road network, in the heart of the city. Every road is named after the node of its departure and arrival. The name of the roads on which the nodes are found is written across the top part of the chart. For example, the first one is "0-31" e the last one is "9-44".
The expected flow, based on collected data, is written across the chart immediately under the name of the roads. Data on flow calculated with the use of computer is shown immediately under the expected flow. On the subsequent row, the percentage differences found between the expected and computer calculated flows are written. On the last row, the absolute values of the difference are written. The total is written on the right end side of the chart; this total was calculated as the average of the absolute values of the differences.
Inside the old city walls of Forlí the total was 26.51% (see figure 6), meanwhile, outside the old city walls, the total was only 4.83%. The reason for this difference was probably because in constructing the graph of the road network we simplified some roads especially in the heart of the city. We did this because in this central area there are many old small lanes with low capacity.

	Road inside the old city wall of Forlì													
	0-31	31-34	34-31	34-44	44-34	60-0	0-60	0-28	28-0	34-35	35-34	44-9	9-44	Totale
Expected flow	1572	898	999	795	803	1218	949	828	782	821	766	301	342	11074
Calculated flo	1669	429	513	554	572	1401	558	645	744	896	307	273	397	8958
% difference	6%	-52%	-49%	-30%	-29%	15%	-41%	-22%	-5%	9%	-60%	-9%	16%	
absolute differ	97	469	486	241	231	183	391	183	38	75	459	28	55	26,51%

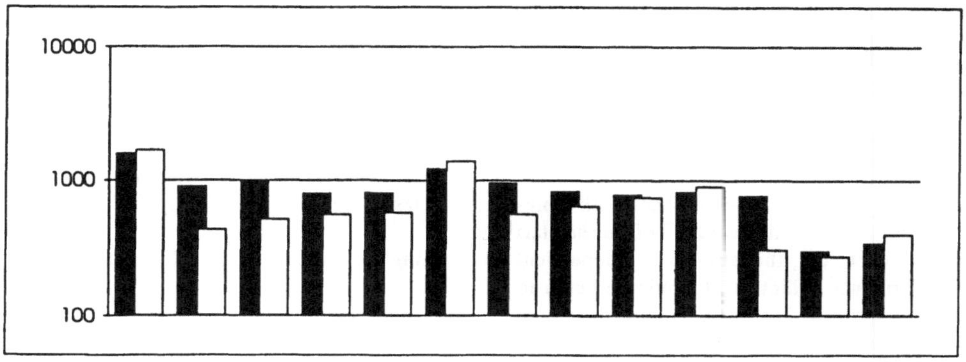

Figure 6; comparison between expected flows and computer calculated flows inside the old city walls of Forlí.

	Road outside the old city wall of Forlì									
	1-0	1-34	35-36	58-59	24-59	58-61	20-29	20-59	23-59	Totale
Expected flow	567	621	1440	1182	821	72	869	520	1263	7355
Calculated flo	490	654	1463	1085	824	68	908	511	1333	7336
% difference	-14%	5%	2%	-8%	0%	-6%	4%	-2%	6%	
absolute differ	77	33	23	97	3	4	39	9	70	4,83%

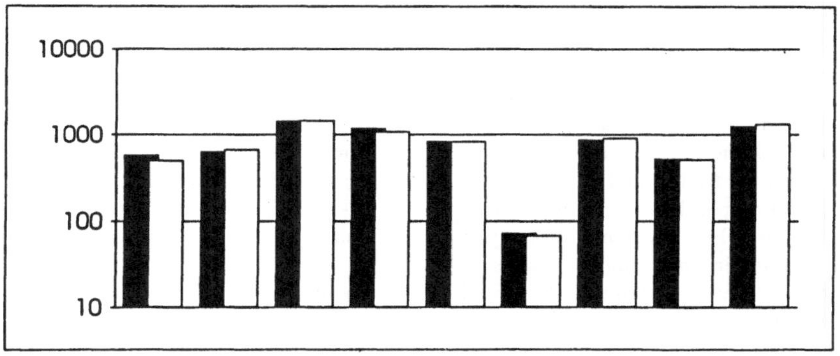

Figure 7; comparison between expected flows and computer calculated flows outside the old city walls of Forlí.

An empirical analysis confirms the above. In fact, the system is capable of drawing up a map of the flow of traffic on the road network. Figure 8 shows a part of such a map. The dark lines, which are often thicker, are the roads that have a superior demand by motorists than their capacity has been calculated at. These roads correspond rather well to the critical points on the road network, in the Forlí area, in the morning rush hour.

Conclusion and possible future developments

In conclusion, we can affirm that the system is able to work out models that correspond rather closely to real life situation, even if only a small amount of up to date data is available.

Even though this system works and produces good results, it should not be the only instrument used to analyse urban traffic. It is only a prototype, with the aim of demonstrating the possibility and the advantage in using the system to confront problems within the limits in the situation given.

In addition to this we intend to develop research further by verifying the conclusions using statistical analysis. Other research to be carried out in this area is as follows.

a. We introduced to our system the attribute "copies" not documented elsewhere for every classifier. This attribute enables us to work not only on singular rules but also on sets of identical rules. The aim of this attribute depends on the way that our system deals with the messages and it increases the stability of the system. Using this attribute, together with the methodology of reproducing rules, helps to produce a more "robust" system, that is less sensitive to the different sequences of pseudo-casual generated numbers. So, in this way, it becomes a more efficient system.

b. The prototype we developed finishes with a pre-set number of generations decided before the start of trail test. Therefore, this number is independent of the Performance achieved. It would be better to have a mechanism to halt the system a soon as a suitable solution is found. This solution happens when the Performance is sufficiently low. The problem is to know exactly what is the Performance threshold to stop the system. A reasonable applicable strategy would be to analyse the trend of the Performance of the last generations. In this way the system can be worked out when the trend is stable. This strategy has not yet been implemented into our system. However, we have introduce the possibility of drawing graphs to show the stability (calculated by using the Kendal test on the sum of ranks). From preliminary studies this technique seems to be successful from the evidence available.

c. There is however a preliminary problem to be resolved. In the actual implementation of the system, the several trials are carried out one after the other, after which the averages are calculated. It would be better if the trials could be carried out at the same time in a "parallel" way. In this way the system would be able to work out an analysis of termination considering not only the results of one trial test but of all the results obtained from all the independent trial tests. Using the above method would mean a more "robust" and efficient system could be guaranteed as result.

Bibliography

Booker L.B. (1982) Intelligent behaviour as an adaptation to the task environment (Doctoral dissertation, Departement of computer and communication sciences, University of Michigan, Ann Arbor).

Booker L.B. (1987) Improving search in genetic algorithms (In L.Davis, Genetic algoritms and simulated annealing, p. 61-73).

Booker L.B. & Goldberg D.E. & Holland J.H. (1987) Classifier system and genetic algorithms (tecnical report 8, University of Michigan).

Booker L.B. (1988) Classifier System that learn internal world models (Machine learning, 3: 161-192).

Booker L.B. & Goldberg D.E. & Holland J.H. (1989) Classifiers systems and genetic algorithms (Artificial Intelligence p. 235-282).

Booker L.B. (1989) Triggered rule discovery in classifier systems (ICGA pp. 265-274).

Buchanam C. (1963) Traffic in towns.

Colorni A. & Dorigo M. & Maniezzo V. (1990) On the use of genetic algorithms to solve the time-table problem (Internal report 90-060).

Cox L.A. & Davis L. & Qiu Y. (1990) Dynamic anticipatory routing in circuit-switched telecomunications networks (Handbook of genetic algoritms, p. 124-143).

Davis L. & Ritter F. (1987) Schedule optimization with probabilistic search (proceedings of the 3rd IEEE conference on artificial intelligence applications).

Dipartimento ambiente-territorio-trasporti (1982) I problemi della viabilità in Emilia Romagna.

Dorigo M. (1989) Genetic algorithms: the state of the art and some research proposal (Internal report 89-058, dipartimento di elettronica, Politecnico di Milano).

Dorigo M. (1990) Machine learning: an approach based on classifier systems and evolutionary algorithms (Internal report 90-043).

Dorigo M. (1991) New perspectives about default hierarchies formation in learning classifier systems (Internal report 91-002).

Dorigo M. (1991) Message based bucket brigade: an algorithm for the apportionment of credit problem (Proceedings of the European working section on learning 91).

Dorigo M. & Schnepf U. (1991) Organization of robot behaviour trough genetic learning processes (to appear in the procedings of the fifth IEEE International conference on advanced robotics).

Filippi F. (1985) [Fil85] Piano del traffico e dei trasporti del comune di Forlì.

Grefenstette J.J. (1985) Proceedings of the first International conference on genetic algorithms.

Grefenstette J.J. & Ramsey C.L. & Schultz A.C. (1990) Learning sequential decision rules using simulation models and competition (Machine learning , 5: p: 355-381).

Goldberg D.E. (1981) Robust learning and decision algorithms for pipeline operations (dissertation proposal, University of Michigan, Ann Arbor).

Goldber D.E. & Lingle R. (1985) Alleles, loci, and the traveling salesman problem (proceedings of an International conference on genetic algorithms and their applications, p. 154-159).

Goldberg D.E. & Samtani M.P. (1986) Enginering optimization via genetic algorithm (procedings of the ninth conference on electronic computation, p. 471-482).

Goldberg D.E. (1987) Genetic algorithms in search, optimalization, and machine learning (Addison - Wesley).

Goldberg D.E. & Kuo C.H. (1987) Genetic algorithms in pipeline optimalization (Journal of computers in civil engineering, 1(2), p.128-141).

Goldberd D.E. (1987) Genetic algorithms in pipeline optimalization (Engineering with computers, 3, p: 35-45).

Goldberg D.E. & Smith R.E. (1987) Non stationary function optimalization using genetic algorithms with dominance and diploidy (genetic algorithms and their applications:proceedings od the 2nd International conference on genetic algorithms).

Goldberg D.E. & Richardson J. (1987) Genetic algorithm with sharing for multimodal function optimalization (proceedings of the second international conference on genetic algorithm, p. 41-49).

Goldberg D.E. & Wilson S.W. (1989) A critical review of classifier systems. (ICGA pp.244-255).

Goldberg D.E. (1990) Probability matching, the magnitude of reinforcement and classifier system bidding (Machine learning, 5, p. 407-425).

Matson & Smith & Hurd (1961) [MSH61] Tecnica del traffico stradale (Padova Cedam).

Michalewicz Z. & Kazemi M. & Janikow C. (1990) On dynamic control problem (29th IEEE conference on decision and control, Honolulu).

Michalewicz Z. & Vignaux G.A. & Hobbs M. (1990) A genetic algoritm for the nonlinear transportation problem.

Michalewicz Z. & Shell J. (1990) Data structures + genetic operators = evolution programs (proceedings of the International conference on tools for AI).

Microsoft corporation (1990) Microsoft C version 6.00 reference manual.

Montana D.J. (1990) Empirical learning using rule threshold optimization for detection of events in synthetic images (Machine learning, 5, p. 427-450).

National Research Council (1985) [NRC85] Highway Capacity Manual (Special report 209, Transportation research board, Washington D.C.)

Orlandi A. (1974) [Orl74] Tecnica della circolazione (Patron editore Bologna).

Proli G. (1992) I classifier systems (Internal report, Università di Bologna).

Regione Emilia Romagna (1982) Piano Regionale Integrato dei Trasporti.

Vose M.D. (1990) Generaliziong the notion of schema in genetic algorithm (Atificial intelligence 50 p: 385-396).

Wilson S.W. & Goldberg D.E. (1990) A critical review of classifier systems. (ICGA pp.244-255).

Zhou H.H. (1990) CSM: computational model of cumulative learning (Machine learning 5, p.383-406).

Genetic Search for Optimal Representations in Neural Networks

Paul W. Munro
Department of Information Science
University of Pittsburgh
Pittsburgh PA 15260
USA

munro@lis.pitt.edu

Abstract

An approach to learning in feed-forward neural networks is put forward that combines gradual synaptic modification at the output layer with genetic adaptation in the lower layer(s). In this "GA-delta" technique, the alleles are linear threshold units (a set of weights and a threshold); a chromosome is a collection of such units, and hence defines a mapping from the input layer to a hidden layer. The fitness is evaluated by measuring the error after a small number of delta rule iterations on the hidden-output weights. Genetic operators are defined on these chromosomes to facilitate search for a mapping that renders the task solvable by a single layer of weights. The performance of GA-delta is presented on several tasks, and the effects of the various operators are analyzed.

Introduction

The power of multilayer feed-forward networks to learn and generalize is well documented but not well understood. Existing techniques for training multilayer feed-forward networks generally rely on backpropagation of error to find appropriate internal representations; this approach has been criticized by biologists for nonconformity with neurophysiological dogma and by engineers for its very long convergence times on nontrivial tasks.

It is known that most mappings cannot be accomplished by letting each component of the target space directly compute a semi-linear function (e.g., linear threshold) across the input components; in general, an intermediate stage is required to transform a pattern space into a representation suitable for the output units. A useful mapping from the input units to the socalled "hidden units", that produces an adequate representation of the pattern space for the task, can sometimes be found by the generalized delta rule (commonly referred to as backward propagation of error) for mulilayered networks, a gradient descent procedure. Gradient descent techniques do not always find a representation that works; they can converge to states that are optimal locally but not globally. Even when a globally optimal state is found, it may require hundreds of thousands or millions of pattern presentations. Despite these problems with local minima and convergence time, backprop is currently the best technique available. This paper presents an alternative approach to finding hidden unit representations using a variant of genetic algorithms.

Background

The Perceptron

Powerful learning algorithms have been developed for feed-forward networks which can perform arbitrary mappings from one space to another; that is, a pattern of activities across a set of input units is transformed to a pattern across a set of output units. Rosenblatt [1] developed a learning rule for his socalled "two layer Perceptron" with output units which each perform linear threshold calculations across the input units directly. Recognizing the inadequacy of two layers, Rosenblatt attempted in vain to extend his learning rule to a three layer architecture.

Let a categorization task be defined in terms of a set of vectors, **A**; the task is to find a function that correctly determines the membership of a given vector. If the membership of a particular vector **x** can be determined according to a rule of the form $\mathbf{w} \cdot \mathbf{x} > \theta$, for some weight vector $\mathbf{w}$ and threshold θ, then the task is termed *linearly separable*. Rosenblatt [1] developed the Perceptron Learning Rule (PLR), which can be used to find a set of weights that solves an arbitrary linearly separable categorization task. The PLR is defined for single layer networks (i.e., networks consisting of a set of input units directly connected to a set of output units). The PLR is a supervised error correction routine. That is, the initial weight components are random; they are incrementally modified by repeatedly presenting input examples, $\mathbf{x}$, and comparing the resulting output with the desired output. If there is a discrepancy, the weights are adjusted.

Unfortunately, most categorization tasks are not linearly separable; however, *they can be made linearly separable by transforming the representation space with a carefully chosen semi-linear mapping*. Hence, a solution to an arbitrary categorization problem generally exists in the form of a multilayered network of semi-linear units, but PLR is not applicable in this case, since it is restricted to single layer networks. Rosenblatt postulated an intermediate layer (A) of associative units between the sensory (S) and response (R) layers, and generated random mappings from S to A using a variety of distributions. PLR was attempted on the A to R connections for each mapping, and statistics were compiled for each random number distribution on several tasks. In Rosenblatt's enumeration of the capabilities and deficiencies of the approach, the former are eclipsed by the latter. A mathematical analysis of the failure of the Perceptron was provided by Minsky and Papert [2] in their book, *Perceptrons*, which was instrumental in diverting the attention of the scientific community away from neural networks for over a decade.

Rosenblatt [1] suggests a "back-propagating error correction procedure" (p.292) to adjust the S-A weights. This entails assigning an "error-indication" to each A unit; the S-A weights would be modified based on this error. Other researchers [3, 4, 5] derived back-propagating error correction procedures from gradient descent assumptions that have been so successful that the study of Multilayered Perceptrons (MLPs) has been revitalized to the extent that hundreds of papers have been published offering analyses of, extensions to, and applications of this approach, commonly referred to as *backprop*.

Backprop requires the computation of partial derivatives of the responses of the output units with respect to the weights in the network, and so the threshold activation functions of Rosenblatt must be replaced by differentiable functions. The usual choice is the sigmoid $f(x)=(1+e^{-x})^{-1}$. The resulting rule is simulataneously applied to all weights in the network. Subject to the appropriate selection of dynamic partamaters, such as the learning rate and momentum [5], backprop generally performs very well relative to competing approaches for tasks in categorization, nonlinear mapping, and function optimization.

While it has emerged as a leading learning algorithm, the performance of backprop is not always satisfactory. The algorithm does not always converge to a solution in an acceptable time; it can sometimes converge to a nonsolution. In terms of the error surface across which the gradient descent process operates, these problems correspond to "shallow regions" and local minima. The poor convergence performance of backprop can also be described in terms of the pattern representations presented to the output units. The preceding layers must transform the input pattern into a representation which can be processed in a single layer; e.g., for a categorization problem (binary valued output), the representation seen by the output unit must be linearly separable.

Genetic Algorithms and Neural Networks

Several techniques for combining GA techniques with backprop have been reported recently; see for example, [6, 7, 8]. These generally use GA techniques to choose optimal architectures or simulation parameters, such as the learning rate, momentum, gain, etc. Chalmers [9] shows how a learning rule can evolve using genetic operators. A recent overview of this hybrid area is given by Schaffer, Whitley, and Eschelman [10]. Yet another approach is under consideration in this paper.

A New Approach

Another approach to the synthesis of the GA and neural network learning methodologies is put forward in this research paper. Here, a GA is used to search for mappings from the input units to the hidden units that give task-appropriate representations at the hidden unit level; i.e., representations that permit solution with a single semi-linear mapping (from the hidden units to the output units). Learning in a single layer system (the simple delta rule) is so much faster than learning a multilayer system with backprop (the generalized delta rule) that it can be used to test each mapping and assign a fitness value.

The *GA-delta* procedure begins with the random generation of a population of input-to-hidden unit mappings (IHMs), where each mapping is characterized as a collection of a fixed number of hidden units. The hidden units are linear threshold units that are each represented as a set of connection strengths weighting the input components and a threshold value. Here, the IHMs are chromsomes and the hidden units are alleles; that is, the alleles are real valued vectors. Several GA operators are applied to the population, including selection of fit individuals, crossover, and mutation. At this level of description, the approach is similar to those of Wilson [11] and Whitley, Starkweather, and Bogart [12], but the techniques for genetic representation, chromosome generation, crossover, and mutation are very different.

Generation of networks

The initial IHMs must be generated using a random process that does not select the weights and thresholds of the hidden units independently. If the weights are chosen randomly, independent of the threshold values, several "trivial" hidden units are created; here, a trivial unit is one that responds identically for every input pattern in the training set. For example, consider a unit receiving two inputs. Let the threshold be 0.5 and the two weights be chosen randomly. The probability of selecting weights that result in a response of 1 to a particular pattern corresponds to the relative areas on each side of a line (Figure 1a). Several patterns break the weight space into regions corresponding to different response characteristics across the set of patterns (Figures 1b, 1c). Trivial units are more likely with smaller pattern sets (Figure 1d).

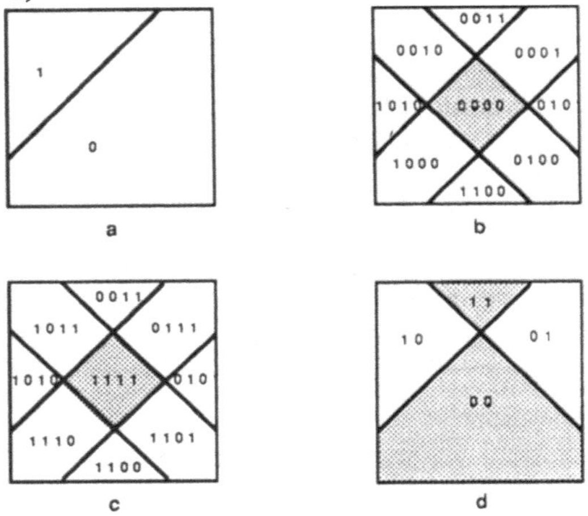

Figure 1. *Two dimensional weight space.* For a given threshold θ (here θ=0.5), each pattern separates the space into two half-planes. (a) The effect of a single pattern (-1,+1); weights in one half plane yield a response of 0 to the pattern, weights in the other half plane yield 1. (b) Four patterns break the space into 9 regions, each with a different set of responses over the training set. Each region is labelled by a string of four bits indicating the responses to each of the four patterns; regions corresponding to trivial units are shaded. (c) The complimentary set of response properties is obtained with here θ=-0.5. (d) Fewer patterns result in a higher probability of randomly selecting trivial units.

In order to generate a set of hidden units with a uniform distribution of response properties and to exclude units with trivial response properties, the weights and thresholds are chosen using the following Hidden Unit Generation (HUG) routine (see Figure 2). For each unit, a set of weights is randomly selected independently from a unform distribution, and the corresponding linear sums are computed for each pattern in the training set. A threshold is then selected such that the unit gives a response of 1 to K of the N input patterns, and a response of 0 to the remaining N-K, where K is randomly selected from the range $1<K<N$; these are strict inequalities to preclude the generation of units which do not discriminate among the patterns. That is, the response properties of all units generated by HUG fall in the unshaded regions of Figure 1.

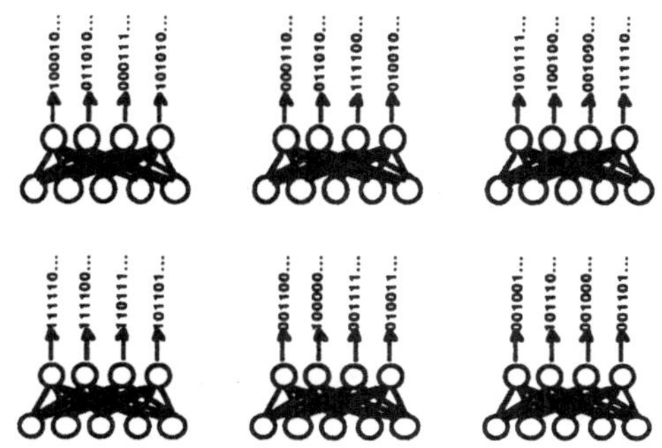

Figure 2. *Generation of the initial population.* Using the Hidden Unit Generation (HUG) procedure, a population of networks is initialized. The HUG procedure initializes the weights and thresholds of each hidden unit for a particular set of patterns such that the response of the unit to the set has some variability; i.e. so the set of responses is not all 1s or all 0s. The figure depicts an initial population of six networks (sets of input units connected to sets of output units); the weights from the hidden units to the output units are initially zero. The numbers above each hidden unit represent its responses to the patterns in the pattern set.

Fitness Calculation

Each IHM must be evaluated for its fitness as a contributor to successive generations. Calculation of the fitness is done by applying gradient descent learning (the delta rule) to the weights from the hidden layer to the output layer for a small number (usually 20) of iterations, and computing the squared error summed over all patterns and all output units. This gradient descent is done deterministically so that comparison of the resulting fitness values is unbiased (the weights from the hidden units to the output units are always initialized at zero). The hidden-to-output weights are initialized to zero and the weight changes are not implemented until the appropriate changes have been calculated for each pattern in the training set. The fitness is computed as a function inversely dependent on the error (high fitness means low error and vice versa). IHMs are randomly selected to continue to the next generation with probabilities proportional to their respective fitnesses.

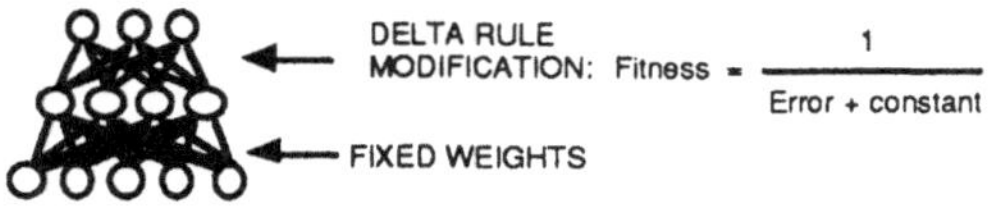

Figure 3. *The GA-delta procedure.* This method incorporates two kinds of learning: genetic search at the lower level (input to hidden) and gradient descent at the upper level (hidden to output).

Crossover

IHMs are selected for successive generations in "parent" pairs. A *crossover operator* is applied to each parent pair is combined to produce two "child" IHMs. This is generally accomplished by randomly selecting a number, h, of the hidden units from one parent IHM, and H-h hidden units from the partner IHM to produce one child IHM. The unused units (H-h from the first parent and h from the second) are used to produce a sibling. If h=0 or h=H, the parents survive to the next generation intact.

Preliminary investigations have shown that random crossover gives very slow progress. A heuristic was designed in response to the observation that random crossover produces IHMs with redundant hidden units [12]. The Hidden Unit Network Comparison Heuristic (HUNCH) consists of two steps: (1) Comparison of the response properties hidden units of each parent network with those of its partner's, and (2) exchange of hidden units which are similar but nonidentical. To avoid the computation of H(H-1)/2 similarity values, HUNCH is currently implemented by performing a single exchange; a hidden unit is randomly selected in one parent and exchanged with the most similar (but nonidentical) hidden unit in its partner.

Mutation

As in most genetic algorithm procedures, a mutation operator is included to promote diversity in the population. Each hidden unit is susceptible with a small probability (about 0.01 to 0.1) to a random perturbation in its weights with each generation. This can be implemented in several different ways. Two methods have been used: (1) a small random change to each weight, or (2) rerandomization of a single weight in the hidden unit. In either case, the unit must be checked in case its new weights render its response properties trivial (i.e. the unit gives responses of all 1s or all 0s to the training patterns). If the mutated unit does not discriminate among the input patterns, a new threshold is assigned to the unit using the technique from the HUG procedure described above.

The formation of a new individual from two parents (using the crossover and mutation operators) is illustrated in figure 4.

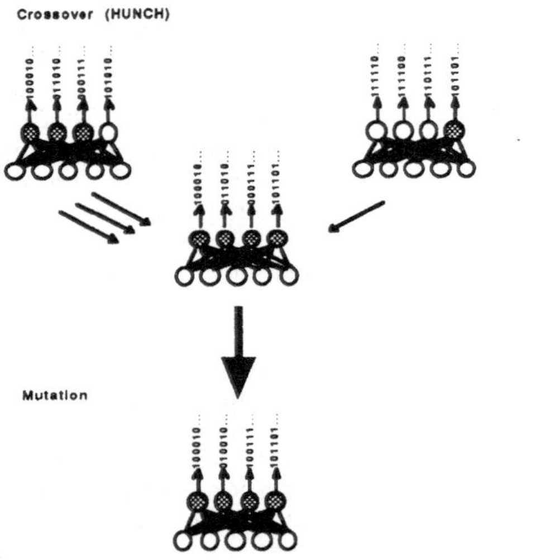

Figure 4. *Crossover and mutation of networks.* The Hidden Unit Network Comparison Heuristic (HUNCH) operates on two "parent" networks to produce a "child" network. One hidden unit is selected in one of the networks (here, the fourth unit in the right parent), and exchanged with the unit in the other parent with the most similar response properties. In the mutation stage, weights are subject, with a very low probability, to rerandomization. Here, two weights have been altered, thereby changing the response properties of the corresponding hidden units.

Results

The GA-delta method, as described above, has been applied to a number of small scale tasks, including parity in 2 (XOR), 3, 4, and 5 bits. Examination of the time course of the populations error profile (Figure 5) shows that while the minimum error descreases, the population maintains a diverse range of individual errors. This phenomenon is probably related to the technique used for selection; e.g., the roulette wheel selection and the particular choice of fitness function.

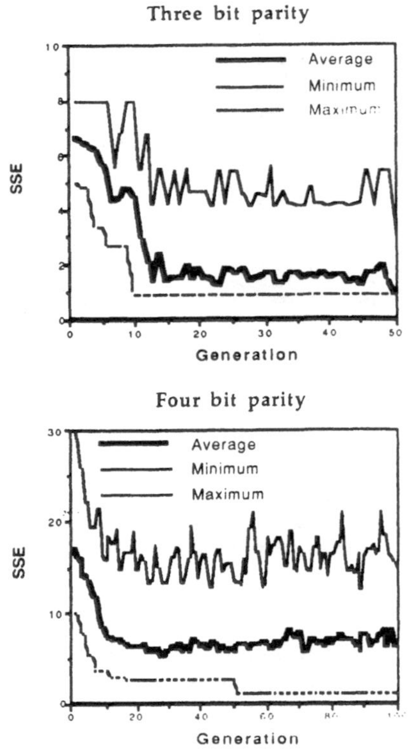

Figure 5. *Error profiles of populations over time.* The time course of error reduction is shown for populations of 100 IHMs trained on the three bit and four bit paity tasks.

Direct comparisons with standard backprop simulations matched by task and architecture (number of hidden units) are difficult due to the platform dependence; the two frameworks have different strengths with different hardware, for example massively parallel hardware will benefit the GA-delta approach, and floating point accelerators are expected to give a relative advantage to backprop. Despite the problems with interpreting such information, Table 1 presents some raw data that contrasts the performance of GA-delta with respect to straight backprop in terms of convergence times for 4 and 5 bit parity tasks. The reader should consider that in addition to the idiosyncrasies of the algorithms, the results are most likely task-sensitive as well.

TABLE I Results of preliminary tests.

Four Bit Parity Task

	BP #iterations	GA #generations
	100000+	28
	35000	50+
	95000	50+
	200000+	11
	200000+	16
	150500	11
	200000+	50+
	300000+	38
	300000+	50+
	200000+	11
success rate	3/10	6/10
Time	93833	19.2

Five Bit Parity Task

	BP #iterations	GA #generations
	7329	27
	15308	22
	181619	37
	200000+	33
	200000+	79
	200000+	37
	200000+	32
	200000+	28
	200000+	24
	200000+	23
success rate	3/10	10/10
Time	68085	34.2

Backprop Parameters	*GA-delta Parameters*
learning rate =0.1	population size = 100
momentum =0.9	crossover rate = 0.5
fitness test uses 20 iterations	mutation rate = 0.01

Discussion

While it is a far cry from a precise model of the interaction of natural selection and synaptic adaptation, the GA-delta procedure is partially motivated by this interaction. The analogy between the model and nature is only put forward at an abstract level, in that natural selection is used to find intermediate representations that facilitate synaptic learning at later stages of the network.

The above results were obtained using only a few parameter sets; a rigorous study of the GA-delta method will require more systematic and thorough investigation of the interaction of crossover rate with mutation rate in the context of several tasks and network architectures. The proposed research will begin with a study of this kind. Specifically, for each of several "small" tasks (e.g. parity, encoders, and symmetry), several simulations are planned, with a variety of crossover and mutation rates. Initially, zero crossover will be analyzed to investigate the role of mutation alone; this effect will be studied for both types of weight mutation described above (at least). Crossover will also be examined in isolation (using zero mutation).

The fitness function must also be examined and improved. The current function is a monotonically decreasing function, f(E), where E is the summed squared error after a fixed number of training epochs; the number of epochs (arbitrarily chosen as 20) could be improved, other function could be examined, or other techniques might be developed that do not require any gradient descent learning. This general approach is the best known to us, that gives a continuous measure of fitness.

References

[1] Rosenblatt, F. *Principles of Neurodynamics.* Spartan: Washington DC, 1962.

[2] Minsky, M. and Papert, S. *Perceptrons* Cambridge: MIT Press, 1968.

[3] Werbos, P. *Beyond regression: new tools for prediction and analysis in the behavioral sciences.* Unpublished doctoral dissertation, Harvard University. November, 1974.

[4] Parker, D. *Learning logic.* TR-47. MIT Center for Computational Economics and Statistics. Cambridge MA, 1985.

[5] Rumelhart D., Hinton G., and Williams R. *Learning representations by back-propagating errors.* Nature **323**:533-536, 1986.

[6] Belew, R. and Gherrity, M. *Evolution, learning, and culture: computational metaphors for adaptive search.* CSE TR CS89-156 UCSD. La Jolla, California, 1989.

[7] Davis, L. *Mapping neural networks onto classifier systems.* In: Proceedings of the Third International Conference on Genetic Algorithms. San Mateo CA: Morgan Kaufmann, 1989.

[8] Miller, G. Todd, P., and Hegde, S. *Designing neural networks using genetic algorithms.* In: Proceedings of the Third International Conference on Genetic Algorithms. San Mateo CA: Morgan Kaufmann, 1989.

[9] Chalmers, D. *The evolution of learning: an experiment in genetic connectionism.* In: Proceedings of the connectionist Summer School. D. S. Touretsky, J. L. Elman, T. J. Sejnowski, and G. E. Hinton, eds. San Mateo CA: Morgan Kaufmann, 1990.

[10] Schaffer, J. D., Whitley, D., and Eschelman, L. J. *Combinations of Genetic Algorithms and Neural Networks: A Survey of the State of the Art.* In: Proceedings of the International Workshop on Combinations of Genetic Algorithms and Neural Networks, J. D. Schaffer and Darrell Whitley, eds. IEEE, 1992.

[11] Wilson, S. *Perceptron Redux: Emergence of Structure.* In: Proceedings of the Conference on Emergent Computation. Los Alamos, 1990.

[12] Whitley, D., Starkweather, T., and Bogart, G. *Genetic algorithms and neural networks: optimizing connections and connectivity.* Parallel Computing **14**:347-361, 1990.

Searching among Search Spaces:
hastening the genetic evolution of feedforward neural networks

Vittorio Maniezzo

Artificial Intelligence and Robotics Project
Dipartimento di Elettronica e Informazione - Politecnico di Milano
20133 Milano, Italy
MANIEZZO@ipmel2.elet.polimi.it

Abstract

The paper introduces a genetic paradigm for evolving feedforward neural network, where both the network topology and the weights distribution are coded in the individuals. The resulting binary coded strings are too long for efficient evolution, thus two novel techniques are employed. The first consists in a coding procedure that allows the genetic algorithm to evolve the length of the coding string along with its content. This goes by the name of granularity evolution procedure. The second technique is inspired by linear programming and yields a fast and accurate fine tuning of the solutions.

These ideas have been implemented in a running system. Computational results show how during evolution the genetic algorithm uses several coding lengths, thus it autonomously identifies the search spaces that lead to more promising solutions.

1. Introduction

Genetic algorithms (GAs) have proved to be a viable alternative for artificial neural networks (ANNs) learning to more commonly used algorithms, notably to backpropagation. Genetically evolved ANNs have usually been tested in the literature over a standard problem test suite relative to Boolean function learning, originally proposed by Rumelhart [1] to evaluate the effectiveness of backpropagation. Several authors (Scholz [8], Kitano [2], Whitley, Hanson [3]; and others) obtained comparable results on those problems using genetically evolved nets.

This research has been partially supported by grants of "Progetto Finalizzato Informatica e Calcolo Parallelo", subproject 2 "Processori dedicati" of the Italian National Research Council.

Moreover, gradient-descent learning algorithms cannot usually learn the topology (number of nodes and their connections) of the net whose weights' distribution they try to optimize. On the contrary, interesting results have been obtained with networks that evolve genetically their own topology along with the weights, distribution, applied to an animat control problem [4].

Search in the space of topologies relieves the network designer from the burden to identify the network structure that allows a problem to be solved: a process that can otherwise be accomplished only through a difficult analysis of the potentialities of a net with a given topology or through the use of oversized networks, that however define huge search spaces for the learning process and therefore make it far slower than it could be.

In this paper it is argued that the main computational advantage of topology learning lies in this possibility to autonomously identify minimal search spaces containing at least one reachable solution of the problem tackled. This capacity is further enhanced by another net coding feature that is let evolve concurrently to the net in the system I present: the bit length of weight coding, i.e., the coding *granularity*.

Granularity is a coding parameter which has always been specified in the design phase and whose computational impact is at least as heavy as that of network topology: too coarse granularities hamper problem solution, too fine-grained granularities lead to huge search spaces, thus slow down learning.

In the paper it is shown how searching in the space of granularities can lead to network coding advantages greater than those due to search in the space of topologies.

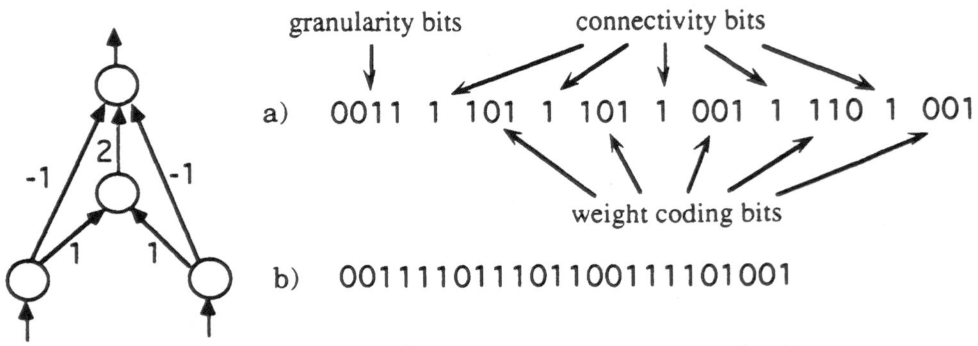

Figure 1. Network coding: a) annotated code b) actual string.

An algorithm is presented that includes both these possibilities. Its implementation has been tested over the standard problem test suite and the results are presented. The system presented in this paper is called ANNA ELEONORA, standing for Artificial Neural Networks Adaptation: Evolutionary LEarning Of Neural Optimal Running Abilities [5].

In particular, it is shown how genetic evolution tries to minimize both weights coding granularities and network topologies. Short coding allows early identification of promising weights distributions, eventually to be refined by subsequent code lengthening. Good topologies lead to a reduction of the number of nodes of the net and to a reduction of the connectivity among the useful nodes: both these processes essentially reduce the number of links of the net thus the length of the coding strings that have to be optimized by the GA.

2. Anna Eleonora

This section presents the evolutionary algorithm which the ANNA ELEONORA system is based on. It consists essentially of a genetic algorithm, used both for topology and for connection weights learning; however, two new and general mechanisms have been added to it. The first one consists of a new genetic operator, so far untested in complex discrete optimization problems, called GA-simplex recently proposed by Bersini and Seront [6]. The second consists of a new dynamic encoding procedure that allows the algorithm to autonomously identify an opportune length of the coding string (therefore the discretization of the potentially real-valued weights, the *coding granularity*). This possibility is ensured by the interpretation of a segment of the string as the value of the length control parameter to be applied to the string itself, a process much alike the evolution of mutation probabilities carried on by Evolution Strategies [7].

The utilization of the evolutionary programming paradigm in a specific application entails two activities: the definition of the coding structure and the definition of the operators to use. In the case of evolution of ANN substantial work has already been presented, allowing some insight in the strengths and weaknesses of the different alternatives.

Our choice has been to let both the net architecture and its weight distribution evolve. However, the representation scheme followed differs from the ones already tested because we put much emphasis on the minimization of the length of the coding string, thus of the search space. We came out with a variable-length binary coding, where also a control parameter is represented. Acting upon these string are potentially standard genetic operators: we modified the probability distribution for mutation to gain efficiency.

2.1 coding scheme

We used an extended direct encoding scheme, where each connection is represented directly by its binary definition. This allows us to achieve better genotype evaluation than what we could achieve by using an indirect encoding scheme and a developmental rule; this reliability in fitness assessment is needed especially in tasks where fitness information are intrinsically fuzzy and computationally hard to get.

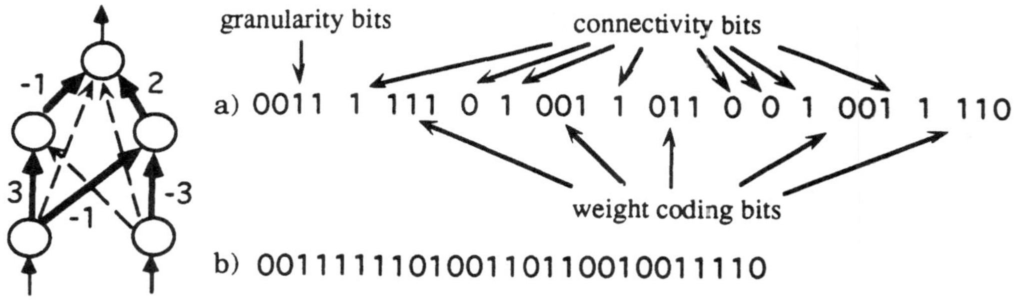

Figure 2. Network coding: a) annotated code b) actual string.

The internal organization of the coding string is decisive for the effectiveness of crossover. The representation chosen wants to identify the networks nodes as basic functional units and keeps together all the information relevant for a node: its input connectivity pattern and the relative weight distribution. Connectivity is coded by presence/absence bits (the *connectivity bits*), that specify whether any possible connection is present or not. Immediately after each connectivity bit there is the binary coding of the (eventual) relative weight. The first byte of the string specifies the number of bits (the *granularity*) according to which the weights of the present connections have been codified.

For example, Fig.1 presents a simple network and its coding; for simplicity, biases have been left out. Fig.2 shows another net coded with granularity 3 but where some connections are declared nonexistent: they are drawn with dashed arcs on the graph on the left and the corresponding connectivity bit is 0 in representation a). Note how no weight follows 0 connectivity bits, thus achieving a variable-length coding even for networks with the same granularity.The possibility of having strings of different length is further emphasized by the granularity mechanism. As already mentioned, coding granularity is a control parameter whose value is coded within the string it is used to interpret.

The importance of a good granularity choice is paramount: if too few bits are chosen, there could be very few or no solution in the search space at all because the useful combinations of real connection weights cannot be translated to integers with the used discretization; if too many bits are used, the strings representing the networks can become very long, thus forcing the learning to be carried out in a huge search space, therefore requiring intolerable amount of time. Since good granularity values are problem-dependent and there is no a priori way to tell which they are, our approach proposes to let the algorithm itself determine them. Note how granularity evolution is an evolutionary mechanism that could be adopted in any problem of optimization of binary-coded functions, not being restricted to the optimization of ANNs.

2.2 Genetic operators

ANNA ELEONORA employs four genetic operators: reproduction, crossover, mutation and GA-simplex. In this section we will briefly get through them and illustrate their characteristics, when they differ from the standard ones.

Reproduction

It is the standard roulette wheel reproduction operator [8], with Montecarlo selection with probabilities based on fitness levels. Since we used no explicit local optimizing operator (but see the discussion on GA simplex later on), we used fitness scaling to improve fine-tuning of the solutions.

Crossover

There has been some debate in the literature about the opportunity of applying crossover to the evolution of ANNs [2], [9]. These considerations have induced several researchers (Fogel, Fogel and Porto, [10]); Parisi, Cecconi and Nolfi [4], among others) to use mutation as the only genetic operator.

	internal representation	network coding
parents	11 1 010 0 001 1 110 1 001 0 111 1 111	11 1 010 0 1 110 1 001 0 1 111
	10 0 110 1 010 0 001 0 010 1 001 1 110	10 0 1 01 0 0 1 00 1 11
offspring	11 1 010 0 001 0 001 0 010 1 001 1 110	11 1 010 0 0 0 1 001 1 110
	10 0 110 1 010 1 110 1 001 0 111 1 111	10 0 1 01 1 11 1 00 0 1 11

Fig.3. Mating two different length strings

We do not want to delve into theoretical considerations here,: in section 3 we present some computational results showing how the use of crossover can benefit the evolution, using our representation scheme and allowing the GA to evolve both architecture and weights.

The crossover we used is totally standard: a single cutting point chosen with uniform probability over the string length and a swap of the genetic material following it. A simple implementational trick allows mating of network with different connectivity and/or different granularity, with no modification of the operator. Coded networks are actually implemented using strings of the same length: the maximal possible length (all connection present and maximal granularity). Crossover is applied to those strings and therefore it has no application problem. Granularity and connectivity bits are actually used to determine *how many* of the weight bits following each granularity one are to be considered for the network definition. Fig.3 exemplifies a crossover of two downscaled different length strings, granularity bits are underlined and connectivity bits are boldface, extra spaces have been added to ease readability.

Obviously the search space is still constrained by the granularity and the connectivity and does not extend to all the possible internal representations.

Mutation

It is the standard mutation operator negating each bit with probability p_m, except that we split it into three: a granularity bits mutation operator, a connectivity bits mutation operator and a weight bits mutation operator. This has been done because of the different interpretation of the bits, which suggests that a unique mutation probability value will be suboptimal with respect to three different values.

GA simplex

This operator has been recently presented by Bersini and Seront [6] as a way of hybridizing the exploration effectiveness of GAs with classical optimization methods for local optimization of the GA individuals. Several authors have already advocated the need of a local optimizing operator [11], [12], which departs from the biological inspiration of the GA paradigm, but which has proved to dramatically increase the effectiveness of the search process.

Local search in the case of feedforward ANNs evolution is usually performed applying backpropagation (which is a gradient-descent algorithm) to the individuals of the GA population [2], [9]. GA allows an efficient search over the whole search space, thus escaping from local minima, BP optimizes locally the weight distribution. There are two main drawbacks in this approach for the kind of application we want to face: it is extremely computational expensive and it requires too many computations of the objective function.

GA simplex on the other hand do not require any extra computation of the objective function, optimizes on the basis of a *set* of solutions, thus promoting cooperation among individuals, and ensures more exploitational power than bare GA, even though it does not guarantee the local optimality of the solutions achieved.

GA simplex is a ternary operator, inspired by the linear programming simplex procedure, that tries to exploit the fitness landscape identified by the already available solutions. It works on three individuals of the population x_1, x_2, x_3 and generates a new one, x_4, on their basis. The simple constructive algorithm is:

Step 1 Rank the three parents by fitness value, (suppose $ff(x_1) \geq ff(x_2) \geq ff(x_3)$).

Step 2 For each i-th bit of the strings

Step 3 if $x_{1i}=x_{2i}$ then $x_{4i}:=x_{1i}$

Step 4 else $x_{4i}:=$ negate(x_{3i})

The original proposal suggested to apply step 4 in probability. We experimentally found to be more effective to apply it deterministically and to subject the use of the operator to a specific probability parameter, p_s.

An example of the application of GA simplex is presented in Fig.4, where we suppose we want to maximize a fitness function given by the interpretation of each individual as a binary coding of an integer value.

individuals	ff(x)=x		
1011101	93		
0110110	54	→	1110101 (117)
0011010	26		

Fig.4. GA simplex

Bersini and Seront applied GA simplex in the context of real function optimization, with promising results. We tested it in the completely different application area of ANN evolution and we found it to provide significant benefit to the search process, thus suggesting that coupling evolutionary and traditional optimization techniques can result in simple, effective and robust algorithms that outperform the parent algorithms in complex tasks.

3. Computational results

In the following are presented the results obtained applying ANNA ELEONORA to boolean function optimization. The set of problems used has originally been proposed by Rumelhart, Hinton and Williams [1] to assess the effectiveness of BP. We used 6 of the proposed problems:

Xor: evolve a 2-inputs, 1-output network that learns to give in output the exclusive or of the input values. It is an easy task, used mainly for comparison. The initial configuration of the network we used to solve it was 2-2-2-1 (2 input neurons, 2 neurons in the first hidden layer, 2 neurons in the second hidden layer and one output neuron).

Parity: a 4-inputs, 1-output problem where the output should be 1 if there is an odd number of 1s in the input pattern, 0 otherwise. The initial configuration of the network we used to solve it was 4-4-4-1.

Encoding: a 8-inputs, 8-outputs problem, requiring a layered net with only 3 neurons in the hidden layer, the output should be equal to the input for any of the 8 combination of seven 0s and a 1. The initial configuration of the network we used to solve it was 8-3-0-8, strictly layered.

Symmetry: a 4-inputs 1-output problem where the output is required to be 1 if the input configuration is symmetrical, 0 otherwise. The initial configuration of the network we used to solve it was 4-4-4-1.

Addition: a 4-inputs, 3-outputs problem, where the output should be the result of the binary sum of two 2-bits input numbers. The initial configuration of the network we used to solve it was 4-4-4-3.

Negation: a 4-inputs, 3-outputs problem, where the three output bits should be equal to the three rightmost input bits if the leftmost one (acting as negation bit) is 0, to their negations if it is 1. The initial configuration of the network we used to solve it was 4-4-4-3.

Since our neurons have a real-valued output ranging on [-1, 1], we scaled -1 to 0 and computed the fitness as the LMS error, with the genetic scaling mechanism superimposed. The network topologies are bigger than the minimal possible ones in order to assess the irrelevancy detection potentiality of ANNA ELEONORA [5].

3.1 Parameter settings

The first set of tests was carried out in order to assess the optimal parameter values. The algorithm has quite a few parameters and there is no way to a priori individuate useful combinations of values. The parameters tested were: n, cardinality of the population, p_c, crossover probability, p_s, GA simplex probability, p_{mg}, granularity mutation probability, p_{mc}, connectivity mutation probability, p_{mw}, weight mutation probability, γ, slope of the neurons sigmoid transfer function in the origin.

The results obtained, averaged over ten runs, are presented in Table 1. All tests have been made using the Addition problem and letting the algorithm run for 3000 generations. The sets of values used for parameter settings were enlarged according to the results obtained. The test sets were initialized with values centered on the setting proposed by de Jong [13] for function optimization and enlarged in case the better performing value turned out to be at an extreme of the considered range.

Table 1. Parameter settings.

p_{mw} = 0.0005	p_{mc} = 0.01
p_{mg} = 0.1	p_c = 0.04
p_s = 0.9	γ = 1
n = 100	

- The best values for the three proposed mutation probabilities as expected are different; in particular, p_{mw} is the order of $1/l$ where l is the string length, in accordance with the prescriptions of Mühlenbein, while the other two mutation probabilities are bigger, testifying a less disruptive impact of the mutations they control.
- Crossover probability is smaller than the commonly adopted values suggested by de Jong (0.6, in [13]) or by Grefenstette (0.95, in [14]). This testifies a less efficient building block decomposition than in the case of function optimization; however, very small crossover probabilities or no crossover at all are less efficient. There is therefore empirical evidence that with the representation used functional (building) blocks can indeed be identified and the search process benefit from their recombination.
- There is evidence of the effectiveness of GA simplex. Its use greatly improves the results obtained, even though applying it too often could lead to early convergence and local minima problems.
- The sigmoid slope is a parameter that affects dramatically the results. A slope of 5 almost makes it a step function, while a slope of 1/5 results in a very mild inclination. Given the boolean discretization of the weights, mild inclination ensures more input combinations to non input neurons and more graded output, but it also entails a correspondingly harder search space. The best slope value corresponds to a rather steep sigmoid, thus to neurons that have a quasi-boolean output behavior. This is correct in this application, but it cannot be generalized to every class of problem [5].
- Bigger populations provide better results, we proposed the value of 100 because it was the biggest tested one. However, the computational cost of evolving big populations does not seem to be rewarded by a comparable improvement of the solutions obtained, for $n \geq 30$. For this reason, we considered n=30 to be a good choice in sequential evolution. Far bigger population cardinalities have been tested on the Connection Machines (n=4096 and n=16384) using a parallelized algorithm, with the results presented in [5].

3.2 Search effectiveness

Having identified a good parameter setting, we started a set of runs to evaluate the effectiveness of Anna Eleonora on the Rumelhart's test suite. Each problem has been tested at least ten times, both on a vax (the sequential model) and on a CM2 (the parallel model). Each run was stopped after 3000 iterations and we present here the results achieved.

For comparison, we implemented also a standard backpropagation learning procedure and tested it (on the vax). BP was allowed to go on for the same number of function evaluations used by the sequential GA (30x3000=90000 evaluations).

Since the neurons we use are real-valued, we implemented a thresholding procedure to binarize the network output values; however, the fitness is computed by a scaled LMS of the error. It is therefore possible to have suboptimal fitnesses for network that have correctly learned the function proposed. In Table 2 we present, for each test problem, the average results of the 10 trials of the sequential version (SGA), of the parallel one (PGA) and of BP.

Table 2. Computational results.

	SGA	PGA	BP
xor	4.00	4	2.24
encoding	8.00	8	4.27
parity	15.30	16	10.34
addition	14.88	16	10.26
symmetry	15.89	16	9.89
negation	14.37	16	9.65

Some considerations are worth being made.

- The xor problem was consistently solved optimally by ANNA ELEONORA in a few dozens generations. The parallel model often had an optimal solution already in the first randomly constructed population. Even the encoding problem has always been solved optimally, but it required many more iterations.
- The parallel version was able to solve to the optimum every instance of every problem, usually within 200 iterations. Tests of our PGA involved populations of 4096 individuals and a radius 2 neighborhood, and lasted about 20' on a CM2 in time-sharing mode.
- BP has a high variance of the results over the 10 runs. Sometimes it moves steadily (though slowly) toward the optimum, sometimes it never modifies much the initial fitness, which is close to the half of the optimum.

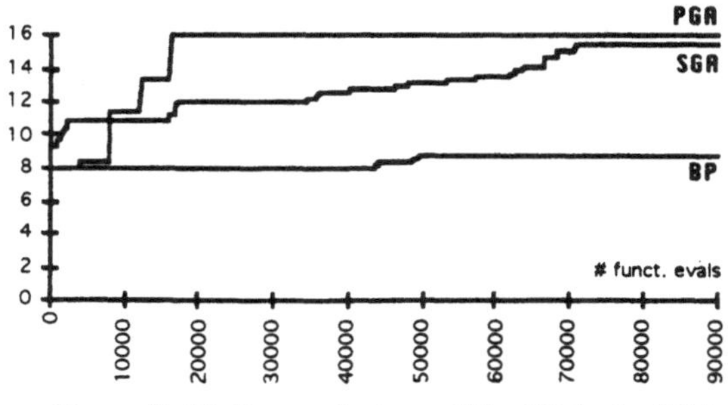

Figure 5. Median evolutions of the PGA, the SGA and BP applied to Neg.

In Fig.5 we present the evolutions that achieved the median result over the trials for the negation problem, plotted against the number of function evaluations.

3.3 Search for smaller spaces

One of the main assumptions of our work was that, given the possibility, a search process tries to identify a promising search space, reducing the length of the coding strings. This has been confirmed for the connection removal mechanism, here we present some results for the granularity updating one.

Table 3. Granularities of best networks

	Granularity	Best value	Iteration
Xor	3	4	877
Enc	3	8	2838
Par	3	16	545
Neg	6	15.42	2795
Sym	3	16	1580
Add	7	15.44	2694

In table 3 are presented the granularities of the best performing networks for each problem. Note how easy problems (Xor, Par) have been solved in few iterations with low granularity, while problem harder to solve (Neg, Add) have been tackled best with high granularities. The high value of p_{mg} ensures that the search considers always all granularity spaces, so it is possible, such as in the case of Enc, to find an optimal solution with low granularity even in a very late phase of the search.

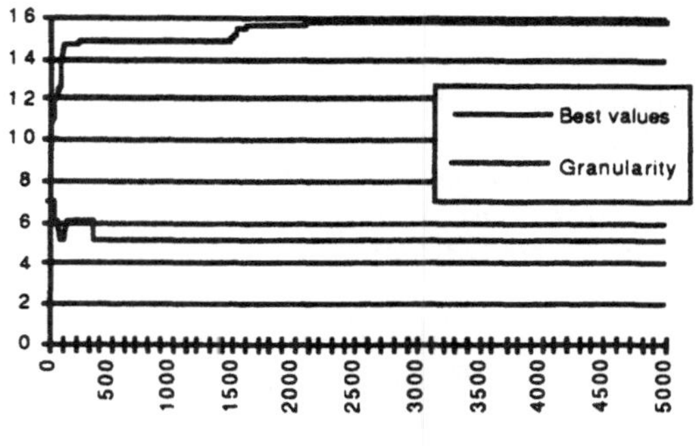

Fig.6. Granularity evolution for negation problem.

The granularity reduction effort is also evident along the search. In Fig.6. we present the evolution of the best performing net and of its granularity for a 5000 iteration long run on a negation problem. Note how in the first hundreds iterations the algorithm tunes in to a promising search space, with granularity 5 in this case, exploring other possibilities but then sticking to one value and getting better and better results with that.

4. Conclusions

In the paper we presented an evolutionary computational approach suitable to maximize very complex functions, such as those that assign a fitness to a neural network on the basis of its topology and its weights distribution.

Search effectiveness is achieved through an encoding procedure for candidate solutions, i.e., through a genotype coding, that allows the algorithm to autonomously identify promising search spaces at evolution time. Specifically, we used different length genotypes codifying also a control parameter, the coding granularity. A cooperative locally optimizing operator adapted from standard LP has also been used.

For computational testing, we used the Rumelhart's problem test suite, comparing our serial algorithm and backpropagation. The comparison was carried out under comparable limited computational constraints and shows the superior efficiency of the evolutionary approach.

Moreover, it is shown how our system is capable of evolving low-connectivity topologies where useless nodes are also discarded. The implicit search for short effective codes, i.e., for small search spaces, is testified by the results achieved.

References

[1] Rumelhart D.E., Hinton G.E. and Williams R.J., "Learning internal representations by error propagation", in *Parallel Distributed Processing: Explorations in the microstructures of cognition*, D.E. Rumelhart, J.L. McLelland Eds, Cambridge: MIT Press, 1986, 318-362.

[2] Kitano H., '*Empirical studies on the speed of convergence of neural network training using genetic algorithms*', in Proc. of the eigth nat. conf. on AI (AAAI-90), 1990, 789-795.

[3] Whitley D. and Hanson T., '*Optimizing neural networks using faster, more accurate genetic search*', Proc. of the Third Int. Conf. on Genetic Algorithms and their Applications, 1989, 391-396.

[4] Parisi D., Cecconi F. and Nolfi S., '*Econets: neural networks that learn in an environment*', Network, 1, 1990, 149-168.

[5] Maniezzo V., '*Anna Eleonora: Genetic Evolution of Feedforward Neural Networks*', Technical Report No. 93-003, Politecnico di Milano, Italy, 1993.

[6] Bersini H. and Seront G., '*In search of a good crossover between evoltion and optimization*', in Parallel Problem Solving from Nature 2, B. Manderick, R.Männer Eds., Amsterdam: Elsevier, 1992, 479-488

[7] Bäck T., Hoffmeister F. and Schwefel H.P., '*A Survey of Evolution Strategies*', Proc. of the fourth int. conf. on Genetic Algorithms, 1991.

[8] Goldberg D.E., *Genetic Algorithms in Search, Optimization and Machine Learning*. Addison Wesley, INDOA, 1989.

[9] Hancock P.J.B., '*Recombination operators for the design of neural nets by genetic algorithm*', in R.Manner, B.Manderick (eds.) Parallel Problem solving from Nature, 2, Amsterdam: Elsevier, 1992, 441-450.

[10] Fogel D.B.,Fogel L.J. and Porto V.W., '*Evolving neural networks*', Biological Cybernetics, 63, 1990, pp.487-493

[11] Mühlenbein H., '*Parallel Genetic Algorithms, Population Genetics and Combinatorial Optimization*', Proc. of the third int. conf. on Genetic Algorithms, 1989, 416-421.

[12] Colorni A., Dorigo M. and Maniezzo V., '*ALGODESK: an experimental comparison of eight evolutionary heuristics applied to the QAP problem*', Technical Report No. 92-052, Politecnico di Milano, Italy, 1992.

[13] de Jong K.A., *Analysis of the behavior of a class of genetic adaptive systems*, Ph.D. dissertation, Univ. of Michigan, 1975.

[14] Grefenstette J.J., '*Optimization of control parameters for genetic algorithms*', IEEE Transaction on System, Man and Cybernetics SMC-16, 1, 1986, 122-128.

Representation and Evolution of Neural Networks

Martin Mandischer
Department of Computer Science VI
University of Dortmund – Germany
e-mail: mam@grete.informatik.uni-dortmund.de

Abstract

An evolutionary approach for developing improved neural network architectures is presented. It is shown that it is possible to use genetic algorithms for the construction of backpropagation networks for real world tasks. Therefore a network representation is developed with certain properties. Results with various application are presented.

1 Introduction

The performance of neural networks highly depends on the architecture of the networks and their parameters. Therefore, determine the architecture of a network (size, structure, connectivity) greatly affects the performance criteria, i.e. learning speed, accuracy of learning, noise resistance and generalization ability. There are only very few mathematical methods to determine the architecture and parameters of networks. Design decisions are normally achieved by rules of thumb and the common approach for designing neural networks is to build a network, test it for the desired function, change structure and parameters if it does not work and repeat this until your evaluation criteria are more or less fulfilled. From a computational and developers viewpoint, this procedure is very expensive and gives no reliable insight into the architecture/performance relation. Recent work of Judd and Lin/Vitter shows that learning in general, as well as choosing an optimal network topology, are NP-complete problems [4, 5]. They also have shown that placing constraints on the topology can help to make learning tractable. This motivates the usage of some evolutionary approach as a heuristic for finding optimal architectures and a network representation which places useful constraints on the architecture.

Several approaches have been made towards the evolutionary design of neural networks [12]. A straight-forward method to determine the architecture of a network has been used Miller et al. The network structure was mapped onto a binary connection matrix where each cell of the matrix determines whether a connection between two units exists or not. A set of matrices were created randomly and the corresponding networks trained. The networks which performed best on the given task, are allowed to produce offsprings for the next generation. New offsprings are created from one or two parent networks by combining or manipulating their connection-matrices. After a fixed number of cycles the best network is chosen [7]. The main problem with this approach is the matrix representation of the network. Incorrect network structure, i.e. feedback connections, and very long codes for larger networks are the result.

A more sophisticated approach has been presented by Harp et al. [3]. They used 'blueprints' for the networks where each network is described by several parameters like, number of layers, layer size, and connections between layers. With this representation it was possible to place constraints on the networks architecture and to reduce the number of incorrect networks. Much less incorrect networks had to be trained an evaluated and the number of evolution cycles needed to get good networks is reduced. Another approach which avoids the representation problem is described by Schiffman et al. To produce the offsprings the best networks are selected and some connections are added or removed from networks and the worst networks are replaced by the offsprings of the best [10].

The next section introduces Genetic Algorithms as a method to search for better networks. Section 3 focuses on a network representation that fulfills some useful properties and the operators to create new networks.

2 Genetic Neural Networks

The basis for evolution is a set of individuals (networks) which establishs a population. The better an individual adapts itself to the given environment, the greater is its chance to survive and produce offsprings. Each individual can be seen as a state in the search space which is, in our case, the space of possible feedforward networks. Each individual is assigned an fitness value which reflects its ability to adapt to the given environment. A genetic algorithms is an iterative procedure, where each iteration is called generation [2]. During each iteration the two natural inspired principles of *selection* and *reproduction* are applied to the population. The selection mechanism determines which individuals are allowed to produce offsprings for the next generation. The probability at which an individual is allowed to reproduce itself and the number of offsprings is based upon its fitness. Supervised learning in networks directly provides a measure of fitness. Therefore we restrict ourselves to the evolution of backpropagation networks. The evolution cycle can now be characterized as follows.

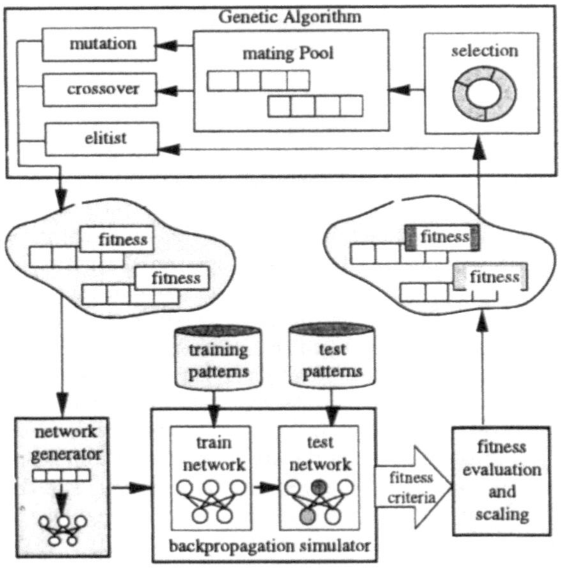

Figure 1: evolution of networks

The use of GA implies a genotypic representation of the individuals. This allows the genetic operators to modify them without using knowledge about the individuals structure. Starting from a randomly generated initial population of genotypic networks a network-generator builds the networks from the genotyp. The backpropagation networks can be interpreted by the network simulator. In the simulator, all networks are trained on the same training data (environment) and their performance on the test-data is evaluated. The simulator delivers several quality criteria (mean-square-error, training time, size of the network) to a device for fitness evaluation and scaling. This device calculates the actual fitness value for each network according to its performance and relatively to the fitness of all other networks. To prohibit that networks with very high fitness values dominate the population in the first generations and to compensate low fitness diversity at the end of the search, we used the sigma-scaling method [9] to rescale the fitness of all networks. The fitness is calculated by several weighted factors:

$$\text{fitness} := \begin{array}{c} w_1 \times \mathit{error} + w_2 \times \mathit{connections} \\ + \\ w_3 \times \mathit{units} + w_4 \times \quad \mathit{epochs} \end{array}$$

With the fitness every network has its expected number of offsprings assigned. The roulette wheel selection method [2] chooses networks with a probability according to their relative fitness, until the population size is reached. The chosen networks are put in the mating pool and allowed to reproduce themselves. Reproduction is achieved by several genetic operators like mutation and crossover which ensure that offsprings are new but somehow related to their parent networks. To prevent the best network from being manipulated by the operators it is automatically put into the new generation. This evolution cycle is iterated for a fixed number of times. During evolution the actual population, best networks and some statistical measures to evaluate the process are collected and saved.

3 Networkrepresentation and Genetic Operators

Weiß distinguished between low- and high-level representations [12]. A *low-level representation* specifies exactly each network parameter (each connection, unit etc.). This causes a large search space for the GA, so that the number of iterations increases dramatically. In contrast a *high-level representation* is an abstract description of the networks where connection patterns and parameter to determine the number and distribution of units characterize the networks structure. The search space is much smaller and networks can be prestructured with the restriction that some architectures are excluded.

The works of Harp et al. and Miller et al. presented representations of both types which are not able to insure the correctness of coded networks [3, 7]. Our representation of backpropagation networks allows only correct networks which prevents incorrect networks from participating in the evolution cycle and saves useless training. Therefore the population size and the number of iterations can be reduced to a number that makes the problem tractable. Excluding incorrect networks and prestructuring gives some evidence that first generation networks can solve the given problem and genetic search can focus on finding improved networks. Additionally training time of the networks can significantly be reduced if layers are sparsely connected rather than fully. The following representation allows only correct networks and the genetic operators do not produce incorrect networks.

The network is specified by describing its main components. A *network-parameter-area* specifies learning rate and momentum term for all connections. Each of the layers has its own layer-area specifying the number of unit in the layer and the connections to other layers. To insure input and output for every unit and therefore correct networks we distinguish projective and receptive connections.

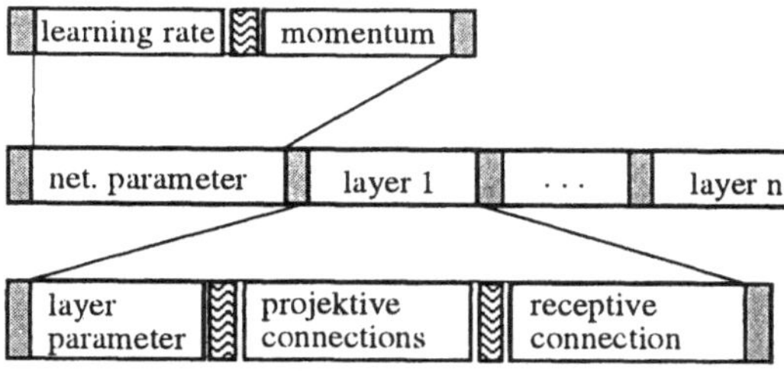

Figure 2: network representation

The receptive connections connect the actual layer with one or more of the preceding layers. Each units receives connections from at least one unit in a precedent layer. The projective connections are used to insure each unit an output and are therefore connected to the next layer. The structure of the network can evolve freely because of the variable number of connection areas.

The connection pattern between two layers is motivated by neurophysiological receptive fields and determined by two parameters. A *radius parameter* specifies the connection radius for each unit of the layer. The size of the radius is given in percent relative to the size of the destination layer. To determine the beginning of a of receptive field an offset relative to the position of the actual unit is added. The second parameter determines the density of the connections within the receptive field and the percentage of units which are connected to an actual unit.

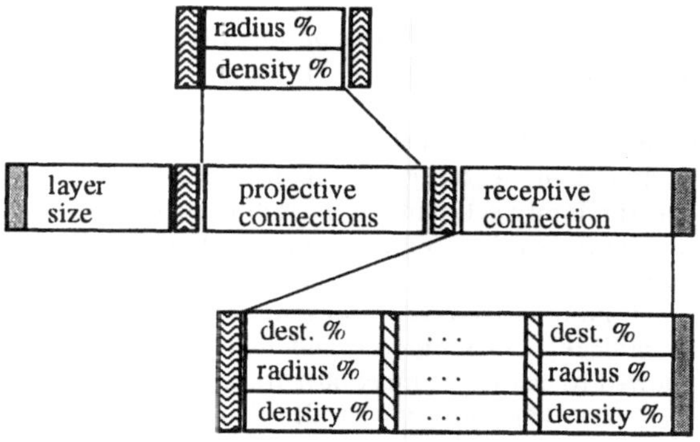

Figure 3: layer representation

Addressing *destination layers* is difficult because of the variable number of layers and connection areas. As a solution relative addresses are used. The projective connections are always related to the following layer, whereas the receptive connections are related to one or more preceding layers. A destination parameter gives the percentage of layers which will be skipped before connections to a preceding layer are established.

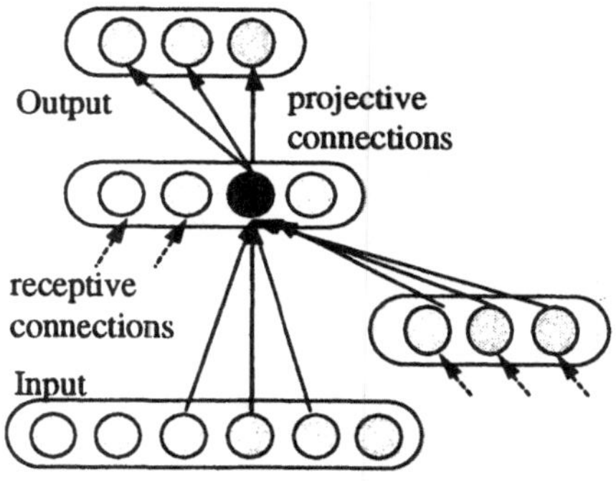

Figure 4: type of connections

The figure illustrates the principles of projective and receptive connections.

This network representation requires a modification of the classical genetic operators. Starting with the idea of mutation should introduce and maintain genetic diversity and crossover should inherit features during reproduction, operators are developed which work on this high level representation and obey the underlying ideas.

Three operators are used: relative mutation, one- and two-point crossover. The *mutation-operator* is applied to those networks in the mating pool which are chosen with the probability given by the mutation-rate ($0 \leq P_m \leq 1$). After a network has been selected for mutation, one of the following parameters in the representation are either increased or decreased: learning-rate, momentum-term, size of one randomly selected layer, radius or density of a randomly selected connection-area. Each of the parameters has an upper bound, a lower bound and a maximum change-rate to prevent changes from being to radical. Learning rate and momentum-term are allowed to change at most 50% from their total range ($0 \leq lr \leq 16$), mutation of all other parameters can vary within 10% of the total range.

The *one-point-crossover* operator exchanges layers between two networks. Networks are again chosen from the mating pool with a probability given by a crossover-rate ($0 \leq P_{c1} \leq 1$). If the representation is viewed a linear ordering of layers, a randomly chosen number (0.100) specifies the crossing-point at which layers are exchanged. The *two-point-crossover* operator exchanges layers between two randomly chosen crossing-points p_1, p_2. Networks are again chosen from the mating pool with a probability given by the crossover-rate ($0 \leq P_{c2} \leq 1$). The following figure illustrates crossover.

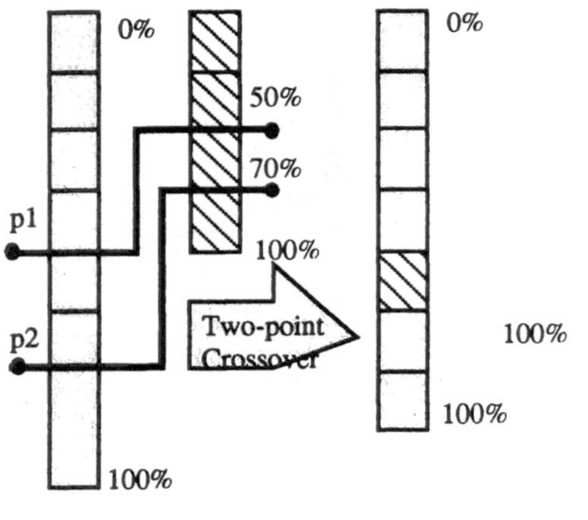

Figure 5: two-point crossover

Due to the relative addressing of destination layers and the usage of percentual parameters instead of fixed values, layers can be exchanged between networks without loosing correctness. It can been shown that the network representation is closed under the given operators and therefore only correct networks participate in the evolution [5].

4 Results

To evaluate our approach we selected four applications for backpropagation networks. The first task is to discriminate edges and corners. Training data consisted of 70 patterns, where each pattern is given as 9x9 binary matrix, which displays either an edge or a corner. A more difficult task is to discriminate 490 handwritten digits from 6 different writers which are given by 8x11 matrices each showing one of 10 possible digits. The third application consists of 279 vectors with 112 real-valued components which encode control-knowledge to improve proofs in PROLOG [11]. It has been proven [13] that feed-forward networks are capable of arbitrary exact function approximation. Despite this theoretical result, Fox showed that it is difficult for a great number of architectures to approximate a Mexican hat function [1]. Therefore we used 841 vectors with x, y, z coordinates building a 3-dimensional Mexican hat function. We developed some evaluation criteria for the networks and the evolution process itself. The func-

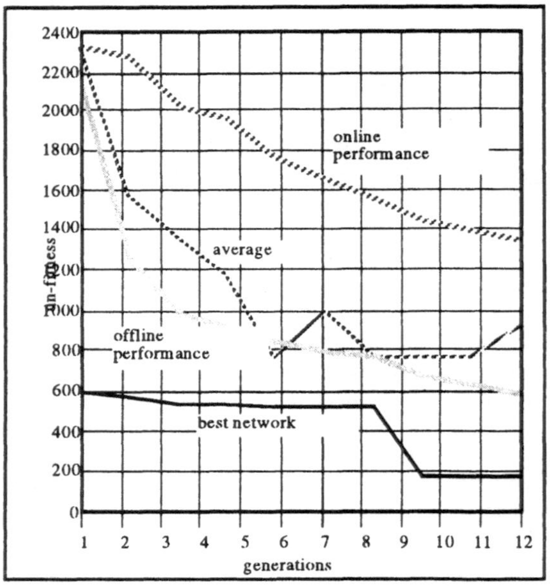

Figure 6: evolution process for function approx.

tionality of our approach is evaluated by its capability to evolve networks which solve a given task and its reliability to reproduce this results with a high probability. A second criteria measures the quality of the evolved networks by evaluating its performance on the given task and a comparison with standard architectures on the same tasks. For all tasks the GA discovered architectures which are capable of solving the given task better or at least as well as all other architectures.

Figure 6 gives an example of the evolutions progress. A population of 10 networks evolves for 12 generations and each network is trained for 500 epochs. The online-performance is the averaged fitness over all generation and the offline-performance averages the best fitness values of all generations.

At the end of the evolution process stands a network with a performance much better than that of comparable standard networks. By standard networks we mean a fully connected input-hidden-output architecture.

For the other applications a population size between 10 and 50 networks and 10 to 100 generations has been typical and yields similar results. To show the reliability of our approach we repeated one experiment 50 times and gained a statistical certainty of 99% that other experiments result in a best network which is within a 5% range from the average best network.

After the successful evolution process we compared the best evolved network for each task to a standard network with the same number of units and a standard network with the same number of connections and, as far as known, results from other publications. The following charts summerizes our results on the applications.

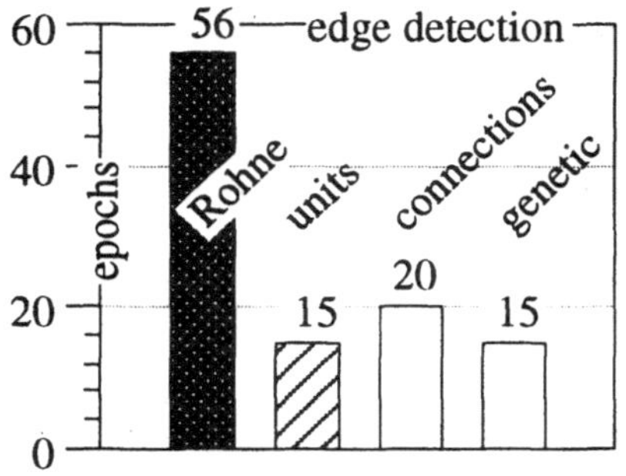

Figure 7: results from edge detection

On the edge detection task the genetic network with 81-45-2 units and 821 connections performed better than a standard network 81-10-2 with 830 connections and as good as a 81-45-2 network with 3735 connections. The classification capability is 100% for all networks. A network (88-52-10) for digit dis-

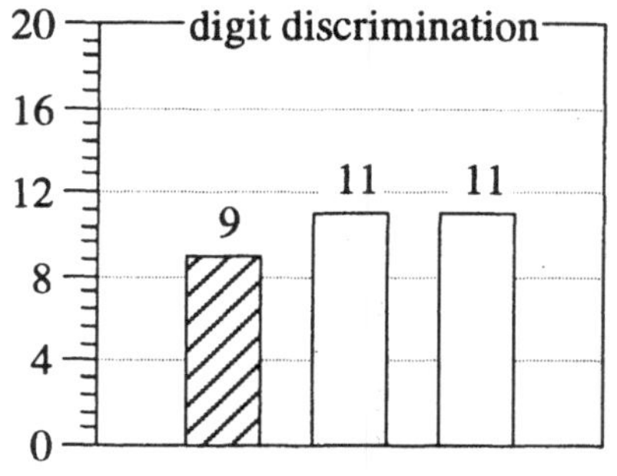

Figure 8: results from digit discrimination

crimination with 2671 connections performed as good as the standard network with the same number of connections and slightly worse than a network with twice as much connections. Classification capability is more than 97% for all networks. Results in learn-

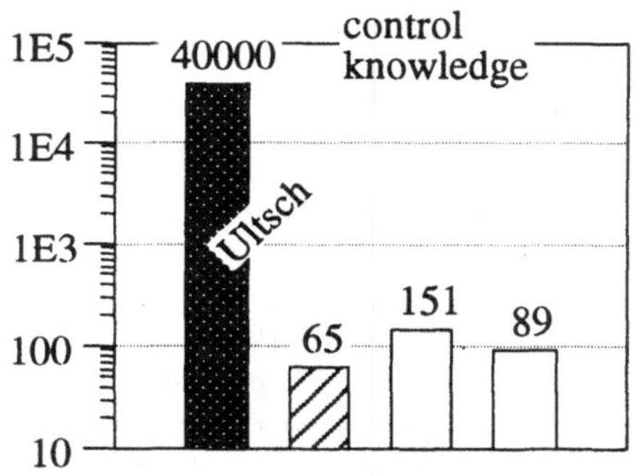

Figure 9: results from proof optimization

ing control-knowledge are comparable to the digit task. A network (112-36-7) with 2103 connections performed better than the standard network with the same number of connections and slightly worse than a network (112-36-7) with twice as much connections. Compared to Ultsch et al. [11] where less than 50% of the control knowledge has been learned, our networks achieved more then 90%. The most astonishing result we gained from the function approximation. We varied the number of hidden units in the standard net-

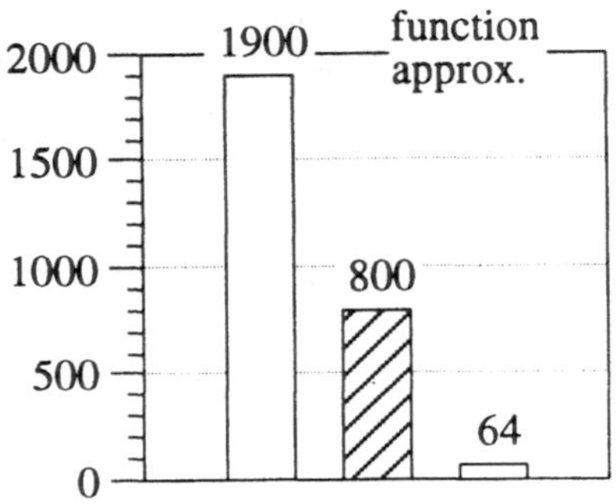

Figure 10: results from function approximation application

works between 40 and 200 but never get below 800 epochs. The genetic network (2-97-1) with 590 connections performed more than 12 times better than the standard network (2-200-1) with 600 connections.

The above results give strong evidence that the underlying structure is essential for the performance of the network. Backpropagation networks can efficiently be developed by GAs and perform better than standard networks. All experiments have been run on a SPARC-station within a couple of hours up to two days.

5 Conclusion

The successful application of GAs to determine the architecture of neural networks shows that an automated design of networks is possible. The resulting networks proved to be better than standard architectures. The best evolutionary networks for all four application solved the tasks correct. Nearly every genetic network showed its superiority over networks with a comparable number of connection, units and learning epochs.

Applied to the function approximation task the genetic network needed 12 times less learning time and showed better approximation quality than standard networks [6]. These results give strong evidence that the underlying structure is essential for the performance of the network. An emerging question in this context is how the structure influences the learning process and how we can gain knowledge about this structure/performance relation from superior genetic neural networks.

References

[1] D. Fox, V. Heinze, K. Möller, S. Thrun, and G. Veenker. Learning by error-driven decomposition. In T. Kohonen, K. Mäkisara, O. Simula, and J. Kangas, editors, *Artificial Neural Networks*, volume 1, pages 207–212, Amsterdam, June 1991. North–Holland.

[2] D. E. Goldberg. *Genetic Algorithms in Search, Optimization and Machine Learning*. Addison-Wesley, 1989.

[3] S. A. Harp, T. Samad, and A. Guha. Designing application-specific neural networks using the genetic algorithm. In D. S. Touretzky, editor, *Proceedings of IEEE conference on Neural Information Processing Systems*, volume 2, pages 447–454, San Mateo, 1990. (Denver 1989), Morgan Kaufmann.

[4] J. S. Judd. *Neural Network Design and the Complexity of Learning*. MIT Press - Bradford Book, Cambridge, MA, 1990.

[5] H. J. Lin and J. S. Vitter. Complexity issues in learning by neural networks. Technical Report TR-CS-90-01, Department of Computer Science, Brown University, Providence, RI, 1990.

[6] M. Mandischer. Genetische Algorithmen zur Optimierung Konnektionistischer Modelle. Master's thesis, Department of Computer Science, University of Dortmund, Dortmund, PO BOX 500500, February 1992.

[7] G. F. Miller, P. M. Todd, and S. U. Hegde. Designing neural networks using genetic algorithms. In J. D. Schaffer, editor, *Proceedings of the Third International Conference on Genetic Algorithms*, pages 379–384, San Mateo, 1989. (Arlington 1989), Morgan Kaufmann.

[8] T. Rohne. Künstliche neuronale netze zur eckendetektion in digitalisierten grauwertbildern. Master's thesis, University of Dortmund, Department of Computer Science X, Dortmund, FRG, 1991.

[9] N. Schaudolph and R. K. Belew. Dynamic parameter encoding for genetic algorithms. Technical Report CSE-TR#CS90-175, University of California, San Diego, CA, 1990.

[10] W. Schiffmann, J. Merten, and W. Randolf. Performance evaluation of evolutionary created neural network topologies. In *First International*

Workshop on Parallel Problem Solving from Nature, pages A–III:1–11, Dortmund, FRG, 1990. University of Dortmund, Department of Computer Science X.

[11] A. Ultsch, R. Hannuschka, U. Hartmann, M. Mandischer, and V. Weber. Optimizing symbolic proofs with connectionist models. In T. Kohonen, K. Mäkisara, O. Simula, and J. Kangas, editors, *Artificial Neural Networks*, volume 1, pages 585–590, Amsterdam, June 1991. North-Holland.

[12] G. Weiß. Combining neural and evolutionary learning: Aspects and approaches. Technical Report FKI-132-90, Technische Universität München, 1990.

[13] H. White and K. Hornik. Universal approximation using feedforward networks with non-sigmoid hidden layer activation functions. In *Proceedings of the International Joint Conference on Neural Networks*. (San Diego, CA), IEEE, 1989.

Using a Genetic Algorithm to tune Potts Neural Networks

W. J. M. Philipsen and L. J. M. Cluitmans
Department of Electrical Engineering, EH 7.34
P.O. Box 513, 5600 MB Eindhoven
The Netherlands
tel +31 40 473352, email wim@es.ele.tue.nl

Abstract

This paper describes the successful application of a Genetic Algorithm to tuning the parameters of a Potts Neural Network.

Potts Neural Networks can be used to solve optimization problems. First the problem is mapped upon the network by generating neurons and connections depending on the instance of the problem at hand. The stable state found with simulation represents a solution for the problem.

There are several parameters that influence correctness and quality of the solutions found. The parameters depend on the problem to be solved. Another optimization technique inspired by nature, Genetic Algorithms, is used to find parameter settings that give the best performance of the network.

The approach has been tested on two applications of Potts Neural Networks. In both cases good results were achieved.

1. Introduction

Our research is mainly focussed on developing heuristics for optimization problems using Potts Neural Networks [1]. An optimization problem is considered to be a black box with several input values, and one output. The problem is to find the optimum input values, that is the set of input values that optimize (minimize or maximize, depending on the formulation) the output. The algorithms that we describe in this paper do not guarantee that the absolute optimum will be found. If the solutions are close enough to the optimum, they will be acceptable for our applications.

First the problem is mapped on the network by generating neurons and connections depending on the instance of the problem at hand. After initialising the network in a random state, it is simulated until it reaches a stable state. The values of the state variables now represent a solution for the problem.

There are several parameters that control the generation of the network and the simulation. The parameters consist of the relative weights of the different groups of connections and a few parameters that control the simulation speed. The values of these parameters depend on the instance of the optimization problem. In general it will be possible to develop a method to calculate these parameters from the instance at hand. This can be more difficult during the development of the mapping of a new problem on a neural network. Changes to the representation of a problem will cause changes in the prediction algorithm too.

This paper describes how a Genetic Algorithm has been used to perform the parameter tuning. Based on [2], a genetic front end has been added to our Neural Network software [3].

The described method will be applied to two different problems (Data Flow Graph scheduling [4] and a partitioning problem [5]).

* This research is part of the ASCIS project sponsored by the European Community under contract BRA 3281.

2. Genetic Algorithms

The technique of Genetic Algorithms is a general technique for solving optimization problems. A Genetic Algorithm tries to let a solution to the problem at hand *evolve* from a set of would-be solutions. Such a would-be solution is called a *genotype*. The set of genotypes in which evolution takes place is called the *genepool*. For a general introduction to Genetic Algorithms see [6].

A Genetic Algorithm performs the following steps:

1. A genepool is created and filled with randomly generated genotypes.
2. For each genotype in the genepool the optimization criterion of the optimization problem is computed. The genotypes in the genepool are sorted according to this value. This divides, in a probabilistic sense, the genepool in 'good' genotypes and 'bad' genotypes.
3. A new genepool is now created based on the old one. It is a copy of the previous one, with some of the 'bad' genotypes removed and some new genotypes inserted (so the number of genotypes in it does not change). These new genotypes are created as crossings or mutations of the existing 'good' genotypes, as will be explained below.
4. Steps 2 and 3 are repeated a few times. For some optimization problems an occurring optimal solution can easily be identified and the Genetic Algorithm stops when it encounters one. For other optimization problems an optimal solution cannot be recognized as such. In these cases one simply stops the algorithm after a given time has elapsed (or when a certain number of new genotypes has been generated). The solution returned by the Genetic Algorithm is the best genotype found so far.

New genotypes can be generated from old ones by two different mechanisms: *crossing* and *mutation*. When crossing, two good *parent* genotypes are chosen from the genepool. Two new *child* genotypes are created from these, usually by a variant of chopping the parents into pieces and reconstructing the children from the remains. In mutation, one parent is chosen, and one child is created by making a small change to the parent genotype. The implementation of these operations depends on the exact structure of the genotypes, and that structure depends in its own turn on the problem at hand.

3. Potts Neural Networks

We will look at an optimization problem as a number of questions that require a yes-or-no answer, with an optimization criterion that gives the quality of solutions. Complex questions, requiring more than just a yes-or-no answer, will have to be split into yes-or-no decisions.

Potts neural networks [1] are an extension of the Hopfield neural network model [7] for optimization. In the Hopfield model an optimization problem is mapped on a neural network by creating a network with one neuron for each possible decision. When the combination of two decisions is illegal or less favorable, their neurons are connected with a negative weight. During the simulation the neurons are forced to become either 0 or 1. After stabilization the states of the neurons give a solution for the problem. If a neuron is on the corresponding decision is taken.

Major problems occur in situations where one out of several options has to be chosen. These complex decisions were split into one yes-or-no decision for each possible answer to the complex decision. These were then mapped onto a vector of neurons. Connections with strong negative weights between each of them were added to ensure that only one of them could become *on* simultaneously. This method is called neuron multiplexing.

As the size of the complex decisions increased, these groups became so large that the optimizing capabilities of the network were gradually destroyed by the large number of connections within the groups. Potts neural networks use a revised updating rule which normalizes the output value within a group. The sum of the output values of the neurons in one group will always be 1. As the neurons are forced to become either 0 or 1, there will be exactly one neuron on in the final state.

An additional advantage of Potts Neural Networks is their increased parameter insensitivity.

Parameter sensitivity is a known problem with Hopfield Neural Networks. The parameter that controls the weight of the connections in a cluster is eliminated, and the network is less sensitive to changes of the other parameters.

3.1 Simulation details

The previous section gave an informal introduction to solving optimization problems with a Potts Neural Network. In this section the practical details of the algorithm will be introduced.

3.1.1 The updating rule

Equations (1) and (2) were introduced by Hopfield [7].

$$E = -\frac{1}{2}\sum_i\sum_j v_i\, v_j\, T_{ij} + \sum_i v_i I_i \tag{1}$$

$$v_i = \tfrac{1}{2} + \tfrac{1}{2}\tanh(u_i) \tag{2}$$

where:

$$u_i = \frac{\sum_{i \neq j} T_{ij} v_j - I_i}{T}$$

Equation (1) gives the energy of the network as a function of the state the network. At initialization, the states v_i of the neurons are all given a random value. During the simulation the updating rule of equation (2) is applied in a random order to all neurons. The value of T is gradually decreased during the simulation. The lower the value for T, the further the states of the neurons are driven to either 0 or 1.

Hopfield showed that during this process, the energy E will be monotonously decreasing. When the energy is a measure for the quality of the solution of an optimization problem, then the network can be used for finding solutions for this problem.

In Potts Neural Networks the neurons are grouped in vectors, and the updating rule of equation (2) is replaced with:

$$v_{ia} = \frac{e^{u_{ia}}}{\sum_b e^{u_{ib}}} \tag{3}$$

where the numeric index runs over the different vectors, and the letter index runs over the neurons within a vector. It easy to see that the sum of the states of the neurons in one vector will always be 1 (equation (4)).

$$\sum_a v_{ia} = 1 \tag{4}$$

3.1.2 The annealing schedule

Figure 1 shows the value of the energy for each value of T during the simulation. There is a state transition for which the energy starts to decrease. The value for which this transition occurs is called the T_c.

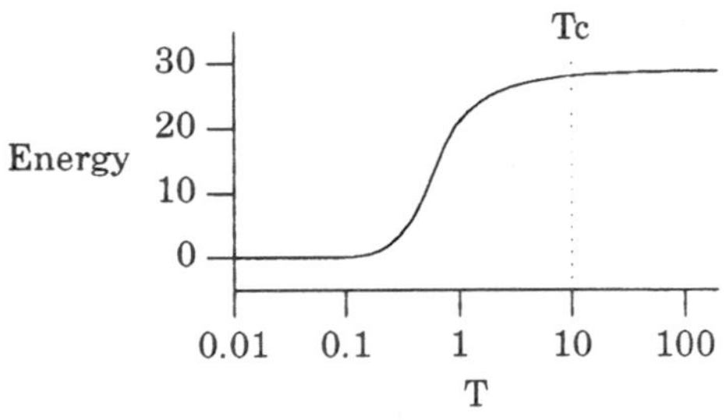

Figure 1. The energy a function of T

The value of T_c depends on the absolute values of the weights of the connections. It can be derived from the the connection matrix [1]. This involves, depending the the type of updating, complicated calculations.

We used the following schedule for the decrease of T.

1. Choose a $T_0 > T_c$
2. iterate over the neurons until all of them reach a stable state
3. Choose a new $T_n = \Delta T \cdot T_{n-1}$, until $\Sigma = 0.9$ (equation (5)).

For values of T well above the T_c, step 2 only takes a few iterations. Therefore it takes only a little extra time to start well above the T_c. It is however important to have a relative small step size round $T = T_c$. Otherwise the quality of the solution will decrease. By changing ΔT we can shift the grain of T.

In [1] the saturation was introduced as an indication of the phase of the simulation.

$$\Sigma = \frac{1}{N} \sum_{ia} v_{ia}^2 \tag{5}$$

It is an indication of the fact that all vectors have already reached a final state. In the final state, in each vector one neuron is 1, and all others are 0. The saturation then becomes 1.

4. The problems

In this section the Potts Neural Network solutions for two optimization problems are introduced..

4.1 ASU clustering problem

The problem addressed in this paper arose in the Cathedral-III design system [8], an architectural design system for irregular high-throughput signal processing applications. The operations are first grouped into clusters. The second step is to decide which clusters will be assigned to the same datapath, called ASU (Application Specific Unit). These ASUs will then be generated. This paper will focus on the assignment of clusters to data paths. At present these problems are transformed into a 0/1 integer linear programming problems [5]. Based on that transformation, we developed a Potts Neural Network algorithm for this problem. The paper will focus on the development of the Potts Neural Networks solution, given the description of the problem.

4.1.1 The cost function

The problem is a special case of the set partitioning problem. We have a set of clusters C which has to be partitioned into N_{ASU} disjoint subsets S_s.

The size of each subset should not exceed N_{cyc}:

$$|S_s| \leq N_{cyc}; \forall s \tag{6}$$

And the total cost should be minimized:

$$\sum_{s=1}^{N_{ASU}} cost(S_s) \tag{7}$$

The cost of the clusters $cost(S_s)$ will be estimated by:

$$cost(S_s) = \sum_{\forall i, j: i, j \in S_s} C_M(i, j) \tag{8}$$

Where similarity of a pair i and j, is given by a compatability measure $C_M(i, j)$. This term expresses the impact on operations, interconnect

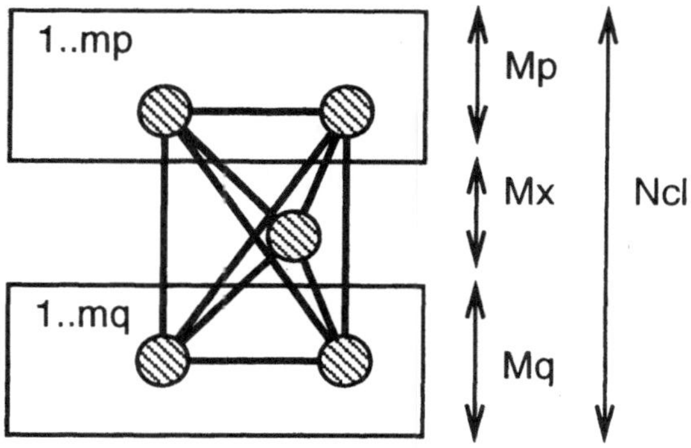

Figure 2. Compatibility Graph

and registers when combining i and j.

4.1.2 Hierarchy

With the introduction of *loops* in the application the constraint of equation (6) needs to be refined. Each cluster will belong to one or more subsets called loops. Loops can be nested, one loop can be entirely part of an other loop. Each loop has a number of cycles N_L associated with it. The number of cycles N_L consumed by a loop L should not exceed the following limit:

$$\sum_{\forall l \in L} m_l \cdot N_l + M_L \leq N_L, \quad \forall L$$

where:

- m_l is the number of iterations of loop l
- $\forall l \in L$ are all sub-loops of L.
- N_l is the number of cycles consumed by loop l.
- M_L is the number of cycles for the cluster belonging to L only.

4.1.3 Mapping on an integer linear programming problem

In this section we describe the method used in [5] to map the problem on a 0/1 integer linear programming problem. The neural network mapping is based on this method. Let $x_{i,s}$ be a binary variable. that indicates if cluster i is assigned to set S_s. Then the cost function becomes:

$$C = \sum_{s=1}^{N_{ASU}} \sum_{\forall i,j} C_M(i, j).\, x_{i,s}.\, x_{j,s}$$

With the following global constraints:

1. To ensure that each cluster is assigned to exactly one subset:

$$\sum_{s=1}^{N_{ASU}} x_{i,s} = 1; \qquad i = 1..N_{cl} \tag{9}$$

2. For each loop L the amount of clusters nested in that loop that can be assigned to a particular ASU is limited by the available number of cycles M_L for the clusters of the loop.

$$\sum_{\forall c \in L} x_{c,s} \le M_L; \qquad s = 1...N_{ASU}, \forall L$$

3. For each loop L the sum of the number of cycles consumed by:
   - nested loops $l \in L$ (N_l)
   - the clusters of that loop (M_L)

   should not exceed the available number of cycles N_L:

$$\sum_{\forall l \in L} m_l \cdot N_L + M_L \le N_{L,} \forall L$$

In [5] these are transformed into a 0/1 integer linear programming problem in the canonical form.

4.1.4 Mapping the problem on a Potts Neural Network

The transformation into a Neural Network is similar to the transformation into the 0/1 programming problem. For each binary variable $x_{i,s}$ we have one neuron. All neurons for one cluster will be combined in one Potts group. The Potts updating rule enforces Equation (4) to be fulfilled.

There will be a bundle of connections from each group to each other group, with connections between neurons representing the same ASU. The weights will be negative, with a value equal to the cost of combining the two clusters. These connections will try to minimize the total cost.

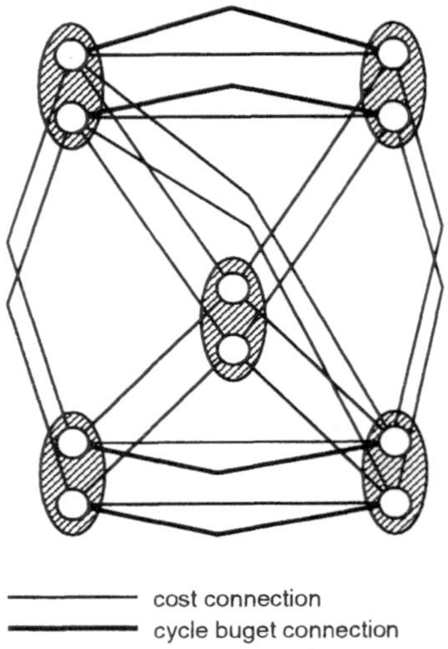

Figure 3. ASU clustering network

The capacity constraint can only be met approximately. By adding a bundle of connections between groups that are in the same group (just like in the previous paragraph) with a negative weight proportional the number of iterations of that loop. In the ideal case the resulting penalty would only depend the maximum number of clusters assigned to one ASU. In this case the contribution of each ASU is quadratic in the number of clusters assigned to it. The nested loops still form a problem. At the moment they are handled as if they were separated, with the number of iterations of the inner loop multiplied by the number of iterations of the outer loop. A more elegant mapping would include the nodes of the inner loop also in the outer loop. First experimental results showed however that this wasn't necessary.

The relative weights of the different types of connections are passed as parameters to the program. They have to be changed when the computed solution does not meet the constraints. If e.g. the cycle budget limit is not met, the value for the connections imposing this limit has to be increased (made more negative). If on the contrary not the whole budget is used, and the cost of the ASU are to high, it can be decreased.

4.2 Data Flow scheduling

Data flow graph scheduling is the assignment of operations to time slots. The input is generally a data flow graph. Figure 4 shows a small example of such a graph. This graph describes a system that adds numbers. The scheduling problem is to

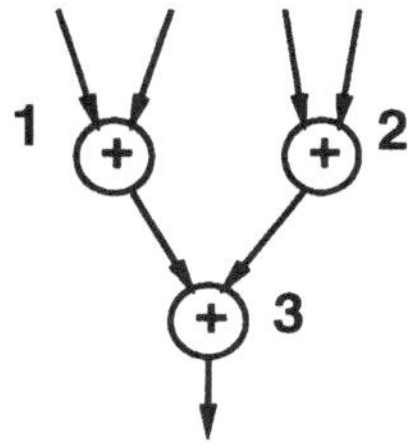

Figure 4. Data flow graph example

assign each of the operations 1 through 3 to a time slot. Figure 5 shows two possible schedules. The graph contains directed edges. When there is an edge from node *i*to node *j* this means that node *j* should be scheduled after *i*. In the first schedule we need two adders in the first time slot. So we have a schedule that takes two time slots, and uses 2 adders. The second schedule

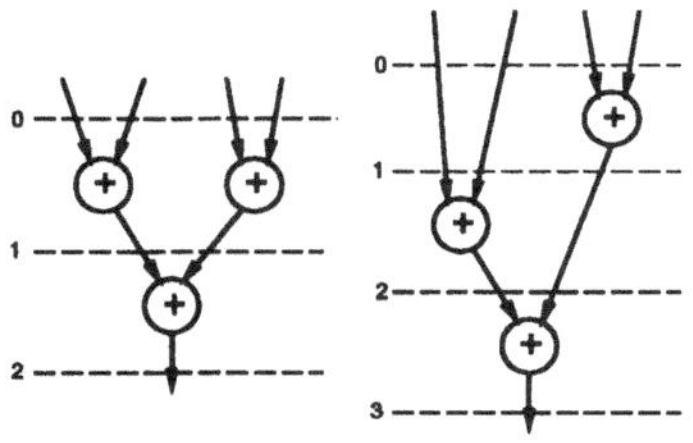

Figure 5. Two possible schedules

uses only one adder, but uses 3 time slots.

4.2.1 Mapping the problem on a Potts Neural Network

We use for each node (operation) in the data flow graph, for each possible time slot one neuron. So if the neuron v_{ij} is *on*, this corresponds to the decision to schedule node *i* at time *j*. All neurons for one node (i.e. for one value of *i*) will be combined in one Potts neuron. In this way we enforce that only one of them can be *on* at a time.

The precedence relations will be enforced by adding connections with a negative weight between neurons which can not be on at the same time. Figure 6 shows the resulting network

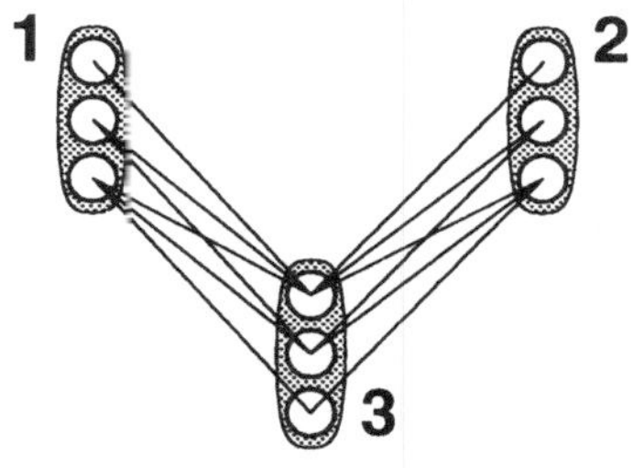

Figure 6. Network for the simple dfg

for the graph of figure 4. The white circles represent the neurons, and those with a grey shaded area will be combined into one Potts neuron. The lines represent connections with a negative weight. For example it is not allowed to schedule both nodes 1 and 3 on time 0 because 3 uses the result of 1. Thus there is a line from node 1 time 0 to node 3 time 0. Inspection of the figure learns that there will be no legal schedule with node 3 at time 0. From this neuron there are connections to every neuron in node 1. The same holds for node 1 and 2 at time 2. Since they are not possible, they need not to be implemented. In literature these times are called *asap* and *alap*.

To enforce the use of as little modules as possible, we now add connections with a negative weight between two neurons if they have the following attributes:

- they represent the same time slot
- they can be implemented by the same module
- they are not mutual exclusive

Figure 7 shows the resulting connections, and figure 8 shows the resulting neural net combining both type of connections. The dark shaded neurons represent times not between the asap and alap time of the node at hand and will not be implemented.

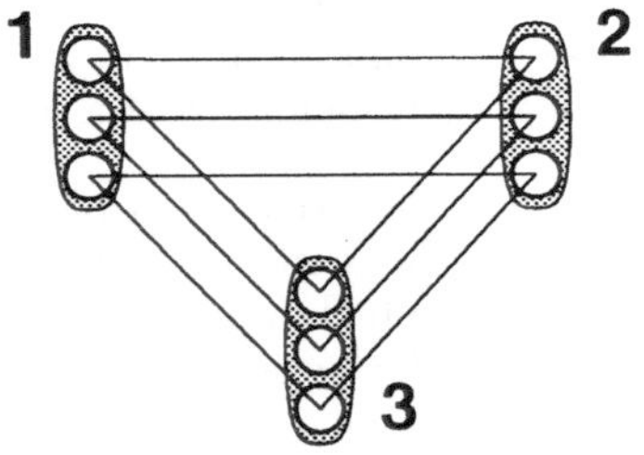

Figure 7. Hardware limiting

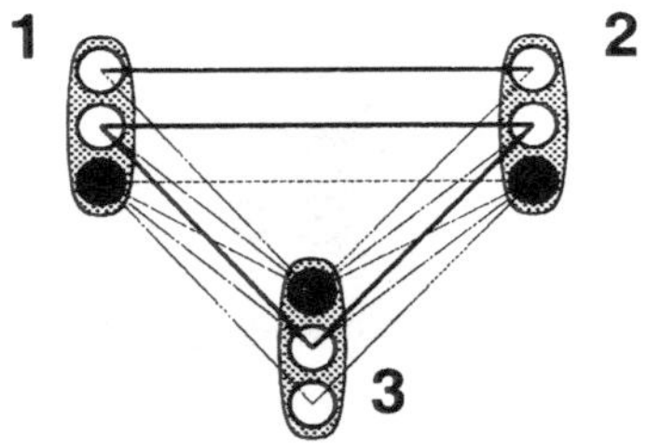

Figure 8. The total neural net

5. Parameter tuning

In the previous sections we described the generation of the networks. Nothing has been said so far about the values of the weights, nor about the parameters that steer the simulation.

5.1 Which parameters are to be tuned?

Generally there are several distinct classes of connections. Connections are of the same class if they have the same "meaning", and will get the same weight. Only their ratios will be of interest. One of the weights can be fixed, and the others can be determined based on that. The number of different weights will in general be rather small. With the data flow scheduling as described here there was only one parameter. We did some additional experiments where two additional parameters were introduced, to steer the solution into a specific direction. With the ASU clustering there are two different weights.

The absolute values do however influence the simulation process. As stated in section 3.1.2 the weights determine the critical gain T_c. By changing the value for ΔT, the grain of T at T_c can be varied. If this grain is not fine enough this will lead to illegal solutions. If the grain is to fine, the simulation will take more time than necessary. The ΔT can take any value between 0 and 1.

5.2 Optimization goal

To determine a fitness function, an optimization goal for the Genetic Algorithm has to be defined.

Tuning the annealing schedule will introduce the possibility of illegal solutions. As illegal solutions are completely useless, there should be a high punishment. The number of constraints violated by the solution given by the neural net, will be used as an indicator.

Next there is the score of the solution that the neural network will find. This is the goal we actually want to optimize. In the case of the ASU-clustering this is the total cost, and in the case of the data flow scheduling this is the number of modules used.

Finally we prefer to get the solutions as fast as possible. As the size of the net and the total number of connections per problem are fixed, the total number iteration steps is a good indication for the time needed.

5.3 The application of the Genetic Algorithm

The total number of parameters is rather small. varying between 2 and 4. However, as it is difficult to predict how the combination will influence the quality of the solution, we tried to tune the parameters using a Genetic Algorithm. The code of a different data flow scheduling approach with a Genetic Algorithm [2] was integrated into the Neural Network simulation package [3].

Our genotypes will consist of the parameters that have to be tuned. These are the value for ΔT and all but one of the different weights.

Only one mutator has been implemented. It multiplies or divides the parameter with a random number between then 2 and 15. Experiments show that this was that finest grain that was needed for the parameters. Smaller parameter changes have no considerable impact in the solution quality. After a mutation it is verified that the value of the parameter is still within the acceptable range.

The fitness function consists of the three components mentioned in the previous paragraph:

- high punishment for illegal or incomplete solutions
- medium punishment for bad scores
- low punishment for large number of sweeps needed in the simulation

6. Results

Each evaluation of a parameter set involves a complete run of the neural network simulator. As quite some evaluations are needed, this can be time consuming. For the ASU clustering this the was no limitation. One evaluation took less then a second. For the data flow scheduling this was a limitation, one evaluation takes between 5 and 600 seconds. In both cases the desired parameter settings were found.

The data flow scheduling problem has often to be solved with several different make spans for the same input graph. The generated networks have a very similar structure and size. Therefore once a setting has been found, it can also be used for a range of related problems. For this problem there were already parameter settings known for many problem instances. All of them were found by Genetic Algorithm too.

Some of the constraints of the ASU clustering, could only be modeled approximately in the Neural Network. The measure of the fitness was however more accurate. This compensated for the shortcomings of the neural model.

7. Conclusions

The application was successful for the two problems it has been tested on. In both cases it eliminated the tuning of the Neural Network parameters by hand. As the running time increases too, the technique will be most useful when developing a Neural Network for a new optimization problem.

The solution space of the tuning problem was *smoother* then we had expected it be. This implies that it may be useful to try other optimization strategies, which are known to take advantage of this characteristic.

References

[1] Carsten Peterson and Bo Söderberg, "A New Method for Mapping Optimization Problems Onto Neural Networks", *International Journal of Neural Systems,* vol. 1, no. 1, pp. 3-22, World Scientific Publishing Company (1989).

[2] L.J.M. Cluitmans, "Using Genetic Algorithms for Scheduling Data Flow Graphs", EUT Report 92-E-266, Eindhoven Unitversity of Technology, Eindhoven, the Netherlands (december 1992).

[3] Wim Philipsen, "Towards a Simulator Dedicated to Solving Optimization Problems Using Potts Neural Networks" in *Proc of the 1993 Int. Symp. on Circuits and Systems* (in press).

[4] Wim Philipsen, "Data Flow Graph Scheduling using Potts neurons" in *Proceedings of the Workshop on Circuits, Systems and Signal Processing,* ed. J.P. Veen, pp. 329-333, STW - Technology Foundation, Houthalen, Belgium (April 8-9, 1992).

[5] Werner Geurts and Wim Philipsen, "Cluster to ASU Assignment: a Global Optimization Problem", ASCIS/CD/m30/C3-D2/1 (February 1992).

[6] L. Davis (ed.), "Handbook of Genetic Algorithms", Van Nostrand Reinhold, New York (1991).

[7] J.J. Hopfield, "Neurons with Graded Response Have Collective Computational Properties Like Those of Two-State Neurons", *Proceedings of the National Academy of Sciences,* vol. 81, pp. 3088-3092 (1984).

[8] S. Note, W. Geurts, F. Catthoor, and H. De Man, "Cathedral-III: Architecture-Driven High-Level Synthesis for High Throughput DSP Applications" in *Proc. of the 1991 Design Automation Conference,* San Francisco, Calif. (1991).

Genetic Weight Optimization of a Feedforward Neural Network Controller

Dirk Thierens*, Johan Suykens, Joos Vandewalle and Bart De Moor

Department of Electrical Engineering
ESAT-lab K.U.Leuven
Kardinaal Mercierlaan 94
B-3001 Leuven
Belgium

Abstract

The optimization of the weights of a feedforward neural network with a genetic algorithm is discussed. The search by the recombination operator is hampered by the existence of two functional equivalent symmetries in feedforward neural networks. To sidestep these representation redundancies we reorder the hidden neurons on the genotype before recombination according to a weight sign matching criterion, and flip the weight signs of a hidden neuron's connections whenever there are more inhibitory than excitatory incoming and outgoing links. As an example we optimize a feedforward neural network that implements a nonlinear optimal control law. The neural controller has to swing up the inverted pendulum from its lower equilibrium point to its upper equilibrium point and stabilize it there. Finding the weights of the network represents a nonlinear optimization problem which is solved by the genetic algorithm.

keywords: genetic algorithm, feedforward neural network, global optimization, nonlinear optimal control.

1 Introduction

Several authors have shown that single hidden layer feedforward neural networks (NNs) are universal approximators for any continuous mapping [1,2,3]. Unfortunately these theoretical proofs give very little guidance on how to obtain a good network for a specific problem. First there is the problem of network architecture: how many hidden layers do we need, how many neurons in each hidden layer, and what connectivity will give us optimal performance? Second there is the problem of weight determination: how do we get the connection weights once we have chosen a particular network topology? In this paper we are mainly concerned with the latter problem. Specifying the weights of a NN is mostly viewed as an optimization process where the goal of the computation is to find an optimal value of an error function. The most commonly used algorithms are backpropagation, conjugate gradient methods and variable metric methods. Although there are considerable differences between these algorithms they do have one significant property in common: they all are local optimization algorithms. Starting from an initial random value a sequence of neighboring points is generated by extracting local information of the search space. Depending on the particular algorithm used this information consists of the function values, the first- and/or second derivative. For non-trivial problems however the error surface is high-dimensional and contains many local optima. By using a local optimization algorithm on such a function there is a significant risk to convergence to some bad value, and in practice people simply do multiple runs with different random starting points. The amount of function evaluations for one single run is usually quite substantial so the ratio between the number of restarts and the number of local function evaluations is typically very low which in fact makes this approach a poor global optimization algorithm.

In this paper we discuss the use of a genetic algorithm (GA) to search the neural network weight space in a global way. In the next section we first look at the functional equivalent symmetries that exist in a wide class of NNs. Section 3 discusses the problems these symmetries cause for the GA, and some changes to the straightforward genotype representation are offered. Section 4 applies the ideas to the optimization of a neural network control function that has to swing up the inverted pendulum and stabilize it in the upper equilibrium point. Experimental results for optimizing the network with and without the proposed recombination modifications are compared.

*email: thierens@esat.kuleuven.ac.be

2 Functional Equivalent Symmetries in Feedforward Neural Networks

The functional mapping implemented by a single hidden layer feedforward network is not unique to one specific set of weights. The same mapping is also obtained by a number of different NNs. What characterizes these networks is that they all are a member of a finite group of symmetries defined by two transformations. Any member of this group can be constructed from any other member by a sequence of these transformations. The first transformation is defined at the single hidden node level. The second is defined at the hidden layer level.

2.1 Hidden Node Redundancy

The most frequently used class of feedforward neural networks consists of hidden nodes that sum their weighted inputs and subsequently apply a transfer function to produce their output value. A number of different transfer functions exist but most of them are odd. Examples are the linear threshold, the logistic and the hyperbolic tangent function. Since all these transfer functions are odd the output of the network does not change if we flip the sign of all the incoming and outgoing weights of a hidden node.

We can choose any combination of the n hidden neurons to flip their weight signs so there are $\sum_{i=0}^{n} \binom{n}{i} = 2^n$ structurally different but functionally identical networks generated by this transformation.

2.2 Hidden Layer Redundancy

A second functional equivalent group of networks is situated at the hidden node layer level. Suppose that we have a network with $h_1 h_2 \ldots h_n$ as hidden nodes. The mapping implemented by the network does not change if a particular hidden node with all its incoming and outgoing weights is exchanged with another neuron and its weights. For instance the networks $h_1 h_2 \ldots h_n$ and $h_2 h_1 \ldots h_n$ are equivalent, even though the first and second neuron have changed their position in the hidden layer. Obviously we can permute any of the n neurons so the total number of functional equivalent networks by this transformation is $n!$.

Since the two transformations are independent of each other, there is a total of $2^n n!$ functional equivalent but structurally different networks. Recently it has also been proven that at least in the case of a single hidden layer, one output neuron and a tangent hyperbolic transfer function the weights within this group of symmetries is unique [4], so there are exactly $2^n n!$ redundant networks for a specific mapping.

For the traditional local weight optimization algorithms this redundancy poses no problem since they only look at a small part of the search space. Global optimization algorithms however will try to explore the whole connection weight search space and this is a factor $2^n n!$ bigger than it really ought to be for the network to function as a universal approximator. For the genetic algorithm the problem is not only one of scale but also of crossover efficiency: functional equivalent near optimal networks often give rise to totally inappropriate networks after recombination because their weight structure is only equivalent up to a certain amount of function invariant transformations.

In the next section we first look at a straightforward genotype representation. The consequences of the functional equivalent groups are discussed and finally a method is offered to eliminate these redundancies.

3 Genotype Representation of Feedforward Neural Networks

3.1 Straightforward Genotype Representation

The standard approach in genetic algorithm practice is to represent the search space simply by concatenating all the parameters in a binary string - the genotype. Parameters close to each other on the genotype are more likely to be processed as a whole because it is less probable that a linkage biased crossover will disrupt them. Whenever a set of parameters form a functional unit it is therefore better to encode them tightly at the genotype. In neural networks the incoming and outgoing weights of a single hidden node form a high dimensional hyperplane, so we want to place them together on the string. The order with which the hidden neurons are placed on the genotype is irrelevant for the mapping performed by the network. Unfortunately for the crossover operator this is not the case: suppose we have a network with two hidden neurons $h_1 h_2$ and $h_{1'} h_{2'}$ with h_i and $h_{i'}$ representing similar hyperplanes. The genotype representation of the first network might be $h_1 h_2$ and for the second $h_{2'} h_{1'}$. When we recombine the two networks by crossing between the two nodes then one offspring inherits h_1 and $h_{1'}$ and the other h_2 and $h_{2'}$. The new neurals nets will almost certainly have a very high error value. Ideally we want to exchange functional similar hidden neurons, and the following two paragraphs discuss a way to achieve this.

3.2 Hidden Layer Redundancy Elimination

The goal of the crossover operator in GAs is to take the partial solutions of two individuals and recombine them to form a better solution. In feedforward NNs with global defined transfer functions, the hidden nodes represent hyperplanes and we want to recombine the good hyperplanes of two NNs to create a better performing network. It is important however that the offspring inherits all the hidden nodes that are necessary to implement the desired mapping. To prevent that recombination would place the functional similar hidden neurons on the same offspring, we rearrange the hidden nodes before crossover is applied such that similar neurons are in the same position. In order to do this we need a way to easily identify the functionality of the hidden neurons and their connecting weights.

The approach we propose here is to look at the signs of the incoming and outgoing weights of every hidden node: the position of the hyperplane is predominantly determined by the weight signs. Hidden neurons that have most of their weight signs in common will be placed at the same position in the genotype before crossover is applied. The reordering of the neurons of one of the two recombining genotypes is done with a simple greedy algorithm. For convenience let us call parent1 the genotype that will be reordered to match the ordering of parent2. First we look for the hidden neuron in parent1 that best matches the first hidden neuron in parent2 and place it at the first position in parent1. Next the best matching neuron of the remaining neurons in parent1 with the second neuron of parent2 is placed at the second position. This matching process is continued for all the hidden neurons. Note that after the reordering the neurons in the last positions will not necessarily match very closely. The greedy reordering algorithm is only suboptimal and is a compromise between optimal matching and computational complexity. The suboptimal reordering should not be a problem however: in fact it introduces some diversity in the recombination process that might counteract premature convergence of the GA.

3.3 Hidden Node Redundancy Elimination

The functional redundancy at the individual hidden neuron level is very easy to eliminate. Whenever the number of positive incoming and outgoing weights of a neuron in the hidden layer is less than the number of negative incoming and outgoing weights, we simply flip their sign. This way we have reduced the 2^n functional equivalent neural networks to just one representative of the group.

4 Example: swinging up the inverted pendulum

4.1 problem description

To test the above ideas we apply the algorithm to find a feedforward neural network controller that has to swing up the inverted pendulum from its lower equilibrium point to its upper equilibrium point and stabilize it there. The design method used to accomplish this is proposed in [5]. The general idea is to use a feedforward neural network as a parametrized control law. The network has as input the four state variables (the position and velocity of the cart, and the position and velocity of the pendulum), and the output is the continuous control force acting on the cart and limited to a maximum allowed force. The control law represented by the neural net is overparametrized and constrained in the following sense: in the neighborhood of the upper equilibrium point, the control law has to coincide with a linear stabilizing controller around the target point. The additional freedom in the parameters is used to enforce the desired swinging up from the lower to the upper point.

Suppose we have a single input nonlinear system

$$\dot{x} = f(x, u)$$

with state vector x, input u and f a vector field. The control task is to bring the state x form the initial state x_o to the target state x_{eq}. We have to determine a nonlinear parametrized static state feedback law

$$u = g(x, w)$$

where w is the parameter vector [6]. In the case of a neural network controller these parameters are the connection weights. For a single layer feedforward network with one output neuron and tangent hyperbolic transfer functions the control law is given by:

$$u = F_{max}.tanh(\sum_{i=1}^{n_h} w_i.tanh(\sum_{j=1}^{n_{in}} v_{ij}.x_j))$$

with F_{max} the maximal allowed control force, n_h the number of hidden neurons, n_{in} the number of inputs, v_{ij} the weights from the input to the hidden layer, and w_i the weights from the hidden layer to the output neuron.

A state-space model of the inverted pendulum can be given by

$$\dot{x} = f(x) + b(x).u$$

with state x, input u and

$$f(x) = \begin{pmatrix} x_2 \\ \frac{\frac{4}{3} m l x_4^2 sin x_3 - \frac{m g}{2} sin(2x_3)}{\frac{4}{3} m_t - m cos^2 x_3} \\ x_4 \\ \frac{m_t g sin x_3 - \frac{m l}{2} x_4^2 sin(2x_3)}{l(\frac{4}{3} m_t - m cos^2 x_3)} \end{pmatrix}$$

$$b(x) = \begin{pmatrix} 0 \\ \frac{4}{3} \cdot \frac{1}{\frac{4}{3} m_t - m cos^2 x_3} \\ 0 \\ \frac{-cos x_3}{l(\frac{4}{3} m_t - m cos^2 x_3)} \end{pmatrix}$$

The states x_1, x_2, x_3 and x_4 are respectively the position and the velocity of the cart, and the position and velocity of the pendulum. The symbol m is the mass of the pendulum, m_t the total mass of cart and pendulum, l is half the pendulum length and G the gravity constant.

The starting point is the lower equilibrium point $x_o = [00\pi0]$ and the target state is the upper equilibrium point $x_{eq} = [0000]$. To swing up the pendulum we take as cost function

$$C = x_N^t x_N$$

with x_N the state vector that we have reached after a certain time period. In optimal control terminology this means that we are performing terminal control. Swinging the pendulum up however is not sufficient; we also want to stabilize it in its upper position. To achieve this the weights of the network are constrained by 4 equations so that the neural controller will coincide with a linear static state feedback controller (LQR) around the target point. The output of the linear controller is given by $u = -k_{lqr}^t.x$. A stabilizing controller around the upper equilibrium point can be achieved with a single neuron with weight vector

$$w = \begin{pmatrix} 0.1000 & 0.2303 & 3.1894 & 0.8178 \end{pmatrix}$$

$F_{max} = 10$ and

$$k_{lqr}^t = -F_{max}.w^t$$

[5].

To let the multilayer neural controller coincide with the linear controller we have to satisfy the four constraints

$$k_{lqr}^t = -F_{max}.w^t.v$$

4.2 experimental results

In the experiments we used a network with 4 hidden neurons. F_{max} and k_{lqr} are known so we can satisfy the constraints by computing the 4 output weights w form the 16 input weights v by simply inverting the input weight matrix. Although the output weights are also represented on the genotype the GA does not actually have to search them: before a newly created network is evaluated the output weights w are first computed from the input weights v. The values of the parameters in the experiments are $m = 0.1$, $m_t = 1.1$, $l = 0.5$ and $F_{max} = 10$. The genetic algorithm used is a steady state GA with a population size $n = 100$. Two parents are randomly selected and recombined following the approach outlined in the previous section - hidden neurons with their incoming and outgoing weights are exchanged as a whole. One of the offspring is evaluated by simulating it for 3 seconds and when it has a better function value than the worst of the current population it replaces this worst network. After every single recombination one individual is randomly picked out for a single hill-climbing step: one of the hidden neurons is selected and its weights are mutated by adding gaussian noise with zero mean and 0.1 variance. When the mutated network is better it replaces its parent, otherwise it is not included.

Figure 1 shows the results for 25 independent optimization runs of the neural controller with and without using the functional redundancy elimination. The curves represent the mean cost function value of the best network in the population after a certain amount of network evaluations. The lower curve is obtained when the hidden neurons are reordered before recombination and with the weight signs flipped. The upper curve is the result obtained when we simply recombine the genotypes without any reordering or sign flipping. The small vertical lines indicate the standard deviation for the lower curve.

Figure 2 shows a simulation of the swinging up of the pendulum by a typical neural controller. This network has a cost function value of 0.052 after 25000 function evaluations which is the median value of the

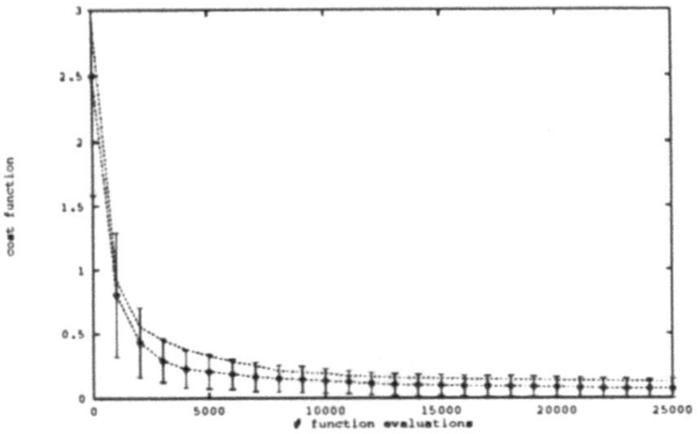

Figure 1: *Curves represent the mean - for 25 runs - cost function value of the best network in the population after the indicated number of network evaluations. The lower curve is obtained when the hidden neurons are reordered before recombination and with the weight signs flipped. The upper curve is the result obtained when we simply recombine the genotypes without any reordering or sign flipping. The small vertical lines indicate the standard deviation for the lower curve.*

25 independent runs when using the modified genotype representation. The weights of this controller are

$$v = \begin{pmatrix} 1.02824 & 1.14631 & 0.58466 & 0.54718 \\ 0.72937 & 2.09564 & 0.04480 & -0.25648 \\ -0.26217 & -0.37379 & 0.11679 & 0.06484 \\ 1.67519 & -1.19728 & 1.22857 & -1.67968 \end{pmatrix}$$

$$w = \begin{pmatrix} 1.78359 & 1.82575 & 13.95555 & 0.35406 \end{pmatrix}$$

The simulation clearly shows how the neural controller smoothly swings up the inverted pendulum and since the output weights w are constrained by the linear LQR controller the pendulum is stable at its upper equilibrium point. Of the 25 optimization trials with the functional redundancy elimination 23 of them were able to swing up the pendulum close enough to the target point so that a stabilizing controller was achieved. For the straightforward recombination only 16 trials were successful.

5 Discussion

The experiments show that the functional redundancy elimination gives better results: the mean value of the straightforward crossover is almost one standard deviation worse than the mean of the modified representation. This might at first seem a very little improvement but one should realize that the neural controller is actually a very small network. Since there are only 4 inputs, 4 hidden neurons and 1 output neuron there are just 16 possible weight sign combinations for a hidden node (1 with no inhibitory link, 5 with 1 inhibitory link and 10 with 2 inhibitory links). It can be expected that the genotype modifications will be more advantageous with increasing network size but this remains to be confirmed.

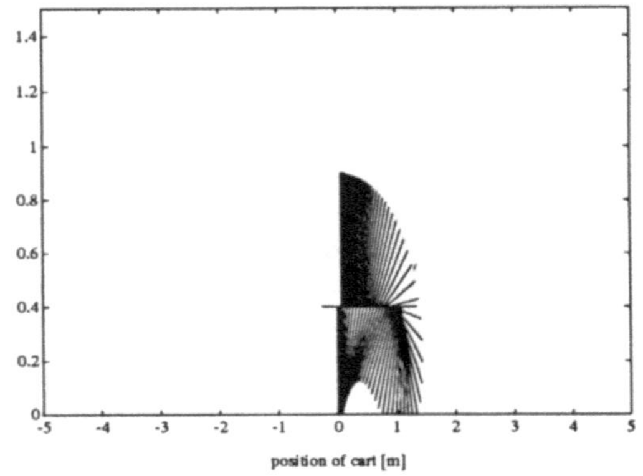

Figure 2: *simulation of the swinging up and stabilization of the inverted pendulum with the genetic optimized neural controller with median cost function value.*

A second remark can be made on the nature of the GA search in general. When inspecting the solutions found it is clear that multiple weight combinations - aside from the symmetrical ones - are able to swing up the pendulum and stabilize it. Since we are using a simple GA with a small population size and a mixed in local hillclimber the search is very rapidly concentrated on one or very few promising regions - a phenomenon called "genetic hillclimbing" [7]. Considering the multimodal character of the cost function, techniques that promote niches to occur (e.g sharing [8]) are no doubt worth investigating.

A last remark concerns the efficiency of the search. The crossover operator as described here only exchanges complete hidden neurons with all their incoming and outgoing weights. The search for good hyperplanes itself is done by the gaussian hillclimber with constant variance. It is well known from the work of Evolution Strategies [9] that this is actually a very bad local search algorithm, so the performance will be much improved if a better local search technique is used, especially when second order gradient information is exploited.

6 Conclusion

The paper discusses the genetic weight optimization of a feedforward neural network. It is argued that the functional equivalent symmetries that exist in such networks cause a problem for the crossover operator. A way to eliminate these representation redundancies is to reorder the hidden neurons on the genotype before recombination according to a weight sign matching criterion and to flip the weight signs of a hidden neuron's connections whenever there are more inhibitory than excitatory incoming and outgoing links.

Experimental results were obtained for optimizing a neural controller that has to swing up the inverted pendulum and stabilize it at the upper equilibrium point.

References

1. Hornik K., *Multilayer Feedforward Networks are Universal Approximators*, Neural Networks, Vol.2, pp.359-366, 1989.
2. Funahashi K.I., *On the Approximate Realization of Continuous Mappings by Neural Networks*, Neural Networks, Vol.2, pp.183-192, 1989.
3. White H., *Connectionist Nonparametric Regression: multilayer feedforward network can learn arbitrary mappings*, Neural Networks, Vol.3, pp.535-549, 1990.
4. Sussmann H.J., *Uniqueness of the Weights for Minimal Feedforward Nets with a Given Input-Output Map*, Neural Networks, Vol.5, pp.589-593, 1992.
5. Suykens J. & De Moor B., *Stabilizing Neural Controllers with Application to the Control of an Inverted Pendulum*, ESAT-SISTA report nr.16, K.U.Leuven, 1992.
6. Bryson A.E. & Ho Y.C., *Applied Optimal Control*, Waltham, Ma. Blaisdel, 1969.
7. Whitley D., Starkweather T. & Bogart C., *Genetic Algorithms and Neural Networks: optimizing connections and connectivity*, Parallel Computing 14, pp.347-361, 1990.
8. Deb K. & Goldberg D.E., *An Investigation of Niche and Species Formation in Genetic Function Optimization*, in Schaffer J.D.(ed.) Proceedings of the Third International Conference on Genetic Algorithms. San Mateo, CA. Morgan Kaufmann Publishers, 1989.
9. Schwefel H.P., *Numerical Optimization of Computer Models*, Wiley, Chichester, 1981.

Using a genetic algorithm to find the rules of a neural network

R.J.Mitchell, J.M.Bishop and W.Low

Neural Network Research Group

Department of Cybernetics, University of Reading, UK.

email: cybrjm@uk.ac.reading.cyber

ABSTRACT

Neural networks can be used in various applications, and there are many forms of neural network. However, all networks have two common factors: they can learn from examples and they can generalise; that is, they can respond suitably both to the data taught and to similar data. Once trained, the network can be used for such tasks as recognition and classification.

One criticism of neural networks is that they are black boxes: appropriate outputs are generated for each set of inputs, but the user has no understanding of how such outputs are generated. The aim of the work here is to generate the rules which determine how a taught network responds.

This is achieved using a genetic algorithm which suggests possible rules for the network. The network itself is then used to evaluate the performance of each rule, and this performance measure is used to determine new rules.

This paper outlines the problem in more detail, describes the proposed technique and discusses the preliminary results obtained so far.

INTRODUCTION

The last few years have witnessed a great increase in interest in neural networks. The basic functions of neural networks are classification, prediction and recognition, and networks can be used for such diverse purposes as control, instrumentation, image processing, speech processing and economic forecasting.

There are many forms of neural network, including the multi-layer perceptron[1], Kohonen networks[2], Hopfield networks[3] and weightless networks[4]. All networks, however, have two important factors in common. They can learn from examples and they can generalise, that is, they can respond suitably both to the data taught and to similar data.

Once trained for a given task, a network can then be used in the application for which it has been trained, by providing suitable data on the network inputs. However, the network remains a black box. Appropriate outputs are generated for any set of inputs, but the user has no understanding of why or how particular results are obtained.

Often this is an acceptable solution, as the user is interested only in these results. Indeed it is often claimed that one advantage of neural networks over, say, expert systems is that they can be used in many different situations without the need for a detailed understanding of the problem. All that is required for training a network, is a suitable training set. The same form of neural network can thus be applied to many diverse situations.

However, a more detailed understanding of the network is sometimes required. For example, in a safety critical application, such as in medicine or, say, a nuclear power plant, people would be very wary of using a neural network control system without knowing how it would respond in all situations. Similarly, although there is some interest in applying neural network techniques in economic systems, there is some concern about applying a neural network black box: speculators are worried about their money!

Thus there is a need for a system which explains the operation of a neural network. The aim of the work described here is to generate the rules which determine how a taught network responds. These rules could then be fed into an expert system, for example.

REQUIRED SYSTEM

A typical result for this process might be that if the data input to the network were in one range, the network would indicate that the input belongs to one class, but a different classification would be made if data were in another range. For a neural network used for control purposes, the system might generate rules indicating that the output would have one value for a particular range of input values. Such rules would be suitable for use in expert systems in applications where detailed explanations of hypotheses are required.

The problem remains, how should such rules be generated? Clearly there are many possible rules, and the system could have some exhaustive search. A better method, it is believed, is to use a suitable search technique, and genetic algorithms seem appropriate.

RULE GENERATION

The method which has been used here for generating the rules is based on a hybrid system developed in the Cybernetics Department for diagnosing faults in the electricity grid[5].

The basic idea there is to use genetic algorithms to suggest rules to explain the operation of the grid. Associated with each rule is a performance measure generated by a suitable critic, and this measure is then used for determining the parent rules from which child rules are bred, in the usual manner[6].

For the grid project, the critic is a model of the grid, but for generating the rules that explain a neural networks, the critic can be the taught network itself.

As an example, a rule might suggest that a network output should be 0.5 for a particular set of inputs. These inputs would then be passed to the network and its output calculated. The difference between the calculated output value and the value generated by the rule, would in part determine the performance measure. Other factors which might affect this measure include the simplicity of the rule. Consider the following examples.

Suppose the network has two inputs, a and b, and one output, y, and the network was trained on the OR function. Thus the output y will be '1' if either a or b is a '1'. The following rules might be generated:

IF a = 1 AND b = 0 THEN y = 1

IF a = 1 AND b = 1 THEN y = 1

IF a = 1 THEN y = 1

The last rule encompasses the first two, and could therefore replace them. Thus the third rule is better than either of the other two, and so it should have a better performance measure than the other two, even though all three rules are equally correct.

In this second example it is assumed that the system is operating on real valued input. For this, the following two rules might be generated by the algorithm:

IF a >= 0.3 AND a <= 0.7 THEN y = 0.5

IF a >= 0.2 AND a <= 0.75 THEN y = 0.5

Assuming that both rules are true, the second rule is better as it encompasses a greater range than the first.

It might be possible to generate the rules by study of the data which are used to train the neural network. However, one of the great advantages of neural networks is that they can generalise. Thus, by using the neural network as the critic, the rules generated should reflect both the training set and the data similar to those in the training set.

Another advantage of the proposed technique is that a trained network can operate very quickly. Thus the performance of the network can be evaluated rapidly by feeding the suggested inputs to the net and calculating the outputs and comparing these with the outputs suggested by the rules.

ALGORITHM

Thus the operation of the system can be defined by the following algorithm.

> Initially the network is trained on a suitable representative data sample.
>
> The first stage of the rule generation process is to produce an initial population of rules. For each rule, the system evaluates its performance measure.
>
> Then the program loops doing the following:
>
> > Selecting parents whose genetic make up contributes to suitable offspring
> >
> > Producing offspring, using suitable genetic operators
> >
> > Evaluating the performance of these offspring
> >
> > Replacing certain parents by new offspring
>
> This process ends when the population has converged or an appropriate number of iterations has occurred.

Reproduction is achieved using the standard genetic operators crossover, inversion and mutation[7]. Various problems, however, need to be considered. These include the frequency with which the different genetic operators are used, the population size, how new offspring should replace their parents and how the rules should be represented.

RULE REPRESENTATION

The method used to represent the rules will vary depending upon the application. Factors to be considered here are the number of inputs and the range of data values for each input.

For example, if the neural network is processing binary data, then each input could be ‘0’, ‘1’ or ‘don’t care’. For a two input problem with inputs a and b and output y, the rules could take the form:

IF a = ‘0’ AND b = don’t care THEN y = 1

If, however, the data in the network were real valued, say having values in the range 0 to 1, then possible rules might be of the form:

IF a >= 0.1 AND a <= 0.3 AND b >= 0.5 AND b <= 0.95 THEN y = 0.3

The equivalent of a ‘don’t care’ would be for a rule to include:

a >= 0.0 AND a <= 1.0

The rules would be represented by strings of binary data. These would consist of the inputs and their associated range of numbers. These numbers would also be represented as binary strings, but the resolution of the numbers must be chosen. For some problems, 8-bit values could be used, representing multiples of 1/255 to give data in the range 0 to 1, but for other problems, 16-bit values might be required.

POPULATION

Various factors need to be considered as regards the population of rules. These include the size of the population, whether the size is fixed, and the means whereby the population is updated.

A suitable size for the population must be chosen. This must clearly be large enough to include the appropriate rules, but this cannot always be known. To cope with this problem, a strategy could be adopted in which the population grows when necessary. This might occur, for example, when a significant number of the rules have a good performance measure, perhaps suggesting that there might be more good rules.

Updating the population is often achieved by breeding as many children as are in the current population, and these children become the new population. An alternative strategy, as has been used when designing a neural network architecture for colour recipe prediction using genetic algorithms[8], is to breed some new children, but only introduce them into the population if they are better than those in the current population. Another possibility is to add new rules to the population.

There is the well known danger at the start of the algorithm that some 'super rules' prevent others from contributing to the next generation. Scaling of the fitness measure is needed to prevent this[9], in which the fitness of weaker rules is increased.

INITIAL TESTS

As stated earlier, there are various forms of neural network. It was decided, however, to initially consider just one type of network, the one most commonly used, the multi-layer perceptron (MLP). Thus the following reports the initial work done in testing the idea on a MLP network.

One of the well known bench marks of MLPs is the XOR problem. That is, the network has two inputs, which can each take the values '0' or '1', and one output which should be a '1' only if the two inputs are different. The aim of these tests was to see if the system could successfully find the rules for this problem, and to investigate suitable reproduction operators, etc., for this system.

There are four rules which define the network, namely

IF $a = 0$ AND $b = 0$ THEN $y = 0$

IF $a = 0$ AND $b = 1$ THEN $y = 1$

IF $a = 1$ AND $b = 0$ THEN $y = 1$

IF $a = 1$ AND $b = 1$ THEN $y = 0$

Thus the network was first trained with data defining the XOR problem, using the back-propagation learning algorithm. Then the genetic algorithm was run in order to see if it could generate these four rules. Various tests were made on the algorithm, varying such factors as population size and probabilities of the genetic operators.

RESULTS

Results obtained from these tests can be summarised as follows. The genetic algorithm can converge and produce a rule population including the four required rules. However, population size and probabilities of genetic operators affect the system.

For example, with a population size of only 20, only three of the four rules are generated unless the mutation rate is high, say 0.3. A lower mutation rate, 0.01, but with a larger population, say 30, can also lead to the generation of the correct rules. For the latter case, the algorithm converged after fewer iterations. These results are depicted in the two graphs shown below.

Figure 1 below shows the algorithm in operation for such a system, depicting the percentage of correct rules found after each new population is bred. Here the population is 20 and the mutation rate is 0.3 and the crossover rate is 0.9.

Figure 2 below shows the algorithm operation when the poluation is 30, the mutation rate is 0.01 and the crossover rate is 0.9.

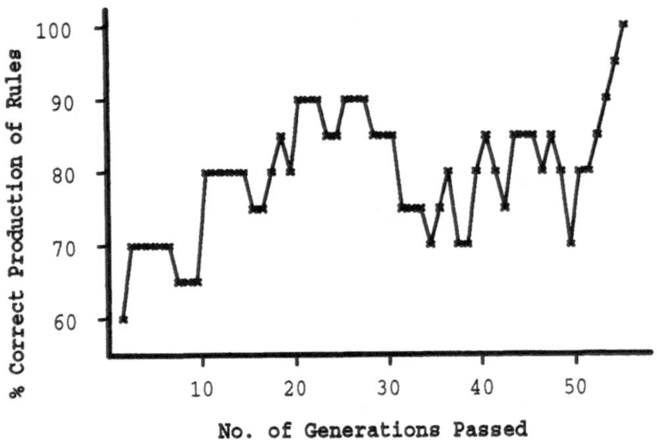

Fig 1. Percentage correct rules vs time

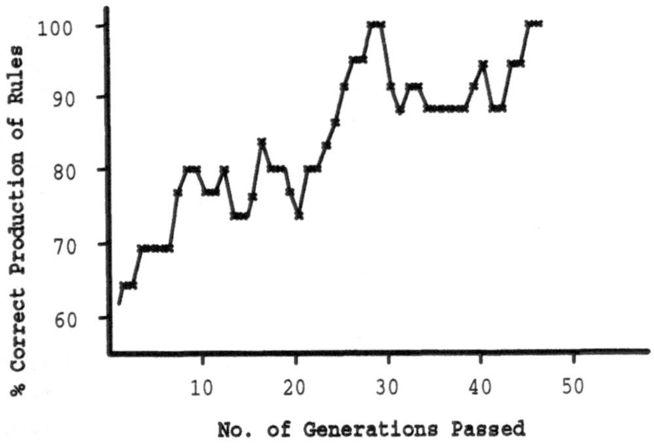

Fig 2. Percentage correct rules vs time

FURTHER WORK

Clearly this work is at an early stage. However, the results have demonstrated the feasibility of the technique on a simple problem. The next stage is to investigate a more complicated problem, say 4-bit parity. Then, some real valued problems will be tested.

CONCLUSION

A technique is proposed by which the rules governing a taught neural network can be discovered. Such a system could then be applied in areas where neural networks are not currently used, because of concern about the 'black box' nature of networks. The feasibility of the technique has been demonstrated, but more work is needed, however, in investigating more complicated networks, and different forms of network.

REFERENCES

1. Rumelhart, D.E. and McClelland, J.L., '*Parallel Distributed Processing*', MIT Press, 1986.

2. Kohonen, T. '*Self Organisation and Associative Memory*', Springer Verlag, 1984.

3. Hopfield, J.J. '*Neural Networks and physical systems with emergent collective properties*', Proc. Nat. Acad. Sci. **79**, 2554- 8, 1982.

4. Aleksander, I., Thomas, W. and Bowden, P. '*WISARD, a radical new step forward in image recognition*', Sensor Rev., 120-4, 1984.

5. Kiernan, L.A. and Warwick, K. '*Learning Systems for Fault Diagnosis in Power Networks*', Proc. UPEC'90, pp591-3, 1990.

6. Holland, J. '*Adaptation in Natural and Artificial Systems*', Ann Arbor, University of Michigan Press, 1975.

7. Goldberg, D.E. '*Genetic Algorithms*', Addison Wesley, 1989.

8. Bishop, J.M., Bushnell, M.J., Usher, A., and Westland, S. '*Genetic Optimisation of Neural Network Architectures for Colour Recipe Prediction*', Proc. International Conference on Neural Networks and Genetic Algorithms, Innsbruk, 1993.

9. De Jong, K.A. '*An analysis of the behaviour of a class of genetic adaptive systems*', Doctoral Thesis, University of Michigan, 1975.

M. L. P. Optimal Topology via Genetic Algorithms

P. Arena, R. Caponetto, L. Fortuna, M.G. Xibilia
Dipartimento Elettrico Elettronico e Sistemistico
Universita' di Catania
Viale A. Doria 6, I-95125 Catania, ITALY
Tel. + 39 95 339535 Telefax + 39 95 338887 EMail dees@dees.unict.it

Abstract
In the paper a Genetic Algorithm in order to select the optimal topology of a Multi Layer Perceptron is adopted. Two different problems are considered. The first one is to select the optimal number of neurons in a structure with one hidden layer. The second one is the choice of the number of layers into which a fixed number of neurons has to be arranged, to solve a given problem. To this aim, a suitable set of genetic operators has been introduced.

I. Introduction

One of the unsolved problems in neural network design consists in the choice of the network topology suitable to solve a given problem. In fact, restricting our attention to the Multi Layer Perceptron (MLP) structure, it is well known that the number of neurons involved, greatly influences the network performance. In fact, too little a number of neurons in the hidden layer could be unable to guarantee good performance during the learning phase, while too large a number of unit leads to a good error index due to a mere association between the training patterns.

To determine the suitable number of neurons to accomplish a given task, a trial-and-error procedure is usually adopted. However, several efforts have been made, including a monitoring approach in order to find the optimal number of neurons by pruning the units that less contribute to the network behavior [2]. Recently, the Genetic Algorithms have been suggested by several authors to approach this problem. In [3] and [4] the genetic algorithms are used to discover the size, structure and learning parameters of a network to be trained by a separate artificial neural network learning algorithm. Here, the great number of variables considered for the optimization, leads to a very complex procedure. It can be argued that, in several cases, an algorithm managing a fewer number of variables could obtain good results being at same time faster.

Our aim is to propose a simpler algorithm by taking into account the following considerations:

-the learning parameters of the network are not included in the optimization algorithm because suitable techniques for their automatic tuning during the training phase have been already developed [5];

-in a first step, only MLPs with only one hidden layer can be considered, in fact it is proved [1] that such a structure can approximate, with an arbitrary degree of accuracy, any non linear function. As a second step, one may think to rearrange the obtained number of neurons in a multi layer structure to reduce the number of interconnections.

A genetic algorithm able to determine the optimal number of hidden neurons in a MLP with only one hidden layer has been presented by the authors in [6]. In this paper an algorithm to choose the multi layer structure into which a fixed number of neurons should be optimally rearranged is proposed.

II. An overview on Genetic Algorithms

Genetic algorithms (GAs') implement optimization strategies, based on the simulation of the natural law of the evolution of the species by natural selection, in order to obtain the "fittest individual" in the evolutionist sense [6]. According to this theory, considering a population which evolves in a particular environment, only the fittest individuals will be able to reproduce, handing down their chromosomes, while the less fit will be doomed to extintion, due to the environmental constraints. Adopting this analogy, the optimal solution of a given problem corresponds to the "fittest individual".

Some important characteristics that make GAs' quite different from traditional optimization strategies are reported below:

- GAs' search for the function optimum starting from a population of points, not from a single point; this feature suggests that they are global search methods, able to climb many peaks in parallel, thus reducing the probability of finding local minima;

- GAs' use information only about the objective function; they do not require knowledge of the first derivatives or other auxiliary informations on the funtion to optimize, allowing a number of problems to be solved without the need to formulate restrictive assumptions on the search space. Moreover, the contraints involved in the problem can easily be taken into account;
-GAs' use probabilistic transition rules during iteration, not deterministic ones.
All these features contribute to GAs' robustness, making them computationally simple and powerful.
The variables involved in the optimization must be codified in particular structures, similar to the chromosomic one: the choice of a suitable representation may condition the performance of the algorithms. By adopting a classical approach, parameters can be translated into binary strings which will be manipulated by appropriate operators to reach the global minimum. However, different codification can be used.
Generally speaking, it is possible to describe GAs' as follows. The first step consists in the creation of an initial population, usually randomly selected in the whole function domain. This population is a dynamic entity; at each iteration some elements of the population will be chosen to reproduce, by using suitable operators, accordind to a fitness value assigned to them by the algorithm. Each generation will tend to contain more of the features that were found useful in the previous generation and an improvement in overall performace can be realized over the previous generation. A condition is addeed to insure that the best individual from one generation is always retained in the next generation. Iterative reproduction leads the population toward the optimal condition.
The function that assigns a fitness value to each string is strictly correlated with the optimization problem considered: a high fitness value is assigned to those strings which evolve toward an optimal condition. The strings will be chosen for reproduction probabilistically, according to their fitness and the genetic operators are applied to the selected strings with fixed probability. The most used operators are reproduction, mutation and cross-over. Their probabilities, "prep", "pmut" and "pcross", must be tuned depending on the application. Other parameters which influence the convergence of the genetic algorithms are the population size and the number of iterations, or in case the criterion adopted to stop the algorithm.

III. Optimal selection of MLP's topology

The fundamental phases of the MLP activities are [8]:

- a learning phase, during which a set of training patterns is presented to the input layer and the weigths of each connection are cyclically updated, according to the Back Propagation learning rule. Back Propagation is a supervised learning technique that performs gradient descent on a quadratic error measure. A network starts with small random weights and is trained by comparing its response to each input pattern of the learning set with the correct output. The error obtained is back-propagated through the network to update the weights. This phase ends when a suitable value of the error is reached;
- a testing phase, in which the network is required, using the aquired experience embedded into the weight values, to answer on-line when a new set of patterns is presented to the input layer. Also in this phase an error value is calculated. If the performance in the testing phase are not satisfactorily, the learning phase has to be taken up again or a new topology has to be considered.

Both in networks with one hidden layer or in multi layer ones, the number of input and output units is fixed and the hidden structure, i.e. the number of layers and the number of neurons in each layer, has to be determined by the genetic algorithm.
Let us briefly discuss the genetic algorithm reported in [6], in which MLP with one hidden layer are considered.
In that case, each member of the population was constitued by a string, representing a network topology by the number of hidden units in binary codification. The strategy adopted to determine the fitness of each string is suitable also for the multi layer case and will be discussed in the following. The genetic operators were in the standard form.
If the multi layer case is considered, the major problem lies in determining a suitable codification and the corresponding genetic operators.
The following notation is adopted:

NN = fixed number of neurons to arrange in the network;
LN = number of layers to be determined by the algorithm;
N(l) = number of neurons in the l-th layer (l = 1...LN);
Lmax = maximum number of layers allowed (Lmax < NN);

In order to carry out the examined problem, a matrix codification is considered; each network topology is represented by a matrix of Lmax rows and NN columns. Each row, representing a layer, contains a number of "1" equals to the number of neurons belonging to the considered layer: the row dimension will be completed by "0". In order to make more efficient the operators, it is very important to place the "1" randomly in the rows. An example of codification is reported in Figure 1.

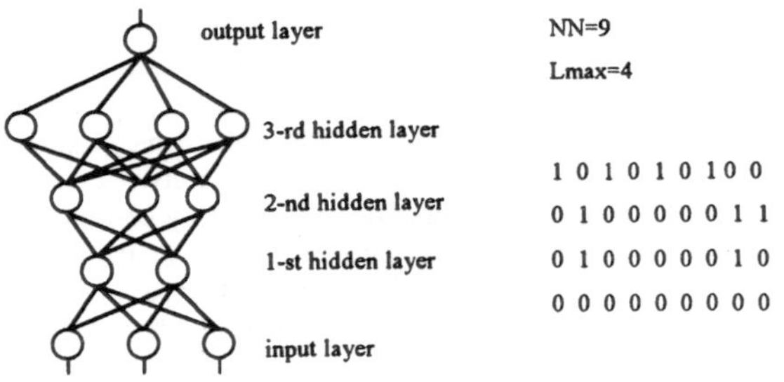

Fig. 1: codification of a neural structure

The entire population consists therefore of a fixed number of matrices, representing neural structures, randomly determined .
Next, it is necessary to define a function which assigns the fitness value to each string. This choice is very important because the genetic algorithms guarantee the convergence to the topology with the greatest fitness value. The fitness function must be consequently a combined measure of the parameters which characterize a "good neural network".
To determine the fitness function, each network belonging to the population is let go through a number of back propagation cycles until an a priori fixed error value Eo is reached.
If some networks are not able to reach the fixed error value in a fixed number Nmax of learning cycle, they are not considered as good networks and an arbitrary low value of fitness is assigned to them.
As well known, a low error value in the learning phase, represents a necessary but not sufficient condition for suitable network performance during the testing phase; a further testing phase with unknown patterns is therefore required for all the networks that have reached the value Eo within Nmax learning cycles.
Taking into account this consideration, a set of testing patterns is considered and a testing phase is performed for each sub-optimal network. Since the target is known for each pattern, the error in the testing phase is evaluated as follows:

$$E_t = \sum_{i=1}^{N} (y_i - t_i)^2$$

where:
t = target;
N = number of testing patterns;
y = output of the network.

Another worth which can be considered in evaluating the performance of the network is the number of connection (NC), which depends on the displacement of the neurons: a good fitness value is assigned to the network with a low value of the testing error and a small number of connections.
The fitness function for each individual i is therefore:

$$F(i) = k_1 * \left(\frac{1}{E_t(i)} \right) + k_2 * \left(\frac{1}{NC(i)} \right)$$

where k_1 and k_2 are suitable constants adopted to tune the worth on the problem.
For the matrix representation adopted, appropriate operators have to be defined. Let us consider, in the matrix representing a network, a window randomly fixed through two points: $P(x_1,y_1)$ and $Q(x_2,y_2)$, with $0 < x_1 < x_2 < NN$ and $0 < y_1 < y_2 < Lmax$.
The genetic operators are defined on the considered windows for each matrix, as shown in Figure 2.and 3.

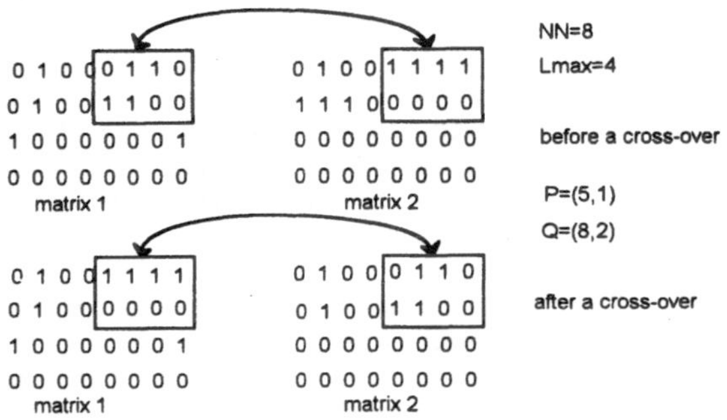

Fig. 2: Cross-over operation

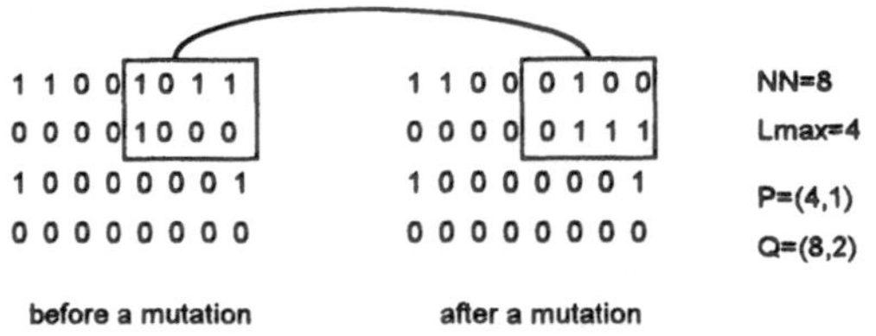

Fig.3: Mutation

Particular attention has to be devoted to assure that the cross-over or mutation operators always yield meaningful blueprints. In fact when applying the cross-over and the mutation operators it is possible that the new offsprings have a number of neurons ("1") not equal to NN. In this case the operation has to be repeated on the same matrices, randomly changing the position and the dimension of the window.
The process of training individual networks, measuring their fitness and applying genetic operators to produce a new population of networks, is repeated until the

maximum value of fitness over the whole population becomes constant with iterations and the mean value approaches the maximum.
This procedure can be viewed as tricky and time consuming: in order to overcome this drawback, a parallel architecture, based on a Transputer array, has been adopted.

IV. Results

The proposed strategy has been applied to the same problem considered in [6] in order to rearrange in a multy layer structure the optimal number of neurons previously found for a structure with only one hidden layer.
The task considered is the identification of a 3-rd order non linear system reported in literature [9]:

$$y_p(k+1) = f[y_p(k), y_p(k-1), y_p(k-2), u(k), u(k-1)]$$

where k is the discrete time and:

$$f(x_1, x_2, x_3, x_4, x_5) = \frac{x_1 x_2 x_3 x_5 (x_3 - 1) + x_4}{1 + x_3^2 + x_2^2}$$

The number of input and output neurons is therefore fixed as follows:
-5 input units
-1 output unit.
The training has been carried out for each network considering 200 patterns from a random input signal uniformly distributed in the interval [-1 1] and the relative output values. The maximum number of training cycles is fixed to Cmax=10000 and the maximum error to Eo=0.05. For the networks which do not reach the value Eo within Cmax cycles, the fictitious value of fitness F= 0.001 has been assigned. The testing phase is therefore performed only for "good" neural networks. The testing phase has been performed with patterns obtained by considering the input signal u(k)=sin(2pk/250) for k<500.
The number of neurons to be rearranged was NN=14, this value has been obtained in [6] and it has been validated by using a monitoring approach [2].
The maximum number of layers has been fixed to the value Lmax = 4.
The parameters of the genetic algorithms are reported in the following:
-population size = 10;
-pcross = 0.7;
-pmut = 0.07;
-prep = 0;
-k_1 = 1;
-k_2 = 10.

The convergence has been reached in about 20 steps to the topology characterized by the following values:
LN = 2;
L (1) = 10;
L (2) = 4;
The trend of the maximum, minimum and of th mean of the fitness values over the whole population versus the step number, are reported in Fig. 4, 5 and 6 respectively.

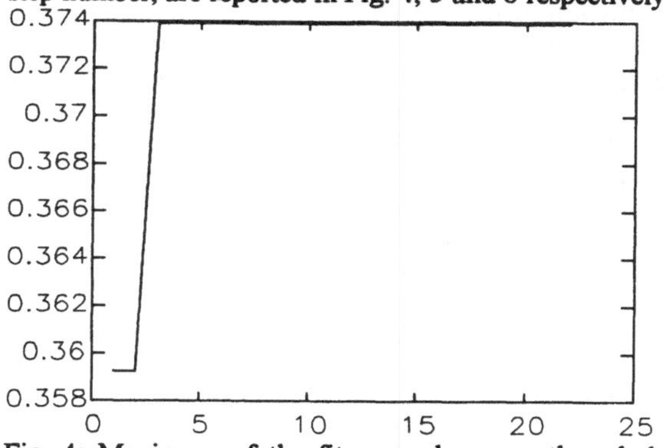

Fig. 4: Maximum of the fitness value over the whole population versus the step number.

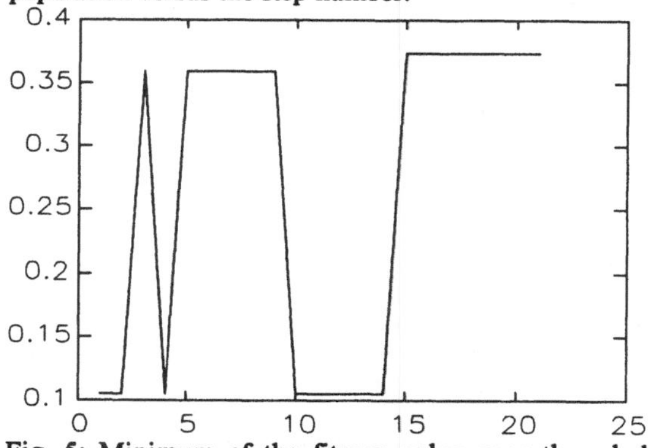

Fig. 5: Minimum of the fitness value over the whole population versus the step number.

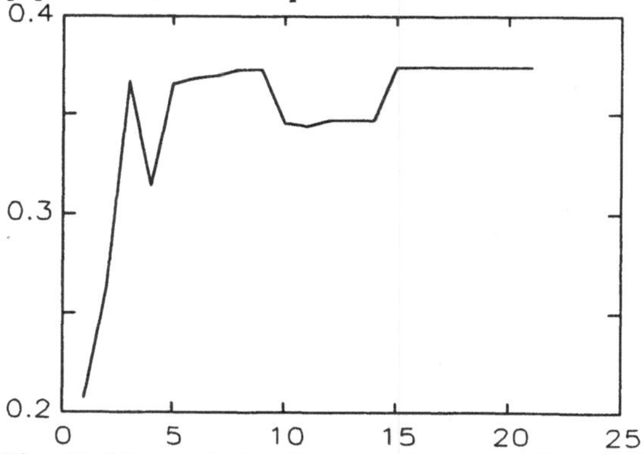

Fig. 6: Mean of the fitness value over the whole population versus the step number.

Also in this case, the suitability of the result has been verified by performing an alternative search with the aid of a monitoring tool which allows to investigate about the strength of each connection of the network as well as the activation value of each neuron.

V. Conclusion

In the paper a genetic algorithm able to determine the optimal topology of a Multy Layer Perceptron is proposed. To this aim, the structure of each neural network has been codified by a suitable matrix on which the genetic operators have been defined. An example is reported in order to show the suitability of the algorithm. The little number of steps spent for the convergence and the number of neural topologies trained by the genetic algorithm have shown the suitability of the proposed approach with respect to the usual "trial-and-error" strategy to select the best network topology.

References

[1] G. Cybenko, "Approximation by superposition of sigmoidal function", Mathematics of control signals and systems, Vol. 2, 1989, pp.303-314

[2] P. Arena, L. Fortuna, S. Graziani, G. Nunnari, "A monitoring approach for the design of a multi-layer neural network", COMADEM '91, Southampton, U:K:

[3] J. D .Schaffer, R. Caruana, L. Eshelman, "Using genetic search to exploit the emergent behavior of neural network", Phillips Laboratories 1989

[4] S. A. Harp, T. Samad "Genetic Synthesis of neural network architecture", Handbook of genetic algorithms, L. Davis, Van Nostrand Reinhold, 1991

[5] P. Arena, L. Fortuna, S. Graziani, G. Muscato, "A real-time implementation of a multi-layer perceptron with automatic tuning of learning parameters" Algorithms and architecture for real-time control, IFAC Workshop Series, N. 4, 1992

[6] P. Arena, R. Caponetto, L.Fortuna, M.G. Xibilia, "Genetic algorithm to select optimal neural network topology" 35-th Midwest Conference on Circuits and Systems, Washington, U.S.A., August 1992

[7] D. E. Goldberg, "Genetic algorithms in search, optimization & machine learning", Addison-Wesley, 1984

[8] D. E. Rumelhart, G. E. Hinton and R.J. Williams, "Learning internal represntation by error propagation", Parallel Distributed Processing: Exploration in the microstructure of cognition, D. E. Rumelhart and J. E. McLelland, Cambridge, Mass: MIT Press, 1986, pp.303-314

[9] K. S. Narendra, "Identification and control of dynamical systems using neural network", IEEE Trans. on Neural Network, No. 1, March 1990

Application of Genetic Algorithms to the Construction of Topologies for Multilayer Perceptrons

W. Schiffmann, M. Joost, R. Werner *

Abstract

In this paper we present a new approach for automatic topology optimization of backpropagation networks. It is based on a genetic algorithm. In contrast to other approaches it allows that two networks with different number of units can be crossed to a new valid "child" network. We applied this algorithm to a medical classification task, which is extremely difficult to solve. The results confirm, that optimization make sence, because the generated network outperform all fixed topologies.

1 Introduction

As Minsky and Papert [1] have shown the XOR–problem cannot be solved without a hidden layer. A learning rule which is able to train this kind of networks was developed by Rumelhart et al. [2]. It is known as *backpropagation* (BP) and it is one of the most often used neural network paradigms. Hornik [3] proved that every function can be approximated by a neural network with just *one* hidden layer. By adding "enough" hidden units the approximation error can be made as small as required. This can be compared with a look–up table which stores one output value/vector per hidden unit. But, this approach doesn't take into account that the network should be able to generalize. It just tries to fit the training data.

2 Automatic Topology Search

BP is working on a given network architecture which is made up by a specific partition of units over layers and by a particular connectivity pattern. Choosing an appropriate topology for a given problem depends on personal experience of the human designer. Almost always a fully interconnected architecture is used. But what is the optimal number of units und their organization into layers?

With the exception of some simple task, e.g. the XOR–problem, humans cannot foresee the optimal network topology. Thumb rules like ***the harder the problem the more units you need*** are of little practical use to the design problem. Manual network design is something of a black art [4] and it can be assumed that most of such network designs are not optimal. In [5] the impact of network topology on the speed and performance of BP trained networks was demonstrated. In order to adapt the network topology to the problem at hand we proposed an *automatic* design procedure which is based on *Genetic Algorithms* (GAs) (see also [3]).

Later on, we evaluated the performance and generalization behaviour of automatic generated network architectures with respect to character recognition tasks [7]. While our first approaches used just mutation to modify the network topology we are now able to cross two architectures as well.

2.1 Possible partitions

Suppose a wizard would tell us that h hidden units and l hidden layers are necessary for a given problem. Now, we have to distribute the hidden units into these layers. We will derive a recursive formula which allows to compute the number of possible partitions $p(h, l)$. It is evident that

$$p(h,l) = 1 \quad \text{if} \quad l = 1 \quad \text{or if} \quad h = l \qquad (1)$$

If $h < l$ then $p(h, l) = 0$ because empty partitions are not reasonable. If $h > l$ we get

$$p(h,l) = p(h-1, l-1) + p(h-1, l) \qquad (2)$$

By this formula we can compute the number of possible partitions from that number of a less complex

*University of Koblenz, Institute of Physics, Rheinau 3-4, D-5400 Koblenz, e-mail: schiff@infko.uni-koblenz.de. This work is supported by the *Deutsche Forschungsgemeinschaft* (DFG) as part of the project *FE–generator* (grant Schi 304/1–1)

h / l	10	20	30	40	50	60	70	80	90	100
1	1	1	1	1	1	1	1	1	1	1
2	9	19	29	39	49	59	69	79	89	99
3	36	171	406	741	1176	1711	2346	3081	3916	4851
4	84	969	3654	9139	18424	32509	52394	79079	113564	156849
5	126	3876	23751	82251	211876	455126	864501	1502501	2441626	3764376

Table 1. Number of possible partitions if h hidden units should be distributed over l (non empty) layers

architecture with $h-1$ units. The first term counts the number of partitions if the boundary of an additional layer seperates the additional unit itself. The second term means that the additional unit is placed into the last hidden layer of a less complex architecture which has already l layers. As Table 1 illustrates the number of possible partitions —even for moderate numbers of hidden units— increases to astronomical values.

2.2 Possible connections

If we have decided for a specific partition we are confronted with the problem of optimizing connectivity. For h hidden units the number of possible connections is limited by two extreme topologies which are fully interconnected from input layer (m units) to output layer (n units):

1. A topology that has as much as possible hidden layers with one hidden unit each. We will refer to that topology as TALL.
2. A topology that has just one hidden layer. It forms a look-up table and we will refer to it as WIDE.

While the TALL-architecture contains

$$C_T = \frac{h^2 - h}{2} + h \cdot (m+n) + m \cdot n \qquad (3)$$

connections, the WIDE-architecture has only

$$C_W = h \cdot (m+n) + m \cdot n \qquad (4)$$

connections. Table 2 compares these two topologies to each other for various values of hidden units. Additionally the fraction of C_W/C_T is given. It can be used as a measure of connectivity.

In most practical applications neither the TALL-nor the WIDE-architecture will be best suited. Thus, we have to find a connectivity pattern which is adapted to a particular task. Because we (humans) cannot comprehend the effects of modifications in topology, we have to provide methods for topology optimization which operate automatically.

h	C_T	C_W	C_W/C_T
1	87	87	1.000000
25	963	663	0.688474
50	2488	1263	0.507637
75	4638	1863	0.401682
100	7413	2463	0.332254
200	24763	4863	0.196382
300	52113	7263	0.139370

Table 2. Number of possible connections and minimum connectivity for the two extreme architectures (m=21, n=3, see also section *Simulation Results*)

3 Applications of GAs to Neural Networks

It should be noted that GAs can be applied to neural networks in two different ways:

1. Optimizing connection weights
2. Optimizing network topology

3.1 GAs for weight adjustment

In the first case the GA works at continuous parameters. Results concerned with this approach can be found in [8] – [16]. An exciting discussion of evolutionary training methods is provided by [17]. It can be summarized that the mentioned approaches differ mainly in three ways:

1. number representation
2. genetic operators used and
3. parent–offspring replacement strategies

If the number of connections is high it takes a lot of time to train a network by means of GAs. As simulations show, just less complex networks (<50 units) can be trained by this method. The speed of convergence can be increased if the rate of mutation is inverse to the diversity in the population of networks

[9]. GAs for weight adjustment implement a parallel search in weight space. Thus, they are able to find approximative solutions in short time. In contrast, GAs have trouble to get an exact solution. In order to solve this problem, one can combine GAs with BP. The starting weight vector for BP is determined by a preceding GA training run. See [18] for details of that approach.

3.2 GAs for topology optimization

Here, learning is done by BP or any other well known learning procedure. Because discrete parameters must be optimized the performance function is undifferentiable. Thus, gradient methods are not applicable. As we have seen the search space of all possible network architectures is vast and noisy. Because it depends on the random initial weights the performance of a specific architecture must be viewed as a random variable. It is also *deceptive* and *multimodal*. These features are concerned with the effects of small changes of the parameters and the objective function. *Deceptive* means that similar network architectures can have different performance. On the other hand, different network architectures show similar performance. Hence, the search space is *multimodal*.

As pointed out in [4] such complex spaces cannot be explored efficiently be enumerative, random or even heuristic knowledge–guided search methods. In contrast, the adaptive features of GAs (building blocks, step–width control by crossover) provide a more robust and faster search procedure. Additionally, it is easy to speed–up the genetic search by means of parallel processing.

Before we discuss the differences between several approaches we want to give a basic GA for topology optimization which is common to all these approaches. It is assumed that two representations of the networks are distinguished:

1. *genotypes* which are modified by the GA's operators (mutation, crossover)
2. *phenotypes* which are trained by a conventional learning procedure (e.g. BP) used for performance evaluation or selection

The basic GA comprises four steps:

1. Initialize a population of random starting architectures
2. Select, cross(over) and mutate these architectures
3. Train the networks for a given number of epochs
4. Stop if the desired performance is achieved; else proceed with step 2

3.3 Representation of genotypes

Roughly two basic representation schemes can be distinguished:

1. low–level genotypes
2. high–level genotypes

While the first one is transparent and easy to use, there are two variants of the second representation scheme. Low–level genotypes directly code the network topology. Each unit and each connection is specified separately. This "blueprint" approach is used by Miller et al. [4], Schiffmann et al. [7] and Dodd [19].

High–level genotypes are more complex coded representations of network architectures. They can be further divided into *parametric* and *recipe* genotypes. Examples for the use of parametric genotypes are found in [20] – [23]. Here, the networks are splited into modules of units which are specified by parameters and which are coupled by parametric connectivity patterns. Even through this representation is more compact and thus well suited to code large network architectures, it is difficult to choose the *relevant* parametric shapes. It should be noted that in using parametric genotypes the search space is confined to a specific subspace.

Similar considerations apply to the recipe genotype representation. Here, the architecture is specified by growth rules [24] and [25] or by sentences of a formal language [26]. The last approach is called *genetic neural networks*. Even though these approaches are mostly biological plausible, one has to commit to primitives used for the rules. Further, the choice of specific primitives hardly influences the application of genetic operators.

Usually neural network applications are limited to network sizes of about 1000 units. Often, research applications are confined to smaller architectures (e.g. XOR– or encoder–problem). Thus, in order to evaluate the properties of GAs for topology optimization the low–level representation is well suited and sufficient. Because it is easy and straightforward to apply genetic operators to the blueprints, we decided to use it to code our network architectures.

3.4 Genetic Operators

As pointed out in [17] the genetic operators must produce *correct* and *complete* offsprings. These requirements are easy to satisfy with blueprints. An efficient blueprint representation can be achieved by using a list where the connectivity of individual units is registered. In order to use this data structure together with BP, the list must be chained in both directions. For the sake of simplicity, this is not shown in Figure 1.

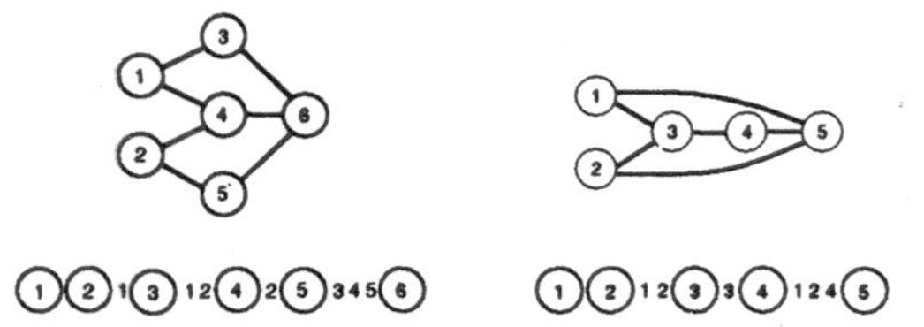

Fig. 1. Two network architectures and its corresponding blueprint representations

Blueprint coding works in two steps. Starting with the input layer, the units are numbered. Then, for each of the units the numbers of its preceding units are registered in the list. Note that topologically equivalent architectures can have different blueprint representations.

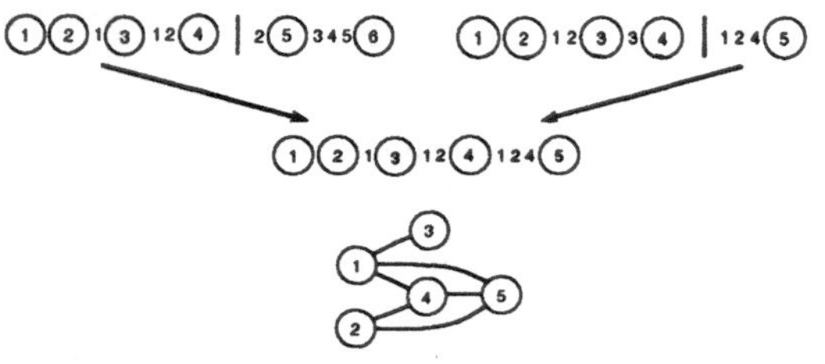

Fig. 2. Application of our crossover operator to the networks in the previous figure. The isolated unit can be removed.

Crossover is done by choosing a common cross point and joining two sub-lists of the blueprints. The cross point must be positioned before the last output unit of that network which has the lower number of units. If this constraint is satisfied it is guaranteed that the produced offspring operates on the same interface units (input and output) as its parents. So we get an *useful* network (see Figure 2). In this way, we have defined a *correct* crossover operator which is able to cross networks of *arbitrary* size.

In order to guarantee that the offspring isn't identical with one of the parent networks we must position the cross point behind the last input unit. Nevertheless we cannot guarantee that always *new* architectures arise. Even though we get useful networks it could happen that isolated or fixed hidden units arise. The output of *isolated* units isn't used by any other unit. *Fixed* units don't have connections to preceding units. Its weighted (constant) output can be substituted by the bias units and its connection weights. Both kinds of useless* units can be eliminated without affecting the networks functionality. In this way the shrunk offspring networks can become smaller than the smallest parent network. On the other side, it isn't possible that the number of units becomes larger than the biggest parent network. However, the number of connections can grow when the described crossover operator is applied.

4 Simulation Results

A description of implementation details can be found in [27] and [28], which are available via ftp†.

XOR–Problem

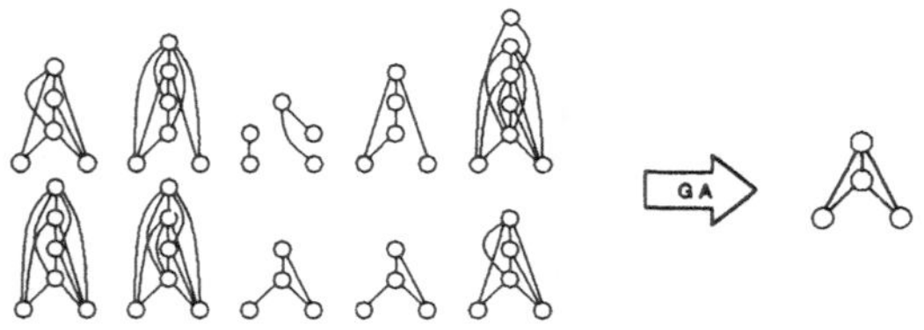

Fig. 3. The starting nets for the XOR Problem. All nets were made by chance.

To test our genetic algorithm, we tried to generate an optimal net for the XOR problem. Ten different nets were made by chance and so-called evolution servers were started on several computers (6 x Sun4, 4 x NeXT). The number of the training epochs per net was fixed on 1000 passes. After a short time of approximately five minutes the genetic algorithm converged to the well known topology for the XOR Problem (Figure 3).

*information is neither combined nor processed

†archive.cis.ohio-state.edu (128.146.8.52) pub/neuroprose/schiff.gann.ps.Z and pub/neuroprose/schiff.bp_speedup.ps.Z.

Thyroid data

The thyroid data are measurements of the thyroid gland. Each measurement vectors consists of 21 values — 15 binary and 6 analog ones. Each pattern is labeled by a class name which corresponds to the hyper–, normal–, and subnormal function of the thyroid gland. Since over 92% of all patients have a normal function, a useful classifier must be significantly better than 92% correct classifications. The training set consists of 3772 measurements and again 3428 measurements are available for testing. This classification problem is difficult to solve by neural nets. Several fixed topologies were tested. Figure 4 shows the course of the error during the training of these fixed topologies. It can be seen that several thousand learning passes are necessary to archieve a reasonable classifier.

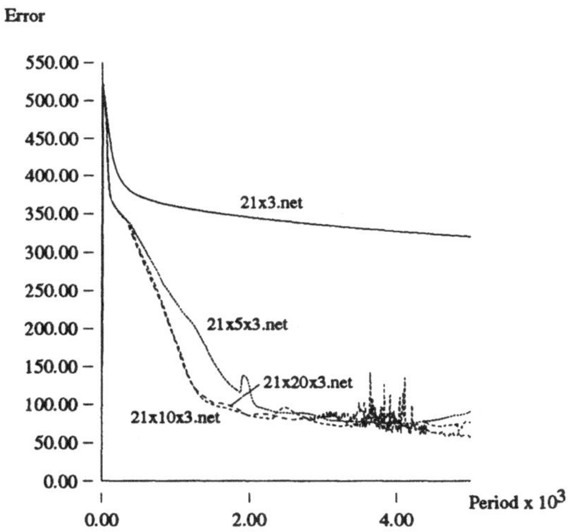

Fig. 4. The error during the training of the thyroid data with fixed topologies. The name of a net specifies its architecture. For example the name 21x10x3 indicates that the net consists of 21 input units, 3 output units and 10 hidden units. All units are fully interconnected.

In order to accelerate the evolution process, we don't use the complete training set. Instead of that, 713 measurements were chosen out of it. These selected patterns are the hardest ones of the training set because they are drawn from the boundary region between two categories. All nets were trained only

	#units	#Weights	Connectivity
n_1.net	54	110	9.0%
n_2.net	34	270	22.4%
n_3.net	49	461	47.9%
n_4.net	34	32	9.2%
n_5.net	39	378	71.6%
n_6.net	54	174	14.3%
n_7.net	24	40	63.5%
n_8.net	39	433	82.0%
n_9.net	31	233	92.5%
n_0.net	34	238	31.6%

Table 3. Characteristics of the starting nets for the thyroid task.

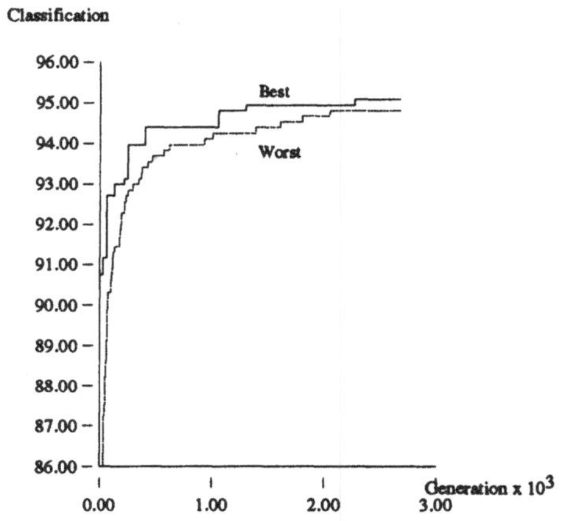

Fig. 5. The achieved classification performance during the evolution process. Only the classification performance of the best and the worst net of the population is shown.

with this reduced set. The training period during the evolution was fixed on 500 passes. Learning rate and momentum were adjusted constantly to 0.01 and 0.9 as well.

To start the evolution process we designed 10 different nets by chance. These nets have various numbers of units and connections (Table 3). The achieved classification performance during the evolution process is shown in Figure 5. Only the values of the best and the worst net of each generation are represented. To get these well adapted networks the evolution needs about 72 hours. It can be seen that the best net reaches a classification performance of more than 95% correct classified patterns.

The development of the network's complexity is de-

	#Units	#Weights	Connectivity	reduced training set	whole training set	test set
fixed nets						
21x5x3.net	29	183	94.8%	76.6%	95.0%	94.3%
21x10x3.net	34	303	87.1%	76.9%	95.5%	94.8%
21x20x3.net	44	543	74.1%	71.1%	94.4%	93.9%
evol. nets						
n_123.net	48	285	31.1%	94.8%	98.3%	97.3%
n_125.net	49	283	29.4%	95.0%	98.1%	96.9%
n_130.net	49	278	28.9%	94.8%	98.3%	97.4%
n_136.net	48	277	30.3%	94.8%	98.2%	97.1%
n_147.net	49	276	28.7%	95.0%	98.4%	97.3%
n_149.net	50	279	27.6%	95.0%	98.1%	97.2%
n_152.net	50	278	27.5%	95.0%	98.2%	97.0%
n_154.net	49	286	29.7%	95.1%	98.3%	97.3%
n_155.net	50	279	27.6%	95.0%	98.6%	97.4%
n_156.net	50	278	27.5%	94.8%	98.4%	97.5%

Table 4. Comparison between the fixed and the generated networks. The evolved architectures archive a better performance although they are smaller.

monstrated in Figure 6. It can bee seen that at first big topologies are preferred because of the better results. Later the algorithm produced smaller architectures.

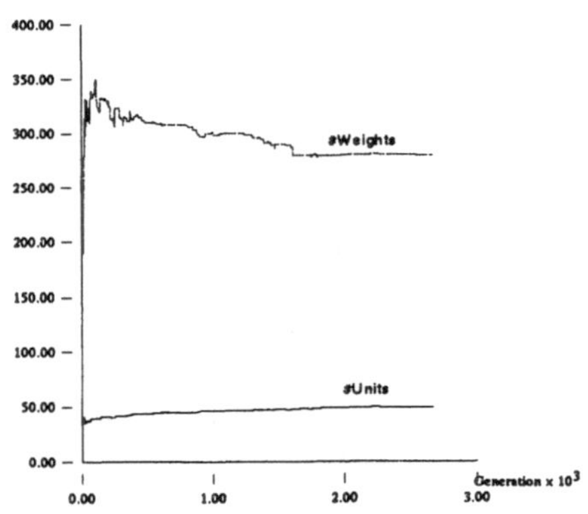

Fig. 6. The development of the net size during the evolution process.

Table 4 compares the results of the nets created by evolution with the results of the fixed architectures. To obtain comparable results, every fixed net was also trained 500 epochs with the reduced training set. The table represents the achieved classification performance with respect to the reduced training set, the whole training set, and the test set. Note, that the nets achieve a better classification performance on the complete training set. This is, because the extracted set consists only of the hardest samples of the training set. Furthermore, the table shows the number of units and connections in order to compare the network's complexity. It should be taken into account that the number of the connections is especially decisive for the size of a net. The adapted nets are more efficient although they are smaller. They train faster and generalize better. To achieve a similar classification performance with the fixed nets, many additional learning passes would be necessary.

Finally, we investigated the training behavior of the created nets with respect to the complete training set. Figure 7 shows the error of selected nets during the learning passes. The illustration demonstrates that the error starts to oscillate. This behavior is typical for all generated nets. Nevertheless, the error of the adapted nets is below the error of the fixed nets.

Table 5 compares the results of the trained nets. It shows that the adapted nets have a some better result concerning the training set. But with respect to the generalization behavior they are, however, superior to the fixed architectures (> 1%). This can be explained by the fact that evolution finds feature extracting cells.

5 Conclusion and further work

Unfortunately, the generated nets show oscillations in the course of error. These oscillations start typically after the number of epochs which was used for the

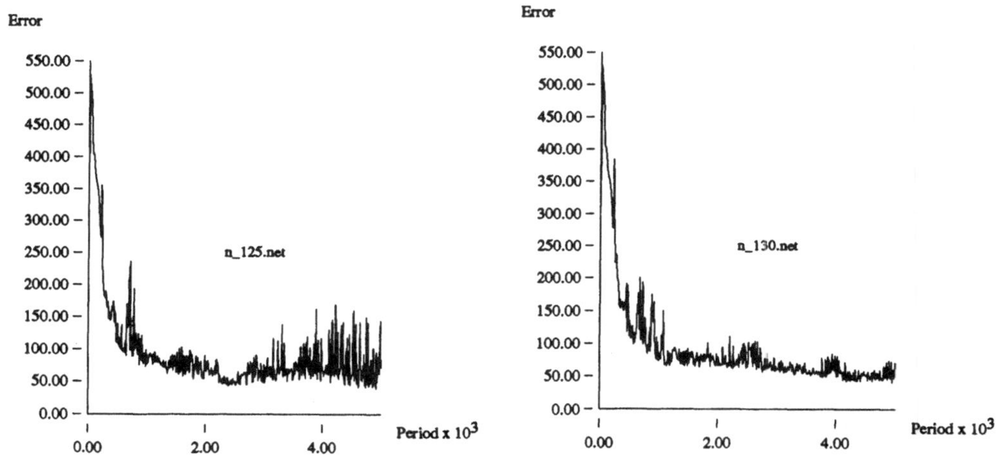

Fig. 7. Error during the training of the complete thyroid data with the evolved networks

	training set	test set
21x5x3.net	99.0%	97.4%
21x10x3.net	99.1%	97.3%
21x20x3.net	99.3%	97.4%
n_123.net	99.5%	98.2%
n_125.net	99.4%	98.6%
n_130.net	99.4%	98.4%
n_147.net	99.4%	98.2%
n_152.net	99.4%	98.5%
n_154.net	99.4%	98.4%
n_155.net	99.4%	98.6%
n_156.net	99.4%	98.4%

Table 5. Comparison between the fixed and the generated networks which are trained with the whole training set.

genetic algorithm (here at 500 epochs). The genetic algorithm generates networks which can be trained quickly up to this epoch. The training behavior of the nets after this number of epoch is uncertain. In order to suppress these oscillations, a learning rate adaption should be used.

Furthermore, we have found that the number of training epochs can be reduced by using quickprop [29]. We plan to implement this training procedure in our algorithm. In order to improve the generalization behavior of the nets, the performance of the produced nets should be determined with respect to the test set. The generalization behavior must then be determined on a third test (test) pattern set. In the case of the thyroid data the nets could be trained with the reduced training set and the quality measure should be determined with the complete training set. Finally, the test set could be used to determine the generalization performance.

6 Acknowledgements

We are grateful to K.-H. Staudt for implementation of the network editor. By means of this tool we are able to visualize and analyse the generated topologies. This work is supported by the *Deutsche Forschungsgemeinschaft* (DFG) as part of the project *FE-generator* (grant Schi 304/1-1).

7 References

[1] Minsky and Papert, *Perceptrons*, MIT Press, 1969

[2] Rumelhart D.E., Hinton G.E. and Williams R.J., *Learning internal representations by error propagation*, in Parallel Distributed Processing: *Explorations in the Microstructures of Cognition*, Vol.I, MIT Press, pp. 318-362, 1986

[3] Hornik K., Stinchcombe M. and White H., *Multilayer Feedforward Networks are Universal Approximators*, Neural Networks, Vol. 2, pp. 359–366, Pergamon Press, 1989

[4] Miller G., Todd P.M. and Hegde S.U., *Designing Neural Networks Using Genetic Algorithms*,Proc. of the third Intern. Conf. on Genetic Algorithms (ICGA), pp. 379–384, San Mateo (CA), 1989,

[5] Schiffmann W.H. and Mecklenburg K., *Genetic Generation of Backpropagation Trained Neural Networks*, Proc. of Parallel Processing in Neural Systems and Computers (ICNC), Eckmiller R. et al. (Eds.), pp. 205–208, Elsevier, 1990

[6] Schiffmann W.H., *Selbstorganisation neuronaler Netze nach den Prinzipien der Evolution*, Fachbericht 7/1989, Universität Koblenz

[7] Schiffmann W.H., Joost M. and Werner R., *Performance Evaluation of Evolutionarily Created Neural Network Topologies*, Proc. of Parallel Problem Solving from Nature, Schwefel H.P. and Maenner R. (Eds.), pp. 274–283, Lect. Notes in Computer Science, Springer, 1991

[8] Whitley D., *Applying Genetic Algorithms to Neural Net Problems*, Neural Networks, Vol. 1, p. 230, 1988

[9] Whitley D. and Hanson T., *Optimizing Neural Networks Using Faster, More Accurate Genetic Search*, Proc. of the third Intern. Conf. on Genetic Algorithms (ICGA), pp. 391–396, San Mateo (CA), 1989

[10] Whitley D. and Bogart C., *The Evolution of Connectivity: Pruning Neural Networks Using Genetic Algorithms*, Proc. of the intern. Joint Conf. on Neural Networks, Vol. I, pp. 134–137, 1990

[11] Whitley D., Starkweather and Bogart C., *Genetic Algorithms and Neural Networks: Optimizing Connections and Connectivity*, Parallel Computing Vol. 14, pp. 347–361, Elsevier, 1990

[12] Whitley D., *The GENITOR Algorithm and Selection Pressure: Why Rank-Based Allocation of Reproductive Trials is Best*, Proc. of the third Intern. Conf. on Genetic Algorithms (ICGA), pp. 116–121, San Mateo (CA), 1989

[13] Heistermann J., *Parallel Algorithms for Learning in Neural Networks with Evolution Strategy*, Parallel Computing, Vol. 12, 1989

[14] Heistermann J., *Learning in Neural Nets by Genetic Algorithms*, Proc. of Parallel Processing in Neural Systems and Computers (ICNC), Eckmiller R. et al. (Eds.), pp. 165–168, Elsevier, 1990

[15] Montana D. and Davis C., *Training Feedforward Neuronal Networks Using Genetic Algorithms*, Technical Report, BBN Systems and Technologies Inc., Cambridge (MA), 1989

[16] de Garis H., *Genetic Programming — Modular Neural Evolution for Darwin Machines*, Proc. of the Intern. Joint Conf. on Neural Networks, Vol I, pp. 194–197, 1990

[17] Weiss, G., *Combining neural and evolutionary learning: Aspects and approaches*, Report FKI–132–90, Technische Universität München, 1990

[18] Kitano H., *Empirical Studies on the Speed of Convergence of Neural Network Training Using Genetic Algorithms*, Proc. of the National Conf. of the American Association of Artificial Intelligence (AAAI), pp. 789–795, 1990

[19] Dodd N., *Optimization of Network Structure Using Genetic Techniques*, Proc. of the Intern. Conf. on Neural Networks, pp. 693–696, Paris, 1990

[20] Lehar S. and Weaver J., *A Developmental Approach to Neural Network Design*, Proc. of the IEEE Intern. Conf. on Neural Networks, Vol. I, pp. 97–104, 1987

[21] Harp S.A., Samad T. and Guha A., *Towards the Genetic Synthesis of Neural Networks*, Proc. of the third Intern. Conf. on Genetic Algorithms (ICGA), pp. 360–369, San Mateo (CA), 1989

[22] Harp S.A., Samad T. and Guha A., *The Genetic Synthesis of Neural Networks*, Technical Report CSDD–89–14852–2, Honeywell, Golden Valley (MN), 1989

[23] Merrill J. and Port R., *Fractally Configured Neural Networks*, Neural Networks, Vol. 4, pp.53–60, 1991

[24] Mjolsness E. and Sharp D.H., *A Prelimiary Analysis of Recursively Generated Networks*, in Denker J. (Eds.): *Neural Networks for Computing*, Snowbird (Utah), 1986

[25] Mjolsness E., Sharp D.H. and Alpert B.K., *Recursively Generated Neural Networks*, Proc. of the IEEE Intern. Conf. on Neural Networks, Vol. III, pp. 165–171, 1987

[26] Mühlenbein H. and Kindermann J., *The Dynamics of Evolution and Learning — Towards Genetic Neural Networks*, in Pfeifer et al. (Eds.): *Connectionism in Perspective*, Elsevier, 1989

[27] Schiffmann W.H., Joost M. and Werner R. *Optimization of the Backpropagation Algorithm for Training Multilayer Perceptrons*, Technical Report 15/1992, University of Koblenz

[28] Schiffmann W.H., Joost M. and Werner R. *Synthesis and Performance Analysis of Multilayer Neural Network Architectures*, Technical Report 16/1992, University of Koblenz

[29] Fahlman S.E., *An Empirical Study of Learning Speed in Back-Propagation Networks*, Technical Report, Carnegie-Mellon University, CMU-CS-88-162, 1988

Genetic Algorithms as Heuristics for Optimizing ANN Design

E. Alba, J.F. Aldana & J.M. Troya

Dpto. de Lenguajes y Ciencias de la Computación
Facultad de Informática
Universidad de Málaga
Pl. El Ejido s/n
29013 Málaga
Spain
aldana@ctima.uma.es

Abstract. The problem of the ANN design is usually thought as residing in solving the training problem for some predefined ANN structure and connectivity. Training methods are very problem and ANN dependent. They are sometimes very accurate procedures but they work in narrow and restrictive domains. Thus the designer is faced to a wide diversity of different training mechanisms. We have selected Genetic Algorithms because of their robustness and their potential extension to train any ANN type. Furthermore we have addressed the connectivity and structure definition problems in order to accomplish a full genetic ANN design. These three levels of design can work in parallel, thus achieving multilevel relationships to build better ANNs. GRIAL is the tool used to test several new and known genetic techniques and operators. PARLOG is the Concurrent Logic Language used for the implementation in order to introduce new models for the genetic work and to attain an intralevel distributed search as well as to parallelize any ANN management and any genetic operations.

1.- Introduction

Artificial Neural Networks (ANNs) represent an important paradigm in AI. They were first proposed as biological models for the human brain dealing with massively parallel information processing. ANNs are widely used to offer human-like skills wherever they are needed, so we can find them in pattern recognition, signal processing, intelligent control and many other applications that can be faced by introducing a network as the heart of the solution system (e.g., see [1]).

Whenever an ANN is to be used it must first be designed. At present, any ANN designer drags along an unstructured, heuristic and arbitrary path in order to reach "the better" structure and connectivity to be trained. Only the training methods are being truly applied, but every ANN type seems to need a different and own training mechanism. Usually, the training mechanism is some kind of hillclimbing prosecution, which is very closely related to (and so dependent on) the problem being solved, the ANN type and/or the pattern set for it. This results in a vast landscape of different multiparameter tuning procedures to be applied for any individual problem and with no warranties for optimum results.

This lack of methodology and the search for a general multifaceted quasioptimum training made Genetic Algorithms (GAs) suitable for our purposes. Defined by J. Holland in 1975, GAs simulate natural evolution. The bases of our GA and ANN approaches can be found in [2], [3] and [4]. In our work chromosomes will encode ANNs, genes will encode the items being optimized (weights, links or hidden layers) and alleles will be gene components (regarding upon the used coding one or more alleles are included to compose a single gene value). Initial strings will be genetically evolved over generations of newly created offsprings searching for an optimum. Our goal is to design the whole ANN through genetic means. We will use a *supervised* genetic training because input/output patterns are to be specified in order to decide the fitness of every genetic string.

Since any ANN must be coded as a chromosome string ANN independence is achieved, just some *evaluation* procedure must be defined to recognize relative fitness of individuals to the problem. GA techniques have a stochastic behavior and so we only can expect **quasioptimum** (very frequent good or optimum) training. Besides local minima avoiding, generality, multiple points parallel search and robustness we can get further in using GAs to complete the ANN design. Since GAs work on some coding of the solution and not on the solution itself, we can code "any" problem as a string of parameters and submit it to genetic optimization.

Thus, it is only necessary to code ANN *connectivity and structure* as strings properly and define an evaluation procedure to get two new levels of ANN design. This way we can bring optimization methodology to these two design stages. This **full three levels design** is thought to help designer's work from the problem specification (patterns) up to a quasi-optimum ANN to solve it (called a **Genetic ANN**). These three levels of design will be fully accomplished by using Genetic Algorithms. An introduction to genetic ANNs can be found in [5] and [6].

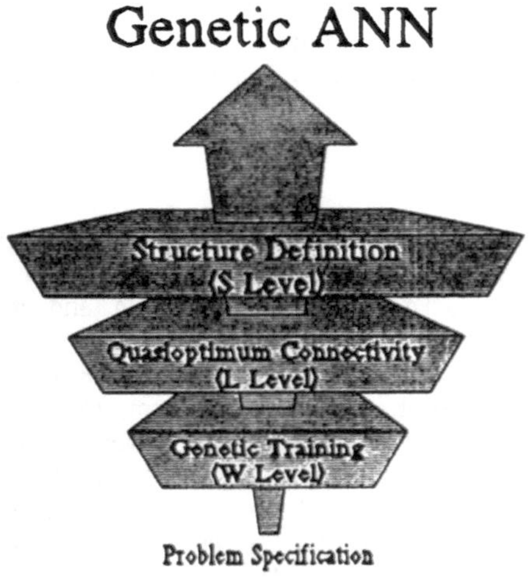

Fig 1.- *Three Levels GANN Design.*

We have designed and built up a genetic tool called **G.R.I.A.L.** (Genetic Research In Artificial Learning) [7] to implement several known and new GA techniques and the three levels of genetic ANN design in order to test their efficacy and properties. We are furthermore concerned with scalability and computational efficiency, thus we have used a Concurrent Logic Language called **PARLOG** to implement GA and ANN behavior in GRIAL as a new kind of computational approach in the aim of profiting from the outcoming parallelism advantages.

The used test suite is made up of four problems. The XOR and the Two-bits Binary Adder (TBA) problem (but with 4 patterns) as presented in [4]. A feedforward network to build Spanish Plurals (three classes' classification) and a Hopfield network to behave as a 9-characters classifier. The XOR and TBA problems are fully tested (training, connectivity and layered structure, either separately and together). The Spanish Plurals ANN has been trained and optimized in structure and the Hopfield network has been trained. We have tested the relative influence of the multiple GRIAL techniques in the aim of trying to get the optimum ANN for every problem.

In this work we have tested in GRIAL the effects of the *traditional selection* procedure versus a *one-at-a-time selection* procedure. We explore the influence of coding on ANN design by using *binary, real and diploid genotypes* (this last never tested before for ANN design). A *migration* scheme and an *adaptive mutation* similar to those used in [4] are tested against sequential single GAs and constant mutation. A smooth bit climber-like operator [8] called *GBit* is tried and a *Mating Restriction* similar to [9] is implemented. Dislike partial genetically defined ANNs (as [4] where genetic training and connectivity are addressed as separate works and [10] where structure is genetically defined but backpropagation is used for training) we have designed a full genetic and automatic ANN designer in order to build the whole ANN.

2.- Parlog

PARLOG [11] is a Concurrent Logic Language that has been developed at the Imperial College. Operationally the computational model of this kind of language consists in a concurrent processes' set which communicate by means of binding logic variables and which synchronize by waiting for unbounded logic variables. The possible behaviors of a process are defined by means of guarded horn clauses: **Head <- Guard : Body.** Head and Guard define the conditions under which a reduction can be made. Body specifies the resulting state of the processes after the reduction.

Parlog is one of these languages that exploit two kinds of parallelism: stream *and* parallelism and *or* parallelism. The first type of parallelism occurs when a goal is reduced to a conjunction of subgoals and they are all tested in parallel. The second type of parallelism appears when a predicate can be solved by more than one clause. In this case, all of them are tested at the same time and, if more than one fulfills its guard, one of them will be selected in an indeterministic way. Parlog also has some primitives that can be used to avoid both types of parallelism (see [12]).

This kind of language fits very well in the parallel programming paradigm. In opposition to sequential logic languages, that present a transformational behavior, concurrent logic languages are well suited for the specification of reactive systems, that is, of open systems which have a high level of interaction with their environment. And this is what a neural network does: it tries to reach a statistical minimum that is environment-dependent.

GRIAL is oriented to provide an easy changing of the GA strategies used to solve a given problem. GRIAL provides a graphic and user-friendly interface in order to allow the user to select the genetic search strategy. *Unix parallelism* is achieved at *interlevel* communications while *Parlog parallelism* appears at *intralevel* searches. Real LAN distribution of intralevel Parlog parallelism can be faced by using Parlog mail-boxes. Parlog allows entering parallelism from string managing genetic operations till ANN neurons' activations. Fine and coarse grained approaches are of a straightforward implementation with Parlog. These advantages along with its lists and symbols processing make Parlog a better language than imperative ones to get good and reliable implementations.

3.- A Complete Genetic ANN Design

In this section a three-levels full genetic ANN design is presented and analyzed by means of GRIAL. New and old existing GA techniques have been tested to get a qualitative understanding on the properties of this kind of design. We envisage the following exposition from bottom (genetic training) to up (genetic structure definition) passing through an intermediate level of genetic connectivity definition.

3.1.- From GAs to ANNs

For a genetic ANN definition we must be aware of several important considerations. All of these considerations present the same underlying problem: since ANN strings are very expensive to manage (think of millions operations to be run on a population of strings), not only the algorithmic implementation of the operations has to be efficient, but the genetic techniques must require a low computational cost.

Since ANN coding/decoding and the evaluation operations are very complex and expensive we look for improved selection mechanisms that minimize wanderings along the search space while still maintaining the GA properties. The **Generations Evolutive Scheme** (a full new generation replaces the old one) using the *Stochastic Remainder Without Replacement* [2] seems to be very expensive in our tests despite its advantages in preserving diversity and genotype exploration. That's why we have designed the **Immediate Effects Evolutive Scheme** to keep strings ranked (from best to worst) and using the *Roulette Wheel* selection to pick up two individuals to be genetically processed and produce offsprings to be inserted in the ranking. This later selection operator is the best choice for ANN design, but population size must be kept large enough to avoid premature convergence due to its more minimum-directed search. A complete-ranked selection as that in [13] could lessen genetic drift because in GRIAL we allow duplicates in the population and the IEES produces a high selection pressure.

In order to evaluate the fitness of a string to the *environment* (problem) strings are *expressed* (decoded) as ANNs. We use *SQE* (squared error between desired and obtained outputs extended to any output neuron and any pattern in the pattern set) as the fitness measurement to help natural selection's work: stochastic selection picks up the best strings to be crossed.

3.2.- Genetic Training

To submit any ANN type to genetic training we must define some proper coding for weights to appear as strings. GA's work needs some local logic meaning to be present in strings, i.e. a chromosome must be coded to include *logic building blocks* with some meaning for the underlying problem [2]. Then we code any ANN string as being a sequence of its input link weights to every neuron in the ANN, from the input layer to the output layer. The genetic crossover of two different strings (encoded ANNs) profits from their best slices to create better trained offsprings. Through natural selection bad *schemata* (bad structures of the solution) are exponentially discarded and good schemata are exponentially reproduced as evolution occurs.

Coding

Weights can be coded in strings attending to several codes. **Binary** code (signed magnitude) is very extended in genetics. **Real, Reordering and Diploid** [2] codings are another known ones we have used to train.

The binary code is very suited for GA's job, but for ANN training it needs too large populations and evolutions, even for very small problems (we want to keep population size on hundreds of individuals). Reordering schemes (genes, PMX and/or Inversion) do not seem to improve binary results, and we think this is because we are using a correct genotype representation of the problem that does not need additional genetic help.

Real codings (one-weight/one-real-gene) seems to be the truly useful genotype because they allow small and quick GAs to solve adequately the trainings, despite they present an undesired low diversity maintenance during evolution that provokes local minima appearances. All these codings are *Haploid* codings (one string encodes one ANN), but *Diploid* chromosomes with triallelic values [2] (and maybe with real values...) have much to say in allowing sophisticated behavior by helping diversity and natural adaptation to traumatic changes in the environmental conditions (they outperform the other codings using half the number of strings). Any feedforward, recurrent, dynamic, static or any other ANN type can be trained by GA means.

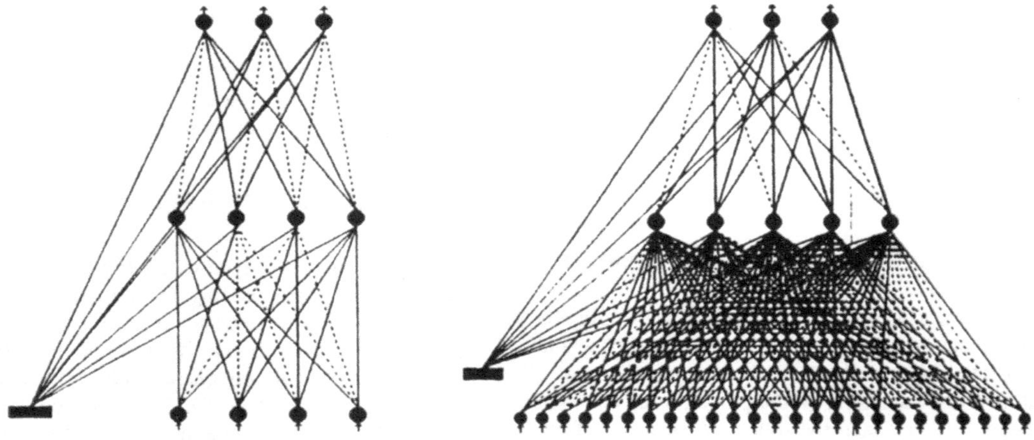

Fig 2.- *The Two-bits Binary Adder problem and the Spanish Plurals problem (24 patterns to determine one of the three Spanish plural types for any word presented to the ANN) solved by using feedforward ANNs with genetic training. In order to solve increasingly complex problems we need a better balance between exploration and exploitation and that's why we state the need of more sophisticated GA strategies and not the use of huge populations (thousands of individuals) and huge evolutions (millions of crossovers). Non-trivial training problems are actually being solved by using GAs.*

Genetic Techniques

We have tried *constant* and *adaptive mutation* operators (this last based on Hamming Distance --hd--) to maintain diversity in population. The results show that mutation is not a secondary operator but an essential technique, very useful in helping to keep population size at a relatively low value.

Probability for adaptive mutation is computed as 1/hd (linear) and we have detected this as a somewhat high value (high mutation has allowed good searches with our one-at-a-time selection), but in GRIAL, a control technique called ***begin-end-frequency*** allows a better pursuit of GA strategies' effects by specifying how and when to apply them.

In order to speed up the search of a quasioptimum weights' set we have designed the **GBit** operator. This hybrid genetic-hillclimbing procedure makes smooth changes in the best-to-now solution string in order to explore its neighborhood.

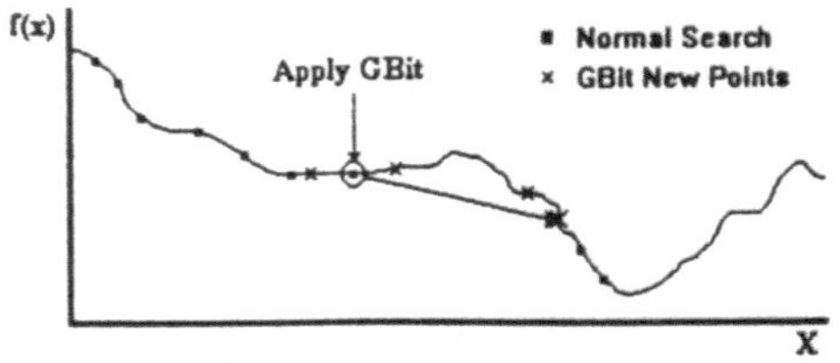

Fig 3.- *Search can be speeded-up by GBit.*

Crossing two strings is not a trivial operation because these two strings may represent different functionality-neurons' distributions for the problem solution and crossover will often yield two new strings (ANNs) that behave much worst than their parents and that are unable of future improvement (called *Lethal* strings).

To solve this problem we have designed a **Restrictive Mating** operator based on hamming distance (similar to the Genotypic Mating Restriction outlined in [9]) to impose a minimum likeness between mates to be ensure that interspecies (very different couples) crossovers do not take place.

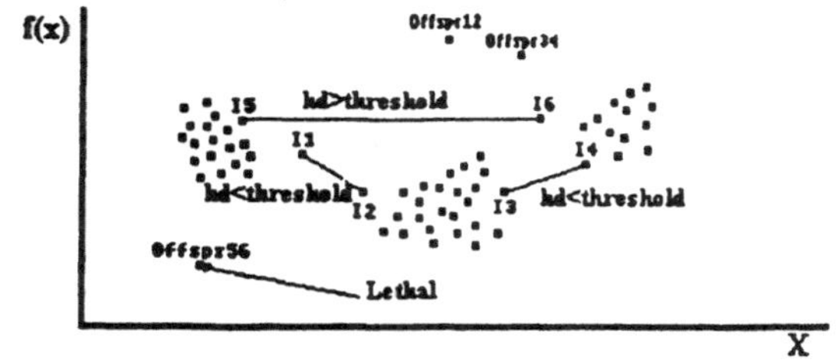

Fig 4.- *How rmating can avoid lethals' appearances.*

This problem is very frequent in big populations where many good, but different, partial solutions often appear. GBit and RMating have shown to turn GAs quick tools of high efficacy.

We have finally tested and stated the superiority of any **Distributed search** using N GAs over any equivalent single GA search for the whole test suite. We have used a **Migration** scheme similar to [4] to define N GAs in parallel evolution with occasional exchanges of their respective best string through Parlog streams. This parallelism improves time, accuracy, diversity, lethals avoiding and is more likely to find a quasioptimum result. Speed-up could actually be brought from the real machine distribution of the N GAs in a LAN or multiprocessor computer to get quick GA evolutions.

3.3.- Connectivity Definition

Every ANN type has traditionally had a link pattern definition associated to the structure being used in the network. Designers modify this connectivity arbitrarily in order to improve the ANN in some way: to make it faster, more accurate, less expensive, easy to train, etc... We expect from GAs a natural definition of any minimum connectivity pattern.

We want to make a *genetic connectivity definition* to accomplish three goals: *(a)* GAs are used to define connectivity and weights (training), *(b)* we want our method to be scalable with the problem size and *(c)* the predefined structure to be connected must be fully used, because we expect this structure to have been selected as quasioptimum by any other means.

Coding

By coding links in strings we want to obtain the GAs advantages not only in that links pruning is tried but links appear and disappear in parallel in strings as evolution occurs. Binary codings as the used in [4] fill every string position to encode presence (1) or absence (0) of a link, but this coding needs excessively long chromosomes that do not scale as the problem grows because, even if the GA discovers the useless links, the strings *must hold room enough* to contain the full link space. To allow the designer decide the magnitude of the search to be carried out we only impose a low and high limit to GA strings length to encode links (regarding the ANN structure).

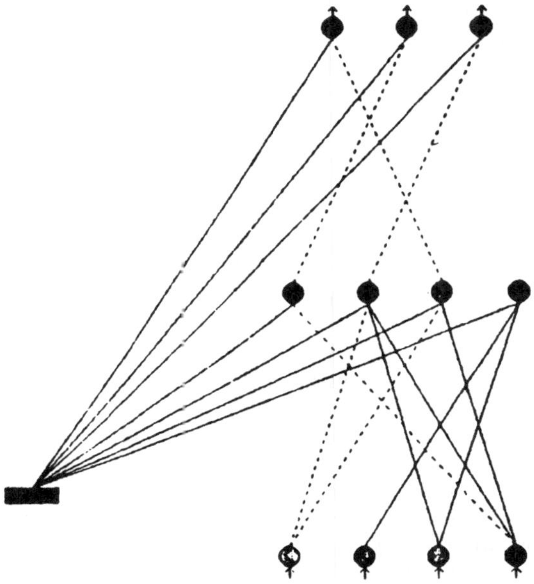

Fig 5.- *GAN version of the feedforward fullconnected TBA ANN. Optimum behavior by using only 40% of the full link space.*

We have designed a **General Artificial Neuralnetwork** sheet (GAN) as an implicit link space for feedforward and/or recurrent ANNs. For a given ANN structure three GAN input link spaces are defined: **FF** (feedforward), **FK** (feedback) and **DIO** (direct-io) input link spaces.

We define the input **FFspace** and the **FKspace** for every neuron belonging to some given layer in a GAN network as *link sets* composed of all the incoming links from the precedent and from the layer being considered itself. For any output neuron, and if any hidden layer exists, the **DIOspace** is defined as being composed of all the input links to the output neuron that can income directly from the input layer.

We associate every link with its weight value by the use of symbolic genes we call **wlink genes** to allow *independent existence* of any link in the ANN. Every string of wlink genes in the GA population encodes a *subset* of the full link space being optimized.

The user of GRIAL can specify the strings' length and the way in which link subsets are encoded. Thus, designer can specify the length of strings in the population, then *pruning* as desired the full link space. Excessive initial pruning has shown to be undesirable because networks are unable to learn the full pattern set. The rational use of the ANN structure brings from the results one can get by pruning links beyond a low limit: many neurons can be present in the final ANN whose inputs and/or output are not being considered when the ANN works. So we have defined **Archetypes** as distinguished string positions respected by the crossover during evolution. Archetypes assure that the ANN structure is being profited by.

Genetic Techniques

This linking (L) level uses the training (W) level to determine weights and fitness for its initial population and evolution is responsible for crossing the subsets of links (strings) looking for a quasioptimum solution. Link's pruning and adding are achieved by a *link's duplication effect*. We have tried penalty schemes attending to links number in order to modify the fitness values, but results indicate that any other penalty schemes (as giving more --W-- learning facilities to strings) should be of greater success. But the real power of this technique becomes from the initial designer-driven pruning and the subsequent GA optimization. Again the best results are these obtained with N distributed GAs working in parallel. At L level we pretend to make a more natural ANN design by interacting with the training lower level and then making it easy as well as more accurate and cheaper to implement (as software or hardware ANNs).

The Mutation, RMating and GBit have been extended to work at this level of connectivity optimization, however, their influence in the results has not been so important as for the genetic training level were. A simple genetic GA along with the IEES and the distributed evolution have been the techniques that produced the best improvements in the goodness of the designed networks. The designer-driven initial pruning has had the greater influence in reducing the connectivity.

3.4.- Structure Optimization

Defining the best ANN structure for a given problem is not a deterministic nor even a well-known process due to the complexity of recognizing every neuron's job in the whole ANN and the relationships among neurons.

There exist many rules based on experience about the number and disposition of neurons, but, as for connectivity definition, designers have not a methodology out of the proof-and-error mechanics.

Coding

The real problem for a genetic structure definition is *to select a good coding* to contain structure information, general --weak-- enough to be combined by crossover and specific --strong-- enough to determine one unique ANN structure when decoded (expressed). We have designed a **binary genotype** to do this job (a *strong* coding of the structure).

Dislike other works like [14], we won't merge in the same string any training or even connectivity parameters, because the two lower levels of design are encharged for these tasks. We want to optimize the number of hidden layers and neurons per hidden layer.

[0,1,0,1, 0,0,1,0, 0,0,0,0]

(a) *String encoding one structure*

GN=3 GL=4

(b) *Genes Number and Length*

[binary, linear, <u>binary</u>]

(c) *Transfer Functions List*

Fig 6.- *Genotype for structure encoding. If a gene (hidden layer) decodes to a non-zero number of neurons its user-associated transfer function will be used for any of the neurons of this hidden layer.*

The strings at the structuring level are **GN*GL** bits length, where **GN** is the max. number of hidden layers to be considered and **GL** is a value such that $2^{GL}-1$ is the max. number of neurons to be included in any hidden layer.

Genetic Techniques

Usually, the kind of desired structure is best known by the designer than the best connectivity pattern or even the weights set to be used, and that's why S level search is accomplished by smaller GAs than those ones needed at the L or W levels. On the other hand S strings are very complex and expensive to evaluate for fitness because they require multiple L and/or W GAs executions. Genetic operators have been extended to this level (e.g., GBit or mutation) but the simple traditional GA has shown to be a good issue for S level.

We encourage the use of the *Immediate Effects Evolutive Scheme* in order to prevent useless wanderings even when we risk to obtain only good structures and not the best ones. For structure definition, hillclimbing-like methods may reduce the overall cost.

For the *XOR* we have duplicated the human proposed structures (2-1-1 with direct IO connections and 2-2-1 without them) and discovered another interesting but unexpected ones. For the *TBA* and *Spanish Plurals* problems we have been sometimes surprised by very complex structures because the GA is only concerned with the strings' fitness values.

We have designed and tested a **penalty scheme** consisting in multiplying the SQE value (only IEES has been used at the S level) by the total number of neurons in the network (including the input and output neurons, that are not being optimized, in order to avoid a zero value to be multiplied when no hidden layers exist). The penalization has been very successfully applied in that we have gotten very reduced structures with optimum training (100% efficacy). Moreover the penalization has helped selection in recognizing the best structures by enlarging the gaps between the new strings' associated error values.

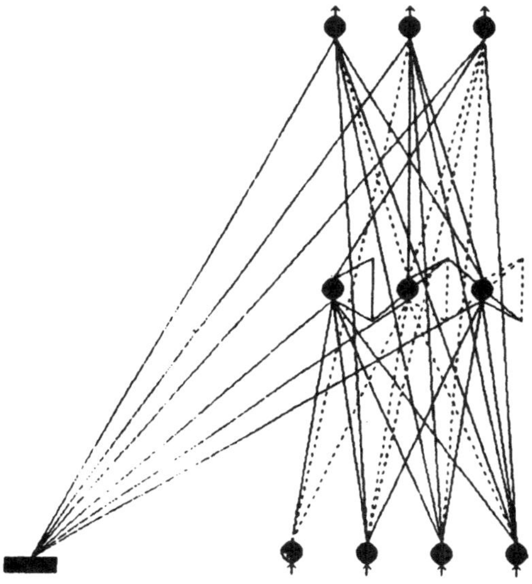

Fig 7.- *Fully automatic genetic TBA-ANN design.*

The final best ANN will have been designed taking account for many aspects to yield a full suited ANN to the proposed environment (the problem).

Every single GA will gain by being allowed to present a higher rate of error due to the multiple level relationships (GA parameters tuning does not need to be of high accuracy to work well). As we can see the parallelism at many computational levels is the best way to bring efficiency and efficacy to this complex Genetic ANN design: three levels run in parallel, several GAs can work in parallel at any level and every genetic operation can be parallelized due to PARLOG software enhancement. PARLOG can be used to define an Object Oriented library of ANN simulators [15] and to implement any kind of GA parallelism as those mechanisms outlined in [16]. GRIAL includes some of these propositions.

4.- Conclusions

There exist many different non GA ways to define the target ANN to solve a problem and many other Evolutive mechanisms but the **Full Three Levels ANN Design** is thought to be the best and more natural way to take designer to quasioptimum ANNs by controlling only some GA strategies. GAs provide the better set of tools to do this work. Resulting ANNs can be later refined by any of the applicable existing mechanisms.

We think that a smart connectivity for initial pruning to be used at L strings (initial link strings' composition) can improve the connectivity definition, because, when fully automatic design is being performed, S level randomly generates this pruning and only a high number of strings at S level can overcome excessive initial pruning (because numerous initial prunings will be tried).

The "best" GA to ANN design is a distributed search with IEES or a similar ranked selection, using a somewhat high mutation rate, two points crossover and some help to avoid lethals and to speed up search (GBit or another bit climber and some niche and species formation technique like RMating if no migration is to be used).

ANNs and GAs are thought to be open and reactive systems which fit very well within the logic and the concurrent programming paradigms. Thus, many advantages regarding parallel implementations and software improvements can be brought from the use of Parlog as the base language. With the correct combination of several GA techniques we can define a general methodology to enter automatic ANN design in a unified fashion while still maintaining a diversity of approaches to solve a problem. The computational requirements of this full genetic ANN design can be reduced by using parallel techniques.

References

[1] K.H. Kim, C.H. Lee, B.Y. Kim & H.Y. Hwang, *'Neural Optimization network for minimum-via layer assignment'*, Neurocomputing 3, 15-27, 1991.

[2] David E. Goldberg, *'Genetic Algorithms in Search, Optimization & Machine Learning'*, Addison-Wesley, 1989.

[3] Darrell Whitley & Thomas Hanson, *'Optimizing Neural Networks Using Faster, More Accurate Genetic Search'*, Proceedings of the Third ICGA, Morgan Kaufmann, 391-396, 1989.

[4] D. Whitley, T. Starkweather & C. Bogart, *'Genetic Algorithms and Neural Networks: Optimizing Connections and Connectivity'*, Parallel Computing 14, 347-361, 1990.

[5] H. Mühlenbein, *'Limitations of multi-layer perceptron networks - steps towards genetic neural networks'*, Parallel Computing 14, 249-260, 1990.

[6] H. Mühlenbein & J. Kindermann, *'The Dynamics of Evolution and Learning - Towards Genetic Neural Networks'*, Connectionism in Perspective, 173-197, 1989.

[7] E. Alba Torres, J.F. Aldana Montes & J.M. Troya Linero, *'Genetic Algorithms as Heuristics for Optimizing ANN Design'*, Technical Report, Dpto Lenguajes y Ciencias de la Computación, Univ. de Málaga, 1992.

[8] Lawrence Davis, *'Bit-Climbing, Representational Bias and Test Suite Design'*, Proceedings of the Fourth ICGA, Morgan Kaufmann, 18-23, 1991.

[9] Kalyanmoy Deb & David E. Goldberg, *'An Investigation of Niche and Species Formation in Genetic Function Optimization'*, Proceedings of the Third ICGA, Morgan Kaufmann, 42-50, 1989.

[10] Steven Alex Harp, Tariq Samad & Aloke Guha, *'Towards the Genetic Synthesis of Neural Networks'*, Proceedings of the Third ICGA, Morgan Kaufmann, 360-369, 1989.

[11] Keith Clark & Steve Gregory, *'PARLOG: Parallel Programming in Logic'*, ACM Trn. on PL & S 1-49, 1986.

[12] Crammond, Davison, Burt, Huntbach & Lam, *'The Parallel Parlog User Manual'*, Imperial College, London, 1-40, 1989.

[13] Darrel Whitley, *'The GENITOR Algorithm and Selection Pressure: Why Rank-Based Allocation of Reproductive Trials is Best'*, Proceedings of the Third ICGA, Morgan Kaufmann, 116-121, 1989.

[14] Steven Alex Harp, Tariq Samad & Aloke Guha, *'Towards the Genetic Synthesis of Neural Networks'*, Proceedings of the Third ICGA, Morgan Kaufmann, 360-369, 1989.

[15] J.M. Troya & J.F. Aldana, *'Extending an Object Oriented Concurrent Logic Language for Neural Network Simulations'*, IWANN, Lecture Notes in Computer Science, Springer-Verlag, 235-242, 1991.

[16] D. Macfarlane & Ian East, *'An investigation of several Parallel genetic algorithms'*, Univ. of Buckingham, MK 18 IEG, 60-67.

Genetic Algorithm Design of Neural Net Based Electronic Nose

Adhanom A. Fekadu, Evor L. Hines and Julian W. Gardner
Department of Engineering
University of Warwick
Coventry CV4 7AL, U.K.
Email: E.L.Hines@uk.ac.warwick.eng
es791@uk.ac.warwick.eng

Abstract

The training of a multi-layer perceptron using the well known back-propagation algorithm normally takes place after the neural network architecture and the initial values of various network parameters have been defined. Since the success of the training process, in terms of a fast rate of convergence and good generalisation, can be affected by the choice of the architecture and the initial network parameters, much time is spent in searching for the optimal neural paradigm. In this paper, results are presented on the use of Genetic Algorithms to determine automatically a suitable network architecture and a set of parameters from a restricted region of design space. The data-set comes from the response of the Warwick Electronic Nose to a set of simple and complex odours.

1. Introduction

The human sense of smell is still the primary faculty upon which many industries rely to monitor the flavour of items such as beers, coffees and foodstuffs. The parallel neural architecture existing in the human nose consists of millions of cells in the olfactory epithelium that act as the primary receptors to odorous molecules. These receptors synaptically link into glomeruli nodes and mitral cells which in turn feed into the brain. This multi-layer architecture suggests an arrangement that could be used in an analogous electronic instrument with which to mimic the biological system. In the Warwick Electronic Nose, the primary olfactory receptors have been replaced by an array of solid-state sensors that respond differentially to a broad range of chemical vapours or odours. This response may be characterised by a change in the electrical resistance of the sensor array which is processed further in order to identify the vapours/odours [1, 2]. The use of a multi-layer perceptron (MLP) to classify the output has already been studied using the back-propagation technique [3, 4].

The most generally used neural network is the MLP usually trained by the standard back-propagation algorithm (SBPA) [5]. Training by SBPA requires a number of user-specified parameters whose values may affect both the training time and the prediction error. Another short-coming associated with SBPA is that the best network architecture to use for a given application is not known a priori. Thus, a considerable time may be spent in finding the appropriate architecture and set of network parameters that produce optimal network, in terms of both the training time and prediction error. Although a number of methods have been proposed to minimize the training time, the ideal architecture and parameters still need to be found by a set of laborious experiments which may not even give the optimum values.

Genetic algorithms (GAs) were first developed by Holland in 1970 to mimic the processes of natural evolution and have been shown since to be useful in a variety of search problems [6].

In this paper, results are presented on the application of GAs in the search for an appropriate network architecture and network parameters for our electronic nose system.

2. Network training

The SBPA of Rumelhart and McClelland [5] is a supervised learning algorithm for a feed-forward MLP. Learning using the SBPA involves two phases. During the first phase, each input vector is fed into the network consecutively producing corresponding output vectors which are compared with target outputs to give the error at each of the output nodes. In the second phase, the error signal is passed backwards through all units and weight changes are made accordingly. Equations (1) and (2) below show the calculations performed during phases 1 and 2 respectively.

$$E = \frac{1}{2}\sum_{p}\sum_{i}(t_{pi}-y_{pi})^2 \quad (1)$$

$$\Delta W(n+1) = -\eta \nabla E(n) + \alpha \Delta W(n) \quad (2)$$

where $W(n)$ is the weight vector at the nth iteration, $\nabla E(n)$ is the gradient of the error, E, evaluated at $W(n)$, η and α are constants referred to as the learning rate and the momentum coefficient, respectively, t_{pi} and y_{pi} are the target and the calculated output of the ith *neuron* corresponding to the pth input vector, respectively. Network training may be stopped when either the number of training cycles reaches the maximum specified or the total sum of squared error (TSSE), E, is below a specified value. SBPA is a gradient descent optimization technique which tries to minimize the TSSE defined by equation (1). The step length or learning rate, η, and the momentum coefficient, α, are predetermined constant parameters of the algorithm and the choice of their value will affect the rate of convergence as well as the prediction error.

In order to increase the rate of convergence, a number of improvements have been suggested. The adaptive back-propagation algorithm (ABPA) [7] which modifies the values of the learning rate and the momentum coefficient has been shown to give the best performance on a number of data-sets [10]. The performance of the training, in terms of the rate of convergence and the network error value, is generally affected by the choice of the initial values of the learning rate, momentum coefficient and weights, and the network architecture. Unfortunately, all these have to be predetermined before applying ABPA or SBPA.

In this paper, GAs are applied to select suitable values for these network parameters, from a given set of range of values, for ABPA.

3. Genetic algorithms

Genetic algorithms are heuristic search algorithms based on the mechanics of natural selection [6]. They use nature's basic philosophy of survival - the fittest survive and the worst die off. Genetic algorithms work on the coding of the parameter rather than the parameters themselves. The structure of the system we are concerned with is mapped into a string of symbols (called chromosomes), usually binary strings consisting of 0s and 1s. Each range of parameter values may be mapped onto a binary string, which when concatenated form a chromosome. GAs are then applied to search populations of chromosomes.

Several genetic operators can be identified, the most commonly used ones in simple GAs being parent selection (artificial version of natural selection), crossover (recombination of parents' genetic material to make children), and mutation (re-introduction of genetic diversity by random modification of chromosomes). Simple GA that uses only the above three operators has been found to be satisfactory in the applications described in this paper, however, several other more sophisticated operators may also be employed [6].

Members of the population are assigned fitness values according to the evaluation of the fitness function to measure how well suited the parameter values encoded in the chromosome are for the application. The higher the fitness value of a chromosome the greater the chance it has to breed.

4. Network representation.

The parameters which need to be searched by GA in our network training are: the range of initial weights (r), initial values of learning rate, (η), and momentum coefficient, (α), and the network architecture. The initial weights are generated from a uniform random number generator in the range [-r, r]. The network architecture is described by the number of hidden layers (*h*), the number of nodes in each layer (n_i, $i=1, \ldots, h$) and by the connectivity of nodes. Since we are using the back-propagation algorithm for training, the neural networks are restricted to a feed-forward MLP. The MLP may be randomly or fully connected. In a fully connected MLP every node in a layer is connected only to all the nodes in the next layer.

Since r, η and α, are real numbers, the mapping of these values to bit strings can easily be done by a linear transformation of the real numbers to binary values using equation (3)

$$b = Integer\left((2^n-1)\frac{(x-x_o)}{(x_m-x_o)}\right) \qquad (3)$$

where *Integer* stands for the 'integer part', x_o and x_m are the minimum and the maximum values of the parameter *x*, *n* is the length of the binary string, *b* is the binary representation of the parameter value *x*. The parameter transformed using equation (3) is discretized to only 2^n possible values in the interval [x_o, x_m]. The integer values, *h* and n_i are represented by their equivalent binary values. Gray-coding is often claimed to be better than binary coding for GAs [6, 8]. Thus, the binary representations are gray-coded in our simulation.

It is the representation of the connectivity that requires much attention since it does not actually have a numerical value. There are only two possible values for a connection from one node to another node: the connection exists or it does not exist. Thus, a single connection can be represented by a binary digit. One way of representing the connectivity is to construct a positional binary string for every possible connection in the network. Assume we have a network with 3 inputs, 2 outputs and a maximum of 1 hidden layer having a maximum of 5 nodes. Then each of the input nodes can be connected to any hidden or output nodes and thus the maximum number of connections from any one of the input nodes is 5+2=7. Similarly, the maximum number of connections from any one of the hidden nodes is 2. The maximum number of connection in the network is (3x7)+(5x2)=31. We now construct a binary string of length 31 representing each possible connection in the network by a single bit. Figure 1 shows a 3-5-2 randomly connected neural network, and its binary representation is shown in Figure 2. The connections from the first node in the input layer to any of the nodes in the hidden or output layers are represented by the first 7 bits (the first 5 bits for the connections to each of the 5 hidden nodes the next 2 bits for the connections to each of the 2 output nodes), from the second input node by the next 7 bits and so on. The position of a bit in the string will determine the connection of which two nodes is represented by the bit. A bit value of 1 means that the connection exists and a bit value of 0 means the connection does not exist. Once the string is initialised, a network can be constructed by examining each bit in the string. One potential problem with this method of representation of connectivity however is that the network constructed by interpreting the string may not be meaningful (for instance, networks with no forward path between the input and output nodes). Therefore, a consistent method of constructing a meaningful network from the representation must be used.

A chromosome is constructed by concatenating the representations of all the parameters giving a long string. A chromosome

Figure 1 : A 3-5-2 randomly connected feed-forward network.

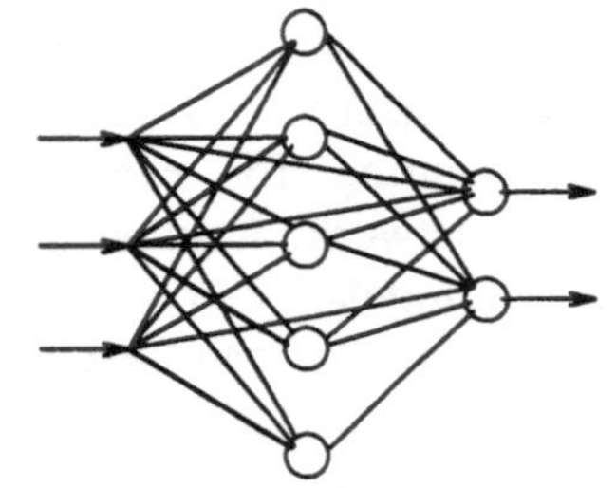

Figure 2: A binary representation for the network in figure 1.

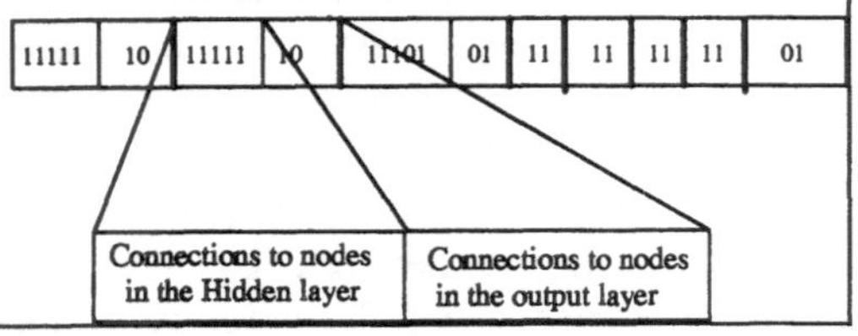

can thus be interpreted to give the initial parameter values and architecture of the network which can be trained by ABPA or SBPA.

5. Network evaluation

The performance of the ith chromosome, or of the network represented by it, is evaluated using the fitness function $f(c_i)$:

$$f(c_i) = au(c_i) + b \tag{4}$$

where f is the fitness function, u is the objective function which we want to optimize, a and b are transformation parameters dynamically adjusted to avoid premature convergence. The performance of a network can be evaluated with respect to different attributes. Different performance measures of a network can be evaluated depending on the relative interest in the attributes. The objective function can be defined as a weighted sum of the various performance measures. The weights can then be adjusted to reflect the interest in any of the performance measures. In our simulation, the objective function, u, for chromosome c_i was defined by (5)

$$u(c_i) = \frac{k}{\sum_j w_j v_j(c_i)} \tag{5}$$

where v_j is a performance measure with respect to the jth network attribute, w_j is a non-negative significance measure attached to the performance of the jth attribute, and k is a constant.

In the solution to our electronic nose problem, the performance measures used in the objective function are based on the network prediction error, the speed of convergence, the size of the network and the level of generalisation achieved. Except for the prediction error, all the other measures are expressed as unitless ratios. The performance of the network with respect to the speed of convergence is expressed as a ratio of the number of epochs the network trained before the training terminates to the maximum number of epochs specified. In terms of size, the number of nodes and the number of weights expressed as fractions of their respective maximum specified values were used. The ratio of the number of misclassification of patterns to the total number of input patterns using both the training and the testing data-sets was used to express generalisation performance. For good generalisation, this ratio has to be small. Since the objective function (equation (5)) is inversely related to the performance measures, larger performance values will result in smaller values of the objective function. The weights, w_j, in (5) are actually cost factors attached to each of the performance measures. Thus, the larger the weight value attached to an attribute's performance measure the greater the emphasis placed on the attribute being small.

6. Simulation results

The procedure was tested by applying it to the standard exclusive-or (XOR) problem and two electronic nose applications: the classification of a set of simple odours (alcohols) and of complex odours (coffees). The alcohol and coffee data-sets were gathered using a version of the Warwick Electronic Nose containing twelve tin oxide gas sensors. The test odours were injected into a glass vessel containing the sensors and the response of each sensor recorded on an IBM PC. Details of the experimental procedure for the alcohols and coffees can be found elsewhere [3, 9]. A series of eight tests were carried out on each of the 5 alcohols : methanol, ethanol, butan-1ol, propan-2-ol, and 2-methyl-1-butanol. The coffee data comprised of 89 samples from each of the 12 sensors for three different coffee types. 30 samples for each of the first two and 29 samples for the third one.

Table 1 shows the specifications for the three problems in the simulation. Although the results are not conclusive yet, it was observed that normalising the input improves performance of the network [10]. The normalisation ranges specified in table 1 were chosen as a result of repeated training for the respective problem.

For the XOR problem, GA yielded a 2-2-1 network after 10 generations of evaluations. Table 2 shows the 10 'best' network parameters selected by GA, ranked in order of performance.

The alcohol classification problem was studied previously [4, 10]. After repeated experiments (using both the SBPA and ABPA methods of network training), a 12-7-5 fully connected feed-forward network with initial weight range [-1,1], learning rate 1.0 and momentum coefficient of 0.7 was suggested. In our simulation, we restricted the network to a maximum of 2 hidden layers each having a maximum of 20 units.

The coffee classification problem has also been studied previously [10]. The network studied was a 12-3-3 fully connected feed forward network trained by the method of ABPA. In our simulation, we allow again a maximum of 2 hidden layers each one having up to 20 units.

The full results obtained are summarised in Table 2, 3 and 4. These tables show the best 10 networks (ranked by performance) that the GA selected for each problem. 100% of classifications were correct during both training as well as testing for the XOR and alcohol problems with an error of less than 0.1 at each output node. In the case of the coffee classification, the networks were able to classify correctly all the coffee training samples, however, one of the test samples was misclassified.

The shift in the fitness curve (Figure 3) and the increasing trend in average population fitness (Figure 4) show that networks with good performance are produced as the number of generations increases. When the algorithm converges, the fitness curve will level out and the change in the average population fitness will be small.

7. Conclusion.

GAs can be applied to find simultaneously the appropriate architecture and set of suitable network parameters. We have applied GAs to the classical XOR problem and two practical odour classification problems. The results presented here are part of a research programme which is concerned with the application of artificial neural networks in practical instrumentation, such as the Warwick Electronic nose. Our results demonstrate that no simplifying assumptions, such as linearity, etc., on the function to be optimised are required, except a search space of possible values with some measure of performance which may be highly complicated and non-linear. The networks generated by the GA gave excellent performance on all three classification problems, with generally 100% success in classifying the training and test samples. However, further improvements may be expected from the extension of this technique to include normalisation procedures within the GA parameters.

Table 1 : Specifications used in the GA simulations.

Specification	XOR	Alcohol	Coffee
Maximum number of hidden layers	2	2	2
Maximum number of units in each hidden layer	5	20	20
Input normalisation range	[-2,1]	[-1,1]	[-1,1]
Maximum number of training cycles	50	500	600
Minimum TSSE to stop training	0.01	0.01	0.01
Generations	10	15	10
Population size	50	100	50
Crossover probability	0.9	0.9	0.9
Mutation probability	0.01	0.01	0.01
Number of training data-sets	4	35	80
Number of test data-sets	_	5	9

Table 2. The best 10 network parameters selected by GA for the XOR problem.

Rank	Network	Links	Epoch	η	α	r
1	2-2-1	7	16	0.69	0.97	0.927
2	2-1-1	5	25	0.94	0.97	0.927
3	2-1-1	5	25	0.80	0.97	0.927
4	2-2-1	6	30	0.91	0.97	0.927
5	2-2-1-1	10	31	0.92	0.97	0.927
6	2-5-1	16	28	0.80	0.80	0.949
7	2-4-1	12	31	0.79	0.79	0.949
8	2-5-3-1	26	24	0.90	0.94	0.948
9	2-3-1-1	13	32	0.42	0.90	0.928
10	2-3-1	10	34	0.10	0.73	0.948

Table 3. The best 10 network parameters selected by GA for the alcohol classification problem.

Rank	Network	Links	Epoch	η	α	r	Error %	Fitness
1	12-3-5	62	65	0.507	0.826	0.763	0.99	14.585
2	12-11-5	129	61	0.310	0.956	0.835	0.91	12.966
3	12-1-11-5	137	61	0.310	0.949	0.084	0.90	12.591
4	12-10-5	114	68	0.295	0.809	0.946	0.95	12.510
5	12-12-5	135	73	0.348	0.980	0.145	0.82	12.180
6	12-14-5	158	63	0.172	0.866	0.794	0.99	12.131
7	12-12-5	134	78	0.652	0.970	0.145	0.94	11.722
8	12-7-8-5	177	69	0.471	0.892	0.061	0.89	11.670
9	12-11-5	123	88	0.530	0.742	0.036	0.81	11.325
10	12-6-15-5	273	59	0.961	0.802	0.038	0.88	10.849

Table 4. The best 10 network parameters selected by GA for the coffee classification problem.

Rank	Network	Links	Epoch	η	α	r	Error %	Fitness
1	12-3-3	45	417	0.937	0.657	0.234	0.99	10.588
2	12-11-3	165	456	0.248	0.897	0.971	1.00	9.087
3	12-3-3	45	538	0.572	0.591	0.008	1.00	8.724
4	12-8-3	99	507	0.412	0.803	0.996	1.00	8.632
5	12-3-3	45	560	0.661	0.682	0.946	0.99	8.454
6	12-1-10-3	120	555	0.394	0.986	0.888	1.00	7.828
7	12-8-3	88	600	0.434	0.803	0.997	5.80	7.351
8	12-15-3	141	600	0.787	0.956	0.005	1.69	7.175
9	12-9-10-3	214	600	0.074	0.771	0.997	1.28	6.971
10	12-11-3	98	600	0.044	0.687	0.756	8.74	6.946

Figure 3. Fitness of chromosomes in alcohol & coffee classification problem

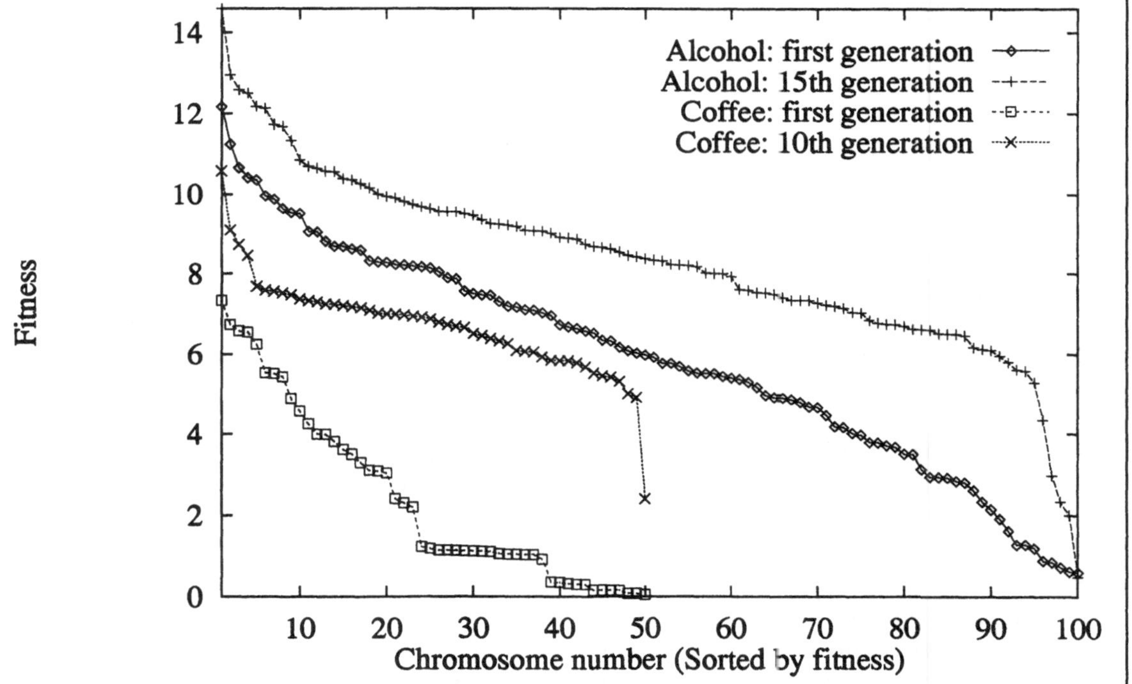

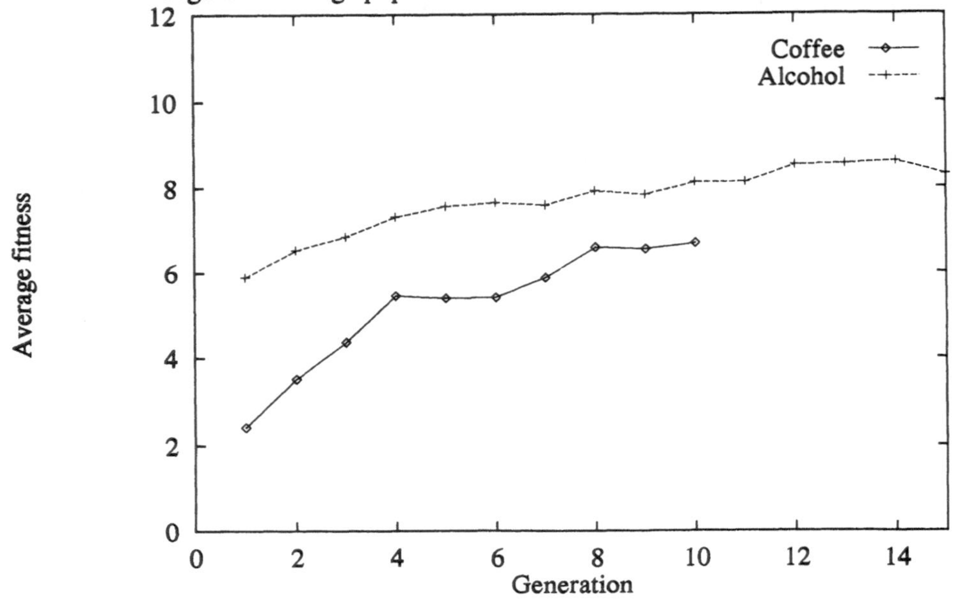

Figure 4.Average population fitness for coffee and alcohol classifications

References.

[1] Shurmer H V, Gardner J W and Chan H T, *'The application of discrimination techniques in alcohols and tobacco using tin oxide sensors'*, Sensors and Actuators, 18, pp 361-371, 1989.

[2] Gardner J W, Bartlett P N, Dodd G H and Shurmer H V, *'Chemosensory information processing'*, ed. D Schild (Berlin: Springer) Vol H39, pp 131, 1990.

[3] Gardner J W, Hines E L and Wilkinson M, *'The application of artificial neural networks to an electronic olfactory system'*, Meas. Sci. Technol. 1, pp 446-451, 1990.

[4] Gardner J W, Hines E L and Tang H C, *'Detection of vapours and odours from a multisensor array using pattern recognition technique. Part 2 : Artificial neural networks'*, Sensors and Actuators, B, pp9-15, 1992.

[5] Rumelhart D E and McClelland J L, *'Parallel distributed processing'*, MIT press, chapter 8, 1986.

[6] Goldberg D E, *'Genetic algorithms, in search, optimization, and machine learning'*, Reading, MA: Addison Wesley, 1989.

[7] Chan L W and Fallside F, *'An adaptive training algorithm for back propagation networks'*, Computer Speech and Language, 2:205-218, 1987.

[8] Davis L, *'Handbook of Genetic Algorithms'*, Van Nostrand Reinhold, 1991.

[9] Leung C W, *'Application of neural networks to the classification of coffee data'*, MSc IT Dissertation, Department of Engineering, University of Warwick, 1991.

[10] Hines E L, Gardner J W, Fung W and Fekadu A A, *'Improved rate of convergence in a MLP based electronic nose'*, 2nd Irish Neural Network Conf., Belfast, June 1992.

CIRCUITS OF PRODUCTION RULE GenNets

The Genetic Programming of Artificial Nervous Systems

Hugo de Garis

Brain Builder Group,
Evolutionary Systems Department,
ATR Human Information Processing
Research Laboratories,
2-2 Hikari-dai, Seiko-cho, Soraku-gun,
Kyoto, 619-02, Japan.
tel : + 81 7749 5 1440,
fax : + 81 7749 5 1408,
email : degaris@hip.atr.co.jp

Keywords :

Artificial Nervous Systems, Genetic Programming (GP), Genetic Algorithms (GAs), GenNets (Genetically Programmed Neural Network Modules), Artificial Creatures, Biots (Biological Robots), Detector GenNets, Decision GenNets, Motion GenNets, Production-Rule-GenNets, Complexity Independence of GAs, ALife, 1000-GenNet Biots, Darwinian Robotics, GenNet Accelerators, Software Programmable Hardware, Darwin Machines.

Abstract :

A year ago, the author evolved a simulated artificial creature (biot, i.e. biological robot), called LIZZY, which consisted of fully connected neural network modules (called GenNets), whose weights were evolved such that each GenNet performed some desired behavior, such as making the biot walk straight ahead, turn clockwise or anticlockwise, peck, or mate [de Garis 1990, 1991, 1993]. Other such GenNets were evolved to detect sinusoidal frequencies, or signal strengths, or signal strength differences, etc. However, the middle layer, between the detector and the motion GenNets, was implemented with traditional symbolic production rules, for reasons of computer simulation speed. Every time a GenNet was added to the system, simulation speed dropped. This paper "completes" the above work, by proposing a model which shows how a circuit of production-rule-GenNets (i.e. GenNets which behave like production rules), can be evolved which implements the middle or "decisional" layer, which takes signals outputted from the detector GenNets, and then decides which motion GenNet should be switched on. This work takes a first step towards the evolution of whole nervous systems, where circuits of GenNet modules (of appropriate types) are evolved to give a biot a total behavioral performance repertoire.

1. Introduction

As mentioned in the abstract, the author was forced to use conventional symbolic production rules to implement the middle decision layer between detector and motion GenNets. This was felt to be unsatisfactory, because the whole point of the exercise was to evolve a biot based on GenNet modules. This paper proposes how the intermediate decision layer can be evolved, so that the incoming signals, as processed by the detector GenNets are able, via the decision layer, to switch on the appropriate motion GenNet. The aim of the exercise is to evolve an artificial nervous system. The artificial nervous system described in this paper, is largely "supervised", i.e. the fitness definition used to drive the evolution was closely linked to a set of predefined and desired targets. However, later research will be devoted to more "open ended" evolution, where artificial nervous systems can be evolved whose fitness definitions will be much less constrained. The ultimate open ended fitness definition would be simply to survive and reproduce. As computer technology progressively allows more elaborate simulations, this kind of "open ended" evolution becomes more feasible.

We begin this paper with a description of the behavioral characteristics that the artificial nervous system that is to be evolved should have. These characteristics are presented in section 2 in the form of symbolic production rules (i.e. a set of preconditions, and an action). To begin with, some background on LIZZY, the biot whose nervous system this is to be, will be presented.

FIG. 1 shows LIZZY, a simulated quadruped biot, whose detectors and motion controllers are GenNets (i.e. Genetically Programmed Neural Network Modules [de Garis 1990, 1991, 1993]. Genetic Programming (GP) is the art of using Genetic Algorithms (GAs) to build/evolve complex systems. Since GAs are driven only by the fitness values of the systems they are evolving, the internal complexities of the evolving systems are irrelevant (provided that the fitness values keep increasing, i.e. that the systems are evolvable). This "complexity independence" of GAs allows one to evolve systems of complexity levels well beyond what is humanly comprehensible, hence one has a tool to "extend the boundary of the buildable". This ability to build/evolve complex systems (e.g. neural network dynamics or artificial embryos [de Garis 1993], etc) is the great attraction of GP.

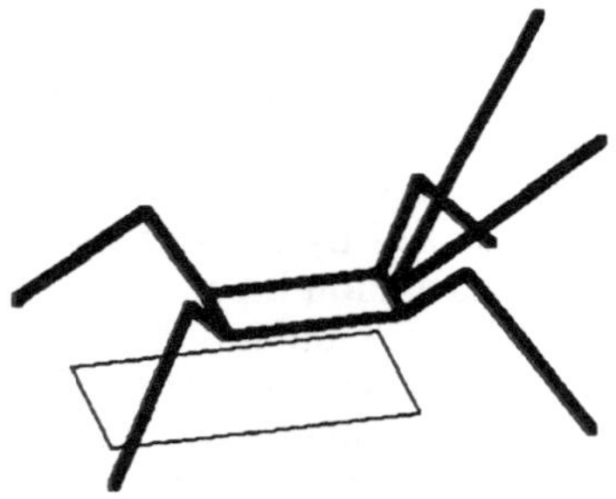

FIG. 1 "LIZZY" the BIOT

One application of GP that the author has investigated over the past few years [de Garis 1990, 1991, 1993], has been the evolution of complex time-dependent behaviors of fully connected neural networks. The traditional neural network training algorithms (e.g. backprop, recurrent backprop) cannot compete with GP in terms of the complexity of behaviors that GP can cope with. GenNets were so versatile, that the author believed it would be possible to build whole artificial nervous systems using them as basic building blocks. Very briefly, a GenNet is made by evolving the weights of a fully connected neural network. If there are N neurons in the GenNet, there will be N^2 connections and hence weights. These weights (usually with an absolute value less than 1.0) can be expressed in binary fraction format (e.g. P = 6 to 8 bits per weight) and one bit for the sign. These weights are concatenated onto a single bit string GA chromosome of total length N*N*(P + 1) bits. These chromosomes are mutated and (uniformly) crossed over in the usual GA fashion, such that their time dependent outputs do what one wants, whatever it is.

LIZZY had an average-signal-strength detector GenNet at each antenna tip. By detecting differences in these two strengths, LIZZY was able to orientate toward or away from the source of the signal. LIZZY also contained a frequency detector, where the assumption was made that the 3 types of creature in its environment (prey, predators, and mates) emitted a characteristic frequency (i.e. low frequency for predators, middle frequency for mates, high frequency for prey). LIZZY would orientate toward a prey or mate, approach, peck at a prey, or mate with a mate, then walk away.

LIZZY would orientate away from a predator and flee. The five behaviors LIZZY possessed were, walk straight ahead, turn clockwise, turn anticlockwise, peck (by pumping with the front feet), and mate (by pumping with the back feet). Each of these behaviors was generated by a separate GenNet. To switch between behaviors A and B, the outputs of GenNet A were input into GenNet B. Since the evolved behaviors (probably) took the form of limit cycles in the phase space of their motions, behavior B always occured, independently of when the behavior switch was made.

2. Production-Rule-GenNets and their Circuits

This section describes the fundamental assumptions made in the model which served as the "evolutionary framework" for the artificial nervous system. An evolutionary framework, is the set of assumptions, parameter ranges, etc which determine the search space within which the Genetic Algorithm used is to operate. The basic unit used to build/evolve the circuit, was the production-rule-GenNet (p-r-GenNet), shown in FIG. 2.

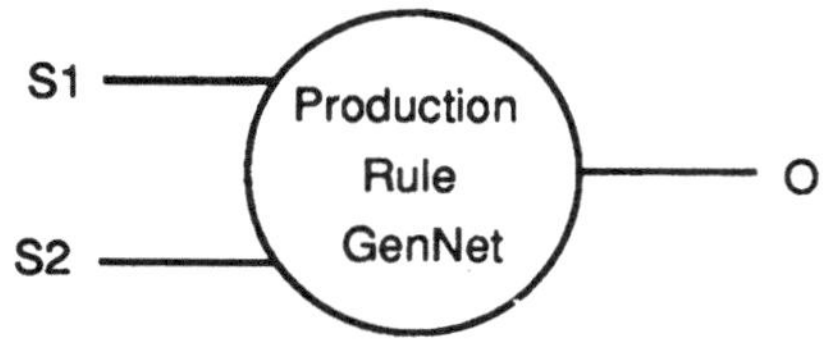

FIG. 2 PRODUCTION-RULE-GenNet

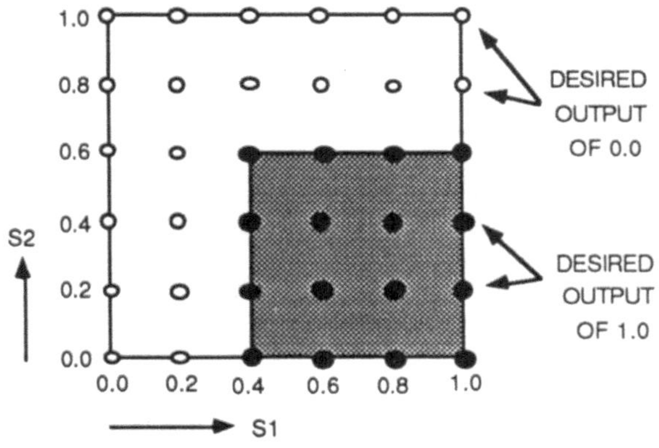

FIG. 3 EVOLVING a P-R-GenNet

Production-rule-GenNets are evolved to obey production rules of the form :-

IF ((input S1 > or < const C1)&
(input S2 > or < const C2))
THEN output O = 1.0,
ELSE output O = 0.0

To evolve such a p-r-GenNet, see FIG. 3 which represents the production rule :-

IF ((input S1 > 0.4)&(input S2 < 0.6))
THEN output O = 1.0,
ELSE output O = 0.0

The fitness definition for this GenNet was :-

$$\text{FITNESS} = \text{Inverse of } \left(\sum_{n=1}^{36} (d_n - a_n)^2 \right)$$

where n ranges over the 36 pairs of inputs (S1, S2). The desired outputs d_n are either 1.0 or 0.0 depending upon whether the (S1, S2) "coordinates" lie inside or outside the shaded region respectively. The a_n represent the actual outputs. In practice, these actual values were very nearly 0.0 in the "white" region, and roughly 0.7 in the "grey" region. Note, it was assumed S1 and S2 were positive reals, with values < 1.0

Experience shows that these production-rule-GenNets can be evolved for any combination of constant pairs and inequality pairs. It was thought that it might be possible to evolve artificial nervous systems by using these p-r-GenNets as building blocks. For example, it might be possible to incrementally evolve a network of these p-r-GenNets, so that they could serve as the middle layer for LIZZY. It might also be possible at a later date to evolve p-r-GenNet circuits which act as production systems, and hence be capable of manipulating symbols. However, since the ideas presented in this paper are still in an early stage of development, only a rather simple artificial nervous system will be proposed here.

FIG. 4 shows an example of a circuit which is built up from production-rule-GenNets (p-r-GenNets). Each p-r-GenNet has two inputs, and a single output which can fan out indefinitely.

Several characteristics can be observed from FIG. 4. In this simple model, each p-r-GenNet (oval shaped in FIG. 4) has its own production rule of the form :-

IF (input1 >< c1)&(input2 >< c2)
THEN out = 1.0,
ELSE out = 0.0

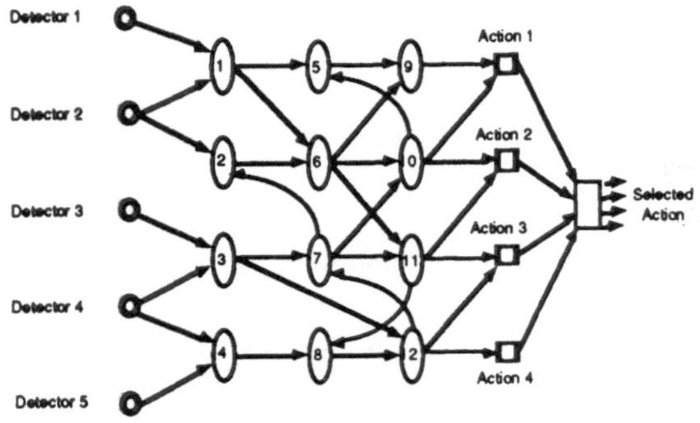

FIG. 4 A Circuit of Production-Rule-GenNets

The outputs from the detectors as well as the inequality constants c_i, are assumed to be real numbered values d_i in the range $0.0 < d_i < 1.0$. The actions are assumed to be mutually exclusive, so that only one action is used to switch on one motion GenNet from a library of motion GenNets. These motion GenNets control the effectors such as the legs of a biot, and hence only one of them can be active at any one time.

For this experiment, the detectors and motions used were :-

Detectors

i) Predator detector (low frequency detection), PRED.
ii) Mate detector (middle frequency detection), MATE.
iii) Prey detector (high frequency detection), PREY.
iv) Signal-strength detector at mouth, Mss.
v) Signal-strength at left antenna tip, LAss.
vi) Signal-strength at right antenna tip, RAss.
vii) Difference of signal-strengths at antenna tips, L-Rss.
viii) Difference of signal-strengths at antenna tips, R-Lss.

Motions

i) Walk straight ahead, WALK.
ii) Turn clockwise, CLK.
iii) Turn anticlockwise, ACLK.
iv) Peck at food (raise and lower front legs), PECKING
v) Mate with mate (raise and lower back legs), MATING

Threshold Constants

i) thrMss = A threshold constant beyond which pecking or mating can begin, once the Mss is sufficiently high.
ii) thrL-R = A threshold constant for rotation.

Examples of Behavioral Rules

i) If ((PREYor MATE)&(L-Rss > thrL-R))
Then ACLK
ii) If ((PREYor MATE)&(R-Lss > thrL-R))
Then CLK
iii) If ((PREYor MATE)&(R-Lss < thrL-R))
Then WALK
iii) If ((PRED)&(L-Rss > thrL-R))
Then CLK
iv) If ((PRED)&(R-Lss > thrL-R))
Then ACLK
iv) If ((PRED)&(R-Lss < thrL-R))
Then WALK
v) If ((PREY)&(Mss > thrMss))
Then PECKING
vi) If ((MATE)&(Mss > thrMss)
Then MATING

The way in which these detectors and actions translate into input/output training vectors to evolve the p-r-GenNet circuit will be discussed in a later section.

Each p-r-GenNet needs to be supplied with the following parameters :-

a) The inequality (> or <) for its input1.
b) The inequality (> or <) for its input2.
c) The constant c1 for its input1.
d) The constant c2 for its input2.

e) The source (i.e. from detector$_i$, or from the output of p-r-GenNet$_j$) for its input1.

f) The source (i.e. from detector$_p$, or from the output of p-r-GenNet$_q$) for its input2.

g) The output to action$_k$ or to the input of another p-r-GenNet$_r$.

From Object (columns) / To Object (rows)

	D1	D2	D3	D4	D5	1	2	3	4	5	6	7	8	9	10	11	12
A1														●	●		
A2															●	●	
A3																●	●
A4																	●
1	1	2															
2		1										2					
3			1	2													
4				1	2												
5						1											
6						1	2										
7								1									2
8									2							1	
9										1	2						
10											1	2					
11											1	2					
12								1					2				

FIG. 5 CONNECTION MATRIX

Note that the structure of the p-r-GenNet network can be recurrent. The above parameters a) to d) are "internal" (i.e. they are concerned with one p-r-GenNet only), whereas the parameters e) to g) are "external" (i.e. they are concerned with connections between two objects, where an object is either a detector, a p-r-GenNet, or an action). These connections can be visualized by means of a connection matrix as shown in FIG. 5, which corresponds to the circuit shown in FIG. 4. Note that for the p-r-GenNets (numbered from 1 to 12) there can only be two entries per row. Note also, that there can be any number of outputs from a detector and any number of inputs to an action. Detectors can be connected directly to actions. A connection to the first (upper) or second (lower) input of a p-r-GenNet is indicated by a **1** or a **2** in FIG. 5.

This connection matrix is useful for defining the rules for the mutation of connections, and other uses. For example, an input at a p-r-GenNet can be disconnected from its output, and replaced by an input from some (arbitrary) other output. This corresponds to moving the appropriate entry within the row of the p-r-GenNet concerned. The question now arises whether we wish our p-r-GenNet circuit to be incremental or not, i.e. whether one can add or delete occasionally, a p-r-GenNet, a detector, or an action. Nature evolved incrementally, so it would be nice to have this capability in our model. However, incrementality creates certain problems :-

a) When an object (i.e. p-r-GenNet, detector, or action) is added or deleted, how does that affect the existing links to that object?

b) The chromosome representation will need to be flexible enough to be able to handle a dynamic population of p-r-GenNets etc in a circuit.

The list below shows the types of incremental modifications considered desirable for a p-r-GenNet circuit. Each of them requires corresponding modifications to the circuit's connection matrix. Unfortunately, there is not enough space in this paper, to go into details, except perhaps to give one example. If a p-r-GenNet is deleted, the corresponding row and column in the circuit's connection matrix needs to be deleted. If a p-r-GenNet is left without an input, then either a detector or the output of another random p-r-GenNet is connected to its vacant input. There are many such details.

<u>Incremental Modifications</u>

i) Adding a Detector
ii) Deleting a Detector
iii) Adding a P-R-GenNet
iv) Deleting a P-R-GenNet
v) Adding an Action
vi) Deleting an Action
vii) Adding a Connection
viii) Deleting a Connection
ix) Changing a Connection

Besides the above changes to the circuit, there are two "standard" mutations :-

a) Changing the value of the inequality constants.
b) Changing the inequalities (e.g. > to <).

3. Chromosome Representation

FIG. 6 shows the chromosome format used for each production-rule-GenNet circuit in this experiment.

FIG. 6 CHROMOSOME FORMAT

INEQs is a binary list of the inequality pairs (i.e. the > or <) for all the p-r-GenNets (where a 1 means >, and a 0 means <), and THRESHs is a binary list of the threshold pairs for all the p-r-GenNets. A fuller explanation of these fields now follows. Assume that there is a maximum number N_{max} of possible p-r-GenNets in any circuit. The INEQs field contains N_{max} subfields, each of length 2 bits, where the ith (2-bit) subfield provides the inequalities for the ith p-r-GenNet. (It is assumed that each p-r-GenNet is labeled by a unique integer). The first bit in the pair specifies the inequality of the first input to the p-r-GenNet, and the second bit in the pair specifies the inequality of the second input. A similar story holds for the threshold values, i.e. N_{max} subfields, each of length 12 bits, where the ith (12-bit) subfield provides the thresholds for the ith p-r-GenNet. The first 6 bits in the pair specify the (positive binary fraction) threshold value of the first input to the p-r-GenNet, and the second 6 bits in the pair specify the (positive binary fraction) threshold value of the second input. Note that the circuit itself, i.e. the specification of the connections, is not included in the chromosome. Associated with each chromosome, is a connection matrix. Whenever a new p-r-GenNet is added to the circuit, it gets a unique integer label, and a correspondingly labelled row and column are added to the circuit's connection matrix. For example, if the circuit's previous new neuron had been labelled "n", then the next neuron would be labelled "n+1", and so would its corresponding row and column. Whenever an existing p-r-GenNet is deleted from the circuit, the correspondingly labelled row and column are deleted from the circuit's connection matrix. Thus an evolved connection matrix will probably be full of "holes" in its row and column labels.

4. Fitness Definition

To define the fitness, a simple supervised learning technique is proposed, i.e. by supplying a set of real valued input/output vector pairs, and letting the fitness definition be

$$\text{PN*Inverse of } \left(\sum_{p=1}^{P} \sum_{n=1}^{N} (d_{np} - a_{np})^2 \right)$$

where the d_{np} are the desired output vector elements, and the a_{np} are the actual measured output vector elements, for a set of N possible output actions (i.e. N is the size of the output vector), for P possible input patterns. Specifically, P detector vectors (of size M, where M is the number of detectors in the nervous system) are applied sequentially. For each (real valued) input vector, there is a desired (real valued) output vector.

The behavior rules mentioned in section 2, can be translated into pairs of input/output training vectors. The PN is a normalization factor for variable P and N. No crossover is proposed for this experiment, because it is thought to be too disruptive of the circuits. Therefore the only modifications to the chromosomes and connection matrices will be by "mutation". For the chromosomes, mutation is the usual bit flip.

Each of the incremental modifications listed in section 2 is given a user defined probability of application, and the appropriate connection matrix is updated accordingly. The connection matrices are used to help calculate the outputs of p-r-GenNets, so that the final action outputs can be used to measure the fitness.

5. Conclusions and Future Work

The incremental evolution of a network of neural networks is an ambitious undertaking. The ideas presented in this paper are only a modest beginning to what is hoped will become an important branch of neural networks and artificial life, namely trying to build/evolve artificial nervous systems for biots (biological robots). Future work will be to implement and test the ideas proposed in this paper. The longer term aim of the author is to build what he calls Darwin Machines, i.e. special hardware designs capable of using "software configurable hardware" techniques **to evolve neural circuits for biots directly in hardware**, to be able to build real world biots with 100, 1000, 10,000 etc behaviors, and their corresponding artificial nervous systems.

6. References

[de Garis 1990] "Genetic Programming : Building Artificial Nervous Systems Using Genetically Programmed Neural Network Modules", Hugo de Garis, Proceedings of the 7th. Int. Conf. on Machine Learning, B.W. Porter and R.J.Mooney (Eds.), Morgan Kaufmann, 1990.

[de Garis 1992] "Genetic Programming", Hugo de Garis, ch. 8 in "Neural and Intelligent Systems Integration", B. Soucek (Ed.), Wiley, 1991.

[deGaris 1993] "Genetic Programming : GenNets, Artificial Nervous Systems, Artificial Embryos", Hugo de Garis, WILEY manuscript.

Note : More complete references can be found in the author's other paper ("Evolvable Hardware : Genetic Programming of a Darwin Machine") in this book.

NEUROMIMETIC ALGORITHMS PROCESSING : TOOLS FOR DESIGN OF DEDICATED ARCHITECTURES

Laurent KWIATKOWSKI
kwiatkow@mimosa.unice.fr

Jean-Paul STROMBONI
strombon@mimosa.unice.fr

Laboratoire Informatique Signaux Systèmes - I3S - URA 1376 du CNRS
Université de Nice-Sophia Antipolis
41, Bd. Napoléon III - 06041 Nice Cedex - Tél : 93.21.79.61 - Fax : 93.21.20.54
250, Av. Albert Einstein, Bât.4 - 06560 Valbonne - Tél : 92.94.27.25 - Fax : 92.94.28.98

Abstract : The finality of the research subject is to derive a new parallel architecture of computer dedicated to connectionist algorithms and especially neuromimetic algorithms. The study of this problem induces us to carry out a *simulation model*. This model allows to highlight difficulties of parallel implementation on multiprocessor architecture of connectionist algorithms, i.e. projection of *application graphs* on the target *machine graph* in order to achieve optimal performances. Implementation is difficult because data transfers between processors induce mapping and scheduling problems. Then, some mapping strategies and optimization heuristics are used, as Node Swapping, Simulated Annealing or Genetic Algorithms. According to the results the model simulation allows to vary several parameters of the studied machine : number of processors, architecture, particular parallel mode (SIMD, SPMD, ...), interconnection network (static or dynamic) and communication mode. It also allows to research a trade off between processing loads and communication loads. Such a way, some basic specifications of *dedicated architecture* will be chosen from the comparison of a set of alternatives.

1 - INTRODUCTION

Many optimization problems use connectionism and more particularly neuromimetic algorithms. Some features of such algorithms and the resulting application graphs are described in the first part. The inherent parallelism of these algorithms suggests an easy implementation on parallel machine dedicated. The second part introduces the model for the machine and the associated graph with different parallelisms to be used in the simulation. Third, the mapping and execution are described. The model is able to estimate the efficiency of mapping optimization strategy in order to test architectural assumptions and analyse the adequation between algorithm and architecture. To wind up, the fourth part presents some results on mapping optimization (using Simulated Annealing and a Genetic Algorithm). The effect on the execution time of neuromimetic algorithms is compared on different parallel architectures (SIMD, SPMD, static or dynamic network, with or without time overlap).

2 - CONNECTIONIST ALGORITHMS

2.1 - Generalities

Many applications and algorithms include an amount of parallelism in processing and communication. The set of data precedence constraints reduces this potential parallelism. The algorithm is devided in successive phases each with some possible parallelism. Connectionist algorithms are a priori parallel : they demonstrate in the formulation few phases doted with high degrees of parallelism. They are described by the interaction of identical cells which execute same processing; initial data are the initial cells states and the results are read in final states. Processing is contained in the weighted cells interactions. The processing of connectionist problem may be described in terms of directed graph G(N,E), called *application graph*, whose N is the set of vertices and E, the edges which are weighted according to the influence between vertices. The variables to be processed are distributed to the different graph vertice : the elementary vertex task is the accumulation of neighbourhood balanced influence (figure1). According to applications, neighbourhoods may be regular or irregular with important or small degree, static or eventually dynamic. Assignment and scheduling of processing and communications on a given machine may arise complex problems.

2.2 - Neuromimetic algorithms

They are a particular case of connectionist algorithms. The application graph is usually stationary. They are characterized by many computational elements (neurons) interconnected with varying weights (synaptic plasticity). Neighbourhoods are regular. The basic neuron sums N weighted inputs (synaptic potential) and applies a non linearity to the result in order to determine the internal activation state. The application graph is then defined by the graph G(N,E) where N is the set of formal neurons and E, the set of links between the neurons. Then the variable computed in every vertex is the activation state of the neuron.

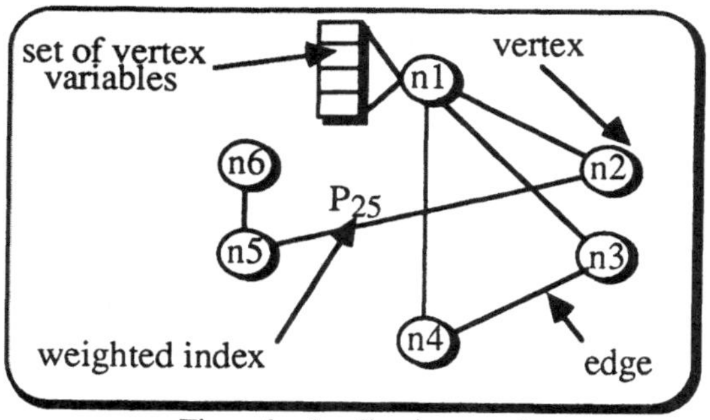

Figure1 : Application graph

Two neuromimetic networks were considered in the study : the Multilayer or direct neural network [11] restricted to the feedforward operation, where the neurons in a layer are connected only to the previous layer. This network has three layers : the input layer with eight neurons, the hidden layer with seven neurons and the output layer with six neurons. Second, a 21 cells Hopfield net fully recurrently interconnected was studied [7]. Regularly mapped on regular architecture, they demonstrate a priori quasi optimal implementation and load balance. So in order to evaluate the performance of different mapping algorithms, a load balanced problem on variable size machine is implemented.

3 - PARALLEL ARCHITECTURES

Parallel architectures are a current concern in the computer sciences development because they offer an alternative for the improvement of performance. The parallel machines contain many processing units of various power (or grain) able to calculate simultaneously (processing parallelism). They use an internal network for exchanging intermediate data simultaneously (communication parallelism). In addition, there are different control parallelisms available in the machine for increasing the execution speed of the program and the memory accesses (control parallelism). For a given algorithm, the mapping, the allocation of communication channels to transfers and the scheduling of operations are a key factor in the final performance. The set of processing units and communication links may also be described by an oriented graph G(P,Ep). The vertices are the processors, including one or more functions : processing, memory, routing and sequencing. An edge leaving a vertex specifies a physical unidirectional communication link towards a neighbour processor. Communication times between processors are associated to the edges. In the same way, executing times are associated to vertices operations (figure2). In existing model, processors are identical (same functions, same execution times), and also the communication links.

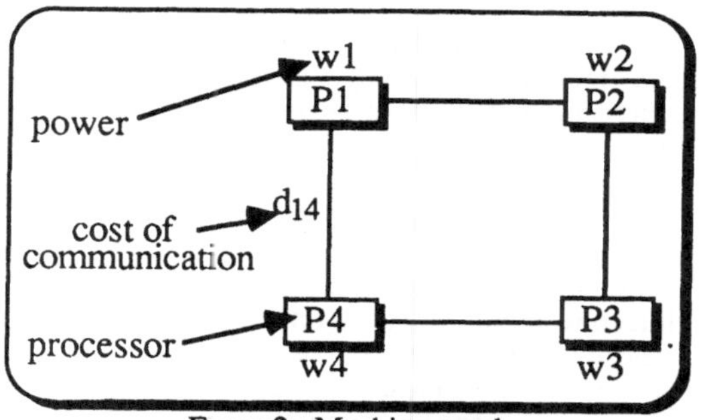

Figure2 : Machine graph

For a given machine, mapping the algorithm consists in allocating every variable with the associated operation to one processor with sometimes a routing information for execution (merging G(N,E) and G(P,Ep)). The regular structure of neural networks suggests efficient simple implementation strategies on multiprocessor machines. However, nowadays available processors are much more powerful but have strongly smaller numbers of input-output ports, than natural neurons. Thus the mapping is not trivial and deserves much attention, since the communication problem is obviously crucial. In this paper, two parallel modes are simulated : SIMD (Single Instruction Multiple Data) and SPMD (Single Program Multiple Data) with or without transfer and processing overlapping. Two interconnection networks are considered in the machine model. First, a *Omega/Benes* reconfigurable network performs the communication by means of successive circular permutations (as the non-linear addressing, scatter and gather used in the parallel machine Opsila [1]). Second, a static *Hypercube* network uses a routing method : transfer then uses a proximity matrix computed from the network topology. The routing algorithm implicitly uses a router in every processor and one buffer at every link origin. Destination address is written in the buffer when transfer is decided. Here, routing is minimal and deterministic [8], links with direct neighbours are prioritary, and if there is no longer some, the messages are sent only when the distance to their destination is reduced.

4 - SIMULATION SOFTWARE

Using the model needs first the definition of the algorithm and the machine by means of graphs (specifed in different files). After a specified mapping initialization, the model will allow to apply a mapping optimization strategy. The parallel processing mode in the machine must be chosen and the communication method, the synchronization between processing and transfer tasks with possible time overlap. Such choices are specified to the model by code writing a procedure which describes the execution schema to be simulated (transfer delays and execution times are specified in the model by means of constants to change at wish, in number of clock cycles). From this, the model delivers some results, execution time in number of clock cycles, parallelization efficiency, speedup gained from parallelism, load repartition obtained from a given mapping, and so on.

4.1 - Mapping optimization

The execution speed of an algorithm on a parallel architecture highly depends on the assignment of the processing tasks to the set of processors and links. Though taking advantage of architectural features to get better performance, the research of optimal mapping remains complex. Two criteria dominate with different weights : the communication load balance for links and the processing load balance for the processors. In our case, tasks are variables with associated operation. The mapping problem may be described by a function of the set of m tasks to the set of q processors. There is q^m possible mappings if one task is assigned a unique processor. Comparing different mappings by means of a *performance criterion* associating a cost to every mapping leads to mapping optimization which tries to minimize the cost. Here, the cost is the sum of one processing load component and one communication load component. The parameters α and β set the relative importance of processing and communicating respectively, if the cost have the same order of magnitude.

$$\text{Cost} = \alpha \cdot \text{Processing Cost} + \beta \cdot \text{Communications Cost}$$

The algorithm is characterized by a application graph G(N,E), where vertices, $N = \{N_1, N_2, \ldots, N_m\}$, represent the tasks in the algorithm, and edges E correspond to data communication dependencies between the tasks. The weight of the edge between i and j, denoted Pij is the number of communication required between the two tasks. The parallel computer is represented as a machine graph G(P,Ep). The vertices, $P = \{P_1, P_2, \ldots, P_q\}$, represent the processors and the edges Ep represent the communication links. The system is assumed to be homogeneous with all processors being equally powerful and all communication links capable of the same rate of communication. The distance d(q,r) between processors q and r is defined as the minimum number of links to be traversed to get from q to r, i.e. the lenght of the shortest path from processor q to processor r. A mapping problem is a function F: N ---->P; the value F(i) is the processor which contains task i. Here, the load balance is the goal and the function cost is a biased variance of the load for processing and transfer.

$$\alpha \cdot \frac{1}{Q} \sum_{k=1}^{Q} (\text{processing load}[k])^2 + \beta \cdot \sum_{k=1}^{Q} \sum_{(i,j) \in C_k} P_{ij}^2 \cdot d_{[F(i),F(j)]}$$

$$\text{avec } C_k = \{(i,j) \mid F(i)=k \text{ et } F(j) \neq k\},\ k = 1..Q$$

An ideal mapping would insure perfect load balance and consequently the maximum speedup with the machine considered. There is no exact solution in the general case, even the optimal mapping determination may rise a non polynomial problem. Existing optimization methods **[10]** start from an initial mapping, iteratively search a mapping which optimizes a criterion selected as representative of the goal i.e. fast execution time and efficient parallel ressources utilization. The actual performance gain depends on the algorithm for optimization and the quality of the criterion. In Node Swapping, the research of the optimum is deterministic but restricted to the swap of all possible pairs of variables and thus non optimal **[4]**. Oppositely, Simulated Annealing is stochastic but without restriction of the solution space [3].

An alternative is offered by Genetic Algorithms. Usually, the drawback of suboptimal methods is the convergence towards local solutions because the search space is reduced. Thus, it is necessary to repeat the algorithm with different initial conditions, then select the best result. Genetic Algorithms scan a more important space of solutions. They result from an analogy with the mechanisms of natural selection and gene reproduction (*crossing-over* and *mutation*) applied to a population of individuals **[6]**. Previous cost function is the selection function. One searches to produce a more performant population. Every individual codes a solution under the form of a string. The algorithm is the following :

```
iteration:=0;
choice of initial population P(0)
repeat
        selection of P(iteration+1) from
        P(iteration)
        crossover to P(iteration+1)
        mutation to P(iteration+1)
        increment [iteration];
until  stop condition (stagnation)
```

Only the three operations of reproduction, crossover and mutation are implemented here, without intervention of the operations of dominance **[5]** and inversion **[2]**. These operations of crossover and mutation use probabilistic decision rules and concern only a part of the population (100 individuals in our case). During the execution of the algorithm, probabilistic values may be modified to obtain a faster convergence to a optimal solution. In our case, the crossover operator uses a mask in order that sons have 50% of the features of their parents and the mutation operator is realized randomly on one allele of all the sons. The objective function (or fitness) F is the same as defined previously. It tends to map the algorithm on the machine in order to reduce the execution time.

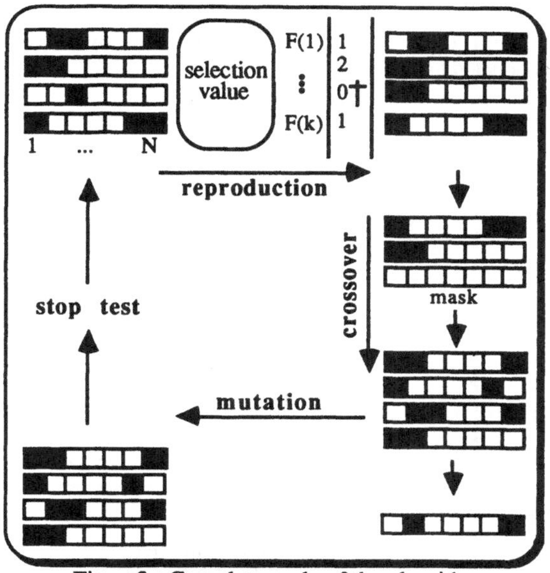

Figure3 : Complete cycle of the algorithm

In order to use genetic algorithms, it is necessary to modelize the research state in terms of strings of values or individuals. The solution of our problem being a particular assignment of m vertices to q processors, a natural way to code it is the previous mapping vector of m integers. However, others representations of problem solution might be considered, for example, an individual of m times q integers. The first m integers allocated to first processor and each integer i indicates the influence of the variable i on the considered processor. Figure3 illustrates the genetic algorithm operation.

4.2 - Performances evaluation of the execution on a parallel architecture model

The analysis of existing machines demonstrates various control and synchronisation alternatives for processing, communication and memory access, and execution. These three kinds of parallelism are some fondamental machine features. The machine description ought to include several processing sites, memory sites, communication links and methods for control and synchronisation of the parallel resources. The model proposes various parallel modes for processing, differing by the control and synchronisation, in particular the start, stop and load conditions (SIMD, SPMD, Dataflow, Systolic). Data are not actually processed, only the resource activity is simulated. For all these modes, it is possible to specify that transfer and processing overlap, or that the whole tranfer occurs first, and then processing only. The model can simulate the routed transfers or communications through reconfigurable networks. Such parallelisms, possibly mixed assist the investigation of actually exploitable parallelism taking into account constraints of the target machine and an implentation method. It is possible to highlight the effect on execution performance and evaluate the adequation between the algorithm and the architecture, or utilization of the parallel ressources in the machine. The model also allows to act on topology, nature of parallel resources and parallel modes used for tranfers and computations in order to modelize some architectural assumptions related to the algorithm.

5 - APPLICATIONS

5.1 - Mapping optimization

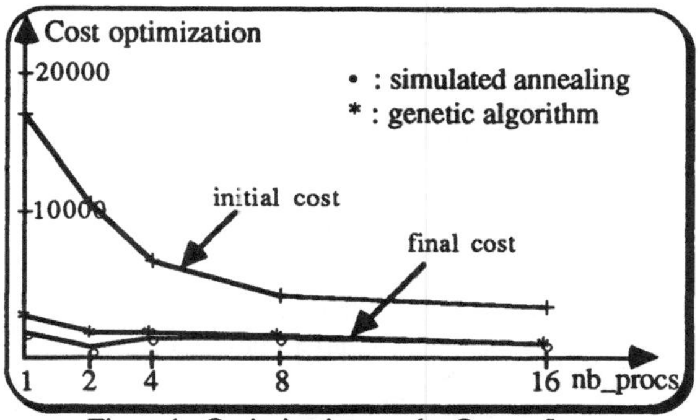

Figure4 : Optimization on the Omega/benes

In order to evaluate the performance of the optimization algorithms, the application graph is regularly mapped on a reduced set of machine processors (1, 2, 4, 8 processors), and the machine is extended to 16 processors.

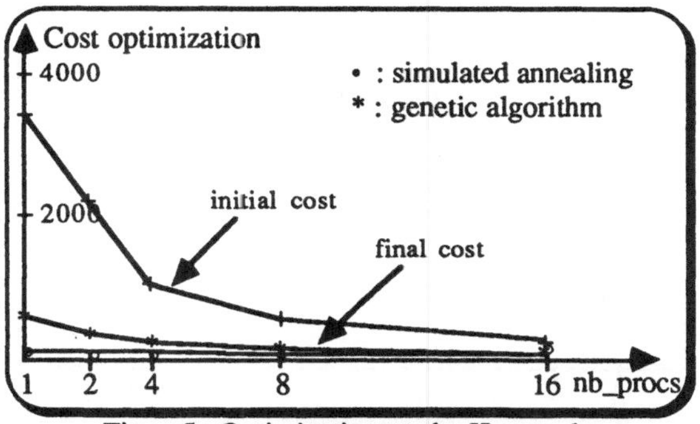

Figure5 : Optimization on the Hypercube

Here, the initial and final costs are compared for two optimization heuristics. The objective function is weighted by $\alpha=1$ (processing load) and $\beta=5$ (communicating load) corresponding to the relative importance of transfer times and the processing times drawn from the TMS320C40 DSP (40ns for multadd 32 bits and 200ns for 32 bit transfer on one communication channel).

The results allow the comparison of the two heuristics for the two previous interconnection networks : the static hypercube network Cube4 (figure5) and the reconfigurable network Omega/benes with 16 processors where the distance between the 16 processors is at most one (figure4). Figures 4 and 5 give the gain from heuristics utilization. Only the results obtained for the Multilayer network are reported, Hopfield network leads to similar results. The application of regular mapping with 16 processors nearly minimizes the cost function, especially in the reconfigurable network case, because of the regularity of the application graph. Moreover, the convergence towards global optimum spends similar lapse of time with the two heuristics. The gain is slightly better for Simulated Annealing when the initial mapping uses 1, 2 or 4 processors. It becomes the same as Genetic Algorithm with 8 processors.

5.2 - Effect on execution time

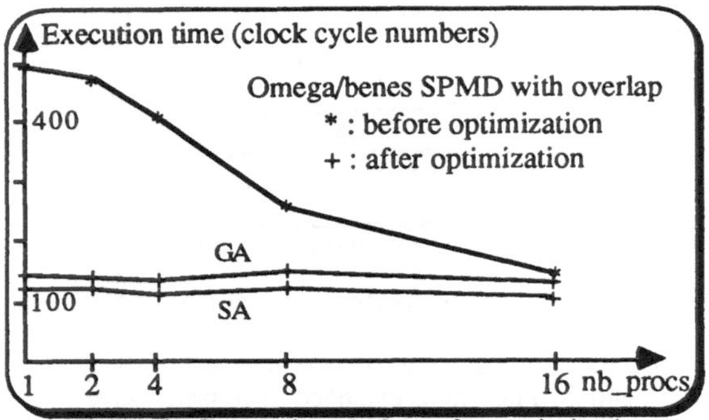

Figure6 : Comparison before-after optimization

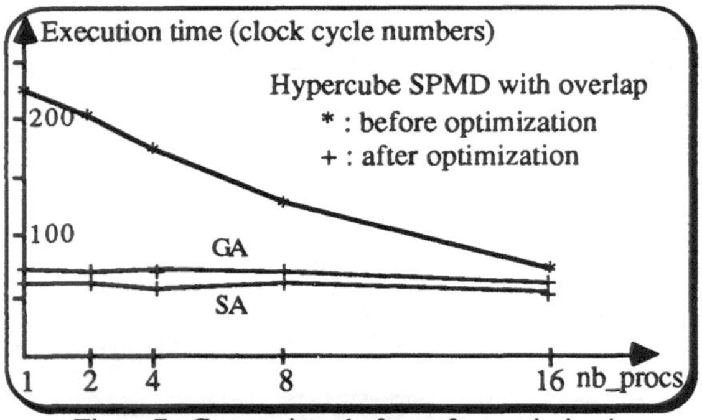

Figure7 : Comparison before-after optimization

Using the mapping obtained from the optimization, the model evaluates the algorithm execution time according to the parallel mode and the interconnection network. Figures 6 and 7 show the execution time improvement, i.e. before and after optimization for the two previous networks. It is a verification of the proper choice of the cost function.

The comparison of SIMD, SPMD parallel modes without time overlap and SPMD with overlap is conducted in figures 8 and 9.

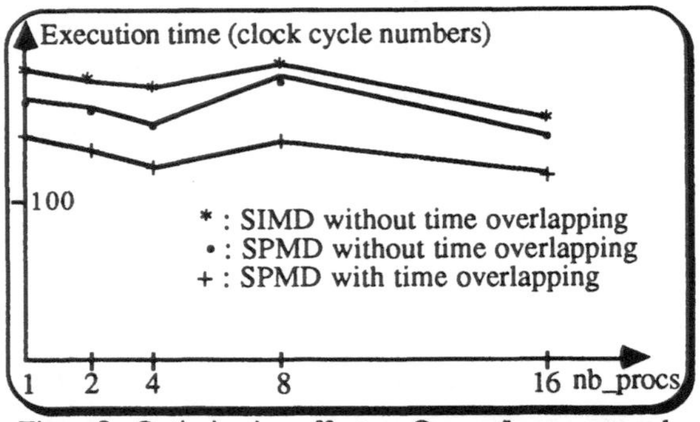

Figure8 : Optimization effect on Omega/benes network

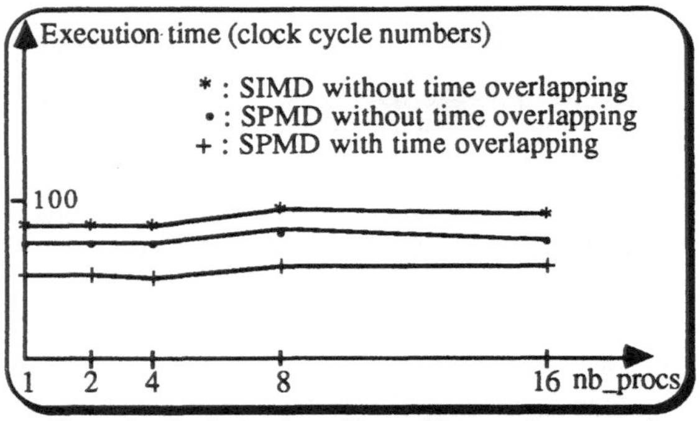

Figure9 : Optimization effet on Hypercube network

Execution times are almost half smaller with the static Hypercube network using a routing algorithm, compared to the reconfigurable network. Finally, confronted with the dynamic problem, these results show that the utilization of optimization heuristics leads to a maximum performance slightly better than regular mapping. Simulated Annealing proves to reach sligtly better minimum than Genetic Algorithm.

6 - CONCLUSION

In the special case where a regular and stationary application graph is implemented on a regular and static architecture, the utilization of assignment optimization heuristics denotes few improvement on execution time. For neuromimetic algorithms case, the synaptic matrix is regular by blocks and the machine is homogeneous : the speed of communication links, also the computational power of processors are the same. This explains that the optimal results are near to those obtained by direct application of regular mapping with 16 processors. However, these heuristics would be useful for static problems in which the algorithm graph is irregular [9], or the machine includes heterogeneous processors and communication links, even in the dynamic case in which the graphs vary in time.

In this paper, we compare two optimization heuristics for mapping neuromimetic algorithms (the Simulated Annealing and a Genetic Algorithm) on a dynamic problem where the machine size increases. A three-layers 8-7-6 neural network is considered, restricted to forward operation and without the data precedence constraints tied to successive layers. The optimization results are nearly identical, considering the obtained gain, the number of iterations necessary and the diminution of execution time. From the study, the Genetic Algorithms is simpler and requires less parameters than the Simulated Annealing algorithm. Unfortunately, more memory is needed in order to store the population and the final performance is slightly less.

In our dynamic problem, one suggestion is that these heuristics could be inserted in the program of a host processor, which would serve the working load according to supply and demand. From the model simulation, it appears that the chosen cost function is well related to the execution time of the neuromimetic algorithm. The SPMD machine doted with time overlapping between transfer and processing gives the best performance.

7 - REFERENCES

[1] M. Auguin, F. Boéri, J.P. Dalban
Synthèse et évaluation du projet OPSILA; Techniques et Sciences Informatiques, Vol 7, N°4, 1988, pp 385-401.

[2] J.D. Bagley
The behavior of adaptative systems wich emply genetic and correlation algorithms; PhD thesis; University of Michigan, 1967.

[3] C. Bonnemoy, S.B. Hamma
La méthode du recuit simulé : Optimisation globale dans R^n; Automatique Productique Informatique Industrielle; AFCET Automatique, Vol 25, N°5, 1991, pp 477-496.

[4] E. Fikret, P. Sadayappan
One to One mapping of process graphs onto a hypercube; Proceedings of the fourth hypercube Conference on Concurrent Computers and Applications, ACM, 1989, pp 91-98.

[5] D.E. Goldberg, R.E Smith
Nonstationary function optimization using genetic algorithms with dominance and diploidy; Genetic Algorithms and their Applications : Proceedings of the second International Conference on Genetic Algorithms, 1987.

[6] D.E. Goldberg
Genetic Algorithms in search, optimization and machine learning; Reading, MA : Addison-Wesley, 1989.

[7] J.J. Hopfield
Neural Network and physical systems with emergent collective computational abilities; Proceedings of the National Academy of Sciences, USA, 79, pp 2554-2558.

[8] S. Konstantinidou, L. Snyder
The chaos router : a practical application of randomize in network rooting; Proceedings of the fifth hypercube Conference on Concurrent Computers and Applications, ACM, 1990, pp 79-88.

[9] L.Kwiatkowski, J.P. Stromboni
Evaluation d'architectures parallèles dédiées aux Algorithmes Connexionnistes; Première conférence AAA, Outils d'aide à la conception et à la programmation d'architectures dédiées au traitement du signal et des images, Gréco TDSI, Lannion, 14-15 Septembre 1992, pp 35-48.

[10] T. Muntean, E. Talbi
Méthodes de placement statique des processus sur architectures parallèles; Techniques et Sciences Informatiques, Vol 10, N°5, 1991, pp 356-373.

[11] D.E. Rumelhart, G.E. Hinton
Learning internal representation by error propagation; Parallel distributed processing : Exploration in the Microstructures of Cognition, Vol 1, Cambridge, M.A.: MITpress, 1986.

• • •

Towards the Development of Cognitive Maps in Classifier Systems

N. R. Ball

Engineering Design Centre
Department of Engineering
University of Cambridge
Trumpington Street
Cambridge
CB2 1PZ
U.K
Facsimile :- 44 - 223 - 332662
E-mail :- nrb@uk.ac.cam.eng

ABSTRACT

Classifier systems are well tested vehicles for implementing genetic algorithms in machine learning environments. This paper presents a novel system architecture that transforms a classifier system's knowledge representation from message-based structures to self-organizing neural networks. These networks have been integrated with a classifier system to produce a Hybrid Learning System (HLS) that exhibits adaptive behaviour when driven by low level environmental feedback. Problems are represented within HLS as objects characterized by environmental features. Objects controlled by the system have preset goals set against a subset of their features and the system has to achieve these goals by developing a behavioural repertoire that efficiently explores and exploits the problem environment. Three types of knowledge structures evolve during this adaptive process : a cognitive map of useful regularities within the environment (encoded in a single self-organizing network); classifier behaviour calibrated against feature states and targets (encoded in a set of self-organizing feature maps); a population of complex behaviours (evolved from a gene pool supplied as part of the initial problem specification).

KEYWORDS

adaptive control, associative networks, classifier systems, feature maps, genetic algorithm, hybrid, self organization.

INTRODUCTION

The development of hybrid systems combining Genetic Algorithms (G.A) and Artificial Neural Networks (A.N.N) is an active part of the research in the machine learning community. Research to date has mainly concentrated on the construction of neural networks (particularly multi-layer perceptrons) using genetic algorithms ((1] [2] [3] [4)). This paper presents a different combination of GAs and ANNs in the development of a Hybrid Learning System (HLS) that embeds network structures within a rule-based Classifier System (CS).

Classifier Systems are message-passing, rule-based, self-organizing systems employing a genetic algorithm as their main learning process ((5] [6] [7] [8)). The architecture of HLS (figure 1) is based around self-organizing network structures that provide the system with its internal 'world model'. This model adapts with experience and the nature of that experience is changed through the activation of external and internal classifiers. The goal of the system is to achieve (fixed) problem domain objectives by learning how classifiers behave during the exploration of the domain and then evolving new classifiers that can optimize search in the domain state space to discover goal states. Two forms of experience-based learning are exhibited by the system -
(1) continual calibration and application of classifiers by matching of feature map activity with current states and predefined targets.
(2) creation of new classifiers by an evolutionary process based on goal convergence rates.

The key concept is to use neural networks to represent the behaviour of each classifier within the population in terms of its effects on the external environment. These effects are encoded within the networks as state vectors with each vector representing the environment state before and after classifier activation. HLS's networks are self-organizing since no external critic is available to provide training examples. The self-organizing process calibrates classifiers by adapting their pre/post activation state vectors and eliminating redundant state data unaffected by classifier activation. Elimination of this redundancy is equivalent to the induction of general rules that describe a classifier's action from a set of specific examples. An example of an HLS classifier is shown in figure 2.

One type of self-organizing network able to extract underlying correlations from large sets of data examples is the Kohonen Feature Map (9] and this network forms the basis for all network structures within HLS. Each classifier is allocated 'F' feature maps where F is the number of features detected in the environment. The resulting multi-layer network self-organizes to represent the relative effect of the classifier across a range of feature states as that classifier is activated and receives post activation feedback from the environment.

An additional feature map provides a long term memory correlating domain regularities (discovered through classifier activation) to the pre-defined domain-dependent goals. Kohonen networks have been used in this manner to perform statistical analysis (10).

1. BASIC ARCHITECTURE

The HLS architecture (figure 1) is an extension of the classifier model which allows for the encoding of multiple classifier condition / fitness values within self-organizing neural network structures ([11] [12]).

The system consists of two modules - the representational interface and the core system. The representational interface isolates the domain-independent core system from the specific problem environment. The detector process extracts and formats pre-activation state data into vectors of features ready for processing by the correlation network. The effector process applies the instantiated classifier actions to the environment creating the post-activation state. The core system is the domain independent part of HLS consisting of the correlation network, classifier population and feature maps associated with each classifier.

The correlation network represents the system's model of the environment. The classifier population encodes the set of actions available to the system. The feature maps provide a representation of a classifier's effect across any domain state given a particular target. Three domain dependent entities are needed for the definition of a new control problem -

- . feature objectives (or targets) which remain fixed;
- . an initial classifier population which evolves over time;
- . a domain state comprised of the collective state of all objects (characterized by feature vectors).

The network subsystem has two functions - to provide an associational memory between internal goal states and external sensory states and to calibrate the current classifier population to produce optimal behaviour dependent on current feature goals. These functions are performed by the correlation network and feature maps respectively. Both networks are based on Kohonen's Self-Organizing Feature Maps that have been modified to enable differentiation between different types of state data from the problem domain -

- . good data points that lead to goal convergence;
- . bad data points that lead to goal divergence;
- . neutral data points that are neither good or bad.

The structure of HLS's classifiers (figure 2) differs from the CS model in two respects. Firstly CS classifier condition data is encoded within the feature maps of HLS classifiers. Secondly the CS strength / fitness measure has been replaced in HLS by a new measure called ranking. Condition data in HLS is represented in the feature maps associated with each classifier as sets of network nodes with multiple weights. Self organization of these networks changes the conditions under which classifiers operate and affects the probability of a classifier's selection (for activation) given a current (pre activation) problem state and a target (post activation) objective state. A classifier's ranking value is proportional to the effectiveness of a classifier in reducing the state difference between the current and target states.

Adaptation of a classifier's feature maps reflects the system's incremental calibration of known behaviours to the problem domain. Production of new classifiers via the adaptive strategy (figure 1) generates new, speculative, behaviours which may improve performance. Three operators are applied across the classifier population as part of this strategy - crossover, inversion and mutation. Classifiers evolved using these operators are built upon the backs of previously successful ones. They displace weaker classifiers within the population and are calibrated in their turn by self-organization of their feature maps. Representation of classifier conditions as network node activity sidesteps the problem of condition-message matching inherent in the CS model. All classifiers in HLS match any problem domain state to a certain degree depending on current goals. Self organization of classifier feature maps improves the accuracy of this match over time.

Classifier activation is based on the CS model but has been simplified by the use of an integer rather than bit-string coding of individual action components ('genes'). This representation permits easy instantiation of classifiers at the expense of an increased 'genetic alphabet' which results in a larger hypothetical search space of possible classifier instances.

HYBRID LEARNING SYSTEM

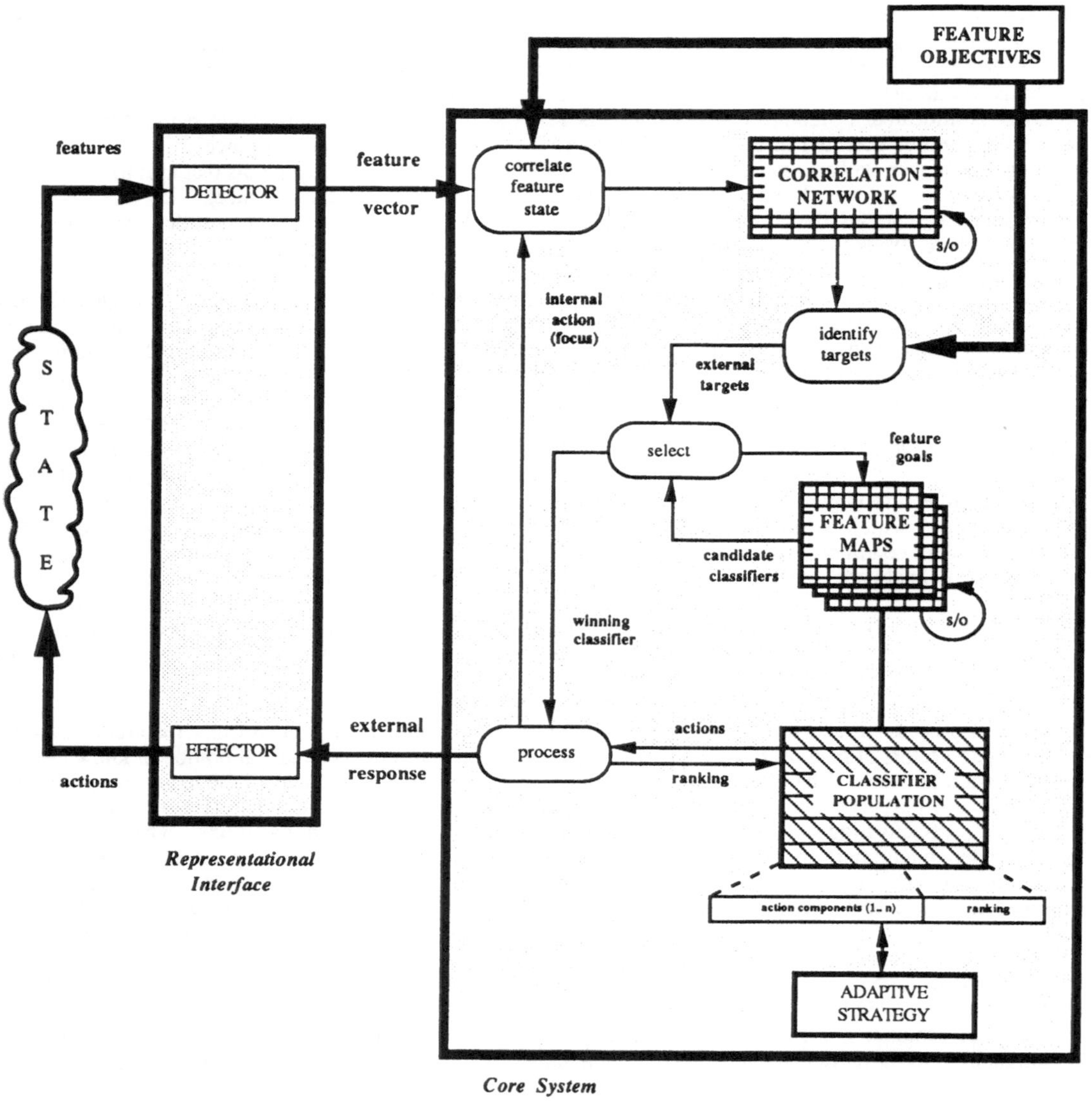

ADAPTIVE STRATEGY - application of G.A operators (based on classifier ranking)
S/O - self organization of neural networks

Figure 1.

HLS Classifier Architecture

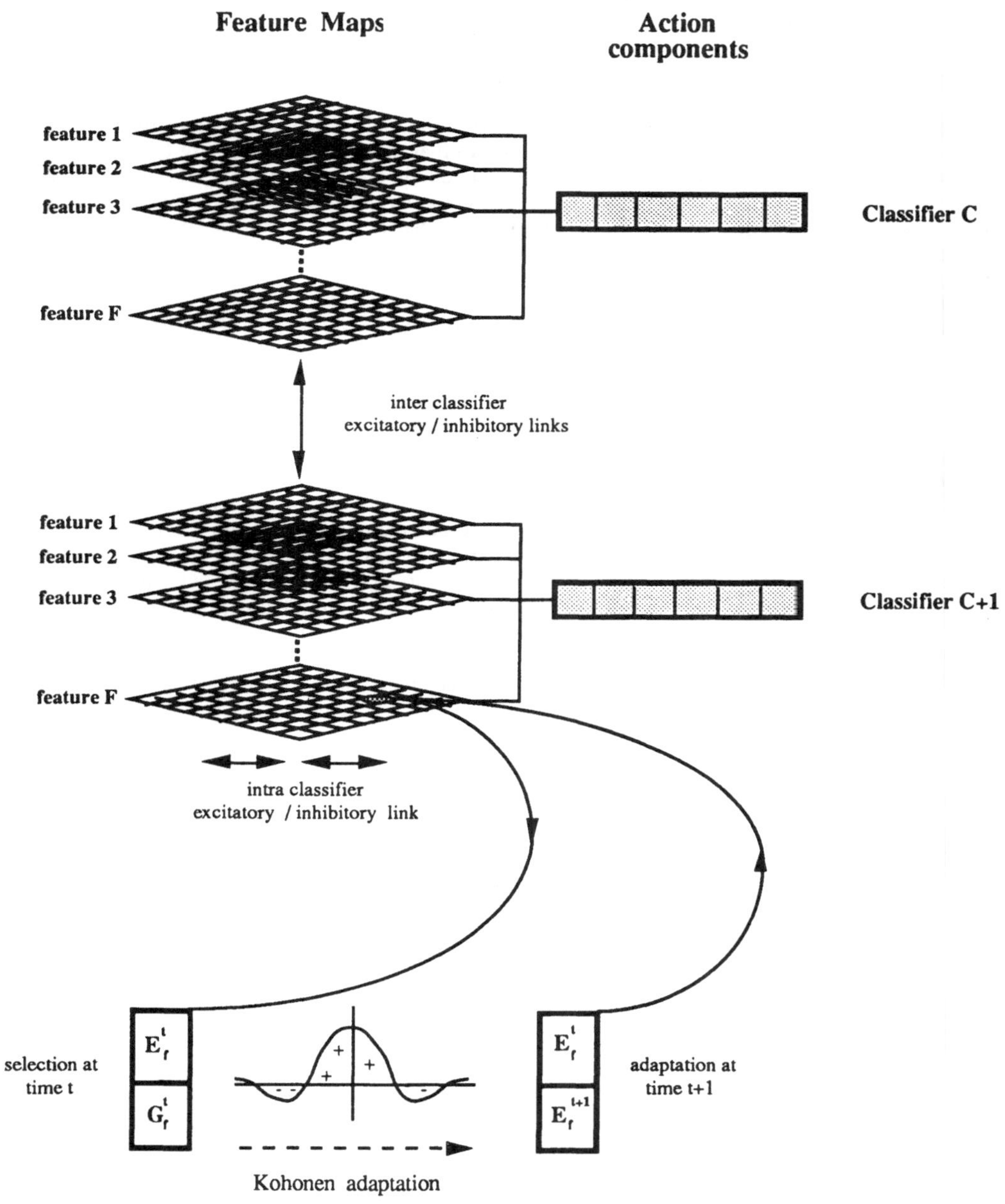

E_f^t = problem domain state for feature F at time t.

G_f^t = predefined / currently correlated goal for feature F.

Figure 2.

2. EXECUTION CYCLE

The iterative process controlling HLS is -
(1) self organization of classifier feature maps based on domain states before and after classifier activation (i.e. low level feedback).
(2) detection of current signals and mapping onto correlation network.
(3) self organization of correlation network to reflect current domain state.
(4) identification of external targets through association of feature objectives with HLS's current internal model.
(5) selection of candidate classifiers for each external target by inspection of nodes on classifier feature maps.
(6 application of selected classifier to the problem domain.
(7) evolution of new classifiers and integration into current population.

Processes (1), (3) and (7) constitute the learning mechanism of the system. Unsupervised network based learning occurs at steps (1) and (3) as the system assimilates low level feedback. Genetic adaptation occurs at step (7) as the classifier population is evolved by application of the genetic operators - crossover, mutation and inversion.

The selection of classifiers for adaptation is based on a ranking vector recorded against each classifier similar to that of the Schaffer's VEGA system [13]. Each element of this vector corresponds to a fitness score for a specific domain feature. Hence in a domain problem characterized by 'F' features, each classifier's fitness is represented by a vector of 'F' real numbers. The measure of fitness is proportional to the classifier's rate of goal convergence and frequency of usage. Removal of classifiers from the (fixed size) population to make way for new individuals is based on a crowding algorithm simila- to that presented by Goldberg (14). The set of possible domain states is clustered into 'niches' and the level of classifier crowding within each niche calculated. Densely populated niches are thinned by identifying pairs of classifiers within the niche that are most genetically similar and then removing the lowest ranking classifier.

3. A SIMPLE CONTROL PROBLEM

Recent research has focussed on the linkage between HLS's network subsystem and the adaptive strategy using a simple control problem domain shown in figure 3. The system to be controlled, S, forms part of an external environment which generates a background disturbance D. The task of the controller is to extract a set of output parameters U from S and to develop a repertoire of external behaviours A to apply to S so that U converges to a pre-defined target G. The relationship between U and G is an unknown transfer function that must be discovered by the adaptive controller. The output and control parameters constitute multi-dimensional signals with each element corresponding to

Simple Control Problem

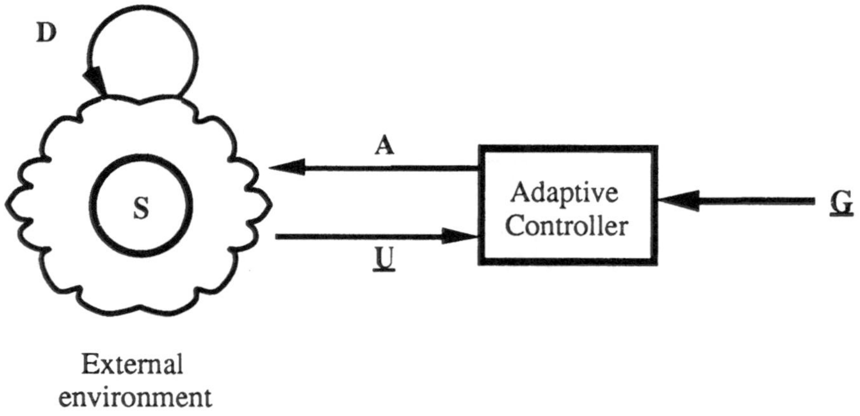

Figure 3.

an 'environmental feature' to be detected by the controller and correlated to other known features.

HLS is currently under test in the following simulated domains -
(1) multidimensional output with no external disturbance;
(2) multidimensional output with external disturbance affecting part or all of U.

Domain (1) is equivalent to a stationary target problem; domain (2) to a moving target problem. Controller convergence for a type (1) domain using a complete classifier population and network-only learning is shown in figure 4(a). Complete in this context means that the population includes a null-action classifier enabling stable behaviour.

Figures 4(b) - (d) show the behaviour of the system with a deficient population (minus the null-action classifier). In 4(b) genetic adaptation is disabled and therefore no new classifiers are generated. In 4(c) genetic adaptation is enabled with a niche strategy generating 25 niches in an environment of 400 states; in 4(d) genetic adaptation is enabled with a 'deletion of the weakest' strategy. All tests were conducted with a population limit of 10 classifiers and crossover / inversion / mutation rates of 0.01 / 0.002 / 0.0001 respectively. (Note that these rates are much lower than in standard CS systems because of the slow rate of classifier feature map calibration).

The results show that the system was able to generate stable behaviour with a multiple niche strategy that maintains diversity in the population but fails using a 'deletion of the weakest' strategy that results in the removal of the lowest ranking classifier regardless of genetic makeup / uniqueness.

CONCLUSION

The development of classifier feature maps within HLS is relevant to research in the fields of Genetic Algorithms (G.A) and Artificial Neural Networks (A.N.N) because it offers a *sub-symbolic* approach to the development of cognitive maps within a CS architecture. The use of low level feedback within HLS makes it applicable to domains where fitness is specified over multiple, discrete, features which can impinge directly upon the classifier population via the feature maps. In this sense the current work continues that on multi-objective optimization pioneered by Schaffer. (13).

REFERENCES

[1] Badii A., Binstead M., Jones A., Stonham T. & Valenzuela C. (1989). *Application of N-tuple sampling and genetic algorithms to speech recognition.* In Aleksander I. (ed) Neural Computing Architectures, pp 172 - 216. North Oxford Academic.

[2] Hancock P. (1990). *GANNET: Design of a Neural Network for Face Recognition by Genetic Algorithm.* Proc. IEEE Workshop on Genetic Algorithms, Simulated Annealing and Neural Nets, University of Glasgow.

[3] Harp S. & Samad T. (1991). *Genetic Synthesis of Neural Network Architecture.* In Davis L. (ed) Handbook of genetic Algorithms, pp 202 - 221, Van Nostrand Reinhold.

[4] DeGaris H. (1990). *BRAIN building with GenNets.* Proc. INNC-90 (Paris), Vol 2, pp 1036 - 1039. Kluwer Academic Publishers.

[5] Holland J. & Reitman J. (1978). *Cognitive Systems based on adaptive algorithms.* In Waterman D. & hayes-Roth F. (eds). Pattern directed inference systems, pp313 - 329. Academic Press.

[6] Booker L. (1982). *Intelligent behaviour as an adaptation to the task environment.* Phd dissertation, Ann Arbor : University of Michigan.

[7] Goldberg D. (1983). *Computer-aided gas pipeline operation using genetic algorithms and rule learning.* Phd dissertation, University of Michigan.

[8] Wilson S. (1985). *Knowledge growth in an artificial animal.* Proc. 1st Int. Conference on Genetic Algorithms and Their Applications, pp 16 - 23. Lawrence Erlbaum Associates.

[9] Kohonen T. (1984). *Self Organization and Associative Memory.* Springer Verlag.

[10] Hecht-Nielsen R. (1988). *Applications of counterpropagation networks.* Neural Networks, Vol 1, No 2, pp 131 -139, Pergamon Press.

[11] Ball N. (1990). *Adaptive signal processing via genetic algorithms and self-organizing neural networks.* Proc. IEEE Workshop on Genetic Algorithms, Simulated Annealing and Neural Nets , University of Glasgow.

[12] Ball N. (1991). *Cognitive Maps in Learning Classifier Systems.* Phd dissertation, University of Reading.

[13] Schaffer J. D. (1985). *Multiple objective optimization with Vector Evaluated Genetic Algorithms.* Proc. 1st Int. Conference on Genetic Algorithms and Their Applications, pp 93 - 100. Lawrence Erlbaum Associates.

[14] Goldberg D. (1989). *Genetic Algorithms in Search, Optimization and Machine Learning.* Addison-Wesley.

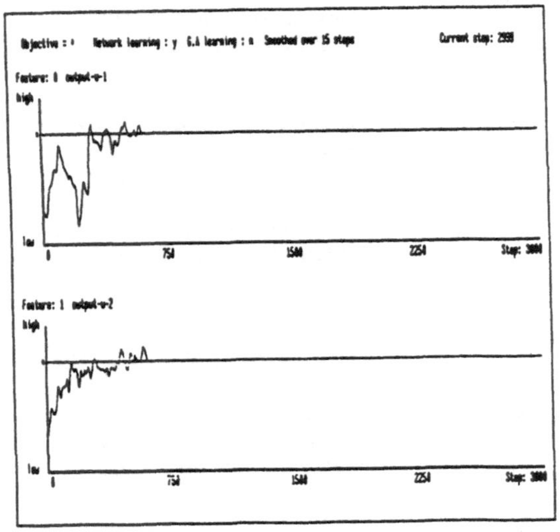

Figure 4(a)

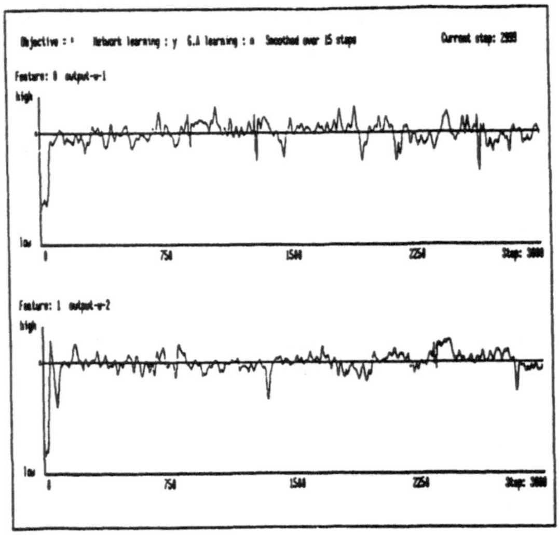

Figure 4(b)

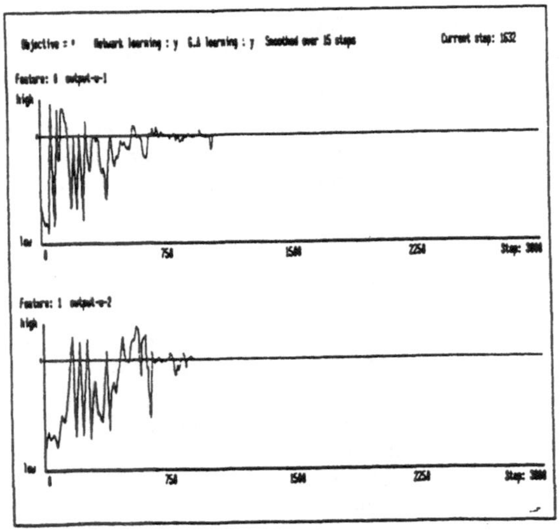

Figure 4(c)

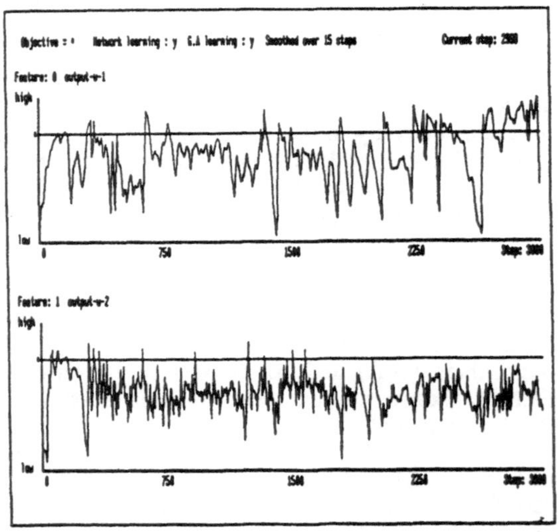

Figure 4(d)

Genetic Optimisation of Neural Network Architectures for Colour Recipe Prediction

J.M.Bishop & M.J.Bushnell
Dept. of Cybernetics, University of Reading, Berkshire, UK.
A.Usher
Courtaulds Research, Spondon, Derby, UK.
S.Westland
Dept. of Communication & Neuroscience, University of Keele, Staffs, UK.

ABSTRACT

Colour control systems based on spectrophotometers and microprocessors are finding increased use in production environments. One of the most important aspects of quality control in manufacturing processes is the maintenance of colour in the product. This involves selecting a recipe of appropriate dyes or pigments which when applied at a specific concentration to the product will render the required colour. This process is known as recipe prediction and is traditionally carried out by trained colourists who achieve a colour match via a combination of experience and trial-and-error. Instrumental recipe prediction was introduced commercially in the 1960's and has become one of the most important industrial applications of colorimetry. The model that is almost exclusively used is known as the Kubelka-Munk theory, however its operation in certain areas of coloration is such as to warrant an alternative approach. The purpose of this paper is to investigate the performance of a Genetically optimised Neural Network applied to this recipe prediction task.

INTRODUCTION

An industrial colour control system will typically perform two primary functions relating to the problems encountered by the manufacturer of a coloured product. Firstly the manufacturer needs to find a means of producing a particular colour. This involves selecting a recipe of appropriate dyes or pigments which when applied at a specific concentration to the product in a particular way, will render the required colour. This process is known as recipe prediction and is traditionally carried out by trained colourists who achieve a colour match via a combination of experience and trial-and-error. The second function of a colour control system is the evaluation of colour difference between a batch of the coloured product and the standard on a pass/fail basis.

The first commercial computer for recipe prediction[1] was an analog device known as the COMIC (COlorant MIxture Computer) but all colour systems on the market today employ digital computers. A typical colour control system consists of a reflectance spectrophotometer connected to a PC-based machine with various peripherals and costs in the region of £20,000 - £50,000. All computer recipe prediction systems developed commercially to date are based on an optical model that relates the concentrations of individual colorants to some measurable property of the colorant in use (e.g. reflectance). The model must also describe how the colorants behave when used in mixtures with each other.

The model that is almost exclusively used is known as the Kubelka-Munk theory[2]. It relates measured reflectance values to colorant concentrations via two terms K and S, which are the Kubelka-Munk version of the absorption and scattering coefficients of the colorant. The Kubelka-Munk theory is a highly simplified version of rigorous radiative-transfer theory[3] whereby only two fluxes of radiation are considered. Attempts have been made to introduce more complex theories by allowing the use of three or more fluxes[4], but the application of these more complex theories is generally not practical[5]. The use of the exact theory of radiation transfer is not of practical interest to the coloration industry.

The use of the conventional two-flux Kubelka-Munk theory has attracted criticism[6]. The popularity of the Kubelka-Munk equations is undoubtedly due to their simplicity and ease of use. The equations give insight and can be used to predict recipes with reasonable accuracy in many cases. In addition the simple principles involved in the theory are easily understood by the non-specialist. However in order for the Kubelka-Munk approximation to be valid a number of restrictions are assumed[6].

There are many applications of the Kubelka-Munk theory in the coloration industry, where these assumptions are known to be false. In particular,the applications to thin layers of colorants, for example, lithographic printing inks[7] and fluorescent dyestuffs[8,9] have generally yielded poor results.

NEURAL NETWORKS AND RECIPE PREDICTION

The performance of the Kubelka-Munk theory in certain areas of coloration is such as to warrant an alternative approach. An empirical colorant mixture model has been suggested[10] using high-order polynomial models but the accuracy of such a model will clearly be influenced by the exact choice of polynomial.

The trained colourist accumulates experience of the behaviour of the colorants and is able to extrapolate and interpolate from this experience to predict recipes for new shades without the use of Kubelka-Munk theory or any other algorithmic model. Neural Networks techniques are now beginning to be able to emulate the performance of human operators or experts in many areas of science and engineering and this paper describes their use on the recipe prediction task.

It was expected that the Neural Network approach would provide a novel and profitable new solution to the recipe prediction problem. It was hoped that a suitable network system would be able to automatically learn relationships between colorants and colour, and hence learn to predict which colorants, and at which concentrations, need to be applied to a particular substrate in order to produce a specified colour. The preliminary results of this work have been published elsewhere[11,12].

NEURAL NETWORK METHODS

Neural Networks consist of collections of connected processing elements that are not individually programmable. Each element usually computes a simple non linear function **f** on the weighted sum of its input (Figure 1).

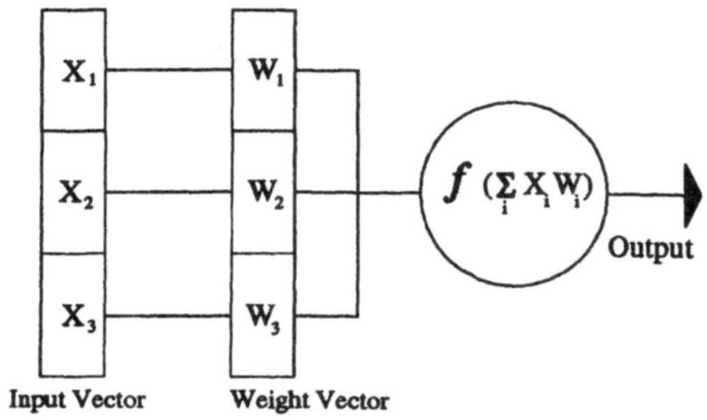

Figure 1. Schematic of Neural Element

The output of this function is defined as the *activation* of the cell. Long term knowledge is stored in the network in the form of interconnection weights linking a network of these cells. Unlike a conventional computer solution to a given problem, a neural network is not explicitly programmed to complete a given task, instead it acquires knowledge over time by adapting the strengths of the interconnections between cells. It is this ability to learn the solution to a problem that enables the use of network models on computationally ill-defined systems.

The network model that has been used with this research is a multi-layer feed forward network architecture, taught using the Generalised Delta rule[13]. This works by performing gradient descent in Error/Weight space. That is, after each pattern has been presented, the resulting error on that pattern is computed, by comparing the actual output with the desired output, and each weight in the network modified by moving down the error gradient towards its minimum for that input/ output pattern pair.

Using a Neural Network for Colour Recipe Prediction

The international standard for colorimetry is the CIE system which takes measured relative reflectance values and converts them into three coordinates in a three dimensional colour space. There are two common ways of representing these numbers, known as *CIELAB* and *CIELCH*. In both methods, the *vertical* axis is the lightness axis, **L**. The other two numbers are either taken as two orthogonal axes in the horizontal plane, **A** and **B**, or polar coordinates **C** and **H**. The **A** and **B** coordinates of the CIELAB system represent the *redness/ greeness* or the *yellowness/blueness* of the colour, whereas the **C** and **H** of the CIELCH system represent the chroma (or saturation) of the colour and its hue. For this work the CIELAB colour specification was used[12].

Information enters the network in the form of a vector of real values, the CIELAB colour coordinates, applied to the input layer of the network. The network output is the vector of real values, the dye concentrations, defined by the activation values of the output layer. The network is trained by the repeated presentation of a list of such input/output vector pairs. The task of the Network is to learn a set of mappings between its input and output, such that the sum of the output errors squared, is minimised. The error, for each pattern in the training set, is defined as the difference between the actual network output and the desired output.

The Network architecture used in these experiments consisted of an input layer where cell inputs are clamped to external values (scaled CIELAB values), a number of hidden layers, where cell inputs are defined by the weighted activation values from the cells in the previous layer that they are connected to, and an output layer connected to the last hidden layer (Figure 2).

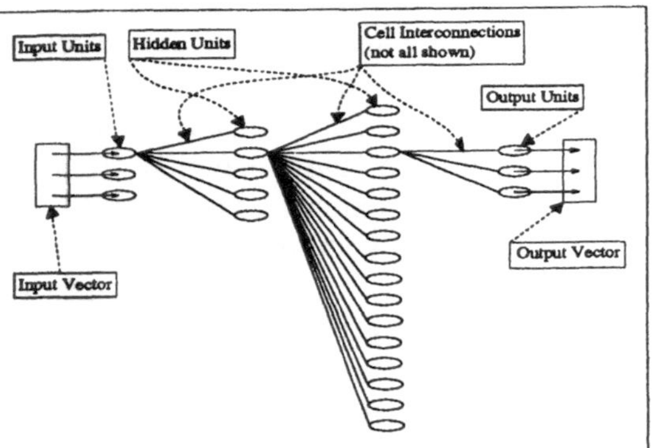

Figure 2. Network Architecture

Historically, the main technique that has been used in network design is trial and error, however even if an adequate network is discovered, there is no means of knowing *if this network is the best network for the task?*

DESIGNING THE NETWORK ARCHITECTURE

One of the key problems in creating a neural network for use on a given problem is to design a suitable architecture. As there are no well defined rules for creating suitable networks, the designer must *guess* a network architecture with the aid of available heuristics, or use an optimisation process.

Designing a network by guesswork is not a profitable design strategy, as it is impossible to evaluate the strength of the chosen design without actually training the network. However there are some rules that can aid the network design process.

It has been shown that any classification task can be solved by a two layer back propagation network using a single layer of hidden units[14]. However, if the number of hidden units exceeds the number of training patterns then the network may store knowledge in a very localised manner, with each hidden unit responding strongly to one training pattern. Clearly, this type of network is impractical when there were many training patterns. In addition, such a network is less able to perform useful generalistion as each unit becomes highly tuned to one training pattern. What is required is a multi-layered network which distributes knowledge throughout the net, such that the training patterns are represented by a pattern of activation across the network.

A second heuristic is that networks with multiple layers and fewer units in the first layer generalise better than shallow networks with many units in each layer[15].

However, even by using such heuristics network design is still largely guesswork and it is often necessary to design many such networks in order to generate one useful net. What is required is an automated method of network design which is capable of optimising network parameters such as the number of layers and the number of neurons needed in each layer, for a given objective function.

Genetic optimisation of network architecture

A programming strategy known as Genetic Algorithms has been used to optimise the design of network architectures[16]. Genetic algorithms use evolutionary techniques to perform an optimising search whose search space is defined by a binary coding of the parameters to be optimised. In the case of neural networks, the genes can represent the numbers of neurons in each layer of the network. An initial population of networks is generated randomly and tested to obtain a measure of the merit of each initial net. The test involves training each network for a short period, in relation to the usual total training time, to obtain a measure of the total error over all the training patterns. This error is then used to calculate a merit function for the net. In the case of the recipe prediction problem, the training period chosen was 20000 epochs, which is short compared to the usual training period of 200000 epochs but long enough for the network to begin to converge if a solution can be found.

Once the initial population has been created and tested, evolutionary techniques are used to improve the population. At each generation, a parent network is chosen from the population using a *weighted roulette wheel* such that each net has a probability of being chosen proportional to its merit after training. The genetic string is then manipulated by one of a number of operations (also chosen on a weighted random basis) to produce a child. The offspring is tested in the same way as the initial population to determine its merit. If the new network is better than the worst network in the population then it survives and the old network is removed from the population.

The possible operations performed on the parent string to produce a daughter string are as follows:

a) **Inversion**: Two sites are chosen randomly within the string to determine how much of the string is to be inverted. Then, the portion of the string defined by these random sites is simply reversed. For example, if the parent string was 0110010110 and the sites chosen were 3 and 7 the offspring created by inversion would be 0101001110 after inverting the 10010 portion of the string.

b) **Crossover**: This operation involves two parent strings, so a second parent is chosen in the same manner as the first, using the weighted roulette wheel approach. A crossing site is selected and the offspring is created by splicing the beginning of the first parents string, up to the crossing site, with the end of the second parents string from the crossing site. For example, if the first parents gene string was 111111 and the second parents string was 000000, using a crossing site of 4, the new offspring would be 111100.

c) **Mutation**: This operation has a very low probability, as it does in nature, but is the simplest of the three operations. A mutation site is chosen at random the the bit value at that site is simply changed: a 1 to a 0 or vice versa.

Although all these operations and selection mechanisms rely on random numbers, a population acted upon in this way soon improves. This is because the parent genes are more likely to be those who gave rise to better networks and that poor genes are killed off by new offspring, where such offspring produce improved networks. Thus the entire population tends to improve.

In this way, after a number of generations the population will consist almost entirely of very similar networks which are all capable of solving the problem. This is a direct result of the optimisation strategy which ensures survival of the fittest and also propagates the best characteristics of those fittest into new generations.

THE EFFECTIVENESS OF A GENETICALLY OPTIMISED NEURAL NETWORK RECIPE PREDICTION SYSTEM

An optimised network was created by running the genetic algorithm program under the following conditions. The initial population consisted of ten genetic strings which were used to create the networks. These networks were each tested by running under the neural network program for 20000 cycles. The genetic algorithm then ran for 20 generations, with each generation creating 2 new offspring and removing the two worst networks from the population. Finally, the best network found by the genetic algorithm was trained for a further 60000 cycles and compared with the best *hand-crafted* network trained for the same total number of cycles (80000).

Although in the initial stages of training the hand-crafted network had a smaller error the the genetically optimised one by the end of the 80000 cycles, the G.A. produced network performed much better than the hand-crafted one. The G.A. network had a final error of 0.000427 while the hand crafted network had an error of 0.003931, almost 10 times worse. Unfortunetely neither network was effective enough when predicting untaught recipes (most ΔE values being significantly greater than 0.8).

Although networks have been able to predict synthesised recipes well[11,12], generalisation still remains a significant problem when dealing with real data. Current research involving adding gaussian noise to the training data[17], is aimed at overcoming this problem.

CONCLUSION

Genetic Algorithms are a useful optimisation tool in the design of artificial neural networks. Results using a simple back propagation network on the problem of colour recipe prediction, have demonstrated that a genetically optimised neural network can significantly out-perform a hand crafted network. The use of Genetic Algorithms has the following advantages and dissadvantages:

i. They are very computer intensive, taking many days of CPU time on a SUN Sparc Station to converge.

ii. They can produce networks that over learn the data - that is they are able to reproduce the training data very accurately but are unable to generalise over unseen data.

iii. Networks designed by GA's are inherently very stable. That is, learning is insensitive to the initial random weights used in a given training run.

Other recent developments that have been reported in network design include the use of networks that dynamically change their topology[18] and the use of Gaussian noise to improve generalisation performance[17]. The use of both these techniques on the recipe prediction problem is ongoing.

REFERENCES

1. Davidson, H.R., Hemmendinger, H. & Landry, J.L.R. A System of Instrumental Colour Control for the Textile Industry, Journal of the Society of Dyers and Colourists, Vol.79, pp. 577, 1963.

2. Judd, D.B. & Wyszecki, G. Color in Business, Science and Industry. 3rd ed., Wiley, New York, 1975, pp. 438-461, 1975.

3. Chandrasekhar, S. Radiative Transfer. Clarendon Press, Oxford, 1950.

4. Mudgett, P.S. & Richards, L.W. Multiple Scattering Calculations for Technology. Applied Optics, Vol.0, pp. 1485-1502, 1971.

5. Mehta, K.T. & Shah, H.S. Simplified Equations to Calculate MIE-Theory Parameters for use in Many-Flux Calculation for Predicting the Reflectance of Paint Films. Color Research and Application, Vol.12, pp. 147-153, 1987.

6. Nobbs, J.H. Review of Progress in Coloration. The Society of Dyers and Colourists, Bradford, 1986.

7. Westland, S. The Optical Properties of Printing Inks. PhD Thesis, University of Leeds, (UK), 1988.

8. Ganz, E. Problems of Fluorescence in Colorant Formulation. Colour Research and Application, Vol.2, pp. 81, 1977.

9. McKay, D.B. Practical Recipe Prediction Procedures including the use of Fluorescent Dyes. PhD Thesis, University of Bradford (U.K), 1976.

10. Alman, D.H. & Pfeifer, C.G. Empirical Research and Application. Vol.12, pp. 210-222, 1987.

11. Bishop, J.M., Bushnell, M.J. & Westland, S. The Application of Neural Networks to Computer Recipe Prediction. Color, Vol.16, No.1, pp.3-9, (USA), 1991.

12: Bishop, J.M., Bushnell, M.J. & Westland, S. Computer Recipe Prediction Using Neural Networks. Proc. Expert Systems '90. (London), 1990.

13. Rumelhart, D.E., Hinton, G.E. & Williams, R.J. Learning Internal Representations by Error Propagation. in D.E.Rumelhart, J.L.McClelland and the PDP Research Group (Eds), Parallel distributed processing: Explorations in the microstructure of cognition: Vol.1, Foundations. pp.318-362. MA: Bradford Books/MIT Press.

14. Funahashi, K. On the approximate realization of continuous mappings by neural networks. Neural Networks, Vol.2, No.3, pp.183-192, 1989.

15. Rumelhart, D.E. Parallel Distributed Processing. Plenary Session, IEEE Int. Conf. Neural Networks, San Diego, CA., 1988.

16. Dodd, N. Optimisation of Network Structure using Genetic Techniques. Proc. INNC. '90. Paris, 1990.

17. Sietsma, J & Dow, R.J.F. Creating Artificial Neural Networks That Generalize. Neural Networks, Vol.4, No.1, pp.67-79, 1991.

18. Hirose, Y., Yamashita, K & Hijiya, S. Back-Propagation Algorithm Which Varies the Number of Hidden Units. Neural Networks, Vol.4, No.1, pp.61-66, 1991.

USE OF GENETIC ALGORITHMS FOR OPTIMAL TOPOLOGY DETERMINATION IN BACK PROPAGATION NEURAL NETWORKS

Philip Robbins[1], Alan Soper[1], Keith Rennolls[2]

[1] School of Computing and Information Technology, University of Greenwich, London, England, SE18 6PF

[2] School of Mathematics, Statistics and Computing, University of Greenwich, London, England, SE18 6PF

A genetic algorithm is applied to evolve neural network topologies suitable for given problem domains. Certain concepts, from the fields of statistics and genetics, are considered with a view to possible future improvements to the genetic algorithm.

Introduction

Neural networks have been used successfully in a variety of applications[1], and yet the definition of a neural network 'topology' (that is, the number of nodes and their arrangement) is problematic. There is currently no algorithmic procedure for defining a best topology for a neural network for a particular problem. Also, no single network topology can be proved the best for any non-trivial problem. Furthermore, networks with the same topologies, but which have been trained from different starting positions (i.e. different initial weight and bias values) will often converge to different solutions, illustrating the multimodality of the criterion being optimised.

In the majority of cases, neural network topologies are designed by trial and error by an experienced expert. This is a time-consuming process, because the task is iterative by nature, and because the amount of time taken to train neural networks is often substantial. Dependence on human control will also lead to a reluctance to search for a better topology once a working topology has been found.

It is clearly desirable to automate the design of the neural network topology. There are a number of possible strategies for doing this. In this paper we consider the use of genetic algorithms restricted to the back propagation network model. In the section on 'Further Work' we will discuss some of the other strategies and some more general issues.

The Genetic Algorithm Paradigm

The basic genetic algorithm paradigm[2,3] is a model inspired by nature (as is the neural network concept), based upon Darwin's theory of natural selection and evolution.

The general genetic algorithm adopted within this paper is :-

1. Generate an initial population of possible solutions. These are generated randomly.

2. Evaluate the fitness of each member of the population.

3. Remove from the population a certain number of the least fit members.

4. Mate randomly selected pairs of the remaining members to produce new members to be added to the population, so that the population size remains constant. Mating involves selecting attributes from each of the two parents which are then inherited by the child.

5. Apply mutations to the new population members. This involves adjusting one or more of the members' attributes. This may shift a solution out of a local minimum. It should be noted that mutation should occur very infrequently.

6. Go back to step 2, until it is decided that the search for good solutions should be allowed to terminate.

The Genetic Algorithm as used to Evolve Neural Network Topologies

A program was written whose purpose is to evolve 'good' neural network topologies for a given problem, given training and test data.

The genetic structure of an individual neural network is characterised by three features (genes). These are :-

1. The number of hidden layers in the network.

2. The number of nodes in each of the hidden layers.

3. The gain term, which dictates the speed of learning. The normal back propagation algorithm for training has been adopted.

It may be noted that the first two features are discrete and correspond to observable 'phenotypic' characteristics in individual neural nets, whilst the last is a continuous parameter representing a 'behaviour' of the individual in the learning situation. Inclusion of the latter may be regarded as a Lamarkian feature of our evolutionary model for neural nets. The fact that Lamarkianism is discredited in the evolution of natural populations does not mean that it will not have a potentially useful role in the evolution of our unnatural population.

The training of a net with given topology by back propagation seeks to minimise the error sum of squares (ESS) on the training data set. It has been well documented that large nets have so many parameters that they can tend to 'over-train'. That is, they are fitting the specific training data set almost exactly and, in doing so, loose their usefulness for prediction for new data sets. Such over-training is a disadvantage, and is, in our competitive model, avoided by use of the ESS on the test data as the fitness measure.

The algorithm uses certain criteria which allow it to terminate training when it appears that either the network is not going to be a good solution, or there is going to be only a very marginal improvement in the ESS as a result of further training.

The reproduction algorithm, which incorporates both mating and mutation (steps 5 and 6 above) can be described as follows :-

1. Select two parents at random from the surviving population members.

2. The number of layers in the new network is inherited from one or other of the parents. Mutation may occur at this point to either increment or decrement this number of layers.

3. From either of the parents, select a random number of layers from the front (input) end of the network. The number of nodes in these layers are inherited by the new network. The number of nodes in each of the remaining layers of the new network is defined by the number of nodes in the back (output) layers of the other parent. Mutation is applied to the number of nodes in each layer, allowing the number of nodes in that layer to be incremented or decremented by one.

4. The gain term from either one of the parents is passed on to the new network. This can also be subjected to mutation.

The Problem Domains

Three simple problem domains were defined to test the ability of the genetic algorithm to evolve good neural network topologies. These domains are :-

The "Exclusive Or" Problem

This is a slight variation on the classic exclusive or problem, which is specified in terms of binary inputs and a binary output. We have used two inputs in the range [0,1] and the exclusive or function is then applied with respect to being in the range [0,0.5] and [0.5,1].

Sphere Containment

The three inputs specify the co-ordinates of a point in three-dimensional space. Each of the co-ordinates is in the range -1 to +1. The neural network should be able to decide whether or not the point is contained within a sphere of unit radius, centred at the origin.

Relationship to a Polynomial Function

The objective is that the network should determine whether a point lies above or below the line defined by a polynomial function. The polynomial selected has roots at $x = 0.25$, $x = 0.1$, $x = -0.3$, $x = -0.8$ and $x = -0.9$. The (x,y) co-ordinates of the input points lay between -1 and +1.

Programs were written to produce training and test data sets for each of the above problems. The training data sets each contain 1000 entries, and the test data sets each contain 100 entries.

Results

Tables 1,2 and 3 show the results in terms of root mean squared error (RMS) and topologies for two simulation runs on each of the three data sets. The population size was 20 in each of the runs. The RMS value is sqrt(ESS/20).

The column labelled 'Best RMS' shows the RMS of the most fit network at that point in the run. When a generation does not produce a network which improves on the previous best network, then this column is left blank. The column labelled 'Hidden Nodes' lists the number of nodes in each of the hidden layers of the most fit network, starting with the layer closest to the input layer.

It is interesting to note that the topologies converge in relatively few generations, possibly indicating the appropriateness of the chosen mating and genetic mechanisms.

Further Work

The problem of determining the best topology of a back propagation neural network is solved by finding the (positive integer) number of layers and the (positive integer) number of nodes in each layer which maximise a fitness criterion. It is therefore a non-linear integer programming method, for which there is already some expertise and algorithms[4]. These techniques may therefore be of use in guiding the development of heuristics for the determination of optimal net topology. Other heuristic features, gained from human experience and expertise, could also be combined with numeric heuristics and it is possible that such methods may produce solutions more quickly than the undirected genetic algorithm described in this paper.

The fact that genetic algorithms are essentially random in operation calls for comparison with the technique of optimisation by simulated annealing[5]. In this technique, an apparently sub-optimal iteration step can be taken, with a probability that decreases as the optimisation process continues. Such random searching behaviour allows the possibility of escaping from local optima. The genetic algorithm has the same benefit by virtue of its random mating and mutation processes.

It is not clear that the genetic algorithm strategy adopted in this paper is the best for the evolution of the neural net species with respect to the fitness criterion adopted. For example, should 'sex' be introduced so that an individual is only allowed to mate with others of a different sex? Genetics determines many behaviour patterns; maybe the adopted social interactions of individuals would have an evolutionary advantage for a neural species adopting these strategies. There is already scope for much further work in this area. It is likely that the concept of evolutionary stable strategies used in current evolution theory will have relevance to neural evolution[2]. An evolutionary stable strategy is one such that small perturbations from it result in a strategy which has a lower average fitness. Such concepts may be of use in eliminating non-viable strategies relatively quickly.

The general mathematical/statistical modelling paradigm consists of three phases :-

Identification of an appropriate family of models (forward propagation nets in this paper).

Calibration of the model; this involves a choice of an objective criterion which is to be optimised by the choice (estimation) of the model's parameters (weights).

Validation, either by the examination of the residuals on the training data set, or by considering performance of the calibrated model on a test data set. The latter practice has been adopted in neural net modelling.

In this paper we have trained the nets by minimising the ESS on the training set, but have judged fitness within the genetic algorithm, which is optimising net topology, with respect to the ESS on the test data set. Such a complex criterion reflects the fact that we have multiple objectives each of which has to play a part in determining the final solution. It is possible that other objective functions might produce better (fitter) solutions. Perhaps the differing variabilities of the output node values should be taken into account. Maybe the variance of the output of a given node depends on the values of the inputs. If so, then a weighted error sum of squares (WESS) has been found to give more precise parameter estimation in traditional statistical modelling. Once one considers differing criteria it is natural to consider maximum likelihood and Baysian methods[6], but these are outside the scope of this paper.

References

1 Marshall, S. J. and Harrison, R. F., 1991, Optimisation and Training of Feedforward Networks by Genetic Algorithms, 2nd IEE International Conference on Artificial Neural Networks, Bournemouth UK, Nov. 91, p39-43.

2 Holland, J. H., 1975, Adaptation in Natural and Artificial Systems, University of Michigan Press.

3 Goldberg, D., 1988, Genetic Algorithms for Search, Optimisation and Machine Learning. Addison-Wesley.

4 Papadimitrion, C. H. and Steiglitz, K., 1982, Combinatorial Optimisation : Algorithms and Complexity. Prentice-Hall.

5 Aarts, E., and Korst, J., 1989, Simulated Annealing and Bolzman Machines (A Stochastic Approach to Combinatorial Optimisation and Neural Computing). John Wiley and Sons.

6 Ripley, B. D., Statistical Aspects of Neural Networks, 1993, Proceedings of SemStat, Sandbjerg, Denmark, April 1992. Chapman and Hall.

Generation	Run #1			Run # 2		
	Best RMS	Hidden Nodes	Avg. RMS	Best RMS	Hidden Nodes	Avg. RMS
0	0.1692	5	0.4525	0.2173	3	0.4762
1			0.4191			0.4613
2			0.3937			0.4612
3	0.1604	7	0.3536	0.1577	5	0.4437
4			0.2877			0.4276
5			0.2743			0.3931
6			0.2996			0.3459
7	0.1457	5	0.2531			0.3077
8			0.2736			0.2492
9			0.1960	0.1566	4	0.2152
10			0.2535			0.2716
11			0.2710	0.1547	4	0.2802
12			0.3292			0.2488
13			0.2303			0.2792
14			0.2777	0.1528	5	0.2944
15			0.2949			0.2683
16			0.2604			0.2269
17			0.1950			0.2709
18			0.2610			0.2432
19			0.2112			0.2479
20			0.2109			0.2202

Table 1 - Exclusive Or Results

Generation	RUN # 1			Run # 2		
	Best RMS	Hidden Nodes	Avg. RMS	Best RMS	Hidden Nodes	Avg. RMS
0	0.2358	20,14	0.4729	0.4457	2,4	0.4937
1			0.4718	0.2051	18	0.4791
2	0.1972	17	0.4504			0.4591
3	0.1886	20	0.3964			0.4464
4			0.3482	0.1927	18	0.4247
5			0.3124			0.4057
6	0.1756	20	0.2729			0.3501
7			0.2676			0.2624
8			0.2301	0.1859	18	0.2468
9			0.2704			0.2375
10			0.2889			0.2323
11			0.2484			0.2653
12	0.1546	20	0.2495	0.1795	18	0.2489
13			0.1972			0.2568
14			0.2501	0.1708	18	0.2125
15			0.2470			0.2657
16			0.2465			0.2892
17			0.2192			0.2385
18			0.2602			0.2760
19			0.2568			0.2906
20			0.2166			0.2449
21						0.2598
22						0.2945
23						0.2733
24						0.3160
25						0.2939
26						0.2624
27				0.1636	18	0.2330

Table 2 - Sphere
Containment Results

Generation	Run # 1			Run # 2		
	Best RMS	Hidden Nodes	Avg. RMS	Best RMS	Hidden Nodes	Avg. RMS
0	0.0211	2	0.2892	0.0137	18,16,16	0.3506
1			0.1568			0.1713
2	0.0161	14,16	0.1029			0.0840
3			0.0378			0.0250
4			0.0597			0.0824
5			0.0546			0.0562
6	0.0146	14,2	0.0238			0.0550
7			0.0749			0.0454
8			0.0483	0.0119	3,16,16	0.0186
9			0.0220			0.0416
10						0.0773
11						0.0387

Table 3 - Polynomial
Function Results

Counting and Naming Connection Islands on a Grid of Conductors

Daniel C. Fielder and Cecil O. Alford
School of Electrical Engineering and Computer Engineering Research Laboratory
Georgia Institute of Technology
Atlanta, GA USA 30332

Abstract

Consider a grid of T horizontal and T vertical information conductors. At each crossing point it is possible to either connect a horizontal to a vertical conductor or just let the connection be open. When connected at a crossing point, both conductors share the same information state throught their lengths.

When several conductors are mutually connected at their crossing points, an island of connections is formed. A requirement for an island is that there be representative conductors of both kinds. All conductors associated with an island share the same information state.

For the case of several islands, each island's conductors are mutually exclusive of any other island's conductors. In fact, an island can be uniquely named by its conductors. As a further requirement on the conductors, the grid is assumed to be completely utilized with no idle conductors.

It is under the above conditions we develop the general combinatorial equations for counting the number of ways s sets of islands can exist on a TxT grid of conductors. The restraints of the problem introduce interesting combinatorial aspects. We also discuss methods of counting connections and naming sets of islands.

Key Words

Information State, Sets of Information States, Information Conductors, Combinatorics, Permutations, Combinations, Partitions of Numbers, Compositions of Numbers.

1. Introduction

Our earlier work [1-3] on counting conversations on the symmetrical TxT crossbar grids used in parallel processing, indicated that there were many unique ways that T horizontal and T vertical busses (hereafter generalized as *information conductors*) can be interconnected. For our conversation work and for this note, the most meaningful connection arrangement occurs when groups of horizontal conductors connect with groups of vertical conductors at their crossing points to form *islands* of connections. In general, there are s islands. Each island must be associated with at least one information conductor of each kind. Because of the connections, all an island's information conductors attain the same information potential. Further requirements are that islands cannot share conductors with other islands (thus avoiding "bus contention") and that there be no idle information conductors. The horizontal conductors are numbered from 1 to T, and the vertical conductors are numbered from T + 1 to 2T.

Figure 1 illustrates a 6 x 6 grid of conductors with one of the ways $s = 3$ islands can exist.

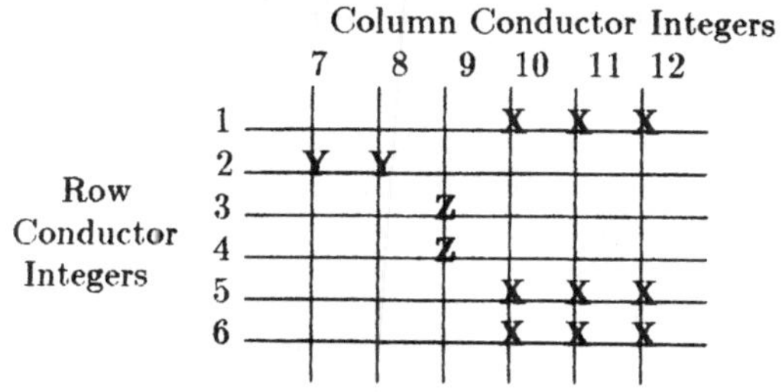

Figure 1. Grid with Islands X, Y, Z

In Figure 1 the integer name of island X is 1 5 6 10 11 12, that of island Y is 2 7 8, and that of island Z is 3 4 9. By following specific rules (developed in next section) which ensure that each valid set of $s = 3$ islands is counted exactly once, the unique name of the set of 3 islands of Figure 1 becomes 2 7 8 3 4 9 1 5 6 10 11 12. To complete the description, island X has 9 connection points, 1,10 1,11 1,12 5,10 5,11 5,12, 6,10 6,11 6,12, island Y has 2 connection points, 2,7 2,8, while island Z has 2 connection points, 3,9 4,9. There are a total of 13 connection points.

The ensuing derivations and formulas address the very interesting questions, "How many ways can sets of s islands of connections be placed on a TxT grid of information conductors? How many configuration patterns are there? How are the number of connection points per island found from an island's name? How many total connections per configuration are there? The answers to these questions are based on work reported in [4,5].

2. Integer Use in Identifications, Part (a)

In this section, we introduce the special uses of partitions of numbers and cascaded combinations in establishing valid sets of island names.

Sets of integers and individual integers are used as names to guide and identify counting procedures throughout this note. Whenever a set of integers is used as an identifier, there is always the question of whether the set is ordered or unordered. If permutation of the set implies different names the set is *ordered*. If the arrangement of the integers is immaterial, the set is *unordered*. For uniformity, unordered sets are listed in non-decreasing order.

Each connection, regardless of island affiliation, must involve conductors of both orientations. Hence, the smallest possible island size is 2. All 2T conductors are used in a set of s names. This means that the the number of conductors per name in a set of s names is a restricted s-part partition of 2T, the restriction being that no member can be less than 2. The example of Figure 1 has the 3-part partition of 12, which being unordered, appears as 3 3 6 . Figure 1's configuration is typical of those with two 3-integer names and one 6-integer name. But, there are other 3-part partitions of 12 which must be included in a complete count. Those restricted 3-part partitions of 12 needed to acccount for all sets of 3 islands on a 6 x 6 grid are 2 2 8, 2 3 7, 2 4 6, 2 5 5, 3 3 6, 3 4 5, and 4 4 4. It should be clear in general that access to the names of the s-partitions of 2T with no member less than 2 is absolutely essential for guiding any total island count calculation. While there are constructive algorithms for listing the partition names, there are, unfortunately, no analytic formulas for the names. We developed a simple computer subroutine to use as needed in getting the partitions.

For any given s and T, the island names are a subset of all the ways the integers from 1 through 2T can be sorted into s sets of integers with the integers within each of the s sets always in ascending order. If any single island candidate fails, the entire set of s candidates fails. Obviously, counting formulas must analytically weed out any s set which does not qualify as an island name.

There is a structure we call a *cascaded combination* [3,6] whose names include both the qualifying and non-qualifying sets of s island names. Suppose, as in Figure 1, we have $2T = 12$ integers from 1 through 12 to be chosen in $s = 3$ consecutive selections. The order of appearance and size of the s island candidates is controlled by the restricted partition 3 3 6. The first selection consists of 3 objects (conductor integers) out of 12, the second selection consists of 3 objects out of the remaining 9. The third selection must consist of the remaining 6 objects. All possible choices are governed by the cascaded combination $\binom{12}{3}\binom{9}{3}\binom{6}{6}$. Since each island name candidate is a name of a combination (an unordered set of integers), the integers within each of the $s = 3$ island name candidates appears in ascending order.

While we do not need to view all the $\binom{12}{3}\binom{9}{3}\binom{6}{6} = 12!/(3!\ 3!\ 6!) = 184{,}800$ candidates for sets of 3 island names in this note, we have developed elsewhere [3] brute-force checking programs which do. We start with the cascaded combination name 1 2 3 4 5 6 7 8 9 10 11 12 and conclude with 10 11 12 7 8 9 1 2 3 4 5 6. The interested reader can find a combinatorial application in [6].

Even if we were to isolate and count the successful candidates, we would find, in general, repeticious sets of s island names. For example, in Figure 1, the name of the set of $s = 3$ island names is the cascaded combination name 2 7 8 3 4 9 1 2 5 6 10 11 12. However, among the cascaded combination names there also appears 3 4 9 2 7 8 1 2 5 6 10 11 12 which has the same island names and conforms to the size-order imposed by partition 3 3 6. To resolve this dilemma and assure uniqeness, we stipulate that for repeated partition integers the *first* integers of the corresponding island names be in ascending order. This was an easy task for a brute-force computer simulation program, but defied the analytic implementation needed for this note. When we realized that the trouble lay in permutations of the island names induced by multiple partition members, we were able to bypass the difficulty by tailoring our derivation to accept the inflated count first and then, as a final step, "unpermute" the count through division by the factorials of the numbers of repeated partition members. This turned out to be a blessing in disguise because the inflated set of cascade combination names is highly symmetrical while the reduced set is not. For Figure 1, the divisor needed is 2! since there are two 3's in the s-part partition of 12. If a partition were 2 3 3 4 4 4, for example, the correcting divisor would be 1! 2! 3!.

3. Integer Use in Identifications, Part (b)

In our brute-force checking programs [4], a count is easy to find because all of the cascaded combination names are available one-by-one for acceptance or rejection. In an analytic derivation, we do not have the names available. However, a combinatorial operation induced by the restricted partition, does lead to the number of ways sets of s islands can exist on a TxT grid. The operation utilizes an association between the restricted partition and all possible row-column integer content of island names *without* knowing the names.

Each of the partitions is associated with many possible sets of s island names. Each set of s island names must include all the row-column integers, with each integer appearing exactly once. In order to exemplify finding the number and type of row-column arrangements corresponding to a given restricted s-part partition, we establish Table 1. for the 3-part partition of 12, 3 4 5. In Table 1, the larger, **bold** integers are **row integers** and the smaller, plain integers are column integers. There must be as many integers per island name as the size of the corresponding partition member. But since an island name must have at least one row and one column integer, the greatest number of either in an island name must be less than the partition member.

Table 1. Possible Row-Column Integer Count per Partition Member

Partition Member	3	4	5
Possible	**1** 2	**1** 3	**1** 4
Numbers of	**2** 1	**2** 2	**2** 3
Rows and Columns		**3** 1	**3** 2
per Partition Member			**4** 1

In the column headed by partition member 5, for instance, we can have **1** row integer and 4 column integers or **2** row integers and 3 column integers, etc. The table construction for any valid partition follows the obvious pattern.

While Table 1 lists all the possible row-column numbers for each individual island, not all combinations of one entry pair from each column can furnish valid row-column name distribution values.

Distributed throughout each set of $s = 3$ island names, there must be exactly $T = 6$ row integers and $T = 6$ column integers. Once the row distribution is fixed, the column distribution must complement it in the sense that the sum must equal the associated partition member size. We can choose trial row count numbers from the boldface entries of the table which total $T = 6$. An exhaustive choice of boldface numbers for use as **row** integers yields Table 2 which follows in the next section

4. Integer Sets Needed for Calculations

To cover all configurations of s islands on a TxT grid, the restricted s-part partitions of 2T with no member less than 2 must be available. For each partition, a set of restricted *compositions* of T must be available. The restriction on the compositions is that each member must be less than its corresponding partition member. Since there are no closed formulas for finding partition and composition names, the partitions and compositions must either be available as look-up tables or be calculated as needed. We used computer programs which produce the restricted partitions first, then the compositions. Because of complementation, only one set of compositions per partion is needed.

Table 2. Choices of Number of Row Integers

Partition Member	3	4	5
	1	1	4
	1	2	3
	2	3	1
	2	1	3
	1	3	2
	2	2	2

The **row** integer selections shown in Table 2 are the $s = 3$-part compositions of $T = 6$ with the restriction that each member be less than the value of the partition member which heads its column.

By including the complementary *column* integer count, Table 3 emerges as

Table 3. Numbers of Row and Column Integers

Partition Member	3	4	5
	1 2	**1** 3	**4** 1
	1 2	**2** 2	**3** 2
	2 1	**3** 1	**1** 4
	2 1	**1** 3	**3** 2
	1 2	**3** 1	**2** 3
	2 1	**2** 2	**2** 3

The entries of Table 3 show all possible distributions of numbers of row-column integers since the sum of an islands **row** integers and column integers must equal an island's partition number. The information of Table 3 's in general is of direct use in calculating the number of ways sets of s islands can appear on a TxT grid. The same information is used in the count of the number of connections per island.

We note that the connections per island is the product of the number of row and the number of column integers. As an example, Table 3 tells us that a set of islands configured as **1** 2 **1** 3 **4** 1 has 2 connection in the first island, 3 in the second, and 4 in the third. The set of 3 islands have 9 connections total.

5. Role of Partitions and Compositions

The previous discussion has indicated that restricted s-part partitions of 2T and restricted s-part compositions of T are indispensable for the calculation approach of this note. Unfortunately, there is no way to choose the names of successive partitions and compositions as analytic functions of s and T. However, our requirements can be satisfied either through use of precalculated look-up tables or by computer generation of the partitions and compositions as needed. In fact, we chose the latter approach.

To summarize, for a given s and T we need access to a list of all s-part partitions of 2T in which no partition member is less than 2. For each of these partitions we need certain s-part compositions of T for row integers. In these compositions a member in a given position cannot exceed the same positioned partition member diminished by 1. There is no need to generate a new set of compositions for use as column integers since these are complements as exemplified in Table 3.

6. A Worked Example

A worked example serves to illustrate how a specific count may be obtained and, more importantly, suggests how general equations evolve.

Consider the example in the earlier discussion in which $s = 3$, $T = 6$, and $2T = 12$. The $s = 3$-part partitions of $2T = 12$ with no member less than 2 is given as

Table 4. $s = 3$-Part Partitions of $2T = 12$

2 2 8
2 3 7
2 4 6
2 5 5
3 3 6
3 4 5
4 4 4

The first distribution of the 3 island name integers has 2 row-column integers for the first name, 2 for the second name, and 8 for the third name. If we were to count by a brute force simulation method (which we are not), the starting cascaded combination name would be

$$1\ 2 \quad 3\ 4 \quad 5\ 6\ 7\ 8\ 9\ 10\ 11\ 12 \tag{1}$$

and the value of the cascaded combination in binomial coefficient form would be

$$\binom{12}{2}\binom{10}{2}\binom{8}{8} \tag{2}$$

Note that the integers in the lower, or *choose*, position of the binomial coefficients are the partition integers.

PARTITION 2 2 8 In order to determine the ways in which integers of individual sets of row and column integers can combine, a table of compositions controlled by partition 2 2 8 is constructed. The set of restricted compositions for use with 2 2 8 is given as the lone entry in Table 5.

Table 5. Compositions for Partition 2 2 8

1 1 4

Out of the $T = 6$ row integers, the first island selects 1 row integer. By complementation, the first island must have 1 integer to be selected out of the $T = 6$ column integers. The second island has $T - 1 = 5$ row integers remaining from which to select 1 and (again by complementation) $T - 1 = 5$ column integers from which to choose 1. The third island has 4 row and 4 column integers from which to choose 4 of each. The number of ways the choices can be made is summarized in the binary coefficient product

$$\binom{6}{1}\binom{6}{1}\ \binom{5}{1}\binom{5}{1}\ \binom{4}{4}\binom{4}{4} = 900 \tag{3}$$

But partition 2 2 8 requires a divisor 2! 1! since there are 2 repeating members. Thus, the corrected number of ways $s = 3$ rook gangs with a 2 2 8 row-column distribution can exist on a 6 x 6 chessboard is

$$\frac{900}{2!\ 1!} = \boxed{450} \tag{4}$$

Since there is only one valid composition, there is only one island distribution.

Since the lower or choose part of the binary coefficients in (3) are the number of rows and columns for each island, the connections per island and the total connections are easily found as

$$(1x1) + (1x1) + (4x4) = 1 + 1 + 16 = 18 \tag{5}$$

PARTITION 2 3 7. Partition 2 3 7 induces compositions 1 1 4 and 1 2 3 which are complementary. 1 1 4 as row integer content and 1 2 3 as column integer content produces a configurations having 1800 counts. 1 2 3 as row integer content and 1 1 4 as column integer content produces *conjugate* configurations with 1800 counts also, giving a total count of 3600. By taking advantage of conjugancy, count calculation can be approximately halved.

The individual and collected connections for composition 1 1 4 are

$$(1x1) + (1x2) + (4x3) = 1 + 2 + 12 = 15 \tag{6}$$

and for composition 1 2 3 are

$$(1x1) + (2x1) + (3x4) = 1 + 2 + 12 = 15 \qquad (7)$$

The product represented by equation (3) is arranged in $s = 3$ sets of two binomial coefficients. The first binomial coefficient of each set pertains to choices of row integers, and the second to choices of column integers. Another way of presenting the product is to group the $s = 3$ binomial coefficients with row information together followed by in kind by the column related binomial coefficients. The rearrangement of equation (3) is

$$\binom{6}{1}\binom{5}{1}\binom{4}{4} \quad \binom{6}{1}\binom{5}{1}\binom{4}{4} = 900 \qquad (8)$$

But each $s = 3$-tuple of binary coefficients is a cascaded combination whose numerical value can be expressed as

$$\frac{6!}{1!1!4!}\ \frac{6!}{1!1!4!} = 900 \qquad (10)$$

The form of (10) highlights the computational aspects and is easily generalized. We prefer this form over that of (3).

PARTITION 2 4 6 The compositions needed for partition 2 4 6 are shown below

Table 6. Compositions for Partition 2 4 6

1 1 4
1 2 3
1 3 2

The computation pattern applied to each composition in order yields

$$\frac{6!}{1!1!4!}\ \frac{6!}{1!3!2!} = 1800 \qquad (11)$$

$$\frac{6!}{1!2!3!}\ \frac{6!}{1!2!3!} = 3600 \qquad (12)$$

$$\frac{6!}{1!3!2!}\ \frac{6!}{1!1!4!} = 1800 \qquad (13)$$

The complementary compositions are evident in the denominators (sans !). It can be seen that (11) and (13) are conjugates, and (12) is self-congugate.
The correcting divisor associated with partition 2 4 6 is 1!1!1!. Thus, the corrected number of ways $s = 3$ islands with a 2 4 6 row-column distribution can exist on a 6 x 6 grids is

$$\frac{7200}{1!\ 1!\ 1!} = \boxed{7200} \qquad (14)$$

There are three compositions and three island distributions. The individual and collected island connections for compositions 1 1 4, 1 2 3, and 1 3 2, respectively, are

$$(1x1) + (1x3) + (4x2) = 1 + 3 + 8 = 12 \qquad (15)$$
$$(1x1) + (2x2) + (3x3) = 1 + 4 + 9 = 14 \qquad (16)$$
$$(1x1) + (3x1) + (2x4) = 1 + 3 + 8 = 12 \qquad (17)$$

7. Summary Of Worked Example

We have sufficient data to deduce the general counting formulas without further calculations. In the interest of completeness, however, we present a complete summary of the example island and connection counts. In the listing below, Prt stands for partition, Cmp for composition, Div for the correcting divisor, Isl for the island count, Con for connection count, and Ttl for totals.

Prt	Cmp	Div	Isl	Ttl	Con	Ttl
2 2 8	1 1 4	2	450	450	1 1 16	18
2 3 7	1 1 4	1	1800		1 2 12	15
	1 2 3		1800	3600	1 2 12	15
2 4 6	1 1 4	1	1800		1 3 8	12
	1 2 3		3600		1 4 9	14
	1 3 2		1800	7200	1 3 8	12
2 5 5	1 1 4	2	450		1 4 4	9
	1 4 1		450		1 4 4	9
	1 2 3		1800		1 6 6	12
	1 3 2		1800	4500	1 6 6	12
3 3 6	1 1 4	2	1350		2 2 8	12
	1 2 3		1800		2 2 9	13
	2 1 3		1800		2 2 9	13
	2 2 2		1350	6300	2 2 8	12
3 4 5	1 1 4	1	1800		2 3 4	9
	1 2 3		5400		2 4 6	12
	1 3 2		3600		2 3 6	11
	2 1 3		3600		2 3 6	11
	2 3 1		1800		2 3 4	9
	2 2 2		5400	21600	2 4 6	12
4 4 4	1 2 3	6	600		3 4 3	10
	1 3 2		600		3 3 4	10
	2 1 3		600		4 3 3	10
	3 1 2		600		3 3 4	10
	2 3 1		600		4 3 3	10
	3 2 1		600		3 4 3	10
	2 2 2		1350	3750	4 4 4	12
Total Ways				48600		

8. General Formula Presentation

The construction of the general formula for the number of ways in which s islands of connections can exist on a fully utilized TxT grid follows as a generalization of the worked example. In order to present the formula in the simplest fashion, abbreviated and simplified symbols are used in the formula. These symbols are described here prior to presentation of the counting formulas.

The summation $\sum_{P}$ means "to sum over all restricted s-part partitions of 2T where no member size is less than 2." The s and T parameters are the number of islands and the size of the grid, respectively. The restricted partitions must be calculated and be available for use.

The summation $\sum_{C \text{ over } P}$ means "to sum over all suitably restricted s-part compositions of T where the suitable restriction requires that no member of a composition equal or exceed its correspondingly positioned partition member." The compositions corresponding to a P must be calculated and be available for use.

The parameters $p_1, p_2, p_3, \ldots, p_s$ are the members of a partition in their order of appearance at the time of use. Similarly, the parameters $c_1, c_2, c_3, \ldots, c_s$ are the members of a composition in their order of appearance at the time of use.

If a partition has repeating members, the repetitions induce multiplicate counts. The multiplicates are the result of permutations induced by the number of repeats in each set of repeating members. To compensate for the permutation effect caused by repeating members, the count calculated under each partition P must be divided by the product of the factorials of the counts of each repeat in the restricted partition P. The value of the compensating divisor for a restricted partition is symbolized by D_P which must be calculated from partition data and be available at the time of use.

N is the number of ways unordered sets of s islands can fully occupy a T x T grid.

For a given partition and an induced composition, I_k is the number of connections for the kth island. $\sum K$ stands for the total number of connections for that configuration of s islands.

The general counting formula follows as

$$N = \sum_{P} \left\{ \sum_{C \text{ over } P} \left[\frac{T!}{c_1!\, c_2! \ldots c_s!} \times \frac{T!}{(p_1 - c_1)!\, (p_2 - c_2)! \ldots (p_s - c_s)!} \right] \frac{1}{D_P} \right\} \tag{18}$$

For a given partition and each of its induced compositions, the number of connections in the kth ordered island is seen to be

$$(c_k)\,(p_k - c_k) \tag{19}$$

and the total number of configurations per each configuration of s islands becomes

$$\sum K = \sum_{i=1}^{s} (c_i)\,(p_i - c_i) \tag{20}$$

9. Conclusion

We have developed equations for finding the number of ways s islands of connections can exist on a fully utilized TxT grid of information conductors with no sharing of conductors between islands. Also, we have given equations for numbers of connections per individual island and total numbers of connections per configurations of s islands. The veracity of the equations has been tested over a wide range against brute-force computer simulation programs [4,5].

While it would be satisfying to have found simple, closed equations in s and T alone, the "no sharing of conductors" makes this impossible. Instead, the equations must depend on known or currently generated names of restricted partitions of numbers and partition-induced restricted compositions of numbers. There are compensations, however, since subsets of total counts can be obtained on a single partition-single compositon level.

For information on executable programs on the general material of this note as well as those which specifically perform counts based on the developed equations, please contact the first author.

10. References

[1] Fielder D C , *'Counting and Naming Private Conversations on a PFP Crossbar System'*, Computer Engineering Engineering Research Laboratory, School of Electrical Engineering, Georgia Institute of Technology, Report # DCF-COUNT-PRI, 8/90 and Addendum #1, 1990.

[2] Fielder D C and Alford C O, *'Counting and Naming Private Conversations on a PFP Crossbar System'*, presented at the ISSM Workshop on Parallel Computing, Trani, Italy, September 10-13, 1991. Also accepted for publication in The International Journal Of Mini and Microcomputers (Acta Press).

[3] Fielder D C, *'Continuing Count of Crossbar Broadcast Conversations'*, Computer Engineering Research Laboratory, School of Electrical Engineering, Georgia Institute of Technology, Interim Report BDCSTCONV 9-91, 1991.

[4] Fielder D C, *'Computer Simulation of and Exact Count Formula Derivation for Non-taking Gangs of Friendly Rooks'*, Computer Engineering Research Laboratory, School of Electrical Engineering, Georgia Institute of Technology, CERL Memorandum Report #DCF-ROOKGNGS-1/92, 1992.

[5] Fielder D C, *'Derivation of Alternate Formulas and Programs for Counting Sets of s Gangs of Friendly Rooks on a TxT Chessboard Without Use of Inclusion-Exclusion'*, Computer Engineering Research Laboratory, School of Electrical Engineering, Georgia Institute of Technology, Memorandum Report, 1992.

[6] Fielder D C and Alford C O, *'Contributions from Cascaded Combinations to the Naming of Special Permutations'*, The Fifth International Conference on Fibonacci Numbers and Their Applications, St Andrews University, Scotland, July 20-24, 1992. Also refereed and accepted for publication in *'Applications of Fibonacci Numbers'*, vol. 5, G. E. Bergum et. al., Eds., Kluwer Academic Publishers, Dordrecht, The Netherlands.

A Min Tjoa, I. Ramos (eds.)

Database and Expert Systems Applications

Proceedings of the International Conference in Valencia, Spain, 1992

1992. 324 figures. XIV, 551 pages.
Soft cover DM 148,–, öS 1036,–
ISBN 3-211-82400-6

Prices are subject to change without notice

The Database and Expert Systems Applications (DEXA) conferences are mainly oriented to establish a state-of-the-art forum on database and expert systems applications. But practice without theory has no sense, as Leonardo said five centuries ago. Therefore, as presented in this book, a compromise has been aimed at these two complementary aspects. Five sessions are application-oriented, ranging from classical applications to more unusual ones in software engineering. Actual research aspects in databases, such as activity, deductivity and/or object orientation are also presented in DEXA '92, as well as the implications of the new "data models" such as OO-model, deductive model, etc. are included in the modelling sessions.

Other areas of interest, such as hypertext and multi-media applications, together with the classical field of information retrieval are also considered. Finally, implementation aspects are reflected in very concrete fields.

Springer-Verlag Wien New York

Evolution and Cognition

Offical Journal of the Konrad Lorenz Institute for Evolution and Cognition Research

Contents Volume 2

Evolution and Cognition is an interdisciplinary forum devoted to all aspects of cognition, be it at the level of animals or at the human level. Particularly, evolutionary explanations of cognitive phenomena are discussed, since organisms are to be considered as products of evolution and their cognitive capacities as results of evolutionary processes. Besides, the impact of the study of cognition on our understanding of evolution is considered of importance.

To this end, **Evolution and Cognition** welcomes papers on different aspects of the relation between evolutionary processes and cognitive phenomena. Original papers as well as review articles will be published; the editors will pay attention to papers reporting empirical research work as well as to articles including important theoretical implications. Each issue of the journal will also contain some critical book reviews.

Subscription Information:

1993. Vol. 3 (4 issues): DM 260,–, öS 1820,–, plus carriage charges

Prices are subject to change without notice

Springer-Verlag Wien New York